INTERNATIONAL BACCALAUREATE

MATHEMATICAL STUDIES STANDARD LEVEL

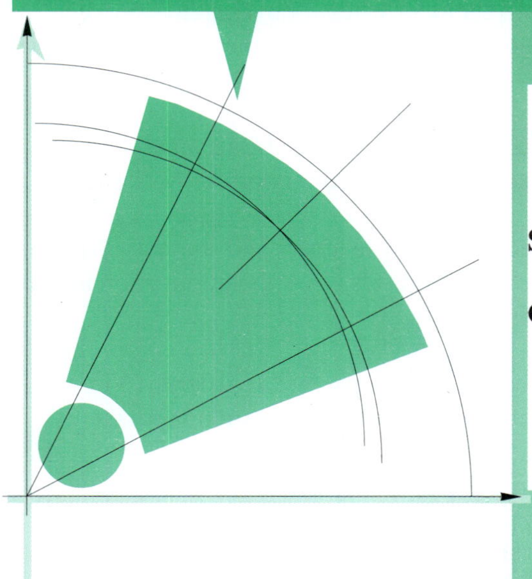

3RD EDITION
2nd Imprint

Series editor: Fabio Cirrito

Contributing authors:
Vicki Strid
Patrick Tobin

First published in 1998 by IBID Press
2nd Edition published in 2000 by IBID Press,
3rd Edition published in 2004 by IBID Press,
3rd Edition, 2nd Imprint published in 2005 by IBID Press,

Published by IBID Press, Victoria.
Library Catalogue:

Cirrito Fabio Editor., Strid. V, Tobin. P
1. Mathematics, 2. International Baccalaureate. Series Title: International Baccalaureate in Detail

ISBN: 1 876659 23 8

We also wish to thank the Incorporated Association of Registered Teachers of Victoria for granting us permission to reproduced questions from their Examination Papers.

Cover design by Adcore Creative.

Published by IBID Press, at www.ibid.com.au
For further information contact fabio@ibid.com.au

Printed by SHANNON Books, Australia.

PREFACE TO 3RD EDITION

The 3rd edition of the Mathematical Studies – Standard Level text has been completely revised and updated. Sections of the previous two editions are still present, but much has happened to improve the text both in content and accuracy.

In response to the many requests and suggestions from teachers worldwide the text was extensively revised. There are more examples for students to refer to when learning the subject matter for the first time or during their revision period. There is an abundance of well-graded exercises for students to hone their skills. Of course, it is not expected that any one student work through every question in this text - such a task would be quite a feat. It is hoped then that teachers will guide the students as to which questions to attempt. The questions serve to develop routine skills, reinforce concepts introduced in the topic and develop skills in making appropriate use of the graphics calculator.

The text has been written in a conversational style so that students will find that they are not simply making reference to an encyclopedia filled with mathematical facts, but rather find that they are in some way participating in or listening in on a discussion of the subject matter.

Throughout the text the subject matter is presented using graphical, numerical, algebraic and verbal means whenever appropriate. Classical approaches have been judiciously combined with modern approaches reflecting new technology - in particular the use of the graphics calculator.

The book has been specifically written to meet the demands of the Mathematical Studies Standard Level course and has been pitched at a level that is appropriate for students studying this subject. The book presents an extensive coverage of the syllabus and in some areas goes beyond what is required of the student. Again, this is for the teacher to decide how best to use these sections.

Sets of revision exercises are included in the text. Many of the questions in these sets have been aimed at a level that is on par with what a student could expect on an examination paper. However, some of the questions do go beyond the level that students may expect to find on an examination paper. Success in examinations will depend on an individual's preparation and they will find that making use of a selection of questions from a number of sources will be very beneficial.

Chapter 3 is dedicated to project work. It includes many topics that students can use, either as they stand or as a source of ideas from which to then produce their own investigation. Each suggestion has been extensively trialled and they have been included in the book because they will address the assessment criteria required for the successful completion of the project work.

I hope that most of the suggestions and recommendations that were brought forward have been addressed in this edition. However, there is always room for improvement. As always, I welcome and encourage teachers and students to contact me with feedback, not only on their likes and dislikes but suggestions on how the book can be improved as well as where errors and misprints occur. There will be updates on the IBID Press website in relation to errors that have been located in the book – so we suggest that you visit the IBID website at www.ibid.com.au. If you believe you have located an error or misprint please email me at fabio@ibid.com.au.

Fabio Cirrito, July 2004

PREFACE TO 1ST EDITION

I am very pleased to finally see in complete form the third of the maths book in this series. Mathematical Studies (Standard Level) has been prepared so as to take into account the needs of students who are studying this subject. At times, the content or depth of the mathematics as gone slightly beyond that required for examination purposes. However, this has been done sparingly and always being conscious of the students' needs. Each (Core) chapter finishes with an '*Examination Style Questions*' section as well as a '*Self Assessment Test*'. These sections are meant to be used for consolidation purposes in the lead up to the final examinations. The questions in both these sections reflect the views of the authors and editor.

A number of people have been involved with the final production of this book. In particular, the contribution made by Vicki Strid has made this book all the better. I thank her for the work she has done in writing a large part of the book as well as her suggestions on how to improve other areas. Also, on behalf of Vicki, I would like to acknowledge the assistance provided by Jill Weitz and Bill Flynn.

This text has been produced independently as a resource to support the teaching of the Mathematical Studies Standard Level Course of the International Baccalaureate. The examples and questions do not necessarily reflect the views of the official senior examining team appointed by the International Baccalaureate Organisation.

The **notation** used is, as far as possible, that specified in the appropriate syllabus guidelines of the IB. The **units** of physical measurements are in S.I. The **language** and **spelling** are U.K. English. **Currency** quantities are specified in dollars, though these could be read as any currency that is decimalised, such as Swiss francs, Lire etc.

The graphic calculators covered directly in the text are the Texas TI/82 and 83. Supplementary material is available from the publisher for students using some other makes and models of calculators. As it is important that students learn to interpret graphic calculator output, the text and answers present a mixture of graphic calculator screens and conventional diagrams when discussing graphs.

The text has been presented in the order in which the topics appear in the syllabus. This does not mean that the topics have to be treated in this order, though it is generally the case that the more fundamental topics appear at the front of the book. Students are reminded that it is the IB Syllabus that specifies the contents of the course and not this text. One of the keys to success in this course is to be thoroughly familiar with the course contents and the styles of questions that have been used in past examinations.

As always, we welcome suggestions, critisms (constructive) or any other comment from teachers, students and those that use this book.
You can call or fax through on + 61 03 9 747 0840.

Fabio Cirrito, July 1998.

CONTENTS

NOTATION

The list below represents the signs and symbols which are recommended by the International Organization for Standardization as well as other symbols that are used in the text.

$\mathbb{N}$	the set of positive integers and zero, $\{0, 1, 2, 3,...\}$
$\mathbb{Z}$	the set of integers, $\{0, \pm 1, \pm 2, \pm 3...\}$
$\mathbb{Z}^+$	the set of positive integers, $\{1, 2, 3,...\}$
$\mathbb{Q}$	the set of rational numbers $\left\{x \mid x = \frac{a}{b}, b \neq 0, a, b \in \mathbb{Z}\right\}$
$\mathbb{Q}^+$	the set of positive rational numbers, $\{x \mid x \in \mathbb{Q}, x > 0\}$
$\mathbb{R}$	the set of real numbers
$\mathbb{R}^+$	the set of positive real numbers $\{x \mid x \in \mathbb{R}, x > 0\}$
$\{x_1, x_2 ...\}$	the set with elements $x_1, x_2 ...$
$n(A)$	the number of elements in the finite set A
$\{x \mid \quad\}$	the set of all x such that...
$\in$	is an element of
$\notin$	is not an element of
$\emptyset$	the empty (null) set
U	the universal set
$\cup$	union
$\cap$	intersection
$\subset$	is a proper subset of
$\subseteq$	is a proper subset of
A'	the complement of set A
$p \wedge q$	conjunction: p and q
$p \vee q$	disjunction: p or q (or both)
$p \veebar q$	exclusive disjunction: p or q (**not** both)
$a \mid b$	a divides b
$\neg p$	negation: not p
$p \Rightarrow q$	implication: if p then q
$p \Leftarrow q$	implication: if q the p
$p \Leftrightarrow q$	equivalence: p is equivalent to q
$a^{1/n}, \sqrt[n]{a}$	a to the power $\frac{1}{n}$ or the nth root of a
$a^{-n} = \frac{1}{a^n}$	a to the power $-n$, reciprocal of a^n

$a^{1/2}$, $\sqrt{a}$	a to the power $\frac{1}{2}$ or the square root of $a \geq 0$
$\equiv$	identity
$\approx$	is approximately equal to
$>$	is greater than
$\geq$	is greater than or equal to
$<$	is less than
$\leq$	is less than or equal to
$\not>$	is not greater than
$\not<$	is not less than
$[a, b]$	the closed interval $a \leq x \leq b$
$]a, b[$	the open interval $a < x < b$
u_n	the nth term of a sequence or series
d	the common difference of an arithmetic sequence
r	the common ratio of an geometric sequence
S_n	the sum of the first n terms of a sequence $u_1 + u_2 + u_3 + \ldots + u_n$
S_∞	the sum to infinity of a sequence $u_1 + u_2 + u_3 + \ldots$
$\sum_{i=1}^{n} u_i$	$u_1 + u_2 + \ldots + u_n$
$f: A \rightarrow B$	f is a function under which each element of set A has an image in set B
$f: x \mapsto y$	f is a function under which x is mapped to y
$f(x)$	the image of x under the function f
$f^{-1}(x)$	the inverse function of the function f
$\lim_{x \rightarrow a} f(x)$	the limit of $f(x)$ as x tends to a
$\frac{dy}{dx}$	the derivative of y with respect to x
$f'(x)$	the derivative of $f(x)$ with respect to x
$\frac{d^2y}{dx^2}$	the second derivative of y with respect to x
$f''(x)$	the second derivative of $f(x)$ with respect to x
sin,cos,tan	the circular functions
A(x, y)	the point A in the plane with Cartesian coordinates x and y
[AB]	the line segment with endpoints A and B
AB	the length of [AB]
(AB)	the line containing points A and B

$\hat{A}$	the angle at A
$C\hat{A}B$	the angle between the lines [CA] and [AB]
ΔABC	the triangle whose vertices are A, B and C
$P(A)$	probability of event A
$P(A')$	probability of the event "not A"
$P(A\|B)$	probability of the event A given B
$x_1, x_2, \ldots$	observations
$f_1, f_2, \ldots$	frequencies with which the observations $x_1, x_2, \ldots$ occur
$\bar{x}$	sample mean
s_n^2	sample variance, $\sigma^2 = \dfrac{\sum_{i=1}^{k} f_i(x_i - \bar{x})^2}{n}$ where $n = \sum_{i=1}^{k} f_i$
s_n	standard deviation of the sample $s_n = \sqrt{\dfrac{\sum_{i=1}^{k} f_i(x_i - \bar{x})^2}{n}}$
s_{n-1}^2	unbiased estimate of the population variance $s_{n-1}^2 = \dfrac{n}{n-1}s_n^2$ or $\dfrac{\sum_{i=1}^{k} f_i(x_i - \bar{x})^2}{n-1}$
$X \sim N(\mu, \sigma^2)$	the random variable X has a normal distribution with mean μ and variance σ^2
Φ	cumulative distribution function of the standardised normal variable: N(0,1)
r	Pearson's product-moment correlation coefficient
χ^2	chi-squared

It is also expected that students possess prior knowledge of mathematics from earlier studies in mathematics. For example, we have not included in Chapter 17, volumes of 3–D solids:

Sphere	Cylinder	Cone
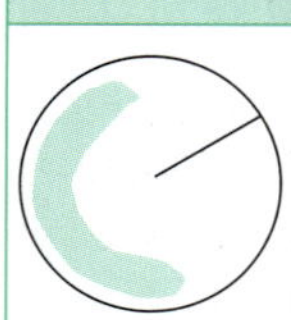 $V = \frac{4}{3}\pi r^3$	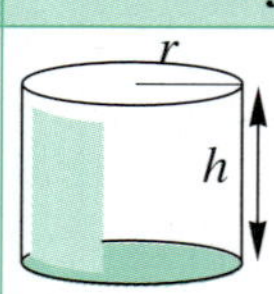 $V = \pi r^2 h$	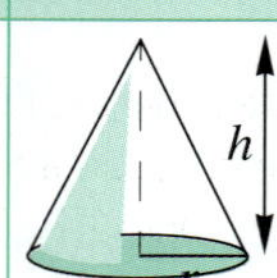 $V = \frac{1}{3}\pi r^2 h$
Rectangular/square Pyramid	**Prisms with uniform cross-section**	
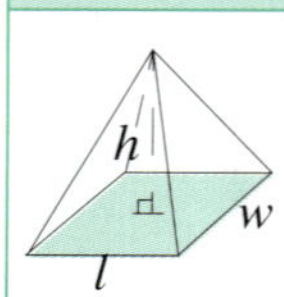 $V = \frac{1}{3}lwh$	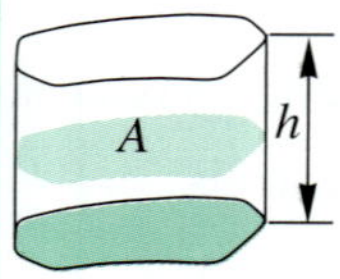 $V = Ah$	

1.1 PURE AND APPLIED MATHEMATICS

Mathematics has clearly played a significant part in the develcpment of many past and present civilisations.

There is good evidence that mathematical, and probably astronomical techniques, were used to build the many stone circles of Europe which are thought to be at least three thousand years old (Thom). It is likely that the Egyptian pyramids and constructions on Aztec and Mayan sites in South America were also built by mathematically sophisticated architects. Similarly, cultures in China, India and throughout the Middle East developed mathematics a very long time ago. It is also the case that there have been very successful cultures that have found little use for mathematics. Ancient Rome, handicapped, as it was, by a non-place value number system did not develop a mathematical tradition at anything like the same level as that of Ancient Greece. Also, the Australian Aborigines who have one of the most long lasting and successful cultures in human history did not find much need for mathematical methods. The same is true of the many aboriginal cultures of Africa, Asia and the Americas. This may well be because these aboriginal cultures did not value ownership in the way that western culture does and had no need to count their possessions. Instead, to aboriginal cultures, a responsible and sustainable relationship with the environment is more important than acquisition and exploitation. Maybe we should learn from this before it is too late!

Mathematics has developed two distinct branches. Pure mathematics, which is studied for its own sake, and applied mathematics which is studied for its usefulness. This is not to say that the two branches have not cross-fertilised each other, for there have been many examples in which they have.

The pure mathematician Pierre de Fermat (1601-1665) guessed that the equation $x^n + y^n = z^n$ has whole numbered solutions for $n = 2$ only. To the pure mathematician, this type of problem is interesting for its own sake. To study it is to look for an essential truth, the 'majestic clockwork' of the universe. Pure mathematicians see 'beauty' and 'elegance' in a neat proof. To pure mathematicians, their subject is an art.

Applied mathematics seeks to develop mathematical objects such as equations and computer algorithms that can be used to predict what will happen if we follow a particular course of action. This is a very valuable capability. We no longer build bridges without making careful calculations as to whether or not they will stand. Airline pilots are able to experience serious failures in commercial jets without either risking lives or the airline's valuable aeroplanes or, indeed, without even leaving the ground.

1.2 AXIOMS

Mathematics is based on axioms. These are 'facts' that are assumed to be true. An axiom is a statement that is accepted without proof. Early sets of axioms contained statements that appeared to be obviously true. Euclid postulated a number of these 'obvious' axioms.

Example:

'Things equal to the same thing are equal to each other';

That is, if $y = a$ and $x = a$ then $y = x$.

Euclid was mainly interested in geometry and we still call plane geometry 'Euclidean'. In Euclidean space, the shortest distance between two points is a straight line. We will see later that it is possible to develop a useful, consistent mathematics that does not accept this axiom.

Most axiom systems have been based on the notion of a 'set', meaning a collection of objects. An example of a set axiom is the 'axiom of specification'. In crude terms, this says that if we have a set of objects and are looking at placing some condition or specification on this set, then the set thus specified must exist. We consider some examples of this axiom.

Example:

Assume that the set of citizens of China is defined. If we impose the condition that the members of this set must be female, then this new set (of Chinese females) is defined.

As a more mathematical example, if we assume that the set of whole numbers exists, then the set of even numbers (multiples of 2) must also exist.

A second example of a set axiom is the 'axiom of powers':

Example:

For each set, there exists a collection of sets that contains amongst its elements all the subsets of the original set. If we look at the set of cats in Bogota, then there must be a set that contains all the female cats in Bogota, another that contains all the cats with green eyes in Bogota, another that contains all the Bogota cats with black tails, etc. A good, but theoretical, account of axiomatic set theory can be found in Halmos, 1960.

Mathematics has, in some sense, been a search for the smallest possible set of consistent axioms. In the section on paradox, we will look further at the notion of axioms and the search for a set of assumptions that does not lead to contradictions. There is a very strong sense in which mathematics is an unusual pursuit in this respect. Pure mathematics is concerned with absolute truth only in the sense of creating a self-consistent structure of thinking.

As an example of some axioms that may not seem to be sensible, consider a geometry in which the shortest path between two points is the arc of a circle and all parallel lines meet. These 'axioms' do not seem to make sense in 'normal' geometry. The first mathematicians to investigate non-Euclidean geometry were the Russian, Nicolai Lobachevsky (1793-1856) and the Hungarian, Janos Bolyai (1802-1860). Independently, they developed self consistent geometries that did not include the so called parallel postulate which states that for every line AB and point C outside AB there is only one line through C that does not meet AB.

Since both lines extend to infinity in both directions, this seems to be 'obvious'. Non-Euclidean geometries do not include this postulate and assume either that there are no lines through C that do not meet AB or that there is more than one such line. It was the great achievement of Lobachevsky and Bolyai that they proved that these assumptions lead to geometries that are self consistent and thus acceptable as 'true' to pure mathematicians. In case you are thinking that this sort of activity is completely useless, one of the two non-Euclidean geometries discussed above has actually proved to be useful; the geometry of shapes drawn on a sphere. This is useful because it is the geometry used by the navigators of aeroplanes and ships.

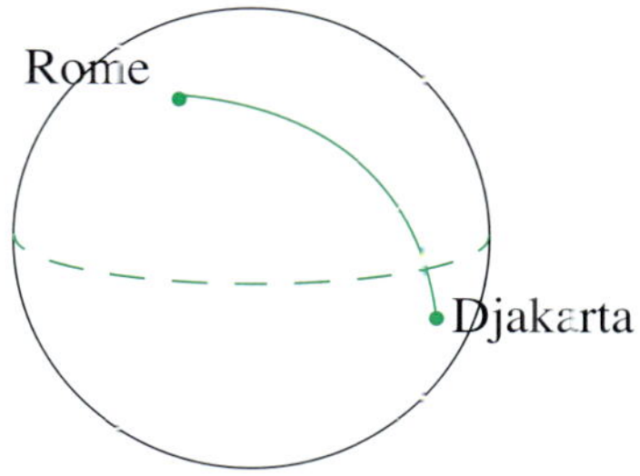

The first point about this geometry is that it is impossible to travel in straight lines. On the surface of a sphere, the shortest distance between two points is an arc of a circle centred at the centre of the sphere (a great circle). The shortest path from Rome to Djakarta is circular. If you want to see this path on a geographer's globe, take a length of sewing cotton and stretch it tightly between the two cities. The cotton will follow the approximate great circle route between the two cities.

If we now think of the arcs of great circles as our 'straight lines', what kind of geometry will we get? You can see some of these results without going into any complex calculations. For example, what would a triangle look like?

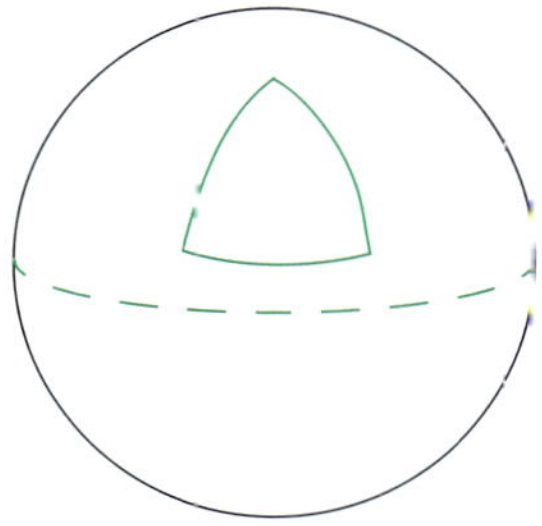

The first point is that the angles of this triangle add up to more than 180°. There are many other 'odd' features of this geometry. However, fortunately for the international airline trade, the geometry is self consistent and allows us to navigate safely around the surface of the globe. Thus non-Euclidean geometry is an acceptable pure mathematical structure.

While you are thinking about unusual geometries, what are the main features of the geometry of shapes drawn on the 'saddle surface'?

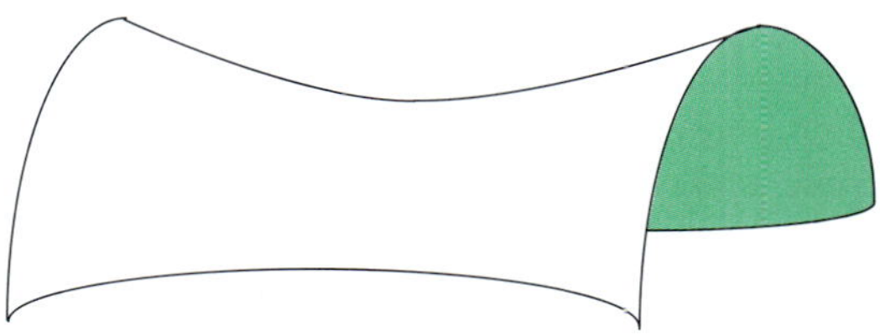

One final point on the subject of non-Euclidean geometries; it seems to be the case that our three dimensional universe is also curved. This was one of the great insights of Albert Einstein (1879-1955). We do not yet know if our universe is bent back on itself rather like a sphere or whether another model is appropriate. A short account of non-Euclidean Geometries can be found in Cameron (pp31-40).

By contrast, applied mathematics is judged more by its ability to predict the future, than by its self-consistency. Applied mathematics is also based on axioms, but these are judged more on their ability to lead to calculations that can predict eclipses, cyclones, whether or not a suspension bridge will be able to support traffic loads, etc. In some cases such mathematical models can be very complex and may not give very accurate predictions. Applied mathematics is about getting a prediction, evaluating it (seeing how well it predicts the future) and then improving the model.

In summary, both branches of mathematics are based on axioms. These may or may not be designed to be 'realistic'. What matters to the pure mathematician is that an axiom set should not lead to contradictions. The applied mathematician is looking for an axiom set and a mathematical structure built on these axioms that can be used to model the phenomena that we observe in nature. As we have seen, useful axiom sets need not start out being 'sensible'.

The system of deduction that we use to build the other truths of mathematics is known as **proof**.

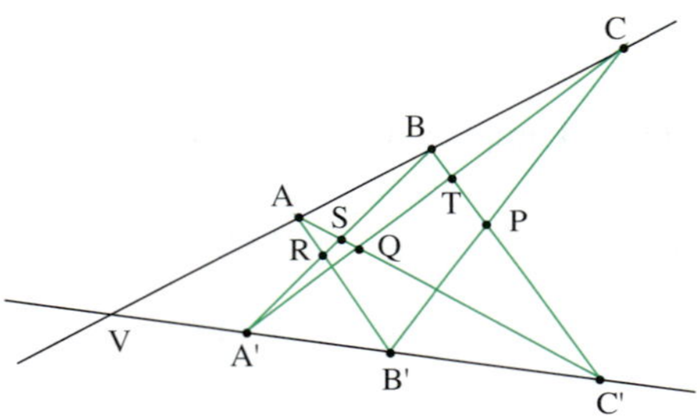

$\Rightarrow P, Q, R$ are collinear

1.3 PROOF

Proof has a very special meaning in mathematics. We use the word generally to mean 'proof beyond reasonable doubt' in situations such as law courts when we accept some doubt in a verdict. For mathematicians, proof is an argument that has *no* doubt at all. When a new proof is published, it is scrutinised and criticised by other mathematicians and is accepted when it is established that every step in the argument is legitimate. Only when this has happened does a proof become accepted.

Technically, every step in a proof rests on the axioms of the mathematics that is being used. As we have seen, there is more than one set of axioms that could be chosen. The statements that we prove from the axioms are known as **theorems**. Once we have a theorem, it becomes a statement that we accept as true and which can be used in the proof of other theorems. In this way we build up a structure that constitutes a 'mathematics'. The axioms are the foundations and the theorems are the superstructure. In the previous section we made use of the idea of consistency. This means that it must not be possible to use our axiom set to prove two theorems that are contradictory.

There are a variety of methods of proof available. This section will look at three of these in detail. We will mention others.

1.3.1 RULES OF INFERENCE

All proofs depend on rules of inference. Fundamental to these rules is the idea of 'implication'.

As an example, we can say that $2x = 4$ (which is known as a **proposition**) implies that $x = 2$ (provided that x is a normal real number and that we are talking about normal arithmetic).

In mathematical shorthand we would write this statement as $2x = 4 \Rightarrow x = 2$.

This implication works both ways because $x = 2$ implies that $2x = 4$ also.

This is written as $x = 2 \Rightarrow 2x = 4$ or the fact that the implication is both ways can be written as

$$x = 2 \Leftrightarrow 2x = 4.$$

The $\Leftrightarrow$ symbol is read as '**If and only if**' or simply as '**Iff**', i.e., If with two fs.

Not every implication works both ways in this manner:

If $x = 2$ then we can conclude that $x^2 = 4$.
However, we cannot conclude the reverse:
i.e., $x^2 = 4$ implies that $x = 2$ is false because x might be -2.

Sothat $x = 2 \Rightarrow x^2 = 4$ is all that can be said in this case.

There are four main rules of inference:

1. The rule of detachment: from a is true and $a \Rightarrow b$ is true we can infer that b is true. a and b are propositions.

Example: If the following propositions are true:

It is raining.
If it is raining, I will take an umbrella.

We can infer that I will take an umbrella.

2. The rule of syllogism: from $a \Rightarrow b$ is true and $b \Rightarrow c$ is true, we can conclude that $a \Rightarrow c$ is true. a, b & c are propositions.

Example: If we accept as true that:
if x is an odd number then x is not divisible by 4 ($a \Rightarrow b$)and,
if x is not divisible by 4 then x is not divisible by 16 ($b \Rightarrow c$)

We can infer that the proposition;

if x is an odd number then x is not divisible by 16 ($a \Rightarrow c$) is true.

3. The rule of equivalence: at any stage in an argument we can replace any statement by an equivalent statement.

Example: If x is a whole number, the statement x is even could be replaced by the statement x is divisible by 2.

4. The rule of substitution: If we have a true statement about all the elements of a set, then that statement is true about any individual member of the set.

Example: If we accept that all lions have sharp teeth then Benji, who is a lion, must have sharp teeth.

Now that we have our rules of inference, we can look at some of the most commonly used methods of proof

1.3.2 PROOF BY EXHAUSTION

This method can be, as its name implies, exhausting! It depends on testing every possible case of a theorem.

Example:

Consider the theorem: Every year must contain at least one 'Friday the thirteenth'.

There are a limited number of possibilities as the first day of every year must be a Monday or a Tuesday or a Wednesday.... or a Sunday (7 possibilities). Taking the fact that the year might or might not be a leap year (with 366 days) means that there are going to be fourteen possibilities.

Once we have established all the possibilities, we would look at the calendar associated with each and establish whether or not it has a 'Friday the thirteenth'. If, for example, we are looking at a non-leap year in which January 1st is a Saturday, there will be a 'Friday the thirteenth' in May. Take a look at all the possibilities (an electronic organiser helps!). Is the theorem true?

1.3.3 DIRECT PROOF

The following diagrams represent a proof of the theorem of Pythagoras described in 'The Ascent of Man' (Bronowski pp 158-161). The theorem states that the area of a square drawn on the hypotenuse of a right angled triangle is equal to the sum of the areas of the squares drawn on the two shorter sides. The method is direct in the sense that it makes no assumptions at the start. Can you follow the steps of this proof and draw the appropriate conclusion?

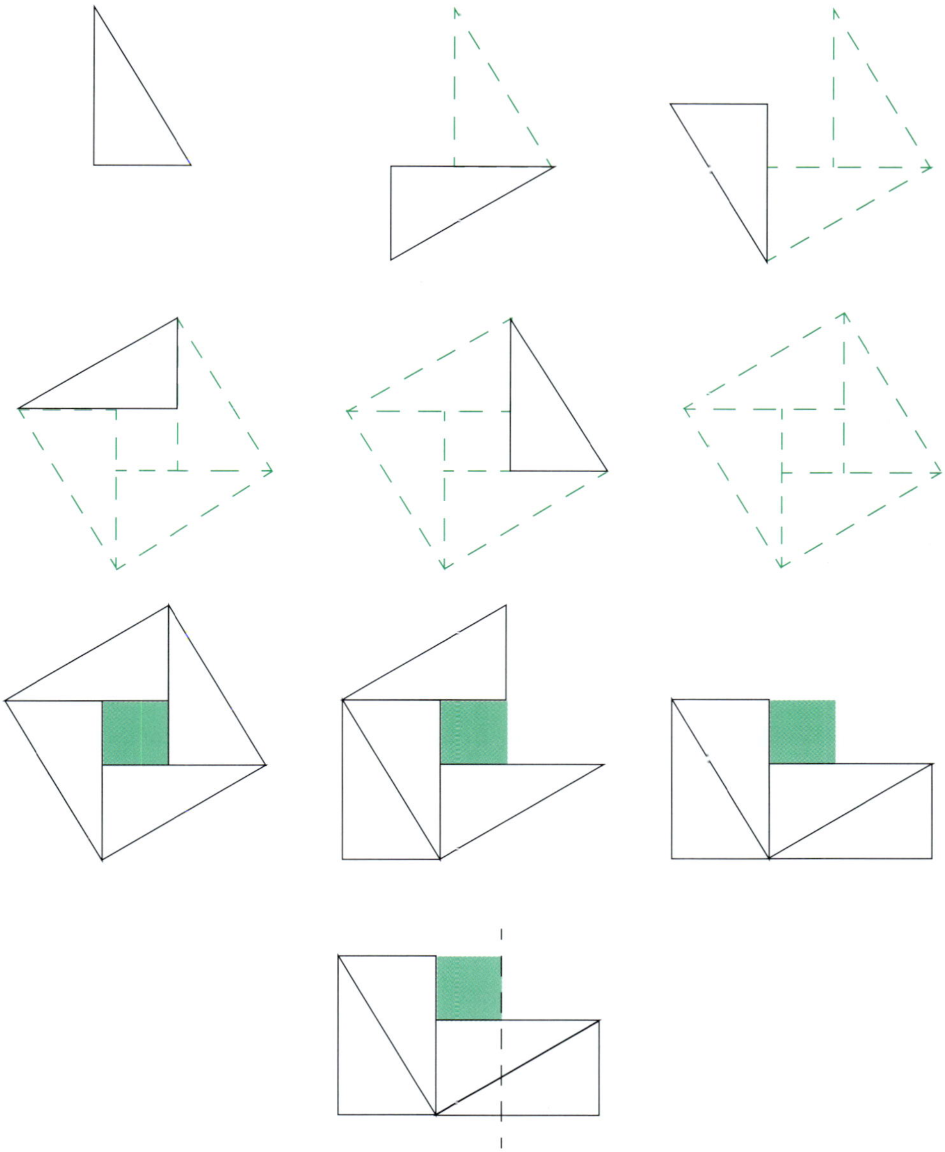

1.3.4 PROOF BY CONTRADICTION

This method works by assuming that the proposition is false and then proving that this assumption leads to a contradiction.

Example:

The number $\sqrt{2}$ greatly interested classical Greek mathematicians who were unable to find a number that, when it was squared, gave exactly 2.

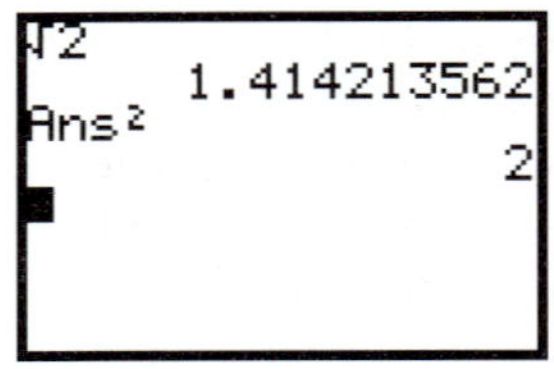

Modern students are often fooled into thinking that their calculators give an exact square root for 2 as when 2 is entered and the square root button is pressed, a result (depending on the model of calculator) of 1.414213562 is produced. When this is squared, exactly 2 results. This is not because we have an exact square root. It results from the way in which the calculator is designed to calculate with more figures than it actually displays. The first answer is stored to more figures than are shown, the result is rounded and then displayed. The same is true of the second result which only rounds to 2. Try squaring 1.414213562, the answer is not 2.

The theorem we shall prove is that there is *no* fraction that when squared gives 2. This also implies that there is no terminating or recurring decimal that, when squared, gives exactly 2, but this further theorem requires more argument.

The method begins by assuming that there *is* a fraction $\frac{p}{q}$ (p & q are integers) which has been cancelled to its lowest terms, such that $\frac{p}{q} = \sqrt{2}$. From the assumption, the argument proceeds:

$$\frac{p}{q} = \sqrt{2} \Rightarrow \frac{p^2}{q^2} = 2 \Rightarrow p^2 = 2q^2 \Rightarrow p^2 \text{ is even} \Rightarrow p \text{ is even}$$

As with most mathematical proofs, we have used simple axioms and theorems of arithmetic.

The most complex theorem used is that if p^2 is even, then p is even. Can you prove this?

The main proof continues with the deduction that if p is even there must be another *integer, r,* that is half p. $p = 2r \Rightarrow p^2 = 4r^2 \Rightarrow 2q^2 = 4r^2 \Rightarrow q^2 = 2r^2 \Rightarrow q^2$ is even $\Rightarrow q$ is even

We now have our contradiction as we assumed that $\frac{p}{q}$ was in its lowest terms so p & q cannot both be even. This proves the result, because we have a contradiction.

This theorem is a very strong statement of *impossibility*.

There are very few other areas of knowledge in which we can make similar statements. We might be virtually certain that we will never travel faster than the speed of light but it would be a brave physicist who would state with certainty that it is *impossible*. Other methods of proof include proof by induction which is mainly used to prove theorems involving sequences of statements. Whilst on the subject of proof, it is worth noting that it is much easier to disprove a statement than to prove it. When we succeed in disproving a statement, we have succeeded in proving its negation or reverse. To disprove a statement, all we need is a single example of a case in which the theorem does not hold. Such a case is known as a **counter-example**.

Example:

The theorem 'all prime numbers are odd' is false. This can be established by noting that 2 is an even prime and, therefore, is the only counter-example we need to give. By this method we have proved the theorem that 'not every prime number is odd'.

This is another example of the way in which pure mathematicians think in a slightly different way from other disciplines. Zoo-keepers (and indeed the rest of us) might be happy with the statement that 'all giraffes have long necks' and would not be very impressed with a pure mathematician who said that the statement was false because there was one giraffe (with a birth defect) who has a very short neck. This goes back to the slightly different standards of proof that are required in mathematics.

Counter-examples and proofs in mathematics may be difficult to find.

Consider the theorem that every odd positive integer is the sum of a prime number and twice the square of an integer. Examples of this theorem that do work are:

$$5 = 3 + 2 \times 1^2,\ 15 = 13 + 2 \times 1^2,\ 35 = 17 + 2 \times 3^2 .$$

The theorem remains true for a very large number of cases and we do not arrive at a counter-example until 5777.

Another similar 'theorem' is known as the Goldbach Conjecture. Christian Goldbach (1690-1764) stated that every even number larger than 2 can be written as the sum of two primes. For example, $4 = 2 + 2$, $10 = 3 + 7$, $48 = 19 + 29$ etc. No-one has every found a counter-example to this 'simple' conjecture and yet no accepted proof has ever been produced, despite the fact that the conjecture is not exactly recent!

Finally, whilst considering proof, it would be a mistake to think that mathematics is a complete set of truths that has nothing which needs to be added. We have already seen that there are unproved theorems that we suspect to be true. It is also the case that new branches of mathematics are emerging with a fair degree of regularity. During this course you will study linear programming which was developed in the 1940s to help solve the problems associated with the distribution of limited resources. Recently, both pure and applied mathematics have been enriched by the development of 'Chaos Theory'. This has produced items of beauty such as the Mandelbrot set and insights into the workings of nature. It seems, for example, that the results of Chaos Theory indicate that accurate long term weather forecasts will never be possible (Mandelbrot).

1.4 PARADOX

1.4.1 WHAT IS A PARADOX?

Pure mathematics is a quest for a structure that does not contain internal contradictions. A satisfactory mathematics will contain no 'nonsense'.

Consider the following proof:

Let $x = 1$	
Then $x^2 - 1 = x - 1$	Try substituting $x = 1$ to check this line.
$(x + 1)(x - 1) = x - 1$	Factorising using the difference of two squares.
$x + 1 = 1$	Dividing both sides by $x - 1$.
$2 = 1$	Substituting $x = 1$.

There is obviously something wrong here as this is the sort of inconsistency that we have discussed earlier in this chapter, but what is wrong? To discover this, we must check each line of the argument for errors or faulty reasoning.

Line 1	must be acceptable as we are entitled to assign a numerical value to a pronumeral.
Line 2	is true because the left hand and right hand sides are the same if we substitute the given value of the pronumeral.
Line 3	is a simple factorisation of the left hand side.
Line 4	is obtained from line 3 by dividing both sides of the equation by $x - 1$ and should be acceptable as we have 'done the same thing' to both sides of the equation.
Line 5	is obtained from line 4 by substituting $x = 1$ and so should give the correct answer.

Obviously we have an unacceptable conclusion from a seemingly watertight argument. There must be something there that needs to be removed as an acceptable operation in mathematics.

The unacceptable operation is dividing both sides by $x - 1$ and then using a value of 1 for x. What we have effectively done is divide by a quantity that is zero. It is this operation that has allowed us to prove that $2 = 1$, an unacceptable result. When a **paradox** of this sort arises, we need to look at the steps of the proof to see if there is a faulty step. If there is, then the faulty step must be removed. In this case, we must add this rule to the allowed operations of mathematics:

Never divide by a quantity that is, or will become, zero.

This rule, often ignored by students, has important implications for Algebra and Calculus.

Some paradoxes are arguments that seem to be sound but contain a hidden error and thus do not contain serious implications for the structure of mathematical logic. An amusing compilation of simple paradoxes can be found in Gardner (1982). An example is the 'elevator paradox'.

Why does it always seem that when we are waiting for an elevator near the bottom of a tall building and wanting to go up, the first elevator to arrive is always going down? Also, when we want to go back down, why is the first elevator to arrive always going up? Is this a real phenomenon or is it just a subjective result of our impatience for the elevator to arrive? Or is it another example of Murphy's Law; whatever can go wrong will go wrong?

This is quite a complex question, but a simple explanation might run as follows:

If we are waiting near the bottom of a tall building, there are a small number of floors below us from which elevators that are going up might come and then pass our floor.

By contrast, there are more floors above us from which elevators might come and then pass our floor going down.

On the basis of this and assuming that the elevators are randomly distributed amongst the floors, it is more likely that the next elevator to pass will come from above and will, therefore, be going down.

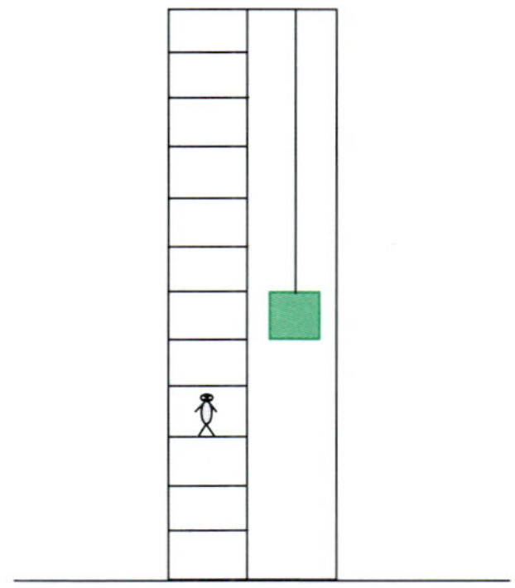

By contrast, if we are waiting near the top of a tall building, there are a small number of floors above us from which elevators that are going down might come and then pass our floor.

Also, there are more floors below us from which elevators might come and then pass our floor going up.

It is more likely that the next elevator to pass will come from below and will, therefore, be going up.

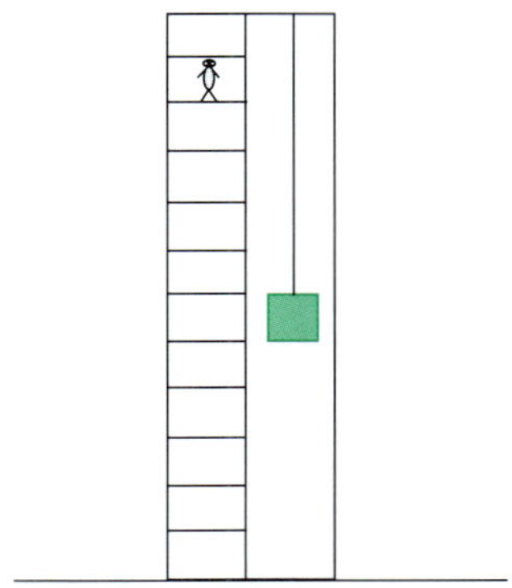

A fuller analysis of this paradox can be found in Gardner (pp96-97).

The elevator paradox does not contain serious implication for the structure of mathematics like our first example. We will conclude this section with a look at a modern paradox that did cause a re-evaluation of one of the basic ideas of mathematics, the set.

1.4.2 RUSSELL'S PARADOX

Bertrand Russell (1872-1970) looked in detail at the basic set axioms of mathematics. We do regard the existence of sets as axiomatic in all mathematical structures. Does this mean that we can make a set that contains 'everything'? There would seem to be no difficulty with this as we just move around the universe and sweep everything that we meet into our set, numbers, words, whales, motorcycles etc. and the result is the set that contains everything.

Russell posed the following question which we will relate in the context of library catalogues. Every library has a catalogue. There are various forms that this catalogue might take; a book, a set of cards, a computer disc etc. Whatever form the catalogue in your local library takes, there is a sense in which this catalogue is a book (or publication) owned by the library and, as such, should appear as an entry in the catalogue:

CATALOGUE	NEWEL LIBRARY
Castle, The. F Kafka 231.72	Catherine the Great A Biography J Nelson 217.42
Catalogue At reception	Catullus The complete works Edited by F Wills 312.42
Catcher in the Rye JD Salinger 123.64	

Of course, many librarians will decide that it is silly to include the catalogue as an entry in the catalogue because people who are already looking at the catalogue know where to find it in the library! It follows that library catalogues can be divided into two distinct groups:

Catalogues that do contain an entry describing themselves.
Catalogues that do not contain an entry describing themselves.

Next, let us make a catalogue of *all* the catalogues of type two, those that do not contain themselves.

This gives us a problem. Should we include an entry describing our new catalogue? If we do, then our catalogue ceases to be a catalogue of all those catalogues that do not contain themselves. If we do not, then our catalogue is no longer a complete catalogue of all those catalogues that do not contain themselves.

The conclusion is that making such a catalogue is impossible. This does not mean that the library catalogues themselves cannot exist. We have, however, defined an impossible catalogue.

In set terms, Russell's paradox says that sets are of two types:

Type 1 Sets that do contain themselves.
Type 2 Sets that do not contain themselves.

The set of all sets of type 2 cannot be properly defined without reaching a contradiction.

The most commonly accepted result of Russell's paradox is the conclusion that we have to be very careful when we talk about sets of everything. The most usual way out is to work within a carefully defined universal set, chosen to be appropriate to the mathematics that we are undertaking. If we are doing normal arithmetic, the universal set is the set of real numbers.

1.5 MATHEMATICS AND OTHER DISCIPLINES

When writing Theory of Knowledge essays, students are required to develop their arguments in a cross disciplinary way. For more details on this, you are strongly advised to read the task specifications and the assessment criteria that accompany the essay title. You are reminded that it is these statements that define what is expected of a good essay, not the contents of this Chapter which have been provided as a background resource. A good essay will only result if you develop your own ideas and examples in a clear and connected manner. Part of this process may include comparing the 'mathematical method' described earlier with the methods that are appropriate to other systems of knowledge.

As we have seen, mathematics rests on sets of axioms. This is true of many other disciplines. There is a sense in which many ethical systems also have their axioms such as 'Thou shalt not kill'.

The Ancient Greeks believed that beauty and harmony are based, almost axiomatically, on mathematical proportions. The golden mean is found by dividing a line in the following ratio:

The ratio of the length AB to the length BC is the same as the ratio of the length BC to the whole length AC. The actual ratio is $1:\frac{1}{2}(1+\sqrt{5})$ or about 1:1.618. The Greek idea was that if this line is converted into a rectangle, then the shape produced would be in perfect proportion:

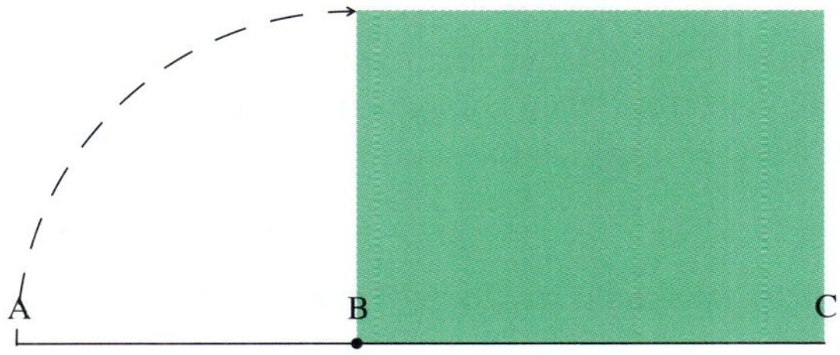

Likewise, the correct place to put the centre of interest in a picture is placed at the golden mean position between the sides and also at the golden mean between top and bottom. Take a look at the way in which television pictures are composed to see if we still use this idea:

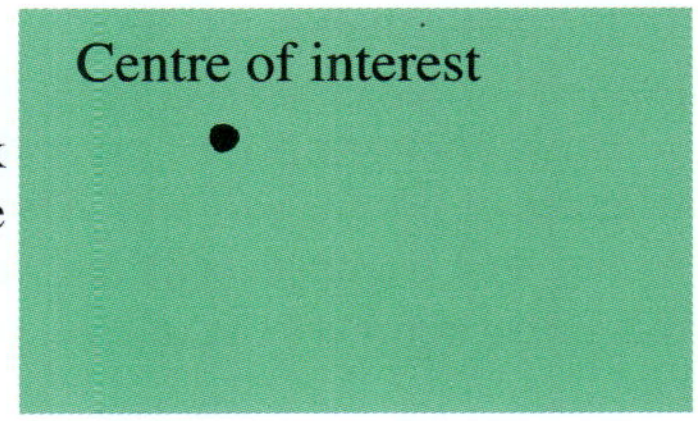

In a similar way, the Ancient Greeks believed that ratio determined harmony in music. If two similar strings whose lengths bear a simple ratio such as 1:2 or 2:3 are plucked together the resulting sound will be pleasant (harmonious). If the ratio of string lengths is 'awkward', such as 17:19, then the notes will be discordant. The same principle of simple ratios is used in tuning musical instruments (in most cultures) today.

The most common connection between mathematics and other disciplines is the use of mathematics as a tool. Examples are: the use of statistics by insurance actuaries, probability by quality control officers and of almost all branches of mathematics by engineers. Every time mathematics is used in this way, there is an assumption that the calculations will be done using techniques that produce consistent and correct answers. It is here that pure mathematical techniques, applied mathematical modelling and other disciplines interface.

In some of these examples, we apply very precise criteria to our calculations and are prepared to accept only very low levels of error. Navigation satellite systems work by measuring the position of a point on or above the Earth relative to the positions of satellites orbiting the Earth.

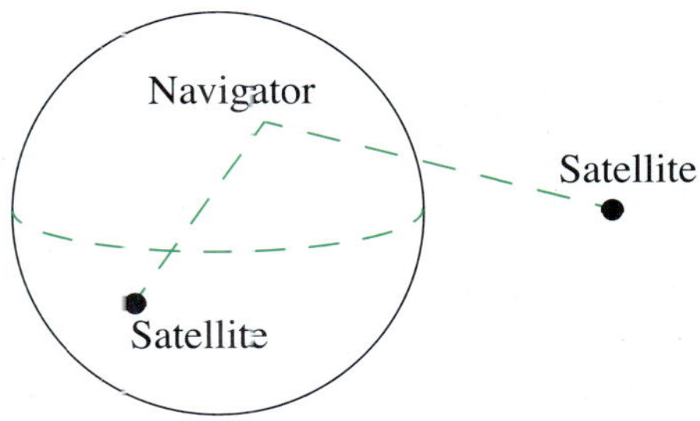

This system will only work if the positions of the satellites are known with very great precision.

By contrast, when calculations are made to forecast the weather, whilst they are done with as much precision as necessary, because the data is incomplete and the atmospheric models used are approximate, the results of the calculations are, at best, only an indication of what might happen. Fortunately, most of us expect this and are much more tolerant of errors in weather forecasting than we would be if airlines regularly failed to find their destinations!

There are, therefore a large number of ways in which mathematics complements other disciplines. In fact, because computers are essentially mathematical devices and we are increasingly dependent on them, it could be argued that mathematics and its methods underpin the modern world.

That is not to say that mathematics is 'everywhere'. Many very successful people have managed to avoid the subject altogether. Great art, music and poetry has been produced by people for whom mathematical ideas held little interest.

In using mathematical ideas in essays, remember that you should produce original examples, look at them in a mathematical context and then compare the ways in which the example might appear to a mathematician with the way in which the same example might appear to a thinker from another discipline.

As a very simple example, what should we think of gambling?

To the mathematician (Pascal was one of the first to look at this activity from the mathematical perspective), a gambling game is a probability event. The outcome of a single spin of a roulette wheel is unknown. If we place a single bet, we can only know the chances of winning, not whether or not we *will* win. Also, in the long run, we can expect to lose one thirtyseventh of any money that we bet every time we play. To the mathematician, (or at least to this mathematician) this rather removes the interest from the game!

Other people look at gambling from a different standpoint. To the politician, a casino is a source of revenue and possibly a focus of some social problems. To a social scientist, the major concern might be problem gamblers and the effect that gambling has on the fabric of society. A theologian might look at the ethical issues as being paramount. Is it ethical to take money for a service such as is provided by a casino? Many of these people might use mathematics in their investigations, but they are all bringing a slightly different view to the discussion.

As we can see, there are many sides to this question as there are many sides to most questions. Mathematics can often illuminate these, but will seldom provide all the answers. When you choose an essay title, you do not have to use mathematical ideas or a mathematical method to develop your analysis. However, we hope that if you do choose to do this, you will find the brief sketch of the mathematical method described in this Chapter helpful.

We will finish with one observation.

Mathematics and mathematicians are sometimes viewed as dry and unimaginative. This may be true in some cases, but definitely not all.

We conclude with some remarks by the mathematician Charles Dodgson (1832-1898), otherwise known as Lewis Carroll:

'The time has come', the Walrus said,
'To talk of many things:
Of shoes and ships and sealing wax,
Of cabbages and kings,
Of why the sea is boiling hot
And whether pigs have wings'.
Through the Looking Glass

References:

Megalithic Sites in Britain. Thom, A. (1967). U.K. Oxford University Press.
Heritage Mathematics. Cameron, M. (1984).U.K. E.J. Arnold.
The Ascent of Man, Bronowski, J. (1973).U.K. BBC.
The Fractal Geometry of Nature, Mandelbrot, B (1977), U.S. W.H. Freeman & Co. 1977.
Gotcha!, Gardner, M. (1977). U.S.A. W.H. Freeman & Co.

2.1 THE GRAPHICS DISPLAY CALCULATOR

The most common tool we will use in this course is the graphics display calculator (GDC). These calculators offer excellent support for traditional scientific calculation with a good display of both problem and solution. However, in addition to this they give a range of graphing capabilities and powerful statistical tools as well as numerical support for calculus procedures and the solution of equations. The examples provided here come from the Texas Instrument TI83 but a similar facility exists within all of the GDCs which are approved for this course.

2.1.1 KEYPAD INSTRUCTIONS

Some operations with the GDC's (of all brands) can be accessed by keypads where all the numbers basics operands and many standard functions such as trigonometric functions are to be found. The keypad is sufficient for all basic arithmetic. In addition to this there are many menus associated with various keys which we will also find very useful in accessing the hidden parts of the GDC. The keypad of the TI83 is shown to illustrate this.

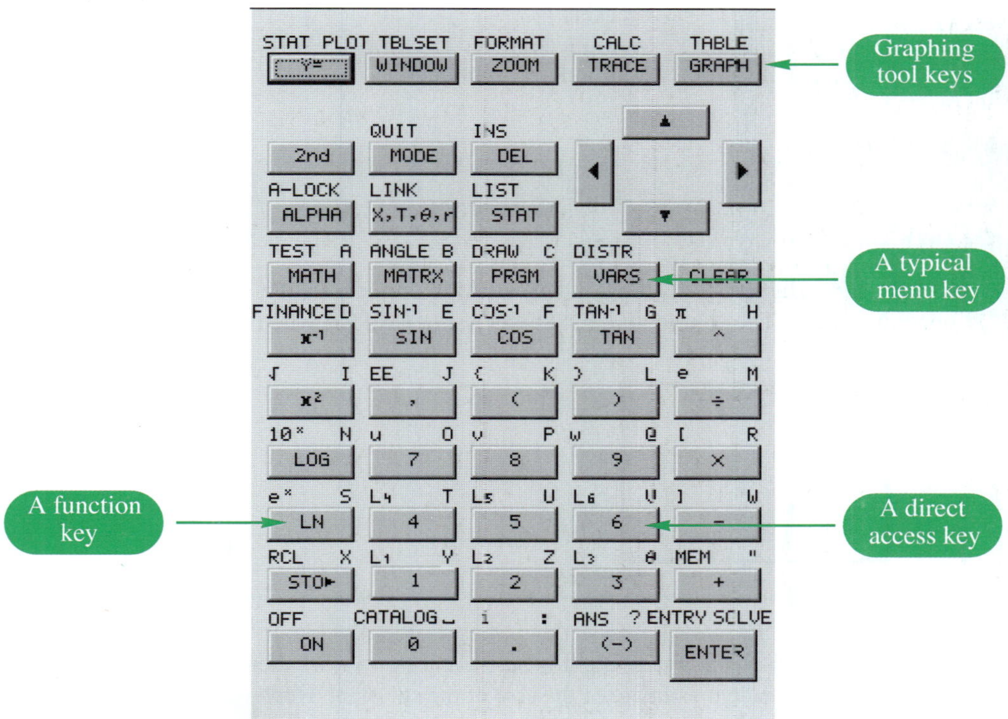

2.1.2 MOVING ABOUT THE CALCULATOR SCREEN

There are many different types of screens which the GDC can produce. Even within the usual operating screen it is necessary to be able to move the cursor. We will look at this briefly as it is important to be able to navigate the calculator. The basic arithmetic and function calculations occur on the **homescreen**. This is always accessible by the **QUIT** which is second function on the **MODE** key. Each time we access a menu we go to a dedicated screen but we can then select menu items for pasting to the homescreen by default.

Note that on the homescreen pressing **CLEAR** will clear up the screen.

Graph screens and function rule entry screens are also used commonly. The arrow keys enable us to navigate on a screen moving up and down and across to various menu items. They also enable us to trace points on graphs. They allow the cursor to move about for deleting or inserting items in mathematical expressions using the **DEL** key.

As the calculator is used many functions, variables, lists, programs and matrices will be stored. The memory can be cleared either entirely or by individual items using the **MEM** which is found at 2nd function on the + key. Note too that if the screen is too faint it can be darkened using the 2nd function and the up arrow in sequence until the contrast is suitable.

2.1.3 COMPUTATION

The calculators provide for traditional arithmetic in a natural setting which means that operations look as they do on your page and the order of operations is as we use in conventional pen and paper calculation, that is

1. Brackets
2. Multiplication and Division
3. Addition and Subtraction

EXAMPLE 2.1

Use a calculator to find $2+\left(5-\frac{6}{3}\right)\times 9$.

Solution

Clearly this is the sort of calculation we can do in our heads but it illustrates the use of operation order. The solution is found by pressing **ENTER** when the problem is written on screen.

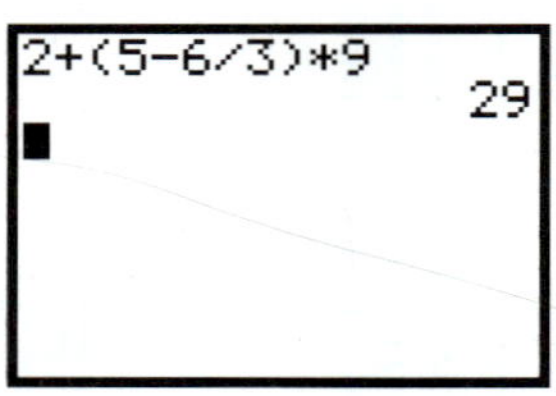

As with any calculator use with arithmetic it is best to check on work mentally with one figure checks to give an independent estimate of the size of the solution. However with a GDC it is at least possible to see both the problem and solution together on screen!

EXAMPLE 2.2

Use a calculator to find $12.315\times\left(19.4-\frac{13.417+46.2}{82.3}\right)$, checking your result with a one figure check.

Solution

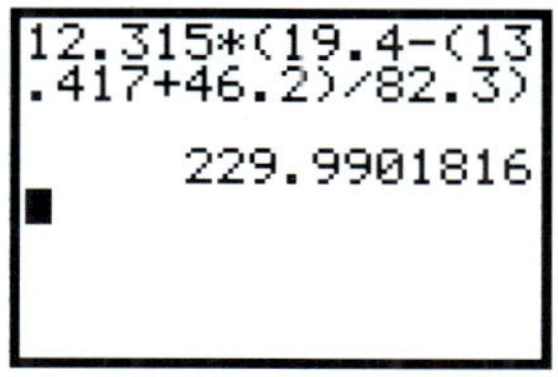

The one figure check replaces each number with a near single digit form. Hence, for example, 12.315 can be replaced by 10, 19.4 by 20, 13.417 by 10, 46.2 by 50 and 82.3 by 80. So the calculation quick estimate is $10 \times \left(20 - \frac{10 + 50}{80}\right)$

which we can see is just 10×19.25 or 192.5. This compares well with the calculated result of 229.9901816 giving us confidence in our entries.

Notice that special numbers like π have their own keys and need not be typed.

The calculator is well suited to editing. The **ENTRY** which is 2nd function on the **ENTER** key brings back the previous entry and enables it to be edited – for example replacing one number by another. This saves time if it would be long to retype.

The answer generated by the **ENTER** key can also be reused by **ANS** which is 2nd function on the (–) key. This is another time saving step.

Finally it is important to note that there is both a binary 'minus' key for which we use in subtraction and a unary 'minus' key (–) which we use for negative numbers.

2.1.4 FUNCTIONS

The basic trigonometric functions, power and root keys reciprocals and log functions are all accessed by the main keypad – occasionally by use of the 2nd function key. Additional special functions can be found on some menus and all functions can be found in the **CATALOG** which is accessed by the Zero key. With trigonometric functions it is important to ensure that you have the correct **MODE** settings to ensure you are using degree measure if that is what you require. We will see that this can be overruled by the **ANGLE** key menus anyway but it is the usual practice. The **MODE** key dictates whether we are in normal number presentation or using scientif or engineering formats where all numbers are expressed using multiples of powers of ten. The **NORMAL** mode is default and the most common.

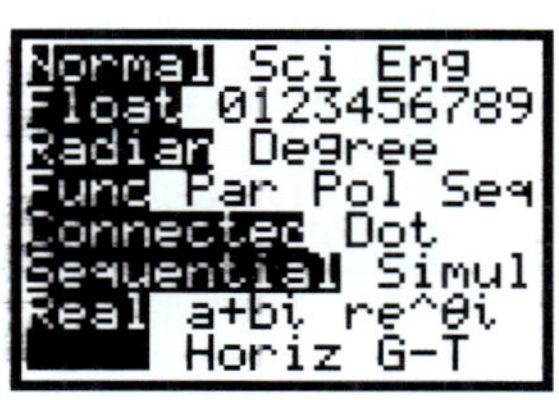

The **MODE** menu tells us the Float which gives the maximum number of decimal points which the calculator will display as well as the angle measure format (degrees or radians). The other **MODE** menu entries dictate features connected with the graphic display, the screen layout and the formats used for complex numbers. We will not concern ourselves with these features yet.

EXAMPLE 2.3

Find $\sin 23°$ using the calculator.

Solution

Ensure calculator is in degree mode and simply use the sin(key and other numerical and bracket keys.

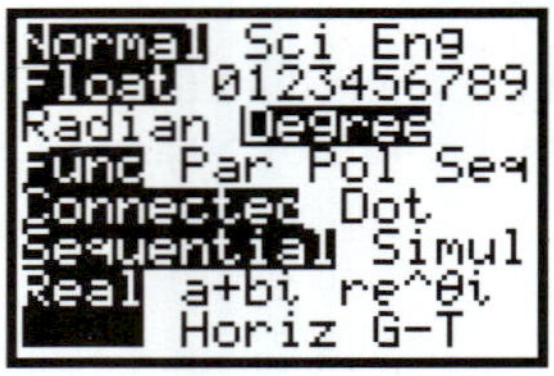

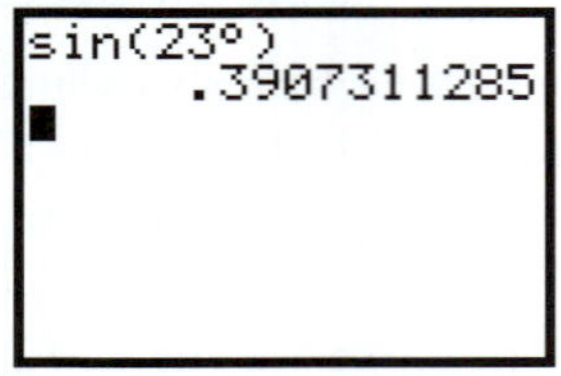

Some functions like powers and roots use unusual forms or may require use of a menu.

EXAMPLE 2.4

Find (a) $2.3^{3.1}$ (b) $\sqrt[3]{13}$

Solution

(a) In this case we use the power key which is shaped like ^.

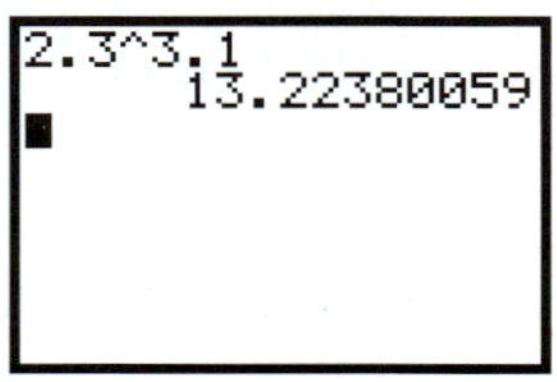

(b) Here we could use a (1/3) power if we recall that roots are fractional powers or we can access the cube root on the **MATH** key on item 4 of the first menu, the **MATH** menu.

Method 1:

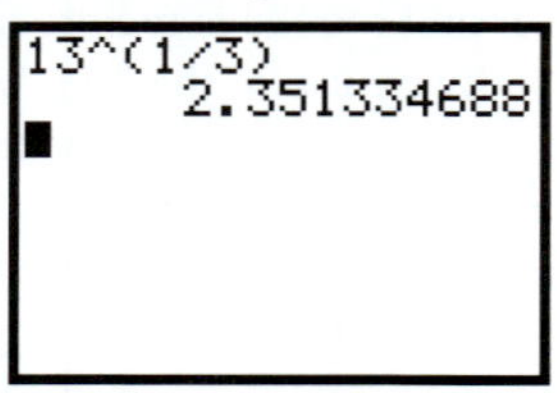

Method 2:

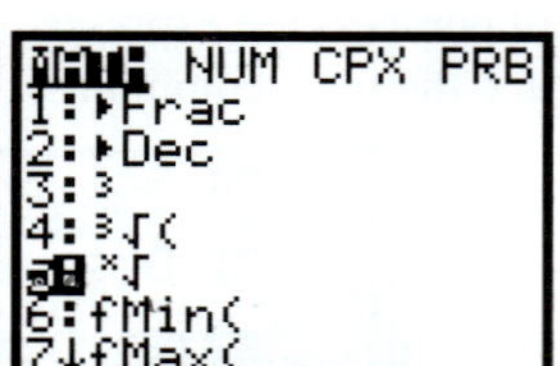

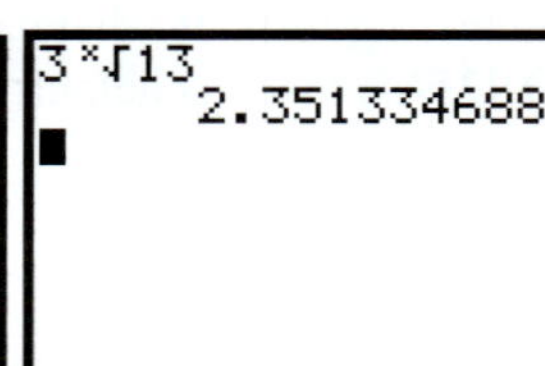

2.1.5 ALPHABET KEYS

The letters of the alphabet are accessed by use of the green **ALPHA** key. In addition to these there is a special key for the main independent variable used in mathematical rules. This is the **X,T,θ, n** key which gives each of these letters depending on the mode being function, parametric, polar or sequence. The normal mode is function and so this gives an X. We can use this to write out expressions which give us rules in equations and graphs. Since these rules often produce a "$y =$ " form in mathematics we can set these up using the Y = key which is part of the set of graph related keys along the top of the keypad.

EXAMPLE 2.5 Write the word TEXT on the screen.

Solution

Use the **ALPHA** key to access the letters shown in green, first locking this in by using 2nd function key to get the A-lock. Pressing clear will have the cursor revert to normal form.

Most commonly in mathematics we use letters to store values we may use repeatedly in calculations. This we can do by using the **STO➧** key.

EXAMPLE 2.6 Store the value 4.7 in the letter A and find $A^2 + A$.

Solution

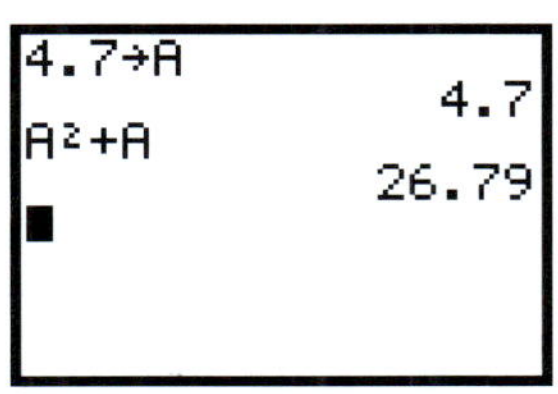

We also find that the cursor turns automatically into an alphabetic one if we enter the catalog. This means we can find any menu item and more besides by scanning down the alphabetic list.

Thus if we want to find **DiagnosticsOn** in the **Catalog** menu we use that key to start the list and press D (note it is already in alphabet mode). This takes us to the start of the D list section and we scroll down using the down arrow to find the **DiagnosticsOn** entry.

2.1.6 GRAPHING KEYS AND TABLES OF FUNCTIONS

The GDC has a powerful graphing facility. In Function mode (the normal one) the Y= key enables us to enter expressions as rules for up to 10 different functions. The function defined can then be used in graphs and tables and even back on the homescreen. For example if we write $X^2 - 3X$ in the $Y_1 =$ position then we can graph this using the **GRAPH** key and change the field of view using the **WINDOW** or **ZOOM** keys. The default window is set at $-10 \le X \le 10$ and $-10 \le Y \le 10$. Note that the resolution limits accuracy in graphing and distorts graphs. There are more pixels across the screen than up and down so the unit is larger on the X-axis.

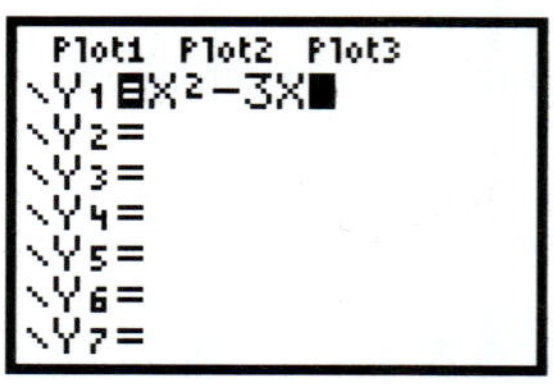

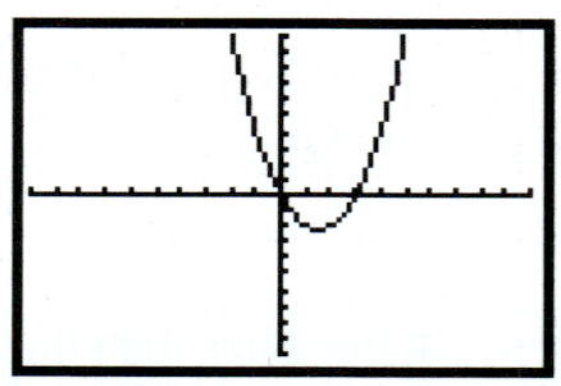

We can trace along the curve using the **TRACE** key, which puts a blinking cursor on the graph that can be moved right or left by the arrow keys. Where there are several graphs on screen the up and down arrows enable us to jump this trace cursor between graphs. The 2nd function on **TRACE** gives **CALC** which enables many operations to be performed on the graph directly such as finding minimum values, or finding intersections when several graphs are drawn.

TRACE:

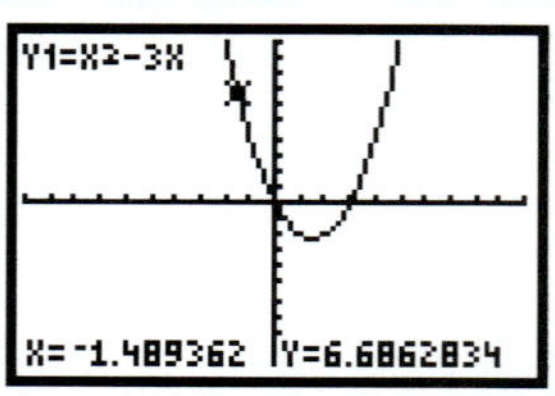

CALC:

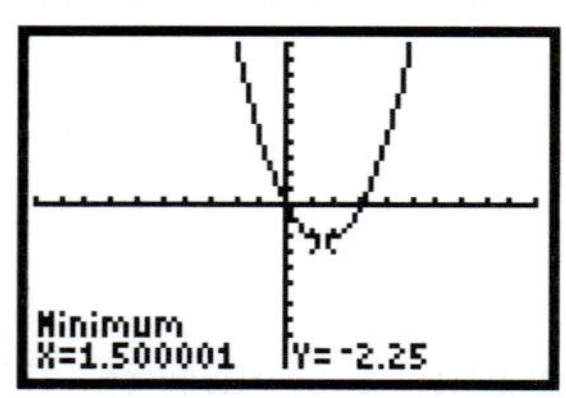

EXAMPLE 2.7

Find the intersection of the curves $y = x^3 + 2x$ and $y = x^2 - 6x + 7$.

Solution

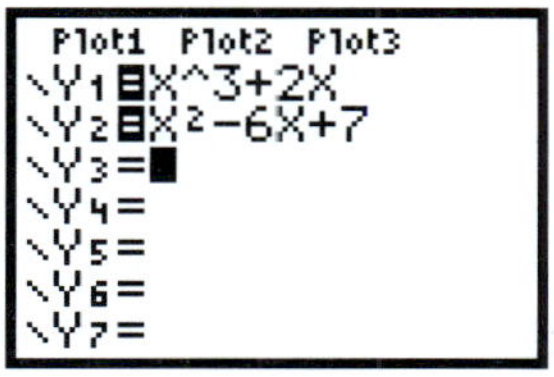

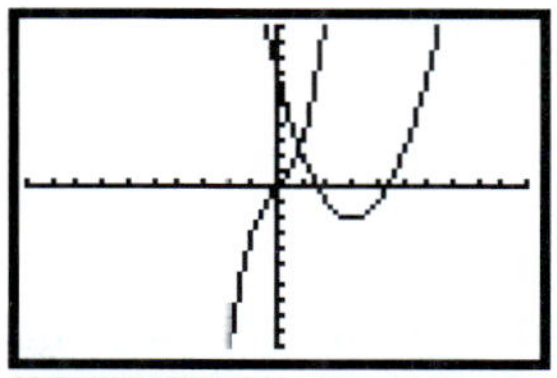

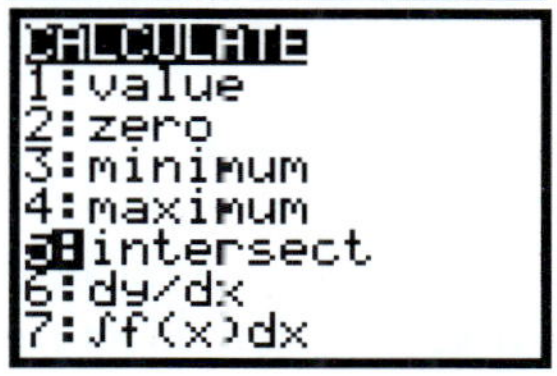

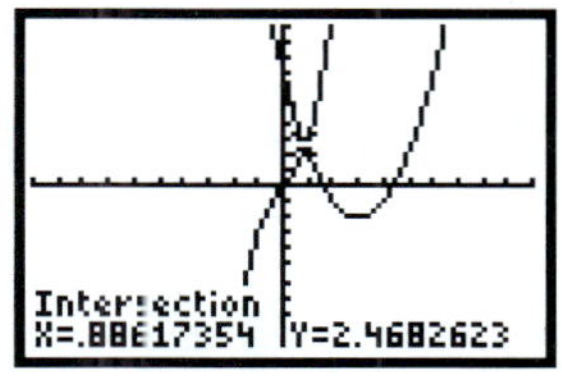

Tables of data can be produced for functions and seen at 2nd function on the **GRAPH** key in **TABLE**. The range of x values and the x step are set in **TBLSET** which is accessed on the **WINDOW** key using 2nd function.

EXAMPLE 2.8

Find the table of values of $y = x^3 + 2x$ and $y = x^2 - 6x + 7$ for x from 0 to 5 in steps of 0.5.

Solution

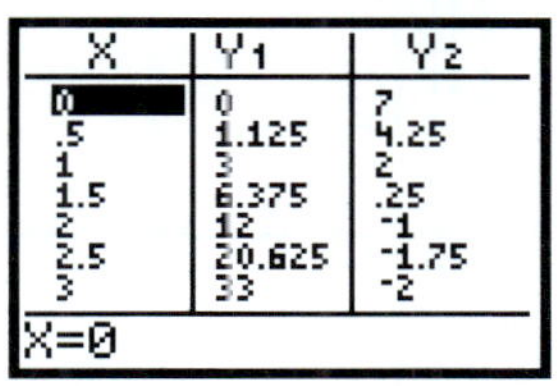

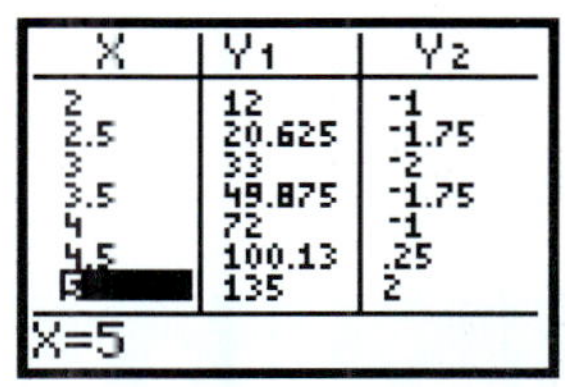

2.1.7 STATISTICS AND MATRICES

The **STAT** key enables us to enter data in table lists and find basic statistics for the data. The **DISTR** which is located at 2nd function on the **VARS** key enables us to do statistical tests. The **PRB** menu on the **MATH** key gives access to basic probability functions.

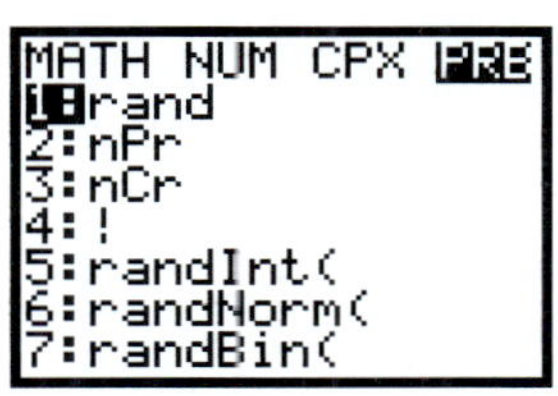

Detailed use of these statistic and probability features will be given in the relevant chapters in examples. Note that creating special statistical plots and graphs are accessed by **STATPLOT** which is 2nd function on the Y= key and these are then seen by using item 9:ZoomStat on the **ZOOM** key to obtain the correct view window.

2.1.8 PROGRAMS

Powerful programs can be written or downloaded to the calculator from other GDCs or the internet. The **PROGRAM** key gives access to any existing programs and the menus to write and execute new programs. These can give additional support in a range of algebraic work like factorisation. However there will not be any specific requirement to create programs in the course.

2.1.9 SOLUTION OF EQUATIONS

We will see that there are many ways we can use the calculator in solving equations. There are graphical procedures accessed on **CALC** as well as direct equation solvers.
The **Solver** is item 0 on the **MATH** menu of the **MATH** key.

We can also use the feature SOLVE which is in the CATALOG.

EXERCISES 2.1

[Answers not provided. You should check with fellow students]

1. Use the calculator to enter and solve $32.715 \times \left(29.4 - \frac{13.417 + \pi}{182.3}\right)$ and perform a one figure check on your answer.

2. Find 3.687×8.715 using
(a) the FLOAT set on default (unrestricted)
(b) the FLOAT set on 2.

3. Store the number 15.9 as B and find $B^4 - \frac{2B}{B+1}$

4. Use the calculator to find (a) $\cos 29°$ (b) 3^7 (c) $\sqrt{23.4}$

5. Enter the function $y = x^3 - 4x + 6$
(a) Graph this using a window of $-5 \le X \le 6$ and $-2 \le Y \le 10$.
(b) Use the trace to estimate the function x-intercept
(c) Set up a table of values for the function starting at $x = -3$ in steps of 0.5

6. Find the intersection of the curves $y = 2x^3 - 3x$ and $y = x^2 - 6x + 4$ using the **CALC** menu.

3.0 INTRODUCING PROJECT WORK

We begin this chapter by observing that the capabilities we are trying to develop are important life skills. We hope that you will use them as you face challenges in your personal and working lives.

Questions that face many of us at some stage in our lives: which job to apply for, which car to buy, which course to take, whether to but or rent a home etc. may seem simple. Some are, but most are not. Many of these carry heavy penalties if you make a poor decision.

We are suggesting that, when faced with an important decision, you should follow the process that we hope you will learn in this part of the course. That said, many of history's most valuable investigations (into, for example, the causes of disease, navigation, atomic structure etc.) have been undertaken more out of interest than in the expectation of personal gain.

The main steps in the process are:

1. Define the problem. Should I buy or rent an flat? How should I finance it? Where should my company buy its paper, cars, iron ore, legal advice...?

2. Identify the parameters that define a 'good decision'. These may be all sorts of things such as price, status value, fast service etc.

3. Research appropriate data. What are the fees? How powerful is the engine? etc.

4. Analyse the data. This is where maths generally comes in!

5. Arrive at a decision and be prepared to justify it to an employer, partner, voter, etc. A good analysis work wonders here! This is known as evaluation. You will probably end up living with the decision (occupying the house, paying for the car, using the software) for some time, so it is very much in your interest to remember this step!

There are many ways to conduct a successful investigation. This chapter is intended to give you ideas. It is not intended to provide a recipe or set of 'how to do it' instructions. Nor does it provide a complete investigation!

We will look at project ideas in the areas of probability, functions and statistics.

To understand these you will need to have covered the relevant chapters:

Idea 1: Monte Carlo methods (Probability, chapter 16).

Idea 2: Damped oscillations (Trigonometric and Exponential Functions, chapter 10 & chapter 11 and calculus, chapters 23 & 24).

Idea 3: **Weights in lifts and transport** (Statistics, chapters 20 & 21))

3.1 CARRYING OUT PROJECTS

3.1.1 CHOICE OF TOPIC (DEFINING THE PROBLEM).

Try not to spend too much time on this! We have provided projects suggestions in two forms:

1. A few suggestions at the end of each chapter, but you should not feel bound by these.

2. A list of project 'themes' at the end of this chapter. We have adopted a 'themes' approach so that the projects would not be prescriptive, but rather, allow you do use them as a springboard to topics of rich mathematical content and of interest to you.

However, some of the best projects that we have seen have arisen from quite simple questions such as:

Is it true that petrol prices go up at the weekend?

OR

Will a coin that is spun on its edge fall equally heads and tails?

Try to avoid questions that are of no interest to you personally. If you are an artist, consider looking at the mathematics of perspective. If you are keen on a sport such as sailing, look at the mathematics of navigation.

Equally, avoid questions that are pointless. Collecting the sizes of the feet of a group of students is pointless. However, imagining that you run a shoe shop and want to know how many of each size of shoe to stock is not!

So, try to pick a simple, relevant question that is in an area that interests you. Then discuss your choice with your teacher.

Of course, your project will also be assessed and graded. So, it is important that you are familiar with the '**Internal assessment criteria**' that will be used to grade your project work. There are seven criteria that will be used to assess your project work. These are based on:

A. **Introduction**

B. **Information/measurement**

C. **Mathematical process**

D. **Interpretation of results**

E. **Validity**

F. **Structure and communication**

G. **Commitment**

Your teacher should provide you with the details of the criteria (i.e., a detailed statement of the Achievemnet level for each criteria). Then, as you proceed with your project work, refer to the criteria and ensure that you are addressing them

3.1.2 WHAT WILL DEFINE A GOOD DECISION?

This will depend heavily on your choice of problem and may be as simple as 'the ship will not run aground', but will probably be complicated by a requirement such as 'will be cheaper than my competitor's ship whilst still not running aground'. Do not get over complicated here as you may make the next step too difficult!

3.1.3 COLLECT DATA

If you want real data (and we suggest that you do!), the internet is an excellent starting point. Our three case studies give examples in which data came from this source as well as other possibilities. There is nothing wrong with collecting your own data. Remember, however, that the data that you collect must be relevant to the problem you are working on.

Also, there are some dangers to the collection of data. Here are a few:

1. It is a general principle that the process of collecting data affects the results that you get. A simple question such as

"How many people are in the room at the moment?"

may seem to be easy to answer exactly.
However, do you know if there is a person hiding in a cupboard?

2. Measurement also affects the quantity that it is trying to measure.

The act of putting a cold thermometer into a hot cup of coffee lowers its temperature. The only way to prevent this is to know the temperature of the coffee and to heat the thermometer to that before putting it into the cup.

But, if we know the temperature to start with, why are we wasting time measuring it?

The size of this error depends on a variety of factors.
It will be very small if we use a thermometer to measure the temperature of a swimming pool, particularly if we guess that it is likely to be about 25°C and warm the thermometer up to this temperature before using it.

Of course, in the case of the swimming pool the 'cooling error' will be much less than the other main source of error: instrument inaccuracy. Thermometers just cannot measure temperature exactly. Some are better than others. The thermometer shown just cannot be relied on to measure to better than 1°C. More accurate instruments are more expensive and none are EXACT.

3. Users can also introduce errors by misreading an instrument.

The photograph shows a view of the thermometer from above. It suggests that the coffee is at about 53°C. However, the scale lies behind the liquid column which contains the possibility of a PARRALAX error.

Parralax error is illustrated next (the diagram exaggerates the problem).

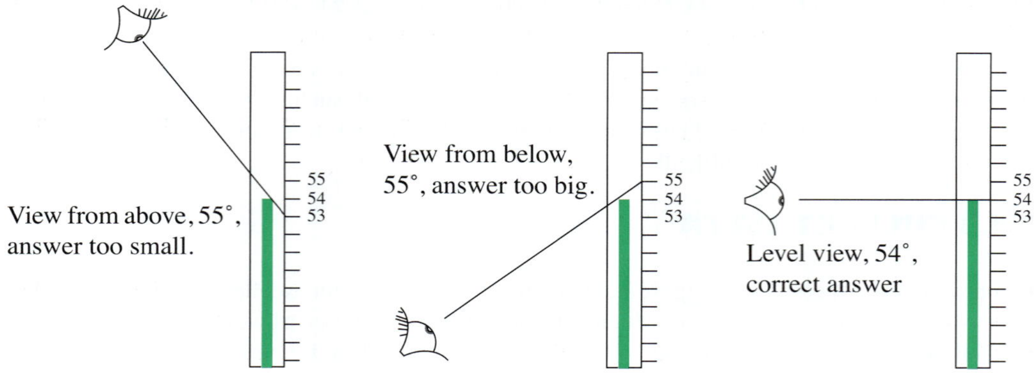

The second of our three project ideas measures the position of an object using a fairly crude method that is subject to the parallax error. It would be important to recognise that this limits the accuracy of the measurements made.

4. Problems can also arise when we ask people questions. There are two main reasons for this. The first is the possibility that people may lie to us.
Suppose you are approached by a researcher in the street and asked,

"*Did you watch the 9 o'clock news on ABC last night*?",

you will probably tell the truth.

If, however, this stranger asks:

"*How much do you weigh*?" or "*How many friends do you have*?",

you may well be less likely to tell the truth by 'massaging' the answer to be closer to what you would like it to be.

The other problem is that your questions may inadvertently affect the answers people give. The question:

"*Will you be going to cheer on the school basketball team in their final on Saturday*?"

may well get a different answer from the question:

"*Will you be going to the Rock-Concert or the basketball on Saturday*?".

5. Sequences of questions can also affect answers:
Suppose you are leading up to the question:

"*Will you be voting for President Clover in the election on Saturday*?"

and you lead up to this question with:

"*Are you aware that President Clover has increased the health care budget by 8%*?"
"*Do you approve of President Clover's new measures to protect the environment*?"
"*Will you be voting for President Clover in the election on Saturday*?"

You will probably get a higher proportion of positive replies than if you use this sequence:

"*Are you aware that President Clover has decreased the police budget by 8%*?"
"*Do you approve of President Clover's frequent trips to resorts in the Caribbean*?"
"*Will you be voting for President Clover in the election on Saturday*?"

If you intend to use a questionnaire, you should be aware of these problems and your report should underline the steps that you have taken to minimise their effects.

Many questionnaires include 'lie detector questions'. One of the best of these is: "Have you ever told a lie?" Would you trust anyone who replied "no" to this question?

Many investigators make use of information derived from the internet. Whilst much of this has been collected by professional researchers who are skilled at minimising the problems described above, the information should not be treated as completely reliable.

3.1.4 ANALYSING YOUR DATA

Following this, you should use mathematics to analyse your data. Analyse means 'pull apart and find how it works' . This is a mathematics project and you should use mathematics contained in the course. Your teacher should help you determine if your problem is beyond the scope of the course.

Remember to **relate the mathematics that you use to your original question**. If you are looking at the question, "Do petrol prices go up at the weekend?", you will have already collected information on 'petrol prices' and you will have noted that since you, presumably, have not visited every petrol retailer in the World, there are limits to the validity of your data. What mathematics is it appropriate to use here? There is a temptation to, for example, work out the average price as this is reasonably easy. But does this help you address your question? Far more relevant would be a price/time graph with the weekends clearly identified. Then, you can start to look at whether you have a pattern of steady increase (a trend) or whether the price is stable and 'spikes' at the weekend (or has both a trend and spiking).

It is also important to know what your mathematical results are telling you. If you, for example, calculate a trend line for petrol price data, make sure that you have understood the implications of the calculation before drawing any conclusions from it.

3.1.5 CONCLUSION

Finally, you should reach a conclusion, related to the question that forms the subject of your investigation. This should be evaluated. In some respects, this is the hardest part of the project.

What do we mean by '**evaluation**'? Some decisions have a right answer. If you are playing 'noughts and crosses' there is always a best move that you should make, whatever the game situation. Chess, and most of life, is different. Some moves are better than others but we do not currently (and probably never will) know the best move in every situation. Your project will probably be like this. Your conclusion may be wrong for a variety of reasons including incomplete information. Evaluation means that you have looked at these possibilities!

Try also to avoid looking for the data to confirm your previous opinion. If you ask a motorist "Do petrol prices go up at the weekend?", you may well get the answer "Sure, everyone knows that!". This means that, if your evidence points to the truth being that prices do not go up at the weekend, you may feel under pressure to go along with what everyone thinks. That is not the scientific method! Whilst you will not be held up to public ridicule because you have written something in a school based investigation that goes against public perceptions, many scientists have been vilified and occasionally imprisoned for insisting on the truth of their (un-popular) findings – the fate of Galileo Galilei (1564–1642) is the best-known case in point.

We will now look at three case studies to expand these ideas. We have taken some ideas further than others and none can be considered as even approaching complete. They are to be taken as skeletons on which a good report might be built!

3.2 APPROACHING YOUR PROJECT

3.2.1 IDEA 1: MONTE CARLO METHODS

(Background theory: Probability).

If you type 'Monte Carlo' into an internet search engine, you will get a lot of information about the small and glamorous Mediterranean state that is almost synonymous with 'high rolling'. The third site listed by my search engine was a mathematical paper on 'Monte Carlo Methods'.

The name is no accident as it is a mathematical method based on gambling.

1. THE QUESTION.

The first question was raised by looking at this aerial photograph. It emphasises that most real objects are not bounded by straight lines, circles etc. as they are in mathematics texts. This coastline is very 'crinkly'.

This suggests the problem: how do we find the areas of irregular shapes?

2. A GOOD ANSWER?

This will be a numerical answer to the problem. Given that there is no 'correct' answer with which we can compare ours, we will be looking at an approximation. Try to set a realistic target such as 'the nearest 100 km^2'.

3. COLLECT INFORMATION.

Even though Monte Carlo methods are not specifically a part of the course, they depend on the concepts of probability that you study. They have many variants. At its simplest, the Monte Carlo method is like throwing darts at a map whilst wearing a blind-fold. You will obviously need to research what this might have to do with area!

It would also be a good idea to set yourself the task of finding the area of a real island such as Hokkaido. You will need to obtain an accurate map. As part of your report, you should note the accuracy and scale of the map. A mariner's chart is made to very different standards from those you see on tourist brochures that encourage you to 'Holiday in Happy Hokkaido' (the ones that have cartoon fish leaping from cartoon lakes and cartoon waves rolling onto cartoon beaches).

4. ANALYSIS.

The simplest version of the method involves copying a scale version of your island into a rectangle of known dimensions.

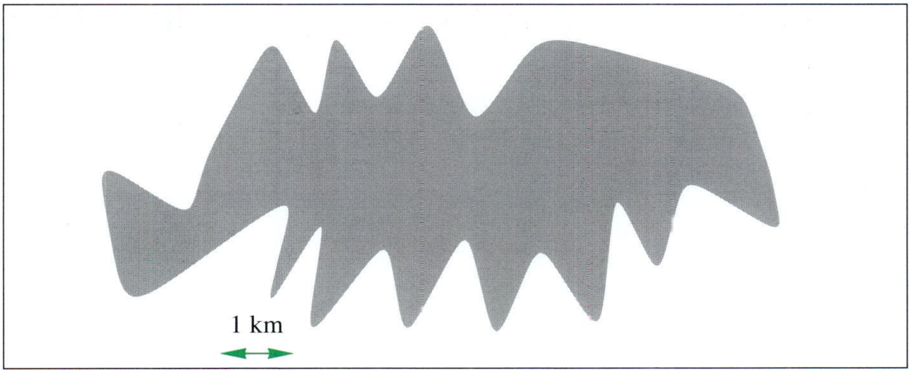

This rectangle measures 12 km by 5 km and so has an area of 60 km^2. But what proportion of this is inside the island?

This is where we don our mathematical blindfold and start throwing darts!

To blindfold ourselves, we can use the random number generator of a computer or calculator. These vary but, commonly, produce a random (what does this mean?) number between 0 &1. If we multiply a random number in this range by 12, we will get a random number in between 0 & 12. If we take a second random number and multiply it by 5, we will get a random number in the range 0 to 5. If we now view these two numbers as a coordinate pair on the map, we will have, effectively, thrown a blind dart at the map.

Here are ten random numbers and the five 'darts' that they throw at the map:

Random 1	Random 2	12×Random1	5×Random2
0.723018492	0.529733397	8.68	2.65
0.310504197	0.673043069	3.73	3.37
0.167302964	0.293733395	2.01	1.47
0.96699594	0.078452756	11.60	0.39
0.695841987	0.075482986	8.35	0.38

When these are transferred to the map (with the origin being at the bottom left of the rectangle) they are:

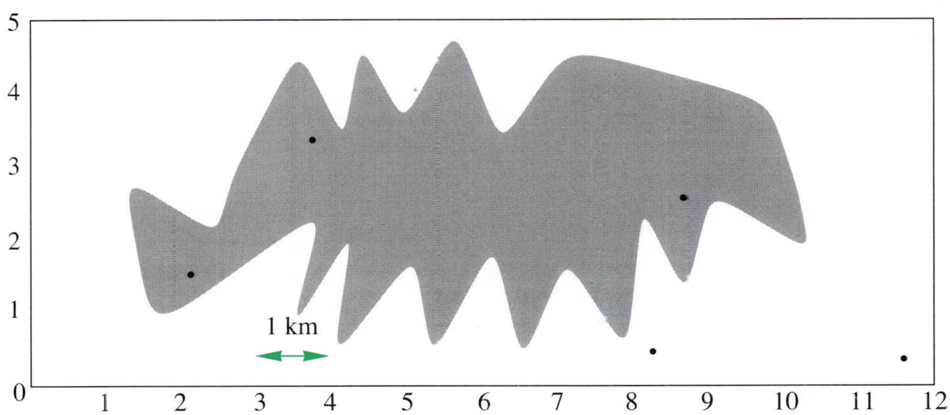

As things stand, 3 out of the 5 'darts' have landed in the island. This suggests that its area is three fifths of that of the rectangle or $\frac{3}{5} \times 12 = 7.2\ \text{km}^2$.

However, if one of the 'darts' that 'hit' was recorded as a 'miss', the area estimate becomes $\frac{2}{5} \times 12 = 4.8\ \text{km}^2$. As this is quite different from the previous answer, we cannot be very confident about our current measure of the area.

How do we improve matters? More darts would seem to be indicated, but just how many do we need to achieve accuracy to a particular level (such as 'nearest 1 km^2')? This is where the 'evaluation' stage of the investigation can begin.

5. EVALUATION AND CONCLUSION.

There are several ways to go about this. If you read further into this subject, you should discover that it is possible to estimate the accuracy of an answer obtained by a Monte Carlo method by using a formula. However, if you calculate your area every 10 'darts' and plot a graph of area estimate against number of darts, you should see that the result starts to stabilise after early fluctuations. The extent of the fluctuations in the estimate can give you a measure of accuracy:

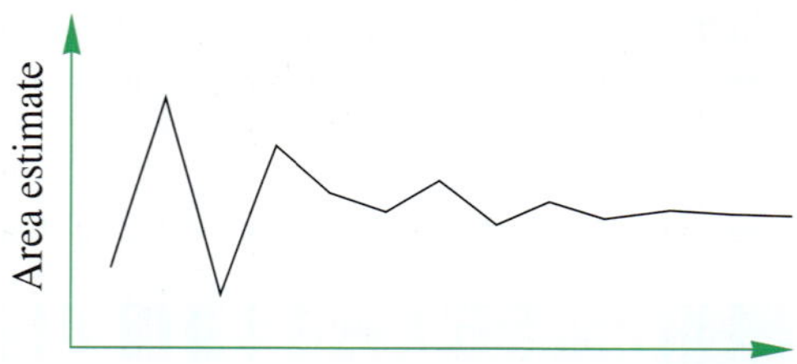

Whatever method you use to assess the accuracy of your method, you should comment on this in your report.

3.2.2 IDEA 2: DAMPED OSCILLATIONS

(Background theory: Trigonometric and Exponential Functions and Calculus).

1. THE QUESTION

This idea came from a very rough ride on the back of a truck along a dirt road! Why do we get a much smoother ride in a modern saloon car than on the back of a truck or tractor?

So why do we 'bounce around' more in some vehicles than others?

There are several answers to this question, not least the different surfaces over which they usually travel! One of the keys is the quality of the wheel springs and what are known as shock absorbers.

The idea also came from the slightly worrying feelings that you get in tall buildings when they 'sway' and in aircraft when the wings 'flap'.

Real vehicle suspensions, buildings and aircraft wings are complex structures that are difficult to observe. What we decided to do was make our measurements on a 'home-built' structure that had similarities to the 'real thing'.

In setting up an experiment to investigate this, we have decided to simplify the problem to look at a simple damped vibrating system using a piece of stiff garden wire, a nut and a digital video camera. The structure is shown in the photograph. Apart from the camera, we used items that were to hand in a domestic garage.

The intention is to move the nut sideways and use pictures taken from the video recording and the grid to measure the position of the nut as it varies over time. To assist in measuring this position, we made and printed a 1 cm square grid using a computer and have stuck a paper arrow on the back of the nut angled backwards to minimise the parallax error (discussed earlier).

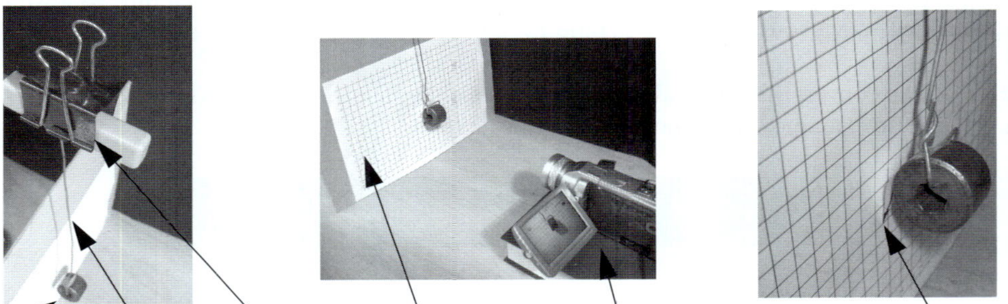

2. DEFINE THE OBJECTIVES AND CRITERIA FOR SUCCESS.

It is our hope that, working from measurements from the 'structure', we can set up a mathematical model that can be used to predict how it will perform under conditions different from the original experiment (a different initial deflection).

3. COLLECT THE DATA.

The complete data set is a digital video.

When the data is imported into a computer editing package, a timing scale is added. As we examined the video frame by frame, we noticed that the times of successive frames were: 0:00, 0:01, 0:02 and so on to 00:24. This was followed by 01:00, 01:00 etc.

To make sure that we had understood this time scale, we took a short video of an electronic stopwatch. Two frames from this video are shown:

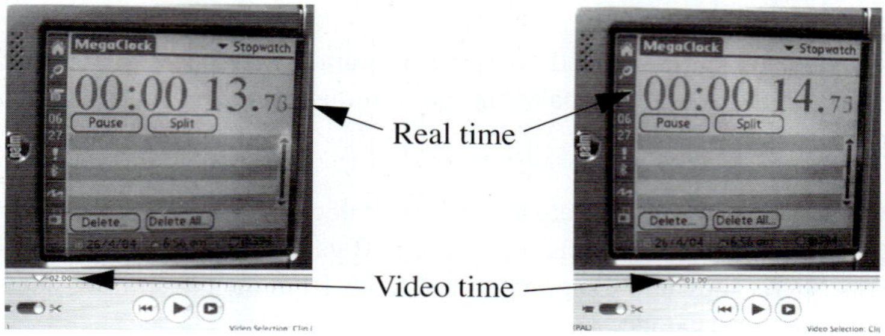

These suggest that when the video camera time goes from 2:00 to 3.00, the real time has advanced from 13.76 to 14.75. This is means that 0:25 video units correspond to 1 second so that the frames are made at the rate of 25 per second.

To convert from a video time to a real time, we need to first express the video time as a whole number. Thus, a video time of 0:15 is time 15, 2:00 is 2×25 = 50 and 3:17 is 3×25+17 = 92.

Next, we can convert these to real times by recognising that they are recorded in 25ths of a second. These examples are summarised in this table:

Video time from computer	Video time as a whole number	Real time (seconds)
0:15	15	$\frac{15}{25} = 0.6$
2:00	50	$\frac{50}{25} = 2$
3:17	92	$\frac{92}{25} = 3.68$

Note that, even before we have started to look at the motion of the structure, we have had to use mathematics to make sure that we have correctly understood the data that we have recorded.

We can now move on to looking at some of the frames from the video of the vibrating structure:

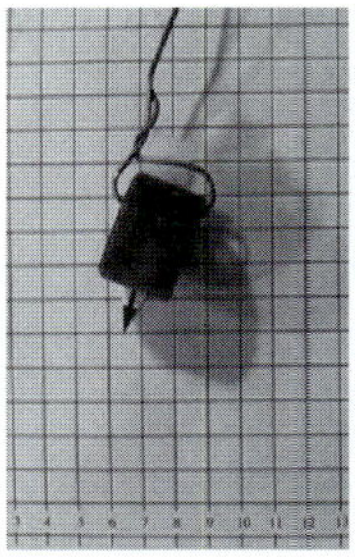

Video time: 00:08
Real time: 0.32 sec
Deflection: 6.5 cm

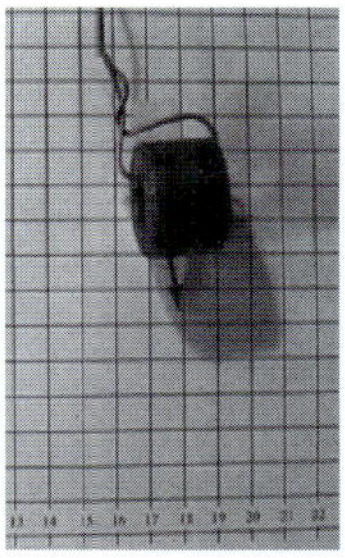

Video time: 00:15
Real time: 0.60 sec
Deflection: 17.9 cm

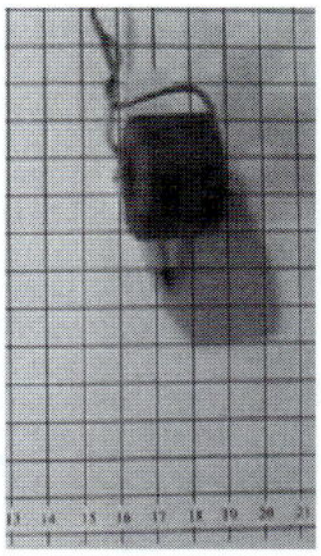

Video time: 01:05
Real time: 1.20 sec
Deflection: 17.4 cm

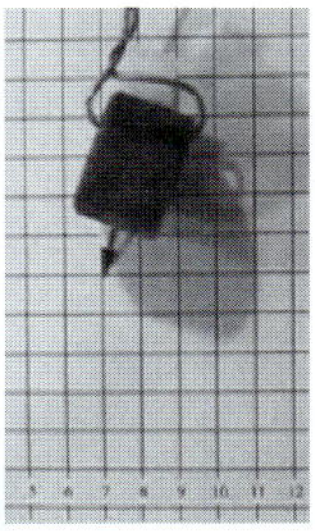

Video time: 01:14
Real time: 1.56 sec
Deflection: 7.0 cm

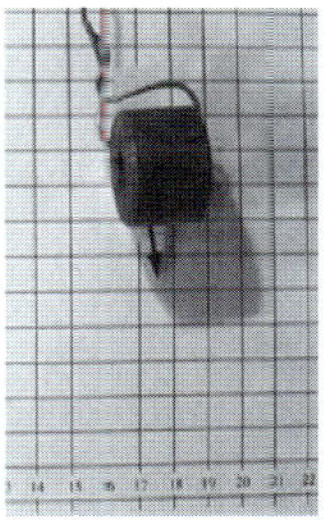

Video time: 01.21
Real time: 1.85 sec
Deflection: 17.5 cm

Video time	Real time (sec)	Deflection (cm)
8	0.32	6.5
9	0.36	7.1
10	0.4	8.8
11	0.44	10.9
12	0.48	13.2
13	0.52	15.3
14	0.56	17.0
15	0.60	17.9
16	0.64	17.9
17	0.68	16.9
18	0.72	15.4
19	0.76	13.3
20	0.80	10.9
21	0.84	8.8
22	0.88	7.4
23	0.92	6.7
24	0.96	6.9
25	1.00	8.1
26	1.04	9.9
27	1.08	12.2
28	1.12	14.4
29	1.16	16.2
30	1.20	17.4

It is an important part of the data collection section of an investigation that you should report on the limitations of your method. In this case, we have used a 1 cm grid and have estimated to one decimal place. It is unlikely that we have done better than an accuracy level of ±0.2 cm, particularly in the frames (such as the one at right) where there is 'motion blur'. Also, there is the effect of parallax, although we have tried to minimise this. Note also that we are ignoring the fact that the arrow moves vertically as well as horizontally (it moves in the approximate arc of a circle). It should also be noted that we are modelling the position of the point of the arrow, not the nut.

We have included an example of a practical investigation to illustrate the fact that these may not be as difficult as they at first appear. However, we strongly advise that you discuss a proposal of this sort with your teacher before undertaking any such experiment. There are many reasons for this, the main ones being practical difficulties and safety. We have also seen proposals for this type of investigation which have been too complex to be carried out in a reasonable time frame.

It is important to avoid investigations that expose people or animals to danger or may produce anti-social side-effects such as loud noise etc.

4. ANALYSIS

As a first step, we have entered the data given in the table above onto a spreadsheet. This enables us to generate a graph of the data set for the early part of the motion.
This is shown below:

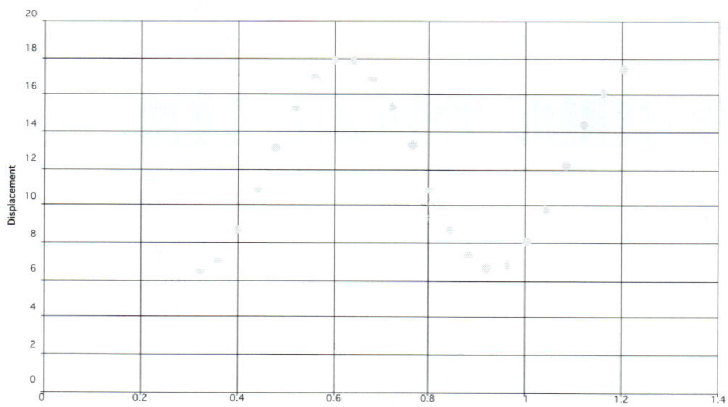

A major objective is to find a mathematical function that will model the motion and which can be used to make predictions about its future. The first step is to note that the graph has the appearance of a sine or cosine graph and we shall, therefore, be looking for a model of this type.

The key features of such a model are amplitude and period. We can add lines to the graph that will enable us to determine approximate values for these:

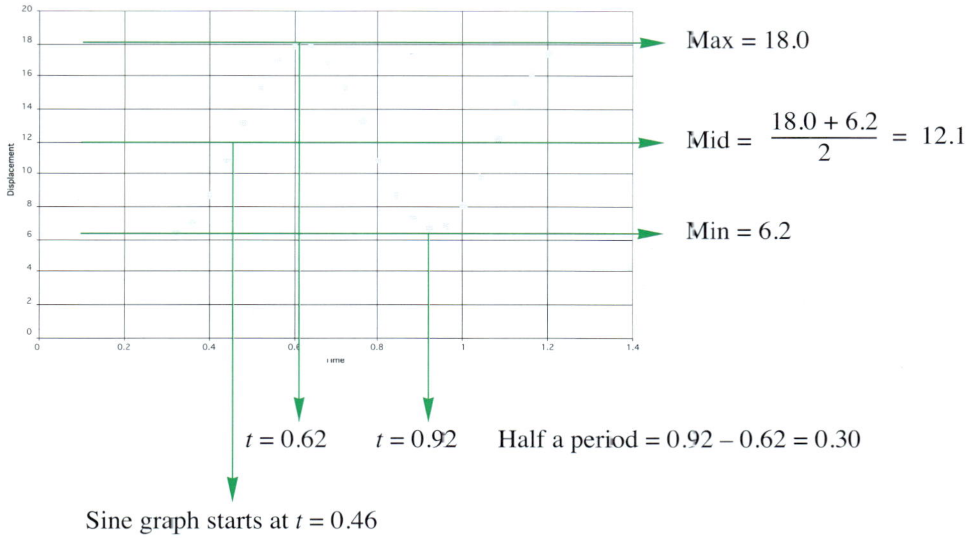

An appropriate modelling function will need to have:

- an amplitude of $\frac{18.0 - 6.2}{2} = 5.9$
- a period of 0.6 which implies that $n = \frac{2\pi}{0.6} \approx 10.5$
- translated 0.46 to the right and 12.1 up.

This suggests that an appropriate modelling function is:

$$D = 5.9 \times \sin(10.5(t - 0.46)) + 12.2 .$$

The first stage in evaluating the model is to see how it fits the existing data set. To do this, we added a column to the spreadsheet that used this function. The resulting graph shows both the data and the model:

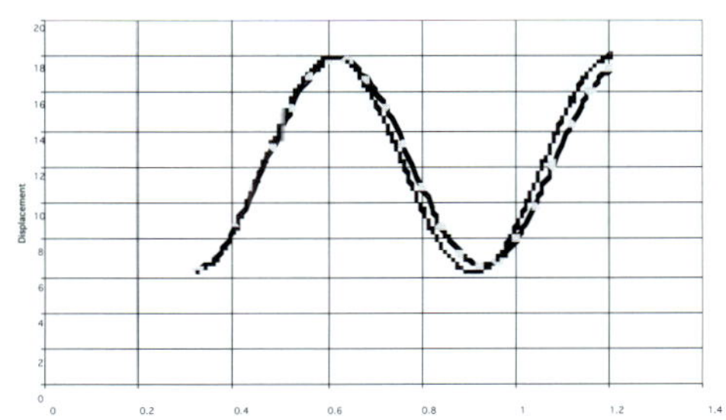

The modelling function (dotted line) seems to be working well but has a period that is a bit too small. After some experimentation, we found that $D = 5.9 \times \sin(10.1(t - 0.46)) + 12.2$ fits the data better:

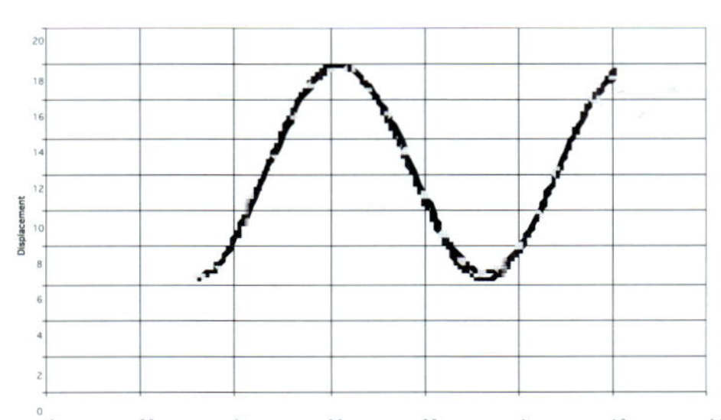

As a further preliminary evaluation, we used the model to predict what will happen in the near future (after $t = 1.2$). For example, the model predicts for:

Time 1.56 (video frame: 01.14) $D = 5.9 \times \sin(10.5(1.56 - 0.46)) + 12.2 = 7.2$

Time 1.85 (video frame: 01.21) $D = 5.9 \times \sin(10.5(1.85 - 0.46)) + 12.2 = 17.5$

The appropriate frames from the video are:

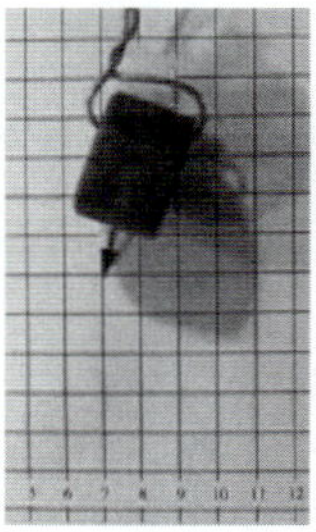

Video time: 01:14
Real time: 1.56 sec
Deflection: 7.0 cm

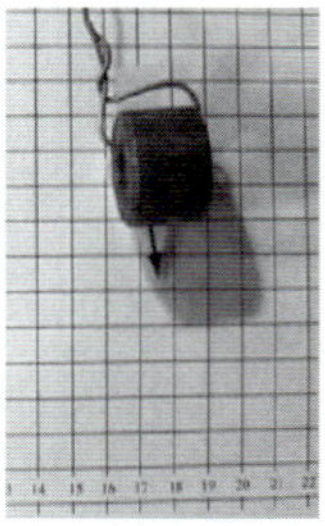

Video time: 01.21
Real time: 1.85 sec
Deflection: 17.5 cm

This seems to be a good result and confirms that we are on the right track with the model.

The model has yet to take into account the fact that the motion dies down to nothing after a minute or so. How are we to model this?

Looking at successive frames that record the rightmost deflections produced the following data set:

Video time	Real time (s)	Deflection	Grid Position
16	0.64	5.8	17.9
31	1.24	5.7	17.8
41	1.64	5.4	17.5
62	2.48	5.3	17.4
77	3.08	5.2	17.3
93	3.72	5.1	17.2
108	4.32	4.9	17
124	4.96	4.8	16.9
139	5.56	4.7	16.8
155	6.2	4.5	16.6
170	6.8	4.5	16.6
185	7.4	4.3	16.4
201	8.04	4.2	16.3
216	8.64	4.1	16.2
232	9.28	4	16.1
247	9.88	3.9	16
262	10.48	3.9	16
278	11.12	3.8	15.9
293	11.72	3.7	15.8

Video time	Real time (s)	Deflection	Grid Position
309	12.36	3.5	15.6
324	12.96	3.5	15.6
339	13.56	3.4	15.5
355	14.2	3.3	15.4
370	14.8	3.2	15.3
385	15.4	3.1	15.2
401	16.04	3	15.1
416	16.64	3	15.1
432	17.28	2.9	15
447	17.88	2.9	15
462	18.48	2.8	14.9
478	19.12	2.7	14.8
508	20.32	2.6	14.7
570	22.8	2.4	14.5
631	25.24	2.2	14.3
708	28.32	2	14.1
784	31.36	1.8	13.9
846	33.84	1.7	13.8
1029	41.16	1.3	13.4
1213	48.52	1	13.1
1350	54	0.9	13
1416	56.64	0.8	12.9

Again, when looking for a model for data, we advise that you plot the associated graph(s). The first is the grid position against time (columns 4 & 1) from the above table):

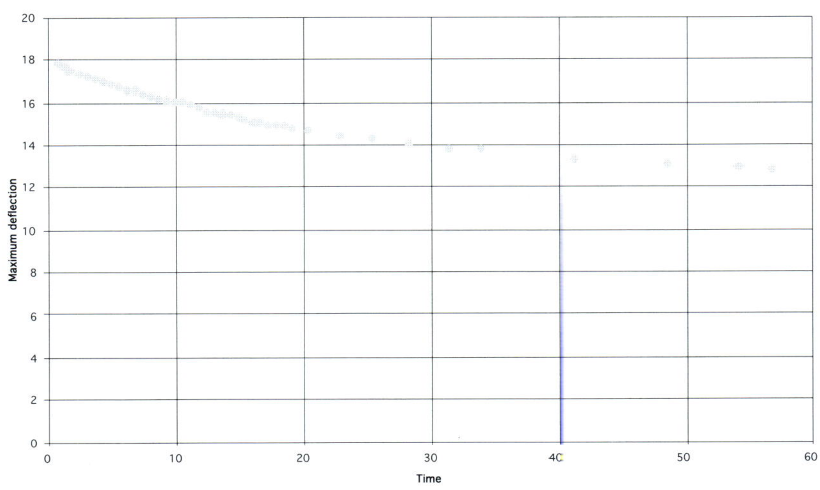

As we found earlier, the central position of the arrow is at about 12.2 cm on the grid. The following graph shows the deflection from this central position as is varies with time (columns 3 & 1) from the above table):

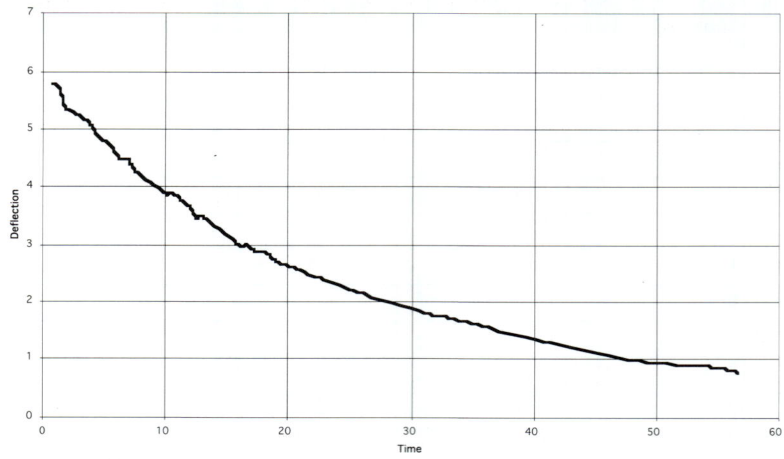

As before, we ask, have we seen a graph of this shape before? Well, it does look somewhat like the graph of $y = e^{-x}$. This means that we should look for a modelling function of the form: $D = Ae^{-kt}$ where D is the maximum deflection at time t. The deflection at time zero is not shown. However, projecting the graph to the left gives a value of about 6 cm. This means that the modelling function is of the form: $D = 6e^{-kt}$ since $e^0 = 1$. The only parameter that remains to be determined is k. This fixes the rate at which the deflection decreases. You can find this out by noticing that the deflection halves (from 6 cm to 3 cm) in about 17 seconds. This suggests that a good value for k might be $k = \left(-\frac{1}{17}\right)\log\left(\frac{1}{2}\right) = 0.0408$ and that the appropriate modelling function for the deflection is: $D = 6e^{(-0.0408t)}$. This model needs to be evaluated by comparing it with the data. This graph shows the raw data and the model.

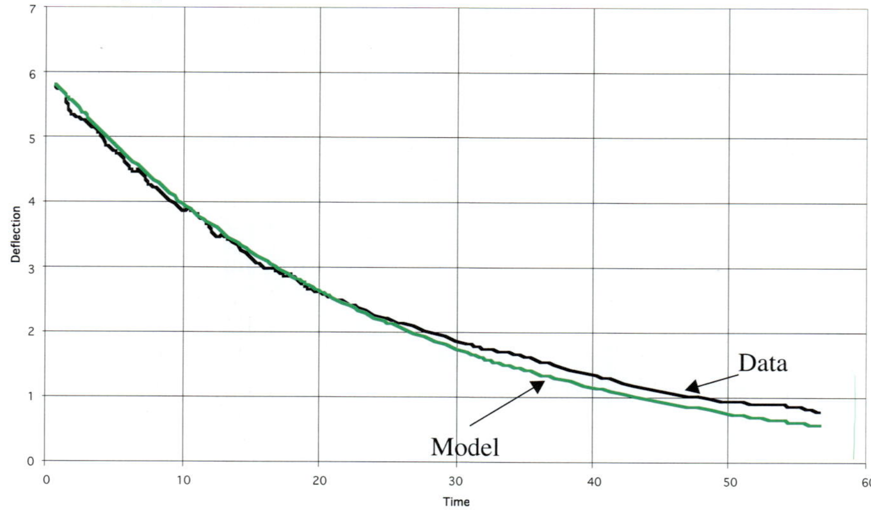

This model seems to fit the data quite well, but it is a bit too big in the early stages and a bit too small later on. How would you change the model to improve the fit?

Next, we can put the two aspects of the motion together to produce a function that models the motion that we recorded on video-tape. The result is the product of the functions that model the periodicity and the damping:

$$D = 6e^{(-0.0408t)} \times \sin(10.1(t - 0.46)) + 12.2$$

Damping part Periodic part 'Zero point'

To see the way the model predicts the motion, we use a graphing program:

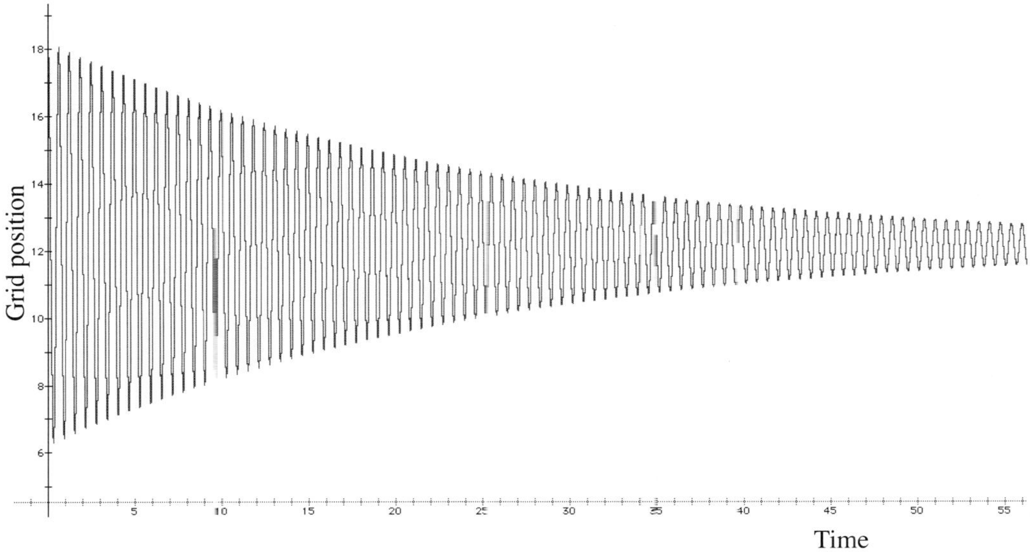

The major purpose of going to the trouble of calculating a model such as this, is to predict what will happen to a 'structure' under a variety of circumstances. Our 'structure' is a length of garden wire and a rusty nut and it is not really important to know what will happen if it is given a big initial deflection as no-one cares if it breaks apart. However, we are taking a very simplified look at what engineers do when they are designing sky-scrapers, aircraft wings etc. These structures have to be able to withstand stresses such as storms and earthquakes in the case of buildings and turbulence in the case of aircraft wings whilst not being unnecessarily large and heavy (for reasons of cost). Most of us consider it important that these structures do not 'fall apart'.

Here is an early frame from a second tape of the structure under a different deflection. It is the first time that the nut moves to its rightmost deflection:

Video time: 0
Real time: 0 sec
Deflection: 19.6 cm

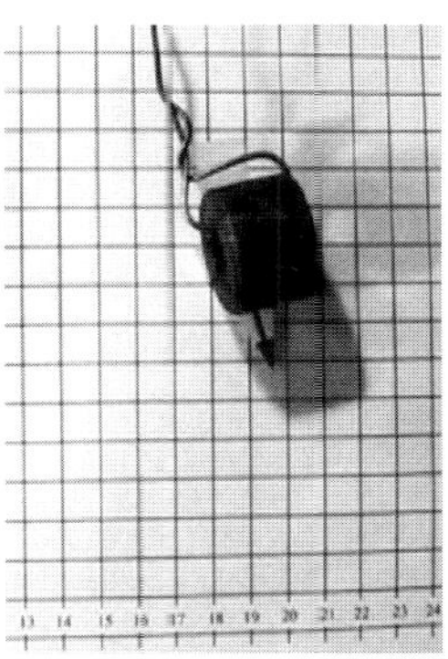

There is a new start time (one quarter of a period = 0.15 sec) and a new initial deflection: $19.6 - 12.2 = 7.4$.

This means that the new modelling function is: $D = 7.4e^{(-0.0408t)} \times \sin(10.1(t+0.15)) + 12.2$. This is found by replacing the time translation (we are now starting at time 0.15 sec) and changing the amplitude from 6 to 7.4.

The important test is how well the model works in predicting what will happen next. Here is frame 22, real time $\frac{22}{25} = 0.88$ sec. The video frame shows a deflection of about 6.1.

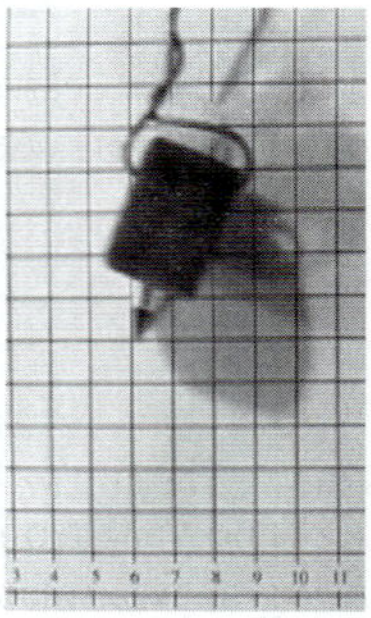

The model predicts:

$$D = 7.4e^{(-0.0408(0.88))} \times \sin(10.1(0.88+0.15)) + 12.2$$
$$= 6.3$$

This is a good result as, remember, we never claimed to be able to make the measurements to better than ±0.2 cm.

How well will the model perform much later in the motion? Here is a second test frame taken at time 12 seconds. The photograph shows a deflection of 9.1.

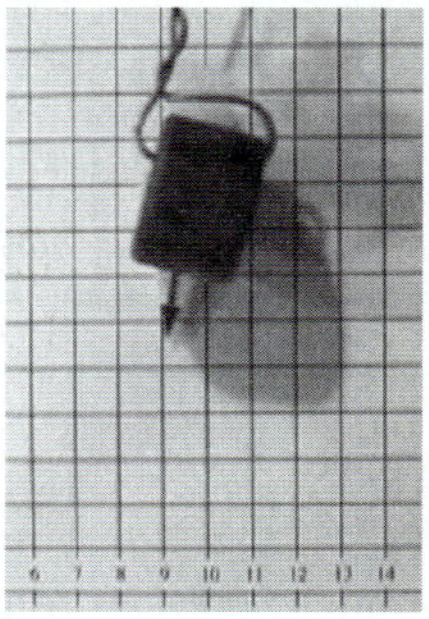

The model predicts:

$$D = 7.4e^{(-0.0408(12))} \times \sin(10.1(12+0.15)) + 12.2$$
$$= 11.3$$

This is a much larger error than for the earlier time. The reason is most probably our assumption that the period of the motion has not been altered by the change in initial deflection.

5. EVALUATION AND CONCLUSION

We have succeeded in taking measurements from a real structure that was set in motion. We have used mathematics to find a model that can be used to predict the behaviour of the structure in circumstances different from those that we actually tried. The model performed much better in the short term than later in the motion. A good investigation will explain that this is much more likely to be due to errors in the trigonometric part of the model than in the exponential part.

3.2.3 IDEA 3: STATISTICS (WEIGHTS IN LIFTS AND TRANSPORT, CHAPTERS 20 & 21)

1. THE QUESTION.

The question here was raised by this safety plaque in an elevator (lift). The manufacturers seem to be assuming an average weight of

$\frac{2030}{30} \approx 68$ kg per person.

Further research on the internet indicated that airlines assume standard weights for passengers. These are under review (people are getting heavier) and vary from country to country (as do people) but at the time of writing approximate values were:

Seating capacity	Adult Male (kg)	Adult Female (kg)	Teenage Male (kg)	Teenage Female (kg)	Child (kg)	Infant (kg)
7-9	86	71	65	56	44	17
40-59	83	68	63	57	42	16
100-149	82	66.9	61.1	55.2	41	16
300-499	81.4	66.3	60.6	54.8	41	16

There seem to be many questions that could be followed here and it is important to define what you choose to investigate.

If you accept the airline figures, what chance is there that the elevator will be over-loaded?

If you accept the implied elevator figures, what chance is there that an aeroplane will be over-loaded?

If you accept none of these figures, you will need to get data on the actual weights of people. This can be a real problem as 'body weight' is very definitely in the area of 'personal sensitivity'. Ask someone their weight and they may well tell you their preferred weight rather than their actual weight. Ask them to get on a set of bathroom scales and they will probably refuse. This seems to be a line of enquiry best avoided!

2. DEFINE THE PROBLEM.

After a variety of internet searches on 'body weight', we failed to come up with any definitive data on this distribution. The outcome of a search for useful data on human weight distributions was to produce a flood of offers for diet programs! You may, of course, be more successful in forming your search questions than were the authors!

This became an investigation in which 'definition' emerged as a major feature.

We have airline data on the 'standard weights' of various classes of humans. We also have a safety plaque that indicates the safe loading of elevators.

One option is to accept the most conservative weights assumed by airlines (the row for 7-9 seat aeroplanes). Why is this the best assumption and why do the airlines assume that everyone gets lighter when they get onto large aeroplanes? There is a good investigation in this question alone!

However, we will accept the most pessimistic data on weights assumed by the airlines and ask ourselves:

What are the chances (mathematicians call this probability) that the elevator whose plaque was shown earlier will accept a load of 30 persons or fewer and yet still be overloaded?

This is now a clearly defined problem. We were not able to arrive at this definition until we looked at the available data. This may well be the case with the investigation that you choose and underlines the fact that the stages identified earlier in the chapter may well over-lap.

3. COLLECT DATA.

Our problem definition has now assumed that we accept these data on human body weight:

Adult Male (kg)	Adult Female (kg)	Teenage Male (kg)	Teenage Female (kg)	Child (kg)	Infant (kg)
86	71	65	56	44	17

The elevator is restricted to 30 persons and a weight of 2030 kg.

If 30 adult males get into the elevator they will weigh: $30 \times 86 = 2580$ kg and the elevator will be overloaded. It appears that the manufacturers are assuming that the normal load will seldom consist of 30 adult males. So what is a normal load?

There are two questions to investigate here:

1. What is the distribution of the numbers of people who enter elevators?

2. What is the distribution of the types of people who enter elevators?

Both these questions can be answered by collecting real data. Observe an elevator and make a frequency distribution of:

The number of people riding in the elevator (not the number who get into it)

AND

The distribution of the types of people (adult females, infants etc. as defined by the weight data) who ride in elevators.

4. ANALYSE THE DATA.

Having collected the data described above, you now need to display it mathematically. This can mean graphs, tables etc.

However, remember that the issue is to look at the probability that the elevator will be overloaded and your analysis should reflect this.

For example, if you observed a family consisting of a mother, father, two female teenagers and an infant riding in the elevator, you can assume that they weigh: $86 + 71 + 2 \times 56 + 17 = 286$ kg which is well inside the weight limit. If, however, you observe a full load of adult males, you will record an overload.

Taking into account both these variables (number of occupants and composition of the load) what is the probability that the elevator will be overloaded?

By the way, you do not need to flee in panic if you find yourself in an elevator full of heavy people. Many have weight sensors that prevent them moving if overloaded. In common with all structures, the designers will have built in a safety factor that they are not keen to advertise. Imagine if the plaque had said '30 persons, but will take 35'. Everyone would work on 35 and it is 'goodbye to the safety factor'. Also, though this will not be a part of a mathematics project, most elevators include a truly ingenious 'fail safe' device that ensures that, even if the cable snaps, the 'cage' will not plummet to the bottom of the shaft.

5. CONCLUSIONS.

This topic has made many assumptions with its data. Your conclusions should reflect on the difficulties that were experienced in acquiring reliable data. These, of course, follow through into the reliance that can be placed on the conclusions.

3.2.4 YES, YES, . . . BUT WHERE DO I START?

The following flowchart highlights the important roles involved in carrying out your project. If you are finding it difficult to start, use this as a guide.

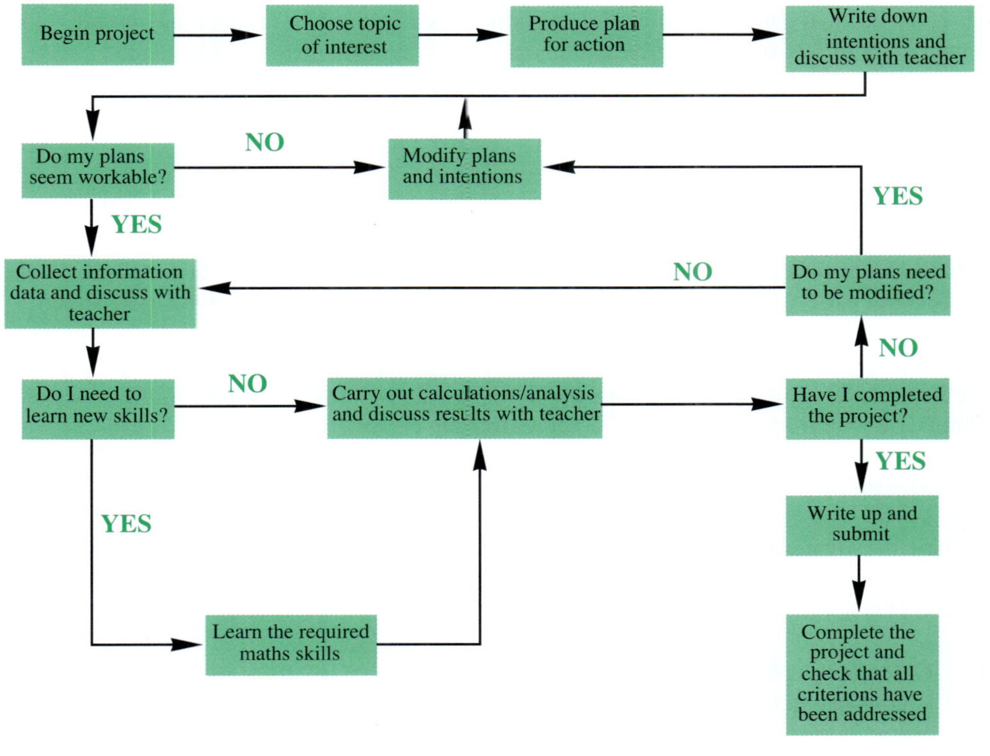

3.3 ASSORTED TOPICS FOR PROJECT WORK

We now provide a number of topics (or themes) that can be used as the embryo of a project. Each project has been carefully selected so that all the criteria used for the internal assessment component of your final grade can be readily addressed.

Each project may require some background knowledge, and so, we provided a (non-exhaustive) list of the relevant chapters that **may** apply to each project. The extent to which these are suitable will depend on the depth with which you carry out your project.

Project 1

THEME: PERIODICITY

Background: Chapters 5, 10, 11, 12, 22, 23

Your project must arise from or incorporate the study of periodicity. You are encouraged to show initiative and be independent in carrying out your project.

1. General Advice

Periodicity is a rich field for mathematical investigation. Many phenomena are periodic. Just consider, for example, the motion of planets, recurring decimals, calendars, gear mechanisms and waves. Such examples offer many opportunities to use equations (for the period), arithmetic (for finding the decimal period of a fraction), trigonometric functions (for modelling waves), geometry (for tilings and decorative patterns) and general mathematical judgement.

The general criteria for assessment should be borne in mind, but here are a few issues which arise specifically in the case of periodicity.

First, you could investigate whether periodicity is present in a specific situation and describe its properties mathematically. Alternatively, you could design and study a mechanism to produce a desired periodic phenomena, for example, a machine, a computer program or a mathematical rule. Second, what mathematics is appropriate ? Algebra, arithmetic, trigonometry, geometry? In some cases a full analysis will not be possible in terms of school mathematics. If so, you may have to approximate or quote appropriate advanced results without proof. Third (or perhaps first), does the problem have a history? Many problems have been solved correctly after a series of attempts, ranging from wildly wrong to approximately correct.

2. Starting points

You may investigate any topic related to the theme. You must discuss your choice of topic and how it relates to the theme with your teacher. The examples below show some starting points for projects. This list is by no means exhaustive.

2.1 **Calendar** - is the calendar exact ? Find the day of the week for a given date in a given year. What about alternative calendars, past and future?

2.2 **Modular arithmetic** - What is it? And how could you use it in investigating periodicity? What is the Chinese remainder Theorem? How does this relate to periodicity?

2.3 **Tides, length of the day, biorhythms.**

2.4 **Economics** - what, if any, periodicity can you find in, for example, the value of the Australian dollar, the stockmarket, the CPI?.

2.5 **Astronomy** - planets, moons, comets, distance of stars, epicycle theories.

2.6 **Waves** - sound, water, light.

2.7 **"Random" sequences** - investigate ways to create such sequences which are never periodic; create rules for repeating number sequences.

2.8 **Periodic patterns in wallpaper, tiling and crystals.**

2.9 **Mechanisms** - investigate mechanisms that produce periodic patterns in materials, carpets, zig-zag and other fancy stitches; in drawings (spirographs, tissajas figures); hammer drills, cogs in a gearing system, music boxes, pianolas.

2.10 **Electric currents and their trigonometric properties.**

2.11 **Sundials** - design a sundial; explain why it works. Where does the sun set?

Project 2

THEME: FRACTALS

Background: Chapters 5, 6, 10, 12

Your project must be based on or incorporate the study of fractals. Some introductory reading is recommended in the general advice below, however you are encouraged to show initiative and be independent in carrying out your project.

1. General advice

The basic concepts of fractals have been summarised in this booklet. The following books are recommended for their general interest: Keith Devlin's Mathematics: The New Golden Age (Pelican 1988), Ian Stewart's The Problems of Mathematics (Oxford 1987) and James Gleick's Chaos (Viking 1987).

The important thing is to have some familiarity with the 'snowflake' curve and to understand the concepts of self-similarity and fractal dimension. Your report should include explanations of these concepts in your own words, and examples of your own which illustrate your explanations.
The mathematical investigation of fractals is likely to involve geometry (length, area, volume), number and algebra (summing infinite series, exponents and logarithms). There is also the opportunity to use computers, particularly if you invent your own fractals.

A likely direction of investigation is in the modelling of nature (which is the reason for most of the current interest in fractals). Here it may also be convenient to use a computer to experiment with different models (though computer print-outs are generally less attractive than pictures carefully drawn by hand).

1. Families of fractals

How does a fractal vary when you vary the rule for making it? Invent examples. Investigate changes in appearance, dimension and so on when the rule is varied. Try including a random element.

2. Geometric properties

Find some properties of the snowflake curve; for example, its length and the area it encloses. Why does it have infinite length? Find out something about the Cantor set, the Sierpinski carpet and/or the Sierpinski sponge.

3. Models of nature

Is the snowflake curve a good model of a coast line? If not, how could it be unproved? Do all coast lines have the same fractal dimension? Investigate or create fractal models of other natural objects; for example, lightning flashes, ferns, the lung and circulatory systems. Can you generate them recursively by simple rules? (Being able to vary rules should be useful here.)

4. Rough and smooth objects

Find examples of smooth, self-similar objects. Do they occur in nature? How rough is the snowflake curve?
(What do tangents have to do with roughness?) How rough should an object be, to be regarded as fractal?

Some background

Fractals: the basic ideas

Fractals are geometric objects having fractional dimension, that is, a dimension between the usual dimensions 1, 2, 3 (for lines, surfaces and solids respectively). This seemingly strange property arises in quite a simple way from basic geometric ideas, which we now describe.

Similarity

Two objects are called similar if they are the same shape, in other words, if one becomes the same as the other under magnification. Figure 1 shows an example

Figure 1

Other examples: any two line segments are similar, any two squares are similar, but a square is not similar to a triangle. Since any segment of a line segment is also a line segment, a line segment is similar to parts of itself. Such 'self-similarity' also occurs in more interesting objects. Here is one.

The snowflake curve

The Koch snowflake (sometimes called the Koch island) is the union of infinitely many triangles, added in stages as indicated in Figure 2.

Figure 2

The rule for going from one stage to the next is: add an equilateral triangle to the middle third of each segment of the boundary. The Koch snowflake is interesting in itself (for example, can you find its area?) but we shall look only at its boundary, called the snowflake curve.

The snowflake curve is the limit of polygons obtained from an equilateral triangle by repeatedly replacing each line segment by a 'spiked line segment' as shown in Figure 3.

Figure 3

At the sixth stage a line segment becomes the polygon shown in Figure 4, which already gives a good impression of the snowflake curve.

Figure 4

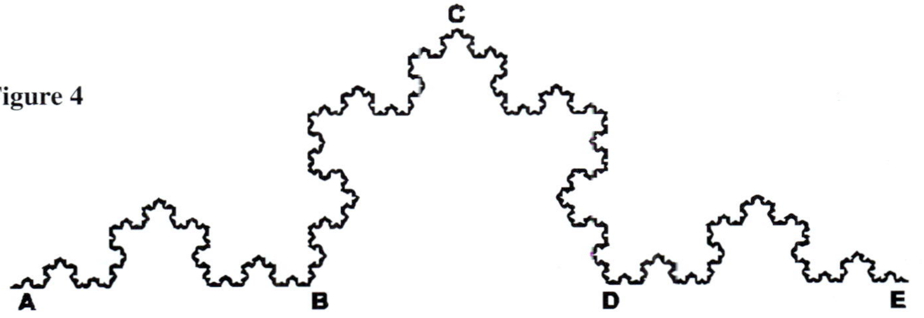

Self-similarity in the snowflake curve

Hold Figure 4 at arm's length so that it begins to blur. Now you can imagine that what you see is not just the sixth stage polygon, but a segment AE of the genuine snowflake curve. If you were to magnify part of it to get a better view - say by magnifying AB by 3, to the size of the whole segment AE - then you would see exactly the same thing! Indeed, it is clear, from the way the snowflake curve is constructed, that any part which begins as a line segment is similar to any other part which begins as a line segment.

Notice that assembling the four equal parts AB, BC, CD, DE, of the snowflake curve gives AE, which is three times the size of AB. In other words, assembling four copies of a segment of the snowflake curve magnifies the segment by 3. This is very strange: just how strange can be seen by looking again at ordinary geometric objects.

Dimension

The dimension of a geometric object is usually defined to be the number of coordinates needed to specify the position of a point in it. Thus the dimensions of a line segment, square and cube are 1,2,3, respectively (Figure 5).

Figure 5

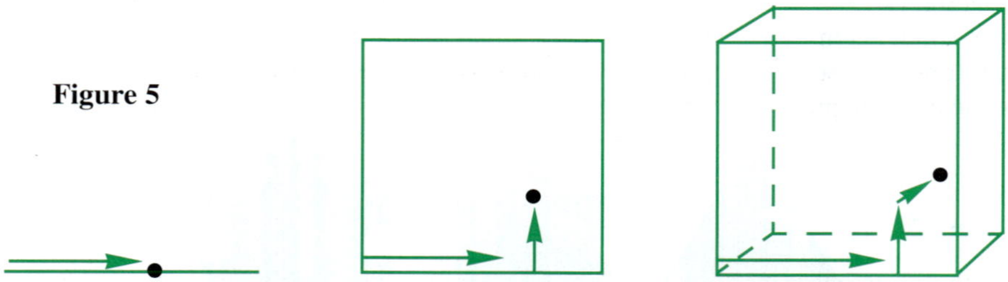

However, the numbers 1,2,3, also turn up as exponents when we calculate how many copies of an object have to be assembled to magnify the object by a given factor. This hints at a different definition of dimension, which we shall arrive at below. Consider, for example, how many copies of a line segment, square and cube have to be assembled to magnify by 3 (that is, to triple all lengths).

$3 = 3^1$ line segments  magnified by 3

$9 = 3^2$ squares 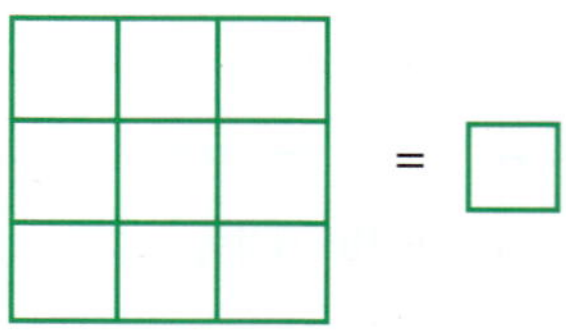 magnified by 3

$27 = 3^3$ cubes 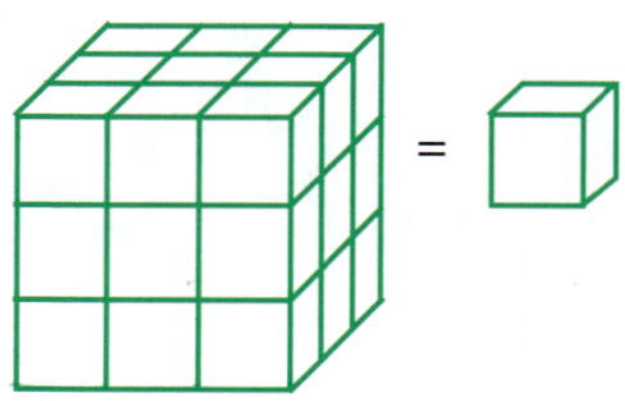 magnified by 3

In other words, another way to define dimension is to call an object d-dimensional if $3d$ copies have to be assembled to magnify the object by 3 (or, more generally, if md copies have to be assembled to magnify the object by m).

Fractal dimension

Now you can see what is strange about the snowflake curve: to magnify a segment of it by 3 one needs 4 copies, which is between the 3 copies needed to magnify a one dimensional object by 3 and the 9 copies needed to magnify a two-dimensional object by 3. Consequently, the snowflake curve has dimension between 1 and 2. In fact, according to the definition, its dimension d satisfies

$$3d = 4.$$

Taking logs gives $\quad d \log 3 = \log 4,$

and hence $\quad d = \log 4/\log 3$

$$= 1.2618\ldots$$

This number is called the fractal dimension of the snowflake curve. Similarly we can find the fractal dimension of other self-similar objects. If n copies of the object form the original, magnified by m, then fractal dimension = $\log n/\log m$

Example 1

Figure 6

Suppose we vary the snowflake rule as follows. Replace each line segment by a 'two-spiked segment' as shown in Figure 6. (Try drawing a few more stages.) Again, any part of the resulting limit curve which begins as a line segment is similar to any other part which begins as a line segment. But now we see that 7 copies of such a curve segment give the original, magnified by 5. (Because the double-spiked segment consists of 7 segments, and it is 5 times the length of each of them.) Consequently, fractal dimension = log 7/log 5.

Example 2 (The 'Sierpinski gasket')

Starting with an equilateral triangle, divide it into four equal equilateral triangles and remove the middle one (Figure 7.) Now repeat the process in each of the three triangles which remain, and continue indefinitely.

Figure 7

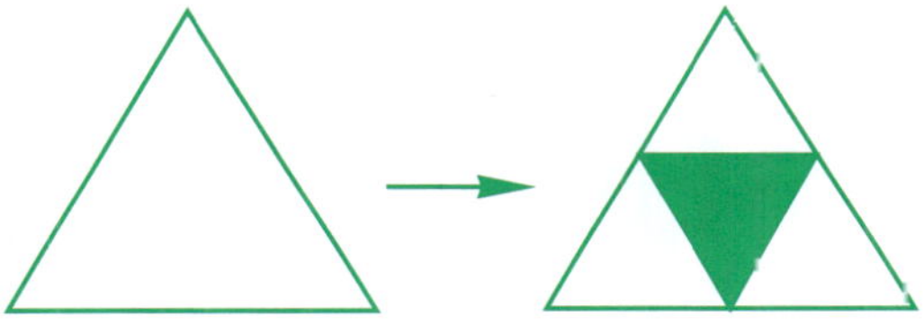

It follows that each small triangle becomes a set similar to the whole. We see from Figure 7 that 3 copies of the set yield the original, magnified by 2. Hence, fractal dimension = log3/log 2.

Objects such as these, with non-integer dimensions, are called fractals. Some books broaden the definition of fractal dimension (which is also called similarity dimensions) by allowing 'approximate or 'statistical' similarity. However, there is ample scope to invent fractal objects using strict similarity, along the lines of the examples above.

An example of a non-fractal, which nevertheless has some self-similarity, is the spiral shown in Figure 8.

Figure 8

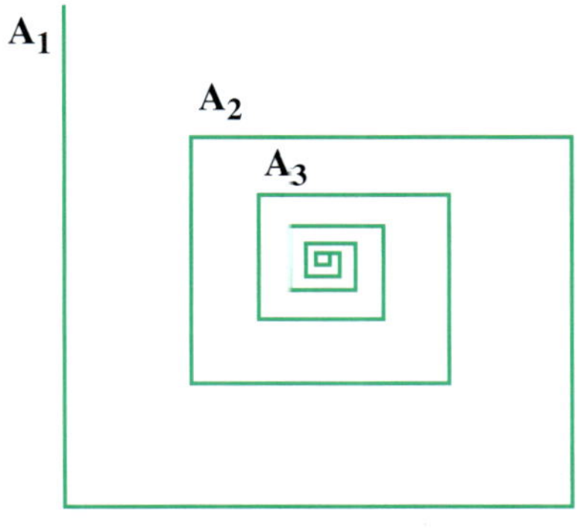

The whole spiral starting at A_1, is similar to the part starting at A_2, but twice as big. It follows that the whole is also similar to the part starting at A_3, but four times as big, and so on. However, one cannot assemble any number of copies of the spiral to make a similar spiral, so there is no way to obtain a fractal dimension. The spiral is in fact one-dimensional, like the line segments from which it is made.

Project 3

THEME: CIRCLES IN DESIGN

Background: Chapters 5, 6, 7, 11, 12, 16, 17

Your project must be based on, or incorporate the study of circles when used in design. You are encouraged to show initiative and be independent in carrying out your project.

1. General advice

Circles are involved in design in various ways ranging from the use of tools, such as a potter's wheel, to the visual effect shown in the design below.

The various characteristics of a circle, including radius and circumference, sector and tangent are well known and, where appropriate, you should refer to these in your project. Reference may be made to geometric properties of circles, their equations and the use of these properties in mensuration and trigonometry.

Acknowledge which area of mathematics you use: arithmetic, geometry, trigonometry, and/or algebra. You may choose to develop a computer program to assist you, or you may use a recognised computer package, but remember to include your own analysis of the problem.

2. Starting points

You may investigate any topic related to the theme: Circles in design. You must discuss your choice of topic and how it relates to the theme with your teacher. The examples listed below are possible starting points although it is not compulsory to use these ones. It is possible to combine related starting points but keep in mind that an investigative project is intended to include depth as well as breadth of research.

2.1 Investigate tools that use a circular motion; for example, lathes producing solids of revolution.

2.2 How does a tool convert circular motion to linear motion? What are the geometric properties involved?

2.3 How do gears and other circle-based mechanisms operate? What is their purpose?

2.4 Investigate patterns drawn by a spirograph. What are the equations of the curves produced?

2.5 How are circles used in sewing patterns? What properties of circles are used in pattern making, for example, altering sixes?

2.6 How are circles used in typefaces for printing or signwriting?

2.7 How are circles used in art? For example, how did Leonardo da Vinci and Albrecht Durer use circles in their art?

2.8 How are circles used in architecture? For example, the use of circles in the design of pneumatic structures, the shells of the Sydney Opera House (why was the original design of the shells changed to one based on circles?).

2.9 How were circles involved in the design of sundials and instruments used in navigation?

2.10 Investigate the use of geometric properties of circles in artistic designs. There are two examples below.

Project 4

THEME: INACCESSIBLE POINTS, MEASUREMENT AND ERRORS

Background: Chapters 4, 5, 16, 17, 18

Your project must be based on, or incorporate, the indirect measurement made necessary because of the inaccessibility of points needed for direct measurement. It must also incorporate an examination of the errors that result from such indirect measurement. You are encouraged to show initiative and be independent in carrying out your project.

1. General Advice

It is not always possible to measure distances or sizes of things outdoors directly For example, the height of Mt Everest, or of a flagpole on the top of a building, or the distance across the entrance to Port Phillip Bay or the area of Port Phillip Bay. These quantities cannot be directly measured with traditional measuring instruments.

Measuring these quantities has to be done indirectly, by combining measurements of directly accessible distances or sizes with measurements of angles. Problems like these led to the development of many parts of geometry (geo = earth, metry = measurement) and trigonometry (tri = three, gon = angle, metry = measurement).

Your task is to use trigonometry or geometry to find lengths or volumes in local outdoor situations where direct measurement is not practical. If necessary you can pretend that there are some obstacles in your local environment that make parts of the area inaccessible. These must be carefully specified when you describe your problem.

Nowadays, some modem instruments can measure the distance to an inaccessible point by using laser technology, but these are not to be used for this project. Methods such as using a barometer, or timing a stone as it falls to the bottom of a deep hole, are also not to be used in this project because they do not use trigonometry or geometry. Whilst the measurements of length and angles that you make should be done carefully, this project does not require you to make highly accurate measurements. NO extra credit will be given for very accurate direct measurement(s). One purpose of the project is for you to discuss the effect on the accuracy of your result that may arise from:

- assumptions you made in setting up your problem and its solutions.
- inaccuracies in each of the measurements you made.

Instructions for Starting Points 1,2,3 and 4

1 Carefully describe the length or volume you are going to measure indirectly and what factor(s) makes the target quantity inaccessible. (A diagram will be useful.)

2 Indirectly measure that length or volume which is not directly accessible for measurement using only one method. You may also wish to try out your indirect method on a length or volume where you already know the answer.

3 You must also make several indirect measurements using the same method. For example, if you were measuring the height of a tree using a directly measured distance from the tree and a directly measured angle of elevation, it would be a good idea to take measurements from several different points around the tree. This will enable you to practise making reasonably accurate measurements with your instruments and it will also help you judge how errors in the measurement of each quantity for a particular situation will effect the result.

4 Discuss the effect on your final answer of small errors in each of the direct measurements you made. For example, consider which measurements of length, angle etc need to be the most accurate for a satisfactory answer. This may vary for different parts of the domain.

5 Use one other method to measure indirectly the same length or volume.

6 Repeat Step 4 for your other method.

7 Compare the answers found using the methods and discuss which method is likely to give the best result.

8 Discuss the mathematical reasons why one method might be preferable and the conditions under which your recommendation is valid.

9 Extending your investigation.
Here are some suggestions for breadth or depth in your investigation. Because of time constraints, you should pursue only one of these suggestions.

- consider an analysis of errors based on data from step (4) or step (6) above.
- consider a mathematical justification of the valid conditions recommended in step (8).
- consider a third method to measure indirectly the same length or volume by repeating steps (5) to (8).

Measuring Instruments you may use

Only simple tools are allowed. These are

i. for distances: measuring tape, ruler, trundle wheel, bicycle, car odometer. Some measurements can be taken from a map (only for very large distances) to assist in calculating inaccessible distances for objects not shown on the map, for example, estimating the distance of a fire, or flare, from known reference points.

ii. for angles: an inclinometer for angles of elevation and depression constructed by yourself, a hiking compass or a protractor, a blackboard protractor with elongated arms.

2. Starting Points

You must investigate ONE of the following Starting Points. You must confirm with your teacher your choice of Starting Point (preferably by the end of the first week of the designated period).

2.1 Height

In the natural and/or the constructed environment, obstructions and conditions sometimes cause difficulty in the indirect measurement of the heights of objects such as trees and towers. Such obstructions include:

- sloping ground

- irregular surface

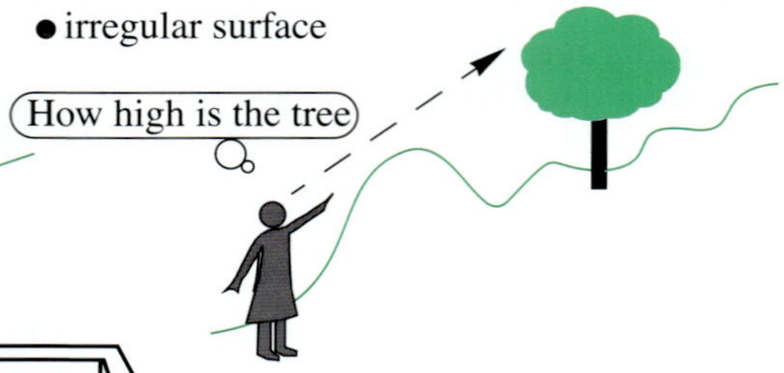

- obstacles preventing measurement from certain positions

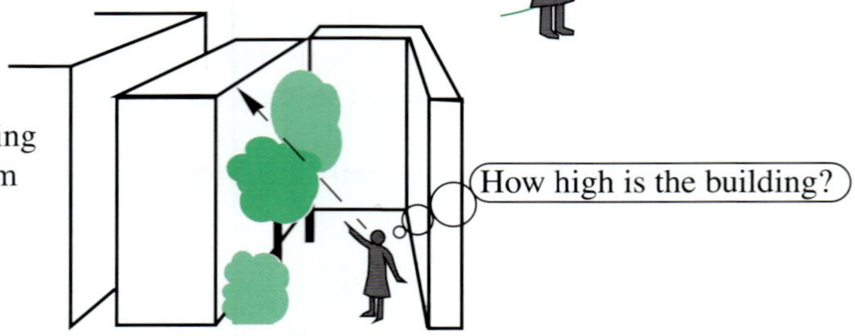

Follow the steps (1) to (9) in the General Advice to find the height of an object preferably in your local environment. where direct measurement is not possible. Provide a scale drawing of the situation, containing all relevant measurements.

2.2 Depth

In the natural and/or the constructed environment, obstructions and conditions sometimes cause difficulty in the indirect measurement of the depths of depressions such as quarries and valleys. Such obstructions include:

- sloping ground

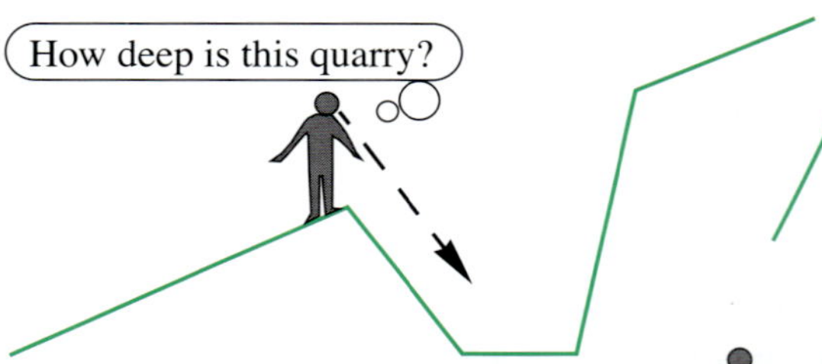

- irregular surface

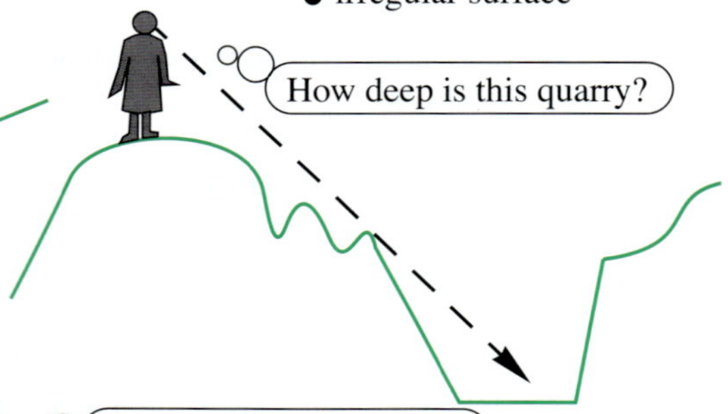

- obscured vision of the bottom of an object.
- obstacles preventing measurement from optimum positions.

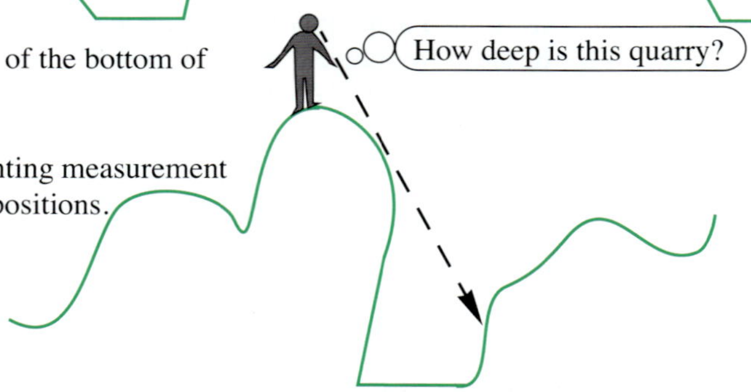

Follow the steps (1) to (9) in the General Advice to find the depth of a depression, preferably in your local environment, where direct measurement is not possible. Provide a scale drawing of the situation, containing all relevant measurements.

2.3 Distance between two points

In the natural and/or the constructed environment obstructions sometimes cause difficulty in measurement of the shortest distance between two points in situations such as those shown below.

- The shortest distance to an inaccessible point.

- The shortest distance between two inaccessible points as measured from a long way off.

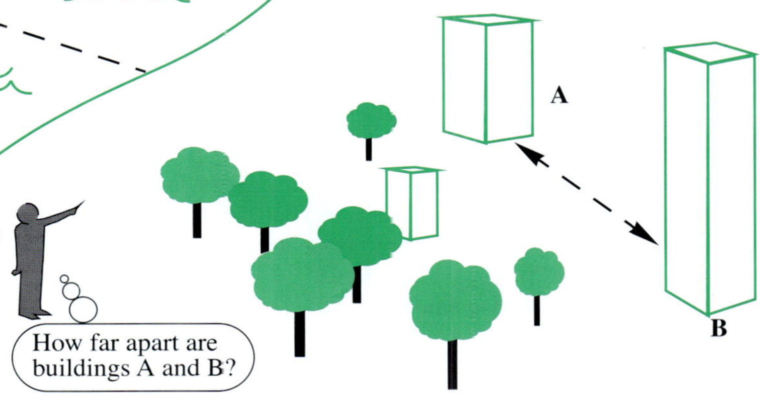

Investigate the indirect measurement of horizontal distances, preferably in your local environment, like those described above. In each case follow the steps (1) to (9) in the General Advice. Provide a scale drawing of the situation, containing all relevant measurements.

2.4 Volume

In the natural and/or the constructed environment, the volume of quantities needs to be found. Consider the situations of

- The need to estimate the volume of material removed from a quarry so that the number of earth carrying trucks required can be calculated.
- The need to estimate the volume of water to be added to a quarry when converting it into a lake.

Indirectly measure the volume of a depression such as a quarry or a valley. As you will need to make a number of indirect measurements, time may not allow a second method of calculation. Accordingly you will need to follow the steps (1) to (4) in the General Advice a number of times to enable indirect measurements of length to be used in the volume calculations. Note: steps (5) to (8) do not need to be followed at this stage. Drawings of the situation are required but need not necessarily be to scale.

Extending your investigation.

Here are some suggestions for breadth or depth in your investigation. You must pursue only one of these suggestions.

- For breadth you may wish to consider one other method to measure indirectly the volume by completing steps (5) to (8).
- for depth you may wish to consider an analysis of errors based on data from step (4).

Project 5

THEME: IS NEAR ENOUGH GOOD ENOUGH? ERRORS AND APPROXIMATIONS

Background: Chapters 4–11, 16–18, 21–23

Your project must be based on or incorporate the study of errors and approximations. You are encouraged to show initiative and be independent in carrying out your project.

1. General advice

Errors arise in different ways and some are more significant than others. You should therefore distinguish between errors in data (due to human error, and limitations of measuring instruments), errors created by calculations with incorrect data, and methods which are inherently approximate; for example, truncation of series, replacement of curves by polygons.

The general criteria for assessment should be borne in mind, but here are some of the issues, which occur specifically when considering errors and approximations.

> What is the effect of the error or approximation on the data and/or results?
> Are the errors acceptable?

Acknowledge which areas of mathematics you use: coordinate geometry, calculus, and algebra. You may choose to develop a computer program to assist you or you may use a recognised computer package but remember to include your own analysis of the problem.

2. Starting points

You may investigate any topic related to the theme. You must discuss your choice of topic and how it relates to the theme with your teacher. The examples below show some starting points for projects. It is not compulsory to use the starting points.

2.1 Errors arising from inaccuracies in data

Some inaccuracies cause larger errors than others. How, for example, does the effect of a one percent error in setting a course for a ship compare with a one percent error in estimating the speed of the ocean?

- Compare the error of angle with error in speed for a tennis serve.
- How accurately does a baseball bat have to be swung compared to a cricket bat?
- Scientific experiments: how do errors in measurement affect results? Which data need to be more accurately measured?
- Compare the effects on predicted population due to errors in estimating birth rates, death rates, and so on.
- Make a mathematical model of a situation that interests you and investigate the effects of inaccuracies in the data.

2.2 Errors due to rounding off and calculations

- Calculators create an error due to rounding off. How does this error affect results when added, multiply, divide, find reciprocals and/or find square roots and so on?

- Devise calculations where the error is catastrophic and/or devise safeguards to avoid such catastrophes

- Find sets of linear equations for which the usual methods of solution by calculator give highly inaccurate results. Explain the inaccuracies.

- Find sets of equations that allow spreadsheets to give inaccurate answers. Explain the inaccuracies.

- How much money can a financial institution make by rounding off interest calculations?

2.3 Errors arising from approximate methods

- How are mathematical tables calculated for sine, cosine and tangent, exponential and square root tables?

- Finding π or e.

- Devise calculations for conversions in your head, for example, Fahrenheit to centigrade, kilograms to pounds, yards to metres.

- Accuracy of calculus methods; for example, approximating a curve by a tangent.

- Series truncation; for example, $\sin x = x - \frac{x^3}{6}$ for two terms. How does an increase in the number of terms increase accuracy?

- Approximating areas using geometry and/or calculus.

- Approximation of curves by polygons.

- Stopping after n iterations; for example, Newtons' method.

- Methods for finding square roots.

Project 6

THEME: EXPONENTIAL AND LOGARITHMIC SCALES

Background: Chapters 4 –10, 22, 23

Your project must be based on, or incorporate the study of exponential or logarithmic scales. You are encouraged to show initiative and be independent in carrying out your project.

1. General advice

In a relationship such as $y = e^x$, where y depends exponentially on x, the exponent x is called the logarithm of y. You should already be familiar with relationships involving exponents, such as the formula for compound interest. The goal of this project is to investigate relationships in which the exponent is used to measure some quantity. Many quantities in science and daily life are actually measured this way. For example, exponents (of energy) make up the scale used to measure earthquake strength. Your task is to explain the exponential relationship in one or more instances and investigate it in depth.

Exponential relationships offer the opportunity to use several parts of mathematics. Explaining the interplay between exponent and logarithm involves the concepts of function and inverse function. Exponential relationships can be illustrated by specific numerical calculations, and by graphing on log-log or semi-log paper. The precise meaning of logarithms can be given in terms of algebra and calculus. You are also invited to make and use actual logarithmic scales in connection with some of the topics suggested for investigation below.

Acknowledge which area of mathematics you use: coordinate geometry, calculus and/or algebra. You may choose to develop a computer program to assist you, or you may use a recognised computer package but remember to include your own analysis of the problem.

2. Starting points

You may investigate any topic related to the theme: Exponential and logarithmic scales. You must discuss your choice of topic and how it relates to the theme with your teacher. The examples listed below are possible starting points, although it is not compulsory to use these ones. You may investigate some or all of the points raised in one of the following topics.

2.1 Music

- How are exponents involved in musical scales?
- Describe mathematically the spacing of frets on a guitar neck.
- Describe the well tempered (or equally tempered) scale and how it departs from ideal harmony.
- Calculate the difference between the ideal fifth and the fifth on the well tempered scale.

2.2 Scientific scales

- Describe some logarithmic scales in science; for example, decibels, pH, the Richter scale, magnitude of stars, colour-coding on resistors.

- Collect and discuss information on these scales; for example, the loudness of typical sounds, the strength of earthquakes, the brightness of stars.
- Why are logarithmic scales used to measure such quantities?
- Suggest other quantities which might be appropriately measured on logarithmic scales.

2.3 The slide rule

A slide rule consists of two scales on which numbers are marked in such a way that numbers in a constant ratio are a constant distance apart. In particular, the powers of 2 appear on each scale as follows.

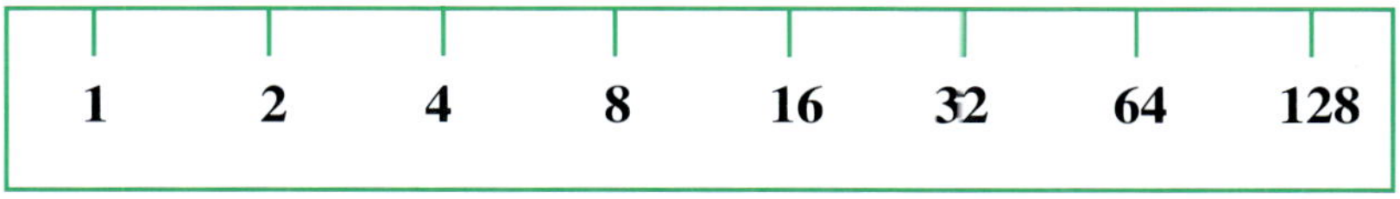

- Explain how a slide rule can be used to do multiplication.
- Using the fact that $5^3 = 125$ is close to 128, estimate the positions of 5 and 25 on the above scale.
- Deduce approximate positions of 10, 20, 40, 50, 80 and then estimate the positions of 3
- Hence make a cardboard slide rule, accurate enough to multiply any two single digit numbers, using only a ruler and elementary arithmetic.
- Discuss different ways of estimating the positions of 3, 5, 7, 11, 13,. . . and ways of increasing accuracy.
- Is it feasible to make a slide rule by hand for multiplying numbers less than 20, or less than 100?

2.4 Logarithms and computation

- Find out about the contribution to the calculation of logarithms made by Napier, Briggs and other mathematicians.
- How were log tables originally calculated?
- Investigate the connection between integration and logarithms.
- Look up, or work out, some infinite series for exponentials, or logarithms.
- Compare the computational performance of different formulas.

2.5 Measurement of information

- The 'length' of a number can be measured by the number of its digits. For example, 27 has length 2, and 17 764 has length 5. If the number carries information (as does, say, a telephone number) then its length is a measure of the amount of information.
- Is this length scale logarithmic? If not, how does it differ from a logarithmic scale?
- How does the number of decimal digits compare with the number of binary digits (bits) in a number?
- If car licence plates were all letters, how long would they need to be? (Similarly for telephone numbers for Melbourne, for Australia and/or the world). How long do bar codes, credit card numbers, car licence plates and/or pin numbers need to be? If binary digits were used, how long would bar codes, credit card numbers, car licence plates, and/or pin numbers need to be?
- Investigate the role of redundancy in information - ways of reducing it, ways of making it useful, (for example, check digits in bar codes).

Project 7

THEME: USE SIMULATION TO SOLVE A PROBLEM INVOLVING CHANCE

Background: Chapters 15, 18

Your project must use simulation to solve a problem involving chance. You must-follow the instructions given below and report on each of the specified steps in the main text of your report.

1. General advice

Simulation is a useful method for investigating problems. When the problem involves chance, the simulation will involve a random experiment. The simulation process is as follows.

i. specify the problem carefully.
ii. identify the important mathematical relationships.
iii. find, use and test a model which represents the important features of the situation.

You need to check that your simulation is realistic, giving answers that agree with a real life situation. By working with a model, you should be able to investigate aspects of the real situation. It is important to evaluate the reliability of the answers obtained from your simulation. Acknowledge which area of mathematics you use: probability, statistics, logic and/or algebra. You may choose to develop a computer program to assist you, or you may use a recognised computer package, but remember to include your own analysis of the problem.

An Example

Imagine you wanted to start a shoe shop and needed to know how many pairs of shoes of each size to keep in stock. You can get data about the percentage of people who wear shoes of each size. This would give you a good idea about the percentage of pairs of shoes in each, size that are needed. If you know how many pairs of shoes you expect to sell, you would then be able to work out how many of each size to stock. However, to be confident that you do not run out of any particular size, you need to know when to order more pairs of shoes of each size. How many pairs should be left before you order more?

The question above can be answered by theoretical statistical methods that are not usually studied at school. However, simulation can be used to help answer the question. The purchase by a customer is simulated by the drawing of a marble from a bag (and afterwards replacing it). Use a different colour for each shoe size and be sure to allocate the right percentage of marbles for each size. Imagine that you had only six pairs left of a certain shoe size-what would be the chance that you would run out in a week if you had not ordered more? To answer this question, conduct some experiments with the marbles, keeping a careful tally of the results. From further experiments, you could also find out how many pairs of shoes of that size should be kept in stock so that you can be reasonably sure to meet the customer's demand. Is it the same quantity for all sizes of shoes? How do the answers vary if the assumptions you have made about the number of customers, frequency of delivery, for instance, are changed?

In the simulation above, physically choosing marbles from a bag is the way in which randomness is generated. The results of each experiment would also have to be tallied by hand. Using a computer to generate the random numbers and also to do the tallying makes the process much quicker, enabling many more trials to be carried out, so that a better idea of the reliability of your answer can be obtained.

Instructions

(Steps v., viii. and ix. are optional.)

i. Choose a real life situation involving chance or uncertainty and describe it carefully. Identify a small number of questions about it that you wish to investigate by simulation.

ii. Identify the factors that will be most important to simulate and describe the factors that you have chosen to disregard.

iii. Describe a way of investigating your questions using a physical simulation. List the materials required and specify the procedure to be carried out carefully.

iv. Clearly list the assumptions that have been made in setting up the simulation and explain why they are reasonable. (To do this, you may need to find some data from books or collect some data of your own, but do not let this take more than a couple of hours.)

v. If possible, describe a way of investigating your questions using a computer simulation instead of a physical simulation. (If you have access to a ready-made computer simulation suitable for answering your questions, you need to give full details of it and describe clearly how it works, including the assumptions that have been built into program and how, the chance element is generated.)

vi. Carry out simulation, either physically or by computer, (that is by writing your own program or using a ready-made program).

vii. Explain the answers clearly and discuss their validity.

viii. You may wish to investigate several different ways of simulating your situation or you may also wish to change some of the assumptions you have made and repeat the simulation. How and why do the answers vary? Which answers best reflect what might happen in the real world?

ix. You may also wish to compare the answers you have obtained with an answer obtained using theoretical probability or statistics, or with what has actually happened in practice.

2. Starting points

You may investigate any topic related to the theme: Use simulation to solve a problem involving chance. You must discuss your choice of topic and how it relates to the theme with your teacher and you must follow the instructions above. The examples listed below are possible starting points, although it is not compulsory to use these ones.

Use simulation to investigate

- How many cereal packets you would have to buy to get a complete set of cards
- The chances of winning a finals series given a particular position in a preliminary competition

- Stock control and inventory control (for example, the shoe shop)
- The chance of getting two identical birthday cards at a child's party

  The likelihood that two people, in a class have the same birthday
- Winners of horse races, or outcomes of bets for punters or bookmakers
- The variation in-time to travel from A to B by public transport
- How much time you should allow to drive from A to B to arrive by 9.00 am
- How much better for clients, in a bank or Medicare office, is a single queue compared with a multiple queuing system
- The likelihood of winning simple games using various strategies
- Winning prizes in a gambling game
- The chance of winning a tennis game after being

Project 8

THEME: PREDICTION OF UNCERTAIN EVENTS

Background: Chapters 5– 11, 15, 18 – 21

Your project must arise from or incorporate the prediction of uncertain events. You are encouraged to show initiative and be independent in carrying out your project.

1. General Advice

There can be significant benefits from being able to make accurate predictions, say for example, on the stockmarket, for trends in population, in games of chance, and for consumer prices. Mathematical techniques often enable us to make predictions and to analyse their limitations. Such techniques might involve finding equations to predict events or explain relationships; or the gathering and statistical analysis of data; or the counting of successful outcomes to determine probability.

The general criteria for assessment should be borne in mind, but here are a few issues, which arise specifically in the prediction of uncertain events.

First, consider in what sense, if any, is the event predictable. In some cases we understand how a system works and we can predict aspects of its behaviour via probability, for example in tattslotto and other games of chance. In other cases it may be necessary to collect data to mathematically model a system, for example, data on life expectancy or on traffic flow. Second what mathematics is appropriate? Algebra, statistics, probability, logical reasoning? Your project should include developing a mathematical method of prediction and discussion of the reliability of the prediction.

Please note that if gathering of data is appropriate for your project, you should not allow this to delay developing a detailed plan or model for the prediction of events using the data.

2. Starting points

You may investigate any topic related to the theme. You must discuss your choice of topic and how it relates to the theme with your teacher. The examples below show some starting points for projects. This list is by no means exhaustive.

2.1 **Trends** - employment trends in particular industries or local companies. How are they related to other economic indicators such as business investments, the value of the dollar, the CPI, et cetera?

2.2 **Trends** - population trends in countries and the world. How is rural population related to urban population?

2.3 **Queues** - compare waiting times in queues. When is it better to wait in an "8 items or less" queue (if you have less than 8 items)? In a bank, how long would you expect to wait if there were a single queue for many tellers, compared with separate queues for each teller?

2.4 **School enrolments** - local, state and national.

2.5 **Consumer prices** - predict the future cost of such items as a compact disk player, home computer, a dozen eggs, et cetera.

2.6 **Economic forecasting** - look for relationships among inflation, interest rates and the local currency which can be used to predict future levels.

2.7 **Games** - investigate games of chance (gambling). Is there a winning strategy? If you play a fair game and you play regularly what is your chance of "breaking even ?" Design a game of chance, modify the game's rules for differing numbers of players, for example, how should tattslotto be modified for half as many players or for twice as many players?

2.8 **Diseases** - predict the decline in the number of cases of measles or polio as a result of an immunization program.

2.9 **Diseases** - predict the spread of AIDS. Is the AIDS death rate exponential?

2.10 **Smoking and alcohol** - what are the long term trends? How are they related to health and life expectancy?

2.11 **Athletic records** - examine the long term trends in athletic performance. Could there be a 40 second 100 metre freestyle? Could there be a 10 metre long jump ? What will the 100 metre sprint record be in the year 2000? Will women's athletic perfomance in track events catch up to those of men?

2.12 **Extinction of species** - kangaroos, koalas, platypus, humans.

2.13 **Opinion polls** - how many people should you survey to gauge public opinion concerning, for example, political leaders, soap powders, perfumes, et cetera?

2.14 **Weather forecasting** - how accurate is your daily weather forecast? How accurate is the long range forecast?

2.15 **Life expectancy** - what are the long term trends? What is the relationship with life insurance premiums?

Project 9

THEME: FITTING FUNCTIONS

Background: Chapters 6 – 10, 23

Your project must centre on the exploration of the 'closeness' of a particular function, class of functions or group of functions to a given curve, a given data set, or another function. You are encouraged to show initiative and originality.

1. General Advice

In this investigation, you are required to examine a problem or series of related problems which involve determination of a function which approximates a given function, or curve, or set of data. The task begins with a relatively straightforward problem. Following this inductory task, a number of more complex variations of the original task are presented which require a greater depth of analysis. You are advised to give careful consideration to domain limitations and to how you measure 'closeness' or 'good fit', as well as other factors which might relate to the reasonableness of your solution. Finally, there is an extension which requires some individual initiative in extending earlier results or placing the problem in a broader, or more realistic, context.

You are at liberty to research a problem that is of interest to you. We provide the following either for you to use or to provide a guide of the type of question questions you should be considering (at the very least) when fitting a curve or curves to your own data.

Topic: Selling ice cream

Those who sell ice cream and gelati are well aware that the amount of ice cream/gelati sold depends both upon the price of the product and how hot the day is, that is, the temperature. Knowledge of both price and temperature and their relationship to consumption are important factors to economists and retailers for the analysis of market trends and sales. In this Starting Point, you will explore how a variety of functions can be used to form models of relationships, and discuss and evaluate their usefulness.

The following table summarises a projected sample of data for a popular retailer in a shopping strip.

price (cents per scoop)	ice cream consumption for different temperatures (litres per hour)			
	0°C	10°C	20°C	30°C
15	12	18	25	50
30	10	12	18	37
45	7	10	13	27
60	5	7	10	20
75	3	5	7	14
90	2	3	5	10
105	1	2	3	6

Data like this can be modelled by equations of two different forms:

1. $f(x) = \dfrac{A}{x-B} + C$, where A, B and C are arbitrary real constants; and

2. $g(x) = De^{kx} + E$, where k, D and E are arbitrary real constants.

a. i. For each temperature, plot consumption as a function of price on the same set of axes.

ii. First take $B = 0$ in $f(x)$ and $E = 0$ in $g(x)$. Select any two appropriate points for the 30°C curve and determine two equations, one of each type, which could model the consumption data.

Discuss how well each function models the data over the entire domain.

b. Now allow B and E to be non-zero and repeat a.ii. using three appropriate points for the 30°C curve. Do these curves fit the data more 'closely'?
You will need to explain how you have measured 'close'.

c. i. Systematically vary the values of the constants in each of the general functions to determine the effect of each constant on the behaviour of the function.

ii. Using your measure of 'closeness', decide which type of function is the better to use to model consumption for all temperatures.

The market cannot tolerate a price of more than 150 cents per scoop (that is, this is the maximum price) and at that price there are no sales. Likewise, no matter how cheap the bottom price is, consumption does not exceed 55 litres per hour on the hottest day shown.

d. Using the type of function that you have chosen to be 'better' in c.ii., find values of the constants which give a good model of the 30°C consumption data, over the domain [0,150]. Discuss any assumptions you have made.

e. It is believed that a better model for the 30°C data over the domain [0, 150] could be provided if it were to consist of a combination of a function of the type you have chosen to be better in c.ii. and two polynomial functions, one for prices over 105 cents and the other for prices below 15 cents.
Explore how such a model could be developed with the curves fitting smoothly together, and using quadratic functions of the form $q(x) = a_0 + a_1x + a_2x^2$, where a_0, a_1 and a_2 are real arbitrary constants, as the two polynomial functions.
Discuss why this model is better than the one developed in part d.

Extension

f. Extend your investigation in part e. in some way.

Project 10

THEME: TRENDS AND RELATIONSHIPS

Background: Chapters 6–10, 18, 19

It will be your task to use statistics to investigate trends in the data and the relationships between variables in the data set provided, and to use these trends or relationships to make some predictions. You are required to carry out your project independently and are encouraged to show initiative and originality where this is possible.

In this investigation you should use only the data provided. Conclusions and interpretations should be made from the given data and the analysis performed in the investigation. You are not expected to carry out any additional data collection and/or research beyond that specified here. You are, however, encouraged to comment on any assumptions and limitations associated with the data provided.

The use of technology for dealing with the data is encouraged. You may choose to develop a computer program to assist you. You may use a statistical or graphics calculator or a recognised computer package, but remember to include your own analysis of the problem. You need to say what technology is being used; what is being calculated, giving the formula and defining the terms; what the purpose of the calculation is and its relevance to the problem.

In using a computer spreadsheet, for example, you must ensure that the appropriate formulation is included and all variables and units are correctly defined. All graphs, including computer generated graphs, must be appropriately scaled and correctly labelled.

Topic: Rabbits, the furry invaders

Rabbits were brought to Australia in 1859 and have since become a pest causing environmental damage and economic losses to farmers. One method of control used is a virus called myxomatosis which can be spread rapidly in various ways, one of which is transmission by mosquitoes and fleas. Many environmental, seasonal and climatic factors affect the number of fleas and/or mosquitoes and hence affect the spread of the virus.

The rabbit population over a five-year period in three different climatic regions is shown in the Data set for Rabbits. Rabbits were injected with the myxomatosis virus in the first year shown (Year 1). At Urana and Chidlow this was done in Spring, sometime during September, October or November. On Ile du Cimetière this was done in Autumn, sometime during March, April or May. Urana is situated inland near a lake in New South Wales. Urana has warm summers and mild winters and the rainfall is typically spread throughout the year. In wet years the region is flooded. Chidlow is situated near the coast close to Perth in Western Australia. Chidlow has warm wet winters and dry summers.

Ile du Cimetière is situated close to Antarctica, and there are no mosquitoes or fleas on this island because of the extreme climatic conditions.

a. i. Randomly choose a region and a three-year period. Explain your method of selection. Graph the rabbit population on a monthly basis for your selection. Describe any trends or patterns in the graph.

ii. Examine the data for the other two regions for the same period to decide if the trends or patterns are repeated. Comment on any similarities or differences between the three regions.

b. i. For either the region of Urana, or the region of Chidlow, use an appropriate method to determine a regression line which can be used to predict the rabbit population. Graph your regression line on the corresponding scatter plot to show the line of best fit. Discuss whether your selected three-year period is representative of the entire five years.

ii. Use your regression line formula to predict the following for your selected region.

- the rabbit population in July of the sixth year
- the rabbit population in December of the tenth year
- the month and year when the population will be half that at the end of the fifth year.

Discuss any assumptions you have made and comment on the reliability of your predictions. Do you believe that myxomatosis will ever cause rabbits to become extinct in your chosen region? Explain your answer using the results of your analyses in part a. or part b. or both.

iii. For your selected region, calculate an appropriate moving average, justifying your choice of length of the moving average. Investigate and discuss any long-term trends in the data. Compare your results with those from b.ii.

c. Randomly select two regions and two years, explaining your method of selection. For each of these years, draw a scatter plot and perform a correlation analysis between the populations for the two regions. Interpret and compare your results.

Rabbit population for the regions of Urana, Chidlow and Ile du Cimitière

URANA:

	YEAR 1	YEAR 2	YEAR 3	YEAR 4	YEAR 5
January	3000	100	25	15	10
February	3500	120	40	35	25
March	3900	140	55	45	35
April	4200	160	65	60	50
May	4500	160	70	65	60
June	4700	200	140	120	100
July	4900	300	250	230	200
Augus	4920	330	290	270	235
September	5000	460	350	330	290
October	5000	530	490	480	350
November	4900	510	480	465	390
December	4500	50	45	30	25

CHIDLOW:

	YEAR 1	YEAR 2	YEAR 3	YEAR 4	YEAR 5
January	1500	3640	1590	1290	200
February	1800	2940	1210	1000	190
March	1980	2720	1210	1000	190
April	2170	2240	800	700	210
May	2380	2000	750	790	240
June	2500	2300	920	830	380
July	3000	2500	1280	1000	490
Augus	3600	2000	1480	1120	500
September	4400	2400	1620	1000	520
October	5250	3000	1850	780	600
November	5210	2500	2000	500	610
December	4500	1860	1620	240	500

ILE DU CIMITIÈRE:

	YEAR 1	YEAR 2	YEAR 3	YEAR 4	YEAR 5
January	940	810	980	980	710
February	940	750	990	720	620
March	900	690	980	640	640
April	975	650	640	700	720
May	760	600	600	630	670
June	775	580	640	1000	680
July	700	590	800	990	800
Augus	1000	630	900	990	950
September	960	800	970	1000	960
October	1000	1000	980	910	1000
November	1000	1010	1010	900	950
December	920	1000	910	800	800

3.4 CHALLENGING PROBLEMS

The project suggestions 1 to 10 in Section 3.3 have been 'Open ended' – i.e., a theme was nominated so that you would be at liberty to investigate any element of that theme. As such, there are no definitive answers (at least for the first 8). The **challenging problems** that follow can be considered as projects that have much more structure associated with them. Because these projects are 'self-contained', there reaches a point where a conclusion can be arrived at. However, to satisfy the requirements of the internal assessment criteria, these structured projects **should be used as the beginings of a broader** investigation rather than providing 'a solution' to set questions. Note also that they might require mathematical concepts and skills that lie outside the scope of this course – which you will therefore need to first acquire

Project 11

PROBLEM: CIRCLES UPON CIRCLES

Background: Chapters 6, 8, 16, 18, 23

Here are two circles of radius 1 with centres A and B, which are 4 units apart.

For a given pair of circles, any circle that intersects each of the original circles twice, with every intersection forming a right angle, we will call perpendicular to that pair.

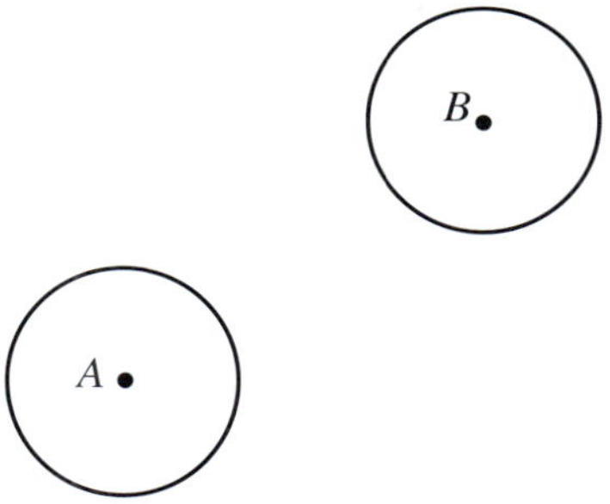

For the pair of circles pictured,

a. i. give a precise description of a circle, perpendicular to the pair, with centre on the line joining A and B.

ii. give a precise description of any other circles perpendicular to the pair.

iii. comment on any noteworthy features of the family of all circles perpendicular to the pair.

b. Consider the circles, perpendicular to the pair, with centres above the line joining A and B. Each such circle has a point furthest from the line joining A and B, which we will call the top. Suppose that the spacing between two such circles is defined to be the distance between their tops. Find and describe a family of 'equispaced' circles, all of which are perpendicular to the original pair of circles.

c. Now consider the following. Each of the circles considered in part b. has a counterpart obtained by reflection about the line joining A and B. If we now take one of the circles from part b. and its reflected counterpart to form a pair of circles, describe the family of circles perpendicular to this pair and comment on any noteworthy features.

Generalise your results in part a. to the case where the original pair of circles have unequal radii.

Project 12

PROBLEM: LATTICE PARALLELOGRAMS

Background: Chapters 6, 8, 16, 18, 23

The diagram shows lattice points from a portion of a triangular lattice which is made up of successive equilateral triangles with sides of unit length.

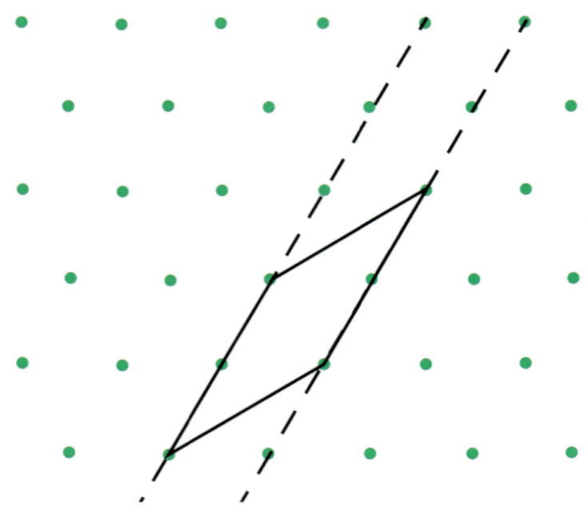

A line is drawn from the origin O to a lattice point P.

Another line is drawn parallel to the line OP, so that there are no lattice points between the two parallel lines.

A parallelogram is formed so that the four comers of the parallelogram are at lattice points lying on the two parallel lines, and no lattice points, other than those at the corners, are on the boundary of the parallelogram.

Calculate the area of any parallelogram which can be drawn in this way.

Project 13

PROBLEM: CABLES

Background: Chapters 6, 8, 16, 18, 23

Part A

Engineers from a telephone company need to provide a telephone link between three towns A, B and C which form an isosceles triangle with AB = BC. The engineers wish to use the shortest possible cable length to join the three towns.

Instead of just solving the problem for the particular towns being considered, the engineers would like to obtain a more general result which could be used if a similar situation arose in the future. To do this they propose the following model: fix the distance between A and C at 2 units and change the isosceles triangle by moving B along a line perpendicular to AC and through the mid-point of AC.

The engineers have put forward three plans which they believe will be useful in solving the problem. These plans, showing the cable links, are shown in Figure 1.

Figure 1: Three plans for joining towns A, B and C

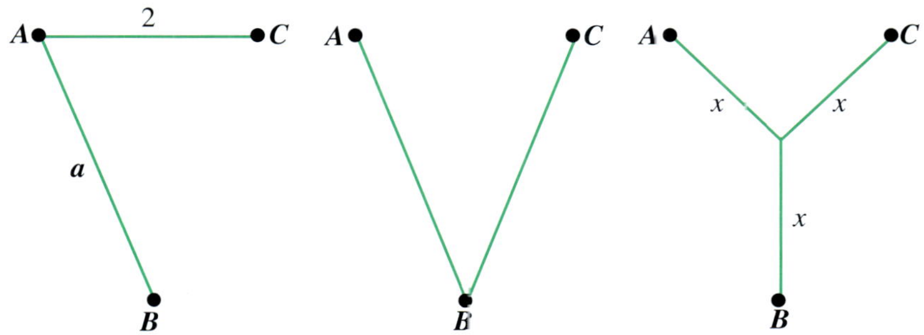

In the C–plan, cable is run from B to A to C. In the V–plan, cable is run from A to B to C. In the Y-plan, cables of equal length are run to A, B and C from a point between the towns.

For different allowable positions of B, when should a given plan be used to obtain the shortest cable length?

Part B

A consulting engineer suggests another option: to link the three towns using a modified Y-plan in which the cables are joined so that the angles between them are each 120°. This plan is shown in Figure 2.

Figure 2

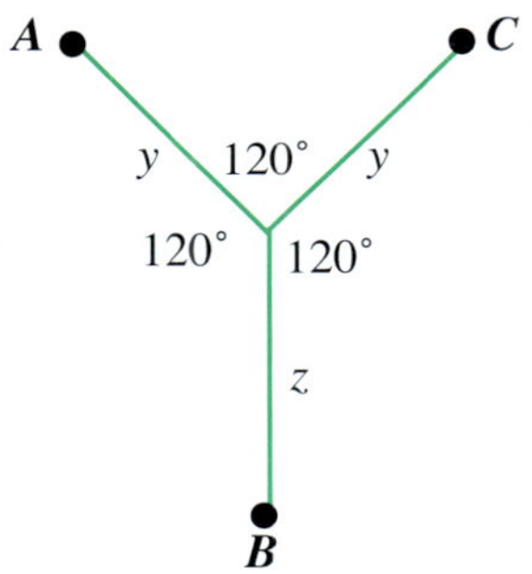

Is this plan better than any of the three plans in part A? Explain your answer fully.

Part C

The consulting engineer is working on the problem of connecting four towns A, B, C and D with telephone cable. She is considering four plans which are shown in Figure 3.

Figure 3: Four plans for joining towns A, B, C and D

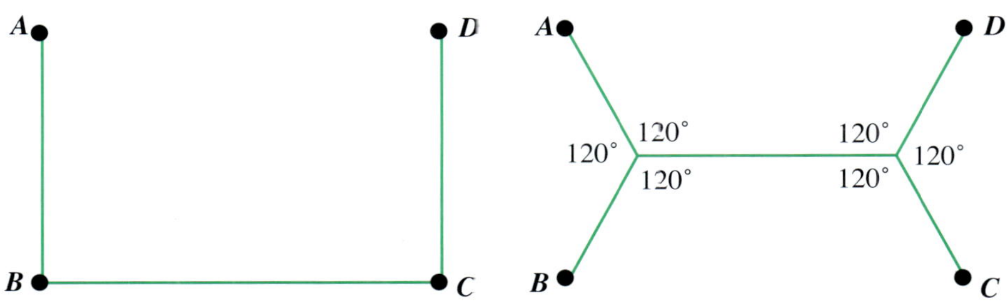

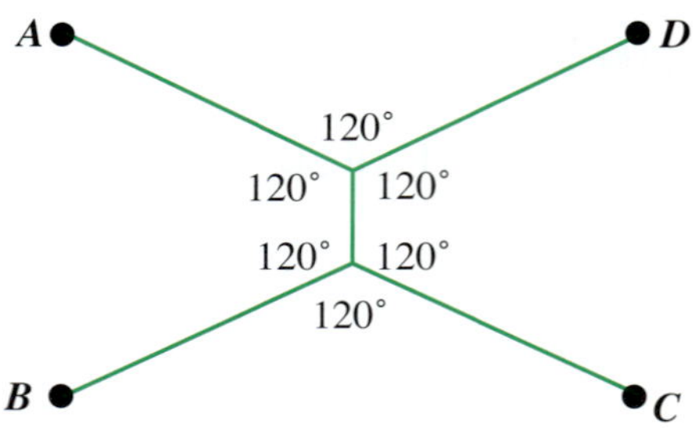

Suppose the distance between A and B is fixed at 2 units and we change the rectangle by adjusting the distance a between B and C.

> If the engineers want to obtain the shortest cable length, which plan should be used (consider all possible values for a)?

Project 14

PROBLEM: LATTICE TRIANGLES

Background: Chapters 6, 8, 16, 18, 23

Consider the upper right hand quadrant of a square lattice with unit spacing between the lattice points. Suppose a point A is chosen such that the straight line OA from the origin O (0,0) to A does not pass through any lattice points between O and A. An example of a particular line OA is given below.

Figure 1

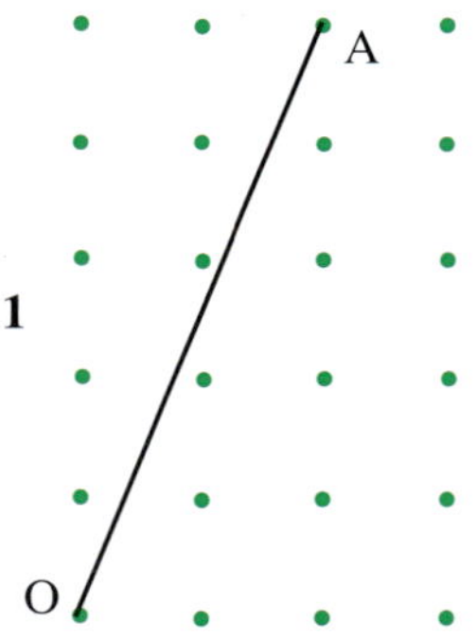

Figure 2

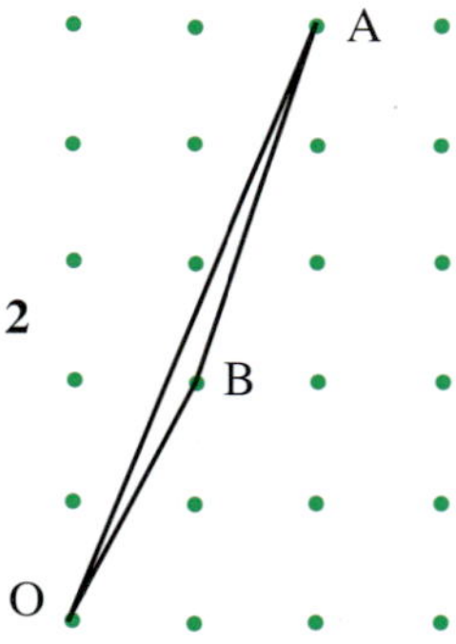

In Figure 1, what are the allowed coordinates of the point A?
For future reference, let N3, N4 and N5 be any positive integers such that (3, N3), (4, N4) and (5, N5) respectively are allowed points A.

Now suppose a triangle OAB is to be formed as in Figure 2, with the line OA as the longest side. The point B is to be chosen such that no lattice point other than the vertices OAB are contained within, or on the boundary of, the triangle.

> Find a formula for an allowed point B when A is the point (3, N3).
> For a general point A, give some criteria which the point B must satisfy and use these criteria to find a formula for an allowed point B when A is the point (4, N4) or (5, N5).

Project 15

PROBLEM: BURIED TREASURE

Background: Chapters 6, 8, 16, 18, 23

You have found a map and instructions to find treasure buried on an island on which are located a water tower, a large gum tree and a peppercorn tree. The instructions include the following directions.

'When you stand at the water tower you should be able to see the large gum tree ahead of you and somewhat to the left, and the peppercorn tree ahead of you and somewhat to the right. Starting from the water tower, walk directly to the large gum tree, counting the number of steps taken. Turn left through an angle of 90° and walk the same distance away from the gum tree. Mark this spot.

Do this again by starting at the water tower, walking in a straight line to the peppercorn tree and counting the number of steps taken, turning right through an angle of 90°, walking the same distance away from the peppercorn tree and marking the spot.

Half-way along a straight line joining the two marked spots lies the treasure.'
Unfortunately, when you get to the island, the gum tree and the peppercorn tree are plainly visible, but the water tower is nowhere to be seen. It appears that the tower has been demolished and all traces removed. Where is the treasure buried?

You calculate the spot where you think the treasure is buried and have been digging unsuccessfully for some time when you remember that your compass was out of alignment and was not measuring angles accurately.

The angle at the gum tree through which you had turned left was only 85° and the angle at the peppercorn tree through which you had turned right was really 95°. Your compass has now broken down completely and you have no way of accurately determining a right-angle. What can you now do in your attempt to find the treasure?

In your concern to find the treasure you become exhausted and confused and you lose the map in a sudden gust of wind. You think that the instructions were as follows

'Starting from the water tower, walk directly to the large gum tree, counting the number of steps taken. Turn right through an angle of 90° and walk twice the same distance away from the gum tree. Mark this spot. Do this again by starting at the water tower, walking in a straight line to the peppercorn tree and counting the number of steps taken, turning left through an angle of 90°, walking twice the same distance away from the peppercorn tree and marking the spot. Half-way along a straight line joining the two marked spots lies the treasure.'

If you were able to find this mid-point, how far from the treasure would you be if the gum tree and the peppercorn tree are 50 metres apart?

Project 16

PROBLEM: EASTER SUNDAY

Background: Arithmetic

In theory, Easter Sunday occurs on the first Sunday after the Paschal full moon, which is the first full moon in Jerusalem after 21 March. In practice, the scheduled date of Easter Sunday in each year is determined by a formula specified in Christian literature.

A simpler formula was derived by the mathematician C F Gauss (1777 - 1855), which gives the same date as the scheduled date for every year this century except for 1954 and 1981.
The formula derived by Gauss involves the use of the symbol $a \bmod b$ which means the remainder when a is divided by b.

For example, 18 mod 7 is equal to 4 since 18 divided by 7 gives 2 with a remainder of 4.
Gauss' calculation of the date of Easter Sunday is as follows.

> For the year which is x years after 1900, for example the year 1931 has $x = 31$, the first full moon occurs c days after 22 March where
>
> $$c = [19\,(x \bmod 19) + 24] \bmod 30$$
>
> The following Sunday, Easter Sunday, occurs d days after the full moon where
> $$d = (2a + 4b + 6c + 3) \bmod 7$$
> with $a = x \bmod 4$, $b = x \bmod 7$, and c defined as before.

You will need to make use of the following information in order to answer the questions below.

1. In 1900 the full moon occurred 24.07 days after 22 March. This was a Sunday.
2. The time between two full moons is 29.53059 days.

Use the Gauss formula to calculate the date of Easter Sunday for each year in the period 1990 - 1999.

Explain the reasoning underlying the formula for c relating it to the full moon cycle.

Now let c be any number of days, from 0 to 29 inclusive, after 22 March. Show that the first Sunday after this date occurs in a further d days, as given by the Gauss formula. In the cases for which this date is already a Sunday, show that the Gauss formula gives $d = 0$.

Project 17

PROBLEM: AREA AND PERIMETER

Background: Chapters 6, 8, 16, 17

The following question was posed to a group of mathematics students.

'Are there shapes for which the numerical value of the length of the perimeter is the same as the numerical value of the area?'

One student quickly saw that a square is a shape with this property because a square which has a side length of 4 units has a perimeter of 16 units and an area of 16 square units. The student could also easily show that there could only be one square with this property.

After looking at families of shapes like triangles, circles, rectangles and other polygons the students made the following conjecture.

'For every family of shapes there is at least one of these shapes for which the numerical value of the area and the numerical value of the perimeter are the same.'

By a 'family of shapes' the students meant all shapes which are similar to a given shape. For example there is only one family of squares, but there is an infinite number of families of rectangles.
You are required to find the following.

For which shapes does the conjecture hold?

For each class of shapes for which the conjecture holds, give a method for finding an actual shape for which the numerical value of the area is the same as the numerical value of the perimeter.

Project 18

PROBLEM: POLYGONS

Background: Chapter 6, 8, 16, 17

Two of the interior angles of the pentagon shown in Figure 1 are right angles.

Figure 1

Figure 2

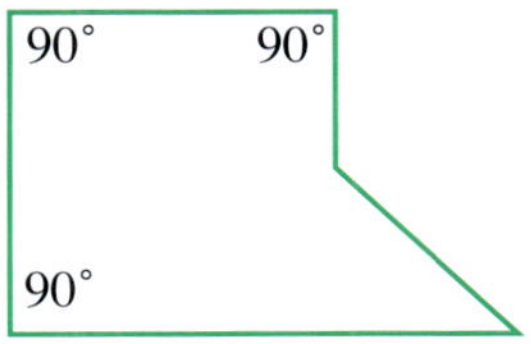

The maximum number of right angles that can occur as interior angles of a pentagon is three. An example is shown in Figure 2.

For any polygon there will be a maximum number of right angles that can occur as interior angles of that polygon.

For example a seven-sided polygon can have at most five interior angles of measure 90°.
Find a way of predicting the maximum number of interior angles of 90° that can occur in any polygon, and show that polygons can be drawn with this maximum number of interior right angles.

Project 19

PROBLEM: TRIANGLES

Background: Chapters 6, 8, 16, 17

Find the number of triangles that can be drawn where all side lengths are whole numbers and where the longest side length is n units.

Explain and justify your solution by examining the number of equilateral, isosceles and scalene triangles that can be drawn for whole number values of n.

Project 20

PROBLEM: LIGHT BULBS

Background: Chapters 6, 8, 9, 18, 21

Several manufacturers of light bulbs are advertising new low-energy bulbs. These bulbs cost a lot more than conventional light bulbs, but they use much less electricity and they last much longer.

One company claims their bulbs
- last as long as eight conventional bulbs
- use 20 per cent of the electricity used by conventional bulbs.

It is claimed that because of the savings possible through use of the low-energy bulbs, spending the money to change over the lights in an average house from conventional bulbs to the new bulbs would be a very good investment.

Use the information provided, and any other information you think relevant, to investigate this claim.

Express the savings as an annual rate of return on the investment in the change over from conventional bulbs to low-energy bulbs.

Information

Electricity costs (domestic rate, as at January 1990)

kWh	cents/kWh
first 120/quarter	24.35*
next 900/quarter	9.79
balance	10.79

(*this means using one kilowatt, or 1000 watts, for one hour costs 24.35 cents)

Note: prices are indexed on July 1 each year, using the CPI (approximately 8 per cent)

Electricity usage

An average household would use 1000 – 1500 kWh in three months.

Comparison of bulbs

	new bulbs	conventional bulbs
Cost per bulb	\$24	\$1
Expected life of bulb	8000 hrs	1000 hrs
Equivalent light strength	7 watts	40 watts
	11 watts	60 watts
	15 watts	75 watts
	20 watts	100 watts

Project 21

PROBLEM: HAMBURGER STALL

Background: Chapters 6, 8, 9, 18, 21

The Student Representative Council (S.R.C.) of Greentrees Secondary College operates a hamburger stall for half an hour each lunchtime. A subcommittee is investigating the most profitable way of running the stall. The following data are used in making their decisions.

1. The rolls used cost \$3.60 per dozen.
2. The patties cost \$3.50 per packet of 10.
3. The hamburgers sell at \$1.50 each.
4. There are overhead costs of \$10.00 per day.
5. Each customer is only allowed to buy one hamburger.
6. The stall is only open for 30 minutes each day, but staff are employed for 2 hours each day to allow for preparation and cleaning.
7. Staff are each paid at \$10.00 per hour.
8. It takes 30 seconds to serve a customer who has the correct money and 60 seconds if change is required. Two thirds of the customers require change.
9. It has been estimated that on average there is a demand for 119 hamburgers per day.

    Find the best way of staffing the stall. That is, what number of workers should be employed so as a make a maximum profit.

    If the price of a hamburger is \$$p$ and the wage of employees is \$$w$ per hour, find a relation between p and w so that the profit is increased if an extra person is employed.

Some of the costs and other variables identified in the data above will tend to change over time while others will not change. How will the likely changes affect your solution to this problem?

Project 22

PROBLEM: RECTANGLES

Background: Chapter 12

You have a supply of 2 × 1 rectangles like this one:

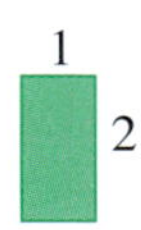

You can use these rectangles to make other rectangles which are 2 units deep and of whatever width you choose.

For example, here are some 2 × 3 rectangles:

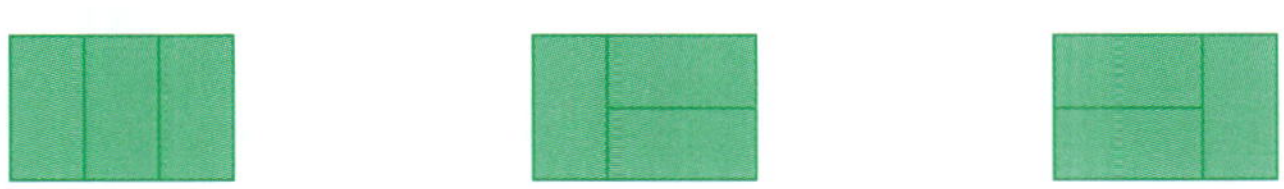

Describe how many $2 \times n$ rectangles it is possible to make from the 2×1 rectangles (where n is a natural number). Justify your conclusions.

Extend your solution describe how many $3 \times n$ rectangles which can be made from 3×1 rectangles.

Extend your solution further to describe how many $m \times n$ rectangles which can be made from $m \times 1$ rectangles (where m and n are natural numbers).

Project 23

PROBLEM: CHOCOLATE WRAPPING

Background: Chapters 5, 16, 17, 23

A triangular slice of chocolate has to be wrapped in a triangular piece of paper, which cannot be cut or torn. The chocolate is sliced so thinly that you can ignore its thickness.

Both the chocolate and paper are equilateral triangles. The paper can be folded along an edge of the chocolate. The edges of the wrapping do not have to overlap, they can just meet.

The chocolate can be positioned in various ways on the paper. Some examples appear in the diagram below:

When the chocolate has a side length of 4 centimetres, what is the smallest side length of the paper that will allow the chocolate to be wrapped?

How does the solution differ if the chocolate and paper are both rectangular (not necessarily similar rectangles)?

What optimisation issues arise when dealing with such a situation?

Project 24

PROBLEM: SLIDING RECTANGLES

Background: Chapters 8, 16, 17, 23

A rectangle ABCD is placed on the cartesian plane as shown in the diagram below.

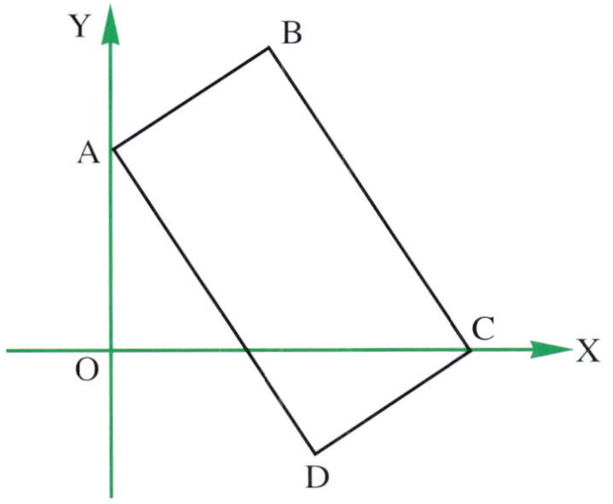

The rectangle is free to slide so that A is always somewhere on the positive Y-axis and C is always somewhere on the positive X-axis.

As the rectangle moves, what path will the point B follow? Explain your results.

What will be the path of the point D?

If the point A can move over both the positive and negative parts of the Y-axis, and the point C can move over both the positive and negative parts of the X-axis, what paths will the points B and D follow?

What effect will changing the dimensions of the rectangle have on the paths of points B and D?

Project 25

PROBLEM: GOAL SHOOTER

Background: Chapters 8, 16, 17, 23

Hockey is played on a rectangular playing field with a goal area at each of the shorter ends of the rectangle. A player on the long boundary, as shown in the diagram below, wants to shoot a ball through the goal.

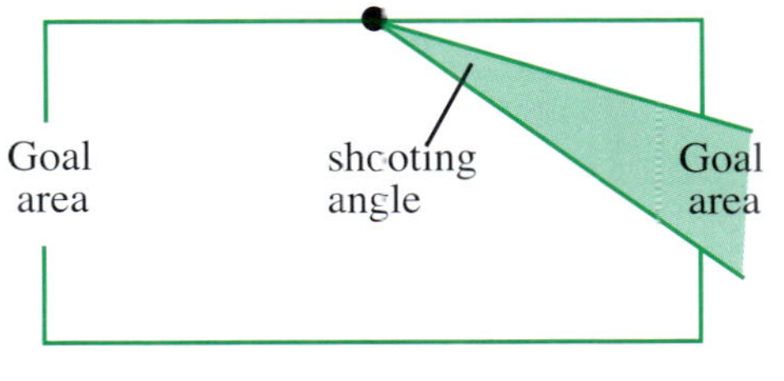

What position on the boundary will give the player the biggest shooting angle through the goal?

If the field was circular instead of rectangular, what position on the boundary would give the player the best shooting angle through the goal?

What is the best position on the boundary to shoot from if the field is elliptical?

Project 26

PROBLEM: JELLY BEANS

Background: Chapters 15, 18

In the diagram below there is a grid that extends infinitely upwards and to the right.
At the bottom left corner there is a start dot (the square dot on the grid).

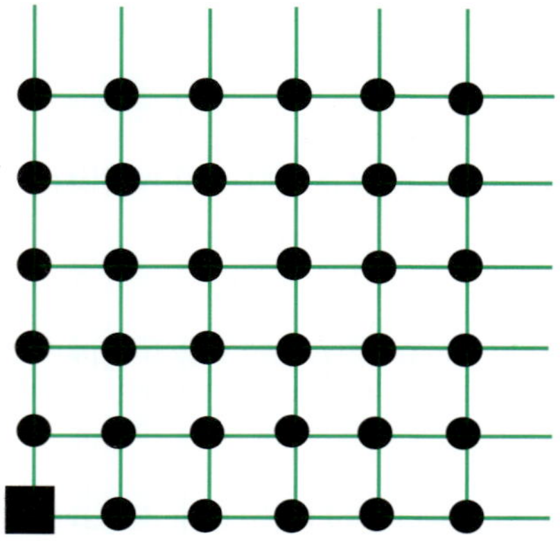

Christine is standing at the start and has a bag containing 3 black jelly beans and 2 purple jelly beans.

She takes a jelly bean from the bag and eats it. If it is a black jelly bean she moves up to the next dot. If it is a purple jelly bean she moves right to the next dot.

Which dots could lie on Christine's path and with what probabilities would she arrive at each of these dots?

What if the bag contained a different number of black and purple jelly beans?

What if Christine did not eat the jelly beans as she took them from the bag, but replaced them instead?

Project 27

PROBLEM: RUNS OF HEADS AND TAILS

Background: Chapters 15, 18

When you toss a fair coin a number of times you get a sequence of heads (H) and tails (T). If you toss the coin a large number of times you might expect the number of heads and the number of tails to be about equal. However when you repeatedly toss a coin you will also notice that you get runs of heads and runs of tails of varying length.

One possible outcome of tossing a coin five times is HHHTH. This sequence contains a run of three heads, a run of one tail and a run of one head, a total of three runs.

Investigate the distribution of runs of different length for the 5 coin situation.

How does the number of coin tosses affect the distribution of runs of heads and tails?

What if there was a different number of equally likely outcomes (for example if you were rolling a die instead of tossing a coin) or if the outcomes were not equally likely (for example if you were using an unfair coin) ?

Investigate the distribution of runs in these situations..

Project 28

PROBLEM: TELEPHONE NUMBERS

Background: Chapters 15, 18

For this problem you should begin by conducting the following experiment.

Obtain a copy of the white pages of a telephone directory and open it to any page in the main body of the directory. From the page you have chosen, take $2N$ consecutive telephone numbers (N must be fairly large, say 120 or more) and place these numbers into N pairs.

For each pair, note the pair of numbers formed by the last two digits of each telephone number. Square these two numbers and add them together.

Assign the pair to one of the four categories A. B, C or D according to the following rule.

A One or both of the numbers being added is equal to 0.
B The sum of the squares is equal to 1002.
C The sum of the squares is less than 1002, and neither number in the pair is zero.
D The sum of the squares is greater than 1002 and neither number in the pair is zero.

For example, a pair of telephone numbers is (3015403, 2133571). The last two digits of each number in the pair form another pair of numbers (03, 71). To obtain the sum of the squares of these numbers, proceed as follows.

$$\begin{aligned}(03)^2 + (71)^2 &= 3^2 + 71^2 \\ &= 9 + 5041 \\ &= 5050\end{aligned}$$

For this pair, the sum of the squares is less than 100^2. Therefore this pair should be assigned to category C.

For the $2N$ telephone numbers you have chosen, count the number of pairs a, b, and c in categories A, B and C respectively.

Now calculate the value of $4\left\{\dfrac{\frac{a}{2} + \frac{b}{2} + c}{N}\right\}$ for the $2N$ numbers you have chosen.

The value generated by the process followed in carrying out this experiment gives a good approximation to a particular irrational number. What is that number?

> Give a mathematical argument which explains why the process gives a good approximation to this number.
>
> Make sure your explanation accounts for the appearance of $\frac{a}{2}$ and $\frac{b}{2}$ in the above expression.
> Why is it that any page in the telephone directory will give a similar result?

Suppose the sum of the squares of the last three digits of each pair of telephone numbers was computed instead of the last two digits.

> How should the process be modified to give an approximation to the same irrational number?
> How would this approximation compare with the previous one?

Project 29

PROBLEM: CRYPTS

Background: Chapters 15, 18

A crypt is found in the human body. It is like a tube of cells which descends from the surface of the colon. Figure 1 shows a highly simplified picture of a crypt which in this case is 11 cells long.

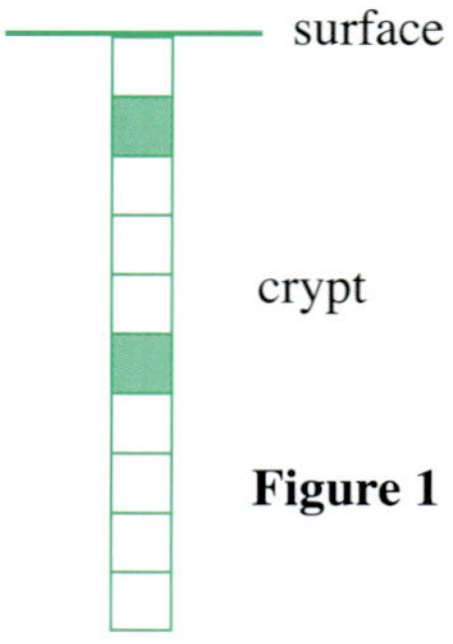

Figure 1

In this picture the shaded boxes show cells which are undergoing cell division. Such cells are found by a staining technique which makes them readily apparent under a microscope. Each clear box shows a cell which is not stained and thus is not undergoing cell division.
It is thought that a high fibre diet may alter the way in which cell division occurs within a crypt.

The diagrams in Figure 2 show 20 crypts selected randomly from the colon of George, who has a normal diet, and 20 crypts selected randomly from the colon of Fred, who has a high fibre diet.

Figure 2: George's and Fred's crypts

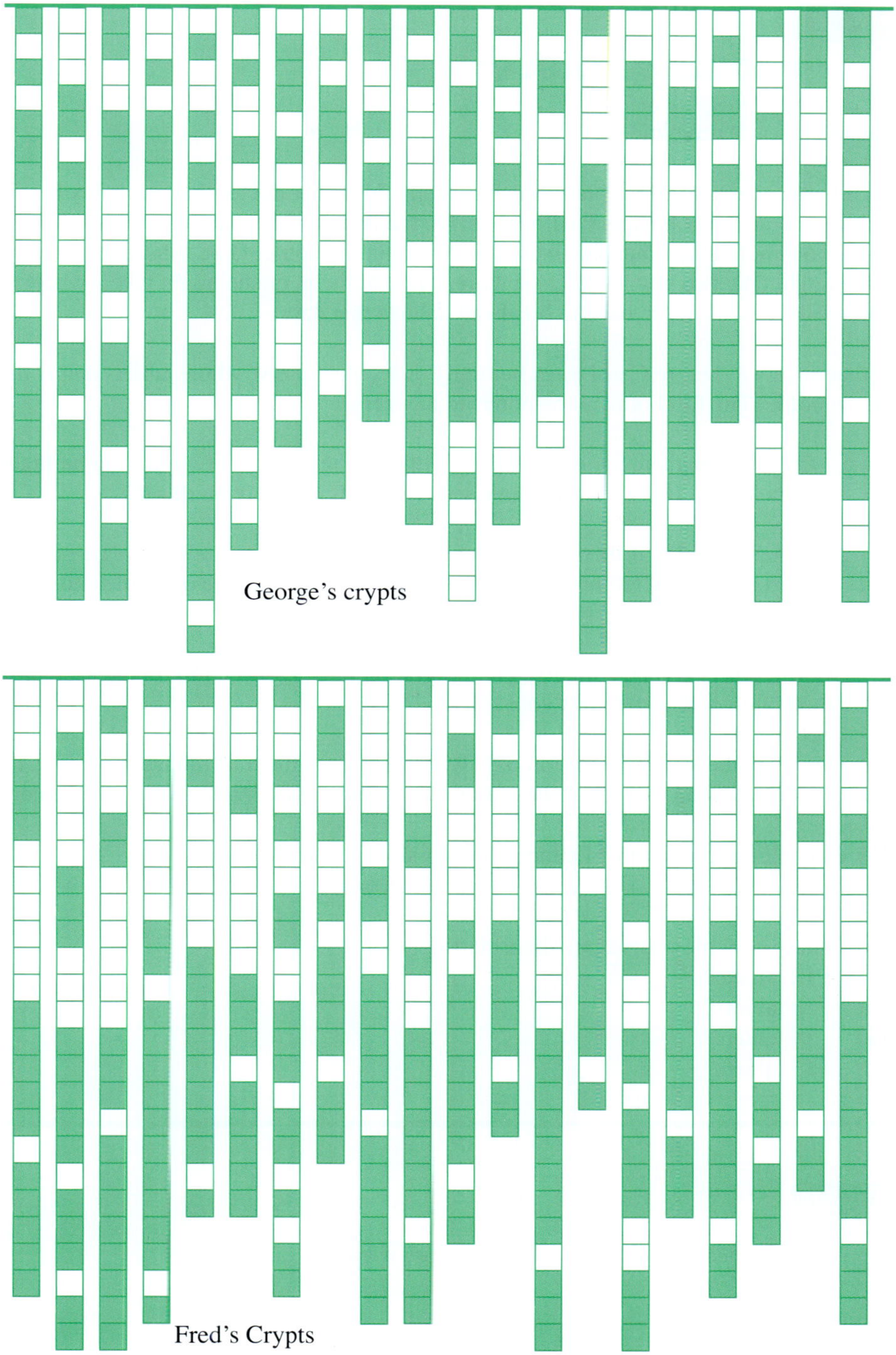

Find a good descriptive variable which indicates that statistically George's crypts are in some way rather different from Fred's crypts. Show also your analysis of other descriptive variables which do not indicate any significant difference between George and Fred.

Project 30

PROBLEM: TABLETS

Background: Arithmetic

Carol has a bottle containing 18 tablets. She has been told by her doctor that, each day before breakfast, she should take one and a half tablets. On the first day she takes two tablets from the bottle, breaks one in half and puts the unused half back in the bottle. On a subsequent day, perhaps she might take two whole tablets and break one of them, or she might take a whole and a half or three halves so as not to have to break any tablets that day.

Suppose that on day d, before Carol takes her daily dose, there are W whole tablets and H half tablets in the bottle. The full bottle corresponds to day zero.

a. i. On day zero there is only one possibility for the mixture of whole and half tablets, namely $W = 18, H = 0$. How many possibilities are there on day d?

ii. More generally, if there were N tablets in the bottle initially, find a rule or set of rules which will tell how many possibilities for the mixture of whole and half tablets there are on day d. The case $N = 18, d = 10$ is a good one to test your rule.

Give the answer in this case, and similarly give several more cases which for one reason or another you think are good ones to test your rule.

b. In going from the starting situation of N whole tablets to the final situation where a daily dose cannot be taken, the sequence of possible values for (W, H) shall be called a course.

For example, starting with 14 tablets, one possible course is
$(14,0) \rightarrow (12,1) \rightarrow (10,2) \rightarrow (9, 1) \rightarrow (7,2) \rightarrow (53) \rightarrow (5,0) \rightarrow (3, 1) \rightarrow (2,0) \rightarrow (0, 1)$

How many possible courses are there starting with
i. 12 tablets?
ii. 13 tablets?
iii. 14 tablets?
iv. 15 tablets?

Project 31

PROBLEM: ACID IN WATER

Background: Chapter 15, 18

The chemical adipic acid has the structure shown in the diagram following.

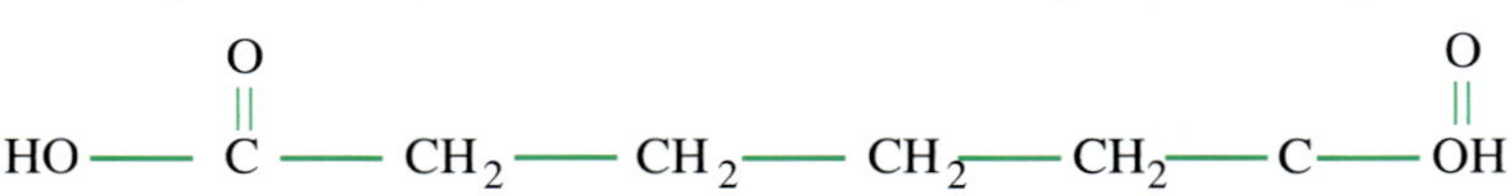

When added to water, the terminal groups (denoted by OH and HO) on either end may dissociate, with the oxygen (O) remaining part of the acid, and the hydrogen (H) becoming free in the water. The acid in water can therefore exist in one of three states, described and shown schematically as follows.

State 1
Both hydrogens still attached. HO — [ ] — OH

State 2
One hydrogen attached. HO — [ ] — O
or O — [ ] — OH

State 3
No hydrogens attached. O — [ ] — O

In each case [ ]

represents

In analysing an experiment, a chemist knows that the total number of hydrogens still attached to the terminal groups of the acid is equal to the number of acid molecules. Of interest are the proportions of molecules in each of the states 1 – 3.

As a first step in studying this problem, consider the case in which 10 molecules originally in state 1 have a total of 10 hydrogens, chosen at random, removed from the terminal groups.

> What is the most probable number of molecules which will remain in state 1?
>
> What is the expected (mean) number of molecules which will remain in state 1?
>
> Suppose the number of molecules is increased from 10 to, say, m. If m hydrogens are removed, how do the quantities calculated above change?

Project 32

PROBLEM: MAD MOUSE

Background: Chapter 7–9, 23

The track of a section of a mad mouse railway at an amusement park is in disrepair and needs replacing. Viewed from the side, the track has the shape shown in the following diagram.

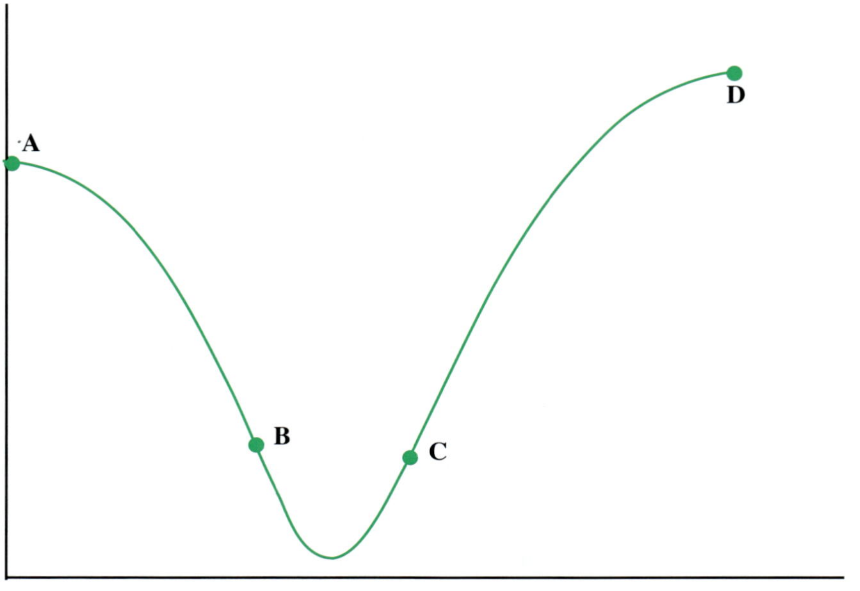

The profile of each of the segments AB, BC and CD is a parabola. The gradients of the profiles at the points B and C, where the segments meet, match so that the track is smooth.

The track at point A is 30 metres above ground level and inclined at an angle of 10° below the horizontal, while the track at point B is 10 metres above ground level and 8 metres horizontally across from A. The point C is at the same height as B and 11 metres across horizontally from A. The track at point D is 35 metres high and has zero slope.

The new track costs \$100 per metre. Devise a method to estimate the cost of the new track.

Extension: Carry out an investigation addressing similar problems as those discussed so far, but this time visit a skateboard circuit.
Draw diagrams to scale.
Consider where the curved ramps meet the ground or other obstacles. How do the designers of this circuits ensure a 'smooth' ride for the skateboarder?

How would you go about designing your own skateboard ramp?

Project 33

PROBLEM: SPREAD OF DISEASE

Background: Chapter 7–10, 12

One Sunday evening, five people with infectious influenza arrive in a large city with a population of about two million. They go to different locations in the city and thus the disease begins to spread throughout the population.

At first, when a person becomes infected he or she shows no sign of the disease and cannot spread it. About one week after first catching the disease this person becomes infective and can spread the disease to other people.

This infective phase also lasts for about one week. After this time the person is free from the influenza, although he or she may catch it again at some later time.

Epidemiologists are trying to model the spread of the influenza. They make a simplifying assumption that the infection progresses in one week units. That is, they assume that everybody who becomes infected does so on a Sunday evening, becomes infective exactly one week later, and is free of infection exactly one week after that. People free of the disease are called **susceptibles**.

The epidemiologists also assume that the city population is large enough so that the population size is constant for the duration of the disease. That is, they ignore births, deaths and other migrations into or out of the population.

Finally, they assume that each infective person infects a fixed fraction f of the number of susceptibles, so that

1. the **number of infectives** at week $n+1$ is equal to
   $f \times$ (the number of susceptibles at week n) $\times$ (the number of infectives at week n)

and

2. the **number of susceptibles** at week $n+1$ is equal to
   (the number of susceptibles at week n) + (the number of infectives at week n)
   – (the number of infectives at week $n+1$).

Why must the number of infectives, plus the number of susceptibles, be constant from week to week?

Choose values of f between $\frac{1}{10^6}$ and $\frac{2}{10^6}$ and use this model to show how the number of infectives changes from week to week from the Sunday when the five infective people arrived.

What limiting values does the model predict for the number of infectives?

Will there always be a limiting value for the number of infectives? If so, how are the limiting values related to the population size and to f?

For what values of f will it be possible to have a situation where the number of infectives eventually oscillates between two values?

Extension: How does the work covered in this project relate to the concepts encountered in Project 2 – Fractals?

What other models for the spread of diseases exist?

4.0 ROMAN & BABYLONIAN NUMBERS

We are used to writing our words using 'LETTERS' (a, b, c, . .) and numbers using 'NUMERALS' (1, 2, 3, . . .).

These are comparatively recent developments. Our alphabet is principally Roman (the Roman alphabet was developed from an Etruscan alphabet around 600BC) and is phonetic.

Our modern number system is known as the Hindu-Arabic system and was developed in the fifth and sixth centuries (AD) by Hindu mathematicians. Later the system was transmitted to Arab mathematicians (possibly by traders using the 'Silk Road') such as Muhammad ibn Musaal-Khwarizmi(c.780–850) who wrote one of the earliest books that used our modern decimal system.

Early methods of writing often used pictographs or pictures to represent objects. Thus 'FISH' might be written:  . Later methods such as hieroglyphs ('sacred carving') used such pictures to represent sounds as well as objects. Much of this type of writing survives on the monuments of ancient Egypt.

The Sumerian civilisation used 'cuneiform'. These were wedge shapes impressed into wet clay tablets using straw pens. The tablets were then fired to provide a durable record.

Try looking at http://www.upennmuseum.com/cuneiform.cgi. At the time of writing, this site has an on-line cuneiform 'translator' which will convert your name to cuneiform.

The best known form of early numerical writing of numbers is the Roman system where letters were used to represent numbers. A few of the basic equivalents are: I=1, V=5, X=10, L=50, C=100 D=500 & M=1,000. The number XXVI means 'ten, ten, five and1' or 26. Note that the system has limited place value. Both Xs mean the same thing. Contrast this with the decimal number 66 in which the first 6 means 'six tens' and the second six means 'six units'. The one exception is that VI in the Roman system means 'one more than five' or 6 and IV means 'one less than five' or 4.

Roman numerals still appear in a few applications such as the dates of films etc.

You might like to research how Roman school students were taught to solve problems such as 'Maximus captured LXI prisoners at Pontum and another CIX prisoners at Vallium. He then sold CV to Governor Bilious as slaves. How many did he take back to Rome?'

Even more challenging would have been: 'The army unit known as the maniple has soldiers arranged in XII files in X ranks. There are IV maniples in each cohort. How many soldiers are there in a cohort?'

It should also be remembered that many cultures developed quite advanced mathematics. As well as those mentioned above, there were working systems in Asia and the Americas.

4.1 NUMBER SYSTEM

We will not be looking at the history of numbers as such, although it could form part of a project, but rather we want to present an awareness of the need to consider how and why the number system needed to expand and evolve over time.

Consider the set of numbers $\{1, 2, 3, 4, 5, \ldots\}$. If we select any two numbers from this set, say 4 and 6, when they are added we obtain another number in this set, i.e., $4 + 6 = 10$. Similarly, when we multiply these two numbers, namely $4 \times 6 = 24$, again we have a number that belongs to this set. Such a set is called a **closed set** under addition and multiplication. But what happens when we divide 6 by 4, does the result belong to this set?

This basic question proved to be quite an issue in the early development of mathematics.We will briefly look at the evolution of the number system which enabled mathematics to develop the different number systems that we so readily use (and take for granted) today.

The set considered so far , i.e., $\{1, 2, 3, 4, \ldots\}$ is known as the set of positive numbers and is denoted by the letter $\mathbb{Z}^+$. This set is also known as the set of **Natural numbers**. Although this appears as if we were born with the ability to possess such 'trivial knowledge', it was only in the nineteenth century that the Italian mathematician Peano (1858–1932) and others like him, were successful in describing the set $\mathbb{Z}^+$ in a way which brought about (in a mathematical way) the nature of the basic properties of $\mathbb{Z}^+$. One of the properties that he formalised became the basis of the so called "Principle of Mathematical Induction" (see HL Mathematics).

Of course, as long as we added or multiplied numbers within this set all was well, we would always end up with another number from that set.

But, what happens when two numbers are subtracted?

If we choose 6 and 4, then $6 - 4 = 2$, which is still a natural number, but $4 - 6 = -2$.
What about $4 - 4$ ($= 0$), where does this fit into this particular number system?

The question then naturally arose (as it did when we considered $6 \div 4$), where do these numbers belong?

Because of questions like these, it was necessary to expand the number system to sets that included negative numbers, zero and fractions. New definitions came about so that we could classify these 'new' numbers. Imagine the difficulties a shepherd would have had when asked to pay in taxes, a total of 12 sheep, when he only had 10! How would he have accounted for (in a mathematical sense) the expression $10 - 12$, if the notion of negative numbers did not exist at that stage?

This did create a problem, and so, the number system was expanded, to allow for expressions such as $3 - 2 = 1$, $3 - 3 = 0$ and $3 - 4 = -1$. This set, called the set of integers and denoted by the letter $\mathbb{Z}$, was defined as $\mathbb{Z} = \{\ldots, -3, -2, -1, 0, 1, 2, 3, \ldots\}$.

This now enabled mathematicians to solve problems like, find x where $x + 6 = 4$ ($x = -2$). This also meant that we now had two sets, one of which was wholly contained within the other. That is, we now had that $\mathbb{Z}^+ \subset \mathbb{Z}$

Times were good, and everyone was happy, they had $\mathbb{Z}$ and they had $\mathbb{Z}^+$ – all questions involving arithmetic could now be addressed. Then, one day, Zackery wrote to Zoi:

Dear Zoi,

I am puzzled somewhat by your question. You make it clear to me that you wish to add one to a number that has been trebled. With five as the outcome.

I have spend many a sleepless nights one this task and I must conclude that I can find no number to satisfy this question.

Zackery.

Zackery was trying to solve the following (mathematical) problem:

Find that value of x, for which $3x + 1 = 5$.

The solution to this problem is $x = \frac{4}{3}$. The problem, however, was that the set $\mathbb{Z}$ that everyone had come to love did not contain such a number.

And so, out of necessity, the set $\mathbb{Z}$ was extended. It was extended to incude the 'gaps' that existed in $\mathbb{Z}$. This new set was called the set of **rational numbers**, and was denoted by the letter $\mathbb{Q}$.

Formally, the set of rational numbers, $\mathbb{Q}$, is given by

$$\mathbb{Q} = \left\{ x \mid x = \frac{a}{b}, b \neq 0 \text{ and } a \text{ and } b \text{ are integers} \right\}.$$

Notice that the restriction on this set is that **division by zero is not allowed**.

We now have the relationship that $\mathbb{Z}^+ \subset \mathbb{Z} \subset \mathbb{Q}$.

Everyone was now truly happy, all seemd to be in order, until Zackery received Zoi's next letter.

Dear Zackery,

I wish to have a square garden patch, so that I may plant, basil.
I've decided that the surface area of this patch is to be 2 sq. metres.

What side lengths should my square have?

Lots of love,
Zoi.

Zackery tried and tried, but there was to be no success! Once more, there was no number in the set $\mathbb{Q}$ that would allow Zackery to give Zoi an answer.

The problem Zackery was trying to solve was, find x, such that $x^2 = 2$.

Numbers generated from such problems could not be found in any of the sets found so far. This is where the set of **irrational numbers** came into play. Irrational numbers became the set of numbers that did not belong to the set $\mathbb{Q}$ and it was denoted by $\overline{\mathbb{Q}}$.

The solution to the equation $x^2 = 2$ is $x = \sqrt{2}$ (assuming that $x > 0$). Numbers that belong to this set are π, e, $\sqrt{2}$ (i.e., surds) and all numbers that cannot be expressed as a fraction.

It is also useful to know where numbers such as $\sqrt{2}$ are relative to the numbers 1 and 2. Is $\sqrt{2}$ closer to 1 or 2? We can make use of a number line and Pythagoras's Theorem to locate $\sqrt{2}$:

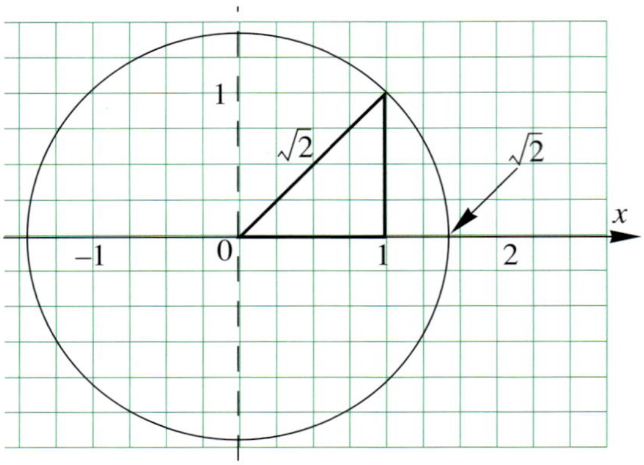

If we use a triangle with side lengths 1 unit then using Pythagoras's Theorem we have that the hypotenuse is $\sqrt{2}$ units.

We then draw a circle having as its radius a length equal to $\sqrt{2}$. Then, where the circle cuts the x-axis will be where $\sqrt{2}$ lies.

How would you go about locating $\sqrt{3}$?

Sometimes we run into numbers like, 0.33333 . . .. What type of number is this? It appears as if it is an irrational number, but in fact, it is a rational number. This can be shown as follows:

Let $x = 0.33333\ldots \therefore 10x = 3.33333\ldots$ [i.e., multiply both sides by 10]
However, the number 3.3333 . . . can be written as 3 + 0.33333 . . .

And so, we have that $\quad 10x = 3 + 0.33333\ldots$

But, $x = 0.33333\ldots$, so, $\quad 10x = 3 + x$

Therefore, $\quad 9x = 3$

i.e., $\quad x = \dfrac{1}{3}$

Therefore, we have that $0.33333\ldots = \dfrac{1}{3}$, which is a rational number.

The **union of** the **rational set** and the **irrational set** produce the **set of real numbers**.

That is, $Q \cup \overline{Q} = \mathbb{R}$. The set of real numbers contains every number that can be thought up (excluding some numbers that belong to a set known as **complex numbers** (see Higher Level course). Also, as the number system has developed over many hundreds of years, it might very well be the case that a new set of numbers, which contains new numbers that cannot be accounted for using our present system is yet to be developed!

We now provide a diagram highlighting the relationship between the sets covered.

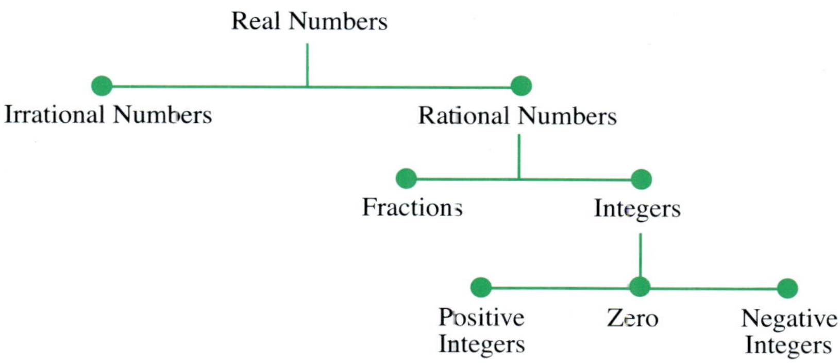

A summary of the sets of numbers is provided below:

Set of positive integers and zero $= \mathbb{N} = \{0, 1, 2, 3, \ldots\}$.

The set of integers $= \mathbb{Z} = \{0, \pm 1, \pm 2, \pm 3, \ldots\}$

The set of positive integers $= \mathbb{Z}^+ = \{1, 2, 3, \ldots\}$ [Also known as Natural numbers]

The set of rational numbers $= \mathbb{Q} = \left\{ x \mid x = \frac{a}{b}, b \neq 0 \text{ and } a \text{ and } b \text{ are integers} \right\}$

The set of positive rational numbers $= \mathbb{Q}^+$

The set of positive real numbers $= \mathbb{R}^+ = \{x \mid x \in \mathbb{R}, x > 0\}$

The empty set $= \varnothing =$ The set with *no* members.

EXERCISES 4.1

1. The following chart gives an idea of the progress of number development, starting from humble beginings, using rudimenatry counting objects such as stones, notches on ropes or markings on a stick to define natural numbers, to the sophistication of the real number set, that involves discussions of infinitesimals.

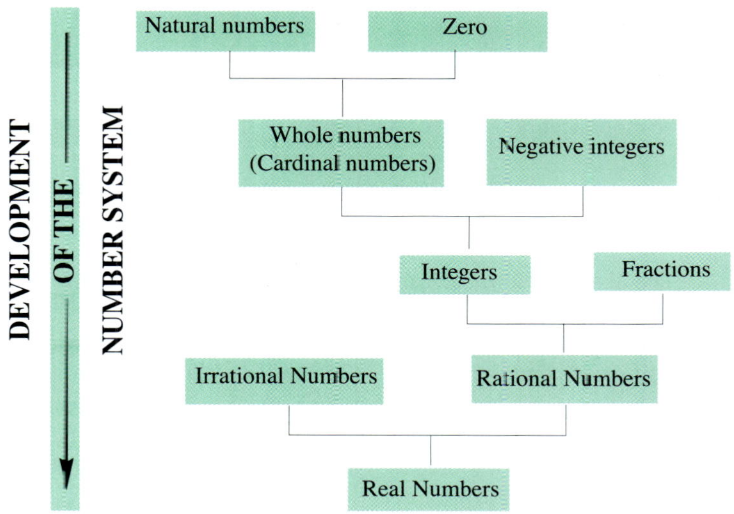

Your task is to research the development of different number systems, looking at systems used on different continents and in different time periods.

4.2 SIGNIFICANT FIGURES

4.2.1 QUOTING SIGNIFICANT FIGURES

Whenever a quantity is measured, there is some error in the measurement. The amount of error usually depends on the quality of the device used to make the measurement. Here are two attempts to measure the length of a rod:

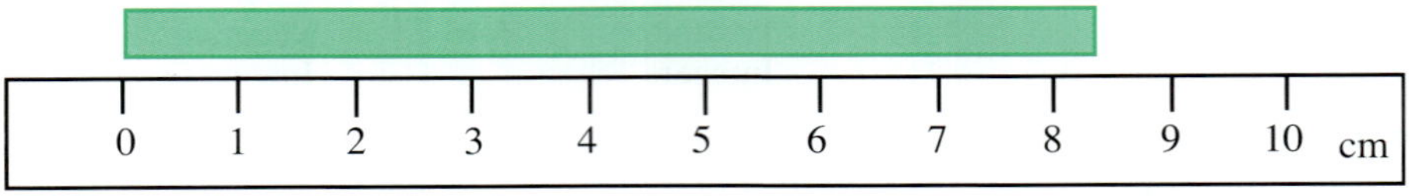

This ruler is only marked with centimetre graduations and suggests that the length of the rod is 8cm.

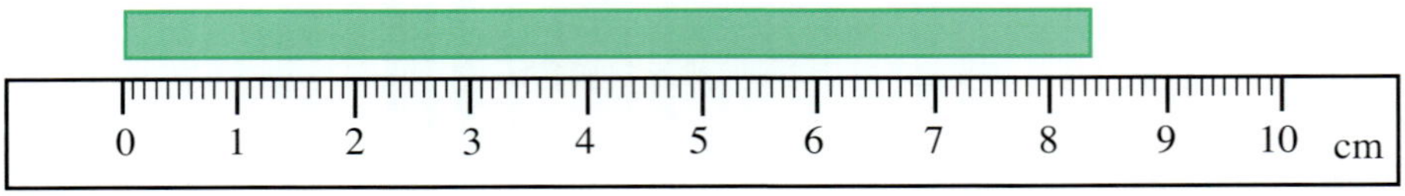

This ruler is marked with millimetre graduations and allows us to estimate the length of the rod as 8.3cm.

In the first case, the length is given as 8cm which means that the length of the rod is somewhere between 7.5 and 8.5cm. In the second case, when using the more accurate ruler, giving the length as 8.3cm means that the true length is between 8.25cm and 8.35cm. In summary:

A measurement of 8cm means that the true value is in the range 7.5cm to 8.5cm and is said to be accurate to **1 significant figure**.

A measurement of 8.3cm means that the true value is in the range 8.25cm to 8.35cm and is said to be accurate to **2 significant figures**.

In most cases, the number of significant figures in a measurement can be found by counting the figures: 6.743gms is 4 significant figures, 45216kg is 5 significant figures etc. Generally, the more significant figures, the more accurate the measurement.

Large and small measurements need to be considered carefully when deciding their level of accuracy. Consider this statement:

"The distance from the Earth to the Sun is 150 000 000km"

How many significant figures is this? There is a sense in which this is not obvious, but it seems likely that the statement means that the distance is 150 million kilometres as opposed to 140 million or 160 million kilometres. This means that the 1 and 5 are the significant figures and the zeros are only there to hold the position of the decimal point and tell us that the units of measurement are tens of millions of kilometres. The safest assumption is that the measurement is only accurate to two significant figures.

If you interpret the statement as meaning that the distance from the Earth to the Sun is 150000000 as opposed to 149999999 or 150000001km (which seems to be highly unlikely!), then the measurement is correct to 9 significant figures. This is seldom the case and you should interpret leading and trailing zeros as not being significant figures.

EXAMPLE 4.1

Find the number of significant figures in the following measurements:

(a)	4035km	(b)	56000kg	(c)	56004kg
(d)	0.001600mm	(e)	2.0016mm	(f)	300m

Solution

The significant figures are shown **bold**. Note that leading and trailing zeros do not count as 'significant'.

(a) **4035**km [4 S.F.]
(b) **56**000kg [2 S.F.]
(c) **56004**kg [5 S.F.]
(d) 0.00**1600**mm [4 S.F.]
(e) **2.0016**mm [5 S.F.]
(f) **3**00m [1 S.F.]

4.2.2 ROUNDING

It is often necessary to round answers to calculations. As a general principle, if you do calculations with numbers that are accurate to, say, 3 significant figures, the answers should be rounded to the same level of accuracy (3 significant figures).

EXAMPLE 4.2

A rectangular field is measured as 91 metres long and 75 metres wide. What is the area of the field?

Solution

The area is $91 \times 75 = 6825\text{m}^2$. This is four significant figures.
It is not appropriate to quote an answer to a much higher level of accuracy than the data used in the calculations. In this case, the data is accurate to 2 significant figures and so the answer should be rounded to this level of accuracy. Rounding the answer to 2 significant figures gives 6800 square metres.

In this case, we should note that the largest possible values of the length and width are 91.5 and 75.5 metres giving a largest possible value for the area of 6908.25m^2. Similarly, the smallest possible area is 90.5 by 74.5 = 6742.25m^2. From these figures, it is evident that we are not able to calculate the area to a greater level of accuracy than 2 significant figures.

EXAMPLE 4.3

Find the area of a circle of radius 2.33cm.

Solution

Area $= \pi \times 2.33^2 \approx 17.05539\text{cm}^2$.
It is appropriate to perform all calculations to a high level of accuracy.
In this case, for example, you should use the 'π' key and not any less accurate approximation such as 3.14.
The data is accurate to 3 significant figures and so the answer should be rounded to a similar level of accuracy. In this case the fourth figure is a 5 and so the result must be rounded up to 17.1cm^2.

EXERCISES 4.2

1. State the number of significant figures in these values:

(a)	34.52	(b)	5673.7	(c)	1200	(d)	4.001
(e)	0.00452	(f)	0.00340	(g)	784520	(h)	0.450
(i)	4503450	(j)	0.004520	(k)	67.4500	(l)	0.56204

2. Round the following values to the number of significant figures given:

(a)	2.526 [2 S.F]	(b)	24650 [3 S.F]	(c)	0.347 [2 S.F]
(d)	45627 [4 S.F]	(e)	0.4523 [2 S.F]	(f)	3.624 [1 S.F]
(g)	56720 [2 S.F]	(h)	0.04537 [3 S.F]	(i)	0.0045 [2 S.F]
(j)	345620 [3 S.F]	(k)	0.0453 [2 S.F]	(l)	89000 [1 S.F]

3. A square has an area of $67\ \text{cm}^2$. Find the length of one of the sides, giving the answer to an appropriate number of significant figures.

4. A rectangle has an area of $56\ \text{cm}^2$ and a length of 5.1cm. Find the width of the rectangle, giving the answer to an appropriate number of significant figures.

5. The angle of elevation of a building is 34° when measured from a distance of 65 metres. Find the height of the building giving the answer to an appropriate number of significant figures.

6. A painting is 782mm wide and 679mm high. What is the length of the diagonal (to an appropriate level of accuracy)?

7. Consider the series of fractions $\frac{1}{2}+\frac{1}{3}+\frac{1}{4}+\frac{1}{5}+\frac{1}{6}$............. . How many terms of this series will you need to add before the total is equal to 3, correct to three significant figures?

8. The fraction $\frac{22}{7}$ is often used as an approximation to π. To how many significant figures is this accurate?

9. The expression $\left(1+\frac{1}{n}\right)^n$ is equal to the irrational number e (press e^x ^1 on a calculator). What is the smallest value of n that will give e correct to 2 significant figures?

4.3 SCIENTIFIC NOTATION

Very large and very small numbers can be written in a convenient, abbreviated, form known as scientific notation. To convert a number into scientific notation, it is first split into two factors, one of which is a power of ten and the other a number in the range 1 to (but not including) 10.

(i.e., 3 'jumps' **to the right**)

$$4000 = 4 \times 1000 = 4 \times 10^3$$

(i.e., 6 'jumps')

$$5603000 = 5.603 \times 1000000 = 5.603 \times 10^6$$

$$372000 = 3.72 \times 100000 = 3.72 \times 10^5$$

5 zeroes

$$62420000 = 6.242 \times 10000000 = 6.242 \times 10^7$$

7 zeroes

Small numbers can be converted to scientific notation in a similar way:

(i.e., 3 'jumps' **to the left**)

$$0.004 = 4 \times \frac{1}{1000} = 4 \times 10^{-3}$$

(i.e., 5 'jumps' **to the left**)

$$0.0000402 = 4.02 \times \frac{1}{100000} = 4.02 \times 10^{-5}$$

Notice that this time, as the **jumps are to the left**, the **power is negative**!

$$0.0002049 = 2.049 \times \frac{1}{10000} = 2.049 \times 10^{-4}$$

4 zeroes

$$0.2501 = 2.501 \times \frac{1}{10} = 2.501 \times 10^{-1}$$

1 zero

It is often desirable to give very large or small answers in scientific notation as it is concise and makes the number of significant figures obvious.

In the case of the number 3.604×10^{-7}, all the figures in the 3.604 part of the number are significant and we can say that the number is accurate to 4 significant figures.

Many scientific quantities are either very large or very small. Here are some commonly used physical constants in scientific notation:

The speed of light: 2.998×10^8 ms^{-1}

The mass of a neutron: 1.67×10^{-24} gm

The charge on an electron: 1.6021×10^{-19} coulombs

You should know how to enter quantities that are in scientific notation on your calculator. Many scientific calculators have an 'EXP' key that allows the direct entry of a power of 10. Thus the speed of light would be entered as 2.998 'EXP'8. Failing this, the number can be entered as $2.998 \times 10x^y8$ using the powers key.

Using a Graphics Calculator

Graphics and other calculators can be set to display answers correct to a fixed number of decimal places.

To display all answers correct to three decimal places, press the **MODE** key and select **Float**.... 3 from the screen. It is necessary to press **ENTER** to confirm this selection.

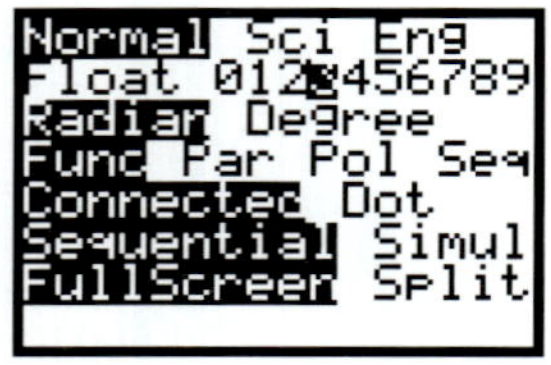

.

All subsequent calculations will now be rounded to three decimal places. This screen shows some examples:

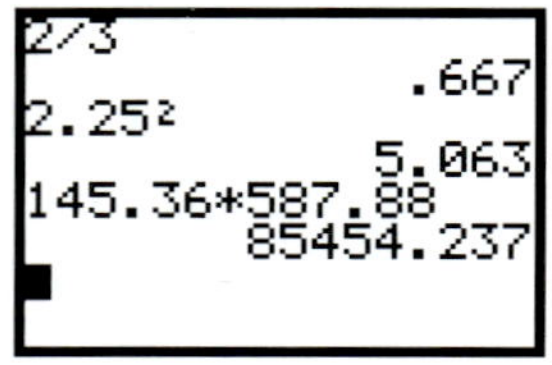

If it is desired to set the calculator to display answers correct to three significant figures, the calculator must be set to display answers in scientific notation and set the display to **Float**... **2** (**ENTER**).

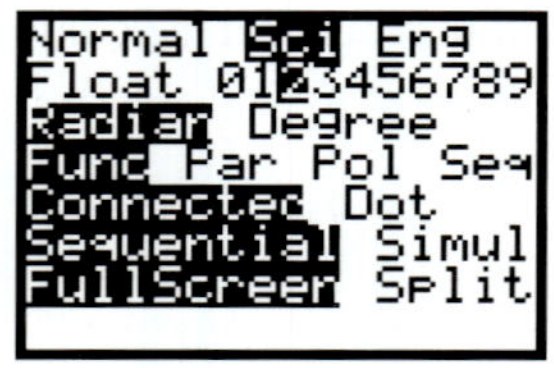

Calculations will now appear in scientific notation:

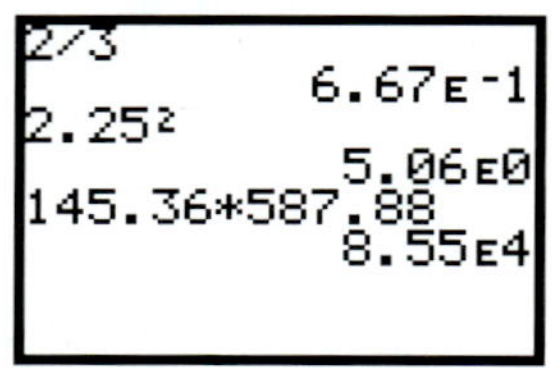

Numbers in scientific notation can be entered using the 2nd **EE** key. Note that the (-) and not the subtraction key must be used for negative indices

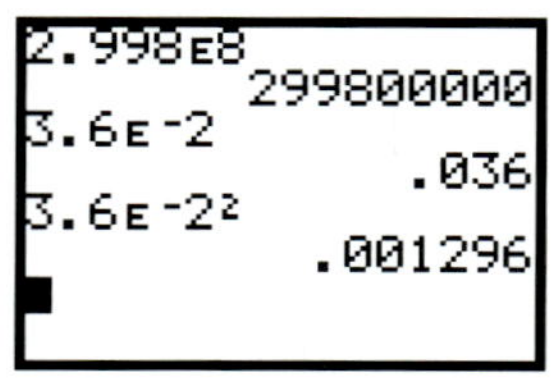

When calculating with numbers in scientific notation, the principle of giving answers to an appropriate level of accuracy discussed earlier, also applies.

EXAMPLE 4.4

If $a = 0.000982$ $\quad b = 3.56 \times 10^{-4}$ $\quad c = 4.56 \times 10^{6}$, evaluate:

(a) $a(b + c)$ (b) ab (c) $\dfrac{b}{a + c}$

Solution

(a) $a(b+c) = 0.000982(3.56 \times 10^{-4} + 4.56 \times 10^{6}) \approx 4.48 \times 10^{3}$ to 3 S.F.

(b) $ab = 0.000982 \times 3.56 \times 10^{-4} \approx 3.50 \times 10^{-7}$ to 3 S.F.

(c) $\dfrac{b}{a+c} = \dfrac{3.56 \times 10^{-4}}{0.000982 + 4.56 \times 10^{6}} \approx 7.81 \times 10^{-11}$ to 3 S.F.

Using a Graphics Calculator

When using a graphics calculator, you should see these, or similar, screens if you first set the display to '**SCI**' '**FLOAT**' **2** format as described earlier. You must be able to interpret such screens and you should give answers in correct scientific notation, not the calculator version depicted.

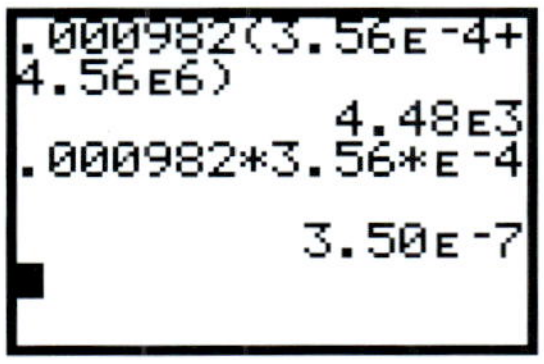

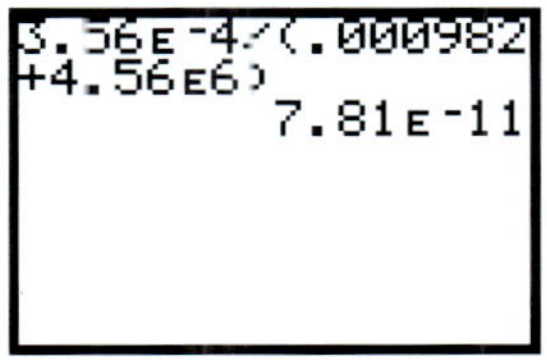

In example (c) the use of brackets for the denominator is essential if your calculator is not to interpret the calculation as $\frac{b}{a} + c$ which gives 4.56×10^{6}, a substantially different and incorrect answer.

It is a good practice when using a calculator to estimate the answer before entering the calculation. This helps in the detection of errors in keying and is a technique that you are advised to practise.

EXERCISES 4.3

1. If $x = 3.8 \times 10^{3}$ $\quad y = 5.7 \times 10^{-3}$ $\quad z = 9.1 \times 10^{5}$, evaluate:

(a) xy (b) $\dfrac{1+y}{z^2}$ (c) $y(x+z)^2$

2. The radius of a planet has been estimated to be 4720000000 metres. Write this in standard form. Find the circumference and the volume of the planet correct to an appropriate level of accuracy.

3. A 2000 metre (to the nearest metre) length of optical fibre cable has a radius of 3.45×10^{-5} metres. Find the volume of the cable.

4. The world's oceans contain about 23 million cubic kilometres of water. If this water were made into a giant ice cube what would be the approximate length of a side? You may ignore the fact that water expands when it freezes.

5. Einstein's formula $E = mc^2$ gives the amount of energy, E, released when m kg of matter is destroyed in a process such as a nuclear reaction. The constant c is the velocity of light, $2.998 \times 10^8 \text{ ms}^{-1}$. The unit of measurement of the energy released is $\text{kgm}^2\text{s}^{-2}$. Find the amount of energy released when 5.000×10^{-2} kg of matter is destroyed.

6. Lindy has an approximate heart rate of 83 beats per minute. How many beats will Lindy's heart make during her expected life-span of 78 years?

7. On average, people have children when they are 27 years old. Thus, a 'generation' lasts 27 years. Every human has two parents, four grand-parents and eight great grand-parents. Jesus Christ was born about 2000 years ago. If you traced your family back to the time when Jesus Christ was alive, how many generations would have passed and how many direct relations would you have?

8. A grain of salt is a cube of side length 0.2mm. Assuming that the grains fit together perfectly (i.e. no space is wasted), how many grains of salt will fit into a cubic box of side 10cm?

9. 2.0 grams of hydrogen contain 6.5×10^{23} molecules. What is the mass of one molecule of the gas?

10. Normal alternating current electricity is produced at 50 cycles per second. How many cycles are delivered in one day?

4.4 ROUNDING ERRORS

Earlier in this chapter we have seen that all measured quantities are subject to some level of error. It is also the case that calculated values are often subject to error. There are several sources of such error. The most common of these arises when we use a calculating device such as a hand held calculator or computer. These devices only calculate to a limited number of significant figures. This is usually quite sufficient for practical calculations. The TI-82 graphic calculator calculates and stores numbers of up to 14 digits with a 2 digit exponent. The display gives answers to a lower level of accuracy. The reason for this will be discussed later in this section. First we must consider the ways in which errors in calculations and measurement are defined:

4.4.1 ABSOLUTE ERROR

Absolute error = the absolute value **of the difference** between an approximation and the true value.

An 'absolute value' is merely the positive value of the error. So, for example, the absolute value of –12.34 is 12.34, or the absolute value of –0.29 is 0.29.

EXAMPLE 4.5

If the true value of a measurement of length is 23.735 metres, what are the absolute errors if three measurements of this length are 24.7 metres, 22.6 metres & 23.733 metres?

Solution

The absolute errors are:

$|23.735 - 24.7| = 0.965\text{m}$

$|23.735 - 22.6| = 1.135\text{m}$

$|23.735 - 23.733| = 0.002\text{m}$

All the errors are positive whether or not the measurement is larger or smaller than the true value. This enables us to make direct comparisons of the three measurements and, for example, rank them in order of accuracy. In this case it is clear that the third measurement is the best and the second is the worst.

Having all the errors positive also allows the averaging of a series of related errors. This would not give meaningful results if some errors were positive and some negative.

EXAMPLE 4.6

If the approximation 1.414 is used for $\sqrt{2}$, what is the absolute error.

Solution

$\sqrt{2}$ cannot be written exactly as either a fraction or a decimal. This disturbing fact was first established by the Pythagorean school of Mathematics in Ancient Greece. Thus, all values of this quantity, including those produced by your calculator are approximate. In this case, we can only estimate the error as:

```
abs (√2-1.414)
     2.135623731E-4
```

A graphics calculator display of the calculation and answer might appear as shown. You would need to interpret the answer as being about 0.00021.

EXAMPLE 4.7

The equatorial diameter of the Earth is 12756km. If the approximation 3.14 is used for π, what is the absolute error in calculating the equatorial circumference of the Earth? What is the error if $\frac{22}{7}$ is used as an approximation to π?

Solution

The ratio π has a similar status to $\sqrt{2}$ in that there is no exact value available to use in calculations.

The π key on calculators is, however, a very good approximation. If using a calculator to evaluate these errors, the screen should appear somewhat like this example. Fewer key-strokes are used if the calculation is entered in factorised form: $\left|\pi - \frac{22}{7}\right| \times 12756$.

```
abs (π*12756-22/
7*12756)
         16.12982509
abs (π*12756-3.1
4*12756)
         20.31588919
```

The answers give an error of 16 km in the second case and about 20 km in the case of the first approximation of 3.14 is used instead of π..

4.4.2 RELATIVE AND PERCENTAGE ERROR:

The second and third examples above illustrate one problem with absolute error as a measure. It appears that an error of 20km in the third case is much worse than the error of 0.00021 in the second. But there is a sense in which this is a false impression because the numbers are of such different size. The question is which is the larger:

An error of 0.00021 in a quantity of about 1.4142

OR:

An error of 20km in a quantity of about 40074km (the approximate equatorial circumference of the Earth.)

To help decide this sort of question, we often use two additional error measures:

$$\text{Relative Error} = \frac{\text{Absolute Error}}{\text{True value}}$$ AND: $$\text{Percentage Error} = \frac{\text{Absolute Error}}{\text{True value}} \times \frac{100}{1}$$

With this definition, it should be noted that neither relative nor percentage error have units.

EXAMPLE 4.8

Find the relative and percentage errors in the previous set of examples, 4.5, 4.6 and 4.7. Which of these are the most accurate?

Solution

1. **Example 4.5:**

The absolute errors are:
$|23.735 - 24.7| = 0.965\text{m}$
$|23.735 - 22.6| = 1.135\text{m}$
$|23.735 - 23.733| = 0.002\text{m}$

The relative errors are:
$$\frac{0.965}{23.735} \approx 0.0407$$
$$\frac{1.135}{23.735} \approx 0.0478$$
$$\frac{0.002}{23.735} \approx 0.0000843$$

The percentage errors are 100 times these figures: 4%, 5% & 0.008% (approx.).

2. **Example 4.6:**

Relative Error $= \dfrac{|1.414 - \sqrt{2}|}{\sqrt{2}} \approx 0.000151$ and percentage error is about 0.0151%

3. **Example 4.7:**

$$\pi \approx \frac{22}{7} \qquad \text{Relative Error} = \frac{\left(\pi - \frac{22}{7}\right)12756}{\pi \times 12756} = \frac{\pi - \frac{22}{7}}{\pi} \approx -0.000402$$

As with absolute error, negative results such as the one above are usually given as the positive (absolute) value. So in this case, we would say that the relative error is approximately 0.000402 and the percentage error about 0.0402%.

$\pi \approx 3.14$ Relative Error $= \frac{(\pi - 3.14)12756}{\pi \times 12756} = \frac{\pi - 3.14}{\pi} \approx 0.000507$

The percentage error is about 0.0507% or a bit worse than the first approximation. It is now possible to make much fairer comparisons of the errors implied in these calculations. For example, if we wanted to compare the errors in the approximation of $\sqrt{2}$ (Ex 4.6.) with those in the approximation of π (Ex 4.7), we can see that the errors in example (Ex 4.7) are about five times larger than those in example (Ex 4.6).

In all cases, when you do calculations with numbers that are in error, your answers should show that you recognise that answers can only be quoted to a limited level of accuracy. You should also recognise that there are several ways of analysing error. Within one set of calculations in which all the results are of a similar size, absolute error is a good measure. When trying to compare the errors in measurements that differ largely in size, relative and percentage errors are probably the best measures to use.

EXERCISES 4.4

1. Find the absolute, relative and percentage errors if 0.33 is used as an approximation to $\frac{1}{3}$.

2. $\sqrt[3]{20} \approx 2.7$. Find the absolute, relative and percentage errors if 2.7 is used as an approximation to $\sqrt[3]{20}$.

3. A digital bathroom scale measures weights to the nearest 0.2kg. Find the absolute, relative and percentage errors involved in using these scales to weigh:

i. Pericles the cat who weighs 7.4kg.
ii. Texas the dog who weighs 19.6kg.
iii. The luggage which weighs 22.4kg.
iv. Leila who weighs 56.8kg.
v. Uncle Bill who weighs 98.8kg.

4. It is possible to use a schoolroom ruler to measure lengths to the nearest millimetre. Find the absolute, relative and percentage errors in using such a ruler to measure lines of length:

i. 0.5cm   ii. 1cm   iii. 12cm   iv. 25cm

5. Before the advent of the calculator, we used to find the trigonometric ratios, logarithms, square roots etc. from books of tables. Find the percentage errors involved in these approximations that come from an old book of tables:

i. $\tan 15^\circ = 0.2679$
ii. $\sin 2^\circ = 0.0349$
iii. $\cos 89^\circ = 0.0175$
iv. $\sqrt{5} = 2.236$

6. A stopwatch can measure time to the nearest tenth of a second. Find the percentage errors involved if this watch is used to measure the times taken for these events:

i. A 100 metre sprint completed in 11.2sec.
ii. A 400 metre race completed in 68.4sec.
iii. A 1500 metre race completed in 4mins 6.7sec.
iv. A 10000 metre race which took 38mins 7.9sec.

7. The Lockheed SR71 'Blackbird' surveillance aircraft was once quoted (in a television documentary) as having flown from New York to London in 2 hours, 7 minutes and 6.8 sec.
What is the percentage error implied by this statement? Is this a reasonable level of accuracy for this measurement?

4.5 COMPUTATION ERRORS

Whenever we calculate with one or more quantities that are in error, the results will also be in error. This section will look at the ways in which such errors propagate through such calculations.

EXAMPLE 4.9

A rectangular field has been measured as being 120metres long and 55metres wide. Both measurements are correct to two significant figures. What are the errors if these figures are used to calculate the perimeter and the area of the field?

Solution

Perimeter:
The length is 120metres to the nearest 10metres (2 significant figures). This means that the smallest possible length is 115metres and the largest is 125metres.

In the same way, the width is correct to 2 significant figures and so is between 54.5 and 55.5metres.

Smallest perimeter = 2(smallest width + smallest length) = 2(115 + 54.5) = 339metres

Perimeter = 2(width + length) = 2(120 + 55) = 350metres

Largest perimeter = 2(largest width + largest length) = 2(125 + 55.5) = 361metres

The perimeter can now be given as: 350 ± 11 metres.

This result is an example of a general principle. We have done a calculation with a length that has an error of 5metres and with a width that has an error of 0.5 metres.

The calculation involved adding the length and the width and then doubling the total.

The errors could have been calculated by adding and then doubling the errors to get:
Error = 2(5 + 0.5) = 11metres.

When calculating the areas, a similar process applies:

Area:

Smallest area = $115 \times 54.5 = 6267.5\text{m}^2$

True area = $120 \times 55 = 6600\text{m}^2$ (the data is in error, so this is not necessarily correct!)

Largest area = $125 \times 55.5 = 6937.5\text{m}^2$

In this case, we can say that the true area is probably best given as being between 6267.5 and 6937.5 square metres.

The situation is much more complicated than the case of the perimeter. The error range is not symmetrical: The difference between the 'true' area and the smallest area is 332.5m^2 and the difference between the largest area and the 'true' area is 337.5m^2.

In these calculations that are more complex than simple addition and subtraction, the only way to work out the effects of the errors is, as we have just done, to find the smallest and largest possible values that the calculation can produce, as well as the 'true' value.

EXAMPLE 4.10

If $a = 3.2 \pm 0.1$ $b = 6.9 \pm 0.1$ $c = 4.7 \pm 0.1$, find the errors that result from evaluating these expressions:

(a) $2a + b$ (c) $a(b + c)$ (d) abc

Solution

(a)

Smallest value	'True' value	Largest value
$2 \times 3.1 + 6.8 = 13.0$	$2 \times 3.2 + 6.9 = 13.3$	$2 \times 3.3 + 7.0 = 13.6$

In this case, the answer can be given as 13.3 ± 0.3. This is a level of error that might be expected with a calculation in which the data is in error by 0.1

(b)

Smallest value	'True' value	Largest value
$3.1(6.8 + 4.6) = 35.34$	$3.2(6.9 + 4.7) = 37.12$	$3.3(7.0 + 4.8) = 38.94$

Answer = 37.12, but we can only say that it is between 35.34 & 38.94.
The error range is not symmetrical, but the average error is about 1.8.

(c)

Smallest value	'True' value	Largest value
$3.1 \times 6.8 \times 4.6$	$3.2 \times 6.9 \times 4.7$	$3.3 \times 7.0 \times 4.8$
$= 96.968$	$= 103.776$	$= 110.88$

The answer is best given as being in the range 96.968 & 110.88. A sensible answer could be 104 ± 7. In this case, it is a borderline decision as to whether or not we should quote three significant figures for the answer. There is a sense in which a better answer might be

100 ± 10. Once again, the data was given to an accuracy a bit better than two significant figures and the detailed error analysis indicates that a similar level of precision is appropriate in the answers.

In all these cases, the relative and percentage errors in the answers are of approximately the same size as those of the data (between about 2% & 7%).

In the following set of examples, you will see that in complex calculations, it is necessary to be much more careful when calculating error bounds. It is also the case that there are some calculations in which the answer can be in error at a much higher level than the data. It is particularly necessary to be aware of this possibility when programming computers to perform calculations with data that contain errors.

EXAMPLE 4.11

If $x = 5.3 \pm 0.1$ $\quad y = 12.2 \pm 0.1$ $\quad z = 5.6 \pm 0.1$, find the errors that result from evaluating these expressions:

(a) $y - z$ (b) $\dfrac{y}{z - x}$ (c) $x^2 - y^2$

Solution

(a)

Smallest value	'True' value	Largest value
$12.1 - 5.7 = 6.4$	$12.2 - 5.6 = 6.6$	$12.3 - 5.5 = 6.8$

In this example the **smallest** possible result comes from subtracting the **largest** z from the **smallest** y.

In the same way, the **largest** possible result comes from subtracting the **smallest** z from the **largest** y.

It is often necessary to think carefully through the logic of these error analyses! The answer can be given as 6.6 ± 0.2. The absolute error is, in this case, equal to the sum of the absolute errors in the data. The percentage errors in the data are about 1%. However, the percentage error in the answer is about three times as much at 3%.

(b)

Smallest value	'True' value	Largest value
$\dfrac{\text{Smallest } y}{\text{Largest } z - \text{Smallest } x}$	$\dfrac{\text{True } y}{\text{True } z - \text{True } x}$	$\dfrac{\text{Largest } y}{\text{Smallest } z - \text{Largest } x}$
$= \dfrac{12.1}{5.7 - 5.2}$	$= \dfrac{12.2}{5.6 - 5.3}$	$= \dfrac{12.3}{5.5 - 5.4}$
$= 24.2$	$= 40.667$	$= 123$

The results of these calculations show that the 'best' answer is about 41, however we can only say that the true answer is in the approximate interval [24,123], a more than 100% error.

Considering that the original data contains errors in the 1% region, the errors in the answer are little short of disastrous! The size of this error results from the nature of the calculation and not from the size of the errors in the data. It is very important to look for calculations that produce this sort of error magnification when programming computers. It is possible for the very small errors that result from the storage of numbers in a computer to be significantly magnified during calculations.

(c) Smallest value $5.2^2 - 12.3^2 = -124.25$ | 'True' value $5.3^2 - 12.2^2 = -120.75$ | Largest value $5.4^2 - 12.1^2 = -117.25$

	Smallest value	'True' value	Largest value
(c)	$5.2^2 - 12.3^2 = -124.25$	$5.3^2 - 12.2^2 = -120.75$	$5.4^2 - 12.1^2 = -117.25$

Note that –117.25 is larger than –124.25.

EXERCISES 4.5

1. If $a = 12.5 \pm 0.2$ $b = 3.4 \pm 0.1$ $c = 56.4 \pm 0.5$. Evaluate the following expressions and give error bounds for the answers:

(a) $\dfrac{a}{b+c}$ (b) $\dfrac{c^2}{a-b^2}$ (c) $\dfrac{a(b-c)}{b}$

2. An engine component consists of a cuboid whose dimensions are 334.7mm long by 126.8mm wide by 12.7mm high. These dimensions are known to the nearest tenth of a millimetre. The circular hole has a radius of 23.7mm also accurate to the nearest tenth of a millimetre.

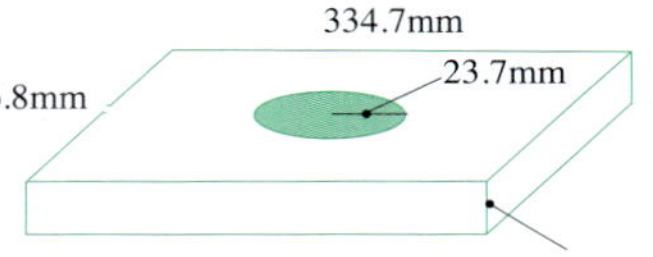

Find the volume of metal used in making this component and the errors that can be expected in this measurement.

3. If we use a value of 3.14 for π to calculate the volume of a sphere of radius 34 ± 1 cm, what is the range of values we might expect to get and what is the percentage error?

$\left[\text{Volume of a sphere} = \frac{4}{3}\pi r^3\right]$

4. The period (time for a complete swing) of a pendulum, T seconds, is given by the formula

$T = 2\pi\sqrt{\frac{l}{g}}$ where l is the length of the pendulum in metres and g is the acceleration due to gravity which is $9.81 \pm 0.01\,\text{ms}^{-2}$. Find the period of a pendulum of length 5.75 ± 0.01 metres.

How many swings will the pendulum complete in one day and what are the approximate error bounds on this number?

5. The angle of elevation of a building is measured as being 31° to the nearest degree from a point 47 metres, to the nearest metre, from the base of the building. What is the height of the building and what level of accuracy can be claimed?

6. A rectangular park has a length of 95 metres and a width of 46 metres. Both measurements are correct to two significant figures. What is the length of the diagonal of the park and what errors are involved in the result?

7. If values of a & b are to be substituted into the formula $Z = \frac{a+b}{a-b}$, what values of a & b will lead to the largest errors?

8. The average weight of a group of netballers is to be calculated. The players have been weighed on digital scales that weigh to the nearest half kilogram. What will be the absolute error in the average weight?

9. An optical lens forms a focussed image when an object is placed in front of it.

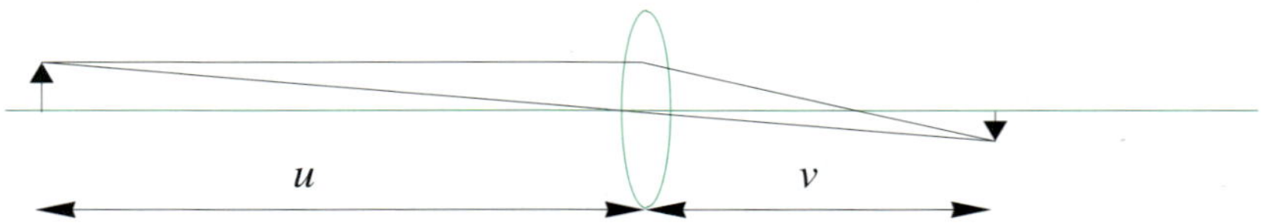

If the object is u metres in front of the lens, the image is v metres behind the lens (and is also upside down). The formula connecting these two values is: $\frac{1}{u} + \frac{1}{v} = \frac{1}{f}$ where f is the focal length of the lens.
A lens has a focal length of 7.5 ±0.1 cm. If an object is placed 14.6 ±0.1 cm in front of the lens, how far behind the lens will the image form and how accurate is this figure?

10. The resistance to motion of an object through a liquid, R, depends upon the velocity of the object, v and is given by the formula $R = 7v^3$. If an object is moving with a velocity of 12 (correct to two significant figures), what is the resistance to motion and what are the errors in this measurement?

11. The area of a cone is given by the formula:

$$A = \pi rs + \pi r^2 .$$

A cone has a base radius (r) of 12cm and **vertical** height (h) of 45cm. Both measurements are correct to two significant figures.

What is the surface area of the cone and what is the percentage error in this figure?

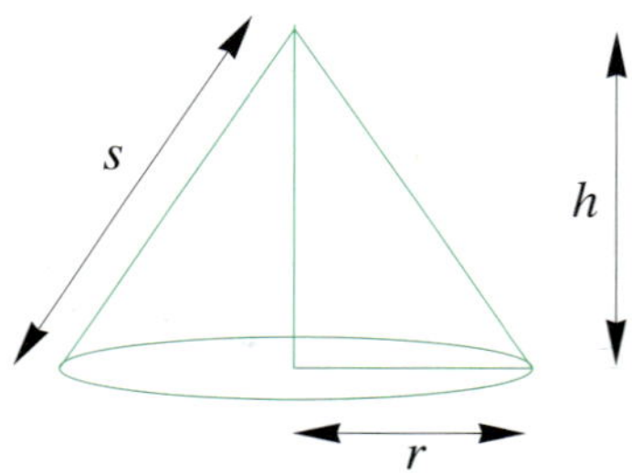

12. If a calculating device approximates all fractions by rounding them to five significant figures, what is the percentage error when calculating: $\frac{1}{2} + \frac{1}{3} + \frac{1}{4} + \frac{1}{5} + \frac{1}{6}$?

4.6 MISCELLANEOUS QUESTIONS

1. If $a = 3.14$, $b = 0.98$ and $c = 1.04$, calculate $a(c-b)(c+b)$, giving the answer
(a) exactly;
(b) correct to three significant figures;
(c) correct to two decimal places;
(d) in the form $a \times 10^k$, where $1 \le a < 10$, $k \in \mathbb{Z}$.

2. If $a = 78.3$, $b = 0.32$ find the value of $4a^2 + ab$, giving the answer
(a) exactly;
(b) correct to three significant figures;
(c) correct to two decimal places;
(d) to the nearest whole number;
(e) in the form $a \times 10^k$, where $1 \le a < 10$, $k \in \mathbb{Z}$.

3. If $x = 12.9$, calculate the value of $\sqrt{\frac{5}{6}\pi x^3}$, giving the answer
(a) i. correct to three significant figures;
ii. correct to two decimal places;
iii. to the nearest whole number.
(b) Write your answer to (a) iii. in the form $a \times 10^k$, where $1 \le a < 10$, $k \in \mathbb{Z}$.

4. The volume, V cm^3, of the cone shown is given by the formula

$$V = \frac{1}{3}\pi r^2 h.$$

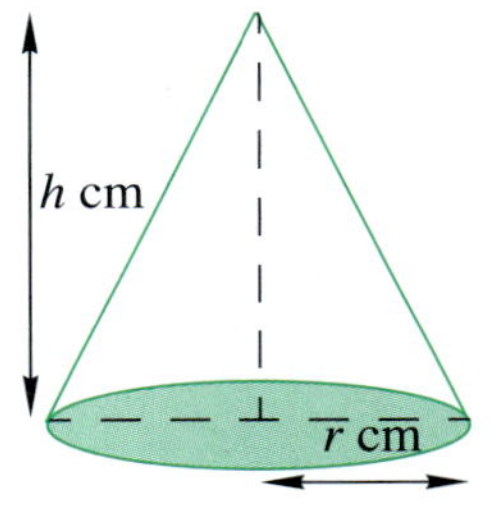

If $r = 7.8$, $h = 12.4$ find its volume, giving the answer
(a) correct to three significant figures;
(b) correct to two decimal places;
(c) to the nearest whole number;

5. If the fraction $\frac{1}{7}$ is approximated to 5 significant figures, find the percentage error in this measurement, correct to 2 significant figures.

6. Express $\frac{\sqrt{12} - \sqrt{5}}{6}$ as a decimal, correct to 3 significant figures.

7. State the number of significant figures in each of these measurements:
(a) 0.00315m (b) 40500gms (c) 0.020kg

8. Evaluate each of these expressions, giving the answers in scientific form, correct to 3 significant figures. (a) 5.27^{30} (b) $\frac{2.6^3}{55.8^{10}}$ (c) $\sqrt{0.045}$

9. Express the following numbers in normal decimal notation:
(a) 3.4×10^4 (b) 7.05×10^{-4}

10. If $a = 6.4 \pm 0.2$, $b = 7.1 \pm 0.4$, evaluate the following expressions, giving bounds on the errors. (a) $2a + b$ (b) $a \times b$ (c) $\frac{1}{b - a}$

11. A method (due to Euler) that has been used to calculate approximations to π is:

$$\frac{\pi^2}{6} = \frac{1}{1^2} + \frac{1}{2^2} + \frac{1}{3^2} + \frac{1}{4^2} + \ldots$$

If the first 5 terms of the series are used to find π, what are the absolute and percentage errors produced by the method?

4.7 GRADED REVISION EXERCISES

LEVEL 1

1. Which of the following equations have solutions that are members of the set $\mathbb{Q}$?

i.	$2x - 3 = 1$	ii.	$\frac{2}{x} = 9$	iii.	$x^2 = 6.25$
iv.	$x^2 = 27$	v.	$x = x + 7$	vi.	$\frac{1}{x} = \frac{x}{7}$

2. Write the following numbers in standard form:

i.	34	ii.	56700	iii.	600056	iv.	0.4
v.	0.00438	vi.	4.0456	vii.	0.0020	viii.	$\frac{3}{4}$
ix.	0.0000054	x.	100^2	xi.	$\frac{6}{100}$	xii.	$\frac{3}{40}$

LEVEL 2

1. If $a = 2.47 \times 10^{-2}$, $b = 7.92 \times 10^{3}$, $c = 4.01 \times 10^{-3}$, evaluate, giving answers in standard form, correct to three significant figures:

i. $a + b$ ii. ab iii. $\frac{1}{bc}$ iv. $\frac{1}{a - c}$ v. a^2c

2. If $x = 27 \pm 1$, $y = 155 \pm 5$, $z = 93 \pm 2$, find the percentage error, correct to two significant figures, that results from calculating:

i. $x + y$ ii. $y - z$ iii. xy iv. $\frac{1}{y - z}$ v. x^2y

3. By writing the following numbers as fractions, prove that they are rational numbers:

i. 1.3 ii. 0.25 iii. $0.\dot{3}$ iv. $0.\dot{5}$ v. $0.8\dot{3}$ vi. $0.\dot{8}\dot{3}$

LEVEL 3

1. When an optical system forms the image of a real object that is u metres away from the lens, the image is formed v metres behind the lens. The image is also formed upside-down.

The formula that relates u to v in terms of the focal length f of the lens is: $\frac{1}{u} + \frac{1}{v} = \frac{1}{f}$.

A lens of focal length 0.4 ± 0.003 metres is used to form the image of an object that is 1.9 ± 0.06 metres from the lens.

(a) Make v the subject of the formula.
(b) Find the largest and smallest possible values of v.
(c) Find the percentage error in the value of v.

2. Einstein's formula for the amount of energy (E Joules) released when matter (m kilograms) is destroyed in a nuclear reaction is $E = mc^2$ where c is the velocity of light 3.00×10^8 metres per second.

(a) Find the energy released when 0.2 kilograms of matter is released.

(b) A neutron weighs 1.67×10^{-27} kg. Find the amount of energy released when five million neutrons are destroyed.

3. The volume, V cm^3, of the square pyramid shown is given by the formula $V = \frac{1}{3}Ah$, where A is the base area.

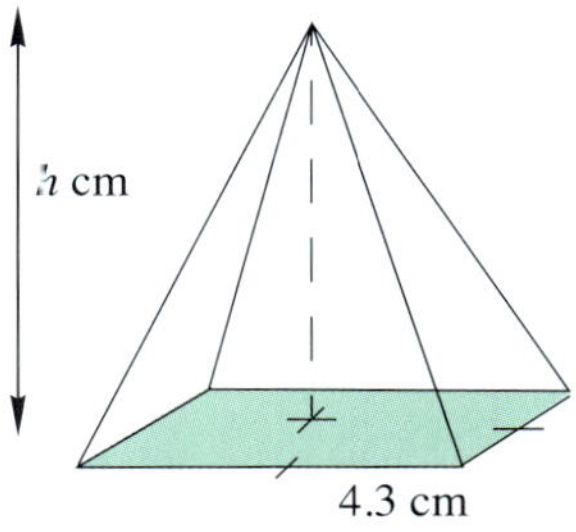

If $h = 12.4$ find its volume, giving the answer

(a) correct to three significant figures.

(b) correct to two decimal places.

(c) to the nearest whole number.

LEVEL 4

1. For what value(s) of k will $kx - 2 = \sqrt{2}$ have a rational solution when solving for x?

2. The function $f(x) = 3x + 6, 0 \le x \le 20$ is to be used to predict values of f throughout the domain. The maximum level of error permitted in these values is 6%. Assuming that the constant is known exactly, find the maximum percentage error allowable in the coefficient of x.

3. The diagram shows part of an engineering component.

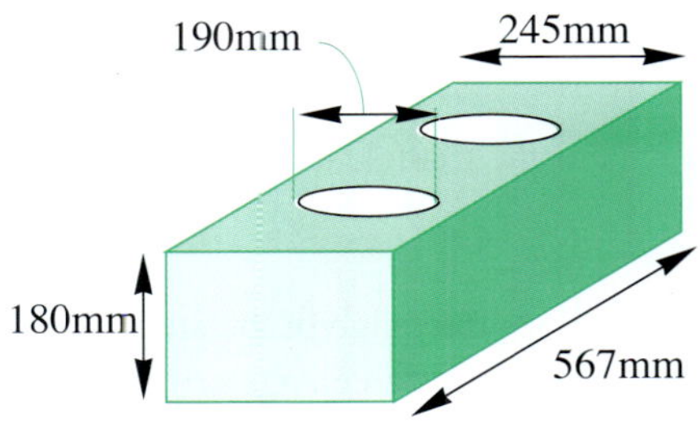

Nominally, the block has dimensions 180 by 245 by 567mm and the holes have diameter 190mm.

The dimensions are known to 0.2% accuracy and the diameters of the circular holes are known to 0.1% accuracy.

Find the percentage errors in the volume of metal used in making the component and in its weight.

4.8 TOPIC TEST

1. If $a = 4.7 \times 10^3$, $b = 9.1 \times 10^{-2}$ & $c = 6.2 \times 10^4$, evaluate, giving answers in scientific form, correct to three significant figures:

i. $\frac{c}{ab}$ ii. $a^3 + c^2$ iii. $\frac{b}{c - a}$

[5 marks]

2. Round the following quantities correct to three significant figures:

i. 4009 ii. 0.0050607 iii. 109056

[3 marks]

3. What is the percentage error in quoting a mass as being 1.50 kg?

[3 marks]

4. If $a = 3.14$, $b = 0.98$ and $c = 1.04$, calculate $a(c^2 - b)$, giving your answer

(a) exactly.
(b) correct to three significant figures.
(c) correct to two decimal places.
(d) in the form $a \times 10^k$, where $1 \le a < 10$, $k \in \mathbb{Z}$.

[5 marks]

5. Express the recurring decimal $2.\dot{2}\dot{3}$ in the form $\frac{a}{b}$, where $a, b \in \mathbb{Z}$ and hence state which of the sets $\mathbb{N}$, $\mathbb{Z}$ or $\mathbb{Q}$, the decimal belongs to.

[3 marks]

6. For the triangle shown, $x = 10.2$, $y = 3.6$ and $\theta = 32°$.

Find
i. the range of the area, A m^2.
ii. the percentage error in the area.

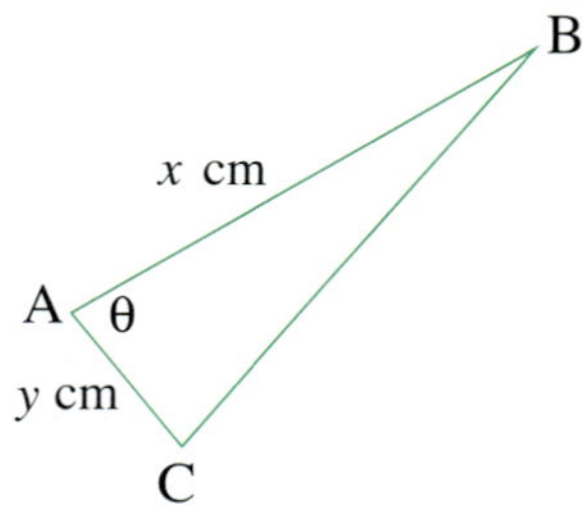

[6 marks]
Total : 25 marks

4.9 PROJECT SUGGESTIONS

Investigate how arithmetic could have been done using the Roman number system. Here is a piece of 'trivia' to get you started:

In 1588 as the English awaited the arrival of the Spanish Armada its naval commander, Howard of Effingham, wrote to Elizabeth's spymaster, Francis Walsingham, complaining that he never received any money to feed his sailors. He referred to the Treasurer, Lord Burghley, as a man...

> '... unable to handle the simplest addition or subtraction sums in any but Roman figures, with the inevitable result of frequent mistakes and faulty account keeping'.

(Quoted in *In the Confident Hope of a Miracle, the True Story of the Spanish Armada*, Neil Hanson, Doubleday)

Next, consider the multiplication 78×26 carried out using the Roman number system:

```
        LXXVIII
           XXVI
    -----------
        LXXVIII
  CCLLLXXVVVVV
       DCCLXXX
       DCCLXXX
  -------------
       MMXXVIII
```

Try this using different numbers.

Perhaps try some simple addition and subtraction first.

What about division?

Some early number systems used base systems other than 10.

The Mayan system was based on 20 and the Sumerian system used 60. The use of 60 as a base is still partly in use. Where?

Computers use base 2 and the old English money system used 12 and 20.
What are the advantages of other number bases?

When does 9 + 4 = 1? What about 9:00 am + 4 hours? You could carry out an investigation into 'Clock-arithmetic' and then extended it to modular arithmetic.

When does 10 + 10 = 100?

Of course there are many other number systems and indeed, number patterns – many of which are found in nature. Any one of these can make a great topic for a project.

A good reference is the book *Mathematics in the Making*, by Lancelot Hogben

5.0 ALGEBRA AND PUZZLES

PUZZLES

Many people enjoy puzzles. The puzzle could be a detective story, a crossword puzzle, a historical mystery or a practical mind-teaser like that in the photograph.

The puzzle is to move no more than four matches so that the pattern has six squares.

There are many puzzles from classical mathematics and it was in part to solve these that algebra and equations were developed.

Here are two of these ancient puzzles. Can you solve them?

This tomb holds Diophantus. This tombstone gives the measure of his life. He was a boy for the sixth part of his life; when a twelfth was added, he acquired a beard; He kindled for him the light of marriage after a seventh, and in the fifth year after his marriage he had a son. Alas! late-begotten and miserable child, when he had reached the measure of half his father's life, illness took him. After consoling his grief by this science of numbers for four years, he reached the end of his life. How old was Diophantus when he died?

Let fall a tear as you pass by; for we are those guests of Antiochus whom his house killed when it collapsed, and God gave us in equal shares this place for a banquet and a tomb. Four of us from Tegea lie here, twelve from Messene, five from Argos, and half of the banqueters were from Sparta, and Antiochus himself. A fifth of the fifth part of those who perished were from Athens, and to thou, Corinth, weep for Hylas alone. How many guests were entombed in the fallen house?

You may not actually use formal algebra to solve these problems ('let the number of guests be x etc.'). However, you will almost certainly use the essential ideas of algebra in arriving at a solution. How to do this well is the subject of this chapter.

5.1 THE REAL NUMBER LINE

In Chapter 4, we briefly looked at the different types of number sets that are available. One particular set that we are interested in is that of the real numbers, denoted by the letter $\mathbb{R}$. A geometric representation of the real numbers may be obtained by associating every real number with a unique point on a straight line. After establishing an origin on the straight line, which will be the number zero, we partition this straight line into three parts:

1. Negative real numbers, $\mathbb{R}^-$.
2. Zero, 0.
3. Positive real numbers, $\mathbb{R}^+$.

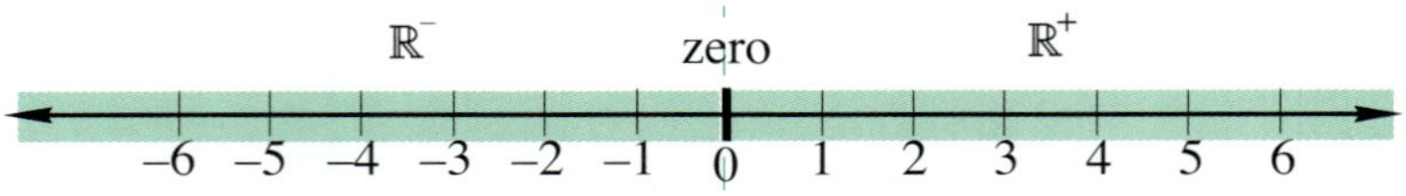

Next consider the task of listing all positive integers greater than 2 but less than 8. This is rather straight forward and is given by the set of numbers 3, 4, 5, 6 and 7. Or, using set notation, we can write it as $\{3, 4, 5, 6, 7\}$. The **braces** are used to denote '*the set of . . .*'.

However, if we wanted to list all real numbers greater than 2 but less than 8, this could not be achieved. For example, what is the first real number greater than 2? Is it 2.1, 2.01, 2.001 etc. . . In this case, the only way that we can list this set is by writting it in **set builder notation**, i.e., if x represents any number satisfying the condition that it must be greater than 2 but less than 8, then we can write this as $2 < x < 8$. Using set builder notation, we write this as $\{x | 2 < x < 8\}$.

There are other ways that this set can be written, for example, we can use **interval notation**, where in this case the set would be written as (2, 8) (NB: do not get this confused with the coordinates (2, 8) used with the Cartesian plane). Notice that as the numbers 2 and 8 are not included as part of the set, we use '*round brackets*'. However, if we were to include the number 2 but still exclude the number 8, then we would write $\{x | 2 \le x < 8\}$ (in set builder notation) and [2, 8) (in interval notation), i.e., we use a square bracket to indicate that the number 2 is included.

There is another form of interval notation that can be used when describing sets. The difference from the one already described is that rather than using round brackets to indicate that we are not including a value, we still use the square brackets but this time we draw them 'opening up' in the opposite direction. For example, the set $\{x | 2 \le x < 8\}$ would be written as [2, 8 [, i.e., the '2' is included whereas the '8' is excluded. While the set $\{x | 2 < x < 8\}$ would be written as] 2, 8 [.

It is also helpful to be able to provide a visual representation of sets such as those discussed above. To do this we use the real number line and the following rules:

If the number is **not included** in the set, you circle the point representing the number (that is, an **open circle**):

$\{x | 2 < x < 8\} = (2, 8)$

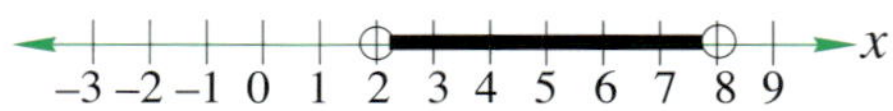

If the number **is included** in the set, you fill–in the circle at the point representing the number (that is, a **closed circle**):

$\{x | 2 \le x < 8\} = [2, 8)$

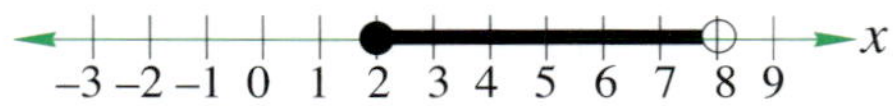

Summary of notation and visual representation on the real number line.

Set Notation	Interval Notation	Real number line	Example
$\{x \mid a \le x \le b\}$	$[a, b]$		$\{x \mid 2 \le x \le 5\} = [2, 5]$ 
$\{x \mid a < x \le b\}$	$(a, b]$ $]a, b]$		$\{x \mid 2 < x \le 5\} = (2, 5]$  $\{x\ \ 2 < x \le 5\} =]2, 5]$
$\{x \mid a \le x < b\}$	$[a, b)$ $[a, b[$		$\{x \mid 2 \le x < 5\} = [2, 5)$  $\{x\ \ 2 \le x < 5\} = [2, 5[$
$\{x \mid a < x < b\}$	(a, b) $]a, b[$		$\{x \mid 2 < x < 5\} = (2, 5)$  $\{x\ \ 2 < x < 5\} =]2, 5[$
$\{x \mid x \ge a\}$	$[a, \infty)$ $[a, \infty[$		$\{x \mid x \ge 2\} = [2, \infty)$  $\{x\ \ x \ge 2\} = [2, \infty[$
$\{x \mid x > a\}$	(a, ∞) $]a, \infty[$		$\{x \mid x > 2\} = (2, \infty)$  $\{x\ \ x > 2\} =]2, \infty[$
$\{x \mid x \le a\}$	$(-\infty, a]$ $]-\infty, a]$		$\{x \mid x \le 2\} = (-\infty, 2]$  $\{x\ \ x \le 2\} =]-\infty, 2]$
$\{x \mid x < a\}$	$(-\infty, a)$ $]-\infty, a[$		$\{x \mid x < 2\} = (-\infty, 2)$  $\{x\ \ x < 2\} =]-\infty, 2[$

Note then that if we wanted to represent the set of **integers** greater than 2 but less than 8 on the real number line, the diagram would be given by:

That is, each individual number would be circled.

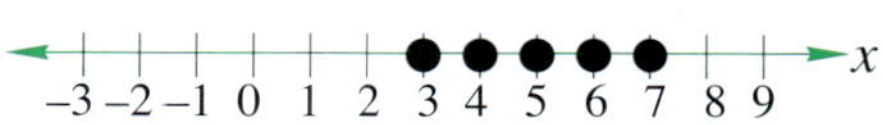

EXERCISES 5.1

1. Represent each of the following sets on the real number line

(a) $\{x|1 \le x \le 4\}$ (b) $\{x|2 < x \le 6\}$ (c) $\{x|-3 \le x < 1\}$

(d) $\{x|-4 < x < 4\}$ (e) $\{x|-5 \le x < -3\}$ (f) $\{x|-2 < x \le 2\}$

2. Represent each of the following sets on the real number line

(a) $\{x|x < 1\}$ (b) $\{x|x \le 6\}$ (c) $\{x|x > -1\}$

(d) $\{x|-4 < x\}$ (e) $\{x|-5 \ge x\}$ (f) $\{x|x \ge 2\}$

3. Convert the following set notations into interval notation.

(a) $\{x|2 < x \le 8\}$ (b) $\{x|x \ge 4\}$ (c) $\{x|0 \le x \le 6\}$

(d) $\{x|x \le 10\}$ (e) $\{x|x < 12\}$ (f) $\{x|3 \le x < 5\}$

(g) $\{x|x > -2\}$ (h) $\{x|-3 < x \le 1\}$ (i) $\{x|x < -7\}$

4. Convert the following interval notations into set notation.

(a) $[4, 6]$ (b) $]-2, 5[$ (c) $]8, 12]$

(d) $[12, 15[$ (e) $[5, \infty[$ (f) $]-\infty, 8[$

(g) $]2, \infty[$ (h) $]-\infty, 3]$ (i) $]-2, \infty[$

5. Represent each of the following sets on the real number line.

(a) $[4, 6]$ (b) $]-2, 5[$ (c) $]8, 12]$

(d) $[12, 15[$ (e) $[5, \infty[$ (f) $]-\infty, 8[$

(g) $]2, \infty[$ (h) $]-\infty, 3]$ (i) $]-2, \infty[$

5.2 SOLVING LINEAR EQUATIONS

All linear equations in one variable take on the form $ax + b = c$, where the variable (in this case) is x. Solving such equations requires the use of basic transposition techniques in order to isolate the varible. This involves the use of subtraction, addition, multiplication and division.

EXAMPLE 5.1

Solve for x, where $2x - 4 = 12$.

Solution

Given that $2x - 4 = 12 \Rightarrow 2x = 12 + 4$. [Adding 4 to both sides]

Therefore, we have that $2x = 16$.

So that $x = \frac{16}{2}$. [Dividing both sides by 2]

$\therefore x = 8$

Note the use of the symbol ‘ $\Rightarrow$ ’. It is a shorthand way of saying, the new statement follows on from the previous equation. That is, starting with the equation $2x - 4 = 12$, then, the equation $2x = 12 + 4$ is a direct consequence (i.e., $2x = 12 + 4$ follows on from $2x - 4 = 12$).

Also, the symbol ‘ $\therefore$ ’ is a shorthand way of writing ‘therefore’.

We next solve a linear equation using words rather than symbols.

EXAMPLE 5.2

Solve for x, where $3 - \frac{1}{2}x = 1$.

Solution

Given that $3 - \frac{1}{2}x = 1$, then $-\frac{1}{2}x = 1 - 3$. [Subtracting 3 from both sides].

Meaning that $-\frac{1}{2}x = -2$

So that, $x = -2 \times -2$. [Multiplying both sides by –2].

Therefore, we have that $x = 4$

Although using mathematical symbols is neater and very precise, either of the approaches shown in the previous two examples can be used – just make sure that the steps follow on from each other.

Using your graphics calculator

Graphics calculators have the capability of solving linear equations. This does not mean that you should now forget how to use classical methods to solve linear equations, but rather use the graphics calculator as a tool to help you check your solution and to develop an in depth understanding of why and how a result is obtained.

We can use the **solve** function from the TI–83. To do this we

1. first need to call up the **CATALOG** option by pressing [2 nd] [0].
2. next, press [LN].
   This will bring up the catalogue listing for those functions or operators that start with the letter ‘s’.
3. then, use the down arrow key until the **solve(** option is located.

The following screen dumps display what you should see as you follow the steps above.

1. **2.** **3.**

Notice that when using the solve function, the information must be written in the form

solve(Equation = 0, variable, initial guess)

where the **initial guess** is a reasonable guess to the value which x might be.

This means that the original equations must be rewritten

Example 5.1: from $2x-4=12$ to $2x-4-12=0$ (or $2x-16=0$),

Example 5.2: from $3-\frac{1}{2}x = 1$ to $3-\frac{1}{2}x-1 = 0$ (or $2-0.5x=0$).

EXAMPLE 5.3

Solve for x, where $5x+4=20$.

Solution

We have that $5x+4 = 20 \therefore 5x = 20-4$ [subtract 4 from both sides]

$$5x = 16$$

$$\therefore x = \frac{16}{5} \quad \text{[divide both sides by 5]}$$

(or $x = 3.2$)

Using the TI–83 (after re-arranging the equation $5x+4=20$ to $5x+4-20=0$) we have:

```
solve(5X+4-20,X,
1)
             3.2
■
```

EXAMPLE 5.4

Solve for x, where $\frac{1}{3}x+1 = -7$.

Solution

Given that $\frac{1}{3}x+1 = -7$, then $\frac{1}{3}x = -8$ [subtract 1 from both sides]

$$\therefore x = 3\times -8 \quad \text{[multiply both sides by 3]}$$

$$= -24$$

Using the TI–83, we have:

```
solve(X/3+1+7,X,
1)
             -24
■
```

Consider the next example where it is not 'obvious' that we are dealing with a linear equation.

EXAMPLE 5.5

Solve for x, where $\frac{2}{x} + 1 = 5$.

Solution

Starting with $\frac{2}{x} + 1 = 5$ and then subtracting '1' from both sides of the equation we have:

$$\frac{2}{x} + 1 = 5 \Rightarrow \frac{2}{x} = 4$$

Then, we multiply both sides by 'x', giving

$$2 = 4x$$

And so we have $x = \frac{1}{2}$

In its original form, the equation $\frac{2}{x} + 1 = 5$ does not represent a linear equation. However, we were able to convert it into a linear equation and then solve as we had done in the previous examples. With the graphics calculator, such equations pose no difficulties. Again we can enter the equation into the calculator and then solve (as in the previous examples). We now display this. Using the TI–83, we have:

```
solve(2/X+1-5,X,
1)
               .5
■
```

There are also situations where there are x-terms located on both sides of the equality sign. We now look at how to solve problems such as these.

EXAMPLE 5.6

Solve the equation $12x - 2 = 3x - 11$.

Solution

This time we need to transpose all x terms to one side and all other terms to the other side.

That is, $12x - 2 = 3x - 11$

$\Rightarrow 12x - 3x = -11 + 2$ [subtract $3x$ from both sides and add 2 to both sides]

$\Rightarrow 9x = -9$

$\therefore x = -1$

Using the TI–83, we have:

```
solve(12X-2-3X+1
1,X,1)
               -1
■
```

Notice how we have consistently used '1' as our guess when using the TI–83. The reason is that even with a guess of 10 we would still obtain the same answer. It might take the calculator a fraction more time, but it would not be noticable. What if we tried 100 as our guess? Try it!

Note that we can also use the **Equation Solver** to solve the equations we have just dealt with. As we have already seen, the TI–83 still needs to have the equation in the form **Equation = 0**.
To call up the **Equation Solver** follow these steps:

1. MATH 0 This brings up the Equation Solver screen:

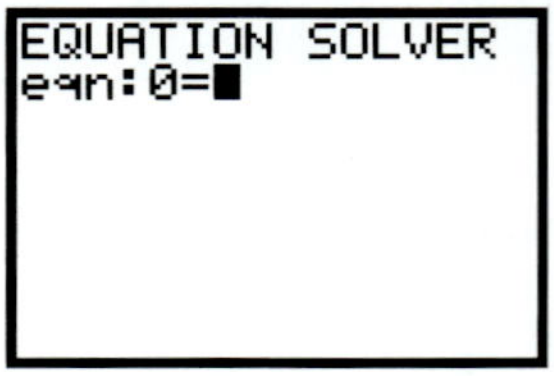

2. Enter the equation directly (after transposing to **Equation = 0**)
(i.e., if solving 12X – 2 = 3X – 11, enter 12X– 2 – 3X +11)

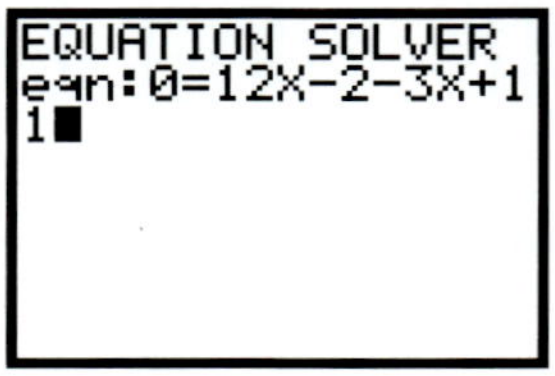

3. Move the cursor so that it covers the variable you are solving for and then press ENTER.

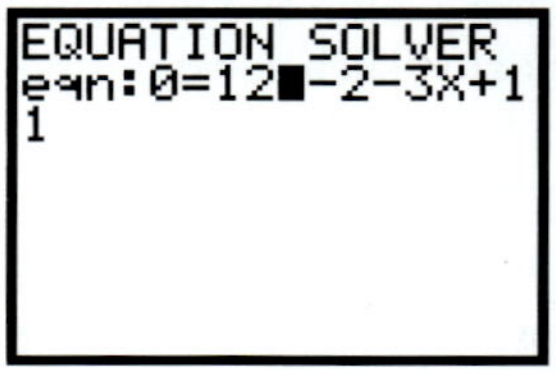

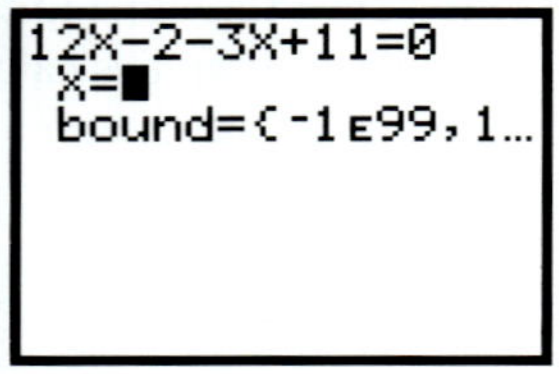

4. Then press ALPHA ENTER (i.e., SOLVE)

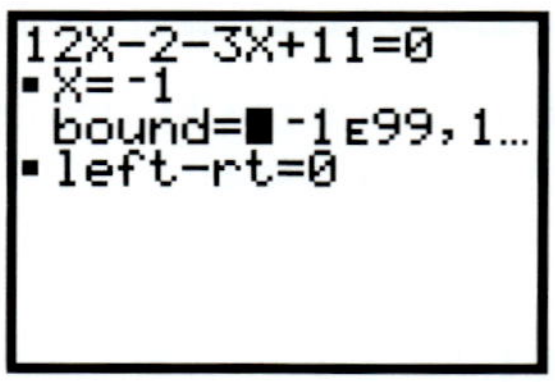

EXAMPLE 5.7

Solve the equation $\frac{3x-1}{2} = \frac{x+4}{3}$.

Solution

We start by 'cross multiplying' by '3' (to transpose the 3 from the denominator on the right hand side) and '2' (to transpose the 2 from the denominator on the left hand side):

$$\frac{3x-1}{2} = \frac{x+4}{3} \Rightarrow 3 \times (3x-1) = 2 \times (x+4)$$

$$\Rightarrow 9x - 3 = 2x + 8$$

$$\Rightarrow 9x - 2x = 8 + 3 \quad \text{[subtract } 2x \text{ and add 3 to b.s]}$$

$$\Rightarrow 7x = 11$$

$$\therefore x = \frac{11}{7} \quad \text{[divide b.s. by 7]}$$

Again, we see that the graphics calculator makes easy work of solving equations such as these:

Note also that we have used the **Frac** option (from **MATH** menu) to convert the answer into a fraction.

```
solve((3X-1)/2-(
X+4)/3,X,1)
        1.571428571
Ans▸Frac
               11/7
```

EXAMPLE 5.8

Solve the equation $\frac{x}{2}+\frac{x}{5} = 3$.

Solution

As the fractions involved have different denominators, we first need to express this sum as a fraction with a common denominator:

$$\frac{x}{2}+\frac{x}{5} = 3 \Rightarrow \frac{5x}{10}+\frac{2x}{10} = 3 \quad \text{[lowest common denominator is 10]}$$

$$\Rightarrow \frac{7x}{10} = 3$$

$$\Rightarrow x = 3\times\frac{10}{7} \quad \left[\text{multiply b.s. by } \frac{10}{7}\right]$$

$$\therefore x = \frac{30}{7}$$

Using the TI-83 we have:

equation solver:

```
X/2+X/5-3=0
▪X=4.2857142857…
 bound=█-1ᴇ99,1…
▪left-rt=0
```

solve(option:

```
solve(X/2+X/5-3,
X,4)
        4.285714286
Ans▸Frac
               30/7
█
```

EXAMPLE 5.9

Solve the equation $\frac{3}{2x-1} = 4$.

Solution

Given that $\frac{3}{2x-1} = 4$ we first transpose the expression in the denominator of the L.H.S by multiplying both sides of the equation by $(2x - 1)$:

$$\frac{3}{2x-1} = 4 \Rightarrow 3 = (2x-1)\times 4$$

$$\Rightarrow 3 = 8x - 4$$

$$\Rightarrow 7 = 8x$$

$$\therefore \frac{7}{8} = x$$

Using the TI–83 we have:

```
solve(3/(2X-1)-4
,X,1)
               .875
Ans▸Frac
                7/8
█
```

EXAMPLE 5.10

Solve the equation $\frac{x}{3} + \frac{2x-5}{5} = 1$.

Solution

This time we need a common denominator of 15, so we have:

$$\frac{x}{3} + \frac{2x-5}{5} = 1 \Rightarrow \frac{x}{3} \times \frac{5}{5} + \frac{2x-5}{5} \times \frac{3}{3} = 1$$

i.e., $$\frac{5x}{15} + \frac{3(2x-5)}{15} = 1$$

$$\Rightarrow \frac{5x + 3(2x-5)}{15} = 1$$

$$\Rightarrow \frac{5x + 6x - 15}{15} = 1 \quad \text{[simplify fraction]}$$

$$\Rightarrow \frac{11x - 15}{15} = 1$$

$$\Rightarrow 11x - 15 = 15 \quad \text{[multiplying b.s by 15]}$$

$$\Rightarrow 11x = 30 \quad \text{[adding 15 to b.s]}$$

$$\therefore x = \frac{30}{11} \quad \text{[dividing b.s by 11]}$$

Again, solving the above equation using the TI–83 is straight forward:

1. Enter equation:

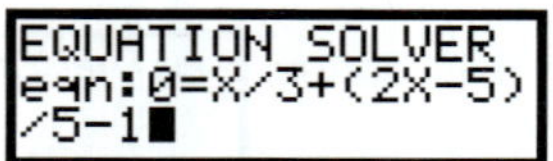

2. Highlight variable:

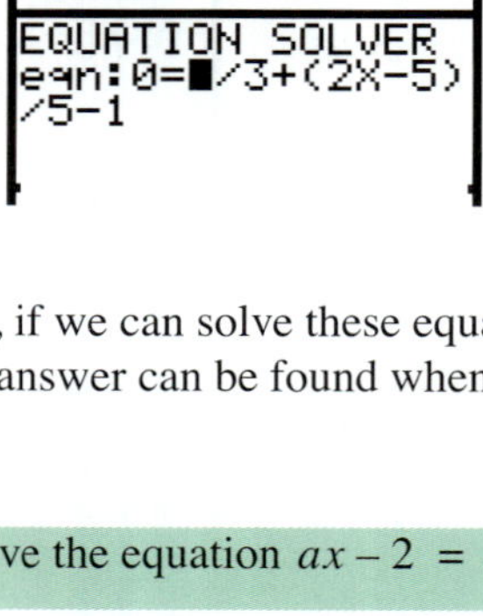

3. Press ENTER:

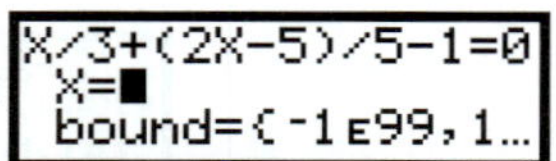

4. Press ALPHA ENTER:

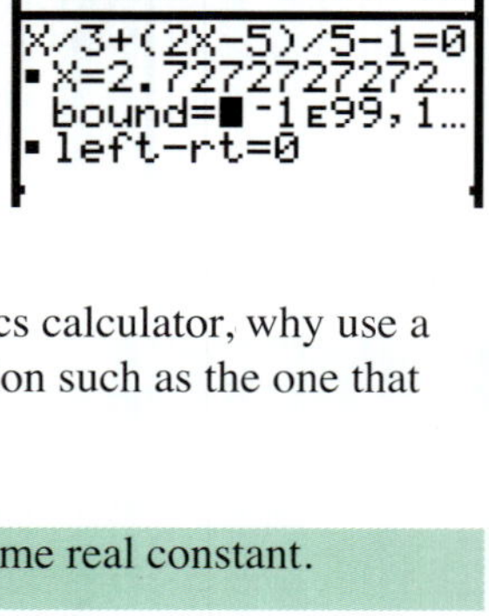

The question then arises, if we can solve these equations with a graphics calculator, why use a classical approach? The answer can be found when addressing a question such as the one that follows.

EXAMPLE 5.11

Solve the equation $ax - 2 = a - 1$, where a is some real constant.

Solution

In this instance, your graphics calculator (with features specified by the I.B.O) will not be able to solve an equation such a this. These are called **literal equations** and solving them requires a classical approach.

So, we have: $$ax - 2 = a - 1 \Rightarrow ax = (a-1) + 2 \quad \text{[adding 2 to b.s]}$$

$$\Rightarrow ax = a + 1$$

$$\therefore x = \frac{a+1}{a} \quad \text{[dividing b.s. by } a\text{]}$$

EXERCISES 5.2

1. Solve the following equations for the unknown

(a)	$4x - 5 = 15$	(b)	$3x + 8 = 23$	(c)	$2 - 5x = 17$
(d)	$7 + 4x = 91$	(e)	$3a + 12 = 40$	(f)	$4 - 6w = 42$
(g)	$8 + 5w = 63$	(h)	$4z - 8 = 36$	(i)	$2y - 3 = 5$

2. Check your answers to Question 1 using a graphics calculator.

3. Solve the following equations for the unknown

(a)	$\frac{x}{3} + 1 = 4$	(b)	$\frac{x}{4} - 3 = 2$	(c)	$3 - \frac{x}{5} = 1$
(d)	$\frac{3}{2}x + 2 = 4$	(e)	$3 - \frac{1}{3}x = 6$	(f)	$\frac{4}{5}w - 2 = 10$

4. Check your answers to Question 3 using a graphics calculator.

5. Solve the following equations for the unknown

(a)	$4 - x = 2 - 3x$	(b)	$5y + 7 = 21 - 2y$
(c)	$3x + 8 = 7x + 28$	(d)	$5 - 8x = 15 + 2x$
(e)	$3(x - 1) + 4 = 6 - 2x$	(f)	$5(2 - 3x) - 4 = 1 - 2(3 + x)$

6. Check your answers to Question 5 using a graphics calculator.

7. Solve the following equations for the unknown

(a)	$\frac{3x + 1}{2} = \frac{x - 5}{3}$	(b)	$\frac{3 - 2x}{4} = \frac{x + 5}{2}$
(c)	$\frac{2x + 1}{5} = \frac{x + 5}{3}$	(d)	$\frac{4 - 5y}{7} = \frac{3y + 1}{2}$
(e)	$\frac{x + 1}{2} + 1 = \frac{x - 5}{3}$	(f)	$\frac{w - 2}{4} - 3 = \frac{1 - 3w}{2}$
(g)	$\frac{x - 3}{3} - 4 = x - 2$	(h)	$5 - \frac{2 - a}{3} = \frac{2a + 1}{4}$

8. Check your answers to Question 7 using a graphics calculator.

9. Solve the following equations for the unknown

(a)	$\frac{2}{2x - 1} = 3$	(b)	$\frac{4}{2 + x} = 6$	(c)	$\frac{5}{1 - x} = 3$
(d)	$\frac{3}{4 - 2x} = -2$	(e)	$\frac{5}{2x - 4} = -1$	(f)	$\frac{1}{y} - 3 = 1$

(g) $\frac{3}{2x-1} = \frac{4}{1-x}$ (h) $\frac{2}{x-4} = \frac{1}{1-2x}$ (i) $\frac{1}{x-1} + 1 = 4$

(j) $\frac{1}{a} + \frac{1}{2a} = 3$ (k) $\frac{3}{a} - \frac{1}{2a} = 1$ (l) $\frac{4}{3p} + \frac{1}{p} = 2$

10. Check your answers to Question 9 using a graphics calculator.

11. Solve the following for the variable in the square bracket.

(a) $6(x-1)+4 = 2(7x+1)$ $[x]$

(b) $3(y+2) = 5y+8$ $[y]$

(c) $6[4-(a-2)] = 2a+10$ $[a]$

(d) $\frac{3}{2}(z-5) - \frac{1}{3}(2-z) = 4$ $[z]$

(e) $\frac{15}{x} - 4 = \frac{6}{x} + 4$ $[x]$

(f) $\frac{5x-4}{2x+3} = 3$ $[x]$

(g) $\frac{1}{y-2} + \frac{2}{y+2} = \frac{5}{y^2-4}$ $[y]$

(h) $\frac{a-2}{3} - \frac{2a+1}{4} = 1$ $[a]$

(i) $\frac{w}{2.15} - 6.55 = 2.05 - \frac{w}{4.30}$ $[w]$

(j) $\frac{2x-1}{2} + 3 = \frac{2(1-x)}{4}$ $[x]$

(k) $\frac{a-4}{5} = 3(a-2)+1$ $[a]$

(l) $2 + \frac{1-v}{3} = 1 + \frac{v}{2}$ $[v]$

12. Check your answers to Question 11 using a graphics calculator.

13. Solve the following literal equations for x.

(a) $ax+2 = a$ (b) $bx-b = 1$ (c) $2-ax = 4$

(d) $3ax-3 = 2ax+2$ (e) $4-ax = ax+8$ (f) $\frac{x}{a} - \frac{x}{2a} = 1$

(g) $2a = 3 - \frac{4}{5}x$ (h) $3a = 2a + \frac{a}{x+a}$ (i) $\frac{2}{a+1} - \frac{3x}{a-1} = 0$

14. Solve the following literal equations for x.

(a) $ax-3 = bx+2$ (b) $ax-b = a+bx$ (c) $ax+b = a+bx$

(d) $ax-2 = 2x-a$ (e) $\frac{x}{a} + 1 = \frac{1}{a}$ (f) $1 - \frac{x}{a} = a-1$

(g) $ax = b(a-x)$ (h) $ax+b^2 = bx+a^2$ (i) $(b+1)x = b-1+2x$

5.3 PROBLEM SOLVING USING LINEAR EQUATIONS

Once the skills for solving linear equations have been established, it then becomes important to be able to translate worded scenarios into mathematical equations. This process is never an easy one, however, there are a number of steps that can provide some guidance as to how to go about setting up equations.

We begin with a couple of examples to highlight this process.

EXAMPLE 5.12 The sum of two consecutive numbers is 27. What are the numbers?

Solution

Two key words in this problem are: **sum** and **consecutive**.
The term *sum* is another way of saying *adding*.
The term *two consecutive* means that we are referring to two numbers, where the second number is one greater than the previous one.

In algebraic terms we can then write:
Let the first number be x.
This means that the next number must be $(x+1)$.

Then, we have that $x+(x+1) = 27$ [i.e., the sum of these two numbers = 27]

Once the equation has been established, we revert back to solving a linear equation.

So, we have
$$x+(x+1) = 27 \Rightarrow 2x+1 = 27$$
$$\Rightarrow 2x = 26$$
$$\therefore x = 13$$

Therefore, the consecutive numbers are 13 ($=x$) and 14 ($=x+1$).

EXAMPLE 5.13 My father is twice as old as I am. If the sum of our ages is 69, what is my age?

Solution

Key terms in this problem are: **twice as** and **sum**.

The term *twice as* is another way of saying *multiply by two*.
The term *sum* is another way of saying *adding*.

In algebraic terms we can then write:
Let my age be x.
This means that twice my age must be $2x$.

That is, this means that my father must be $2x$ years old (i.e., he is twice as old as I am).

However, the sum of our ages is 69, meaning that $x + 2x = 69$.
This is the mathematical equation that represents the worded problem.

What now remains is to solve the linear equation:

$$x + 2x = 69$$
$$\Rightarrow 3x = 69$$
$$\therefore x = 23$$

Therefore, I am 23 years old.

EXAMPLE 5.14 Five times the difference between a number and 3 is the same as the difference between 4 times the number and 3.

Solution

Key terms in this problem are: **times**, **difference** and **same as**.

The term *times as* is another way of saying *multiply*.
The term *difference* is another way of saying *subtract*.
The term *same as* is another way of saying *is equal to*.

In algebraic terms we can then write:

Let the number be x.
This means that the difference between the number and 3 must be $x - 3$.
Therefore, five times this difference must be $5(x - 3)$ – (1)

Next, we have: 4 times the number must be $4x$.
Therefore, the difference $(4x - 3)$ – (2)

However, statement (1) and (2) are the same, therefore we have that $5(x - 3) = (4x - 3)$.

Having established the algebraic equation we now solve it.

$$5(x-3) = 4x - 3 \Rightarrow 5x - 15 = 4x - 3$$
$$\Rightarrow 5x - 4x = 15 - 3$$
$$\therefore x = 12$$

Therefore, the number is 12.

There are many ways of wording a problem, for example, rather than using the term *difference* we could say *is subtracted from*. Equally, rather than using the term *sum* we could say *is added to* and so on.

There also exist standard types of problems in the areas of

– money
– motion
– productivity
– mixture

Each area has its own particularities, and so there will be some recurring terminology and approaches to solving such problems. So, before we look at more examples we examine some of the key steps involved in tackling such problems. The following problem solving steps provide a starting point when dealing with worded problems.

Step 1: Recognising key terms:

Worded expression	Algebraic expression
The sum of two numbers.	$x + y$ [i.e., x plus y]
The difference of two numbers.	$x - y$ [i.e., x minus y]
The product of two numbers.	$x \times y$ [i.e., x times y]
The quotient of two numbers.	$\frac{x}{y}$ [i.e., x divided by y]

Step 2: Task: Interpret the verbal problem

Carefully read the problem – then, re-read it again. Underline key information. If necessary (and it is often useful) draw a rough diagram to help you visualize the situation. For example, use a flow chart, a table, anything that will help you get a better understanding of the situation.

Step 3: Task: Assign labels

Introduce a variable (or more than one variable) to identify what it is you are looking for.

Step 4: Task: Write an algebraic equation

Construct an equation that relates the worded problem into an algebraic expression. This is the most difficult step in the process

Step 5: Task: Solve the algebraic equation

Solve the equation for the variable from the equation in Step 3.

Step 6: Task: Answer the question

Once you have solved the problem, **check** you solution and see if you have actually answered the question.

EXAMPLE 5.15 The sales tax on shoes is 12%. If you have \$25.70 what is the most expensive pair of shoes that you can buy?

Solution

We want the cost of shoes, so let \$ x be the cost of the most expensive pair you can afford.

As there is 12% tax on the shoes, the tax amount for the shoes is $0.12 \times x$.
This amount must then be added to the advertised price.
Therefore, the total cost of the shoes is x (advertised price) + $0.12 \times x$ (tax amount).

This must equal your \$25.70, i.e., $x + 0.12x = 25.70$.

We can now solve for x: $1.12x = 25.70$

$$\therefore x = \frac{25.70}{1.12}$$

$$= 22.946\ldots$$

Therefore, the most expensive pair of shoes that you can buy costs \$22.95.

EXAMPLE 5.16 Melissa has invested some money in a biotech company, which pays a dividend of 18% p.a., as well as in a food export company which pays a dividend of 12% p.a. In all, at the end of the financial year Melissa, recieved \$2395 from her investments. If she invested \$2000 more in the biotech company than in the food export company, how much money did she invest in the biotech company?

Solution

Let the amount of money that Melissa invests in the food export company be \$ x.
This means that she must have invested \$ $(x + 2000)$ in the biotech company.

The return on her investments are: 18% on biotech
12% on food export

This means that Melissa will receive \$ $\frac{18}{100} \times (x + 2000)$ from the biotech shares

\$ $\frac{12}{100} \times x$ from the food export shares.

Then, as the sum of these two amounts make up the \$2395 she received we have that

$$\frac{18}{100} \times (x + 2000) + \frac{12}{100} \times x = 2395$$

We can now solve for x. However, rather than labouring through the algebra, seeing as the difficult part of the problem has been resolved, we can make use of the TI–83:

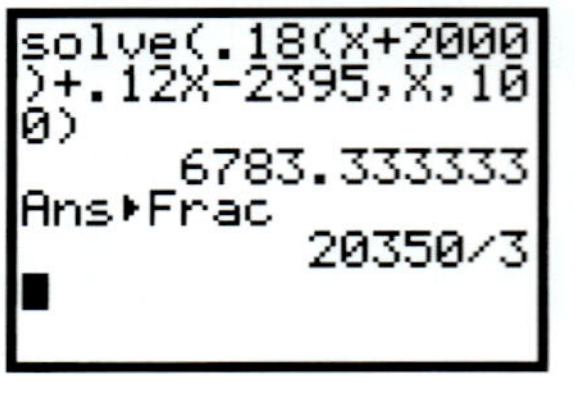

That is, Melissa invested \$(6783.33 + 2000) = \$ 8783.33 in the biotech company.

When dealing with problems like Example 5.15, constructing an information table can be helpful. For Example 5.15 we would have:

Share	Melissa's investment	dividend	Amount received
biotech company	\$ $(x + 2000)$	18%	$\frac{18}{100} \times (x + 2000)$
food export company	\$ x	12%	$\frac{12}{100} \times x$

EXAMPLE 5.17 Patrick has gone on a bike ride. For the first 40 minutes he maintains a steady speed. Then, he increases his speed by 5 kmh^{-1} for the next 30 minutes. If Patrick has travelled a total of 45 km, how fast was he travelling during the first 40 minutes of his ride?

Solution

This is a motion problem and so we will most certainly need the relationship between speed, time and distance. The equation that relates these three quantities is given by

$$\text{distance} = \text{speed} \times \text{time}$$

Now, let Patrick's speed be x kmh^{-1} for the first 40 minutes.

This then means that he was travelling at $(x + 5)$ kmh^{-1} for the next 30 minutes.

Distance travelled during the first 40 minutes = $\frac{40}{60} \times x$ km.

Distance travelled during the next 30 minutes = $\frac{30}{60} \times (x+5)$ km.

Therefore, as Patrick has travelled a total of 45 km, we have that

$$\frac{40}{60} \times x + \frac{30}{60} \times (x+5) = 45$$

We can now solve for x:

$$\begin{aligned} \frac{40}{60} \times x + \frac{30}{60} \times (x+5) &= 45 \\ \Rightarrow \frac{4}{6}x + \frac{3}{6}(x+5) &= 45 \\ \Rightarrow 4x + 3(x+5) &= 270 \\ \Rightarrow 7x + 15 &= 270 \\ \Rightarrow 7x &= 255 \\ \therefore x &= \frac{255}{7} \end{aligned}$$

That is, $x = 36.43$ and so Patrick was riding his bike at 36.43 kmh^{-1}.

Of course, we could have made use of the TI–83 to solve the equation in the example above. Using an information table can also be helpful when dealing with motion problems. Example 5.16 could have been set up as follows:

Path	Patrick's speed	Time	Distance travelled
1st 40 minutes	x kmh^{-1}	40 min = $\frac{40}{60}$ hr	$\frac{40}{60} \times x$
next 30 minutes	$(x+5)$ kmh^{-1}	30 min = $\frac{30}{60}$ hr	$\frac{30}{60} \times (x+5)$

EXAMPLE 5.18 The two rectangles shown below have the same perimeter. Find the value of a.

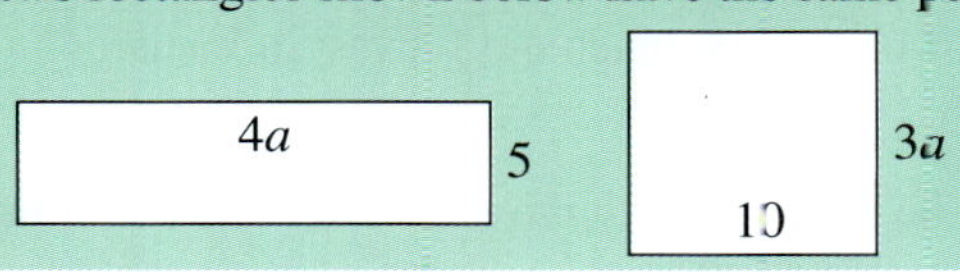

Solution

The perimeter of the first rectangle is $4a + 5 + 4a + 5 = 8a + 10$

Next, the perimeter of the second rectangle is $3a + 10 + 3a + 10 = 6a + 20$.

Because they have the same perimeter, we then have that

$$\begin{aligned} 8a + 10 &= 6a + 20 \\ \Rightarrow 8a - 6a &= 20 - 10 \\ \Rightarrow 2a &= 10 \\ \therefore a &= 5 \end{aligned}$$

EXERCISES 5.3

1. If y is subtracted from 12 the result is 5.
(a) Set up a linear equation satisfying the above statement.
(b) Solve for y.

2. Two more than four times a number is 30. If the number is x,
(a) set up a linear equation satisfying the above statement.
(b) solve for x.

3. Five less than twice a certain number is 19. Set up an equation and use it to find the number.

4. The sum of two consecutive numbers is 25. What are the numbers?

5. The sum of three consecutive numbers is 96. What are the numbers?

6. Three people, Fabio, Juan and Isabelle divide £84 amongst themselves. Isabelle receives twice as much as Juan and Fabio receives three times as much as Juan.
(a) If Juan receives £x, set up an equation in x.
(b) How much will each person get?

7. The sum of the perimeter of two squares, one having side length x cm and the other 4 cm is 96 cm. Find x.

8. My mother is three times as old as I am. If the sum of our ages is 112, how old is my mother?

9. I have 4 times as many two–cent coins as I have five–cent coins. All up I have \$6.50. How many two–cent coins do I have?

10. I invest in two types of shares, X and Y. X shares pay a dividend of 10% p.a. and Y shares pay a dividend of 8%. If I invest \$4000 more in X than in Y and my total income from both investments is \$1010, how much did I invest in Y shares?

11. What number must be added to both the numerator and denominator of $\frac{2}{3}$ to give $\frac{10}{13}$?

12. Tony has invested money in two different companies. Bio-hazard Pty Ltd pays a dividend of 8% p.a., whereas Wheat Products Pty Ltd pays a dividend of 15% p.a. Tony recieves \$3000 from his investments. If he invested \$2000 less in Bio-hazard Pty Ltd than in Wheat Products Pty Ltd, how much money did he invest in Wheat Products Pty Ltd?

13. A 10 ml solution made up of 70% water and 30% acid is added to a second 20 ml solution consisting of 80% water and 20% acid.
(a) How many ml of acid is there in the combined solution?
(b) What percentage of the solution is made up of acid?

14. Tim is half as old as his sister Claire and two years younger than his brother Thomas. The sum of their ages is 34. How old is each child?

15. Joanna either walks or rides to school. Riding is 10 kmh^{-1} faster than walking. If it takes her 15 minutes to ride to school and 45 minutes to walk to school, how far is her school?

16. A team has won 50 of the 75 games it has played so far. There are still 45 games to be played. If they are to win 60% by the end of the season, how many more games must they win?

17. Miko has \$14,000 to invest. She invests in two enterprises which together yield an income of \$742 p.a. She invests \$8,400 in the enterprise that yields 5.5%. What is the other enterprise's yield?

18. A train travels at a steady speed fom town A to town B. If the train's speed had been 5 kmh^{-1} faster, it would have taken 24 minutes less. However, had the train's speed been 4 kmh^{-1} slower, the trip would have taken 24 minutes more.
(a) How far apart are the towns?
(c) What was the train's speed?

19. Francesco invested \$18,000 in two research companies, Ab Pty Ltd and Ba Pty Ltd. Ab gives a return 4% p.a. while Ba gives a return of 5% p.a. Had Francesco invested twices as much in Ba Pty Ltd, his total return would have been \$55 more. How much did Francesco invest in each company?

20. Simar wishes to mix two types of tea. Type A costs \$ 0.36 per kg while type B costs \$ 0.50 per kg. He then sells the mixture at \$ 0 46 per kg and makes a profit of 20% of the total cost. What percentage of the total mixture is made up of Type A?

21. Mee Yee walked d km from Ballarat to Creswick at a constant speed of m kmh^{-1}. On her return trip she walked at $(m + 1)$ kmh^{-1}.
(a) Find an expression for the total time that she walked.
(b) What was her average speed?

22. The length of a rectangle is increased by 20% while its width by 10%. What percentage increase is there in its area?

23. A number consists of two digits followed by the digit 4. A second number starts with the digit 4 followed by the same two digits as used in the first number.
(a) If x is the two digit number, express in terms of x
i. the first number.

ii. the second number.

(b) If the second number is as much greater than 400 as the first number is smaller than 400, what is the first number?

24. Sea water contains 5% of salt (by weight). How many kg of fresh water should be added to 40 kg of sea water for the latter to contain 2% of salt?

5.4 LINEAR INEQUATIONS

The methods used in solving linear inequalities in one variable are identical to those used when solving linear equalities. The only difference is when we divide or multiply both sides of the equation by a negative number.

For example, the statement '$5 > 2$' is true. However, should we multiply both sides of the inequality by '-1' and leave the direction of the inequality as is, we would have '$-5 > -2$', which is not true! Because of this, we have the following rule when dealing with inequalities:

> when **dividing or multiplying** by a **negative number** the **direction of the inequality reverses**.

For example, if we have $-2x > 8$ then diving both sides by '-2' implies that $x < -4$. That is, the direction of the inequality has been reversed.

EXAMPLE 5.19

Solve for x, where $2x - 4 \geq 12$. Represent the solution set on the real number line.

Solution

The method is the same as when solving equations:

$$2x - 4 \geq 12$$
$$\Rightarrow 2x \geq 12 + 4$$
$$\Rightarrow 2x \geq 16$$
$$\therefore x \geq 8$$

In this case, as we did not divide (or multiply) by a negative number, so the inequality sign was not affected. The inequality can then be displayed on the real number line as shown:

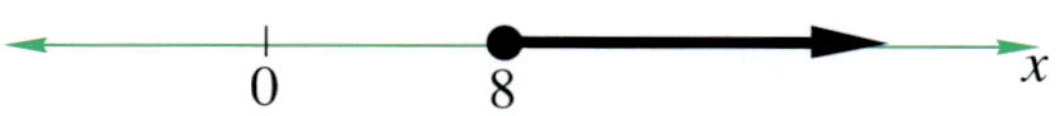

EXAMPLE 5.20

Represent, on the real number line, those values of x for which $3 - \frac{1}{2}x > 1$.

Solution

$3 - \frac{1}{2}x > 1 \Rightarrow -\frac{1}{2}x > -2$ [i.e., subtract 3 from both sides]

$\Rightarrow x < -2 \times -2$ [× by –2, so reverse direction of inequality sign]

$\therefore x < 4$

On the real number line we have:

Note that the inequality could have also been solved as follows:

$3 - \frac{1}{2}x > 1 \Rightarrow 3 - 1 > \frac{1}{2}x$ [i.e., subtract 1 and add $\frac{1}{2}x$ to both sides]

$\Rightarrow 2 > \frac{1}{2}x$

$\therefore 4 > x$ [× by 2, but this time leave the inequality sign as is]

We still obtain the same solution.

EXAMPLE 5.21

Represent the following on the real number line $\frac{3x-1}{2} \leq \frac{x+4}{3}$.

Solution

$\frac{3x-1}{2} \leq \frac{x+4}{3} \Rightarrow 3 \times (3x-1) \leq 2 \times (x+4)$

$\Rightarrow 9x - 3 \leq 2x + 8$

$\Rightarrow 9x - 2x \leq 8 + 3$

$\Rightarrow 7x \leq 11$

$\therefore x \leq \frac{11}{7}$

On the real number line we have:

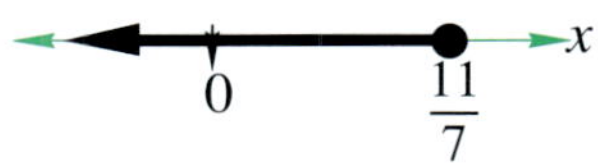

EXAMPLE 5.22

Express the set in interval notation, $\left\{x \mid \frac{x+1}{2} - x \leq \frac{x}{3} - 4\right\}$.

Solution

$\frac{x+1}{2} - x \leq \frac{x}{3} - 4 \Rightarrow \frac{x+1}{2} - x - \frac{x}{3} \leq -4$

$\Rightarrow \frac{x+1}{2} - \frac{3x}{3} - \frac{x}{3} \leq -4$

$\Rightarrow \frac{x+1}{2} - \frac{4x}{3} \leq -4$

$\Rightarrow \frac{3(x+1)}{6} - \frac{8x}{6} \leq -4$ [common denominator]

$$\Rightarrow \frac{3(x+1)-8x}{6} \le -4$$

$$\Rightarrow \frac{3x+3-8x}{6} \le -4$$

$$\Rightarrow \frac{3-5x}{6} \le -4$$

$$\Rightarrow 3-5x \le -24 \qquad \text{[multiplying b.s. by 6]}$$

$$\Rightarrow 3+24 \le 5x \qquad \text{[adding 24 and } 5x \text{ to b.s]}$$

$$\Rightarrow 27 \le 5x$$

$$\therefore \frac{27}{5} \le x$$

That is, $\left\{x \mid \frac{x+1}{2} - x \le \frac{x}{3} - 4\right\} = \left\{x \mid x \ge \frac{27}{5}\right\}$. Using interval notation, $x \in \left]\frac{27}{5}, \infty\right[$.

EXERCISES 5.4

1. Find the value(s) of x for which

(a)	$x+5>9$	(b)	$x-2\le 7$	(c)	$5-x>1$
(d)	$x-2<-1$	(e)	$4-x\ge -2$	(f)	$x+9\le 8$
(g)	$4x-5\ge 15$	(h)	$3x+8<23$	(i)	$2-5x\ge 17$
(j)	$7+4x>91$	(k)	$3x+12\ge 40$	(l)	$4-6x<42$
(m)	$8+5x\le 63$	(n)	$4x-8<36$	(o)	$4+2x<1$

2. Represent the solutions to Question 1 on the real number line.

3. Solve the following inequations for the unknown

(a)	$\frac{x}{3}+1<4$	(b)	$\frac{x}{4}-3\ge 2$	(c)	$3-\frac{x}{5}<1$
(d)	$\frac{3}{2}x+2\le 4$	(e)	$3-\frac{1}{3}x<6$	(f)	$\frac{4}{5}x-2\ge 10$

4. Represent the solutions to Question 3 on the real number line.

5. Solve the following inequations for the given unknown

(a)	$4-x>2-3x$	(b)	$5y+7<21-2y$
(c)	$3x+8\ge 7x+28$	(d)	$5-8x\le 15+2x$
(e)	$3(x-1)+4<6-2x$	(f)	$5(2-3x)-4>1-2(3+x)$

6. Represent the solutions to Question 5 on the real number line.

7. Solve the following inequations for the given unknown

(a) $\frac{3x+1}{2} < \frac{x-5}{3}$ (b) $\frac{3-2x}{4} \geq \frac{x+5}{2}$

(c) $\frac{2x+1}{5} \geq \frac{x+5}{3}$ (d) $\frac{4-5y}{7} \leq \frac{3y+1}{2}$

(e) $\frac{x+1}{2} + 1 \leq \frac{x-5}{3}$ (f) $\frac{w-2}{4} - 3 > \frac{1-3w}{2}$

8. Represent the solutions to Question 7 on the real number line.

9. Solve the following inequalities, expressing your answers using interval notation.

(a) $\left\{x \mid \frac{2x}{3} - \frac{x+1}{4} > 2 + \frac{x}{2}\right\}$ (b) $\left\{x \mid \frac{3-x}{5} + \frac{x}{2} \leq 1 + \frac{x}{3}\right\}$

(c) $\left\{x \mid \frac{2+x}{7} - \frac{x}{5} < 2 - \frac{1}{3}x\right\}$ (d) $\left\{x \mid \frac{5-3x}{3} + 2x \geq \frac{5-3x}{4} - 2x\right\}$

10. Find the set of integers that satisfy the inequalities $x + 4 > \frac{3x+10}{4} \geq 2x + 3$.

5.5 PROBLEM SOLVING USING LINEAR INEQUATIONS

In the same way that we looked at key terms when constructing an algebraic equation from a worded description, there are also some key terms to note when dealing with problems involving inequalities.

Some key words (and their algebraic equivalents) are shown in the table below.

Term	Example	Algebraic expression
At most	x is at most 10	$x \leq 10$
At least	x is at least 20	$x \geq 20$
The minimum	The minimum is –3	$x \geq -3$
The maximum	The maximum is 12	$x \leq 12$
Cannot exceed	a cannot exceed 15	$a \leq 15$
Less than	y is less than one	$y < 1$
Greater than	z is greater than –5	$z > -5$

EXAMPLE 5.23

A rectangle has a perimeter of a least 40 cm. Given that the length of the rectangle is 12 cm, what must the width of this rectangle be?

Solution

We start by drawing a diagram to help us visualise the situation:
Let the width of the rectangle be x cm.

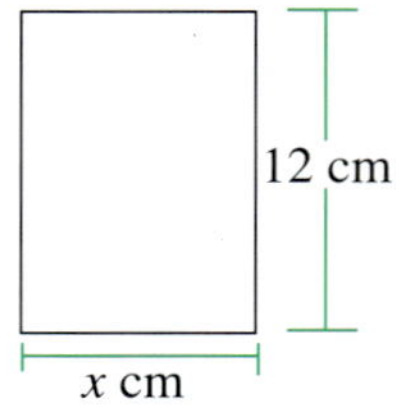

Therefore, the perimeter is $12 + x + x + 12 = 24 + 2x$.
However, the perimeter is at least 40, i.e., perimeter ≥ 40.

Therefore, we have $24 + 2x \geq 40$. [This is the algebraic representation of the problem]
We now solve the inequality:

$$24 + 2x \geq 40 \Rightarrow 2x \geq 16$$

$$\therefore x \geq 8$$

That is, the width must be at least 8 cm.

EXAMPLE 5.24

The sum of two integers, one of which is 10, is less than three times the unknown number. Given that the unknown number is at least 1, what values can it take?

Solution

Let the unknown number be y.
Then, we have that
1. the sum of y and $10 = y + 10$
2. three times the unknown number $= 3y$

But, (the sum of y and 10) is less than (three times the unknown number)
That is, $y + 10 \quad < \quad 3y$
Solving this inequality we have $y + 10 < 3y \Rightarrow 10 < 2y$

$$\therefore 5 < y$$

However, we are also told that the unknown is at least 1, i.e., $y \geq 1$.

Therefore, combining our two results we have that $1 \leq y < 5$.
Then, as y is an integer, the only possible values for y are, 1, 2, 3, 4.

EXERCISES 5.5

1. A rectangle has a perimeter of a least 60 cm. Given that the width of the rectangle is 15 cm, what must the length of this rectangle be?

2. The sum of two integers, one of which is 20, is greater than twice the unknown number. Given that the unknown number is at least 3, what values can it take?

3. The mean of 20, 35 and x must exceed 31. What must x be if $x < 50$?

4. What is the minimum amount of premium tea, which costs \$8.0 per kg, that must be mixed with 12 kg of a lesser quality tea, costing \$4.50 per kg, so that the mixture costs at least \$7.20 per kg.

5. Determine the values of x for each of the triangles shown.

(a)

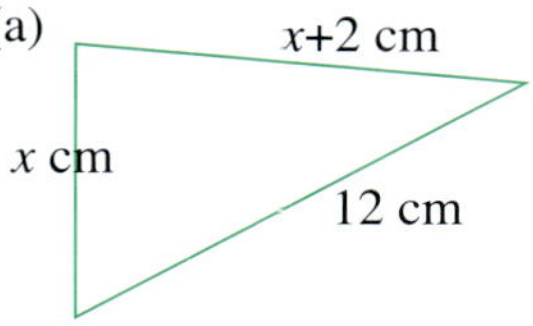

(b)

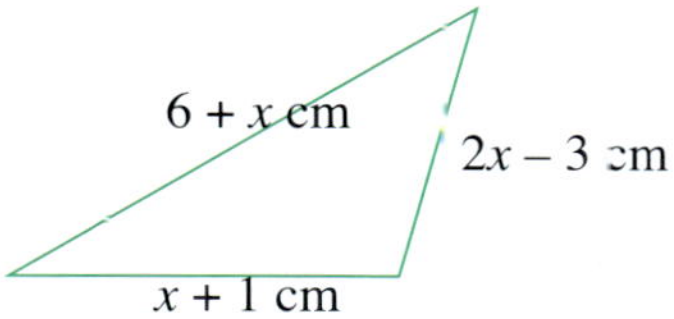

6. Jones has \$15,000 to invest. He invests in two enterprises which together must yield an income of no less than \$780 p.a. He invests in an enterprise that yields 6.5%. and in one that yields 5%. What is the minimum that he must invest in the enterprise yielding 6.5%?

7. Jennifer and her brother Michael enjoy racing each other. One day, Michael challenges his sister to a race. Michael is to set off on his bike at a constant speed of 15 kmh^{-1}. Twenty minutes later, Jennifer sets off in pursuit. What is the minimum speed that Jennifer must travel at in order to catch Michael 30 minutes after she sets off?

8. Peter has scored 60 and 62 and 65 in his last three biology tests. To ensure an overall B grade for the semester, his average mark must be at least 70. What is the minimum score that Peter needs in his fourth test in order to be awarded a B grade for the semester?

5.6 MISCELLANEOUS QUESTIONS

1. Solve for x, where $3(2x-1) = 12$.

2. Find the value(s) of x for which

(a) $7x+5 = 3x-10$ (b) $\frac{4x-3}{2} < 1$.

3. Represent the following on the real number line

(a) $\frac{3-2x}{4} \geq 3$ (b) $\frac{4-x}{2} + \frac{3-x}{4} > 2$

4. Solve for x, where

(a) $\frac{2}{3}(x-1) - \frac{1}{3} = 8$ (b) $\frac{9}{5-2x} = 3$

5. Find the value(s) of x for which $2 + \frac{1-x}{3} = 4$. Hence, represent the inequality

$2 + \frac{1-x}{3} < 4$ on the real number line.

6. Solve the equations

(a) $\frac{x-2}{3} - 1 = x$ (b) $\frac{x-2}{3} - \frac{3-x}{2} = 1$

7. What number when added to both the numerator and denominator of $\frac{1}{4}$ gives $\frac{8}{9}$?

8. Sally has kept her 2–cent and 5–cent coins from last week. In total she has \$1.78. There are two more 5–cent coins than 2–cent coins. How many 5–cent coins does Sally have?

9. Solve the equations

(a) $2a - 4 = 10$ (b) $3y - 6 = 8y + 6$ (c) $7(3x - 2) = 21$

10. Find the value(s) of x for which

(a) $7(x-3) = 3(2x-5)$ (b) $\frac{3-5x}{2} > 8$

11. Represent the following on the real number line

(a) $2x - 1 \leq 4 - 3(1-x)$ (b) $\frac{y-1}{5} - 3 > \frac{y-2}{4}$.

12. Solve for x, where

(a) $ax - a = ba, a \neq 0$ (b) $\frac{x}{a} - b = \frac{x}{b} - a$

13. How much water must be added to 3 litres of 10% salt–water solution to produce a 5% salt–water solution?

5.7 GRADED REVISION EXERCISES

LEVEL 1

1. Write down, the sets represented by the number lines

i. using set notation. ii. using interval notation.

2. Represent the following sets on a real number line.

i. $[1, \infty[$ ii. $]-10,3[$ iii. $]-\infty,9[$

LEVEL 2

1. Solve the following equations for the unknown.

i. $2x = 8$ ii. $3s + 2 = 11$ iii. $9 - 7x = 2$

iv. $4 - \frac{1}{2}t = 2$ v. $2(x - 1) = 7$ vi. $1 - 3(y + 2) = 5$

2. Solve the following inequations for the unknown.

i. $1 + x > 0$ ii. $5 - x \leq 9$ iii. $4 + \frac{1}{2}y < 1$

iv. $2 - 3(1 - x) > 5$ v. $\frac{2}{3}(x - 4) + 1 < 3$ vi. $3 - 2\left(1 + \frac{2}{5}x\right) < 1$

LEVEL 3

1. Solve the following for the unknown.

i. $\frac{x-1}{3} + 1 = \frac{x}{2}$ ii. $4 - \frac{2}{5}(2 - x) = \frac{2x+3}{4}$

iii. $\frac{5-2s}{3} - 1 < \frac{s}{6}$ iv. $\frac{3y-1}{2} > \frac{y+4}{3} + 1$

v. $\frac{2}{1-x} + 1 = \frac{x}{1-x}$ vi. $\frac{1}{5}(5 - 2a) \geq 2 - \frac{3}{2}(1 + a)$

2. Joseph's mother is three time as old as he is. How old is Joseph, given that their combined age is 88.

LEVEL 4

1. Find x: i. $ax + a = ba, a \neq 0$ ii. $\frac{a}{b}x - b > a - \frac{b}{a}x$, $a \neq 0, b \neq 0$

2. A train departs at 1:00 pm and travels at 80 km h^{-1}. An aeroplane departs from the same area at 4:00 pm and travels at 200 km h^{-1}. How far will the aeroplane travel before it overtakes the train? What assumption have you made?

5.8 TOPIC TEST

1. On a real number line, represent the set $\{x | 3 < x \leq 8\}$.

[2 marks]

2. Solve the following equations

i. $8h - 2 = 14$ ii. $5y - 10 = 2y + 5$ iii. $3(2x - 1) = 21$

[2 + 2 + 2 marks]

3. (a) Find $\{x | 4 - 2(x - 3) > x - 5\}$.

(b) Represent the solution set found in (a) on the real number line.

[3 + 2 marks]

4. Solve the following

i. $\frac{a-1}{2} - 2 = a + 4$ ii. $\frac{4x+1}{3} \leq \frac{2-x}{5}$

[3 + 3 marks]

5. i. Express DC in terms of x.

ii. Find the value of x.

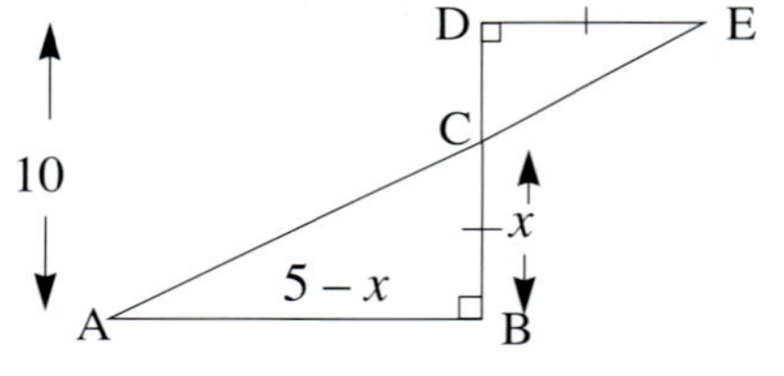

[1 + 3 marks]

6. Linley purchased some stock which, after making a profit of $y\%$, was worth \$6400. A week later she sold it for \$6000 making a loss of $y\%$.

i. Show that $\frac{100 + y}{100 - y} = \frac{16}{15}$.

ii. Find y.

iii. How much did Linley pay for the stock?

[4 + 2 + 1 marks]

7. Sea water contains 8% of salt (by weight). How many kg of fresh water should be added to 50 kg of sea water for the latter to contain 3% of salt?

[5 marks]

Total 35 marks

5.9 PROJECT SUGGESTIONS

The two problems given at the start of this chapter are known as diophantine problems - their solution has to be a whole number. These are named after Diophantus (~200-280AD). Little is known about this pioneer of Algebra, but the problems named after him require skill in their solution.

Hypatia (hy-pay-shuh) was an Egyptian philosopher who was the first notable woman in mathematics. She was head of the school of philosophy at Alexandria in about 400 where she attracted a large number of pupils. Hypatia played an important part in the development of Algebra and could be the start of an interesting project. Her life, which ended in a brutal death, was full of intrigue, with involvement in politics and religion.

> Can you find and solve some of Hypatia's original problems that deal with Diophantine Algebra?

A starting point could be working on the solution to the following problem:

> Your teacher ordered white chalk to last for the academic year. When the box arrived it was found that the chalk had been packed in 17 blue containers and 5 red containers. However, the pieces of chalk (none of which had broken) were no longer in their containers. All together, your teacher counted 234 pieces of chalk and was now faced withthe task of repacking them into their containers. How should your teacher fill each container?
>
> Consider the linear equation $ax + by = c$, where a, b and c are integers.
>
> Start by investigating integer solutions to this equation.
>
> For example, the equation $5x + 6y = 28$, has an integer solution $x = 2$ and $y = 3$.
> Are there other integer solutions satisfying $5x + 6y = 28$?
>
> What about the equation $5x + 3y = 100$ or $5x + 3y = 1000$ or $5x + 3y = 10000$?
> How many distinct pairs of **positive integers** (x, y) satisfy these equations?
> Should you consider negative integers as well?
>
> Next restrict the equation $ax + by = c$, where a, b and c are integers, with the following conditions: $b > a$, with a and b relatively prime
> and $1 \leq c \leq b$

6.0 CARBON DIOXIDE

We read almost daily about the 'Greenhouse Effect'. What is this and how can mathematics help in disentangling fact from fiction?

Some years ago, researchers in Antarctica realised that when they drilled into the icecap, they could see layers in the ice that resulted from each year's snowfall. This meant that they could date each layer by counting back from the surface much in the way we can find the age of a tree by counting its growth rings. In addition, each layer contained trapped air bubbles that were like time capsules of the Earth's past atmosphere. By testing the composition of these bubbles, the scientists found that the concentration of carbon dioxide in the air had increased steadily over the centuries preceding 1900.

Some of the figures are shown in the table below:

Year	1760	1774	1790	1804	1820	1836	1848	1855	1866	1873	1877	1889	1898
CO_2 (ppm)	277	278	280	284	284	286	287	288	290	290	292	293	295

When these figures are plotted on a graph, the result is as shown.

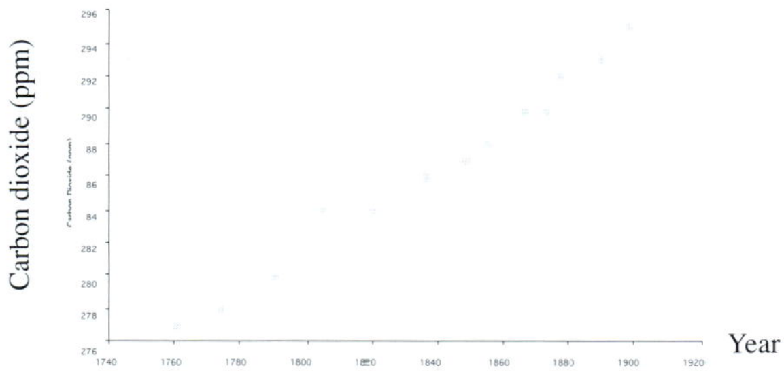

As you can see, the points lie on an approximate straight line. Handling data of this sort is the subject of this chapter. In it you will learn how to find a formula that can help predict the carbon dioxide into the future. In this case, an approximate model is: CO_2 (ppm) = 0.126×Year + 55.

The prediction for 2004 is CO_2 (ppm) = 0.126×2004 + 55 = 308.

Why do we care about the concentration of carbon dioxide in the atmosphere? The short answer is that the gas acts as a blanket trapping the sun's heat. More carbon dioxide means a warmer Earth. Sounds nice? Not really as warming is predicted to melt the polar ice caps (flooding low lying land), disrupt agriculture (threatening food supplies) and increase the frequency of extreme storms.

What has some scientists worried is that the actual value of the CO_2 concentration at present is over 350 ppm, a figure much larger that the linear model's prediction. The major reason is our extensive use of fossil fuels. This could make a good starting point for a project.

6.1 GRAPHS

6.1.1 INTRODUCING LINEAR GRAPHS

Graphical representation of data is one of the most powerful means by which information can be interpreted and analysed. Consider the following set of data, representing the cost involved in the production of an item (in batches of five):

No. of items	0	5	10	15	20	25	30
Cost ($)	25	50	75	100	125	150	175

From the table it appears that as the number of items increase, so too does the cost. However, at a glance, we do not get an overall feel for how the cost is increasing. On the other hand, if we were to represent this data graphically, we would get a better feel for how the two **variables**, that is the number of items and the cost, are related.

In other words, we could more easily 'see' the relationship between these two variables.

As the cost is dependent on the number of items, we say that the cost is the **dependent variable** and that the number of items the **independent variable**.

Next we **plot a graph** of the above data with the **independent variable along the horizontal axis** and the **dependent variable along the vertical axis**. To plot the relationship between these two variables we use a set of ordered pairs:

When 5 items are sold, we have an associated cost of $50 and so this would correspond to the ordered pair (5, 50).

When 15 items are sold, we have a cost of $100, giving the ordered pair (15,100).

We continue in this manner until we have a complete set of ordered pairs:

$$\{(0, 25), (5, 50), (10, 75), (15, 100), (20, 125), (25, 150), (30, 175)\}.$$

We are now in a position to plot these points on a set of axes:

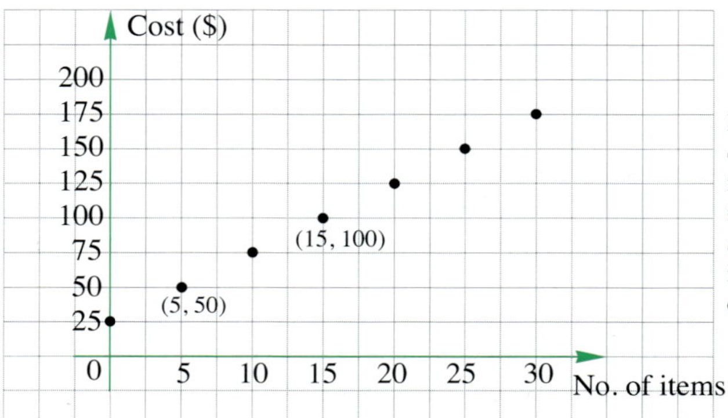

From the graphical representation of the data, it is clear that not only is the cost increasing as the number of items increases, but that the relationship appears to be linear. That is, the cost increases by the same amount of $25 for every extra five items produced.

Using the Graphics Calculator

We can make use of the TI–83 to reproduce the table of values and then plot the points. To do this we need to create two lists, L1 and L2. There are a number of ways this can be done.

Clearing all lists

First we clear all existing lists from the claculator.
Press **2nd +**, and then choose **4:ClrAllLists**

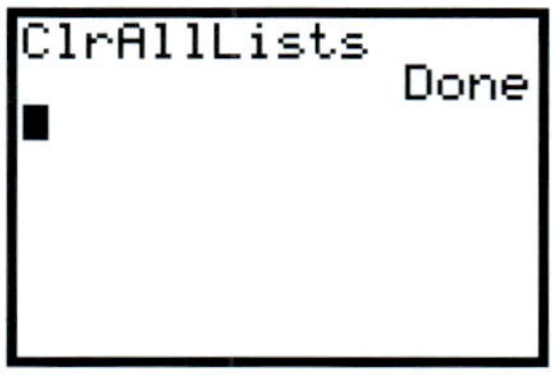

Clearing selected lists

To clear some lists, use the **STAT** option.
Press **STAT**, choose option **4:ClrList**
then enter the lists you wish to delete by pressing **2nd 1** (for List1), **2nd 2** (for List 2) and so on.

To enter the data we then press **STAT** and choose option **1:Edit**. This will send us to a Table window where we can then enter the data:

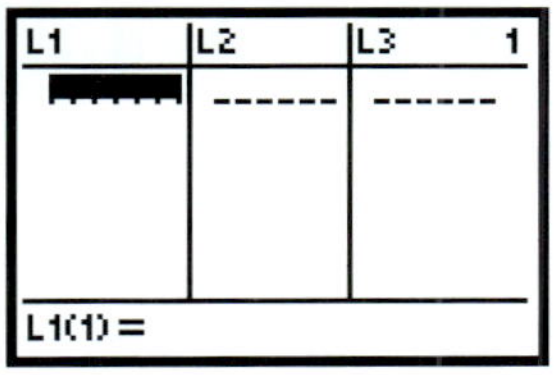

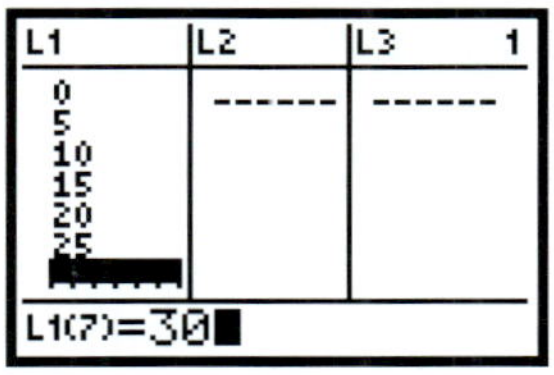

First enter the idependent variable, type in the number and then press **ENTER** (repeat this until all the data has been entered).

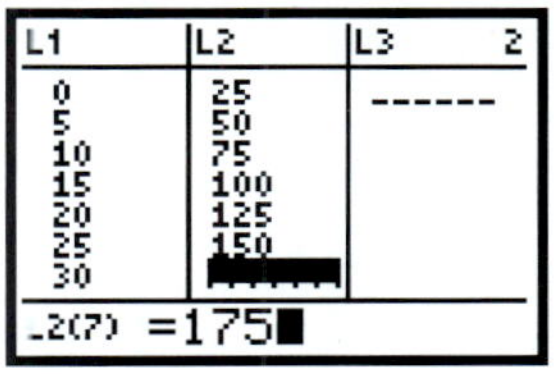

Once that is done, use the arrow key to move to the L2 column and do the same as for L1.

Once the data is entered, we must have the **STAT PLOT** option on:
Press **2nd Y=**, select option **1:Plot 1. . .**, place cursor over the **On** option and then press **ENTER**.

Once that is done, press **2nd MODE** (to quit). Then set your **WINDOW** so that all the data points can be displayed. In this case, the x-values (independent variable) range from 0 to 30 and the y-values (dependent variable) range from 0 to 175. Then press **GRAPH**:

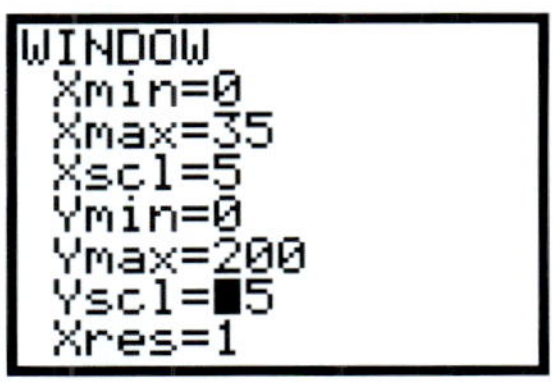

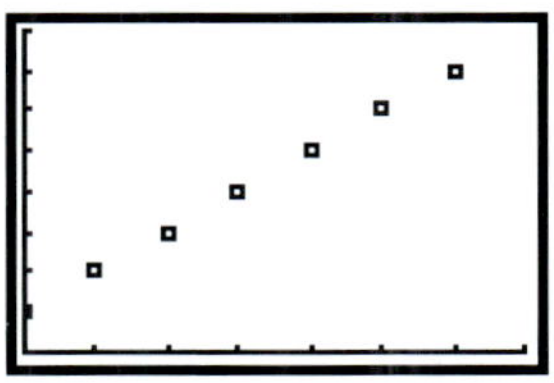

Another way that the relationship between two quantities can be presented is through the use of a formula (or **equation**).

Consider the relationship describing the conversion from degrees Fahrenheit (°F) to degrees Celsius (°C), $F = 1.8C + 32$. It would be possible, once more, to set up a table of values, however, because we are given the relationship in an algebraic form, we could simply use the graphics calculator to visualise this relationship.

This time we use the equation editor screen.
We enter the equation, set the window ranges and then press **GRAPH**

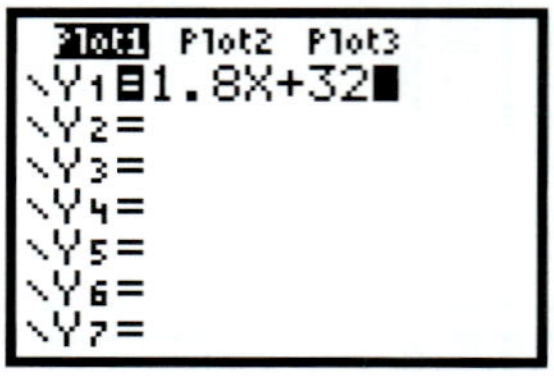

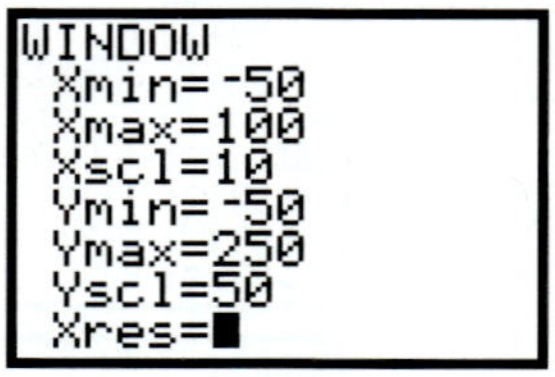

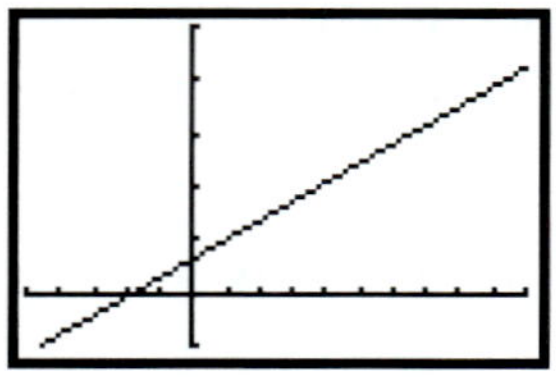

From the display we can see that a linear relationship exists.

At this point we make note of the following. With the example involving cost, we produced with a set of points that aligned themselves in a straight line. Such a graph is known as a **discrete linear relationship** – that is, a disconnected set of points. However, with the rule for converting from degrees celsius to degrees Fahrenheit, we found a **continuous** line.

EXAMPLE 6.1

For the continuous relation, $T = 1.2x + 4$, construct a table of values for values of x from -1 to 5 and use it to sketch the graph of this relation. Give a brief description of this graph.

Solution

The table is constructed by substuting values of x into the equation and determining the corresponding values for T. For example, when $x = -1$, $T = 1.2 \times -1 + 4 = 2.8$ Similarly, when $x = 0$, $T = 1.2 \times 0 + 4 = 4$ and so on. So, we have:

x	-1	0	1	2	3	4	5
T	2.8	4	5.2	6.4	7.6	8.8	10

We can now plot the points on a set of axes, however, as we are told that it is a continuous relation, we can also draw a straight line through the points.

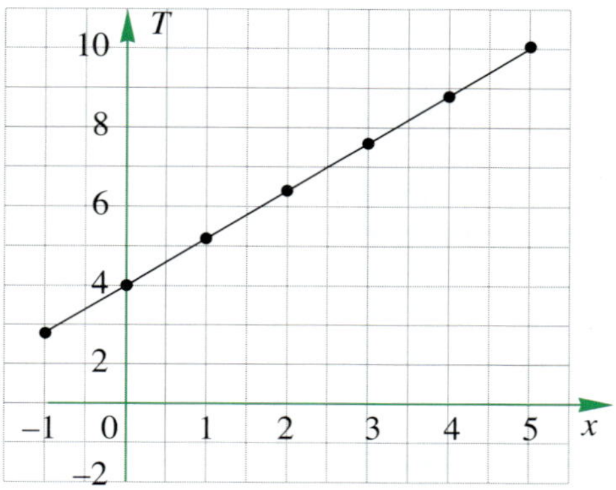

The following observations can be made:

1. The relation is **linear**.

2. It cuts the T–axes at the point (0, 4). i.e., line has a T-intercept of 4.

3. It increases at a **constant rate**. T increases by 1.2 for every increase of 1 unit in x.

   i.e., $\frac{\text{rise in } T}{\text{run in } x} = 1.2$

EXAMPLE 6.2

Using tables of values for the following relations, draw their graphs.

(a) $y = 2x + 1$ (b) $x + y = 2$ (c) $\frac{2t}{3} - s = 4$

Solution

(a) An equation of this type in which one of the variables is the subject can be tabulated by choosing values of x and using the rule to calculate the corresponding values of y. Such a relation is sometimes called '**explicit**'. On this occasion, there are no instructions as to what values of x are allowed, so any values can be chosen:

If $x = 5$ then $y = 2 \times 5 + 1 = 11$ and if $x = -4$, $y = 2 \times -4 + 1 = -8 + 1 = -7$. A table of values for this relation could be:

x	–1	–0.5	0	0.5	1	1.5
y	–1	0	1	2	3	4

These table entries can be converted to ordered number pairs: (–1, –1), (0, 1), (1, 3), (2, 5) etc.

Next, these number pairs can be plotted onto a cartesian graph:

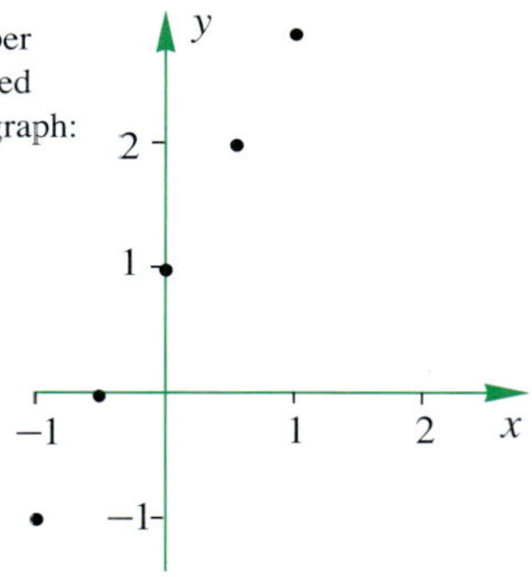

The points of the graph lie on a straight line, hence the description, 'linear'. It should also be noted that only a sample of the points has been plotted on the graph. If all the points are added, they will form a continuous straight line:

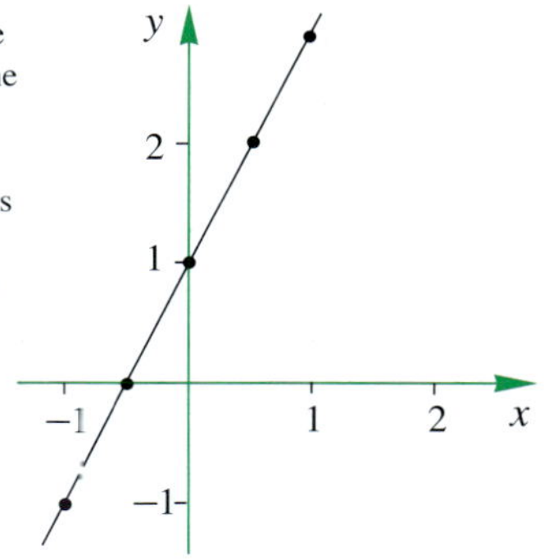

This line passes through the y-axis at 1 and has a slope of 2 (the line rises by 2 units for each unit moved to the right).

(b) $x + y = 2$ is known as an **implicit relation**. It is possible to make y the subject of the relation (to get $y = 2 - x$), but this is not necessary. All that is required is to find pairs of numbers that add up to 2.

Some examples are:

x	–1	0	1	2	3	4
y	3	2	1	0	–1	–2

As before, the entries in this table are only examples. If the complete set of such points is plotted, the result is:

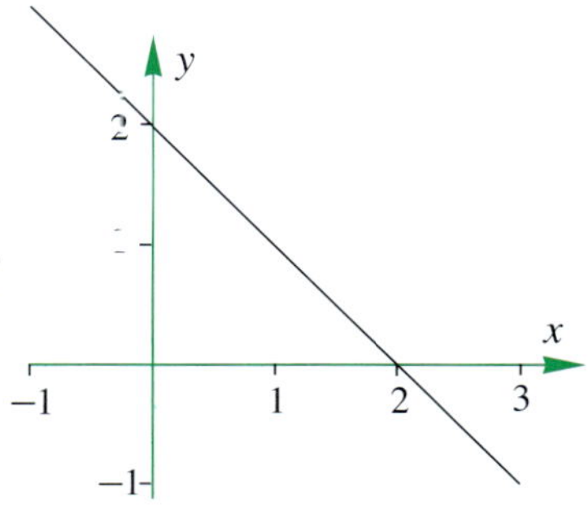

(c) The previous examples used the variables x and y. In these cases it is conventional to plot x values on the horizontal axis and y values on the vertical axis. When other variables are used, if the relation is explicit (one of the variables is on its own on one side of the equation), this variable is plotted on the vertical axis.

In this case, it is not clear which variable should be plotted on each axis. Also, it is probably easiest to make one of the variables the subject of the equation:

$$\frac{2t}{3} - s = 4 \Rightarrow \frac{2t}{3} = s + 4 \Rightarrow s = \frac{2t}{3} - 4$$

This can now be used as in part (a) to calculate values of s after choosing values for t. The table could be:

t	–3	0	3	6	9	12
s	–6	–4	–2	0	2	4

The graph can then be completed as shown:

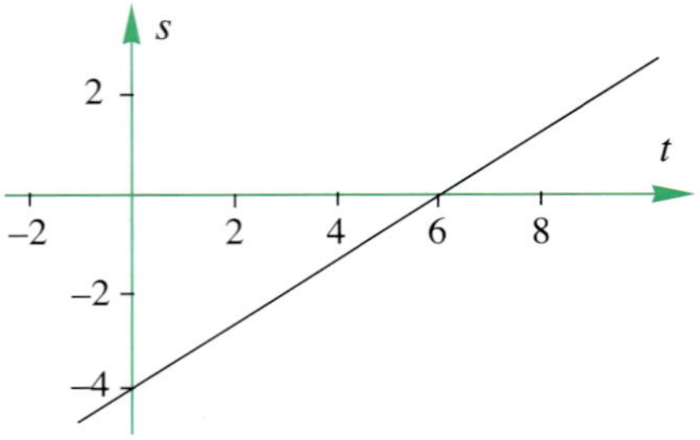

Examples 6.1 and 6.2 bring to light key features of a straight line.

Observation 1: The equation $T = 1.2x + 4$ represents a linear form.

Observation 2: The intercept along the vertical axis is 4, which corresponds to the constant term in the equation.

Observation 3: T increases at a constant rate of 1.2, which happens to be the number in front of the x-term. This is known as the **gradient** (or **slope**) of the straight line.

We can then make the following generalisations about linear graphs.

1. General equation of a **linear graph** is $y = mx + c$.
2. The **gradient** (or slope) of the line $y = mx + c$ is given by $\boldsymbol{m}$.
3. The ***y*-intercept** for the line $y = mx + c$ is $\boldsymbol{c}$, i.e., the straight lines cuts the vertical axes at **(0, *c*)**.

This means that given an equation such as $y = 2x + 3$, we can immediately tell that

1. It is a straight line.
2. It has a gradient of 2.
3. It cuts the y-axes at (0, 3)

Which can be verified using the graphics calculator:

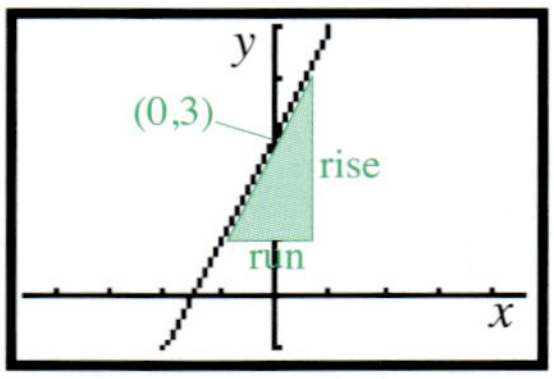

$\frac{\text{rise}}{\text{run}} = 2$

Sometimes it is possible that a linear relation is 'hidden' within a given rule. For example, the equation $-4x + y - 1 = 0$ might not immediately appear to be a linear relation, however, after some rearranging of the variables we have: $-4x + y - 1 = 0 \therefore -4x + y = 1$

$$\therefore y = 4x + 1$$

Which is of the form, $y = mx + c$, with $m = 4$ and $c = 1$, and thus, telling us that the straight line has a gradient of 4 ($m = 4$) and a y-intercept of 1 ($c = 1$).

EXAMPLE 6.3

For each of the following linear relations, identify its gradient, y-intercept and then, sketch its graph.

(a) $y = 3x + 8$ (b) $y = x - 2$ (c) $4x - 8y = 24$

Solution

(a) Comparing the equation $y = 3x + 8$ with that of $y = mx + c$, we have that $m = 3$ and $c = 8$. Therefore, the gradient is 3 (= m) and the y-intercept is 8 (= c).

The graph needs to 'show' that the line has a slope of 3 and that it passes through the point (0, 8). Also, there are no restrictions on the x-values that can be used, and so we can assume that we can use the set of real values for x. The set of values of x that can be used is called the **domain**.

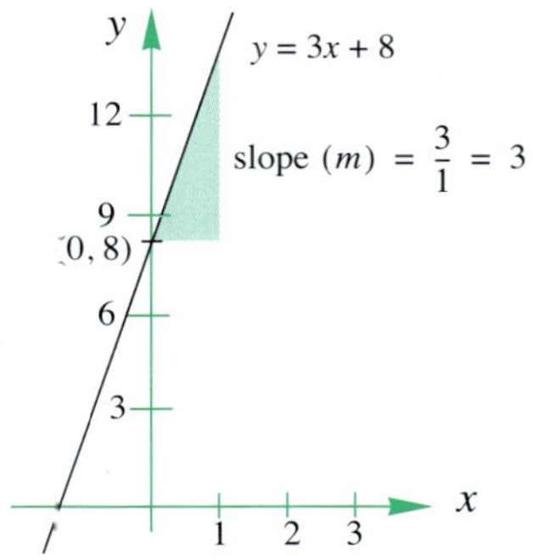

Notice that although the graph seems to 'stop', it does in fact continue. To indicate that a graph does not go beyond a particular point, we place a closed circle at the end point(s).

(b) Comparing the equation $y = x - 2$ with that of $y = mx + c$, we have that $m = 1$ and $c = -2$.
Therefore, the gradient is 1 (= m) and the y-intercept is –2 (= c).

The graph should display a line passing through the point (0, –2) and with a gradient of 1.

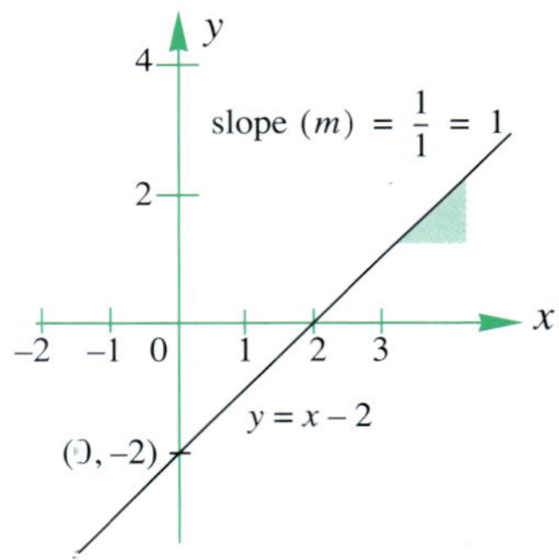

(c) The equation $4x - 8y = 24$ is not in the form $y = mx + c$, and so, this time, we need to first carry out a little algebra. We need to rearrange the equation $4x - 8y = 24$ and make y the subject.

$$4x - 8y = 24 \therefore -8y = -4x + 24$$

Then, dividing both sides by –8 gives:
$$y = \frac{-4}{-8}x + \frac{24}{-8}$$
$$y = \frac{1}{2}x - 3$$

We can now compare $y = \frac{1}{2}x - 3$ with $y = mx + c$.

In this case we have a gradient of $\frac{1}{2}(= m)$ and a y-intercept of –3.

The straight line will then have to pass through the point (0, –3) and have a gradient of 1.

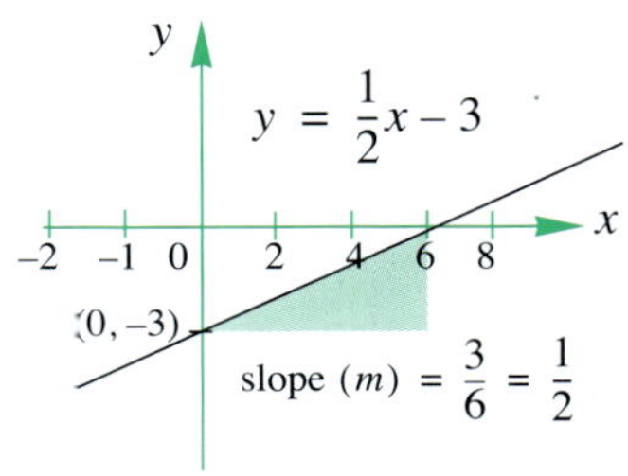

EXAMPLE 6.4

State the gradient and y-intercepts of the graphs related to these equations:

(a) $y = 2x - 3$ (b) $y = \frac{x-1}{3}$ (c) $y = 2 - \frac{3x}{4}$

Solution

(a) By comparing the equation $y = 2x - 3$ to the standard form $y = mx + c$, the gradient $(m) = 2$ and the y-intercept $(c) = -3$ can be read directly from the equation.

(b) The equation can be written in the standard form: $y = \frac{x-1}{3} \Rightarrow y = \frac{1}{3}x - \frac{1}{3}$.

The gradient is $\frac{1}{3}$ $(= m)$ and the y-intercept is $-\frac{1}{3}$ $(= c)$.

(c) Similarly: $y = 2 - \frac{3x}{4} \Rightarrow y = \left(-\frac{3}{4}\right)x + 2$.

The gradient is $-\frac{3}{4}$ $(= m)$ and the y-intercept is 2 $(= c)$.

6.1.2 SUMMARY OF LINEAR GRAPHS

We can summarise our observations of linear graphs as follows:

For the linear relationship, $y = mx + c$, where m is the **gradient** of the straight line and c the **y-intercept**, passing through the point (0, c), the following outcomes are possible:

Equation	$m > 0$	$m < 0$	$m = 0$
$y = mx + c$ with $c > 0$	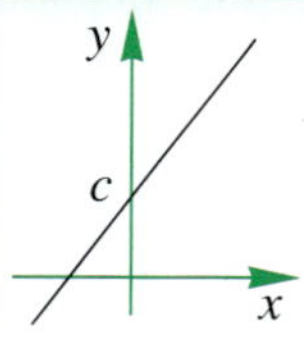	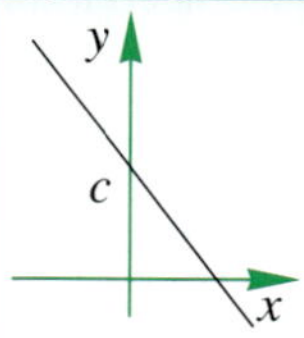	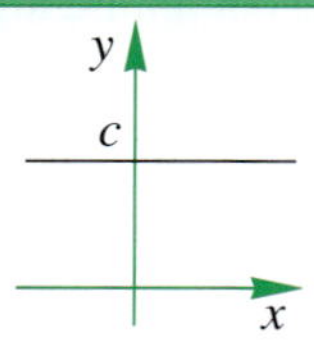
$y = mx + c$ with $c < 0$	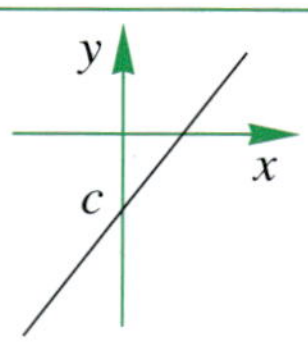	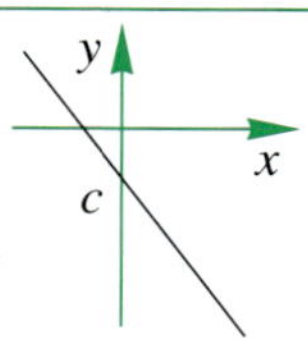	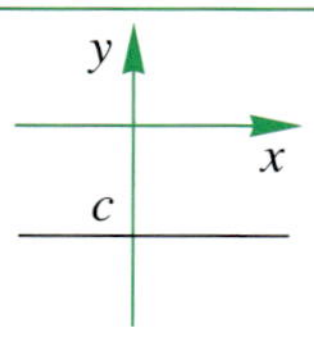

There is also a special case to consider, that is, the case where the gradient is undefined, In such a situation we have a **vertical straight line**. Such a graph is shown alongside:

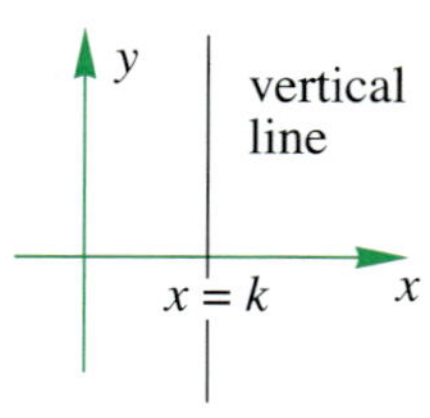

EXERCISES 6.1

1. For each table of values (you may wish to use a graphics calculator)
(a) plot the set of points on a set of axes.
(b) draw a continuous line through the points in (a).

i.

x	0	1	2	3	4	5
y	3	5	7	9	11	13

ii.

x	–3	–1	1	3	5	7
y	1	3	5	7	9	11

iii.

x	–4	–2	0	2	4	6
y	6	4	2	0	–2	–4

iv.

x	–4	–2	0	2	4	6
y	0	3	6	9	12	15

2. Using tables of values for the following relations, draw their graphs.
(a) $y = 3x - 1$ (b) $y = 4x + 2$ (c) $y = -2x + 4$
(d) $s = -\frac{1}{2}t + 3$ (e) $q = 1.2r - 3.5$ (f) $v = -\frac{2}{7}t + 1$

3. For each of the following linear relations, identify its gradient, y-intercept and then, sketch its graph.
(a) $y = -x + 1$ (b) $x - y = 2$ (c) $2x + 3y = 12$
(d) $3x + y = 6$ (e) $x - 3y + 1 = 0$ (f) $\frac{1}{2}s = 2t + 1$

4. Sketch the graphs of the following
(a) $x - 3y = 6$ (b) $2x = 3y + 9$ (c) $6s = -2t + 12$

5. Sketch the graphs of the following
(a) $4x - 2y = 3$ (b) $x + y = 7$ (c) $y = -1.5x + 1.5$

6. Sketch the graphs of the following
(a) $y = 5$ (b) $x = 2$ (c) $y = 3 - \frac{1}{3}x$

7. Sketch the graphs of the following
(a) $y = 2 - \frac{2}{3}x$ (b) $v = \frac{1}{5}u + 1.2$ (c) $5x + 3y = 0$

8. Sketch the graphs of the following

(a) $y = \frac{2}{3}x$ (b) $s = \frac{1}{2}(t-1)$ (c) $\frac{2y-x}{3} = 1$

9. Draw the straight line that
(a) has a gradient of 2 and passes through the point (0, 5).
(b) has a gradient of –1 and passes through the point (0, 3).
(c) has a gradient of –2 and passes through the point (4, 0).
(d) passes through the points (0, 3) and (4, 0).
(e) passes through the points (0, –2) and (1, 0).

10. A computer is purchased for \$900 at the start of 2002 and losses \$150 of its value every year.
(a) Construct a table showing the value of the computer, \$ V, t years since the start of 2002.

t (years)	0	1	2	3	4	5	6
V (\$)	900						

(b) Graph the relationship between the computer's value \$ V and the number of years, t, since the start of 2002.
(c) Why did we stop at $t = 6$? What implications does this have for your graph?

11. A tank, containing 250 litres of water has its tap left open and water is leaking at a constant rate of 25 litres per hour.
(a) Construct a table of values for the amount of water, V litres, left in the tank every hour.

t (hours)	0	1	2	3	4	5	6	7	8	9	10
V (Litres)	250										

(b) Plot your results on a set of axes with V litres representing the volume of water left in the tank and t hours the time that water has been leaking from the tank.
(c) Graph the relationship between the volume V litres and the time t hours.
(d) Why did we stop at $t = 10$? What implications does this have for your graph?

12. A door-to-door salesperson receives \$750 per week plus \$75 for every set of books sold. Let \$I represent the salesperson's total weekly income and n the number of sets of books sold.
(a) Construct a table of values for the the salesperson's total weekly income.
(b) Plot your results on a set of axes.
(c) Graph the relationship between the total weekly income \$I and the number of sets of books sold.

13. The relationship between a father's height, H cm, and the average of that of his sons' heights, h cm, can be approximated by the linear relation $h = \frac{1}{2}H + 87$.
(a) Construct a realistic table of values for the relationship between h and H.

H											
h											

(b) Plot your results on a set of axes.
(c) From your graph, determines a son's height if the fathers is 187 cm tall.

6.2 GRADIENT FORM OF A LINE

6.2.1 THE GRADIENT OF A STRAIGHT LINE

We have defined the gradient of a straight line, $y = mx + c$, as $m = \frac{\text{rise}}{\text{run}}$. We can also use a more formal definition for the gradient of a straight line, which is sometimes called the two point form definition.

We still make reference to the rise and run of a straight line, but this time we express these in terms of two points that lie on the straight line.

Consider two points, $A(x_1, y_1)$ and $B(x_2, y_2)$ on the line straight line with equation $y = mx + c$. We then have:

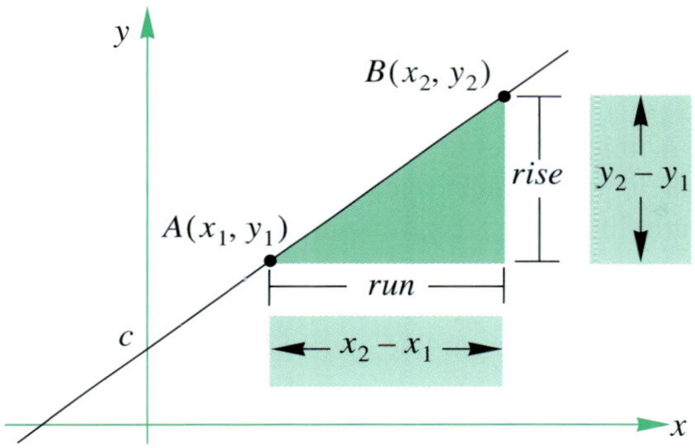

The gradient, m is given by

$$m = \frac{\text{rise}}{\text{run}}.$$

However, the $\text{rise} = y_2 - y_1$

and $\text{run} = x_2 - x_1$

therefore, we have

$$m = \frac{y_2 - y_1}{x_2 - x_1}$$

EXAMPLE 6.5

Find the gradient of the straight line pasing through the points

(a) (3, 7) and (5, 9) (b) (–3, 2) and (4, 0) (c) (2.6, 5.5) and (3.8, 3.8)

Solution

(a) We can label the points A as (3, 7) and B as (5, 9).

Then, using the definition, $m = \frac{y_2 - y_1}{x_2 - x_1}$ we have

$$m = \frac{9-7}{5-3} = \frac{2}{2} = 1$$

A quick sketch of the line passing through A and B validates our result (at the very least it shows that the line has a positive gradient).

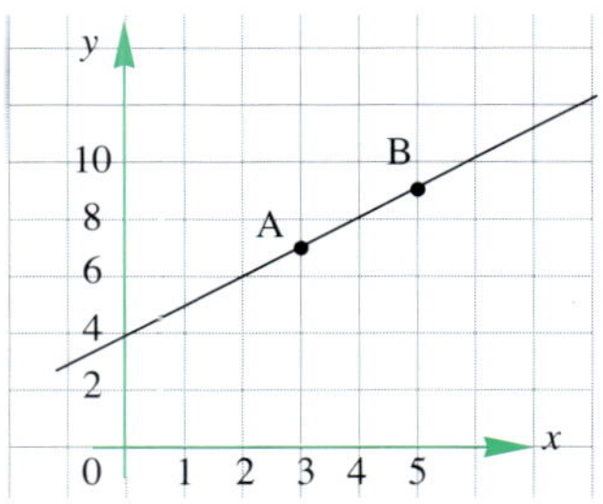

(b) Labelling the points as A(–3, 2) and B(4, 0), we have $m = \frac{y_2 - y_1}{x_2 - x_1} = \frac{0-2}{4-(-3)} = \frac{-2}{7}$.

That is, the gradient $m = -\frac{2}{7}$.

A sketch confirms that we have a negative gradient.

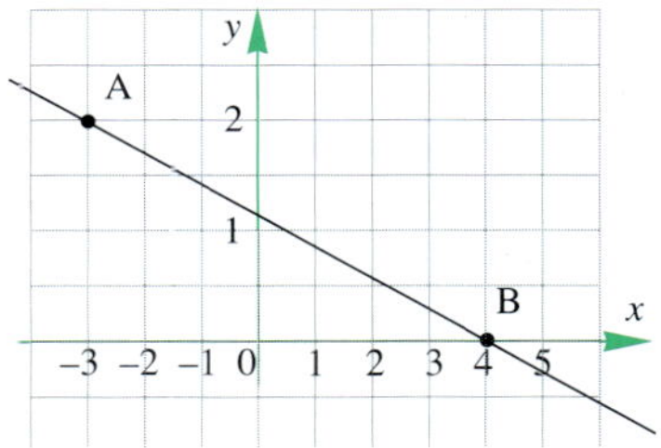

(b) Labelling the points as A(2.6, 5.5) and B(3.8, 3.8), we have

$$m = \frac{y_2 - y_1}{x_2 - x_1} = \frac{3.8 - 5.5}{3.8 - 2.6} = \frac{-1.7}{1.2}$$

$$= -\frac{17}{12}.$$

A note about labelling points when finding the gradient. Consider Example 6.5 (b), had we reversed the points and labelled them as A(4, 0) and B(–3, 2), would this have changed our answer? Let's see: $m = \frac{y_2 - y_1}{x_2 - x_1} = \frac{2-0}{-3-4} = \frac{2}{-7}$.

That is, the gradient $m = -\frac{2}{7}$.

Giving us the same result as before. That is, it does not matter which point you label A and which point you label B.

6.2.2 EQUATION OF STRAIGHT LINE FROM TWO POINTS

What we have observed is that a straight line is uniquely defined by two different points. That is, as long as we have two points we can draw one unique straight line passing through those points. Why is it not possible to define a unique straight line if we are given only one point?

Having drawn a line passing through two points, it is very useful to know its equation. In this section we will see how to determine the equation of a straight line given any two points on the line.

Consider the straight line passing through the points (1, 3) and (4, 9). We start by plotting these points on a set of axes and drawing the straight line through these points:

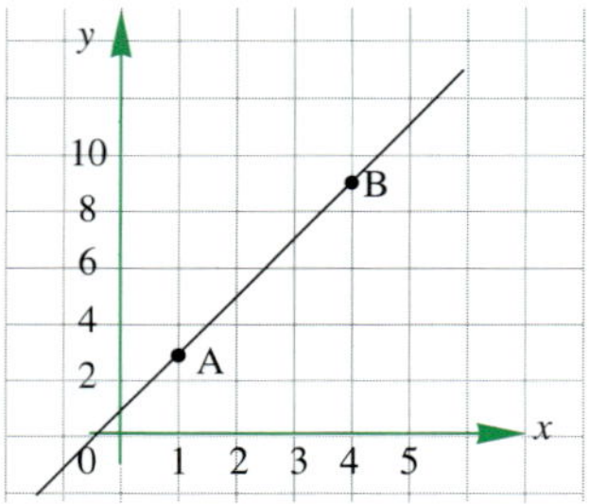

From the given information we can determine the gradient of the line:

$$m = \frac{9-3}{4-1} = \frac{6}{3} = 2$$

Next, we know that a straight line can be written in the form $y = mx + c$, where c is the y-intercept.

So far, we have that $y = 2x + c$ (as $m = 2$). The question then remains, 'How do we determine the value of c?'

One possible way is to look at the graph and see where the line cuts the y-axes. From our graph it appears that the line cuts at $y = 1$, meaning that $c = 1$, however, we need to be absolutely sure.

To do this, we use the fact that the straight line passes through A(1, 3), that is, when $x = 1$, $y = 3$. So, we can substitute these values into the equation $y = 2x + c$ giving $3 = 2 \times 1 + c \Rightarrow c = 1$.

Therefore, we have the equation $y = 2x + 1$. It is reassuring that this result is consistent with our visual observation.

EXAMPLE 6.6 Find the equation of the straight line passing through the points (2, 3) and (6, –3).

Solution

Step 1: Find the gradient of the straight line using $m = \frac{y_2 - y_1}{x_2 - x_1}$:

$$m = \frac{-3-3}{6-2} = -\frac{6}{4} = -\frac{3}{2}$$

Step 2: Use the equation $y = mx + c$ with the value of m found in step 1:

$$y = -\frac{3}{2}x + c$$

Step 3: Use any one of the points that the line passes through to solve for c:

Using the point (2, 3), i.e., $x = 2, y = 3$, we have, $3 = -\frac{3}{2} \times 2 + c$

$$3 = -3 + c$$

$$\therefore c = 6$$

Therefore, $y = -\frac{3}{2}x + 6$

Would it have made a difference if we had used the point (6, –3) in Example 6.6? The answer is no, it makes no difference which point is used. In our case, had we used the point (6, –3) we would have: $-3 = -\frac{3}{2} \times 6 + c \Rightarrow -3 = -9 + c \therefore c = 6$

giving the same result.

6.2.3 OTHER FORMS OF EQUATIONS FOR STRAIGHT LINES

The other common forms for linear equations are: $ax + by + c = 0$ and $px + qy = r$. These forms of the linear equation do not give the gradient and y-intercept of the line. When sketching or plotting linear equations in this form it is probably best to calculate the intercepts on both axes.

For $ax + by + c = 0$ this is done as follows:

For x–intercept, let $y = 0$, so that $ax + c = 0 \Rightarrow x = -\frac{c}{a}$

For y–intercept, let $x = 0$, so that $by + c = 0 \Rightarrow y = -\frac{c}{b}$

EXAMPLE 6.7

Find the axial intercepts and sketch the graphs of these linear relations.

(a) $2x - 3y = 12$ (b) $\frac{2}{3}x - \frac{y}{4} = 4$ (c) $2x + 3y - 4 = 0$

Solution

(a) For y-intercept set $x = 0$: $2 \times 0 - 3y = 12 \Rightarrow -3y = 12 \therefore y = -4$. i.e., $(0, -4)$

For x-intercept set $y = 0$: $2x - 3 \times 0 = 12 \Rightarrow 2x = 12 \therefore x = 6$. i.e., $(6, 0)$

Then, to sketch the graph, first plot the points $(0, -4)$ and $(6, 0)$ and then draw a line passing through both points.

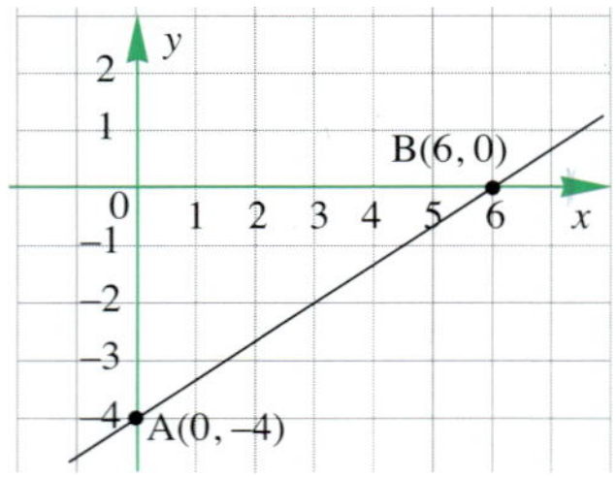

(b) y-intercept set $x = 0$: $\frac{2}{3}x - \frac{y}{4} = 4 \Rightarrow \frac{2}{3} \times 0 - \frac{y}{4} = 4 \Rightarrow -\frac{y}{4} = 4 \therefore y = -16$. i.e., $(0, -16)$

x-intercept set $y = 0$: $\frac{2}{3}x - \frac{0}{4} = 4 \Rightarrow \frac{2}{3}x = 4 \Rightarrow x = \frac{3 \times 4}{2} \therefore x = 6$. i.e., $(6, 0)$

To sketch the graph, first plot the points $(0, -16)$ and $(6, 0)$ and then draw a line passing through both points.

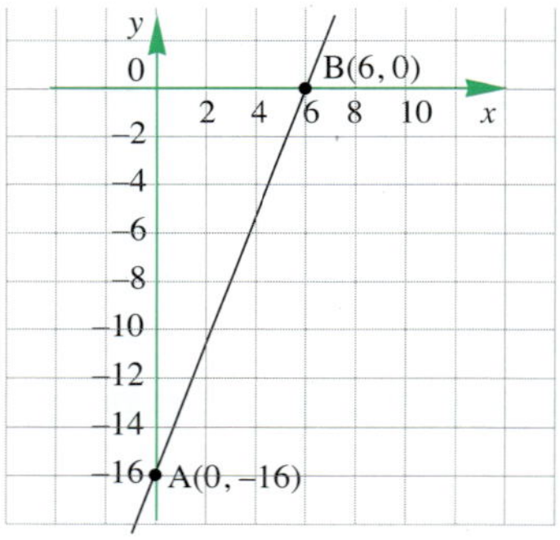

(c) y-intercept set $x = 0$: $2x + 3y - 4 = 0 \Rightarrow 3y - 4 = 0 \Rightarrow 3y = 4 \therefore y = \frac{4}{3}$. i.e., $\left(0, \frac{4}{3}\right)$.

x-intercept set $y = 0$: $2x + 3y - 4 = 0 \Rightarrow 2x - 4 = 0 \Rightarrow 2x = 4 \therefore x = 2$. i.e., $(2, 0)$

To sketch the graph, first plot the points $\left(0, \frac{4}{3}\right)$ and $(2, 0)$ and then draw a line passing through both points.

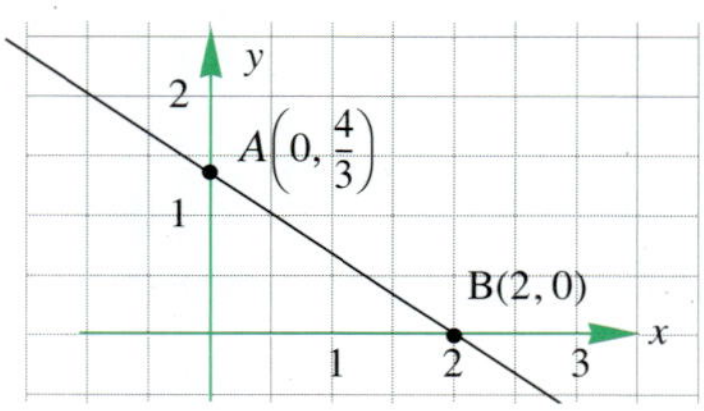

EXERCISES 6.2

1. Find the gradient of the straight line passing through the points

(a) (5, 6) and (9, 12)
(b) (2, 5) and (5, 8)
(c) (2.3, 5.8) and (6.5, 14.2)
(d) (–3, 4) and (–6, –4)
(e) (5, –3) and (–2, 6)
(f) (2, 12) and (–3, 9)
(g) (–2, 5) and (2, 4)
(h) (1.5, 3.7) and (3.0, 8.2)
(i) $(1, 3a)$ and $(4, 6a)$
(j) $(b + c, 4b)$ and $(a, 6b)$

2. Find the gradients of the following straight lines

(a)

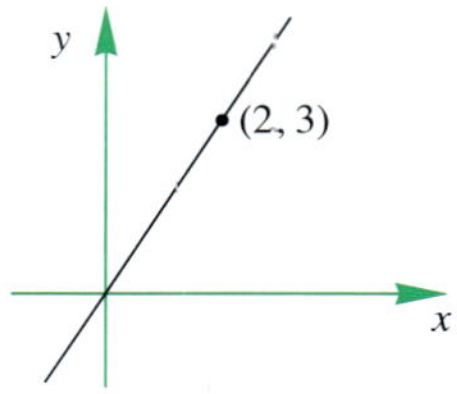

(b)

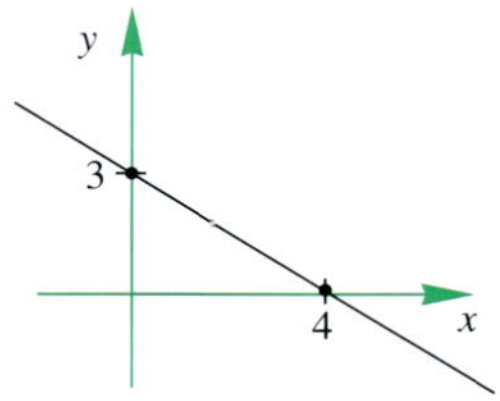

(c)

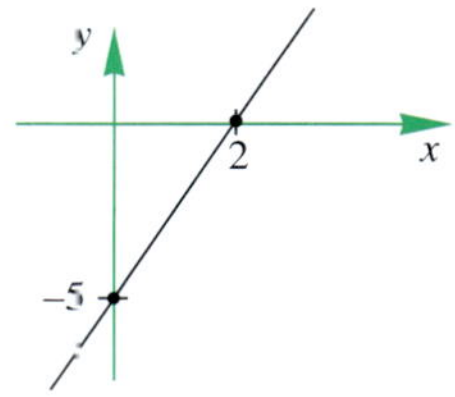

(d)

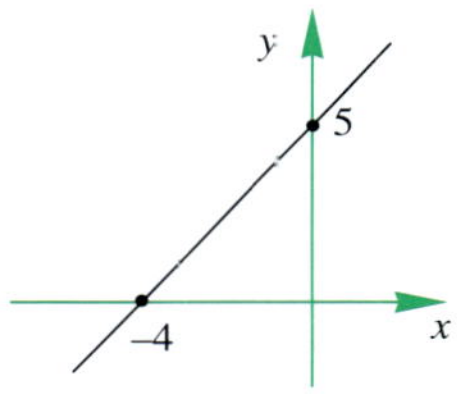

(e)

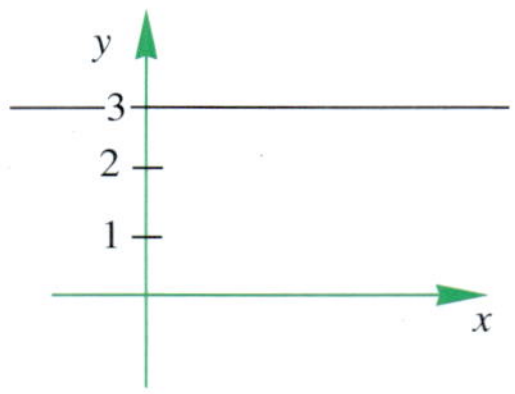

(f)

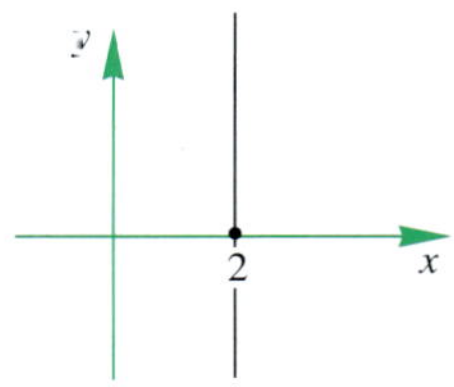

3. Find the equation of the straight line that

(a) has a gradient of 2 and passes through the point (0, 5).
(b) has a gradient of –1 and passes through the point (0, 3).
(c) has a gradient of –2 and passes through the point (4, 0).
(d) has a gradient of –0.5 and passes through the point (4, 1)
(e) has a gradient of $\frac{2}{3}$ and passes through the point (6, 4)

4. Find the equation of the straight line that

(a) passes through the points (0, 3) and (4, 0).
(b) passes through the points (0, –2) and (1, 0).
(c) passes through the points (2, –1) and (5, 5).
(d) passes through the points (–5, 1) and (4, –8).
(e) passes through the point (2.1, 3.7) and (4.3, –2.9)

5. Find the equation of each of the straight lines in question 2.

6. For each of the following, find the x-intercept and y-intercept and hence sketch its graph.

(a) $2x + 4y = 8$
(b) $-x + 3y = 6$
(c) $y - 4x = 9$
(d) $x - y + 2 = 0$
(e) $2y - 3x - 6 = 0$
(f) $5x + 2y + 5 = 0$

7. For each of the following, find the x-intercept and y-intercept and hence sketch its graph.

(a) $\frac{1}{2}x - \frac{3}{4}y = 1$ (b) $\frac{3}{5}x + \frac{2}{3}y + 6 = 0$ (c) $x + \frac{1}{2}y - 7 = 0$

8. It costs \$4 to set the type to print business cards. Each card costs \$0.02 to print.
(a) Write an equation that could be used to calculate the cost (\$$C$) of printing n business cards.
(b) Sketch the graph of the relationship between the \$$C$ and the number of cards, n.
(c) What cost is involved in printing 2000 cards?

9. Joe's telephone account is made up of a fixed rental charge of \$44 per three months, plus \$0.22 per local call.
Assume that Joe has not made any non-local calls.
(a) Write down a relationship between Joe's telephone bill, \$$B$, and the number of calls, n, he makes in a three month period.
(b) Sketch the relationship in (a).
(c) What was the amount shown on Joe's last telephone bill if he made 168 calls during the last three-month period?
(d) How many calls would Joe have made during a three month period if his telephone bill amounted to \$86.24 for that three month period?

10. It costs 35 cents per kilometre to run a car on petrol. If the car is converted to liquefied petroleum gas, the cost becomes 27 cents per kilometre.
(a) Write equations that give the cost of running the car on
  i. petrol (P) for k kilometres.
  ii. gas (G) for k kilometres.
(b) On the same set of axes, sketch both graphs in (a).
(c) If it costs \$1000 to convert a car from petrol to gas, how far would the owner have to drive before recovering the conversion costs?

6.3 SIMULTANEOUS EQUATIONS

To solve **simultaneously**, when given two linear equations, means to determine the set of points that belong to both equations. For example, given two equations $y = x + 2$ and $y = 2 - x$, then as both lines pass through the point (0, 2) – i.e., when $x = 0$, $y = 2$, a solution to this **system of equations** would be $x = 0$, $y = 2$.

We now consider solving simultaneous equations in some detail.

6.3.1 GRAPHICAL METHODS

We begin by solving the system of equations $y = -x + 8$ – (1) and $y = 2x - 1$ – (2)

To sketch both these straight lines we need to set up a table of values.

For $y = -x + 8$:

x	0	2
y	8	6

For $y = 2x - 1$:

x	0	2
y	–1	3

That is, the straight line $y = -x + 8$ passes through the points with coordinates $(0, 8)$ & $(2, 6)$.
The straight line $y = 2x - 1$ passes through the points with coordinates $(0, -1)$ & $(2, 3)$.
We can now sketch both these straight lines on the cartesian plane.

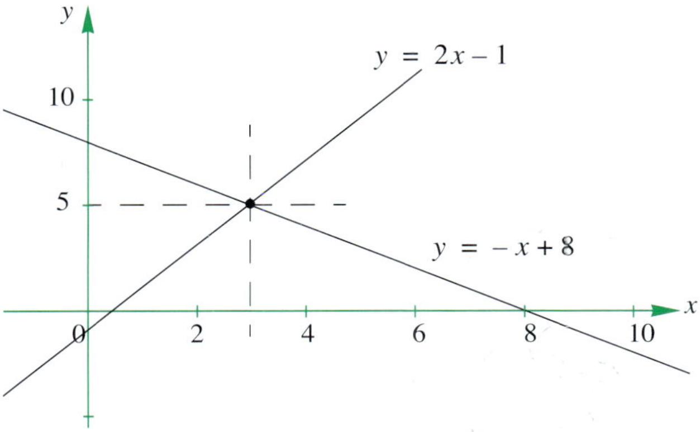

The point where the two lines meet is called the **point of intersection**. From the graphs we can see that the lines intersect at the point with coordinates $(3, 5)$. We say that the solution to the simultaneous equations (or to the system of equations) is $x = 3$ and $y = 5$.

Notice that we have obtained this solution by looking at our graph of the two straight lines.

Using a Graphics Calculator

There are a number of methods that can be employed when using a graphics calculator. We consider three methods.

1. Using a TABLE of values

This is done by entering the equations into the equation editor screen and then calling up the **Table** (of values). From this table we can then identify where the two lines have the same x and y values simultaneously. In this instance it occurs when $x = 3$ and $y = 5$.

$x = 3, y = 5$

2. Using the TRACE option

Using the same set of equations, $y = -x + 8$ and $y = 2x - 1$, we enter their equations into the equation editor screen and then press **GRAPH**. Once that is done press **TRACE** and move the cursor along until it is placed where the two lines meet. The coordinates will be shown at the bottom of the screen.

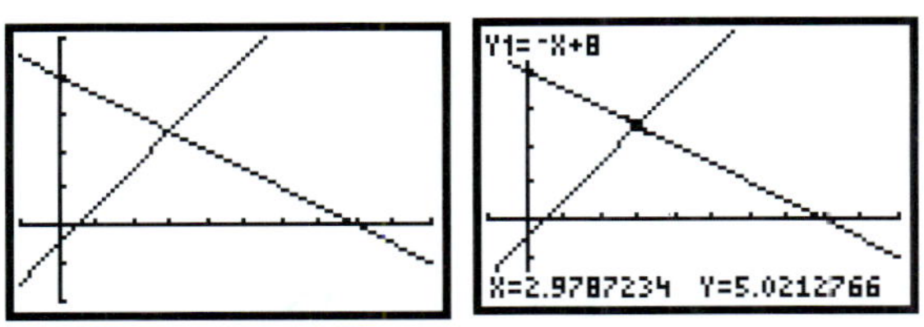

Notice that the screen shows that the two lines intersect at $x = 2.987\ldots$ and $y = 5.021\ldots$
However, this is not accurate enough.

Obtaining a more accurate answer will require that we use the **ZOOM** feature. Use either option **1:ZBox** or **2:Zoom In**. The **ZBox** option focuses directly on the region of interest as it will allow you to 'box-in' the region where the two graphs meet.

Using the **ZBox** option:

Step 1: Press **ZOOM** and select option 1

Step 2: Move cursor to the left-upper corner (that will be the start of your box). Then press **ENTER** (this will anchor the corner of your box).

Step 3: Then use the arrow keys to drag the box over the required region.

Step 4: Press **ENTER** – this will produce a screen that includes the region you have just boxed-in.

Step 5: Move the cursor to the point where the lines meet. In this case we obtain the solutions $x = 2.9986 \ldots$ and $y = 4.9885 \ldots$.

Step 1:

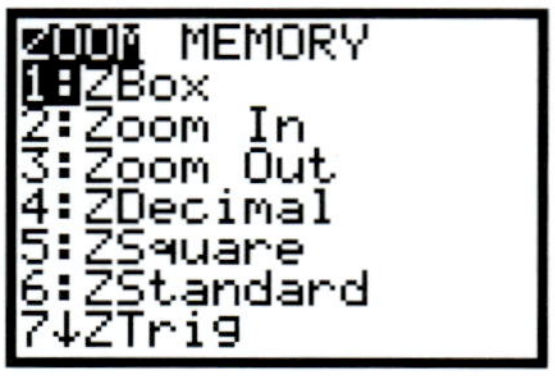

Step 2:

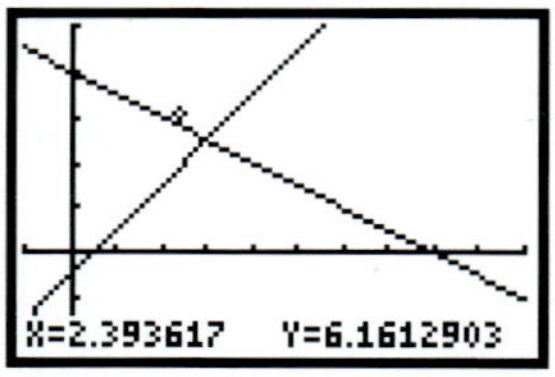

Step 3:

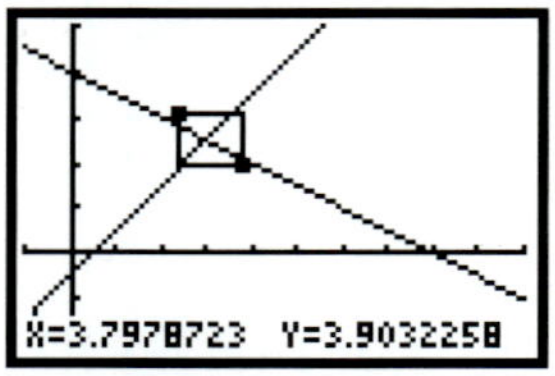

Steps 4 &5:

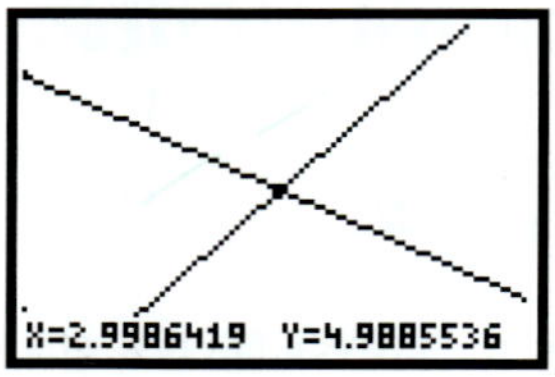

However, we still do not obtain the exact solutions.

3. Using the intersect option from the CALC menu

Of the visual alternatives, this method is probably the most efficient.

Step 1: Press **2nd TRACE** to call up the **CALC** menu, then select option **5:intersect**.

Step 2: You are then asked to identify your equations, starting with First curve. At this point, first move the cursor close to where the lines meet, and then press **ENTER**.

Step 3: Do the same as for Step 2, but this time for the second curve.

Step 4: When asked to guess, move your cursor to the point where the graphs meet and then press **ENTER**.

Step 5: The calculator will then provide you with the coordinates of the point of intersection.

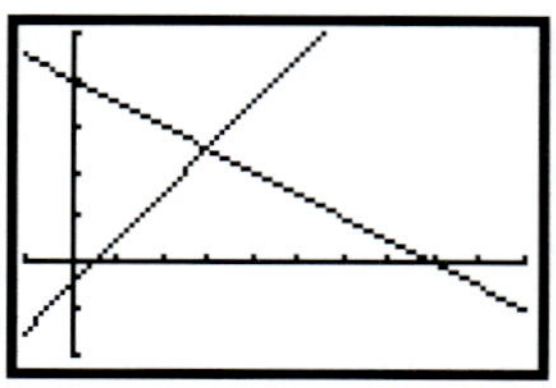

Step 1:

Step 2:

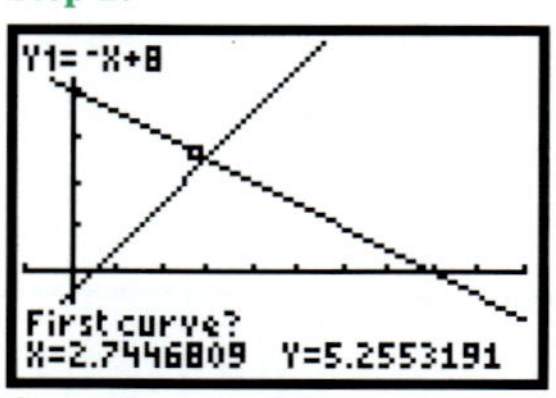

Step 3:

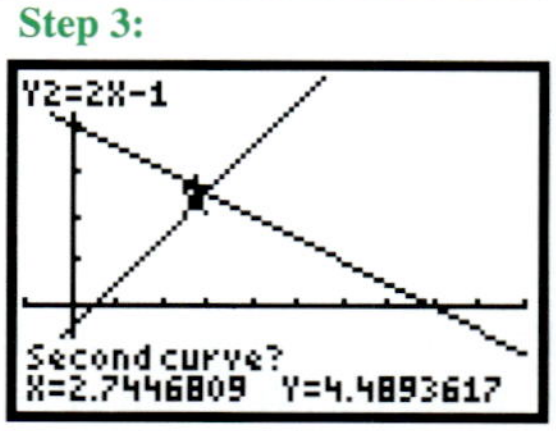

Step 4:

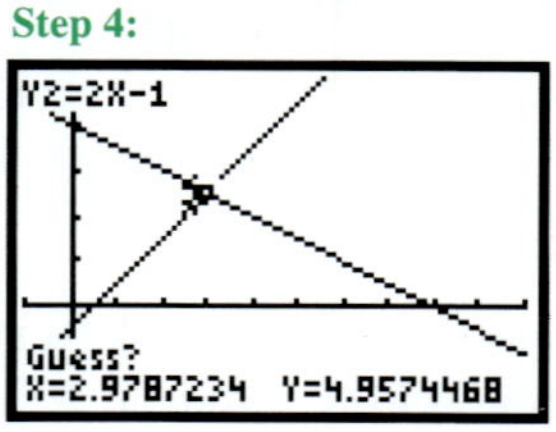

Step 5:

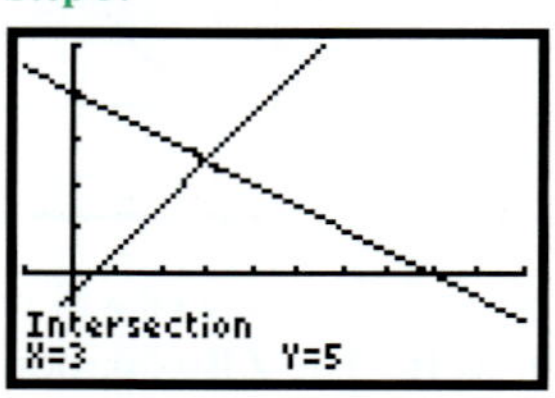

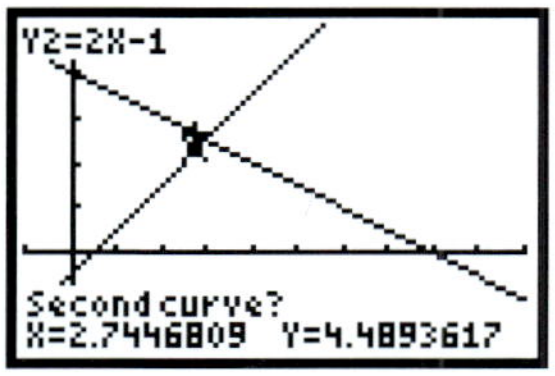

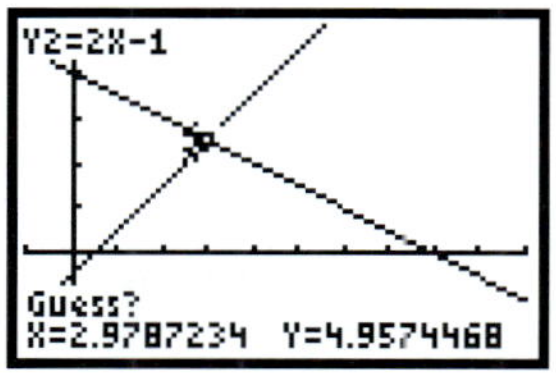

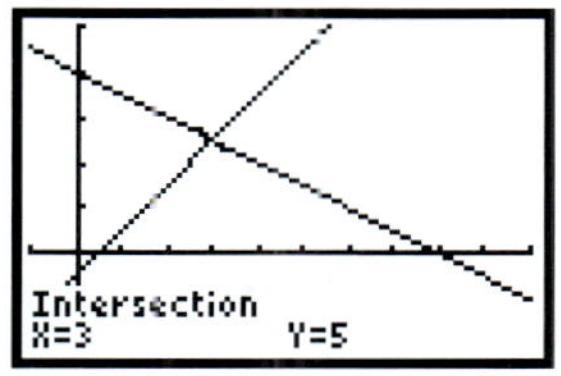

Notice that the linear equations can only be entered into the graphics calculator if they are in the form $y = mx + c$. So that if the equations had been $x - 2y = -7$ – (1) and $2x + 3y = 0$ – (2), then we would first need to rewrite them as:

$$x - 2y = -7 \Rightarrow -2y = -x - 7 \therefore y = \frac{1}{2}x + \frac{7}{2} \quad \text{– (1)}$$

$$2x + 3y = 0 \Rightarrow 3y = -2x \therefore y = -\frac{2}{3}x \quad \text{– (2)}$$

EXERCISES 6.3.1

1. For the graphs shown below, state the coordinates of the points of intersection.

(a)

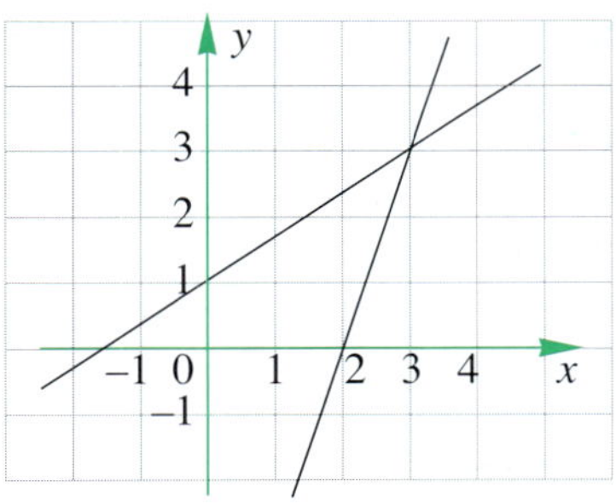

(b)

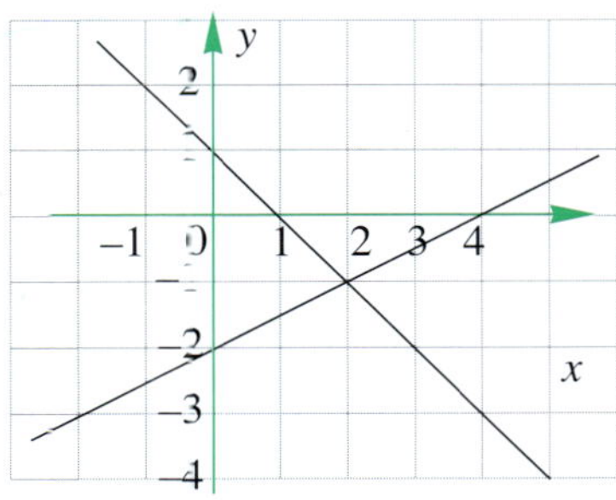

(c)

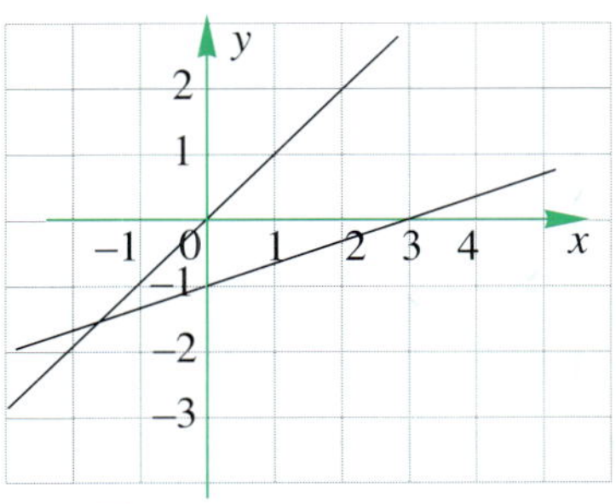

(d)

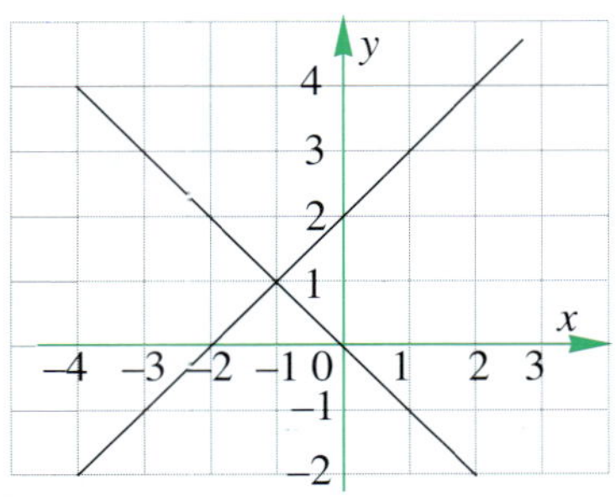

2. Sketch each pair of equations on a set of axes. Then use these graphs to solve simultaneously for the given equations.

(a) $x + y = 4$
$y = x$

(b) $y = 2x + 1$
$y + x = 2$

(c) $2y + x = 6$
$-x - 2 = y$

3. By making use of a table of values, solve each of the following systems of equations.

(a) $y = 10 - x$, $y = 5 + x$ (b) $y = 4x + 3$, $y = 12 - x$ (c) $2x + 7 = y$, $5y + 3x + 4 = 0$

4. Use either a table of values or a graphical method to solve each of the following simultaneous equations.

(a) $2y - x = 2$, $8y + 1 = x$ (b) $3u + v = 2$, $v - 5u = 8$ (c) $2n + 5m = 5$, $4n + 3m = 17$

(d) $3a + 2b = 2$, $a - b = 4$ (e) $x + y = 1$, $y + x = 3$ (f) $2y - x = 4$, $\frac{1}{2}x - y = 6$

5. Solve by any means discussed in this section, the system of equations

(a) $x + 2y - 9 = 0$, $5x - y - 1 = 0$ (b) $2x + y = 8$, $x - y = 1$ (c) $y = -2x + 3$, $y = -x + 1$

(d) $2y = 3x - 2$, $y = 2x - 3$ (e) $3x + 2y = 6$, $5x + 3y = 11$ (f) $x - 2y = -7$, $2x + 3y = 0$

6.3.2 ALGEBRAIC METHOD

Unfortunately, when using graphs or even the graphics calculator, we might not always obtain exact answers. Now, in most cases it might not be necessary to have exact values for our answers, however, because of this 'inconvenience' we need to look towards an algebraic approach. Having said this, it is sometimes the case that the only possible method that can be used to solve simultaneous equations is via the use of numerical or graphical methods.

There are two algebraic methods that we can use:

1. **Elimination Method** and
2. **Substitution Method**.

Method 1 Elimination Method

The **key step** in using the elimination method is to obtain, for one of the variables (in both equations), coefficients that are the same (or only differ in sign).

Then:

1. If the **coefficients are the same you substract** one equation from the other (this will eliminate one variable) leaving you with only one unknown.

2. If the **coefficients only differ in sign then you add** the two equations (this will again eliminate one variable), leaving you with only one unknown.

EXAMPLE 6.8

Solve the system of equations $2x + 7y = 9$, $3x - 7y = 6$

Solution

We start by identifying the equations: $2x + 7y = 9$ – (1)
$3x - 7y = 6$ – (2)

The coefficient of the y-variable in (1) & (2) only differ in sign, in this case we need to add (1) and (2):

$$2x + 7y = 9 \quad -(1)$$
$$3x - 7y = 6 \quad -(2)$$
$$(1) + (2): \quad 5x + 0 = 15$$
$$\therefore x = 3$$

We can now substitute the value for $x = 3$ into either (1) or (2).
Substituting into (1) gives $2 \times 3 + 7y = 9 \Rightarrow 7y = 3$

$$\therefore y = \frac{3}{7}$$

Therefore the solution is $x = 3$ and $y = \frac{3}{7}$.

This means that the two lines meet at the point $\left(3, \frac{3}{7}\right)$.

Visually this is represented as shown:

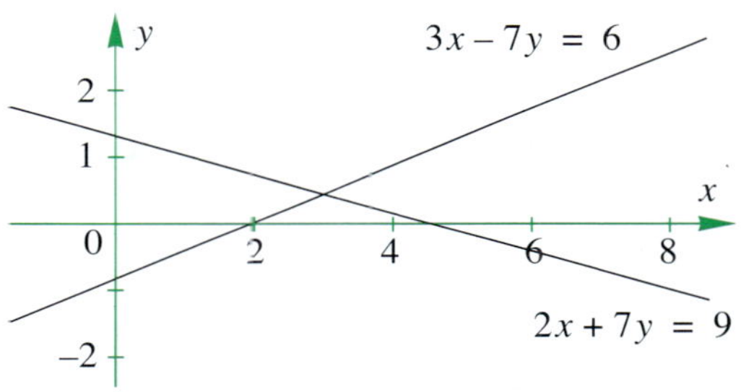

Most often you will need to do some work before you can add or subtract equations.

EXAMPLE 6.9

Solve the system of equations $3x + y = 4$
$2x + 3y = 2$

Solution

First let's label the equations: $3x + y = 4$ – (1)
$2x + 3y = 2$ – (2)

This time all of the coefficients are different and of the same sign. However, this does not create too much of a difficulty. To solve for such a system of equations there are a number of steps to follow:

Step 1: **Choose the variable you wish to eliminate.**
This time we'll choose y.

Step 2: **We now need to get the coefficients of y to be the same.**

This means multiplying (1) by 3.
This will provide a new equation:

$$9x + 3y = 12 \quad -(3)$$

Our system of equations is now:

$$2x + 3y = 2 \quad -(2)$$
$$9x + 3y = 12 \quad -(3)$$

Step 3: Subtract and solve:

(3) – (2): $7x + 0 = 10$

$$\therefore x = \frac{10}{7}$$

Step 4: Solve for the other variable (in this case, *y*).

Substitute $x = \frac{10}{7}$ into (1): $3 \times \frac{10}{7} + y = 4 \Rightarrow y = 4 - \frac{30}{7}$

$$\therefore y = -\frac{2}{7}$$

The solution is $x = \frac{10}{7}$ and $y = -\frac{2}{7}$.

This can be visualised as in the diagram alongside.

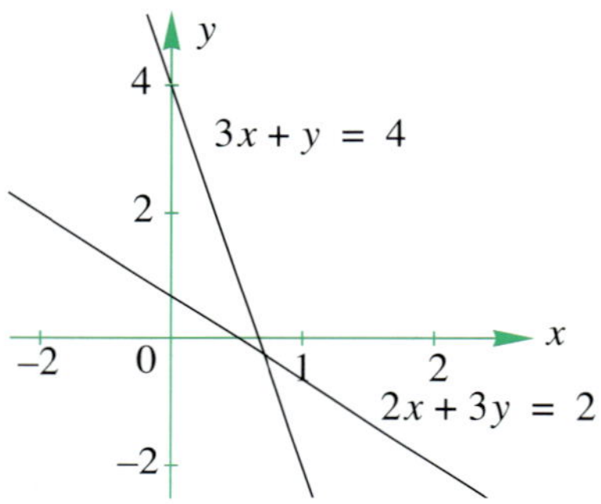

Method 2 Substitution Method

This process can be summarised as follows:

Step 1	Rewrite (i.e., transpose) one of the equations so that you have one of the variables in terms of the other.
Step 2	Substitute this equation into the other equation and solve for the variable under consideration.

EXAMPLE 6.10

Solve the system of equations $3x - y = 4$
$2x + 5y = 14$

Solution

Label the equations: $3x - y = 4 - (1)$
$2x + 5y = 14 - (2)$

Use (1) to express y in terms of x: $y = 3x - 4 - (3)$

Now substitute (3) into (2): $2x + 5(3x - 4) = 14$
[i.e., replace the y-term in (2) with $(3x - 4)$ from (3)]

We can now solve for x:

$$\begin{aligned} 2x + 15x - 20 &= 14 \\ 17x &= 34 \\ \therefore x &= 2 \end{aligned}$$

Substitute $x = 2$ into (3), giving $y = 3\times2 - 4 = 2$.
The solution is therefore $x = 2$ and $y = 2$. That is, the lines intersect at the point (2,2).

EXAMPLE 6.11

Solve the system of equations: $\frac{x}{3} - \frac{y}{2} = 4$ and $\frac{x}{4} + y = 1$

Solution

Label the equations:

$$\frac{x}{3} - \frac{y}{2} = 4 \quad -(1)$$

$$\frac{x}{4} + y = 1 \quad -(2)$$

A rule of thumb is, to 'remove' all the fractions – this will make the arithmetic a little easier. In this case we have:

Multiply (1) by 6: $2x - 3y = 24$ – (3)
Multiply (2) by 4: $x + 4y = 4$ – (4)
Now, from (4) we get that $x = -4y + 4$ – (5)

Substituting (5) into (3), we have:

$$\begin{aligned} 2(-4y + 4) - 3y &= 24 \\ -8y + 8 - 3y &= 24 \\ -11y &= 16 \\ \therefore y &= -\frac{16}{11} \end{aligned}$$

Substituting this value into (4): $x = -4 \times -\frac{16}{11} + 4 = \frac{108}{11}$.

The solution is then $x = \frac{108}{11}$, $y = -\frac{16}{11}$.

That is, the point of intersection of the two straight lines is $\left(\frac{108}{11}, -\frac{16}{11}\right)$.

Not all simultaneous equations have solutions. Also, some pairs of equations have infinite solution sets. If equations are of these types, you will need to be able to recognise the 'problem' in the processes of both algebraic and graphical solution.

The following examples illustrate these possibilities:

EXAMPLE 6.12

Solve: (a) $\begin{aligned} 2x + 6y &= 8 \\ 3x + 9y &= 12 \end{aligned}$ (b) $\begin{aligned} 2x + 6y &= 8 \\ 3x + 9y &= 15 \end{aligned}$

Solution

(a) **Algebraic solution:**

$$2x + 6y = 8 \quad -(1)$$
$$3x + 9y = 12 \quad -(2)$$

$3\times(1)$: $6x + 18y = 24 \quad -(3)$

$2\times(2)$: $6x + 18y = 24 \quad -(4)$

$(4) - (3)$: $0 + 0 = 0$

At this point there is nothing else that we can do.
The algebraic method produces an equation that is always true. This means that any pair of numbers that satisfy either equation will satisfy both and are, therefore, solutions to the problem.
Examples of solutions are: $x = 4, y = 0, x = 1, y = 1$ & $x = 7, y = -1$ and the list goes on.

Graphical Solution:

If we sketch both equations (1) and (2), we end up with only one graph! That is, the lines lie on each other.
This tells us that there are an infinite number of solutions (as the lines are in fact the same).

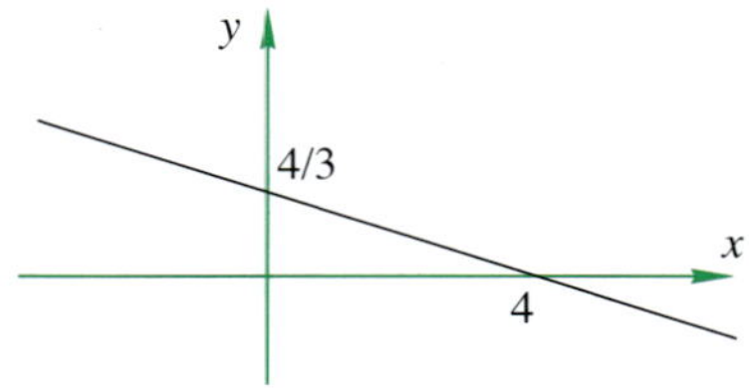

(b) **Algebraic solution:**

$$2x + 6y = 8 - (1)$$
$$3x + 9y = 15 - (2)$$

$3\times(1)$: $6x + 18y = 24 - (3)$

$2\times(2)$: $6x + 18y = 30 - (4)$

$(4) - (3)$: $0 + 0 = 6$

In this case the results are **inconsistent**! Therefore, there are no real solutions for this system of equations.

Graphical Solution:

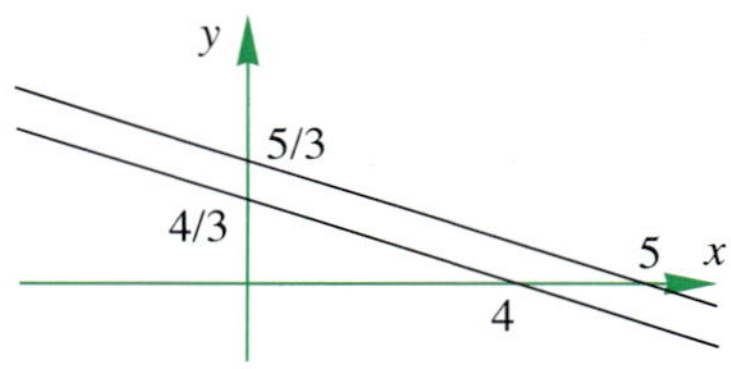

Graphically, the two lines are parallel and produce no points of intersection.

EXERCISES 6.3.2

1. Solve these simultaneous equations, giving exact answers:

(a) $3x - 2y = -1$, $5x + 2y = 9$

(b) $3x + 5y = 34$, $3x + 7y = 44$

(c) $2x + 4y = 6$, $4x - 3y = -10$

(d) $3x + 2y = 2$, $2x - 6y = -6$

(e) $5x + 4y = -22$, $3x - y = -3$

(f) $5x - 9y = -34$, $2x + 3y = -7$

2. Solve these simultaneous equations, giving fractional answers where appropriate:

(a) $3x - y = 2$, $5x + 2y = 9$ (b) $4x + 2y = 3$, $x - 3y = 0$ (c) $-3x + y = 0$, $2x - 4y = 0$

(d) $\frac{x}{2} - 3y = 4$, $4x + \frac{3y}{2} = -1$ (e) $5x + \frac{2y}{3} = -4$, $4x + y = 2$ (f) $\frac{3x}{5} - 4y = \frac{1}{2}$, $x - 2y = \frac{1}{3}$

3. Solve the system of equations

(a) $y + 9 = 2x$, $y + 3x = 6$ (b) $2x - y = 5$, $y + 7 = 3x$ (c) $7x - 3y = 18$, $2x + 5y = 11$

(d) $2x + 3y = 3$, $x - y = 4$ (e) $3x - y - 1 = 0$, $x + y + 5 = 0$ (f) $2x - y - 7 = 0$, $3(3y - x) + 1 = 0$

4. Solve the system of equations for x and y.

(a) $x + y = a$, $x - y = a$ (b) $x + y = a$, $x - y = b$ (c) $x + y = 0$, $x - y = a$

5. Graph the set of equations and hence solve the system of equations

(a) $x + y = 2$, $y = -x$ (b) $2x - y = 3$, $\frac{1}{2}y = x + 1$ (c) $2x + y = 0.5$, $4x + 2y - 1 = 0$

6. Find the values of m such that these equations have no solutions:

(a) $3x - my = 4$, $x + y = 12$ (b) $5x + y = 12$, $mx - y = -2$ (c) $4x - 2y = 12$, $3x + my = 2$

7. Find the values of m and a such that these equations have infinite solution sets:

(a) $4x + my = a$, $2x + y = 4$ (b) $5x + 2y = 12$, $mx + 4y = a$ (c) $3x + my = a$, $2x - 4y = 6$

8. Find the a and b for each of the following

(a)

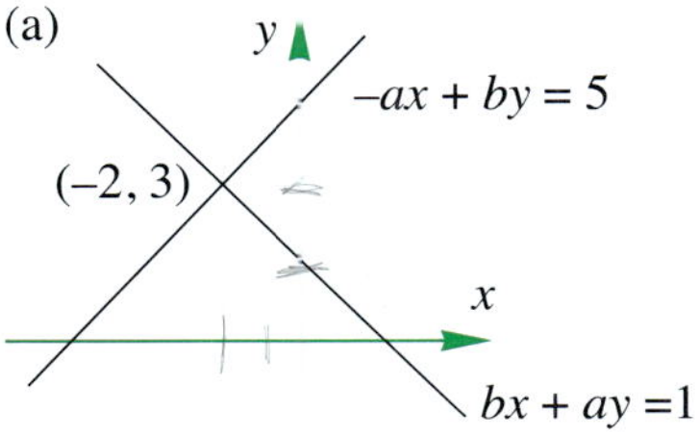

(b)

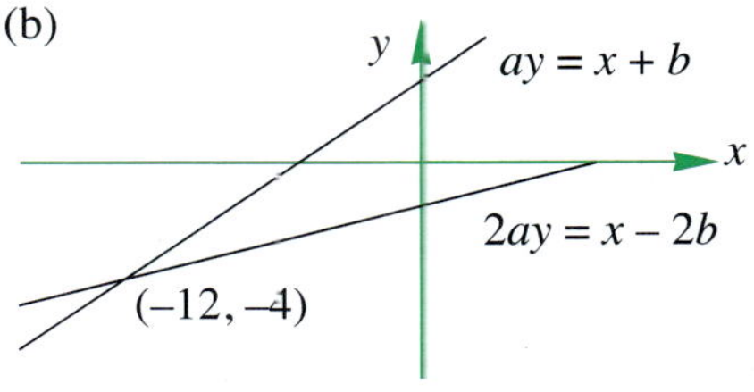

9. Find the solution set of

(a) $\{(x, y)|ax + by = 1\} \cap \{(x, y)|ax - by = 1\}$.

(b) $\{(x, y)|x + y = a + b\} \cap \{(x, y)|bx + ay = 2ab\}$.

(c) $\{(x, y)|ax + y = ab\} \cap \{(x, y)|bx - ay = b^2\}$.

(d) $\{(p, q)|ap - bq = a\} \cap \{(p, q)|bp + aq = b\}$.

6.3.3 APPLICATIONS OF SIMULTANEOUS EQUATIONS

Many situations require the ability to use information simultaneously. In such circumstances it will be necessary to construct sets of equations that can then be solved simultaneously. Using a systematic approach often provides a pathway towards the solution to such problems. Whenever such a situation arises, we suggest the following approach as a guide.

Step 1. Read the problem very carefully (at least twice).

Step 2. (Optional) Draw a diagram (or use a table of values) to try and visualise the problem.
We do not mean an artist's impression of the situation, simply something that gives you a feel for what is going on.

Step 3. Define two variables and clearly state what these variables represent.
(Note: Variables can only take on numerical values, e.g., $x = 2$ (say). It would be incorrect to write $x = 2$ metres! For this reason your variable x should be defined with units 'attached':
e.g., 'Let x metres denote the length of . . .'.).

Step 4. Form two equations in two unknowns. (This is the most difficult step!)

Step 5. Solve simultaneously by any method you feel is appropriate (usually this will be an algebraic method).

Step 6. Clearly write the solution to the question. (This might involve a sentence or two).

Step 7. Check your final answer by substituting your values into the original equations.

EXAMPLE 6.13

The sum of two numbers is 2 while their difference is 6. What are the numbers?

Solution

Let the numbers be x and y.

We are told that their sum is 2, therefore we can write $x + y = 2$ – (1)
We are also told that their difference is 6, therefore $y - x = 6$ – (2)

All that remains is to solve.
In this instance we use the elimination method:

(1) + (2): $2y = 8$
$\therefore y = 4$

Substituting into (1): $x + 4 = 2 \therefore x = -2$
Checking our results, we have (using (2)): L.H.S $= 4 - (-2) = 4 + 2 = 6$ ($\surd$)

EXAMPLE 6.14

While searching in the attic for some old belongings I found a bag consisting of only five-cent and two-cent coins. In all, there were 37 coins making up \$1.49 in total. How many of each coins were there in the bag?

Solution

Let x be the number of five-cent coins and y be the number of two-cent coins.
Therefore, if I have x five-cent coins, the amount made up from the five-cent coins is $5x$.
Similarly the amount made up from the y two-cent coins is $2y$.
In all, the sum $5x + 2y$ equals 149 (\$1.49 is equal to 149 cents).

We have our first equation: $5x + 2y = 149$ – (1)

We also know that we have 37 coins in all. This means that $x + y$ must be 37.
This gives the second equation: $x + y = 37$ – (2)

We can now use the substitution method (or any other appropriate method) to solve.
From (2) we get $y = 37 - x$ – (3)

Substituting (3) into (1) we get

$$\begin{aligned} 5x + 2(37 - x) &= 149 \\ 5x - 74 - 2x &= 149 \\ \Rightarrow 3x &= 149 - 74 \\ \Rightarrow 3x &= 75 \\ \therefore x &= 25 \end{aligned}$$

Substituting back into (3) we get: $y = 37 - 25 = 12$.

That is the bag contained 25 five-cent coins and 12 two-cent coins.
We now need to check our answer: $25 \times 5 + 12 \times 2 = 125 + 24 = 149$ (√).

EXAMPLE 6.15

An equilateral triangle has sides measuring 5, $x + y$ and $4x - y$ units. Find the value of x and y.

Solution

This time we might find a diagram useful (to start with anyway).

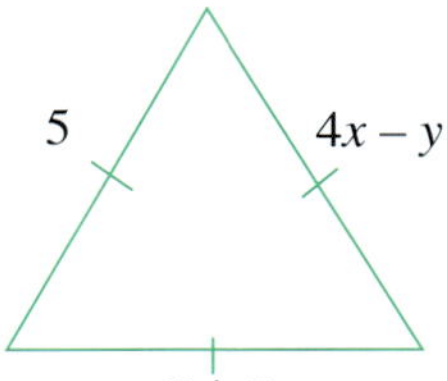

The relationship between each side is that they are all of the same length (property of an equilateral triangle). We can therefore set up our equations as follows:

$4x - y = 5$ – (1)

$x + y = 5$ – (2)

Adding (1) + (2): $5x = 10$

Therefore, $x = 2$

Substitute into (2): $2 + y = 5$ so that $y = 3$.

Check (using (1)): L.H.S $= 4\times2 - 3 = 8 - 3 = 5 =$ R.H.S.(√)

EXERCISES 6.3.3

1. The difference of two numbers is 15 and their sum is 57. What are the numbers?

2. Timothy purchased 2 CDs and 3 blank video tapes for \$35 from a music store. Dianne on the otherhand purchased 2 CDs and one blank video tape for \$23. How much does a CD and a blank video tape cost?

3. Rory purchased a shirt and a pair of socks for \$11.00 while Brett purchased 3 shirts and a pair of socks from the same store for \$29.00. How much does each item cost?

4. A rectangle's length is 3 m longer than its width. If the perimeter is 54 m, find the dimensions of this rectangle.

5. Joseph's test score for maths was 28 more than his test score for physics. His aggregate score for both subjects was 116. How much did Joseph score on each test?

6. Attendance at a local pantomine show costs \$16.00 for 2 adults and 3 children. While for 3 adults and 2 children it costs \$19.00. How much does an adult ticket and a child's ticket cost?

7. Three brown boxes and four yellow boxes cost \$78.00 while two brown boxes and three yellow boxes cost \$56.00. How much does each box cost?

8. The cost, $\$C$, of making two types of invitation cards, deluxe and standard, is dependent on the numbers of cards produced. The relationship between the cost and the number of cards, n, is shown.

How many deluxe and standard cards need to be produced so that the same cost is involved?

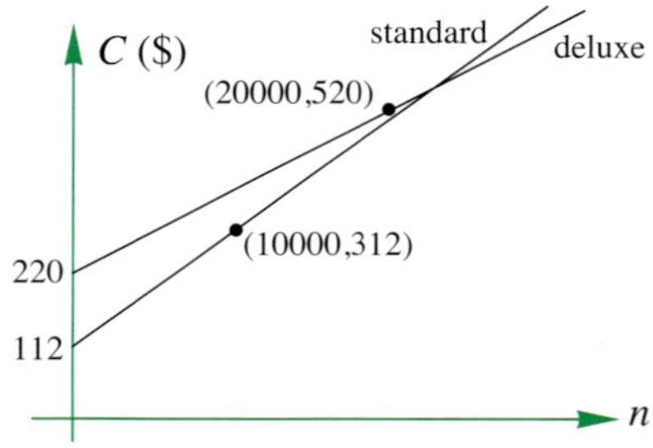

9. Three compact discs and two tapes cost \$114 and two compact discs and five tapes cost \$142. Find the cost of one compact disc.

10. The cost of entry to a football match is \$87 for 3 adults and 4 children. For a family of two adults and two children, the entry charge is \$52. What are the costs of entry for adults and for children?

11. Consider the equations:
$$\begin{aligned} 3x + 2y &= 4 \\ 9x + 6.1y &= 5 \end{aligned}$$

(a) If all the coefficients are known exactly, what is the solution of these equations?

(b) If the coefficient of y in the second equation has been incorrectly measured and should be 6.05, what is the new solution of the equations?

(c) Explain why the solutions to these equations, whose coefficients are so similar, are so different.

6.4 MISCELLANEOUS EXERCISES

1. Sketch the systems of equations and use the elimination method to solve.

(a) $\begin{aligned} 7x+2y &= 12 \\ 3x-2y &= 8 \end{aligned}$ (b) $\begin{aligned} 5x-2y &= 7 \\ 3x+4y &= 25 \end{aligned}$

(c) $\begin{aligned} 2x-4y &= 14 \\ 4x+2y &= -27 \end{aligned}$ (d) $\begin{aligned} 4y-3x &= 6 \\ 2y-x &= 4 \end{aligned}$

2. Sketch the systems of equations and make use of the substitution method to solve.

(a) $\begin{aligned} r+2s &= 4 \\ r-3s &= 2 \end{aligned}$ (b) $\begin{aligned} 3x-y &= 8 \\ 2x+3y &= 9 \end{aligned}$

(c) $\begin{aligned} x+2y &= 1 \\ 5x-4y &= -23 \end{aligned}$ (d) $\begin{aligned} 2x+y &= 11 \\ 3x-2y &= -8 \end{aligned}$

3. Use an appropriate method to solve simultaneously for each of the following:

(a) $\begin{aligned} 2x-5y &= 11 \\ 7x+3y &= 18 \end{aligned}$ (b) $\begin{aligned} 0.02x-0.05y &= -0.38 \\ 0.03x+0.04y &= 1.04 \end{aligned}$

(c) $\begin{aligned} \frac{x}{4}+\frac{y}{6} &= 1 \\ x-y &= 3 \end{aligned}$ (d) $\begin{aligned} \frac{2}{3}u+\frac{1}{6}v &= \frac{2}{3} \\ u+\frac{1}{4}v &= 1 \end{aligned}$

4. Find the solution set to each of the following:

(a) $\{(x, y)|2y+x=0\} \cap \{(x, y)|y-2x=5\}$

(b) $\{(p, q)|2p-q+3=0\} \cap \{(p, q)|2p+q-5=0\}$

(c) $\left\{(x, y)|\frac{x}{9}-\frac{y}{3}=-1\right\} \cap \left\{(x, y)|\frac{x}{2}+\frac{y}{3}=1\right\}$.

(d) $\left\{(u, v)|u+\frac{v}{3}=1\right\} \cap \{(u, v)|u=v-1\}$

(e) $\{(m, n)|m+2n=1\} \cap \{(m, n)|3m-8n=-2(4m+n)\}$

5. Find (a) $\{(x, y)|ax+y=a\} \cap \{(x, y)|ax-y=b\}$

(b) $\{(p, q)|p-q=a\} \cap \{(p, q)|ap-bq=b\}$

6. Find $\left\{(x, y)|\frac{x}{a}+\frac{y}{b}=2\right\} \cap \left\{(x, y)|\frac{x}{b}+\frac{y}{a}=2\right\}$

7. The difference between two numbers is 15 and their sum is 57. What are the numbers?

8. Find two numbers such that twice the first increased by three times the second equals 51 while four times the first decreased by 5 times the second is 3.

9. The sum of two numbers is 42, whilst their difference is 28. Find the two numbers.

10. Sally bought four pens and three pencils for a total of \$13.60, whereas Ted bought five pens and two pencils for a total of \$14.90. How much would it cost to buy 2 pens and one pencil?

11. When emptying my jar of old coins, I found 39 coins made up of one-cent and two-cent coins. In all I had forty-six cents. How many of each type of coin do I have ?

12. According to Kirchoff's Law, the current i and j flowing through two resistors in the electrical circuit shown, must satisfy the system of equations:

$$10i - 50j = 0$$
$$i + j = 6.$$

Find the current flowing trough each resistor.

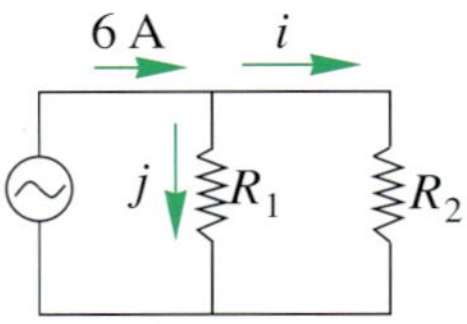

13. Rates for long-distance telephone calls to a certain city are based on a fixed charge of \$$p$ for the first minute and another charge of \$$q$ for each additional minute. The cost of a 9 minute call is \$ 11.60 while that of a four minute call is \$5.20.
What are the rates involved for calls to this city?

14. Two social evenings have been organised by the local scout group. During the first night 25 children and 20 adults attended and the takings for that evening was \$150.00. On the second night there were 30 children and 22 adults, with takings of \$170.00.
How much would each adult have paid to attend this social ?

15. The points (1,3) and (–2,10) lie on the line with equation $ax + by = 16$. Find the values of a and b.

16. Joanne is seven years older than Felicity. Fifteen years ago, Felicity was three–quarters of Joanne's age. Find their ages today.

17. Two men, A and B, 11 kilometres apart, set out at the same time in opposite directions. They end up meeting 1 hr and 20 minutes later. On the other hand, if they had set out and travelled in the same direction, they would have meet after 14 hours and 40 minutes. Find their speeds (you may assume that their speeds are uniform and that they travel along a straight line).

18. A number, made up of two digits, is 8 times the sum of its digits. If the digits are reversed, the number is 45 less than the original number. What is this number ?

19. A rectangular grazing area has been subdivided into 8 equal sections. The external perimeter of the fencing is 80 m and the total length of fencing used is 144 m. Find the total grazing area enclosed by the fence.

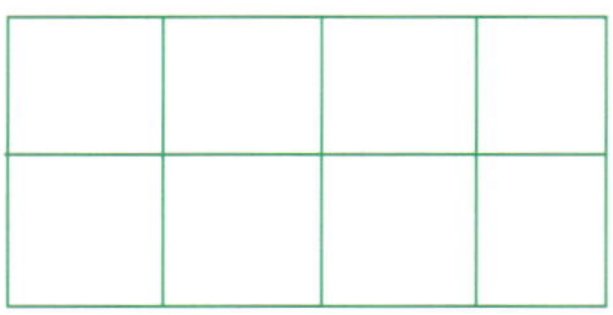

20. A riverboat departs from point A and travels 30 km upstream to a point B in five hours. On its return trip (coming downstream) the boat is moving along with the current and so it only takes four hours to return to point A.

Find (a) the speed of the current (assume it to be constant)
(b) the speed of the riverboat (in still waters).

6.5 GRADED REVISION EXERCISES

LEVEL 1

1. State the gradient and y-intercept of the following linear relations

(a) $y = 2x + 5$ (b) $y = -0.3x - 0.4$ (c) $5y = 8 - 2x$

2. Determine the point of intersection in each of the following

(a)

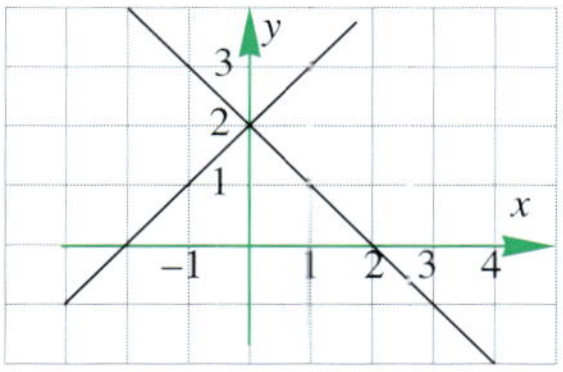

(b)

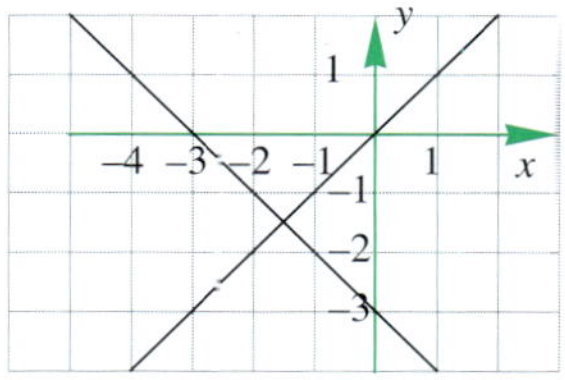

(c)

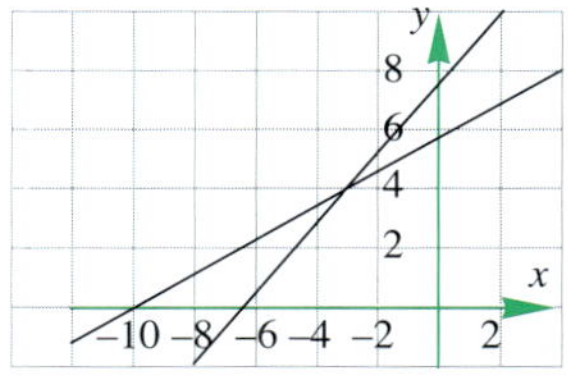

3. Find the value of a in each of the following.

(a)

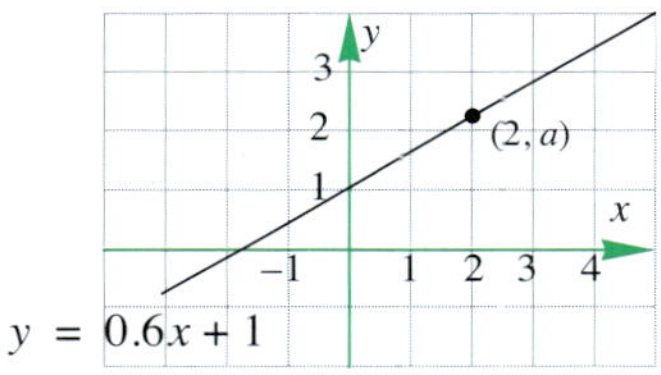

(b)

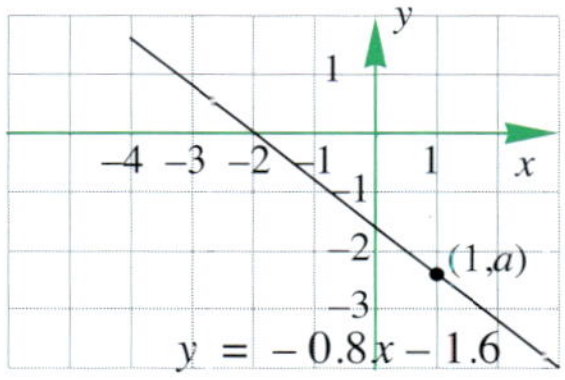

(c)

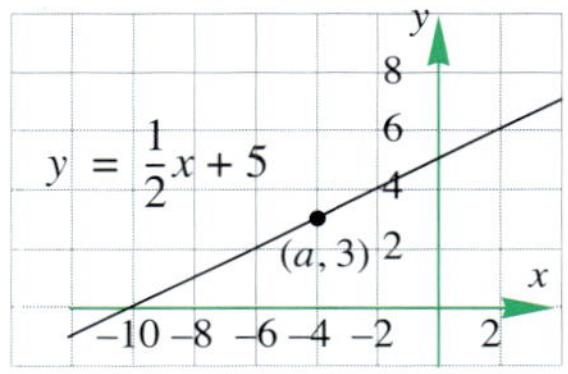

LEVEL 2

1. Find the gradient of the straight line passing through the points

(a) (3, 7) and (5, –2) (b) (–1.2, 3.4) and (2.8, 11.4)

2. The straight line $y = 4x + c$ passes through the point (–1, 6). Find the y-intercept.

3. On different sets of axes, sketch the graphs of the straight lines

(a) $y = -x - 2$ (b) $2x + y = 4$ (c) $y = 5x - 1$

4. Find the equation of the following straight lines

(a)

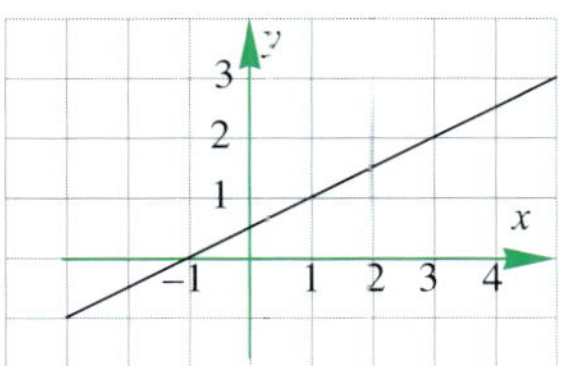

(b)

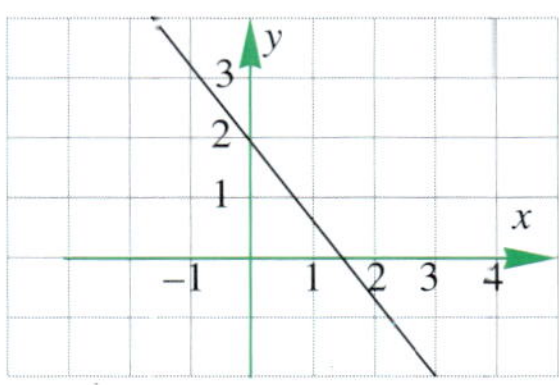

(c)

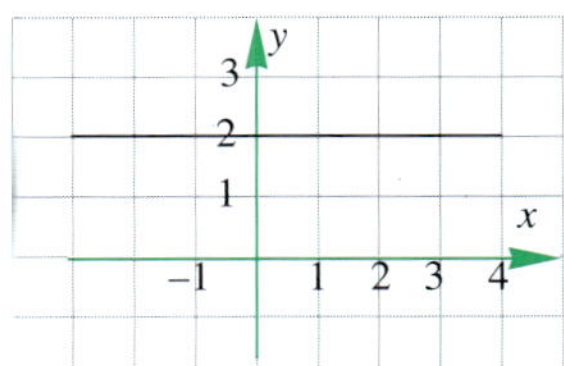

5. Tickets for the local primary school fair are \$0.40 for adults and \$0.10 for children. By the close of the day 550 tickets were sold for a total of \$100. How many adults attended the fair?

6. Solve the system of equations $y = x - a$, $ay = -x + a$

LEVEL 3

1. Solve the system of equations

(a) $\begin{aligned} 3x + 2y &= 17 \\ 2x - y &= 2 \end{aligned}$ (b) $\begin{aligned} 2x + y &= 11 \\ 2y &= 3x + 1 \end{aligned}$ (c) $\begin{aligned} x - 3y &= -2 \\ 4x - 9y &= 7 \end{aligned}$

2. Find the equation of the straight line passing through the points
(a) (3, 7) and (5, –2) (b) (–2, 3) and (0, 1) (c) $(a, 2b + a)$ and $(1, 2b + 1)$

3. For each of the following
(a) Sketch the system of equations on a set of axes.
(b) Determine the point of intersection of the two lines.

i. $\frac{1}{2}x + 2y = 1, -x + \frac{1}{2}y = 1$ ii. $0.1x - y = 0.2, x + y = 0.1$

4. The line $4x + 5y = 20$ crosses the x-axis at M and the y-axis at N. Find the distance between M and N.

5. The point $(7, b)$ lies on the line $2y = 8 - 5x$. Find the value of b

6. The relationship between the market demand equation, Q_d, and the market supply equation, Q_s, for a particular product is shown alongside:

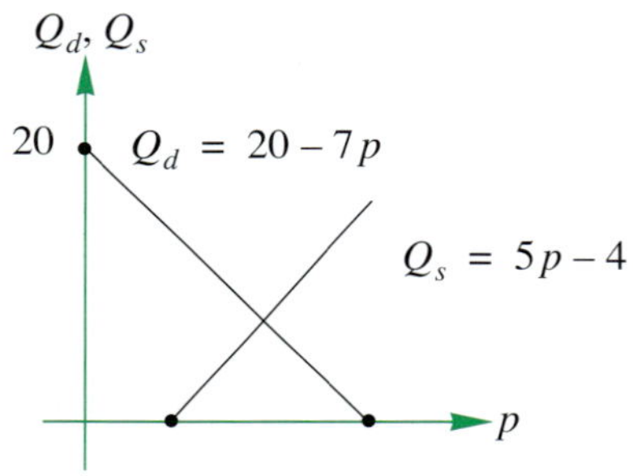

Find the equilibrium price (p).
What significance can be attached to this equilibrium price ?

LEVEL 4

1. The lines $y = \frac{3}{2}ax - b$ and $y = -ax + 3b$ intersect at the point (2, 4). Find a and b.

2. $\left\{(x, y) | \frac{a}{x} + \frac{b}{y} = 1\right\} \cap \left\{(x, y) | \frac{2a}{x} + \frac{b}{y} = 2\right\}$ $\left[\text{Hint: Let } p = \frac{1}{x} \text{ and } q = \frac{1}{y}\right]$

3. Solve the system of equations $\begin{aligned} 5x + 3y + 2z &= 1 \\ 2x - y + z &= -1 \\ -2x + 2y - z &= 2 \end{aligned}$.

4. (a) For the line $3y = k - 2x$, $k > 0$, find
i. the x-intercept
ii. the y-intercept.

(b) Draw this straight line on a set of axes.

(c) A triangle is formed by the x-axis, the y-axis and the line $3y = k - 2x$, $k > 0$. The area of this triangle is 6 sq. units. Find the value of k.

6.6 TOPIC TEST

1. State the gradient and y-intercept for the straight line with equation $y = -3x + 7$.

[2 marks]

2. Find the gradient of the straight line passing through the points $(5, 8)$ and $(9, 16)$.

[2 marks]

3. (a) Find the equation of the straight line shown.
(b) Find the value of a.

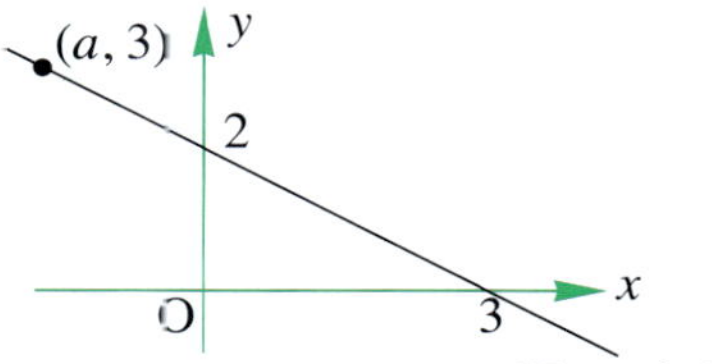

[5 marks]

4. Solve the system of equations $\begin{aligned} 3x - y &= 6 \\ x + 2y &= 9 \end{aligned}$.

[4 marks]

5. Find the coordinates of the point of intersection of the two straight lines shown.

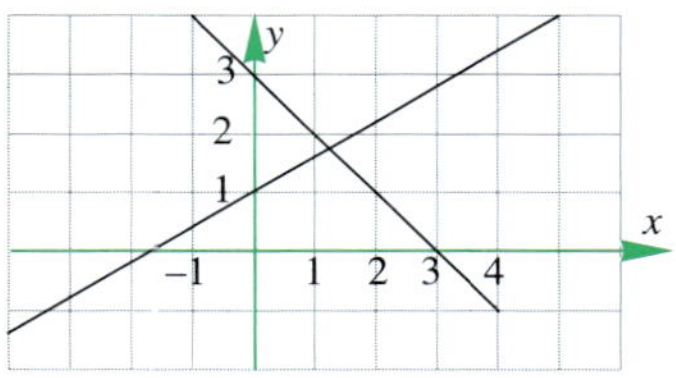

[6 marks]

6. A contractor employed 5 adults and 3 teenagers for one day and paid them a total of \$224 per hour. The following day he employed 3 adults and 5 teenagers for \$160 per hour. What was the hourly rate paid to each adult and teenager?

[6 marks]

7. A door-to-door salesperson selling vacuum cleaners receives a fixed income of \$200 per week plus \$40 commission for every vacuum cleaner, n, sold.

(a) Copy and complete the following table.

n	0	1	2	3	4	5	10
Total income (\$$T$)							

(b) Plot the data on a set of axes.
(c) i. What is the salesperson's total income if 8 vacuum cleaners are sold.
ii. The salesperson receives a total income of \$720. How many vacuum cleaners did the salesperson sell?
(d) i. What values can n have?
ii. Find an equation to describe the relationship between T and n.

[10 marks]
Total 35 marks

6.7 PROJECT SUGGESTIONS

Develop the discussion at the start of this chapter on the greenhouse effect.

Many athletic records have been improving in a linear manner. Investigate a question such as: 'Will women ever run the marathon faster than the best male runner?'.

What has been the history of the World's tallest building?

What will be human life expectancy at the turn of the next millennium?

Consider the following problem:

A petstore owner can sell two particular food mixes. Mix A at a profit of £ 0.20 per kg and mix B at a profit of £ 0.40 per kg. The nutrients, N1 (1000 units) and N2 (1000 units) present in each mix is tabulated below:

	Nutrients	
	N1	N2
Mix A	3	1
Mix B	4	4

The weekly requirements for this pet store are 36,000 units of N1 and 20,000 units of N2.

How much of each mix should the owner purchase in order to maximise profits?

Such a problem belongs to an area of mathematics known as linear programming. You could carry out a project that produces a proposal to a business about how they could **optimise** an aspect of their business dealings.

7.0 CANNON BALLS

History's most famous artillery officer was Napoleon Bonaparte (born Aug. 15, 1769, Corsica, died 1821, St. Helena Island). After a series of military victories, Napoleon became leader and later Emperor of France. He made many changes, not least to secondary education. Mathematics became a key subject in the new Lycées because, it has been said, Napoleon understood its value in organising effective artillery regiments (as well as other civil projects).

Gunpowder is thought to have been invented by the Chinese around 900AD though they probably used it in fireworks and signalling rather than in war. The gun is said to have been an Arab invention.

From the invention of the explosive cannon, people argued about the path taken by the cannonballs they fired they moved too fast to follow with the eye. Various 'trajectories' were suggested two of which are shown below:

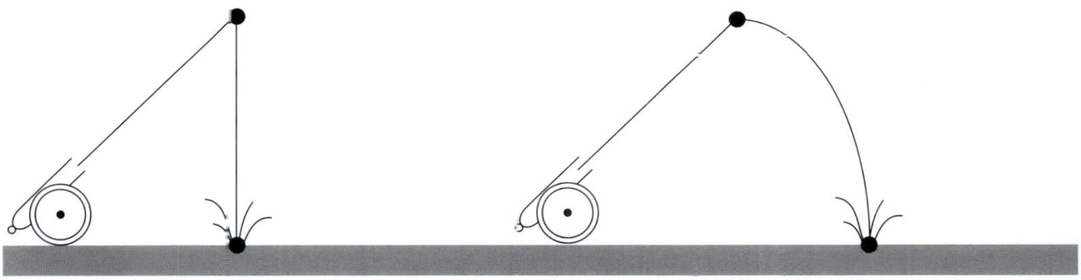

The problem was not sorted out until Galileo Galilei (1564 -1643) both experimented with and used mathematics to demonstrate that the correct model is quadratic, the subject of this chapter.

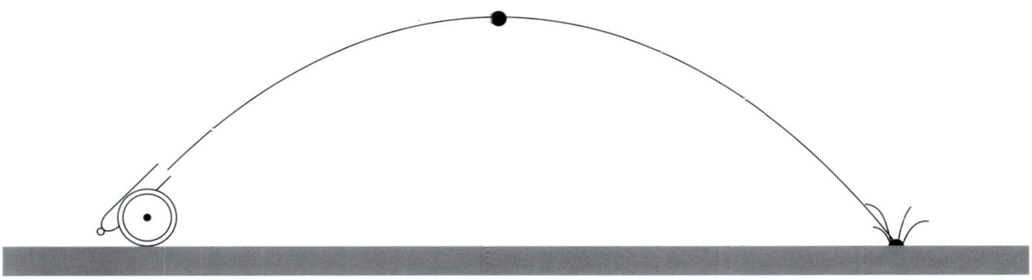

A project that looks at the paths of trajectories is full of interesting mathematics. Do not only consider the path of a cannon ball.

What about the path of objects at a venue like the Olympic Games – the javelin, the shot-putt, the discus. What about the pole vault or the gymnastic events, diving, the triple jump? The list goes on.

What about bridges? Look at suspension bridges. Do any of them contain a parabolic curve within their structure?

7.1 FACTORISATION

7.1.1 WHAT IS FACTORISATION?

Factorisation is a process which enables us to express a number as a product of its **factors**. For example, the number 12 can be expressed as the product of 2 and 6, i.e., $12 = 2 \times 6$. Or it can also be expressed as the product of 3 and 4, i.e., $12 = 3 \times 4$. So the factors are 1, 2, 3, 4, 6, & 12.

On the otherhand, the expression $3xy^2$ has the following factors:

$$3, 3x, 3xy, 3y, 3y^2, 3xy^2, x, y, xy, y^2, xy^2 \text{ and of course, } 1.$$

However, when dealing with expressions that consist of more than one term, and the need to factorise we look for the highest common factor from all terms.

EXAMPLE 7.1

Factorise the following

(a) $5x^2 - 3x$ (b) $4x^3 + 9x^2$ (c) $3xy^2 - xy + x^2y$

Solution

(a) To factorise this expression we need to determine the factors of each term.

The factors of $5x^2$ are $1, 5, x, 5x$ and $5x^2$

$3x$ are $1, 3, x$ and $3x$.

In this case we see that the highest common factor is x. This means that we can 'take out' the x term:

$$5x^2 - 3x = x \times 5x - x \times 3$$
$$= x(5x - 3)$$

(b) The highest common factor in the expression $4x^3 + 9x^2$ is x^2. Therefore, we have

$$4x^3 + 9x^2 = x^2 \times 4x + x^2 \times 9$$
$$= x^2(4x + 9)$$

(c) The highest common factor in the expression $3xy^2 - xy + x^2y$ is xy. Therefore, we have

$$3xy^2 - xy + x^2y = 3xy \times y - xy + x \times xy$$
$$= yx(y - 1 + x)$$

7.1.2 FACTORISATION BY GROUPING

The above examples relied on a relatively straight forward observation process – i.e., looking at each term and identifying the highest common factor. However, more often than not we are given expressions that require a little work before we can identify the common factors.

Consider the expression $2x + xy + 2y + y^2$ there is no one common factor that can be applied to every term. For example, x is common to $2x$ and xy but not to $2y$ and y^2. Similarly, y is common to xy, $2y$ and y^2 but not to $2x$. Therefore there appears to be no factor that can be used for every term – the operative word is 'appears'. However, let's look at the expression once more and carry out a little rearranging of the terms:

First we group the terms as $2x + xy$ and $2y + y^2$.

Next, we factorise each of these groups: $2x + xy = x(2 + y)$

$$2y + y^2 = y(2 + y)$$

This means we can rewrite the expression $2x + xy + 2y + y^2$ as:

$$\begin{aligned} 2x + xy + 2y + y^2 &= (2x + xy) + (2y + y^2) \\ &= x(2 + y) + y(2 + y) \\ &= (x + y)(2 + y) \end{aligned}$$

That is, we have managed to factorise an expression that at first sight appeared as if it could not be factorised. The key to factorising expressions of this form is to identify the correct grouping that will enable us to factorise.

The general approach to factorisation by grouping follows the following pattern:

$$\begin{aligned} ax + ay + bx + by &= a(x + y) + b(x + y) \\ &= (x + y)(a + b) \end{aligned}$$

That is, we first group the $ax + ay$ and then 'take out' the 'a', giving $a(x + y)$ and then group the $bx + by$, to give $b(x + y)$. This means that we have $(x + y)$ as a common factor.

EXAMPLE 7.2

Factorise the following

(a) $3(x - y) + a(x - y)$ (b) $ab - cb + xa - xc$ (c) $x^3 + ax + 4x^2 + 4a$

Solution

(a) In the expresssion $3(x - y) + a(x - y)$ we see that the term $(x - y)$ is the highest common factor.

So, we have that $3(x - y) + a(x - y) = 3 \times (x - y) + a \times (x - y)$

$$= (3 + a)(x - y)$$

(b) This time we need to first set up the grouping. It seems reasonable to group the first two terms and then the second two terms.

Meaning that $ab - cb + xa - xc = b(a - c) + x(a - c)$.

Then, we have $= (b + x)(a - c)$.

(c) This time we 'observe' that the two groups are $x^3 + ax$ and $4x^2 + 4a$.

The reason is that *thinking ahead*, we 'see' that the $x^3 + ax$ can be written as $x(x^2 + a)$ and the $4x^2 + 4a$ can be written as $4(x^2 + a)$. Producing a common factor of $(x^2 + a)$.

Therefore, we have $x^3 + ax + 4x^2 + 4a = x(x^2 + a) + 4(x^2 + a)$

$$= (x + 4)(x^2 + a)$$

EXERCISES 7.1

1. Factorise the following

(a)	$2ax + 4$	(b)	$9y - ay$	(c)	$4t + st$
(d)	$x^2 + x^3$	(e)	$3xy - xy^2$	(f)	$r^3s + rs^3$
(g)	$2z^2 - 3zy$	(h)	$ab^2 - a^4b^3$	(i)	$3x^2y^2 - 8xy$
(j)	$3xy^3 + 9y^2$	(k)	$2wt + 8w^2t$	(l)	$3ps - 12(ps)^2$

2. Factorise the following

(a)	$5ax + 10ay - 5axy$	(b)	$2x^2u - 4xu^2 + xu$
(c)	$3mn - 2n^2m + 6mn^2$	(d)	$ab^2c - a^3b^2 + a^2b^3c$
(e)	$3xy^3 + 6x^2y - 12xy$	(f)	$12uv - 9u^2v - 6u^2v^2$

3. Factorise the following

(a)	$2(x + y) + z(x + y)$	(b)	$3(t - s) - r(t - s)$
(c)	$x(x + 2) - y(x + 2)$	(d)	$st(a - b) + 2(a - b)$
(e)	$r^2(xy + 1) + 4(xy + 1)$	(f)	$y^3(2 + z) + 5(z + 2)$

4. Factorise the following

(a)	$4(s + 1) + xs + x$	(b)	$a(2 - c) + 2b - bc$
(c)	$10x + 2y + t(5x + y)$	(d)	$3x + 6 - 5k(x + 2)$
(e)	$ab^2 - b + 4(ab - 1)$	(f)	$y(a + b) - az - bz$

5. Factorise the following

(a)	$a^4 - a^3 + 4a - 4$	(b)	$x + 1 - yxz - yz$
(c)	$mn - m - n + 1$	(d)	$y^3 - y^2 + 3y - 3$
(e)	$(x - b)^2 + x - b$	(f)	$(a + b)^2 - a - b$
(g)	$2a - 2b - xa + xb$	(h)	$xz - yz + 3x - 3y$

7.2 FACTORISATION OF QUADRATICS

A quadratic expression is one which takes on the form $ax^2 + bx + c$. Expressions such as these can (where possible) be factorised into the form $(Mx + N)(Px + Q)$.

That is, we can write the quadratic as a product of its factors:

$$ax^2 + bx + c \equiv (Mx + N)(Px + Q)$$ [1]

Unlike the previous section, it might not be as obvious as to what the factors of a quadratic are. However, one thing is for certain, the more practice you have, the easier it will be for you to work out the factors.

1. Notice the use of the symbol '≡' instead of '='. Discuss this notation with your teacher.

We now proceed in search for a relationship between M, N, P and Q and the coefficients, a, b and c of the quadratic.

We first expand the brackets on the right–hand side:

$$\begin{aligned}(Mx+N)(Px+Q) &\equiv MNx^2 + MQx + NPx + NQ \\ &\equiv MNx^2 + (MQ+NP)x + NQ\end{aligned}$$

That is, we have that $ax^2 + bx + c = MNx^2 + (MQ+NP)x + NQ$

Then, comparing the coefficients on the left hand side with those on the right hand side, we have with the following relationships:

$$MP = a, \qquad MQ + NP = b, \qquad \text{and} \qquad NQ = c$$

Our task then, is to determine the numbers M, N, P and Q for which this can occur.

7.2.1 QUADRATICS OF THE FORM $x^2 + bx + c$ (i.e., $a = 1$)

When we have an expression of the form $x^2 + bx + c$ and wish to factorise it, we only need to consider the values of N and Q!

That is, $\quad x^2 + bx + c \equiv (x+N)(x+Q) \quad$ [Because $M = 1$ and $P = 1$]

This therefore means that the numbers N and Q must satisfy the conditions

$$N + Q = b \quad \text{and} \quad NQ = c.$$

That is, we need to find two numbers such that their sum is 'b' and their product is 'c'.

For example, to factorise the quadratic $x^2 + 5x + 6$ we need to decide on two numbers so that their sum is 5 and their product is 6. We see that $2 + 3 = 5$ and $2 \times 3 = 6$, meaning that the two numbers we are looking for are '2' and '3'.
Therefore, we have that $x^2 + 5x + 6 = (x+2)(x+3)$.
NB: We could expand the right hand side to verify this result.

Put simply, we have used a trial and error method to determine the values that will satisfy the conditions. Of course, our guesses were educated guesses.

Some observations:

1. We note that the x^2 term can only be arrived at if both brackets start with 'x' i.e., we must have an expression of the form

$$(x \;\ldots\;)(x \;\ldots\;)$$

2. We need to take into account the numbers **and** their signs, meaning that we would need to consider expressions of the form:

$$(x + \ldots)(x + \ldots) \text{ or } (x + \ldots)(x - \ldots) \text{ or } (x - \ldots)(x - \ldots)$$

EXAMPLE 7.3

Factorise the following (a) $x^2 + 7x + 10$ (b) $x^2 - x - 12$

Solution

(a) We need to come up with two numbers so that they add up to 7 and when multiplied equal 10. The only possible combination is '5' and '2', i.e., $5 + 2 = 7$ & $5 \times 2 = 10$.
Therefore, we have $x^2 + 7x + 10 = (x+2)(x+5)$.

(b) This time we need to find two numbers that add to '–1' and have a product of '–12'.
The following pairs are available (using trial and error):
–12 and 1, 12 and –1, 3 and –4, –3 and 4, 6 and –2, –6 and 2

Numbers	Product	Sum
–12 and 1	–12	–11
12 and –1	–12	11
3 and –4	–12	–1
–3 and 4	–12	1
6 and –2	–12	4
–6 and 2	–12	–4

From the table we see that the only correct combination is 3 and –4.
Therefore, $x^2 - x - 12 = (x-4)(x+3)$.

As we said earlier, there will be a stage that you will go through where trial and error will be the only means available to you, however, as you do more factorisation, you will start to recognise pairs of numbers that are appropriate and the 'guessing game' will feature less. Having said this, we now consider some more observations - to reduce the amount of time guessing.

A. Factorising $x^2 + bx + c$ where c is positive and b is positive:

In such cases we have $N + Q = b > 0$ and $NQ = c > 0$
That is, the product and the sum of the two numbers is positive.
The only way this can occur is if **both Q and N are positive**.

e.g. Factorise $x^2 + 5x + 4$; $x^2 + 5x + 4 = (x+Q)(x+N)$

The only two positive numbers that add to 5 and multiply to 4 are, '1' and '4'.
Therefore, $x^2 + 5x + 4 = (x+1)(x+4)$.

B. Factorising $x^2 + bx + c$ where c is positive and b is negative:

In such cases we have $N + Q = b < 0$ and $NQ = c > 0$
That is, the product of the two numbers is positive while their sum is negative.
The only way this can occur is if **both Q and N are negative**.

e.g. Factorise $x^2 - 5x + 4$; $x^2 - 5x + 4 = (x+Q)(x+N)$

The only two negative numbers that add to –5 and multiply to 4 are, '–1' and '–4'.
Therefore, $x^2 - 5x + 4 = (x-1)(x-4)$.

c. Factorising $x^2 + bx + c$ where c is negative and b is positive:

In such cases we have $N + Q = b > 0$ and $NQ = c < 0$
That is, the product of the two numbers is negative while their sum is positive.
The only way this can occur is if **Q is negative and N is positive and $N > |Q|$**.

e.g. Factorise $x^2 + 3x - 4$; $x^2 + 3x - 4 = (x + Q)(x + N)$

The only two numbers that add to 3 and multiply to –4 are, '4' and '–1'.
Therefore, $x^2 + 3x - 4 = (x - 1)(x + 4)$.

D. Factorising $x^2 + bx + c$ where c is negative and b is negative:

In such cases we have $N + Q = b < 0$ and $NQ = c < 0$
That is, the product of the two numbers is negative and their sum is negative.
The only way this can occur is if **Q is negative and N is positive and $N < |Q|$**.

e.g. Factorise $x^2 - 7x - 30$; $x^2 - 7x - 30 = (x + Q)(x + N)$

The only two numbers that add to –7 and multiply to –30 are, '–10' and '3'.
Therefore, $x^2 - 7x - 30 = (x - 10)(x + 3)$.

SUMMARY:

To factorise $x^2 + bx + c$ into $(x + Q)(x + N)$			
Case	Conditions	Example	Signs of Q and N
1.	$b > 0, c > 0$	$x^2 + 5x + 4 = (x + 1)(x + 4)$	Both positive
2.	$b < 0, c > 0$	$x^2 - 5x + 4 = (x - 1)(x - 4)$	Both negative
3.	$b > 0, c < 0$	$x^2 + 3x - 4 = (x - 1)(x + 4)$	1 positive, 1 negative
4.	$b < 0, c < 0$	$x^2 - 7x - 30 = (x - 10)(x + 3)$	1 positive, 1 negative

EXERCISES 7.2.1

1. Factorise the following

(a)	$x^2 + 9x + 8$	(b)	$x^2 + 11x + 10$	(c)	$x^2 + 13x + 12$
(d)	$x^2 + 10x + 24$	(e)	$x^2 + 8x + 15$	(f)	$x^2 + 13x + 36$
(g)	$x^2 + 10x + 16$	(h)	$x^2 + 11x + 24$	(i)	$x^2 + 10x + 25$
(j)	$x^2 + 12x + 36$	(k)	$x^2 + 14x + 33$	(l)	$x^2 + 19x + 78$

2. Factorise the following

(a)	$x^2 - 9x + 8$	(b)	$x^2 - 11x + 10$	(c)	$x^2 - 13x + 12$
(d)	$x^2 - 10x + 16$	(e)	$x^2 - 9x + 20$	(f)	$x^2 - 12x + 32$
(g)	$x^2 - 13x + 36$	(h)	$x^2 - 21x + 110$	(i)	$x^2 - 15x + 56$
(j)	$x^2 - 12x + 27$	(k)	$x^2 - 12x + 35$	(l)	$x^2 - 12x + 36$

3. Factorise the following

(a)	$x^2 + 3x - 28$	(b)	$x^2 + 6x - 27$	(c)	$x^2 + 4x - 12$
(d)	$x^2 + 8x - 48$	(e)	$x^2 + 3x - 40$	(f)	$x^2 + 6x - 72$
(g)	$x^2 + 11x - 42$	(h)	$x^2 + 10x - 75$	(i)	$x^2 + 2x - 99$
(j)	$x^2 + x - 72$	(k)	$x^2 + 10x - 39$	(l)	$x^2 + 9x - 52$

4. Factorise the following

(a)	$x^2 - 4x - 21$	(b)	$x^2 - 6x + 16$	(c)	$x^2 - 5x - 36$
(d)	$x^2 - 10x - 24$	(e)	$x^2 - 10x + 56$	(f)	$x^2 - 6x - 72$
(g)	$x^2 - 10x + 75$	(h)	$x^2 - 8x - 105$	(i)	$x^2 - 9x - 52$
(j)	$x^2 - 6x - 91$	(k)	$x^2 - 12x - 64$	(l)	$x^2 - x - 42$

7.2.2 QUADRATICS OF THE FORM $ax^2 + bx + c$ (i.e., $a \neq 1$)

We deal with quadratics of the form $ax^2 + bx + c$ in exactly the same way as we dealt with quadratics of the form $x^2 + bx + c$ — except this time we have an extra number to take into account. Nonetheless, we still make use of educated guesses and trial and error.

case 1: $a > 0$

EXAMPLE 7.4

Factorise $2x^2 + 7x + 3$

Solution

The coefficient of the x^2 term is 2, so, the first thing that we do is check if 2 is a common factor of the other terms. As this is not the case in this example, we will have a factored expression of the form $(2x\ldots\ )(x\ldots\ )$.

That is, the $2x$ term and x term will multiply to give the $2x^2$ term. What remains then, is to find the appropriate constants.

The last term in the quadratic is 3. This number can only be obtained by multiplying 3 and 1.

We then have two options:

Option 1: $2x^2 + 7x + 3 = (2x + 3)(x + 1)$

Option 2: $2x^2 + 7x + 3 = (2x + 1)(x + 3)$

For option 1, the 'cross–product' term is

$$\begin{aligned} 2x^2 + 7x + 3 &= (2x+3)(x+1) \\ &= 2x^2 + (2x \times 1 + 3 \times 1x) + 3 \\ &= 2x^2 + 5x + 1 \end{aligned}$$

Which is not correct.

For option 2, the 'cross–product' term is

$$\begin{aligned} 2x^2 + 7x + 3 &= (2x+1)(x+3) \\ &= 2x^2 + (2x \times 3 + 1 \times 1x) + 3 \\ &= 2x^2 + 7x + 1 \end{aligned}$$

Which is the required term!

Therefore, we have that $2x^2 + 7x + 3 = (2x+1)(x+3)$

Example 7.4 highlighted the 'extra work' involved in factorising a quadratic whose leading coefficient (i.e., the coefficient of x^2) is no longer '1'. Consider then the next example.

EXAMPLE 7.5

Factorise $6x^2 - 5x - 4$.

Solution

In this instance, the leading coefficient is 6 and so, the possible factors are, 1, 2, 3 and 6.

This means that we could have factors that start with

$(x \ \ldots\ )(6x \ \ldots\ )$ or $(2x \ \ldots\ )(3x \ \ldots\ )$

Next, the constant term is '–4' and so, the factors could end with

$(\ \ldots\ 4\ )(\ \ldots\ -1\ )$ or $(\ \ldots\ -4\ )(\ \ldots\ 1\ )$ or $(\ \ldots\ 2\ )(\ \ldots\ -2\ )$ or $(\ \ldots\ -2\ )(\ \ldots\ 2\ )$

What remains now is to consider the possible combinations:

$(x+4)(6x-1)$, $(x-4)(6x+1)$, $(6x+4)(x-1)$, $(6x-4)(x+1)$

$(6x-2)(x+2)$, $(6x+2)(x-2)$

$(2x+4)(3x-1)$, $(2x-4)(3x+1)$, $(3x+4)(2x-1)$, $(3x-4)(2x+1)$

$(2x-2)(3x+2)$, $(3x-2)(2x+2)$

All of this is a lot of work!

Nonetheless, the process is still the same – trial and error. How quickly you determine the factors relies on how quickly you can add/subtract and multiply. For, at the end of the procees, all that has happened is that we have considered a number of factors so that when multiplied and added/subtracted in the correct combination, you end up with the coefficients '6', '–5' and '–4'.

In our case, we have that the correct combination leads to $6x^2 - 5x - 4 = (3x-4)(2x+1)$.

There are a number of 'systematic' approaches to factorising quadratic trinomials. One such method is known by a host of different names, two of which are 'the cross-product method' the 'the gate method'. It still requires that we use one pair of factors for the x^2 term and one pair for the constant term. For Example 7.5 it would look like this:

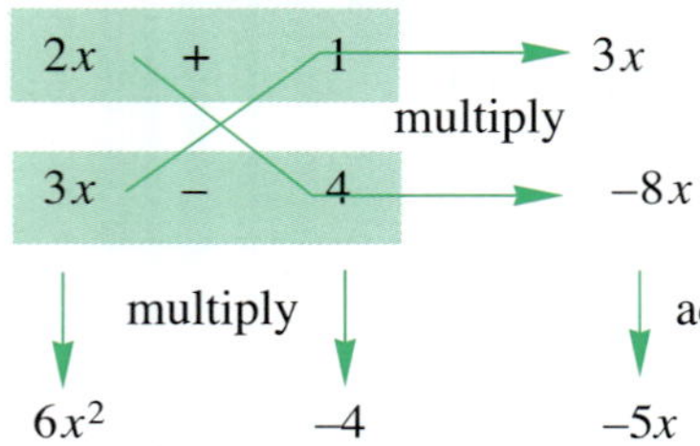

All the correct terms are produced.

Now, although this might appear as a systematic approach, you still need to decide which combination to multiply and add. The method is systematic in the sense that you do not need to write down the factored form until you obtain the correct combination.

EXAMPLE 7.6

Factorise $2x^2 - x - 3$.

Solution

Possible factors for the leading coefficient, '2' are: 1 and 2
Possible factors for the constant term, '–3' are: (1 and –3) or (–1, 3)
Using the cross-product method we have:

Attempt 1.

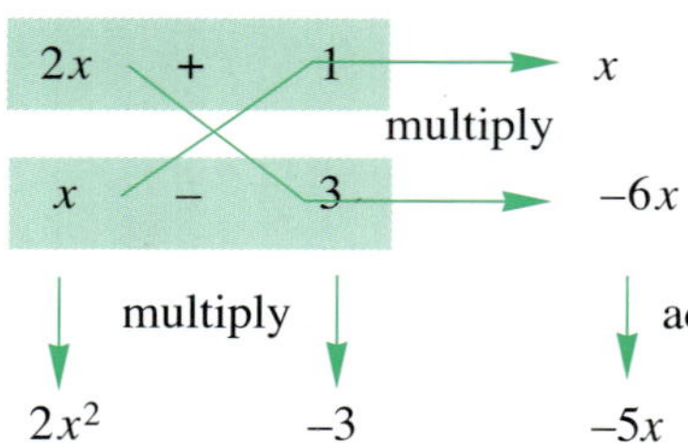

Which doesn't work.

Attempt 2.

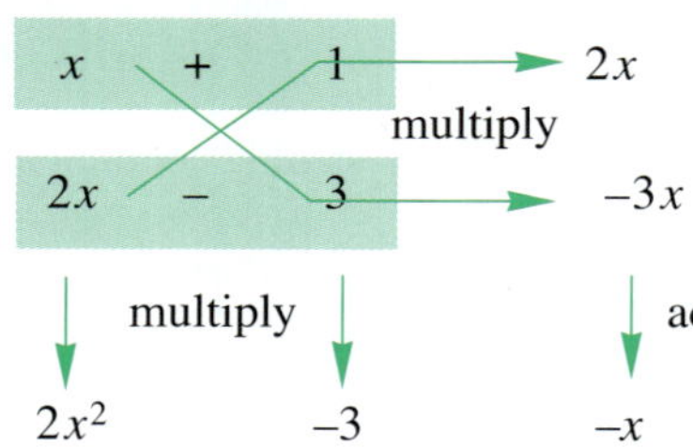

Which gives the correct coefficients.
Therefore, $2x^2 - x - 3 = (x + 1)(2x - 3)$

When factorising expressions such as those just encountered, try not to overlook the possibility of a common factor throughout the expression. For example, given $3x^2 - 21x + 36$ to factorise, it will be much easier to first express it as $3(x^2 - 7x + 12)$ and then concentrate on factorising $(x^2 - 7x + 12)$. So that

$$3x^2 - 21x + 36 = 3(x^2 - 7x + 12) = 3(x-3)(x-4) .$$

case 2: $a < 0$

There is not much to say about this case, except that it will be easier to first factorise the negative in the trinomial. e.g., $-x^2 + 7x + 18 = -(x^2 - 7x - 18) = -(x-9)(x+2)$.

EXERCISES 7.2.2

1. Factorise the following

(a)	$2x^2+3x+1$	(b)	$2x^2+5x+2$	(c)	$2x^2+7x+3$
(d)	$3x^2+4x+1$	(e)	$3x^2+7x+2$	(f)	$3x^2+10x+3$

2. Factorise the following

(a)	$2x^2-3x+1$	(b)	$2x^2-5x+2$	(c)	$2x^2-7x+3$
(d)	$3x^2-4x+1$	(e)	$3x^2-7x+2$	(f)	$3x^2-10x+3$

3. Factorise the following

(a)	$2x^2-x-1$	(b)	$2x^2-3x-2$	(c)	$2x^2-5x-3$
(d)	$3x^2-2x-1$	(e)	$3x^2-5x-2$	(f)	$3x^2-8x-3$
(g)	$6x^2+x-1$	(h)	$6x^2-x-1$	(i)	$4x^2-4x+1$

4. Factorise the following

(a)	$-3x^2-3x+6$	(b)	$3x^2+9x+6$	(c)	$5x^2-7x+2$
(d)	$5x^2-13x-6$	(e)	$-4x^2-10x+6$	(f)	$-3x^2-11x+4$
(g)	$9x^2+15x+6$	(h)	$-2x^2+2x+12$	(i)	$10x^2+29x+10$
(j)	$8x^2-2x-3$	(k)	$-6x^2-5x+6$	(l)	$-9x^2+17x+2$

7.2.3 PERFECT SQUARES AND DIFFERENCE OF SQUARES

A. Quadratics of the form $x^2+2bx+b^2$

Quadratics that take on the form $x^2+2bx+b^2$ (e.g., x^2+6x+9, where $b=3$) are known as **perfect squares**. The process for factorising these quadratics remains the same as before, the only difference is that the factors are repeated. That is,

$$x^2+2bx+b^2 = (x+b)(x+b) = (x+b)^2$$

EXAMPLE 7.7

Factorise the following (a) $x^2+8x+16$ (b) $2x^2+8x+8$

Solution

(a) For the quadratic $x^2+8x+16$ we need to come up with a number so that when it is doubled it equals 8 (i.e., $2b=8$) and when squared it equals 16 (i.e., $b^2=16$).

In this case, we have that $b=4$.
Therefore, we have that $x^2+8x+16 = (x+4)(x+4) = (x+4)^2$.

(b) The first thing that we notice about the quadratic $2x^2 + 8x + 8$ is that there is a common factor of '2', therefore we 'take out' the '2' and then proceed as before:

$$\begin{aligned} 2x^2 + 8x + 8 &= 2(x^2 + 4x + 4) \\ &= 2(x+2)(x+2) \\ &= 2(x+2)^2 \end{aligned}$$

B. Quadratics of the form $x^2 - 2bx + b^2$

Again, we have a situation where we could use trial and error to factorise an expression such as $x^2 - 6x + 9$, or we could recognise it as a perfect square, where $b = 3$. That is, $2b = 6$ and $b^2 = 9$, meaning that it would fit the form $x^2 - 2bx + b^2$ and therefore could be factorised immediately as $(x-3)^2$. Notice that we have $(x-3)$ this time and not $(x+3)$ – otherwise we would not produce a negative for the middle term.

That is, we have that

$$x^2 - 2bx + b^2 = (x-b)(x-b) = (x-b)^2$$

EXAMPLE 7.8

Factorise the following (a) $x^2 - 10x + 25$ (b) $3x^2 - 18x + 27$

Solution

(a) The first thing we notice is that $5^2 = 25$ and that $2 \times 5 = 10$ and so it is of the form $x^2 - 2bx + b^2$.

Then, because the middle term (coefficient of x) is negative we would use $(x-5)$ as the factor to be repeated.

That is, $x^2 - 10x + 25 = (x-5)^2$.

(b) With $3x^2 - 18x + 27$ we observe that '3' is a common factor, so we first 'take that out', leaving us with

$$3x^2 - 18x + 27 = 3(x^2 - 6x + 9)$$

Next, we recognise $x^2 - 6x + 9$ as a perfect square. That is, $x^2 - 6x + 9 = (x-3)^2$. And so, we have

$$\begin{aligned} 3x^2 - 18x + 27 &= 3(x^2 - 6x + 9) \\ &= 3(x-3)^2 \end{aligned}$$

C. Quadratics of the form $x^2 - b^2$

Quadratics that take on the form $x^2 - b^2$ (e.g., $x^2 - 9$, where $b = 3$) are known as **difference of perfect squares (i.e., D.O.P.S)**. The process for factorising these quadratics remains the same as before, the only difference is that there is no middle term. That is, we need to determine two numbers that add to zero and multiply to give '$-b^2$'.

For the quadratic $x^2 - 9$, choosing the numbers '3' and '–3' gives a sum of '3 + (–3) = 0' and a product of '$3 \times -3 = -3^2 = -9$'. Therefore we have that $x^2 - 9 = (x-3)(x+3)$.

In general we have that

$$x^2 - b^2 = (x+b)(x-b)$$

EXAMPLE 7.9

Factorise the following (a) $x^2 - 16$ (b) $yx^3 - xy^3$

Solution

(a) The only two numbers satisfying the conditions that their 'sum = 0' and their 'product = – 16' are '4' & '–4'.
Therefore, we have that $x^2 - 16 = (x+4)(x-4)$.

(b) First we observe that in the expression $yx^3 - xy^3$ there is a common factor of xy, meaning that $yx^3 - xy^3 = xy(x^2 - y^2)$.
Then, the term $x^2 - y^2$ is a D.O.P.S., i.e., $x^2 - y^2 = (x+y)(x-y)$.
Therefore, we have that

$$\begin{aligned} yx^3 - xy^3 &= xy(x^2 - y^2) \\ &= xy(x+y)(x-y) \end{aligned}$$

EXAMPLE 7.10

Factorise the following (a) $16x^2 - 25b^2$ (b) $9 - (a-b)^2$

Solution

(a) We first observe that $16x^2 = (4x)^2$ and that $25b^2 = (5b)^2$.
Therefore we can rewrite $16x^2 - 25b^2$ as $(4x)^2 - (5b)^2$.

We can then apply the D.O.P.S rule:

$$\begin{aligned} 16x^2 - 25b^2 &= (4x)^2 - (5b)^2 \\ &= (4x - 5b)(4x + 5b) \end{aligned}$$

(b) We notice that $9 = 3^2$, so that $9 - (a-b)^2 = (3)^2 - (a-b)^2$.
Then, using the D.O.P.S rule, we have

$$\begin{aligned} 9 - (a-b)^2 &= [3 - (a-b)][3 + (a-b)] \\ &= (3 - a + b)(3 + a - b) \end{aligned}$$

The key to factorising expressions such as these in Examples 7.7 – 7.10 is **recognition**. If you can recognise the form of the expression that needs to be factorised, then you are more than half way there. As always, practice makes perfect – or at least, it will make your task easier.

We consider one more example requiring the use of D.O.P.S.

EXAMPLE 7.11

Factorise the following (a) $y^2 - 10$ (b) $(a-b)^2 - 3$

Solution

(a) The expression $y^2 - 10$ represents a difference of perfect squares, namely, because we can write 10 as $(\sqrt{10})^2$ so that $y^2 - 10 = y^2 - (\sqrt{10})^2$

$$= (y - \sqrt{10})(y + \sqrt{10})$$

(b) Again, we have the difference of perfect squares, as

$$(a-b)^2 - 3 = (a-b)^2 - (\sqrt{3})^2$$
$$= (a - b + \sqrt{3})(a - b - \sqrt{3})$$

EXERCISES 7.2.3

1. Factorise the following

(a)	$x^2 + 4x + 4$	(b)	$x^2 + 14x + 49$	(c)	$y^2 + 16y + 64$
(d)	$16x^2 + 8x + 1$	(e)	$25a^2 + 10a + 1$	(f)	$36z^2 + 12z + 1$
(g)	$4x^2 + 12x + 9$	(h)	$9x^2 + 24x + 16$	(i)	$25x^2 + 20x + 4$
(j)	$x^2 + 2x + 1$	(k)	$x^2 + x + \frac{1}{4}$	(l)	$x^2 + 3x + \frac{9}{4}$

2. Factorise the following

(a)	$x^2 - 4x + 4$	(b)	$x^2 - 14x + 49$	(c)	$y^2 - 16y + 64$
(d)	$4u^2 - 12u + 9$	(e)	$9a^2 - 24a + 16$	(f)	$25v^2 - 20v + 4$
(g)	$y^2 - y + \frac{1}{4}$	(h)	$a^2 - 5a + \frac{25}{4}$	(i)	$x^2 - 7x + \frac{49}{4}$

3. Factorise the following

(a)	$a^2 - y^2$	(b)	$y^2 - 1$	(c)	$100 - a^2$
(d)	$64 - b^2$	(e)	$81 - z^2$	(f)	$r^2 - 49$
(g)	$16x^2 - 49$	(h)	$25 - 4a^2$	(i)	$1 - 9y^2$
(j)	$4a^2 - 36$	(k)	$2m^2 - 8$	(l)	$3y^2 - 27$
(m)	$(x+2)^2 - 9$	(n)	$(ab)^2 - (a+b)^2$	(o)	$(x-2y)^2 - (x+y)^2$

4. Factorise the following

(a)	$x^2 - 3$	(b)	$y^2 - 5$	(c)	$b^2 - 7$
(d)	$(a+b)^2 - 5$	(e)	$a^2 - 10b^2$	(f)	$(x+y)^2 - 2$
(g)	$2x^2 - 4(x-y)^2$	(h)	$4x^2 - 20$	(i)	$36a^2 - 18b^2$
(j)	$(a+2)^2 - 10$	(k)	$(x-3)^2 - 6$	(l)	$4(x-y) - (x-y)^3$

7.2.4 FACTORISING BY COMPLETING THE SQUARE

Sometimes we have quadratic expressions for which 'obvious factors' are not immediately recognisable. For example, looking at the quadratic $x^2 + 6x - 1$, it is difficult (if not nearly impossible) to determine the two factors. However, we notice that had the quadratic been $x^2 + 6x + 9$ then, it would have been a perfect square, giving us $(x+3)^2$.

So, how different are the expressions $x^2 + 6x - 1$ and $x^2 + 6x + 9$?

We observe that $x^2 + 6x - 1 = x^2 + 6x + 9 - 10$. That is we have 'replaced' the '–1' with a 9 (thereby creating a perfect square) and then, to make sure that we haven't altered the original expression, we have subtracted 10 (as 9 – 10 = –1).

So, we now have that
$$\begin{aligned} x^2 + 6x - 1 &= x^2 + 6x + 9 - 10 \\ &= (x+3)^2 - 10 \\ &= (x+3)^2 - (\sqrt{10})^2 \\ &= (x+3+\sqrt{10})(x+3-\sqrt{10}) \end{aligned}$$

It would have been extremely difficult to use a trial and error approach to factorise this quadratic.

EXAMPLE 7.12

Factorise the following (a) $y^2 + 10y + 20$ (b) $a^2 - 4a + 2$

Solution

(a) Looking at the first two terms in the expression $y^2 + 10y + 20$, i.e., $y^2 + 10y$, we observe that had the quadratic been $y^2 + 10y + 25$, then we would have a perfect square, i.e., $y^2 + 10y + 25 = (x+5)^2$.
However, we have $y^2 + 10y + 20$.
So, let's rewrite it as
$$\begin{aligned} y^2 + 10y + 20 &= y^2 + 10y + 25 - 5 \\ &= (y+5)^2 - 5 \\ &= (y+5)^2 - (\sqrt{5})^2 \\ &= (y+5+\sqrt{5})(y+5-\sqrt{5}) \end{aligned}$$

(b) Again, there are no obvious factors of $a^2 - 4a + 2$.
Looking at the first two terms, i.e., $a^2 - 4a$, we observe that had it been $a^2 - 4a + 4$, we would have had a perfect square, i.e., $a^2 - 4a + 4 = (a-2)^2$.
We then rewrite $a^2 - 4a + 2$ as $a^2 - 4a + 4 - 2$. (4 – 2 = 2 so the quadratic is unaltered)
i.e., $a^2 - 4a + 2 = (a^2 - 4a + 4) - 2$
$$\begin{aligned} \therefore a^2 - 4a + 2 &= (a-2)^2 - 2 \\ &= (a-2)^2 - (\sqrt{2})^2 \\ &= (a-2-\sqrt{2})(a-2+\sqrt{2}) \end{aligned}$$

This method is known as **completing the square**. The basis of this method lies in two steps:

Step 1: Create a perfect square (by adding and subtracting an appropriate number).
Step 2: Use the difference of two squares (i.e., D.O.P.S.) to factorise.

We illustrate this with the quadratic $x^2 + 6x - 1$ (which we have already factorised)

Step 1: To create a perfect square, we set aside the '–1' term momentarily and concentrate on the $x^2 + 6x$ term.
To convert this term into a perfect square we would need to have a '+9' attached to it, i.e., $x^2 + 6x + 9$. However, this would actually alter the original expression, and so, we also need to subtract '9' (so that we do not alter the $x^2 + 6x$ term).

That is we have $x^2 + 6x = x^2 + 6x + 9 - 9 = (x+3)^2 - 9$
Now, we mustn't forget that we still have that '–1' that we momentarily set aside.
So, putting this back into the equation we have

$$(x+3)^2 - 9 - 1 = (x+3)^2 - 10\,.$$

Step 2: We now have a difference of two squares (note that $(\sqrt{10})^2 = 10$), so that

$$\begin{aligned}(x+3)^2 - 10 &= (x+3)^2 - (\sqrt{10})^2\\ &= (x+3+\sqrt{10})(x+3-\sqrt{10})\end{aligned}$$

One question that does arise, is 'How do we know what to add?". The number that needs to be added (to complete the square of a quadratic whose leading coefficient is one) is the square of half the coefficient of the x-term.

In the example above, the coefficient of x is 6, then, a half of 6 is 3, and squaring 3 we get 9.

Now, this might look rather lengthy, however, it will often only take a few lines of work.

EXAMPLE 7.13

Factorise (a) $x^2 + 8x + 10$ (b) $x^2 - 4x + 1$

Solution

(a) As there are no obvious factors, we use the completing the square method.
Placing aside the '+10' term momentarily, we see that to convert the $x^2 + 8x$ term into a perfect square we need to add 16. [That is, a half of 8 is 4 and 4 squared is 16.]

Therefore, we have
$$\begin{aligned}x^2 + 8x &= x^2 + 8x + 16 - 16\\ &= (x^2 + 8x + 16) - 16\\ &= (x+4)^2 - 16\end{aligned}$$

However, we mustn't forget the '10'. So, we now re–instate the '+10' term to get:

$$\begin{aligned}x^2 + 8x + 10 &= (x+4)^2 - 16 + 10\\ &= (x+4)^2 - 6\\ &= (x+4)^2 - (\sqrt{6})^2\\ &= (x+4+\sqrt{6})(x+4-\sqrt{6}) \quad \text{(Voila!)}\end{aligned}$$

(b) $x^2 - 4x + 1 = (x^2 - 4x + 4) - 4 + 1$ [i.e., add and subtract 4 to complete a square]
$= (x-2)^2 - 3$ [Factorise the quadratic and create a D.O.P.S.]
$= (x - 2 + \sqrt{3})(x - 2 - \sqrt{3})$

So, what happens if the leading coefficient is not one?

First, we need to factorise the coefficient of the x^2 from the quadratic,

$$\text{e.g., } 2x^2 + 4x - 8 = 2(x^2 + 2x - 4).$$

Then, we concentrate on the expression inside the brackets, i.e., $x^2 + 2x - 4$.
We factorise this expression by completing the square:

$$\begin{aligned} x^2 + 2x - 4 &= (x^2 + 2x + 1) - 1 - 4 \\ &= (x+1)^2 - 5 \\ &= (x + 1 + \sqrt{5})(x + 1 - \sqrt{5}) \end{aligned}$$

Finally, we multiply by the term we factored out:

$$\begin{aligned} 2x^2 + 4x - 8 &= 2(x^2 + 2x - 4) \\ &= 2(x + 1 + \sqrt{5})(x + 1 - \sqrt{5}) \end{aligned}$$

That is, the process is still the same except that we first 'remove' the coefficient of x^2. Depending on the numbers involved, the arithmetic can be straight forward (as in this example) or it can become arithmetically challenging! For example, what if the expression had been $3x^2 + 4x - 8$? The answer in this case is $3\left(x - \frac{2}{3}\sqrt{7} + \frac{2}{3}\right)\left(x + \frac{2}{3}\sqrt{7} + \frac{2}{3}\right)$!

EXERCISES 7.2.4

1. Factorise the following

(a)	$x^2 + 2x - 2$	(b)	$x^2 + 4x - 2$	(c)	$x^2 - 6x + 2$
(d)	$x^2 - 8x - 3$	(e)	$x^2 - 2x - 1$	(f)	$x^2 - 4x - 4$
(g)	$2x^2 - 6x - 2$	(h)	$3x^2 + 12x - 9$	(i)	$2x^2 + x - 2$

EXERCISES 7.2.5 – MISCELLANEOUS QUESTIONS

1. Factorise the following

(a)	$x^2 + 3x + 2$	(b)	$x^2 + 7x + 6$	(c)	$x^2 + 6x + 8$
(d)	$x^2 + 9x + 20$	(e)	$z^2 + 9z + 18$	(f)	$x^2 + 7x + 10$

2. Factorise the following

(a) $x^2 - 3x + 2$ (b) $x^2 - 7x + 6$ (c) $x^2 - 6x + 8$
(d) $x^2 - 9x + 20$ (e) $z^2 - 9z + 18$ (f) $x^2 - 7x + 10$

3. Factorise the following

(a) $x^2 - 2x - 3$ (b) $y^2 + 3y - 10$ (c) $s^2 - 3s - 10$
(d) $x^2 - x - 12$ (e) $y^2 - 5y - 14$ (f) $r^2 - 4r - 45$

4. Factorise the following

(a) $2x^2 - 6x + 4$ (b) $3x^2 + 21x + 30$ (c) $4s^2 - 12s - 40$
(d) $3y^2 + 9y - 30$ (e) $5x^2 + 30x + 40$ (f) $6x^2 - 12x - 18$

5. Factorise the following

(a) $y^2 - 16xy + 15x^2$ (b) $z^2 + zw - 42w^2$ (c) $a^2 + 2ab - 35b^2$
(d) $2y^2 + 18xy + 36x^2$ (e) $3x^2 - 3yx - 36y^2$ (f) $5a^2 - 25ab + 30b^2$

6. Factorise the following

(a) $2x^2 + 5x + 2$ (b) $3x^2 + 5x + 2$ (c) $3x^2 + 7x + 2$
(d) $2x^2 - 5x + 2$ (e) $2x^2 + 3x - 2$ (f) $7s^2 + 2s - 5$
(g) $5x^2 + 2x - 3$ (h) $7x^2 - 36x + 5$ (i) $3y^2 - 5y - 2$
(j) $3z^2 - 10z + 8$ (k) $5w^2 - w - 4$ (l) $2x^2 - 7x - 15$
(m) $2y^2 + 7y + 6$ (n) $5x^2 + 9x + 4$ (o) $3z^2 + 8z + 4$
(p) $2 - 6x - 8x^2$ (q) $3 - 2x - 5x^2$ (r) $4 + x - 5x^2$

7. Factorise the following

(a) $x^2 + 8x + 16$ (b) $y^2 + 10y + 25$ (c) $z^2 + 6z + 9$
(d) $x^2 - 10x + 25$ (e) $b^2 - 12b + 36$ (f) $x^2 - 14x + 49$
(g) $y^2 - 25$ (h) $x^2 - 36$ (i) $z^2 - 16$
(j) $4x^3 - 36x$ (k) $3s^2 - 48$ (l) $4x^2 - 9y^2$
(m) $4x^2 + 4x + 1$ (n) $9z^2 + 12z + 4$ (o) $9z^2 + 6z + 1$

8. Factorise the following by completing the square

(a) $x^2 + 2x - 4$ (b) $x^2 - 2x - 2$ (c) $x^2 + 4x + 2$
(d) $x^2 + 4x - 3$ (e) $x^2 - 6x + 3$ (f) $x^2 + 6x + 2$
(g) $z^2 + 8z + 13$ (h) $y^2 - 8y + 14$ (i) $z^2 + 10z + 20$
(j) $x^2 - 10x + 20$ (k) $z^2 - 4z + 1$ (l) $a^2 + 14a + 30$

7.3 QUADRATIC EQUATIONS

Quadratic equations such as $2x^2 + 5x = 3$ can be solved by several methods. It is usually necessary to rearrange the equation so that it has a zero on one side and all non-zero terms on the other side. In the present case, the equation would need to be rearranged to give $2x^2 + 5x - 3 = 0$. Once the equation is in this form, we can proceed using a number of different methods.

7.3.1 THE FACTORISATION METHOD

To solve a quadratic, once it has been rearranged so that one side is zero, requires the use of the **Null Factor Law**. The Null Factor Law states the following:

If $a \cdot b = 0$ then either $a = 0$ or $b = 0$ or both $a = 0 = b$

Although this might seem rather trivial, it makes solving equations less of a guessing game and more of a guarranteed outcome.

For example, consider the equation $x(x-5) = 0$. The only way that the product of x and $(x-5)$ can be 0 is if either $x = 0$ **or** $(x-5) = 0$ [Note: we use the **inclusive 'or'** – meaning that when solving equations, '**or**' stands for one, or the other, or both]. We then have that $x = 0$ or $x = 5$.

The sequence would then be: $x(x-5) = 0 \Leftrightarrow x = 0 \text{ or } x-5 = 0$

$\Leftrightarrow x = 0 \text{ or } x = 5$

The symbol: $\Leftrightarrow$ (i.e., if and only if, which is abbreviated to 'iff' is discussed in Chapter 1 – The Theory of Knowledge).

So, at the end of the process there has been no guessing as to what the **solutions** to the equation are. The solutions are also known as the **zeros** or **roots** of the equation.

Solving quadratic equations by first factorising the quadratic requires that you are able to factorise expressions such as those found in § 7.2. Once you have successfully factorised the quadratic you can make use of the Null Factor Law and then solve. That is:

		Example
Step 1:	Transpose all terms to one side of the equality sign. So that one side is left with zero.	$x^2 + 2x = 8$ $\therefore x^2 + 2x - 8 = 0$
Step 2:	Factorise the quadratic completely into its two factors.	$(x+4)(x-2) = 0$
Step 3:	Apply the Null Factor Law.	$x + 4 = 0$ or $x - 2 = 0$
Step 4:	Solve for the variable.	$x = -4$ or $x = 2$

EXAMPLE 7.14

Solve the quadratic equations:

(a) $x^2 - 2x = 0$ (b) $3x^2 - 27 = 0$ (c) $2x^2 + 5x = 3$

Solution

(a) $x^2 - 2x = 0 \Leftrightarrow x(x-2) = 0$

The factorisation technique used is the 'single common factor' method.

Now that the equation is factorised (a product of terms) we can use the Null Factor Law.

That is, $x(x-2) = 0 \Leftrightarrow x = 0$ or $x - 2 = 0$

$\Leftrightarrow x = 0$ or $x = 2$

Therefore, there are two solutions to the equation, $x = 0$ or $x = 2$.

(b) $3x^2 - 27 = 0 \Leftrightarrow 3(x^2 - 9) = 0$

$\Leftrightarrow 3(x+3)(x-3) = 0$

The Null Factor Law can now be used to solve the problem:

$3 = 0$, which is never true, or $x + 3 = 0 \Leftrightarrow x = -3$ or $x - 3 = 0 \Leftrightarrow x = 3$

Therefore, we have that $x = -3$ or 3.

(c) We transpose the equation so that we have zero on one side:

That is, $2x^2 + 5x = 3 \Leftrightarrow 2x^2 + 5x - 3 = 0$

Next, we factorise: $\Leftrightarrow (2x-1)(x+3) = 0$

Make use of the N.F.L: $\Leftrightarrow 2x - 1 = 0$ or $x + 3 = 0$

Solve for x: $\Leftrightarrow x = \frac{1}{2}$ or $x = -3$

Not every quadratic equation leads to an expression that will factorise and the method can generally only be applied to a minority of examples. There are also many quadratic equations that have no real solutions. For example, the equation $x^2 + 1 = 0 \Rightarrow x^2 = -1$ has no real solution! (Because the square of any real number is always positive or zero).

EXERCISES 7.3.1

1. Use the factorisation method to solve these quadratic equations.

(a)	$x^2 - 7x = 0$	(b)	$x^2 - 7 = 0$	(c)	$x^2 + 3x = 0$
(d)	$x^2 + 4 = 0$	(e)	$x^2 + x - 12 = 0$	(f)	$2x^2 = 4x$
(g)	$x^2 + 2x - 8 = 0$	(h)	$x^2 = 3x + 10$	(i)	$10 + 3x - x^2 = 0$
(j)	$30 = x + x^2$	(k)	$6 - 5x - x^2 = 0$	(l)	$x^2 - 3x - 28 = 0$
(m)	$2x^2 + x - 28 = 0$	(n)	$11x = 28 + x^2$	(o)	$x^2 - 12x + 27 = 0$
(p)	$x^2 = 6x + 27$	(q)	$x^2 + 7x + 12 = 0$	(r)	$3x^2 + 13x + 12 = 0$
(s)	$2x^2 + 18x = -28$	(t)	$4x^2 = 8 + 14x$	(u)	$3x^2 - 11x - 4 = 0$
(v)	$11x = 5 + 2x^2$	(w)	$6x^2 - 13x + 5 = 0$	(x)	$12x^2 - 23x + 5 = 0$
(y)	$11x = 2 + 15x^2$	(z)	$-2 + 9x - 4x^2 = 0$		

7.3.2 EQUATIONS SOLVED BY COMPLETING THE SQUARE

If the quadratic expression will not factorise by observation (i.e., trial and error), it could very well be the case that we need to factorise it by 'completing the square'.

EXAMPLE 7.15 Solve the quadratic equations:

(a) $x^2 + 6x - 5 = 0$ (b) $x^2 - 5x = 3$

Solution

(a) As there are no obvious factors, we try to factorise it by completing the square:

$$x^2 + 6x - 5 = (x^2 + 6x + 9) - 9 - 5$$

$$\left(\frac{6}{2}\right)^2 = 9$$

Therefore, we have that $x^2 + 6x - 5 = 0 \Leftrightarrow [(x+3)^2 - 9] - 5 = 0$

$$\Leftrightarrow (x+3)^2 - 14 = 0$$

$$[(x+3) + \sqrt{14}][(x+3) - \sqrt{14}] = 0 \quad \text{[Using D.O.P.S.]}$$

$$\therefore x + 3 + \sqrt{14} = 0 \Leftrightarrow x = -3 - \sqrt{14} \quad \text{or} \quad x + 3 - \sqrt{14} = 0 \Leftrightarrow x = -3 + \sqrt{14}$$

An alternative is to rearrange the equation and take square roots of both sides.

That is, $(x+3)^2 - 14 = 0 \Leftrightarrow (x+3)^2 = 14$

$$\Leftrightarrow x + 3 = \pm\sqrt{14}$$

$$\Leftrightarrow x = -3 \pm \sqrt{14}$$

The '±' informs us that we have two solutions to consider; $x = -3 - \sqrt{14}$ or $x = -3 + \sqrt{14}$.

(b) Again, as no factors are obvious, we try factorising by completing the square:

$$x^2 - 5x = 3 \Leftrightarrow x^2 - 5x - 3 = 0$$

$$\left(x - \frac{5}{2}\right)^2 - \left(\frac{5}{2}\right)^2 - 3 = 0$$

$$\Leftrightarrow \left(x - \frac{5}{2}\right)^2 - \frac{37}{4} = 0$$

$$\Leftrightarrow \left(x - \frac{5}{2} - \sqrt{\frac{37}{4}}\right)\left(x - \frac{5}{2} + \sqrt{\frac{37}{4}}\right) = 0$$

$$\Leftrightarrow x = \frac{5}{2} + \sqrt{\frac{37}{4}} \text{ or } x = \frac{5}{2} - \sqrt{\frac{37}{4}}$$

That is, $x = \frac{5}{2} + \frac{\sqrt{37}}{2}$ or $x = \frac{5}{2} - \frac{\sqrt{37}}{2}$

EXAMPLE 7.16 Solve the quadratic equations:

(a) $2x^2 + 4x - 12 = 0$ (b) $3x^2 + x + 7 = 0$

Solution

(a) In this case, remove a common factor so that the coefficient of x^2 is 1. In the case of an equation, it is possible to simplify the problem by dividing both sides by this factor.

$$2x^2 + 4x - 12 = 0 \Leftrightarrow 2(x^2 + 2x - 6) = 0 \Leftrightarrow x^2 + 2x - 6 = 0$$
$$\Leftrightarrow [(x+1)^2 - 1] - 6 = 0$$
$$\Leftrightarrow (x+1)^2 = 7$$
$$\Leftrightarrow x + 1 = \pm\sqrt{7}$$
$$\Leftrightarrow x = -1 \pm \sqrt{7}$$

Notice that in this example we took a short-cut!
That is, rather than factorising the equation, we took the square root of both sides, producing the $\pm\sqrt{7}$ term.

(b) $3x^2 + x + 7 = 0 \Leftrightarrow x^2 + \frac{1}{3}x + \frac{7}{3} = 0$

$$\left(x + \frac{1}{6}\right)^2 - \left(\frac{1}{6}\right)^2 + \frac{7}{3} = 0$$
$$\left(x + \frac{1}{6}\right)^2 = -\frac{7}{3} + \frac{1}{36}$$
$$= -\frac{83}{36}$$

In this case, the solution cannot proceed since it is not possible to take the square root of a negative number, i.e., $\sqrt{-\frac{83}{36}}$ does not produce a real number.
We therefore say that there are **no real solutions**.

It is not uncommon for a quadratic equation to have no real solutions. There will also be occasions on which there is exactly one solution to an equation.

EXERCISES 7.3.2

1. Use the method of completing the square to solve the following quadratic equations:

(a)	$x^2 + 5x + 2 = 0$	(b)	$x^2 + 4x - 3 = 0$
(c)	$x^2 - 2x - 5 = 0$	(d)	$x^2 - 2x + 3 = 0$
(e)	$x^2 + 5x = 7$	(f)	$x^2 + 7x - 9 = 0$
(g)	$x^2 + 11x = 9$	(h)	$x^2 + 3x - 12 = 0$
(i)	$x^2 + 5x = 12$	(j)	$-x^2 + 4x - 12 = 0$
(k)	$-x^2 + 4x + 7 = 0$	(l)	$x^2 = 6x + 11$
(m)	$-x^2 - 4x + 20 = 0$	(n)	$2x^2 + 20 = 4x$
(o)	$2x^2 - 4x - 9 = 0$	(p)	$2x^2 - 5x - 3 = 0$
(q)	$2x^2 - 7x - 4 = 0$	(r)	$2x^2 - 3x - 4 = 0$
(s)	$3x(x-1) = 4$	(t)	$3x^2 - 5x - 4 = 0$

7.3.3 EQUATIONS SOLVED BY USING THE FORMULA

The process of completing the square can be applied to the general quadratic equation

$$ax^2 + bx + c = 0.$$

When this is carried out, we obtain a formula that will enable us to find all real solutions that correspond to any quadratic (as long as there are real solutions).

The resulting formula – known as The Quadratic Formula – (which we will not prove here) is given by

$$ax^2 + bx + c = 0 \Leftrightarrow x = \frac{-b \pm \sqrt{b^2 - 4ac}}{2a}$$

Success in using this formula will depend on your ability to identify the correct values of a, b and c. The following examples illustrate the use of the formula.

EXAMPLE 7.17

Use the formula to solve the quadratic equations:

(a) $x^2 - x - 4 = 0$ (b) $2x^2 = 4 - x$ (c) $x^2 - 3x + 7 = 0$

Solution

(a) For the quadratic $x^2 - x - 4 = 0$ we have that $a = 1, b = -1$ & $c = -4$. These values can now be substituted into the quadratic formula.

$$x = \frac{-b \pm \sqrt{b^2 - 4ac}}{2a} = \frac{1 \pm \sqrt{(-1)^2 - 4 \times 1 \times -4}}{2 \times 1}$$

$$= \frac{1 \pm \sqrt{1 + 16}}{2}$$

$$= \frac{1 \pm \sqrt{17}}{2}$$

(b) The equation $2x^2 = 4 - x$ must first be rearranged into the form $ax^2 + bx + c = 0$. So, $2x^2 = 4 - x \Leftrightarrow 2x^2 + x - 4 = 0$. This means that $a = 2, b = 1$ and $c = -4$. Therefore, using the quadratic formula we have that

$$x = \frac{-b \pm \sqrt{b^2 - 4ac}}{2a} = \frac{-1 \pm \sqrt{(-1)^2 - 4 \times 2 \times -4}}{2 \times 2}$$

$$= \frac{-1 \pm \sqrt{33}}{4}$$

(c) In this case, the equation $x^2 - 3x + 7 = 0$ implies that $a = 1, b = -3$ and $c = 7$.

Then, $x^2 - 3x + 7 = 0 \therefore x = \frac{-b \pm \sqrt{b^2 - 4ac}}{2a} = \frac{3 \pm \sqrt{(-3)^2 - 4 \times 1 \times 7}}{2}$

$$= \frac{3 \pm \sqrt{-19}}{2}$$

However, the negative quantity in the square root, $\sqrt{-19}$ has no real value and so the equation has no real solutions.

EXERCISES 7.3.3

1. Use the quadratic formula to solve these equations:

(a)	$x^2 - 3x - 7 = 0$	(b)	$x^2 - 5x = 2$	(c)	$x^2 - 3x - 6 = 0$
(d)	$x^2 = 7x + 2$	(e)	$x(x+7) = 4$	(f)	$x^2 + 2x - 8 = 0$
(g)	$x^2 + 2x - 7 = 0$	(h)	$x^2 + 5x - 7 = 0$	(i)	$x^2 - 3x - 7 = 0$
(j)	$x^2 - 3x + 9 = 0$	(k)	$x^2 + 9 = 8x$	(l)	$4x^2 - 8x + 9 = 0$
(m)	$4x^2 = 8x + 9$	(n)	$5x^2 - 6x - 7 = 0$	(o)	$5x^2 - 12x + 1 = 0$
(p)	$7x^2 - 12x + 1 = 0$				

7.4 QUADRATIC GRAPHS

7.4.1 THE BASICS – DILATIONS AND TRANSLATIONS

In the same way that we produced a graphical representation of linear relations, we now produce a graphical representation of a quadratic relation. The simplest quadratic relation is $y = x^2$. The table of values is:

x	–3	–2	–1	0	1	2	3
y	9	4	1	0	1	4	9

When plotted on a Cartesian plane, the result is a curve known as a parabola.

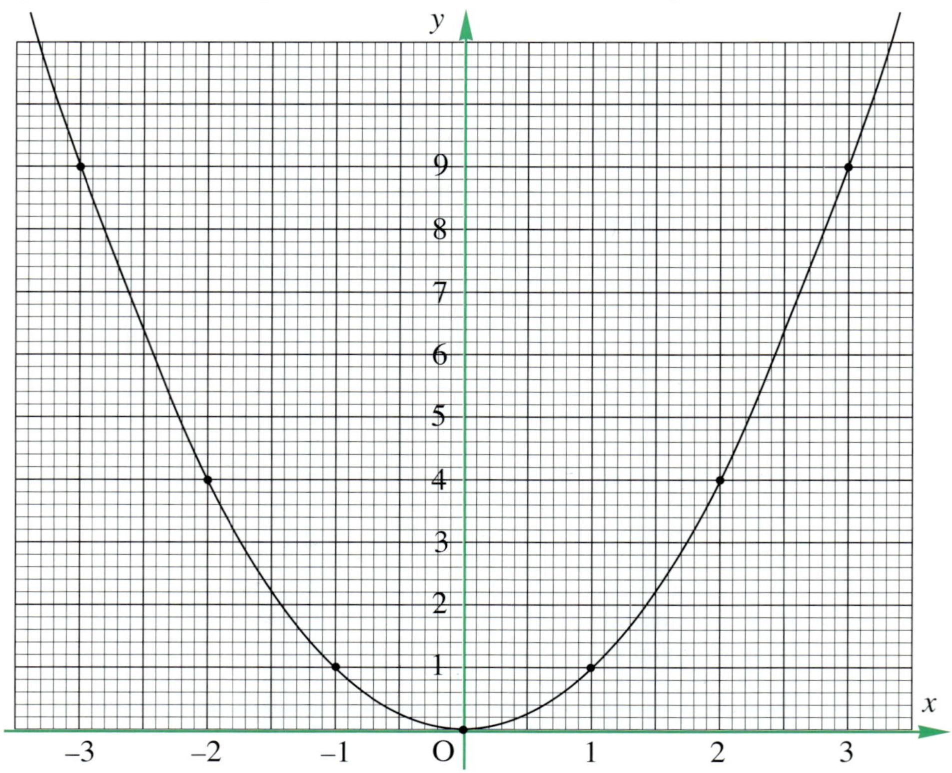

This diagram is a ''**plot**' of the graph. It is common to be asked to 'sketch' graphs of quadratics and other functions. A sketch must show the correct shape of the graph and other key features such as intercepts and turning points. There are two main techniques for sketching parabolas:

1. The turning point method
2. The intercept method

However, before we begin with these methods we need a little ground work on the features of a parabolic curve and how they relate back to the equation of the quadratic relation.

Also, because functions is the domain of Chapter 8, it will suffice at this point in time to refer to to quadratic relations as quadratic functions, so that rather than always writing $y = x^2$ we can substitute $f(x) = x^2$, where the notation $f(x)$ reads as 'f of x' – i.e., 'a function of x' – meaning that the expression on the right hand side is in terms of x. This also has the added advantage that when we wish to determine the y–value corresponding to a particular value of x, say $x = 3$, then we use the following setting out:

$f(3) = 3^2 = 9$. So that when $x = 3$, $y = 9$.
Therefore curve passes through the point (3, 9)

Also, the use of $f(x)$ is arbitrary. We could have used the notation, $g(x)$ 'read as g of x' or $h(x)$ 'read as h of x' or . . . It is simply a labelling system.

1. DILATIONS ALONG THE Y-AXIS – $y = ax^2$

Case 1. $y = ax^2, a > 0$

Consider the graphs of the following functions,

Figure 1: $f(x) = x^2$, $h(x) = 2x^2$, $k(x) = 3x^2$.

Figure 2: $f(x) = x^2$, $g(x) = \frac{1}{2}x^2$, $p(x) = \frac{1}{3}x^2$.

We sketch each of these and then comment on the results.

Figure 1:

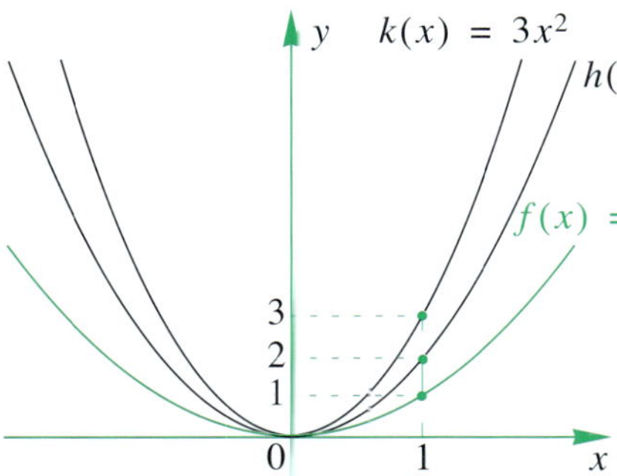

Figure 2:

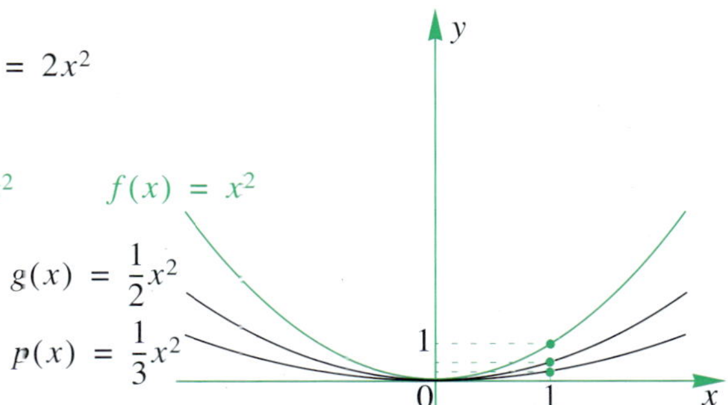

Observations:

Figure 1
As the coefficient, 'a', increases, the parabola is **stretched** along the y–axis. Often, people say that the graph looks more narrow as 'a' increases.

Figure 2
As the coefficient, 'a', decreases, the parabola '**shrinks'** along the y–axis. Often, people say that the graph looks 'fatter' as 'a' decreases.

Case 2. $y = ax^2, a < 0$

With the coefficient of x^2 being negative, the parabola is now **inverted**. That is, it maintains the same properties as for Case 1, but is reflected about the x–axis:

Figure 3:

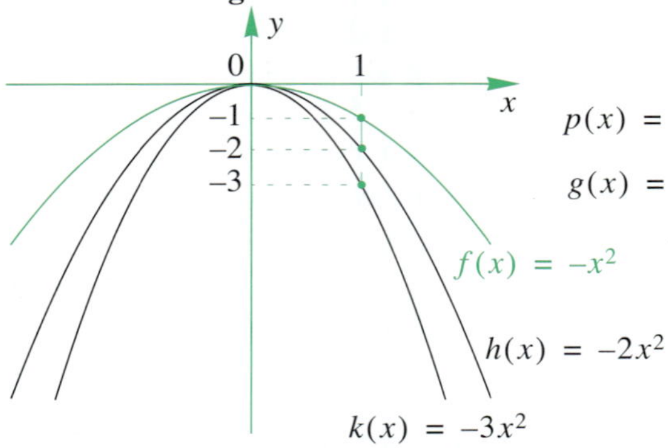

Figure 4:

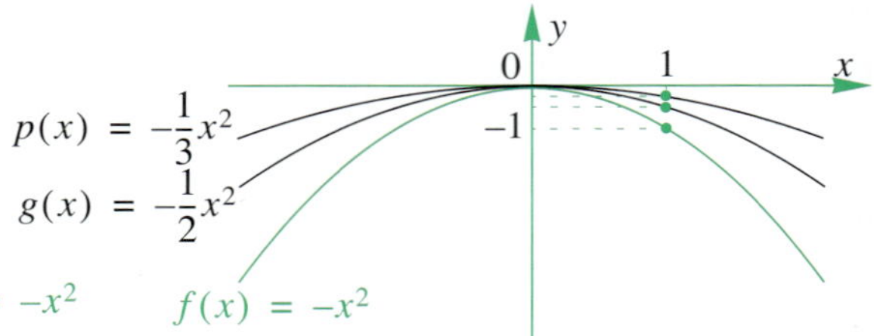

Observations: Both Figure 3 and Figure 4 are identical to Figures 1 and 2 respectively, except that this time they are inverted (or reflected about the x-axis).

2. TRANSLATIONS ALONG THE X-AXIS – $y = (x \pm h)^2$

Case 1. $y = (x-h)^2, h > 0$

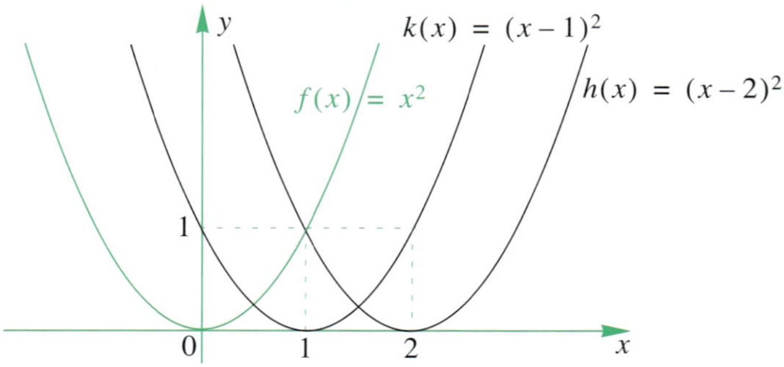

Observation:
The curve of $y = x^2$ has been moved (translated) **to the right** by the value of h.

That is,
if $h = 1$, then the curve is moved accross by 1 unit to the right.
if $h = 2$, then the curve is moved accross by 2 units to the right.

Case 2. $y = (x+h)^2, h > 0$

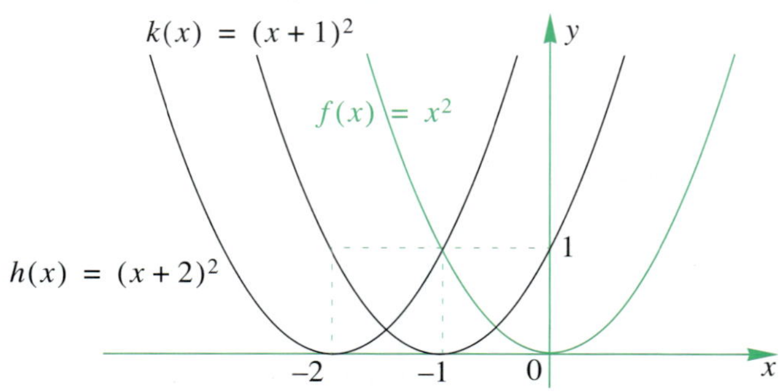

Observation:
The curve of $y = x^2$ has been moved (translated) **to the left** by the value of h.

That is,
if $h = 1$, then the curve is moved accross by 1 unit to the left.
if $h = 2$, then the curve is moved accross by 2 units to the left.

Note that there is no need to consider the case where $h < 0$. For example, if $h = -2$, then we still have either $y = (x-2)^2$ or $y = (x+2)^2$ – which are dealt with by $h > 0$.

3. TRANSLATIONS ALONG THE Y-AXIS – $y = x^2 \pm k$

Case 1. $y = x^2 + k, k > 0$

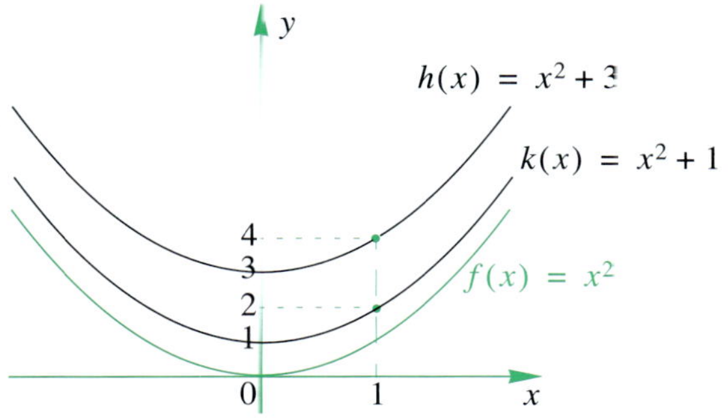

Observation:
The curve of $y = x^2$ has been moved (translated) **up along the y-axis** by the value of k.

That is
if $k = 1$, then the curve is moved up by 1 unit.
if $k = 3$, then the curve is moved up by 3 units

Case 2. $y = x^2 - k, k > 0$

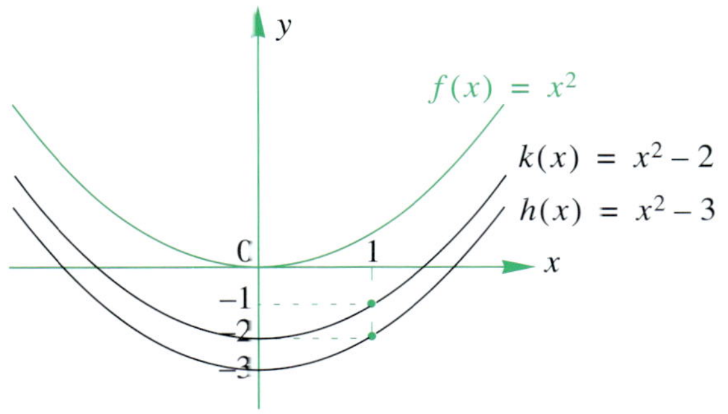

Observation:
The curve of $y = x^2$ has been moved (translated) **down along the y-axis** by the value of k.

That is,
if $k = 2$, then the curve is moved down by 2 unit.
if $k = 3$, then the curve is moved down by 3 units.

EXERCISES 7.4.1

1. Sketch the following pairs on the same set of axes

(a) $f(x) = x^2, h(x) = 4x^2$ (b) $f(x) = x^2, h(x) = 1.5x^2$

(c) $f(x) = x^2, h(x) = \frac{1}{2}x^2$ (d) $f(x) = x^2, h(x) = \frac{3}{4}x^2$

(e) $f(x) = x^2, h(x) = -2x^2$ (f) $f(x) = x^2, h(x) = -\frac{1}{2}x^2$

(g) $f(x) = 3x^2, h(x) = -3x^2$ (h) $f(x) = 2x^2, h(x) = -\frac{1}{4}x^2$

2. Sketch the following pairs on the same set of axes

(a) $f(x) = x^2, g(x) = (x-3)^2$ (b) $f(x) = x^2, g(x) = (x+2)^2$

(c) $f(x) = x^2, g(x) = \left(x+\frac{1}{3}\right)^2$ (d) $f(x) = x^2, g(x) = \left(x-\frac{1}{2}\right)^2$

(e) $f(x) = x^2, g(x) = x^2 - \frac{9}{4}$ (f) $f(x) = x^2, g(x) = x^2 + \frac{1}{2}$

(g) $f(x) = x^2, g(x) = x^2 + 4$ (h) $f(x) = x^2, g(x) = x^2 - 9$

7.4.2 TURNING POINT FORM

We can now combine the transformations from §7.4.1 to produce a general expression of the form $y = a(x \pm h)^2 \pm k$. This is known as the turning point form of the parabola. It is important that when sketching graphs of this form, you apply each transformation in the following order.

Order	Observation	Graph	
1.			
Look at the sign of 'a'	If $a > 0$, parabola is upright:	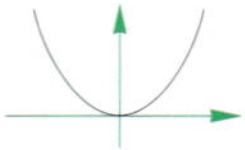	
	If $a < 0$, parabola is inverted:	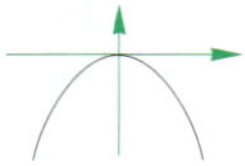	
2.		$(a > 0)$	$(a < 0)$
Look at the magnitude of 'a'	If *the magnitude of a is* > 1, parabola is 'thin':	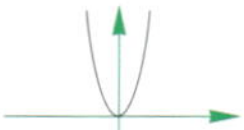	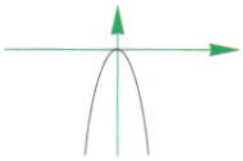
	If *the magnitude of a is* greater than 0 but less than 1, parabola is 'wide':	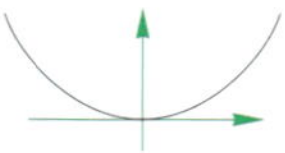	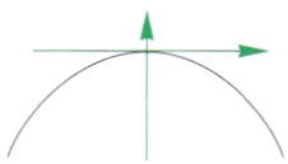
3.			
Look at $(x \pm h)^2$ term	If it contains the term $(x - h)^2$, move curve to the right by h units.	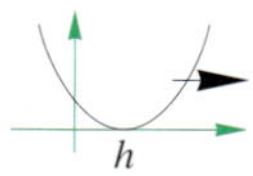	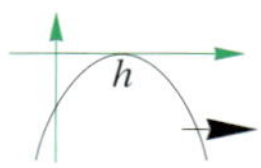
	If it contains the term $(x + h)^2$, move curve to the left by h units.	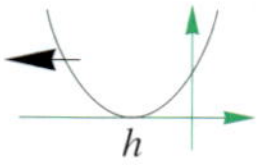	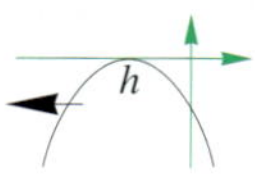
4.			
Look at '$\pm k$' term	If it contains the term '+ k', move curve up by k units .	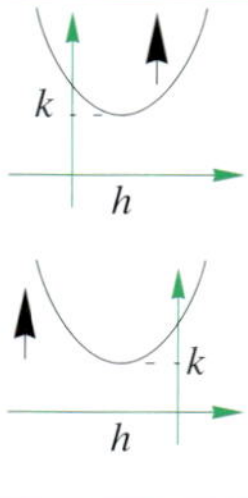	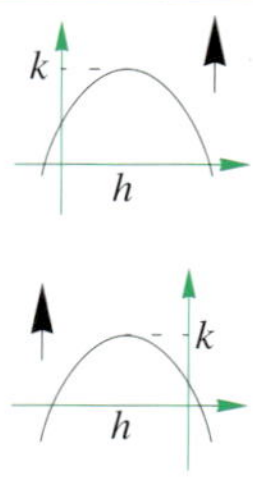
	If it contains the term '– k', move curve down by k units.	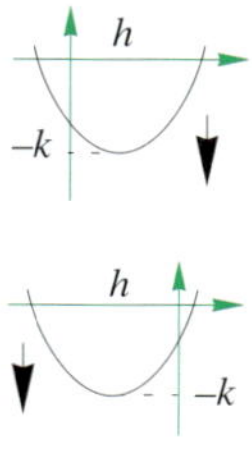	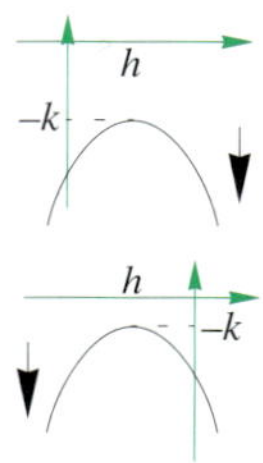

We are not suggesting that you memorise the summary table just provided – that would defeat the whole purpose of the observations. The table is provided simply to show the consistency in applying the transformations. Furthermore, this consistency applies to many other curves, not just parabolas.

Notice then, that the graph with equation

$$y = a(x-h)^2 + k$$

is a parabola

1. with turning point at (h, k).
2. that is symmetrical about the line $x = h$.

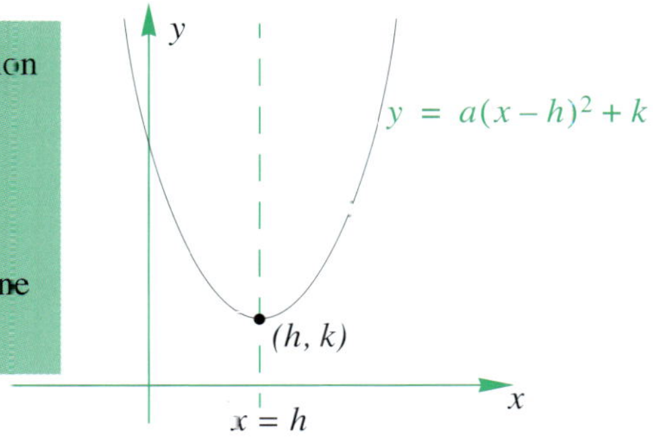

EXAMPLE 7.18

Sketch the graphs of the following functions

(a) $f(x) = 2(x-3)^2$ (b) $y = -\frac{1}{2}x^2 + 2$ (c) $y = 3(x+1)^2 - 2$

Solution

(a) We first identify the features of the function:

$f(x) = 2(x-3)^2$

3. Translated 3 units **to the right**.

1. Positive coefficient, so parabola is **upright**
2. Magnitude is greater than 1, so parabola is '**thinner**' than that of $y = x^2$.

Note: Turning point at (3, 0)

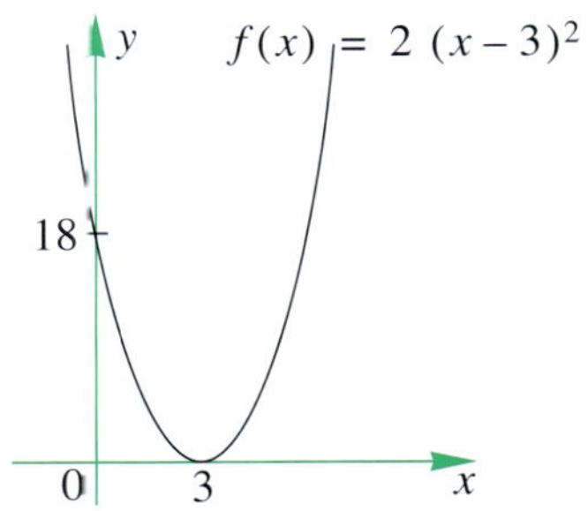

It is always good practice to find where the curve cuts the axes. For now, we will only concern ourselves with finding the y-intercept. To do this we determine the value of y when $x = 0$.
In our case we have, $f(0) = 2(0-3)^2 = 2 \times 9 = 18$ and so the curve cuts the y-axis at (0, 18).

(b) Again we identify the features of the function:

$y = -\frac{1}{2}x^2 + 2$

3. Translated 2 units **up**.

1. Negative coefficient, so parabola is **inverted.**
2. Magnitude is less than 1, so parabola is '**wider**' than that of $y = x^2$.

Note: Turning point at (0, 2)

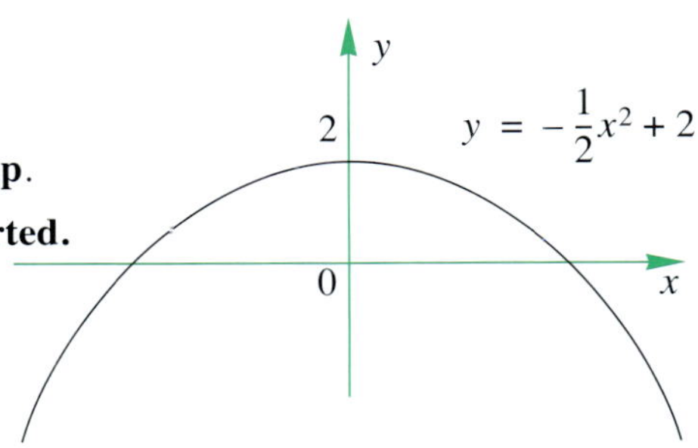

(c) For this equation we have 3 transformations to consider:

$y = 3(x+1)^2 - 2$

4. Translated 2 units **down**.

3. Translated 1 units **to the left**.

1. Positive coefficient, so parabola is **upright.**
2. Magnitude is greater than 1, so parabola is '**thinner**' than that of $y = x^2$.

Note: Turning point at $(-1, -2)$

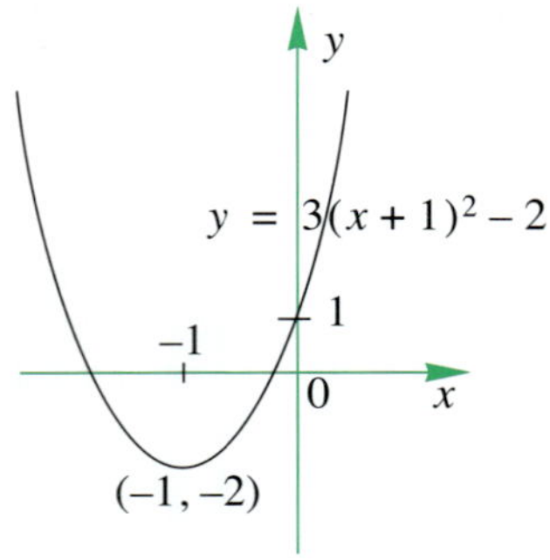

To find the y-intercept, set $x = 0$ so that $y = 3(0+1)^2 - 2 = 1$. i.e., curve cuts y-axis at (0, 1).

One approach that simplifies the sketching of graphs is to first pencil-in each stage of the process and then conclude by using a biro on the final graph. For example, if asked to sketch the graph of the function $f(x) = 2(x-1)^2 - 4$ then 'break-up' the process as follows:

1. Pencil-in the graph of $y = x^2$
2. Stretch the curve in 1., by a factor of 2 and so, pencil-in $y = 2x^2$.
3. Move the curve in 2., one unit to the right and pencil-in $y = 2(x-1)^2$.
4. Move the graph in 3., down 4 units, and so pencil-in $y = 2(x-1)^2 - 4$.

Then, as the graph of $y = 2(x-1)^2 - 4$ is the last graph to be penciled-in, we go over it in black biro. We see how this works in practice:

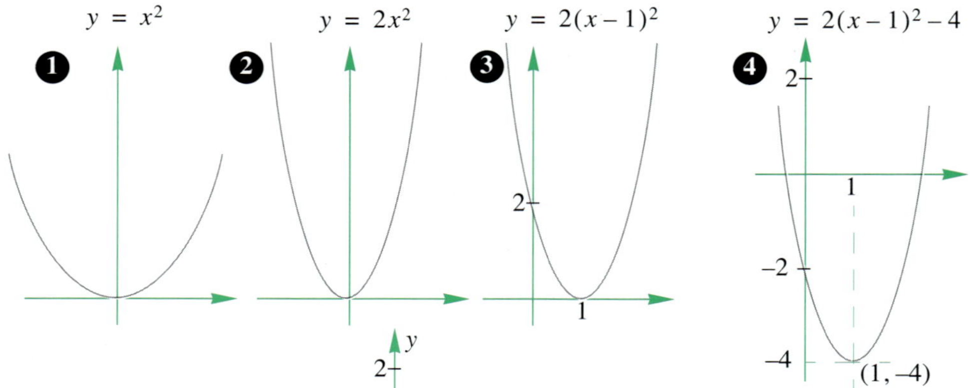

Final Graph:

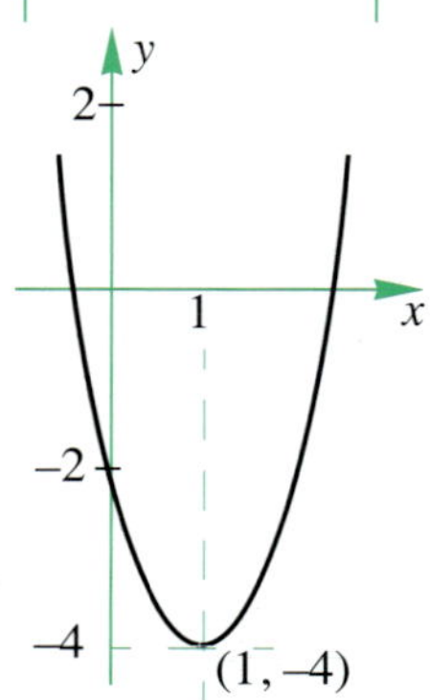

EXERCISES 7.4.2

1. Sketch the graphs of the following

(a) $y = (x+1)^2 - 1$ (b) $y = (x+2)^2 - 4$

(c) $y = (x-2)^2 + 1$ (d) $y = (x-1)^2 + 2$

(e) $y = \left(x - \frac{1}{2}\right)^2 + \frac{3}{4}$ (f) $y = \left(x + \frac{1}{2}\right)^2 - \frac{1}{4}$

2. Sketch the graphs of the following

(a) $y = -(x+1)^2 - 1$ (b) $y = -(x+2)^2 - 4$

(c) $y = 1 - (x-2)^2$ (d) $y = 2 - (x-1)^2$

(e) $y = \frac{3}{4} - \left(x - \frac{1}{2}\right)^2$ (f) $y = -\left(x + \frac{1}{2}\right)^2 + \frac{1}{4}$

3. Sketch the graphs of the following

(a) $y = 2(x-3)^2 + 2$ (b) $y = \frac{1}{2}(x-1)^2 + 1$

(c) $y = 3(x-2)^2 + 3$ (d) $y = \frac{1}{4}(x-4)^2 + 2$

(e) $y = 4 + \frac{1}{3}(x+2)^2$ (f) $y = 1 + \frac{2}{3}(x+2)^2$

4. Sketch the graphs of the following

(a) $y = -2(x+1)^2 + 3$ (b) $y = -\frac{1}{3}(x-2)^2 + 4$

(c) $y = -3(x-2)^2 + 1$ (d) $y = -\frac{1}{4}(x+3)^2 + 2$

(e) $y = 5 - \frac{1}{2}(x+1)^2$ (f) $y = 1 - \frac{2}{3}(x+2)^2$

7.4.3 EXPRESSING FUNCTION IN TURNING POINT FORM

In §7.4.2 the functions were already given in turning point form and so all that remained was to invert, stretch, squash and translate the curve of $y = x^2$. However, often we are not provided with the equation already in turning point form and so, we first need to convert the equation into the form $y = a(x \pm h)^2 \pm k$.

EXAMPLE 7.19

Use the method of completing the square to sketch the graphs of

(a) $y = x^2 + 2x + 3$ (b) $y = x^2 - 3x - 4$ (c) $y = 3x^2 - 6x + 4$

Solution

(a) Completing the square gives:

$$\begin{aligned} y &= x^2 + 2x + 3 \\ &= (x+1)^2 - 1 + 3 \\ &= (x+1)^2 + 2 \end{aligned}$$

In this case $a = 1$, $h = -1$ and $k = 2$.
There is no vertical dilation.
The curve is translated one unit to the left and two units up.
The turning point (vertex) is at the point (–1,2).
The y-intercept (found by substituting $x = 0$ into the equation) is (0,3).

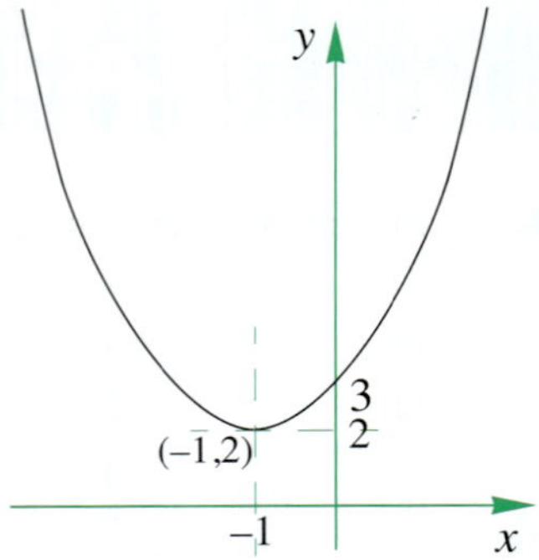

(b) We start by completing the square:

$$\begin{aligned} y &= x^2 - 3x - 4 \\ &= \left(x - \frac{3}{2}\right)^2 - \left(\frac{3}{2}\right)^2 - 4 \\ &= \left(x - \frac{3}{2}\right)^2 - \frac{25}{4} \end{aligned}$$

Here, $a = 1$, $h = \frac{3}{2}$ (moved $\frac{3}{2}$ units to the right)

and $k = -\frac{25}{4}$ (moved $\frac{25}{4}$ down).

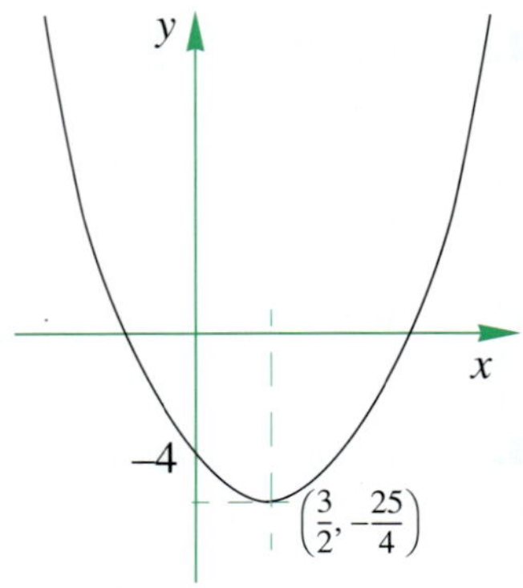

There is no dilation and the turning point has coordinates $\left(\frac{3}{2}, -\frac{25}{4}\right)$.

The y–intercept occurs at (0, –4).

(c) We start by completing the square:

$$\begin{aligned} y &= 3x^2 - 6x + 4 \\ &= 3[x^2 - 2x]^2 + 4 \\ &= 3[(x-1)^2 - 1] + 4 \\ &= 3(x-1)^2 - 3 + 4 \\ &= 3(x-1)^2 + 1 \end{aligned}$$

As $a = 3$, there is a dilation of 3 along the y-axis.
Then a translation of 1 to the right followed by a translation of 1 up.
The turning point is located at (1, 1) and the y-intercept occurs at (0, 4).

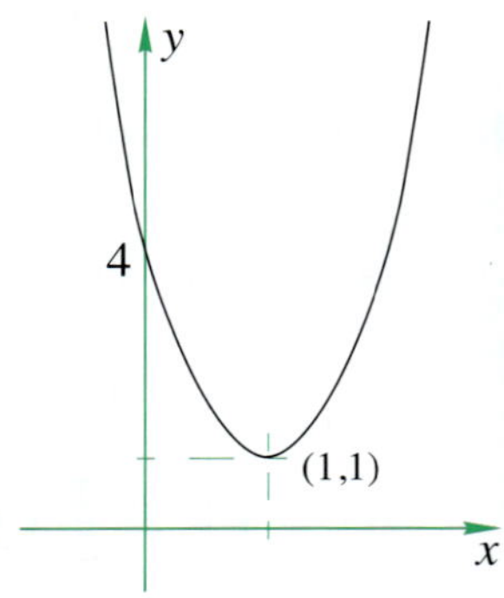

EXERCISES 7.4.3

1. Express the following functions in turning point form and hence sketch their graphs.

(a)	$y = x^2 - 2x + 1$	(b)	$y = x^2 + 4x + 2$	(c)	$y = x^2 - 4x + 2$
(d)	$y = x^2 + x - 1$	(e)	$y = x^2 - x - 2$	(f)	$y = x^2 + 3x + 1$

(g) $y = -x^2 + 2x + 1$ (h) $y = -x^2 - 2x + 2$ (i) $y = 2x^2 - 2x - 1$

(j) $y = \frac{-x^2}{2} + 3x - 2$ (k) $y = -\frac{x^2}{3} + x - 2$ (l) $y = 3x^2 - 2x + 1$

7.4.4 THE INTERCEPT METHOD

As an alternative (or in addition to) completing the square, many parabolas can be sketched by finding all their intercepts.

The y-intercept was discussed in the previous section, i.e., we substitute $x = 0$ into the equation and determine the corresponding y-value. This will give us the coordinates of the point C.

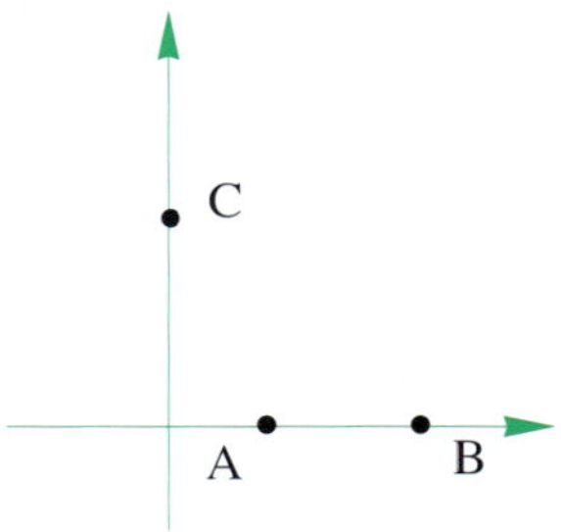

To find the x-intercepts it is necessary to solve a quadratic equation, i.e., we set $y = 0$ and solve for x.
If there are two solutions, we have two points of intersection, points A and B.

Then, we sketch the parabola making sure that it passes through each of the points A, B and C. An important result is that the parabola will be symmetrical about the the vertical line that lies half-way from A to B. This line is known as the axis of symmetry. Note too then that the turning point also lies on this line! This is shown in the diagram alongside:

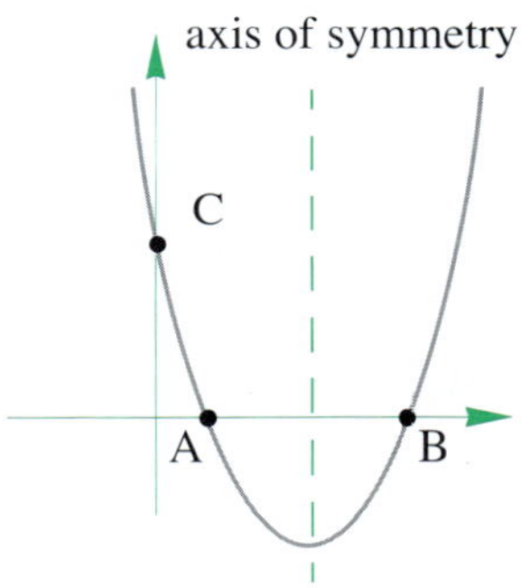

EXAMPLE 7.20

Use the intercept method to sketch the graphs of:

(a) $y = x^2 - x - 2$ (b) $y = x^2 - 3x - 7$ (c) $y = -3x^2 - x + 5$

Solution

(a) $y = x^2 - x - 2$

The y-intercept:

When $x = 0$, $y = 0^2 - 0 - 2 = -2$

i.e., y-intercept occurs at $(0, -2)$.

The x-intercepts are found by solving:

$$x^2 - x - 2 = 0$$
$$(x-2)(x+1) = 0$$
$$x = 2, -1$$

That is, x-intercepts occur at $(-1, 0)$ and $(2, 0)$.

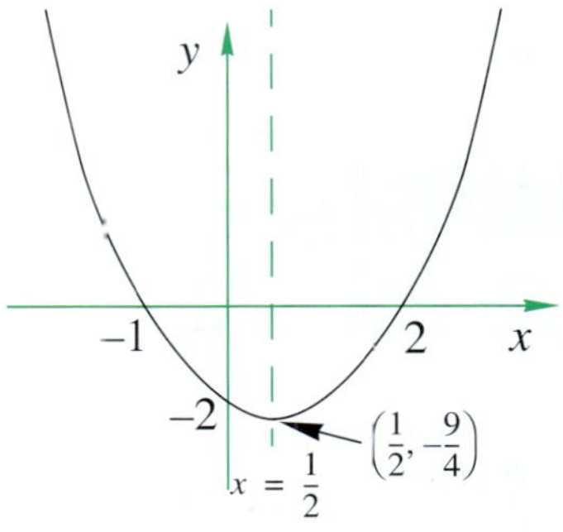

The axis of symmetry runs through the midpoint of $(-1, 0)$ and $(2, 0)$, i.e., $x = \frac{1}{2}(-1+2) = \frac{1}{2}$.

Also, when $x = \frac{1}{2}$, $y = \left(\frac{1}{2}\right)^2 - \frac{1}{2} - 2 = -\frac{9}{4}$. i.e., turning point occurs at $\left(\frac{1}{2}, -\frac{9}{4}\right)$.

(b) $y = x^2 - 3x - 7$

The y-intercept:

When $x = 0$, $y = 0^2 - 3 \times 0 - 7 = -7$.

The x-intercepts:

When $y = 0$, $0 = x^2 - 3x - 7$

However, factors are not immediately obvious, so we make use of the quadratic formula:

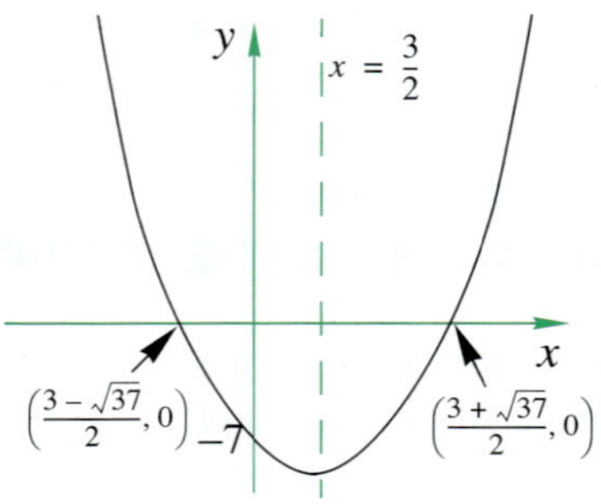

$$x^2 - 3x - 7 = 0$$

$$x = \frac{3 \pm \sqrt{(-3)^2 - 4 \times 1 \times (-7)}}{2}$$

$$= \frac{3 \pm \sqrt{37}}{2}$$

That is, the curve cuts the x-axis at $\left(\frac{3-\sqrt{37}}{2}, 0\right)$ and $\left(\frac{3+\sqrt{37}}{2}, 0\right)$.

The axis of symmetry is given by $x = \frac{1}{2}\left(\frac{3-\sqrt{37}}{2} + \frac{3+\sqrt{37}}{2}\right) = \frac{1}{2}\left(\frac{6}{2}\right) = \frac{3}{2}$.

(c) $y = -3x^2 + 2x + 1$

The y-intercept:

When $x = 0$, $y = -3 \times 0^2 + 2 \times 0 + 1 = 1$

i.e., cuts at (0, 1).

The x-intercepts:

When $y = 0$, $0 = -3x^2 + 2x + 1$

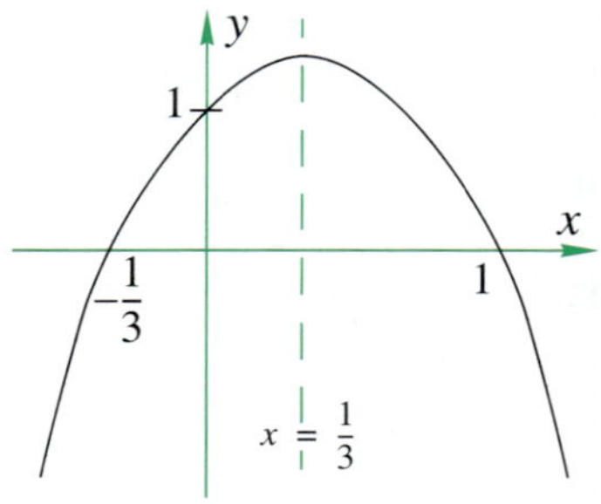

We therefore have, $-(3x^2 - 2x - 1) = 0$

$$\therefore -(3x + 1)(x - 1) = 0$$

$$\therefore x = -\frac{1}{3} \text{ or } x = 1$$

Axis of symmetry:

$$x = \frac{1}{2}\left(-\frac{1}{3} + 1\right) = \frac{1}{2} \times \frac{2}{3} = \frac{1}{3}$$

EXAMPLE 7.21

Where possible, use the intercept method to sketch the graphs of:

(a) $y = x^2 - 2x + 1$ (b) $y = x^2 - 4x + 6$

Solution

(a) ***y*-intercept:**

Set $x = 0$, so that $y = 0^2 - 2 \times 0 + 1 = 1$

i.e., parabola cuts y-axis at (0, 1).

***x*-intercept:**

Set $y = 0$, so that $0 = x^2 - 2x - 1$

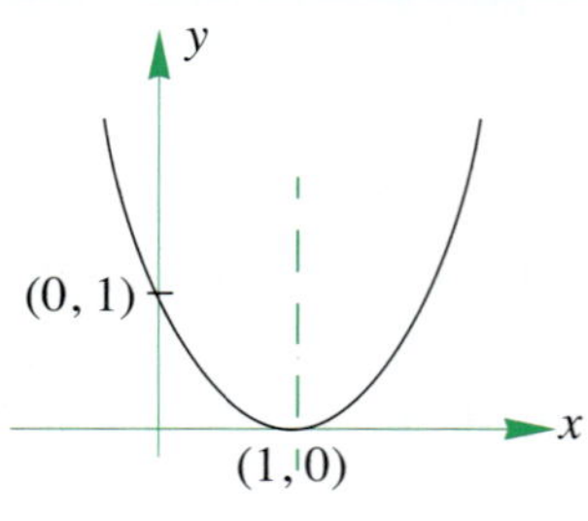

Solving, we have: $0 = (x - 1)^2$

i.e., $x = 1$.

That is, the parabola 'cuts' at $(1, 0)$. In fact, we notice that the graph actually **touches** the x-axis, so that its turning point lies on the x-axis at the point $(1, 0)$.

(b) **y-intercept:**

Set $x = 0$, so that $y = 0^2 - 4 \times 0 + 6 = 6$

i.e., parabola cuts y-axis at $(0, 6)$.

x-intercept:

Set $y = 0$, so that $0 = x^2 - 4x + 6$

However, there are no solutions to this quadratic equation. This means that the parabola will not cut (or touch) the x-axis.

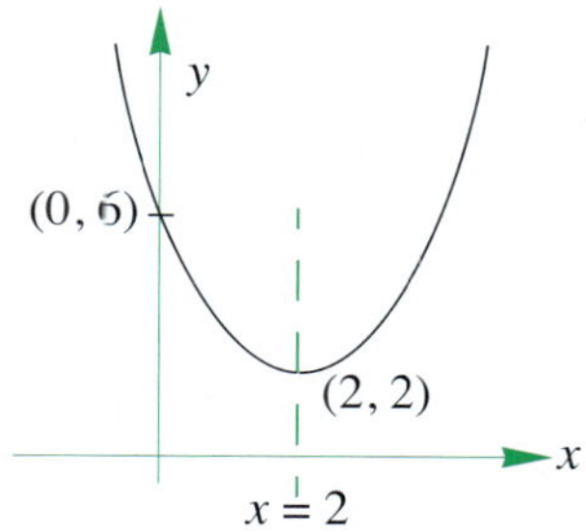

To sketch the graph we will need to complete the square and use the turning point form.

So, $y = x^2 - 4x + 6 = (x^2 - 4x + 4) + 2$

$= (x - 2)^2 + 2$

i.e., it will have a turning point at $(2, 2)$ and an axis of symmetry given by $x = 2$.

Summary of the general properties of the graph of $f(x) = ax^2 + bx + c, a \neq 0$

We consider the case that $a > 0$ (the case that $a < 0$ is identical except that the curve is inverted)

1. **y–intercept:**
   This occurs when $x = 0$, so that $y = f(0) = a(0)^2 + b(0) + c = c$.
   That is, the curve passes through the point $(0, c)$

2. **x–intercept(s):**
   This occurs where $f(x) = 0$.
   Therefore we need to solve $ax^2 + bx + c = 0$.
   To solve we either factorise and solve or use the quadratic formula, which would provide the solution(s) $x = \frac{-b \pm \sqrt{b^2 - 4ac}}{2a}$.

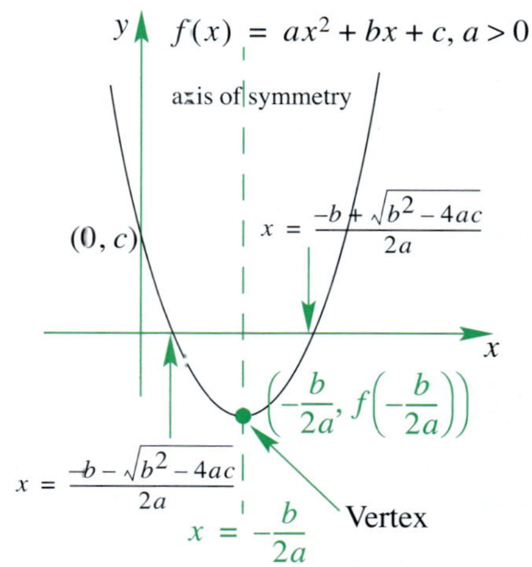

3. **Axis of symmetry:**
   This occurs at $x = -\frac{b}{2a}$.

4. **Vertex (turning point)**
   The vertex occurs when $x = -\frac{b}{2a}$. Then, to find the y–value, find $f\left(-\frac{b}{2a}\right)$

Also, it might be the case that there are no x-intercepts (i.e., no roots) or only one x-intercept (i.e., one root) so that the parabola only touches the x-axis at that point.

EXERCISES 7.4.4

1. Find the axial intercepts of these quadratic functions (correct to 2 decimal places) and hence sketch their graphs.

(a)	$y = x^2 + 3x + 2$	(b)	$y = x^2 - x - 6$	(c)	$y = 2x^2 - 5x - 3$
(d)	$y = x^2 - 4$	(e)	$y = x^2 + x - 5$	(f)	$y = -x^2 + x + 6$
(g)	$y = -x^2 + x + 1$	(h)	$y = -2x^2 - 3x + 5$	(i)	$y = 2x^2 + 5x - 3$
(j)	$y = \frac{x^2}{3} - 2x + 3$	(k)	$y = -\frac{x^2}{2} + x + 4$	(l)	$y = 3x^2 - 2x - 4$

7.4.5 USING A GRAPHICS CALCULATOR

Graphics calculators can be used to check that parabola sketches are correct. The stages are similar to those needed to produce other graphs. To plot $y = x^2 + x - 5$

Step1: **Set the equation using the Y= menu**

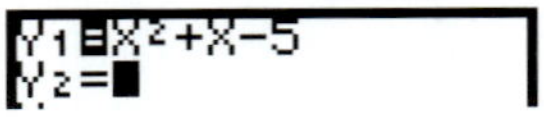

Step 2: **Set an appropriate viewing window.**
It can be a good idea to use the decimal window (**ZOOM4**) as this sets scales that are multiples of the numbers of screen pixels. In this case this window is not large enough to see all the graph.

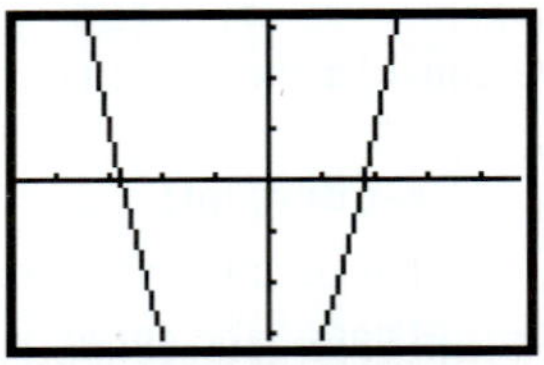

Step 3: **Adjust the viewing window.**
In this case the x scale is correct. If the y-scale is changed (using the **WINDOW** command) to exactly twice the range set by **ZOOM4**, there will still be an integral relationship between the scale and the number of pixels on the screen.

Step 4: **Display the graph**.

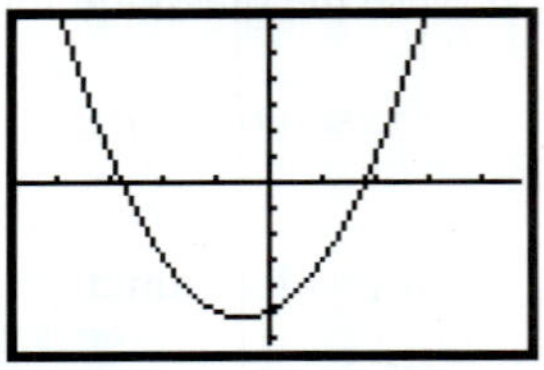

Step 5: If necessary, use **TRACE** to display the coordinates of points on the curve. The diagram shows the coordinates of the minimum point, (–0.5,–5.25). Using **ZOOM4** tends to give better results from the **TRACE** facility. Other viewing windows will give long decimals for the coordinates.

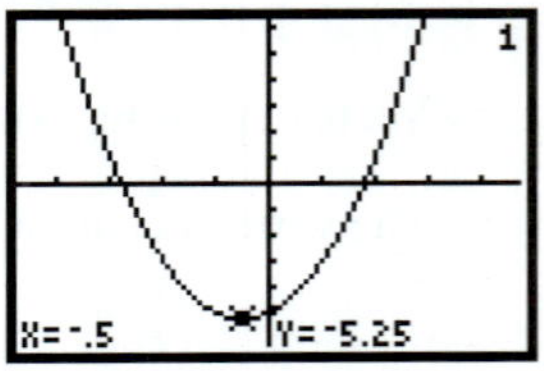

Of course there are short cuts to sketching a parabola on a graphics calculator. For example, let's consider the parabola defined by the equation $y = x^2 - 4x + 6$:

We first enter the equation and select an appropriate window setting (it might take a couple of tries to get an appropriate window):

The final screen dump provides a reasonable result. We can now determine the turning point by using the **CALC** menu and then selecting option **3:minimum**, then moving the cursor about so that we can select a '**Left Bound**', a '**Right Bound**' and eventually a '**Guess**' we obtain:

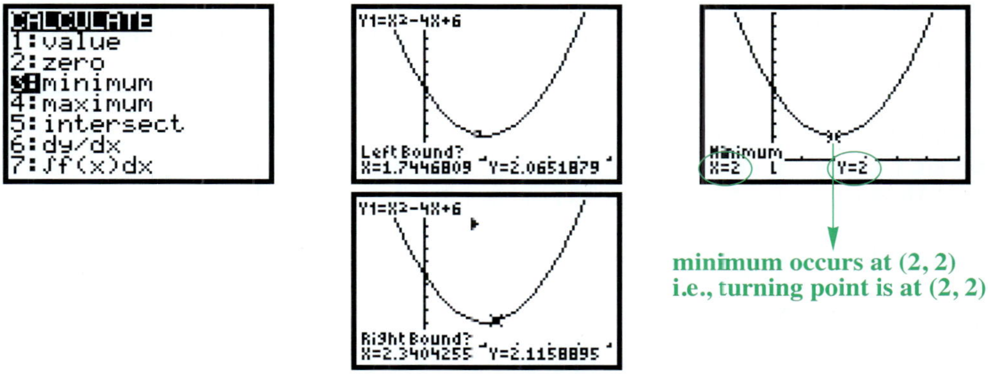

What about finding the x-intercept?

Let's consider the parabola defined by the equation $y = 2x^2 + x - 6$. As in the previous example we enter the equation and select an appropriate window setting:

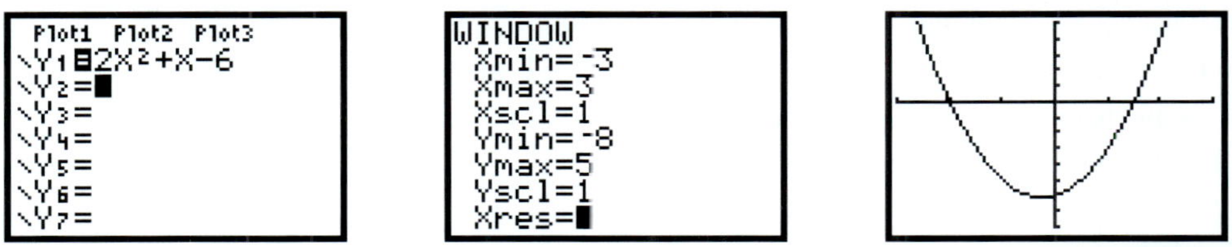

To find the x–intercepts, again we use the **CALC** menu but this time we select option **2:zero**. This will enable us to determine when $y = 0$. Remember, the **zeros** of a quadratic equation correspond to the x-intercepts of the curve defined by that equation. As before, we 'guess' the solutions, first for the negative intercept and then for the positive intercept:

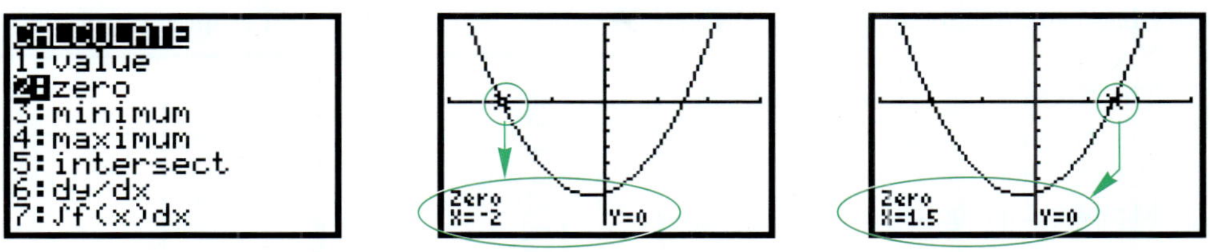

From the screen we see that one zero is $x = -2$ and the other zero is $x = 1.5$ and so the intercepts occur at $(-2, 0)$ and $(1.5, 0)$.
To find the y-intercept, use the CALC menu, select option **1:value** and enter '0' at the prompt.

EXERCISES 7.4.5

1. Select a number of questions from Exercises 7.4.3 and 7.4.4 and use your graphics calculator to sketch their graphs. Make sure that you are able to locate the turning points and x-intercepts (where they exist) for each graph

7.5 USING A GRAPHICS CALCULATOR TO FACTORISE QUADRATICS

Because of the relationship between a quadratic equation, its roots (or zeros) and its factors, it is possible to factorise a quadratic given its graph.

For example, given the graph of $y = x^2 + 2x - 3$, we can see that its roots are $x = 1$ and $x = -3$.

Meaning that the factors are $(x - 1)$ and $(x + 3)$.

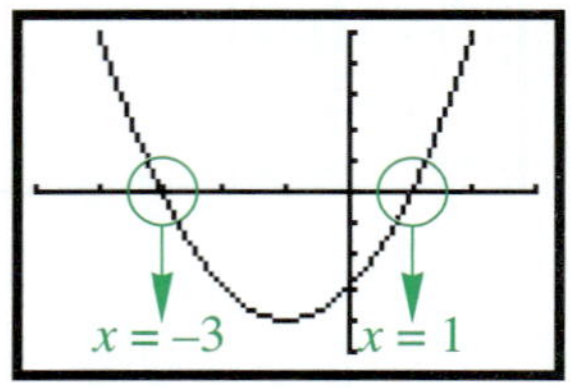

Looking at the solving process in reverse, we can see why this works.

Consider the solution to $x^2 + 2x - 3 = 0$:

The first step is to factorise this expression: $x^2 + 2x - 3 = (x-1)(x+3)$.

Next we use the null factor theorem: $x^2 + 2x - 3 = 0 \Leftrightarrow (x-1)(x+3) = 0$

$\Leftrightarrow x - 1 = 0$ or $x + 3 = 0$

$x = 1$ or $x = -3$.

Now, reversing the process means that we would start with $x = 1$ or $x = -3$,

which leads to the previous step, i.e., $\Leftrightarrow x - 1 = 0$ or $x + 3 = 0$,

which in turn leads to $(x-1)(x+3) = 0$

and finally, back to the original equation $x^2 + 2x - 3 = 0$.

That is, from the zeros, we obtained the equation.

EXAMPLE 7.22

Factorise the quadratic $x^2 - 2x - 8$.

Solution

We start by sketching the graph of $y = x^2 - 2x - 8$ using the graphics calculator. Make sure that an appropriate **WINDOW** setting is selected as well as an appropriate **Xscl**.

The **Xscl** setting here is 1, therefore we see that the graph cuts the x–axis at $x = -2$ and $x = 4$.

Meaning that the factors are

$(x + 2)$ and $(x - 4)$.

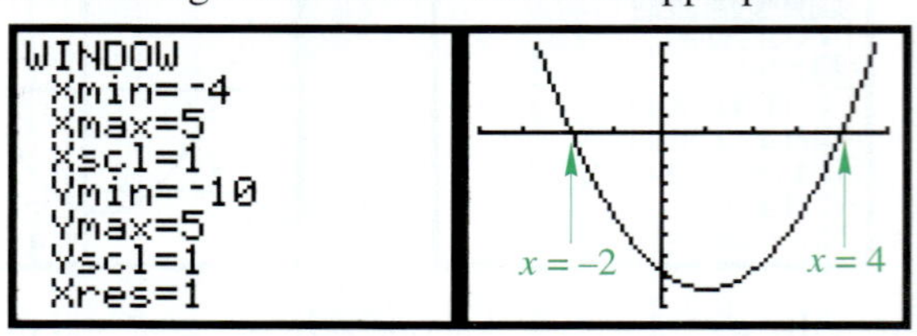

i.e., $x^2 - 2x - 8 = (x+2)(x-4)$.

Having said that this approach works, we need to realise that this might not always be the case.

A word of caution!

1. Using the graphics calculator to factorise a quadratic such as $x^2 - x - 1$ would not be possible. Sketching the graph of $y = x^2 - x - 1$ is easy enough, however, the zeros are irrational numbers – i.e., they aren't numbers that we can recognise.
So, unless you can tell from the graph, that the zeros are $\frac{1+\sqrt{5}}{2}$ and $\frac{1-\sqrt{5}}{2}$, then, coming findings the factors is not possible.

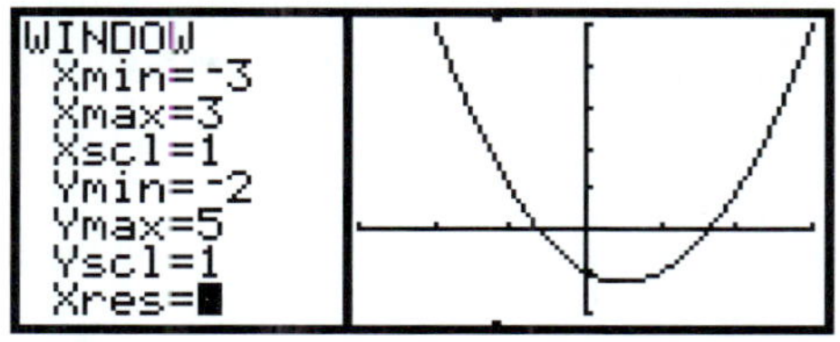

2. Sometimes we need to be extra careful. For example, consider the equation $2x^2 - 4x - 16$. Using the same WINDOW as in Example 7.22, we would have produced the graph shown below:

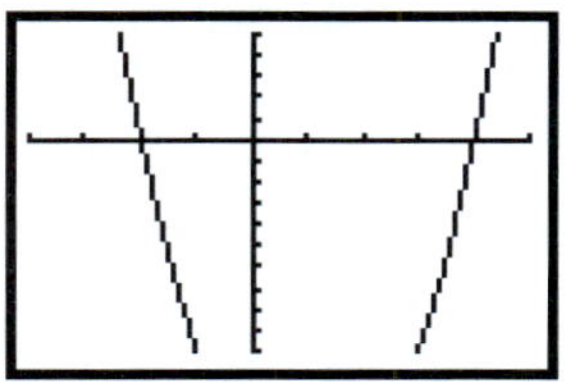

This graph still has intercepts at $x = -2$ and $x = 4$.
It would then be easy to think that the quadratic factorised to $(x+2)(x-4)$.
However, this is clearly not the case.
The first thing that should be done is factorise the '2' and then proceed as in Example 7.22, so that

$$\begin{aligned} 2x^2 - 4x - 16 &= 2(x^2 - 2x - 8) \\ &= 2(x+2)(x-4). \end{aligned}$$

EXERCISES 7.5

1. Use a graphics calculator to factorise the following

(a)	$x^2 - 3x + 2$	(b)	$x^2 - 7x + 6$	(c)	$x^2 - 6x + 8$
(d)	$x^2 - 9x + 20$	(e)	$z^2 - 9z + 18$	(f)	$x^2 - 7x + 10$
(g)	$x^2 - 2x - 3$	(h)	$y^2 + 3y - 10$	(i)	$s^2 - 3s - 10$
(j)	$x^2 - x - 12$	(k)	$y^2 - 5y - 14$	(l)	$r^2 - 4r - 45$
(m)	$2x^2 - 6x + 4$	(n)	$3x^2 + 21x + 30$	(o)	$4s^2 - 12s - 40$
(p)	$3y^2 + 9y - 30$	(q)	$5x^2 + 30x + 40$	(r)	$6x^2 - 12x - 18$

7.6 SYSTEM OF EQUATIONS

7.6.1 LINEAR-QUADRATIC SYSTEM

Consider the equation $x^2 = 2x + 3$. What geometrical interpretation can we give to this equation? As it stands, and without further algebraic manipulation, we have, on the left hand side, a quadratic that represents the parabola with equation $y = x^2$. And, on the right hand side, we have a linear equation that represents the straight line with equation $y = 2x + 3$.

We can sketch both these functions on the same set of axes:

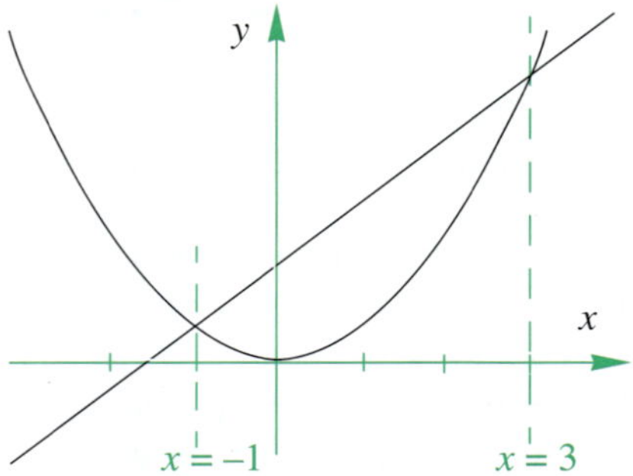

We see that the parabola and the straight line intersect when $x = -1$ and $x = 3$.

So that the solution to the original quadratic equation $x^2 = 2x + 3$ is in fact given by the x-values of where the two graphs intersect.

The solution can be verified as follows:

$$x^2 = 2x + 3 \Leftrightarrow x^2 - 2x - 3 = 0$$
$$\Leftrightarrow (x + 1)(x - 3) = 0$$
$$\therefore x = -1 \text{ or } x = 3$$

This result can be generalised to include many different types of equations. For example, solving the equation $x - 2 = \frac{1}{x}$ would amount to finding the x-values of where the graph of the straight line defined by the equation $y = x - 2$ intersects the hyperbola defined by the equation $y = \frac{1}{x}$. Similarly, solving the equation $5x = 3^x$ amounts to finding the x-values of where the straight line defined by the equation $y = 5x$ and the exponential curve defined by the equation $y = 3^x$ intersect. The graphs below show these graphical representations.

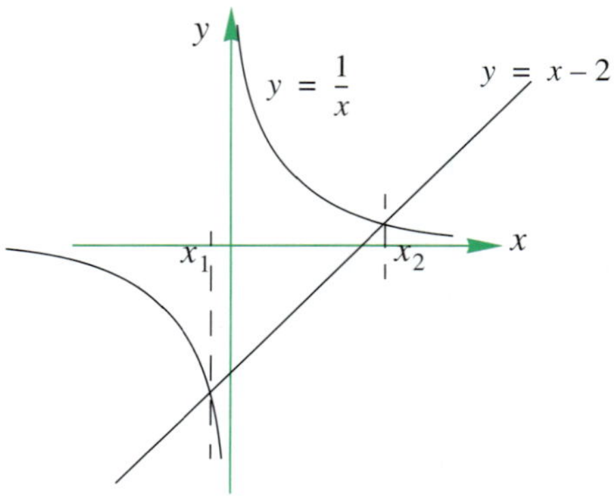

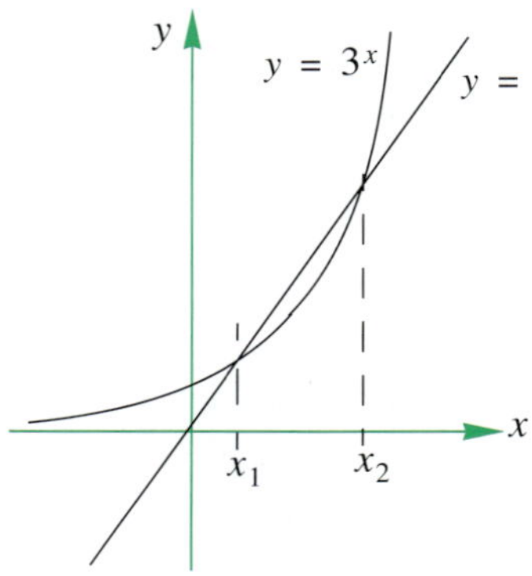

We will look into solutions of unfamiliar equation such as these in Chapter 8. For now we will concentrate on quadratic-linear systems of equations.

Going back to our original problem, i.e., $x^2 = 2x + 3$, we now have more than one way to interpret the solution (as well as how to solve for x).

Established approach: $x^2 = 2x + 3 \Leftrightarrow x^2 - 2x - 3 = 0$

Interpretation: Finding the x-intercepts of the parabola with equation $y = x^2 - 2x - 3$.

Solution method: Factorisation

Alternate approach: $x^2 = 2x + 3$

Interpretation: Finding the x-values of where the two graphs, $y = x^2$ and $y = 2x + 3$ intersect.

Solution method: Read straight off the graphs or an algebraic method, which could be the same as for the established method.

It should also be noted that there is a difference between solving

the system of equations $\begin{matrix} y = x^2 \\ y = 2x + 3 \end{matrix}$ – (1)

and the equation $x^2 = 2x + 3$ – (2)

When solving (1), not only do we need to provide the x-values, but we also need to provide the y-values. So that once we have established that $x = 3$ and $x = -1$, we would then need to determine the corresponding y-values. That is, when $x = 3$, $y = 3^2 = 9$ and when $x = -1$, $y = (-1)^2 = 1$. Meaning that the graphs intersect at the points (3, 9) and (–1, 1).

However, when solving (2), our solutions are made up entirely of x-values. i.e., $x = 3$ & $x = -1$.

EXAMPLE 7.23

Solve the system of equations $\begin{matrix} y = x^2 - 4x - 5 \\ y = 3 - 2x \end{matrix}$.

Solution

As the original question states that we need to solve the system of equations we will need to provide both x- and y-values.

We set up our system of equations using the following labels:

$y = x^2 - 4x - 5$ – (1)

$y = 3 - 2x$ – (2)

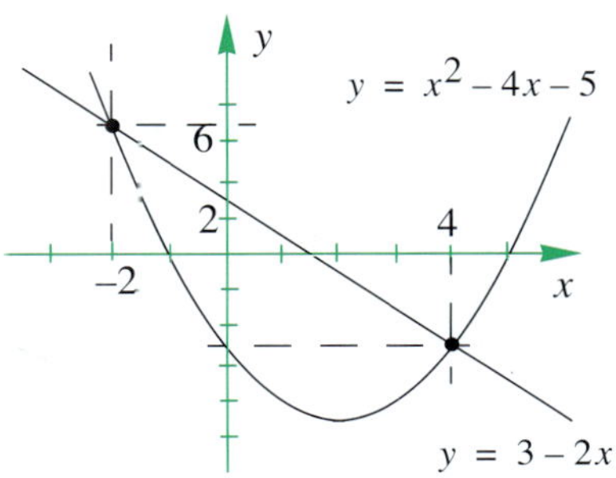

Equating (1) and (2) we have:

$$x^2 - 4x - 5 = 3 - 2x$$

$$\Leftrightarrow x^2 - 2x - 8 = 0$$

$$\Leftrightarrow (x-4)(x+2) = 0$$

Therefore, $x = 4$ or $x = -2$

When $x = 4$, $y = 3 - 2 \times 4 = -5$.

When $x = -2$, $y = 3 - 2 \times -2 = 7$.

That is, the line $y = 3 - 2x$ and the parabola $y = x^2 - 4x - 5$ intersect at (4, –5) and (–2, 7).

EXERCISES 7.6.1

1. (a) On the same set of axes, draw accurate graphs of $y = 5 - x$ and $y = 5x - x^2$.
(b) Using part (a), give a geometrical interpretation of the equation $5 - x = 5x - x^2$.
(c) Solve $5 - x = 5x - x^2$ using part (a).

2. (a) On the same set of axes, draw accurate graphs of $y = x + 6$ and $y = x^2$.
(b) Using part (a), give a geometrical interpretation of the equation $x^2 = x + 6$.
(c) Solve $x^2 = x + 6$ using part (a).

3. (a) On the same set of axes, draw accurate graphs of $y = 5$ and $y = x^2 + 1$.
(b) Using part (a), give a geometrical interpretation of the equation $x^2 + 1 = 5$.
(c) Solve $x^2 + 1 = 5$ using part (a).

4. (a) On the same set of axes, draw accurate graphs of $y = x + 1$ and $y = x^2 - x - 2$.
(b) Interpret the equation $x + 1 = x^2 - x - 2$ using part (a).
(c) Solve $x + 1 = x^2 - x - 2$ using part (a).

5. (a) On the same set of axes, draw accurate graphs of $y = 2 - x$ and $y = x^2 - x - 2$.
(b) Interpret the equation $2 - x = x^2 - x - 2$ using part (a).
(c) Solve $2 - x = x^2 - x - 2$ using part (a).

6. Solve the system of equations

(a) $\begin{array}{l} y = x^2 \\ y = x + 6 \end{array}$ (b) $\begin{array}{l} y = x^2 + 1 \\ y = 5 \end{array}$ (c) $\begin{array}{l} y = -2x + 4 \\ y = x^2 - 4x + 5 \end{array}$

(d) $\begin{array}{l} y = x^2 - 4 \\ y = x + 2 \end{array}$ (e) $\begin{array}{l} y = -x + 4 \\ y = x^2 - 2 \end{array}$ (f) $\begin{array}{l} y = 4x - 3 \\ y = x^2 - 2x + 6 \end{array}$

7. Solve the system of equations $\begin{array}{l} y = 2ax + a^2 \\ y = x^2 + 3ax - a^2 \end{array}$, where a is a real number (except zero).

7.6.2 QUADRATIC-QUADRATIC SYSTEM

In the same way that we solved linear-quadratic systems of equations, we can also solve for quadratic-quadratic systems. Equally, we can also provide the same type of geometrical interpretation.

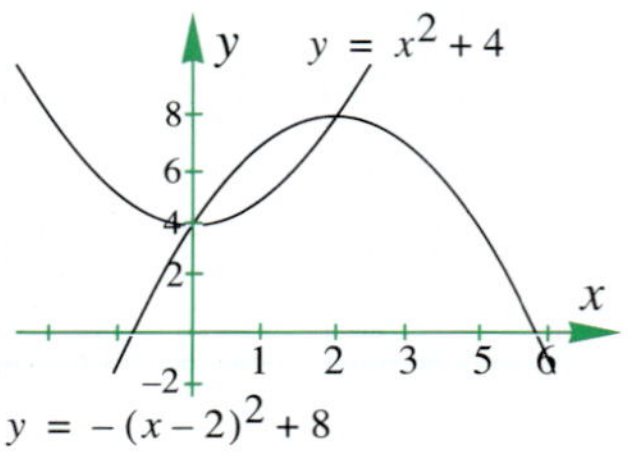

For example, the equation $x^2 + 4 = -(x-2)^2 + 8$ can be interpreted geometrically as representing the x-value of where the parabola with equation $y = x^2 + 4$ meets the parabola with equation $y = -(x-2)^2 + 8$. As shown in the diagram alongside. From where we obtain the solutions, $x = 0$ and $x = 2$.
The algebraic method of solving quadratic-quadratic systems is

the same as that for solving linear-quadratic systems. This approach can be summarised as follows:

Step 1:	Set up (i.e., label) the two equations by writing them both as $y = \ldots$
Step 2:	Equate the two expressions.
Step 3:	Transpose equation so that one side is left with 0. (This will produce a quadratic on the other side.)
Step 4:	Solve the equation in Step 3.
Step 5:	Once you have the x-values, substitute them back into either of the original equations and find the corresponding y-values. This will provide you with solution pairs.

We now illustrate this process.

EXAMPLE 7.24

Solve the system of equations $\begin{aligned} y &= x^2 - 4x + 7 \\ y &= 2x^2 + 2x \end{aligned}$

Solution

Label the equations as follows: $y = x^2 - 4x + 7 \quad -(1)$ **Step 1**

$y = 2x^2 + 2x \quad -(2)$

As they are already in the form $y = \ldots$ we equate (1) and (2):

That is, $2x^2 + 2x = x^2 - 4x + 7$ **Step 2**

$\Leftrightarrow x^2 + 6x - 7 = 0$ **Step 3**

$\Leftrightarrow (x + 7)(x - 1) = 0$ **Step 4**

Therefore, we have that $x = -7$ or $x = 1$

Substituting back into (2) we have: **Step 5**

$x = -7,\ y = 2(-7)^2 + 2(-7) = 84$.

$x = 1,\ y = 2(1)^2 + 2(1) = 4$

Therefore, the solution pairs are: $(-7, 84)$ and $(1, 4)$.

We can obtain or confirm the solution above using a graphics calculator. However, we need to be careful on the window selected.

For the TI–83 a good choice would be $-10 \le x \le 10$ and $-10 \le y \le 120$. Solutions can then be obtained using the **CALC** menu and option **5:intersect** :

Left intersection:

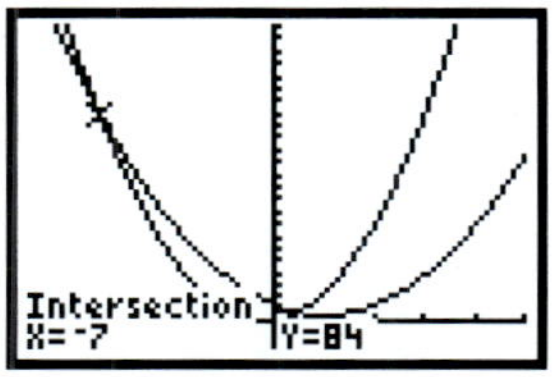

Right intersection:

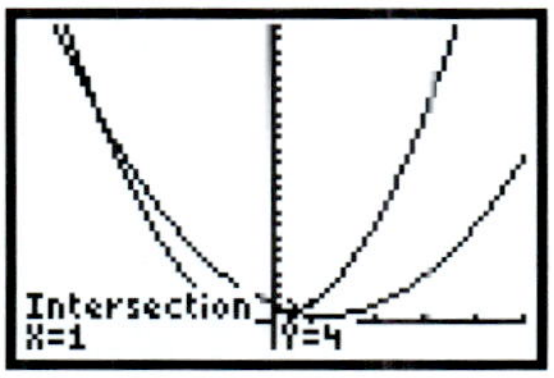

EXERCISES 7.6.2

1. (a) On the same set of axes, accurately draw the graphs of the curves whose equations are $y = x^2$ and $y = 8 - x^2$.
(b) Hence, state the coordinates of the points of intersection.

2. (a) On the same set of axes, accurately draw the graphs of the curves whose equations are $y = 2x^2$ and $y = 3 - x^2$.
(b) Hence, state the coordinates of the points of intersection.

3. (a) On the same set of axes, accurately draw the graphs of the curves whose equations are $y = x(x-2)$ and $y = (2-x)(2x+3)$.
(b) Hence, state the coordinates of the points of intersection.

4. Solve the following systems of equations

(a) $y = 2x^2 + x - 4$, $y = x^2 + 2x + 2$ (b) $y = 2x^2 + x + 3$, $y = x^2 + 4x + 1$ (c) $y = 2x^2 + 4x + 3$, $y = 3x^2 + 4x + 2$

(d) $y = -x^2 - x + 3$, $y = x^2 + 2x + 1$ (e) $y = 2x^2 + 5x - 6$, $y = 3x^2 - 4x + 14$ (f) $y = 4x^2 + 3x + 2$, $y = 3x^2 - 2x - 4$

7.7 MISCELLANEOUS EXERCISES

1. Factorise the following

(a) $4xy - 12y^2$ (b) $6a^2b + 4ab^2$ (c) $3uv^3 - 9u^2v^2 - 12u^3v$

2. The graph of $y = x^2 - 19x - 42$ cuts the x-axis at M and N. Find the distance from M to N.

3. Find the coordinates of the turning point of the parabola $y = (x+2)(x-6)$.

4. How many times will the graphs of $y = x^2 + 5x + 2$ and $y = x - 3$ meet?

5. Give the equation of the axis of symmetry for the parabola $y = x^2 - 2x$. Hence, determine the coordinates of its turning point.

6. If $3x + 2y = 1$ and $x - y = 7$, find the value of $x^2 + y^2$.

7. If $(x^2 - 5)(x^2 - 9) = 0$ determine
(a) the number of rational solutions.
(b) the number of real solutions.

8. The quadratic equation $x^2 + 4x - 21 = 0$ has 2 solutions. Find the product of the two solutions.

9. Find the larger solution to $x^2 + 17x + 30 = 0$.

10. Factorise the expression $3x - 12 + xy - 4y$

11. If $xy = 5$ and $\frac{1}{x^2} + \frac{1}{y^2} = 3$, find the value of $x^2 + y^2$.

12. How many real solutions are there to the quadratic equation $3x^2 - 5x - 2 = 0$?

13. Find the sum of the solutions to the quadratic equation $0.1x^2 - 7.2x + 4.8 = 0$.

14. Find the solution(s) to the equation $x^2 + 24 = 10x$.

15. For what value of x will the number N, where $N = (2x + 3)^2 + 11$, be a minimum. What is the minimum value of N?

16. Give the solution sets of the following equations:

(a) $(x - 3)^2 = x^2$

(b) $(x - 3)^2 = -6x$

(c) $(x - 3)^2 = 9$

(d) $(x - 3)^2 = x^2 - 6x + 9$

17. If $f(x) = ax^2 - 3x + 4$ and $f(2) = 6$, find the value of a.

18. Solve for x: (a) $x^2 + 5x = 24$ (b) $2x^2 - 5x - 4 = 0$

19. The line $y = x - 1$ and the curve $y = x^2 + bx + c$ intersect at $(-2, -3)$ and $(3, 2)$. Find the values of b and c.

20. Find the minimum value of $y = x^2 - 2x + 3$.

21. (a) Factorise the expression $x^2 + 2x - 8$.

(b) Determine the integer values for which $x^2 + 2x - 8$ is less than 0.

22. (a) Factorise the quadratics (i) $x^2 + 6x + 5$ (ii) $x^2 - 5x - 6$

(b) Hence sketch the graphs of (i) $y = x^2 + 6x + 5$ (ii) $y = x^2 - 5x - 6$

23. Factorise each of the following

(a) $x^2 - x - 12$ (b) $6x^2 - 13x + 6$ (c) $(x - 1)^2 - y^2$

24. Sketch the graphs of (a) $y = -x^2 + 9$ (b) $y = -x^2 + 6x - 9$

25. The graph of $y = x^2 - ax + b$ is shown.

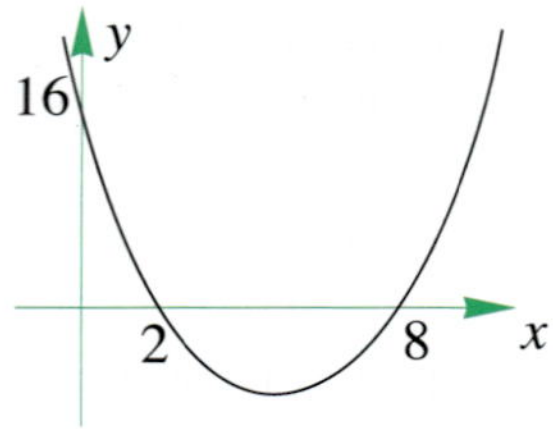

(a) Determine the values of a and b.

(b) Hence, factorise the quadratic $x^2 - ax + b$

26. Write the quadratic $y = x^2 - 3x + 7$ in turning point form and hence find the coordinates of the vertex of the parabola.

27. The graph shows a parabola with equation $y = (x - a)^2 + b$.

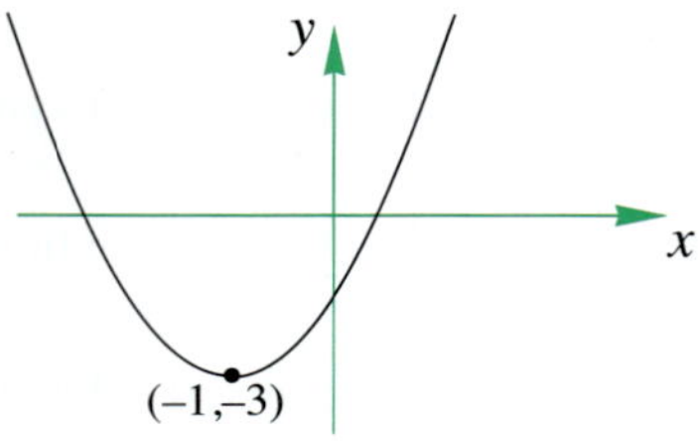

Find the values of a and b.

28. Solve $x(2x + 5) = 3$

29. Find the coordinates of the axis intercepts of the parabola $y = -x^2 + 2x + 7$

30. Find the values of b for which the quadratic equation $x^2 + bx + 9 = 0$ has two real solutions.

31. The graph of $y = ax^2 + bx + c$ is shown.

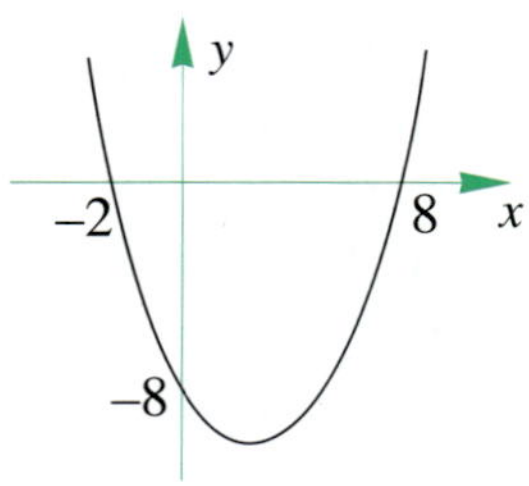

(a) Determine the values of a and b and c.

(b) Hence, factorise the quadratic $ax^2 + bx + c$.

(c) i. Find the range of integers for which $ax^2 + bx + c$ is less than 0.

ii. Find the minimum value of $ax^2 + bx + c$

32. Solve the following systems of equations

(a) $y = 2x^2 + 6x - 2$
$y = -x^2 + 2x + 2$

(b) $y = x^2 - 3x + 2$
$y = 4 - x^2$

(c) $y = x^2 - 2x - 1$
$y = -x + 1$

33. (a) On the same set of axes, sketch the graphs of functions $f(x) = (x + 2)^2 - 2$ and $g(x) = -(x + 2)^2 + 2$. Label the intercepts with the x-axis as X and Y.

(b) Let A and B be the points whose coordinates are the turning points of the functions $f(x)$ and $g(x)$ respectively.

i. How far apart are the points A and B?

ii. A quadrilateral is formed by joining the vertices whose coordinates are those of the points A, B, X and Y. Find the area of this quadrilateral.

34. Find the product of the zeros of the function $f(x) = x^2 - (m + n)x + mn$.

7.8 GRADED REVISION EXERCISES

LEVEL 1

1. Which of the following are quadratic equations?

(a) $3x + x - 7 = 0$ (b) $5x^2 - 5x + 1 = 0$ (c) $16 - y^2 = 0$

2. Factorise the following.

(a) $x^2 + x$ (b) $2x - x^2$ (c) $4ax + ax^2$

(d) $6y^2 + 12y$ (e) $9b^2 - 3b$ (f) $12ay^2 + 4ay$

3. Factorise the following.

(a) $2(a+b) + a(a+b)$ (b) $z(z-1) - 3(z-1)$ (c) $x(x+1) - 2(x+1)$

LEVEL 2

1. Factorise the following.

(a) $x^2 + 3x - 4$ (b) $y^2 - 6y + 5$ (c) $x^2 + 5x - 14$

(d) $a^2 - 2a + 1$ (e) $a^2 - 2a - 3$ (f) $y^2 - 16y + 63$

2. Factorise the following.

(a) $2x^2 + 5x - 3$ (b) $3x^2 + 4x + 1$ (c) $2y^2 + y - 3$

(d) $6z^2 + 5z + 1$ (e) $12x^2 + 25x + 12$ (f) $5x^2 - 14x - 3$

3. Solve the following equations:

i. $x^2 + 3x - 4 = 0$ ii. $x^2 - 6x + 5 = 0$ iii. $x^2 + 5x + 6 = 0$

iv. $x^2 - 2x + 1 = 0$ v. $x^2 + 5x - 14 = 0$ vi. $x^2 - 16x + 63 = 0$

4. Use the quadratic formula to solve these equations (give answers in exact form):

i. $x^2 + x - 7 = 0$ ii. $2x^2 + 3x - 4 = 0$ iii. $-x^2 - 4x + 9 = 0$

iv. $5x^2 - 3x - 4 = 0$ v. $2x^2 - 5x + 7 = 0$ vi. $3x^2 + 5x - 9 = 0$

LEVEL 3

1. Find the turning points of the graphs of the following functions:

i. $y = x^2 - 6x + 13$ ii. $y = x^2 + 2x + 4$ iii. $y = 2x^2 - 4x - 1$

iv. $y = 3x^2 - 6x + 4$ v. $y = x^2 - x + \frac{13}{4}$ vi. $y = 5x^2 - 10x + 5$

2. Give equations for the functions whose graphs are shown:

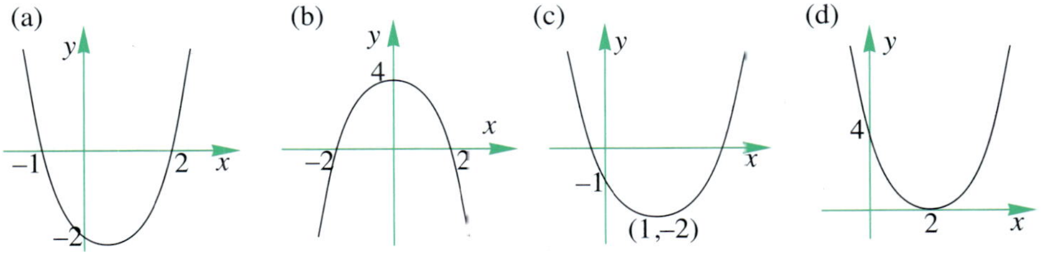

3. By making use of completing the square, find the roots of the following quadratic equations.

(a) $x^2 + x - 4 = 0$ (b) $2x^2 - 2x - 7 = 0$

(c) $x^2 - 6x - 1 = 0$ (d) $3x^2 - 12x + 3 = 0$

(e) $2x^2 - x - 2 = 0$ (f) $x^2 - 6x + 2 = 0$

4. Sketch the graph of the following functions. On your graphs indicate clearly the coordinates of the turning point and the intercepts with the y–axis.

(a) $y = (x-1)^2 + 3$ (b) $y = (x+2)^2 - 1$

(c) $y = 4 - (x-2)^2$ (d) $y = 2(x+1)^2 - 4$

(e) $y = 9 - (x+3)^2$ (f) $y = \frac{1}{2}(x+4)^2 + 2$

5. Sketch the graph of the following functions. On each graph indicate clearly all intercepts with the axes.

(a) $y = x^2 - 2x$ (b) $y = x^2 - 4x - 5$

(c) $y = 9 - x^2$ (d) $y = -x^2 + 6x - 5$

(e) $y = x^2 - 6x + 12$ (f) $y = x^2 - 6x + 9$

6. Solve the system of equations

(a) $y = 2x^2 + 5x - 10$, $y = x^2 + 3x + 5$ (b) $y = x^2 + 4x - 6$, $y = 2(x+1)$

7. The factors of the function $f(x) = ax^2 - 7x + c$ are $(2x+1)(x-k)$.

(a) Determine the values of a, c and k.

(b) Find the x-intercepts for the graph of $y = f(x)$.

(c) Find the equation of the axis of symmetry.

(d) Find the minimum value of $f(x)$.

LEVEL 4

1. Find the coordinates of the turning point on the graph of the function $y = x^2 + kx + 4$ in terms of the parameter k, where k is a real number.

2. (a) Show that $N = a^2 - 3a + 3$ is positive for all real values of a.

(b) Find the minimum value of N.

3. (a) On the same set of axes, sketch the graphs of $y = x^2 - 3x + 2$ and $y = 4 - x^2$, showing all intercepts with the axes.

(b) Solve the quadratic equation $2x^2 - 3x - 2 = 0$.

(c) Hence, find the coordinates of the points of intersection of the curves in (a).

4. (a) Solve for x: $\frac{x}{a} + \frac{1}{x} + \left(1 + \frac{1}{a}\right) = 0$.

(b) Hence, solve the quadratic equation $0.5x^2 + 1 + 1.5x = 0$.

7.9 TOPIC TEST

1. Solve the equation $x^2 - 6x - 7 = 0$.

Hence, sketch the graph of $y = x^2 - 6x - 7$, showing all axial intercepts.

[4 marks]

2. Factorise the following.

(a) $x^2 - 12x + 32$ (b) $x^2 - 12x + 36$ (c) $x^2 - 12x + 30$

[6 marks]

3. Find the values of x such that $5 - 3x - 3x^2 = 0$, correct to three significant figures.

[2 marks]

4. The diagram shows a parabola.

Find an appropriate equation for this graph.
Hence find the coordinates of the maximum point.

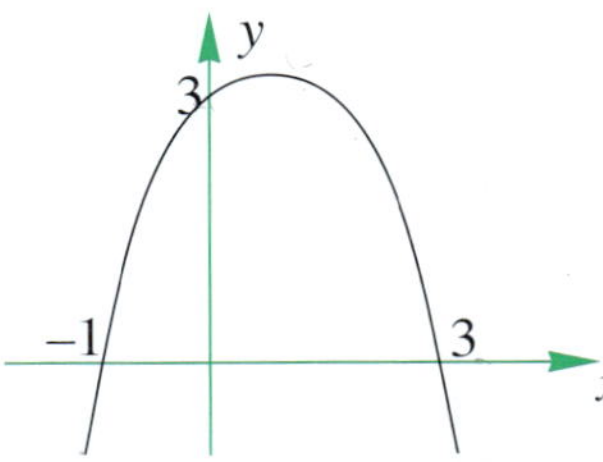

[5 marks]

5. Solve the equation $x = 1 + \frac{6}{x}$.

[3 marks]

6. The parabola $y = x^2 + 6x - 9$ passes through the point $(k, 2k^2)$. Find the value(s) of k.

[4 marks]

7. Parts of the graphs of $y = -x^2 + 14x - 40$ and $y = 8x - x^2$ are shown below.

(a) Factorise i. $-x^2 + 14x - 40$

ii. $8x - x^2$

Hence, find the coordinates of A, B and C.

(b) Find the x-coordinate of the point X.

(c) Find the area of the triangle BCX.

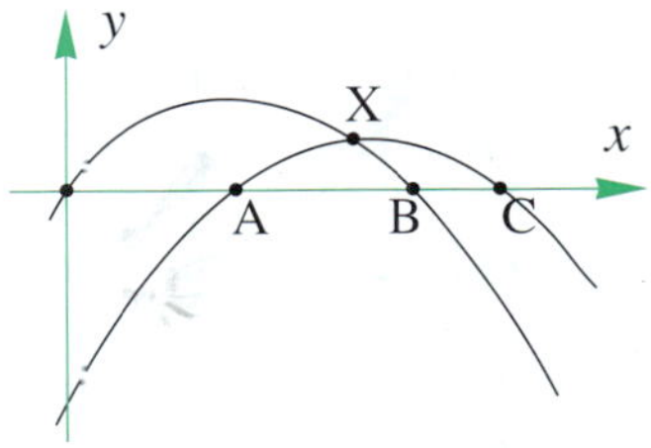

[10 marks]

8. Consider the diagram shown below.

(a) Find the equation of the straight line

(b) Find the equation of the parabola.

(c) Find the coordinates of the point A.

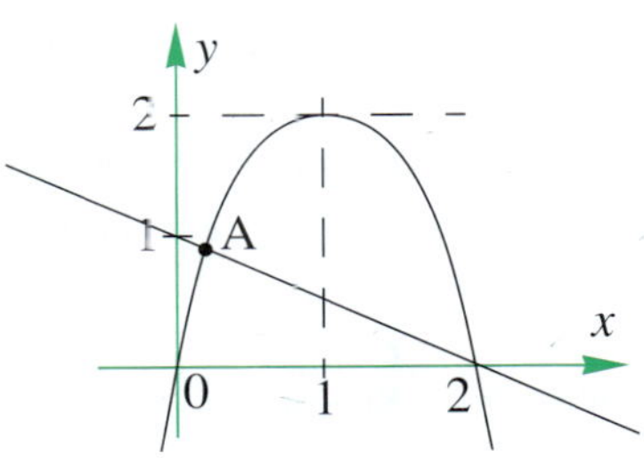

[8 marks]
Total 42 marks

7.10 PROJECT SUGGESTIONS

Look at the art of origami - i.e., paper folding.

Can you construct mathematical rules that involve quadratic equations for the folding process involved in some of the paper folding sequences.

For example, using a sheet of A4 paper, which has a width of a cm and height $a\sqrt{2}$ cm, fold it as shown in the diagram, so that the bottom right-hand corner touches the left-hand side at point X, x cm from the bottom of the page.

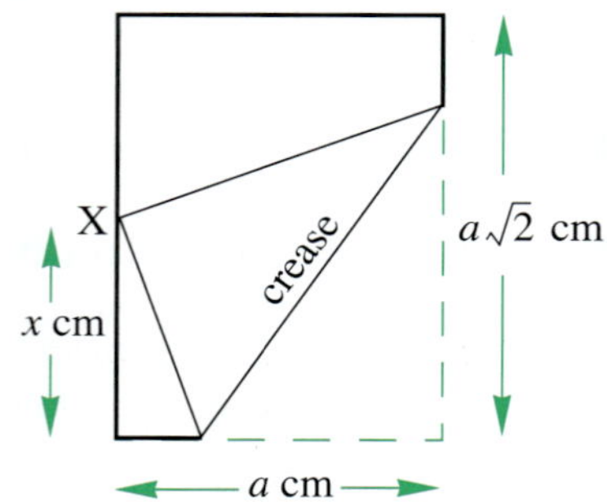

Investigate how the length of the crease varies as the point X varies its position along the edge.

What other properties can you investigate?

How can the graph of $y = x^2$ be use to multiply numbers?
Will it work for all numbers? If not, which ones and why?

Consider the following problem:

Two students were discussing equations they were expected to be able to solve in mathematics. One student argued that as most quadratic equations do not have real solutions, students could only be expected to learn a few examples off by heart. The other student said this must be wrong because most quadratic equations do have real solutions so it is important to learn the formula for the solution of a quadratic equation.

Start by determining the proportion of cases in which the equation $Ax^2 + Bx + C = 0$ has a real solution where the numbers A, B and C are restricted in various ways, as follows.

a. $A = 1$ and B and C are chosen from the list $(0, 0.1, 0.2, 0.3, 0.4, 0.5, 0.6, 0.7, 0.8, 0.9)$.
b. $A = 1$ and B and C are chosen from real numbers between 0 and 1.
c. $A = 1$ and B and C are chosen from real numbers between 0 and 5.
d. $A = 1$ and B and C are chosen from real numbers greater than or equal to 0.

Explain how you interpret the phrase 'proportion of cases' in (b), (c) and (d).

Describe how you could go about finding the proportion of cases for which the equation $Ax^2 + Bx + C = 0$ has a real solution when A, B and C are chosen from real numbers between 0 and 1.

What would you say to these two students?

8.0 RELATIONSHIPS

The classic car shown in the photograph has registration plate 'number' 5604 KV. There is a direct relationship between the car and its 'number'. This is a useful relationship that is stored as a (computer) list by the regulating authority. The complete list of this relation includes all the cars in the region (the *domain*), and the list of actual numbers used (the *range*). The relation is useful in the sense that it enables stolen cars to be returned to their owners and parking fines to be correctly addressed.

A second example is the relationship between each person and their blood group. The domain is the set of all people. The range is the set of all blood groups. There are several ways of grouping blood. A common method is the ABO system which gives a domain of type A, type B, type O and type AB. Do you know which group you belong to? Does it matter?

In both these examples there should be no doubt as to which domain element is related to which range element. Given a car, there should be no doubt as to its registration 'number' and given patients there should be no doubt as to their blood groups. This is an important feature of relations - they need to be well defined. There is however a difference between these two relations. The car example is 'one to one' in the sense that each car has a unique registration 'number' and each number belongs to a single car. The blood group example is 'many to one' in the sense that many people have one blood group. Such distinctions are known as 'classifications'.

Note that the domain and range in these examples are not numbers. Many examples of relations that you meet in maths courses relate sets of numbers to other sets of numbers. The idea of a relation is, as you have seen, much more general. You have almost certainly been using 'relations' without calling them by that name for much of your life. This chapter is about the formal definition of relations and their classifications.

When you read the chapter title 'relations' you may have thought of 'cousins', 'aunts' etc. Do these form relations in the mathematical sense? If you look at the relation 'are my parents', to how many people are you related? What is the answer for the relation 'are my great-great grand parents'? What methods are available for 'graphing' family relationships?

8.1 RELATIONS

So far we have looked at linear and quadratic functions. In this chapter we formalise the notion of a function and take an indepth look at terms like domain, range. We will also look at relations other than those you have already encountered. Some relations that you come across in this chapter might be familiar to you while others may not. When a sketch of a relation that is not familiar to you is required you should make use of your graphics calculator as a guide.

8.1.1 RELATIONS

Consider the relationship between the weight of five students and their ages as shown below.

We can represent this information as a **set of ordered pairs**. An age of 10 years would correspond to a weight of 31kg. An age of 16 years would correspond to a weight of 53kg and so on.

Age (years)	Weight (kg)
10	31
12	36
14	48
16	53
18	65

This type of information represents a **relation** between two sets of data. This information could then be represented as a set of ordered pairs,

$$\{(10, 31), (12, 36), (14, 48), (16, 53), (18, 65)\}$$

The **set of all first elements** of the ordered pair is called the **domain** of the relation and is referred to as the **independent variable**. The **set of all second elements** is called the **range** and is referred to as the **dependent variable**.

For the above example, the domain $= \{10, 12, 14, 16, 18\}$
and the range $= \{31, 36, 48, 53, 65\}$.

Notice that $(10, 31)$ and $(31,10)$ are not the same! This is because the ordered pair $(10, 31)$ provides the correct relation between age and weight, i.e., at age 10 years the weight of the student is 31kg. On the other hand, the ordered pair $(31,10)$ would be informing us that at age 31 years the weight of the student is 10kg!

Summary:

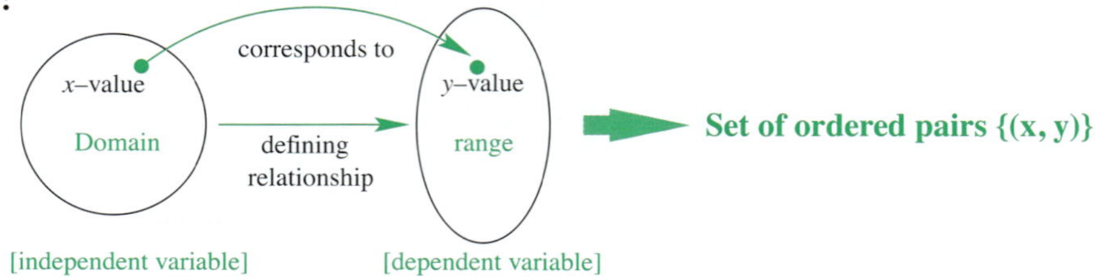

EXAMPLE 8.1

Determine the domain and range for each of the following relations:

(a) $\{(0,0), (1,1), (2,4), (3,9), (4,16), (5,25)\}$.

(b) $\{(-3,4), (-1,0), (2,-2), (-2,2)\}$.

Solution

(a) The domain is the set of all first elements, i.e., $\{0, 1, 2, 3, 4, 5\}$.
The range is the set of all second elements, i.e., $\{0, 1, 4, 9, 16, 25\}$.

(b) The domain is the set of all first elements, i.e., $\{-3, -1, 2, -2\}$.
The range is the set of all second elements, i.e., $\{4, 0, -2, 2\}$

The letter 'X' is often used to denote the domain and the letter 'Y' to denote the range. For part (a) this means that we could write X = $\{0, 1, 2, 3, 4, 5\}$ and Y = $\{0, 1, 4, 9, 16, 25\}$ and for (b) we could write X = $\{-3, -1, 2, -2\}$ and Y = $\{4, 0, -2, 2\}$.
This is a convention, nothing more.

Rather than giving a verbal description of how the independent variable and the dependent variable are related, it is much clearer to provide a **mathematical rule** that shows how the elements in the range relate to the elements in the domain.

EXAMPLE 8.2

A relation is defined by the rule $y = x + 2$, where $x \in \{0, 1, 2, 3, 4\}$.

(a) Determine the range of this relation.
(b) Express this relation as a set of ordered pairs.

Solution

(a) The domain of this relation is given by the x–values, i.e., $\{0, 1, 2, 3, 4\}$. We can therefore substitute these values into the equation $y = x + 2$ and determine their corresponding y–values. This will provide the range of the relation.
Substituting we have,

$$x = 0 \Rightarrow y = 0 + 2 = 2$$
$$x = 1 \Rightarrow y = 1 + 2 = 3$$
$$x = 2 \Rightarrow y = 2 + 2 = 4 \text{, and so on.}$$

This produces a set of y–values $\{2, 3, 4, 5, 6\}$ that defines the range.

(b) The set of ordered pairs would be $\{(0, 2), (1, 3), (2, 4), (3, 5), (4, 6)\}$.

Notice that we can describe the set of ordered pairs more formally as:

$$\{(x,y) : y = x + 2, x \in \{0, 1, 2, 3, 4\}\}.$$

which is read as;
"The set of ordered pairs x and y, such that $y = x + 2$, where x is an element of the set of values $\{0, 1, 2, 3, 4\}$."

The information in Example 5.2 can be displayed in different ways. Both those shown below are visual displays – they show the mappings in different ways:

Mapping diagram

The mapping diagram below displays which y–value corresponds to a given x–value.

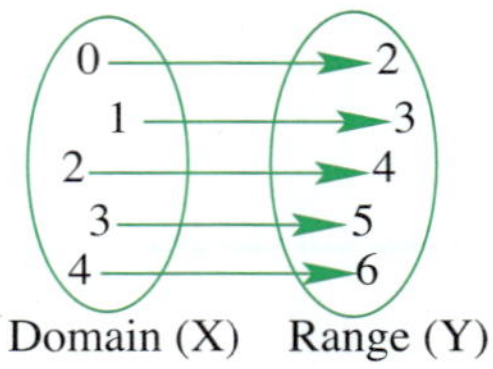

However it is often not easy to see the 'pattern' between the variables with this style of diagram.

Cartesian Plane

The Cartesian plane is made up of a horizontal axis (independent variable, X) and a vertical axis (dependent variable,Y).

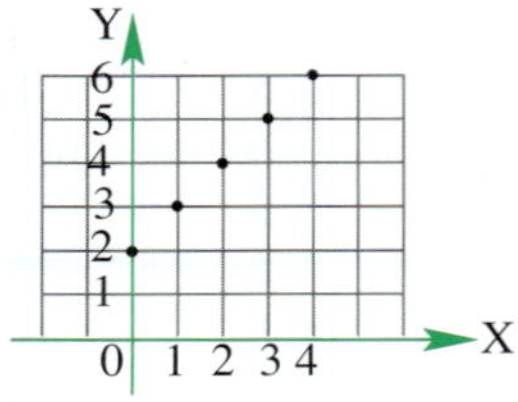

We plot the points on the grid, so that (3, 5) is 3 units to the right and 5 units up.

Notice that in the mapping diagram that uses the Cartesian plane, we have not joined the points together in a straight line. This is because the domain specifies that the only values of x that can be used must be from the set $\{0, 1, 2, 3, 4\}$, and so a value such as $x = 2.4$ cannot be used.

Both these visual representations are useful in displaying which values in the domain generate a given value in the range. However, the Cartesian plane more readily gives a quick overview of what the underlying relationship between the two variables is. It is very easy (and quick) to see that as the x–values increase, so too do the y–values. We can do this by simply looking at the points on the graph and observing the 'trend' without really concerning ourselves with what the actual values are.

We now provide a formal definition of the Cartesian plane and a relation.

8.1.2 THE CARTESIAN PLANE

The **Cartesian plane** is formed by constructing two real lines that intersect at a right-angle where the point of intersection of these two lines becomes the **origin**. The horizontal real line is usually referred to as the x–axis and the vertical real line is usually called the y-axis. This also implies that the plane has been divided into four **quadrants**. Each point on this plane is represented by an ordered pair (x, y) where x and y are real numbers and are the **coordinates** of the point.

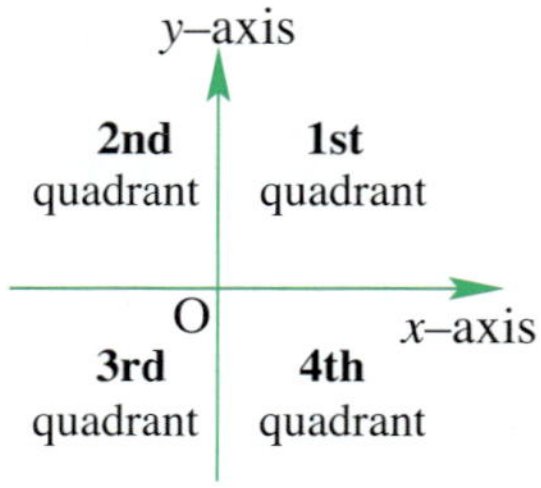

Relation

A **relation** is any subset of the Cartesian plane and can be represented by a set of ordered pairs $\{(x, y)\}$.

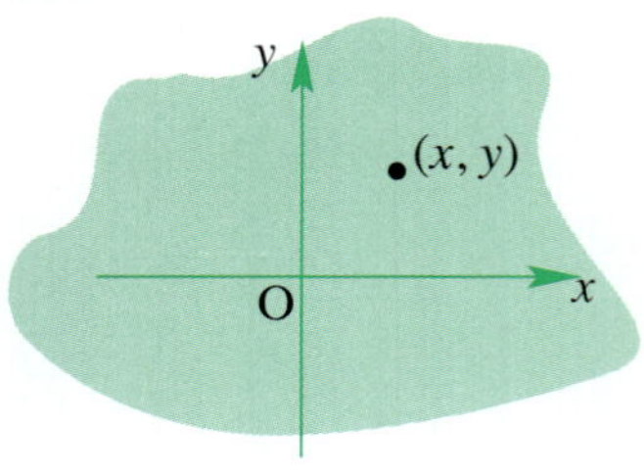

We now consider some further examples.

EXAMPLE 8.3

Determine the domain and range of the following relations

(a) $y = 2 + x, x \geq 0$ (b) $y = 4 - x, -1 \leq x \leq 2$ (c) $y = 1 - x^2, x \geq 0$

Solution

Note: When finding the range of a relation, it is always a good idea to sketch its graph.

(a) The equation $y = 2 + x, x \geq 0$ represents a straight line with the restriction that $x \geq 0$. So in this case, the domain is $[0,\infty[$.
From the graph, the range is given by $[2, \infty[$.

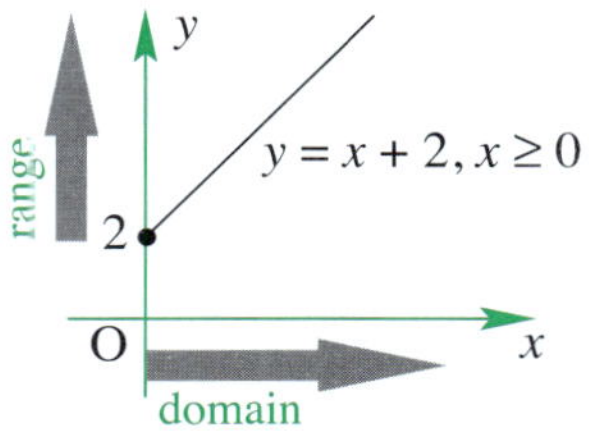

(b) The equation $y = 4 - x, -1 \leq x \leq 2$ represents a straight line with the restriction that $-1 \leq x \leq 2$. So in this case, the domain is $[-1, 2]$.
when $x = -1, y = 5$ and when $x = 2, y = 2$
From the graph, the range is given by $[2,5]$.

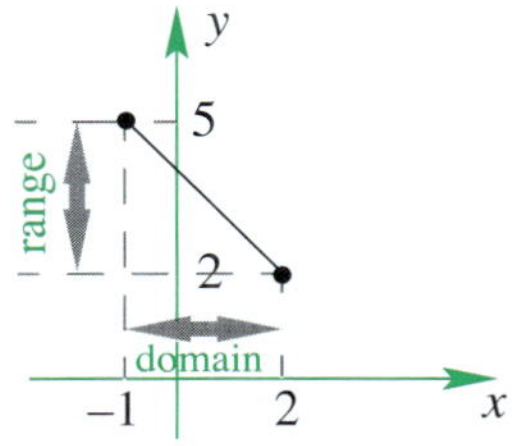

(c) For this relation the domain is specified as $\{x : x \geq 0\}$ or simply $[0, \infty)$. So we can only sketch the graph of $y = 1 - x^2$, for these values of x. Using the graph we can see that the range is $\{y : y \leq 1\}$ or simply $]-\infty, 1]$.

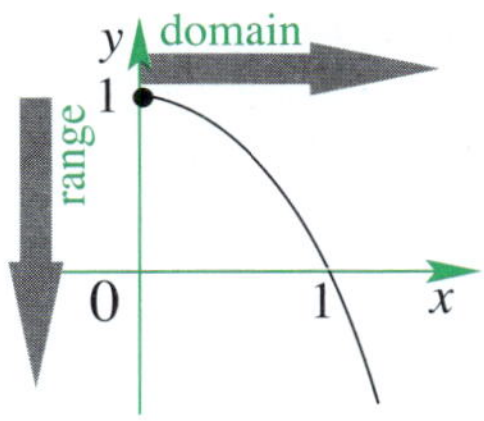

8.1.3 IMPLIED DOMAIN

So far we have looked at examples for which a domain has been specified. For example, if asked to find the range of the relation $y = 1 + x^2, x \geq 3$. Then, after sketching its graph, we would determine its range to be $[10, \infty[$. However, what if we only wanted to know the range of the relation $y = 1 + x^2$? In this case, because we have not been provided with any restriction on the x–values, we will need to assume that we can **use the largest possible set of x–values for which the relation is defined** – this domain is known as the **implied domain** (or **maximal domain**) – in this case that would be the real number set, $\mathbb{R}$. Then, after sketching the graph of $y = 1 + x^2$ for all real values of x we would have a range defined by $[1, \infty[$.

EXAMPLE 8.4

Determine the domain and range of the following relations

(a) $y = \sqrt{x-3}$ (b) $y = \dfrac{2}{\sqrt{x-3}}$ (c) $y = \dfrac{3}{2-x}$

Solution

(a) Using the TI–83 to sketch the graph of $y = \sqrt{x-3}$ (i.e., the square root relation) we observe that its domain is $[3, \infty[$.

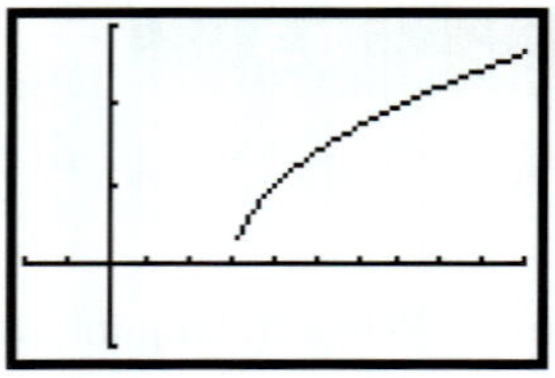

Now, let's take a closer look at why that is the case. Because we are dealing with an expression that involves a square root, then, the term 'inside' the square root must be greater than or equal to zero (as we cannot take the square root of a negative number).

So, we must have that $x - 3 \geq 0 \Leftrightarrow x \geq 3$. Therefore, the implied domain is $\{x: x \geq 3\}$. From the graph, the range can be seen to be $[0, \infty[$.

It should be noted that the TI–83 uses the implied domain when graphing. Also realise that from the sketch, we could be misled into thinking that there is a 'gap' at the point $(3, 0)$. Be careful with this – use the graphics calculator as an aid, then, double check to make sure.

(b) The equation $y = \dfrac{2}{\sqrt{x-3}}$ represents the reciprocal of a square root relation. As in (a), we must have that $x - 3 \geq 0 \Leftrightarrow x \geq 3$.

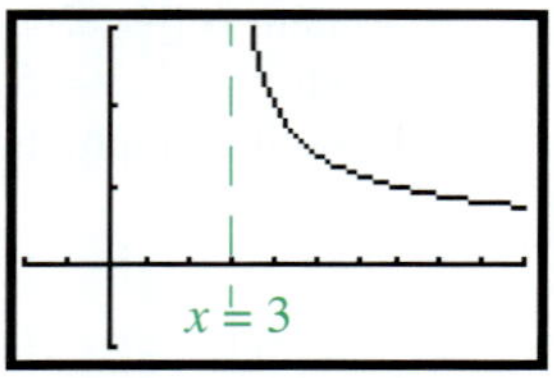

However, this time we have another restriction – we cannot divide by zero and so we cannot include $x = 3$ in our domain. So , at $x = 3$, we draw an **asymptote**.

We then have $x - 3 > 0 \Leftrightarrow x > 3$. This leads to a range of $]0, \infty[$.

(c) The only restriction that can be readily seen for the relation $y = \dfrac{3}{2-x}$ is that we cannot divide by zero and so, we must have that $2 - x \neq 0$. That is, $x \neq 2$.

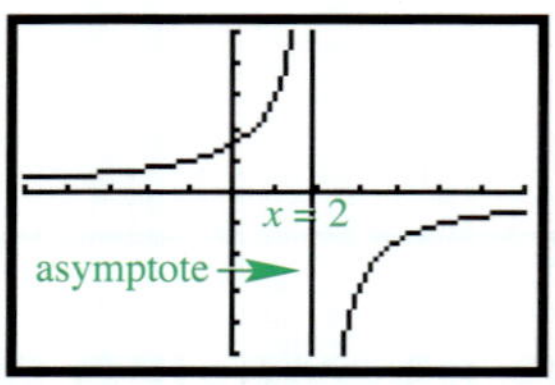

As it is a reciprocal relation, we have an asymptote at $x = 2$. So, the domain is given by $]-\infty, 2[\ \cup\ ]2, \infty[$ or simply, $\mathbb{R}\backslash\{2\}$.
The range can then be seen to be $\mathbb{R}\backslash\{0\}$

8.1.4 TYPES OF RELATIONS

Relations fall into one of four categories. These are:

1. One to one relations [one x to one y]

For any one value of x, there will be only one corresponding value of y.

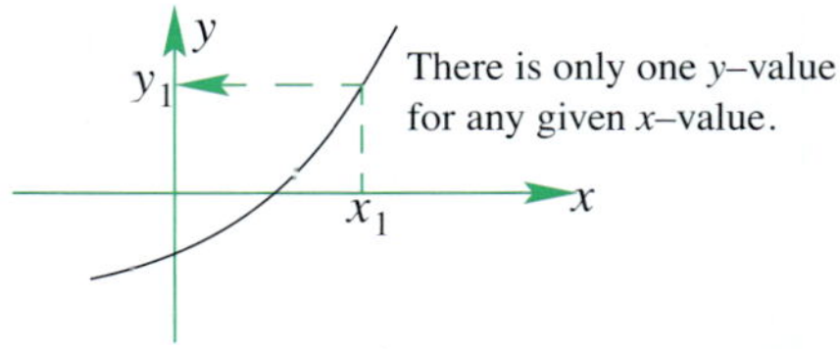

2. One to many relation [one x to many y]

There is at least one value of x for which there exists more than one corresponding value of y.

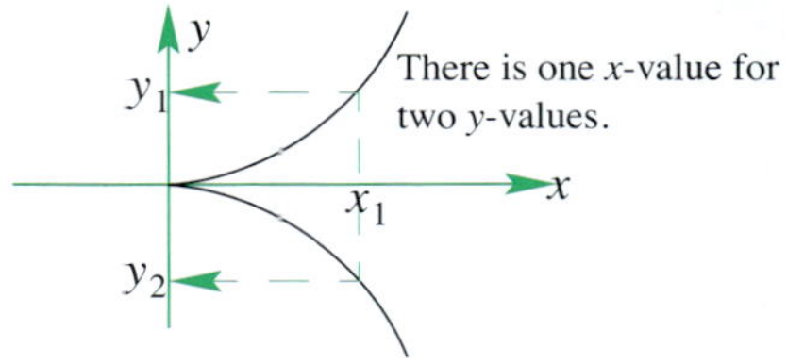

3. Many to one relation [many x to one y]

There are at least two different values of x that will correspond to only one value of y.

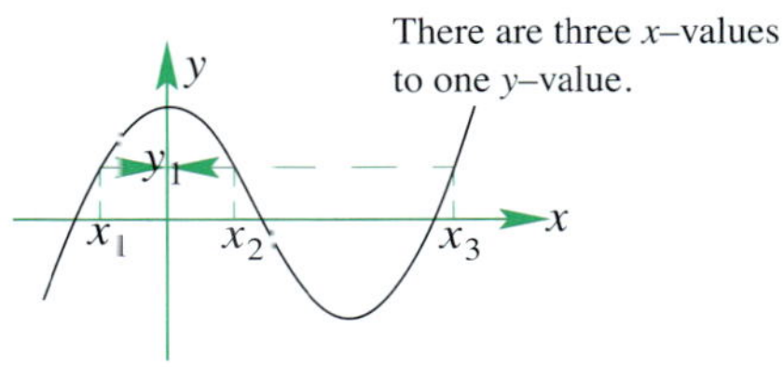

4. Many to many [many x to many y]

There are at least two different values of x that will correspond to at least two different values of y.

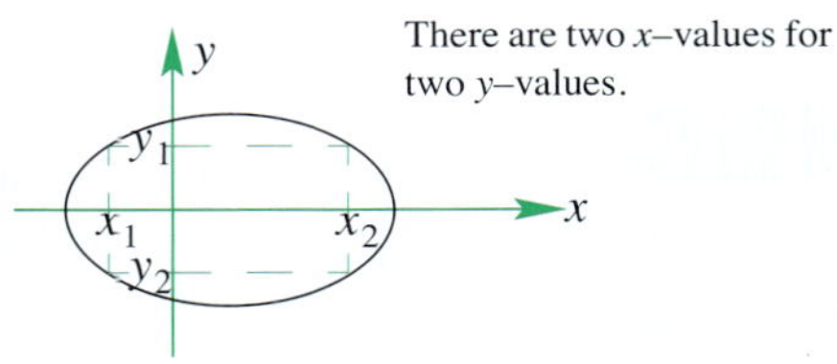

8.1.5 SKETCHING WITH THE GRAPHICS CALCULATOR

Why do graphs sometimes appear distorted on the graphics calculator display screen?

On most graphics calculators, the display screen is two thirds as high as it is wide. Because of this, when we graph a relation that describes a circle we obtain a diagram that is not a true geometric representation. To obtain a true geometric representation we need to use a window that produces a square setting. This is done by using the fact that $\frac{Y_{max} - Y_{min}}{X_{max} - X_{min}} = \frac{2}{3}$.

For example, the window alongside shows such a setting:

i.e., $\frac{Y_{max} - Y_{min}}{X_{max} - X_{min}} = \frac{4-(-4)}{6-(-6)} = \frac{8}{12} = \frac{2}{3}$

These settings enable us to obtain a true geometric representation of the circle with equation $x^2 + y^2 = 16$.

Re–arranging the equation to make y the subject, we have, $x^2 + y^2 = 16 \Rightarrow y = \pm\sqrt{16 - x^2}$.

We can now graph the equations $y = \sqrt{16 - x^2}$ and $y = -\sqrt{16 - x^2}$.

We enter both the positive and the negative equations [we do this using the **VARS** option]. Then we use the above window settings. Finally press the **GRAPH** command.

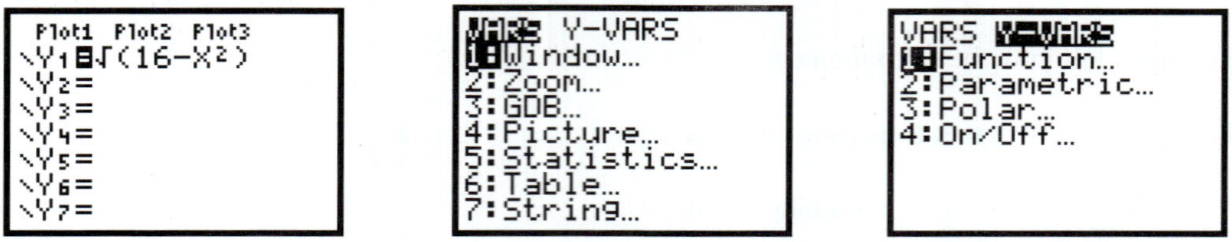

The final output is:

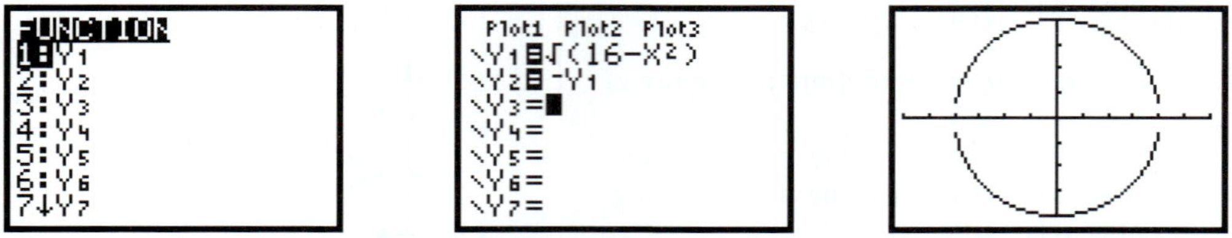

Using the window setting shown below, we obtain a distorted graph, as shown below:

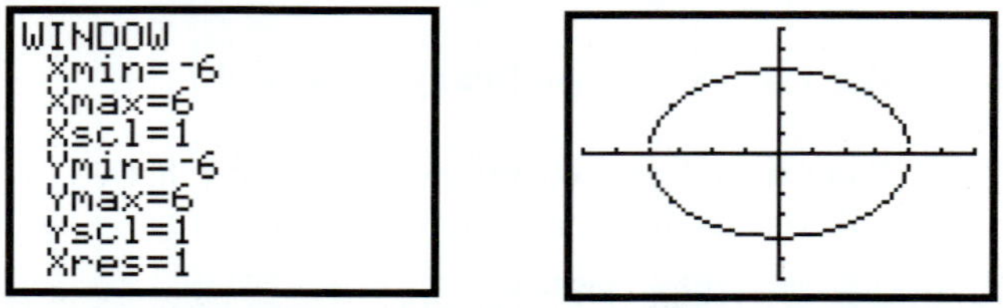

NB: You can also use the **ZOOM** options **4:ZDecimal** or **5:ZSquare** to generate correct graphs.

EXAMPLE 8.5

Determine the domain and range of the following relations.

(a) $x^2 + y^2 = 4$.

(b) $y > 2 - x, x < 0$.

(c) $y = \dfrac{2}{x^2+1}, -1 \le x < 2$

Solution

(a) The relation $x^2 + y^2 = 4$ represents a circle of radius 2 units with its centre at the origin.

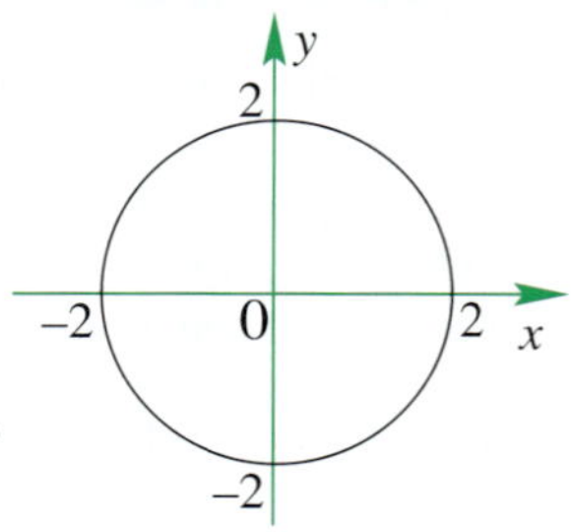

Note that we can only use values of x between –2 and 2. For example, if we have $x = 3$, then we must have $9 + y^2 = 4 \Rightarrow y^2 = -5$ for which there are no real solutions. Therefore, the domain is $-2 \le x \le 2$ (or [–2, 2]). The range is $-2 \le y \le 2$ (or [–2, 2]).

(b) This time we have a subset of the cartesian plane. That is, we are defining a region in the plane as opposed to a set of points lying on a curve.

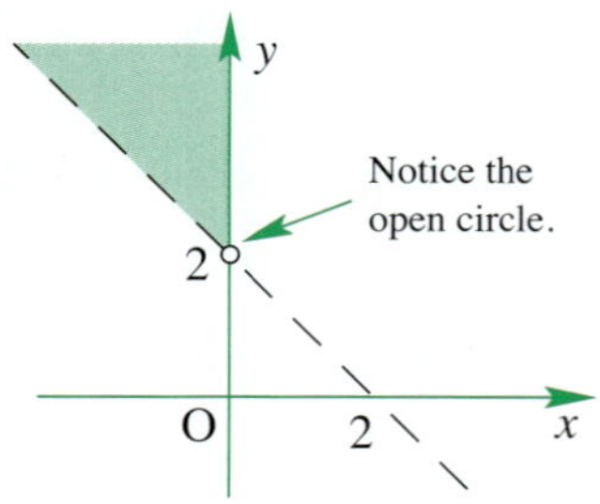

The domain has already been set as $\{x : x < 0\}$ (or $]-\infty, 0[$). From our sketch, we see that the range is $\{y : y > 2\}$ (or $]2, \infty[$).

(c) This time we make use of the TI–83 to sketch the graph:

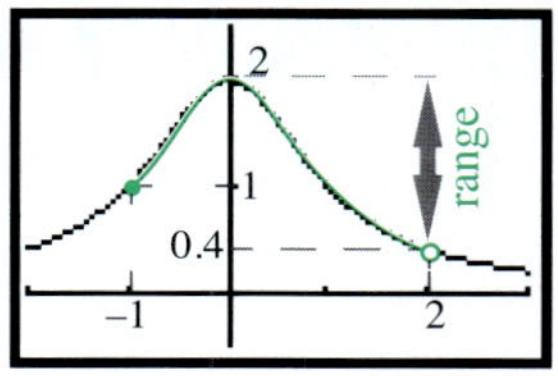

The domain has already been specified as [–1, 2[, meaning that we include the value $x = -1$ (which generates a y–value of 1) but exclude the value $x = 2$ (which would have produced a y–value of $\frac{2}{5} = 0.4$).

From the sketch, the range is therefore]0.4, 2].

NB: Had we been given the relation $y = \frac{2}{x^2+1}$ without a specified domain, then its implied domain would be]–∞, ∞[. This time, unlike Example 5.4 (b), as $x^2 + 1$ will always be greater than or equal to one, there is no danger of ever dividing by zero (for any value of x). This would have then produced a range]0, 2].

The fact that you have access to a graphics calculator will make it easier for you to sketch graphs, however, it is still important that you have a 'feel' for the shape of some standard relations. We will deal with a number of relations in detail throughout this book, so, at this stage we will provide a reference table for some standard relations.

Relation	Properties
$y = mx + c$ 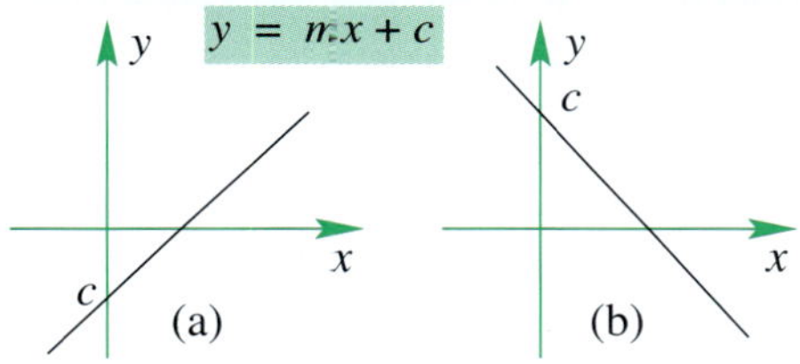	**1.** Linear relation [straight line] **2. Case (a):** $m > 0$, y–intercept at (0, c) increasing **Case (b):** $m < 0$ y–intercept at (0, c) decreasing
$y = ax^2 + bx + c$ 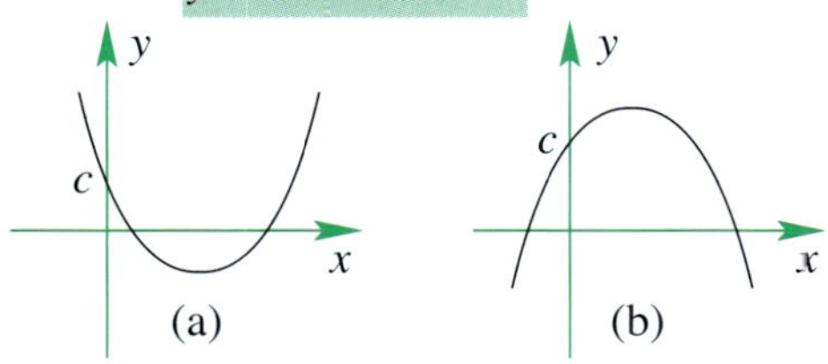	**1.** Quadratic relation [parabola] **2. Case (a):** $a > 0$, y–intercept at $(0, c)$ axis of symmetry at $x = -\frac{b}{2a}$. **Case (b):** $a < 0$ y–intercept at $(0, c)$ axis of symmetry at $x = -\frac{b}{2a}$.
$y = ax^3 + bx^2 + cx + d$ 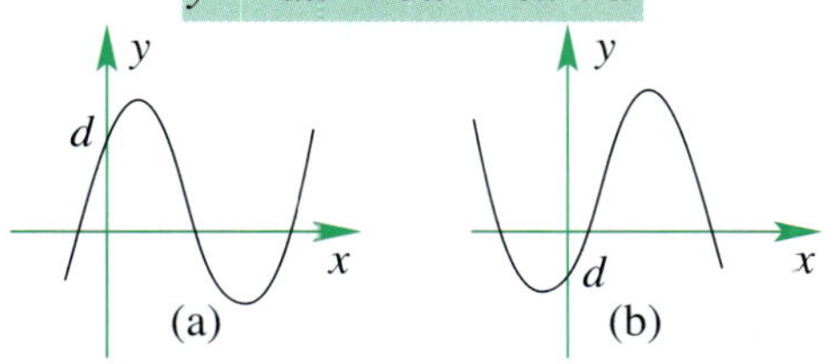	**1.** Polynomial relation (of order 3) [cubic] **2. Case (a):** $a > 0$, y–intercept at $(0, d)$ **Case (b):** $a < 0$ y–intercept at $(0, d)$

Relation	Properties
$y = a \times b^{kx} + c$ 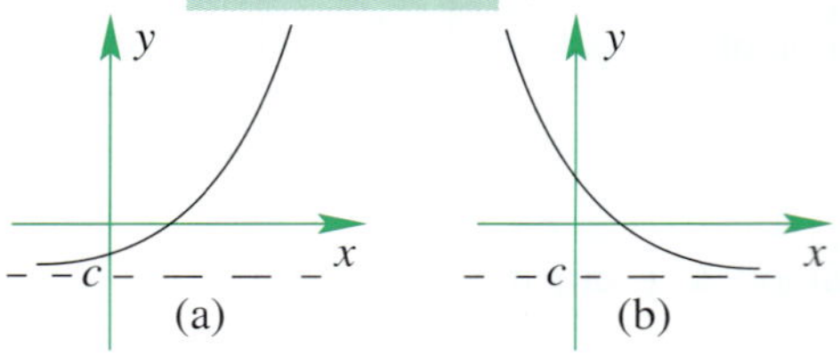 	**1.** Exponential relation [exponential] **2. Case (a):** $a > 0, k > 0$ *asymptote* at $y = c$ increasing **Case (b):** $a < 0$ *asymptote* at $y = c$ decreasing NB: If $a < 0$, then graphs of $y = a \times b^{kx}$ is first reflected about the x – axis.
$y = \frac{a}{x-b} + c$ 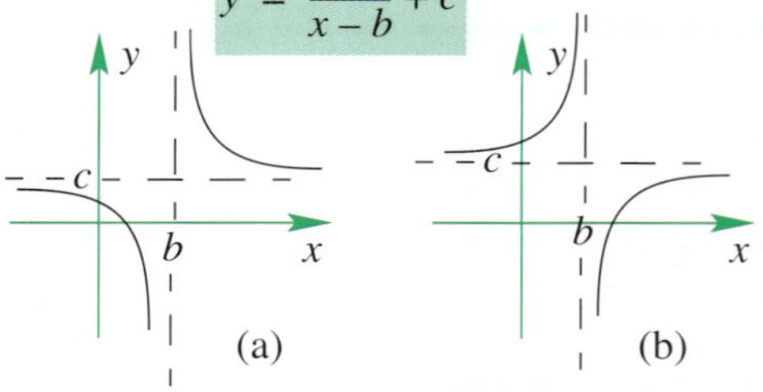 	**1.** Reciprocal relation [rectangular hyperbola] **2. Case (a):** $a > 0$, *asymptote* at $y = c$ and $x = b$ **Case (b):** $a > 0, k < 0$ *asymptote* at $y = c$ and $x = b$
$y = a\sqrt{x} + b$ 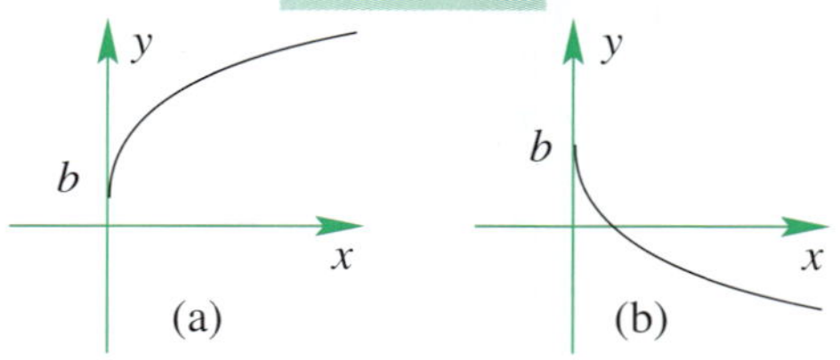 	**1.** Square root relation [sideways parabola] **2. Case (a):** $a > 0$, y–intercept at $(0, b)$ increasing **Case (b):** $a < 0$ y–intercept at $(0, b)$ decreasing NB: $y^2 = x$ is a sideways parabola, $y = \sqrt{x}$ is only 'half' of a sideways parabola.
$y = \frac{a}{(x-b)^2} + c$ 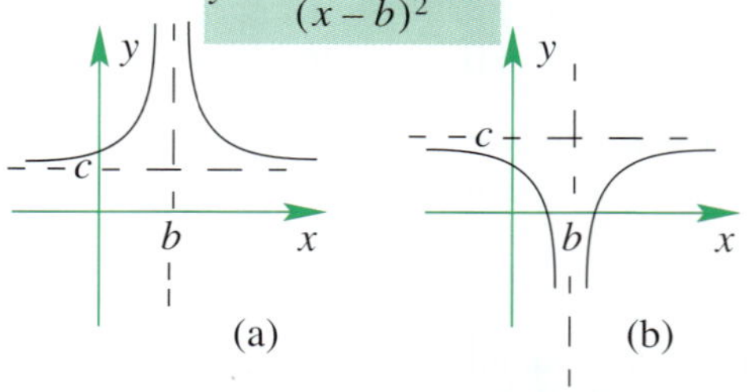 	**1.** Reciprocal squared relation [truncus] **2. Case (a):** $a > 0$, *asymptote* at $y = c$ and $x = b$ **Case (b):** $a < 0$ *asymptote* at $y = c$ and $x = b$
$(x-a)^2 + (y-b)^2 = r^2$ 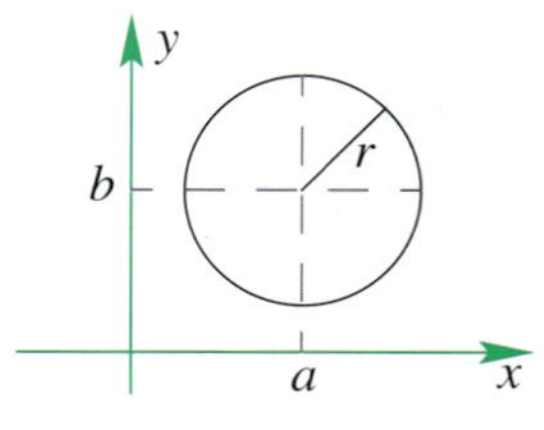 	**1.** Circular relation [circle] **2.** Radius r **3.** Centre at (a, b) NB: The equation $(x-a)^2 + (y-b)^2 = r^2$ can be rewritten as $y = \pm\sqrt{r^2 - (x-a)^2} + b$ with the positive root representing the top half and the negative root the bottom half.

Remember, this 'bank' of relations is a mere sample of possible relations and are not necessarily given in their most general form.

EXERCISES 8.1

1. State the domain and range of the following relations.

(a) $\{(2,4), (3,-9), (-2,4), (3,9)\}$

(b) $\{(1,2), (2,3), (3,4), (5,6), (7,8), (9,10)\}$

(c) $\{(0,1), (0, 2), (1,1), (1,2)\}$

2. Find the range for each of the following.

(a) $\{(x,y):y = x + 1, x \in \mathbb{R}^+\}$ (b) $\{(x,y):y \geq x, x \geq 0\}$

(c) $y = x^2 + 2x + 1, x > 2$ (d) $y = 2x - x^2, x \in \mathbb{R}$

(e) $x^2 + y^2 = 9, -3 \leq x \leq 3$ (f) $x^2 - y^2 = 9, x \geq 3$

(g) $y = x - 1, 0 < x \leq 1$ (h) $y = 4 - x^2, -2 \leq x < 1$

(i) $y = \sqrt{x}, x \geq 0$ (j) $y = \sqrt{x}, 1 \leq x \leq 25$

(k) $y = \dfrac{4}{x+1}, x > 0$ (l) $\{(x,y):y^2 = x, x \geq 1\}$

3. State the range and domain for each of the following relations.

(a)

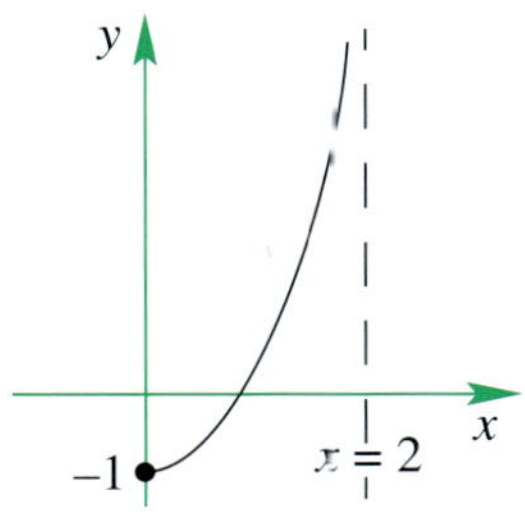

(b)

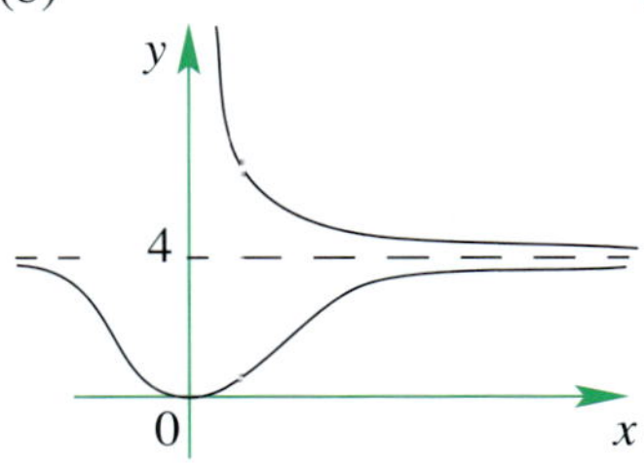

(c)

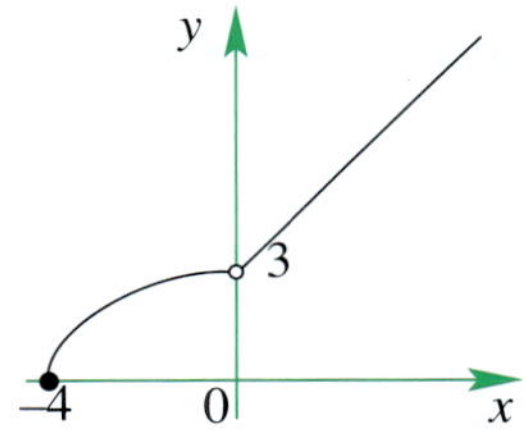

(d)

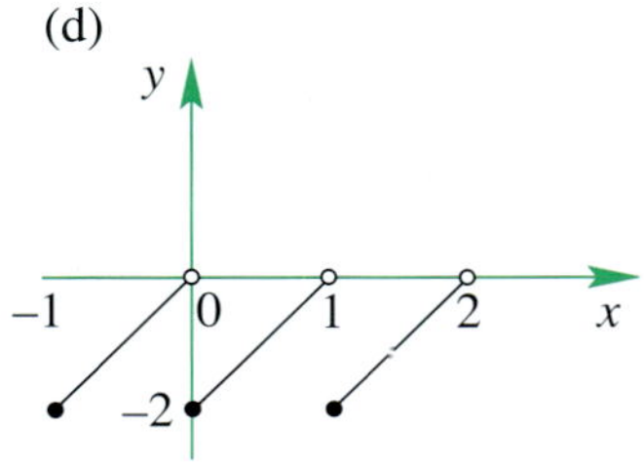

(e)

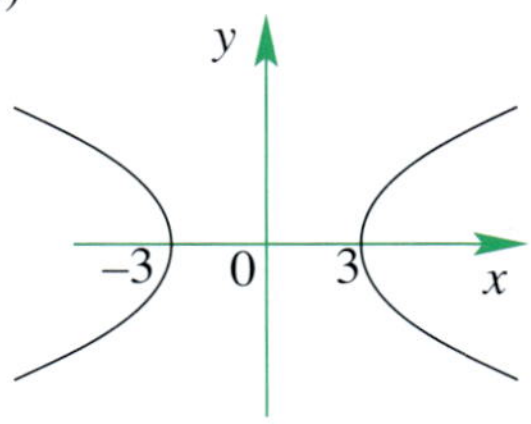

(f)

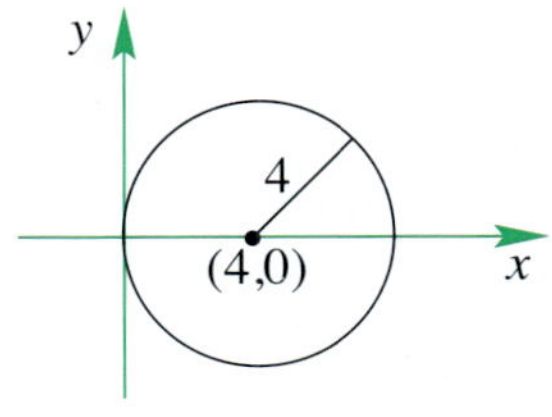

4. What types of relation are the following?

(a)

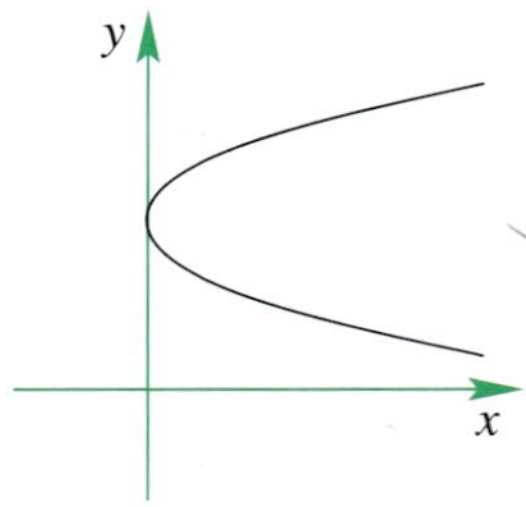

(b)

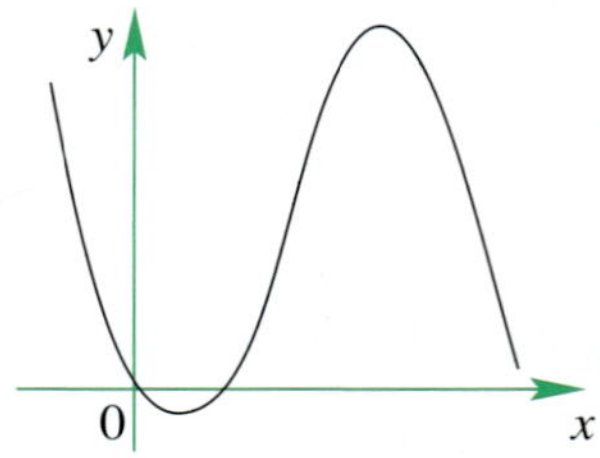

(c)

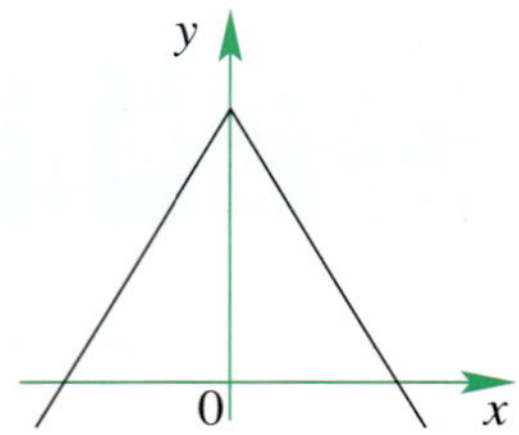

(d)

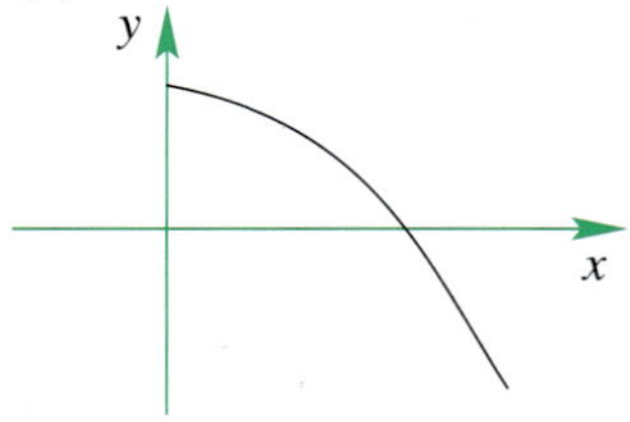

(e)

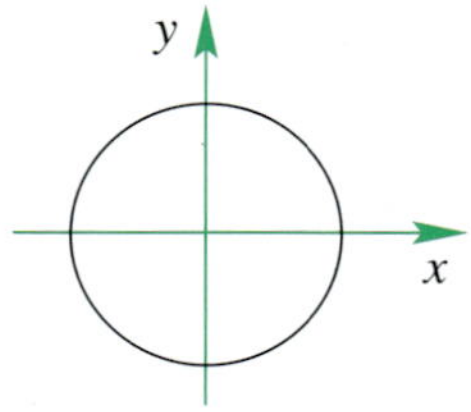

(f)

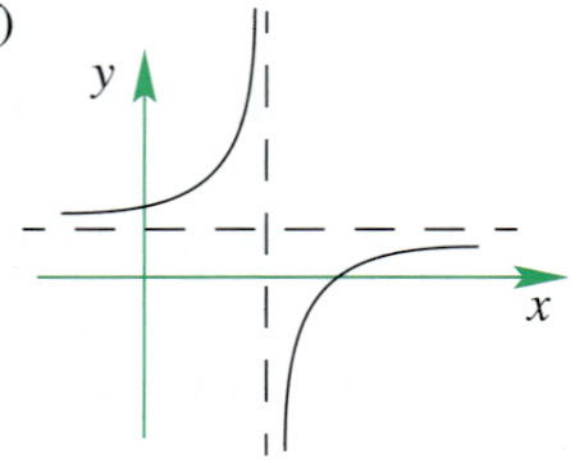

5. Determine the implied domain for each of the following relations.

(a) $y = \dfrac{2x}{x+2}$ (b) $y = \dfrac{3}{\sqrt{9-x}}$ (c) $y = \sqrt{16-x^2}$

(d) $y = \sqrt{x^2-4}$ (e) $xy - x = 3$ (f) $y = \dfrac{2}{x^2+1}$

8.2 FUNCTIONS

8.2.1 DEFINITIONS

There is a special group of relations which are known as **functions**. This means that every set of ordered pairs is a relation, but **every relation is not a function**. Functions then make up a subset of all relations.

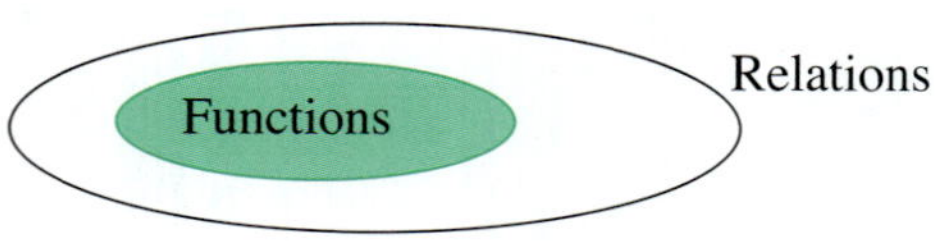

A function is defined as a relation that is either one to one or many to one. That is a function is a relation for which no ordered pairs have the same first element.

There are two ways to determine if a relation is a function.

Method 1: Algebraic approach

For Method One we use the given equation and determine the number of y–values that can be generated from one x–value.

EXAMPLE 8.6

Determine which (if any) of the following are functions.

(a) $y^3 - x = 2$ (c) $y^2 + x = 2$

Solution

(a) From $y^3 - x = 2$ we have that $y = \sqrt[3]{2+x}$, then for any given value of x, say $x = a$, we have that $y = \sqrt[3]{2+a}$ which will only ever produce one unique y–value.
Therefore, the relation $y^3 - x = 2$ is a function. In fact, it is a one to one function.

(b) From $y^2 + x = 2$ we have that $y^2 = 2 - x \Leftrightarrow y = \pm\sqrt{2-x}$. Then, for any given value of x, say $x = a$ (where $a \le 2$), we have that $y = \pm\sqrt{2-a}$, meaning that we have two different y–values; $y_1 = \sqrt{2-a}$ and $y_2 = -\sqrt{2-a}$, for the same x–value. Therefore, this relation is not a function. In fact, it is a one–to–many relation.

Method 2: Vertical line Test

The method is quite simple:

Step 1 Sketch the graph of the relation.
Step 2 Make a visual check of the number of times a vertical line would cut the graph.
Step 3 If the vertical line only ever cuts at one place for every value in the domain the relation is a function.

Function

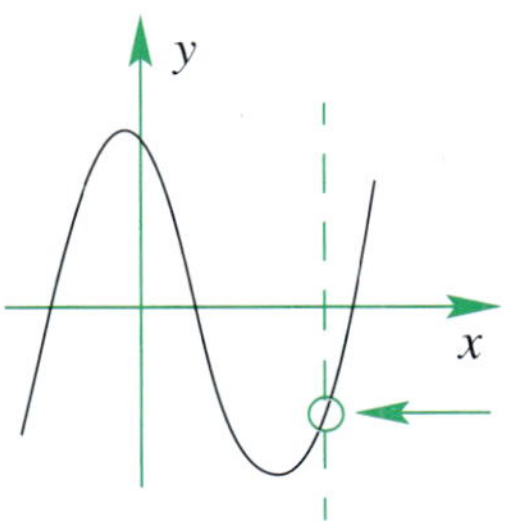

No matter where along the graph we draw a vertical line, the graph will only ever be cut once. Therefore this relation is a function.

Only one cut (anywhere)

Not a function

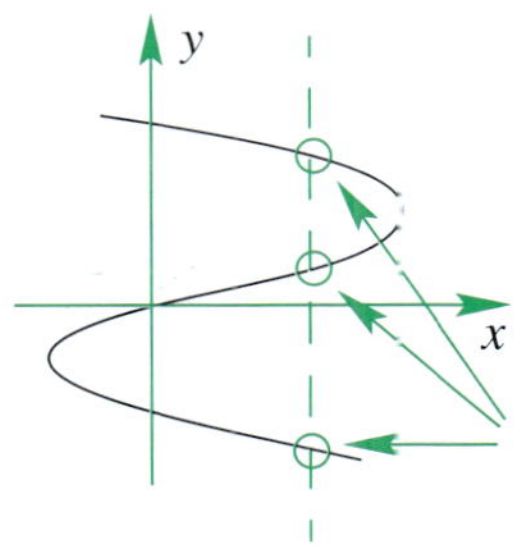

The vertical line has cut the graph at more than one point for a given value of x. Therefore this relation is a not function.

Vertical has cut the graph at three different points for the same x–value.

EXAMPLE 8.7

Which of the following defines a function?

(a) $\{(0,2), (1,2), (2,1)\}$ (b) $\{(x, y): y = x^3 + 1, x \in \mathbb{R}\}$

(c) $y^2 = x, x \ge 0$ (d) $\{(x,y): x^2 + y^2 = 16\}$

Solution

(a) Clearly, we have every first element of the ordered pairs different. This means that this relation is also a function.

(b) Using the TI–83 to provide a visual check:
From the graph shown, a vertical line drawn anywhere on the domain for which the relation is defined, will cut the graph at only one place.
This relation is therefore a function.

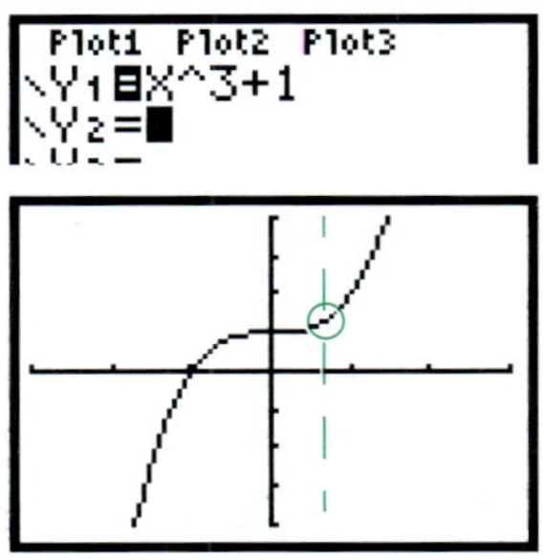

(c) Again we make use of a visual approach to determine if the relation is a function.
First we write the relation in a form that will enable us to enter it into the TI–83: $y^2 = x \Rightarrow y = \pm\sqrt{x}$
We can therefore define the relation $Y_1 = \sqrt{X}$ and $Y_2 = -\sqrt{X}$ and sketch both on the same set of axes.

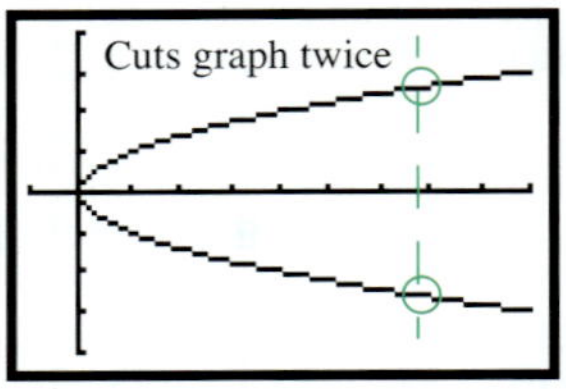

Placing a vertical line over sections of the domain shows that the line cuts the graph in two places (except at the origin). Therefore this relation is not a function.

Algebraic proof
We can also determine if a relation is a function by using algebraic means.
Begin by choosing a value of x that lies in the domain. For example $x = 4$.
This gives the following equation: $y^2 = 4 \Rightarrow y = \pm\sqrt{4}$
From which we can say that when $x = 4$, $y = 2$ **and** $y = -2$, so that there are two ordered pairs, (4,2) and (4, –2). As we have two different y-values for one x-value this relation is not a function.

(d) This relation describes the equation of a circle with radius 4 units and centre at the origin.
The graph of this relation is shown alongside. The graph fails the vertical line test, and so is not a function.

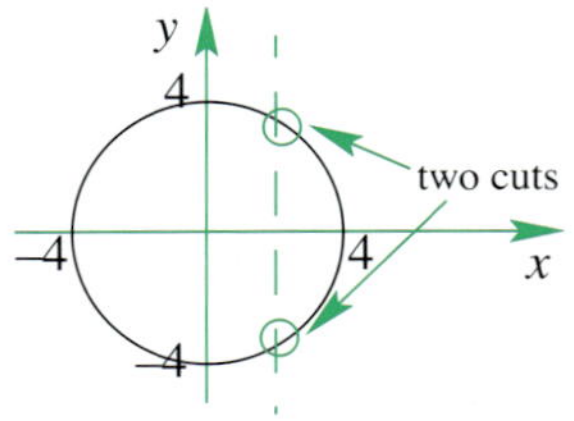

We now provide a formal definition together with commonly used notation for a funcion.

> A function f, (or a mapping f), from a set X to a set Y is a relation that assigns to each element x in the set X a unique element y in the set Y.

The set X is called the **domain** of f and the set Y the **co-domain**. The element y is called the **image** of x under f and we denote this image by $f(x)$, the value of the function f at x (read as; f of x).

We write this mapping as: $f: x \mapsto f(x)$

We can also write this mapping as follows

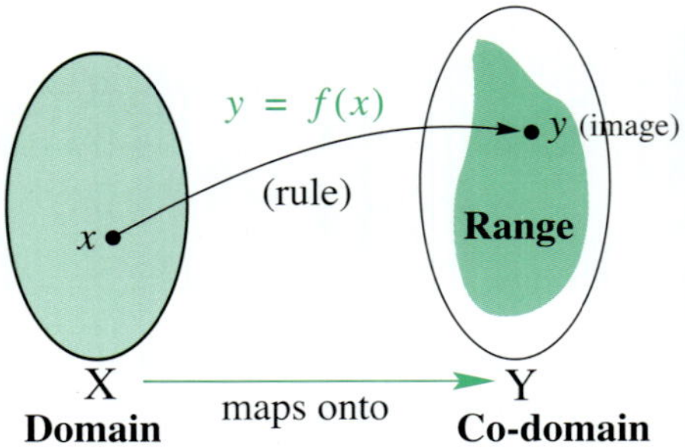

1. $f: X \mapsto Y$, where $y = f(x)$
2. $f: X \mapsto Y, y = f(x)$
3. $y = f(x), x \in X$

It is important to realise that the range of f is not necessarily the set Y. The **range** of f is actually a subset of Y (sometimes it can be equal toY). Set Y, i.e., the co-domain, describes the **types** of numbers that will be produced when f is applied to different x-values — not necessarily which numbers will result! The range of f is given by the values of $f(x)$.

Translating the mathematical notation into English we have the following:

$f:X \mapsto Y$, where $f(x)$ = rule in terms of x

"f is such that the set X maps onto the set Y where f of x is equal to . . ."

Notice that f describes not only the rule, $f(x)$, but also the domain, X.

EXAMPLE 8.8

For the function $f(x) = x^3 + 1, x \in \mathbb{R}$ find

(a) $f(-1)$, $f(2)$,. (b) the element of the domain that has an image of 28.

Solution

(a) $f(x) = x^3 + 1, x \in \mathbb{R} \Rightarrow f(-1) = (-1)^3 + 1 = -1 + 1 = 0$.

Similarly, $f(2) = 2^3 + 1 = 9$.

(b) If the image is 28, then we want the value of x for which $f(x) = 28$.

We then have, $f(x) = 28 \Leftrightarrow x^3 + 1 = 28$

$\Leftrightarrow x^3 = 27$

$\Leftrightarrow x = 3$ [Taking the cube root of both sides]

Therefore, the element of the domain that has an image of 28 is 3.

EXAMPLE 8.9

Determine the range of the function $f:\{x:x \geq 0\} \mapsto \mathbb{R}$, $f(x) = x + 1$.

Solution

First note that the co-domain is given by $\mathbb{R}$, the set of real numbers – meaning that all image values will be real. To determine the actual range of this function we sketch its graph.

From the graph, the only possible values of y are those for which $y \geq 1$. In this case, because $x = 0$ is included in the domain, we also include the value y = 1 in the range. Therefore, we have a closed circle at the end point.

The range of f is then given by $\{y:y \geq 1\}$ (or $[1, \infty[$).

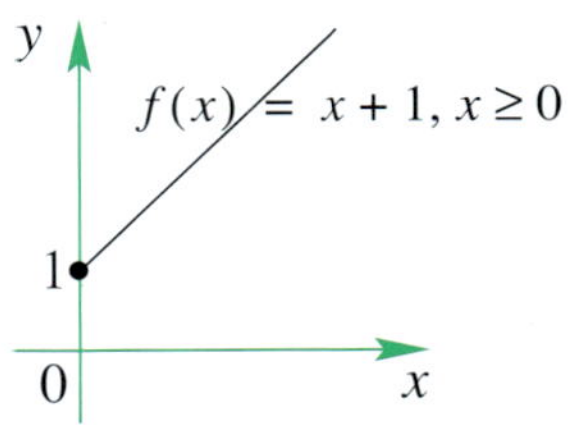

EXAMPLE 8.10

For the function $f(x) = x^2 - 3, x \in \mathbb{R}$ find

(a) $f(6)$ (b) $f(x+1)$ (c) $f(x+h) - f(x)$

Solution

(a) To determine the value of $f(6)$, we 'replace' the x–term in the rule of $f(x)$ with the number '6' i.e., $f(6) = (6)^2 - 3 = 36 - 3 = 33$.

(b) This time we 'replace' the x–term in the rule of $f(x)$ with '$x - 1$':

So, $f(x+1) = (x+1)^2 - 3 = x^2 + 2x + 1 - 3 = x^2 + 2x - 2$.

(c) $f(x+h)-f(x) = (x+h)^2-3-(x^2-3)$

$$= (x^2+2xh+h^2)-3-x^2+3$$

$$= 2xh+h^2$$

EXAMPLE 8.11

Consider the function $g:\{x: -1 \le x \le 2\} \mapsto \mathbb{R}$, where $g(x) = x^2+2$,

(a) Find $g(-1)$, $g(0)$, and $g(2)$. (b) Determine the range of g.

Solution

(a) $g(-1) = (-1)^2+2 = 3$,

$g(0) = (0)^2+2 = 2$,

$g(2) = (2)^2+2 = 6$

(b) To determine the range of the function g, we first need to sketch its graph.

We begin by sketching the graph of $y = x^2+2$ for all real values of x. Then, as the domain is restricted to $x \in [-1, 2]$. This means that we 'remove' the parts of the graph that lie outside this domain. This leaves the required part of the graph.

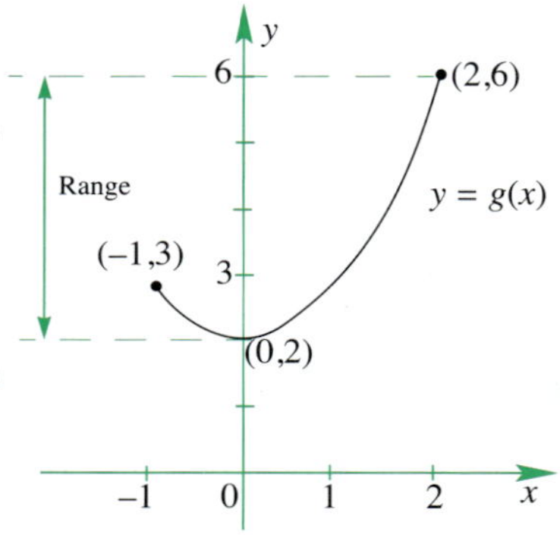

From our graph of g, the range of this function is given by $\{y: 2 \le y \le 6\}$ (or $[2, 6]$).

EXAMPLE 8.12

Consider the function $f:\mathbb{R} \mapsto \mathbb{R}$, where $f(x) = 2-x^2$.

(a) Find i. $f(5)$ ii. $\{x : f(x) = -7\}$ iii. $\{x : f(1-x) = -7\}$

(b) What type of function is f?

Solution

(a) i. $f(5) = 2-5^2 = 2-25 = -23$

ii. $f(x) = -7 \Leftrightarrow 2-x^2 = -7 \Leftrightarrow x^2 = 9$

$$\therefore x = \pm 3$$

So, the solution set is $\{-3, 3\}$.

iii. $f(1-x) = -7 \Leftrightarrow 2-(1-x)^2 = -7$

$$\Leftrightarrow (1-x)^2 = 9$$

$$\Leftrightarrow 1-x = \pm 3$$

Therefore, the solution set is $\{-2, 4\}$.

(b) From (a) we obtained two values of x for one value of y and so $f(x)$ is a many–to –one function.

EXERCISES 8.2

1. A function is defined as follows, $f:x \mapsto 2x+3, x \geq 0$.

(a) Find the value of $f(0)$, $f(1)$.

(b) Evaluate the expressions i. $f(x+a)$ ii. $f(x+a)-f(x)$

(c) Find $\{x:f(x) = 9\}$.

2. If $f(x) = \dfrac{x}{x+1}, x \in [0, 10]$, find

(a) $f(0), f(10)$.

(b) $\{x:f(x) = 5\}$.

(c) the range of $f(x) = \dfrac{x}{x+1}, x \in [0, 10]$.

3. For the mapping $x \mapsto 2-\frac{1}{2}x^2, x \in \mathbb{R}$, find

(a) $f(x+1)$, $f(x-1)$.

(b) a, given that $f(a) = 1$.

(c) b, given that $f(b) = 10$.

4. A function is defined as follows, $y = x^3 - x^2, x \in [-2, 2]$.

(a) Find the value(s) of x such that $y = 0$.

(b) Sketch the graph of $y = x^3 - x^2, x \in [-2, 2]$ and determine its range.

5. The function f is defined as $f:]-\infty, \infty[\mapsto \mathbb{R}$, where $f(x) = x^2 - 4$.

(a) Sketch the graph of i. f

ii. $y = x+2, x \in]-\infty,\infty[$

(b) Find i. $\{x:f(x)= 4\}$ ii. $\{x:f(x)= x+2\}$

6. Which of the following relations are also functions?

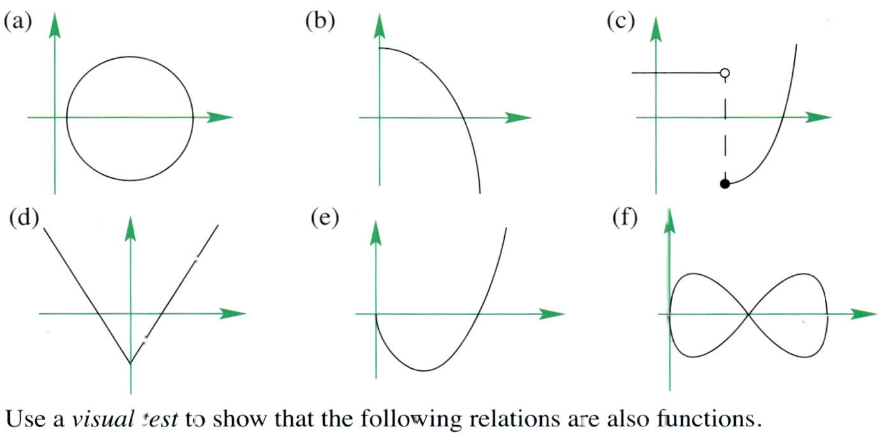

7. Use a *visual test* to show that the following relations are also functions.

(a) $x \mapsto x^3+2, x \in]0,5[$

(b) $x \mapsto \sqrt{x}+1, x \in [0, 9[$

(c) $\{(x,y):y^3= x+1, x \in \mathbb{R}\}$

(d) $\{(x,y):y= x^2+1, x \in \mathbb{R}\}$

8. Decide which of the following relations are also functions.

(a) $f:x \mapsto \frac{1}{x}, x \in \mathbb{R}\backslash\{0\}$

(b) $\{(x,y):y^2 - x = 9,\ x \geq -9\}$

(c) $\{(x,y):y^2 - x^2 = 9,\ x \geq -9\}$

(d) $f(x) = \frac{1}{x^2} + 1, x \neq 0$

(e) $f(x) = 4 - 2x^2, x \in \mathbb{R}$

(f) $f:x \mapsto \frac{4}{x+1}, x \in \mathbb{R}\backslash\{-1\}$

9. Sketch the graph of $f: \mapsto \frac{x^2}{x^2+2}, x \in \mathbb{R}$ and use it to

(a) show that f is a function.
(b) determine its range.

10. A function is defined by $f:x \mapsto \frac{x+10}{x-8}, x \neq 8$ and $x \geq 0$.

(a) Determine the range of f.
(b) Find the value of a such that $f(a) = a$.

11. Find the the largest possible subset X of $\mathbb{R}$, so that the following relations are one to one increasing functions

(a) $f : X \to \mathbb{R}$, where $f(x) = x^2 + 6x + 10$
(b) $f : X \to \mathbb{R}$, where $f(x) = \sqrt{9 - x^2}$
(c) $f : X \to \mathbb{R}$, where $f(x) = \sqrt{x^2 - 9}$

8.3 SOME STANDARD FUNCTIONS

8.3.1 HYBRID FUNCTIONS AND CONTINUITY

A hybrid function is a relation that consists of more than one function, where each function is defined over a mutually exclusive domain.

EXAMPLE 8.13

Sketch the graph of $f(x) = \begin{cases} -2 & x \leq 1 \\ x+1 & x > 1 \end{cases}$, stating its doman and range.

Solution

We first look at the domain of the function. In this instance we have that $x \geq 1$ and $x < 1$, so in fact we have that $x \in]-\infty, \infty[$ (or simply, $x \in \mathbb{R}$).

To determine the range we will need to sketch the graph of f.

Hybrid functions can be sketched using the TI–83 (use the **TEST** menu)

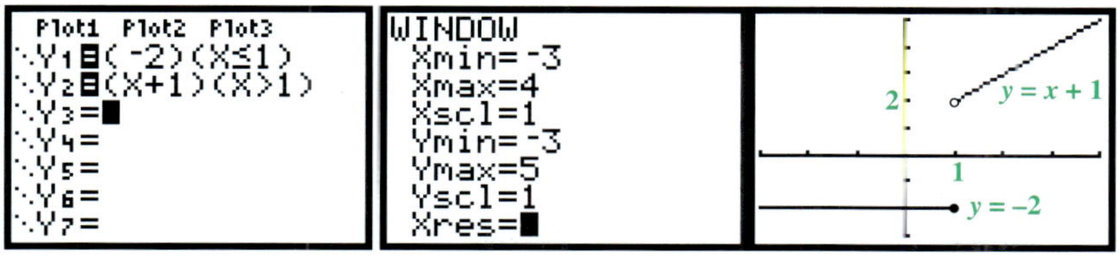

From the graph the range is given by the set $\{y : y > 2\} \cup \{-2\}$.

A few important points to note:

1. Even though $x = 1$ is not in the domain of the $y = x + 1$ part of the function, we have still used $x = 1$ to find the value of $y = 2$.
2. At $x = 1$, we have what is known as a **discontinuity** – there exists a 'gap'.

Example 8.14

Sketch the function $y = \begin{cases} x^2 + 1, & x > 1 \\ -x + 4, & x \le 1 \end{cases}$.

Solution

We first consider that part of the function where $x > 1$:

Here we need to sketch part of the curve of $y = x^2 + 1$

If we could have used the value $x = 1$, then we would have had an image of $y = 1^2 + 1 = 2$.

However, as $x > 1$, we place an open circle at the point with coordinate $(1, 2)$.

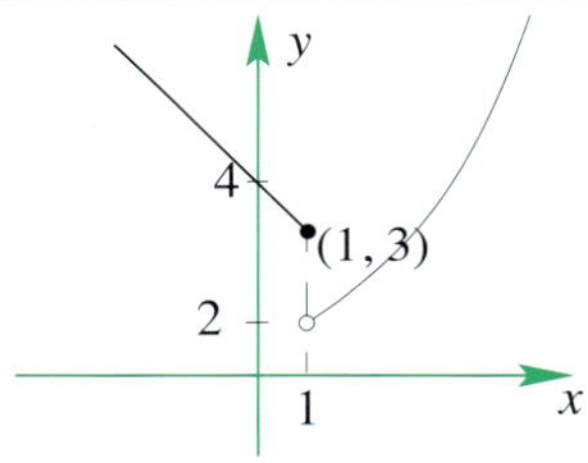

Similarly, for $x \le 1$ we sketch part of the line $y = -x + 1$. However, this time when we use $x = 1$ (to produce the image $y = -1 + 4 = 3$) we include the coordinate $(1, 3)$.

Exercises 8.3.1

1. Sketch the graphs of the following functions

(a) $f(x) = \begin{cases} -x + 2, & x > 3 \\ 1, & x \le 3 \end{cases}$

(b) $f(x) = \begin{cases} x^2 + 2, & x > -1 \\ 3, & x \le -1 \end{cases}$

(c) $f(x) = \begin{cases} \sqrt{x}, & x \ge 4 \\ 6 - x, & x < 4 \end{cases}$

(d) $f(x) = \begin{cases} \dfrac{1}{x}, & x > 1 \\ 2 - x^2, & x \le 1 \end{cases}$

2. Sketch the graph of the function $f(x) = \begin{cases} -2, & x < 0 \\ x - 2, & 0 \le x \le 4 \\ 2 & x > 4 \end{cases}$

3. Sketch the graphs of the following functions

(a) $h(x) = \begin{cases} x^3 + 1, & x > 0 \\ -1, & x \le 0 \end{cases}$ (b) $g(x) = \begin{cases} x + 2, & x > 1 \\ x^2 - 1, & x \le 1 \end{cases}$

(c) $f(x) = \begin{cases} \dfrac{x}{x+1}, & x \ge 0 \\ 1, & x < 0 \end{cases}$ (d) $f(x) = \begin{cases} 2 - \sqrt{x}, & x > 0 \\ x + 3, & x \le 0 \end{cases}$

4. Sketch the graph of the function $f(x) = \begin{cases} -4, & x < -2 \\ x^2 - 4, & -2 \le x \le 2 \\ 4 & x > 2 \end{cases}$

8.3.2 THE ABSOLUTE VALUE FUNCTION

The absolute value function is defined as

$$y = |x| = \sqrt{x^2} = \begin{cases} x & \text{if } x \ge 0 \\ -x & \text{if } x < 0 \end{cases}$$

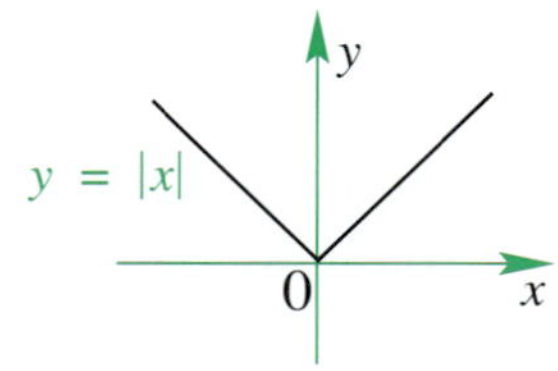

That is, sketch the graph of $y = x$ for $x \ge 0$, and then sketch the graph of $y = -x$ for $x < 0$.

Similarly, the function $x \mapsto |ax + b|$, represents the absolute value of the linear function $y = ax + b$.

EXAMPLE 8.15

Sketch the graph of (a) $y = |x - 2|$

(b) $y = |x| + 1$

Solution

Parts (a) and (b) are best done by considering the functions as translations of the basic absolute value function. That is, the graph of $y = |x - 2|$ is in fact the graph of $y = |x|$ translated two units to the left. While the graph of $y = |x| + 1$ is in fact the graph of $y = |x|$ translated one unit vertically up. So, we have:

(a)

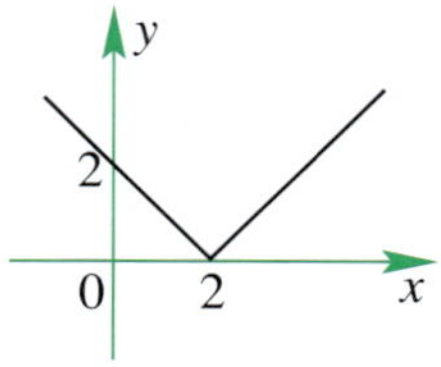

(b)

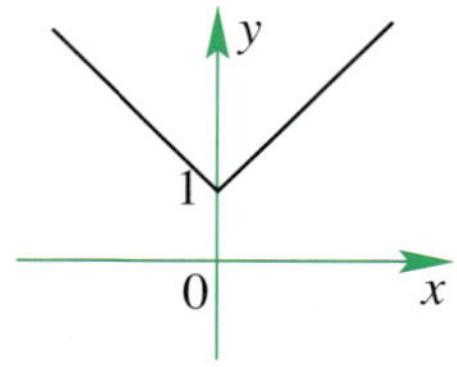

We can also make use of the TI–83 to sketch graphs of absolute value functions:

EXAMPLE 8.16

Find the range of the following functions

(a) $x \mapsto |3 - 2x|, x \in \mathbb{R}$ (b) $y = |x - 4| - 2, x \in \mathbb{R}$

Solution

(a) Use the **MATH** menu and then selecting the **NUM** option, we can choose the **abs(** option. After 'pasting' the **abs(** command, enter the equation as shown on the screen.

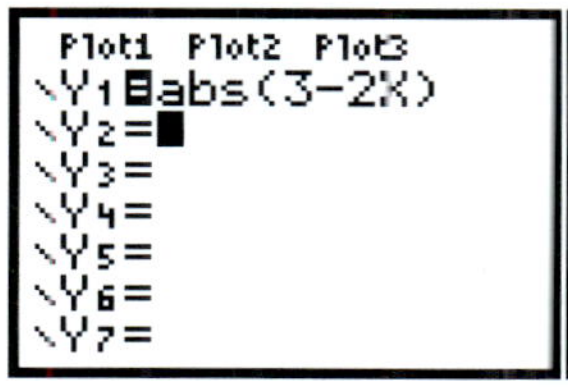

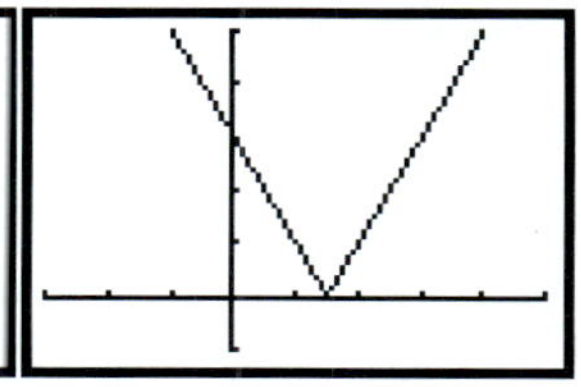

From the given graph, the range is defined as $\{y : y \geq 0\}$.

(b)

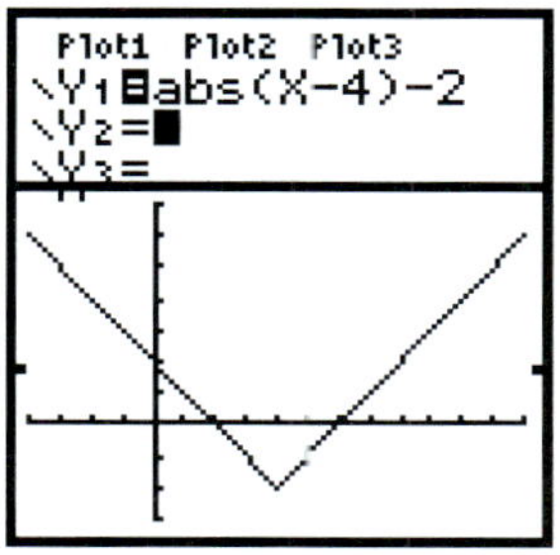

range = $[-2, \infty[$

EXERCISES 8.3.2

1. On separate sets of axes, sketch the graphs of the following functions for $x \in \mathbb{R}$.

(a) $f(x) = |2x|$ (b) $f(x) = |x| + 2$ (c) $f(x) = |x + 1|$

(d) $f(x) = |4x - 2|$ (e) $f(x) = |2x| - 4$ (f) $f(x) = -|x| + 1$

(g) $g(x) = |3 - x|$ (h) $f(x) = \left|5 - \frac{1}{2}x\right|$ (i) $g(x) = \left|\frac{4 - x}{2}\right|$

8.4 MORE EQUATIONS

8.4.1 SOLVING EQUATIONS INVOLVING UNFAMILIAR FUNCTIONS

As we have discussed, there are many types of functions, many of which are unfamiliar to you. As such, this also means that there you will be required to solve equations that involve unfamiliar functions. That is, you will need to solve equations of the form $f(x) = g(x)$ where either one or both are unfamiliar. For example,

$x - 2 = \frac{1}{x}$ i.e., a straight line and a reciprocal function

$5x = 3^x$ i.e., straight line and an exponential.

$2^{0.5x} = 2 + \sin(\pi x)$ i.e., exponential and circular trigonometric function

Many of these are best attempted using your graphics calculator. We consider each of these next.

EXAMPLE 8.17

Solve each of the following

(a) $x-2 = \frac{1}{x}$ (b) $5x = 3^x$ (c) $2^{0.5x} = 2+\sin(\pi x)$

Solution

(a) We start by sketching both curves on the same set of axes.

Let $f(x) = x-2$ (i.e., a linear function) and $g(x) = \frac{1}{x}$ (i.e., a hyperbola).

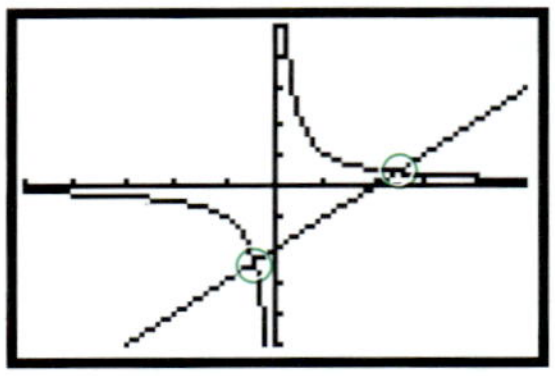

The graph is a great way to see how many solutions there are.
To determine the solutions we can use either

1. **intersect** option from the **CALC** menu

or 2. the **solve** function through the **CATALOG**

or 3. **Solver** option from the **MATH** menu

Make sure that you are familiar with all three methods.
For this problem we will use the **intersect** option from the **CALC** menu.

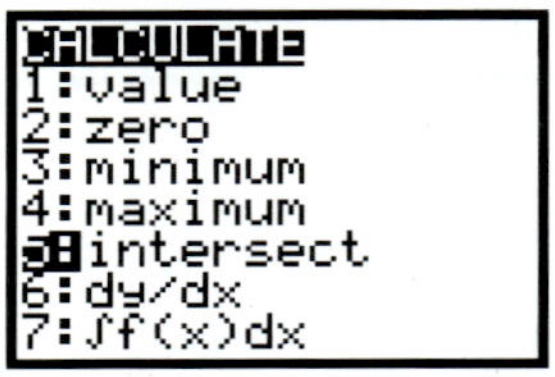

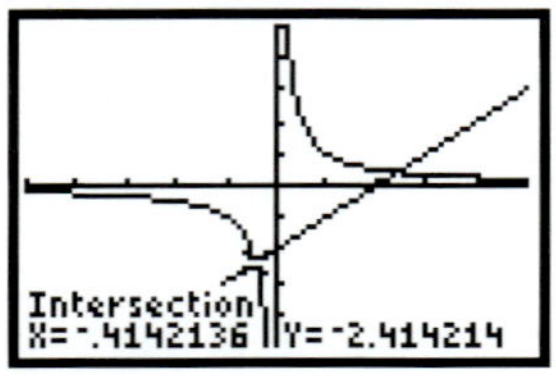

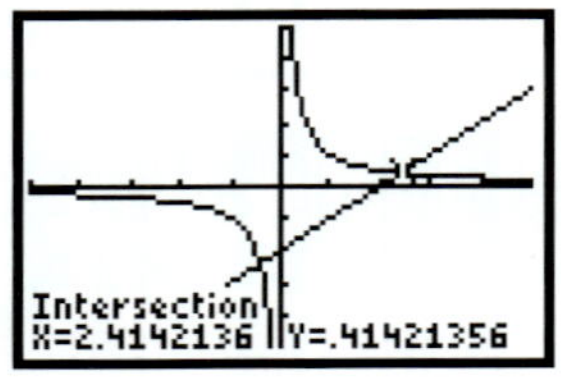

That is, we have that $x = -0.41$ and $x = 2.41$ (to 2 d.p).

Note that in this instance we could have solved this algebraically:

$$x-2 = \frac{1}{x} \Leftrightarrow x(x-2) = 1 \Leftrightarrow x^2-2x-1 = 0$$

$$\therefore x = \frac{2 \pm \sqrt{4-4\times 1\times -1}}{2}$$

$$x = \frac{2\pm\sqrt{8}}{2}$$

$$= 1\pm\sqrt{2}$$

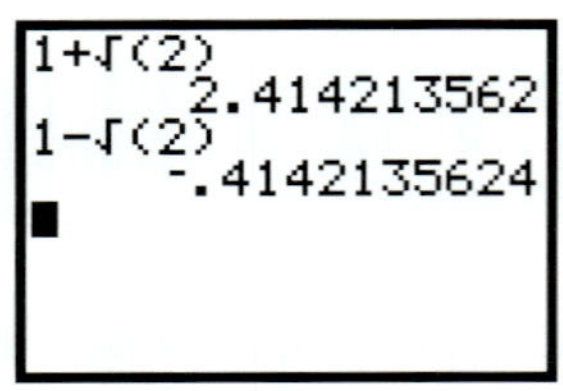

(b) Here we make use of the solve option from the catalog.
However, we first sketch the curves to identify the number of solutions. We observe that there are two solutions:

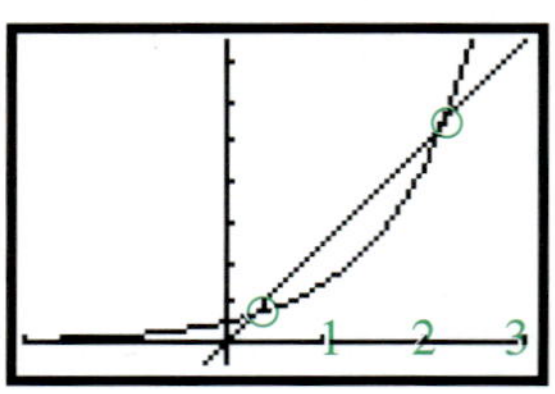

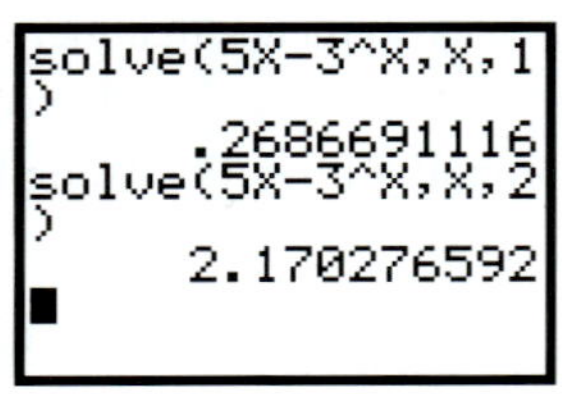

When using solve function we must first enter the equation as $f(x)-g(x)$.
That is, first write $f(x) = g(x)$ as $f(x)-g(x) = 0$ and then use the expresion $f(x)-g(x)$.
Also, we had to use two different guesses – these were determined from the graph.
Therefore, we have that $x = 0.27$ and $x = 2.17$ (to 2 d.p).

(c) Once again we start by sketching the curves on the same set of axes. This will show that there is only one point of intersection and therefore only one solution.

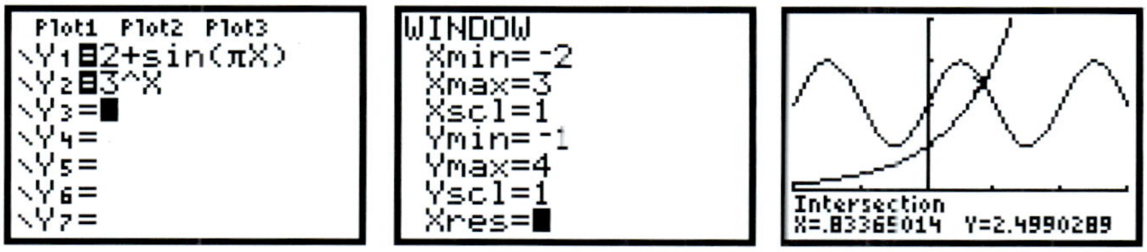

Therefore, $x = 0.83$ to 2 d.p.

EXERCISES 8.4

1. Solve each of the following, giving your answer to 2 decimal places.
Nb: You should use a graphical display to confirm the number of solutions in each case.

(a) $x + 1 = \dfrac{2}{x}$ (b) $2x = 2^x$ (c) $\dfrac{1}{x-1} = 3^x$

(d) $x^3 = x + 2$ (e) $\sqrt{x} = 2 - x$ (f) $2 + \sin x = x$

(g) $x^4 = 4 - x^2$ (h) $\dfrac{1}{x^2} = 9 - x$ (i) $2^2 = 1 - 3^x$

(j) $\sqrt{x-1} = \sqrt{5-x}$ (k) $\sqrt{x-1} = 5 - x$ (l) $1 - \dfrac{3}{x} = x^2$

8.5 MISCELLANEOUS EXERCISES

1. A relation is given by $f : x \mapsto \dfrac{2x}{3} + 1$, for $0 \le x \le 3$.

(a) Is this relation a function?
(b) Sketch the graph of f.
(c) For f, state i. the domain ii. the range.
(d) For what value of x will the image of f be 9?

2. The displacement, s m, of an object from a fixed point O after being in motion for t seconds is given by the function $s : t \mapsto t^2 + 2$, where $t \ge 0$.

(a) Find the object's initial displacement.
(b) How far from the origin will the object be after travelling for 8 seconds?
(c) Sketch the graph representing the object's displacement at any time t seconds.

3. Sketch the function defined as $f(x) = \begin{cases} \dfrac{1}{2}x + 1 & \text{if } x > 2 \\ x & \text{if } x \le 2 \end{cases}$

4. The function $f:x \mapsto a(x-3)^2 - b$ is shown in the figure alongside.

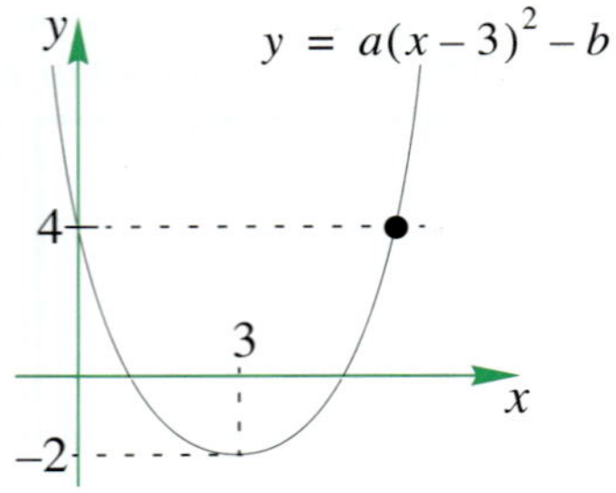

(a) State the range of the function.
(b) State the coordinates of
  (i) the y-intercept(s).
  (ii) the x–intercept(s).
(c) Find the values of a and b.

5. The temperature, T°C of kettle, t minutes after it is removed from the stove is thought to to be given by $T(t) = 28 + \frac{64}{t}, t > 0$.

(a) Complete the table below

t min	2	4	8	16	32
T °C					

(b) Use your results of (a) to plot the graph of $T(t) = 28 + \frac{64}{t}, t > 0$.
(c) What will the eventual temperature of the kettle be?
(d) How long will it take the kettle to reach a temperature of 40°C?

6. If $f:x \mapsto x^2 - x + 1$, i. find $f(2)$ ii. a given that $f(a) = 3$

7. The point $(1, 6)$ and $(-1, 0)$ lie on the curve $y = x^2 + bx + c$.
(a) Show that $b + c = 5$ and $-b + c = -1$.
(b) Using (a) solve for b and c.

8. A function is defined by $f:\mathbb{R} \mapsto \mathbb{R}, x \mapsto 7 - x^2$. Find the range of this function. Classify the function as one to one or many to one.

9. Sketch the graph of the linear piecewise function $f(x) = \begin{cases} x-1 & \text{if } x \geq 3 \\ -x+5 & \text{if } x < 3 \end{cases}$

10. A function is defined by $f(x) = 2x + 1, x \in \mathbb{R}$. Find the value(s) of x for which $f(2x) = f(x+1)$.

11. A sheet of paper of dimension 12 cm by 8cm has four equal squares of side length x cm cut out from its corners.

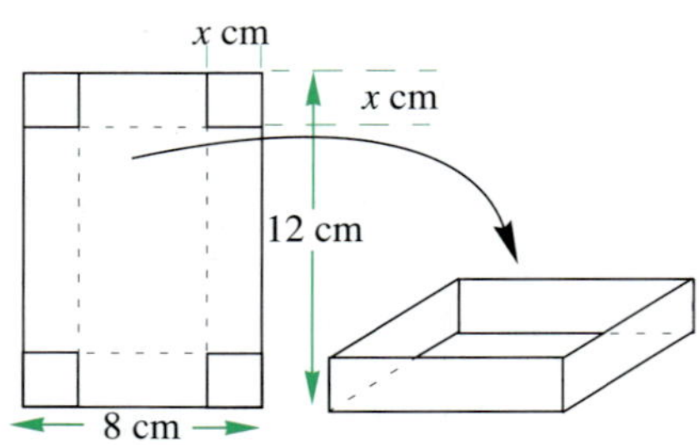

(a) Show that the volume, $V(x)$ cm^3 of the box is given by

$$V(x) = 4x(6-x)(4-x)$$

(b) Complete the table below

x cm	**0**	**1**	**2**	**3**	**4**
V cm^3					

(c) Estimate, to one decimal point, the maximum volume this box can have.

8.6 GRADED REVISION EXERCISES

LEVEL 1

1. State the domain and range of the following relations.

(a)

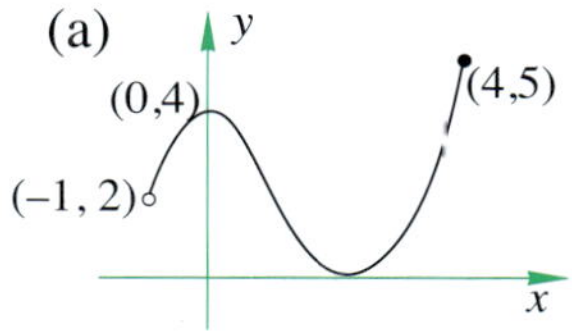

(b)

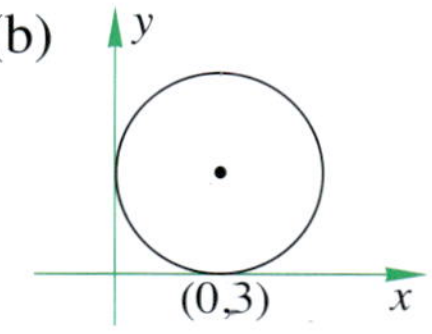

(c)

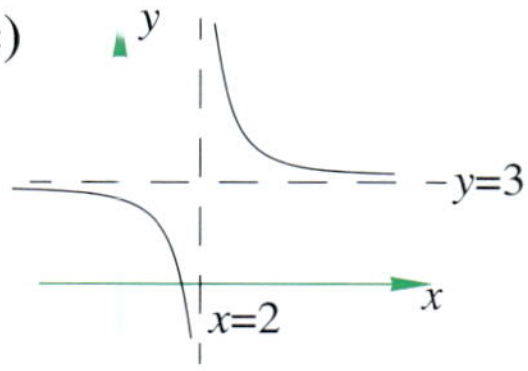

2. Which of the relations in question 1 are also functions?

3. Evaluate each of the following at $x = 2$.

(a) $y = x + 2$ (b) $y = (x+2)^2 + 2$ (c) $f(x) = 2^x + 2$

LEVEL 2

1. Given that $f(x) = \dfrac{x}{x+2}$, find $\{x : f(x) = 4\}$.

2. Find the largest possible set of x–values for which the following are defined.

(a) $x \mapsto \sqrt{2-x}$ (b) $x \mapsto \sqrt{x-3}$ (c) $x \mapsto \dfrac{1}{\sqrt{1+x}}$

3. Given that $f(x) = 2x - 1$ and $g(x) = \dfrac{1}{x+1}$ find

(a) $g(f(x))$ (b) $f(g(x))$

4. Sketch the graphs of the functions

(a) i. $x \mapsto x^2 + 1, x \in \mathbb{R}$ ii. $x \mapsto x^2 + 1, -1 \le x \le 2$

(b) State the range in each case.

5. For the function $f(x) = x^2 + 2x, x \in \mathbb{R}$, evaluate:

(a) $f(-1)$ (b) $f(2x)$ (c) $f(x-2)$ (d) $f(x) - f(-x)$

6. Sketch the graphs of the following, clearly showing intercepts with all axes.

(a) $f : x \mapsto 4x - 2$ (b) $f : x \mapsto 5 - \frac{1}{2}x$ (c) $f : x \mapsto 2(3-x)$

7. For the function $f(x) = x(x-3), x \in \mathbb{R}$ find

(a) i. $f(3)$

ii. $f(h)$

(b) $f(3+h) - f(3)$

LEVEL 3

1. Sketch the graphs of

(a) $f(x) = \begin{cases} x+1 & \text{if } x > 1 \\ 2x & \text{if } x \le 1 \end{cases}$ (b) $f(x) = \begin{cases} 2 & \text{if } x \ge 0 \\ 1-x & \text{if } x < 0 \end{cases}$

2. Given the function, $f: x \mapsto 3 + \dfrac{1}{x-2}$, $x \ne 2$, find i. $f(3)$ ii. $f\left(\dfrac{2x+1}{x}\right)$

3. For the functions $f(x) = \dfrac{1}{x+1}$, $x \ne -1$ and $g(x) = \dfrac{1}{x} - 1$, find $g(f(x))$ and $f(g(x))$.

4. Using a graphics calculator, sketch each of the functions in question 1.

5. For the function $f(x) = \dfrac{x}{x-1}$, $x \ne 1$, find $\{x : f(x) = k\}$.

6. Sketch the graph of $f(x) = \begin{cases} \sqrt{x} & \text{if } x > 4 \\ \sqrt{8-x} & \text{if } x \le 4 \end{cases}$.

LEVEL 4

1. If $f: \mathrm{S} \mapsto \mathbb{R}$, where $f(x) = 4x - x^2$, find S, the largest subset of the positive real numbers set, such that f is an increasing function.

2. Find $\left\{x: f(x) = \dfrac{1}{f(x)}\right\}$ if $f: [0, \infty) \mapsto \mathbb{R}$, where $f(x) = x^2 - 2$.

3. Find the domain of the function $f : x \mapsto \dfrac{3}{\sqrt{9-x^2}}$. Sketch the graph of f.

4. The function f is defined by $f: x \mapsto 2x^2 - 2$, $x \in \mathbb{R}$.

(a) Find the integers a, b and c given that $f(x+2) = ax^2 + bx + c$.

(b) Find the values of x that satisfy the equation $f(x) + 2 = f(x+2)$.

5. (a) Sketch the graph of $f(x) = a + \dfrac{b}{x}$, $a > 1$, $b > 0$.

(b) On the same set of axes, sketch the graphs of

i. $y = f(x)$ ii. $y = (f(x))^2$

8.7 TOPIC TEST

1. If $f : x \mapsto x^2 - \frac{1}{x} + 1$ find i. $f(1)$ ii. $\{a : f(a) = 1\}$

[4 marks]

2. Sketch the graph of the linear piecewise function $f(x) = \begin{cases} 2x - 1 & \text{if } x > 4 \\ 4 & \text{if } x \le 4 \end{cases}$

[3 marks]

3. (a) Find the range of $f : x \mapsto 8x - x^2$, where $x \in \mathbb{R}$.

(b) Hence, find the range of $f : x \mapsto k + 8x - x^2$, where $x \in \mathbb{R}$.

[4 marks]

4. A function is defined as $f(x) = \begin{cases} bx & 0 \le x < 1 \\ 2^{-kx} & x \ge 1 \end{cases}$, where $k > 0$.

(a) If f is continuous in the interval $[0, \infty[$, show that $b \times 2^k = 1$.

(b) Sketch the graph of f and state its range.

[3 + 3 marks]

5. The temperature, T°C of kettle, t minutes after it is removed from the stove is thought to be given by $T(t) = 30 + \frac{60}{t+1}, t \ge 0$.

(a) Complete the table below

t min	1	3	7	14	29	59
T °C						

(b) Use your results of (a) to sketch the graph of $T(t) = 30 + \frac{60}{t+1}, t \ge 0$.

(c) What will the eventual temperature of the kettle be?

(d) How long will it take the kettle to reach a temperature of 50°C?

[5 marks]

6. The minimum charge for a telegram is $4.80 which includes the use of up to 10 words. Each additional word is charged at a rate of 40 cents. Show, on a graph, the cost of telegrams up to 14 words.

[4 marks]

7. Give the equations which define the graphs shown below.

(a)

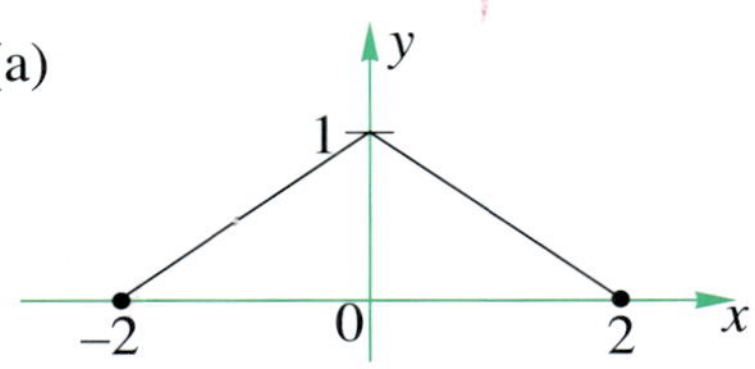

(b)

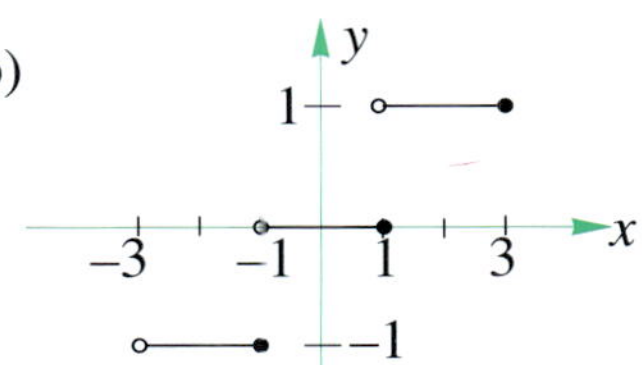

[6 marks]

Total 32 marks

8.8 PROJECT SUGGESTIONS

As we saw in the chapter introduction, the idea of a relation is quite general. It underlies much classification. Here are some examples of classification systems that you might investigate and explain in terms of relations.

Biological classification.

> Biologists relate the 'Great White Shark' to the genus *Carcharodon* and the species *carcharias*.
> How do the various methods of classification work?

The world-wide web:

> The internet consists of a huge number of 'pages' that are related to each other by a web of 'links'. Mostly, we now use search engines to find the pages we want.
> How do these work and how do they exploit web links?

Chapter 3 contains many examples of projects that deal with functions. Take your time and spend some time considering the different tasks (involving function work) that are available.

9.0 FIBONACCI

Leonardo of Pisa (Leonardo Pisano) known as Fibonacci was born around 1170 and died after 1240. He is best known for the sequence that bears his name. He heads this chapter because he was an early exponent of 'modelling' - the process of using mathematics to explore the real world. Fibonacci is also remembered for popularising our modern decimal system of writing numbers.

The modelling problem first given in *Liber Abaci* can be translated as:

'A man put a pair of rabbits in a place surrounded on all sides by a wall. How many pairs of rabbits can be produced from that pair in a year if it is supposed that every month each pair begets a new pair which from the second month on becomes fertile?'

Can you see how this leads to the sequence: 1,1,2,3,5,8...?

Fibonacci also stated that $x^2 + y^2$ and $x^2 - y^2$ cannot both be squares. Is this true?

The photograph shows the famous *Torre Pendente di Pisa*. The bell tower was begun in 1173 and is the third and final building of the city's cathedral complex. Fibonacci may well have seen the problem with the foundations as it developed!

Look at the straight line drawn along the right hand side of the tower. It suggests that the tower is not linear. Why is this?

9.1 MODELLING

This Chapter is dedicated to modelling. In particular, modelling that requires the use of work covered in Chapters 5 through to Chapter 7 – that is, work based on Linear Functions and Quadratic Functions as well as some work from Chapter 8.

However, before we start on the topic of modelling, we will consider a precursor to modelling and look at worded problems – which in themselves incorporate elements of modelling. We have already dealt with (in Chapters 5 and 6) worded problems based solely on linear equations and functions, so, we begin this chapter with worded problems that involve work based on the theory of quadratics.

We will take a three phase approach towards modelling.

Phase 1:	Constructing equations based on verbal descriptions
Phase 2:	Constructing functions/equations based on applications and verbal descriptions
Phase 3:	Constructing models based on applications and data.

9.1.1 MODELLING – PHASE 1

In this phase of our process we will encounter problems that are presented via some form of description of the situation at hand. The problems or situations can be described in one sentence or could entail an elaborate account of the problem. Whatever the details of the problem presented, the task at hand is to translate english into mathematics.

Much of what is required for solving these problems, you already possess. The rest is based on your experiences and general knowledge. For example, if the problem contains elements that refer to areas, then you will need to recall facts that allow you to work out areas of different shapes. On the other hand, if the problem involves a situation that is referring to time and distances then you will need to recall facts about the relationship between speed, distance and time. Once you have recalled all necessary knowledge, you can go about constructing an equation that can then be solved.

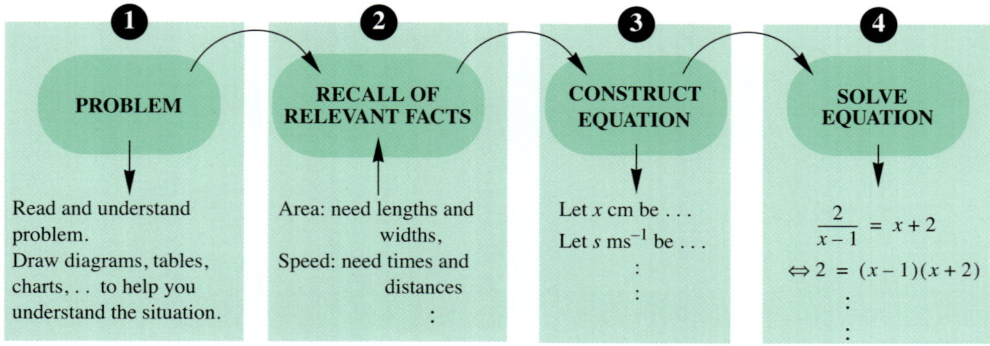

Let's consider some examples to highlight this process.

EXAMPLE 9.1

Find two positive numbers that differ by 2 and have a product of 63.

Solution

Having read the problem, it seems that we are in search of two numbers that possess particular features:

1. their difference is 2
2. their product is 63

Let these numbers be x and y, so that from 1. $y - x = 2$

2. $xy = 63$

That is we have formulated two equations with two unknowns. As such, it would then seem appropriate to consider this a system of equations that requires to be solved using one of the methods encountered in earlier chapters.
We label them as follows:

$$y - x = 2 \quad -(1)$$
$$xy = 63 \quad -(2)$$

From (1) we have $y = 2 + x \quad -(3)$
Substituting (3) into (2) gives:

$$x(2 + x) = 63 \Rightarrow 2x + x^2 = 63$$

We have obtained a quadratic equation. $2x + x^2 = 63$ which we can now solve.

$$2x + x^2 = 63 \Leftrightarrow x^2 + 2x - 63 = 0$$
$$\Leftrightarrow (x + 9)(x - 7) = 0$$
$$\Leftrightarrow x = -9 \text{ or } x = 7$$

If $x = -9$, from (3) we have $y = -7$
If $x = 7$, from (3) we have $y = 9$.
Therefore we have two pairs of numbers satisfying the conditions.
First pair: $x = -9, y = -7$ and the second pair is $x = 7, y = 9$.

However, the problem stipulates that the numbers are positive numbers and so we must discard the pair $x = -9, y = -7$. Leaving the solution as $x = 7, y = 9$.

It is also a good idea to check you answer: $9 - 7 = 2$ √ and $9 \times 7 = 63$ √.

EXAMPLE 9.2

Eighteen times a certain number exceeds the square of the number by forty five. What is this number?

Solution

Let this number be a. Then, eighteen times this number is written as $18a$.

We are told that $18a$ exceeds the square of the number by forty five.
The key terms here are:

1. the square of the number $\Rightarrow a^2$.
2. exceeds $\Rightarrow$ is greater than a^2 by 45 $\therefore 18a = a^2 + 45$.

We have formulated the equation $18a = a^2 + 45 \quad -(1)$
All that remains is to solve for a:

$$18a = a^2 + 45 \Leftrightarrow a^2 - 18a + 45 = 0$$
$$(a - 15)(a - 3) = 0$$

Therefore, $a = 15$ or $a = 3$.

We still need to check our answers.

Case 1. $a = 15$, L.H.S = $18a = 18 \times 15 = 270$ [substituting into (1)]
R.H.S = $a^2 + 45 = 15^2 + 45 = 270$
Therefore, as L.H.S = R.H.S, $a = 15$ is a solution.

Case 2. $a = 3$, L.H.S = $18a = 18 \times 3 = 54$ [substituting into (1)]
R.H.S = $a^2 + 45 = 3^2 + 45 = 54$
Therefore, as L.H.S = R.H.S, $a = 3$ is a solution.

Therefore, there are two such numbers, 3 and 15.

EXAMPLE 9.3

Florence purchased a number of hand-held electrical games for \$180. She decided to keep one for herself and sell the rest. She sold the remaining games for \$1.00 more than she bought them for. After keeping one game for herself, Florence made a profit of \$10.00. How many games did she purchase?

Solution

Let n be the number of games Florence bought, meaning that each game cost her $\$\frac{180}{n}$.

She then sold each game for \$1.00 more than she paid for.

i.e., she sold them for $\$\left(\frac{180}{n} + 1\right)$.

As she kept one for herself, she had $(n - 1)$ to sell.

Therefore, the revenue from selling the games is $\$\left(\frac{180}{n} + 1\right)(n - 1)$.

We also know that Profit = revenue – cost

Therefore, $$10 = \left(\frac{180}{n} + 1\right)(n - 1) - 180$$

Or, $$\left(\frac{180}{n} + 1\right)(n - 1) = 190. \text{ – (1)}$$

Now that we have our equation we proceed to solve it:

$$\left(\frac{180}{n} + 1\right)(n - 1) = 190 \Leftrightarrow \left(\frac{180 + n}{n}\right)(n - 1) = 190$$
$$\Leftrightarrow (180 + n)(n - 1) = 190n$$
$$\Leftrightarrow n^2 + 179n - 180 = 190n$$
$$\Leftrightarrow n^2 - 11n - 180 = 0$$
$$\Leftrightarrow (n - 20)(n + 9) = 0$$
$$\therefore n = 20 \text{ or } n = -9$$

As $n > 0$, $n = 20$ is the only plausible solution.
Checking with equation (1) shows that $n = 20$ is a solution.
Therefore, Florence bought 20 games.

EXAMPLE 9.4

Francesco walks 24 km everyday. He always maintains a constant speed. If he had walked 2 kmh^{-1} faster than he usually does, he would have completed his walk one hour earlier. At what speed does Francesco usually walk?

Solution

The problems is looking for a speed and so, we need to have a variable that defines this speed. Let x kmh^{-1} be Francesco's usual speed.
We are given information about time and speed and distance. These quantities are related by the following equation; $\text{speed} = \dfrac{\text{distance}}{\text{time}}$.
As we are comparing to what ususally happens it is worthwhile finding out the time that it usually takes Francesco to walk 24 km:

Let t_1 hr be the time it usually takes to walk 24 km. $\therefore t_1 = \dfrac{24}{x}$. – (1)

Let t_2 hr be the time it takes if he walks 2 kmh^{-1} faster. $\therefore t_2 = \dfrac{24}{x+2}$. – (2)

However, it takes him 1 hour less if he walks 2 kmh^{-1} faster, i.e., $t_2 = t_1 - 1$

Using equations (1) and (2) we have

$$\frac{24}{x+2} = \frac{24}{x} - 1 \quad \text{– (3)}$$
$$\Leftrightarrow 24x = 24(x+2) - x(x+2)$$
$$\Leftrightarrow 24x = 24x + 48 - x^2 - 2x$$
$$\Leftrightarrow x^2 - 2x - 48 = 0$$
$$\Leftrightarrow (x+8)(x-6) = 0$$
$$\therefore x = -8 \text{ or } x = 6$$

As $x > 0$, discard $x = -8$.

Check $x = 6$: Using (3), we have, L.H.S $= \dfrac{24}{6-2} = \dfrac{24}{8} = 3$. R.H.S $= \dfrac{24}{6} - 1 = 4 - 1 = 3$.

Therefore $x = 6$ is the solution.
That is, Francesco's usual speed is 6 kmh^{-1}.

EXAMPLE 9.5

A man $18x$ years old has a son who is $2x^2$ years old. When the man was $3x^2$ years old, the son was only $(x+4)$ years old. How old would the man be today?

Solution

In this scenario we have a mixture of english and mathematics within the question. The variable has already been defined for us. So, we begin by setting up a table to display the information:

	Age today	Age then	Difference in age
Father	$18x$	$3x^2$	$18x - 3x^2$
Son	$2x^2$	$(x+4)$	$2x^2 - (x+4)$

The number of years that has passed from 'then' to 'now' is the same for both the father and the son. Therefore, the difference in their respective ages must be the same.

Therefore, we have that $18x - 3x^2 = 2x^2 - (x+4)$ – (1)

$$\Leftrightarrow 5x^2 - 19x - 4 = 0$$

$$\Leftrightarrow (5x+1)(x-4) = 0$$

$$\text{i.e., } x = -\frac{1}{5} \text{ or } x = 4$$

Now, as $x > 0$, the only possible solution is $x = 4$.
Therefore the father must be $18 \times 4 = 72$ years old.
Checking with (1) will show this to be the case.

EXERCISES 9.1.1

1. What positive number, when subtracted from its own square gives an answer of 56?

2. Eight times the sum of a number and its reciprocal results in 34. Find this number.

3. The square of the difference between a number and 12 is twice the number. Find it.

4. The product of two numbers is 88. What are the numbers if one number is 3 more than the other?

5. The sum of four times a number and its square is 60. What is the number?

6. Twice the square of Graham's height is 6 metres more than his height. How tall is he?

7. Rachelle has taken $(4x-6)$ hours to travel 102 km at a speed of $(25x+1)$ kmh^{-1}. How fast was she travelling?

8. The product of two consecutive odd numbers is 143. If x is the smaller of the two numbers, show that $x^2 + 2x = 143$. Find the numbers

9. It will take an apprentice 15 days longer than a qualified painter to complete a job. However, if 3 apprentices and two qualified painters work together the job can be completed in $3\frac{1}{8}$ days. How long would it take one apprentice to do the job?

10. At a local screening of a movie, 450 people are seated in rows such that every row contains an equal number of people. However, if there were three more people in each row the number of rows needed would be reduced by 5. How many rows are there?

11. Derek sold walking sticks. Had he charged \$10 extra for each stick he would have made \$1,800. However, had he sold 10 more sticks at the original price he would have made \$2,000.
(a) How many sticks did he sell?
(b) How much was each stick?

12. Bella set off on a 96 km journey, all the while she maintained a constant speed. On her return she again kept a constant speed but this time increased it by 2 kmh^{-1}, reducing her travel time by 4 hours. What was her original speed?

13. The hiring cost for transportation for a class of students was \$144. If two students from a different class were to attend, the cost per student would have decreased by \$1.00. How many students are there in the class?

14. A person walked a distance of 4 km at a constant speed of v kmh^{-1}. If this person increased his speed by 2 kmh^{-1}, the journey would have been reduced by 10 minutes. Find v.

15. A train travels at a constant speed for 120 km. Had the train been travelling 5 kmh^{-1} faster, it would have completed the journey 20 minutes earlier. How long would it take the train to travel 120 km?

16. A group of students are to pay equal amounts to a total of £72. However, Leanne, Josh and Andrew have successfully argued that they should not pay, in which case the others would each have to pay an extra £4. How many people were there in the group?

[Questions 17 – 20 are more challenging. These are often referred to as productivity problems.]

17. Two taps having different rates of flow are used to fill a large water tank. If tap A is used on its own it will take 5 hours longer to fill the tank than it would tap B to fill it on its own. Together, the taps would fill the tap in 6 hours. Assuming that the taps are running at full capacity, find
(a) how long will it take for tap A to fill the tank.
(b) how long will it take for tap B to fill the tank.

18. Two chair manufacturers have been in competition for many years. To protect the innocent in this story we will refer to these companies as Chair-one and Chair-two. Chair-one had to fulfill an order for 810 chairs within a set time, while Chair-two had to fulfill an order for 900 chairs in the same amount of time. Competition led Chair-one and Chair-two to complete the task 3 days and 6 days in advance respectively. We know that Chair-one makes 4 chairs less, per day, than Chair-two. How many chairs does each manufacturer produce per day, assuming that they are working at maximum capacity.

19. The local council has sent out two teams to complete work on a stretch of road. Team A works on the left lane while team B works on the right lane. Although Team B started one day after Team A, each team did manage to repair a 10 km stretch of the road. The teams daily joint quota was 4.5 km per day. How many kilometres per day did
(a) Team A repair?
(b) Team B repair?

20. Two workers need to dig a trench. One worker (working alone) can do the job three hours quicker than the second worker. The total time taken by the workers if they were to work separately on this job is $\frac{144}{35}$ times longer than the time it would take them to do the job if they had worked together.
How long would it take each worker to do the job if they were to do it on their own?

9.1.2 MODELLING – PHASE 2

There are no real changes in moving from Phase 1 to Phase 2. The changes, where they exist, are rather subtle. They might involve the use of diagrams, an aspect that did not feature in Phase 1. They might involve the use of trial and error or the construction of a function as opposed to an equation.

For example, if the problem is referring to the area of a square, then after defining the side length of the square as having a length x cm, you might decide that it is appropriate to intoduce the function $A(x)$ cm^2 to define the area of this square.

One difference, however, can occur in the way information is provided and therefore, how that information and/or data is used. Here is an example:

A common task that is peformed by farmers is setting up enclosures on their farm. A particular farmer has 30 metres of fencing with which to enclose an area of 100 m^2 . The farmer uses a wall to act as one side of the area to be enclosed. What are the possible dimensions of this enclosure?

We begin by using a diagram that includes the information provided:

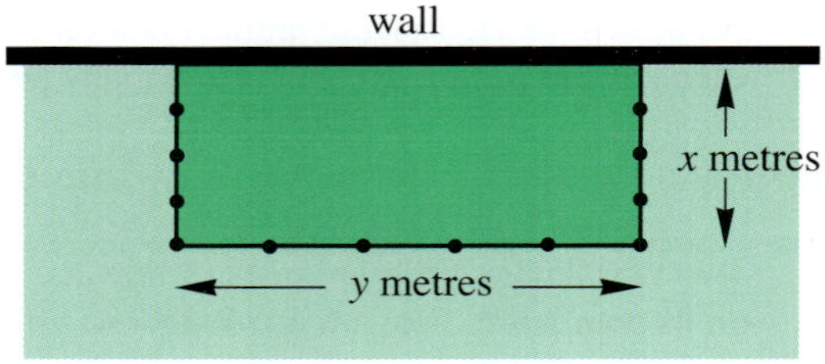

Next we introduce variables to define the length and width, so,

let the length be y m and the width x m.

Then, as we are given 30 metres of fencing, we must have that $x + y + x = 30$

i.e., $2x + y = 30$ – (1)

Also, we are told what the area is to be, 100 m^2 .
As the area is a feature of this problem we also define it by introducing a function to represent the area:
Let the area enclosed be defined by A m^2 , so that $A = x \times y$.

Now, at this point in time we observe that there are three variables, two independent ones, x and y and the dependent variable A.

Somehow we need to reduce this to one independent variable and one dependent variable. We keep A and now look to keep one of x or y. The choice as to which one is kept is not important (in this example), so we choose to keep x, meaning that we need an expression for y in terms of x.
From (1) we have that $2x + y = 30 \Leftrightarrow y = 30 - 2x$.

This means that we can now express A in terms of one variable, namely

$$A = x \times y = x \times (30 - 2x)$$

However, we know that the area must be 100 m^2 , therefore,

$$100 = x(30 - 2x)$$

$$100 = 30x - 2x^2$$

That is, we now have a quadratic equation to solve.

$$2x^2 - 30x + 100 = 0$$
$$\Leftrightarrow 2(x-5)(x-10) = 0$$
$$\therefore x = 5 \text{ or } x = 10$$

When $x = 10$, the length is $30 - 2(10) = 10$, giving an area of $10 \times 10 = 100 \text{ m}^2$. √
When $x = 5$, the length is $30 - 2(5) = 20$, giving an area of $5 \times 20 = 100 \text{ m}^2$. √
It is always a good idea to check your answer.
So that the possible dimension satisfying the restictions are:10 m by 10 m and 5 m by 20 m

Comparing this solution with those in the previous section, we see that there is a build-up to the solution, in the sense that we had three variables to start with and then somehow had to express one in terms of the other. Although we did solving an equation, we first created a function. Now, these are subtle differences, but differences nonetheless. This does not mean that all problems in this section will have three variables to start with. Each question will need to be dealt with on its own merit.

Before moving onto some examples, lets consider a slight alteration to the question above.
This time, we want:

Of all possible rectangular enclosures, which one will have the maximum area?

Of course we could use the method of trial and error. Some possible scenarioes are show below.

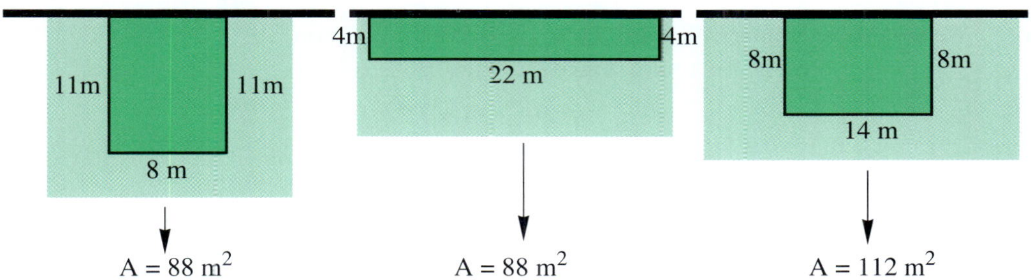

As we can see, different dimensions produce different areas (in some cases, the same area!).

To determine the dimensions that will provide the largest area we make use of the area function, $A = x(30-2x)$. In fact, because A is a function of x only, we can write it as

$$A(x) = x(30-2x)$$

All that remains is to sketch the graph and use it to determine its maximum value.
Notice that if $x = 0$, $A = 0$ and if $x = 15$, $A = 0$.

So, we need to restrict the values of x to $0 < x < 15$.
Our definition of A now becomes, $A(x) = x(30-2x)$, $0 < x < 15$.

Using the **CALC** menu and option **4:maximum** we were able to obtain the maximum area as 112.5 m^2.

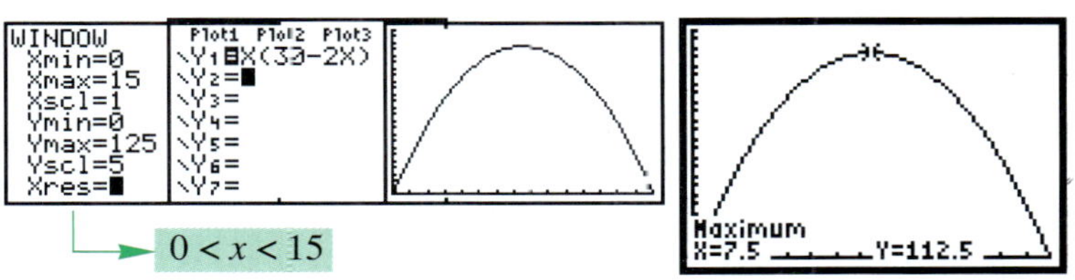

EXAMPLE 9.6 The height, h m, above ground level, reached by a ball, t seconds from when it is thrown, is given by the equation $h = 20t - 5t^2 + 1$.

(a) How long did it take for the ball to first reach a height of 16 metres?
(b) What is the maximum height reached by the ball?
(c) For how long is the ball in flight?

Solution

Unlike previous problems, we are supplied with all the variables and the equation! In fact, there is no need to define or introduce any new variables. However, here is a problems where our task is to interpret the information that has been given.

We start by visualising and interpreting the situation:

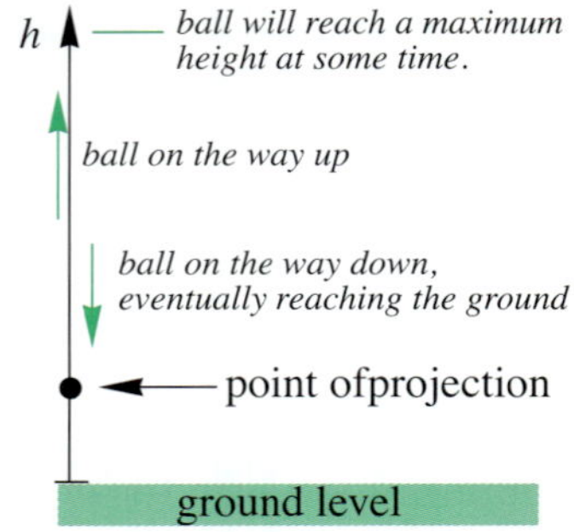

As t represents time, then we know that $t \geq 0$.
Also, when $t = 0$, we have that

$$h = 20 \times 0 - 5 \times 0^2 + 1 = 1$$

This tells us that initially, the ball is 1m above ground level (the point of projection).

Having obtained some 'feel' for what is happening, we now start answering the questions.

(a) We wish to determine t when $h = 16$, so, we solve the equation

$$16 = 20t - 5t^2 + 1 \Leftrightarrow 5t^2 - 20t + 15 = 0$$
$$\Leftrightarrow 5(t^2 - 4t + 3) = 0$$
$$\Leftrightarrow 5(t-3)(t-1) = 0$$
$$\therefore t = 3 \text{ or } t = 1$$

As we are asked for the first time, then $t = 1$. That is, it took 1 sec to reach 16 m.

But what of $t = 3$, what does it represent?
As the ball must come down, $t = 1$ represents the time when the ball is 16 m above ground level on its way up. While $t = 3$ represents the time taken for the ball to be 16 m from the ground, but this time it is on its way back down. This also means that it took 2 seconds for the ball come back to a height of 16 m from the time it first reached that height on its way up.

(b) We recognise the equation as a quadratic, so we can sketch its graph and use it to determine the maximum height by locating its turning point.

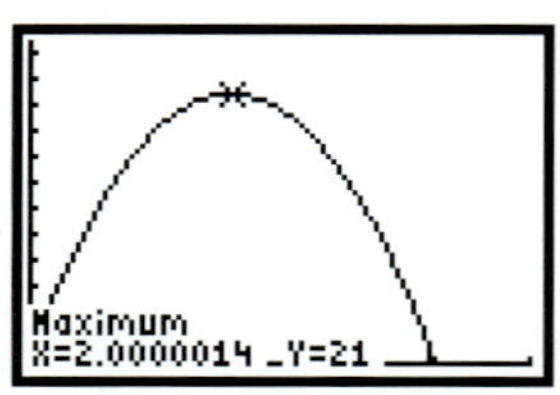

Using the **CALC** menu and option **4:maximum** we have that the maximum height reached is 21 metres, 2 seconds after it was projected.

(c) When $h = 0$, the ball will have reached ground level and so, we need to find that value of t for which $h = 0$.

i.e., $0 = 20t - 5t^2 + 1$

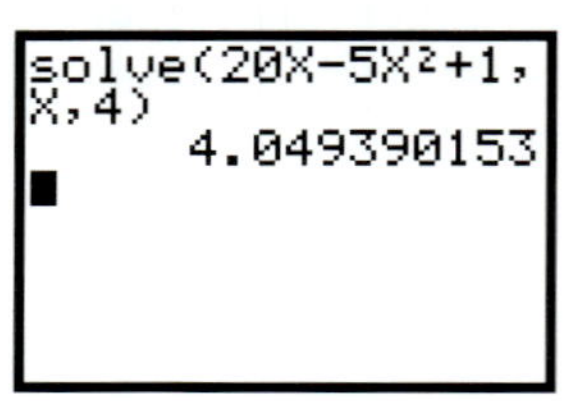

There are a number of options that can be used.
e.g., use **CALC** menu and select option **2:zero** or the **TRACE** option or as we have done, use solve from the catalogue menu.
Giving $t = 4.05$ (to 2 d.pl.).

Had we used the quadratic formula in part (c), we would have obtained the following solutions,

$$0 = 20t - 5t^2 + 1 \therefore t = \frac{-20 \pm \sqrt{20^2 - 4 \times -5 \times 1}}{2 \times -5}$$

$$= 4.0494, -0.0494$$

However, as $t \geq 0$ we would have rejected $t = -0.0494$.

We find that linear models are used in many facets of life as was seen in Chapter 6 and will be seen in Chapter 24. One area where this is very common is that of economics and business. Examples can be found in equations that model company profits and revenue as well as the costs incurred during the running of a business. There are a number of terms that are often referred to when dealing with economic and business situations. Some of these terms are; profit, revenue, cost, depreciation, market demand, market supply, market equilibrium and break-even point. Analysing economic behaviour through mathematical interpretation is a skill which is sought by many of the leading firms around the world. Modelling such economic behaviour displays the power inherent in mathematics.

EXAMPLE 9.7

The total cost, $\$C$, to a manufacturer of calculators is given by the equation

$$C = 12n + 40000$$

where n is the total number of calculators produced. The manufacturer sells these calculators to suppliers at a fixed price of \$32 each.

(a) Find the manufacturer's cost when 1000 calculators are produced.

(b) Will the manufacturer make a profit when 1000 calculators are produced?

(c) How many calculators must this manufacturer produce in order to break-even?

Solution

(a) When $n = 1000$, $C = 12(1000) + 40000$
$= 52000$
That is, it will cost the manufacturer \$52,000.

(b) Let R, dollars, denote the revenue. If the manufacturer has sold 1000 calculators, the revenue is given by $R = 32 \times 1000 = 32000$.

Therefore the manufacturer makes a profit of \$32,000 – \$52,000 = – \$20,000.
A negative profit means that the manufacturer has made a loss.
Therefore the manufacturer has lost \$20,000!

(c) The break-even point occurs when no gain or loss is made, that is, when $R = C$.

Because the **profit** is defined as the **revenue – cost**, i.e., $\boldsymbol{P = R - C}$, the **break-even** point occurs when $\boldsymbol{P} = 0$.

Therefore, if n calculators are produced, we have

$$32n = 12n + 40000$$
$$\Leftrightarrow 20n = 40000$$
$$\Leftrightarrow n = 2000$$

That is, to break-even, the manufacturer needs to produce (and sell) 2000 calculators.
Although we have answered the questions, it will be of great use if we can visualise this situation

by sketching the relationships involved:

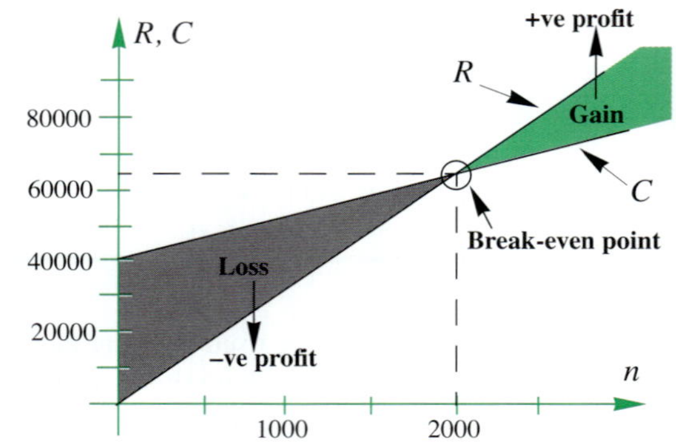

From the graphs, we see that the break-even point corresponds to the point of intersection of the two straight lines representing

A. the cost function, $C = 12n + 40000$

and

B. the revenue function, $R = 32n$

Notice how we have only sketched the graphs for values of n that are greater than (or equal to) zero.

The above example illustrates that the amount of items a manufacturer sells and the price of the item determine the total amount of money earned by selling the items. If the item is more expensive the number sold commonly decreases. The cost of producing items often depends on a fixed (set-up) cost and a price per item (resource cost). Three other terms that are of importance in economic modelling are:

Market demand The quantity of a product (goods or services) that consumers are willing to buy at various prices.

Market supply The quantity of a product (goods or services) that producers are prepared to offer for sale at various prices.

Market equilibrium This occurs when the quantity demanded equals the quantity supplied.

EXAMPLE 9.8

Suppose that we have found that x units can be sold daily for a price of p dollars per unit where $x = 800 - p$ and the cost of making these x units is given by the function

$$C(x) = \begin{cases} 500 + 20x & \text{if } x > 0 \\ 0 & \text{if } x = 0 \end{cases}$$

Find (a) the daily revenue function, $R(x)$.

(b) the daily profit function $P(x)$.

Assuming the production capacity is at most 400 units per day, find

(c) how many units should be made and sold to maximize profit.

(d) the maximum profit per day.

(e) the unit price to be charged to maximize the profit.

Solution

(a) The revenue is $R(x) = x \times p = x(800 - x)$.

(b) Profit is $P(x) = R(x) - C(x)$

$$= x(800 - x) - (500 + 20x)$$

$$= 780x - 500 - x^2$$

(c) The profit function is given by a quadratic function, meaning that it will form an arc of a parabola (or at least part of an arc). The reason we state '*part of an arc*' is that we need to consider the restrictions on x.
First of all $x \geq 0$. Then, we are told that the production capacity is 400 units, so we have that $x \leq 400$. That is, the profit function has a domain given by $0 \leq x \leq 400$.

We can then use the turning-point form of the parabola:

$$\begin{aligned} P(x) &= 780x - 500 - x^2 \\ &= -(x^2 - 780x + 500) \\ &= -[(x-390)^2 - 152100 + 500] \\ &= -(x-390)^2 + 151600 \end{aligned}$$

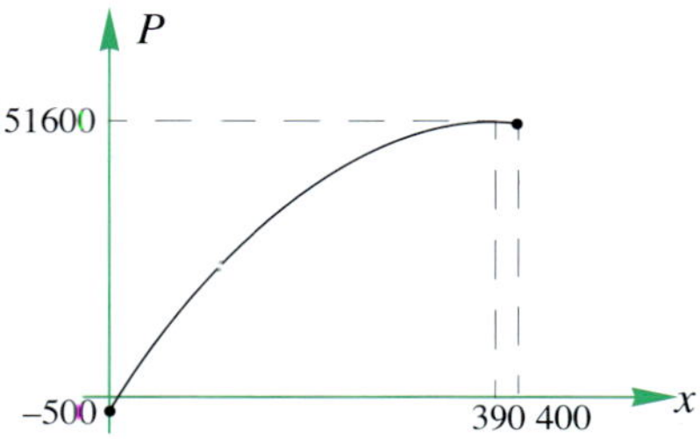

From this equation we can see immediately that the maximum occurs at $x = 390$.

(d) The maximum profit is then \$151600.

(e) The unit price is $800 - 390 = 410$ dollars.

Note: In this case we have not used the full capacity. If our production limit were 350 units per day then the optimum would have occurred at $x = 350$ which would not be at the vertex of the arc but at an endpoint.

EXAMPLE 9.9

The graph shows the cost function, $C(x)$, and the revenue function, $R(x)$, for a transistor manufacturer, where x is the number of units sold.

(a) Determine the number of transistors that need to be sold for the manufacturer to break-even.

(b) For what values of x will the manufacturer be
  i. in the black?
  ii. in the red?

(c) Determine the maximum profit that the manufacturer can make.

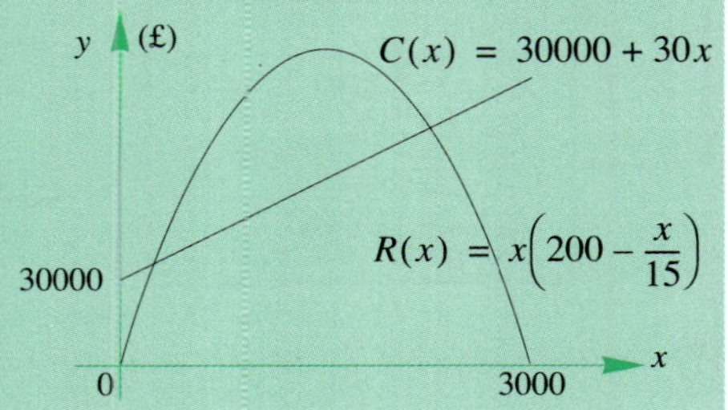

Solution

(a) The manufacturer will break-even when $R(x) = C(x)$.

That is, $x\left(200 - \frac{x}{15}\right) = 30000 + 30x \Leftrightarrow 200x - \frac{x^2}{15} = 30000 + 30x$

$$\Leftrightarrow \frac{x^2}{15} - 170x + 30000 = 0$$

$\therefore x \approx 190.74$ or $x \approx 2359.26$ [using the quadratic formula]

Therefore, as x is an integer, we can say that at $x = 191$ and $x = 2359$, the manufacturer will break-even.

(b) i. To be in the black means to be making a positive profit –i.e., a gain.
This occurs when $R(x) > C(x)$ or simply, $P(x) > 0$.
From part (a) and the graph, this would occur when $191 \le x \le 2359$.

ii. To be in the red means to be making a negative profit – i.e., a loss.
This occurs when $R(x) < C(x)$ or simply, $P(x) < 0$.
From part (a) and the graph, this would occur when $0 \le x \le 190$ or $2360 \le x \le 3000$.

These results can be shown on the graph below:

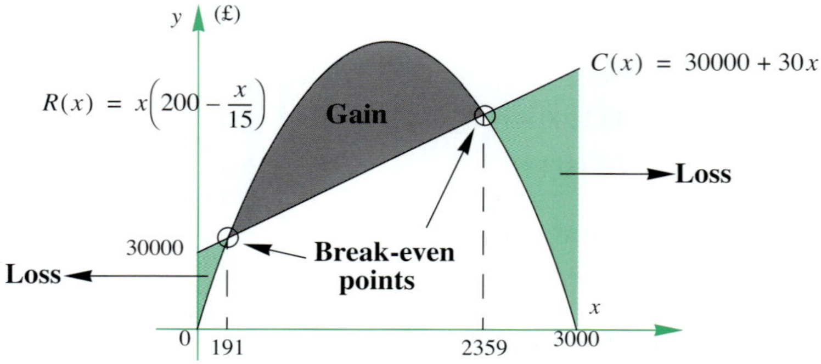

(c) The profit equation is given by, $P(x) = R(x) - C(x)$.

$$\therefore P(x) = x\left(200 - \frac{x}{15}\right) - (30000 + 30x)$$

$$= -\frac{x^2}{15} + 170x - 30000,\ 0 \le x \le 3000$$

We determine the maximum value of this quadratic function using the graphics calculator:

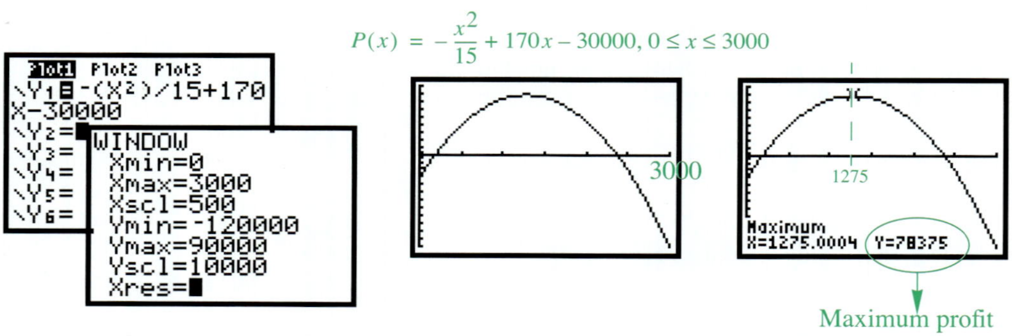

We can then say that the maximum profit is \$78,375 and occurs when 1275 items are sold.

Note the importance of setting the appropriate window in order to obtain a good visual display of the behaviour of the profit function.

As we have seen in this section, there are times when information, including equations is given. This information has been obtained by some means and is presented as a summary – which, as we have seen can come in different forms. However, it is then up to you to use this information and from it be able to analyse and provide findings that are consistent with the information presented.

We conclude this section with a quick note on aspects of linear functions that have not been dealt with specifically in previous chapters but have a number of applications in many standard practices that involve the usage of utilities such as water, electricity and so forth as well as payments for taxi fares, postage and the like.

PIECEWISE LINEAR RELATIONS

Some functions are made up of several different rules for different subsets of their domains. Such functions are often referred to as hybrid functions. A special case of hybrid functions is when each rule is a linear function, in this case we have what is known as a **piecewise linear function**. An example of such a function is

$$f(x) = \begin{cases} x-2 & \text{if } x > 4 \\ 2x-6 & \text{if } x \le 4 \end{cases}$$

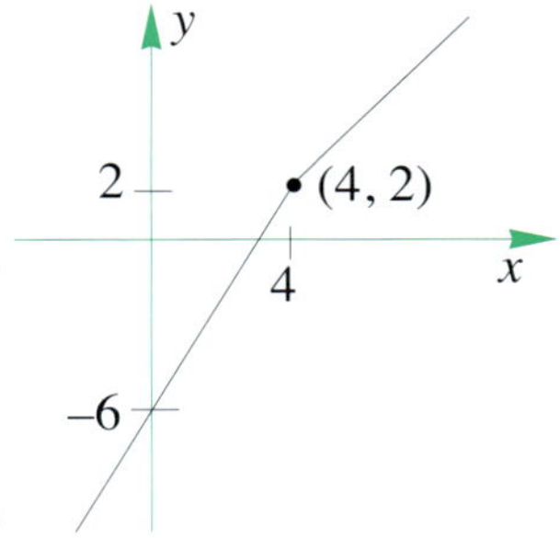

To sketch this graph we sketch the graph of $y = x-2$ for $x > 2$ and then sketch the graph of $y = 2x-6$ for $x \le 4$:

Notice then, that because the two lines meet at the point (4, 2), the resulting function is called a **piecewise continuous linear function**.

Hint: The easiest way to sketch these graphs (other than using a graphics calculator) is to sketch each straight line as if there were no restrictions and then rub out those sections that are not required.

On the TI–82 or TI–83 graphics calculator, we can sketch the above function as follows:
After entering the equation editor, i.e., **Y =** screen, type the following:

Y = (X – 2)(X > 4) + (2x – 6)(X ≤ 4)

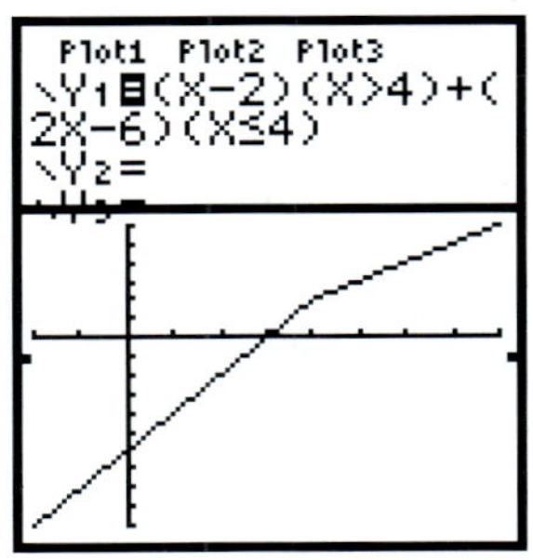

The expressions (**X > 4**) and (**X ≤ 4**) are found under the **TEST/ LOGIC** window. To access this window press **2nd MATH**.
i.e., type **(** **X** **2nd MATH** [then use the arrow key to select **>**, press ENTER] **4** **)** + **(** **2** **X** **–** **6** **)(** **X** 2nd MATH [then use the arrow key to select ≤, press ENTER] **4**) and then press **GRAPH**.

EXAMPLE 9.10

A Water Supply company decides to change its rates to increase the user-pays component of water supply and it phases in the procedure. The new method included a fixed charge (possibly different for each property depending on its assessed value) and a rate of 45 cents per kilolitre for the first part used and 65 cents per kilolitre for the second part.

In a particular bill, the charge for a quarter is \$150 plus a charge for up to 50kl at the lower rate and the rest at the higher rate. If the water used is 63kl what will be the total bill?

Solution

Let the volume of water used be denoted by v kL, then, based on the information given, we have that

1. for $v \le 50$, the total cost is given by $150 + 0.45v$

2. for $v > 50$, the total cost is given by $172.5 + 0.65(v - 50)$

As there are two linear graphs that make up the function, we have a piecewise linear function. The graph then has the form shown.

Here for $v = 63$ we have the second rule applying and therefore, $C = 180.95$

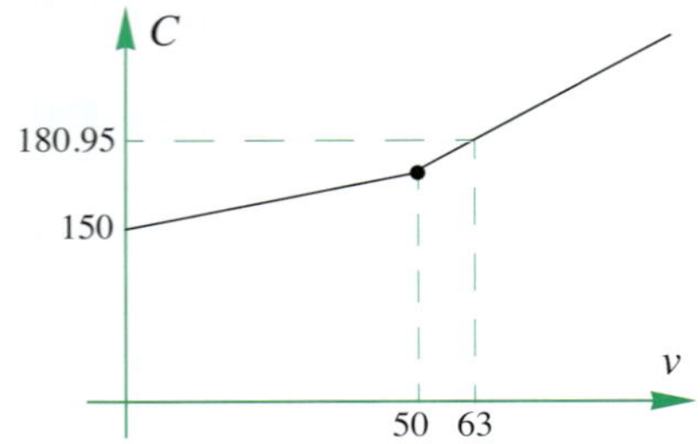

GREATEST INTEGER FUNCTION

Another type of piecewise linear function is the **greatest integer function** (y = [x]) or the **step function**.

These graphs are made up of discontinuous horizontal lines (segments) that resemble a stair case.

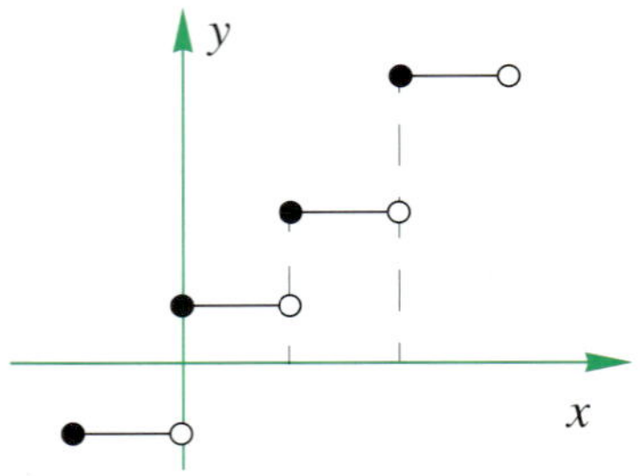

Such graphs are useful when modelling the cost of mailing parcels of different weights.

For example, if it costs \$3.00 to send parcels weighing between at least 1kg but less than 3 kg, \$ 7.00 for parcels weighing at least 3 kg but less that 5 kg, \$ 13.00 for parcels weighing at least 5 kg but less that 8 kg and so on, then, we can summarise that information using the graph shown:

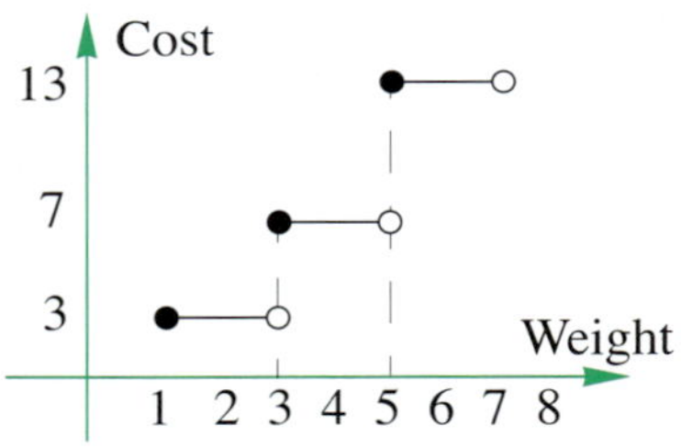

EXERCISES 9.1.2

1. Bobby has decided to start a vegetable garden so that he may sell his produce at the local Sunday market. He buys 100 metres of wirenetting to fence off a rectangular area using an existing fence for one side of the rectangle.

Bobby wishes to fence off the largest possible area. Let the length of one side of the rectangle be x m as shown in the diagram.

(a) i. Find the length of the other dimension in terms of x.
ii. What values can x take?

(b) Let A m^2 denote the area of the enclosure.
i. Find an expression for A in terms of x.
ii. What are the dimensions of the enclosure if it has an area of 800 m^2 ?
iii. What is the largest possible area that the enclosure can have?
iv. What are the dimensions of the enclosure that produce the largest area?

2. A small rectangular enclosure is to be placed in a paddock. There is 24 m of fencing available for the enclosure.

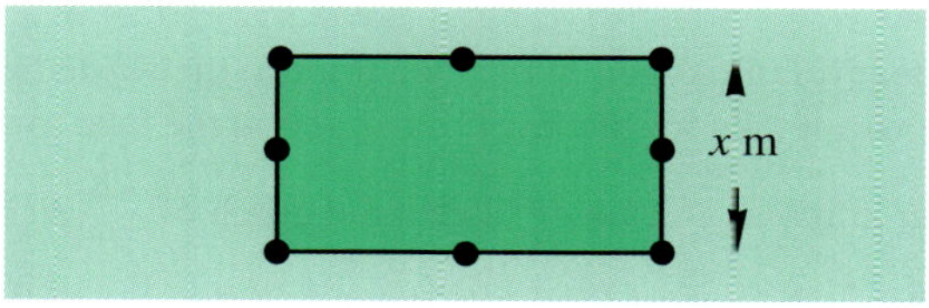

(a) i. If x m is the length of one of the sides, show that the area, A m^2, is given by $A = x(12 - x)$.

ii. What restrictions need to be placed on x?

(b) Find the area of this enclosure when

i. $x = 2$ ii. $x = 4$ iii. $x = 8$

(c) Sketch the graph of the function $A(x)$ for its specified domain.

(d) Find the dimensions of the enclosure that will have the largest possible area.

3. The cost, $\$C$, of new small cars in a particular state have been increasing according to the relation $C(t) = 12,000 + 25t^2$, where t is the time in years since 2000.

(a) Sketch a graph of this relation from 2000 to 2008.

(b) How much would you expect to pay for a new small car from this state in 2006?

4. The demand equation for a certain product is given by the equation $p = 40 - 0.0004x$, where p is the price per unit and x is the number of units sold.

(a) Find the equation for the total revenue, $\$R$, when x units are sold.

(b) i. What is the revenue when 40000 are sold?

ii. How many units must be sold to produce a revenue of \$600,000?

iii. What is the maximum revenue that the product will return?

5. Kow Boy has decided to use the 200 m of left over fencing to create a rectangular enclosure that consists of two adjacent sections as shown below.

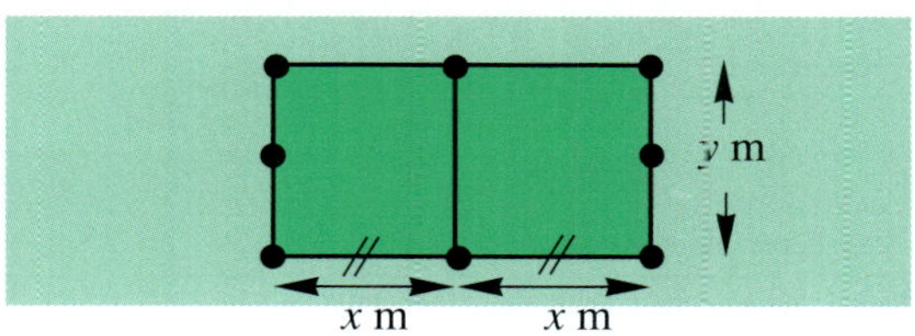

(a) Find an expression for y in terms of x.

(b) Let the area of the combined enclosures be A m^2.

i. Show that $A = \frac{8}{3}x(50 - x)$.

ii. What restrictions need to be placed on x?

(c) i. What is the maximum area that the combined enclosures have?

ii. What are the dimensions of the enclosure with the maximum area?

6. The stopping distance, d metres, of a heavily weighted vehicle travelling at v kmh^{-1} can be approximated by the formula $d = v + 0.05v^2$. At what speed was the vehicle travelling if it took the vehicle 100 m to stop? (Assume $v > 0$)

7. The percentage of normal level oxygen, $P\%$, in a lake, t weeks after organic waste is dumped into it is approximated by the model $\frac{1}{100}P = \frac{t^2 - 2t + 3}{t^2 + 3}, t \geq 0$.

(a) What is the level of oxygen in the lake prior to the waste being dumped?

(b) How long will it take for the level of oxygen in the lake to reduce by 15%?

(c) Using your graphics calculator, sketch the graph of $\frac{1}{100}P(t), t \geq 0$.

(d) i. What is the lowest level of oxygen that the lake will reach?

ii. How long after the waste has been dumped will this level occur?

(e) Describe the behaviour of the level of oxygen in the lake over time. Assume that waste is dumped only once.

8. A Gas company charges a fixed connection fee of $20 and 0.7 cents per MegaJoule (MJ) of gas up to 5000 MJ usage and thereafter charges 0.9 cents per MJ.

(a) Sketch a graph of the total bill as a function of usage (in MJ).

(b) Determine the bill for a household using 9473 MJ in a particular billing period.

9. Following the construction of two new roads, traffic lights need to be put in place where the roads meet.

One of the roads, along O, B, C, is modelled by the equation $y = \frac{3}{5}x$, while the other, along the path A, B, C, is modelled by the equation $y = -(x-5)^2 + 5$, where x and y are measured in kilometres.

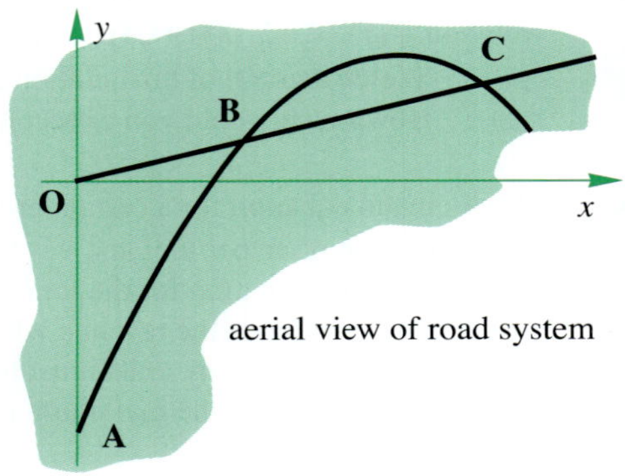

The section of the road has been set up on a set of axes as shown in the diagram. Where must the traffic lights, relative to O, be placed?

10. Tele-Kom use the following scale of charges for a 5 minute S.T.D call, between the hours of 7:30 am and 6:30 pm

Distance	Cost ($)
Less than 50 km	0.25
As of 50 km but less than 80 km	0.65
As of 80 km but less than 160 km	1.30
As of 160 km but less than 320 km	1.95
As of 320 km but less than 500 km	2.50
500 km or more	3.40

(a) Construct a function that will give the cost, $\$C$, in terms of the distance, x km, up to a distance of 1000 km for each 5 minute call.

(b) Sketch the graph of your function.

(c) Peter makes three 5 minute S.T.D calls in succession. The first to Town A, 250 km away, the next to Town B, 45 km away and the third to Town C, 450 km away. How much did these three calls cost all together.

(d) If the cost is calculated on a sliding scale relative to the time of the call. How much would Peter have paid for the three calls if the calls only lasted 2 minutes each?

11. ABCD is a trapezium where [AB] and [CD] are parallel and 4 m apart. Angle DAB is a right angle. The points X and Y lie on [AD] and [BC] respectively with [XY] being parallel to [AB]. The lines [XY] and [AB] are x m apart.

(a) What values can x take?

(b) Given that AB = 3 and that CD = 5, find an expression for the area, A m^2, of the quadrilateral ABYX, in terms of x.

(c) i. Sketch the graph of $A(x)$.

ii. Find the area of the quadrilateral ABYX when $x = 1$.

iii. Find the value of x so that the quadrilateral ABYX has an area of 3 m^2.

12. The height, $h(t)$ metres, of a stone above ground level, t seconds, after it has been projected vertically upwards is given by the equation

$$h(t) = 120t - 10t^2, t \geq 0.$$

(a) Calculate its height

i. after 2 seconds ii. after 4 seconds

(b) i. At what times will the stone reach a height of 40 metres?

ii. How long was it between the times when the stone was 40 metres from ground level?

(c) How long did it take for the stone to come back to ground level?

(d) What was the maximum height reached by the stone?

13. A stone is projected vertically upwards from the top of a building 250 m high. The height of the stone, $h(t)$ metres, above ground level, t seconds after it is thrown is given by the equation $$h(t) = 250 + 100t - 10t^2, t \geq 0.$$

(a) How long did it take for the stone to reach a height of 50 m above the building?

(b) How long did it take for the stone to be back at the point of projection?

(c) What is the maximum height that the stone reached (above ground level).

(d) The stone is allowed to continue down to ground level. How long did it take for the stone to reach the ground?

(e) What distance did the stone travel by the time it reached the ground?

14. The total cost, $\$C$, to manufacturer A, of electrical parts is given by the equation

$$C = 15n + 50000$$

where n is the total number of electrical parts produced. Manufacturer A sells these parts to manufacturer B at a fixed price of \$35 each.

(a) Find manufacturer A's cost when 1500 electrical parts are produced.

(b) Will manufacturer A make a profit when 1500 parts are produced?

(c) How many parts must manufacturer A produce to break-even?

15. A company manufactures and sells x cheap radios per month.
The cost, $\$C$, involved in producing x radios per month is given by the equation

$$C = 60x + 70000, 0 \leq x \leq 6000.$$

The revenue equation, $\$R$, based on the sales of x radios per month, is given by

$$R = -\frac{1}{30}x^2 + 200x, 0 \leq x \leq 6000.$$

(a) Accurately draw the graphs of the cost and revenue functions on the same set of axes.

(b) What is i. the minimum cost involved?
ii. the maximum revenue?

(c) Why is there a cost involved when no radios are produced?

(d) On your graph, identify the break-even points.

(e) What profit does the company make when 2000 radios are produced and sold?

(f) i. Find an expression in terms of x for the profit, $\$P$, this company makes on the sales of their radios.
ii. What is the maximum profit the company can hope to make?
iii. How many radios would they need to sell to achieve this maximum profit.

(g) For what values of x will the company be
i. in the red?
ii. in the black.

(h) Clearly identify each of the regions in part (g) on your graph from part (a).

16. The market research department of a company recommends that the company manufacture and market a new transistor radio. After extensive surveys, the research department submits the following demand and cost equations respectively:

$$x = 6000 - 30p \qquad (x \text{ is the demand at } \$p \text{ per radio})$$
$$C(x) = 72000 + 6x .$$

The revenue equation, $\$R$, is given by the product of the number of units sold the price per unit.

(a) Show that the revenue equation is given by $R(x) = x\left(200 - \frac{x}{30}\right)$

(b) i. Find the profit equation, $\$P$, where $P(x) = R(x) - C(x)$.
ii. Define the domain for the function $P(x)$.

(c) On the same set of axes, sketch the graph of the revenue, $\$R$ and cost $\$C$ functions.

(d) Provide a brief analysis based on the graph in part c., making sure to discuss issues such as i. break-even points.
ii. regions of gain and loss.

(e) How many units need to be sold in order to maximize their profit?

9.1.3 MODELLING – PHASE 3

And so we arrive at the last phase of our process – or, sometimes known as 'Are we there yet?' The question then is, how does this phase differ from the previous two? If we look at the previous two stages we progressed from constructing equations based on some verbal account, which then had to be solved, to constructing functions (or being presented with functions) which then had to be interpreted and/or analysed. In the next stage we still require to construct a model or a function, but this time we do so based on information presented in the form of data.

On many occasions the form of a function may be derived from data, say, from an experiment perhaps. The character of a polynomial function can be easily determined from data which is sufficiently orderly. To see this clearly for a linear function let's look at the following examples.

EXAMPLE 9.11

The results of an experiment (a rather precise experiment) are shown in the table below. What type of model best describes these results?

l (cm)	5.0	5.1	5.2	5.3	5.4
Q (cm)	13.6	14.9	16.2	17.5	18.8

Solution

We start by plotting the data and try to establish the type of trend displayed by the data. After turning the **STAT PLOT** feature on, we create two lists, **L1** (for l) and **L2** (for Q): To create the lists:

❶ select **STAT**, press **ENTER** and select the scatter-diagram icon, then **QUIT.**

❷ choose **4:ClrList** (to identify which lists you wish to clear), then press **ENTER.**

❸ select **STAT**, choose **1:Edit...**, and then enter the data, first **L1** (after every input press **ENTER**) and then use the right arrow key to move to the **L2** column and repeat process as for **L1**.

❶ 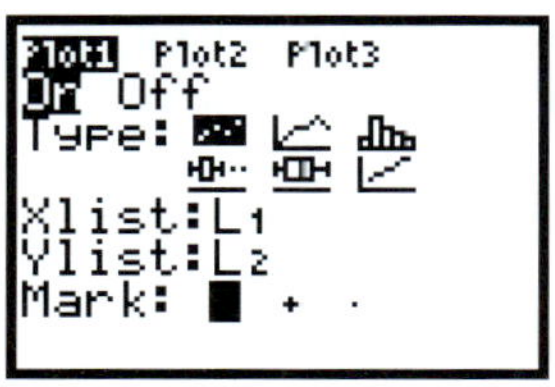

❷ 

❸ 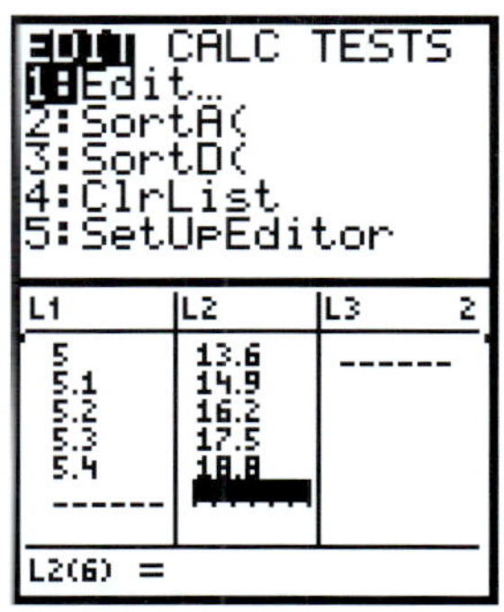

We can now plot the data and look for a trend. Using two different windows sometimes helps.

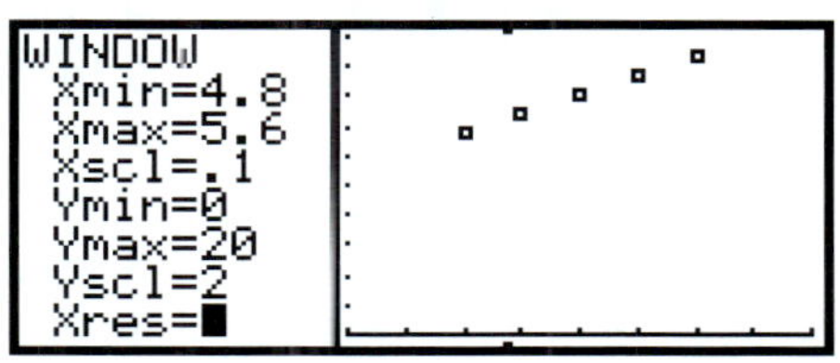

Although the first plot indicated a linear trend, after altering the window, as in the second plot, it became more obvious that there was a linear trend. We could now suggest that the variables l and Q could be **modelled** by the equation $Q = al + b$, where a and b are constants to be determined.

To solve for a and b,we use two points from our data, let's choose (5.0, 13.6) and (5.1, 14.9). We can then set up a system of equations:

$$\text{From } (5.0, 13.6) \text{ we have } 13.6 = 5a + b \quad -(1)$$

$$\text{From } (5.1, 14.9) \text{ we have } 14.9 = 5.1a + b \quad -(2)$$

$$(2)-(1): \quad 1.3 = 0.1a$$

$$\text{Therefore,} \quad a = 13$$

$$\text{Sub. into } (1): \quad 13.6 = 5 \times 13 + b \Rightarrow b = -51.4$$

Therefore we have the data modelled by the equation $Q = 13l - 51.4$.

There is another method that can be used to find the equation of a straight line for which a linear trend in the data is suspected. This method is known as the **least squares regression line**. This is actually the domain of Chapter 22 and there are some cautions that need to be taken into account when using this method. However, in this case, as the data seems to fit a straight line perfectly, we use it for demonstrative purposes.

From the **STAT** menu use the right arrow key and select **CALC** then use the down arrow key and select option **4:Linreg(ax+b)**: then enter **L1, L2** in that order, followed by , **Y1**:

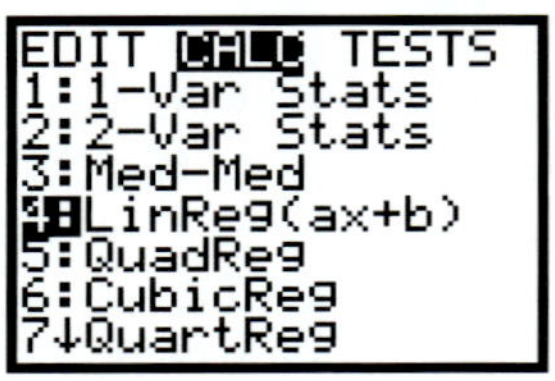

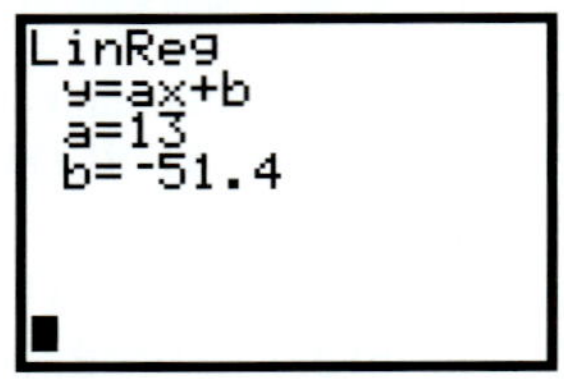

As we can see, the last screen provides the same results as before.

Now, the reason why caution needs to be taken when using the least squares regression equation is that this method uses a process whereby a great many points can be used to obtain the **line of best fit**, even when the points from the data set do not lie perfectly on a straight line.

We can also sketch the line on the same screen as the scatter diagram. The second screen above tells us that the linear regression line has been stored as **Y1**. If we then press **GRAPH** we have:

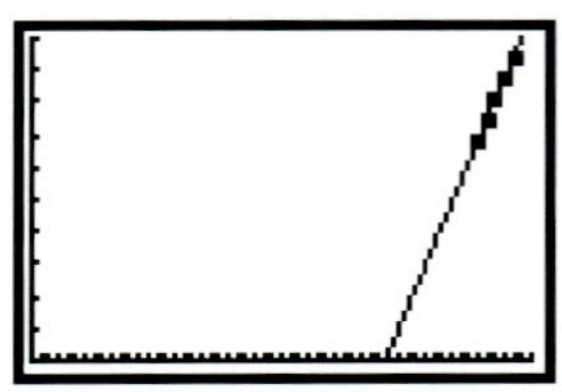

A glance at the table also reveals a constant first differences:

l (cm)	5.0	5.1	5.2	5.3	5.4
Q (cm)	13.6	14.9	16.2	17.5	18.8
First difference	14.9 – 13.6 = 1.3	16.2-14.9 = 1.3	17.5 – 16.2 = 1.3	18.8 – 17.5 = 1.3	

As the first difference is constant we then know that the data fits a linear model. Then, using substitution of data solves the details of the model.

This practice can be extended to higher polynomial functions and in particular to quadratic functions. The quadratic functions exhibit constant second differences - that is, the differences between successive first differences are constant. To see this, consider the function defined by

$f(x) = ax^2 + bx + c$. We will tabulate the function for the first few non-negative integer values.

x	$f(x)$	1st difference	2nd difference
0	c		
	↦	$a + b$	
1	$a + b + c$	↦	$2a$
	↦	$3a + b$	
2	$4a + 2b + c$	↦	$2a$
	↦	$5a + b$	
3	$9a + 3b + c$	↦	$2a$
	↦	$7a + b$	
4	$16a + 4b + c$		

EXAMPLE 9.12

The following results where obtained from a medical laboratory. What appropriate model would fit the data?

r	3	3.2	3.4	3.6	3.8
T	3	3.84	4.76	5.76	6.84

Solution

Our first step is to visualise the data. So, as in Example 9.11 we plot the data points. Using a window of [0,4] by [0,7] with a scale of 1 we have:

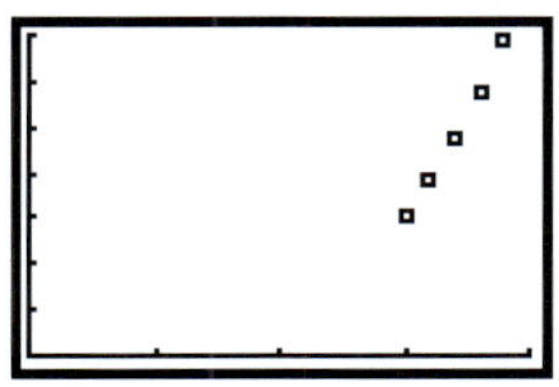

It appears that the data displays a linear relationship. However, closer inspection reveals that the first differences of the T-values are not constant.

First difference: $3.84 - 3.00 = 0.84$
$4.76 - 3.84 = 0.92$
$5.76 - 4.76 = 1$
$6.84 - 5.76 = 1.08$

Therefore a linear model is not supported by the data.

On the otherhand, we see that the second differences are constant.

Second difference: $0.92 - 0.84 = 0.08$
$1.00 - 0.92 = 0.08$
$1.08 - 1.00 = 0.08$

And so, the data supports a quadratic model!

We can therefore write the model as $T = ar^2 + br + c$ and use the raw data to determine a, b and c by substitution. As we have three constants to evaluate we need three data pairs to solve the problem which will then be a set of three linear equations in the unknowns, a, b, c. We have the first three data pairs giving:

$$3 = 9a + 3b + c \quad -(1)$$

$$3.84 = 10.24a + 3.2b + c \quad -(2)$$

$$4.76 = 11.56a + 3.4b + c \quad -(3)$$

We can solve this system of linear equations in the same way that we solved systems of linear equations in Chapter 5. However, we first eliminate c using (1).

$$c = 3 - 9a - 3b \quad -(4)$$

Then, we substitute this into (2) and (3):

$$3.84 = 10.24a + 3.2b + (3 - 9a - 3b) \Leftrightarrow 0.84 = 1.24a + 0.2b \quad -(5)$$

$$4.76 = 11.56a + 3.4b + (3 - 9a - 3b) \Leftrightarrow 1.76 = 2.56a + 0.4b \quad -(6)$$

Then, (6) – 2×(5) gives:

$$0.08 = 0.08a$$

$$\Leftrightarrow a = 1$$

By substitution in (5), $b = -2$. Finally substitution of these into (4) gives $c = 0$.
Therefore, our quadratic model is given by, $T = r^2 - 2r$.

We can fit a least squares quadratic equation by selecting **5:QuadReg**, saving it as **Y1** and then sketching it. From the last screen, it appears to be a pretty good fit!

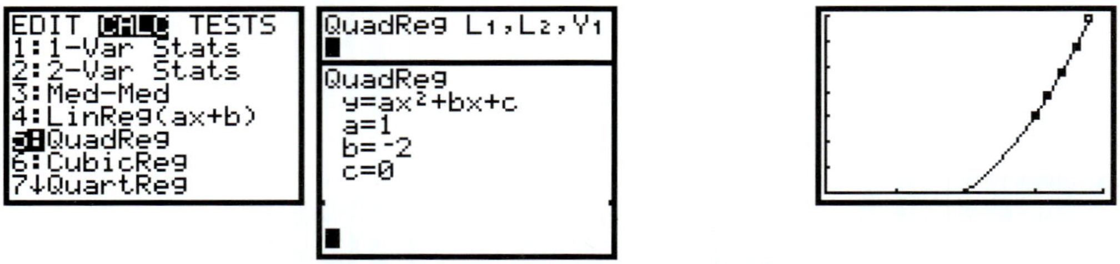

Again, caution needs to be taken when using the least squares method. In this case, because the data was a perfect fit, the equation we obtained using simultaneous equations and that obtained using the method of least squares are the same. On the otherhand, if our data looked like the one shown below, the least squares approach would still find the equation of best fit, whereas, depending on which three points we decide to use to obtain 3 equations and then solve, our result and that of the TI–83 will most probably not be the same (but hopefully not too different).

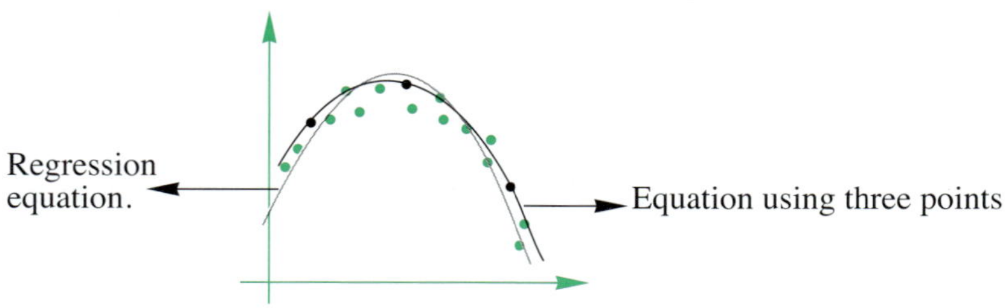

Information and data also come in forms other than tabular – although once we extract the data from the information presented, we can then organise it in tabular form. Consider the following example.

EXAMPLE 9.13

A piece of machinery was bought for \$12,000 new and was then depreciated uniformly to an estimated scrap value of \$1200 after 6 years of use.

(a) Use the information to sketch a graph of the machines values against time.
(b) How much does the machine depreciate by annually?
(c) Construct a model for the value of the machine during its operating life.

Solution

In this problem there are probably some terms that are unfamiliar. When this occurs, make sure that you research the meaning of the terms that are not familiar to you.
We are given that at the begining, the machine is worth $12,000 and that in 6 years time it will be worth $1200. The key term is that it depreciates (i.e., losses its value) uniformally, that is, each year its value will decrease by the same amount. A linear model displays this attribute, and so we will use that as the base for building our model.

(a) We start by setting up a table of values.
Let t years denote the numbers of years since the machine was bought and $\$V$ the value of the machine at time t.

Time (t years)	0	6
Value ($\$V$)	12,000	1200

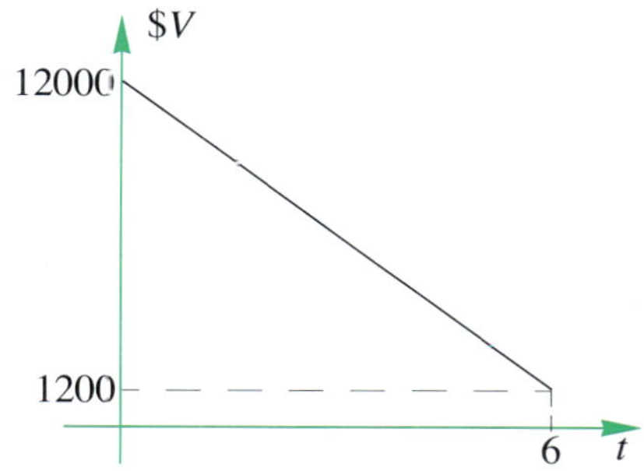

We can use the above information to sketch the graph.

(b) Over a period of 6 years the machine has depreciated by $10,800.

Therefore, the rate of **depreciation** is $\frac{10800}{6} = 1800$. (Why don't we write '–1800'?)

Meaning that every year the machine depreciates by $1800

(c) We can now use our work in Chapter 5 to determine the equation of the straight line.
We have a gradient of –1800 and a V-intercept of 12,000 therefore the equation becomes

$$V = -1800t + 12000$$

However, to fully specify the model we need to take into account any restrictions. In this case we have a restriction on the time, i.e., $0 \le t \le 6$.

Therefore, our model becomes: $V = -1800t + 12000,\ 0 \le t \le 6$

Of course, data can be presented in different formats, but we still need to extract all (and only) relevant information so that we can proceed with constructing a model. Consider the following example.

EXAMPLE 9.14

A suspension bridge, one very much the Golden Gate Bridge or the Severn Suspension Bridge, has its roadway supported by parabolic cables hanging from support towers. The cost involved in constructing such a bridge is a complicated affair, however, our task is to provide a cost for the suspender cables, running vertically from the road level to the parabolic cable, assuming that they are 2 metres apart. All measurements are shown on the plan below. A marker at A is known to be 55 m above the road level and 40 m from the right support tower.

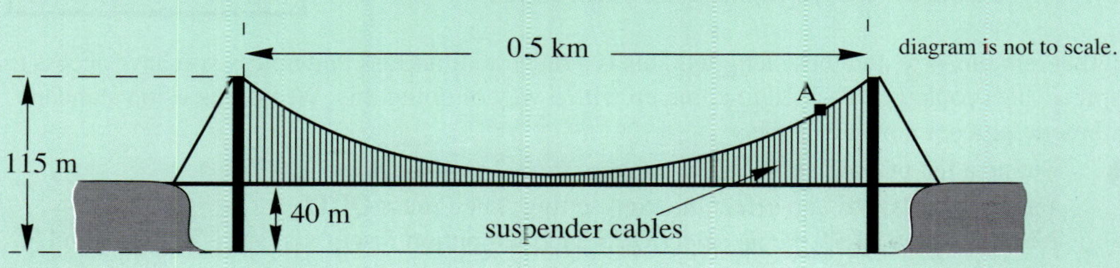

Solution

Part of the modelling process is to make certain assumptions. One assumption we will make is that the cost of the suspender cable is only dependent on the material used and the length of cable used.

So, we let the cost of the suspender cable per unit length (for a particular material) be $\$k$. Then, once we know the total length of suspender cable required, L m say, the total cost for the cable would be $\$kL$.

Notice that we have made no mention of waste. Clearly there will have need to be extra cable necessary so that the cable can be 'attached' to the parabolic cable as well as to the road level. However, this is all part of the simplifying assumptions we are making (among others).

We now need to come up with a value for L. What needs to be found is the length of each suspender cable and sum their lengths –this will provide us with the value of L. We start by 'redrawing' the bridge (without concerning ourselves with accurate scales):

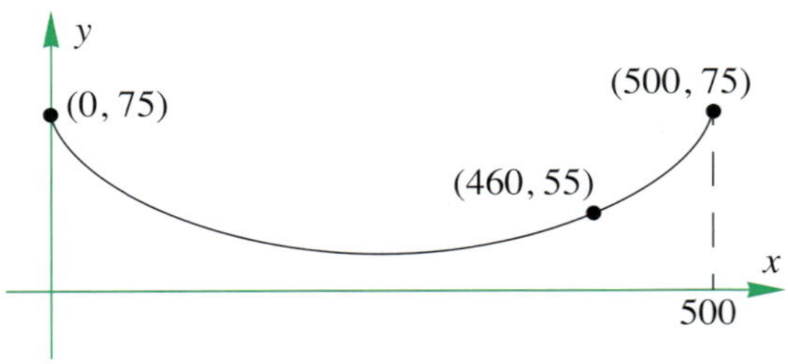

We can make use of the function

$$y = ax^2 + bx + c$$

to model the parabolic cable and from it find the lengths of the suspending cables.

The data consists of the three points lying on this parabola. As we know that we are dealing with a parabola and we have three points on the parabola, we can obtain the equation by using the least squares method (as was done in Example 9.12). Otherwise we can make use of simultaneous equations. So, using the TI–83, we enter our data and obtain the equation:

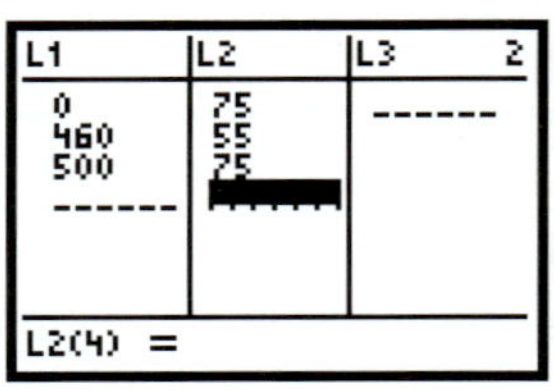

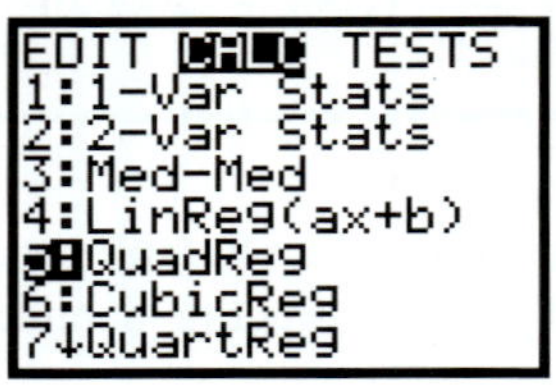

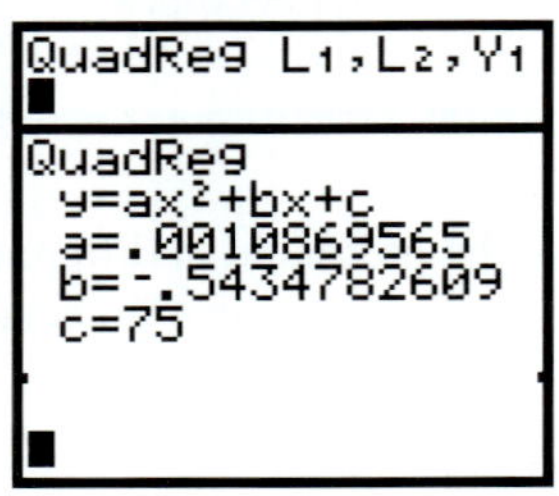

This gives the equation $y = (5 \times 10^{-4})x^2 - 0.25x + 50$.

So that when $x = 2$ (the first suspender cable) it will have a length of 49.502 m

when $x = 4$ (the second suspender cable) it will have a length of 49.00 m

Obviously, a table of values will make life much easier.

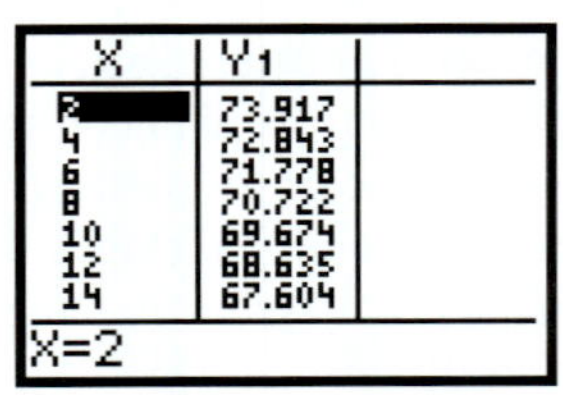

All that remains is to add these lengths. This is a rather tedious task and unless you have access to a spreadsheet package, we need to come up with a way of doing this. We do this using standard arithmetic procedures on our lists:

❶ Set up a list of even consecutive integers from 2 to 498. (Store it as L3).
First press **MODE** and select the **Seq** setting. Then press **QUIT**.
Next go to the **LIST** menu, select **OPS** and then option **5:seq(** and store sequence in L3.

❷ Evaluate the y-value (using the quadratic regression equation) for each number in L3. Store these values in L4.

❸ Go back to **LIST** select **MATH** and then option **5:sum(**
This will add the values in L4 and hence give the sum of the lengths of the cables)

❶
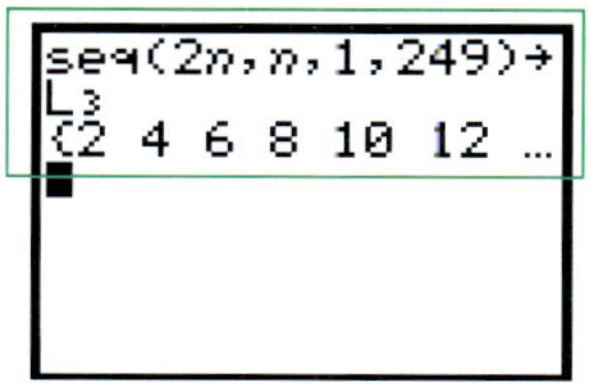

❷
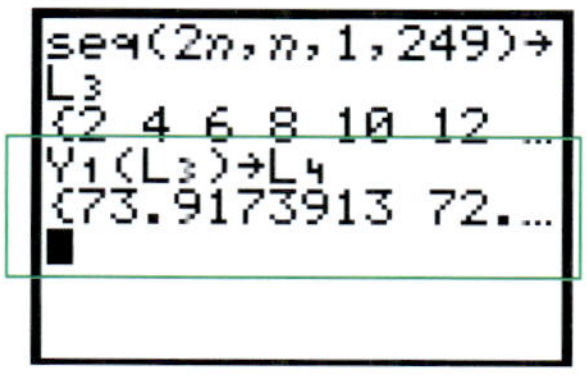

❸
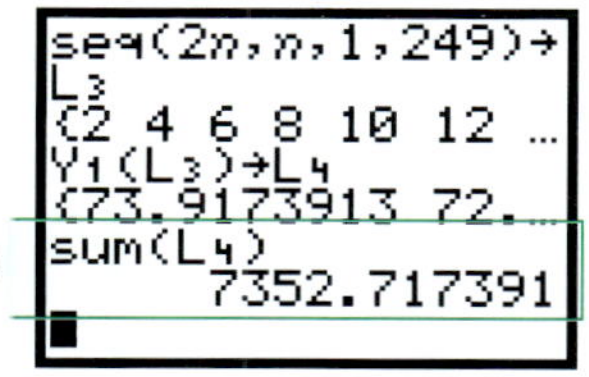

That is, we will need 7352.72 m of suspender cable for one side of the suspension bridge which will then need to be doubled (as we need an equal amount for the other side).

Therefore, the cost for the suspender cable is \$14705.43k. (\$14706k)

Of course, we should always question our findings and then validate them. It is possible to obtain a model that acurrately represents as set of data and yet, the model might not be realistic.

A somewhat simplified approach to phase 3 of our modelling process can be summarised using the following diagram:

Process

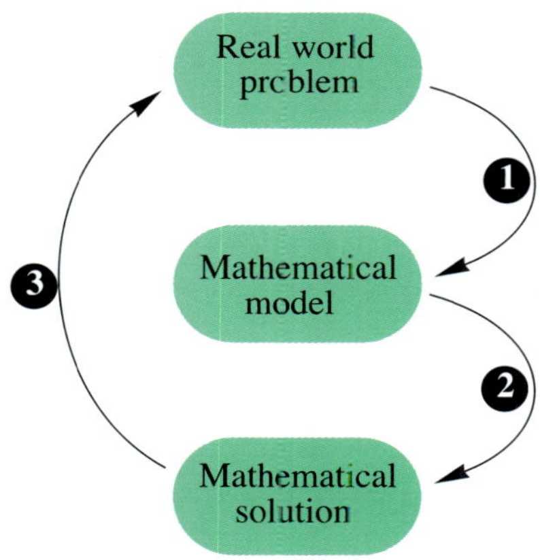

Stage ❶ Collect information and analyse data to construct a mathematical statement about the problem. Based on this statement, build a mathematical model.

Stage ❷ Solve the mathematical model.

Stage ❸ Interpret your solution in light of the original problem. Validate your solution. Then, if necessary refine your mathematical model and repeat the process.

EXERCISES 9.1.3

1. (a) Show that the following data supports a linear relationship between the variables x and y.
(b) Find the linear relationship in each case.

i.

x	2	3	4	5	6
y	8	8.4	8.8	9.2	9.6

ii.

x	0	2.4	4.8	7.2	9.6
y	6	1.2	–3.6	–8.4	–13.2

iii.

x	–6	–4	–2	0	2
y	0.2	1.2	2.2	3.2	4.2

2. Show that the following data supports a quadratic relationship between x and y.

x	2	2.4	2.8	3.2	3.6
y	6	9.12	12.88	17.28	22.32

3. (a) Show that the following data supports a quadratic relationship between x and y.

x	0	1	2	3	4
y	2	7	14	23	34

(b) Determine the rule for this quadratic.

4. A parabolic curve passes through the points (3, 16), (–1, 12) and (2, 9).
(a) Plot these points on a set of axes.
(b) Find the equation of this curve.
(c) Draw the graph of the curve found in (b) on the set of axes in (a).

5. Find the equation of the parabola passing through the points A(1, 4), B(2, 9) and C(–1, 6).

6. The tables below show the demand and supply data for a particular product.
Demand:

q	5	10	15	20
p	3	2	1	0

Supply:

q	5	10	15	20
p	1.11	1.46	1.81	2.16

Where q represents the quantity in thousands and p the price per item.

(a) On the same set of axes, plot the
i. data representing the demand.
ii. data representing the supply.

(b) Show that both demand and supply can be modelled by a linear relationship.
(c) Find the equation for
i. demand. ii. supply.
(d) Describe the optimum scenario.

7. The graph shows the temperature readings taken by a student on a winter's day.

The readings were taken at hourly intervals after 9:00 am. Unfortunately a number of the readings were smudged and therefore could not be transferred onto the graph.

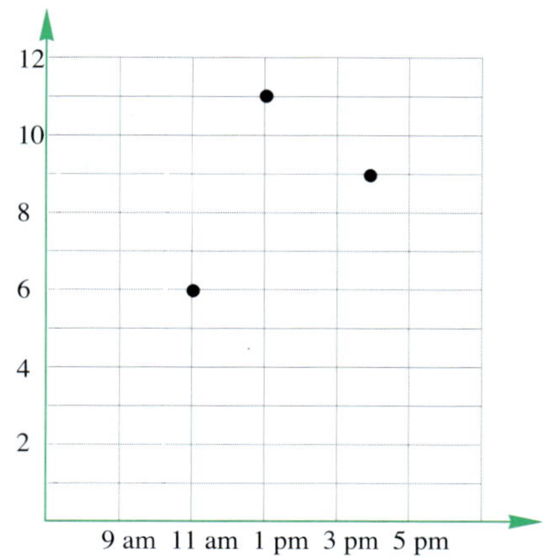

(a) What type of graph would most probably fit the set of data shown?

(b) What would you expect the temperature to be at 11:00 pm?

8. During a baseball game the ball is hit 1 metre above ground level with a bat. After having travelled a horizontal distance of 30 m, it is 20 m above ground level. An outfielder runs to meet the ball and catches it as he dives to the ground. The outfielder is 80 metres from home base when he catches the ball. Asuming that the path of the ball is parabolic, what is the greatest height that the ball reaches?

9. A spring is attached to a bar. The extension, x m, of this spring when different masses, M kg, are attached at the free end, is record and tabulated:

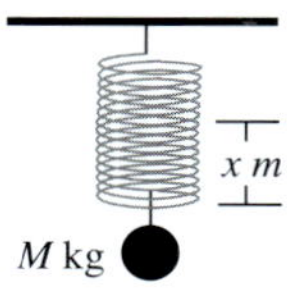

M (kg)	1	2	4	6	10	12
x (cm)	1.5	2.0	3.0	4.0	6.0	7.0

(a) Plot the graph of x versus M.
(b) What model would appear to suit the data?
(c) i. Determine the equation that best fits the data.
ii. What is the spring's natural length?

10. An object is dropped from a building, 125 m high. During its descent, the distance, x m, above ground level is recorded as tabulated:

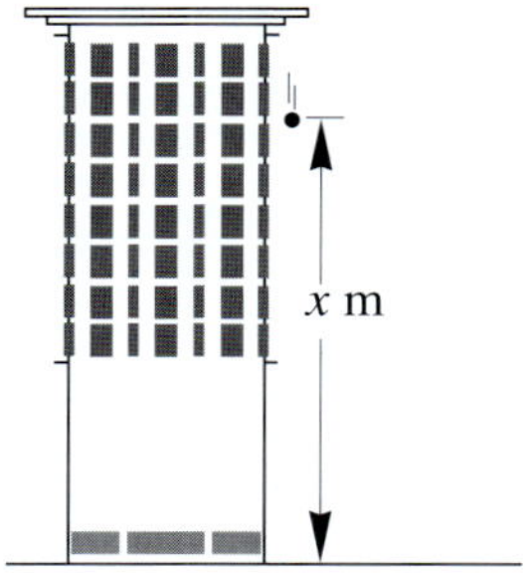

t (sec)	0	1	2	3	4	5
x (m)	125	80	45	20	5	0

(a) Plot the data on a set of axes.
(b) Fit, by eye, a curve that fits the data.
(c) i. What type of curve would fit this data?
ii. Use a difference table to verify your statement.
(d) Find the equation of the curve that best models the data.

11. (a) A parabolic curve passes through the points (1, 2), (–1, 0) and (2, –3). Determine the equation of this curve.

(b) Find the *family* of quadratics that pass through the points (1, 2) and (–1, 0).

12. Over the years 2000 to 2004, the sales of a company have increased at a constant rate. The table below shows the variation over this period.

Year	2000	2004
Sales	\$500 000	\$680 000

Assuming that a linear model of the form $S(t) = at + b, k_1 \le t \le k_2$ exists, since 2000

(a) i. find the values of a and b.

ii. determine the values of k_1 and k_2 for which this model is assumed to hold.

(b) Sketch the graph showing the relationship between the value of sales \$$S$ and the time, t years, since 2000.

13. In 1988, Natalya Lisovskaya (ot the then USSR) won the gold medal for the shot putt at the Olympic Games.

The height of the shot putt from ground level, relative to the horizontal distance it travelled from the point where it left Natalya's hand was recorded and tabulated.

Horizontal distance, x m	12.19	18.29	21.34
Height, h m	6.58	3.74	1.01

(a) Plot these points on a set of axes.

(b) i. What type of curve would best model this situation?

ii. Determine the equation of such a curve.

(c) i. What is the maximum height reached by the shot putt?

ii. How far was did Natalya throw the shot putt?

14. An archway is constructed as shown in the diagram below.

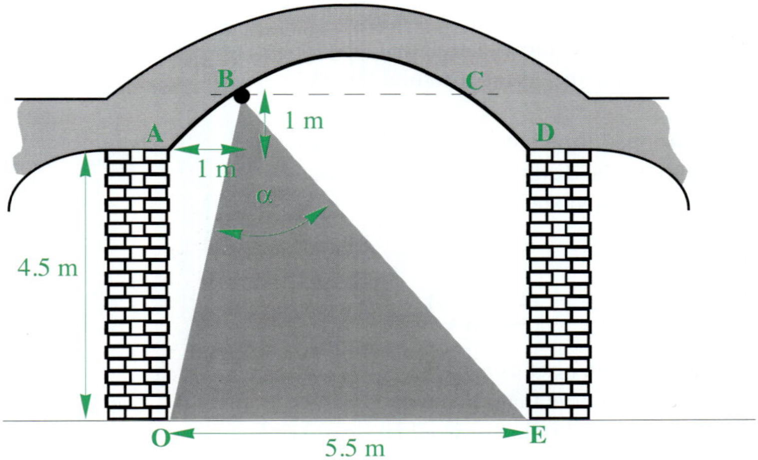

A light source is positioned at B as shown. The archway can be modelled by a parabolic curve.

(a) Using O as the origin, determine the quadratic equation that produces a parabolic curve passing through the points A, B and D.
(b) Find the equations of the straight lines that form the 'edges' of the light beams
 i. [BO].
 ii. [BE].
(c) Find the tilting angle, α, so that light can spread from O to E, where the pillars meet the floor.

15. The total cost, $C(x)$ in thousands of dollars, for manufacturing specialised cars was recorded so that it could then be analysed.

Number of cars produced (x)	Cost ('000) $C(x)$
0	580
10	805
20	980
30	1105
40	1180
50	1205

(a) On a set of axes, plot the data points.
(b) Verify that a quadratic model is appropriate in this situation.
(c) Determine the equation of the quadratic that satisfies this data.
(d) What is the average rate of increase in production costs for the first 10 cars?
(e) i. How much did it cost to produce 11 cars?
 ii. How much did it cost to produce the eleventh car?

16. Stella has just been employed as Marketing Manager at Nattel Toys Group. During the first few weeks of her employment she has undertaken to research the sales of one of their product, know as 'Talking Tutu". Stella provided the following data:

Price, $\$p$, per Talking Tutu:

x	0	10	100	1000
p	10	9.99	9.9	9

Cost, $\$C$, for producing x Talking Tutus:

x	0	10	100	1000
C	7000	7020	7200	9000

Revenue, $\$R$, for selling x Talking Tutus:

x	0	10	100	1000
R	0	99.9	990	9000

(a) Show that the relations between
 i. p and x is linear
 ii. C and x is linear
 iii. R and x is quadratic.
(b) i. On the same set of axes, plot the graphs of C and R versus x.
 ii. Determine the equations that best model each of p, C and R in terms of x.
(c) Determine the maximum profit Stella can predict based on expected future sales.

9.2 PROJECT SUGGESTIONS

Investigate and model the actual shape of the Leaning Tower of Pisa.

Modern digital video cameras allow you to investigate events that happen quite rapidly such as balls falling under the influence of gravity.
Isolate every tenth frame (or use a stopwatch in the frame) to work out the position of the ball at equal time intervals and then find a model that describes the motion.
Remember to evaluate the model. An example of this is discussed in Chapter 3.

At the other end of the spectrum of speed, you can use time lapse photography (one photo a day) to investigate the growth of a plant seedling.

Are there any suspension bridges close to you? If so, there is a wealth of modelling problems that can be carried out.

Investigate the problem of Example 9.14 for a suspension bridge you know of – make sure you obtain true data.

How would you calculate the length of the parabolic cable as opposed to the suspender cables?

What about curves in art?

Churches, cathedrals, mosques, and so on, all have interesting curves that form part of their designs.

What about curve stitching?

What is the diffeence between a curve of pursuit and a tractix? How can they be modelled?

Reseach a company's cost–revenue structure. Analyse their data and produce a report to optimise their revenue.

REVISION SET A - PAPER 1 & PAPER 2 STYLE QUESTIONS

1. Which of the following equations have solutions that are members of the set $\mathbb{Q}$?

i. $x^2 + 2 = 0$ ii. $\sqrt{x} = 7$ iii. $x^3 = -1$

2. Write the following numbers in standard form:

i. $\left(\frac{5}{200}\right)^2$ ii. $\sqrt{\frac{1}{10000}}$ iii. 0.0009^2

3. If $a = 2.47 \times 10^{-2}$, $b = 7.92 \times 10^3$, $c = 4.01 \times 10^{-3}$, evaluate, giving answers in standard form, correct to three significant figures:

i. $\frac{a+c}{a-c}$ ii. $a(b+c)$ iii. $a + \frac{1}{b}$

iv. $\frac{1}{a^2} - b^2$ v. a^c

4. If $x = 27 \pm 1$, $y = 155 \pm 5$, $z = 93 \pm 2$, find the percentage error, correct to two significant figures, that results from calculating $\frac{x+z}{y-z}$

5. By writing the number $0.\dot{4}2857\dot{1}$ as a fraction, prove that it is a rational number.

6. The temperature (O°C) of an industrial oven t hours after it is started is modelled by the function $O(t) = -0.9t^2 + 66t + 25, 0 \le t \le 50$.

(a) Copy and complete the table of values for the function giving answers in scientific form, correct to one decimal place.

t	0	10	20	30	40	50
O						

(b) Sketch the graph of the function.

(c) Find the average rate of change of temperature between $t = 20$ and $t = 30$.

(d) If the coefficient of t (i.e., 66) has a 10% error, find the largest and smallest possible values of $O(20)$ and $O(30)$ and hence find the largest and smallest possible values of the average rate of change of temperature between $t = 20$ and $t = 30$.

7. The voltage, V volts, in an electrical circuit is given by the product RI, where R ohms is the total resistance of the circuit and I amps is the current flowing in the circuit. If $R = 6$ and $I = 4$, each being correct to 1 significant figure, find the range of V.

8. The dimensions of the triangle shown are known to an accuracy of 5%.

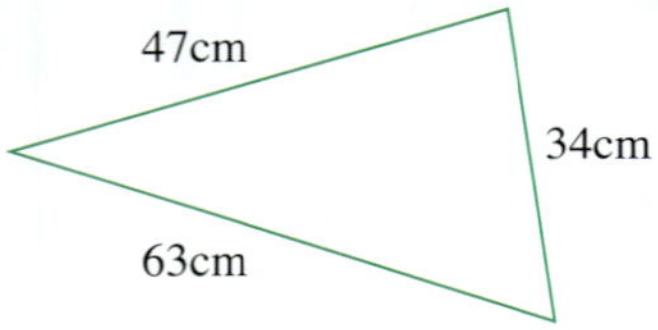

Find the percentage error in calculating:
(a) The perimeter.
(b) The largest angle.
(c) The area.

9. A function is defined by $f(x) = 35 - ax^3, 0 \le x \le 2$. The value of the coefficient a is 3.6 ± 0.2.

Find the percentage error in calculating $f(1)$.

10. (a) Sketch the graph of $f(x) = x(4 - x), x \in \mathbb{R}$, showing all intercepts with the axes and giving the coordinates of the turning point.
(b) Define, I as the largest subset of $\mathbb{R}^+$ for which the function $f(x)$ will be a one–to–one function.
(c) i. Sketch the graph of $g: I \mapsto \mathbb{R}$ where $g(x) = f(x)$.
ii. State the range of $g(x)$.

11. Let the function f be defined by, $f(x) = \begin{cases} 4 & x < 0 \\ 4 - x & 0 \le x \le 2 \\ \dfrac{4}{x} & x > 2 \end{cases}$

(a) Sketch the graph of f.
(b) State the range of f.
(c) Find the values of x for which $f'(x)$ does not exist.

12. A piecewise linear function is defined by $f(x) = \begin{cases} \frac{1}{2}x + k \text{ if } x > 2 \\ 2 - x \text{ if } x \le 2 \end{cases}$.

(a) Sketch the graph of $f(x)$ if
i. $k = 0$
ii. $k = 2$
iii. $k = -2$
(b) Find k if the graph is to be continuous for all real values of x.

13. For the two functions, $f(x) = 9 - x^2$ and $g: [0, \infty) \mapsto \mathbb{R}$, where, $g(x) = \sqrt{x}$,
(a) Find $f(g(x))$.
(b) Let $h(x) = f(g(x))$
i. State the domain of $h(x)$.
ii. Sketch the graph of $h(x)$.
iii. Find the range of $h(x)$.

14. A relation is given by $f : x \mapsto 4 - \frac{3x}{4}$, for $0 \le x \le \frac{16}{3}$.

(a) Is this relation a function?
(b) Sketch the graph of f.
(c) For f, state
  i. the domain
  ii. the range.
(d) For what value of x will the image of f be 2?

15. Consider the function $f(x) = \frac{2}{3}x + 2$.

Find (a) the value of $f(9)$.
(b) the value of x for which $f(x) = x$.
(c) the value of x having an image of 6.

16. (a) Solve the following i. $15x^2 - 2x - 8 = 0$
ii. $5x = 9x^2$

(b) Use the quadratic formula to solve $3x^2 + 2x - 2 = 0$, giving your answer to two decimal places.

17. Given that $y = 2(x-4)^2 - 2$,
(a) state i. the coordinates of the turning point.
ii. the equation of the axis of symmetry.
iii. whether the function has a maximum or minimum value and give this value.
iv. the value of the y-intercept.
(b) find the value(s), if any, of the x-intercepts.

18. Sketch the graph of the following functions
(a) $f(x) = -x^2 + 4x + 5$
(b) $f(x) = x^2 + 5x - 3$

19. A piece of metal sheet, 40 cm long and 30 cm wide has equal-sized squares cut out of each corner so that the remaining metal can be folded into a baker's cake tin which has a base area of 600 cm^2. Find the dimensions of the tin.

20. The velocity, v ms^{-1}, of object A at time t seconds is given by

$$v = 17.2 + 2.8t, 0 \le t \le 40$$

and the velocity, v ms^{-1}, of object B at time t seconds is given by

$$v = 50.5 - 1.2t, 0 \le t \le 40.$$

(a) Find the velocity of each object at $t = 5$.
(b) Find the time taken, correct to one decimal place, for each object to reach a velocity of 30 ms^{-1}
(c) i. At what point in time do both objects have the same velocity?
ii. What is that velocity?

21. (a) In a certain country the present income tax system requires payment of tax under the following scheme

Income	Tax Payable
Up to \$10, 000	No tax payable
\$10,000 and above	30% of income in excess of \$10,000

i. Calculate the tax payable on an income of \$30,000.

ii. Calculate the percentage of income that is paid in tax.

iii. Calculate the income of a person who pays \$4,800 in tax.

iv. If $\$y$ is the tax payable on an income of $\$x$, construct a rule of the form

$$y = \begin{cases} mx + c, & x \geq 10000 \\ 0 & , x < 10000 \end{cases}$$

v. Sketch the graph of y against x.

(b) An alternative system has been suggested by the opposition in this country. Their scheme is as follows:

Income	Tax Payable
Up to \$38,750	20% of income
\$38,750 and above	\$7,750 + 40% of income in excess of \$38,750

i. Calculate the tax payable on an income of \$30,000.

ii. If $\$y$ is the tax payable on an income of $\$x$, construct a rule of the form

$$y = \begin{cases} m_1x + c_1, & x \geq 38750 \\ m_2x + c_2, & x < 38750 \end{cases}$$

iii. Sketch the graph of y against x on the same set of axes as (a) v.

(c) Which incomes would pay less tax under the proposed new system?

22. The height reached by a ball thrown into the air is modelled by the function

$$h = -5t^2 + 12t$$

where t is the time in seconds after the ball is thrown and h is its height in metres.

i. Find the height of the ball after 2 seconds.

ii. Complete the square on the function and hence find the maximum height of the ball and the time at which this is achieved.

iii. Find the times at which the stone is 5 metres from the ground, correct to three significant figures.

iv. Find the total time that the ball is in the air.

23. A square, ABCD, of side length 5 cm is shown below.

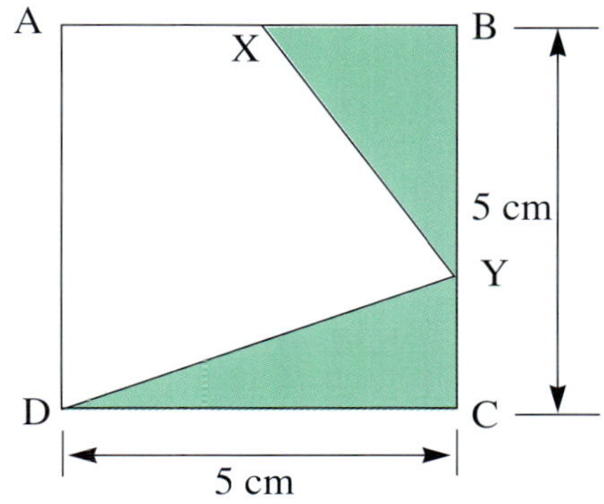

The points X and Y are on the sides AB and BC respectively and are such that BX = CY.

(a) If CY = x cm, write an expression for the shaded areas in terms of x.

(b) Given that the ratio of the area of ΔDCY to the area of ΔXBY is 4:3, find x.

24. The graph of $y = x^2 + ax + b$ is shown.

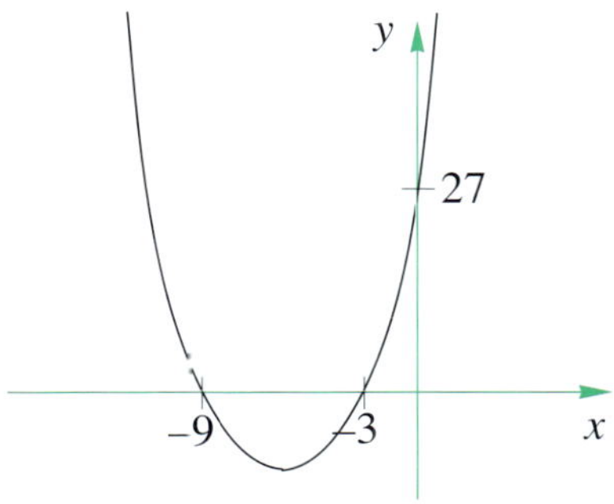

(a) Determine the values of a and b.

(b) What is the minimum value of y?

25. A train travels for 3 hours at a constant speed and then for the next 4 hours at 10 km h^{-1} faster than the first three hours. The train travels a total of 650 km.

i. If the train is travelling at s km h^{-1} during the first three hours, how far did it travel during the next four hours? Give your answer in terms of s.

ii. How fast did the train travel during the first three hours?

26. Find the perimeter of the triangle shown.

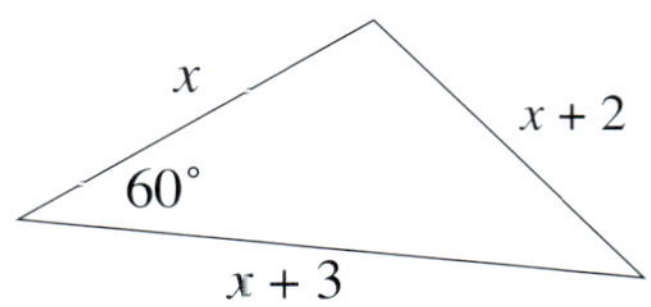

27. A bag contains 10-cent and 50-cent coins. In all, there are 26 coins. The bag contains more than $ 4.50. There are x 50-cent coins and y 10-cent coins in the bag.

i. Write down an inequation involving x and y.

ii. Express y in terms of x.

iii. Find the minimum number of 50-cent coins in the bag.

28. A rectangular dog run is to be built on flat ground from a length of 20 metres of wire fencing which forms three sides, the fourth side being part of a straight wooden fence. If the width of the dog run is x m, then the area, A m^2, is given by the rule $A = x(20 - 2x)$, for $a < x < b$.

(a) State the values of a and b.

(b) Find the maximum area of the dog run.

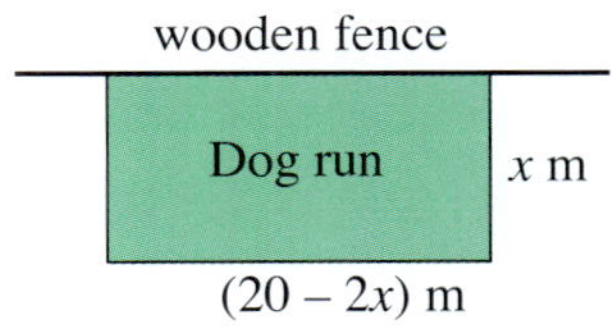

29. A typical fully grown Eucalyptus Regnans tree drops up to half a million seeds each year. We have developed the following model for where the seeds land.

- All seeds land within 80 m of the base or foot of the tree.
- The base of the tree is a circle of radius 1 metre. Variable x represents the distance from the base of the tree, in metres.
- Variable N represents the number of seeds per square metre on the ground.
- $N = b(x-80)^2$, where b is a constant.
- The largest concentration of seeds, found right at the base of the tree, is 3200 seeds per square metre.

(a) Calculate the value of the constant b.

(b) Of the seeds that fall each season, only a small fraction germinate to become seedlings. The graph shows how the value of this fraction F, depends on x, the distance from the base of the tree. Write down the equation connecting F and x.

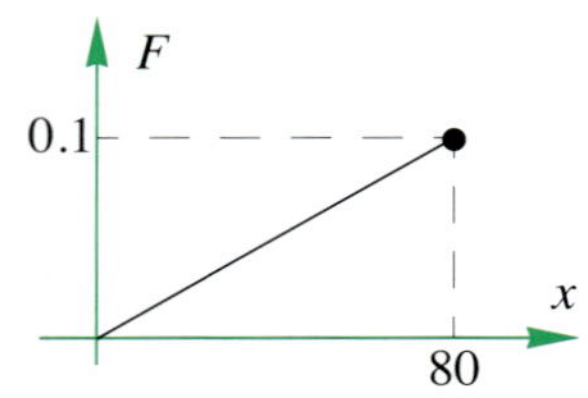

(c) Let the variable y = the number of seedlings per square metre. Find the equation connecting y and x.

(d) Copy and then complete the following table and sketch a graph showing the relationship between y and x.

x	1	20	40	60	80
y					

(e) Calculate the distance from the base of the tree where the model predicts there will be the highest concentration.

10.0 EXPONENTIALS

The four photographs show structures.

Look first at the top two photographs. The left hand picture is of a traditional pylon bridge and the right hand one is of a suspension bridge. Whilst there are obvious differences in the design of the two bridges (the first is supported from below and the second from above), there is a very important difference in the way the two structures react to stress. Stress on bridges comes from the traffic that passes over them, wind, earthquakes etc. A pylon bridge deals with these stresses by being very strong and rigid. Look at the number and size of the pylons the builders used. A suspension bridge is designed to be much thinner (and hence lighter) than the pylon bridge. It deals with stresses by 'flexing and bending'. If you have ever walked across a suspension bridge, you may have felt the bridge deck 'bouncing' as traffic passes over it. This helps the bridge survive these stresses. The amount of 'bouncing' that is allowed is carefully calculated as the bridge is designed. If the engineers get their calculations wrong, the bridge may bounce too much and collapse. This happened to the Tacoma Narrows Bridge (Washington State, USA) in 1940. The full story of this collapse is told on the website: http://www.wsdot.wa.gov/TNBhistory/default.htm which also contains information on bridge design.

The two aircraft shown have different purposes and the same sort of differences in their designs as we saw in the bridges. The first is a small aerobat. It is designed to be very strong. The second picture shows a modern heavy passenger jet. Whilst this is also very strong, it is also designed to be quite flexible. If you have every travelled in a jet such as this, you may have noticed the wings 'flapping'. They do so for a different reason from birds, who also flap their wings. The reason is to help the airframe distribute the stresses that result from atmospheric turbulence.

If engineers design structures that they will allow to vibrate, they will want these vibrations to die away rapidly. This is known as damping. A common model for the way damping works is 'exponential decay'. You will see some of the mathematics that describes this in this chapter.

10.1 EXPONENTIAL FUNCTIONS

10.1.1 THE EXPONENTIAL FUNCTION

The **exponential function** takes the form $f(x) = a^x, x \in \mathbb{R}, a > 0, a \neq 1$, where the independent variable is the exponent.

Graphs with a > 1

An example of an exponential function is $f(x) = 2^x, x \in \mathbb{R}$. So, how does the graph of $f(x) = 2^x$ compare to that of $f(x) = x^2$?

We know that the graph of $f(x) = x^2$ represents a parabola with its vertex at the origin, and is symmetrical about the y–axis. To determine the properties of the exponential function we set up a table of values and use these values to sketch a graph of $f(x) = 2^x$.

x	–3	–2	–1	0	1	2	3	4	5
$y = x^2$	9	4	1	0	1	4	9	16	25
$y = 2^x$	2^{-3}	2^{-2}	2^{-1}	2^0	2^1	2^2	2^3	2^4	2^5
	$= \frac{1}{8}$	$= \frac{1}{4}$	$= \frac{1}{2}$	= 1	= 2	= 4	= 8	= 16	= 32

We can now plot both graphs on the same set of axes and compare their properties:

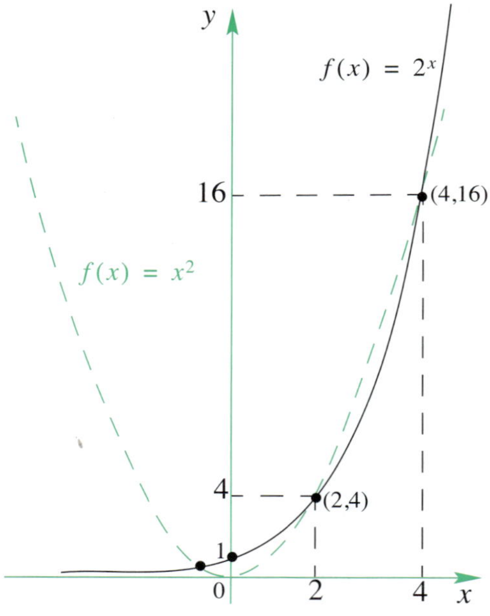

Properties of $f(x) = 2^x$

1. The function increases for all values of x (i.e., as x increases so too do the values of y).

2. The function is always positive (i.e., it lies above the x-axis).

3. As $x \to \infty$ then $y \to \infty$
   $x \to -\infty$ then $y \to 0$.
   i.e., the x-axis is an asymptote.

4. When
   i. $x > 0$ then $y > 1$,
   ii. $x = 0$ then $y = 1$
   iii. $x < 0$ then $0 < y < 1$.

Notice how different the graphs of the two functions are, even though their rules appear similar. The difference being that for the quadratic function, the variable x is the base, whereas for the exponential, the variable x is the power.

We can now investigate the exponential function for different bases. Consider the exponential functions $f(x) = 3^x$ and $g(x) = 4^x$:

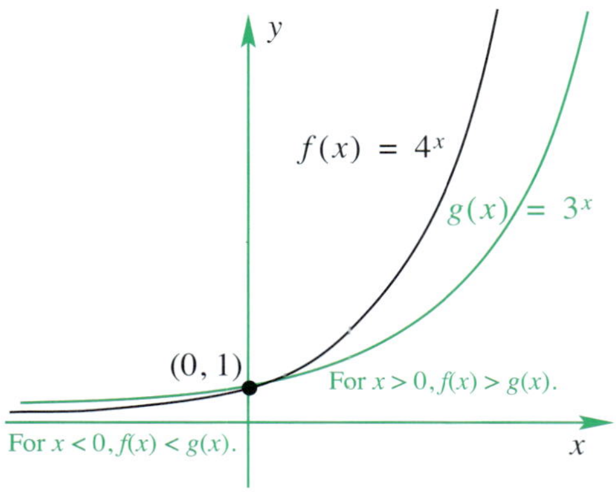

From the graphs we can see that $f(x) = 4^x$ increases much faster than $g(x) = 3^x$ for $x > 0$. For example, at $x = 1$, $f(1) = 4$, $g(1) = 3$ and then, at $x = 2$, $f(2) = 16$, $g(2) = 9$.

However, for $x < 0$ we have the opposite, $f(x) = 4^x$ decreases faster than $g(x) = 3^x$.

Notice then that at $x = 0$, both graphs pass through the point (0, 1).

From the graphs we can see that for values of x less than zero, the graph of $f(x) = 4^x$ lies below that of $f(x) = 3^x$. Whereas for values of x greater than zero, then the graph of $f(x) = 4^x$ lies above that of $f(x) = 3^x$.

Exponential functions that display these properties are referred to as **exponential growth** functions.

What happens when 0 < a < 1?

We make use of the TI–83 to investigate such cases. Consider the case where $a = \frac{1}{2}$.

Rather than using a table of values we provide a sketch of the curve. The graph shows that the function is decreasing – such exponential functions are referred to as **exponential decay**. In fact, from the second screen we can see that the graph of $y = \left(\frac{1}{2}\right)^x$ is a reflection of $y = 2^x$ about the y–axis.

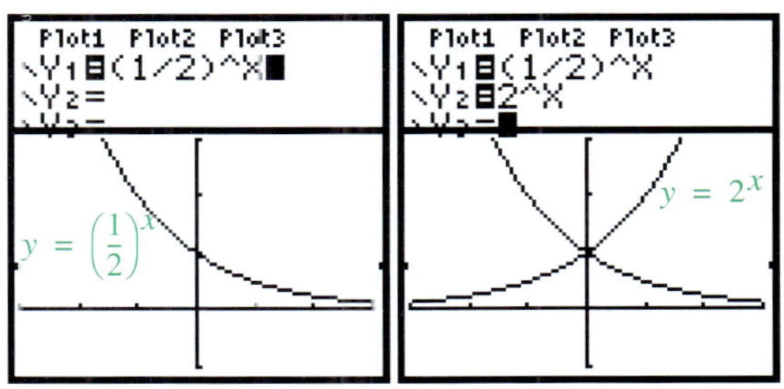

We note that the function $y = \left(\frac{1}{2}\right)^x$ can also be written as $y = (2^{-1})^x = 2^{-x}$. Meaning that there are two ways to represent an **exponential decay** function;

either as $f(x) = a^x, 0 < a < 1$

or $f(a) = a^{-x}, a > 1$.

For example, the functions $f(x) = \left(\frac{1}{4}\right)^x$ and $g(x) = 4^{-x}$ are identical.

We can summarise the exponential function as follows:

$x \mapsto a^x, a > 1, x \in \mathbb{R}$

(e.g., $f(x) = 2^x$, $f(x) = 3^x$)

Properties

Domain	:	$\mathbb{R} = (-\infty, \infty)$
Range	:	$\mathbb{R}^+ = (0, \infty)$
Asymptote	:	$y = 0$ (or x–axis)
Intercepts	:	Cuts y–axis at (0,1)
Other	:	Increases **[growth]** Continuous

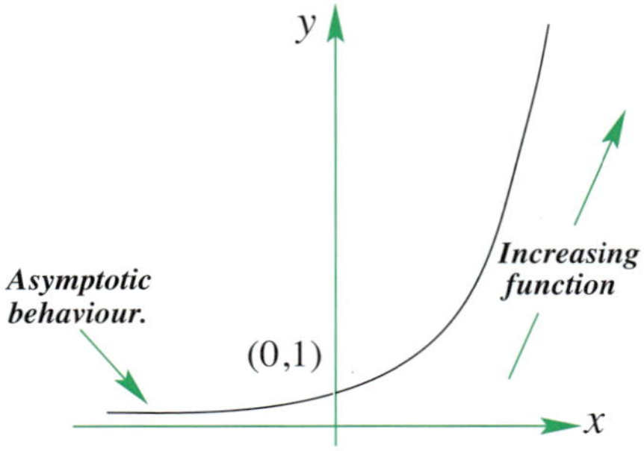

$x \mapsto a^x, 0 < a < 1, x \in \mathbb{R}$

(e.g., $f(x) = \left(\frac{1}{2}\right)^x$, $f(x) = \left(\frac{1}{3}\right)^x$)

Properties

Domain	:	$\mathbb{R} = (-\infty, \infty)$
Range	:	$\mathbb{R}^+ = (0, \infty)$
Asymptote	:	$y = 0$ (or x–axis)
Intercepts	:	Cuts y–axis at (0,1)
Other	:	Decreases **[decay]** Continuous

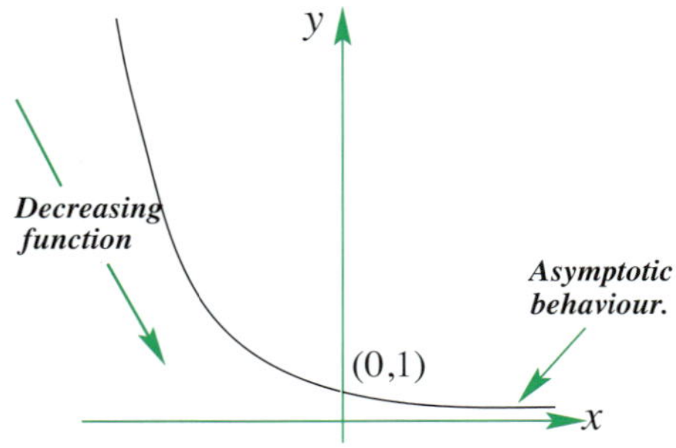

EXAMPLE 10.1

Sketch the graphs of (a) $f(x) = 4^x$ (b) $f(x) = 4^x + 2$

Solution

(a) We've already seen what the graph of $f(x) = 4^x$ looks like. However, should we need to sketch this graph from scratch, we would simply set up a table of values

x	–2	–1	0	1	2	3
$f(x)$	$\frac{1}{16}$	$\frac{1}{4}$	1	4	16	64

From which we would then plot the points on a set of axes:

(a)

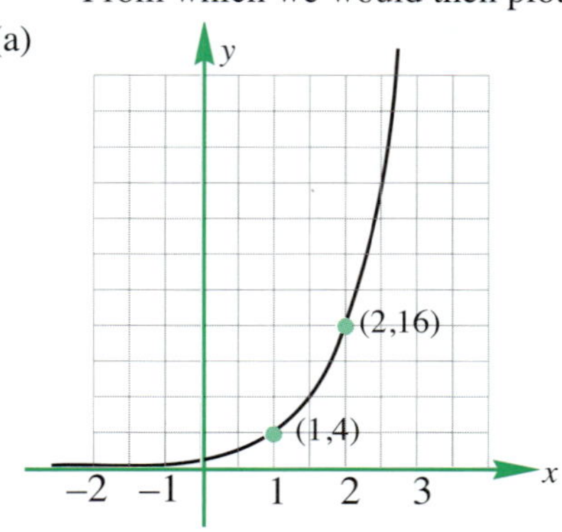

(b)

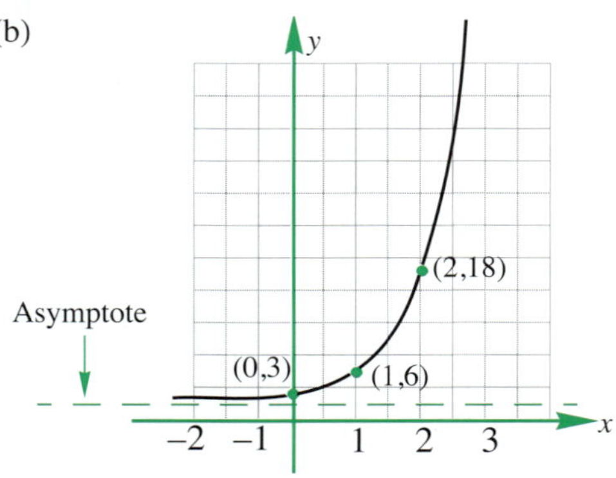

(b) The graph of $f(x) = 4^x + 2$ is simply the same as that of $f(x) = 4^x$ except this time the graph has been moved up 2 units. Notice also that the asymptote moved 2 units upwards.

EXAMPLE 10.2

On the same set of axes sketch the following.

(a) $f(x) = 2^x, g(x) = 2^x - 1$ (b) $f(x) = 3^x, g(x) = 3^x + 2$

Solution

(a)

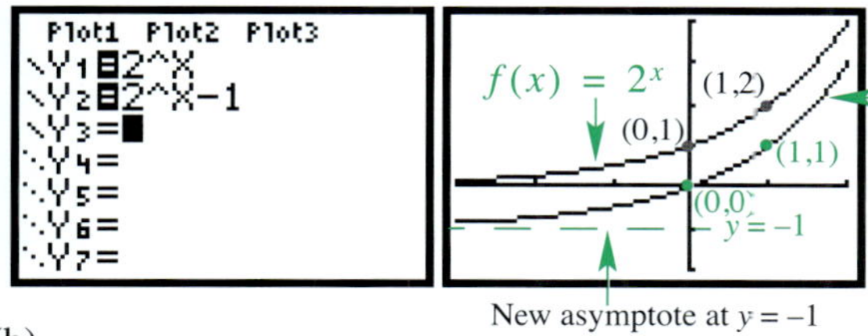

Observations:
The graph of $g(x)$ is the graph of $f(x)$ moved (i.e., translated) one unit down. Notice too that the asymptote has also moved down one unit, from $y = 0$ to $y = -1$.

(b)

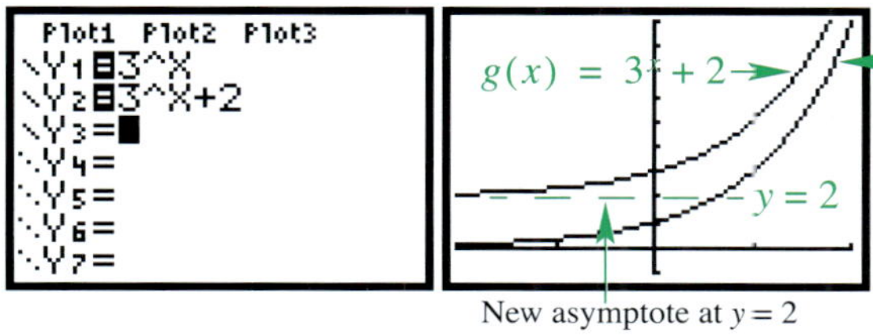

Observations:
The graph of $g(x)$ is the graph of $f(x)$ moved (i.e., translated) two units up. Notice too that the asymptote has also moved up two units, from $y = 0$ to $y = 2$.

From our observations we can make the following generalisation about the graph of $y = a^x \pm k$.

1. The graph of $y = a^x - k, k > 0$ is identical to $y = a^x$ but moved 'k' units down.
2. The graph of $y = a^x + k, k > 0$ is identical to $y = a^x$ but moved 'k' units up.

EXAMPLE 10.3

On the same set of axes sketch the following.

(a) $f(x) = 2^x, g(x) = 3 \times 2^x$ (b) $f(x) = 3^x, g(x) = -\left(\frac{1}{2}\right) \times 3^x$

Solution

(a)

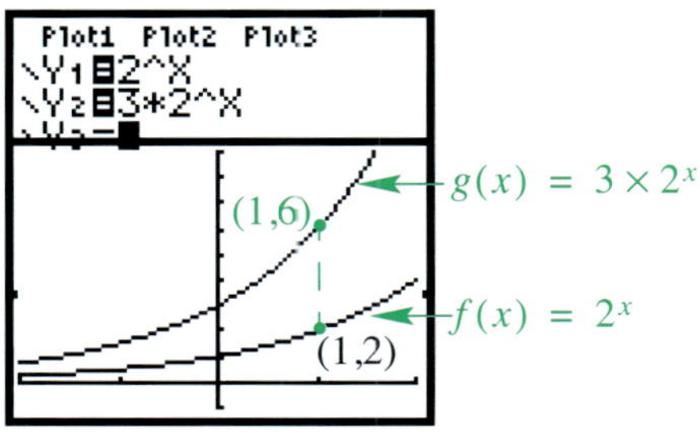

Observations:
The graph of $g(x)$ is the graph of $f(x)$ stretched by a factor of 3 along the y–axis. Notice too that the asymptote has not changed.

(b)

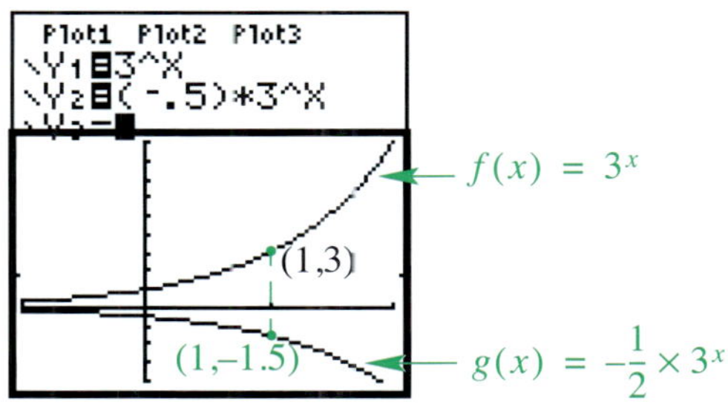

Observations:
The graph of $g(x)$ is the graph of $f(x)$ shrunk by a factor of 2 along the y–axis and reflected about the x–axis. Notice too that the asymptote has not changed.

From our observations we can make the following generalisation about the graph of $y = k \times a^x$.

1. The graph of $y = k \times a^x, k > 0$ is identical to $y = a^x$ but
   i. stretched along the y–axis if $k > 1$.
   ii. shrinks along the y–axis if $0 < k < 1$.

2. The graph of $y = k \times a^x, k < 0$ is identical to $y = a^x$ but
   i. Reflected about the x–axis and stretched along the y–axis if $k < -1$.
   ii. Reflected about the x–axis and shrunk along the y–axis if $-1 < k < 0$.

We consider one more transformation on the graph of $f(x) = a^x$, i.e., we look at the graph of the function $f(x) = a^{\lambda x}$, where λ is a rational number.

We do this by comparing the graphs of $f(x) = 2^x$, $g(x) = 2^{2x}$ and $h(x) = 2^{0.5x}$.
Using the TI–83 we have:

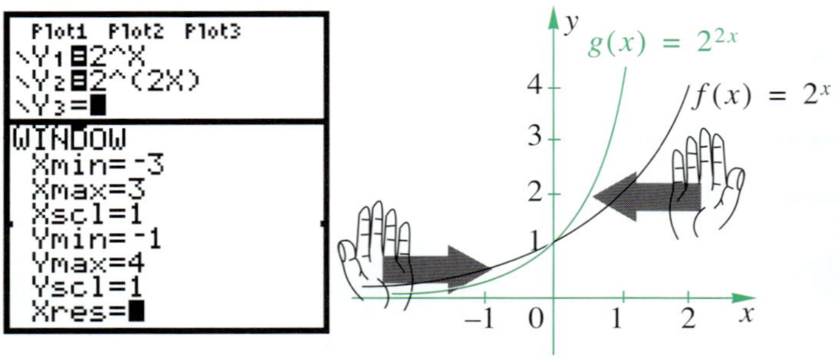

We observe that the graph of $g(x) = 2^{2x}$ has the same shape as that of $f(x) = 2^x$, except that it has been shrunk along the x-axis by a factor of 2.

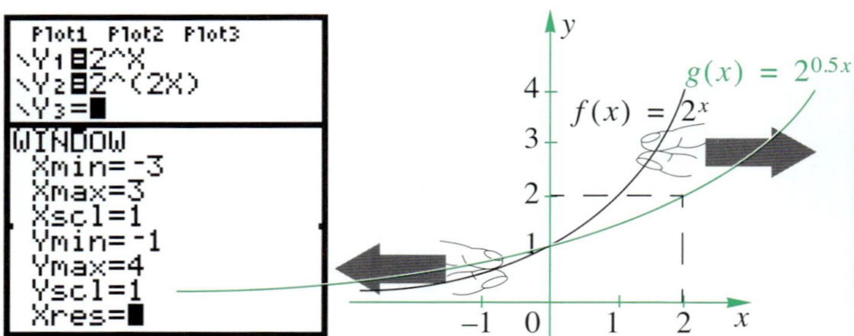

We observe that the graph of $h(x) = 2^{0.5x}$ has the same shape as that of $f(x) = 2^x$, except that it has been stretched along the x-axis by a factor of 2.

Notice that in the last case although the exponent was '$0.5x$' we still say that the graph is stretched by a factor of 2 and not 0.5. The reason is that o.5 can be written as $\frac{1}{2}$, so that

$h(x) = 2^{0.5x}$ can be written as $h(x) = 2^{\frac{1}{2}x}$ and so we stretch the graph by the value of the denominator (of the exponent).

Obviously we can use a combination of these 'effects' on the basic exponential function.

The general form of the exponential function that you will encounter in this course is

$f(x) = ka^{\lambda x} + c$, where k, a, λ and c are all rational numbers.

EXAMPLE 10.4

Sketch the graphs of $g(x)$ where $g(x) = 0.3 \times 3^{-x} + 2$.

Solution

We consider each part of $g(x)$:

$g(x) = 0.3 \times 3^{-x} + 2$ the '$-x$' tells us that it is an exponential decay function.

$g(x) = 0.3 \times 3^{-x} + 2$ the 0.3 will shrink the graph of $f(x) = 3^{-x}$ along the y-axis.

$g(x) = 0.3 \times 3^{-x} + 2$ the '+2' tells us that the graph of $f(x) = 3^{-x}$ is moved up by 2 units. That is, there is an asymptote at $y = 2$.

Using the above information we can now sketch the graph. We do this in stages, so that we 'pencil in' each stage and then use a pen for the final graph.

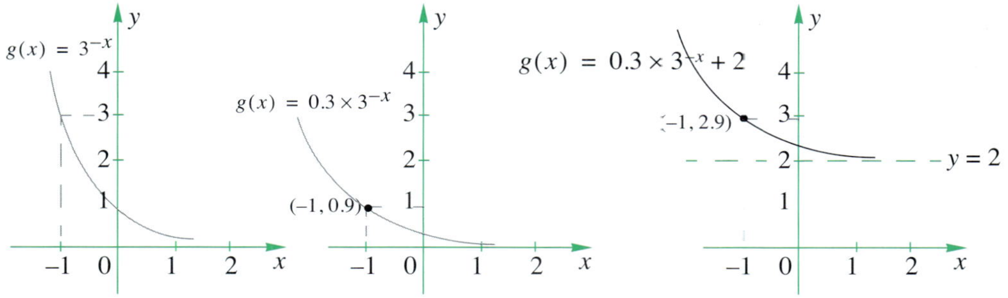

EXERCISES 10.1

1. On separate sets of axes sketch the graphs of the following functions and determine the range of each function.

(a)	$f(x) = 4^x$	(b)	$f(x) = 3^x$	(c)	$f(x) = 5^x$
(d)	$f(x) = (2.5)^x$	(e)	$f(x) = (3.2)^x$	(f)	$f(x) = (1.8)^x$
(g)	$f(x) = \left(\frac{1}{2}\right)^x$	(h)	$f(x) = \left(\frac{1}{3}\right)^x$	(i)	$f(x) = \left(\frac{1}{5}\right)^x$
(j)	$f(x) = \left(\frac{3}{4}\right)^x$	(k)	$f(x) = \left(\frac{5}{8}\right)^x$	(l)	$f(x) = (0.7)^x$

2. Sketch the following on the same set of axes, clearly labelling the y–intercept.

(a)	$f(x) = 3^x + c$	where	i.	$c = 1$	ii.	$c = -2$
(b)	$f(x) = 2^{-x} + c$	where	i.	$c = 0.5$	ii.	$c = -0.5$

3. Sketch the following on the same set of axes, clearly labelling the y–intercept.

(a)	$f(x) = b \times 3^x$	where	i.	$b = 2$	ii.	$b = -2$
(b)	$f(x) = b \times \left(\frac{1}{2}\right)^x$	where	i.	$b = 3$	ii.	$b = -2$

4. On the same set of axes, sketch the following graphs

(a) $f(x) = 3^x$ and $f(x) = 3^{-x}$ (b) $f(x) = 5^x$ and $f(x) = 5^{-x}$

(c) $f(x) = 10^x$ and $f(x) = 10^{-x}$ (d) $f(x) = \left(\frac{1}{3}\right)^x$ and $f(x) = \left(\frac{1}{3}\right)^{-x}$

5. Find the range of the following functions:

(a) $f:[0,4] \mapsto \mathbb{R}, y = 2^x$ (b) $f:[1,3] \mapsto \mathbb{R}, y = 3^x$

(c) $f:[-1,2] \mapsto \mathbb{R}, y = 4^x$ (d) $f:[-1,2] \mapsto \mathbb{R}, y = 2^x$

(e) $f:[2,3] \mapsto \mathbb{R}, y = 2^{-x}$ (f) $f:[-1,1] \mapsto \mathbb{R}, y = 10^{-x}$

6. (a) Solve for x, if $x^2 - 4x - 5 = 0$.

(b) On the same set of axes, sketch the graphs of $f(x) = 5 \times 5^{-x}$ and $g(x) = 5^x - 4$.

(c) Find i. $\{(x, y) : f(x) = g(x)\}$.

ii. $\{x : f(x) > g(x)\}$.

7. Sketch the graphs of the following functions and find their range.

(a) $f(x) = \begin{cases} 2^x, & x < 1 \\ 3, & x \geq 1 \end{cases}$ (b) $f(x) = \begin{cases} 3 - e^x, & x > 0 \\ x + 3, & x \leq 0 \end{cases}$

10.2 EXPONENTIAL EQUATIONS

10.2.1 SOLVING EQUATIONS

Solving exponential equations can carried out by

(a) plotting an accurate graph and reading solution from the graph.
(b) using an algebraic solution.
(c) using a graphics calculator.

We illustrate methods (a) and (c) by looking at two specific examples. Method (b) can lead to situations that require mathematics that is beyond the scope of this course and so we leave this method out.

Of course, equations such as $2^x = 8$ for which is is clear that the solution is $x = 3$ or $5^x = 625$ for which, by observation, we obtain $x = 4$, can be considered as having been solved using an algebraic method. However, these are rather straight forward equations and can be dealt with accordingly. It is for equations such as $2^x = 3$ that we will not consider using an algebraic solution.

EXAMPLE 10.5

Solve for x: (a) $2^x = 10$ (b) $2 \times 3^x + 1 = 40$

Solution

(a) To start with we need to plot a graph of the function $f(x) = 2^x$, from where we will then simply read the solution from the graph (i.e., along the x–axis).

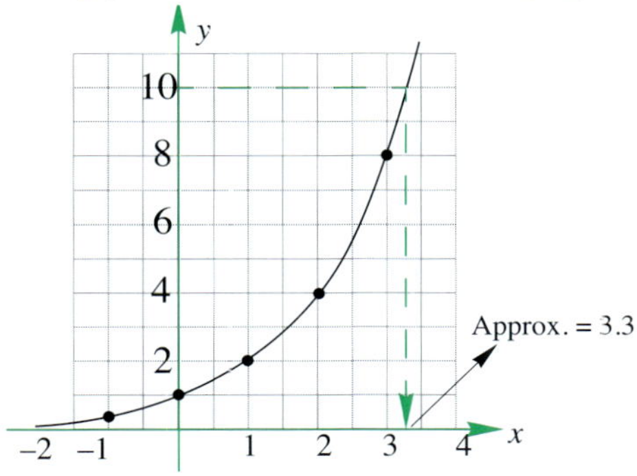

From our graph we therefore have that if $2^x = 10$ then $x \approx 3.3$.

(b) To solve the equation $2 \times 3^x + 1 = 40$ using the TI–83, all we need to do is use the **solve(** function:

Press **2nd 0** [CATALOG] **LN**

Then use the down arrow key to move cursor to **solve(** function:

Press **ENTER**, type the equation in the form **Equation = 0**,

(this means that the equation $2 \times 3^x + 1 = 40$ must first be transposed into the equation $2 \times 3^x + 1 - 40 = 0$) next type X, then provide an initial guess (2 in this case), close the parenthesis and then press ENTER:

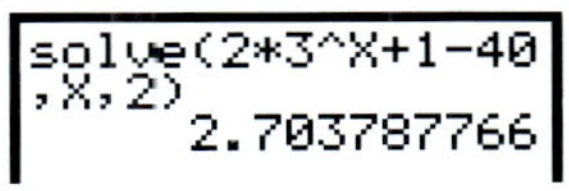

Note, applying this method to the first example gives:

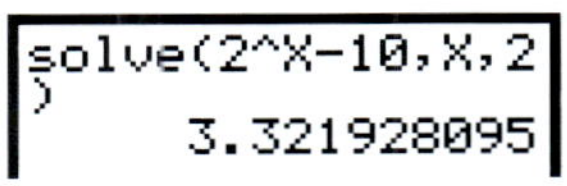

EXERCISES 10.2

1. Solve the following equations using
(a) a graphical solution. (b) a graphics calculator.

i.	$2^x = 8$	ii.	$3^x = 27$	iii.	$4^x = 16$
iv.	$2^x = 15$	v.	$3^x = 25$	vi.	$4^x = 10$
vii.	$3^x - 2 = 12$	viii.	$5^x + 2 = 19$	ix.	$2^x - 3 = 15$
x.	$2^{-x} = 8$	xi.	$3^{-x} = 27$	xii.	$4^{-x} = 16$
xiii.	$2^{-x} = 10$	xiv.	$3^{-x} + 2 = 3$	xv.	$8 - 2^{-x} = 3$

2. Solve the following equations using
(a) a graphical solution. (b) a graphics calculator.

i.	$2 \times 3^x + 4 = 16$	ii.	$3 \times 2^x - 1 = 9$
iii.	$5 \times 2^x - 4 = 21$	iv.	$8 \times 3^x + 4 = 68$

10.3 APPLICATIONS

There are many situations and examples where an exponentional function is an appropriate function to model a particular **growth** or **decay** process. For example:

1. When looking at the bacteria count of an experiment, the growth in the number of bacteria present in the colony is accurately represented by an exponential growth model.

If there are initially 100 bacteria in a colony and the size doubles every day, we model this situation by making use of the exponential function,

$$f: [0, a[\mapsto\mathbb{R}\text{, where } f(t) = 100\times 2^t, a\in\mathbb{R}.$$

The graph of such a model is given below.

Exponential growth

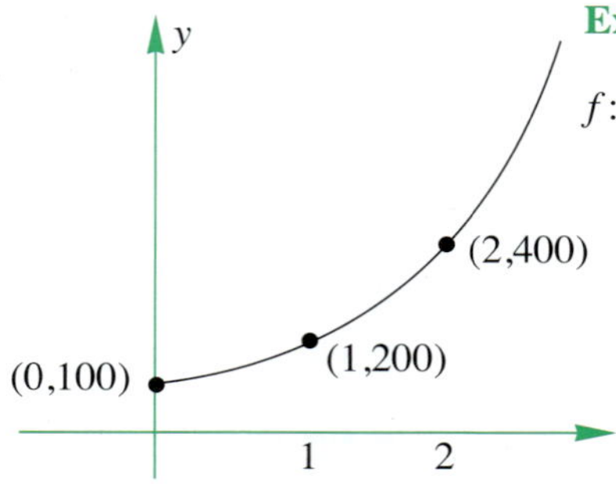

$f: [0, a[\mapsto\mathbb{R}$, where $f(t) = 100\times 2^t, a\in\mathbb{R}$

2. Certain physical quantities decrease exponentially, for example, the decay of a radioactive substance, or isotope. Associated with this is the half–life, that is, the time that it takes for the substance to decay to one half of its original amount.

A radioactive bismuth isotope has a half–life of 5 days. If there is 100 milligrams initially, then we can model this situation by making use of the exponential function,

$$f : [0, \infty)\mapsto\mathbb{R}\text{, where } f(t) = 100\times\left(\frac{1}{2}\right)^{t/5}$$

Exponential decay

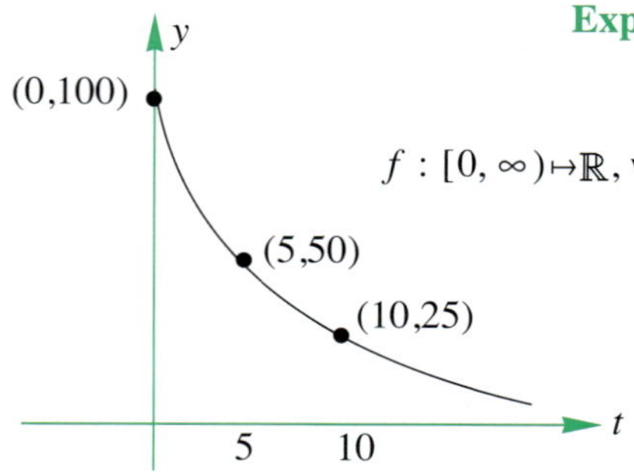

$f : [0, \infty)\mapsto\mathbb{R}$, where $f(t) = 100\times\left(\frac{1}{2}\right)^{t/5}$

Other areas where the use of exponential modelling appears include, medicine (drug dosage), economics (compound interest), oceanography (light penetration in an ocean), environment (endangered species) and many more. We shall look at a few examples of exponential modelling in detail.

Notice that whenever making use of an exponential function to model a real life situation, the domain of consideration is always restricted to $[0, \infty[$. Corresponding to time, $t = 0$ (or $x = 0$), there exists an initial amount. This initial amount is usually denoted by a capital letter with a subscript of '0'. For example, if we are referring to the population size of bacteria, N or the number of radioactive particles P, then their initial amounts would be represented by N_0 and P_0 respectively, so that when $t = 0$, $N = N_0$ and $P = P_0$

Such equations would then be given by

1. $N = N_0 \times a^t, t \geq 0, a > 1$ [growth]

2. $P = P_0 \times a^{-t}, t \geq 0, a > 1$ [decay]

EXAMPLE 10.6

During the chemical processing of a particular type of mineral, the amount M kg of the mineral present at time t hours since the process started, is given by

$$M(t) = M_0(2)^{kt}, t \geq 0, k < 0$$

where M_0 is the original amount of mineral present. If 128 kilograms of the mineral are reduced to 32 kilograms in the first six hours of the process find,

(a) i. the value of k.

ii. the quantity of the mineral that remains after 10 hours of processing.

(b) Sketch a graph of the amount of mineral present at time t hours after the process started.

Solution

(a) i. We have that when $t = 0$, $M = 128 \Rightarrow M_0 = 128$ (the initial amount of mineral).

The equation then becomes

$$M(t) = 128 \times (2)^{kt}, t \geq 0, k < 0 .$$

Next, when $t = 6$, $M = 32$, so that when we substitute this information into the equation, we have, $32 = 128 \times (2)^{6k} \Leftrightarrow 2^{6k} = \frac{1}{4}$

$$\Leftrightarrow 2^{6k} = 2^{-2}$$

$$\Leftrightarrow 6k = -2$$

$$\Leftrightarrow k = -\frac{1}{3}$$

Therefore, the equation is given by, $M(t) = 128 \times (2)^{-\frac{1}{3}t}, t \geq 0$

ii. After 10 hours, we have, $M(10) = 128 \times (2)^{-\frac{1}{3} \times 10}$

$$= 12.699$$

That is, there is approximately 12.70 kg of mineral left after 10 hours of processing.

(b) We notice that the equation is of the form $f : [0, \infty) \mapsto \mathbb{R}$,where $f(t) = a^{-x}, a > 1$ i.e., an exponential decay.

Hence, we have a decreasing function:

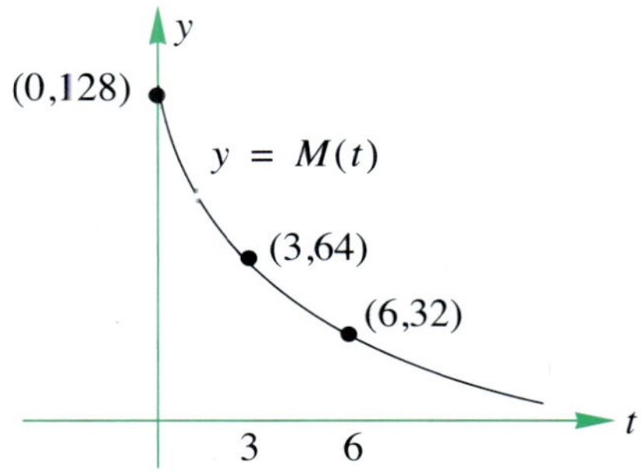

EXAMPLE 10.7

The scrap value , $V, of some machinery after t years is given by

$$V(t) = 50000(0.58)^t, t \geq 0 .$$

(a) What was the initial cost of the machine?
(b) What is the scrap value of the machine after 4 years?
(c) How long would it be before the scrap value reaches $20000?
(d) The machine needs to be sold at some time when the scrap value of the machine lies somewhere between 10000 and 15000. What time–frame does the owner have?

Solution

(a) When $t = 0$, we have $V(0) = 50000(0.58)^0 = 50000$.
That is, the machine initially cost $50000.

(b) After 4 years, we have $V(4) = 50000(0.58)^4 = 5658.25$
That is, after 4 years, the scrap value of the machine would be $5658.25.

(c) We need to determine the value of t when $V = 20000$:

$$20000 = 50000(0.58)^t \Leftrightarrow 0.4 = 0.58^t$$

Then, using the TI–83 we have (using the solve facility):
That is, $t = 1.68$ (to 2 d.p)

```
solve((.58)^X-.4
,X,2)
          1.682109454
```

(d) This time we want to solve for t where $10000 \leq V(t) \leq 15000$.

Now, $10000 \leq V(t) \leq 15000 \Leftrightarrow 10000 \leq 50000(0.58)^t \leq 15000$

$$\Leftrightarrow 0.2 \leq (0.58)^t \leq 0.3$$

Solving the corresponding equalities, we have:
Giving $\quad 2.21 \leq t \leq 2.95$.

```
solve(.58^X-.2,X
,2)
          2.954576135
solve(.58^X-.3,X
,2)
          2.210230843
```

Notice that the graph helped in guessing the values of t that were used in determining the solutions.

Using the TI–83, we can easily sketch the graph of $y = (0.58)^t, t \geq 0$:

```
Plot1 Plot2 Plot3
\Y1=(.58)^X
\Y2=
\Y3=
\Y4=
\Y5=
\Y6=
\Y7=
```

```
WINDOW
 Xmin=0
 Xmax=5
 Xscl=1
 Ymin=0
 Ymax=1
 Yscl=.1
 Xres=
```

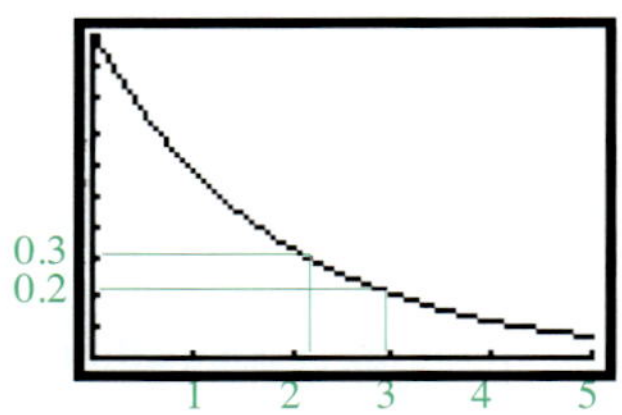

EXAMPLE 10.8

A radioactive element decays in such a way so that the amount present each year is 0.95 of the amount present the previous year. At the start of 1990, there were 50 mg of the element present.

(a) Produce a table of values showing how much of the element is present at the start of each year from 1990 to 1998.
(b) The rule for this situation is given by $N = k \times a^t$, where N mg is the amount of element present t years after the start of 1990. Find the value of a and k.
(c) Sketch the graph of $N = k \times a^t$.
(d) How long will it be before there is 25 mg of this element remaining?

Solution

(a) Initially there is 50 mg of the element present. At the start of the following year there will be 95% of that left over, i.e., there will be $0.95 \times 50 = 47.5$ mg left over. At the start of the year after there will be 95% of 47.5 mg left over. i.e., the will be $0.95 \times 47.5 = 45.125$ mg, and so on. Continuing in this manner we can complete the table of values:

t	0	1	2	3	4	5	6	7	8
N	50	47.5	41.13	42.87	40.73	38.69	36.75	34.92	33.17

(b) The initial amount is 50 mg, therefore, $k = 50$. The quantity is decreasing by 95% each time, therefore, $a = 0.95$.

(c) Using the TI–83, we have:

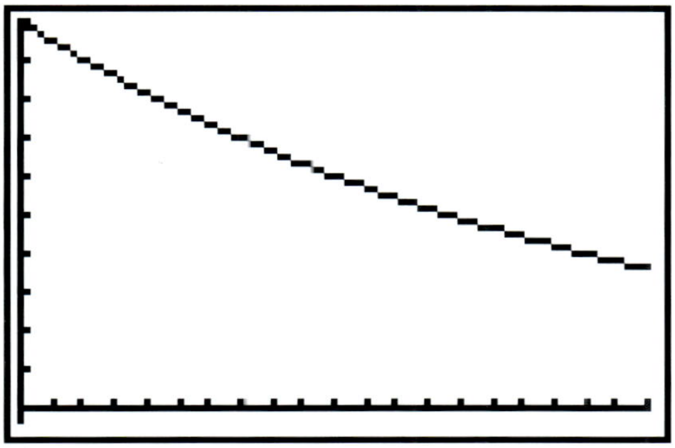

(d) Using the TI–83 we are solving the equation $25 = 50 \times (0.95)^t$ or (as is required for the TI–83), we need to solve the equation

$$50 \times 0.95^t - 25 = 0:$$

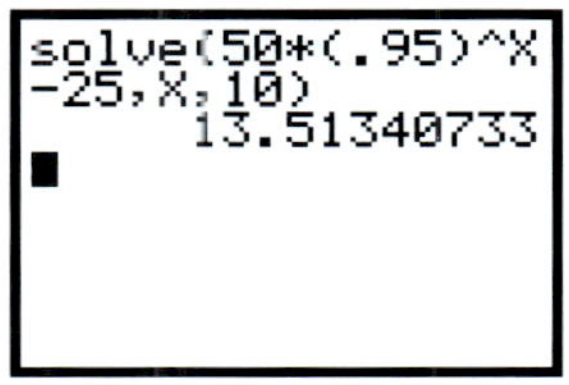

That is, it takes 13.51 years for this amount of radioactive element to halve.

EXERCISES 10.3

1. The charge Q coulombs stored at the plates of a leaky capacitor after time t seconds is determined by the equation $Q(t) = Q_0 \times (1.122)^{-t}, t \geq 0$, where Q_0 is the initial charge.

(a) If the charge on the plates is 250 coulombs after 5 seconds, determine the original charge.
(b) What proportion of the original charge remains after 10 seconds?
(c) How long will it take the charge to reduce by half?
(d) Sketch the graph of the charge remaining on the plates for $t \geq 0$.

2. The number of bacteria in a culture, N, has been modelled by the exponential function $N(t) = 1200 \times (1.148)^t, t \geq 0$, where t is measured in days.

(a) Find the initial number of bacteria in this culture.
(b) How many bacteria (to the nearest hundred) will there be after i. 5 days? ii. 10 days?
(c) How long will it take for the number of bacteria to reach 5000?
(d) Sketch the graph showing the number of bacteria for $t \geq 0$.

3. A population of 120 mammals from an endangered species were introduce onto an enclosed reserve at the start of 1990. By the start of 1995 the population had increased to 300.

It is believed that a model of the form $N(t) = N_0(2.718)^{kt}, t \geq 0$ (where $N(t)$ represents the number of mammals present at time t years since their introduction onto the reserve) adequately approximates the population size of this mammal.

(a) State the value of N_0. What does it represent?

(b) Determine the value of k to 4 decimal places.

(c) Find the population size i. at the start of the year 2000.
ii. at the end of the year 2000.

(d) When will the population reach a size of 1000?

(e) Sketch a graph showing the behaviour of the mammal population on the reserve for $0 \leq t \leq 20$.

4. The number of bacteria in a culture, N, is modelled by the exponential function

$$N = 1000 \times 2^{0.2t}, t \geq 0$$

where t is measured in days.

(a) Find the initial number of bacteria in this culture.

(b) Find the number of bacteria after i 3 days.
ii. 5 days.

(c) How long does it takes for the number of bacteria to grow to 4000?

5. The 'growth' of crystals, measured in kilograms, in a chemical solution, has been approximately modelled by the exponential function $W = 2 \times 10^{kt}, t \geq 0$, where W is measured in kilograms and t in years. After 1 year in a chemical solution, the amount of crystal in the chemical increased by 6 grams.

(a) Find the value of k.

(b) Find the amount of crystal in the chemical solution after 10 years.

(c) How long does it takes for this crystal to double in 'size'?

(d) Sketch the graph showing the amount of crystal in the chemical solution at time t.

6. An endangered species of animal is placed into a game reserve. 150 such animals have been introduced into this reserve. The number of animals, $N(t)$, alive t years after being placed in this reserve is predicted by the exponential growth model $N(t) = 150 \times 1.05^t$.

(a) Find the number of animals that are alive after
i. 1 year ii. 2 years iii. 5 years

(b) How long will it take for the population to double?

(c) How long is it before there are 400 of this species in the reserve?

(d) Sketch a graph depicting the population size of the herd over time.
Is this a realistic model?

7. The processing of a type of mineral in a chemical solution has been found to reduce the amount of that mineral left in the solution. Using this chemical process, the amount W kg of the mineral left in the solution at time t hours is modelled by the exponential decay function $W = W_0 \times 10^{-kt}$, where W_0 kg is the original amount of mineral.

It is found that 50 kilograms of mineral are reduced to 30 kilograms in 10 hours.

(a) Write down the value of W_0.

(b) Find the value of k (to 4 decimal places).

(c) How much of the mineral will be in the solution after 20 hours?
(d) Sketch the graph representing the amount of mineral *left* in the solution.
(e) Sketch the graph representing the amount by which the mineral is *reduced*.

8. The temperatures of distant dying stars have been modelled by exponential decay functions. A distant star known to have an initial surface temperature of 15000°C, is losing heat according to the function $T = T_0 \times 10^{-0.1t}$, where T_0 °C is its present temperature, and T °C the temperature at time t (in millions of years).

(a) Determine the value of T_0.
(b) Find the temperature of this star in i. one million years, ii. 10 million years.
(c) How long will it be before the star reaches a temperature that is half its original surface temperature?
(d) Sketch a graph representing this situation.

9. The amount of radioactive material, Q grams, decays according to the model given by the equation $Q = 200 \times 10^{-kt}, t \geq 0$, where t is measured in years. It is known that after 40 years, the amount of radioactive material present is 50 grams.

(a) Find the value of k (to 4 d.p.).
(b) Find the amount of radioactive material present after 80 years.
(c) What is the half life for this radioactive substance ? *The half–life is the time taken for the radioactive material to reach half its original amount.*
(d) Sketch the graph representing the amount of radioactive material present as a function of time, t years.

10. The resale value, V dollars, of a structure, decreases according to the function

$$V = 2000000(10)^{-0.01t}, t \geq 0$$

where t is the number of years since the structure was built.

(a) How much would the structure have sold for upon completion?
(b) How much would the structure have sold for 10 years after completion?
(c) How long will it take for the structure to lose half its value? (Answer to 1 d.p)
(d) Sketch the graph of the structure's value since completion.

11. The population number N in a small town in northern India is approximately modelled by the equation $N = N_0 \times 10^{kt}, t \geq 0$, where N_0 is the initial popluation and t is the time in years since 1980.
The population was found to increase from 100,000 in 1980 to 150,000 in 1990.

(a) Show that $N_0 = 100000$ and that $1.5 = 10^{10k}$.
(b) Hence find the value of k (to 5 d.p.).
(c) Find the population in this town in 1997.
(d) How long (since 1980) will it be before the population reaches 250,000?

10.4 MISCELLANEOUS EXERCISES

1. On the same set of axes, sketch the graphs of

(a) $f(x) = 3^x + 1$

(b) $g(x) = \left(\frac{1}{2}\right)^x$

Hence, find the value of x, to one decimal place, where $f(x) = g(x)$.

2. The graph with equation $f(x) = 2^x + c$ passes through the point $(1, 3)$.

(a) Find the value of c.

(b) Hence, solve for x, where $f(x) = 9$.

3. A bacteria culture is placed under a strict controlled environment. Under these conditions, the bacteria is allowed to double in number every hour. Initially there were 8 bacteria.

(a) Complete the following table

t, time in hours	0	1	2	3	4	5
N, number of bacteria	8					

(b) Using an appropriate scale, draw a graph with N, the number of bacteria, on the vertical axis and t, the number of hours, on the horizontal axis.

(c) Using your graph in (b), estimate

i. the number of bacteria after 2.5 hours.

ii. after how many hours would there be 100 bacteria.

4. The population of a nest of ants, t weeks after it was located is thought to follow the exponential model defined by

$$N = 400 \times (1.12)^t, t \geq 0.$$

(a) How many ants were there when the nest was discovered?

(b) How many ants were there in the nest after

i. 5 weeks.

ii. 10 weeks.

(c) Using an appropriate scale, draw a graph with N, the number of ants, on the vertical axis and t, the number of weeks, on the horizontal axis.

(d) Using your graph in (c), estimate after how many weeks would there be 2000 ants.

5. The current, I amps, flowing through a circuit, once switched off, dies away according to the rule

$$I(t) = 20 \times (0.25)^t,$$

where t is measured in seconds.

(a) Draw a graph of I, for $0 \leq t \leq 4$.

(b) Estimate how long it takes for the current to reach 10 amps.

6. On the same set of axes, sketch the graphs of

(a) $f(x) = 2^x + 3$, $-1 \le x \le 3$, stating its range.

(b) $g(x) = \left(\frac{1}{3}\right)^x - 1$, $-1 \le x \le 3$, stating its range.

7. Solve the equations

(a) $3^x + 2 = 11$ (b) $4 - \left(\frac{1}{2}\right)^x = 2$

8. The exponential function $f(x) = k \times 2^x + c$ is shown in the diagram.
Find k and c.

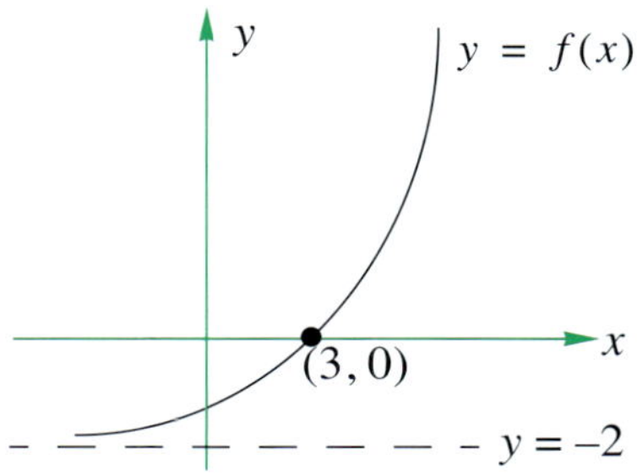

9. A bacteria culture is placed under a strict controlled environment. Under these conditions, the bacteria is allowed to double in number every hour. Initially there were 10 bacteria.

(a) Complete the following table

t, time in hours	0	1	2	3	4	5	6	7
N, number of bacteria	10							

(b) Using a scale of 1 cm to represent 1 hour and 1 cm to represent 100 bacteria, draw a graph with N, the number of bacteria, on the vertical axis and t, the number of hours, on the horizontal axis.

(c) Using your graph in (b), estimate
i. the number of bacteria after 2.5 hours.
ii. after how many hours would there be 500 bacteria.

10. The results from one annual reproduction of Pacific halibut provides that the number of halibut, $N(t)$ still alive after t years is approximated by the equation

$$N(t) = N_0 \times (0.82)^t,$$

where N_0 is the initial size of the Pacific halibut.

(a) Sketch the graph of $y = N(t)$.

(b) Approximate the percentage of the original number still alive after 12 years.

(c) How long will it take for half of the halibut to disappear?

The lengths, L cm, of the halibut from this cohort have lengths modelled by the equation $L(t) = a[1 - b \times (0.85)^t]$.

(d) Using $a = 250$ and $b = 0.95$, estimate the length of a 6 year old halibut.

(e) A halibut measuring 20 cm is caught by the fisheries department. Estimate the age of this halibut.

(f) Sketch the graph of $y = L(t)$.

10.5 GRADED REVISION EXERCISES

LEVEL 1

1. Sketch the following functions

i. $f(x) = 2^x$ ii. $x \mapsto 0.3^x$ iii. $g(x) = 4^{-x}$

2. Evaluate the following for the given value of x:

i. $y = 2 + 3 \times 5^{x-1}$ at $x = 2$. ii. $y = 3 - \frac{1}{3} \times 2^{x+1}$ at $x = 1$

LEVEL 2

1. Solve, to one decimal place, the following equations using graph paper.

i. $2^x = 12$ ii. $3^x = 15$ iii. $4^x = 12$

LEVEL 3

1. The graph with equation $f(x) = 2^{-x} + k$ passes through the point (1,1). Sketch the graph of $y = f(x)$. Hence, solve for x, where $f(x) = 2$.

2. The exponential function $y = k \times 3^x + c$ has an asymptote at $y = 2$ and passes through the point (1,0). Find the values of k and c.

3. Under laboratory conditions, the number of bacteria, N, in a particular culture doubles every 15 seconds. Initially there were 2 such bacteria in the culture.

i. Construct a table of values of N for the first minute of growth using 15 sec intervals.
ii. Sketch the graph of N against time t seconds for the first minute.
iii. Calculate the number of bacteria in the culture after 5 minutes.

LEVEL 4

1. i. On the same set of axes sketch the graphs of $f(x) = 2^x$ and $g(x) = 4 \times (0.5)^x$. Hence sketch the graph of $y = f(x) + g(x)$.

ii. Show that if $f(x) + g(x) = 5$, then $(2^x)^2 - 5 \times 2^x + 4 = 0$.

iii. Solve for x, if $f(x) + g(x) = 5$.

2. The rate R at which the drug enters the bloodstream t minutes after it was administered is approximated by the equation

$$R(t) = 5 \times (0.95)^t \text{ mg/min},$$

while the amount A mg of the drug in the bloodstream t minutes after being administered is approximated by

$$A(t) = 98(1 - (0.95)^t).$$

i. What is the maximum rate at which the drug enters the bloodstream?

ii. Sketch, on separate axes, the graphs of (a) $y = R(t)$.
(b) $y = A(t)$.

iii. How much of the drug is in the bloodstream when the drug is entering at a rate of 2.5 mg/min?

10.6 TOPIC TEST

1. Sketch the graphs of $f(x) = 2 + 3^x, -1 \le x \le 2$. State its range.

[5 marks]

2. The exponential function $y = a - k \times b^{x+1}$ is shown in the diagram.
Find the values of a, k and b.

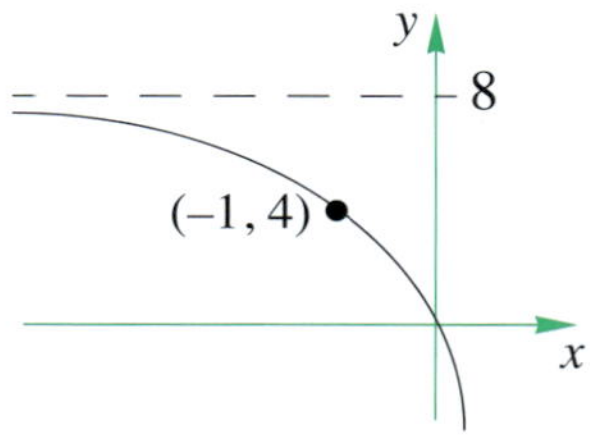

[5 marks]

3. Solve the equations (a) $2^x + 3 = 12$ (b) $\frac{1}{2} - 4^x = \frac{1}{8}$.

Give your answers correct to three decimal places.

[4 marks]

4. A radioactive substance decomposes in such a way that its mass, M gm, present t years after some initial time is given by $M = 200 \times a^t$. After 2 years, its mass is calculated to be 150 gm.

i. Find the value of a, giving your answer to three decimal places.
ii. Determine how long it takes for the mass to decompose by 60%.
iii. Sketch the graph of $y = M(t), 0 \le t \le 10$.
iv. Sketch the graph of the amount of radioactive substance that has decomposed.

[2 + 2 + 3 + 3 marks]

5. On the same set of axes, sketch the graphs of $g(x) = 3^x$ and $f(x) = 12 - 2^x$. Hence find $\{x | 3^x + 2^x = 12\}$, giving your answer to one decimal place.

[4 marks]
Total 28 marks

10.7 PROJECT SUGGESTIONS

You will have seen many ideas for exponential growth and decay in the chapter. Many of these can lead to simple experiments that form the basis of good projects.

If you choose to investigate cooling or another physical phenomenon, you may be able to use one of the data collection devices offered by the various calculator companies. Your school may own one or more of these devices.

Examples are the Texas CBL 2™ & CBR™ and the Casio Data Analyzer.

If you have access to one of these devices, you may be able to investigate a damped oscillation of the type described in the chapter header.

Also, you are encouraged to model a slightly more complex cooling problem.

> Imagine, for example, that you have five greasy plates to wash, a bowl of hot soapy water and a bowl of hot rinsing water.
>
> What happens to the temperatures of these seven items?

11.0 TIDES

Most of us are familiar with things that are repeated at regular intervals. Such events are said to be 'periodic' and are the subject of this chapter.

The photograph shows the fortified town of Mont-Saint-Michel which is built on a granite outcrop that rises sharply out of Mont-Saint-Michel Bay (between Brittany and Normandy, France). The bay is famous for its large tides. These were feared by invading armies and kept the small number of townspeople safe from attack for centuries.

We now know that tides are predictable and the site is protected not by its natural advantages but by its UNESCO World Heritage listing.

We probably think that tides are a nuisance. They are, however, an essential part of many ecosystems. The periodic flooding and drying of mudflats and sandbanks provide feeding grounds for many seabirds. There are also some practical applications. A short distance to the west of Mont-Saint-Michel is the site of the World's first tidal power station (La Rance). The station uses the power of the tides to generate electricity without producing any greenhouse gases. However, the other environmental effects of the project (silting etc.) may outweigh these advantages.

11.1 TRIGONOMETRIC FUNCTIONS

11.1.1 THE SINE AND COSINE FUNCTIONS

There exists a category of functions, known as the circular trigonometric functions, that possess an interesting property – that of periodicity. That is, they behave in such a way that their values are repeated over and over. This is known as **periodic** behaviour. Many real world phenomena are periodic in the sense that the same patterns repeat over time. The **circular trigonometric functions** (or simply trigonometric functions) are often used to model such phenomena which include sound waves, light waves, alternating current electricity and other more approximately periodic events such as tides and even our moods (i.e., biorythms).

When the trigonometric functions are used for these purposes, the angles are almost always measured in **radians** (which is a different way of measuring angles). However there is no reason why we cannot use degrees. It will always be obvious from the equation as to which mode of angle we are using, an expression such as $\sin x$ will imply (by default) that the angle is measured in radians, otherwise it will be written as $\sin x°$, implying that the angle is measured in degrees. In fact, for the purposes of this course we will only use degrees. The two functions thatwe will consider are known as the **sine function** and the **cosine function** (or cos function for short).

What do trigonometric functions look like?

The most appropriate way to determine the shape of these graphs is to plot them. We start with the sine function. Make sure that you calutor is in degree mode.

The sine function:

θ	0	30	45	60	90	120	135	150	180	. . .	330	360
$\sin\theta°$	0.0	0.5	0.71	0.87	1.0	0.86	0.71	0.5	0.0	. . .	–0.5	0.0

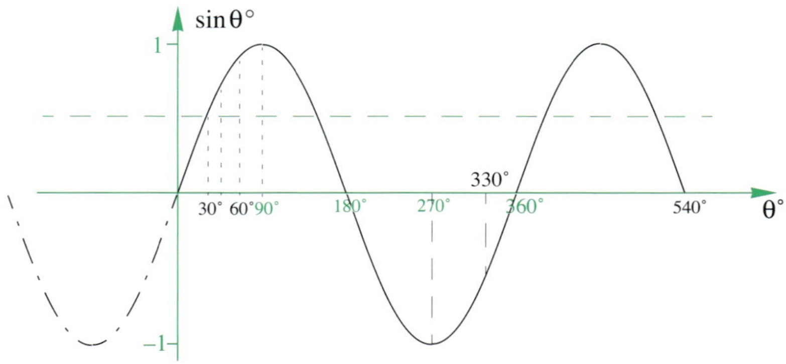

We make note of the following observations:

Feature of sine graph:

1. Maximum value = 1, Minimum value = –1
2. Period = 360° (i.e., graph repeats itself every 360°)
3. We can use both positive and negative angles.
4. Function is not a one to one function.

The cosine function:

θ	0	30	45	60	90	120	135	150	180	...	330	360
$\cos\theta°$	1.0	0.87	0.71	0.5	0.0	–0.5	–0.71	–0.87	–1.0	...	0.87	1.0

Plotting these points on a $\cos\theta°$ versus θ–axis, we have:

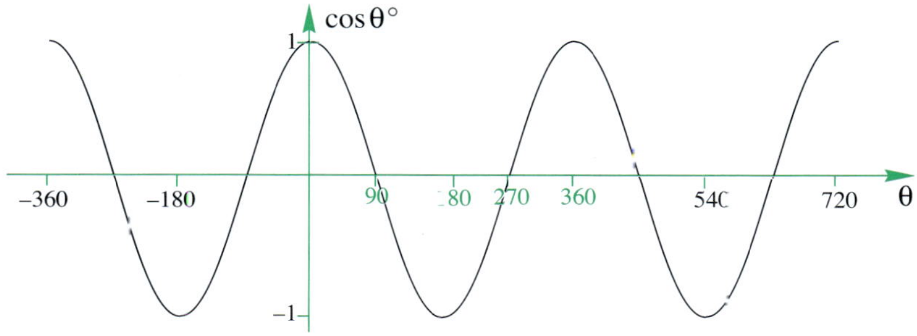

Feature of cosine graph:
1. Maximum value = 1, Minimum value = –1
2. Period = 360° (i.e., graph repeats itself every 360°)
3. We can use both positive and negative angles.
4. Function is not a one to one function.

There are some common observations to be made from these two graphs;

1. We have that the **period** of each of these functions is 360°.
   This is the length that it takes for the curve to start repeating itself.

2. The **amplitude** of the function is the distance between the centre line (in this case the θ -axis) and one of the maximum points. In this case, the amplitude is 1.

The sine and cosine functions are useful for modelling wave phenomena such as sound, light, water waves etc.

When sketching these graphs using the TI–83, be sure that the **WINDOW** settings are appropriate for the **MODE** setting. In the case of degrees we have:

Step 1. Make sure that the calculator is in degree mode. Failure to do this could be disastrous!
Step 2. Select an appropriate range
Step 3. Enter the function rule.

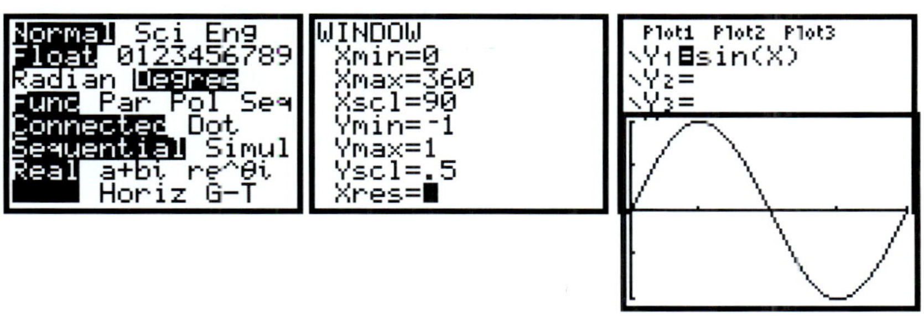

We now provide both graph, one above the other, to show their similarities.

$f:\mathbb{R}\mapsto\mathbb{R}, f(x) = \sin x$

period 360°
amplitude 1

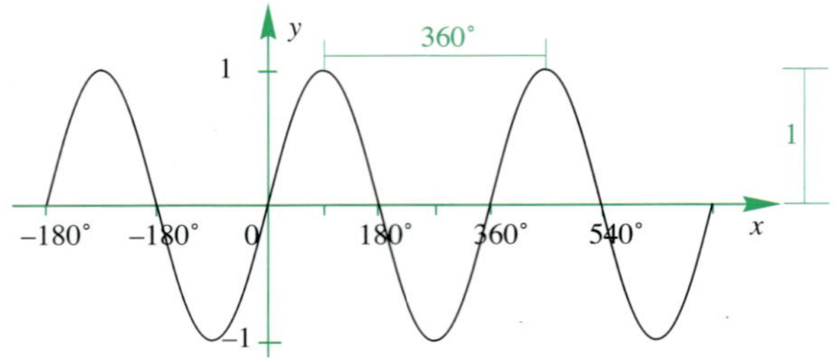

$f:\mathbb{R}\mapsto\mathbb{R}, f(x) = \cos x$

period 360°
amplitude 1

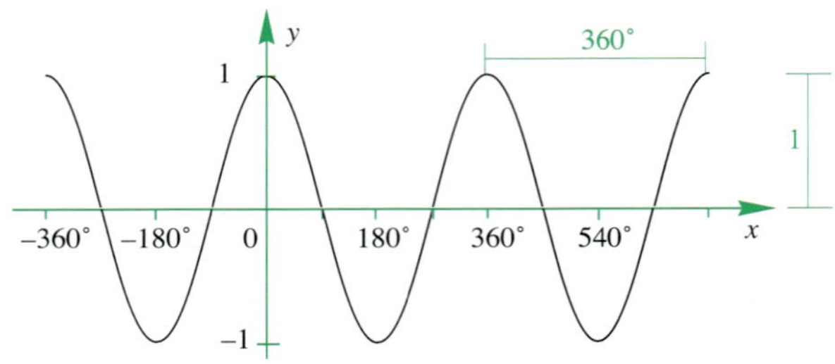

In fact, the cosine graph is the same as the sine graph except that it has been shifted to the left by 90°.

11.1.2 TRANSFORMATIONS OF TRIGONOMETRIC FUNCTIONS

We now consider some of the possible transformations that can be applied to the standard sine and cosine function and look at how these transformations affect the basic properties of both these graphs

1. Vertical translations

Functions of the type $f(x) = \sin(x) + c$ and $f(x) = \cos(x) + c$ represent vertical translations of the curves of $\sin(x)$ and $\cos(x)$ respectively. If $c > 0$ the graph is moved vertically up and if $c < 0$ the graph is moved vertically down.
That is, adding or subtracting a fixed amount to a trigonometric function translates the graph parallel to the y-axis.

EXAMPLE 11.1

Sketch the graphs of the functions for x values in the range -2π to 4π.

(a) $y = \sin(x) + 1$ (b) $y = \cos(x) - 2$

Solution

(a) The '+1' indicates that the graph of $\sin(x)$ must be translated 1 unit up.

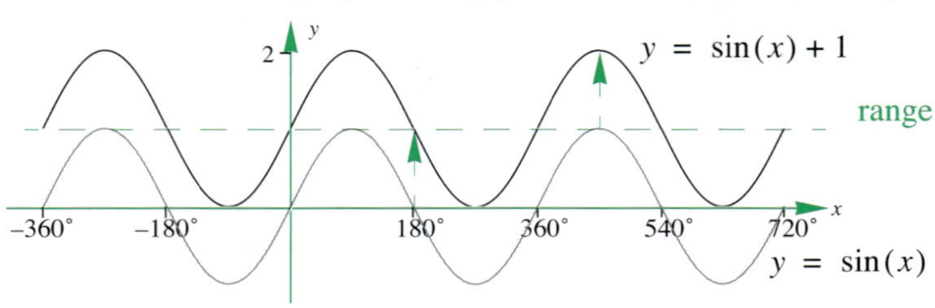

A graph sketch should show all the important features of a graph. In this case, the axes scales are important and should show the correct period (360°) and range [0,2].

(b) This time the graph of $\cos(x)$ needs to be translated down by 2 units.

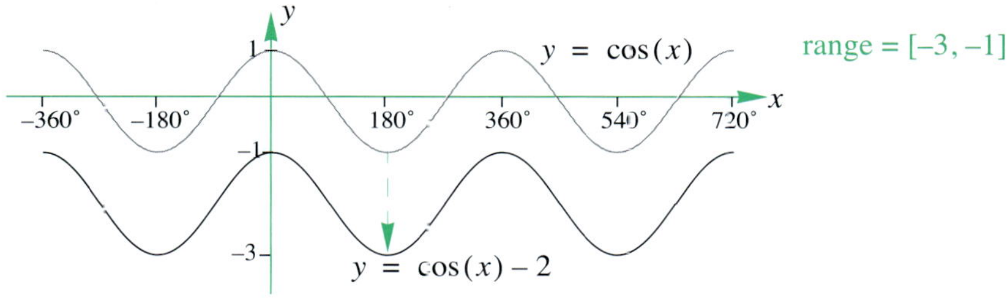

A graph sketch should show all the important features of a graph. In this case, the axes scales are important and should show the correct period (360°) and range [–3,–1].

2. Horizontal translations*

Functions of the type $f(x) = \sin(x \pm \alpha)$ and $f(x) = \cos(x \pm \alpha)$ where $\alpha > 0$ are horizontal translations of the curves $\sin x$ and $\cos x$ respectively.

So that $f(x) = \sin(x - \alpha)$ and $f(x) = \cos(x - \alpha)$ are translations to the **right**

while that of $f(x) = \sin(x + \alpha)$ and $f(x) = \cos(x + \alpha)$ are translations to the **left**.

EXAMPLE 11.2 For $-360° \le x \le 720°$, sketch the graphs of the curves with equations

(a) $y = \cos(x - 45°)$ (b) $y = \cos(x + 60°)$

Solution

(a)

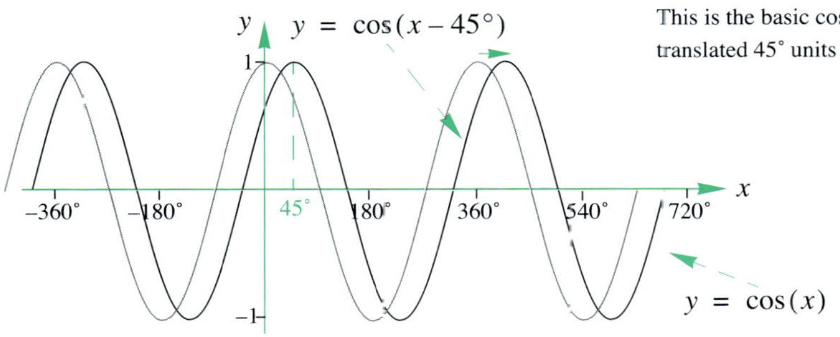

This is the basic cosine graph translated 45° units to the **right**.

(b)

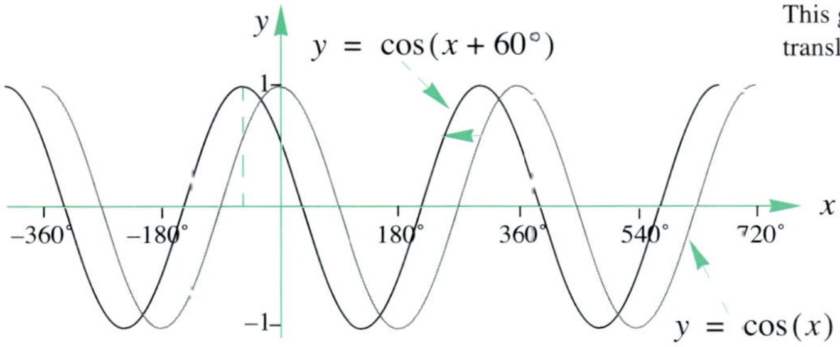

This graph is the basic cosine graph translated 60° units to the **left**.

*Horizontal translations are not on the syllabus. However, we have included them here for the sake of completion.

Of course, it is also possible to combine vertical and horizontal translations, as the next example shows.

EXAMPLE 11.3

Sketch the graph of the function $f(x) = \sin(x - 45°) + 2$, for $-360° \leq x \leq 720°$.

Solution

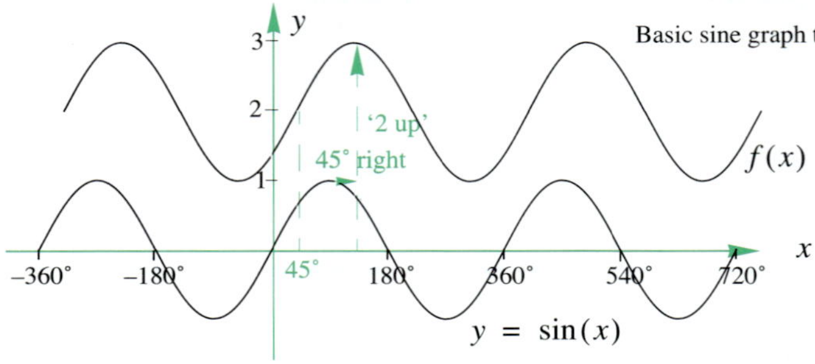

Basic sine graph translated 45° units right and 2 up.

3. Dilations

Functions of the form $f(x) = a\sin x$ and $f(x) = a\cos(x)$ are **dilations** of the curves $\sin(x)$, and $\cos(x)$ respectively, **parallel to the *y*-axis**. The **amplitude** becomes $|a|$ and not 1.

EXAMPLE 11.4

Sketch the graphs of the following functions for $-360° \leq x \leq 720°$.

(a) $f(x) = 2\cos x$ (b) $f(x) = \frac{1}{3}\sin x$

Solution

(a)

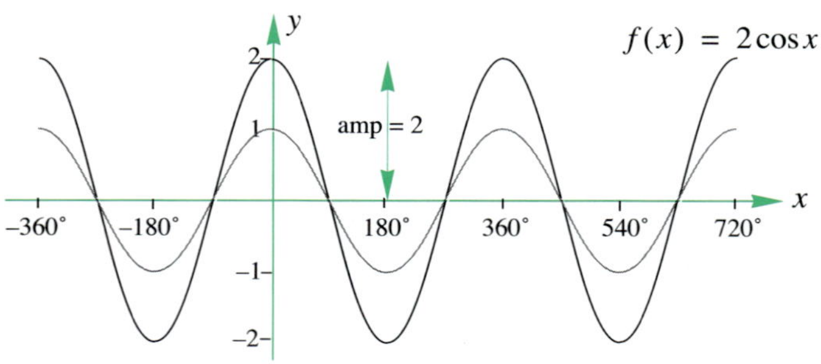

i.e., the cosine graph with an amplitude of 2.

(b)

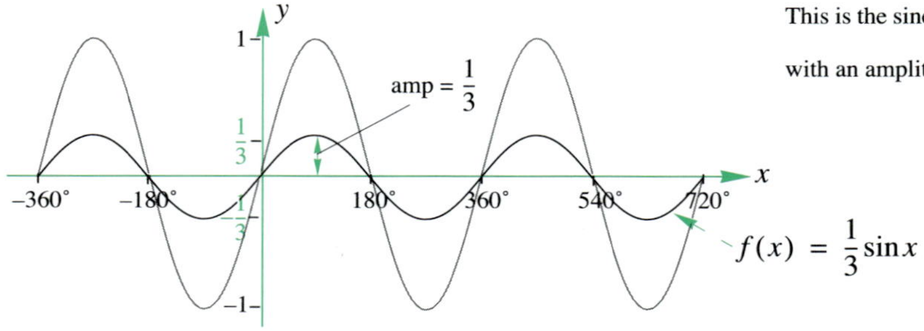

This is the sine graph (broken line) with an amplitude of $\frac{1}{3}$.

Functions of the form $f(x) = \cos(bx)$ and $f(x) = \sin(bx)$ are **dilations** of the curves $\sin(x)$ and $\cos(x)$ respectively, **parallel to the x-axis**.

This means that the period of the graph is altered. It can be valuable to remember and use the formula that relates the value of b to the period τ of the dilated function:

1. The graph of $f(x) = \cos(bx)$ will show b cycles in 360°, meaning that its period will be given by $\tau = \frac{360°}{b}$.

2. The graph of $f(x) = \sin(bx)$ will show b cycles in 360°, meaning that its period will be given by $\tau = \frac{360°}{b}$.

EXAMPLE 11.5

Sketch graphs of the following functions for $x \in [-360°, 720°]$.

(a) $f(x) = \sin(2x)$ (b) $f(x) = \cos\left(\frac{x}{2}\right)$

Solution

(a) $f(x) = \sin(2x)$. The value of n is 2 so the period is $\tau = \frac{360°}{n} = \frac{360°}{2} = 180°$. Note that this means that the period is **half** that of the basic sine function.

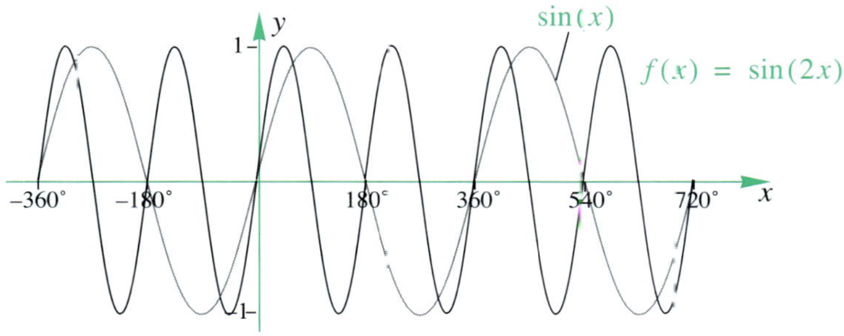

(b) $f(x) = \cos\left(\frac{x}{2}\right)$. This time $n = \frac{1}{2}$ and the period $\tau = \frac{360°}{n} = \frac{360°}{1/2} = 720°$.

The graph is effectively stretched to **twice** its original period.

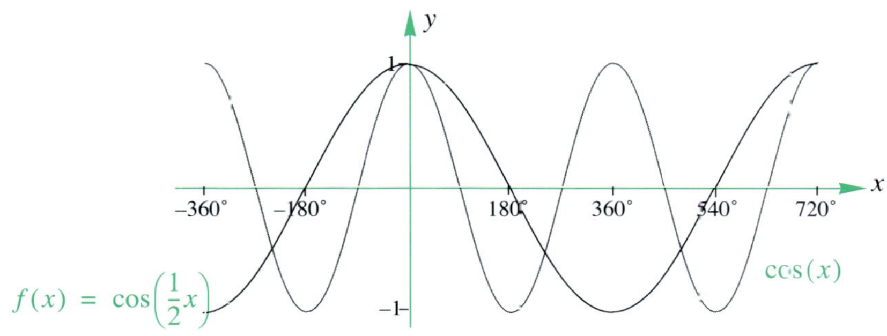

4. Reflections

Recall that the graph of $y = -f(x)$ is the graph of $y = f(x)$ reflected about the x–axis, while that of $y = f(-x)$ is the graph of $y = f(x)$ reflected about the y–axis.

$f(x) = -\cos x$ is the basic cosine graph (broken line) **reflected in the x-axis**.

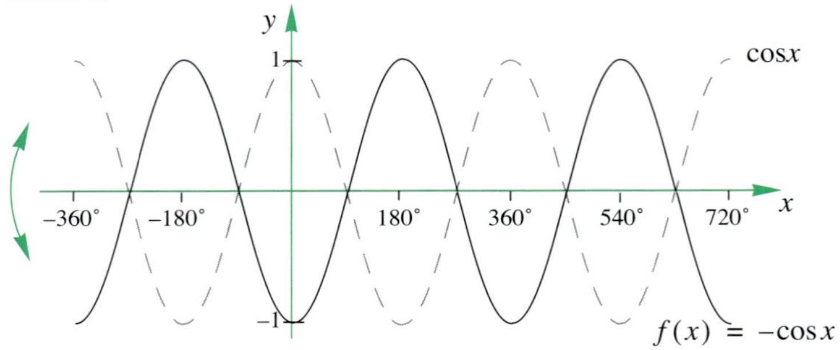

$f(x) = \sin(-x)$ is the basic sine graph **reflected in the y-axis**.

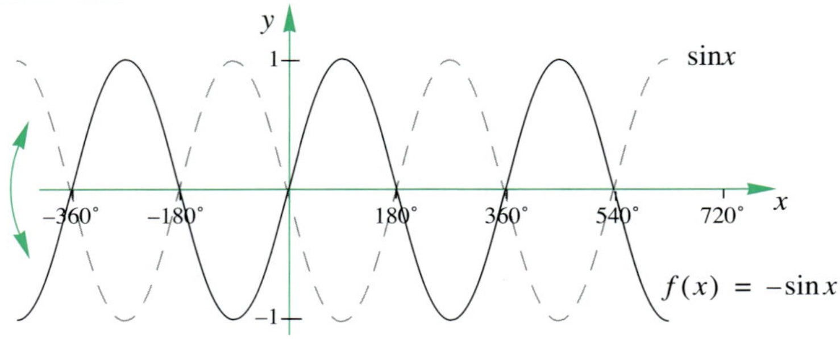

5. Combined transformations

You may be required to combine some or all of these transformations in a single function. The general function of the type

$y = a\sin(bx) + c$ & $y = a\cos(bx) + c$ has:

an amplitude of $|a|$

a period of $\dfrac{360°}{b}$

a vertical translation of 'c' units.

EXAMPLE 11.6

Sketch graphs of the following functions for x values in the range –360° to 720°.

(a) $y = 2\sin(2x) + 1$ (b) $y = -\cos\left(\frac{1}{2}x\right) + 2$

Solution

(a) $y = 2\sin(2x) + 1$ This graph has an amplitude of 2, a period of π and a vertical translation of 1 unit up.

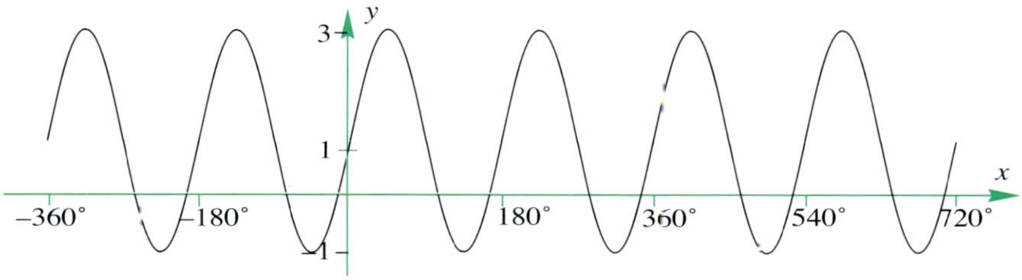

(b) $y = -\cos\left(\frac{1}{2}x\right) + 2$. The transformations are a reflection in the x-axis, a dilation of factor 2 parallel to the x-axis and a translation of 2 up.

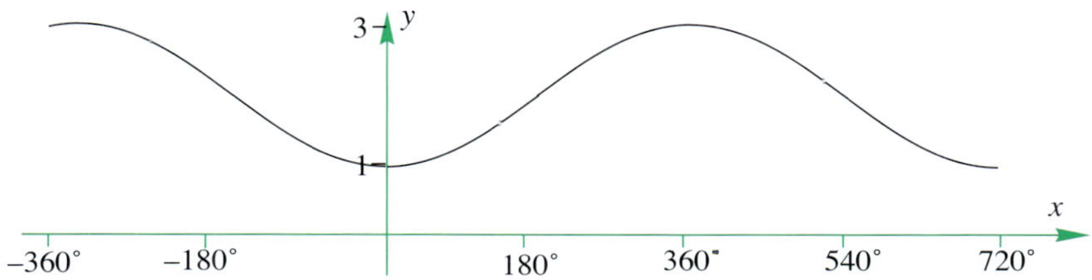

Using a Graphics Calculator

EXAMPLE 11.7

Sketch graph of $f(x) = 3\sin(2x) + 4$ for $x \in [-360°, 720°]$. State its amplitude and range.

Solution

We start with the period: The value of n is 2 so the period is $\tau = \frac{360°}{n} = \frac{360°}{2} = 180°$.

Next, there is a '3' in front of the sine function, this is the amplitude.
Finally, because we add '4' to $3\sin(2x)$, this means that the graph is moved up 4 units.

This time we make use of the TI–83 to sketch the graph. However, we still need to set the window to make sure we see the whole graph over the interval $[-360°, 720°]$.
Now, because the **amplitude is 3** and the graph is moved up 4 units, the maximum value it will have is 3 + 4 = 7. Similarly, the minimum value it will have is 4 – 3 = 1.
This gives us the **range of [1, 7]**.

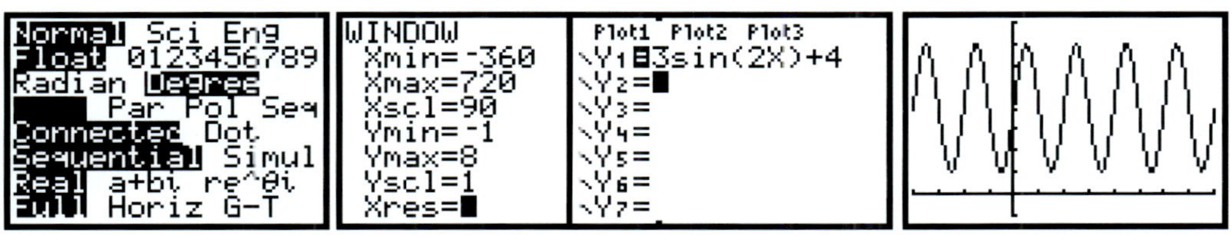

EXERCISES 11.1

1. Find the period and the amplitude of the following functions.

(i) $y = 2\sin x$ (ii) $y = 3\cos\frac{x}{3}$ (iii) $y = -4\sin(2x) + 1$

(iv) $y = 2 - 3\cos(2x)$ (v) $y = \frac{1}{4}\cos(3x) + 5$ (vi) $y = -\frac{2}{3}\cos\left(\frac{3}{4}x\right) + 5$

2. Sketch graphs of the following functions for x values in the interval [-360°,360°].

(i) $y = \sin(2x)$ (ii) $y = -\cos\left(\frac{x}{2}\right)$ (iii) $f(x) = 2\sin(x)$

(iv) $y = 1 - 2\sin(2x)$ (v) $y = -2\cos\left(\frac{x}{2}\right)$ (vi) $y = 3\cos(x)$

(vii) $y = 2\cos\left(\frac{1}{3}x\right) - 1$ (viii) $f(x) = 4\sin\left(\frac{x}{2}\right)$ (ix) $y = 2 - \sin\left(\frac{x}{3}\right)$

(x) $y = 5 + 15\sin(2x)$ (xi) $f(x) = 8 - 4\cos\left(\frac{1}{2}x\right)$

(xii) $f(x) = 12.5 + 2.5\sin(4x)$

3. Sketch graphs of the following functions for x values in the interval [0,4].

(i) $y = 2\cos(360x)$ (ii) $f(x) = 2 + \cos(180x)$

(iii) $y = \sin(360x) - 1$ (iv) $y = 12 + 1.5\sin(180x)$

(v) $f(x) = 8 - 4\sin(60x)$ (vi) $f(x) = 4.5 - 1.5\cos(360x)$

4. Find the values of a, b and c in each of the following

(a)

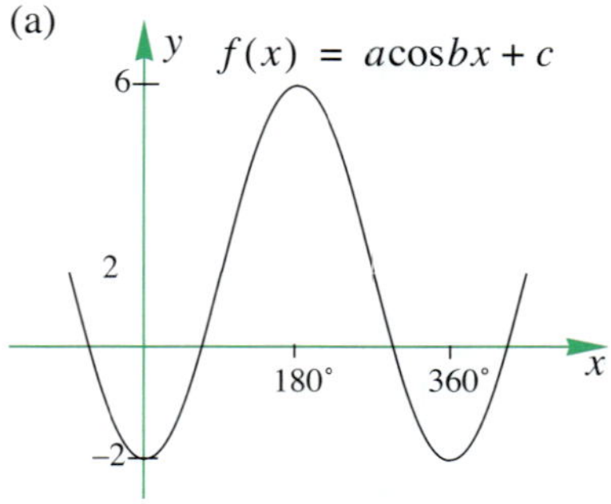

(b)

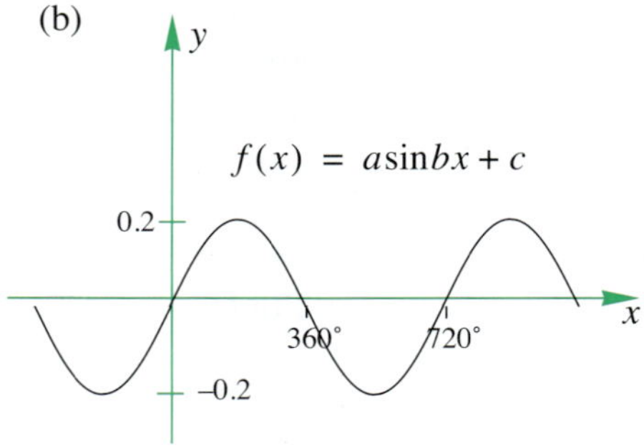

5. (a) Sketch the graph of $g(t) = 3 + 2.5\cos(45t°)$, $0 \le t \le 12$

(b) State i. the amplitude of $g(t)$.

ii. range of $g(t)$.

(c) On a different set of axes sketch the function $f(t) = 0.4g(t) - 1.2$, $0 \le t \le 12$.

(d) For what value(s) of k will the equation $g(t) = k$ have

i. no solution ii. two solutions iii. three solutions?

11.2 APPLICATIONS

Functions of the type considered in the previous section are useful for modelling periodic phenomena. These sorts of applications usually start with data that has been measured in an experiment. The next task is to find a function that 'models' the data in the sense that it produces function values that are similar to the experimental data. Once this has been done, the function can be used to predict values that are missing from the measured data (interpolation) or values that lie outside the experimental data set (extrapolation).

EXAMPLE 11.8

The table shows the depth of water at the end of a pier at various times (measured, in hours after midnight on the first day of the month.)

t (hr)	0	3	6	9	12	15	18	21	24	27	30	33
d (m)	16.20	17.49	16.51	14.98	15.60	17.27	17.06	15.34	15.13	16.80	17.42	15.89

Plot the data as a graph. Use your results to find a rule that models the depth data. Use your model to predict the time of the next high tide.

Solution

We start by entering the data as lists and then plotting them using the TI–83:

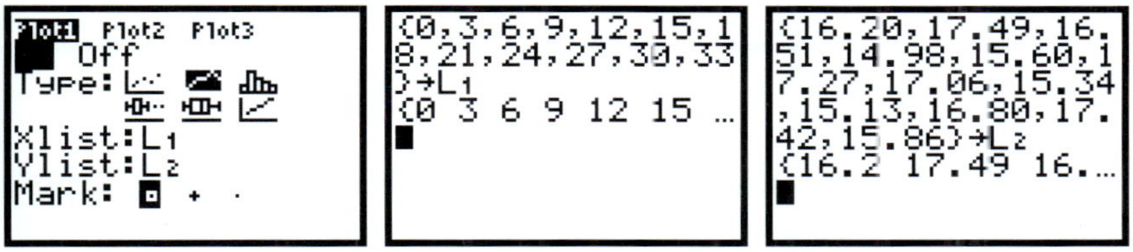

Using two different scales can provide graphs that do not really resemble each other.

Although both display a periodic nature, the second seems to reflect the need for a circular trigonometric function to model the behaviour.

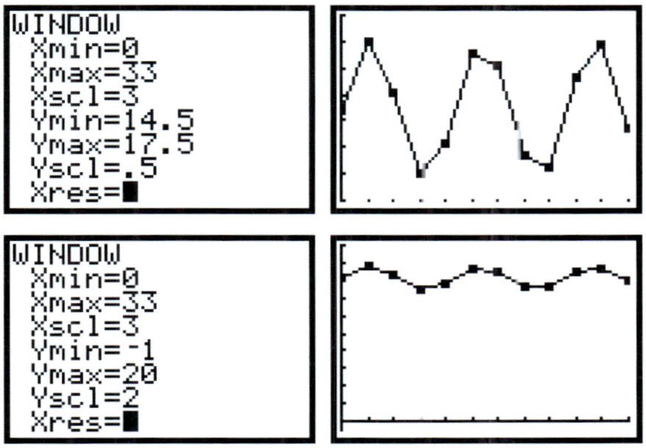

This does suggest that the depth is varying periodically. It appears that the period is approximately 13 hours. This is found by looking at the time between successive high tides. This is not as easy as it sounds as the measurements do not appear to have been made exactly at the high tides. This means that an estimate will need to be made based upon the observation that successive high tides appear to have happened after 3, 16 and 32 hours. Next, we look at the amplitude and vertical translation. Again, because we do not have exact readings at high and low tides, these will need to be estimated. The lowest tide recorded is 14.98 and the highest is 17.49.

A first estimate of the vertical translation is $\dfrac{17.49 + 14.98}{2} = 16.235$ and the amplitude is

$17.7 - 16.235 = 1.465$. Since the graph starts near the mean depth and moves up it seems likely that the first model to try might be: $y = 1.465 \times \sin\left(\frac{360t°}{13}\right) + 16.235$

Notice that the dilation factor (along the x–axis) is found by using the result that if

$$\tau = 13 \Rightarrow \frac{360}{n} = 13 \therefore n = \frac{360}{13}.$$

The model should now be 'evaluated' which means testing how well it fits the data.

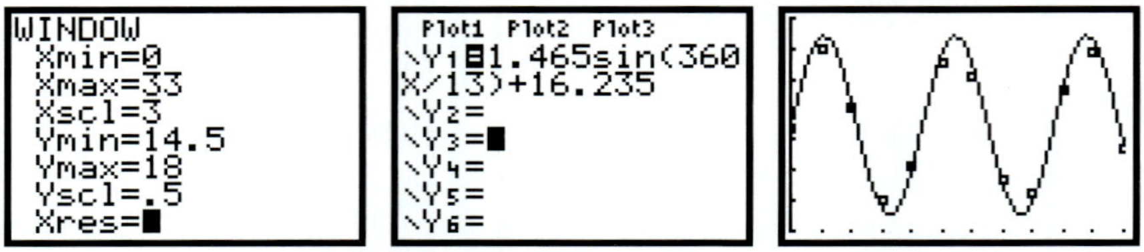

The model shown is quite good in that the errors are small with some being positive and some being negative. The function used is $d = 1.465 \times \sin\left(\frac{360t}{13}\right) + 16.235$ and this can now be used to predict the depth for times that measurements were not made. Also, the graph of the modelling function can be added to the graph of the data (as shown).

The modelling function can also be used to predict depths into the future (extrapolation). The next high tide, for example can be expected to be 13 hours after the previous high tide at about 29.3 hours. This is after 42.3 hours.

EXAMPLE 11.9

During the summer months, a reservoir supplies water to an outer suburb based on the water demand, $D(t) = 120 + 60\sin(2t°), 0 \le t \le 90$, where t measures the number of days from the start of Summer (which lasts for 90 days).

(a) Sketch the graph of $D(t)$.

(b) What are the maximum and minimum demands made by the community over this period?

Solution

(a) The features of this function are:

Period $= \frac{360}{2} = 180$ days

Amplitude $= 60$

Translation $= 120$ units up.

We 'pencil in' the graph of $y = 60\sin(2t°)$ and then move it up 120 units:

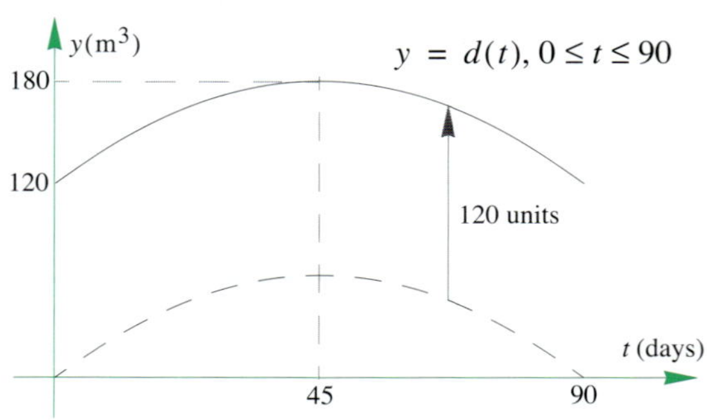

(b) The minimum is 120 m^3 and the maximum is 180 m^3 .

EXAMPLE 11.10

The height of the tide is h metres at time t hours, where

$$h = 4.5 + 2.5\sin 30t°$$

(a) Copy and complete the table below

t hours	0	4	8	12	16	20	24
h metres							

(b) On a set of axes draw the graph of $h = 4.5 + 2.5\sin 30t°$.

(c) Find the height of the tide at the following times
i. 3 hours ii. 7.5 hours.

(d) Find all the times in the first 24 hours when the height is 6.5 m.

Solution

(a) When $t = 0$, $h = 4.5 + 2.5\sin(30 \times 0)° = 4.5 + 0 = 4.5$

When $t = 4$, $h = 4.5 + 2.5\sin(30 \times 4)° = 4.5 + 2.5\sin 120° \approx 6.67$

Continuing in this manner we can complete the table of values:

t hours	0	4	8	12	16	20	24
h metres	4.5	6.67	2.33	4.5	6.67	2.33	4.5

Or we could use the TI–83: Notice that it is important to have the TABLE SET UP correctly, so that in this case we have that the time t starts at 0 and it increases by 4 hours each time.

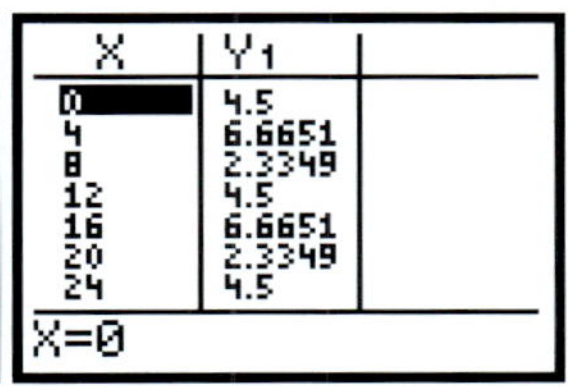

(b) Although such a graph can be sketched on graph paper accurately, we will make use of the TI–83 in this instance:

ter equation:

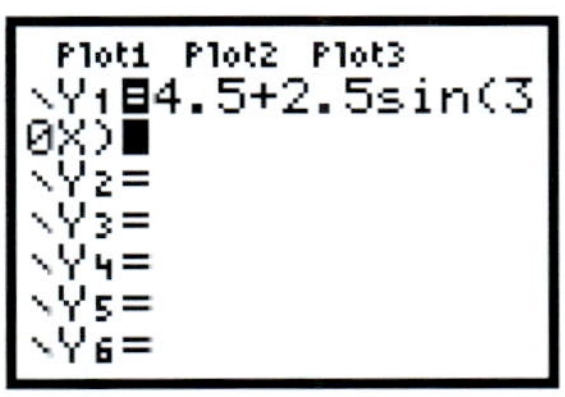

Use an appropriate WINDOW setting:

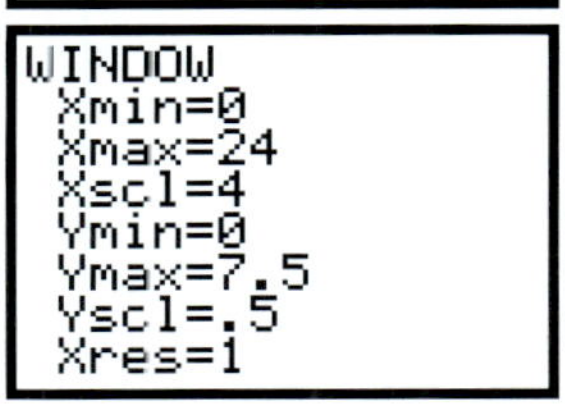

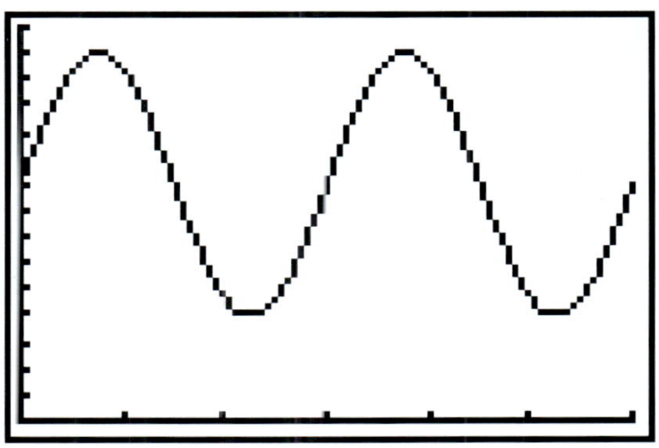

(c) i. When $t = 3$, $h = 4.5 + 2.5\sin(30 \times 3)° = 4.5 + 2.5 \times 1 = 7.0$

ii. When $t = 7.5$, $h = 4.5 + 2.5\sin(30 \times 7.5)° = 2.73$

Again, we can make use of the CALC function on the TI–83, from which these values can then be obtained:

i.e., press **2nd TRACE**, **ENTER** and then type in the value of t, (at the X = prompt) this will produce a graph, where the value is on the graph and what the value of h is.

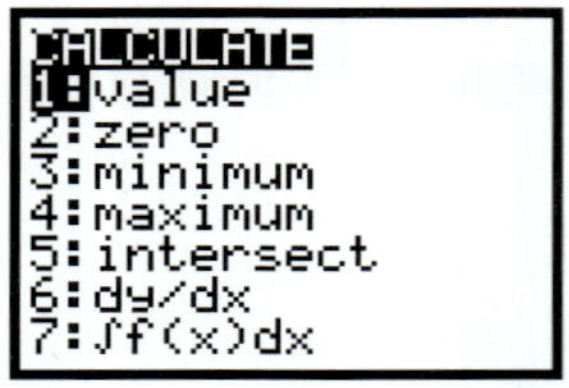

i.

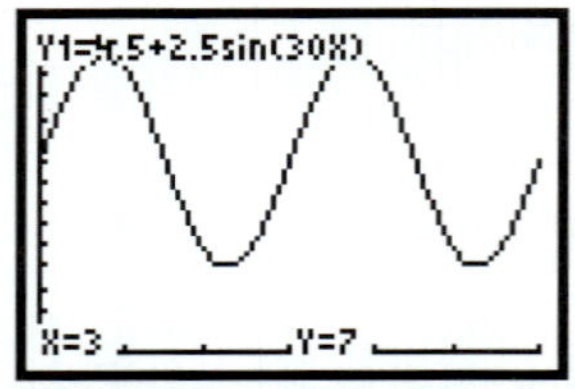

ii.

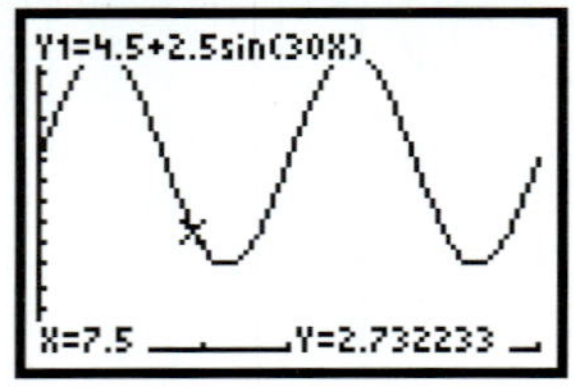

(d) In this case we simply use the graph we have and read the values from it:

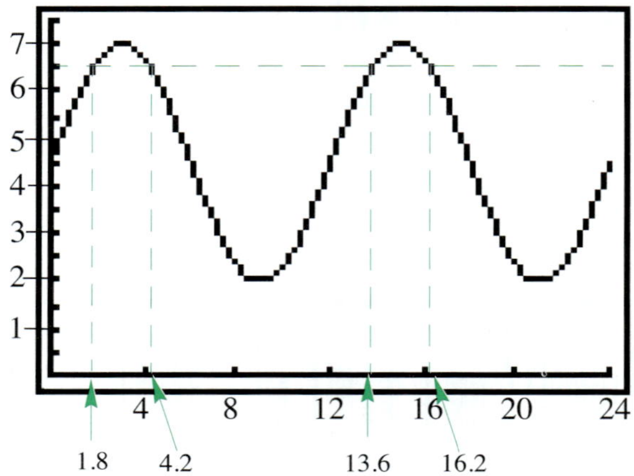

From the graph, the first time $h = 6.5$ is approximately when $t = 1.8$. The second time $t \approx 4.2$. Similarly, the next two times are, $t \approx 13.8$ and 16.2

EXERCISES 11.2

1. The table shows the temperature in an office block over a 36 hour period.

t (hr)	0	3	6	9	12	15	18	21	24	27	30	33	36
T ˚C	18.3	15.0	14.1	16.0	19.7	23.0	23.9	22.0	18.3	15.0	14.1	16.0	19.7

(a) Estimate the amplitude, period, horizontal and vertical translations.
(b) Find a rule that models the data.
(c) Use your rule to predict the temperature after 40 hours.

2. The table shows the light level (L) during an experiment on dye fading.

t (hr)	0	1	2	3	4	5	6	7	8	9	10
L	6.6	4.0	7.0	10.0	7.5	4.1	6.1	9.8	8.3	4.4	5.3

(a) Estimate the amplitude, period, horizontal and vertical translations.
(b) Find a rule that models the data.

3. The table shows the value in \$s of an industrial share over a 20 month period.

Month	0	2	4	6	8	10	12	14	16	18	20
Value	7.0	11.5	10.8	5.6	2.1	4.3	9.7	11.9	8.4	3.2	2.5

(a) Estimate the amplitude, period, horizontal and vertical translations.
(b) Find a rule that models the data.

4. The table shows the population (in thousands) of a species of fish in a lake over a 22 year period.

Year	0	2	4	6	8	10	12	14	16	18	20	22
Pop	11.2	12.1	13.0	12.7	11.6	11.0	11.6	12.7	13.0	12.1	11.2	11.2

(a) Estimate the amplitude, period, horizontal and vertical translations.
(b) Find a rule that models the data.

5. The table shows the average weekly sales (in thousands of \$s) of a small company over a 15 year period.

Time	0	1.5	3	4.5	6	7.5	9	10.5	12	13.5	15
Sales	3.5	4.4	7.7	8.4	5.3	3.3	5.5	8.5	7.6	4.3	3.6

(a) Estimate the amplitude, period, horizontal and vertical translations.
(b) Find a rule that models the data

6. The table shows the average annual rice production, P, (in thousands of tonnes) of a province over a 10 year period.

t (yr)	0	1	2	3	4	5	6	7	8	9	10
P	11.0	11.6	10.7	10.5	11.5	11.3	10.4	11.0	11.6	10.7	10.5

(a) Estimate the amplitude, period, horizontal and vertical translations.
(b) Find a rule that models the data.

7. The population (in thousands) of a species of butterfly in a nature sanctuary is modelled by the function:

$$P = 3 + 2\sin(45t°), 0 \le t \le 12$$

where t is the time in weeks after scientists first started making population estimates.

(a) What is the initial population?
(b) What are the largest and smallest populations?
(c) When does the population first reach 4 thousand butterflies?

8. A water wave passes a fixed point. As the wave passes, the depth of the water (D metres) at time t seconds is modelled by the function:

$$D = 7 + \frac{1}{2}\cos(72t°), t > 0$$

(a) What are the greatest and smallest depths?
(b) Find the first two times at which the depth is 6.8 metres.

9. The rabbit population, $R(t)$ thousands, in a northern region of South Australia is modelled by the equation $R(t) = 12 + 3\cos(30t°), 0 \le t \le 24$, where t is measured in months after the first of January.

(a) What is the largest rabbit population predicted by this model?
(b) How long is it between the times when the population reaches consecutive peaks?
(c) Sketch the graph of $R(t)$ for $0 \le t \le 24$.
(d) Find the longest time span for which $R(t) \ge 13.5$.
(e) Give a possible explanation for the behaviour of this model.

10. The depth, $d(t)$ metres, of water at the entrance to a harbour at t hours after midnight on a particular day is given

$$d(t) = 12 + 3\sin(30t°), 0 \le t \le 24$$

(a) Sketch the graph of $d(t)$ for $0 \le t \le 24$.
(b) For what values of t will
i. $d(t) = 10.5$, $0 \le t \le 24$. ii. $d(t) \ge 10.5$, $0 \le t \le 24$.

Boats requiring a minimum depth of b metres are only permitted to enter the harbour when the depth of water at the entrance of the harbour is at least b metres for a continuous period of one hour.

(c) Find the largest value of b, correct to two decimal place, which satisfies this condition.

11. When a person is at rest, the blood pressure, P millimetres of mercury at any time t seconds can be approximately modelled by the equation

$$P(t) = -20\cos(300t°) + 100, t \ge 0$$

(a) Determine the amplitude and period of P.
(b) What is the maximum blood pressure reading that can be recorded for this person?
(c) Sketch the graph of $P(t)$, showing one full cycle.
(d) Find the first two times when the pressure reaches a reading of 110 mmHg.

11.3 MISCELLANEOUS EXERCISES

1. On the same set of axes sketch the graphs of
(a) $f(x) = 2\sin x°, 0 \le x \le 360$. (b) $g(x) = \cos(3x), 0° \le x \le 360°$.

2. (a) Sketch the graph of $f: x \mapsto \sin 2x$ where $-180° \le x \le 180°$.
(b) Hence, find the two smallest positive values of x for which $f(x) = 0.5$.

3. The height of the tide is h metres at time t hours, where $h = 5 + 2\sin 30t°$.
(a) Copy and complete the table below

t hours	0	4	8	12	16	20	24
h metres							

(b) On a set of axes draw the graph of $h = 5 + 3\sin 30t°$.
(c) Find the height of the tide at the following times
(i) 3 hours (ii) 5.5 hours.

(d) Find all the times in the first 24 hours when the height is 3.0 m.

4. The height, h metres, above ground of a chair in a big wheel is found to be modelled by the equation

$$h(t) = 5.8 - 4.6\cos 6t°,$$

where the chair is initially at its lowest point.

(a) How long does it take for the chair to come back to its initial position?

(b) Copy and complete the table below

t seconds	0	12	24	36	48	60	72
h metres							

(c) Sketch the graph showing the height of the chair against time.

(d) When will the chair first reach a height of 8.0 m?

5. For the function with equation $f: x \mapsto 5\cos 3x° - 2$

(a) state

(i) its amplitude (ii) its period (iii) its maximum value.

(b) Find the smallest positive value for which $f(x) = 0.5$

6. State (a) the period (b) the amplitude

of the functions (i) $2\cos 3x°$. (ii) $-4\sin\frac{1}{2}x°$

7. Determine the maximum value of the function $x \mapsto 4 - 2\sin(2x)$

8. Sketch the graphs of the functions:

(i) $y = 2\sin\left(\frac{x}{2}\right), 0° \le x \le 360°$ (ii) $y = 2 - 3\cos(90x°), 0 \le x \le 8$

9. When a person is at rest, the blood pressure P millimetres of mercury at any time t seconds can be modelled approximately by the equation

$$P(t) = -20\cos(300t)° + 100, t \ge 0.$$

(a) Determine the amplitude and period of P.

(b) What is the maximum blood pressure reading that can be recorded?

(c) Sketch the graph of $P(t)$, showing two full cycles.

(d) Find the first two times when the pressure reaches a reading of 110 mmHg.

(e) For what percentage of time will the pressure be in excess of 110 mmHg?

10. The table shows the depth of water (D metres) over a 5 second period as waves pass the end of a pier.

t (sec)	0	0.5	1	1.5	2	2.5	3	3.5	4	4.5	5
D	11.3	10.8	10.3	10.2	10.4	10.9	11.4	11.7	11.8	11.5	11.0

(a) Estimate the amplitude, period, horizontal and vertical translations.

(b) Find a rule that models the data.

11.4 GRADED REVISION EXERCISES

LEVEL 1

1. Give values, correct to four signficant figures for:

i. $\sin 23°$ ii. $\sin 145°$ iii. $\cos -10°$

iv. $\sin 560°$ v. $\cos(-500°)$ vi. $\sin 15°47'$

LEVEL 2

1. State the maximum values reached by the following functions:

i. $f(x) = 2\sin 3x$ ii. $f(x) = 1 - \cos x$ iii. $f(x) = 3 - 4\sin 2x$

iv. $f(x) = \dfrac{1 + \cos x}{2}$ v. $f(x) = \sin\left(\dfrac{x}{4}\right)$ vi. $f(x) = 3 + \cos 4x$

2. State the period of these functions:

i. $f(x) = 2\sin x$ ii. $f(x) = \cos\dfrac{x}{2}$ iii. $f(x) = \cos 3x$

iv. $f(x) = \cos x$ v. $f(x) = \dfrac{1}{2}\sin\dfrac{x}{3}$ vi. $f(x) = 3\sin\dfrac{3x}{4}$

3. Find the y-intercepts of the following graphs:

i. $y = \sin(x) + 1$ ii. $y = 2\cos x$ iii. $y = 2\sin(x) - 3$

LEVEL 3

1. Sketch the graphs of the following functions:

i. $f(x) = 1 + \cos x, x \in [-360°, 360°]$

ii. $f(x) = \cos(2x), x \in [-180°, 360°]$

iii. $f(x) = 1 + 2\sin x, x \in [-360°, 360°]$

iv. $f(x) = 1 - \dfrac{1}{2}\cos(2x), x \in [0°, 180°]$

v. $f(x) = 3 + 2\sin(4x), x \in [0°, 90°]$

vi. $f(x) = 2 - 3\cos\left(\dfrac{1}{2}x\right), x \in [0°, 360°]$

2. Find the smallest positive x-intercept of these graphs:

i. $y = \cos 2x$ ii. $y = \sin 3x$ iii. $y = \cos\dfrac{x}{3}$

LEVEL 4

1. Give rules for the functions whose graphs are shown:

i.

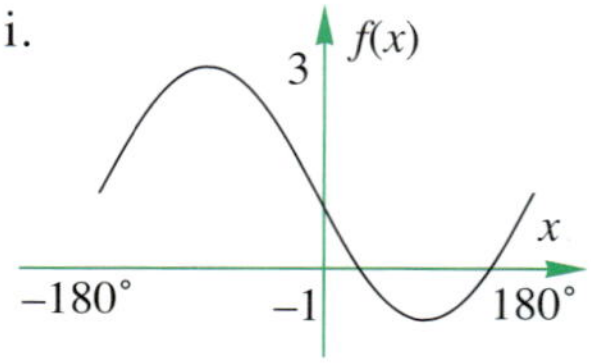

ii.

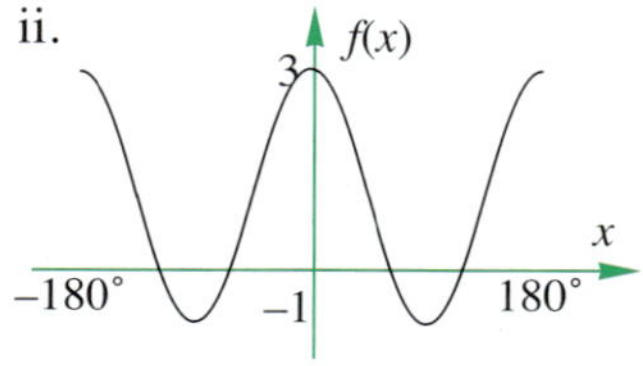

iii.

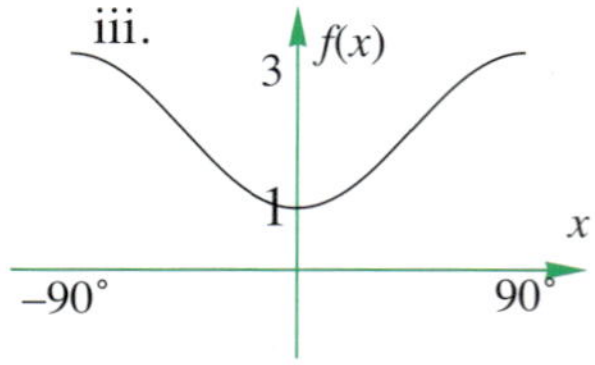

2. The temperature (C°) over a 24 hour day in Montaville is modelled by the function:

$$C(t) = 18 - 6\cos(15t°), 0 \le t \le 24$$

where t is the time in hours after midnight.

i. What is the temperature at
   (a) midnight?
   (b) 2 pm?
ii. (a) What are the maximum and minimum temperatures reached?
   (b) When do the maximum and minimum temperatures occur?
iii. Sketch the graph of the function.
iv. At what times of the day is the temperature warmer than 20°C?
v. When is the temperature rising most rapidly. What is this largest rate of change?

3. A function is defined by: $f(x) = 3\sin(360x°)$ for $0 \le x \le 2$.
i. What is the function's (a) period? (b) amplitude?
ii. Sketch the graph of the function.
iii. Find the average rate of change of the function over the interval $\left[\frac{1}{3}, \frac{1}{2}\right]$.

Give your answer in exact form.

4. The height of the tide is h metres at time t hours, where $h = 5.5 + 1.5\sin(45t°)$.
i. Copy and complete the table below.

t hours	0	1	2	3	4	5	6	7	8
h metres									

ii. On a set of axes sketch the graph of $h = 5.5 + 1.5\sin(45t°)$, $0 \le t \le 24$.
iii. Find the height of the tide at the following times
   (a) 12 hours. (b) 18 hours.
iv. Find all the times in the first 24 hours when the height is 4.0 m.

5. A particle moves in a straight line according to the rule $x = 3 + 2\cos(120t°)$, where x metres is the distance from the origin O, t seconds after it started its motion.
i. Find the (a) greatest distance reached by the particle from O.
   (b) least distance reached by the particle from O.
ii. For $0 \le t \le 6$, find when the particle is 4 metres from O.
iii. Sketch the graph of $x = 3 + 2\cos(120t°)$ for $0 \le t \le 6$.
iv. For what percentage of the time is the particle at least 4 m from O during the first 6 seconds of its motion?

6. Florence is seated on one of the carriages at the local fair's Big Wheel. Her height, H metres above the ground after t seconds is given by the equation

$$H(t) = a\cos(kt°) + b, 0 \le t \le 12.$$

i. The time it takes Florence to undergo one revolution is 12 seconds. Find k.
ii. When Florence steps into the carriage, she is 1metre above the ground and six seconds later she is 21 metres above the ground. Find a and b.
iii. Sketch the graph of H.
iv. Find when Florence is 16 metres above the ground during one revolution.
v. For what percentage of time is Florence at least 16 metres above the ground?

11.5 TOPIC TEST

1. Give an exact value for i. $\sin(-270°)$.

ii. $\cos(135°)$.

[2 marks]

2. State i. the amplitude

ii. the period

iii. the maximum value

of the function $f(x) = 5 - 2\sin(3x°)$.

[3 marks]

3. Find the smallest positive value of x such that $\sin 3x = \frac{1}{2}$. Give an exact answer.

[3 marks]

4. Sketch the graph of $y = 3 + 6\cos(60x°), x \in [0, 6]$. Your sketch must include all intercepts with the axes.

[7 marks]

5. Find a function that has the graph shown:

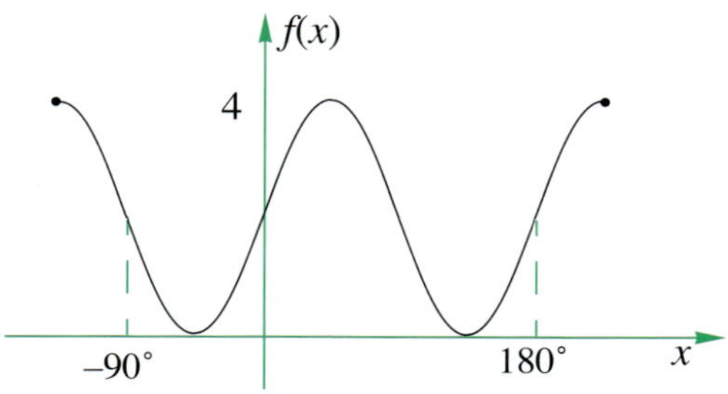

[4 marks]

6. The population, N, of a particular species of insect varies with time according to the model defined by the equation

$$N(t) = 2500\sin\left(\frac{90}{13}t°\right) + 4500,$$

where t is the time in weeks from the 1st of January 2000.

i. State the period of $N(t)$.

ii. State the largest and smallest population sizes reached by this species.

iii. Sketch the graph of $y = N(t), 0 \le t \le 143$.

iv. When will the species first reach

(a) its maximum size?

(b) 6000?

v. For the years 2000, during which months will the population be less than 3000?

[1 + 2 + 4 + 4 + 5 marks]
Total 35 marks

11.6 PROJECT SUGGESTIONS

As suggested at the start of this chapter, the study of tides could be a good topic for a project.

Being able to predict tides is essential for sea captains and others who work in and on the sea. It may be, for example, that there are times at which the entrance to a harbour is too shallow for a large cargo ship to enter without running aground.

The topic is, therefore, important and the project can be set up to answer a well defined question such as 'When will it be high tide at the start of next month?'

After defining the question, you will need to source data. Below are some tidal data obtained by typing 'tide table' into an internet search engine.

Minnie Bay, Alaska

16 April 2004 - 17 April 2004
54.7267° N, 132.3016° W

2004-04-16	05:42	AKDT	Sunrise	
2004-04-16	05:51	AKDT	1.76 feet	Low Tide
2004-04-16	11:43	AKDT	11.64 feet	High Tide
2004-04-16	18:00	AKDT	0.45 feet	Low Tide
2004-04-16	19:56	AKDT	Sunset	
2004-04-17	00:15	AKDT	12.53 feet	High Tide
2004-04-17	05:40	AKDT	Sunrise	
2004-04-17	06:33	AKDT	0.79 feet	Low Tide
2004-04-17	12:31	AKDT	11.85 feet	High Tide
2004-04-17	18:38	AKDT	0.69 feet	Low Tide
2004-04-17	19:57	AKDT	Sunset	
2004-04-18	00:45	AKDT	12.91 feet	High Tide

Source: WWW Tide and Current Predictor.

The next stage is to make sure that you understand the data.

What, for example, is the meaning of 'AKDT'?

The key to a good project will be to find a mathematical model (in this case a function) that will enable you to predict what will happen into the future. A particular problem here is how you are going to handle time as days, hours and minutes are not decimal.

> If you set 05:42 on 2004-04-16 to be $t = 0$, what value of t is appropriate for the high tide at 00:15 on 2004-04-17?

When you have a function that models the tides, you should 'evaluate' it.

> Begin by checking that it predicts the tides given in the existing table. Next, use the function to predict the tides into the future. If you are using the internet source quoted above you will need to wait a few days to get the information needed to check your predictions.

Tidal bores. These are destructive waves in river estuaries. How are they predicted and their destructive effects minimised?

12.0 CRASH DATA

Road accident's are one of the World's most serious problems in both social and economic terms. Much work has been done to try to reduce the accident rate. Mathematicians have played their part in analysing the causes of crashes and in forming the measures (such as seatbelts, speed cameras etc.) that various governments have introduced to deal with the problem. In some areas, the actual accident rate (taking into account the increases in traffic) are one tenth of their level of 30 years ago underlining the success of these measures. Here are some examples of these figures.

In a 60 kph speed zone, the risk of a casualty crash doubles for every 5 kph over the speed limit.

In average conditions, a car travelling at 60km/h will take about 45 metres to stop in an emergency braking situation. A car braking from 65km/h will still be moving at close to 32km/h after 45 metres travelled.

Imagine that you are travelling in a car and are 45 metres away from a road junction. Suddenly a car pulls out and the driver performs an emergency stop. Is a crash inevitable?
The Australian Transport Safety Bureau quote the following data

Initial speed (kph)	Impact speed (kph)	
	Dry road	Wet road
50	stops	stops
55	stops	14
60	touches	32
65	32	44
70	46	53
75	57	63
80	66	70

Look at the wet road data. How much speed does the driver succeed in losing before impact? The table below gives these values:

	Impact speed (kph)	
Initial speed (kph)	Impact speed (kph)	Speed lost (kph)
50	stops	50–0 = 50
55	14	55–14 = 41
60	32	60–32 = 28
65	44	65–44 = 21
70	53	70–53 = 17
75	63	75–63 = 12
80	70	80–70 = 10

Each 5 kph added to the initial speed results in an approximately 20% decrease in the amount of speed the driver manages to lose by braking. In increasing initial speed from 65 to 70 kph the speed lost in braking is reduced from 21kph to $0.8 \times 21 = 16.8 \approx 17$ kph. Why is this?

This is an example of an (approximate) geometric sequence and these are the subject of this chapter.

12.1 ARITHMETIC SEQUENCES & SERIES

12.1.1 ARITHMETIC SEQUENCES

A **sequence** is a set of quantities arranged in a definite order.

$1, 2, 3, 4, 5, 6, \ldots\ldots\ldots\ldots$ $-1, 2, -4, 8, -16\ldots.1,$ $1, 2, 3, 5, 8, 13\ldots.$

are all examples of sequences. When the terms of a sequence are added, we obtain a series. Sequences and series are used to solve a variety of practical problems in, for example, business.

There are two major types of sequence, **arithmetic** and **geometric**. This section will consider arithmetic sequences (also known as arithmetic progressions, or simply A.P). The characteristic of such a sequence is that there is a common difference between successive terms. For example:

$1, 3, 5, 7, 9, 11, \ldots$ (the odd numbers) has a first term of 1 and a common difference of 2.
$18, 15, 12, 9, 6, \ldots$ has a first term of 18 and a common difference of –3 (sequence is decreasing).

The terms of a sequence are generally labelled $u_1, u_2, u_3, u_4, \ldots u_n$. The '$n$th term' of a sequence is labelled u_n. In the case of an arithmetic sequence which starts with a and has a **common difference** of d, the **nth term** can be found using the formula:

$$u_n = a + (n-1)d \quad \text{where } d = u_2 - u_1 = u_3 - u_2 = \ldots$$

EXAMPLE 12.1

(a) Is the sequence 9, 12, 15, 18, . . . an arithmetic sequence?
(b) Find the 8th term for the sequence described in part a.

Solution

(a) To decide if the sequence is arithmetic we need to show that consecutive terms increase by a fixed amount, i.e., that there is a common difference.
We have that $u_1 = 9, u_2 = 12, u_3 = 15, u_4 = 18, \ldots$, so that the difference between the first two terms is $u_2 - u_1 = 12 - 9 = 3$, the difference between the next two terms is $u_3 - u_2 = 15 - 12 = 3$, and so on.
Therefore, we have a common difference, $d = 3$, and so we have an arithmetic sequence.

(b) To find the 8th term we could keep adding 3 (another 4 times), to get the sequence $9, 12, 15, 18, 21, 24, 27, 30, \ldots$ So that the 8th term, $u_8 = 30$.
We could also make use of the formula $u_n = a + (n-1)d$, with $a = 9(= u_1)$, $d = 3$ and $n = 8$. So that
$$\begin{aligned} u_8 &= 9 + (8-1) \times 3 = 9 + 7 \times 3 \\ &= 9 + 21 \\ &= 30 \end{aligned}$$

EXAMPLE 12.2 An arithmetic sequence has 10 as its first term and a common difference of –4. Find its 12th term.

Solution

We have that $u_1 = 10$ and $d = -4$, so that the general term, given by $u_n = a + (n-1)d$, becomes, $u_n = 10 + (n-1) \times -4$.

Therefore, the 12th term, u_{12}, is given by $u_{12} = 10 + (12-1) \times -4 = 10 + (-44)$
$= -34$

EXAMPLE 12.3 The 15th term of an arithmetic sequence is 45. If the first term is –11, find the common difference.

Solution

We are given that $n = 15$, $u_{15} = 45$ and $a = -11$. Using the formula for the general term of an arithmetic sequence, $u_n = a + (n-1)d$, we have

$$45 = -11 + (15-1) \times d$$
$$\Leftrightarrow 45 = -11 + 14d$$
$$\Leftrightarrow 56 = 14d$$
$$\Leftrightarrow 4 = d$$

Therefore, the common difference is 4.

Notice that if the nth term of an arithmetic sequence is u_n and the common difference is d, then the next term, $u_{n+1} = u_n + d$, or, $d = u_{n+1} - u_n$.

EXAMPLE 12.4 For the sequence 7, 11, 15, 19, . . . , find the 20th term.

Solution

In this case $a = 7$ and $d = 4$ because the sequence starts with a 7 and each term is 4 bigger than the one before it, i.e., $d = 11 - 7 = 4$. Therefore the nth term is given by

$$u_n = 7 + (n-1) \times 4$$

That is, $u_n = 4n + 3$

$\therefore u_{20} = 4 \times 20 + 3 = 83$ [$n = 20$ corresponds to the 20th term]

EXAMPLE 12.5 An arithmetic sequence has a first term of 120 and a 10th term of 57. Find the 15th term.

Solution

The data is: $a = 120$ and when $n = 10$, $u_{10} = 57$ [i.e., 10th term is 57].

This gives, $u_{10} = 120 + (10 - 1)d = 57 \Leftrightarrow 120 + 9d = 57$

$$\therefore d = -7$$

Using $u_n = a + (n - 1)d$, we then have $u_n = 120 + (n - 1) \times (-7) = 127 - 7n$.

Therefore, when $n = 15$, $u_{15} = 127 - 7 \times 15 = 22$.

EXAMPLE 12.6

An arithmetic sequence has a 7th term of 16.5 and a 12th term of 24. Find the 24th term.

Solution

In this instance we know neither the first term nor the common difference and so we will need to set up equations to be solved simultaneously.

The data is: $u_7 = a + 6d = 16.5 \quad -(1)$

$u_{12} = a + 11d = 24 \quad -(2)$

We first solve for 'd': $(2) - (1)$: $5d = 7.5 \Leftrightarrow d = 1.5$

Substituting into (1): $a + 6 \times 1.5 = 16.5 \Leftrightarrow a = 7.5$

To find the 24th term we use the general term $u_n = a + (n - 1)d$ with $n = 24$:

$$u_{24} = 7.5 + (24 - 1) \times 1.5 = 42$$

EXAMPLE 12.7

A car whose original value was \$25600 decreases in value by \$90 per month. How long will it take before the car's value falls below \$15000?

Solution

The values can be seen as a sequence: \$25600, \$25510, \$25420 etc.

In this case $a = 25600$ and $d = 25510 - 25600 = -90$ so that:

$$u_n = 25600 + (n - 1) \times (-90)$$
$$= 25690 - 90n$$
$$\therefore 15000 = 25690 - 90n$$
$$\Leftrightarrow 90n = 25690 - 15000$$
$$\Leftrightarrow n = 118.777$$

The car will be worth less than \$15000 after 119 months

Using a Graphics Calculator

Most graphic calculators have an automatic memory facility (usually called **Ans**) that stores the result of the last calculation as well as an ability to remember the actual calculation. This can be very useful in listing a sequence.

EXAMPLE 12.8

List the arithmetic sequence 5, 12, 19, 26,......

Solution

The sequence has a first term of 5. Enter this and press **ENTER** or **EXE**.
The common difference of the sequence is 7 so enter + 7.
The display will show Ans + 7 which means 'add 7 to the previous answer.
From here, every time you press **ENTER** (or **EXE**), you will repeat the calculation, generating successive terms of the sequence.

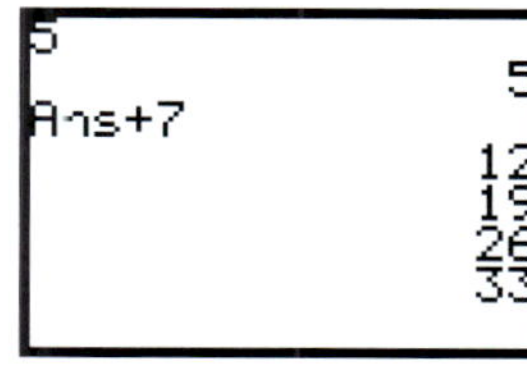

However, the TI–83 is much more sophisticated than this. It is possible to set up a sequence rule on the TI–83. To do this we use the **MODE** key to switch to **Seq** mode and this changes the Equation editor screen from **Y=** to a sequence version (instead of the usual function form).

There are three sequence forms; $u(n)$, $v(n)$ and $w(n)$, which can be accessed on the home screen using the 2nd function key with 7, 8 and 9 respectively. Once these equations are defined we can plot their sequence graph.

We now consider Example 8.2, where we obtained the sequence $u_n = 127 - 7n$ and wished to determine the 15th term.

Setting into sequence mode

Define sequence equation

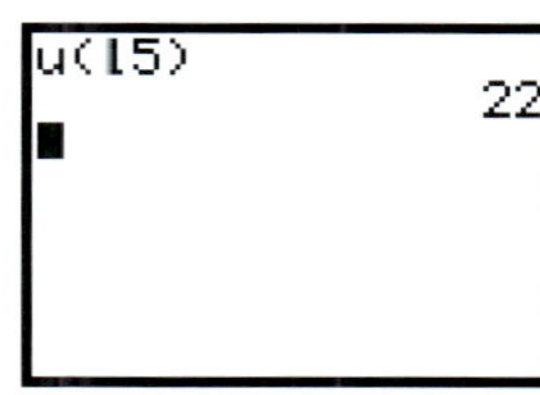

Use 2nd key '7' to call up *u*.

We can also use other features of the TI–83. For example, set up the sequence in a table format:

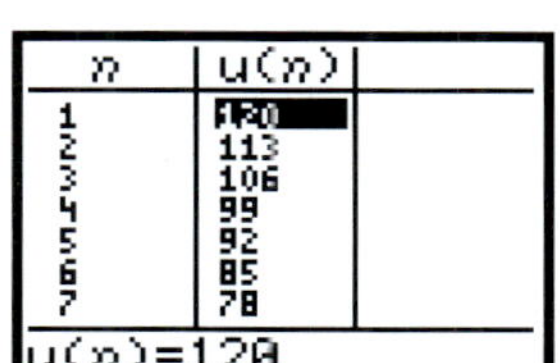

We can plot the sequence:

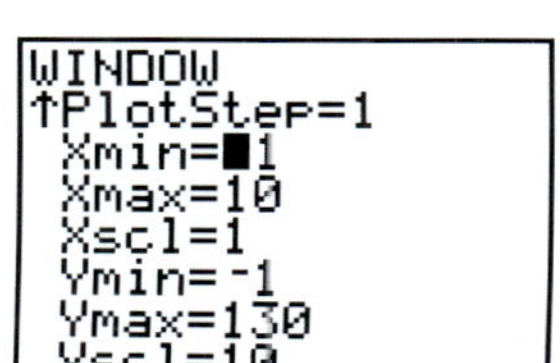

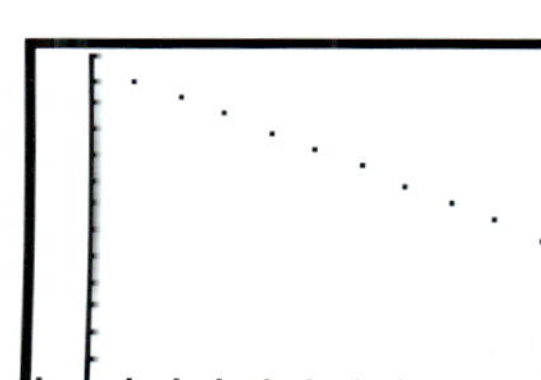

The TI–83 has many features that can be used with sequences. Become familiar with all of them.

EXERCISES 12.1.1

1. (a) Show that the following sequences are arithmetic.
(b) Find the common difference.
(c) Define the rule that gives the nth term of the sequence.

i. $\{2, 6, 10, 14, \ldots\}$ ii. $\{20, 17, 14, 11, \ldots\}$
iii. $\{1, -4, -9, \ldots\}$ iv. $\{0.5, 1.0, 1.5, 2.0, \ldots\}$
v. $\{y+1, y+3, y+5, \ldots\}$ vi. $\{x+2, x, x-2, \ldots\}$

2. Find the 10th term of the sequence whose first four terms are $8, 4, 0, -4$.

3. Find the value of x and y in the arithmetic sequence $\{5, x, 13, y, \ldots\}$.

4. An arithmetic sequence has 12 as its first term and a common difference of –5. Find its 12th term.

5. An arithmetic sequence has –20 as its first term and a common difference of 3. Find its 10th term.

6. The 14th term of an arithmetic sequence is 100. If the first term is 9, find the common difference.

7. The 10th term of an arithmetic sequence is –40. If the first term is 5, find the common difference.

8. If $n+5$, $2n+1$ and $4n-3$ are three consecutive terms of an arithmetic sequences, find n.

9. The first three terms of an arithmetic sequence are $1, 6, 11$.
(a) Find the 9th term.
(b) Which term will equal 151?

10. Find x and y given that $4-\sqrt{3}, x, y$ and $2\sqrt{3}-2$ are the first four terms of an arithmetic sequence.

11. For each of the following sequences
(a) $u_n = -5+2n, n \geq 1$. (b) $u_n = 3+4(n+1), n \geq 1$.
determine
i. its common difference
ii. its first term

12. The third and fifth terms of an A.P are $x+y$ and $x-y$ respectively. Find the 12th term.

13. The sum of the fifth term and twice the third of an arithmetic sequence equals the twelth term. If the seventh term is 25 find an expression for the general term, u_n.

14. For a given arithmetic sequence, $u_n = m$ and $u_m = n$. Find
(a) the common difference.
(b) u_{n+m}.

12.1.2 ARITHMETIC SERIES

If the **terms of a sequence are added**, the result is known as a **series**.

The	sequence:	$1, 2, 3, 4, 5, 6, \ldots$
gives the	series:	$1 + 2 + 3 + 4 + 5 + 6 + \ldots$
and the	sequence:	$-1, -2, -4, -8, -16 \ldots$
gives the	series:	$(-1) + (-2) + (-4) + (-8) + (-16) - \ldots$
		[or $-1 - 2 - 4 - 8 - 16 - \ldots$

The sum of the terms of a series is referred to as S_n, the **sum of n terms of a series**.
For an arithmetic series, we have

$$S_n = u_1 + u_2 + u_3 + \ldots + u_n$$
$$= a + (a + d) + (a + 2d) + \ldots\ldots\ldots\ldots + a + (n - 1)d$$

For example, if we have a sequence defined by $u_n = 6 + 4n, n \geq 1$, then the sum of the first 8 terms is given by

$$S_8 = u_1 + u_2 + u_3 + \ldots + u_8$$
$$= 10 + 14 + 18 + \ldots + 38$$
$$= 192$$

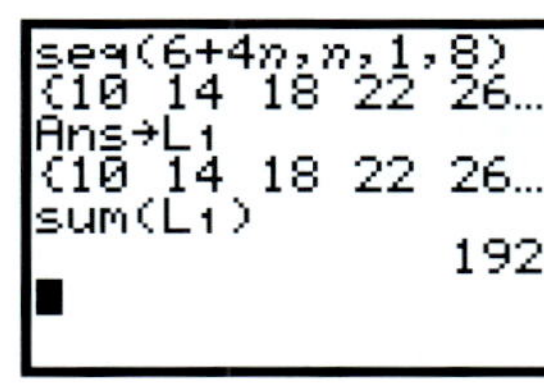

Again, the screen display of the TI–83 shows how readily we can obtain the sum. Once the sequence has been stored as a **List**, use the **sum(** operation to obtain the answer.

There will be many cases in which we can add the terms of a series in this way. If, however, there are a large number of terms to add, a formula is more appropriate.

There is a story that, when the mathematician Gauss was a child, his teacher was having problems with him because he always finished all his work long before the other students. In an attempt to keep Gauss occupied for a period, the teacher asked him to add all the whole numbers from 1 to 100.

'5050' Gauss replied immediately.

It is probable that Gauss used a method similar to this:

1	2	3	4	5	6	......,,	96	97	98	99	100
<u>100</u>	<u>99</u>	<u>98</u>	<u>97</u>	<u>96</u>	<u>95</u>	......,,	<u>5</u>	<u>4</u>	<u>3</u>	<u>2</u>	<u>1</u>
101	101	101	101	101	101	......,,	101	101	101	101	101

Adding each of the pairings gives 100 totals of 101 each. This gives a total of 10100. This is the sum of two sets of the numbers $1 + 2 + 3 + \ldots + 98 + 99 + 100$ and so dividing the full answer by 2 gives the answer 5050, as the young Gauss said, 5050.

It is then possible to apply the same approach to such a sequence, bearing in mind that the sequence of numbers must be arithmetic.

Applying this process to the general arithmetic series we have:

a	$a+d$	$a+2d$	...	$a+(n-3)d$	$a+(n-2)d$	$a+(n-1)d$
$a+(n-1)d$	$a+(n-2)d$	$a+(n-3)d$	...	$a+2d$	$a+d$	a

Each of the pairings comes to the same total.

Here are some examples:

1st pairing: $a+[a+(n-1)d] = 2a+(n-1)d$

2nd pairing: $(a+d)+[a+(n-2)d] = 2a+(n-1)d$

3rd pairing: $(a+2d)+[a+(n-3)d] = 2a+(n-1)d$

$\vdots$

There are n such pairings so $S_n + S_n = n\times[2a+(n-1)d]$

That is, $2S_n = n[2a+(n-1)d]$

Giving the formula, for the sum of n terms of a sequence

$$S_n = \frac{n}{2}[2a+(n-1)d]$$

This formula can now be used to sum large arithmetic series:

EXAMPLE 12.9

Find the sum of 20 terms of the series $-2+1+4+7+10+\ldots$

Solution

We have the following information: $a = u_1 = -2$ and $d = u_2 - u_1 = 1-(-2) = 3$.

Then, the sum to n terms is given by $S_n = \frac{n}{2}[2a+(n-1)d]$

So that the sum to 20 terms is given by

$$\begin{aligned} S_{20} &= \frac{20}{2}[2\times(-2)+(20-1)\times 3] \\ &= 10[-4+19\times 3] \\ &= 530 \end{aligned}$$

EXAMPLE 12.10

Find the sum of 35 terms of the series $-\frac{3}{8}-\frac{1}{8}+\frac{1}{8}+\frac{3}{8}+\frac{5}{8}+\ldots\ldots$

Solution

We have the following information: $a = u_1 = -\frac{3}{8}$ and $d = u_2 - u_1 = -\frac{1}{8}-\left(-\frac{3}{8}\right) = \frac{1}{4}$.

Then, with $n = 35$ we have $S_{35} = \frac{35}{2}\left[2\times-\frac{3}{8}+(35-1)\frac{1}{4}\right] = 17.5\left[-\frac{3}{4}+34\times\frac{1}{4}\right]$

$$= 135\frac{5}{8}$$

EXAMPLE 12.11

An arithmetic series has a third term of 0. The sum of the first 15 terms is –300. What is the first term and the sum of the first ten terms?

Solution

From the given information we have: $u_3 = a + 2d = 0$ – (1)

&: $S_{15} = \frac{15}{2}[2a + 14d] = -300$

i.e., $15a + 105d = -300$

$\therefore a + 7d = -20$ – (2)

The pair of equations can now be solved simultaneously:

(2) – (1): $5d = -20 \Leftrightarrow d = -4$

Substituting result into (1) we have : $a + 2 \times -4 = 0 \Leftrightarrow a = 8$

This establishes that the series is $8 + 4 + 0 + (-4) + (-8) + \ldots$

So the first term is 8 and the sum of the first ten terms is $S_{10} = \frac{10}{2}[16 + 9 \times -4] = -100$.

Using the TI–83 we have, using the general term $u_n = 8 + (n-1) \times -4 = 12 - 4n$

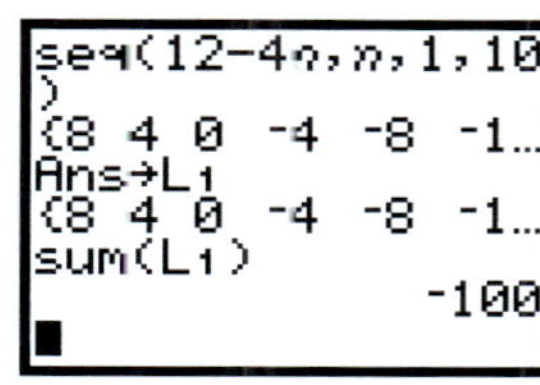

EXAMPLE 12.12

A new business is selling home computers. They predict that they will sell 20 computers in their first month, 23 in the second month, 26 in the third and so on, in arithmetic sequence. How many months will pass before the company expects to sell their thousandth computer.

Solution

The series is: 20 + 23 + 26 + ...

The question implies that the company is looking at the **total** number of computers sold, so we are looking at a series, not a sequence.

The question asks how many terms (months) will be needed before the total sales reach more than 1000. From the given information we have: $a = 20$, $d = 23 - 20 = 3$.

Therefore, we have the sum to n terms given by $S_n = \frac{n}{2}[2 \times 20 + (n-1) \times 3]$

$$= \frac{n}{2}[3n + 37]$$

Next, we determine when $S_n = 1000$: $\frac{n}{2}[3n + 37] = 1000 \Leftrightarrow 3n^2 + 37n = 2000$

$$\Leftrightarrow 3n^2 + 37n - 2000 = 0$$

We solve for n use either of the following methods:

Method 1: Quadratic formula

$$n = \frac{-37 \pm \sqrt{37^2 - 4 \times 3 \times -2000}}{2 \times 3}$$

$$= 20.37 \text{ or } (-32.7)$$

Method 2: Graphics Calculator **Solve** function

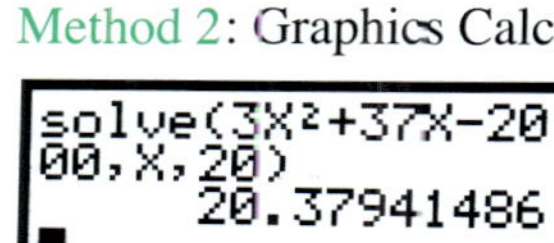

Method 3: Table of values

Notice that we have entered the expression for S_n as the sequence rule for $u(n)$. In fact, the series itself is made up of terms in a sequence of so-called **partial sums**, often called a **sum sequence**. That is, we have that $\{S_1, S_2, S_3, \ldots\} = \{15, 33, 54, \ldots\}$ forms a sequence.

The answer then, is that the company will sell its thousandth computer during the 20th month.

EXERCISES 12.1.2

1. Find the sum of the first ten terms in the arithmetic sequences
(a) $\{1, 4, 7, 10, \ldots\}$ (b) $\{3, 9, 15, 21, \ldots\}$ (c) $\{10, 4, -2, \ldots\}$.

2. For the given arithmetic sequences, find the sum, S_n, to the requested number of terms.
(a) $\{4, 3, 2, \ldots\}$ for $n = 12$
(b) $\{4, 10, 16, \ldots\}$ for $n = 15$
(c) $\{2.9, 3.6, 4.3, \ldots\}$ for $n = 11$

3. Find the sum of the following sequences:
(a) $\{5, 4, 3, \ldots, -15\}$
(b) $\{3, 9, 15, \ldots, 75\}$
(c) $\{3, 5, 7, \ldots, 29\}$

4. The weekly sales of washing machines from a retail store that has just opened in a new housing complex increases by 2 machines per week. In the first week of January 1995, 24 machines were sold.
(a) How many are sold in the last week of December 1995?
(b) How many machines did the retailer sell in 1995?
(c) When was the 500th machine sold?

5. The fourth term of an arithmetic sequence is 5 while the sum of the first 6 terms is 10. Find the sum of the first nineteen terms.

6. Find the sum of the first 10 terms for the sequences defined by
(a) $u_n = -2 + 8n$ (b) $u_n = 1 - 4n$

12.1.3 SIGMA NOTATION

There is a second notation to denote the sum of terms. This other notation makes use the Greek letter $\sum\ldots$ as the symbol to inform us that we are carrying out a summation.
In short, $\sum\ldots$ stands for '**The sum of . . .**'.

This means that the expression $\sum_{i=1}^{n} u_i = u_1 + u_2 + u_3 + \ldots + u_{n-1} + u_n$.

For example, if $u_i = 2 + 5(i-1)$, i.e., an A.P with first term $a = 2$ and common difference $d = 5$, the expression $S_n = \sum_{i=1}^{n} [2 + 5(i-1)]$ would respesent the sum of the first n terms of the sequence. So, the sum of the first 3 terms would be given by

$$S_3 = \sum_{i=1}^{3} [2 + 5(i-1)] = \underbrace{[2 + 5(1-1)]}_{i=1} + \underbrace{[2 + 5(2-1)]}_{i=2} + \underbrace{[2 + 5(3-1)]}_{i=3}$$
$$= 2 + 7 + 12$$
$$= 21$$

Properties of $\sum$

1. $\sum$ is distributive. That is, $\sum_{i=1}^{n} [u_i + v_i] = \sum_{i=1}^{n} u_i + \sum_{i=1}^{n} v_i$.

2. $\sum_{i=1}^{n} ku_i = k\sum_{i=1}^{n} u_i$, for some constant value k.

3. $\sum_{i=1}^{n} k = kn$, i.e., adding a constant term, k, n times is the same as multiplying k by n.

EXAMPLE 12.13

Given that $u_i = 5 + 2i$ and that $v_i = 2 - 5i$ find

(a) $\sum_{i=1}^{5} u_i$ (b) $\sum_{i=1}^{5} [2u_i - v_i]$ (c) $\sum_{i=1}^{1000} [5u_i + 2v_i]$

Solution

(a) $\sum_{i=1}^{5} u_i = u_1 + u_2 + u_3 + u_4 + u_5 = [5 + 2] + [5 + 4] + [5 + 6] + [5 + 8] + [5 + 10]$

$= 7 + 9 + 11 + 13 + 15$
$= 55$

(b) $\sum_{i=1}^{5} [2u_i - v_i] = \sum_{i=1}^{5} (2u_i) + \sum_{i=1}^{5} (-v_i) = 2\sum_{i=1}^{5} u_i - \sum_{i=1}^{5} v_i$

Now, $2\sum_{i=1}^{5} u_i = 2 \times 55 = 110$

and $\sum_{i=1}^{5} v_i = \sum_{i=1}^{5} (2-5i) = \sum_{i=1}^{5} (2) - 5\sum_{i=1}^{5} i = 2 \times 5 - 5[1+2+3+4+5]$ [Using properties]

$= -65$

Therefore, $\sum_{i=1}^{5} [2u_i - v_i] = 110 - (-65) = 175$

(c) $\sum_{i=1}^{1000} [5u_i + 2v_i] = \sum_{i=1}^{1000} [5(5+2i) + 2(2-5i)]$

$= \sum_{i=1}^{1000} [25 + 10i + 4 - 10i]$

$= \sum_{i=1}^{1000} 29$

$= 29\,000$ [i.e., 29×1000]

In this example we have tried to show that there are a number of ways to obtain a sum. It is not always necessary to enumerate every term and then add them. Often, an expression can first be simplified.

EXERCISES 12.1.3 – MISCELLANEOUS QUESTIONS

1. Find the twentieth term in the sequence 9, 15, 21, 27, 33, . . .

2. Fill the gaps in this arithmetic sequence: –3, _, _, _, _, _, 12.

3. An arithmetic sequence has a tenth term of 17 and a fourteenth term of 30. Find the common difference.

4. If $u_{59} = \frac{1}{10}$ and $u_{100} = -1\frac{19}{20}$ for an arithmetic sequence, find the first term and the common difference.

5. Find the sum of the first one hundred odd numbers.

6. An arithmetic series has twenty terms. The first term is –50 and the last term is 83, find the sum of the series.

7. Thirty numbers are in arithmetic sequence. The sum of the numbers is 270 and the last number is 38. What is the first number?

8. How many terms of the arithmetic sequence: 2, 2.3, 2.6, 2.9, . . . must be taken before the terms exceed 100?

9. Brian and Melissa save \$50 in the first week of a savings program, \$55 in the second week, \$60 in the third and so on, in arithmetic progression. How much will they save in ten weeks? How long will they have to continue saving if their target is to save \$5000?

10. A printing firm offers to print business cards on the following terms: \$45 for design and typesetting and then \$0.02 per card.

(i) What is the cost of 500 cards from this printer?

(ii) How many cards can a customer with \$100 afford to order?

11. A children's game consists of the players standing in a line with a gap of 2 metres between each. The child at the left hand end of the line has a ball which s/he throws to the next child in the line, a distance of 2 metres. The ball is then thrown back to the first child who then throws the ball to the third child in the line, a distance of 4 metres. The ball is then returned to the first child, and so on until all the children have touched the ball at least once.

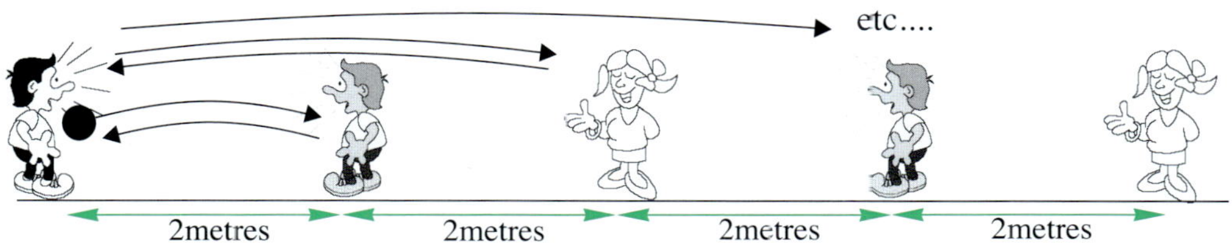

(a) If a total of five children play and they make the least number of throws so that only the leftmost child touches the ball more than once:

i. What is the largest single throw?

ii. What is the total distance travelled by the ball?

(b) If seven children play, what is the total distance travelled by the ball?

(c) If n children play, derive a formula for the total distance travelled by the ball.

(d) Find the least number of children who need to play the game before the total distance travelled by the ball exceeds 100 metres.

(e) The children can all throw the ball 50 metres at most

i. What is the largest number of children that can play the game?

ii. What is the total distance travelled by the ball?

12. Find each sum,

(a) $\sum_{k=1}^{100} k$ (b) $\sum_{k=1}^{100} (2k+1)$ (c) $\sum_{k=1}^{51} (3k+5)$

13. If $u_i = -3 + 4i$ and $v_i = 12 - 3i$ find

(a) $\sum_{i=1}^{10} (u_i + v_i)$ (b) $\sum_{i=1}^{10} (3u_i + 4v_i)$ (c) $\sum_{i=1}^{10} u_i v_i$

14. (a) Show that for an arithmetic sequence, $u_n = S_n - S_{n-1}$, where u_n is the nth term and S_n is the sum of the first n terms.

(b) Find the general term, u_n, of the A.P given that $\sum_{i=1}^{n} u_i = \frac{n}{2}(3n-1)$.

12.2 GEOMETRIC SEQUENCES & SERIES

12.2.1 GEOMETRIC SEQUENCES

Sequences such as $2, 6, 18, 54, 162, \ldots$ and $200, 20, 2, 0.2, \ldots$ in which each term is obtained by multiplying the previous one by a fixed quantity are known as **geometric sequences**.

The sequence: $2, 6, 18, 54, 162, \ldots$ is formed by starting with 2 and then multiplying by 3 to get the second term, by 3 again to get the third term, and so on.

For the sequence $200, 20, 2, 0.2, \ldots$, begin with 20 and multiplied by 0.1 to get the second term, by 0.1 again to get the third term and so on.

The constant multiplier of such a sequence is known as the **common ratio**.

The common ratio of 2, 6, 18, 54, 162,.... is 3 and of 200, 20, 2, 0.2,...... it is 0.1.

The **nth term of a geometric sequence** is obtained from the first term by multiplying by n–1 common ratios. This leads to the formula for the

> *n*th term of a geometric sequence: $u_n = a \times r^{n-1}$ where $r = \dfrac{u_2}{u_1} = \ldots = \dfrac{u_n}{u_{n-1}}$
> and n is the term number, a the first term and r is the common ratio.

EXAMPLE 12.14

Find the tenth term in the sequence $2, 6, 18, 54, 162, \ldots$

Solution

The first term is $a = 2$. The common ratio $r = 3 = \dfrac{6}{2} = \dfrac{18}{6}$ and n, the required term, is 10.

Use the formula to solve the problem:

$$\begin{aligned} u_n &= a \times r^{n-1} \\ u_{10} &= 2 \times 3^{(10-1)} \\ &= 2 \times 3^9 \\ &= 39366 \end{aligned}$$

EXAMPLE 12.15

Find the fifteenth term in the sequence $200, 20, 2, 0.2, \ldots$

Solution

In this case, $a = 20$, $r = \dfrac{20}{200} = \dfrac{1}{10} = 0.1$ and $n = 15$.

Using the general term $u_n = a \times r^{n-1}$, the 15th term is given by

$$\begin{aligned} u_{15} &= 200 \times 0.1^{(15-1)} \\ &= 200 \times 0.1^{14} \\ &= 2 \times 10^{-12} \end{aligned}$$

EXAMPLE 12.16

Find the eleventh term in the sequence $1, -\frac{1}{2}, \frac{1}{4}, -\frac{1}{8}, \frac{1}{16}, \ldots$

Solution

The sequence $1, -\frac{1}{2}, \frac{1}{4}, -\frac{1}{8}, \frac{1}{16}, \ldots$ has a common ratio of $r = \frac{-1/2}{1} = -\frac{1}{2}$.

Using the general term $u_n = a \times r^{n-1}$, we have

$$u_{11} = 1 \times \left(-\frac{1}{2}\right)^{(11-1)}$$

$$= \left(-\frac{1}{2}\right)^{10}$$

$$\approx 0.000977$$

Many questions will be more demanding in terms of the way in which you use this formula. You should also recognise that the formula can be applied to a range of practical problems.

Many of the practical problems involve growth and decay and can be seen as similar to problems studied in Chapter 7.

EXAMPLE 12.17

A geometric sequence has a fifth term of 3 and a seventh term of 0.75. Find the first term, the common ratio and the tenth term.

Solution

From the given information we can set up the following equations:

$$u_5 = a \times r^4 = 3 \quad -(1)$$

$$\& \qquad u_7 = a \times r^6 = 0.75 \quad -(2)$$

As with similar problems involving arithmetic sequences, the result is a pair of simultaneous equations. In this case these can best be solved by dividing (2) by (1) to get:

$$\frac{a \times r^6}{a \times r^4} = \frac{0.75}{3} \Leftrightarrow r^2 = 0.25 \Leftrightarrow r = \pm 0.5$$

Substituting results into (1) we have: $a\left(\pm\frac{1}{2}\right)^4 = 3 \Leftrightarrow a = 48$

Therefore, the 10th term is given by $u_{10} = 48 \times (\pm 0.5)^9 = \pm\frac{3}{32}$

There are two solutions: 48, 24, 12, 6, . . . (for the case $r = 0.5$) & 48, –24, 12, –6, . . . ($r = -0.5$).

EXAMPLE 12.18

Find the number of terms in the geometric sequence:
0.25, 0.75, 2.25, . . . , 44286.75.

Solution

The sequence $0.25, 0.75, 2.25, \ldots, 44286.75$ has a first term $a = 0.25$ and a common ratio $r = \frac{0.75}{0.25} = 3$. In this problem it is n that is unknown. Substitution of the data into the formula gives: $u_n = 0.25 \times 3^{(n-1)} = 44286.75$

The equation that results can be solved using a graphics calculator.

EXAMPLE 12.19

A car originally worth \$34000 loses 15% of its value each year.

(a) Write a geometric sequence that gives the year by year value of the car.

(b) Find the value of the car after 6 years.

(c) After how many years will the value of the car fall below \$10000?

Solution

(a) If the car loses 15% of its value each year, its value will fall to 85% (100% – 15%) of its value in the previous year. This means that the common ratio is 0.85 (the fractional equivalent of 85%). Using the formula, the sequence is: $u_n = 34000 \times 0.85^{(n-1)}$
i.e. \$34000, \$28900, \$24565, \$20880.25, . . .

(b) The value after six years have passed is the **seventh** term of the sequence. This is because the first term of the sequence is the value after **no** years have passed.
$u_7 = 34000 \times 0.85^6 \approx 12823$ or \$12823.

(c) We need to solve the equation $10000 = 34000 \times 0.85^n$.
We solve this equation by making use of the graphics calculator:

This means that the car's value will fall to \$10000 after about 7 years 6 months.

EXAMPLE 12.20

The number of people in a small country town increases by 2% per year. If the population at the start of 1970 was 12500, what is the predicted population at the start of the year 2010?

Solution

A quantity can be increased by 2% by multiplying by 1.02. Note that this is different from finding 2% of a quantity which is done by multiplying by 0.02.

The sequence is: $12500, 12500 \times 1.02, 12500 \times 1.02^2$ etc. with $a = 12500$, $r = 1.02$.

It is also necessary to be careful about which term is required. In this case, the population at the start of 1970 is the first term, the population at the start of 1971 the second term, and so on. The population at the start of 1980 is the **eleventh** term and at the start of 2010 we need the forty-first term:

$$u_{41} = 12500 \times 1.02^{40}$$
$$\approx 27600$$

In all such cases, you should round your answer to the level given in the question or, if no such direction is given, round the answer to a reasonable level of accuracy. In this question, the original population appears to have been given to the nearest 100 and so it is hardly reasonable to give a higher level of accuracy in the answer.

Using a Graphics Calculator

As with arithmetic sequences, geometric sequences such as 50, 25, 12.5, . . . can be listed using a graphics calculator. For this sequence we have $a = 50$ and $r = 0.5$, so, $u_n = 50(0.5)^{n-1}$

We first set the **MODE** to **Seq** and then enter the sequence rule:

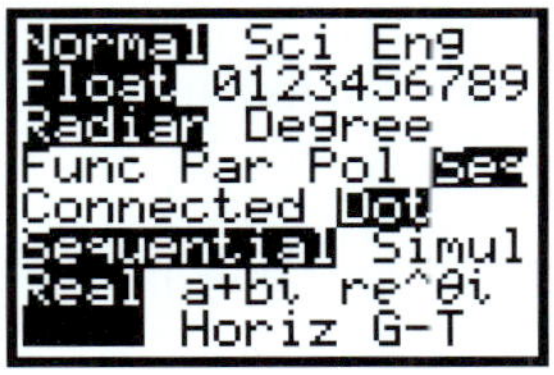

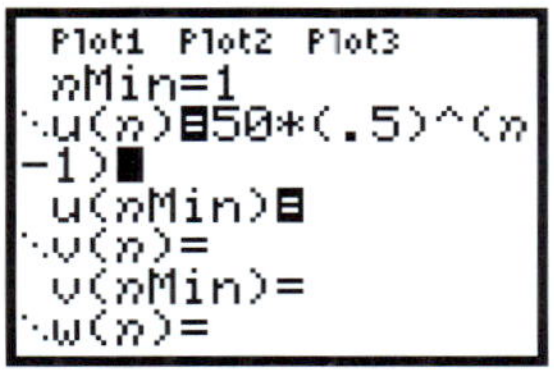

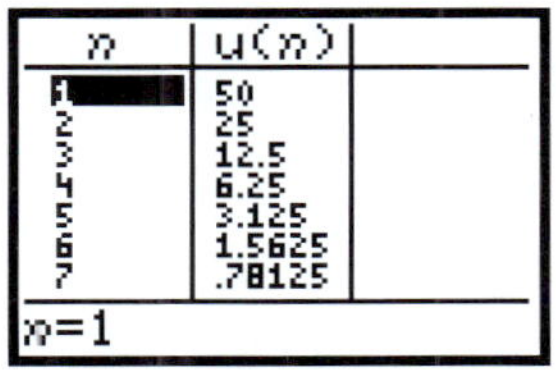

Plot of sequence:

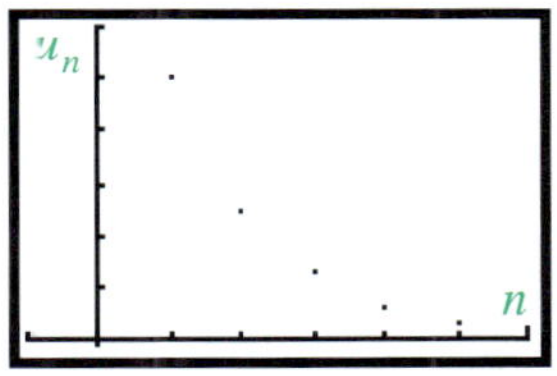

EXERCISES 12.2.1

1. Find the common ratio, the 5th term and the general term of the following sequences

(a) 3, 6, 12, 24, . . . (b) $3, 1, \frac{1}{3}, \frac{1}{9}, \ldots$ (c) $2, \frac{2}{5}, \frac{2}{25}, \frac{2}{125}, \ldots$

(d) $-1, 4, -16, 64, \ldots$ (e) $ab, a, \frac{a}{b}, \frac{a}{b^2}, \ldots$ (f) $a^2, ab, b^2, \ldots$

2. Find the value(s) of x if each of the following are in geometric sequence

(a) $3, x, 48$

(b) $\frac{5}{2}, x, \frac{1}{2}$

3. The third and seventh terms of a geometric sequence are $\frac{3}{4}$ and 12 respectively.

(a) Find the 10th term.

(b) What term is equal to 3072?

4. A rubber ball is dropped from a height of 10 m and bounces to reach $\frac{5}{6}$ of its previous height after each rebound. Let u_n is the ball's maximum height **before** its nth rebound.
(a) Find an expression for u_n.
(b) How high will the ball bounce **after** its 5th rebound.
(c) How many times has the ball bounced by the time it reaches a maximum height of $\frac{6250}{1296}$ m.

5. The terms $k + 4, 5k + 4, k + 20$ are in a geometric sequence. Find the value(s) of k.

6. A computer depreciates each year to 80% of its value from the previous year. When bought the computer was worth \$8000.
(a) Find its value after
i. 3 years
ii. 6 years
(b) How long does it take for the computer to depreciate to a quarter of its purchase price.

7. The sum of the first and third terms of a geometric sequence is 40 while the sum of its second and fourth terms is 96. Find the sixth term of the sequence.

8. The sum of three successive terms of a geometric sequence is $\frac{35}{2}$ while their product is 125. Find the three terms.

9. The population in a town of 40,000 increases at 3% per annum. Estimate the town's population after 10 years.

10. Following new government funding it is expected that the unemployed workforce will decrease by 1.2% per month. Initially there are 120,000 people unemployed. How large an unemployed workforce can the government expect to report in 8 months time.

11. The cost of erecting the ground floor of a building is \$44,000, for erecting the first floor it costs \$46,200, to erect the second floor costs \$48,510 and so on.
Using this cost structure
(a) How much will it cost to erect the 5th floor?
(b) What will be to total cost of erecting a building with six floors?

12.2.2 GEOMETRIC SERIES

When the terms of a geometric sequence are added, the result is a **geometric series**.
For example:
The sequence 3, 6, 12, 24, 48, . . . gives rise to the series: $3 + 6 + 12 + 24 + 48 + \ldots$
and, the sequence $24, -16, 10\frac{2}{3}, -7\frac{1}{9}, \ldots$ leads to the series $24 - 16 + 10\frac{2}{3} - 7\frac{1}{9} + \ldots$

Geometric series can be summed using the formula that is derived by first multiplying the series by r:

$$S_n = a + ar + ar^2 + ar^3 + \ldots\ldots\ldots + ar^{n-3} + ar^{n-2} + ar^{n-1}$$

$$r \times S_n = ar + ar^2 + ar^3 + \ldots\ldots\ldots + ar^{n-3} + ar^{n-2} + ar^{n-1} + ar^n$$

$$S_n - r \times S_n = a - ar^n \quad \text{(subtracting the second equation from the first)}$$

$$S_n(1-r) = a(1-r^n)$$

$$S_n = \frac{a(1-r^n)}{1-r}$$

This formula can also be written as: $S_n = \frac{a(r^n-1)}{r-1}, r \neq 1$. It is usual to use the version of the formula that gives a positive value for the denominator. And so, we have:

The sum of the first n terms of a geometric series, S_n, where $r \neq 1$ is given by

$$S_n = \frac{a(1-r^n)}{1-r}, |r| < 1 \text{ or } S_n = \frac{a(r^n-1)}{r-1}, |r| > 1$$

EXAMPLE 12.21

Sum the following series to the number of terms indicated.

(a)	$2 + 4 + 8 + 16 + \ldots$	9 terms.
(b)	$5 - 15 + 45 - 135 + \ldots$	7 terms.
(c)	$24 + 18 + \frac{27}{2} + \frac{81}{8} + \ldots$	12 terms.
(d)	$20 - 30 + 45 - 67.5 + \ldots$	10 terms.

Solution

(a) In this case $a = 2, r = 2$ and $n = 9$.
Because $r = 2$ it is more convenient to use:

$$S_n = \frac{a(r^n-1)}{r-1}$$

$$S_9 = \frac{2(2^9-1)}{2-1}$$

$$= 1022$$

Using this version of the formula gives positive values for the numerator and denominator. The other version is correct but gives negative numerator and denominator and hence the same answer.

(b) $a = 5, r = -3$ and $n = 7$.

$$S_n = \frac{a(1-r^n)}{1-r} \qquad \text{or} \qquad S_n = \frac{a(r^n-1)}{r-1}$$

$$S_7 = \frac{5(1-(-3)^7)}{1-(-3)} \qquad\qquad S_7 = \frac{5((-3)^7-1)}{(-3)-1}$$

$$= 2735 \qquad\qquad = 2735$$

(c) $a = 24, r = 0.75$ and $n = 12$. Using $S_n = \frac{a(1-r^n)}{1-r}$ we have:

$$S_{12} = \frac{24\left(1-\left(\frac{3}{4}\right)^{12}\right)}{1-\left(\frac{3}{4}\right)} \quad \text{This version gives the positive values.}$$

$$= 92.95907$$

(d) $a = 20, r = -1.5$ and $n = 10$.

$$S_n = \frac{a(1-r^n)}{1-r}$$

$$S_{10} = \frac{20(1-(-1.5)^{10})}{1-(-1.5)}$$

$$= -453.32031$$

When using a calculator to evaluate such expressions, it is advisable to use brackets to ensure that correct answers are obtained. For both the graphics and scientific calculator, the negative common ratio must be entered using the +/– or (–) key.

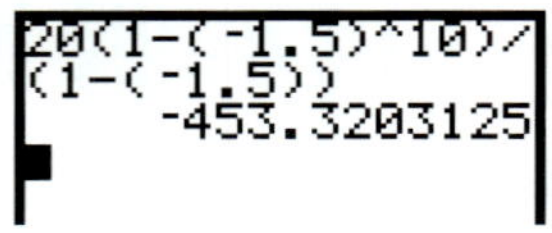

Other questions that may be asked in examinations could involve using both formulas. A second possibility is that you may be asked to apply sequence and series theory to some simple problems.

EXAMPLE 12.22

The 2nd term of a geometric series is –30 and the sum of the first two terms is –15. Find the first term and the common ratio.

Solution

From the given information we have: $u_2 = -30 \therefore ar = -30$ – (1)

$$S_2 = -15 \therefore \frac{a(r^2-1)}{r-1} = -15 \text{ – (2)}$$

The result is a pair of simultaneous equations in the two unknowns. The best method of solution is substitution:

From (1): $a = \frac{-30}{r}$. Substituting into (2): $\frac{\frac{-30}{r}(r^2-1)}{r-1} = -15 \Leftrightarrow \frac{(-30)(r^2-1)}{r(r-1)} = -15$

$$\therefore \frac{-30(r+1)(r-1)}{r(r-1)} = -15$$

$$\Leftrightarrow -30(r+1) = -15r$$

$$\Leftrightarrow -30r - 30 = -15r$$

$$\Leftrightarrow r = -2$$

$$\therefore a = \frac{-30}{r} = \frac{-30}{-2} = 15$$

The series is 15 – 30 + 60 – 120 + 240 – . . . which meets the conditions set out in the question.

EXAMPLE 12.23 A family decide to save some money in an account that pays 9% annual compound interest calculated at the end of each year. They put \$2500 into the account at the beginning of each year. All interest is added to the account and no withdrawals are made. How much money will they have in the account on the day after they have made their tenth payment?

Solution

The problem is best looked at from the last payment of \$2500 which has just been made and which has not earned any interest.

The previous payment has earned one lot of 9% interest and so is now worth 2500×1.09.

The previous payment has earned two years' worth of compound interest and is worth 2500×1.09^2.

This process can be continued for all the other payments and the various amounts of interest that each has earned. They form a geometric series:

Last payment ... First payment

$$2500 + 2500 \times 1.09 + 2500 \times 1.09^2 + \ldots\ldots + 2500 \times 1.09^9$$

The total amount saved can be calculated using the series formula:

$$S_n = \frac{a(r^n - 1)}{r - 1}$$

$$S_{10} = \frac{2500(1.09^{10} - 1)}{1.09 - 1}$$

$$= 37982.32$$

The family will save about \$37,982.

EXERCISES 12.2.2

1. Find the common ratios of these geometric sequences:

(a) $7, 21, 63, 189, \ldots$ (b) $12, 4, \frac{4}{3}, \frac{4}{9}, \ldots$ (c) $1, -1, 1, -1, 1, \ldots$

(d) $9, -3, 1, -\frac{1}{3}, \frac{1}{9}, \ldots$ (e) $64, 80, 100, 125, \ldots$ (f) $27, -18, 12, -8, \ldots$

2. Find the term indicated for each of these geometric sequences:

(a) $11, 33, 99, 297, \ldots$ 10th term. (b) $1, 0.2, 0.04, 0.008, \ldots$ 15th term.

(c) $9, -6, 4, -\frac{8}{3}, \ldots$ 9th term. (d) $21, 9, \frac{27}{7}, \frac{81}{49}, \ldots$ 6th term.

(e) $-\frac{1}{3}, -\frac{1}{4}, -\frac{3}{16}, -\frac{9}{64}, \ldots$ 6th term.

3. Find the number of terms in each of these geometric sequences and the sum of the numbers in each sequence:

(a) $4, 12, 36, \ldots, 236196$ (b) $11, -22, 44, \ldots, 704$

(c) $100, -10, 1, \ldots, -10^{-5}$ (d) $48, 36, 27, \ldots, \frac{6561}{1024}$

(e) $\frac{1}{8}, -\frac{9}{32}, \frac{81}{128}, \ldots, \frac{6561}{2048}$ (f) $100, 10, 1, \ldots, 10^{-10}$

4. Write the following in expanded form and evaluate.

(a) $\sum_{k=1}^{7} \left(\frac{1}{2}\right)^k$ (b) $\sum_{i=1}^{6} 2^{i-4}$ (c) $\sum_{j=1}^{4} \left(\frac{2}{3}\right)^j$

(d) $\sum_{s=1}^{4} (-3)^s$ (e) $\sum_{t=1}^{6} 2^{-t}$

5. The third term of a geometric sequence is 36 and the tenth term is 78,732. Find the first term in the sequence and the sum of these terms.

6. A bank account offers 9% interest compounded annually. If \$750 is invested in this account, find the amount in the account at the end of the twelfth year.

7. When a ball is dropped onto a flat floor, it bounces to 65% of the height from which it was dropped. If the ball is dropped from 80 cm, find the height of the fifth bounce.

8. A computer loses 30% of its value each year.

(a) Write a formula for the value of the computer after n years.

(b) How many years will it be before the value of the computer falls below 10% of its original value?

9. A geometric sequence has a first term of 7 and a common ratio of 1.1. How many terms must be taken before the value of the term exceeds 1000?

10. A colony of algae increases in size by 15% per week. If 10 grams of the algae are placed in a lake, find the weight of algae that will be present in the lake after 12 weeks. The lake will be considered 'seriously polluted' when there is in excess of 10000 grams of algae in the lake. How long will it be before the lake becomes seriously polluted?

11. A geometric series has nine terms, a common ratio of 2 and a sum of 3577. Find the first term.

12. A geometric series has a third term of 12, a common ratio of $-\frac{1}{2}$ and a sum of $32\frac{1}{16}$. Find the number of terms in the series.

13. A geometric series has a first term of 1000, seven terms and a sum of $671\frac{7}{8}$. Find the common ratio.

14. A geometric series has a third term of 300, and a sixth term of 37500. Find the common ratio and the sum of the first fourteen terms (in scientific form correct to two significant figures).

15. A \$10000 loan is offered on the following terms: 12% annual interest on the outstanding debt calculated monthly. The required monthly repayment is \$270. How much will still be owing after nine months.

16. As a prize for inventing the game of chess, its originator is said to have asked for one grain of wheat to be placed on the first square of the board, 2 on the second, 4 on the third, 8 on the fourth and so on until each of the 64 squares had been covered. How much wheat would have been the prize?

12.2.3 COMBINED A.PS AND G.PS

There will be occasions on which questions will be asked that relate to both arithmetic and geometric sequences and series.

EXAMPLE 12.24 A geometric sequence has the same first term as an arithmetic sequence. The third term of the geometric sequence is the same as the tenth term of the arithmetic sequence with both being 48. The tenth term of the arithmetic sequence is four times the second term of the geometric sequence. Find the common difference of the arithmetic sequence and the common ratio of the geometric sequence.

Solution

When solving these sorts of question, write the data as equations, noting that a is the same for both sequences. Let u_n denote the general term of the A.P and v_n the general term of the G.P. We then have:

$$u_{10} = a + 9d = 48, \; v_3 = ar^2 = 48,$$

i.e.,

$$a + 9d = ar^2 = 48 \quad -(1)$$

$$u_{10} = 4v_2 \Rightarrow a + 9d = 4ar \quad -(2)$$

(1) represents the information 'The third term of the geometric sequence is the same as the tenth term of the arithmetic sequence with both being 48'.
(2) represents 'The tenth term of the arithmetic sequence is four times the second term of the geometric sequence'.

There are three equations here and more than one way of solving them. One of the simplest is:

From (1) $a + 9d = 48$ and so substituting into (2) : $48 = 4ar \Leftrightarrow ar = 12 \quad -(3)$

Also from (1) we have: $ar^2 = 48 \Leftrightarrow (ar)r = 48 \quad -(4)$

Substituting (3) into (4): $12r = 48 \Leftrightarrow r = 4$

Substituting result into (1): $a \times 16 = 48 \Leftrightarrow a = 3$

Substituting result into (1): $3 + 9d = 48 \Leftrightarrow d = 5$

The common ratio is 4 and the common difference is 5.

It is worth checking that the sequences are as specified:

Geometric sequence: 3, 12, 48
Arithmetic sequence: 3, 8, 13, 18, 23, 28, 33, 38, 43, 48

EXERCISES 12.2.3

1. Consider the following sequences:
Arithmetic: 100, 110, 120, 130, . . .
Geometric: 1, 2, 4, 8, 16, . . .
Prove that:
The terms of the geometric sequence will exceed the terms of the arithmetic sequence after the 8th term.
The sum of the terms of the geometric sequence will exceed the sum of the terms of the arithmetic after the 10th term.

2. An arithmetic series has a first term of 2 and a fifth term of 30. A geometric series has a common ratio of –0.5. The sum of the first two terms of the geometric series is the same as the second term of the arithmetic series. What is the first term of the geometric series?

3. An arithmetic series has a first term of –4 and a common difference of 1. A geometric series has a first term of 8 and a common ratio of 0.5. After how many terms does the sum of the arithmetic series exceed the sum of the geometric series?

4. The first and second terms of an arithmetic and a geometric series are the same and are equal to 12. The sum of the first two terms of the arithmetic series is four times the first term of the geometric series. Find the first term of each series if the A.P has $d = 4$.

5. Bo-Youn and Ken are to begin a savings program. Bo-Youn saves \$1 in the first week \$2 in the second week, \$4 in the third and so on, in geometric progression. Ken saves \$10 in the first week, \$15 in the second week, \$20 in the third and so on, in arithmetic progression. After how many weeks will Bo-Youn have saved more than Ken?

6. Ari and Chai begin a training program. In the first week Chai will run 10km, in the second he will run 11km and in the third 12km, and so on, in arithmetic progression. Ari will run 5km in the first week and will increase his distance by 20% in each succeeding week.
   (a) When does Ari's weekly distance first exceed Chai's?
   (b) When does Ari's total distance first exceed Chai's?

7. The Fibonacci sequence: 1, 1, 2, 3, 5, 8, 13, 21, . . . in which each term is the sum of the previous two terms is neither arithmetic nor geometric. However, after the eighth term (21) the sequence becomes approximately geometric. If we assume that the sequence is geometric:
   (a) What is the common ratio of the sequence (to four significant figures)?
   (b) Assuming that the Fibonacci sequence can be approximated by the geometric sequence after the eighth term, what is the approximate sum of the first 24 terms of the Fibonacci sequence?

12.2.4 CONVERGENT SERIES

If a geometric series has a common ratio between –1 and 1, the terms get smaller and smaller as n increases.

The sum of these terms is still given by the formula: $S_n = \dfrac{a(1-r^n)}{1-r}, r \neq 1$

For $-1 < r < 1$, $r^n \to 0$ as $n \to \infty \Rightarrow S_n = \frac{a}{1-r}$.

If $|r| < 1$, the infinite sequence has a sum given by $S_\infty = \frac{a}{1-r}$.

This means that if the common ratio of a geometric series is between –1 and 1, the sum of the series will approach a value of $\frac{a}{1-r}$ as the number of terms of the series becomes large i.e., the **series is convergent**.

EXAMPLE 12.25

Find the sum to infinity of the series

(a) $16 + 8 + 4 + 2 + 1 + \ldots$

(b) $9 - 6 + 4 - \frac{8}{3} + \frac{16}{9} - \ldots$

Solution

(a) $16 + 8 + 4 + 2 + 1 + \ldots$

In this case $a = 16$, $r = \frac{1}{2} \Rightarrow S_\infty = \frac{a}{1-r} = \frac{16}{1-\frac{1}{2}} = 32$

(b) $9 - 6 + 4 - \frac{8}{3} + \frac{16}{9} -$

$a = 9$, $r = -\frac{2}{3} \Rightarrow S_\infty = \frac{a}{1-r} = \frac{9}{1-\left(-\frac{2}{3}\right)} = 5.4$

There are a variety of applications for convergent geometric series. The following examples illustrate two of these.

EXAMPLE 12.26

Use an infinite series to express the recurring decimal $0.\dot{4}6\dot{2}$ as a rational number.

Solution

$0.\dot{4}6\dot{2}$ can be expressed as the series: $0.462 + 0.000462 + 0.000000462 + \ldots$

or $\frac{462}{1000} + \frac{462}{1000000} + \frac{462}{1000000000} + \ldots$

This is a geometric series with $a = \frac{462}{1000}$, $r = \frac{1}{1000}$

It follows that $S_\infty = \frac{a}{1-r} = \frac{\frac{462}{1000}}{1-\frac{1}{1000}} = \frac{\frac{462}{1000}}{\frac{999}{1000}} = \frac{462}{999}$

EXAMPLE 12.27

A ball is dropped from a height of 10 metres. On each bounce the ball bounces to three quartes of the height of the previous bounce. Find the distance travelled by the ball before it comes to rest (if it does not move sideways).

Solution

The ball bounces in a vertical line and does not move sideways. On each bounce after the drop, the ball moves both up and down and so travels twice the distance of the height of the bounce.

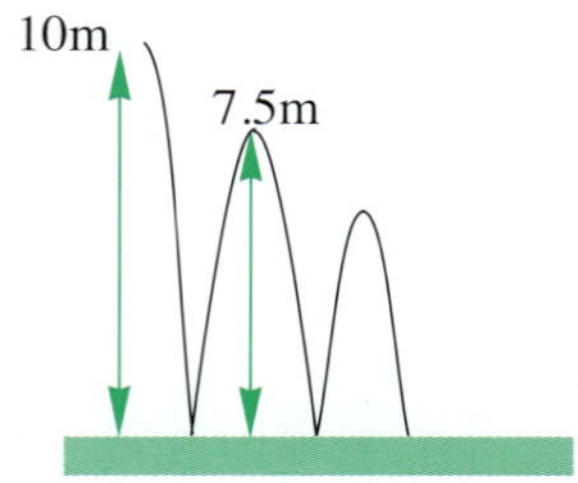

$$\text{Distance} = 10 + 15 + 15 \times \frac{3}{4} + 15 \times \left(\frac{3}{4}\right)^2 + \ldots$$

All but the first term of this series are geometric $a = 15, r = \frac{3}{4}$

$$\text{Distance} = 10 + S_\infty = 10 + \frac{15}{1 - \frac{3}{4}} = 70\text{m}$$

EXERCISES 12.2.4

1. Evaluate:

(i) $27 + 9 + 3 + \frac{1}{3} + \ldots$ (ii) $1 - \frac{3}{10} + \frac{9}{100} - \frac{27}{1000} + \ldots$

(iii) $500 + 450 + 405 + 364.5 + \ldots$ (iv) $3 - 0.3 + 0.03 - 0.003 + 0.0003 - \ldots$

2. Use geometric series to express the recurring decimal 23.232323… as a mixed number.

3. Biologists estimate that there are 1000 trout in a lake. If none are caught, the population will increase at 10% per year. If more than 10% are caught, the population will fall. As an approximation, assume that if 25% of the fish are caught per year, the population will fall by 15% per year. Estimate the total catch before the lake is 'fished out'. If the catch rate is reduced to 15%, what is the total catch in this case? Comment on these results.

4. Find the sum to infinity of the sequence 45, –30, 20, . . .

5. The second term of a geometric sequence is 12 while the sum to infinity is 64. Find the first three terms of this sequence.

6. Express the following as rational numbers

(a) $0.3\dot{6}$ (b) $0.\dot{3}\dot{7}$ (c) $2.1\dot{2}$

7. A swinging pendulum covers 32 centimetres in its first swing, 24 cm on its second swing, 18 cm on its third swing and so on. What is the total distance this pendulum swings before coming to rest?

8. The sum to infinity of a geometric sequence is $\frac{27}{2}$ while the sum of the first three terms is 13. Find the sum of the first 5 terms.

9. Find the sum to infinity of the sequence $1 + \sqrt{3}, 1, \frac{1}{\sqrt{3}+1}, \ldots$

EXERCISES 12.2.5 – MISCELLANEOUS QUESTIONS

1. $2k + 2$, $5k + 1$ and $10k + 2$ are three successive terms of a geometric sequence. Find the value(s) of k.

2. Evaluate $\dfrac{1+2+3+\ldots+10}{1+\frac{1}{2}+\frac{1}{4}+\ldots+\frac{1}{512}}$

3. Find a number which, when added to each of 2, 6 and 13 gives three numbers in geometric sequence.

4. Find the fractional equivalent of

(a) $2.3\dot{8}$ (b) $4.\dot{6}\dot{2}$ (c) 0.41717...

5. Find the sum of all intergers between 200 and 400 that are divisible by 6.

6. Find the sum of the first 50 terms of an arithmetic progression given that the 15th term is 34 and the sum of the first 8 terms is 20.

7. Find the value of p so that $p + 5$, $4p + 3$ and $8p - 2$ will form successive terms of an arithmetic progression.

8. For the series defined by $S_n = 3n^2 - 11n$, find t_n and hence show that the sequence is arithmetic.

9. How many terms of the series $6 + 3 + \frac{3}{2} + \ldots$ must be taken to give a sum of $11\frac{13}{16}$?

10. If $1 + 2x + 4x^2 + \ldots = \frac{3}{4}$, find the value of x.

11. Logs of wood are stacked in a pile so that there are 15 logs on the top row, 16 on the next row, 17 on the next and so on. If there are 246 logs in total,

(a) how many rows are there?

(b) how many logs are there in the bottom row?

12. The lengths of the sides of a right angled triangle form the terms of an arithmetic sequence. If the hypotenuse is 15 cm in length, what is the length of the other two sides?

13. The sum of the first 8 terms of a geometric series is 17 times the sum of its first four terms. Find the common ratio.

14. Three numbers a, b and c whose sum is 15 are successive terms of a G.P, and b, a, c are successive terms of an A.P. Find a, b and c.

15. The sum of the first n terms of an arithmetic series is given by $S_n = \frac{n(3n+1)}{2}$.

(a) Calculate S_1 and S_2.

(b) Find the first three terms of this series.

(c) Find an expression for the nth term.

16. An ant walks along a straight path. After travelling 1 metre it stops, turns through an angle of 90° in an anticlockwise direction and sets off in a straight line covering a distance of half a metre. Again, the ant turns through an angle of 90° in an anticlockwise direction and sets off in a straight line covering a quarter of a metre. The ant continues in this manner indefinitely.

(a) How many turns will the ant have made after covering a distance of $\frac{63}{32}$ metres?

(b) How far will the ant 'eventually' travel?

12.3 COMPOUND INTEREST AND SUPERANNUATION

12.3.1 COMPOUND INTEREST

We have already come across some practical examples of the use of G.Ps in the area of finance. In this section we further develop these ideas and look at the area of compound interest and superannuation.

EXAMPLE 12.28 Find what \$600 amounts to in 20 years if it is invested at 8% p.a. compounding annually.

Solution

End of year 1 value $= \$600 + 8\% \times \$600$

$= \$600(1.08)$

End of year 2 value $= \$600(1.08) + 8\% \times \$600(1.08)$

$= \$600(1.08) + 0.08 \times \$600(1.08)$
$= \$600(1.08)[1 + 0.08]$

$= \$600(1.08)^2$

End of year 3 value $= \$600(1.08)^2 + 8\% \times \$600(1.08)^2$

$= \$600(1.08)^2 + 0.08 \times \$600(1.08)^2$
$= \$600(1.08)^2[1 + 0.08]$

$= \$600(1.08)^3$

↓

End of year 20 value $= \$\,600(1.08)^{20}$

Thus, after 20 years the \$600 amounts to \$2796.57.

Looking closely at the terms of the sequence, they form a G.P:

$$600(1.08), 600(1.08)^2, 600(1.08)^3, \ldots, 600(1.08)^{20}$$

where $a = 600$ and $r = 1.08$.

Developing a formula for compound interest

In general, if \$$P$ is invested at $r\%$ p.a. compound interest, it grows according to the sequence

$$P\left(1+\frac{r}{100}\right), P\left(1+\frac{r}{100}\right)^2, P\left(1+\frac{r}{100}\right)^3, \ldots, P\left(1+\frac{r}{100}\right)^n$$

where $a = P\left(1 + \frac{r}{100}\right)$ and $r = \left(1 + \frac{r}{100}\right)$ so that

$$A_n = P\left(1 + \frac{r}{100}\right)^n$$ where A_n is the amount after n time periods.

12.3.2 SUPERANNUATION*

EXAMPLE 12.29 A woman invests \$1000 at the beginning of each year in a superannuation scheme. If the interest is paid at the rate of 12% p.a. on the investment (compounding annually), how much will her investment be worth after 20 years?

Solution

t_1 = the 1st \$1000 will be invested for 20 years at 12% p.a.

t_2 = the 2nd \$1000 will be invested for 19 years at 12% p.a.

t_3 = the 3rd \$1000 will be invested for 18 years at 12% p.a.

⋮

t_{20} = the 20th \$1000 will be invested for 1 year at 12% p.a.

Finding the amount compounded annually using $A = P\left(1 + \frac{r}{100}\right)^n$, we have:

$$t_1 = 1000\left(1 + \frac{12}{1000}\right)^{20} = 1000(1.12)^{20}$$

$$t_2 = 1000\left(1 + \frac{12}{1000}\right)^{19} = 1000(1.12)^{19}$$

$$t_3 = 1000\left(1 + \frac{12}{1000}\right)^{18} = 1000(1.12)^{18}$$

⋮

$$t_{20} = 1000\left(1 + \frac{12}{1000}\right)^{1} = 1000(1.12)^{1}$$

To find the total of her investment after 20 years, we need to add the separate amounts:

$$\begin{aligned}\text{Total} &= 1000(1.12)^{20} + 1000(1.12)^{19} + 1000(1.12)^{18} + \ldots + 1000(1.12)^{1} \\ &= 1000[(1.12)^{20} + (1.12)^{19} + (1.12)^{18} + \ldots + (1.12)^{1}] \\ &= 1000\left[\frac{1.12(1-(1.12)^{20})}{1-1.12}\right] \qquad \text{Using } S_n = \frac{a(1-r^n)}{1-r} \\ &= \$80,698.74\end{aligned}$$

Thus her total investment amounts to \$80,698.74

---

* This section lies outside the syllabus. It has been included as extension material.

EXAMPLE 12.30 Linda borrows \$2000 at 1% per month reducible interest. If she repays the loan in equal monthly instalments over 4 years, how much is each instalment?

Solution

Amount borrowed = \$ 2000, $r = 1\%$ per month = 0.01 and $n = 4 \times 12 = 48$ months

Let the monthly instalment be = \$ M and the amount owing after n months = \$ A_n

Our aim is to find \$$M$ i.e., the amount of each instalment.

After 1 month (after paying the 1st instalment), we have:

$$A_1 = 2000 + \text{ interest } - M = 2000 + 2000 \times 0.01 - M$$
$$= 2000(1.01) - M$$

After 2 months

$$A_2 = A_1 \times 1.01 - M = [2000(1.01) - M] \times 1.01 - M$$
$$= 2000(1.01)^2 - 1.01 \times M - M$$
$$= 2000(1.01)^2 - M(1.01 + 1)$$

After 3 months

$$A_3 = A_2 \times 1.01 - M = [2000(1.01)^2 - M(1.01 + 1)] \times 1.01 - M$$
$$= 2000(1.01)^3 - M(1.01 + 1) \times 1.01 - M$$
$$= 2000(1.01)^3 - M[1.01^2 + 1.01 + 1]$$

After 4 months

$$A_4 = A_3 \times 1.01 - M = [2000(1.01)^3 - M(1.01^2 + 1.01 + 1)] \times 1.01 - M$$
$$= 2000(1.01)^4 - M(1.01^3 + 1.01^2 + 1.01) - M$$
$$= 2000(1.01)^4 - M[1.01^3 + 1.01^2 + 1.01 + 1]$$

$\vdots \qquad\qquad \vdots$

After n months, we then have

$$A_n = 2000(1.01)^n - M[1 + 1.01 + 1.01^2 + 1.01^3 + \ldots + 1.01^{n-1}]$$

thus, $A_{48} = 2000(1.01)^{48} - M[1 + 1.01 + 1.01^2 + 1.01^3 + \ldots + 1.01^{47}]$.

Now, the loan is repaid after 48 months, meaning that $A_{48} = 0$, therefore, solving for M we have

$$0 = 2000(1.01)^{48} - M[1 + 1.01 + 1.01^2 + 1.01^3 + \ldots + 1.01^{47}]$$
$$\Leftrightarrow 2000(1.01)^{48} = M[1 + 1.01 + 1.01^2 + 1.01^3 + \ldots + 1.01^{47}]$$
$$\Leftrightarrow M = \frac{2000(1.01)^{48}}{[1 + 1.01 + 1.01^2 + 1.01^3 + \ldots + 1.01^{47}]}$$

The denominator is a G.P with $a = 1$, $r = 1.01$ and $n = 48$, so that

$$[1 + 1.01 + 1.01^2 + 1.01^3 + \ldots + 1.01^{47}] = S_{48} = \frac{1(1 - 1.01^{48})}{1 - 1.01} = 61.22261$$

Therefore, $M = \dfrac{2000(1.01)^{48}}{61.22261} = 52.67$

That is, each instalment must be \$52.67.

The total paid = 52.67 × 48 = 2528.16 so that the interest paid = 2528.16 – 2000 = 528.16
That is, she ends up paying $ 528.16 in interest.

Although it is important that you understand the process used in the examples shown, it is also important that you can make use of technology. We now look at how the TI–83 can help us ease the pain of long calculations.

The TI–83 has a number of financial functions which enable computational ease. In particular, it has a **TVM Solver**. The **TVM** Solver displays the time-value-of-money (TVM) variables. In short, given four variable values, the **TVM Solver** solves for the fifth variable. To access the finance screen simply press

2nd [FINANCE] ENTER

Then, it is a matter of entering the 4 known pieces of information and then letting the **TVM Solver** do the rest.

Note: When using the TI–83 financial functions, you must enter cash inflows (cash received) as positive numbers and cash outflows (cash paid) as negative numbers.

Once you have entered your data, there are two ways in which you can then obtain the value of the unknown variable.

Method 1: Place the cursor (using the arrow keys) on the TVM variable for which you want to solve. Press ALPHA [SOLVE]. The answer is computed, displayed in the **TVM Solver** screen and stored to the appropriate TVM variable.

Method 2: You need to leave this window by pressing 2nd QUIT , and return again to the finance menu, by pressing 2nd [FINANCE]. Select the variable you wish to solve for and then press ENTER .

We now illustrate this process using the previous example. In this example we have the known quantities: Linda borrowed $2000, $\therefore PV = -2000$

$I\% = 1$ and $N (= 4 \times 12) = 48$

$FV = 0$ (i.e., loan is fully repaid)

$PMT = ?$ (the monthly repayment required)

That is, once we have entered the information, we then make use of method 1 while at the TVM Solver screen:

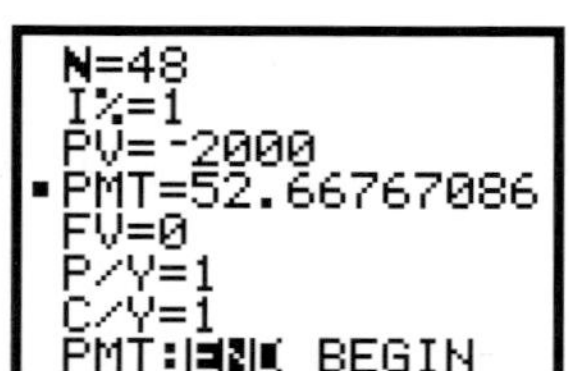

Notice that in the second screen there is a square next to the PMT variable. This is to indicate which variable has just been calculated.

EXERCISES 12.3

1. To how much will $1000 grow to if it is invested at 12% p.a. for 9 years, compounding annually?

2. A bank advertises an annual interest rate of 13.5% p.a. but adds interest to the account monthly, giving a monthly interest rate of 1.125%. Scott deposits $3500 with the bank. How much will he withdraw in 20 months time?

3. To what amount will $900 grow to if it is invested at 10% p.a. for 7 years, compounding every 6 months?

4. A man borrows $5000 at 18% p.a. over a period of 5 years, with the interest compounding every month. Find to the nearest dollar the amount owing after 5 years.

5. Find the total amount required to pay off a loan of $20,000 plus interest at the end of 5 years if the interest is compounded half yearly and the rate is 12%.

6. A man invests $500 at the beginning of each year in a superannuation fund. If the interest is paid at the rate of 12% p.a. on the investment (compounding annually), how much will his investment be worth after 20 years?

7. A woman invests $2000 at the beginning of each year into a superannuation fund for a period of 15 years at a rate of 9% p.a. (compounding annually). Find how much her investment is worth at the end of the 15 years.

8. A man deposits $3 000 annually to accumulate at 9% p.a. compound interest. How much will he have to his credit at the end of 25 years? Compare this to depositing $750 every three months for the same length of time and at the same rate. Which of these two options gives the better return?

9. A woman invests $200 at the beginning of each month into a superannuation scheme for a period of 15 years. Interest is paid at the rate of 7% p.a. and is compounded monthly. How much will her investment be worth at the end of the 15 year period?

10. Peter borrows $5000 at 1.5% per month reducible interest. If he repays the loan in equal monthly instalments over 8 years, how much is each instalment, and what is the total interest charged on the loan? Compare this to taking the same loan, but at a rate of 15% p.a. flat rate.

11. Kevin borrows $7500 to be paid back at 12.5% p.a. monthly reducible over a period of 7 years. What is the amount of each monthly instalment and what is the total interest charged on the loan. Find the equivalent flat rate of interest.

12.4 MISCELLANEOUS EXERCISES

1. The terms of an arithmetic sequence are given by the formula $u_n = 5 + (n-1) \times 3$.

(a) State
 i. the first term and
 ii. the common difference
(b) Find the first 5 terms.
(c) Find the 20th term.
(d) Which term would 302 be?

2. The maximum that a part time worker can expect to earn in a year is £7200. The starting salary is £5100 and it is increased by £75 per year. How long will it take a part time worker to earn the maximum available amount?

3. The first term of an arithmetic sequence is 10 and the sixth term is 70.

(a) Show that the general term is given by $u_n = 12n - 2$.
(b) Find
 i. the first four terms.
 ii. the 20th term.
(c) Find the sum of the first 10 terms.

4. The first four terms of a geometric sequence are 5, 15, 45, 135 .

(a) Find the common ratio.
(b) Find the 10th term.
(c) What is the sum of the first 10 terms?

5. A principal of \$2500 is invested with a return rate of 12% per annum. Assume that no money was ever taken out of this account and that the interest is compounded annually

(a) What will the earnings (due to interest) be by the 7th year since the principal was deposited?

(b) How much will the investor have in the account after 7 years.

6. A small car costing \$12 000 new depreciates at a rate of 15% per year, calculated on its value at the beginning of a year.

(a) What is the annual decay factor (i.e., the common ratio)?
(b) What will its value be in 4 years time?
(c) i. If the value of the car n years after it was first purchased is $\$u_n$, find a formula for u_n.
 ii. How long will it be for the car to be worth just under \$3000?

7. A ball rebounds to 0.85 of the height from which it was dropped. Given that the ball was allowed to fall 8 m before hitting the ground, find

(a) the height to which the ball rebounds after 2 bounces.
(b) how many bounces it takes before it first rebounds to a height of less than 1 m.
(c) the total distance the ball travels after 5 bounces

8. Evaluate: $2 + 6 + 10 + 14, \ldots + 46$

9. If: $3 - 6 + 12 - 24 + \ldots + x = -63$, then x is equal to:

10. Find the number of terms in the sequence $27, -9, 3, -1, \ldots, \frac{1}{27}$.

11. How many odd numbers (beginning with 1) must be added before the total reaches one million?

12. An investment, originally worth \$1250 grows at the rate of 12% per year, compounded annually.

Find:
(a) The value of the investment after 5 years.
(b) The number of years that must pass before the investment is worth more than \$10000.

13. Two separate species of mice live in a national park.

The number of species A was 12000 at the start of 1990 and increased by 200 per month after that date. The number of species B was 8000 at the start of 1990 and increase by 5% per month after that date.

(a) If n is the number of months after the start of 1990 (when $n = 0$), write formulas for the populations of the two species A_n and B_n.

(b) After how many months will the population of species B exceed that of species A?

12.5 GRADED REVISION EXERCISES

LEVEL 1

1. Classify the following sequences as arithmetic, geometric or neither:

i.	2	–1	–4	–7	–10	–13	–16	–19
ii.	$\frac{1}{8}$	$\frac{1}{4}$	$\frac{3}{8}$	$\frac{1}{2}$	$\frac{5}{8}$	$\frac{3}{4}$	$\frac{7}{8}$	1
iii.	4	9	16	25	36	49	64	81
iv.	3	–6	12	–24	48	–96	192	–384

2. Find the missing term in each of the following.
i. $4, 7, 10, x, 16, 19$
ii. $1, -4, -9, -14, x, -24$
iii. $2, 4, x, 8, 10, 12$
iv. $2, 4, 8, x, 32, 64$

3. The general term of an arithmetic sequence is given by $u_n = 8 + (n-1) \times 10$,
find i. u_1 ii. u_5

4. Find the fifteenth term for each of the sequences in question 1.

5. Find the sum of the first ten terms for each of the sequences in question 1.

LEVEL 2

1. How many terms are there in the arithmetic sequence –12,–7, –2, ... , 208?

2. Find the number of terms in the geometric sequence 1, -5, 25, ... ,15625.

3. Find the sum of the geometric series: 2, 6, 18, 54, ... , 3188646

4. \$2500 is invested at 7% annual interest with the interest compounded annually. How many years must pass before there is more than \$10 000 in this account?

5. The first four terms of an A.P are given by a, b, c, d. Express b in terms of a and c.

6. Roy has a monthly salary of \$5000 during his first year of employment which will increase by \$500 after each year of service. Find the total amount Roy will have earnt after 5 years of service.

7. The numbers $126, x, 56$ form 3 consecutive terms of a G.P. Find x.

8. Find the sum of the first 5 terms in the geometric sequence 9, 3, 1, ...

9. The first three terms of a geometric sequence are 3, 6 and 12. The sum of the first N terms is 1533. Find N.

LEVEL 3

1. An arithmetic series has first term 3, 11 terms and a sum of 528. Find:
 (a) the common difference.
 (b) the last term.
 (c) the middle term.

2. A geometric series and an arithmetic series both have first term 3 and 11 terms. The 4th term of the arithmetic series and the 6th term of the geometric are both 96. Find:
 (a) the last term of the arithmetic series.
 (b) the sum of the arithmetic series.
 (c) the common ratio of the geometric series.
 (d) the sum of the geometric series.

3. An arithmetic series has first term –50 and common difference 4. How many terms must be taken before the sum of the series first exceeds 100?

4. The first 4 terms of an A.P are a, b, c, d. Express b in terms of a and d.

LEVEL 4

1. Leo wants to save for a holiday. He needs to save a total of 1000 Euros. In the first week, Leo saves 30 Euros. In the next week he saves 32 Euros and so on in arithmetic progression. For how many weeks will Leo need to save?

2. A colony of rodents initially consists of 2700 animals. The population grows by 7% per month. How many months must pass before the population first passes ten thousand?

3. A car costs \$32 000 new. After 10 years its value has fallen to \$6 300. If it depreciates at d% per annum, find the value of d correct to the nearest one percent.

4. A radioactive source decays at the rate of 1% of its activity per day. On the first day, a source registered 5 000 disintegrations.
 (a) How many disintegrations could be expected on the second day?
 (b) On which day do the number of disintegrations first fall below 4 000?
 (c) How many disintegrations in total could be expected to be recorded if this source is observed for 30 days?

5. Carlo saves \$100 at the beginning of every month for a period of 50 months. The money is saved in an account that pays 6% annual interest calculated monthly.
 (a) Prove that the first amount of \$100 deposited becomes $\$100 \times 1.005^{50}$ by the end of Carlo's saving period (the end of the 50th month).
 (b) Prove that the second amount deposited becomes $\$100 \times 1.005^{49}$ by the end of Carlo's saving period.
 (c) Express the amounts saved and the interest received as a series and hence find the total amount Carlo has saved at the end of the 50 months.

12.6 TOPIC TEST

1. Find the number of terms in the sequence –12, –7, –2, 3, ..., 123.

[2 marks]

2. Find the sum of the geometric series –5 + 10 – 20 + ... 640.

[3 marks]

3. A car costs $30,000 when new and is expected to lose 6.5% of its value each year.

(a) How much is the car worth after 2 years?

(b) Complete the table below

Years since purchase (t)	1	2	3	4
Value of car ($ V)				

(c) Find an expression for V in terms of t.

[6 marks]

4. Chocolates are made in the shape of a sphere. They are to be marketed in triangluar boxes that contain one layer. The diagram shows a box with three rows of chocolates.

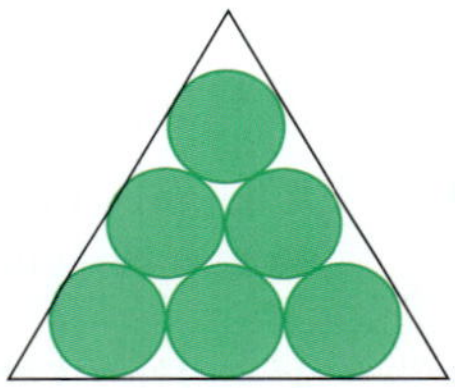

(a) Find the number of chocolates in a box with four rows of chocolates.

(b) Find a formula for the number of chocolates in a box with n rows.

(c) Find the number of rows needed if the box is to contain 55 chocolates.

[5 marks]

5. A newly formed production company expects to manufacture 12 000 ratchets in their first year of operation. This is expected to grow by 8% annually.

(a) Find the number of ratchets manufactured in the fifth year of operation.

(b) Find the total number of ratchets manufactured by the end of the fifth year of operation.

(c) Find the first year in which production will be more than twice that in the first year.

(d) Find the year in which the company can expect to celebrate the manufacture of its 'millionth ratchet'.

[6 marks]
Total 22 marks

12.7 PROJECT SUGGESTIONS

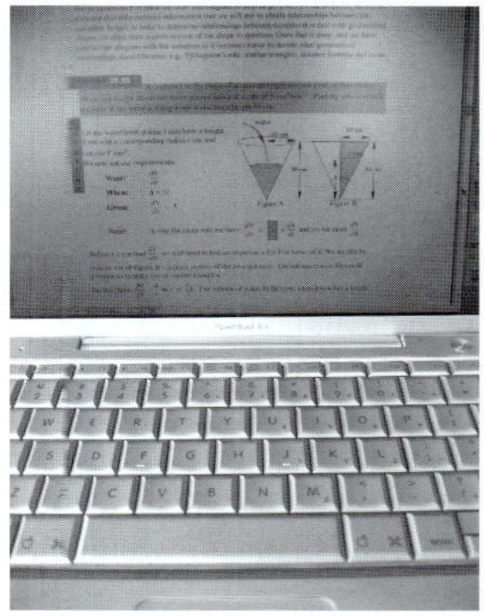

This computer depreciates. It loses 20% of its value each year. What sequence is implied here?

How do businesses handle the depreciation of their equipment?

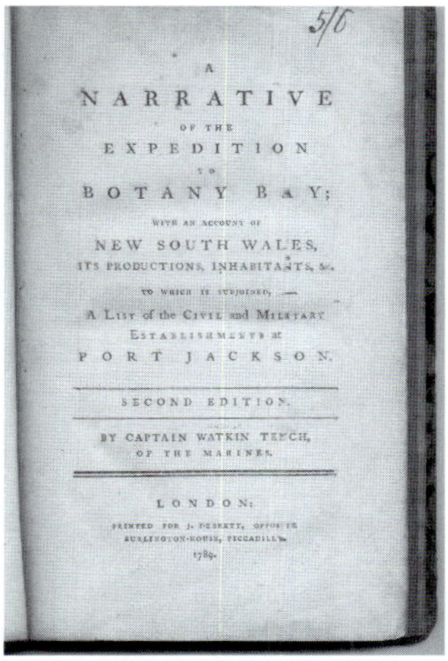
A NARRATIVE OF THE EXPEDITION TO BOTANY BAY; WITH AN ACCOUNT OF NEW SOUTH WALES, ITS PRODUCTIONS, INHABITANTS, &c. TO WHICH IS SUBJOINED, A LIST of the CIVIL and MILITARY ESTABLISHMENTS at PORT JACKSON.

SECOND EDITION.

BY CAPTAIN WATKIN TENCH, OF THE MARINES.

LONDON: PRINTED FOR J. DEBRETT, OPPOSITE BURLINGTON-HOUSE, PICCADILLY. 1789.

Not all goods decrease in value over time.

This is the title page of an old book (published in 1789).

It appears that it was once sold for 5/6. What does this mean?

Not only do antiques increase in value, the value of the money against which their values are measured decreases (due to inflation).
Investigating these two effects could make a good project.

The following photographs show some situations that could lead to a project based on arithmetic sequences. Try to find one of your own!

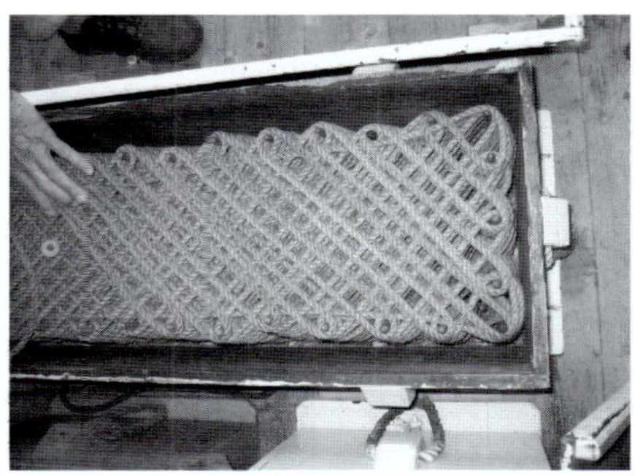

This is a photograph of a 'rope box' that used to be used in rescuing the victims of shipwrecks. The rope is wound in this way so that its loose end could be attached to a rocket. The rocket was fired with the hope that it would carry the end of the rope to a ship in distress. This could then be used to haul more substantial cables to link the ship to the shore with a device known as a 'bosun's chair. The folding pattern of the rocket rope is designed so that it will not tangle as the rocket tows it to the ship.

What number patterns can you see here and are any of them arithmetic?

This photograph shows 'geological strata' (layers).
If you count the layers from the surface, what does this tell you about the age of the stratum?

This city square in Santiago, Chile shows both classical and modern buildings. Architecture can offer many examples of the use of mathematics in a practical context. Look at the modern building on the right.
How would you calculate the number of panes of glass used in cladding buildings of this design?

If you prefer the more classical building to the right of the photograph, what mathematics is evident here?

This diver carries a fixed amount of compressed air in the cylinder on his back. The length of time that the air lasts depends on a number of factors (experience, fitness etc.). These factors are different for each diver. After that, the major factor that determines the time that the dive can last is its depth. If the diver descends to 10 metres below the surface, he will be subjected to twice the pressure at the surface and his air will last half as long. At 20 metres, the pressure is three times that at the surface.

What sequences are implied here?

The other factor that determines the time that the diver can stay underwater is decompression sickness or 'the bends'.
What factors affect this?

13.0 RUSSELL'S PARADOX

Bertrand Russell born May 18, 1872, died Feb. 2, 1970, acquired the impressive name of Bertrand Arthur William Russell, 3rd Earl Russell of Kingston Russell, Viscount Amberley of Amberley and of Ardsalla during his long and full life.

Russell was a philosopher and mathematician as well as being an outspoken social campaigner. He was a prolific author and was awarded the Nobel Prize for Literature in 1950.

Most of his work is complex and deals with the foundations of mathematics. He did, however, pose a comparatively simple question that has a very complex answer.

Russell's 'Barber Paradox' goes like this:

There is a small town with one barber who claims:

"I shave every man in the town who does not shave himself".

On the face of it there is nothing wrong with this claim until one asks the question,

"Who shaves the barber?"

His claim is wrong whether or not he shaves himself. Can you see why? Try to avoid answers such as "The barber is bald, a woman etc." as these do not actually solve the problem!

The answer to this 'paradox' (a self-contradictory statement that at first seems to be true) is that the barber cannot make this claim.

In mathematical term this translates to

"Do not talk about the set of all sets that do not contain themselves".

The conclusion is that this means we need to be careful when talking about sets of things! That is the subject of this chapter.

13.1 DEFINITIONS

13.1.1 SETS

The concept of number sets builds many of the structures that are in mathematics today. Although at times it appears overwhelming, it serves to provide precise definitions that can be used in many branches of mathematics. The notation and definitions are transportable to all these areas. However, because of its precise language and notation, we need to provide some background language and theory so that we can accurately use the concept of sets in this and future studies (in particular, the work in Chapter 15).

A **set** is any well–defined list or collection of objects.

These objects can be anything: letters, shapes, cars, people, numbers etc. We call these objects the **elements** or **members** of the set.

Examples of sets are

1. The letters; a, e, i, o, u (The set of vowels)
2. The numbers 2, 3, 5, 7 (The set of the first four prime numbers)
3. The people in your class
4. The numbers 2, 4, 6, 8, .. (The set of even numbers)

13.1.2 NOTATION

We use **braces** (curly brackets) '{ . . . }' to denote '**The set of . . .** '.

These are usually labelled by capital letters, A, B, X, Y, ...

So that , **A = {a, e, i, o, u}** would be read as '**The set of vowels.**'

This type of set is known as the **Tabular form** of a set. This is because the **contents** of set A can clearly be displayed.

Another form that can be used to display the contents of a set is that of a **Venn–Euler** diagram – in this case we list the elements of the set in a labelled region as shown alongside.

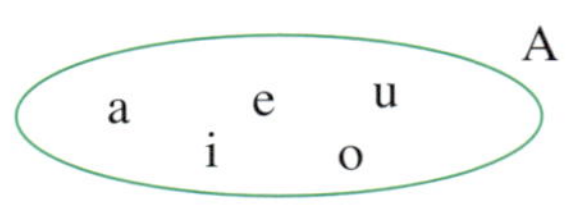

The following are examples of sets, expressed in the correct format:

Example 1 A = {Jack, Alison, Ismael, Baljeet, Yoshi}

Example 2 B = {Brisbane, Sydney, Melbourne, Canberra, Darwin, Adelaide, Perth}

Example 3 C = {Colombia, Peru, Ecuador, Brazil, Chile, Argentina}

Example 4 D = {Clinton, Reagan, Ford, Kennedy}

Example 5 E = {2, 3, 9, 14, 20}

Each item, name, object or number in the set is called an **element** of the set.

As we have seen, a Venn diagram is extremely useful when trying to visualize the elements of a set. In the example used, we say that a, e, i, o and u are elements of the set A.

The symbol for **'is an element of'** is '$\in$ '.

So that, if A = {a, e, i, o, u} then 'a $\in$ A' is read as 'a *is an element of* A'.

Similarly; 'e $\in$ A' is read as " e is an element of A".

We can also write e $\in$ {a, e, i, o, u}. **However**, we **cannot** write {e} $\in$ {a, e, i, o, u}. Nor can we write {e} $\in$ A. That is, you cannot have a set on both sides of the ' $\in$ ' symbol.

That is

The expressions	'object $\in$ { }' and 'object $\in$ A ' are **correct**.
The expression	{ } $\in$ { } is **not correct**.

Using the examples on the previous page we have

Example 1 Jack $\in$ A i.e., Jack is an element of set A

Example 2 Sydney $\in$ B i.e., Sydney is an element of set B

Example 3 Colombia $\in$ C i.e., Colombia is an element of set C

Example 4 Ford $\in$ D i.e., Ford is an element of set D

Example 5 9 $\in$ E i.e., 9 is an element of set E

As the letter b does not belong to the set A = {a, e, i, o, u}, we write
'b $\notin$ A' read as 'b is *not an element of* A'.

The symbol ' $\notin$ ' is read as 'is **not an element** of'.

Using some of our previous sets we have

Example 1 Tom $\notin$ A i.e., Tom is not an element of set A

Example 2 Ballarat $\notin$ B i.e., Ballarat is not an element of set B

Example 3 Mexico $\notin$ C i.e., Mexico is not an element of set C

Example 4 Thatcher $\notin$ D i.e., Thatcher is not an element of set D

Often it is not possible to display all of the elements that belong in a set using the tabular form.

For example, the set E = {2, 4, 6, ...} indicates that we are looking at the set of even numbers, even though it has only listed the first three elements.

Such sets are usually written as $\{x \mid x \text{ is even}\}$ – this is known as **set–builder notation**. This is read as 'E is equal to the set of values x, such that x is even'. The x is known as a *dummy variable* (simply because we could have written $\{z \mid z \text{ is even}\}$ or $\{y \mid y \text{ is even}\}$ without changing the meaning).

The 'bar' symbols ' | ' is read as **'such that'**.

This form can be divided into three parts: { part 1 part 2 part 3 }

For example, the set $F = \{x \mid x \text{ is odd}\}$ is broken up as follows:

Part 1: In the first part a **variable is identified**, in this case the variable is x.

Part 2: In the second part the symbol '|' which means '**such that**' is used to introduce the conditions that applies to the variable.

Part 3: The final part is the **conditions** that the variable(s) follow. In this case there is only one condition, that x is odd.
When more than one condition is used they are separated by commas.

The following examples illustrate the three parts involved when dealing with set builder notation.

Example 1 $G = \{x \mid x \in \mathbb{N}, 4 < x < 12\}$
The set G equals x such that x is an element of $\mathbb{N}$, and x is between 4 and 12.
G is the set of natural numbers between 4 and 12. G = {5, 6, 7, 8, 9, 10, 11}

Example 2 $H = \{y \mid y \in \mathbb{Z}, y < 3\}$
The set H equals y such that, y is an element of $\mathbb{Z}$ and y is less than 3. Or simply H is the set of integers less than 3.

Example 3 $M = \{x \mid x \text{ is a capital city in Australia}\}$
In this instance, the set builder notation is used to describe a set containing names of places.

There also exists another commonly used notation for 'such that', and that is the colon ':'. So a set such as $B = \{x : x \text{ is odd}\}$ (and by odd, we mean 1, 3, 5, . . . not strange!) would read as 'B is equal to the set of values x, such that x is odd'

EXAMPLE 13.1

Consider the sets A = {2, 3, 5, 7} and B = {2, 4, 6, 8}.

(a) Which of the following are true?
i. $3 \in A$ ii. $4 \in A$ iii. $\{5\} \in A$

(b) Write the sets A and B in set builder notation.

Solution

(a) i. As 3 does belong to the set A, then '$3 \in A$' is true.

ii. As 4 does not belong to the set A, then '$4 \in A$' is false.

iii. Now, although 5 does belong to the set A, that is '$5 \in A$' is true, the statement $\{5\} \in A$ is not true (as the set A does not contain the element $\{5\}$).

(b) $A = \{x \mid x$ is a prime number less than or equal to seven$\}$
$B = \{y \mid y$ is an even number less than nine$\}$

EXAMPLE 13.2

List the elements of the following sets

(a) $A = \{b|2b+1 = 3\}$ (b) $X = \{x|(x-1)(x+2) = 0\}$

Solution

(a) The elements of set A must satisfy the condition that $2b+1 = 3$. The only value of b that satisfies this condition is 1.
Therefore the list consists of one element, namely, $\{1\}$.

(b) The elements of set X that satisfy the condition that $(x-1)(x+2) = 0$ are 1 and –2.
Check: when $x = 1$, L.H.S $= (1-1)(1+2) = 0 \times 3 = 0$
when $x = -2$, L.H.S $= (2-1)(-2+2) = 1 \times 0 = 0$
Therefore the set is $\{1, -2\}$.

Now consider the set $A = \{x|x^2 = -4$, where x is an integer$\}$. Can we list the elements of A? In this case we want integer values such that when they are squared produce the answer –4. There are no such integer values that satisfy this condition and so there is nothing to list. In such cases, where no elements exist we write the braces with nothing in them.
That is, A = { }. This is known as the **empty set**.

The **empty set** = { }, also called the **null set** is a set with no elements and is written as $\varnothing$.

EXAMPLE 13.3

List the elements of the following sets

(a) A = {The days of the week that end in the letter "m" when spelt in English}.
(b) B = {The natural numbers less than zero}.
(c) $C = \{x|x+9 > 12$ and $x^2 < 9$ and x is an integer$\}$

Solution

(a) Since no day of the week ends in the letter "m" when spelt in English, A = { } = $\varnothing$

(b) Since there are no natural numbers less than zero, B = { } or $\varnothing$

(c) From the first condition, $x+9 > 12 \Rightarrow x > 3$. The second condition implies that $x = 1, 2$.
And so there are no possible values of x satisfying both conditions. That is $C = \varnothing$.

EXERCISES 13.1

1. Write the following in words.

(a) hammer $\in$ A (b) axe $\in$ C (c) Tuesday $\in$ B

(d) Tuesday $\notin$ A (e) January $\in$ B (f) Sunday $\notin$ C

2. Write the following using set notation.

(a) 32 is an element of set C (b) 45 is an element of set N

(c) green is not an element of set K (d) Mary is not an element of set P

(e) Horse is not an element of set M (f) Banana is an element of set H

3. Which of the following statements are true?

(a) If $A = \{2, 4, 6, 8, 10\}$, then i. $2 \in A$

ii. $\{4, 6\} \in A$

(b) If $B = \{y | y \text{ is an odd number}\}$ i. $5 \in B$

ii. $2k + 1 \in B$, k is a positive integer

4. Write in set builder notation the following sets.

(a) A, the set of even numbers less than 10.

(b) B, the set of odd perfect squares between 40 and 169.

(c) C, the set of positive integers less than 12.

(d) D, the set of prime numbers less than 25.

(e) E, the set of factors of 28.

5. List the elements of the set

(a) $A = \{x | 1 < x < 40, \text{ where } x \text{ is a perfect square}\}$.

(b) $B = \{(x, y) | 2x + 3y = 20, \text{ where } x \text{ and } y \text{ are positive integers}\}$.

(c) $C = \{n | n \text{ is an even number and } n^2 \text{ is an odd number}\}$

6. Given that $X = \{2, 4, 5, 7, 8, 10\}$, which of the following statements are correct?

(a) $5 \in X$ (b) $6 \notin X$ (c) $\{2, 4\} \in X$

7. Which of the following are empty sets? For those that are not, list the elements.

(a) $\{x | x \text{ is an odd integer and } x^2 < 25\}$.

(b) $\{x | x \text{ is an even integer and } x^2 < 25\}$.

(c) $\left\{x | \frac{x}{6} = 15\frac{1}{3}, \text{ where } x \text{ is an integer}\right\}$

(d) $\{x | x^2 + 12 = 7x \text{ where } x \text{ is an integer}\}$

(e) $\{x | x^2 + 9 = 4x \text{ where } x \text{ is an integer}\}$

8. Given that $X = \{1, \{2, 3\}, 4, 5, 7, \{8\}, 10\}$, which of the following statements are correct?

(a) $1 \in X$ (b) $2 \in X$ (c) $\{2, 3\} \in X$ (d) $\{2\} \in X$

(e) $8 \in X$ (f) $\{5\} \in X$

13.2 SETS OF NUMBERS

13.2.1 NUMBER SYSTEMS

The set of real numbers can be broken down into two subsets, namely, the set of **rational numbers** and the set of **irrational numbers**. The set of rational numbers can itself be broken down into two sets, the set of **integers** and the set of fractions. The set of integers can then be broken down into the set of positive integers, the set of negative integers and the set that includes the number zero. Each of these sets can be represented by a different symbol.

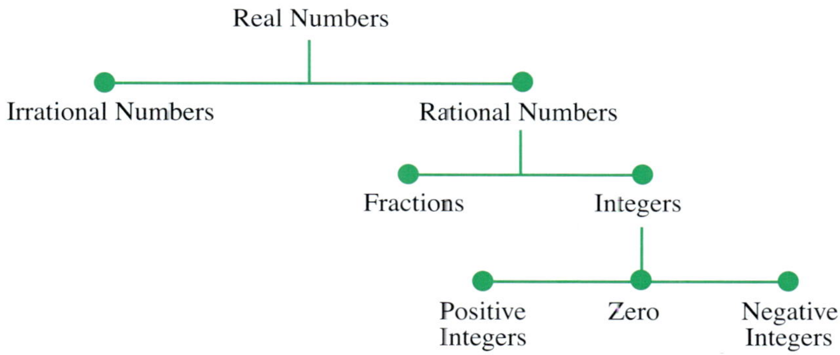

In this book we will use the following notation and definitions:

Set of **positive integers and zero** $= \mathbb{N} = \{0, 1, 2, 3, \ldots\}$. [Also known as counting numbers]

The set of **integers** $= \mathbb{Z} = \{0, \pm 1, \pm 2, \pm 3, \ldots\}$

The set of **positive integers** $= \mathbb{Z}^+ = \{1, 2, 3, \ldots\}$ [Also known as Natural numbers]

The set of **rational numbers** $= \mathbb{Q}$

Definition: $\mathbb{Q} = \left\{ x \mid x = \frac{a}{b}, b \neq 0 \text{ and } a \text{ and } b \text{ are integers} \right\}$

That is, numbers that can be expressed as a fraction with both numerator and denominator being integers, where the denominator is not zero.

Notice then that an integer is also a rational number. For example, the number 2 can be expressed as a fraction, e.g., $2 = \frac{6}{3}$ (which satisfies the conditions above).

The set of positive rational numbers $= \mathbb{Q}^+$

The set of positive real numbers $= \mathbb{R}^+ = \{x \mid x \in \mathbb{R}, x > 0\}$

Notice that the set of integers, $\mathbb{Z}$ can be broken into 3 distinct parts.

The set of positive integers $\mathbb{Z}^+ = \{1, 2, 3 \ldots\}$
The set containing Zero $\{0\}$
The set of negative integers $\mathbb{Z}^- = \{\ldots -3, -2, -1\}$

We have already seen that because integers can be expressed as fractions, they are also rational numbers. Similarly, **terminating decimals are rational numbers** because they can be expressed as the ratio of two integers.

For example, we have $0.2 = \frac{2}{10}$, $0.345 = \frac{345}{1000} = \frac{69}{200}$ and $2.87 = \frac{287}{100}$

Recurring decimals are also rational numbers because they can be expressed as the ratio of two integers.

Convert recurring decimals to fractions is not required at this level. However the process has been shown in the first example for interest's sake. It should also be noted that a treatment of recurring decimals is given in Chapter 13.)

EXAMPLE 13.4

Show that 0.235235 . . . is a rational number.

Solution

Let $x = 0.235235\ldots$

There are three recurring digits, so multiply both sides by 1000

That is, $1000x = 235.235235\ldots$
$= 235 + 0.235235\ldots$

However, the term $0.235235\ldots$ is actually x (again).
Therefore we can write

$$1000x = 235 + x$$

Solving for x, we now have

$$\Rightarrow 1000x - x = 235.235235\ldots$$

$$\Rightarrow 999x = 235$$

$$\therefore x = \frac{235}{999}$$

Check this result on your calculator.

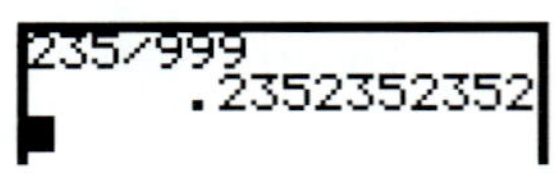

Notice then that we could also have used the conversion into fraction capabilities of the graphics calculator. To do this,

1. type in the recurring decimal
2. call up the MATHS menu
3. select Frac (option 1:)
4. press enter twice

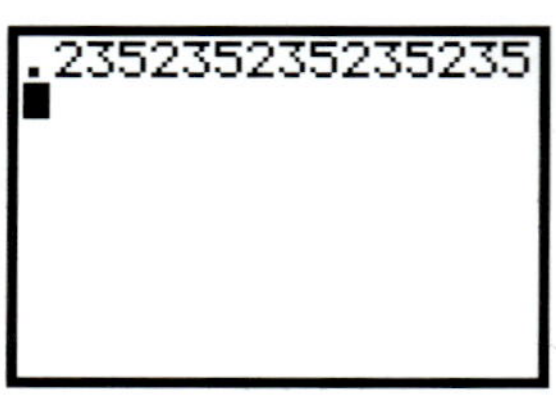

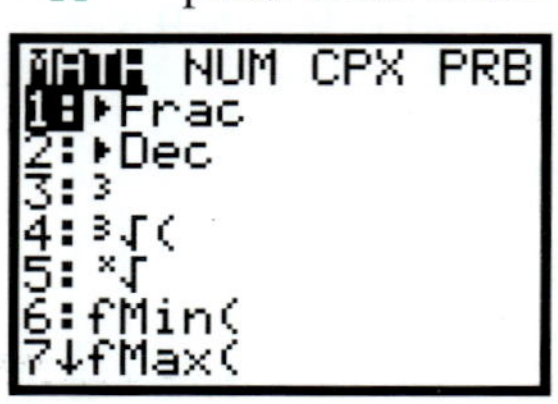

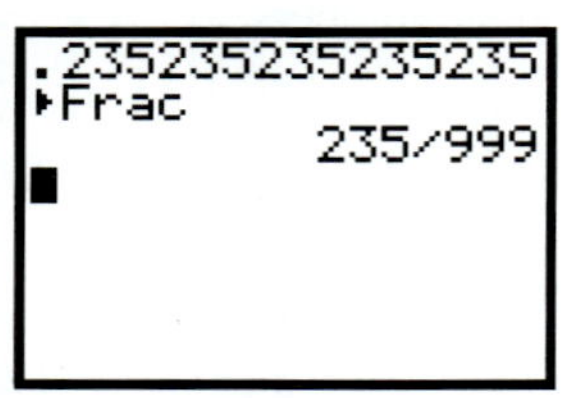

In the same way, we can convert the following into rational numbers:

$$0.3434\ldots = \frac{34}{99}$$

$$0.33\ldots = \frac{33}{99} = \frac{1}{3}$$

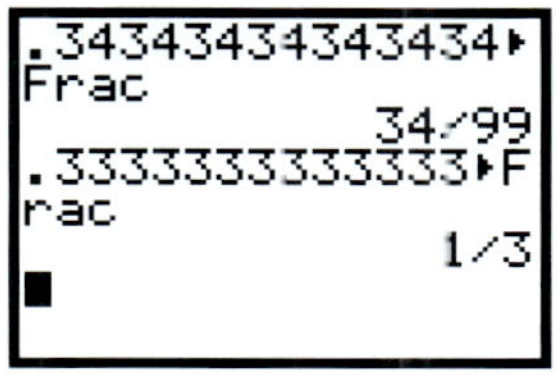

We saw that real numbers are made up of rational and non-rational numbers (or irrational numbers). An **irrational number** is a number that cannot be expressed as the ratio of two integers. Because of this, non terminating, non recurring decimals are irrational. For example, the square root of numbers that are not square numbers are irrational, so that $\sqrt{2}$ is an irrational number. In this case we would write $\sqrt{2} \in Q'$. Other examples of irrational numbers are π and $\sqrt{3}$ and 2.3256487456239088 . . . and so on.

Last, but not least, we come to the set of real numbers. The set of **real numbers** encompasses all of the sets of numbers discussed so far. A visual method to represent the set of real numbers, $\mathbb{R}$, is to use a straight line. This geometrical representation is known as the **real number line**.

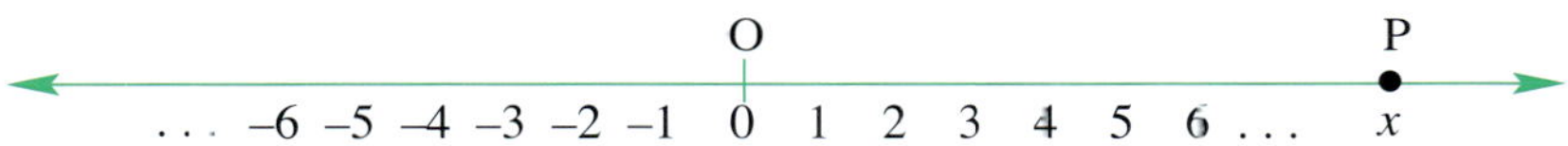

In set buider notation we define the set of real numbers as $\mathbb{R} = \{x : -\infty < x < \infty\}$

EXERCISES 13.2

1. State whether the following are true or false.

(a) $-3 \in \mathbb{N}$ (b) $\frac{4}{5} \in Q$ (c) $0.3333\ldots \in Q$ (d) $\sqrt{2} \in Q$

(e) $-4.7 \in \mathbb{Z}$ (f) $2\frac{1}{4} \in \mathbb{R}$ (g) $\pi \in \mathbb{R}$ (h) $\pi \in Q$

(i) $\sqrt{3} \in Q$ (j) $0 \in \mathbb{Z}^+$ (k) $0.3456723\ldots \in Q$

2. Express the recurring decimals as fractions (without using a graphics calculator).

(a) 0.454545 . . .
(b) 3.191919 . . .
(c) 5.3424242 . . .
(d) 12.78939393 . . .

3. (a) Prove that if $x \in \{x | 2x = 8 - 3x\}$ then $x \in Q$.

(b) List the elements of $\{(x, y) | x + y = 5, x \in \mathbb{N}, y \in \mathbb{N}\}$.

(c) List the elements of $\{(x, y) | x^2 + y^2 \leq 36, x \in \mathbb{N}, y \in \mathbb{N}\}$.

(d) Find x if $x \in A$ and $x \in B$ where $A = \{x | 2 < x \leq 6\}$ and $B = \{x | x > 3\}$

13.3 CARDINAL NUMBER OF A SET

13.3.1 CARDINAL NUMBER OF A SET

The **cardinal number** of a set A, $n(A)$, is the **number of elements** in that set.

$n(A)$ and is read as 'the number of elements in set A'.

Example 1 If $A = \{\text{apple, pear, orange, pineapple, banana}\}$ then $n(A) = 5$.

Example 2 If $B = \{1, 2, 4, 6, 8, 10\}$ then $n(B) = 6$.

EXAMPLE 13.5

Find the number of elements in each of the following sets.

(a) $A = \{2,3,5,7\}$.

(b) $X = \{x|(x-1)(x+2) = 0\}$.

(c) $C = \{x|x^2 + 9 = 0, x \in \mathbb{R}\}$.

Solution

(a) As the set A is in tabular form, we can count the number of elements in the set. In this instance, $n(A) = 4$.

(b) This time we need to first determine the values of x that satisfy the equation. Here we have that $(x-1)(x+2) = 0 \therefore x = 1$ or $x = -2$. Then, as there are only two answers, $n(X) = 2$.

(c) As it is impossible to solve the equation $x^2 + 9 = 0$ when x is a real number, then there are no values of x satisfying the equation $x^2 + 9 = 0$. Therefore, $n(C) = 0$.

13.3.2 FINITE AND INFINITE SETS

A **finite** set has a **countable** number of elements.

Examples of finite sets are:

C = {The people in Austria}
D = {The grains of sand on the beach}
E = {The fish in the sea}

An **infinite** set has an infinite number of elements.

Examples of finite sets are:

$\mathbb{N}$ = {The natural numbers}
$\mathbb{Z}$ = {The integers}
$\mathbb{R}$ = {The real numbers between 1 and 2}

Technically we can have countable and uncountable infinte sets. That is, it is possible to use the natural numbers to count the set of even numbers (an infinite set); 2 is the 1st even number, 4 is the 2nd, 6 the 3rd etc. The integers are also 'countable' in this sense, though the real numbers are not. The proofs of these statements are beyond the scope of this course.

We summarise three important sets that have been discussed so far.

Type of set	Definition	Example
Null set	A set that has no elements. It is denoted by the symbol ∅ or { } (that is, braces with nothing in them)	$\{x \mid x^2 + 1 = 0, x \in \mathbb{R}\}$ {people who are 4 m tall}
Finite set	A set whose elements can be counted.	The set {1, 2, 4, 5, 10, 22} contains 5 elements.
Infinite set	A set for which it is impossible to list the elements.	The set $\{y \mid y$ is a real number$\}$ cannot be counted, as we cannot write a list of all its elements.

Note: We do not use '0' to represent the null set. The reason is that the symbol ∅ **is** in fact something. It is the set that has nothing in it. An analogy is, an empty classroom is still a classroom!

EXERCISES 13.3

1. Given the following sets
A = {hammer, axe, mower, nail}
B = {Months of the year}
C = {Days of the week}
D = {0, 1, 2, 3, 4, 5, 6, 7, 8, 9, 10}

State whether the following are true or false.

(a) hammer ∈ A (b) axe ∈ C (c) Tuesday ∈ B
(d) Tuesday ∉ A (e) January ∈ B (f) Sunday ∉ C
(g) 4.5 ∉ D (h) 3 ∈ D (i) 10 ∈ D

2. For the sets described in Q.1., evaluate

(a) $n(A)$ (b) $n(B)$ (c) $n(C)$ (d) $n(D)$

3. State whether each of the following is a finite or infinite set.

(a) A = {Letters of the English alphabet} (b) B = {Natural Numbers}
(c) C = {Capital cities in Europe} (d) D = {Planets in our solar system}
(e) E = {Fractions between 4 and 10} (f) F = {people in China}

4. List the elements in the following sets, if possible.

(a) A = {Letters that come before "a" in the English alphabet}.
(b) B = {Vowels in the word "zebra"}

(c) C = {Days of the week}
(d) $D = \{x \mid x \in \mathbb{N}, 4 \le x \le 5\}$
(e) $E = \{y \mid y \in \mathbb{N}, -4 \le y \le 5\}$
(f) $F = \{m \mid m \in \mathbb{Z}, -4 \le m \le 5\}$
(g) $G = \{x \mid x \in \mathbb{Q}, -4 \le x \le 5\}$
(h) $H = \{x \mid x \in \mathbb{R}, 4 \le x \le 5\}$

5. Evaluate the following for each set in Q.4.

(a) $n(A)$ (b) $n(B)$ (c) $n(C)$ (d) $n(D)$
(e) $n(E)$ (f) $n(F)$ (g) $n(G)$ (h) $n(H)$

6. Which of the sets listed above are infinite?

7. Express the following using set builder notation.

(a) {The natural numbers between 3 and 18}
(b) {Real numbers less than 12}
(c) {Integers from –56 to 45}
(d) {Rational numbers greater than or equal to –5 but less than or equal to 5}

13.4 THE ALGEBRA OF SETS

13.4.1 EQUALITY OF SETS

Set A is said to be **equal** to set B if both sets are identical, that is, they contain the same elements. We write this as A = B.

EXAMPLE 13.6

Which of the following are identical sets?

(a) $A = \{1, 9, 12\}$, $B = \{9, 1, 12\}$ and $C = \{3, 5, 6\}$.
(b) $A = \{x \mid (x-2)^2 = 0\}$, $B = \{2\}$ and $C = \{1\}$.
(c) $A = \{1, 2\}$ and $B = \{1, 1, 2, 2\}$.

Solution

(a) As every element in A is in B and vice–versa, we can write that A = B.
However, the elements in set C differ from those in both sets A and B and so set C does not equal either set A or set B.

(b) As the elements in B and C are not the same, then we can write $B \neq C$.
Next, we need to determine the elements in A. To do this we need to solve the equation $(x-2)^2 = 0$. The answer is $x = 2$.
So, in fact, set A could be written as $A = \{x \mid (x-2)^2 = 0\}$ or simply $\{2\}$.
We can now say that A = B (and that $A \neq C$).

(c) Every element in A is found in B.
Now, consider each element in B, is that element also in A?
Sorting through every element in B we find that each element in B is also in A.

So, by our definition, A must be equal to B, that is, $\{1, 2\} = \{1, 1, 2, 2\}$.

There are some important points to be made at this stage.

1. **Notice** that it makes no difference as to how the elements are ordered, as long as every element in one set is found in the other and vice-versa – see part (a).

2. In part (c), because A = B, then $n(B) = n(A) = 2$, **not 4**! Even though $B = \{1, 1, 2, 2\}$.

EXAMPLE 13.7

List all the elements of $C = \{x | x^2 - 4x + 4 = 0\}$

Solution

In this case we start by solving the quadratic:

$$x^2 - 4x + 4 = 0 \Rightarrow (x-2)(x-2) = 0$$
$$\therefore x = 2 \text{ or } x = 2$$

And so we write $C = \{x | x^2 - 4x + 4 = 0\} = \{2, 2\}$
$= \{2\}$

Because when we sketch the graph of $y = x^2 - 4x + 4$ it touches the x–axis at only one point, $x = 2$.
This also means that $n(C) = 1$.

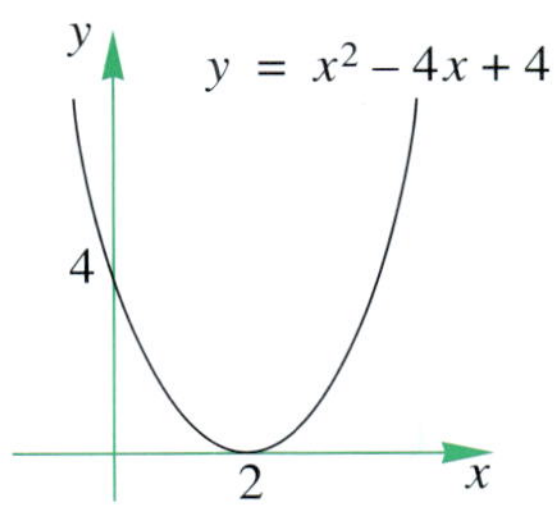

13.4.2 EQUIVALENT SETS

Set A is said to be **equivalent** to set B if $n(A) = n(B)$.

That is, if the sets A and B have the **same number** of elements, then we have that $n(A) = n(B)$ – regardless of what the elements are. In this case we write $A \leftrightarrow B$.

For example, if $A = \{10, 20, 30, 40, 50\}$ and $B = \{a, b, c, d, e\}$ then, because $n(A) = 5$ and $n(B) = 5$, the sets A and B are equivalent. That is $\{10, 20, 30, 40, 50\} \leftrightarrow \{a, b, c, d, e\}$, or $A \leftrightarrow B$.

The reason is that we can create a one to one correspondence between the elements in both sets:

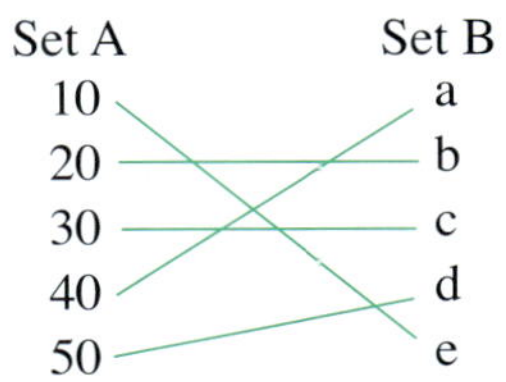

The order in which the elements are matched is not important, simply that every element in both sets has been accounted for (once only).

13.4.3 SUBSETS

A set B is a **subset** of set C if all the elements in set B are in set C.
We write this as $B \subseteq C$, the symbol $\subseteq$ means : 'is a subset of'.

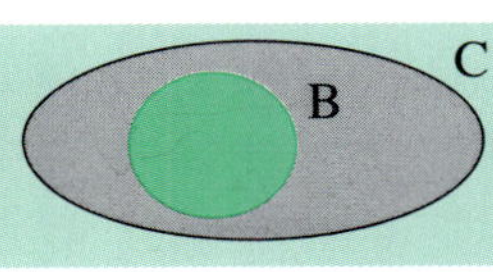

If, on the otherhand, none of the elements in set B are in set C, then set B is not a subset of C, we then write this as $B \not\subseteq C$. In summary we have:

$\subseteq$ denotes is a subset of

$\not\subseteq$ denotes is not a subset of

13.4.4 SPECIAL SUBSETS

1. The null set, $\varnothing$, is a subset of every set.
2. Any set is a subset of itself.

EXAMPLE 13.8 List all the possible subsets of A = {apple, pear, orange}.

Solution

The subsets of A are:

{} [*The empty set*].
{apple}
{pear}
{orange}
{apple, pear}
{apple, orange}
{pear, orange}
{apple, pear, orange} [*Every set is a subset of itself*].

That is, there are 8 subsets.

A note about using the symbols $\in$ and $\subseteq$. The table below provides a summary of the syntax for the use of the symbols $\in$ and $\subseteq$:

Symbol	Correct usage	Incorrect usage
$\in$	$2 \in \{1, 2, 4\}$	$\{2\} \in \{1, 2, 4\}$
$\subseteq$	$\{2\} \subseteq \{1, 2, 4\}$	$2 \subseteq \{1, 2, 4\}$

13.4.5 PROPER SUBSETS

A set B is said to be a **proper subset** of a set C if all the elements of set B are in set C, but the two sets are not the same.

$\subset$ denotes is a proper subset of
$\not\subset$ denotes is not a proper subset of

In Example 14.8, all the subsets listed are proper subsets except for {apple, pear, orange}, since this subset contains the same elements as set A it is not a proper subset.

13.4.6 THE NUMBER OF SUBSETS

Consider the following 4 sets, {a}, {a, b}, {a, b, c} and {a, b, c, d}. We can provide a list of all the possible subsets by being systematic:

Set	Subsets	Number of subsets
{α}	{α}, { }	2
{α, β}	{ },{α}, {β}, {α, β}	4
{α, β, γ}	{ }, {α}, {β}, {γ}, {α, β}, {α, γ}, {β, γ}, {α, β, γ}	8
{α, β, γ, δ}	{ }, {α}, {β}, {γ}, {δ}, {α, β}, {α, γ}, {α, δ},{β, γ}, {β, δ}, {γ, δ}, {α, β, γ}, {α, β, δ}, {α, γ, δ}{β, γ, δ}, {α, β, δ, γ}	16

Realise that for the null set (where there is no element) there still is a subset, namely, itself.

We can summarise the results in table form to compare the number of subsets with the number of elements in each set.

n	N =Number of subsets
0	$1 = 2^0$
1	$2 = 2^1$
2	$4 = 2^2$
3	$8 = 2^3$

Let n be the number of elements in the set and N be the number of subsets. From the table we can deduce a relationship between n and N.

In this case we have $N = 2^n$

That is, if a set A has n elements, then the total number of subsets, N, is 2^n.

If a set A has p elements $n(A) = p$, there are 2^p **subsets** and $2^p - 1$ **proper subsets.**

13.4.7 INTERSECTION OF SETS

The **intersection** of two sets is the elements that are **common** to both sets.

∩ denotes 'the intersection of'

The shaded region in the diagram would then be written as

A ∩ B Reads 'A intersection B'

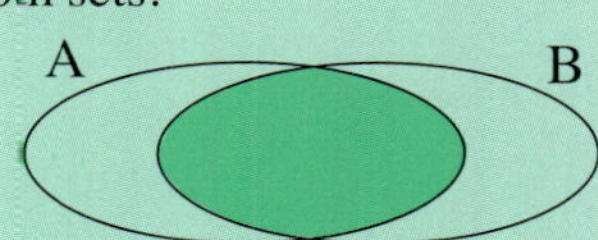

For example, if A = {2, 4, 6, 8, 10, 12, 14, 16} and B = {3, 6, 9, 12, 15} then A ∩ B = {6, 12}. As 6 and 12 are the only elements that belong to both sets, then {6, 12} is the intersection.

EXAMPLE 13.9

If $C = \{x \mid x \in \mathbb{Z}, x < 10\}$ and $D = \{x \mid x \in \mathbb{Z}, x > 5\}$, find $C \cap D$.

Solution

Now, $C = \{x \mid x \in \mathbb{Z}, x < 10\} = \{\ldots, -2, -1, 0, 1, 2, 3, 4, 5, 6, 7, 8, 9\}$
and $D = \{x \mid x \in \mathbb{Z}, x > 5\} = \{6, 7, 8, 9, 10, 11, 12, \ldots\}$

Then, the only elements common to both are $\{6, 7, 8, 9\}$.
So, $C \cap D = \{6, 7, 8, 9\}$
or, $C \cap D = \{x \mid x \in \mathbb{Z}, 5 < x < 10\}$ [using set-builder notation]

13.4.8 UNION OF SETS

The **union** of two sets is the set of **all** the elements in both sets listed once.

$\cup$ denotes 'the union of'
The shaded region in the diagram would then be written as
$A \cup B$ reads 'A union B'.

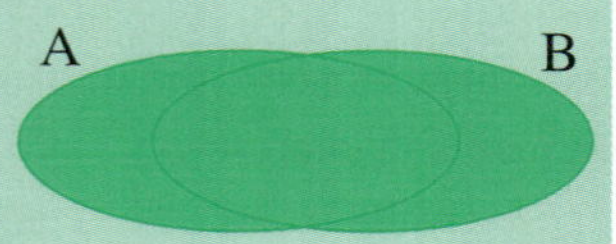

For example, if $A = \{2, 4, 6, 8, 10, 12, 14, 16\}$ and $B = \{3, 6, 9, 12, 15\}$ then
$A \cup B = \{2, 3, 4, 6, 8, 9, 10, 12, 14, 15, 16\}$.

Notice that we have included the elements 6 and 12 only once in the set $A \cup B$. All that has happened is that whatever elements from set B that were not in set A were added to set A. Starting with set B and looking for those elements in set A that are not in set B would lead to the same result.

Notice then that although $n(A) = 8$ and $n(B) = 5$, $n(A \cup B) = 11$ and not $8 + 5 = 13$! This observation leads to the following discussion.

The number of elements in the union of two sets is not necessarily the sum of the numbers of elements in each set. Look at the people in a cinema audience and count the number of women and the number of people with fair hair. Is it true that the number of people who are women or who have fair hair is the sum of the two numbers? Not necessarily as this sum counts the fair haired women twice. This is the intersection of the set of women and the set of fair haired people.

So, going back to our previous example we have that $n(A) = 8$, $n(B) = 5$ and $n(A \cup B) = 11$. We also note that $n(A \cap B) = 2$. Then, adding the number of elements from both sets means that we will have counted those elements common to both sets twice. It then seems appropriate to subtract that amount once from the total.

That is,

$$\begin{aligned} n(A \cup B) &= n(A) + n(B) - n(A \cap B) \\ &= 8 + 5 - 2 \\ &= 11 \end{aligned}$$

In short, we have that

$$\boldsymbol{n(A \cup B) = n(A) + n(B) - n(A \cap B)}$$

EXAMPLE 13.10

Find the union of the sets

(a) M = {apple, pear, orange. pineapple} and N ={nut, carrot, pear, banana}.

(b) $P = \{t \mid t \in \mathbb{R}, -5 < t < 7\}$ and $Q = \{t \mid t \in \mathbb{R}, -3 < t < 10\}$.

Solution

(a) The only element that occurs twice is 'pear' and so we need to make sure that it is included only once in our final answer.
So, in this case we have that M ∪ N = {apple, pear, orange, pineapple, nut, carrot, banana}

(b) We can make use of a number line to help out in this instance. We start by showing each set on the real number line:

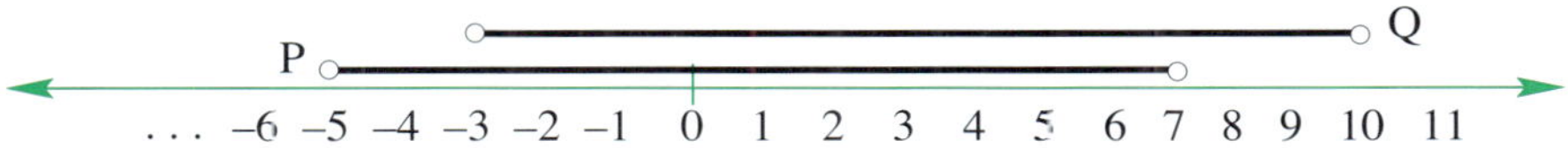

As the sets range from –5 to 10, we then have that $P \cup Q = \{t \mid t \in \mathbb{R}, -5 < t < 10\}$.

Notice that $P \cap Q = \{t \mid t \in \mathbb{R}, -3 < t < 7\}$

13.4.9 THE UNIVERSAL SET

The **universal** set is the set from which the sets of a particular situation are drawn.
The capital letter 'U' **denotes the universal set**

For example, if we want the set of students that have brown hair in your classroom, then the set of all students in your classroom would make up the universal set, of which those with brown hair form a subset.

13.4.10 THE COMPLEMENT

The **complement** of a set B is all the elements that are not in the set B but are in the universal set.

B' denotes the complement of set B (read as 'B-dash')

For example, if the universal set is given by U = {a, d, e, f, r, g, t} and set B = {d, e, f, r, g}, then the set B' = {a, t}. That is, it is made up of all elements that are not in B but still in U.

We note the following results:

$B \cup B' = U$	The **union** of a set and its complement is the universal set.
$B \cap B' = \varnothing$	The **intersection** of a set and its complement is the empty set.
$n(B) + n(B') = n(U)$	The sum of the elements in the set and its complement equals the total number of elements in the universal set.

The following examples illustrate complementary sets.

Example 1 If $U = \{x \mid x \in \mathbb{Z}, 0 < x < 10\}$ and $A = \{2, 4, 6, 8\}$ then $A' = \{1, 3, 5, 7, 9\}$

Example 2 If $U = \{x \mid x \in \mathbb{Z}\}$ and $B = \{x \mid x \in \mathbb{Z}, x \geq -6\}$ then $B' = \{x \mid x \in \mathbb{Z}, x < -6\}$

Example 3 If U = {Students in year 11} and C = {Students with red hair} then
C′ = {Students in year 11 without red hair}

EXERCISES 13.4

1. If A = {Days of the week}, B = {Months of the year}, C = {Letters in the alphabet} and D = {Letters in the word "Standard"}, state which the following are true and which false.

(a) {Saturday, Sunday} $\subseteq$ A (b) {January} $\subseteq$ A (c) February $\in$ B
(d) {a, e, i, o, u} $\subseteq$ C (e) $D \subseteq C$ (f) $C \not\subset D$
(g) {Monday, Tuesday, Wednesday, Thursday, Friday, Saturday, Sunday} $\not\subset$ A
(h) $\varnothing \subseteq D$ (i) $\varnothing \not\subset A$

2. State which the following are true and which false.

(a) $\mathbb{N} \subset \mathbb{R}$ (b) $\mathbb{R} \subseteq \mathbb{Q}$ (c) $\mathbb{Z} \subseteq \mathbb{Q}$
(d) $\mathbb{N} \subseteq \mathbb{Z}$ (e) $\mathbb{N} \subseteq \mathbb{Z} \subseteq \mathbb{R}$ (f) $\mathbb{Q} \supseteq \mathbb{N}$
(g) $\{\pi, \sqrt{2}\} \subseteq \mathbb{Q}$ (h) $\{\pi, \sqrt{2}\} \subseteq \mathbb{R}$ (i) $\left\{\frac{1}{2}, \frac{3}{4}\right\} \supseteq \mathbb{Q}$

3. Write all the subsets of $\{\alpha, \beta, \mu, \sigma\}$

4. How many subsets does the set {1, 2, 3, 4, 5} have?

5. Find the number of proper subsets of the sets
(a) {1, 3, 5, 7, 9}
(b) {1, 3, 3, 4, 5}

6. If A = {quadrilaterals}, B = {rectangles} and C = {squares}, which of the following statements are true?
(a) $A \subset B$ (b) $B \subset C$ (c) $B \subset A$ (d) $C \subseteq A$

7. In each of the following determine if set B is a subset of set A.

(a) A = {–2, –3, 0, 5, 7}, B = {–3, 5}
(b) A = {2, 4, 6, 8}, B = {0}
(c) A = {1, 5, 9, 10, 12, 13}, B = { }
(d) A = {1, 5, 9, 10, 12, 13}, B = {10}

8. List all subsets of
(a) {1, 2} (b) { } (c) { {1}, {1, 2}, 3}

9. Given that $X = \{x\}, Y = \{x, y\}, Z = \{x, y, z\}, P = \{y, z\}, Q = \{w, x, y\}$, state the truth of

(a) $X \subset Y$ (b) $X \subseteq Y$ (c) $X \in Y$ (d) $P \subset Q$

(e) $P \subseteq Z$ (f) $z \subset P$ (g) $Q \subseteq Y$ (h) $z \in P$

10. If $\mathrm{H} = \{(x, y) | 3x + 4y = 80, x, y \in \mathbb{N}\}$, find the number of subsets of H.

11. If $\mathrm{A} = \{3, 6, 9, 12, 15, 18\}$, $\mathrm{B} = \{2, 4, 6, 8, 10, 12, 14, 16, 18\}$ and $\mathrm{C} = \{1, 4, 9, 16\}$
List the elements in

(a) $\mathrm{A} \cap \mathrm{B}$ (b) $\mathrm{B} \cap \mathrm{C}$ (c) $\mathrm{A} \cap \mathrm{C}$
(d) $\mathrm{A} \cap \mathrm{B} \cap \mathrm{C}$ (e) $\mathrm{A} \cup \mathrm{B}$ (f) $\mathrm{A} \cup \mathrm{C}$
(g) $\mathrm{B} \cup \mathrm{C}$ (h) $\mathrm{A} \cup \mathrm{B} \cup \mathrm{C}$ (i) $(\mathrm{A} \cup \mathrm{B}) \cap \mathrm{C}$
(j) $(\mathrm{B} \cap \mathrm{C}) \cup \mathrm{A}$ (k) $(\mathrm{A} \cap \mathrm{C}) \cup \mathrm{B}$

12. Evaluate, for the sets in Q.11.,

(a) $n(\mathrm{A} \cap \mathrm{B})$ (b) $n(\mathrm{B} \cap \mathrm{C})$ (c) $n(\mathrm{A} \cap \mathrm{C})$
(d) $n(\mathrm{A} \cap \mathrm{B} \cap \mathrm{C})$ (e) $n(\mathrm{A} \cup \mathrm{B})$ (f) $n(\mathrm{A} \cup \mathrm{C})$
(g) $n(\mathrm{B} \cup \mathrm{C})$ (h) $n(\mathrm{A} \cup \mathrm{B} \cup \mathrm{C})$

13. If $\mathrm{D} = \{x \mid x \in \mathbb{Z}, -3 \leq x < 4\}$, $\mathrm{E} = \{x \mid x \in \mathbb{Z}, -5 \leq x \leq -2\}$ and $\mathrm{F} = \{x \mid x \in \mathbb{Z}, -3 \leq x \leq 5\}$
List the elements in

(a) $\mathrm{D} \cap \mathrm{E}$ (b) $\mathrm{E} \cap \mathrm{F}$ (c) $\mathrm{D} \cap \mathrm{F}$
(d) $\mathrm{D} \cap \mathrm{E} \cap \mathrm{F}$ (e) $\mathrm{D} \cup \mathrm{E}$ (f) $\mathrm{D} \cup \mathrm{F}$
(g) $\mathrm{E} \cup \mathrm{F}$ (h) $\mathrm{D} \cup \mathrm{E} \cup \mathrm{F}$ (i) $(\mathrm{D} \cup \mathrm{E}) \cap \mathrm{F}$
(j) $(\mathrm{E} \cap \mathrm{F}) \cup \mathrm{D}$ (k) $(\mathrm{D} \cap \mathrm{F}) \cup \mathrm{E}$

14. Evaluate, for the sets in Q.13.,

(a) $n(\mathrm{D} \cap \mathrm{E})$ (b) $n(\mathrm{E} \cap \mathrm{F})$ (c) $n(\mathrm{D} \cap \mathrm{F})$
(d) $n(\mathrm{D} \cap \mathrm{E} \cap \mathrm{F})$ (e) $n(\mathrm{D} \cup \mathrm{E})$ (f) $n(\mathrm{D} \cup \mathrm{F})$
(g) $n(\mathrm{E} \cup \mathrm{F})$ (h) $n(\mathrm{D} \cup \mathrm{E} \cup \mathrm{F})$

15. If $\mathrm{G} = \{x \mid x \in \mathbb{R}, -3 \leq x < 4\}$, $\mathrm{H} = \{x \mid x \in \mathbb{R}, -5 \leq x \leq -2\}$ and $\mathrm{J} = \{x \mid x \in \mathbb{R}, -3 < x \leq 5\}$
Write in set builder notation.

(a) $\mathrm{G} \cap \mathrm{H}$ (b) $\mathrm{H} \cap \mathrm{J}$ (c) $\mathrm{G} \cap \mathrm{J}$
(d) $\mathrm{G} \cap \mathrm{H} \cap \mathrm{J}$ (e) $\mathrm{G} \cup \mathrm{H}$ (f) $\mathrm{G} \cup \mathrm{J}$
(g) $\mathrm{H} \cup \mathrm{J}$ (h) $\mathrm{G} \cup \mathrm{H} \cup \mathrm{J}$ (i) $(\mathrm{G} \cup \mathrm{H}) \cap \mathrm{J}$
(j) $(\mathrm{H} \cap \mathrm{J}) \cup \mathrm{G}$ (k) $(\mathrm{G} \cap \mathrm{J}) \cup \mathrm{H}$

16. If U = {Letters of the English alphabet}, A = {Letters in the word "mathematics"}, B = {The consonants in the alphabet} and C = {a, b, d, m, n, o, p, l, j, z}, list the elements of

(a) A' (b) B' (c) C'

(d) $(\mathrm{A} \cup \mathrm{B})'$ (e) $(\mathrm{A} \cup \mathrm{B} \cup \mathrm{C})'$ (f) $\mathrm{A}' \cap \mathrm{B}'$

(g) $\mathrm{A}' \cap \mathrm{B}' \cap \mathrm{C}'$ (h) $(\mathrm{B}' \cap \mathrm{C}') \cup \mathrm{A}'$

17. If $U = \{x \mid x \in \mathbb{Z}, -10 \le x \le 10\}$, $D = \{x \mid x \in \mathbb{Z}, -5 \le x \le 2\}$, $E = \{x \mid x \in \mathbb{Z}, -1 \le x \le 6\}$ and $F = \{x \mid x \in \mathbb{Z}, 0 < x < 3\}$, list the elements of

(a) D' (b) E' (c) F'
(d) $(D \cup E)'$ (e) $(D \cup E \cup F)'$ (f) $D' \cap E'$
(g) $(E' \cap F') \cup D'$ (h) $(E' \cap F') \cup D'$

18. Using the sets in Q. 17., write the following using set builder notation.
(a) D' (b) E' (c) F' (d) $(D \cup E)'$ (e) $(D \cup E \cup F)'$

19. Using the sets $U = \{x \mid x \in \mathbb{R}, -10 \le x \le 10\}$, $G = \{x \mid x \in \mathbb{R}, -5 < x \le -3\}$, $H = \{x \mid x \in \mathbb{R}, 3 < x \le 10\}$ and $J = \{x \mid x \in \mathbb{R}, -3 < x < 4\}$, write the following using set builder notation.
(a) G' (b) H' (c) J' (d) $(G \cup H)'$ (e) $(G \cup H \cup J)'$

20. $n(U) = 40$, $n(A) = 12$ $n(B) = 21$ $n(C) = 15$, find
(a) $n(A')$ (b) $n(B')$ (c) $n(C')$

13.5 VENN DIAGRAMS

13.5.1 DEFINITION

Sets can be represented using **Venn diagrams**.
Each individual set is represented by a circle (or an oval) and the universal set is represented by a rectangle which encloses all of the sets under investigation.

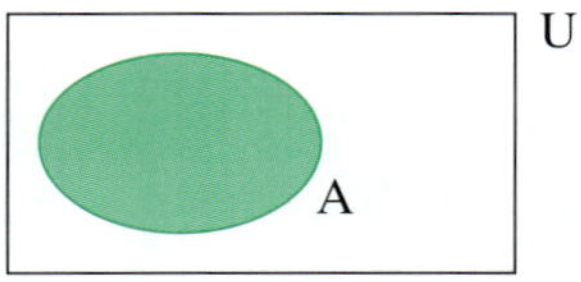

Of course, there can be more than one set inside the rectangle. Then, depending on how the sets relate to each other (i.e., if they have elements in common, or have no elements in common) we place each set inside the rectangle.

13.5.2 DISJOINT SETS

Sets are **disjoint** if they have no elements in common.

Therefore, we have that $A \cap B = \varnothing$.

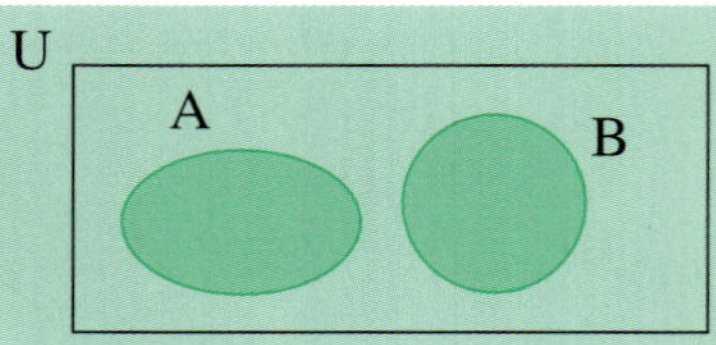

The following example illustrates a Venn diagram with two circles (i.e., two sets) that do not overlap and are enclosed by a rectangle representing the universal set.

Given the set $U = \{1, 2, 3, 4, 5, 6, 7, 8, 9, 10\}$ with $A = \{2, 4, 6, 8, 10\}$ and $B = \{1, 3, 5\}$, the Venn diagram shows that the sets A and B are disjoint (i.e., have no elements in common).

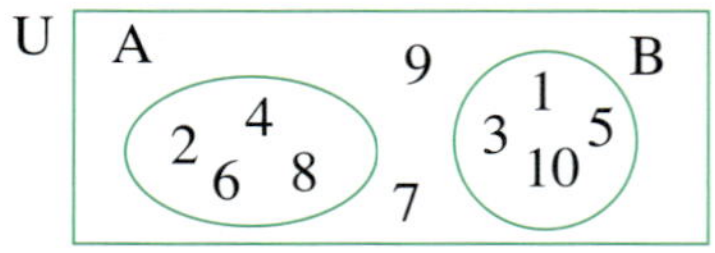

We now provide an algebraic and diagrammatic representation of some possible relationships between two sets A and B where it is known that there exists an intersection.

Operation	Symbol	Venn diagram	Algebraic
Union The set of all elements that belong to A or B or both.	$A \cup B$	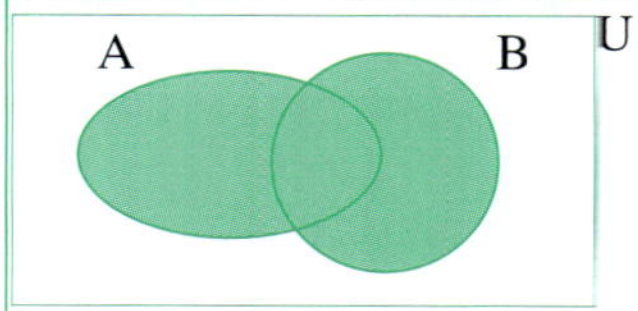	$\{x\ x \in A \text{ or } x \in B \text{ or both}\}$
Intersection The set of all elements that belong to both A and B	$A \cap B$	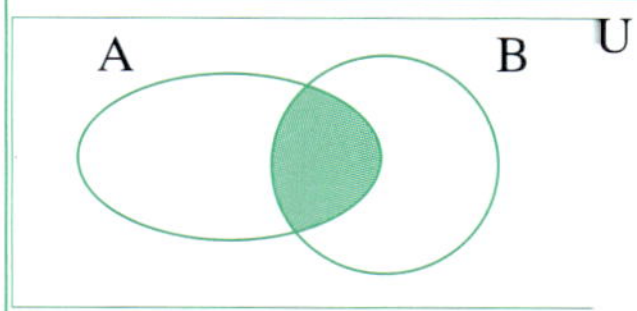	$\{x \mid x \in A \text{ and } x \in B\}$
Complement The set of all elements in U that are not in A.	A'	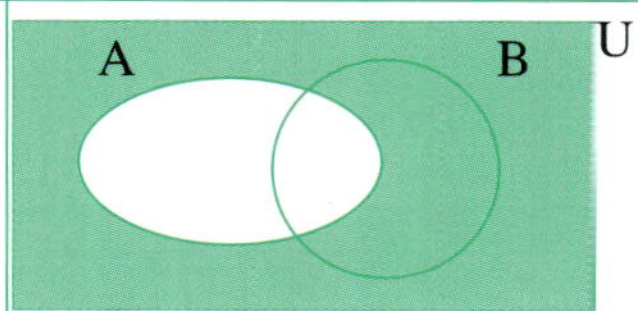	$\{x \mid x \in U \text{ and } x \notin A\}$
Difference The set of all elements in A and not in B.	$A \backslash B$ (A slash B)	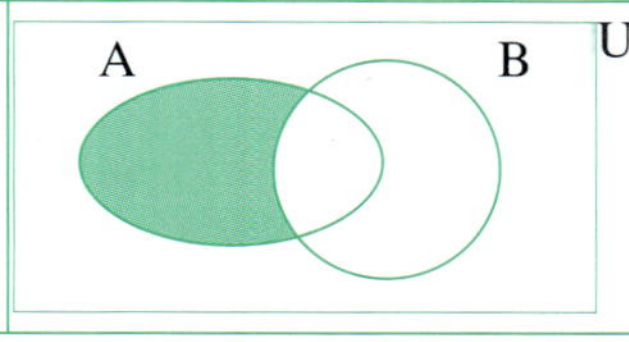	$\{x \mid x \in A \text{ and } x \notin B\}$

EXERCISES 13.5

1. On a Venn diagram, show, by using shaded regions, each of the following sets, clearly labelling set A, set B and the universal set, U.

(a) The intersection of the sets A and B given that

i. $A \cap B = \varnothing$ ii. $n(A \cap B) > 1$ iii. $B \subset A$

(b) The union of the sets A and B given that

i. $A \cap B = \varnothing$ ii. $n(A \cap B) > 1$ iii. $B \subset A$

(c) The complement of A given that

i. $A \cap B = \varnothing$ ii. $n(A \cap B) > 1$ iii. $B \subset A$

(d) i. $A' \cap B'$ given that $A \cap B = \varnothing$.

ii. $A' \cap B$ given that $n(A \cap B) > 1$.

iii. $A \cap B'$ given that $B \subset A$.

2. Given that U = {Natural numbers}, A = {Multiples of 3 less than 40} and B = {Multiples of 4 less than 40}

(a) Draw a Venn diagram to show this relationship, listing all the elements.

(b) Explain in words, the significance of $A \cap B$.

(c) Find

i. $n(A)$ ii. $n(B)$ iii. $n(A \cap B)$ iv. $n(A \cup B)$

(d) Show that $n(A \cup B) = n(A) + n(B) - n(A \cap B)$

(e) Shade the region of the Venn diagram that represents A'.

3. Given that U = {Natural numbers}, A = {Factors of 36} and B = {Factors of 60}

(a) Draw a Venn diagram to show this relationship, listing all the elements.

(b) Explain in words, the significance of $A \cap B$.

(c) Find

i. $n(A)$ ii. $n(B)$ iii. $n(A \cap B)$ iv. $n(A \cup B)$

(d) Show that $n(A \cup B) = n(A) + n(B) - n(A \cap B)$

(e) Shade the region of the Venn diagram that represents B'.

4. Given that U = {Natural numbers}, A = {Prime numbers less than 30} and B = {Odd numbers less than 30}

Note: Prime numbers only have 1 and themselves as factors. 1 is neither prime nor composite.

(a) Draw a Venn diagram to show this relationship, listing all the elements.

(b) Explain in words, the significance of $A \cap B$.

(c) Find

i. $n(A)$ ii. $n(B)$ iii. $n(A \cap B)$ iv. $n(A \cup B)$

(d) Show that $n(A \cup B) = n(A) + n(B) - n(A \cap B)$

(e) Shade the region of the Venn diagram that represents $A \cap B$.

5. Given U = {Fruits}, A = {Apple, pear, banana, pineapple, water melon} and B = {Banana, apple, pear}

(a) Draw a Venn diagram to show this relationship, listing all the elements.

(b) Explain in words, the significance of $A \cap B$.

(c) Find

i. $n(A)$ ii. $n(B)$ iii. $n(A \cap B)$ iv. $n(A \cup B)$

(d) Show that $n(A \cup B) = n(A) + n(B) - n(A \cap B)$

(e) Shade the region of the Venn diagram that represents $A \cup B$.

6. Given that U = {Natural numbers}, A = {Factors of 36}, B = {Multiples of 6 less than 61} and C = {Even number less than 21}

(a) Draw a Venn diagram to show this relationship, listing all the elements.

(b) Explain in words, the significance of $A \cap B \cap C$.

(c) Find

i. $n(A)$ ii. $n(B)$ iii. $n(C)$

iv. $n(A \cap B \cap C)$ iv $n(A \cup B \cup C)$

(d) Shade the region of the Venn diagram that represents C'.

7. (a) For each of the following copy the Venn diagram and shade the region that represents:

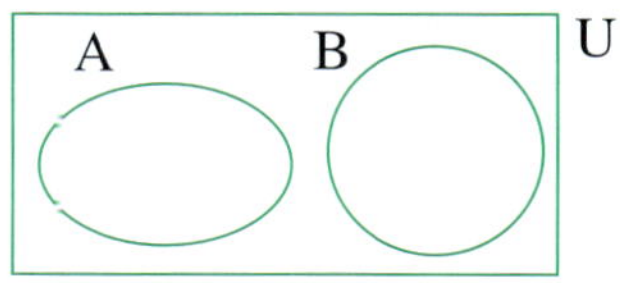

i. A ii. A′ iii. A ∪ B
iv. A ∩ B v. (A ∪ B)′
vi. A′ ∩ B vii. A′ ∩ B′

8.

(a) For each of the following copy the Venn diagram and shade the region that represents:

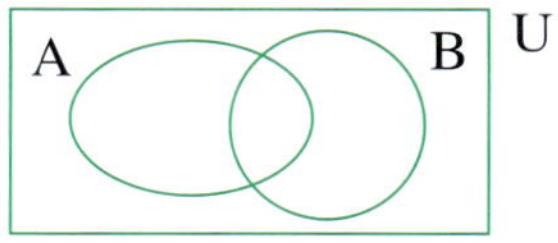

i. A ii. A′ iii. A ∪ B
iv. A ∩ B v. (A ∪ B)′
vi. A′ ∩ B vii. A′ ∩ B′

9.

(a) For each of the following copy the Venn diagram and shade the region that represents:

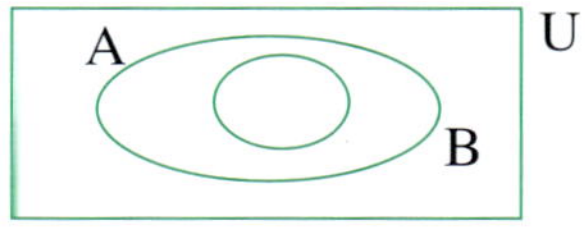

i. A ii. A′ iii. A ∪ B
iv. A ∩ B v. (A ∪ B)′
vi. A′ ∩ B vii. A′ ∩ B′

10. (a) For each of the following copy the Venn diagram and shade the region that represents:

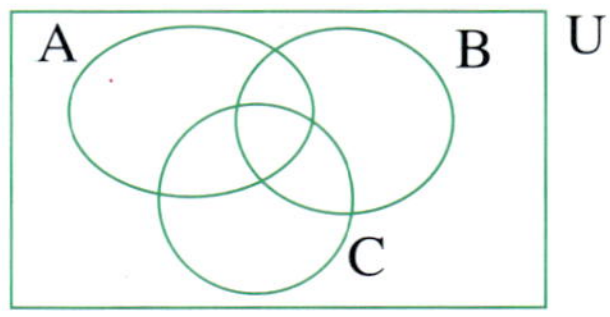

i. A ii. A′
iii. B′ iv. C′
v. A ∪ B ∪ C vi. A ∩ B ∩ C
vii. (A ∪ B ∪ C)′ viii. (A ∩ B ∩ C)′

11. For the universal set U and two sets A and B, write the following in the simplest possible form, where $A \cap B \neq \varnothing$.

(a) $A \cap U$ (b) $A \cap \varnothing$ (c) $A \cup U$
(d) $A \cup \varnothing$ (e) $A \cap A$ (f) $A \cup A$
(g) $A \cap (A \cup B)$ (h) $A \cup (A \cap B)$ (i) $A \cup (A' \cap B)$
(j) $B \cap (B' \cup A)$

13.6 APPLICATIONS

13.6.1 THE NUMBER OF ELEMENTS IN TWO REGIONS

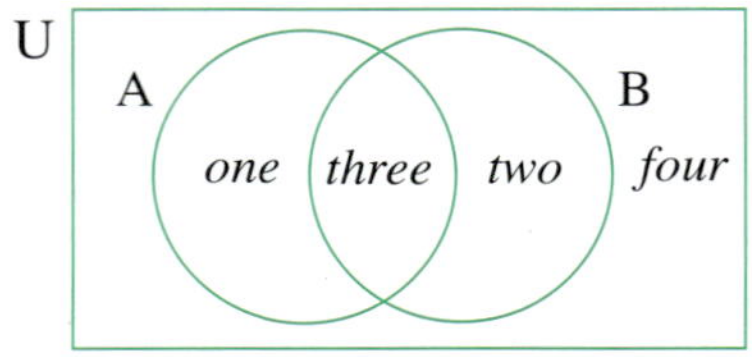

Region	Description of Region	Number of Elements
One	The elements that belong to **only** set A	$n(A) - n(A \cap B)$
Two	The elements that belong to **only** set B	$n(B) - n(A \cap B)$
Three	The elements that belong to A **and** B.	$n(A \cap B)$
Four	The elements that **do not** belong to set A or set B	$n(U) - n(A \cup B)$

EXAMPLE 13.11

Draw a Venn diagram to show the number of elements in each region if, $n(U) = 50, n(A) = 32, n(B) = 25, n(A \cap B) = 11$

Solution

The number of elements in A only $= n(A) - n(A \cap B)$
$= 32 - 11 = 21$

The number of elements in B only $= n(B) - n(A \cap B)$
$= 25 - 11 = 14$

The number of elements in U $= n(U) - n(A \cup B)$
$= 50 - (32 + 25 - 11) = 4$

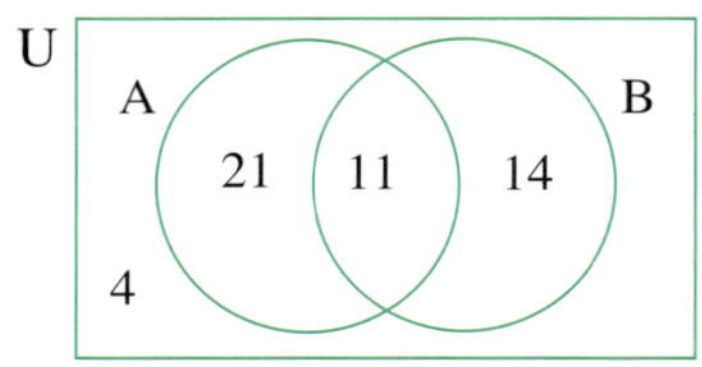

EXAMPLE 13.12

Out of a class of 30 IB students, 22 study Mathematical Studies (M), 15 study Biology (B) and 7 study both. Draw a Venn diagram to show the number of students in each region of the Venn diagram.

Solution

The number of students who only study Maths $= n(M) - n(M \cap B) = 22 - 7 = 15$
The number of students who only study Biology $= n(B) - n(M \cap B) = 15 - 7 = 8$
The number of students who study neither $= n(U) - n(M \cup B) = 30 - (22 + 15 - 7) = 0$

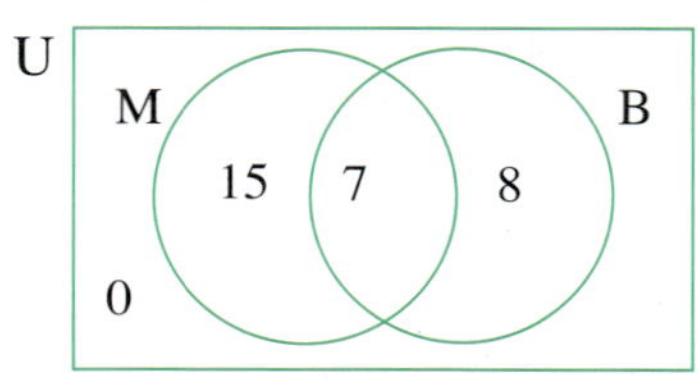

EXERCISES 13.6.1

1. Draw Venn diagrams to show the number of elements in each separate region for the following.

(a)	$n(\text{U}) = 25$	$n(\text{A}) = 10$	$n(\text{B}) = 12$	$n(\text{A} \cap \text{B}) = 5$
(b)	$n(\text{U}) = 55$	$n(\text{A}) = 34$	$n(\text{B}) = 27$	$n(\text{A} \cap \text{B}) = 16$
(c)	$n(\text{U}) = 64$	$n(\text{A}) = 23$	$n(\text{B}) = 41$	$n(\text{A} \cap \text{B}) = 13$
(d)	$n(\text{U}) = 30$	$n(\text{A}) = 19$	$n(\text{B}) = 17$	$n(\text{A} \cap \text{B}) = 8$
(e)	$n(\text{U}) = 100$	$n(\text{A}) = 67$	$n(\text{B}) = 50$	$n(\text{A} \cap \text{B}) = 34$

2. In an athletics club with 20 members, 15 compete in the 100m, 12 compete in the 400m and 9 compete in both. Draw a Venn diagram to show this.

3. 50 people where surveyed as to whether they liked tea, coffee or both. Every person surveyed chose one of the three. If 38 people liked tea and 32 people liked coffee, how many people liked both. Draw a Venn diagram to show this information.

4. Out of 30 students, 19 play hockey and 15 play tennis. How many play both? Draw a Venn diagram to show this.

5. The number of elements in sets A and B are shown on the Venn diagram below.
If $n(\text{A}) = 2 \times n(\text{B})$, find the value of p.

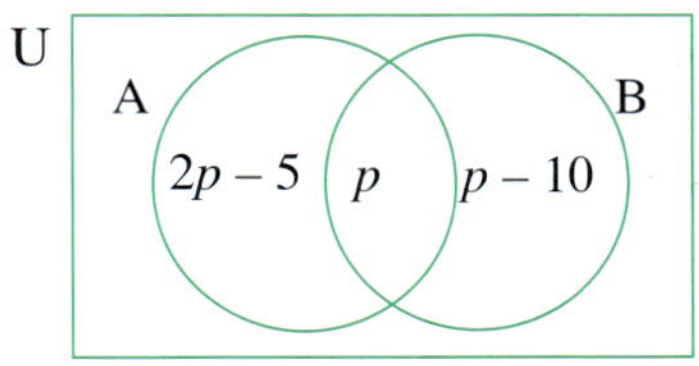

6. A small primary school consists of 300 students. Of these, 250 drink plain milk while 150 students have chocolate biscuits for recess. 50 students drink plain milk and eat chocolate biscuits. Is this information consistent? Use a Venn diagram to justify your answer.

7. A girl scout group consists of 125 girls. In a fundraising event 60 girls sell biscuits while 30 of the girls bake and sell the biscuits. If all 125 girls are involved in either selling or baking the biscuits, how many of the girls

(a) only bake the biscuits?

(b) bake biscuits?

8. In a town that borders France and Italy, 60% of the residents speak French while 40% speak Italian. If there are 1200 residents how many of them speak both Italian and French given that 20% of the residents speak neither language.

9. We are given that $n(U) = 200$, $n(A) = 100$ and $n(B) = 150$ and $n(A \cap B) = x$.

(a) Draw a Venn diagram for this situation.

(b) What are the restrictions on $n(A \cap B)$?

13.6.2 THE NUMBER OF ELEMENTS IN THREE REGIONS

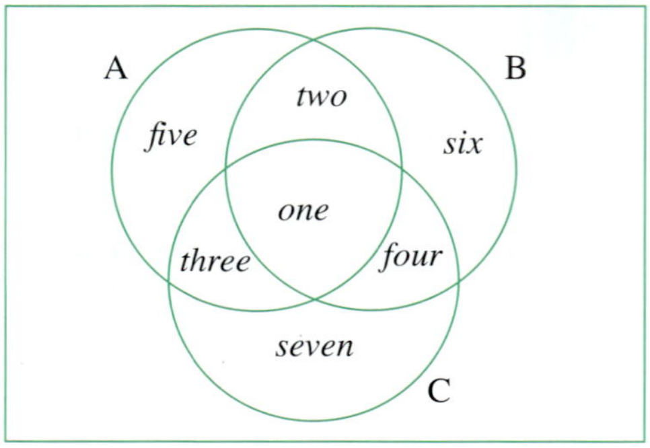

Region	Description of the region	Number of elements
One	The elements that belong to A **and** B **and** C	$n(A \cap B \cap C)$
Two	The elements that belong to A **and** B **but not** C	$n(A \cap B) - n(A \cap B \cap C)$
Three	The elements that belong to A **and** C **but not** B	$n(A \cap C) - n(A \cap B \cap C)$
Four	The elements that belong to C **and** B **but not** A	$n(C \cap B) - n(A \cap B \cap C)$
Five	The elements that belong to A **only**	$n(A) - [n(A \cap B) + n(A \cap C) - n(A \cap B \cap C)]$
Six	The elements that belong to B **only**	$n(A) - [n(A \cap B) + n(B \cap C) - n(A \cap B \cap C)]$
Seven	The elements that belong to C **only**	$n(A) - [n(A \cap C) + n(B \cap C) - n(A \cap B \cap C)]$

EXAMPLE 13.13

From a survey of 50 customers at a video store, the following was found.
29 enjoyed adventure videos, 22 enjoyed comedies, 18 enjoyed horror videos of these
8 liked both adventure and comedy, 10 liked both comedies and horror
5 liked both adventure and horror and 4 liked all three types of videos
Draw a Venn diagram to show the number of people in each region of the Venn diagram.

Solution

Let A = {people who liked adventure videos}, C = {people who liked comedy videos} and H = {people who liked horror videos}

Region	Method	Number of elements
One	4 liked all three types of videos	4
Two	8 liked both adventure and comedy but of these 4 liked all three	$8 - 4 = 4$
Three	5 liked both adventure and horror but of these 4 liked all three	$5 - 4 = 1$
Four	10 liked comedies and horror but of these 4 liked all three	$10 - 4 = 6$

Put this information in the Venn diagram. Then proceed with the second part of the table.

Five	29 enjoyed adventure but some of these already are accounted for on the Venn diagram	$29-(4+4+1)=20$
Six	22 enjoyed comedies but some of these already are accounted for on the Venn diagram.	$22-(4+4+6)=8$
Seven	18 enjoyed horror but some of these are already accounted for on the Venn diagram.	$18-(4+1+6)=7$

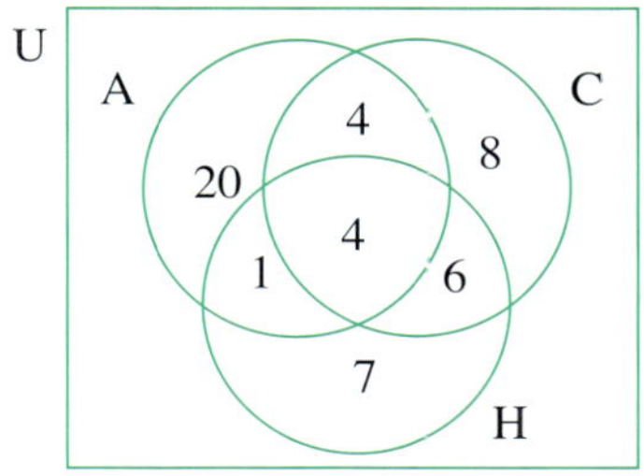

EXAMPLE 13.14

Of 40 students, 22 play soccer, 19 play basketball and 13 play tennis. Six of the students play tennis and basketball only while 7 play tennis and soccer only, and 10 play basketball and soccer only. If there are 9 students that do not participate in any of these sports, how many play all three sports?

Solution

Let S denote the set of students that play soccer,
B denote the set of students that play basketball,
T denote the set of students that play tennis,

Let the number of students that play all three sports be x, i.e., $n(S \cap B \cap T) = x$.

Next, we draw a Venn diagram with as much information as possible:

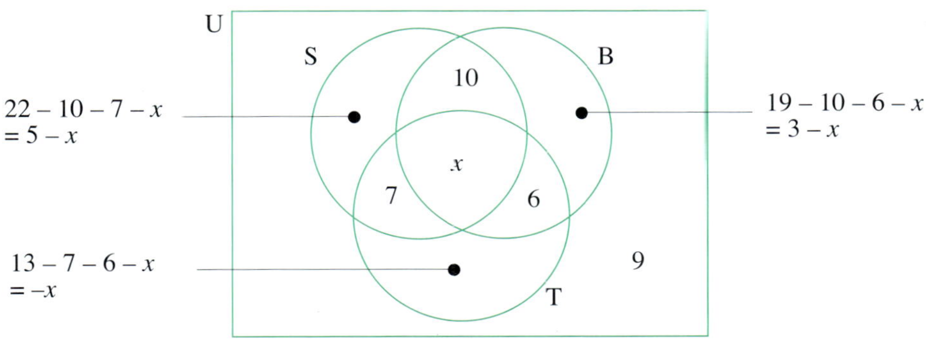

We see that the number of students that play only play tennis is $-x$. Then, as it is not possible to have a negative number of players, the only possible value for x is 0, i.e., $x = 0$. So, no student plays all three sports.

EXERCISES 13.6.2

1. 45 students were asked which of the following activities they enjoyed, roller blading, surfing or BMX racing. Each student listed at least one activity and the results are summarised below.
24 enjoy roller blading, 24 enjoy surfing, 15 enjoy BMX racing
11 enjoy both roller blading and surfing, 6 enjoy both roller blading and BMX racing
4 enjoy both surfing and BMX racing, 3 enjoy all three activities.
Draw a Venn diagram to show the number of students in each region.

2. 18 senior citizens all participate in at least one of the following activities, lawn bowls, backgammon or cards.
11 participate in lawn bowls, 4 participate in lawn bowls and backgammon
3 only participate in backgammon, 2 participate only in cards
1 participates in all three activities
Draw a Venn diagram to show the number of senior citizens in each region.

3. In a class of 50 IB students, each student studies at least one of the following subjects, computing, geography and/or biology.
19 study computing, 23 study geography, 31 study biology
8 study computing and geography, 9 study geography and biology
11 study computing and biology, 5 students study all three subjects
Draw a Venn diagram to show the number of students in each region.

4. 131 children went to a school assembly and all had to wear the correct school uniform, which was to include a school hat, school blazer and grey socks. Of these students 84 wore their hat, 108 wore their blazer, 108 wore grey socks, 62 wore both a hat and blazer, 66 wore grey socks and a hat, while 85 wore a blazer and grey socks.
You may assume that all students wore at least one item.
(a) How many wore a hat, blazer and grey socks?
(b) How many wore a hat but neither grey socks nor a blazer?

5. 104 students from an IB school attended a seminar on how to cope with their mathematics programme. Of these students, 80 studied Maths SL, 60 studied Biology and 64 Art. Of these, 51 studied both Maths SL and Art, 45 studied both Biology and Maths SL while 48 studied Biology and Art.
(a) How many students studied all three subjects?
(b) How many studied Art only?

6. Of 450 people interviewed, 135 contributed to Red House, 180 contributed to Blue House and 180 contributed to Yellow House. 27 contributed to Red House and Yellow House, 54 contributed to Yellow House and Blue House while 45 contributed to Red House and Blue House. 63 of the people interviewed contributed to none of the houses. How many contributed to all three houses?

7. 80 students at Ballarat Grammar School study musical art. Of the 80 students 56 like popular music with 8 liking popular music only. 22 like popular but not classical while 38 like classical and techno. 40 like both techno and popular and 56 like techno.
(a) How many of the students like classical music?
(b) How many liked classical and techno but not popular music?

13.7 MISCELLANEOUS EXERCISES

1. Given that sets A = {1, 2, 3, 4, 5}, B = {2, 4, 6, 8, 10} and C = {white, black, red} state which of the following statements are true and which are false:

(a)	$3 \in B$	(b)	$4 \in B$	(c)	$3 \subset A$	(d)	$\{\text{white}\} \in C$
(e)	$\text{white} \subset C$	(f)	$1 \in \{1, 2, 3\}$	(g)	$\{1, 2\} \subset A$	(h)	$\{1, 2, 3\} \in A$
(i)	$5 \notin B$	(j)	$\text{white} \in C$	(k)	$\text{green} \notin C$	(l)	$\{\text{black}\} \subset C$

2. If A = { 1, 2, 3, 4, 5}, B = {1, 3, 5, 7} and C = {2, 4, 6} find

(a) $A \cap B$ (b) $A \cap C$ (c) $B \cap C$ (d) $A \cap (B \cap C)$

3. If N = {a, b, c, d, e, f, g, h}, A = {a, e, i, o, u} , B = {b, c, d, f, g, h} and C = {o, g, u, h}, find:

(a) $A \cap B$ (b) $A \cup B$ (c) $(A \cap C) \cap N$ (d) $(A \cup C) \cap N$

(e) $(A \cap B) \cup (N \cap C)$

4. If A = {1, 2, 3 ,4 ,5, 6}, B = {2, 4, 6, 8, 10} and C = {3, 5, 7, 9, 11} find:

(a)	i.	$Q = \{x \mid x < 3, x \in A\}$	ii.	$n(Q)$
(b)	i.	$R = \{x \mid x > 4, x \in B\}$	ii.	$n(R)$
(c)	i.	$S = \{x \mid x = 2a + 1, a \in B\}$	ii.	$n(S)$
(d)	i.	$T = \{x \mid x = a + b, a \in A, b \in B\}$	ii.	$n(T)$
(e)	i.	$U = \{(x,y) \mid x = y, x \in A, y \in B\}$	ii.	$n(U)$

5. State which of the following sets are finite and which are infinite

(a) {inhabitants of China} (b) {counting numbers}

(c) {Grains of sand in the Simpson Desert} (d) {1, 2, 3, 4, . . ., 5001}

(e) {1, 2, 3, 4, ... }

6. List the elements of the sets defined below

(a) $P = \{x \mid x > 2, x \in \mathbb{N}\}$ (b) $Q = \{x \mid x \leq 8, x \in \mathbb{N}\}$

(c) $S = \left\{y \mid 1 \leq y \leq \frac{9}{2}, y \in \mathbb{N}\right\}$ (d) $X = \{x \mid x^2 \leq 25, x \in \mathbb{N}\}$

7. (a) If $n(A) = n(B)$, must A = B? (b) If A = B, must $n(A) = n(B)$?

Justify your answer in each case.

8. Write down the sets that are equal to

(a) $\{y \mid 2y - 6 = 10, y \in \mathbb{R}\}$ (b) $\{x \mid x^2 - 4 = 0, x \in \mathbb{R}\}$

(c) $\{x \mid x^2 - 4 = 0, x \in \mathbb{N}\}$

9. (a) If $S = \{x \mid x = 2n, n = 0, 1, 2, 3\}$ and $R = \{y \mid y = x^2, x \in S\}$, find:

i. $n(R)$ ii. $S \cap R$ iii. $n(S \cap R)$

(b) If $T = \{x \mid (x-2)(x-3) = 0\}$ and $U = \{y \mid 2y - 1 = 2\}$, find

i. $n(T)$ ii. $T \cup U$ iii. $V = \{v \mid v = y, y \in T \cap U\}$

10. If A = {1, 2, 3, 4, 5, 6, 7} and B = {3, 4, 7,9} and U = {1, 2, 3, 4, 5, 6, 7, 8, 9} find
i. $A \cup B$ ii. $A \cap B$ iii. $A' \cap B$

11. If $A \subset B$, what can be said about $A \cap B$?

12. If A = {1, {3, 4}, 5}, which of the following are correct
i. $\{1, 5\} \in A$ ii. $\{3, 4\} \in A$ iii. $\{3, 4\} \subset A$ iv. $\{\{3, 4\}\} \subset A$

13. For each of the Venn diagrams shown, shade the region defined by
(a) $A \cup B$ (b) $A \cap B$ (c) $B \backslash A$

i. ii. iii.

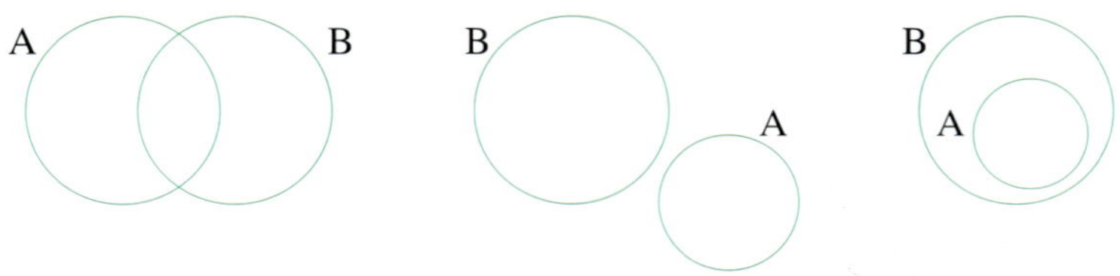

14. Consider the universal set U = {a, b, c, d, e, f, g} with associated sets A = {a, b, c), B = {a, c, e, f} and C = {d, e, f, g}. Find:
(a) $n(A \cap C)$ (b) $n(B \backslash C)$ (c) $n(A' \cap C)$ (d) $n((A \backslash C)')$

15. If $U = \{0, 1, 2, 3, 4, 5, 6, 7, 8\}$, $A = \{0, 1, 2, 3, 4\}$ and $B = \{3, 4, 5, 6\}$, find
(a) $A \cup B$ (b) $(A \cup B)'$ (c) $(A \cap B)'$
(d) $A \cap U$ (e) $A' \cup B$ (f) $A' \cap B$

16. Let $A = \{1, 3, 5\}$, $B = \{2, 3, 4, 5\}$ and $C = \{4, 6, 7, 8, 9\}$ be such that $U = A \cup B \cup C$.
Find: (a) $A \cup B$ (b) $A \backslash B$ (c) $A \backslash B'$
(d) $A' \backslash B$ (e) $A \cap C$ (f) $(A \cap B) \cap (B \cap C)$

17. Using the Venn diagram shown, draw the corresponding Venn diagram representing each of the following
(a) $A \cup B'$
(b) $A \cap B'$
(c) $(A \cup B')'$
(d) $A \cup (A' \cap B)$

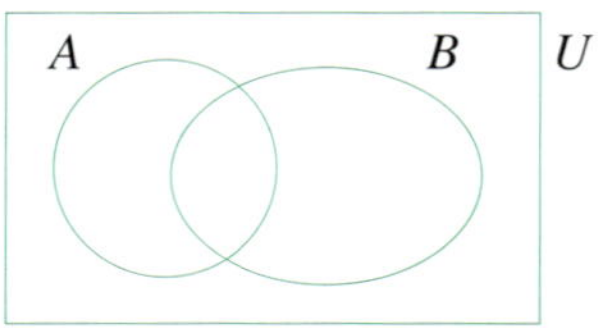

18. If P = {Quadrilaterals}, Q = {Rectangles}, R = {Squares}, which of the following are true ?
(a) $R \in Q$ (b) $R \subset Q$ (c) $Q \subset S$

19. A universal set consists of 100 objects which belong to three sets, A, B and C. Due to some common attributes within the sets it is necessary to display the information using the Venn diagram shown alongside:
Find the number of elements that belong to
(a) $B \cap C$ (b) $(A \cap B) \cup C$
(c) $(A \cap B)' \cap C$ (d) $A' \cap (B \cup C)$

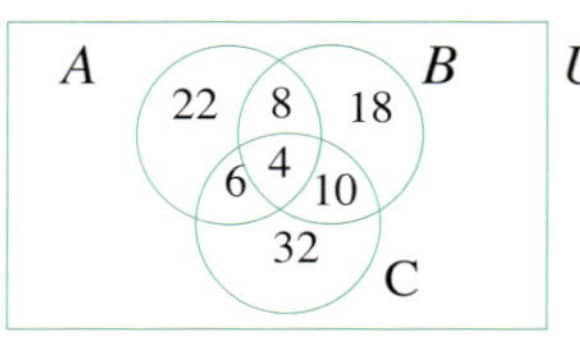

20. If $A = \{x | 1 \le x \le 100$ and x is divisible by $5\}$ and

$B = \{y | 1 \le y \le 50$ and y is divisible by $5\}$

Find: (a) $n(A)$ (b) $n(B)$ (c) $n(A \cap B)$

21. The sets A, B and C satisfy the following three conditions:

1. $A \cap B = A$ 2. $A \cap C = \varnothing$ 3. $n(B \cap C) > 0$

Represent this information on a Venn diagram.

22. (a) If $U = \{\text{Positive integers}\}$, $A = \{x | 21 \le x \le 42\}$ and $B = \{x | x \text{ is a prime}\}$, write down the value of

i. $n(A)$ ii. $n(A \cap B)$

(b) Given that $A = \{s | 1 < 5s < 26\}$, $B = \left\{s | \frac{1}{2}s < 5 < s + 1\right\}$ and $s \in \{1, 2, 3, \ldots\}$,

list all the elements of $A \cup B$ and $A \cap B$.

23. A and B are two sets with the number of elements in each set shown in the Venn diagram.

Given that $n(A) = n(B)$, find

(a) x (b) $n(A \cup B)$

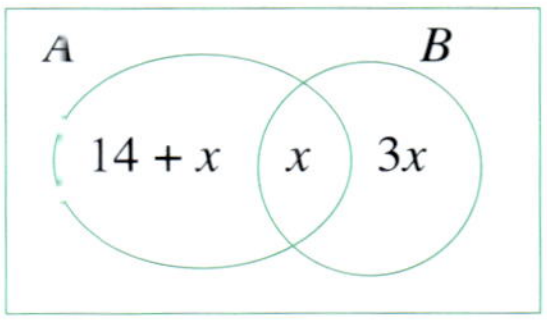

24. A is the set of all integers from 20 to 70 inclusive. Sets B, C and D are defined as follows:

$B = \{x | x \in A$ where x is a perfect square$\}$

$C = \{x | x \in A$ where x is a prime number$\}$

$D = \{x | x \in A$ where the first digit of $x >$ the second digit of $x\}$

(a) Express the sets $B \cap C$ and $B \cap D$ in set builder notation.

(b) List the elements of $B \cap D$ and $C \cap D$.

25. In a large school, 400 of the students play tennis or hockey or both. 65% play hockey and 55% play tennis. Let x denote the number of students who play hockey and tennis.

(a) Represent this information on a Venn diagram.

(b) Determine the value of x.

26. 30 students sit for an examination in both French and English. 25 pass French, 24 pass English and 3 fail both. Determine the number of students who

(a) passed French also passed English (b) failed English and passed French.

27. In a class of 40 students, 23 have dark hair, 18 have brown eyes, and 26 have dark hair, brown eyes or both. Determine how many children have

(a) dark hair and brown eyes.

(b) neither dark hair nor brown eyes.

(c) dark hair but not brown eyes.

13.8 GRADED REVISION EXERCISES

LEVEL 1

1. List the elements of the set $X = \{x \mid x \in \mathbb{N}, 3 \le x < 8\}$.

2. If $n(A) = 20$, $n(B) = 12$ and $n(A \cup B) = 28$, find $n(A \cap B)$.

3. Copy the Venn diagram and shade the region
i. $A \cap B$ ii. $A \cap B'$ iii. $A' \cap B'$

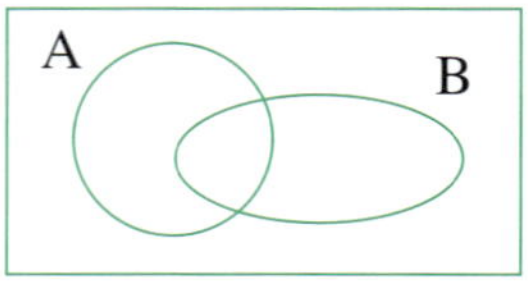

4. List all the possible subsets of $\{4, 8, 12\}$.

5. If $n(X) = 4$, how many possible subsets of X are there?

LEVEL 2

1. In a town of 420 homes it is known that every household has a car or a motorbike or both. In 300 of the households, you will find a car and 70 of these households will also have a motorbike. How many households only have a car?

2. Draw a Venn diagram to display the following information:
$n(U) = 60$, $n(A) = 28$, $n(B) = 18$ and $n(A \cap B) = 10$.

3. Using the Venn diagram alongside, clearly show each of the regions indicated:
(a) $A \cap (B \cap C)$ (b) $A \cap (B \cap C')$
(c) $(A \cup C) \cap B'$

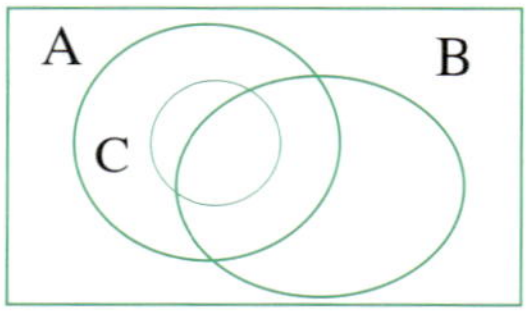

4. A group of 40 students, is made up in such a way that 25 study Italian, 15 study French and 10 study both languages.
(a) Draw a Venn diagram to describe this situation.
Using your Venn diagram, find the number of students that
(b) study either language. (c) study neither language.

5. Consider the sets $A = \{1, 3, 5, 7\}$, $A \cap B = \{3, 5\}$ and $A \cup B = \{1, 3, 4, 5, 6, 7, 8\}$
(a) Find $n(B)$ (b) List the elements of B.

LEVEL 3

1. The sets L, P and C are the sets of all points on a given straight line, a given plane and the circumference of a given circle respectively.
(a) Given that $n(L \cap P) > 1$ and $n(C \cap P) < 2 \times n(L \cap C)$, write down the possible values of $n(L \cap C)$.
(b) If L_1 is the set of all points on a second straight line such that $n(L_1 \cap P) = 1$ write down the possible value(s) of $n(L_1 \cap C)$.

LEVEL 4

1. Simplify the expression $A' \cup B' \cup (A \cap B \cap C')$ using the laws of sets.

13.9 TOPIC TEST

1. Use the Venn diagram to shown shade the region

i. $A' \cap B$ ii. $(A \cap B)'$

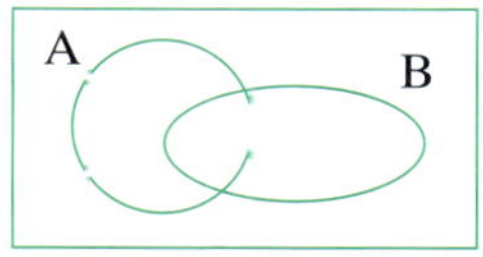

[4 marks]

2. The universal set is defined as the set of positive integers less than 16, with subsets A and B defined as;

A = {integer numbers that are multiples of 4}, B = {integer numbers that are factors of 40}

(a) List the elements of

i. set A ii. set B

(b) Copy and complete the following Venn diagram showing the relevant elements.

[5 marks]

3. Represent the following information, $n(U) = 50$, $n(A) = 18$, $n(B) = 8$ and $n(A \cap B) = 5$ using a Venn diagram and then find $n(A \cap B')$.

[4 marks]

4. A group of 42 children have been discussing their favourite action heroes. For 25 of the children, their favourite action hero is The Fanom, for 12 of them it is Justecia while 5 of them find that they cannot decide between The Fanom and Justecia.

(a) Copy and complete the table below

	The Fanom is their favourite	The Fanom is not their favourite	Total
Justecia is their favourite			
Justecia is not their favourite			
Total			42

(b) Using U as the universal set, F to represent those children whose favourite action hero is The Fanom and J the set of children whose favourite action hero is Justecia, Draw a Venn diagram to illustrate the information in the table above, making sure to label fully all relevant information.

(c) Find $n(F' \cap J)$

[6 marks]

5. In a class of 37 students, 20 have played volleyball and 22 have played basketball. If 25 have played at least one of these two sports, how many students

i. played neither sport?

ii. have not played volleyball?

iii. played volleyball but not basketball?

[8 marks]

Total 27 marks

13.10 PROJECT SUGGESTIONS

There are a number of set based paradoxes that are well described on a variety of websites.

Here are a few examples that you might enjoy investigating:

The Omnipotence Paradox:

"Can an all-powerful being create something that is greater than itself?"

Epimenides's Liar Paradox:

"I am a Cretan and I tell you that all Cretan's are liars."

The Voting Paradox:

What dangers await societies that use referenda to make decisions?

The Truth Value Paradox:

Sentence A: Sentence B is False
Sentence B: Sentence A is True

14.0 LEWIS CARROLL

Charles Lutwidge Dodgson, born 1832, died 1898, wrote two of the most popular children's books ever written in any language: *Alice's adventures in wonderland* (1865) and *Through the looking glass* (1872) under the pseudonym 'Lewis Carroll'. Dodgson was not a full time author. Rather, he was a mathematician at Christ Church, Oxford.

Dodgson's main interest was mathematical logic. He tried to increase understanding by treating logic as a game, publishing *The Game of Logic* in 1887 and *Symbolic Logic Part I* in 1896. He invented an alternative to Venn Diagrams described as Dodgson's gameboard method.

If you read the *Alice* books from the perspective of their logic, you may see how, by removing our ability to answer questions by using commonsense. Dodgson forces us to use logic.

Can a cat's smile still be there after the cat has gone? Is a magnetic field still there after you remove the compass that detects it? Dodgson wrote a number of these 'no common-sense logic problems'. Here are a few examples

A.
1. Things sold in the street are of no great value
2. Nothing but rubbish can be had for a song
3. Eggs of the great Auk are very valuable
4. It is only what is sold in the street that is really rubbish

Can you see how this leads to the conclusion: An egg of the Great Auk is not to be had for a song?

B.
1. No interesting poems are unpopular among people of real taste
2. No modern poetry is free from affectation
3. All your poems are on the subject of soap-bubbles
4. No affected poetry is popular among people of real taste
5. No ancient poetry is on the subject of soap-bubbles

Conclusion: All your poems are uninteresting.

C.
1. Animals, that do not kick, are always unexcitable
2. Donkeys have no horns
3. A buffalo can always toss one over a gate
4. No animals that kick are easy to swallow
5. No hornless animal can toss one over a gate
6. All animals are excitable, except buffalo

Conclusion: Donkeys are not easy to swallow.

As I am sure that many of you cannot follow the reasoning here, I feel that you will enjoy the contents of this chapter!

If you are bothered by this short introduction to the wonderful world of logic, I can only leave the final word to Dodgson:

Beware the Jabberwock, my son!
The jaws that bite, the claws that catch!
Beware the Jubjub bird, and shun
The frumious Bandersnatch!

14.1 DEFINITIONS

14.1.1 LOGIC AND PROPOSITIONS

Logic is the study of correct reasoning.

A proposition is a statement that is either true or false.

The following are examples of propositions:

1. Today is Saturday.
2. All animals respire.
3. Mary is 18 years old.
4. $4 > 5$
5. All mammals are warm blooded.
6. Jose is sick.
7. Frederick studies French.
8. Jock plays football.
9. The moon orbits the earth.
10. All roses are red.
11. $4 + 8 = 13$
12. All swans are white.

Each of the above statements are **either true or false** and **therefore** can be defined as **propositions**.

EXAMPLE 14.1

Consider the the statements:
(a) Go get the book.
(b) Have you seen my new shirt?
(c) The dog is behind the shed.

Which of these represents a proposition?

Solution

(a) This is a command and it does **not** have either a true or a false response.

However if it were worded Anne said "Go get the book" then this is a proposition as the truth or falsity is connected to whether or not Anne **said** "Go get the book".

(b) Have you seen my new shirt?
This is a question and therefore not a proposition.

(c) The response to the statement "The dog is behind the shed." is either it is (i.e., true) or it isn't (false). Therefore, as the statement is **either true or false**, it can be defined as a **proposition**.

EXERCISES 14.1.1

1. Identify which of the following are propositions.

(a) Juan studies French.
(b) The cooker is on.
(c) Is it raining?
(d) All sheep are white.
(e) Make the tea.
(f) Emus can fly.
(g) What is the date today?
(h) $5 > 2$
(i) The glass is full.
(j) Have a good weekend.
(k) All cows are herbivores.
(l) The sun is shining.
(m) Take out the rubbish.
(n) Put your hat on.

14.1.2 NOTATION

Propositions are usually represented by the letters p, q and r. The truthfulness or falsity of a proposition is called its **truth value**.

1. Negation Not $\neg p$

The negation of a proposition can be formed by inserting in the proposition "not" or "do not" as appropriate.

The negation of a proposition p is not p and is denoted by $\neg p$

EXAMPLE 14.2

Find the negations of the propositions:

(a) Today is Saturday
(b) All mammals respire
(c) The glass is full

Solution

(a) Proposition p: Today is Saturday
Negation $\neg p$: Today is not Saturday

(b) Proposition p: All mammals respire
Negation $\neg p$: All mammals do not respire

(c) Proposition p: The glass is full
Negation $\neg p$: The glass is not full

In Example 14.2 part (c), there might be a temptation here to write the negation as the glass is empty. However, if the glass is not completely empty it is still not full.

EXERCISES 14.1.2

1. Write the negation of the following propositions.

(a) Brutus is sick.
(b) The cup is empty.
(c) Monday is a holiday.
(d) There are twelve months in a year.
(e) $4 < 6$
(f) Ismael studies Geography.
(g) Yoshi plays football.
(h) All roses are red
(i) March has 31 days.
(j) All quadrilaterals are rectangles.
(k) Carlos lives in Japan.

14.1.3 TRUTH VALUES

A proposition is a statement that can either be true or false. The truth value of a proposition is T (true) or F (false).

Truth Values of Negation

p	$\neg p$
T	F
F	T

If the original proposition is true then the negation of the proposition would be false.

If the original proposition is false then the negation of the proposition is true.

EXAMPLE 14.3

Find the negation of:

(a) p: x is a prime number (b) p: y is odd (c) p: z is greater than 100

Solution

(a) $\neg p$: x is not a prime number.

(b) $\neg p$: y is not odd. (That is, y is even.)

(c) $\neg p$: p: z is not greater than 100. (That is, z is at most 100)

Connectives

Propositions can be combined to form compound statements. The truth or falsity of a compound statement is dependent on the truth or falsity of the individual propositions.

Conjunction 'and', $\wedge$

A conjunction is formed when propositions are joined by the word 'and'

The conjunction of two propositions p and q is denoted by $p \wedge q$.

EXAMPLE 14.4

Find the conjunctions of the propositions:

(a)	p:	John studies English.	q:	Juan studies Spanish
(b)	p:	Rebecca has red hair.	q:	Carla has black hair
(c)	p:	Liz is a teacher.	q:	Anne is a teacher.

Solution

(a) p: John studies English.
q: Juan studies Spanish
$p \wedge q$: John studies English **and** Juan studies Spanish.

(b) p: Rebecca has red hair.
q: Carla has black hair
$p \wedge q$: Rebecca has red hair **and** Carla has black hair.

(c) p: Liz is a teacher.
q: Anne is a teacher.
$p \wedge q$: Liz **and** Anne are both teachers.

Truth Values of Conjunction

p	q	$p \wedge q$
T	T	T
T	F	F
F	T	F
F	F	F

The truth value of a conjunction is only true when the truth values of all of the individual conjunctions are true. Otherwise the truth value is false.

EXAMPLE 14.5

Find the truth values of the propositions:

p: London is the capital of England
q: Cardiff is the capital of Wales
$p \wedge q$: London is the capital of England **and** Cardiff is the capital of Wales.

Solution

p: London is the capital of England. T

q: Cardiff is the capital of Wales. T

$p \wedge q$: London is the capital of England **and** Cardiff is the capital of Wales. T

The "**and**" means that all parts of the conjunction must be true for the conjunction to be true.

EXAMPLE 14.6

Find the truth values of the propositions:

p: London is the capital of England
q: Cardiff is the capital of France
$p \wedge q$: London is the capital of England **and** Cardiff is the capital of France.

Solution

p: London is the capital of England T

q: Cardiff is the capital of France F

$p \wedge q$: London is the capital of England **and** Cardiff is the capital of France. F

The "**and**" means that all parts of the conjunction must be true for the conjunction to be true. It is false that Cardiff is the capital of France therefore the conjunction is false.

Disjunction 'or' $\vee$

A disjunction is formed when propositions are joined by the word "or".

The disjunction of two propositions p and q is denoted by $p \vee q$.

EXAMPLE 14.7

Find the disjunctions of these propositions:

(a) p: London is the capital of England
q: Cardiff is the capital of Wales

(b) p: Enrique is a mathematician
q: Bill is a pilot

(c) p: Paris is in France
q: Paris is in England

Solution

(a) $p \vee q$: London is the capital of England **or** Cardiff is the capital of Wales.

(b) $p \vee q$: Enrique is a mathematician **or** Bill is a pilot.

(c) $p \vee q$: Paris is in either France **or** England.

Truth Values of Disjunctions

p	q	$p \vee q$
T	T	T
T	F	T
F	T	T
F	F	F

The truth value of a disjunction is true when **at least one** of the propositions is true.

EXAMPLE 14.8

Find the truth values of p, q and $p \vee q$:

p: Bogota is the capital of Colombia
q: Santiago is the capital of Wales

and

p: Bogota is the capital of Sweden
q: Santiago is the capital of Wales

Solution

p: Bogota is the capital of Colombia T

q: Santiago is the capital of Wales F

$p \vee q$: Bogota is the capital of Colombia **or** Santiago is the capital of Wales. T

The "**or**" in the statement means that only one of the propositions has to be true for the whole statement to be true.

The truth value of a disjunction is false when all of the individual propositions are false.

p: Bogota is the capital of Sweden F

q: Santiago is the capital of Wales F

$p \vee q$: Bogota is the capital of Sweden **or** Santiago is the capital of Wales. F

Since no part of the disjunction is true the disjunction is false.

Relationship Between Sets And Logic

Much of the language that is used when describing sets can also be found in logic statements. Because of this, it is possible to represent logic statements using set notation, or, as we will see, using Venn diagrams.

We now look at how the two relate to each other.

Negation and Sets $\neg p$, P′

EXAMPLE 14.9

Find the negation of: p: x is a prime number

Solution

$\neg p$: x is not a prime number
Let P be the truth set of the proposition p

Then P = {2, 3, 5, 7...}
Therefore P′ = {1, 4, 6, 8, 9...} which is the truth set for $\neg p$.

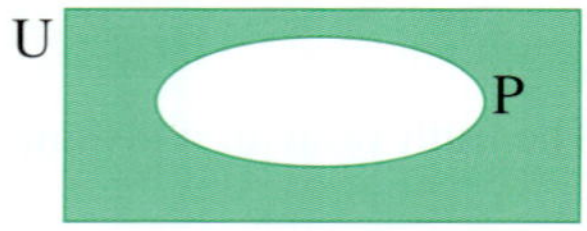

Conjunction and Sets $\wedge$, $\cap$

EXAMPLE 14.10

Find the truth sets of p, q and $p \wedge q$.
p: x is a multiple of 3 between 1 and 20
q: x is a factor of 45

Solution

Let P be the truth set of p
and Q be the truth set of q
Therefore P = {3, 6, 9, 12, 15, 18} and Q = {1, 3, 5, 9, 15, 45}
$p \wedge q$: x is a multiple of 3 between 1 and 20 and is a factor of 45

Therefore the truth set of $p \wedge q$ is {3, 9, 15} which is P ∩ Q.

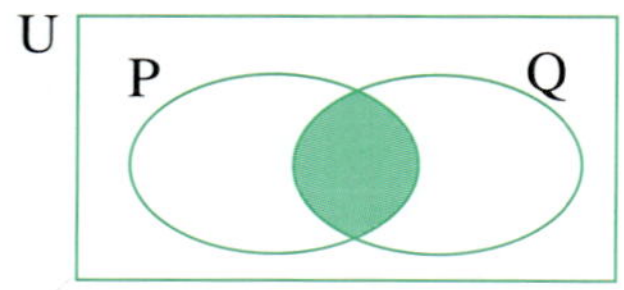

Disjunction and Sets $\vee$, $\cup$

EXAMPLE 14.11

Find the truth set of the disjunction of p and q.
p: x is a multiple of 3 between 1 and 20
q: x is a factor of 45

Solution

Let P be the truth set of p and Q be the truth set of q

Therefore P = {3, 6, 9, 12, 15, 18} and Q = {1, 3, 5, 9, 15, 45}
$p \vee q$: x is a multiple of 3 between 1 and 20 or is a factor of 45

Therefore the truth set of $p \vee q$ is {1, 3, 5, 6, 9, 12, 15, 45} which is P ∪ Q.

Exclusive Disjunction $\veebar$

So far we have illustrated the *inclusive* use of the word *or*. By this, we mean that when we consider the sentence

Millicent shall go home if Rory is running late or it is raining

we mean that if either Rory is late, or it is raining or indeed if it is raining and Rory is running late, then, Millicent shall go home. In this case, the *or* is an *inclusive* 'or' because Millicent will still go home if both events occur, that is, one event need not occur at the exclusion of the other for Millicent to go home.

However, now consider the sentence

You will go to Tasmania by plane or by ferry.

In this instance, the *or* is an *exclusive* 'or' because you cannot be on the plane and on the ferry at the same time. That is, you can travel by plane, or by ferry, but it is not possible for you to travel on both plane and ferry at the same time.

The symbol $\veebar$ is used to define 'or' in the exclusive sense. The Venn diagram representation of $p \veebar q$ is shown alongside. Notice then that this time the area corresponding to the region $P \cap Q$ is unshaded. In set notation we would have $(P \cup Q) \backslash (P \cap Q)$

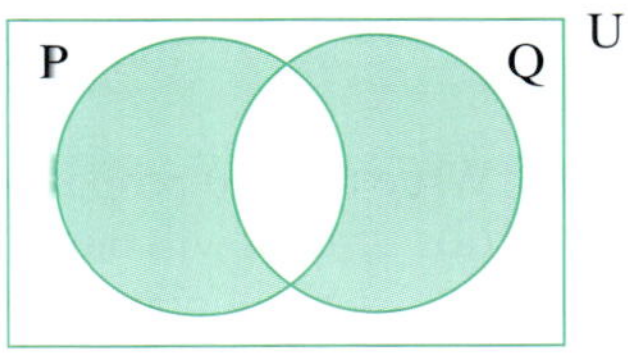

EXERCISES 14.1.3

1. Write the conjunctions of the following propositions

(a) p: All trees are green.
q: All mammals are warm blooded.
(b) p: Josh studies French
q: Anne studies French
(c) p: Abdul plays football
q: Abdul plays squash
(d) p: The moon orbits the earth
q: The earth orbits the sun
(e) p: It is raining
q: Today is Saturday

2. Write the two propositions that make up the following conjunctions

(a) Jennifer studies both physics and chemistry
(b) All mammals are warm blooded and all roses are red
(c) Ruth is 16 years old and Janet is 17 years old
(d) Today is Sunday and it is fine
(e) Ronnie plays football and Renee plays tennis

3. i. Write the truth sets P and Q for the propositions p and q.
ii. Draw a Venn diagram and show all the elements.
iii. Write the truth set for $p \wedge q$.
(a) p: x is a multiple of 4 between 1 and 20
q: x is a multiple of 6 between 1 and 20
(b) p: x is a square number less than 100
q: x is a number that ends in 6 less than 40
(c) p: x is a prime number less than 20
q: x is a multiple of 3 less than 20

4. Write the disjunctions of the following pairs of propositions
(a) p: All tree are green.
q: All mammals are warm blooded.
(b) p: Josh studies French
q: Anne studies French
(c) p: Abdul plays football
q: Abdul plays squash
(d) p: The moon orbits the earth
q: The earth orbits the sun
(e) p: It is raining
q: Today is Saturday

5. Write the two propositions that make up the following disjunctions.
(a) John will buy either a TV or a stereo
(b) Yoshi plays tennis or Mohammed plays squash
(c) The moon orbits the earth or the sun
(d) Emus are birds or goats are horses
(e) Ruth likes watching the theatre or watching sports

6. i. Write the truth sets P and Q for the propositions p and q.
ii. Draw a Venn diagram and show all the elements.
iii. Write the truth set for $p \vee q$.
(a) p: x is a multiple of 4 between 1 and 20
q: x is a multiple of 6 between 1 and 20
(b) p: x is a square number less than 100
q: x is a number that ends in 6 less than 40
(c) p: x is a prime number less than 20
q: x is a multiple of 3 less than 20

14.1.4 IMPLICATIONS

An implication is joined when propositions p and q are linked in the following ways.

If p then q
p only if q
p is sufficient for q
q is necessary for p

In the cases above the proposition p is called the **antecedent** and the proposition q is called the **consequent**.

The implication of two propositions p and q is denoted by $p \Rightarrow q$.

EXAMPLE 14.12

Find $p \Rightarrow q$ in each of these cases:

(i) p: John is not at work
q: John is sick

(ii) p: My watch is slow
q: I will be late

(iii) p: The radio works
q: The power is on

Solution

(a) $p \Rightarrow q$:
If John is not at work **then** he is sick

(b) $p \Rightarrow q$:
If my watch is slow **then** I will be late

(c) $p \Rightarrow q$:
The radio works **only if** the power is on

Truth Values for Implication

p	q	$p \Rightarrow q$
T	T	T
T	F	F
F	T	T
F	F	T

Consider the following propositions and the implication

p: You get an A on your maths test
q: Your mum buys you a pair of jeans
$p \Rightarrow q$: If you get an A on your maths test then your mum will buy you a pair of jeans.

p: You get an A on your maths test T
q: Your mum buys you a pair of jeans T
$p \Rightarrow q$: If you get an A on your maths test then your mum will buy you a pair of jeans. T

The implication is true because you got the A and your mum kept her promise.

p: You get an A on your maths test T
q: Your mum buys you a pair of jeans F
$p \Rightarrow q$: If you get an A on your maths test then your mum will buy you a pair of jeans. F

The implication is false because you got the A but your mum did not keep her promise.

p: You get an A on your maths test F
q: Your mum buys you a pair of jeans T
$p \Rightarrow q$: If you get an A on your maths test then your mum will buy you a pair of jeans. T

The implication is true because even though you did not get an A there is no information given about what your mum will do if you didn't get an A. You may have only missed getting an A by a very small margin.

p: You get an A on your maths test F
q: Your mum buys you a pair of jeans F
$p \Rightarrow q$: If you get an A on your maths test then your mum will buy you a pair of jeans. T

As mentioned previously there is no information given about what your mum will do if you do not get an A. In this example your mum has a choice of buying the jeans or not buying the jeans. She decided not to buy the jeans but the important thing to remember is that in this case as in the previous case your mum has not lied.

Implication and Sets $\Rightarrow \subset$ (Is a subset of)

EXAMPLE 14.13

If p: x is a multiple of 4 greater than 1 but less than 20
q: x is an even number greater than 1 but less than 20

Find $p \wedge q$ and $P \cap Q$

Solution

Let P be the truth set of p and Q be the truth set of q
Therefore $P = \{4, 8, 12, 16\}$
$Q = \{2, 4, 6, 8, 10, 12, 14, 16, 18\}$

$p \Rightarrow q$: If x is a multiple of 4 between 1 and 20 then it is an even number between 1 and 20
Therefore $p \wedge q$ is $P \cap Q$.
(Which is P).

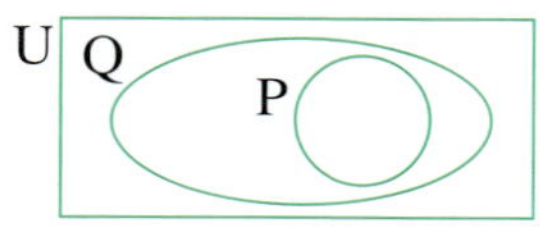

EXERCISES 14.1.4

1. Write implication statements using the following pairs of propositions. Take p as the antecedent and q as the consequent.

(a) p: x is a multiple of 9
q: x is divisible by 3
(b) p: Yoshi is sick
q: Yoshi is not at work
(c) p: Paris is in France
q: The sky is blue
(d) p: This is the right time for an argument
q: This is the right place for and argument
(e) p: John works hard
q: John earns money

2. Identify the antecedent and the consequent in the following implication statements.

(a) If a number is divisible by 10 then it ends in a zero
(b) I will earn more money only if I work hard
(c) The flowers will grow only if there is enough rain
(d) If a number is divisible by four then it is even
(e) If I think then I am

3. i. Write the truth sets P and Q for the propositions p and q.
ii. Draw a Venn diagram showing the relationship between P and Q

(a) p: x is a square number between 10 and 50
q: x is a number between 10 and 50 that ends in either 5,6 or 9
(b) p: Fuzzy is a bear
q; Fuzzy is a mammal

14.1.5 Converse, Inverse and Contrapositive of Implications

For the statements:

p: A quadrilateral is a rectangle
q: A quadrilateral is a parallelogram

The implication $p \Rightarrow q$ is:
If a quadrilateral is a rectangle then it is a parallelogram.

The converse $p \Leftarrow q$ is:
If a quadrilateral is a parallelogram then it is a rectangle.
[Swap the antecedent and the consequent]

The inverse $\neg p \Rightarrow \neg q$ is:
If a quadrilateral is not a rectangle then it is not a parallelogram.
[Negate both the antecedent and the consequent]

The contrapositive $\neg p \Leftarrow \neg q$ is:
If a quadrilateral is not a parallelogram then it is not a rectangle.
[Swap the antecedent and the consequent and negate both]

EXERCISES 14.1.5

1. Write the converse, inverse and contrapositive statements for the following implication statements.

(a) If a number is divisible by 10 then it ends in a zero
(b) I will earn more money only if I work hard
(c) The flowers will grow only if there is enough rain
(d) If a number is divisible by four then it is even
(e) If I think then I am

14.1.6 EQUIVALENCE 'IF AND ONLY IF', IFF, $\Leftrightarrow$

If $p \Rightarrow q$ is true and its converse $p \Leftarrow q$ is true then $p \Leftrightarrow q$ is true

Equivalence is formed when the propositions are joined by the phrase 'if and only if' which has the abbreviation 'iff'.

The equivalence of two propositions p and q is denoted by $p \Leftrightarrow q$.

EXAMPLE 14.14

For p and q defined below, describe $p \Leftrightarrow q$.

(a) p: We will play tennis
q: It is fine
$p \Leftrightarrow q$: We will play tennis **if and only if** it is fine

(b) p: Mary will pass maths
q: The exam is easy
$p \Leftrightarrow q$: Mary will pass maths **iff** the exam is easy

(c) p: Madrid is in Spain
q: Spain is in Europe
$p \Leftrightarrow q$: Madrid is in Spain **iff** Spain is in Europe

Solution

(a) $p \Leftrightarrow q$:
We will play tennis **if and only if** it is fine

(b) $p \Leftrightarrow q$:
Mary will pass maths **iff** the exam is easy

(c) $p \Leftrightarrow q$:
Madrid is in Spain **iff** Spain is in Europe

Truth Values for Equivalence

p	q	$p \Leftrightarrow q$
T	T	T
T	F	F
F	T	F
F	F	T

The truth value of equivalence is true only when all of the propositions have the same truth value.

Consider the following propositions and the implications.

p: I will buy you an ice-cream
q: You get an A on your maths exam
$p \Leftrightarrow q$: I will buy you an ice-cream if and only if you get an A on your maths exam

The implication statement is very clear that the only way you are going to get an ice-cream is if you get an A on your maths exam.

p: I will buy you an ice-cream F
q: You get an A on your maths exam T
$p \Leftrightarrow q$: I will buy you an ice-cream if and only if you get an A on your maths exam. F

In this case you got an A but I didn't buy you an ice-cream so I lied therefore the equivalence statement is false.

p: I will buy you an ice-cream T
q: You get an A on your maths exam F
$p \Leftrightarrow q$: I will buy you an ice-cream if and only if you get an A on your maths exam. F

In this case I bought you an ice-cream even though you didn't get an A which means I lied also so the equivalence statement is false.

Equivalence and Sets $\Leftrightarrow$ = (Equal Sets)

EXAMPLE 14.15

If p: A polygon has 4 sides
q: A polygon is a quadrilateral
Show that P and Q are equal sets.

Solution

$p \Rightarrow q$: If a polygon has 4 sides then it is a quadrilateral

$p \Leftarrow q$: If a polygon is a quadrilateral then it has 4 sides

$p \Leftrightarrow q$: A polygon has 4 sides iff it is a quadrilateral

The propositions p an q have the same truth sets.
Therefore P and Q are equal sets.

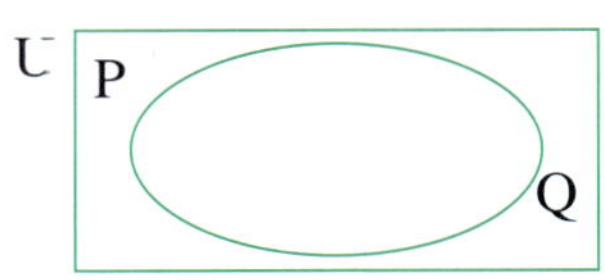

Truth Tables

You have already seen tables being used to show the truth values of compound propositions formed by two propositions and a connective. This idea can be extended to find the truth values of compound propositions that are formed by a number of propositions and a number of connectives as well as negation.

EXAMPLE 14.16

Construct a truth table to show the truth values of the compound proposition: $\neg p \wedge q$

Solution

Step One: Identify the propositions and the connectives: $p, q, \neg p, \neg p \wedge q$

These become the headings of the truth table.

Step Two: Construct a table using the propositions above as headings and enter all combinations for the truth values of p and q.

p	q	$\neg p$	$\neg p \wedge q$
T	T		
T	F		
F	T		
F	F		

Step Three: Complete the truth table using the rules previously outlined.

p	q	$\neg p$	$\neg p \wedge q$
T	T	F	
T	F	F	
F	T	T	
F	F	T	

p	q	$\neg p$	$\neg p \wedge q$
T	T	F	F
T	F	F	F
F	T	T	T
F	F	T	F

Note the compound proposition is only true when the original proposition p is F and the proposition q is T.

EXERCISES 14.1.6

1. Write equivalence statements for the following pairs of propositions.

(a) p: I will paint the house
q: It is fine

(b) p: We will go to Disneyland
q: There are exciting rides

(c) p: I will take my umbrella
q: It is raining

(d) p: John will hit a home run
q: The pitcher is useless

2. i. Construct truth tables for each of the following compound propositions.
ii. State the truth value of the original propositions when the truth value of the compound proposition is true.

(a) $\neg p \vee q$
(b) $\neg p \vee \neg q$
(c) $\neg p \wedge \neg q$
(d) $\neg p \Rightarrow q$
(e) $(p \vee \neg q) \Rightarrow q$
(f) $\neg p \Leftrightarrow q$
(g) $(p \vee \neg q) \Rightarrow (p \wedge \neg q)$
(h) $(\neg p \wedge q) \vee (q \wedge p)$
(i) $\neg(p \Rightarrow \neg q)$
(j) $(p \wedge q) \Rightarrow (p \vee q)$
(k) $\neg(p \wedge q) \vee \neg(q \Leftrightarrow p)$

14.1.7 TAUTOLOGIES AND CONTRADICTIONS

A tautology is a compound proposition that is **always true** regardless of the truth values of the individual propositions.

EXAMPLE 14.17

Show that the compound proposition p or not p is a tautology.

Solution

p	$\neg p$	$p \vee \neg p$
T	F	T
F	T	T

All the final entries are 'T' so the proposition is a tautology.

EXAMPLE 14.18

Show that the compound proposition 'If p, then p or q' is a tautology.

Solution

p	q	$p \vee q$	$p \Rightarrow (p \vee q)$
T	T	T	T
T	F	T	T
F	T	T	T
F	F	F	T

All the final entries are 'T' so the proposition is a tautology.

A contradiction is a compound proposition which is always false regardless of the truth values of the individual propositions.

EXAMPLE 14.19

Show that the compound proposition 'p and not p' is a contradiction.

Solution

p	$\neg p$	$p \wedge \neg p$
T	F	F
F	T	F

All the final entries are 'F' so the proposition is a contradiction.

Truth tables involving 3 propositions

Truth tables can be extended to deal with a number of propositions. 3 propositions is the limit required for this course.

The only difference in constructing truth tables for 3 propositions is that you must ensure that you have all the combinations of original truth values for p, q and r.

Your truth table for 3 propositions should start with the following three columns.

p	q	r
T	T	T
T	T	F
T	F	T
T	F	F
F	T	T
F	T	F
F	F	T
F	F	F

EXAMPLE 14.20

Construct a truth table to show all the truth values of the compound proposition: $(\neg p \vee q) \Rightarrow r$

Solution

p	q	r	$\neg p$	$\neg p \vee q$	$(\neg p \vee q) \Rightarrow r$
T	T	T	F	T	T
T	T	F	F	T	F
T	F	T	F	F	T
T	F	F	F	F	T
F	T	T	T	T	T
F	T	F	T	T	F
F	F	T	T	T	T
F	F	F	T	T	F

EXERCISES 14.1.7

1. i. Construct truth tables for the following compound propositions.
ii. State the truth value of the original propositions when the compound proposition is true.

(a) $(p \vee q) \Rightarrow r$
(b) $\neg p \vee \neg q \wedge r$
(c) $\neg p \wedge \neg q \wedge \neg r$
(d) $\neg p \Rightarrow (q \vee r)$
(e) $(p \vee \neg q) \Rightarrow r$
(f) $(\neg p \Leftrightarrow q) \vee r$
(g) $(p \vee \neg r) \Rightarrow (r \wedge \neg q)$
(h) $(\neg p \wedge q) \vee (q \wedge r)$
(i) $(p \wedge r) \Rightarrow (r \vee q)$
(j) $\neg(p \wedge q) \vee \neg(q \Leftrightarrow r)$

14.1.8 TESTING THE VALIDITY OF SIMPLE ARGUMENTS USING TRUTH TABLES

An argument is made up of one or more premises that lead to a conclusion. The premises and the conclusion are propositions. If the premises provide support for the conclusion then the argument is said to be valid. The conclusion of an argument can be identified as it is introduced by words such as therefore, hence, so, it follows.

EXAMPLE 14.21 Prove: If Fuzzy is a bear then Fuzzy is a mammal. Fuzzy is a bear. Therefore Fuzzy is a mammal.

Solution

The premises are the propositions: p: Fuzzy is a bear.
q: Fuzzy is a mammal.

The conclusion is the proposition: q: Fuzzy is a mammal.

The argument can be written in logical form in the following way.
If Fuzzy is a bear then Fuzzy is a mammal. $p \Rightarrow q$
Fuzzy is a bear p

These two propositions combined $(p \Rightarrow q) \wedge p$

The premises logically imply the conclusion: $[(p \Rightarrow q) \wedge p] \Rightarrow q$

The argument is valid if the truth values of the logic statement form a tautology

p	q	$p \Rightarrow q$	$(p \Rightarrow q) \wedge p$	q	$[(p \Rightarrow q) \wedge p] \Rightarrow q$
T	T	T	T	T	T
T	F	F	F	F	T
F	T	T	F	T	T
F	F	T	F	F	T

Therefore the argument is valid because the compound proposition is a tautology.

EXAMPLE 14.22

Discuss the validity of the argument:
If Julia is sick she will not go to work. Julia is not sick. Therefore Julia will go to work.

Solution

The premises are the propositions
p: Julia is sick
$\neg p$: Julia is not sick [Notice that $\neg$ is used because of 'not' in the proposition]

The conclusion is the proposition: q: Julia will go to work
$\neg q$: Julia will **not** go to work
The argument can be written in logical form in the following way.

If Julia is sick then she will not go to work $p \Rightarrow \neg q$
Julia is not sick $\neg p$
These premises combined $(p \Rightarrow \neg q) \wedge \neg p$

The premises logically imply the conclusion: $[(p \Rightarrow \neg q) \wedge \neg p] \Rightarrow q$

p	q	$\neg q$	$p \Rightarrow \neg q$	$\neg p$	$(p \Rightarrow \neg q) \wedge \neg p$	q	$[(p \Rightarrow \neg q) \wedge \neg p] \Rightarrow q$
T	T	F	F	F	F	T	T
T	F	T	T	F	F	F	T
F	T	F	T	T	T	T	T
F	F	T	T	T	T	F	F

Since the compound proposition is not a tautology the argument in invalid.

EXERCISES 14.1.8

1. Express the following arguments symbolically and use truth tables to determine whether or not each argument is valid.

(a) If it rains then the concert will be cancelled. If the concert is cancelled the money for the tickets will be refunded. It rains. Therefore the money for the tickets is refunded.

(b) If Linda doesn't study then she will not pass her exam. Linda studies. Therefore Linda passes her exam.

(c) If Hank is elected class captain then Martin will be vice captain. If Julia is elected class captain then Hank will not be vice captain. Julia is not elected class captain. Therefore Hank is elected vice captain.

14.2 MISCELLANEOUS EXERCISES

1. State whether each of the following is a proposition.
(a) The moon is made of blue cheese
(b) Can you see the moon?
(c) Look at the moon
(d) What is the time?
(e) Yesterday was Tuesday.
(f) Margaret is 17

2. Write the implication, the converse, the inverse and the contrapositive of the following pairs of propositions in both words and logical form.

(a) p: Fuzzy is a bear
q: Fuzzy is cute

(b) p: Mary lives in Spain
q: Mary loves fish

(c) p: It is fine
q: We will go to the concert

(d) p: I work hard
q: I pay taxes

(e) p: John loves fishing
q: John lives by the sea

3. Which one of the following statements is a tautology?
(a) $((p \wedge q) \wedge p) \Rightarrow q$ (b) $(p \wedge q) \vee \neg p$ (c) $p \Rightarrow (q \vee r)$

4. Consider the truth table shown below:

p	q	$\neg q \Rightarrow \neg p$
T	T	T
T	F	T
F	T	F
F	F	T

Which row has an incorrect entry in the column $\neg q \Rightarrow \neg p$?

5. Which of the following are valid arguments?
(a) $(p \Rightarrow q) \wedge (q \Rightarrow r) \therefore (p \Rightarrow r)$
(b) $(p \Leftrightarrow q) \wedge p \therefore q$
(c) $(p \Rightarrow q) \wedge (\neg q) \therefore (\neg p)$

6. Use a truth table to comment on the statement $(p \Rightarrow q) \Rightarrow (\neg p \vee q)$.

7. Which of the following statements are false?

(a) If $n^2 = n$ then $n = 0$.

(b) If $2n$ is an even number, then n is an odd number.

(c) If two of the internal angles of the triangle XYZ are equal, then the triangle is an isosceles triangle.

8. Copy and complete the following truth table

p	q	$p \wedge q$	$\neg p$	$\neg q$	$(\neg p \wedge \neg q)$	$(p \wedge q) \vee (\neg p \wedge \neg q)$
F	F					
F	T					
T	F					
T	T					

9. (a) Use a Venn diagram to show the region defined by $A \cup (A' \cap B')$.

(b) Simplify the statement $p \vee (\neg p \wedge \neg q)$.

10. Write the equivalent statement in set notation for each of the following logic statements.

(a) $p \wedge q$ (b) $p \vee (q \vee r)$ (c) $p \wedge (q \vee r)$

11. Rewrite the logic statement $[p \vee (q \wedge \neg p)] \wedge (q \vee p)$ as an expression involving sets. Hence, simplify the statement $[p \vee (q \wedge \neg p)] \wedge (q \vee p)$.

12. Simplify the following logic statements

(a) $p \wedge (q \vee \neg q)$ (b) $(p \vee \neg p) \wedge p$ (c) $(p \wedge \neg p) \vee p$

(d) $q \vee (\neg p \wedge p)$ (e) $\neg p \vee \neg q \vee p$ (f) $(p \wedge q) \vee (p \wedge \neg q)$

13. Which one of the following statements should occupy the last column in the truth table?

(a) $p \wedge q$ (b) $\neg(p \vee q)$ (c) $\neg(p \Rightarrow q)$ (d) $p \Leftrightarrow q$

p	q	
T	T	F
T	F	T
F	T	F
F	F	F

14. Complete the truth table below

p	q	$\neg p$	$\neg p \wedge q$
T	T		
T	F		
F	T		
F	F		

15. In a proof, using the contrapositive, of the statement "If $x^2 > x$, then $x > 1$" which of the following valid arguments can be used

(a) assume $x^2 > x$ is true and show $x > 1$.

(b) assume $x^2 < x$ is true and show $x < 1$.

(c) assume $x > 1$ is true and show $x^2 > x$.

(d) assume $x > 1$ is true and show $x^2 < x$.

(e) assume $x < 1$ is true and show $x^2 < x$.

16. Identify the individual propositions in each of the following statements and write the compound propositions in logical form.

(a) Today is not Tuesday.
(b) x is either an even number or a prime number.
(c) Mary and John study French.
(d) If it is raining then the concert will be cancelled.
(e) Yoshi will go to the concert iff it is not raining.
(f) Birgit likes ice-cream and cake.
(g) Jessica will go to the concert if and only if Mary goes to the concert and it is not raining.
(h) If it is fine and the temperature is between 20°C and 30°C then Paul will play tennis.
(i) If I work hard then I will pass my exams
(j) If Bill wins this race then he will make the final.
(k) If Bill does not win this race then he will not make the final.

17. Express the following arguments in logical form then use a truth table to determine whether or not the argument is valid.

(a) If I study then I will pass my exams. I study. Therefore I pass my exams.
(b) If I study then I will pass my exams. If I pass my exams I will go to university. I study. Therefore I go to university.
(c) If Fuzzy is a bear then Fuzzy is a mammal. Fuzzy is a mammal. Therefore Fuzzy is a bear.
(d) I study hard iff I go to university. I go to university. Therefore I study hard.
(e) All accountants are boring people. Mary is not an accountant. Therefore Mary is not boring.

18. Consider the Venn diagram below which is composed of three sets namely W = {well fed cats}, F = {fat cats} and H = {hungry cats}.
The universal set, U, is the set of all cats and no region is the null set.
Which statement below is false?

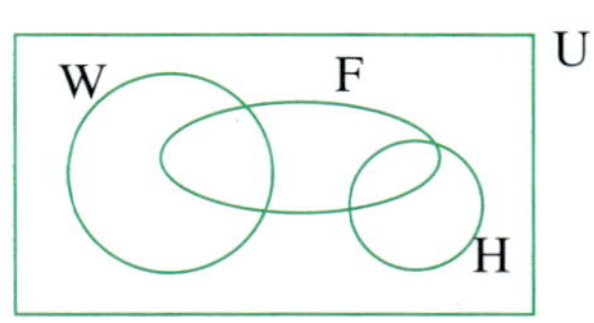

(a) all hungry cats are not well fed.
(b) some fat cats are hungry and some are well fed.
(c) some fat cats are hungry and well fed.
(d) well fed cats cannot be hungry.
(e) some well fed cats are fat cats.

19. Given the statements p: $x^2 = 2x, x \in \mathbb{R}$, q: $x = 2$

(a) Is p a sufficient condition for q?
(b) Is q a sufficient condition for p?
(c) Write down a valid logic statement between p and q.

14.3 GRADED REVISION EXERCISES

LEVEL 1

1. Write the contrapositive statement of the sentence "All flibs are blets".

2. Let p represent " It is hot" and q represent "I will go swimming". Using p and q, write in symbolic form: "If it is hot, then I will *not* go swimming".

3. For the propositions p: Physics is difficult q: I don't study hard
write verbal expressions for

(a) $p \Rightarrow q$ (b) $q \Rightarrow p$ (c) $p \Leftrightarrow q$

(d) the contrapositive of $p \Rightarrow q$

LEVEL 2

1. i. What is the converse of $\neg p \Rightarrow q$?

ii. If x represents "She is tall" and y represents "She is beautiful", write in symbolic form: "She is not tall and she is beautiful".

2. Which of the following statements is a logical contradiction?

(a) $x \wedge \neg y$ (b) $(x \Rightarrow y) \wedge (y \Rightarrow x)$ (c) $(x \Rightarrow y) \vee (y \Rightarrow x)$

(d) $x \Rightarrow \neg x$ (e) $(x \vee y) \wedge (\neg x \wedge \neg y)$

LEVEL 3

1. i. Complete the following truth table with these headings.

p	q	$\neg p$	$\neg p \Rightarrow q$	$p \vee q$	$(\neg p \Rightarrow q) \Leftrightarrow (p \vee q)$

ii. Is $(\neg p \Rightarrow q) \Leftrightarrow (p \vee q)$ a tautology?

iii. Let p represent: "I do my training" and q represent: "I get into trouble".
Write down a sentence which is equivalent to $(p \vee q)$.

2. If $x + y$ stands for x **or** y and xy stands for x **and** y, write down the following statements as simply as possible. i. $x + xy$ ii. $xy' + x'y + xy$

3. Use a truth table to determine the validity of the following argument:

p: If I miss the bus, I will be late for school
q: If I am late for school, the teacher will not be happy
r: Therefore, if I miss the bus, the teacher will not be happy

LEVEL 4

1. Using the propositions

p: the line drawn from the vertex of a triangle to the midpoint of the opposite side does not intersect this side at a right angle.

q: the triangle is not equilateral.

Prove that "If a median does not intersect the third side at a right angle, the triangle cannot be equilateral".

2. Show that $(p \Rightarrow q) \wedge (q \Rightarrow r) \Rightarrow (p \Rightarrow r)$ is a tautology.

14.4 TOPIC TEST

1. Identify the word statements that must be attributed to p and q given that

$p \Rightarrow q$ reads: If the movie is finished, we can go home

[2 marks]

2. Let x represent the statement "The number is an even integer" and y the statement "Three times the number is 18". Analyse each statement and then decide which statement(s) is(are) true if the number is 4?

A. $x \wedge y$ B. $x \vee y$ C. y D. $x \Rightarrow y$

[5 marks]

3. Consider the statements p: x is prime q: x is odd r: $x > 2$

(a) Write the statement defined by $(p \wedge q) \Rightarrow r$.

(b) Write into symbolic form, the sentence:

If x is a prime number greater than 2, then x is odd.

[5 marks]

4. (a) Using these headings, complete the truth table below.

p	q	$p \Rightarrow q$	$\neg q$	$p \wedge \neg q$	$\neg(p \wedge \neg q)$	$(p \Rightarrow q) \Leftrightarrow \neg(p \wedge \neg q)$
T	**T**					
T	**F**					
F	**T**					
F	**F**					

(b) Let p represent the statement "Nora lives in Sydney" and let q represent the statement "Nora lives in New South Wales".
With the aid of the table in part i., write the statement which is equivalent to
"*If Nora lives in Sydney, then Nora lives in New South Wales*."

[10 marks]

5. Given the propositions

p: he eats too much
q: he is healthy
r: he is happy

Write each of the following symbolic statements in verbal form.

(a) $\neg p$ (b) $\neg p \wedge r$ (c) $q \vee r$ (d) $p \Rightarrow \neg r$
(e) $(\neg p \wedge q) \Rightarrow r$ (f) $p \Leftrightarrow \neg r$ (g) $p \vee q$ (h) $(\neg p \vee q) \Rightarrow r$
(i) $\neg p \Rightarrow (q \wedge r)$ (j) $r \Leftrightarrow (\neg p \wedge q)$

[10 marks]

6. (a) Illustrate, using the Venn diagram shown, the equivalent representation of the logic statements

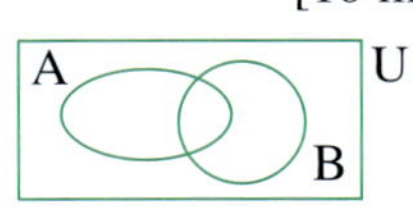

i. $(p \wedge \neg q)$ ii. $(p \wedge q) \vee (\neg p \wedge \neg q)$

(b) Give the equivalent expressions in set notation for part (a).

[6 marks]

7. Given the statements p: $x^2 = 3x, x \in \mathbb{R}$

q: $x = 3$

(a) Is p a sufficient condition for q?

(b) Is q a sufficient condition for p?

(c) Write down a valid logic statement between p and q.

[7 marks]

Total 45 marks

14.5 PROJECT SUGGESTIONS

The chapter introduction posed some questions that you might choose to investigate. What are the conclusions to be drawn from the logical problems given and can you research more?

Logic is widely used in practical applications that could lead to good projects. Here are some questions that you might choose to research.

How are houses wired for electricity? Most switches are 'up-off, down-on', at least in most countries. Houses with stairs often have a safety light that can be switched on or off by altering the position of the switch at either end of the house. How is this achieved?

Trains are assembled by 'shunting' in specially built 'yards' How are these designed to make the process as efficient as possible?

George Boole (1815–1864), is recognised for the invention of a branch of mathematics known as Boolean algebra, which is dependent entirely on True or False (or Yes/No or On/Off) outcomes. Boolean algebra played an important part in the electronic revolution. Investigate the relationship that exists between Logic Theory and Boolean Algebra

There has been only one abstract mathematical problem mentioned in a major feature film (*Enigma*, Columbia/Tristar/Miramax), at least as far as I know! This was the *Entscheidungsproblem* (German: decision problem). The challenge is to find a logical algorithm which decides for given first-order statements whether they are valid or not. Alonzo Church and, independently, Alan Turing (who seems to have contributed a part of one character in *Enigma*) showed in 1936 that this was impossible.

The above logical problem was at the heart of the development of the algorithm and, hence, the computer. Aside from its historical significance, why did this 'decision problem' prove so interesting?

15.0 GAMBLING

The playing of sports and games is as old as civilisation. People play them for fun - but they also like to bet (or gamble) on the outcome. Sports such as soccer and baseball require fitness and skill. Most also have an element of chance such as the weather, field conditions etc. Many games however, depend entirely on the skill of the players. Examples are:

Go originated either in India or in China as early as 2500 BC and was brought to Japan about 500 AD.

Chess first appeared in India about the 6th century AD and by the 10th century had spread from Asia to the Middle East and Europe.

Other games are a blend of skill and chance (the deal of cards, fall of dice etc.)

The Chinese game **Mah-Jongg** is probably of 19th-century origin though it seems likely to have been derived from much older games.

The **whist** and **bridge** family of games is of English origin, having evolved from other games, principally one called **triumph**, a name that became corrupted to trump. Whist developed into **bridge whist** in the 1890s. This evolved into auction bridge in the early 20th century. **Contract bridge** was developed in the 1920s.

Poker developed from a 16th-century European three-card game called **primera** in Spain or **primero** in England.

Blackjack (also called Twenty-one, Vingt-et-un, Van John, Pontoon) is related to the French **Ferme** and **Chemin de Fer** and the Italian **Seven and a Half**.

Other games depend solely on chance. The best known of these is probably **Roulette**. Roulette emerged in the late 18th century. It became a glamorous attraction in the casinos of Europe, such as that at Monte-Carlo. It was to analyse these games that mathematicians developed probability, the subject of this chapter. Two of the leading figures were:

The Italian Jerome Cardano (described by one biographer as 'an ambitious, dishonest, hot-tempered, quarrelsome, conceited and humorless man'). His 15 page booklet on probability, *Liber de Ludo* (1663) is primarily a practical guide to gambling, including cards and dice and cheating. In the mathematical sections Cardano talks about probability, compound events, expectation, frequency tables for dice, and the 'law of large numbers'.

The Frenchman Blaise Pascal born 1623, died 1662, formulated much of the theory of probability as it exists today. He is said to have advised the casino owners as to how their games should be organised so that they appeared fair whilst in fact guaranteeing steady losses for gamblers. It is Pascal who should be remembered by the lines of disappointed punters leaving the World's casinos. If there is a lesson from this chapter it is that, if you want to gamble, do not do so in the expectation of winning. Instead, view it as a recreation that, like most recreations, comes at a cost!

15.1 PROBABILITY

We are often faced with statements that reflect an element of likelihood, For example, "It is likely to rain later in the day" or "What are the chances that I roll a six?". Such statements relate to a level of uncertainty (or indeed, a level of certainty). It is this element of likelihood in which we are interested. In particular, we need to find a measure of this likelihood — i.e., the associated probability of particular events.

15.1.1 PROBABILITY AS A LONG-TERM RELATIVE FREQUENCY

An experiment is repeated in such a way that a series of independent and identical trials are produced, so that a particular event A is observed to either occur or not occur. We let N be the total number of trials carried out and $n(A)$ (or $|A|$) be the number of times that the event A was observed.

We then call the ratio $\frac{n(A)}{N}$ $\left(\text{or } \frac{|A|}{N}\right)$ the **relative frequency** of the event A. This value provides some indication of the likelihood of the event A occuring.

In particular, for large values of N we find that the ratio $\frac{n(A)}{N}$ tends to a number called the **probability** of the event A, which we denote by $p(A)$ or $P(A)$.

As $0 \leq n(A) \leq N$, this number, $P(A)$, must lie between 0 and 1 (inclusive), i.e., $0 \leq P(A) \leq 1$.

A more formal definition is as follows:

If a random experiment is repeated N times, in such a way that each of the trials is identical and independent, where $n(A)$ is the number of times event A has occured after N trials, then

$$\text{As } N \to \infty, \frac{n(A)}{N} \to P(A).$$

It is possible to provide a graph of such a situation, which shows that as N increases, the ratio $\frac{n(A)}{N}$ tends towards some value p, where in fact, $p = P(A)$.

Such a graph is called a **relative frequency graph**.

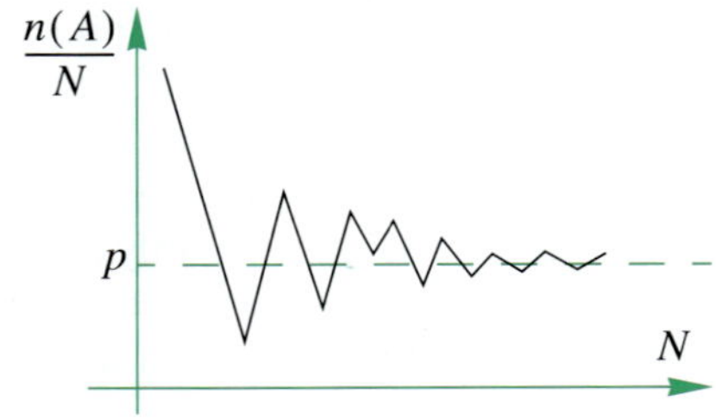

15.1.2 THEORETICAL PROBABILITY

When the circumstances of an experiment are always identical, we can arrive at a value for the probability of a particular event by using mathematical reasoning, often based on an argument reflecting some form of symmetry (i.e., without the need to repeatedly perform the experiment). This type of probability is called **theoretical probability**.

For example, when we roll a die, every possible outcome, known as the **sample space**, can be listed as $U = \{1, 2, 3, 4, 5, 6\}$ (sometimes the letter ε is used instead of U). The probability of obtaining a "four" (based on considerations of **symmetry of equal likelihood**) is given by $\frac{1}{6}$.

Such a probability seems obvious, as we would argue that;

"Given there are six possible outcomes and each outcome is equally likely to occur (assuming a fair die), then the chances that a "four" occurs must be one in six, i.e., $\frac{1}{6}$."

15.1.3 LAWS OF PROBABILITY

We will restrict our arguments to **finite sample spaces**. Recall, that a **sample space** is the set of every possible outcome of an experiment, and that an **event** is any subset of the sample space. This relationship is often represented with a Venn diagram:

The Venn diagram shows the sample space U, with the event A, as a subset.

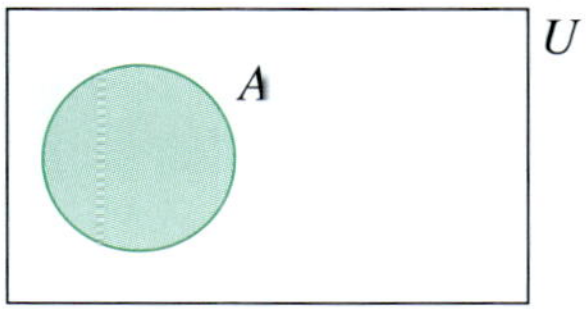

15.1.4 DEFINITION OF PROBABILITY

If an experiment has equally likely outcomes and of these the event A is defined, then the **theoretical probability of event A** occuring is given by

$$P(A) = \frac{n(A)}{n(U)} = \frac{\text{Number of outcomes in which } A \text{ occurs}}{\text{Total number of outcomes in the sample space}}$$

Where $n(U)$ is the total number of possible outcomes in the sample space, U, (i.e., $n(U) = N$).

As a consequence of this definition we have what are known as the **axioms of probability**:

1. $0 \le P(A) \le 1$

2. $P(\varnothing) = 0$ and $P(\varepsilon) = 1$

That is, if $A = \varnothing$, then the event A can never occur.

$A = U$ implies that the event A is a certainty.

3. If A and B are both subsets of U and are mutually exclusive, then

$$P(A \cup B) = P(A) + P(B).$$

Note:

Two events A and B are said to be **mutually exclusive** (or disjoint) if they have no elements in common, i.e., if $A \cap B = \varnothing$.

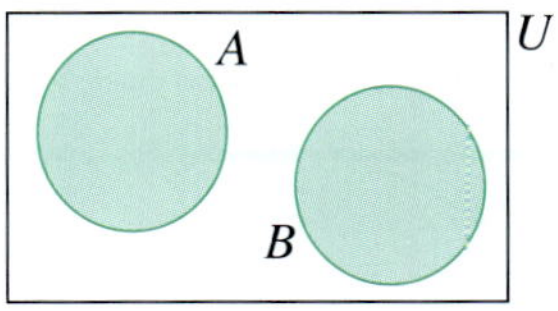

EXAMPLE 15.1

A fair die is thrown. List the sample space of the experiment and hence find the probability of observing:

(a) a multiple of 3

(b) an odd number.

Are these events mutually exclusive?

Solution

(a) The sample space is $U = \{1, 2, 3, 4, 5, 6\}$.
Let A be the event 'obtaining a multiple of 3'.

We then have that $A = \{3, 6\}$. Therefore, $P(A) = \frac{n(A)}{n(U)} = \frac{2}{6} = \frac{1}{3}$.

(b) Let B be the event 'obtaining an odd number'.

Here $B = \{1, 3, 5\}$ and so $P(B) = \frac{n(B)}{n(U)} = \frac{3}{6} = \frac{1}{2}$.

In this case, $A = \{3, 6\}$ and $B = \{1, 3, 5\}$, so that $A \cap B = \{3\}$. Therefore, as $A \cap B \neq \varnothing$ A and B are not mutually exclusive.

EXAMPLE 15.2

Two coins are tossed. Find the probability that:

(a) two tails are showing (b) a tail is showing.

Solution

Let H denote the event a head is showing and T the event a tail is showing. This means that the sample space (with two coins) is given by $U = \{HH, HT, TH, TT\}$.

(a) The event that two tails are showing is given by the event $\{TT\}$, therefore, we have that $P(\{TT\}) = \frac{n(\{TT\})}{n(U)} = \frac{1}{4}$.

(b) The event that one tail is showing is given by $\{HT, TH\}$, therefore, we have that

$P(\{HT, TH\}) = \frac{n(\{HT, TH\})}{n(U)} = \frac{2}{4} = \frac{1}{2}$.

EXAMPLE 15.3

A card is drawn from a standard deck of 52 playing cards. What is the probability that a diamond card is showing?

Solution

Let D denote the event "a diamond card is selected".

This means that $n(D) = 13$ as there are 13 diamond cards in a standard deck of cards.

Therefore, $P(D) = \frac{n(D)}{n(U)} = \frac{13}{52} = \frac{1}{4}$.

15.1.5 PROBLEM SOLVING STRATEGIES IN PROBABILITY

When dealing with probability problems it is often useful to use some form of diagram to help 'visualize' the situation. **Diagrams** can be in the form of

1.	Venn diagrams.
2.	Tree diagrams.
3.	Lattice diagrams.
4.	Karnaugh maps (probability tables).
5.	As a last resort, any form of diagram that clearly displays the process under discussion (e.g. flow chart).

It is fair to say that some types of diagrams lend themselves well to particular types of problems. These will be considered in due course.

EXAMPLE 15.4

Find the probability of getting a sum of 7 on two throws of a die.

Solution

In this instance, we make use of a lattice diagram to display all possible outcomes. From the diagram, we can list the required event (and hence find the required probability):

Let S denote the event "A sum of seven is observed". From the lattice diagram, we see that there are 6 possibilities where a sum of seven occurs.
In this case we have

$S = \{(1, 6), (2, 5), (3, 4), (4, 3), (5, 2), (6, 1)\}$.

Therefore, we have that $P(S) = \frac{n(S)}{n(U)} = \frac{6}{36} = \frac{1}{6}$

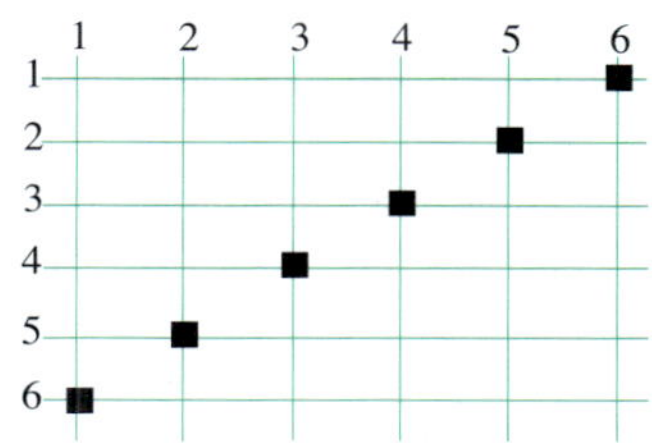

EXERCISES 15.1

1. From a bag containing 6 white and 4 red balls, a ball is drawn at random. What is the probability that the ball selected is
(a) red (b) white (c) not white

2. From an urn containing 14 marbles of which 4 are blue and 10 are red, a marble is selected at random. What is the probability that
(a) the marble is blue (b) the marble is red

3. A letter is chosen at random from the letters of the alphabet. What is the probability that
(a) the letter is a vowel (b) the letter is a consonant

4. A coin is tossed twice. List the sample space and find the probability of observing
(a) two heads.
(b) at least one head.

5. A coin is tossed three times. List the sample space and find the probability that
(a) two heads, show uppermost.
(b) at least two heads show uppermost.
(c) three heads or three tails are showing.

6. A letter is chosen at random from the word FERTILITY. Find the probability that the letter chosen is
(a) a T (b) an I (c) a consonant (d) a vowel.

7. A bag has 20 coins numbered from 1 to 20. A coin is drawn at random and its number is noted.
What is the probability that the coin drawn
(a) has an even number on it?
(b) has a number that is divisible by 3?
(c) has a number that is divisible by 3 or 5?

8. A die is rolled twice. Use a lattice diagram to illustrate the sample space. What is the probability of observing
(a) at least one five (b) a four and a three
(c) a pair (d) a sum of eight

9. A family has three children. List the sample space and hence find the probability that
(a) there are 3 boys.
(b) there are 2 boys and 1 girl.
(c) there are at least two girls.

10. A card is selected from a pack of 52 cards. Find the probability that the card is
(a) red (b) a heart (c) red and a heart

11. From an urn containing 16 cubes of which 6 are red, 4 are white and 6 are black, a cube is drawn at random.
Find the probability that the cube is
(a) red (b) white (c) black (e) red or black

12. A coin and a die are tossed simultaneously. Draw a lattice diagram to depict this situation.
(a) Using your lattice diagram, list the sample space.
(b) What is the probability of observing a tail and an even number?

13. A die is thrown three times. Find the probability of observing
(a) three sixes.
(b) three even numbers.
(c) two odd numbers.
Hint: You might need to draw a three dimensional lattice diagram.

15.2 PROBABILITY & VENN DIAGRAMS

From the axioms of probability we can develop further rules to help solve problems that involve chance. We illustrate these rules with the aid of Venn diagrams.

Event	Set language	Venn diagram	Probability result
The **complement** of A is denoted by A'.	A' is the complement to the set A, i.e., the set of elements that do not belong to the set A.	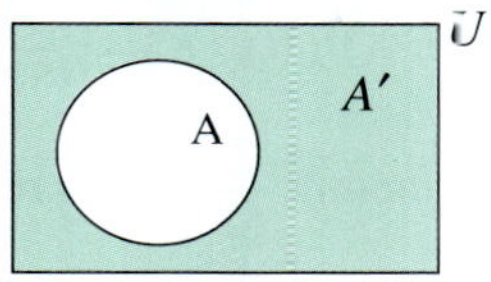	$P(A') = 1 - P(A)$ $P(A')$ is the probability that event A does not occur.
The **intersection** of A **and** B : $A \cap B$	$A \cap B$ is the intersection of the sets A and B, i.e., the set of elements that belong to **both** the set A **and** the set B.	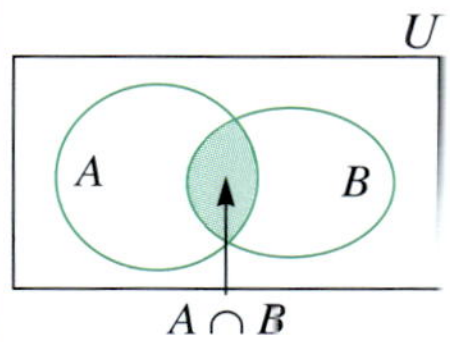	$P(A \cap B)$ is the probability that both A and B occur.
The **union** of events A and B : $A \cup B$	$A \cup B$ is the union of the sets A and B, i.e., the set of elements that belong to A **or** B or both A **and** B.	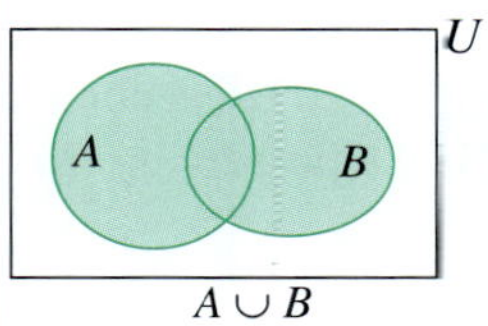	$P(A \cup B)$ is the probability that either event A or event B (or both) occur. From this we have what is known as the '**Addition rule**' for probability: $P(A \cup B) = P(A) + P(B) - P(A \cap B)$
If $A \cap B = \varnothing$ the events A and B are said to be **disjoint**. That is, they have no elements in common.	If $A \cap B = \varnothing$ the sets A and B are **mutually exclusive**.	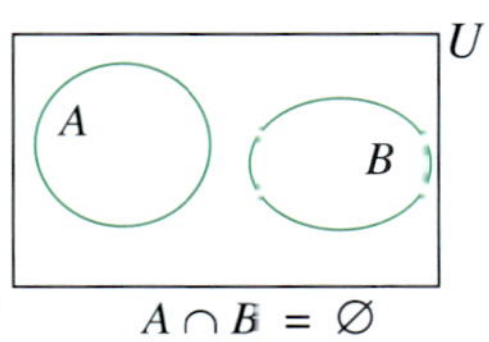	If A and B are mutually exclusive events then event A and event B cannot occur simultaneously, i.e., $n(A \cap B) = 0$ $\Rightarrow P(A \cap B) = 0$ Therefore: $P(A \cup B) = P(A) + P(B)$

Although we now have a number of 'formulas' to help us solve problems that involve probability, using other forms of diagrams to clarify situations and procedures should not be overlooked.

EXAMPLE 15.5

A card is randomly selected from an ordinary pack of 52 playing cards. Find the probability that it is either a 'black card' or a 'King'.

Solution

Let B be the event 'A black card is selected.' and K the event ' A King is selected'.
We first note that event B has as its elements the Jack of spades(J♠), the Jack of clubs (J♣), the Queen of spades(Q♠), the Queen of clubs(Q♣) and so on.
This means that
B ={K♠,K♣,Q♠,Q♣,J♠,J♣,10♠,10♣,9♠,9♣,8♠,8♣,7♠,7♣,6♠,6♣,5♠,5♣,4♠,4♣,3♠, 3♣,2♠,2♣,A♠,A♣} and
K = {K♠, K♦, K♥, K♣}, so that $B \cap K$ = {K♠, K♣}.
Using the addition rule, $P(B \cup K) = P(B) + P(K) - P(B \cap K)$

we have $P(B \cup K) = \frac{26}{52} + \frac{4}{52} - \frac{2}{52} = \frac{7}{13}$.

Note the importance of subtracting $\frac{2}{52}$ as this represents the fact that we have included the event {K♠, K♣} twice when finding B and K.

We now consider one of the problems from Exercise 15.1, but this time we make use of the addition rule.

EXAMPLE 15.6

A bag has 20 coins numbered from 1 to 20. A coin is drawn at random and its number is noted. What is the probability that the coin has a number that is divisible by 3 or by 5?

Solution

Let T denote the event "The number is divisible by 3" and S, the event "The number is divisible by 5".
Using the addition rule we have $P(T \cup S) = P(T) + P(S) - P(T \cap S)$
Now, $T = \{3, 6, 9, 12, 15, 18\}$ and $S = \{5, 10, 15, 20\}$ so that $T \cap S = \{15\}$.

Therefore, we have $P(T) = \frac{6}{20}$ and $P(S) = \frac{4}{20}$ and $P(T \cap S) = \frac{1}{20}$.

This means that $P(T \cup S) = \frac{6}{20} + \frac{4}{20} - \frac{1}{20} = \frac{9}{20}$.

EXAMPLE 15.7

If $p(A) = 0.6$, $p(B) = 0.3$ and $p(A \cap B) = 0.2$, find

(a) $p(A \cup B)$ (b) $p(B')$ (c) $p(A \cap B')$

Solution

(a) Using the addition formula we have, $p(A \cup B) = p(A) + p(B) - p(A \cap B)$

$$\Rightarrow p(A \cup B) = 0.6 + 0.3 - 0.2 = 0.7$$

(b) Using the complementary formula, we have $p(B') = 1 - p(B) = 1 - 0.3 = 0.7$.

(c) To determine $p(A \cap B')$, we need to use a Venn diagram:

Using the second Venn diagram we are now in a position to form a new formula:

$p(A \cap B') = p(A) - p(A \cap B)$.

$\therefore \ p(A \cap B') = 0.6 - 0.2 = 0.4$.

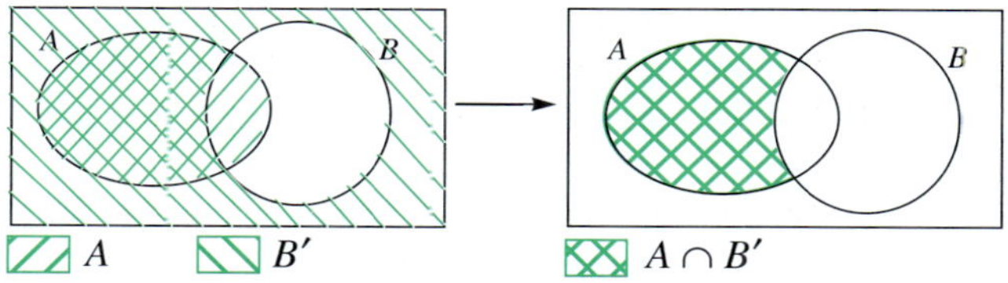

EXAMPLE 15.8

A coin is tossed three times. Find the probability of

(a) obtaining three tails (b) obtaining at least one head

Solution

We begin by drawing a tree diagram to describe the situation:

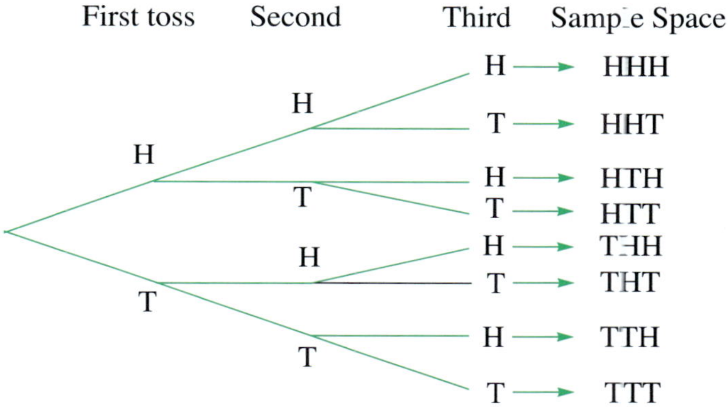

From the tree diagram we have a sample space made up of eight possible outcomes:

{HHH, HHT, HTH, HTT, THH, THT, TTH, TTT}

(a) Let X be the event "Obtaining three tails.", so $X = \{TTT\}$. Therefore $P(X) = \frac{1}{8}$.

(b) Although we can answer this question by using the tree diagram, we make use of complementary events to solve this problem.
Notice that "At least one head." is the complement of no heads.

Therefore, p(At least one head) $= P(X') = 1 - P(X) = 1 - \frac{1}{8} = \frac{7}{8}$.

EXERCISES 15.2

1. A letter is chosen at random from the letters of the word TOGETHER.

(a) Find the probability of selecting a T.
(b) Find the probability of selecting a consonant.
(c) Find the probability of not selecting an E.

2. A card is drawn at random from a standard deck.

(a) Find the probability that the card is an ace.

(b) Find the probability that the card is black.
(c) Find the probability that the card is an ace and black.
(d) Find the probability that the card is an ace or black.

3. A letter is selected at random from the alphabet. Find the probability that the letter is a vowel or comes from the word 'helpful'.

4. The events A and B are such that $P(A) = 0.5$, $P(B) = 0.7$ and $P(A \cap B) = 0.2$. Find

(a) $P(A \cup B)$ (b) $P(B')$ (c) $P(A' \cap B)$

5. The events A and B are such that $p(A) = 0.35$, $p(B) = 0.5$ and $p(A \cap B) = 0.15$. Using a Venn diagram (where appropriate) find:

(a) $p(A')$ (b) $p(A \cup B)$ (c) $p(A \cup B')$.

6. The events A and B are such that $p(A) = 0.45$, $p(B) = 0.7$ and $p(A \cap B) = 0.20$. Using a Venn diagram (where appropriate) find:

(a) $p(A \cup B)$ (b) $p(A' \cap B')$ (c) $p((A \cap B)')$.

7. A coin is tossed three times.

(a) Draw a tree diagram and from it write down the sample space.

Use the results from part (a) to find the probability of obtaining

(b) only one tail.
(c) at least 2 tails.
(d) 2 tails in succession.
(e) 2 tails.

8. In a class of 25 students it is found that 6 of the students play both tennis and chess, 10 play tennis only and 3 play neither. A student is selected at random from this group.

Using a Venn diagram, find the probability that the student.

(a) plays both tennis and chess.
(b) plays chess only.
(c) does not play chess.

9. A blue and a red die are rolled together (both numbered one to six).

(a) Draw a lattice diagram that best represents this experiment.
(b) Find the probability of observing an odd number.
(c) Find the probability of observing an even number with the red die.
(d) Find the probability of observing a sum of 7
(e) Find the probability of observing a sum of 7 or an odd number on the red die.

10. A card is drawn at random from a standard deck of 52 playing cards. Find the probability that the card drawn is

(a) a diamond.
(b) a club or spade.
(c) a black card or a picture card.
(d) a red card or a queen.

11. A and B are two events such that $\text{P}(A) = p$, $\text{P}(B) = 2p$ and $\text{P}(A \cap B) = p^2$.

(a) Given that $\text{P}(A \cup B) = 0.4$, find p.

Use a Venn Diagram to help you find the following:

(b) $\text{P}(A' \cup B)$

(c) $\text{P}(A' \cap B')$.

12. In a group of 30 students 20 hold an Australian passport, 10 hold a Malaysian passport and 8 hold both passports. The other students hold only one passport (that is neither Australian nor Malaysian). A student is selected at random.

(a) Draw a Venn diagram which describes this situation.
(b) Find the probability that the student has both passports.
(c) Find the probability that the student holds neither passport.
(d) Find the probability that the student holds only one passport.

15.3 CONDITIONAL PROBABILITY

15.3.1 INFORMAL DEFINITION OF CONDITIONAL PROBABILITY

Conditional probability works in the same way as simple probability. The only difference is that we are provided with some prior knowledge (or some extra condition about the outcome). So rather than considering the whole sample space, ε, given some extra information about the outcome of the experiment, we only need to concentrate on part of the whole sample space, ε^*. This means that the sample space is reduced from ε to ε^*. Before formalising this section, we use an example to highlight the basic idea.

EXAMPLE 15.9

(a) In the toss of a die, find the probability of obtaining a 'Two'.
(b) After tossing a die, it is noted that an even number appeared, what is the probability that it is a 'Two'?

Solution

(a) This part is straight forward: $U = \{1, 2, 3, 4, 5, 6\}$, and so P('Two') $= \frac{1}{6}$.

(b) This time, because *we know that an even number has occured*, we have a new sample space, namely $U^* = \{2, 4, 6\}$. The new sample size is $n(U^*) = 3$.

Therefore, P('Two' *given that an even number showed up*) $= \frac{1}{3}$.

15.3.2 FORMAL DEFINITION OF CONDITIONAL PROBABILITY

If A and B are two events, then **the conditional probability of event A given event B** is found using

$$\text{P}(A|B) = \frac{\text{P}(A \cap B)}{\text{P}(B)}, \text{P}(B) \neq 0.$$

Note **1.** If A and B are mutually exclusive then $P(A|B) = 0$.

2. From the above rule, we also have the general **Multiplication rule**:

$$P(A \cap B) = P(A|B) \times P(B)$$

It should also be noted that usually $P(A|B) \neq P(B|A)$.

EXAMPLE 15.10

Two dice numbered one to six are rolled onto a table. Find the probability of obtaining a sum of five given that the sum is seven or less.

Solution

We first draw a lattice diagram:

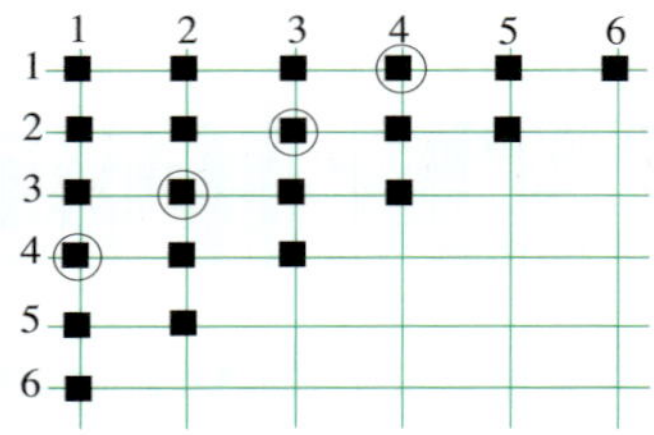

From the diagram we see that the new sample space is made up of 21 outcomes (black boxes) and the event we want (circled) consists of 4 outcomes.

Then, $P((X=5) \cap (X \leq 7)) = \frac{4}{36}$ and $P(X \leq 7) = \frac{21}{36}$.

Therefore, $P(X=5|X \leq 7) = \dfrac{\frac{4}{36}}{\frac{21}{36}} = \frac{4}{21}$.

EXAMPLE 15.11

A box contains 2 red cubes and 4 black cubes. If two cubes are chosen at random, find the probability that both cubes are red given that

(a) the first cube is not replaced before the second cube is selected.

(b) the first cube is replaced before the second cube is selected.

Solution

Let A be the event "the first cube is red" and B be the event "the second cube is red". This means that the event $A \cap B$ must be "both cubes are red".

Now, $p(A) = \frac{2}{6} = \frac{1}{3}$ (as there are 2 red cubes from a total of 6 cubes in the box).

The value of $P(B)$ depends on whether the selection is carried out with or without replacement.

(a) If the first cube selected is red and it is not replaced, then we only have 1 red cube left (in the box) out of a total of five cubes.

So, the probability that the second cube is red given that the first is red is $\frac{1}{5}$.

That is $p(B|A) = \frac{1}{5} \Rightarrow P(A \cap B) = P(B|A) \times P(A) = \frac{1}{5} \times \frac{1}{3} = \frac{1}{15}$.

(b) This time, because the cube is replaced, the probability that the second cube is red given that the first one is red is still $\frac{1}{3}$.

So that, $P(B|A) = \frac{1}{3} \Rightarrow P(A \cap B) = P(B|A) \times P(A) = \frac{1}{3} \times \frac{1}{3} = \frac{1}{9}$.

EXAMPLE 15.12

Two events A and B are such that $P(A) = 0.5$, $P(B) = 0.3$ and $P(A \cup B) = 0.6$.

Find (a) $P(A|B)$. (b) $P(B|A)$ (c) $P(A'|B)$

Solution

(a) $P(A|B) = \frac{P(A \cap B)}{P(B)}$, therefore we need to find $P(A \cap B)$.

Using the addition rule we have $P(A \cup B) = P(A) + P(B) - P(A \cap B)$

$$0.6 = 0.5 + 0.3 - P(A \cap B)$$

$$\therefore P(A \cap B) = 0.2$$

Therefore, $P(A|B) = \frac{P(A \cap B)}{P(B)} = \frac{0.2}{0.3} = \frac{2}{3}$

(b) $P(B|A) = \frac{P(B \cap A)}{P(A)} = \frac{0.2}{0.5} = 0.4$.

(c) $P(A'|B) = \frac{P(A' \cap B)}{P(B)} = \frac{P(B) - P(A \cap B)}{P(B)} = \frac{0.3 - 0.2}{0.3} = \frac{1}{3}$

15.3.3 INDEPENDENCE

The events *A* and *B* are said to be statistically independent if the probability of event *B* occuring is not influenced by event *A* occuring.

Therefore we have the mathematical definition:

> Two events ***A* and *B* are independent if, and only if,**
> $P(A|B) = P(A)$ and $P(B|A) = P(B)$

However, a more convenient definition for independence can be given as follows:

> ***A* and *B* are independent if, and only if**
> $P(A \cap B) = P(A) \times P(B)$

This definition can be used as a test to decide if two events are independent. However, as a rule of thumb, if two events are 'physically independent' then they will also be statistically independent.

There are a few points that should always be considered when dealing with independence:

1. Never assume that two events are independent unless you are absolutely certain that they are independent.

2. How can you tell if two events are independent ?
A good rule of thumb is:
If they are physically independent, they are mathematically independent.

3. Make sure that you understand the difference between mutually exclusive events and independent events.
Mutually exclusive means that the events A and B have nothing in common and so there is no intersection. i.e., $A \cap B = \varnothing \Rightarrow P(A \cap B) = 0$.
Independent means that the outcome of event A will not influence the outcome of event B i.e., $P(A \cap B) = P(A) \times P(B)$.

4. Independence need not only be for two events, it can be extended, i.e., if the events A, B and C are each independent of each other then

$$P(A \cap B \cap C) = P(A) \times P(B) \times P(C)$$

5. Showing that two events, A and B are independent, requires three steps:

Step 1 Evaluate the product $P(A) \times P(B)$.

Step 2 Determine the value of $P(A \cap B)$ using any means (other than step 1) i.e., use grids, tables, Venn diagrams, . . .
That is, you must not assume anything about A and B.

Step 3 If the answer using Step 1 is equal to the answer obtained in Step 2, then and only then will the events be independent. Otherwise, they are not independent.
Notice that not being independent does not therefore mean that they are mutually exclusive. They simply aren't independent. That's all.

6. Do not confuse the multiplication principle with the the rule for independence;
Multiplication principle is $P(A \cap B) = P(A|B) \times P(B)$
Independence is given by $P(A \cap B) = P(A) \times P(B)$

EXAMPLE 15.13

Two fair dice are rolled. Find the probability that two even numbers will show up.

Solution

Let the E_1 and E_2 denote the events "An even number on the first die." and "An even number on the second die." respectively.
In this case, the events are physically independent, i.e., the outcome on one die will not influence the outcome on the other die, and so we can confidently say that E_1 and E_2 are independent events.

Therefore, we have $P(E_1 \text{ and } E_2) = P(E_1 \cap E_2) = P(E_1) \times P(E_2) = \frac{1}{2} \times \frac{1}{2} = \frac{1}{4}$.

EXAMPLE 15.14 Debra has a chance of 0.7 of winning the 100 m race and a 60% chance of winning the 200 m race.

(a) Find the probability that she only wins one race.
(b) Find the probability that she wins both races.

Solution

Let W_1 denote the event "Debra wins the 100 m race." and W_2, the event "Debra wins the 200 m race.".

(a) If Debra wins only one race she must either
win the 100 m **and** lose the 200 m **or**
win the 200 m **and** lose the 100 m.

That is, we want $P(W_1 \cap W_2') = P(W_1) \times P(W_2') = 0.7 \times 0.4 = 0.28$ or
we can multiply the probabilities because the events are independent (why?):
$P(W_2 \cap W_1') = P(W_2) \times P(W_1') = 0.6 \times 0.3 = 0.18$.
Therefore, the required probability is $0.28 + 0.18 = 0.46$

Notice that if W_1 and W_2 are independent, then so too are their complements.

(b) Winning both races means that Debra will win the 100 m **and** 200 m race.
Therefore, we have $P(W_1 \cap W_2) = P(W_1) \times P(W_2) = 0.7 \times 0.6 = 0.42$.

Notice how we have made repeated use of the word '**and**', this emphasizes the fact that we are talking about the intersection of events.

EXAMPLE 15.15 Four seeds are planted, each one having an 80% chance of germinating. Find the probability that

(a) all four seeds will germinate.
(b) at least one seed will germinate.

Solution

(a) Let G_i denote the event that the i th seed germinates.
This means that $P(G_1) = P(G_2) = P(G_3) = P(G_4) = 0.8$
It is reasonable to assume that each seed will germinate independently of the other.

Therefore, $P(\text{All four seeds germinate}) = P(G_1 \cap G_2 \cap G_3 \cap G_4)$
$= P(G_1) \times P(G_2) \times P(G_3) \times P(G_4)$
$= (0.8)^4$
$= 0.4096$

(b) Now, $p(\text{At least one seed will germinate}) = 1 - p(\text{No seeds germinate})$.
$P(\text{Any } \textbf{one} \text{ seed does not germinate}) = P(G_i') = 0.2$
Therefore, $P(\text{At least one seed will germinate}) = 1 - (P(G_i'))^4 = 1 - (0.2)^4 = 0.9984$.

EXAMPLE 15.16

A bag contains 5 white balls and 4 red balls. Two balls are selected in such a way that the first ball drawn is not replaced before the next ball is drawn.
Find the probability of selecting exactly one white ball.

Solution

We begin by drawing a diagram of the situation:

From our diagram we notice that there are two possible sample spaces for the second selection.

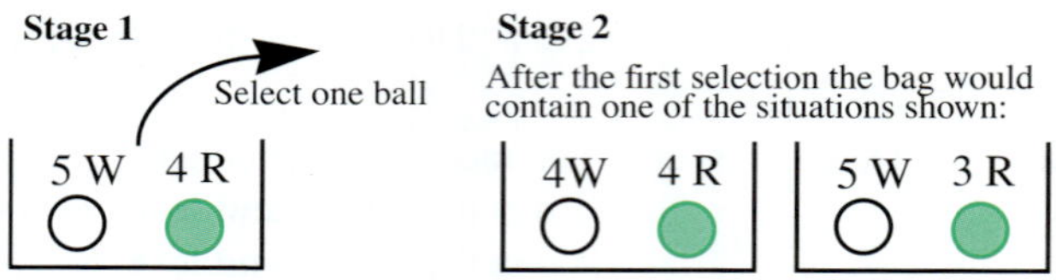

As an aid, we make use of a tree diagram, where W_i denotes the event "A white ball is selected on the ith trial" and R_i denotes the event "A red ball is selected on the ith trial".

The event "Only one white" occurs if the first ball is white **and** the second ball is red, **or** the first ball is red **and** the second ball is white.

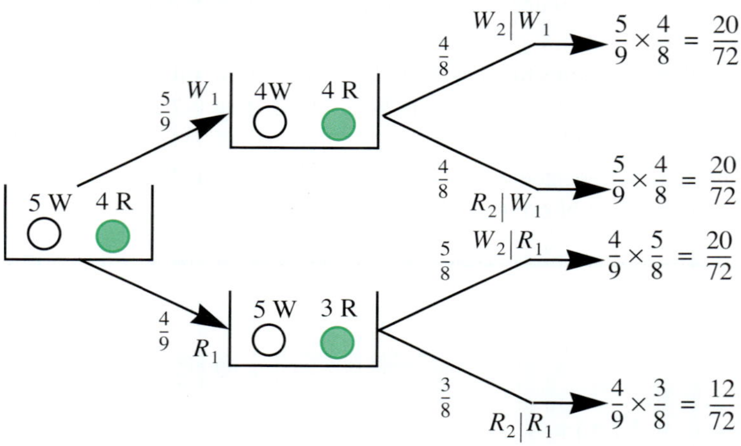

$$\begin{aligned}
\text{P(One White ball)} &= \text{P}(W_1 \cap R_2) + \text{P}(R_1 \cap W_2) \\
&= \text{P}(R_2|W_1)\times\text{P}(W_1) + \text{P}(W_2|R_1)\times\text{P}(R_1) \\
&= \frac{4}{8}\times\frac{5}{9}+\frac{5}{8}\times\frac{4}{9} \\
&= \frac{5}{9}
\end{aligned}$$

EXERCISES 15.3

1. Two events A and B are such that $p(A) = 0.6, p(B) = 0.4$ and $p(A \cap B) = 0.3$. Find the probability of the following events:

(a) $A \cup B$ (b) $A|B$ (c) $B|A$ (d) $A|B'$

2. A and B are two events such that $p(A) = 0.3$, $p(B) = 0.5$ and $p(A \cup B) = 0.55$
Find the probability of the following events:
(a) $A|B$ (b) $B|A$ (c) $A|B'$ (d) $A'|B'$

3. Urn A contains 9 cubes of which 4 are red. Urn B contains 5 cubes of which 2 are red. A cube is drawn at random and in succession from each urn.
(a) Draw a tree diagram representing this process.
(b) Find the probability that both cubes are red.
(c) Find the probability that only 1 cube is red.
(d) If only 1 cube is red, find the probability that it came from urn A.

4. A box contains 5 red, 3 black, and 2 white cubes. A cube is randomly drawn and has its colour noted. The cube is then replaced, together with 2 more of the same colour. A second cube is then drawn.
(a) Find the probability that the first cube selected is red.
(b) Find the probability that the second cube selected is black.
(c) Given that the first cube selected was red, what is the probability that the second cube selected is black?

5. A fair coin, a double-headed coin and a double-tailed coin are placed in a bag. A coin is randomly selected. The coin is then tossed.
(a) Draw a tree diagram showing the possible outcomes.
(b) Find the probability that the coin lands with a tail showing uppermost.
(c) In fact, the coin falls "heads", find the probability that it is the "double-headed" coin.

6. Two unbiased coins are tossed together. Find the probability that they both display heads given that at least one is showing a head.

7. A money box contains 10 discs, 5 of which are yellow, 3 of which are black and 2 green. Two discs are selected in succession, with the first disc not replaced before the second is selected.
(a) Draw a tree diagram representing this process.
(b) Hence find the probability that the discs will be of a different colour.
(c) Given that the second disc was black, what is the probability that both were black?

8. Two dice are rolled. Find the probability that the faces are different given that the dice show a sum of 10.

9. Given that $p(A) = 0.6$, $p(B) = 0.7$ and that A and B are independent events.
Find the probability of the event
(a) $A \cup B$ (b) $A \cap B$ (c) $A|B'$ (d) $A' \cap B$

10. The probability that an animal will still be alive in 12 years is 0.55 and the probability that its mate will still be alive in 12 years is 0.60.
Find the probability that
(a) both will still be alive in 12 years.
(b) only the mate will still be alive in 12 years.
(c) at least one of them will still be alive in 12 years.
(d) the mate is still alive in 12 years given that only one is still alive in 12 years.

11. Tony has a 90% chance of passing his maths test, whilst Tanya has an 85% chance of passing the same test. If they both sit for the test, find the probability that
(a) only one of them passes.
(b) at least one passes the test.
(c) Tanya passed given that at least one passed.

12. The probability that Rory finishes a race is 0.55 and the probability that Millicent finishes the same race is 0.6. Because of team spirit, there is an 80% chance that Millicent will finish the race if Rory finishes the race.
Find the probability that
(a) both will finish the race.
(b) Rory finishes the race given that Millicent finishes.

13. If A and B are independent events, show that their complementary events are also independent events.

14. A student runs the 100 m, 200 m and 400 m races at the school athletics day. He has an 80% chance of winning any one given race. Find the probability that he will
(a) win all 3 races.
(b) win the first and last race only.
(c) win the second race given that he wins at least two races.

15. Dale and Kritt are trying to solve a physics problem. The chances of solving the problem are Dale—65% and Kritt— 75%. Find the probability that
(a) only Kritt solves the problem.
(b) Kritt solves the problem.
(c) both solve the problem.
(d) Dale solves the problem given that the problem was solved.

16. A coin which is weighted in such a way that there is a 70% chance of it landing heads. The coin is tossed three times in succession, find the probability of observing
(a) three tails.
(b) two heads.
(c) two heads given that at least one head showed up.

15.4 MISCELLANEOUS EXERCISES

1. The frequency distribution for how long (in minutes) customers stayed at a bistro is shown below:

Time customer stayed in bistro	Frequency
1-10	11
11-20	8
21-30	9
31-40	7
41-50	13
51-60	2

(a) What is the probability that a customer stayed between 11–20 minutes.
(b) What is the probability that a customer stayed no more than 40 minutes
(c) What is the probability that a customer stayed between 21–50 minutes.

2. A box containing small cardboard squares numbered 1 to 10 has one of the cardboarb pieces randomly selected.

What is the probability that the number selected is

(a) even?
(b) greater than 3?
(c) a prime number?
(d) a prime number and at least 5.

3. A class of 40 students consist of 24 studying History, 22 studying Economics and 15 studying both these subjects.

(a) Illustrate this information using a Venn diagram.

(b) Use your diagram to find the probability that a student studies
i. both subjects.
ii. History only.
iii. Exactly one subject.
iv. neither subject.

4. A student runs the 100 m, 200 m and 400 m races at the school athletics day. He has an 70% chance of winning any one given race.
Find the probability that he will

(a) win all 3 races,
(b) win the first and last race only.
(c) win the second race given that he wins at least two races.

5. In a group of 70 people, 42 have fair hair, 34 do not have blue eyes and 23 have fair hair but do not have blue eyes.
(i) Find the number of students that have both blue eyes and fair hair.

(ii) Find the probability that a randomly chosen student has neither blue eyes nor fair hair.

(iii) If a person with blue eyes is chosen at random from the group, what is the probability that this person has fair hair?

(iv) Is the characteristic of having blue eyes independent of the characteristic of having fair hair? Justify your answer.

6. A bowl contains 8 red balls and 4 green balls. A ball is selected at random from the bowl, its colour noted and it is then replaced. A second ball is drawn and its colour is also noted.

(i) What is the probability that both balls are green?
(ii) What is the probability that the balls are of different colours?
(iii) If the first ball drawn is red, what is the probability that the second is also red?

7. Two events A & B are exclusive. Prove that if $p(\mathrm{A}) > 0$ and $p(\mathrm{B}) > 0$ then the events are dependent.

8. The following data represents the number of eggs laid each year by a species of sea bird.

6	11	4	10	5	5	5	9	2	8
2	4	7	2	5	5	9	7	4	9
9	6	7	5	6	11	9	3	11	9
10	2	8	7	11	4	2	5	5	4

i. Tally the data using class intervals of one.
ii. What is the probability that a randomly chosen bird will lay more than 9 eggs?
iii. Find the mean number of eggs laid.
iv. What is the probability that a randomly chosen bird will lay more than the average number of eggs?
v. Find the standard deviation of the data, correct to four significant figures.
vi. Find the number of eggs laid that lie within 1 standard deviation of the mean.
vii. What is the probability that a randomly chosen bird will lay a number of eggs that is within 1 standard deviation of the mean?

9. A biassed coin has a probability of 0.6 of falling heads. The coin is spun three times in succession and the number of heads counted.

i. Explain why each spin is independent of the others.
ii. Complete the following probability table:

Heads	Probability
0	
1	
2	
3	

iii. If the first spin was a head, what is the probability that the next two throws will also be heads?

The coin is spun repeatedly until a tail is obtained.

iv. What is the probability that the coin is spun ten times before the first tail in the sequence is obtained?

15.5 GRADED REVISION EXERCISES

LEVEL 1

1. What is the probability that a card drawn from a normal pack is a red ace?

2. A bag contains three red balls, four green balls and two white balls. If a ball is chosen at random, what is the probability that the ball is green?

3. The diagram shows a target. If projectiles are fired at this target at random, what is the probability that a single projectile hits the 3 sector?

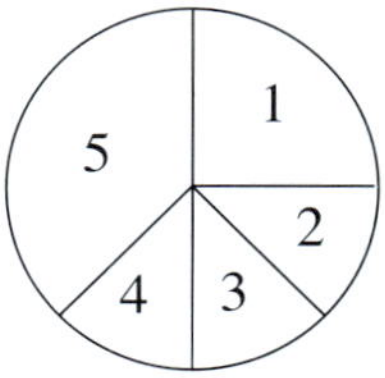

4. If a letter is chosen at random from that word RANDOM, what is the probability that this letter is a vowel?

5. If a coin and a die are thrown together, what is the probability that the result is a head and an even number?

6. If a coin is spun twice, what is the probability that the two results are different?

LEVEL 2

1. The events A and B are such that $p(A) = 0.3$, $p(B) = 0.5$ and $p(A \cap B) = 0.1$
Find:

i. $p(A \cup B)$ ii. $p(B')$ iii. $p(A' \cap B)$

2. A coin is tossed four times.

i. How many outcomes are there in the sample space?
ii. What is the probability of getting four heads?
iii. What is the probability of getting the same number of heads as tails?
iv. What is the probability of getting more heads than tails?

3. Two letters are randomly chosen, one after the other, from the word ARTICHOKE. What is the probability that these letters spell AT if:

i. the same letter can be chosen twice
ii. the same letter cannot be chosen twice?

4. In a group of 80 students, 35 play baseball, 25 play football and 30 play neither.

i. Draw a Venn diagram to show these data.
ii. What is the probability that a randomly selected student plays both baseball and football?
iii. What is the probability that a randomly selected student plays baseball but not football?
iii. What is the probability that a randomly selected student plays neither sport?

LEVEL 3

1. Two events A and B are such that $p(A) = 0.5$, $p(B) = 0.4$ and $p(A \cap B) = 0.3$
Find:

i. $p(A|B)$

ii. $p(B|A)$

iii. Are the events A & B independent?

2. Two cards are drawn from a normal pack without replacement.

i. What is the probability that the second card is an ace given that the first card was not an ace?

ii. Are the two draws independent?

iii. If we do not look at the first card, what is the probability that the second card is an ace?

3. Two letters are selected at random from the word ATTITUDE.

(a) If the letters are selected with the first one selected being replaced before the second one is selected:

i. are the events independent?

ii. what is the probability that both letters are Ts?

iii. what is the probability that both letters are vowels?

(b) If the letters are selected without replacement:

i. are the events independent?

ii. what is the probability that both letters are Ts?

iii. what is the probability that both letters are vowels?

LEVEL 4

1. Given that $p(A) = 0.6$, $p(B) = 0.4$ and that A and B are independent events.
Find the probability of the events:

i. $A \cup B$ ii. $A \cap B$ iii. $A|B'$

2. A group of 200 language students, all of whom speak at least one language have the following languages:
Italian 99, japanese 115, german 87, italian and japanese 44, german and japanese 38 and italian and german 31.

i. Draw a Venn diagram to show this information.

ii. If a student is chosen at random, what is the probability that s/he speaks all three languages?

iii. If a student is chosen at random, what is the probability that s/he speaks exactly two languages?

iv. If a student is chosen at random, what is the probability that s/he speaks italian given that she also speaks japanese?

v. If two students are randomly chosen to represent the group on a student council, are the selections independent of each other? Give a brief reason.

15.6 TOPIC TEST

1. If a letter is chosen at random from the word SUSTENANCE, what is the probability that the letter is:

i. S.
ii. a vowel.

[3 marks]

2. If two cards are drawn from a normal pack:

i What is the probability that the first card is a 'club'?
ii. What is the probability that both cards are 'clubs'?
iii. If the first card is black, what is the probability that the second is also black?

[5 marks]

3. Given that $p(A) = 0.2$, $p(B) = 0.8$ and that A and B are independent events.

i. Show the data as a Venn diagram.

Find the probability of the events:

ii. $A \cup B$
iii. $A \cap B$
iv. $A|B$

[7 marks]

4. 'Weston Athletic' soccer club have a $\frac{3}{7}$ chance of winning each game. The result of each game is independent of any other games the club plays.

i. The club has won its last two games. What is the probability that it will win its next game?

ii. If the club plays four games in January, what is the probability that it will win exactly half of them?

iii. If the club plays three games in May and has lost the first game, what is the probability that it will win the other two?

[5 marks]
Total 20 marks

15.7 PROJECT SUGGESTIONS

The introduction gave some information about probability and gambling. This remains a good area for projects. For, example, how do bookmakers fix the odds on horse races?

However, there are many other areas of activity that depend on probability.

The principle of independence is at the heart of much engineering design. Aircraft are designed so that, if one component fails, there is another ready to take over. This famous bridge over the Avon Gorge in Bristol, UK (designed by Brunel) will survive the failure of one of the support cables. If this happens, the failed component can be replaced before the structure collapses. This is known as 'damage tolerance'.

The entire insurance industry is based on the experimental determination of the probability that damaging events (storms, illness, accidents etc.) will happen. This work is completed by 'actuaries' who generally have mathematical backgrounds.

Why is it that a 21 year old driver pays more for insurance than a 31 year old?

REVISION SET B - PAPER 1 & PAPER 2 STYLE QUESTIONS

1. The first term of an arithmetic sequence is 5 and the eight term is twice the fourth.
(a) Find the common difference, d.
(b) Find the sum of the first twelve terms.

2. (a) In the sequence $2, 2\frac{1}{2}, 3, 3\frac{1}{2}, \ldots$, what term will the number 96 be?
(b) In a city there was on average a crime rate every 90 minutes. This represented an increase of 10% over the previous year. In the previous year there was on average a crime every x minutes. Find x.
(c) Find x given that $3 - 6 + 12 - 24 + \ldots + x = -63$.

3. The mass M in grams of a radioactive substance is given by the equation $M = A \cdot 10^{-kt}$, where t is the time in years and k is a constant. At time $t = 0$, $M = 10$.
(a) Find the value of A.
(b) After 10 years, the mass is 8 grams. Find the value of k.

4. P, Q and R are subsets of a universal set U, and $n(P)$ represents the number of members in P. Given that $n(P) = 8, n(Q) = 27, n(R) = 13, n(P \cup Q \cup R) = 37$, $n(P \cap Q \cap R') = 4$, $n(Q \cap R \cap P') = 2$ and $n(P \cap R \cap Q') = 3$, find $n(P \cap Q \cap R)$.

5. A formula for estimating tidal heights at Grey Rock is given by

$$h(t) = 3\cos(30t°) + 4,$$

where h = the height of the tide in metres on the Grey Rock pier maker and t = the time in hours (measured from midnight)

(a) Find an estimate of the height at
i. midnight ii. 3 a.m. iii. 6 a.m.
(b) Accurately draw the graph of this function for a whole day.
(c) Find the times when the height at the pier marker is 5 metres.
(d) The captain of a boat knows that he can safely dock at the pier when the height of the tide is at least 5 metres. At what times is it safe for him to dock the boat?
(e) It is found that the times between high tides should be 13 hours. State a modified formula which will agree with this information.

6. (a) A clerk is employed at an initial salary of \$8320 per annum and, after each year of service, he receives an increase of \$20 per week.
i. What is his salary in the tenth year of service?
ii. What will be his total earnings for the first ten years?
(b) Another clerk had the same initial salary of \$8320 per annum and, after each year of service, his salary increased by 10% of the salary of the preceding year.
i. Calculate his salary in the tenth year.
ii. Calculate his total earnings for the first ten years.
(c) After how many years will they both be earning at least \$12,000 per year?

7. 50 students go bushwalking. 23 get sunburnt, 22 get bitten by ants and 5 are both sunburnt and bitten by ants.

(a) Using the labels, S for student getting sunburnt
A for student getting bitten by ants
illustrate the above using a Venn diagram.

(b) How many students
i. escaped being bitten?
ii. were either bitten or sunburnt?
iii. were neither bitten nor sunburnt?

8. Twin sisters, Sue and Debbie, are both good maths students. Their Mathematics teacher calculated that Sue has a probability of 0.7 of getting an A at the end of the year and that Debbie has a probability of 0.6. Find the probability that:

(a) neither gets an A.
(b) if only one gets an A, it is Sue.

9. If $P(A) = 0.4$, $P(B) = 0.7$ and $P(A|B) = 0.3$. Find

(a) $P(A \cup B)$ (b) $P(B|A)$ (c) $P(A|A \cup B)$

10. In a certain town, 60% of the people are females and 40% are males. In an opinion poll taken, 30% of the females approved of the work done by the Town Mayor while 70% of the males approved of his work.

(a) What is the probability that a person chosen at random in the town will approve of the work done by the Town Mayor?

(b) i. Find the probability that a person chosen at random will be a female.
ii. Also, if the person chosen did approve of his work, what is the probability that person is a female?

11. (a) A geometric sequence has a first term 2 and a common ratio 1.5. Find the sum of the first 11 terms.

(b) Find the number of terms of the sequence 3, 7, 11, . . . required to sum to 820.

12. The number of cancer cells in a solution is believed to increase in such a way that at the end of every hour, there are α % more cells than at the end of the previous hour. Dr. Bac Teria, who is in charge of this experiment, has been recording the cell counts. Unfortunately, because of his preoccupation with another experiment he has been rather neglectful. The only available readings are shown in the table below:

t	1	2	3	4	5	6
$N(t)$			2420			3221

(a) The number of cells, $N(t)$, in the solution at the end of every hour is thought to be modelled by a geometric progression. That is, $N(t) = N_0 \times r^{t-1}$, $t = 1, 2, \ldots$

i. Show that $r = 1 + \dfrac{\alpha}{100}$.

ii. Find the values of N_0 and α .

(b) Copy and complete the table above.

(c) When is the first time that the number of cells in the solution exceeds 258259?

(d) At the end of every hour a new identical solution is set up in the same way that the first one was. How many cells will there be altogether at the end of a 24 hours run of introducing a new solution at the end of every hour.

13. In a class of 30 Year 12 students, 8 play both tennis and football, 15 play tennis only and 4 play no sport at all. Find the number of students that

(a) play tennis. (b) play football only.

(c) do not play tennis. (d) play tennis or football.

14. (a) On the same set of axes, sketch the graph of the following functions

i. $f(x) = 3 \times 2^{0.5x} + 1$

ii. $g(x) = 1 - 0.5x$

(b) i. Using your graph in (a), what does the value a, in the equation $a + 6 \times 2^{0.5a} = 0$ represent?

ii. Find, to two decimal places, the value of a for which $a + 6 \times 2^{0.5a} = 0$.

15. Drainage pipes are stored at a hardware shop in a stack such that the top row consists of one pipe, the second row two, the third row three and so on. Find the number of rows which would be required to store 210 pipes

16. Given the propositions

p: x is a multiple of 3 between 0 and 40

q: x is an even number between 0 and 40

r: x is a factor of 36

(a) List the elements of the truth sets P, Q and R.

(b) For each of the following compound propositions

(i) Write in words

(ii) Construct a truth table to show the truth values

(iii) Draw a Venn diagram and shade the region that represents the truth set of the compound proposition

(iv) List the elements of the truth set of the compound propositions:

(a) $\neg p$ (b) $p \wedge q$

(c) $p \wedge q \wedge r$ (d) $p \vee q \vee r$

(e) $\neg p \wedge \neg q$ (f) $\neg p \wedge \neg q \wedge \neg r$

(g) $\neg(p \wedge q \wedge r)$ (h) $\neg p \wedge (q \vee r)$

17. A and B are events with $P(A) = x$, $P(B) = 2x$ and $P(A \cup B) = \frac{3}{4}$. Using the expansion for $P(A \cup B)$, find x in each of the following cases:

(a) A and B are mutually exclusive.

(b) $A \subset B$.

(c) A and B are independent.

18. Three cards are drawn (together) at random from a full pack of 52. Find the probability that at least two of these are 'spades'.

19. A pack of cards contains 4 red and 5 black cards. A hand of 5 cards is drawn without replacement. What is the probability of there being 2 red and 3 black cards in the hand?

20. On the same set of axes, sketch the graphs of $f(x) = 5 - 2^{-x}$ and $g(x) = 2 + 5^x$.
Hence show that there will only ever be two real values of x for $10^x - 3 \times 2^x + 1 = 0$

21. Sketch the graph of

(a) $f(t) = 3\sin(45t°) + 3, 0 \le t \le 16$.

(b) For what value(s) of k will the equation $f(t) = k, 0 \le t \le 16$, have

i. two solutions.

ii. five solutions.

22. (a) In Figure 1., a square having side length 2 cm has a quarter of a circle drawn within it as shown in the diagram below.
Find the area of the shaded region in Figure 1.

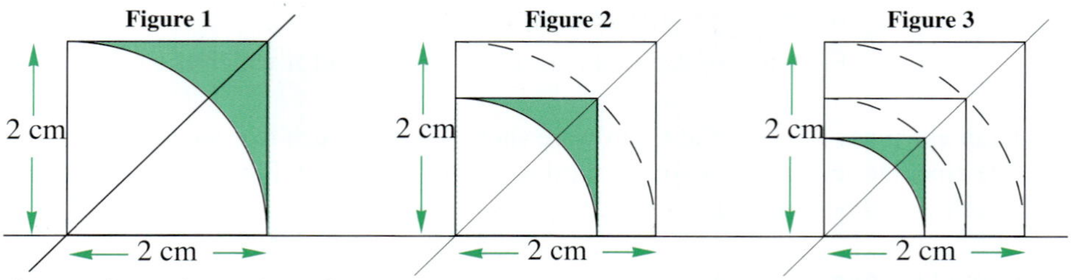

(b) Figure 2 consists of the same diagram as Figure 1 but this time a second square is drawn so that its diagonal line coincides with that of the original square and its corner meets the quarter circle. Once again, a quarter of a circle is drawn within the smaller square as shown.
Find the area of the shaded region in Figure 2.

The process is then continued, as in Figure 3, so that the shaded areas in each of the figures, form a sequence, $A_1, A_2, A_3, \ldots$, where A_1 is the measure of the shaded area in Figure 1, A_2 is the measure of the shaded area in Figure 2, and so on.

(c) i. Show that the sequence $A_1, A_2, A_3, \ldots$ is in fact geometric.

ii. Find the common ratio.

iii. Write down an expression for the nth term, A_n.

iv. What would the area of 5th shaded region be if we continued drawing figures using the same method?

(d) i. Find the sum of the first 5 shaded regions.

ii. If the process was to continue indefinitely, what would the areas of the shaded regions add to?

(e) What type of a sequence would the arc lengths of the quarter circle form? Explain.

23. Construct truth tables for the following and state whether the compound proposition is a tautology, a contradiction or neither.

(a) $(p\wedge q)\vee(\neg p\wedge\neg q)$ (b) $\neg(p\vee q)\wedge p$

(c) $(p\vee q)\Rightarrow\neg(p\vee q)$ (d) $(p\Rightarrow q)\Leftrightarrow(p\wedge\neg q)$

(e) $\neg p\Rightarrow(p\Rightarrow q)$ (f) $[p\wedge(p\Rightarrow q)]\Rightarrow p$

(g) $(p\Rightarrow q)\Leftrightarrow(p\wedge\neg q)$ (h) $\neg(p\wedge q)\vee(p\vee q)$

24. In a test match, the probability of New Zealand defeating the Australian cricket team is 0.2, of a draw is 0.3 and of an Australian win is 0.5. For a three math test series, what is the probability of

i. New Zealand winning the first two matches only?

ii. New Zealand winning two matches?

iii. Australia winning two matches, the other being drawn?

iv. Australia winning the series (that is, winning more matches than New Zealand)?

25. Students at Bankruik Secondary College study Maths or History or both. 12 students study Mathematics only, whilst 30 study only one of Maths or History. We also know that there are 24 students that study History.

(a) Show this information on a Venn diagram.

(b) Find the number of students that study both Maths and History.

(c) Find the number of students that study History only.

26. Sketch an accurate graph of

(a) $f(t) = 4\sin(90t°) + 2, 0 \leq t \leq 6$.

(b) Find i. the values of x for which $f(t) = 0, 0 \leq t \leq 6$

ii. a and b such that $\{t | f(t) < 0, 0 \leq t \leq 6\} = \{x | a < t < b\}$

(b) For what value(s) of k will the equation $f(x) = k , 0 \leq t \leq 6$, have

i. one solutions.

ii. two solutions.

27. A man who works in Melbourne drives home, either along Road A or Road B. He varies his route, choosing Road A $\frac{2}{3}$ of the time. If he drives along Road A, he arrives home before 6:00 pm 90% of the time, while by the more attractive route, along Road B, he gets home before 6:00 pm 60% of the time. What is the probability

i. that he gets home before 6:00 pm?

ii. that he travelled along Road B, if he gets home before 6:00 pm?

28. A builder wishes to tile a section of a roof which has the symmetric shape indicated in the diagram. In order to fix the tiles, it is necessary to lay battens across the roof, parallel to each other, centred $\frac{1}{3}$ m apart. Battens are required at both AB and CD.

(The first four battens are shown on the diagram.)

(a) Find by how much adjacent battens differ by in length.

(b) Find the total length of the battens required.

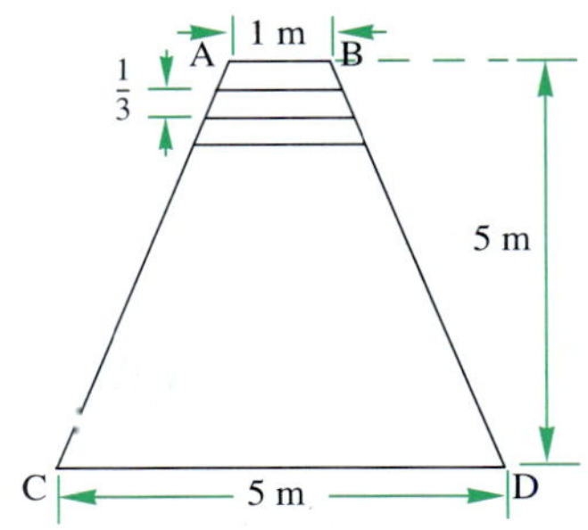

29. (a) Sketch the graph of $f(x) = k \times 2^{\lambda x} + c$ for the case that

i. $k = 3, c = -3, \lambda > 0$. ii. $k = -3, c = 3, \lambda > 0$.

(b) The graph of the function $g(x) = \frac{1}{2} \times 3^{2x} + c$ is shown alongside.

Find the values of a, b, c and d.

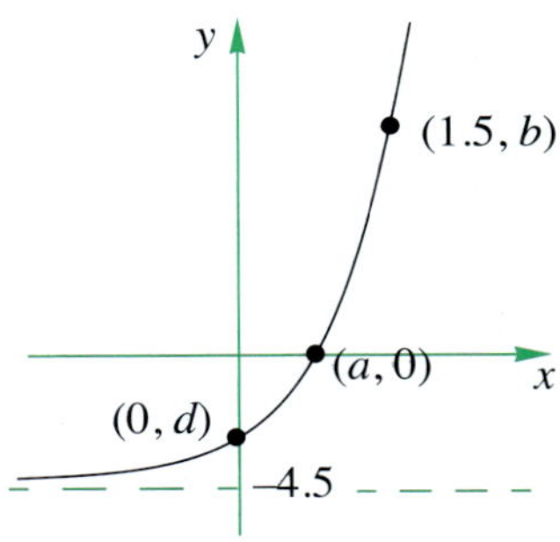

30. The graph below shows the heights of tides, h metres, at time t hours after midnight, for an island near the coast of Australia.

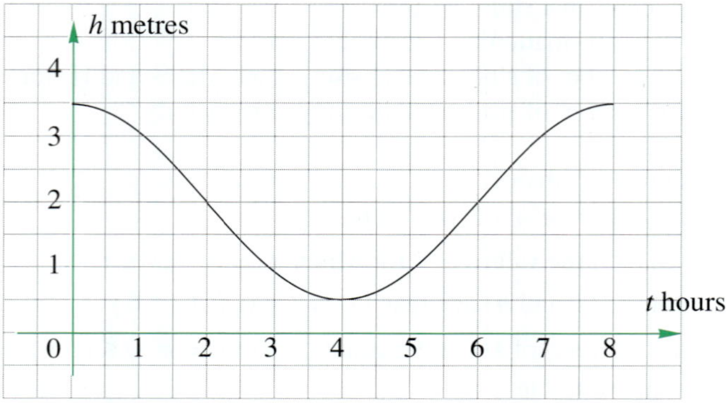

(a) Using the graph above, find
i. when the height of the tide reaches 3 metres.
ii. the height of the tide at 3 am.

(b) For safety purposes, you are only allowed to swim when the tide is less than 2.5 m. Assuming that the behaviour of the tide remains the same over every eigth hour cycle, how long can you swim for in any 24 hour period?

(c) The function that models the height h shown in the graph above is of the form

$$f(t) = a\cos(bt°) + c .$$

It is also found that this model can be extended for values of $t > 8$.
i. Determine the values of a, b and c.
ii. Find the height of the tide at 10 am.

(d) i. In a 24 hour period, how often will the height of the tide reach 2.25 metres?
ii. For what percentage of the time will the height of the tide be at least 2.25 metres?

31. Let A and B be events such that $P(A) = \frac{1}{3}$, $P(B) = \frac{1}{4}$ and $P(A \cup B) = \frac{5}{12}$.

i. Find $P(A|B)$ ii. $P(A|B')$ iii. Are A and B independent? Why?

32. The scrap value, $\$V$, of some machinery after t years is given by

$$V(t) = 50000(0.65)^t, t \geq 0 .$$

(a) i. What was the initial cost of the machine?
ii. What is the scrap value of the machine after 5 years?
iii. How long would it be before the scrap value reaches \$20000?
iv. The machine needs to be sold at some time when the scrap value of the machine lies somewhere between 10000 and 15000. What time–frame does the owner have?

(b) It is suggested that a different model is used to calculate the scrap value of the machine after t years. The new model for the scrap value, $\$V_2$, is given by,

$$V_2(t) = 50000 - kt$$

where k is a constant.
Find the value of k that will give the same scrap value after 5 years as the one obtained when using the first model.

16.0 TRIGONOMETRIC RATIOS

The word TRIGONOMETRY is derived from the Greek words trigonon (a triangular harp) and metron (to measure) and means 'measurement of triangles'.

A major use of trigonometry is in surveying and map making. An early example of this technique was the expedition to India led by William Lambton and George Everest. This was conceived in 1800 and took over half a century to complete. The full story is told in the excellent book *The Great Arc* by *John Keay* (ISBN: 0 00 257062 9).

The expedition began by setting out an accurately measured 'baseline' of 7.5 miles in length at Madras (now Chennai) in the south of the country. The team then used a theodolite (a telescope and accurate protractors mounted on a tripod) to measure the angles A & B between the baseline and a hilltop visible from both places. You will occasionally see surveyors with modern versions of this equipment making measurements around city streets. The theodolites used by the India survey weighed about half a tonne and had to be carried to the tops of hills, temple towers and scaffolding structures to extend the survey.

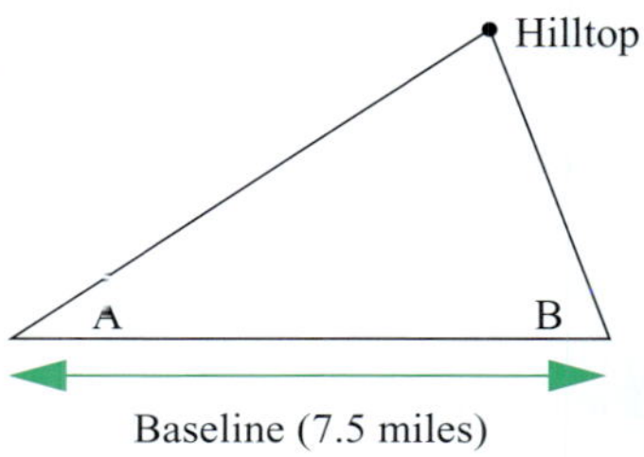

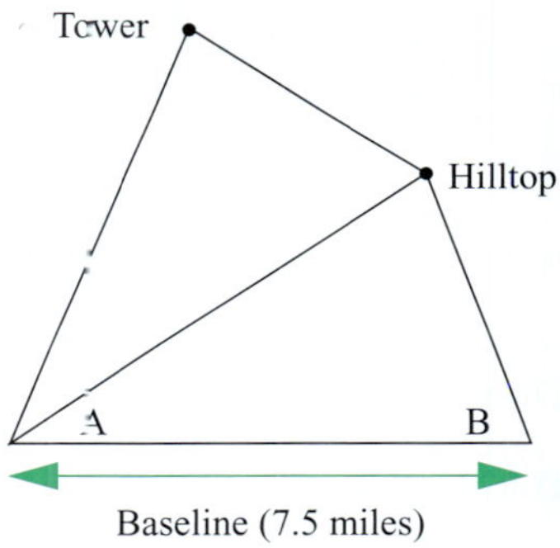

The surveyors then used the mathematics that you will learn in this chapter to calculate the position of the hilltop to great accuracy.

Having done this, the team were able to repeat the process for a second triangle to another landmark. In this way, the survey proceeded northwards through central India measuring hundreds of linked triangles in a connected web. On the way they had to battle tropical diseases, scorpions and the occasional tiger.

Sometimes, teams would be sent east or west to extend the survey to the coasts. By 1834, the surveyors were in sight of the Himalayas. In 1856 the team were able to announce that the peak called Himalaya XV was 29,002 feet above sea level and was the highest point surveyed. The currently accepted value is 29,035 feet (obtained using GPS), an error of one tenth of one percent!

George Everest is said to have wanted the peaks to receive their correct local names. Many of the Himalayas have names such as Kanchenjunga (Five Treasuries of the Great Snow) and Ama Dablam (Mother and her Necklace). Others await correct naming and are still called K2 etc.

Mount Everest is an exception. Its local name is said to be Mi-thik Dgu-thik Bya-phur Long-nga (You cannot see the summit from nearby and birds who fly this high go blind) which probably explains why it is known by its present name, even though Everest himself would not have approved. George Everest called himself 'Eve-rest' not 'Ever-est' and so we are all pronouncing it incorrectly as well!

16.1 TRIGONOMETRIC RATIOS

16.1.1 REVIEW OF TRIGONOMETRIC FUNCTIONS FOR RIGHT-ANGLED TRIANGLES

The trigonometric functions are defined as **ratio functions** in a right-angled triangle. As such they are often referred to as the **trigonometric ratios**.

The trigonometric ratios are based on the right-angled triangle shown alongside. Such right-angled triangles are defined in reference to a nominated angle. In the right-angled triangle ABC the longest side [AB] (opposite the right-angle) is the **hypotenuse**. Relative to the angle $\angle BAC$ of size $\theta°$, the side BC is called the **opposite** side while the side AC is called the **adjacent** side.

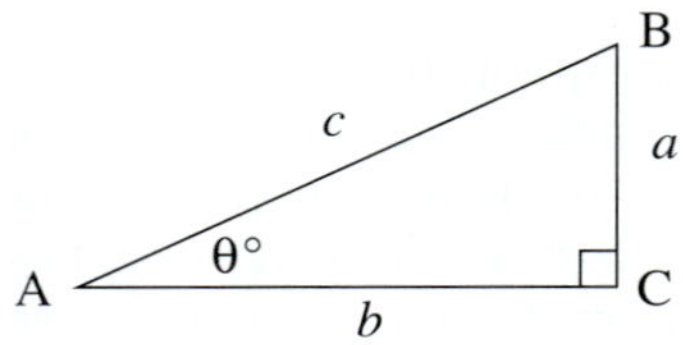

The trigonometric ratios are defined as

$$\sin\theta° = \frac{\text{opposite}}{\text{hypotenuse}} = \frac{a}{c}, \quad \cos\theta° = \frac{\text{adjacent}}{\text{hypotenuse}} = \frac{b}{c}, \quad \tan\theta° = \frac{\text{opposite}}{\text{adjacent}} = \frac{a}{b}$$

Note then, that $\tan\theta° = \frac{\sin\theta°}{\cos\theta°}, \cos\theta \neq 0$.

There also exists another important relation between the side lengths of a right-angled triangle.

This relationship, using **Pythagoras's Theorem** is $a^2 + b^2 = c^2$

Do not forget to adjust the mode of your calculator to degree mode when necessary. On the TI–83, this is done by pressing **MODE** and then selecting the **Degree** mode. As angles can be quoted in degrees '°', minutes '′' and seconds '″' we make use of the **DMS** option under the **ANGLE** menu (accessed by pressing **2nd APPS**) to convert an angle quoted as a decimal into one quoted in degrees, minutes and seconds.

16.1.2 EXACT VALUES

There are a number of special right-angled triangles for which exact values of the trigonometric ratios exist. Two such triangles are shown below:

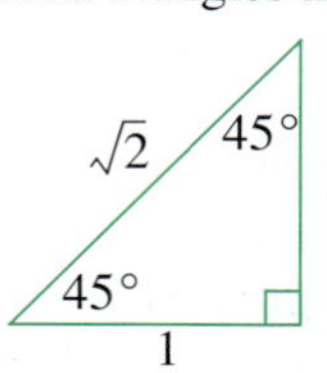

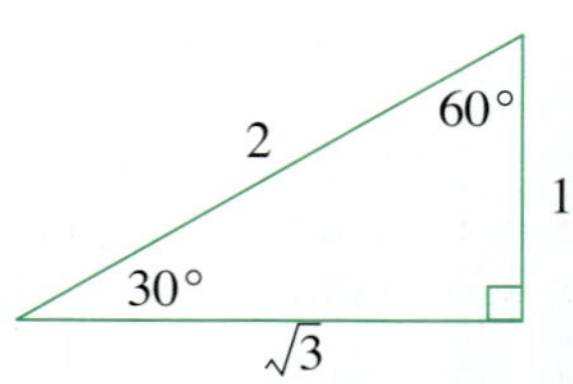

From these triangles we can tabulate the trigonometric ratios as follows:

θ	sinθ°	cosθ°	tanθ°
30	$\frac{1}{2}$	$\frac{\sqrt{3}}{2}$	$\frac{1}{\sqrt{3}}$
45	$\frac{1}{\sqrt{2}}$	$\frac{1}{\sqrt{2}}$	1
60	$\frac{\sqrt{3}}{2}$	$\frac{1}{2}$	$\sqrt{3}$

EXAMPLE 16.1

Find x in each of the following triangles (correct to 4 d.p)

(a)

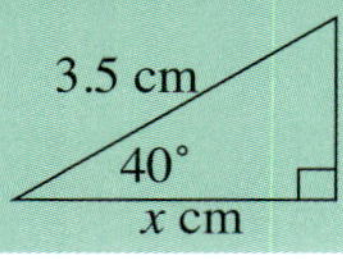

(b)

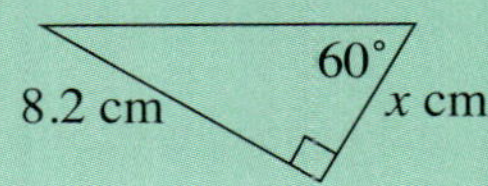

(c)

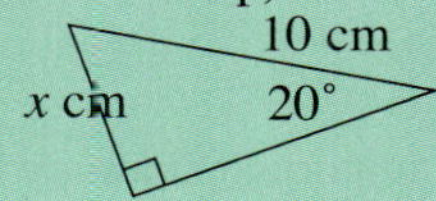

Solution

(a) We label the sides relative to the given angle:
As the sides involved are the adjacent (adj) and the hypotenuse (hyp). The appropriate ratio is the cosine ratio, i.e., $\cos\theta = \frac{\text{adj}}{\text{hyp}}$. Then, substituting the information into the expression we can solve for x:

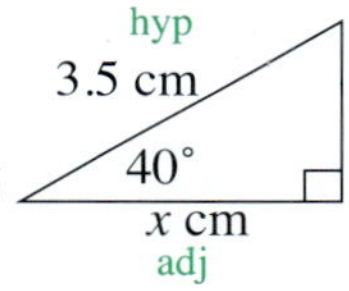

$$\cos 40° = \frac{x}{3.5} \Leftrightarrow 3.5 \times \cos 40° = x$$

$$\therefore x = 2.6812 \text{ (to 4 d.p)}$$

A quick word about using the TI–83. Below we show that depending on the mode setting we obtain different values. In particular, note that in Case B, even though the mode setting was in radians, we were able to over ride this by using the degree measure, '°', under the **ANGLE** menu.

Case A:

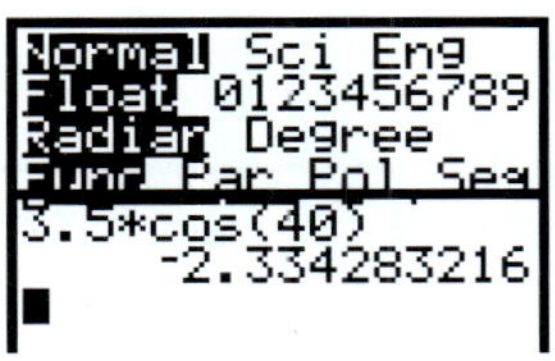

Case B:

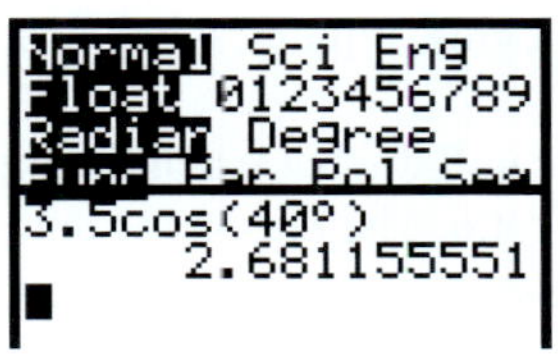

Case C:

(b) We label the sides relative to the given angle.
The sides involved are the adjacent (adj) and the opposite (opp) the appropriate ratio is the tangent ratio, i.e., $\tan\theta = \frac{\text{opp}}{\text{adj}}$. Then, substituting the information into the expression we can solve for x:

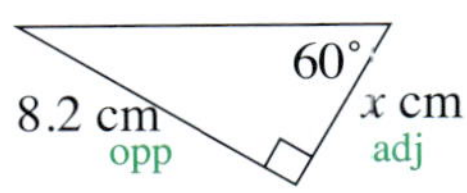

$$\tan 60° = \frac{8.2}{x} \Leftrightarrow x\tan 60° = 8.2 \Leftrightarrow x = \frac{8.2}{\tan 60°}$$

$$\therefore x = 4.7343 \text{ (to 4 d.p)}$$

(c) We label the sides relative to the given angle.
The sides involved are the opposite (opp) and the hypotenuse (hyp).
The appropriate ratio is the sine ratio, i.e., $\sin\theta = \frac{\text{opp}}{\text{hyp}}$. Then, substituting the information into the expression we can solve for x:

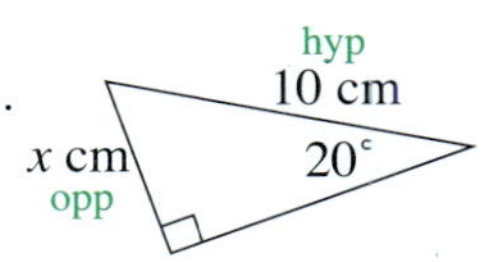

$$\sin 20° = \frac{x}{10} \Leftrightarrow 10 \times \sin 20° = x$$

$$\therefore x = 3.4202 \text{ (to 4 d.p)}$$

EXAMPLE 16.2

Find x and θ in the following triangles

(a) 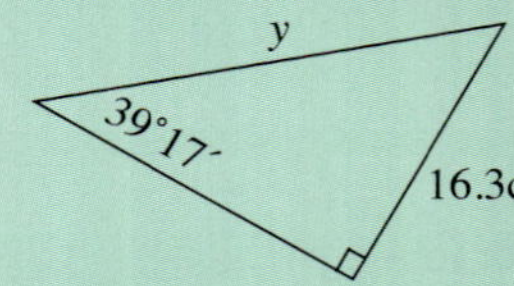

(b) 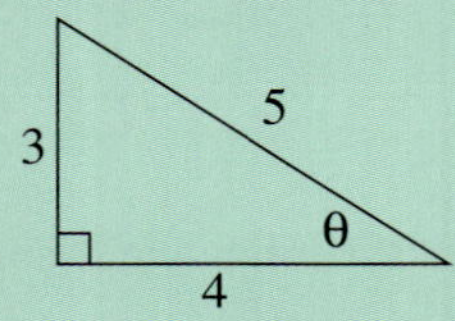

Solution

(a) The important sides are the opposite and hypotenuse. So,

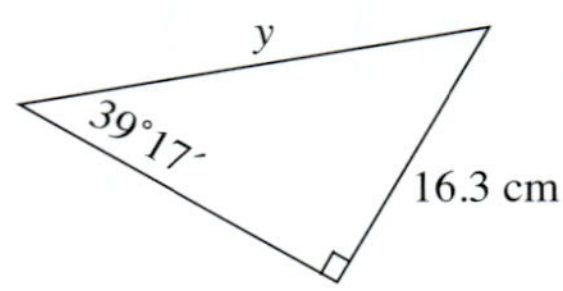

$$\sin 39°17' = \frac{16.3}{y} \Leftrightarrow y \times \sin 39°17' = 16.3$$

$$\Leftrightarrow y = \frac{16.3}{\sin 39°17'}$$

$$\therefore y = 25.7 \text{ cm}$$

The TI–82/83 calculators accept angle inputs using the **2nd ANGLE** menu. **Option 1** allows entry of angles in degrees irrespective of the **MODE** setting of the calculator. **Option 2** allows the entry of degrees, minutes, seconds.

The problem would be solved using the keying sequence

16.3÷sin 39 **2nd ANGLE 1** 17 **2nd ANGLE 2** ENTER.

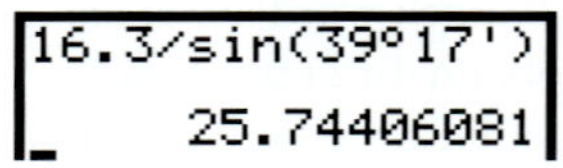

(b) When using a calculator to find an angle, **option 4** of the **2nd ANGLE** menu will allow you to display an answer in degree, minute, second format.

Any of the three trigonometric ratios will do, but when finding angles, it is generally best to use the cosine ratio. The reason for this should become apparent as this chapter progresses.

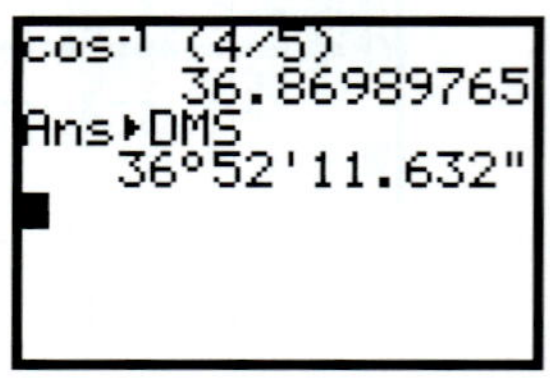

$$\cos\theta = \frac{4}{5} \Rightarrow \theta = \cos^{-1}\left(\frac{4}{5}\right) \therefore \theta \approx 36°52'12''$$

rounded to the nearest second.

EXAMPLE 16.3

Using the triangle shown, find

(a) i. AB

ii. $\cos\alpha$

iii. $\tan\alpha$

(b) If $\cos\alpha = 0.2$ find $\sin(90° - \alpha)$.

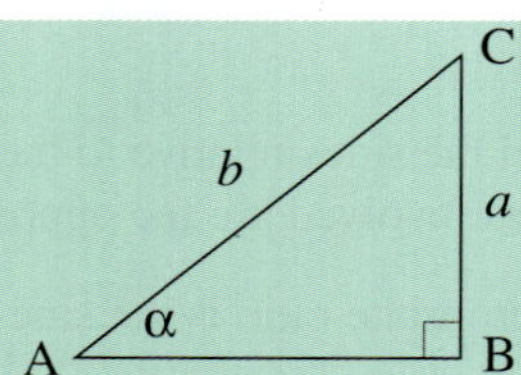

Solution

(a) Using Pythagoras's Theorem we have $AC^2 = AB^2 + BC^2 \therefore b^2 = AB^2 + a^2$

$$\Leftrightarrow AB^2 = b^2 - a^2$$

$$\Rightarrow AB = \sqrt{b^2 - a^2}$$

(b) $\cos\alpha = \dfrac{AB}{AC} = \dfrac{\sqrt{b^2 - a^2}}{b}$

(c) $\tan\alpha = \dfrac{CB}{AB} = \dfrac{a}{\sqrt{b^2 - a^2}}$

(d) $\sin(90° - \alpha) = \dfrac{AB}{AC}$, but $\cos\alpha = \dfrac{AB}{AC} \therefore \sin(90° - \alpha) = \cos\alpha$.

i.e., $\sin(90° - \alpha) = 0.2$

Often we are dealing with non right-angled triangles. However, these can be 'broken up' into at least two right-angled triangles, which then involves solving simultaneous equations. This is illustrated in the next example.

EXAMPLE 16.4

Find the angle θ in the diagram shown.

Note that $\angle ACB \neq 90°$.

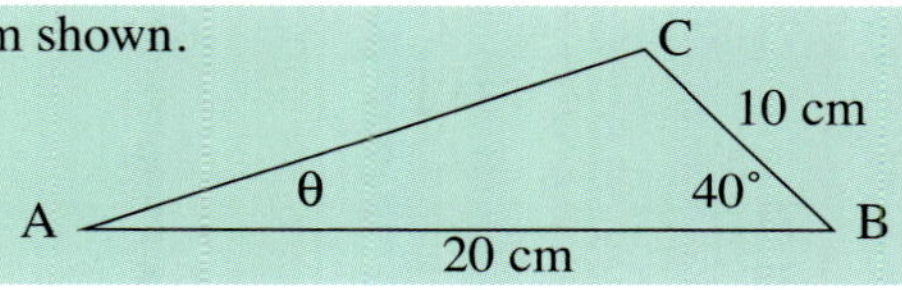

Solution

We start by 'breaking-up' the triangle into two right-angled triangles as follows:

Using ΔACP:

$$\tan\theta = \frac{PC}{AP} = \frac{y}{20 - x} \quad -(1)$$

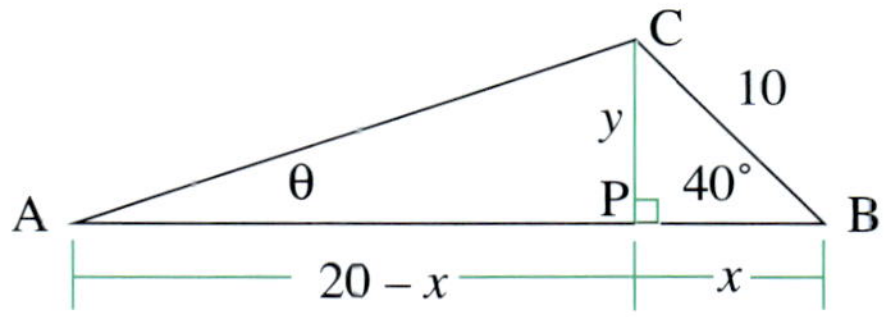

We now need to determine x and y.

Using ΔBPC:

$$\sin 40° = \frac{PC}{BC} = \frac{y}{10}$$

$$\Leftrightarrow y = 10\sin 40° \quad -(2)$$

and $$\cos 40° = \frac{BP}{BC} = \frac{x}{10}$$

$$\Leftrightarrow x = 10\cos 40° \quad -(3)$$

Therefore, substituting (3) and (2) into (1) we have:

$$\tan\theta = \frac{10\sin 40°}{20 - 10\cos 40°}$$

$$= 0.5209$$

$$\therefore \theta = \tan^{-1}(0.5209)$$

$$= 27.5157$$

$$= 27°31'$$

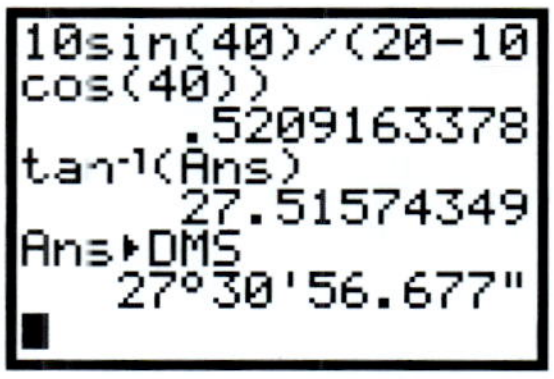

Note that we have not rounded down our answer until the very last step.

EXERCISES 16.1

1. The parts of this question refer to the triangle shown. Complete the blank spaces in this table, giving lengths correct to three significant figures and angles correct to the nearest degree.

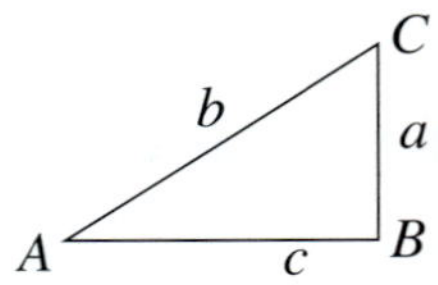

	a cm	*b* cm	*c* cm	*A*	*B*	*C*
1.			1.6		90°	23°
2.		98.3			90°	34°
3.			33.9		90°	46°
4.		30.7			90°	87°
5.	2.3				90°	33°
6.		77			90°	51°
7.	44.4		68.4		90°	57°
8.			12.7	13°	90°	
9.		94.4		52°	90°	

	a cm	*b* cm	*c* cm	*A*	*B*	*C*
10.	71.8		64.6	48°	90°	
11.		34.1		43°	90°	
12.			2.3	87°	90°	
13.	71.5			63°	90°	
14.	33.5		6.5		90°	
15.	6.1	7.2			90°	
16.		30	7.3		90°	
17.	29.0		2.0		90°	
18.	34.5	88.2			90°	
19.	24.0	29.7			90°	
20.		46.2			90°	27°
21.	59.6		41.8		90°	35°
22.		6.8			90°	37°
23.			14.9	41°	90°	49°
24.			16.1	41°	90°	49°
25.			33.3	68°	90°	22°

2. Find the exact value of x in each of the following

(a)

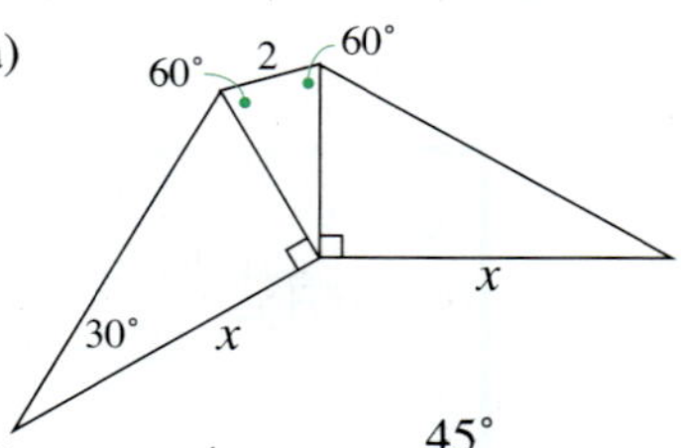

(b)

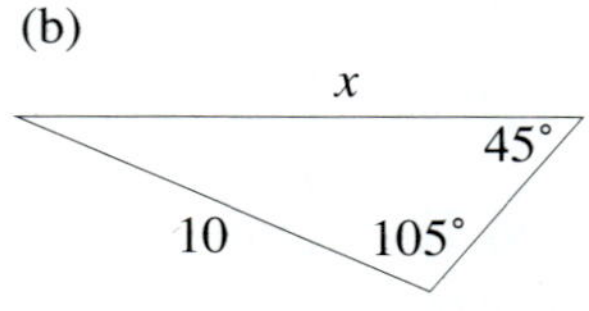

(c)

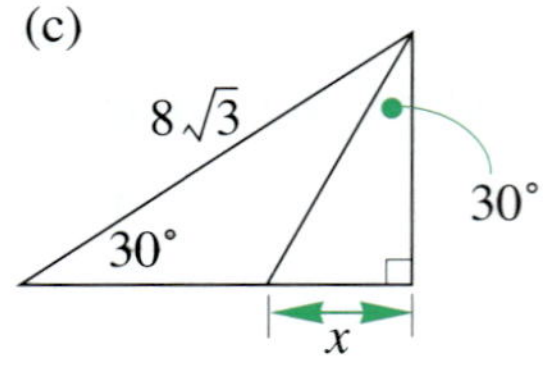

(d)

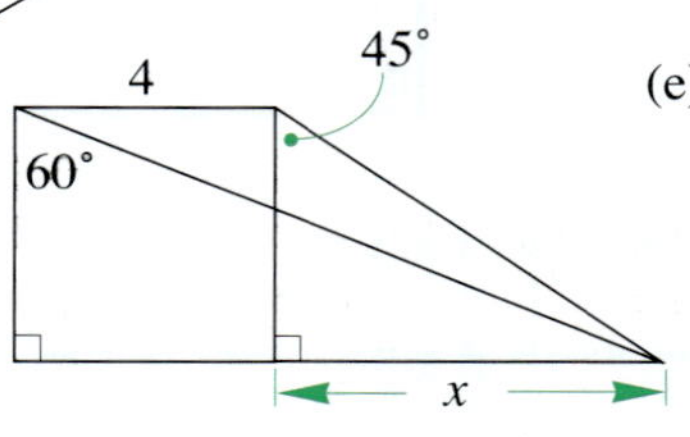

(e)

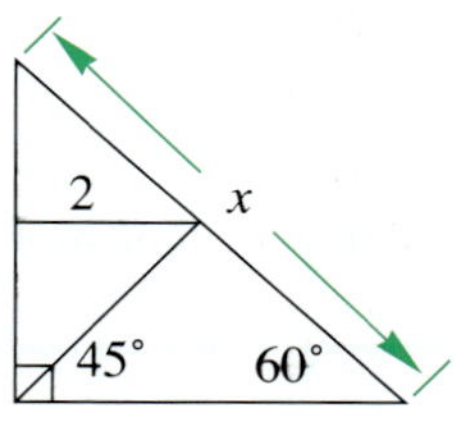

(f)

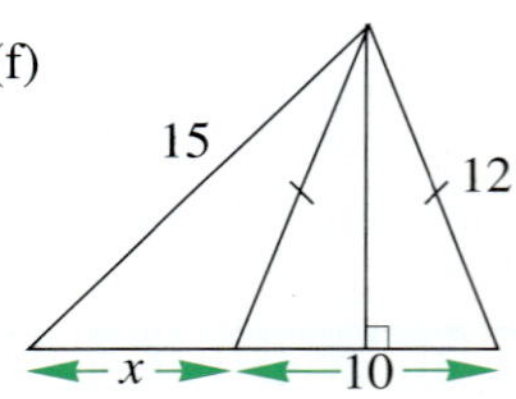

3. Using the triangle on the right, show that

(a) $\sin(90° - \theta) = \cos\theta$

(b) $\cos(90° - \theta) = \sin\theta$

(c) $\tan(90° - \theta) = \dfrac{1}{\tan\theta}$

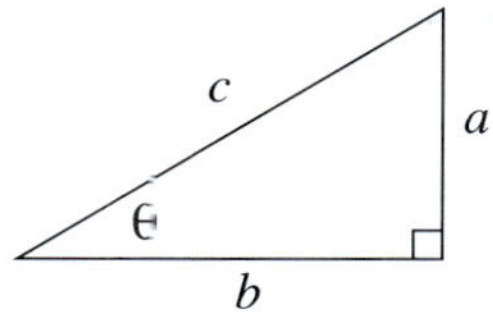

4. Find the exact value of x in each of the following.

(a)

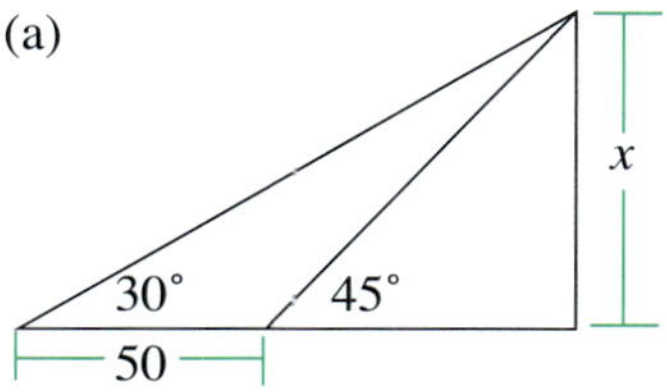

(b)

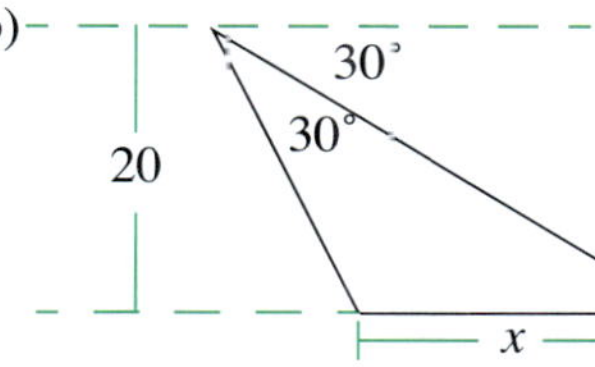

5. Show that $OZ = \dfrac{x\tan\theta(\sin\alpha + \cos\alpha\tan\beta)}{\tan\theta - \tan\beta}$

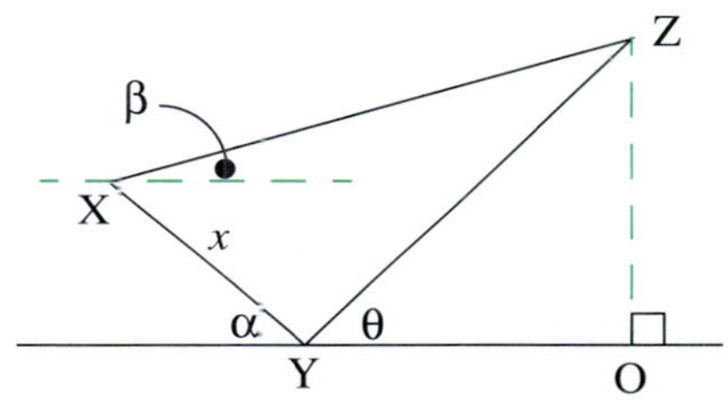

16.2 APPLICATIONS

Applications that require the use of trigonometric ratios and right-angled triangles are many and varied. In this section we consider a number of standard applications to highlight this.

16.2.1 ANGLE OF ELEVATION AND DEPRESSION

The **angle of elevation** is the angle of the line of sight **above** the horizontal of an object seen above the horizontal

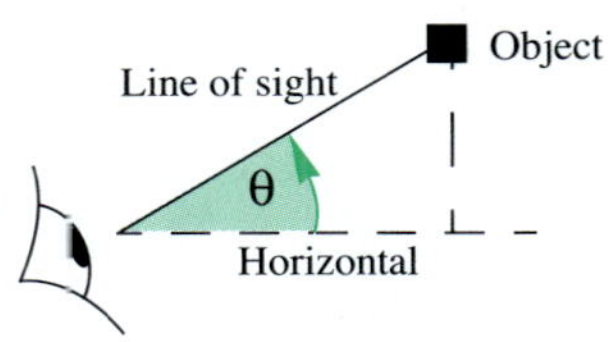

The **angle of depression** is the angle of the line of sight **below** the horizontal of an object seen below the horizontal

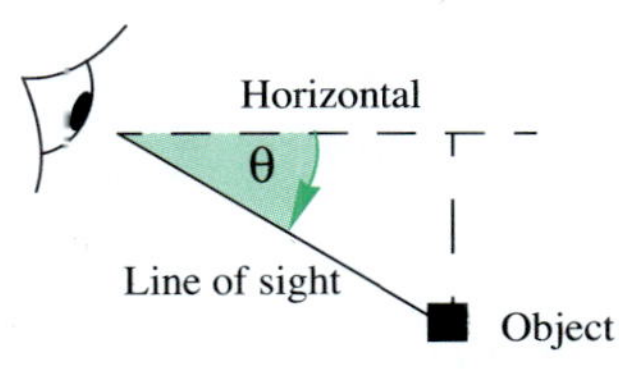

Note that the angle of depression and elevation for the same line of sight are **alternate angles**.

EXAMPLE 16.5

An observer standing on the edge of a cliff 82 m above sea level sees a ship at an angle of depression of 26°. How far from the base of the cliff is the ship situated?

Solution

We first draw a diagram to represent this situation:
Let the ship be at point S, x metres from the base of the cliff, B, and let O be where the observer is standing.
Using the right-angled triangle OBS we have:

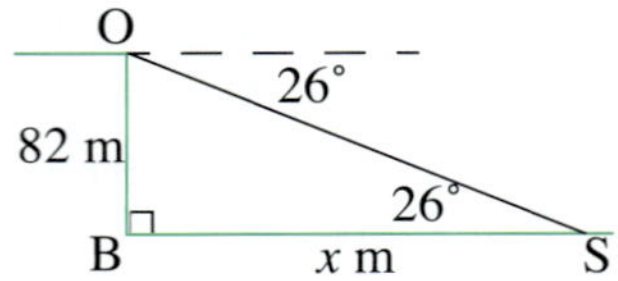

$$\tan 26° = \frac{82}{x} \Leftrightarrow x\tan 26° = 82$$

$$\Leftrightarrow x = \frac{82}{\tan 26°}$$

$$= 168.1249\ldots$$

Therefore, the ship is 168 m from the base of the cliff.

EXAMPLE 16.6

The angle of depression from the roof of building A to the foot of a second building, B, across the same street and 40 metres away is 65°. The angle of elevation of the roof of building B to the roof of building A is 35°. How tall is building B?

Solution

Let the height of building B be x m and that of building A be y m.

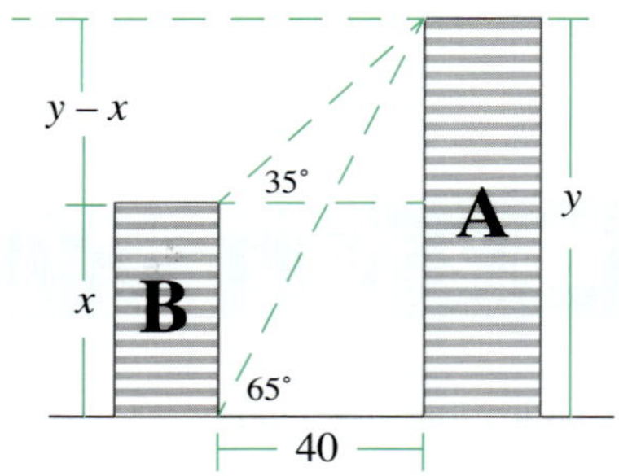

Note that we are using the fact that for the same line of sight, the angle of depression is equal to the angle elevation.

The height difference between the two buildings must then be $(y - x)$ m.

We now have two right-angled triangles to work with:

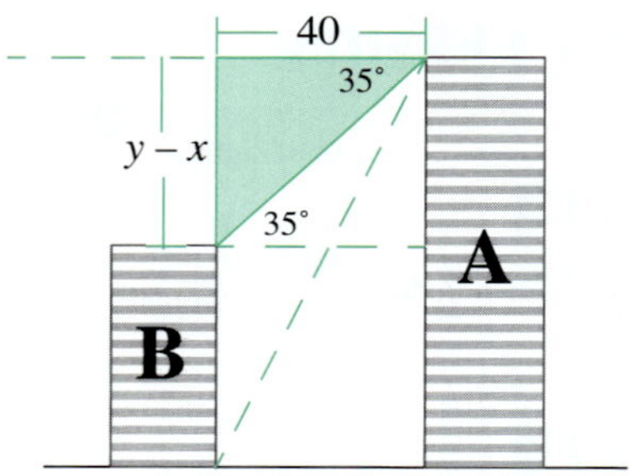

$$\tan 35° = \frac{y-x}{40}$$

$$\Leftrightarrow y - x = 40\tan 35° \quad -(1)$$

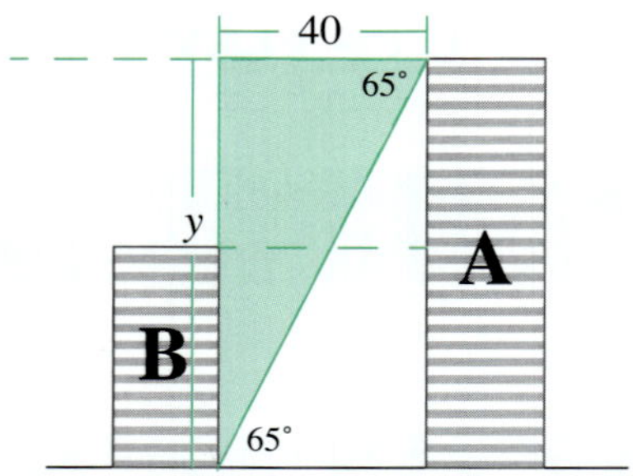

$$\tan 65° = \frac{y}{40}$$

$$\Leftrightarrow y = 40\tan 65° \quad -(2)$$

Substituting (2) into (1) we have:

$$40\tan 65° - x = 40\tan 35°$$

$$\Leftrightarrow x = 40\tan 65° - 40\tan 35°$$

$$\therefore x = 57.7719\ldots$$

That is, building B is 57.77 m

16.2.2 BEARINGS

In the sport of orienteering, participants need to be skilled in handling bearings and reading a compass. Bearings can be quoted by making reference to the North, South, East and West directions or using true bearings. We look at each of these.

Compass bearings

These are quoted in terms of an angle measured East, West, North or South, or somewhere in between. For example, North 30° East, expressed as N30°E, informs us that from the North direction we rotate 30° towards the East and then follow that line of direction. The following diagrams display this for a number of bearings.

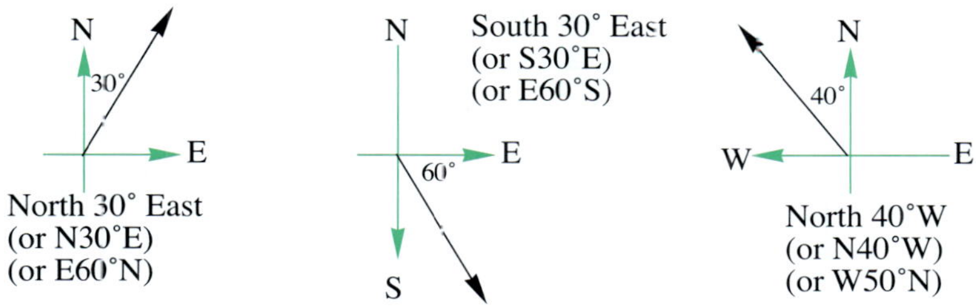

True bearings

These are quoted in terms of an angle measured in a clockwise direction from North (and sometimes a capital T is attached to the angle to highlight this fact). So, for example, a bearing of 030°T would represent a bearing of 30° in a clockwise direction from the North – this corresponds to a compass bearing of N30°E. Using the above compass bearings we quote the equivalent true bearings:

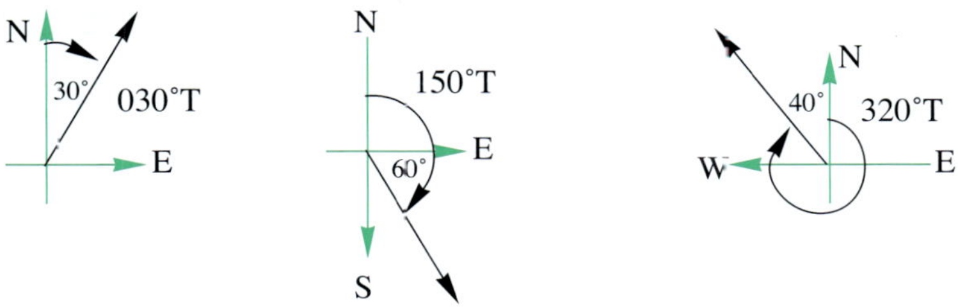

EXAMPLE 16.7

Janette walks for 8 km North-East and then 11 km South-East. Find the distance and bearing from her starting point.

Solution

First step is to draw a diagram.

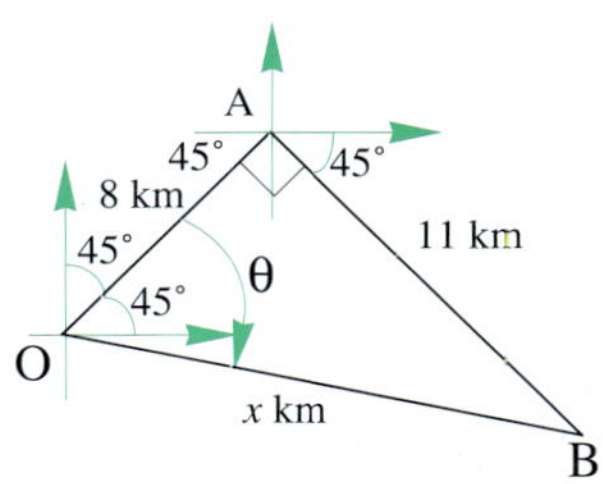

As $\angle OAB = 90°$ we can make use of Pythagoras's Theorem:

$$x^2 = 8^2 + 11^2$$

$$\therefore x = 13.60 \text{ [taking +ve square root]}$$

Let $\theta = \angle AOB$ so that $\tan\theta = \frac{11}{8}$. $\therefore \theta = \tan^{-1}\left(\frac{11}{8}\right)$

$$= 53.97°$$

Therefore, bearing is $45° + \theta = 45° + 53.97° = 98.97°$

That is, B has a bearing of 98.97°T from O and is (approx) 13.6 km away.

EXAMPLE 16.8

The lookout, on a ship sailing due East, observes a light on a bearing of 056˚T. After the ship has travelled 4.5 km, the lookout now observes the light to be on a bearing of 022˚T. How far is the light source from the ship at its second sighting?

Solution

As always, we start with a diagram.

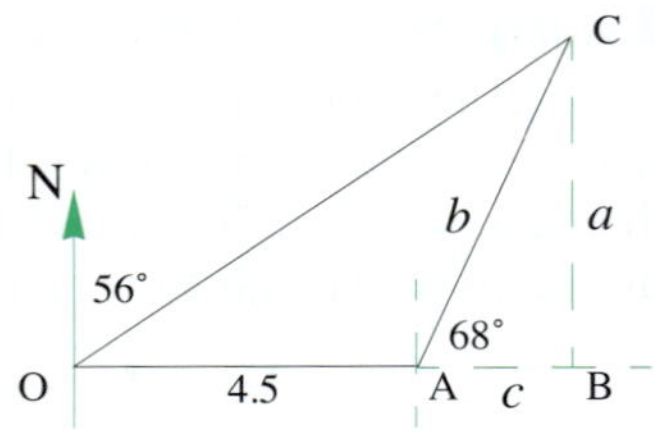

Using ΔOBC we have, $\tan 34° = \frac{BC}{OB} = \frac{a}{4.5 + c}$

$\therefore a = (4.5 + c)\tan 34°$ – (1)

Using ΔABC we have, $\tan 68° = \frac{BC}{AB} = \frac{a}{c}$

$\therefore a = c\tan 68°$ – (2)

Equating (1) and (2) we have,

$$\begin{aligned} c\tan 68° &= (4.5 + c)\tan 34° \\ c\tan 68° &= 4.5\tan 34° + c\tan 34° \\ \Leftrightarrow c(\tan 68° - \tan 34°) &= 4.5\tan 34° \\ \Leftrightarrow c &= \frac{4.5\tan 34°}{(\tan 68° - \tan 34°)} \\ \therefore c &= 1.6857 \end{aligned}$$

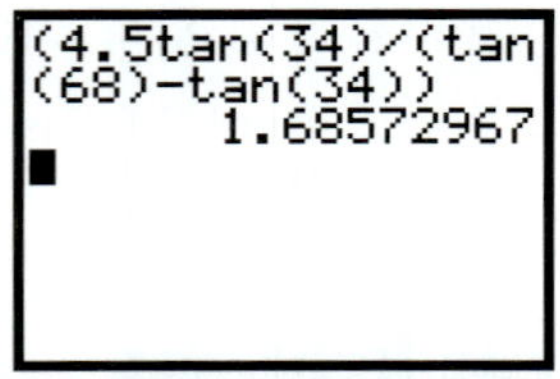

Substituting this result into (2) we have,

$$a = \frac{4.5\tan 34°}{(\tan 68° - \tan 34°)} \times \tan 68°$$

$$\therefore a = 4.1723$$

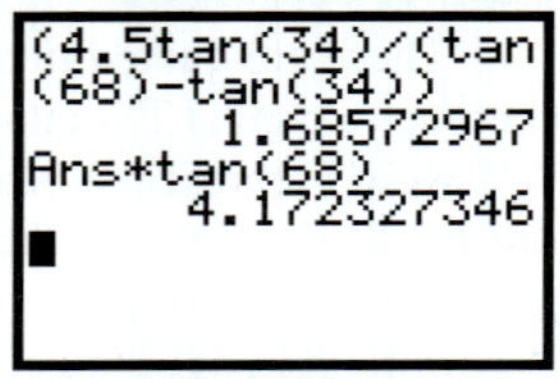

Then, using ΔABC and Pythagoras's Theorem, we have

$$\begin{aligned} b^2 &= a^2 + c^2 \\ &= 4.1723^2 + 1.6857^2 \\ \therefore b &= \sqrt{20.2496} \\ &= 4.4999 \end{aligned}$$

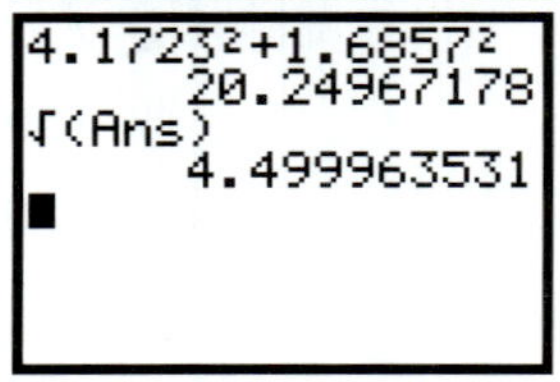

That is, the light is 4.5 km from the ship (at the second sighting).
Can you see a much quicker solution? Hint – think isosceles triangle.

EXERCISES 16.2

1. (a) Change the following compass bearings into true bearings

i. N30˚E ii. N30˚W iii. S15˚W iv. W70˚S

(b) Change the following true bearings into compass bearings

i. 025˚T ii. 180˚T iii. 220˚T iv. 350˚T

2. The angle of depression from the top of a building 60 m high to a swing in the local playground is 58°. How far is the swing from the foot of the building?

3. From a point A on the ground, the angle of elevation to the top of a tree is 52°. If the tree is 14.8 m away from point A, find the height of the tree.

4. Find the angle of elevation from a bench to the top of an 80 m high building if the bench is 105 m from the foot of the building.

5. Patrick runs in a direction N60°E and after 45 minutes finds himself 3900 m North of his starting point. What is Patrick's average speed in ms^{-1}.

6. A ship leaves Oldport and heads NW. After covering a distance of 16 km it heads in a direction of N68°22'W travelling a distance of 22 km where it drops anchor. Find the ship's distance and bearing from Oldport after dropping anchor.

7. From two positions 400 m apart on a straight road, running in a northerly direction, the bearings of a tree are N36°40'E and E33°22'S. What is the shortest distance from the tree to the road?

8. A lamp post leaning away from the sun and at 6° from the vertical, casts a shadow 12 m long when the sun's angle of elevation is 44°. Assuming that the level of the ground where the pole is situated is horizontal, find its length.

9. From a window, 29.6 m above the ground, the angle of elevation of the top of a building is 42°, while the angle of depression to the foot of the building is 32°. Find the height of the building.

10. Two towns P and Q are 50 km apart, with P due west of Q. The bearing of a station from town P is 040°T while the bearing of the station from town Q is 300°T. How far is the station from town P?

11. When the sun is 74° above the horizon, a vertical flagpole casts a shadow 8.5 m onto a horizontal ground. Find the shadow cast by the sun when it falls to 62° above the horizontal.

12. A hiker walks for 5km on a bearing of 053° true (North 53° East). She then turns and walks for another 3km on a bearing of 107° true (East 17° South).

(a) Find the distance that the hiker travels North/South and the distance that she travels East/West on the first part of her hike.

(b) Find the distance that the hiker travels North/South and the distance that she travels East/West on the second part of her hike.

(c) Hence find the total distance that the hiker travels North/South and the distance that she travels East/West on her hike.

(d) If the hiker intends to return directly to the point at which she started her hike, on what bearing should she walk and how far will she have to walk?

13. A surveying team are trying to find the height of a hill. They take a 'sight' on the top of the hill and find that the angle of elevation is 23°27′. They move a distance of 250 metres on level ground directly away from the hill and take a second 'sight'. From this point, the angle of elevation is 19°46′. Find the height of the hill, correct to the nearest metre.

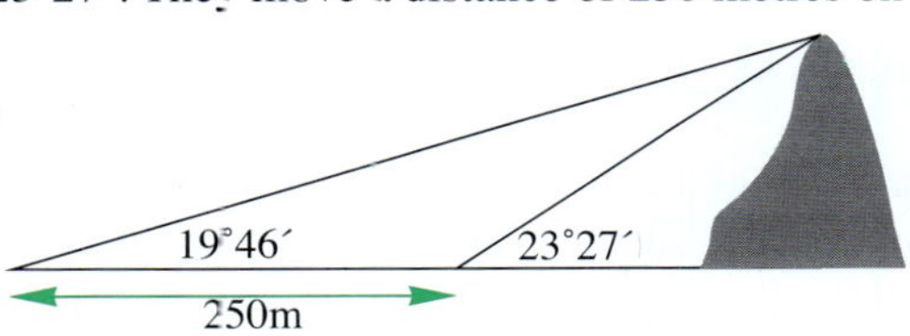

16.3 RIGHT ANGLES IN 3–DIMENSIONS

When dealing with problems in three dimensions, we draw the figures in perspective, so that a model can be more accurately visualised. This does not mean that you must be an artist, simply that you take a little time (and a lot of practice) when drawing such diagrams. The key to many 3–D problems is locating the relevant right–angled triangles within the diagram. Once this is done, all of the trigonometric work that has been covered in the previous two sections can be applied. As such, we will not be learning new theory, but rather developing new drawing and modelling skills. Some typical examples of solids that may be encountered are:

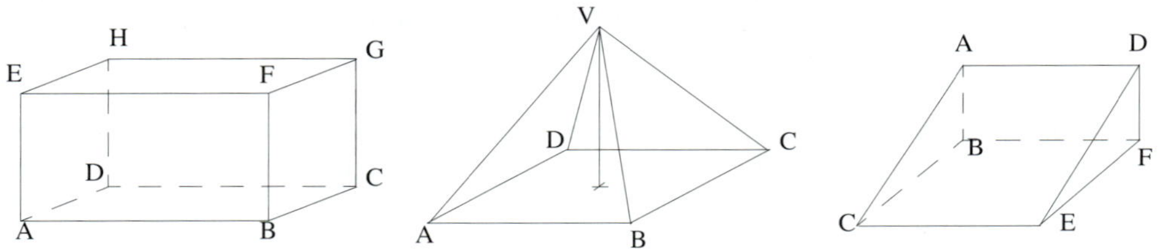

Cuboid ABCD, EFGH   Right Pyramid V, ABCD   Wedge ABC, DEF

We look at two basic concepts and drawing techniques to help us.

1. A line and a plane:

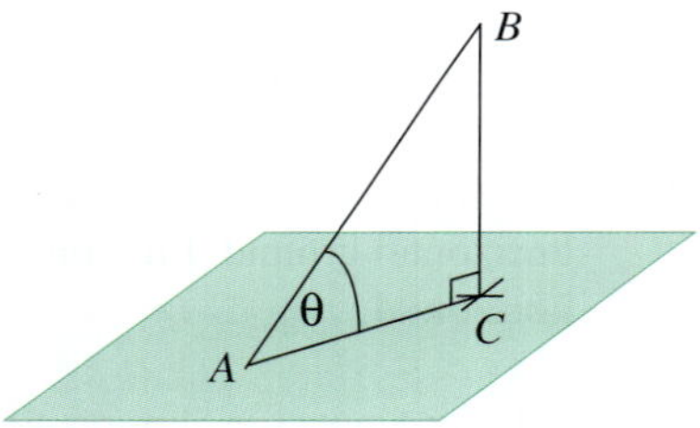

A line will always cut a plane at some point (unless the line is parallel to the plane). To find the angle between a line and a plane construct a perpendicular from the line to the plane and complete a right–angled triangle. In our diagram, we have that the segment $[AB]$ is projected onto the plane. A perpendicular, $[BC]$ is drawn, so that a right–angled triangle, ABC is completed. The angle that the line then makes with the plane is given by θ (which can be found by using the trig–ratios).

2. A plane and a plane:

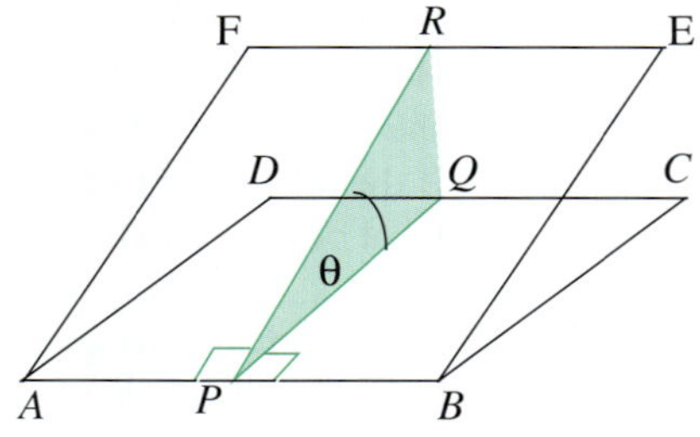

To find the angle between two planes ABCD and ABEF (assuming that they intersect), take any point P on the intersecting segment $[AB]$ and draw $[PQ]$ and $[PR]$ on each plane in such a way that they are perpendicular to $[AB]$. Then, the angle between $[PQ]$ and $[PR]$ (θ) is the angle between the two planes.

EXAMPLE 16.9

A cube ABCD, EFGH has a side length measuring 6 cm.

(a) Find the length of the segment $[AC]$.

(b) The length of the diagonal $[AG]$.

(c) The angle that the diagonal $[AG]$ makes with the base.

Solution

First we need to draw a cube:

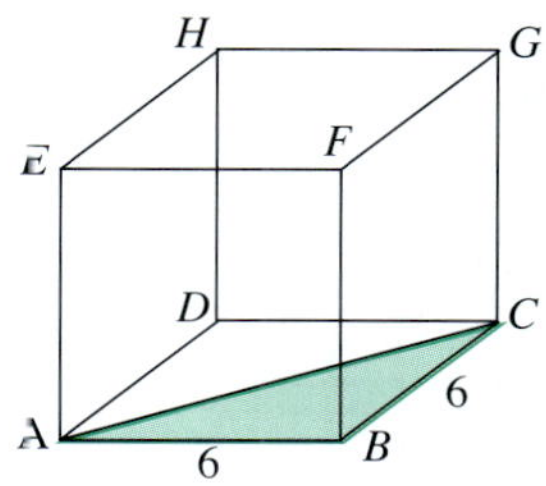

(a) Now the base of the cube is a square, so that $\angle ABC = 90°$, i.e., we have a right–angled triangle, so we can use Pythagoras's Theorem:

$$AC^2 = AB^2 + BC^2$$
$$= 6^2 + 6^2$$
$$= 72$$
$$\therefore AC = \sqrt{72} \approx 8.49$$

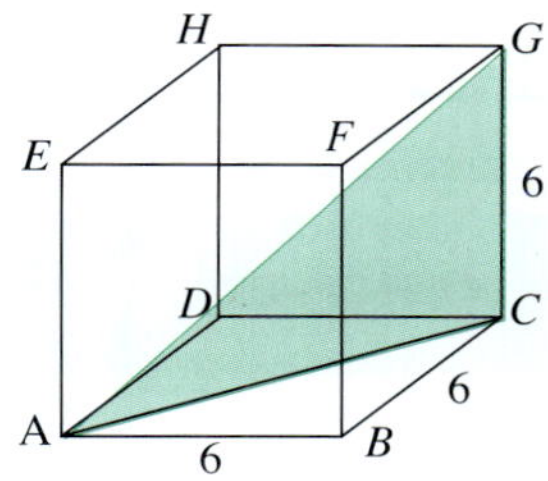

(b) This time we have that $\angle ACG = 90°$, therefore,

$$AG^2 = AC^2 + CG^2$$
$$= (\sqrt{72})^2 + 6^2$$
$$= 108$$
$$\therefore AC = \sqrt{108} \approx 10.39$$

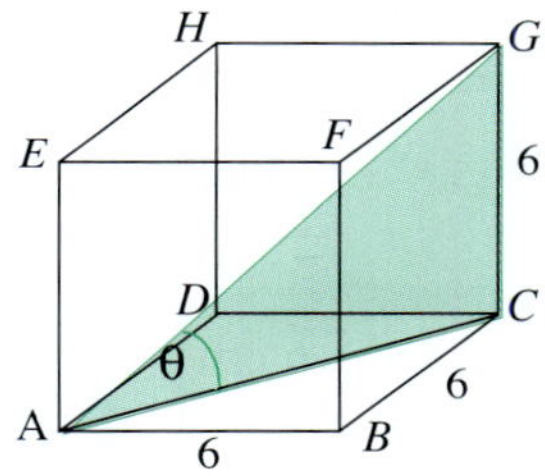

(c) Using triangle ACG, $\tan\theta = \frac{CG}{AC} \therefore \tan\theta = \frac{6}{\sqrt{72}}$

$$\theta = \tan^{-1}\left(\frac{6}{\sqrt{72}}\right)$$
$$= 35.26°$$
$$= 35°16'$$

EXAMPLE 16.10

From a point X, 200 m due South of a cliff, the angle of elevation of the top of the cliff is 30°. From a point Y, due East of the cliff, the angle of elevation of the top of the cliff is 20°. How far apart are the points X and Y?

Solution

We start by illustrating this information on a 3–D diagram (Note that North–South and West–East are drawn on a plane. It is necessary to do this otherwise the diagram will not make sense).

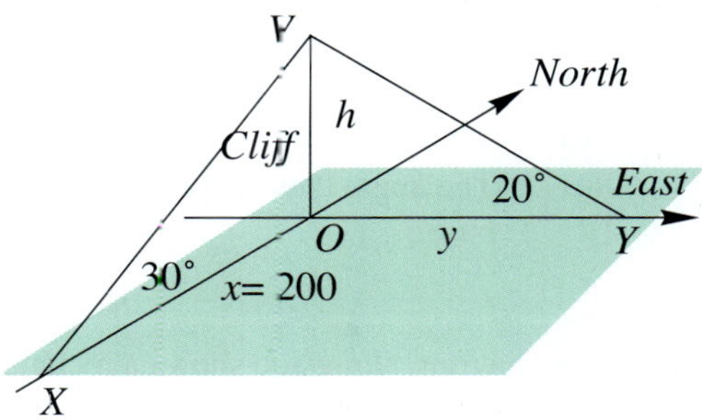

Let the cliff be h metres high. The distance from X to the base of the cliff be x metres and the distance from Y to the base of the cliff be y metres.

As $\angle XOY = 90°$, then $XY^2 = x^2 + y^2$
$$= 200^2 + y^2$$

But, $\tan 20° = \frac{h}{y}$, of which we know neither h or y.

However, using triangle XOV, we have that $\tan 30° = \frac{h}{200} \Rightarrow h = 200 \times \tan 30°$.

Therefore, we have that $\tan 20° = \dfrac{200 \times \tan 30°}{y} \Leftrightarrow y = \dfrac{200 \times \tan 30°}{\tan 20°}$

That is, $y = 317.25$

Therefore, $XY^2 = x^2 + y^2 = 200^2 + \left(\dfrac{200 \times \tan 30°}{\tan 20°}\right)^2$

$= 140648.4289$

$XY = 375.0312$.

That is, X and Y are approximately 375 m apart.

```
200²+((200tan(30
))/tan(20))²
        140648.4289
√(Ans
        375.0312373
■
```

EXERCISES 16.3

1. For the diagram shown, determine the angle of inclination between the plane
(a) ABCD and the base, EABH (Figure 1).
(b) ABC and the base EBFA (Figure 2).

Figure 1.

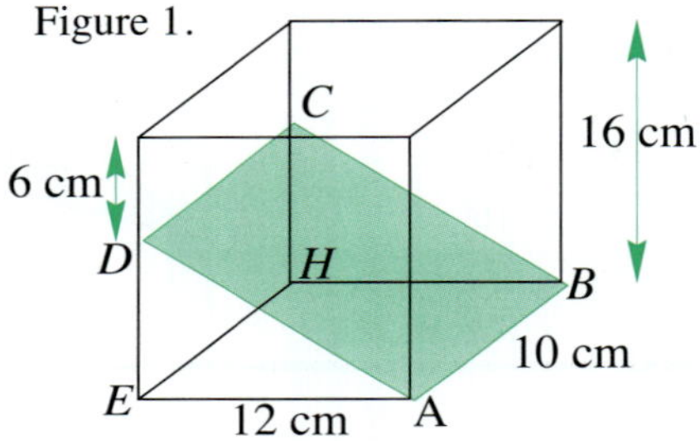

Figure 2.

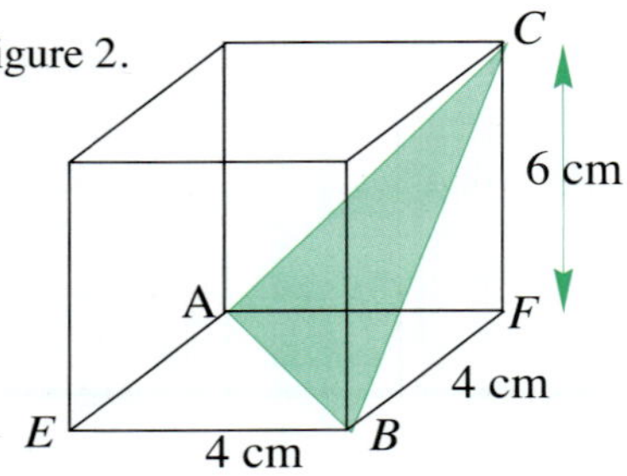

2. A right pyramid with a rectangular base and a vertical height of 60 cm is shown in the diagram alongside. The points X and Y are the midpoints of the sides [AB] *and* [BC] *respectively*
Find
(a) the length, AP.
(b) the length of the edge [AV].
(c) the angle that the edge AV makes with the base ABCD.
(d) the length, YV .
(e) The angle that the plane BCV makes with the base.

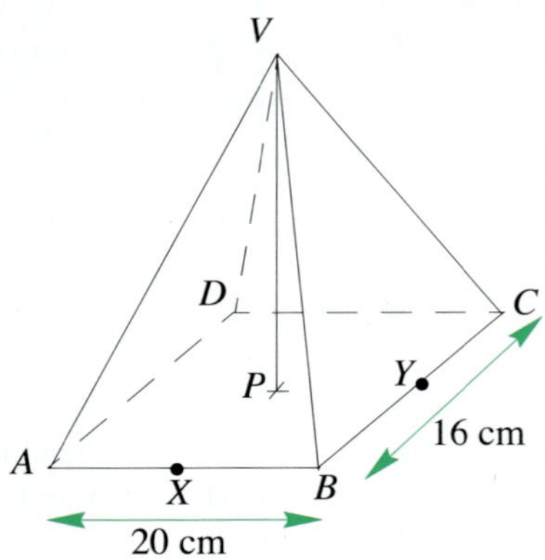

3. The diagram alongside shows a rectangular box with side lengths AB = 8 cm, BC = 6 cm and CG = 4 cm. Find the angle between

(a) the line [BH] and the plane $ABCD$.
(b) the lines [BH and [BA].
(c) the planes $ADGF$ and $ABCD$.

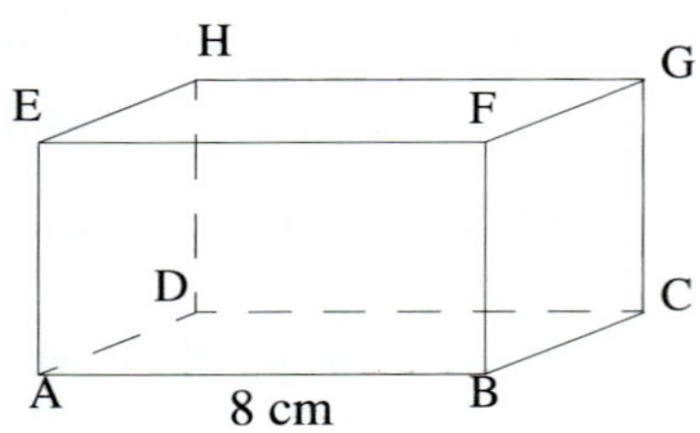

4. For the wedge shown alongside, given that the angle between the lines EA and ED is 50°find

(a) the length of [AE].

(b) the $\angle AEB$.

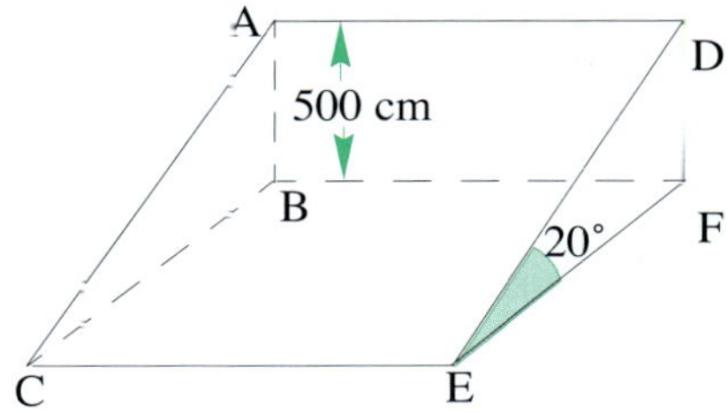

5. From a point A, 100 m due South of a tower, the angle of elevation of the top of the tower is 40°. From a point B, due East of the tower, the angle of elevation of the top of the tower is 20°. How far apart are the points A and B?

6. For the triangular prism shown alongside find

(a) the value of h

(b) the value of α

(c) the angle that the plane ABV makes with the base ABC.

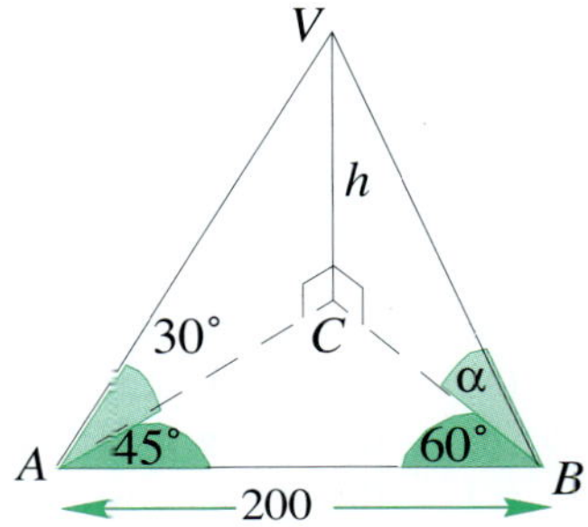

7. The angle of depression from the top of a tower to a point X South of the tower, on the ground and 120 m from the foot of the tower is 24°. From point Y due West of X the angle of elevation to the top of the tower is 19°.

(a) Illustrate this information on a diagram.

(b) Find the height of the tower.

(c) How far is Y from the foot of the tower?

(d) How far apart are the points X and Y?

8. A mast is held in a vertical position by four ropes of length 60 metres. All four ropes are attached at the same point at the top of the mast so that their other ends form the vertices of a square when pegged into the (level) ground. Each piece of rope makes an angle of 54° with the ground.

(a) Illustrate this information on a diagram.

(b) How tall is the mast?

9. A symmetrical sloping roof has dimensions as shown in the diagram. Find

(a) the length of [FM].

(b) the angle between the plane BCEF and the ground.

(c) the angle between the plane ABF and the ground

(d) the total surface area of the roof.

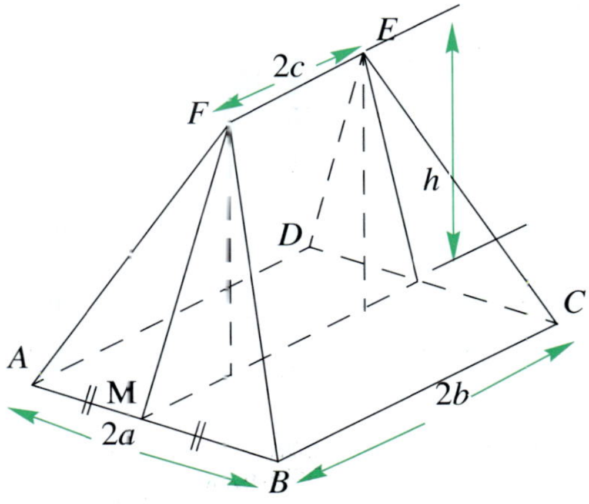

10. The angle of elevation of the top of a tower from a point A due South of it is 68°. From a point B, due East of A, the angle of elevation of the top is 54°. If A is 50 m from B, find the height of the tower.

11. A tower has been constructed on the bank of a long straight river. From a bench on the opposite bank and 50 m downstream from the tower, the angle of elevation of the top of the tower is 30°. From a second bench on the same side of the river and 100 m upstream from the tower, the angle of elevation of the top of the tower is 20°. Find
(a) the height of the tower.
(b) the width of the river.

12. A right pyramid of height 10 m stands on a square base of side lengths 5 m. Find
(a) the length of the slant edge.
(b) the angle betwen a sloping face and the base.
(c) the angle between two sloping faces.

13. A camera sits on a tripod with legs 1.5 m long. The feet rest on a horizontal flat surface and form an equilateral triangle of side lengths 0.75 m. Find
(a) the height of the camera above the ground.
(b) the angles made by the legs with the ground.
(c) the angle between the sloping faces formed by the tripod legs.

14. From a point A due South of a mountain, the angle of elevation of the mountaintop is α. When viewed from a point B, x m due East of A, the angle of elevation of the mountaintop is β. Show that the height, h m, of the mountain is given by $h = \dfrac{x\sin\alpha\sin\beta}{\sqrt{\sin^2\alpha - \sin^2\beta}}$.

16.4 AREA OF A TRIANGLE

Given **any** triangle with sides a and b, and the included angle θ, the area, A, is given by

$$A = \frac{1}{2}bh$$

However, $\sin\theta = \frac{h}{a} \Leftrightarrow h = a \times \sin\theta$ and so, we have that

$$A = \frac{1}{2}b \times a \times \sin\theta$$

where θ is the angle between sides a and b.

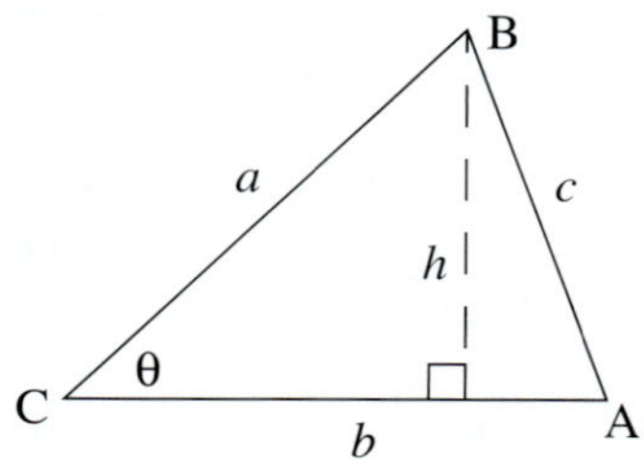

Note that the triangle need not be a right-angled triangle.

Because of the standard labelling system for triangles, the term $\sin\theta$ is often replaced by $\sin C$, given the expresion Area $= \frac{1}{2}ab\sin C$.

A similar argument can be used to generate the formulas: Area $= \frac{1}{2}bc\sin A = \frac{1}{2}ac\sin B$

EXAMPLE 16.11

Find the area of the triangle PQR given that PQ = 9 cm, QR = 10 cm and $\angle PQR = 40°$.

Solution

Based on the given information we can contruct the following triangle:

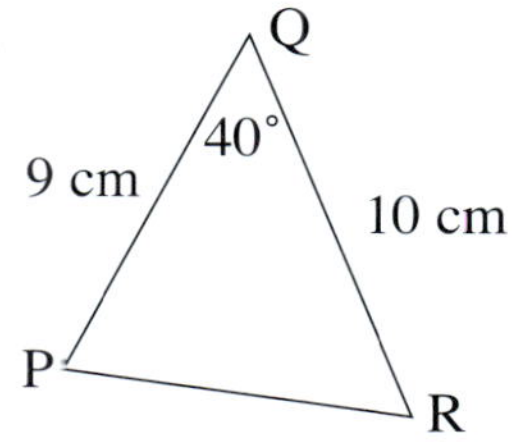

Meaning that, the required area, A, is given by

$$A = \frac{1}{2}ab\sin\theta = \frac{1}{2}\times 9\times 10\times \sin 40°$$
$$= 28.9$$

That is, the area is 28.9 cm^2.

EXAMPLE 16.12

The diagram shows a triangular children's playground. Find the area of the playground

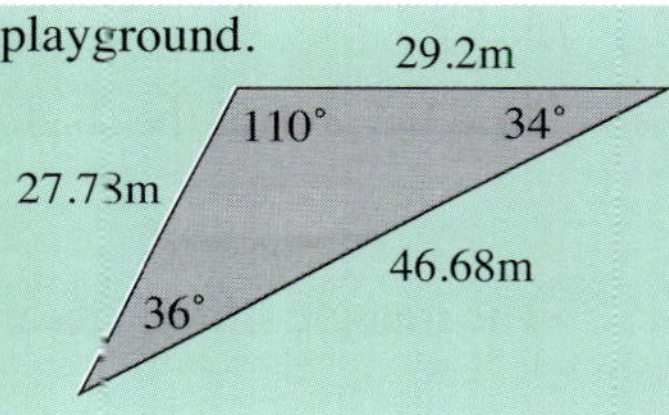

Solution

Since all the measurements of the triangle are known, any one of the three formulas could be used. Many people remember the formula as 'Area equals half the product of the lengths of two sides times the sine of the angle between them'.

$$\text{Area} = \frac{1}{2}\times 27.78\times 46.68\times \sin 36° = 381\,m^2$$

$$\text{Area} = \frac{1}{2}\times 27.78\times 29.2\times \sin 110° = 381\,m^2$$

$$\text{Area} = \frac{1}{2}\times 29.2\times 46.68\times \sin 34° = 381\,m^2$$

EXERCISES 16.4

1. Find the areas of these triangles that are labelled using standard notation.

	a cm	*b* cm	*c* cm	*A*	*B*	*C*
(a)	35.94	128.46	149.70	12°	48°	120°
(b)	35.21	54.55	81.12	20°	32°	128°
(c)	46.35	170.71	186.68	14°	63°	103°
(d)	33.91	159.53	163.10	12°	78°	90°
(e)	42.98	25.07	48.61	62°	31°	87°
(f)	39.88	24.69	34.01	84°	38°	58°
(g)	43.30	30.26	64.94	34°	23°	123°
(h)	12.44	2.33	13.12	68°	10°	102°

	a cm	*b* cm	*c* cm	*A*	*B*	*C*
(i)	43.17	46.44	24.15	67°	82°	31°
(j)	23.16	32.71	24.34	45°	87°	48°
(k)	50.00	52.91	38.64	64°	72°	44°
(l)	44.31	17.52	48.77	65°	21°	94°
(m)	12.68	23.49	22.34	32°	79°	69°
(n)	42.37	42.37	68.56	36°	36°	108°
(o)	40.70	15.65	41.26	77°	22°	81°

2. A car park is in the shape of a parallelogram. The lengths of the sides of the car park are given in metres.

What is the area of the car park?

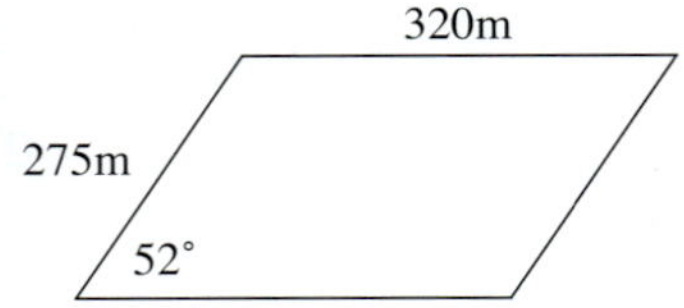

3. The diagram shows a circle of radius 10 cm. AB is a diameter of the circle. AC = 6 cm.

Find the area of the shaded region, giving an exact answer.

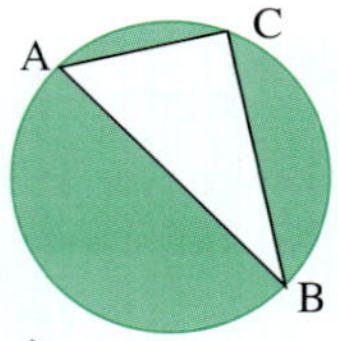

4. The triangle shown has an area of 110 cm^2. Find x.

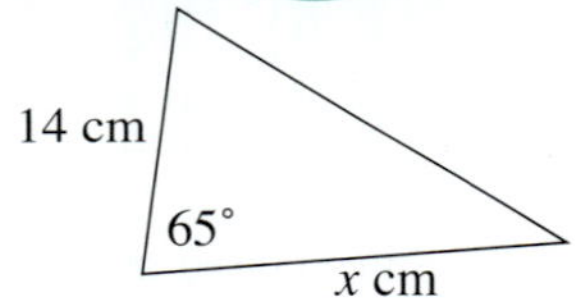

5. Find the area of the following

(a) (b) (c)

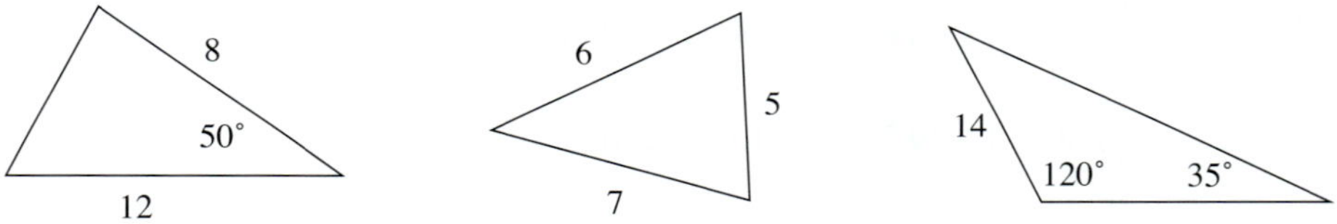

6. A napkin is in the shape of a quadrilateral with diagonals 9 cm and 12 cm long. The angle between the diagonals is 75°. What area does the napkin cover when laid out flat?

7. A triangle of area 50 cm^2 has side lengths 10 cm and 22 cm. What is the magnitude of the included angle?

8. A variable triangle OAB is formed by a straight line passing through the point $P(a, b)$ on the Cartesian plane and cutting the x–axis and y–axis at A and B respectively.
If $\angle OAB = \theta$, find the area of ΔOAB in terms of a, b and θ.

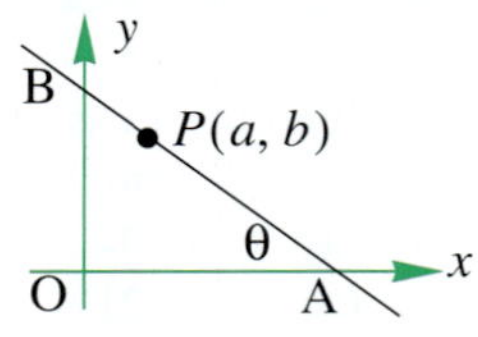

9. Find the area of ΔABC for the given diagram.

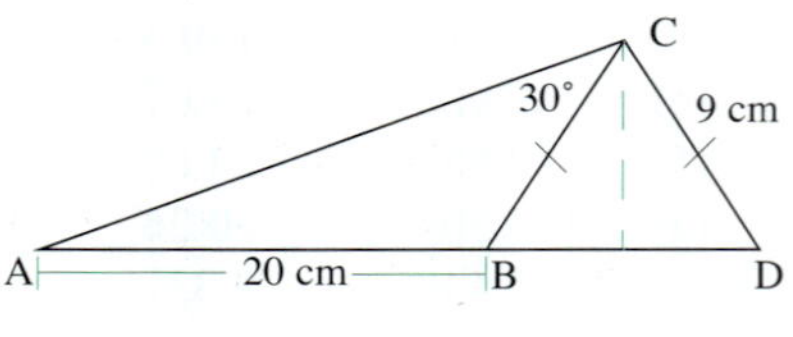

16.5 NON RIGHT-ANGLED TRIANGLES

16.5.1 THE SINE RULE

Previous sections dealt with the trigonometry of right-angled triangles. The trigonometric ratios can be used to solve non right-angled triangles. There are two main methods for solving non right-angled triangles, the **sine rule** and the **cosine rule** (which we look at later in this chapter). Both are usually stated using a standard labelling of the triangle. This uses capital letters to label the vertices and the corresponding small letters to label the sides opposite these vertices.

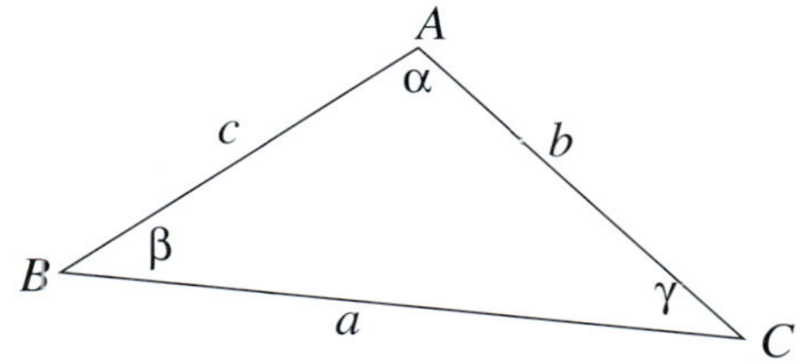

$$\sin A = \sin\alpha$$
$$\sin B = \sin\beta$$
$$\sin C = \sin\gamma$$

Using this labelling of a triangle, the sine rule can be stated as:

$$\frac{a}{\sin A} = \frac{b}{\sin B} = \frac{c}{\sin C} \quad \text{or} \quad \frac{\sin A}{a} = \frac{\sin B}{b} = \frac{\sin C}{c}$$

Note: the sine rule can only be used in a triangle in which an angle and the side **opposite** that angle are known.

So, why does this work?

Using the results of the last section and labelling a triangle in the standard manner we have:

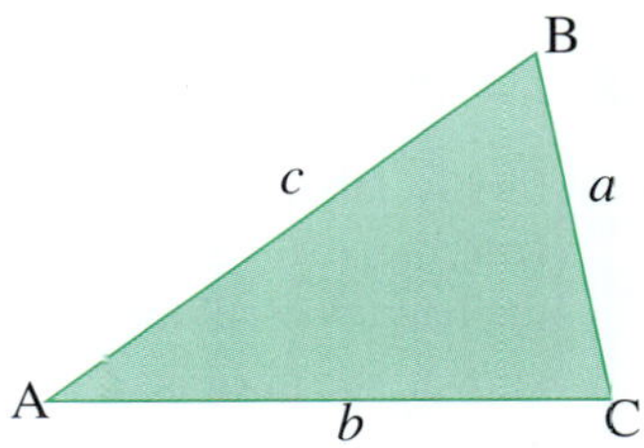

$$Area = \frac{1}{2}bc\sin A,$$

$$Area = \frac{1}{2}ac\sin B$$

and $$Area = \frac{1}{2}ab\sin C$$

However, each of these are equal, meaning that

$$\frac{1}{2}ac\sin B = \frac{1}{2}bc\sin A \Leftrightarrow a\sin B = b\sin A \Leftrightarrow \frac{a}{\sin A} = \frac{b}{\sin B}$$

Similarly,

$$\frac{1}{2}ac\sin B = \frac{1}{2}ab\sin C \Leftrightarrow c\sin B = b\sin C \Leftrightarrow \frac{c}{\sin C} = \frac{b}{\sin B}$$

Combining these results we have that

$$\frac{a}{\sin A} = \frac{b}{\sin B} = \frac{c}{\sin C}$$

So, when should/can we make use of the sine rule?

Although the sine rule can be used for right-angled triangles, it is more often used for situations when we do not have a right-angled triangle, and when the given triangle has either of the following pieces of information:

(a) Two angles and one side

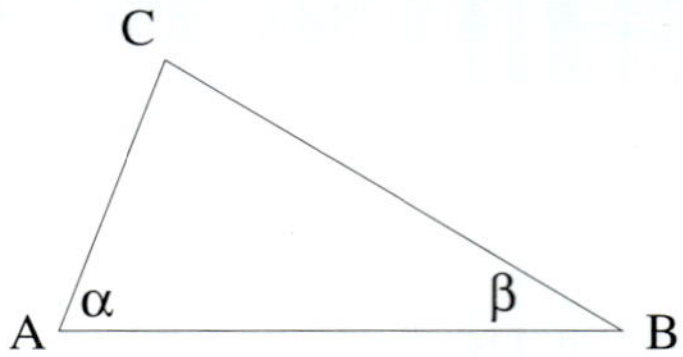

Either the length CB, AB or AC can be given and the triangle can be 'solved'. i.e., we can find the length of every side and every angle.

In this case, if we are give the length AB, we need $\angle ACB$, which can be found using $\angle ACB = 180° - \alpha - \beta$

(b) Two sides and a non-included angle

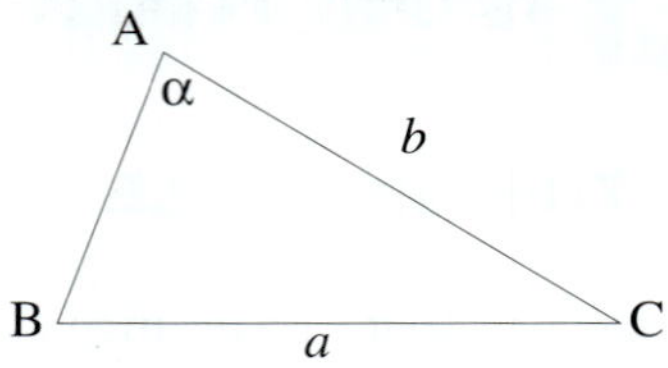

In this case, we first need to determine angle B using $\frac{\sin\alpha}{a} = \frac{\sin B}{b}$.

Once we have angle B, we can the find angle C and then the length AB.

EXAMPLE 16.13

Solve the following triangles giving the lengths of the sides in centimetres, correct to one decimal place and angles correct to the nearest degree.

(a)

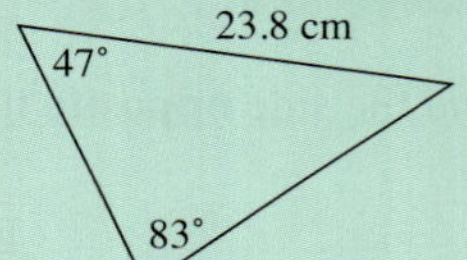

(b)

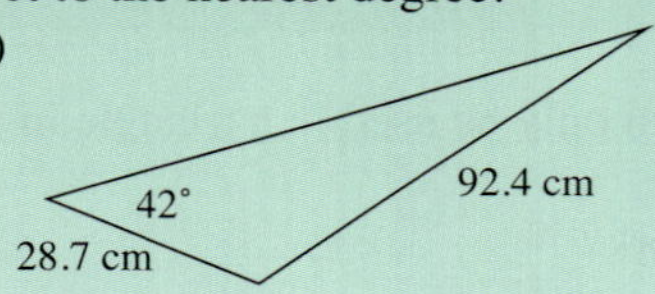

Solution

(a) Firstly, label the triangle using the standard method of lettering. 'Solve the triangle' means find all the angles and the lengths of all the sides. Since two of the angles are known, the third is; $C = 180° - 47° - 83° = 50°$.
The lengths of the remaining sides can be found using the known pairing of side and angle, b and B.

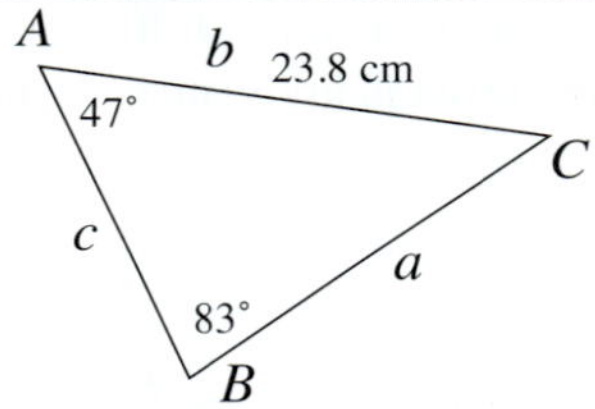

$$\frac{a}{\sin A} = \frac{b}{\sin B} \Leftrightarrow \frac{a}{\sin 47°} = \frac{23.8}{\sin 83°}$$

$$a = \frac{23.8 \times \sin 47°}{\sin 83°}$$

$$= 17.5369\ldots$$

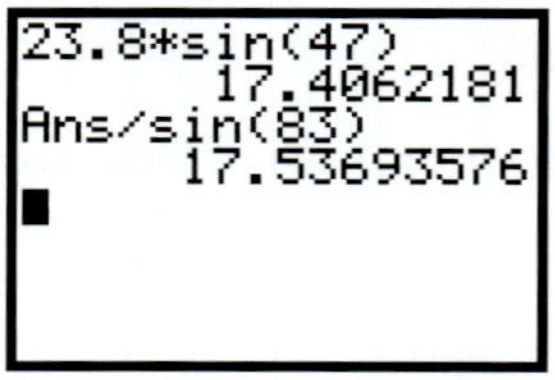

That is, BC is 17.5 cm (correct to one d.p).

Similarly, the remaining side can be calculated: $\frac{c}{\sin C} = \frac{b}{\sin B} \Leftrightarrow \frac{c}{\sin 50°} = \frac{23.8}{\sin 83°}$

$$\therefore c = \frac{23.8 \times \sin 50°}{\sin 83°}$$

$$= 18.3687\ldots$$

That is, AB is 18.4 cm (correct to one d.p).

(b) This triangle is different from the previous example in that only one angle is known. It remains the case that a pair of angles and an opposite side are known and that the sine rule can be used. The angle A must be found first.

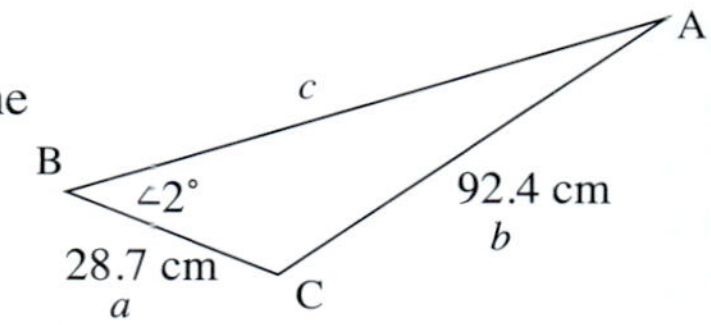

$$\frac{\sin A}{a} = \frac{\sin B}{b} \Leftrightarrow \frac{\sin A}{28.7} = \frac{\sin 42°}{92.4}$$

$$\Leftrightarrow \sin A = \frac{28.7 \times \sin 42°}{92.4}$$

$$= 0.207836$$

$$\therefore A = \sin^{-1} 0.207836$$

$$= 11.9956°$$

$$= 11°59'44''$$

The answer to the first part of the question is 12° correct to the nearest degree. It is important, however, to carry a much more accurate version of this angle through to subsequent parts of the calculation. This is best done using the calculator memory.

The third angle can be found because the sum of the three angles is 180°.
So, $C = 180° - 12° - 42° = 126°$

An accurate version of this angle must also be carried to the next part of the calculation. Graphics calculators have multiple memories labelled A, B, C etc. and students are advised to use these in such calculations.

The remaining side is: $\frac{c}{\sin 126°} = \frac{28.7}{\sin 12°} \Leftrightarrow c = \frac{28.7 \sin 126°}{\sin 12°}$

$$\therefore c = 111.6762\ldots$$

That is, AB is 111.7 cm (correct to one d.p)

EXERCISES 16.5.1

Use the sine rule to complete the following table, which refers to the standard labelling of a triangle.

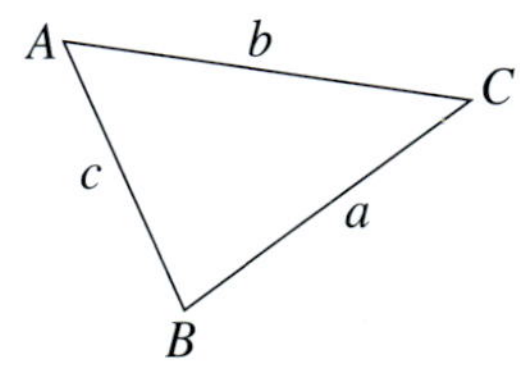

	a cm	*b* cm	*c* cm	*A*	*B*	*C*
1.			48.2		29°	141°
2.		1.2		74°	25°	
3.			11.3	60°		117°
4.			51.7	38°		93°
5.	18.5	11.4		68°		
6.	14.6	15.0			84°	
7.		7.3			16°	85°
8.			28.5	39°		124°
9.	0.8		0.8	82°		
10.			33.3	36°		135°
11.	16.4			52°	84°	

	a cm	*b* cm	*c* cm	*A*	*B*	*C*
12.			64.3		24°	145°
13.	30.9	27.7		75°		
14.			59.1	29°		102°
15.		9.8	7.9		67°	
16.			54.2	16°		136°
17.	14.8		27.2			67°
18.			10.9		3°	125°
19.			17.0		15°	140°
20.			40.1	30°		129°

EXAMPLE 16.14

For the triangle shown, find the angle ABC.

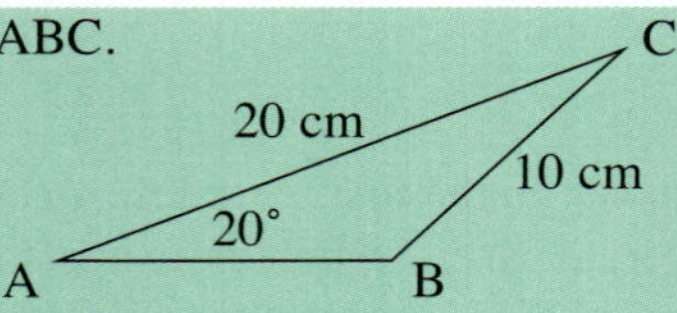

Solution

Making use of the sine rule we have:

$$\frac{\sin A}{a} = \frac{\sin B}{b} \Leftrightarrow \frac{\sin 20°}{10} = \frac{\sin B}{20}$$

$$\Leftrightarrow \sin B = \frac{20\sin 20°}{10}$$

$$\therefore B = \sin^{-1}(2\sin 20°)$$

$$= 43.1601\ldots$$

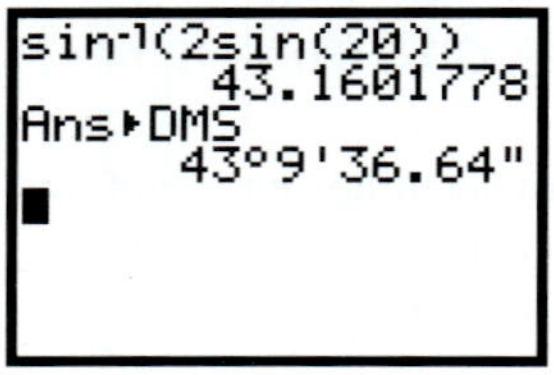

That is, $B = 43°10'$

However, from our diagram, the angle ABC should have been greater than 90°! That is, we should have obtained an **obtuse angle** (90° < B < 180°) rather than an **acute angle** (0° < B < 90°).

So, what went wrong?

This example is a classic case of what is known as the **ambiguous case**, in that, from the given information it is possible to draw two different diagrams, both having the same data. we show both these triangles:

Case 1

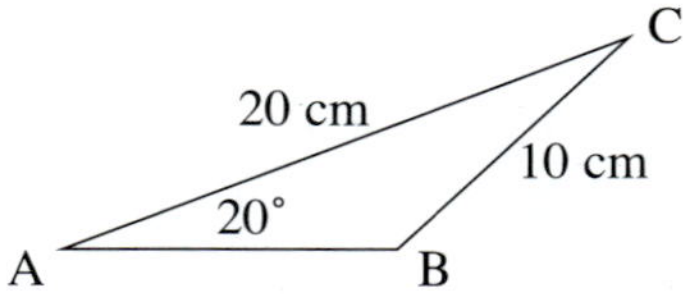

Case 2

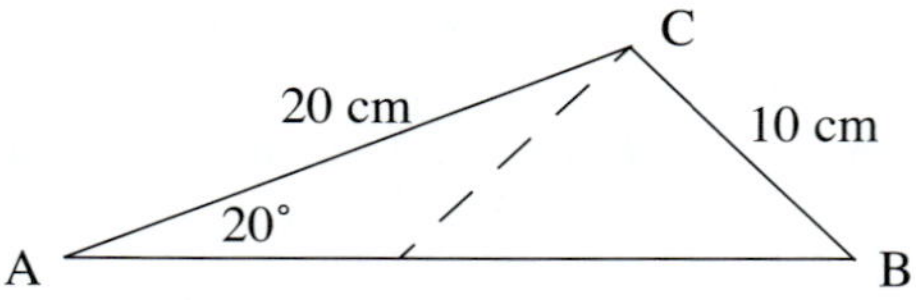

Notice that the side BC can be pivoted about the point C and therefore two different triangles can be formed with BC = 10. This is why there are two possible triangles based on the same information.

In the solution above, $B = 43°10'$ – representing Case 2. However, our diagram is represented by Case 1! Therefore, the correct answer is $180 - 43°10' = 136°50'$.

16.5.2 THE AMBIGUOUS CASE

From Example 9.14, it can be seen that an ambiguous case can arise when using the sine rule. In the given situation we see that the side CB can be pivoted about its vertex, forming two posible triangles.

We consider another such triangle.

EXAMPLE 16.15

Draw diagrams showing the triangles in which AC = 17 cm, BC = 9 cm and A = 29° and solve these triangles.

Solution

Applying the sine rule to the triangle gives:

$$\frac{\sin B}{17} = \frac{\sin 29^\circ}{9} \Leftrightarrow \sin B = \frac{17 \times \sin 29^\circ}{9}$$

$$= 0.91575$$

$$\therefore B = 66^\circ$$

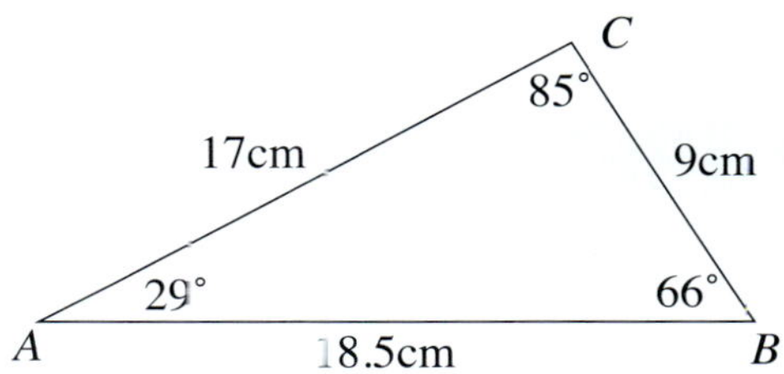

Next, we have,

$$C = 180^\circ - 29^\circ - 66^\circ = 85^\circ$$

$$\frac{c}{\sin 85^\circ} = \frac{9}{\sin 29^\circ} \Leftrightarrow c = 18.5$$

There is, however, a second solution that results from drawing an isosceles triangle BCE. This creates the triangle AEC which also fits the data. The third angle of this triangle is 37° and the third side is: $\frac{AE}{\sin 37^\circ} = \frac{9}{\sin 29^\circ} \Leftrightarrow AE = 11.2$

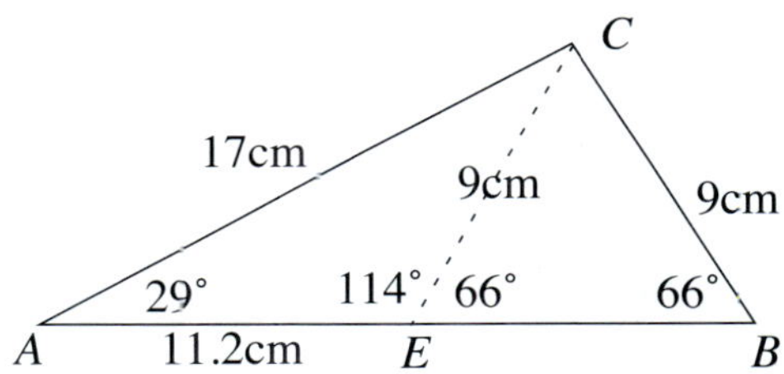

The original data is ambiguous in the sense that there are two triangles that are consistent with it.

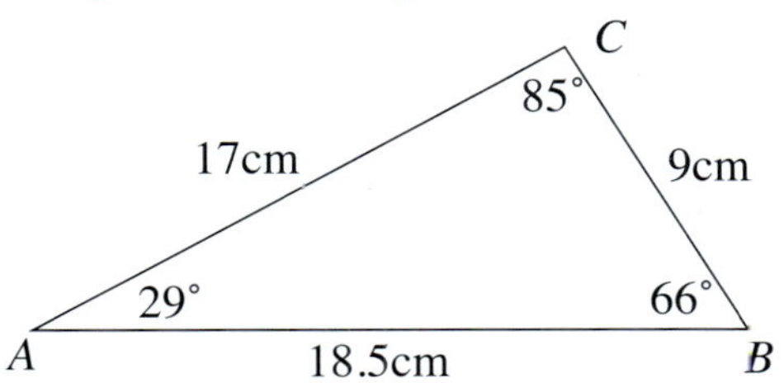

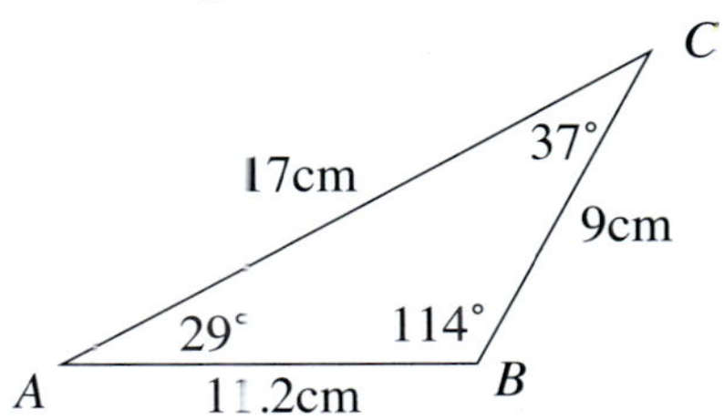

You should also notice that the two angles in the solution are 66° and 114° and that sin66° = sin114°. (That is, sin 66° = sin(180° – 66°) = sin 114°. This will be developed further in Chapter 10).

In fact, we can go one step further and make the following statement:

If we are given two sides of a triangle and the magnitude of an angle opposite one of the sides, there may exist one, two or no possible solutions for the given information.

We summarise our findings:

Type 1

Given the **acute angle** α and the side lengths a and b, there are four possible outcomes:

Notice that the number of solutions **depends on** the length a relative to the perpendicular height, h, of the triangle as well as the length b. Where the height h is based on the right-angled triangle formed in each case, i.e., $\sin\alpha = \frac{h}{b} \Leftrightarrow h = b\sin\alpha$.

So that

– if $a < h$, then the triangle cannot be completed.
– if $a = h$, then we have a right-angled triangle.
– if $a > b$, then we have a triangle that is consistent with the given information.
– if $h < a < b$, then the side BC can be pivoted about the vertex C, forming two triangles.

<table>
<tr><th>Number of Δ s</th><th>Necessary condition</th><th colspan="2">Type of triangle that can be formed</th></tr>
<tr><td>None</td><td>$a < h$</td><td>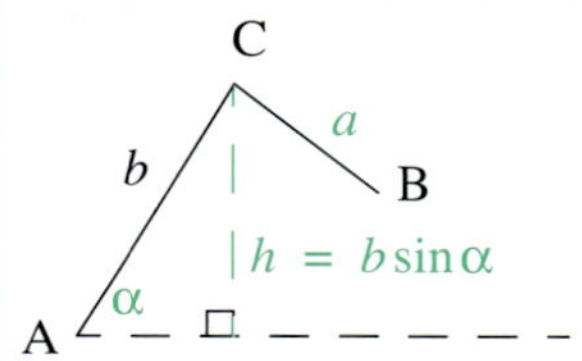
</td><td>In this case, the triangle cannot be constructed.</td></tr>
<tr><td>One</td><td>$a = h$</td><td>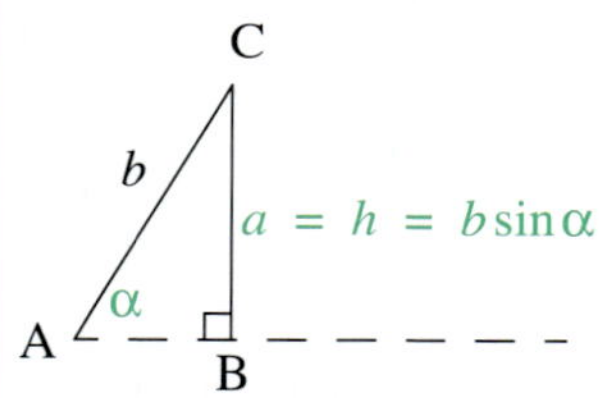
</td><td>In this case we have formed a right-angled triangle.</td></tr>
<tr><td>One</td><td>$a > b$</td><td>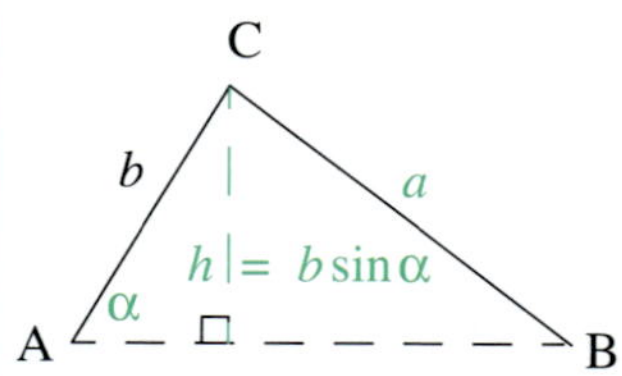
</td><td>In this case there can be only one triangle that is consistent with the given information.</td></tr>
<tr><td>Two</td><td>$h < a < b$</td><td>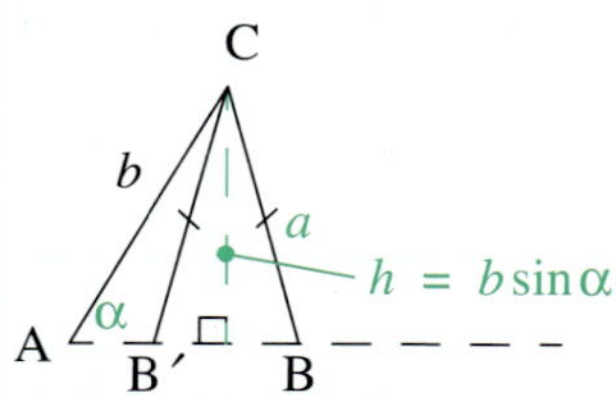
</td><td>In this case there are two possible triangles, ΔABC and $\Delta AB'C$. This is because BC can be pivoted about C and still be consistent with the given information.</td></tr>
</table>

The table above reflects the case where α is acute. What if α is obtuse?

Type 2

Given the **obtuse angle** α and the side lengths a and b, there are two possible outcomes:

Number of Δ s	Necessary condition	Type of triangle that can be formed	
None	$a \le b$	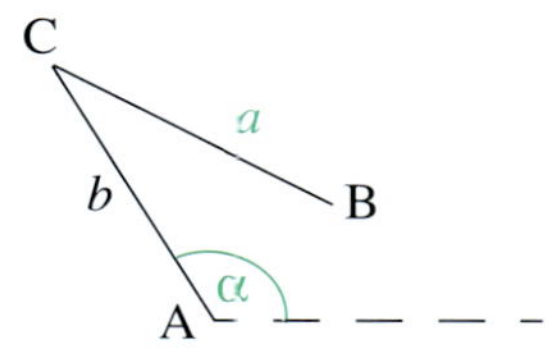	In this case, the triangle cannot be constructed.
One	$a > \text{b}$	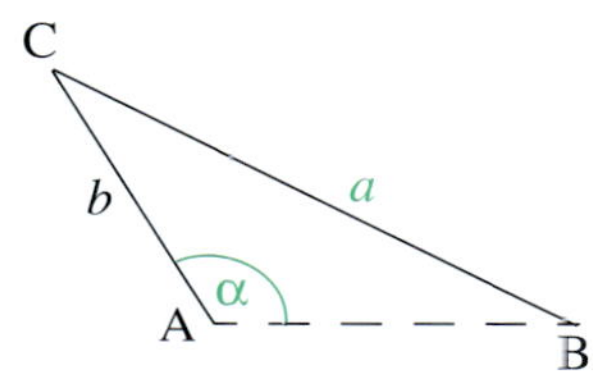	In this case there can be only one triangle that is consistent with the given information.

EXAMPLE 16.16

Find $\angle ABC$ for the triangle ABC given that $a = 50$, $b = 80$ and $A = 35°$.

Solution

We first determine the value of $b\sin\alpha$ and compare it to the value a:

Now, $b\sin\alpha = 80\sin 35° = 45.89$

Therefore we have that $b\sin\alpha\ (= 45.89) < a\ (= 50) < b\ (= 80)$

meaning that we have an ambiguous case.

Case 1:

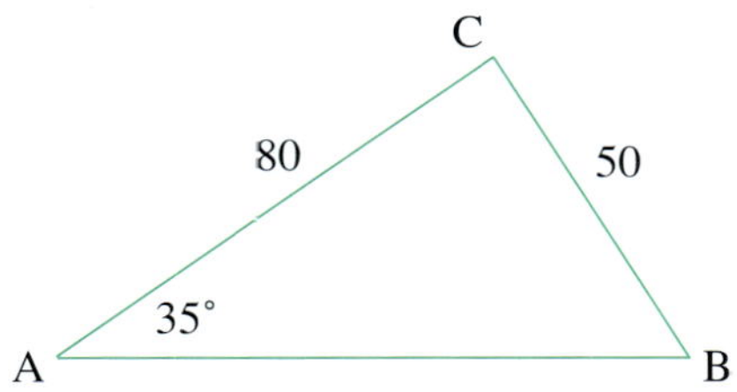

Using the sine rule, $\dfrac{\sin A}{a} = \dfrac{\sin B}{b}$, we have

$$\frac{\sin 35°}{50} = \frac{\sin B}{80} \Leftrightarrow \sin B = \frac{80\sin 35°}{50}$$

$$\therefore B = 66°35'$$

Case 2:

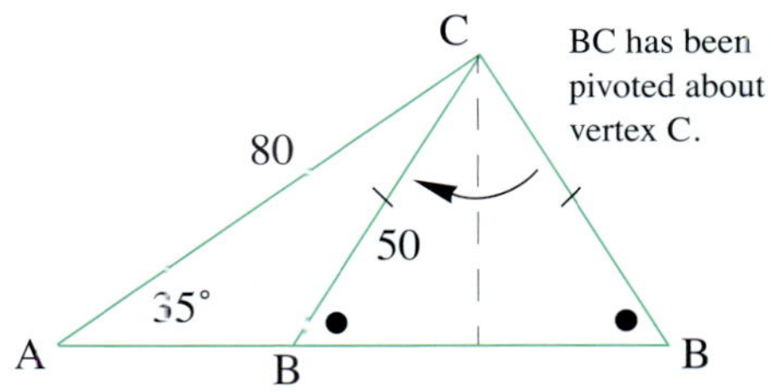

From case 1, the obtuse angle B' is given by $180° - 66°35' = 113°25'$.

This is because $\Delta B'CB$ is an **isosceles** triangle, so that $\angle AB'C = 180° - \angle CB'B$

EXAMPLE 16.17

Find $\angle ACB$ for the triangle ABC given that $a = 70$, $c = 90$ and $A = 75°$.

Solution

We start by drawing the triangle with the given information:

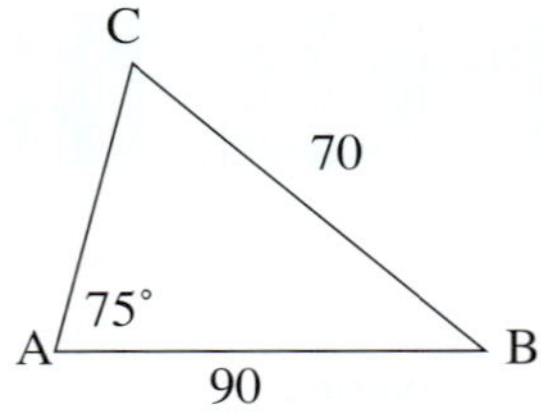

Using the sine rule we have

$$\frac{\sin C}{90} = \frac{\sin 75}{70} \Leftrightarrow \sin C = \frac{90 \sin 75}{70}$$

$$\therefore \sin C = 1.241\ldots$$

Which is imposible to solve for as the sine of an angle can never be greater than one.
Therefore no such triangle exists.

EXERCISES 16.5.2

Find the two solutions to these triangles which are defined using the standard labelling:

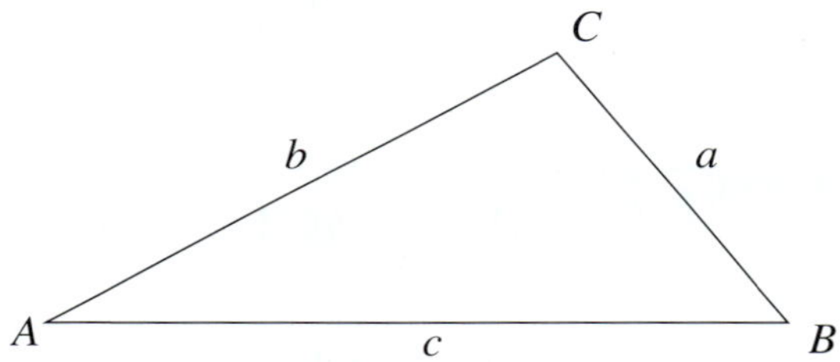

	a cm	*b* cm	*A*
1.	7.4	18.1	20°
2.	13.3	19.5	14°
3.	13.5	17	28°
4.	10.2	17	15°
5.	7.4	15.2	20°
6.	10.7	14.1	26°
7.	11.5	12.6	17°
8.	8.3	13.7	24°
9.	13.7	17.8	14°
10.	13.4	17.8	28°
11.	12.1	16.8	23°
12.	12	14.5	21°
13.	12.1	19.2	16°
14.	7.2	13.1	15°
15.	12.2	17.7	30°
16.	9.2	20.9	14°
17.	10.5	13.3	20°
18.	9.2	19.2	15°
19.	7.2	13.3	19°
20.	13.5	20.4	31°

21. Solve the following triangles

(a) $\alpha = 75°, a = 35, c = 45$

(b) $\alpha = 35°, a = 30, b = 80$

(c) $\beta = 40°, a = 22, b = 8$

(d) $\gamma = 50°, a = 112, c = 80$

16.5.3 APPLICATIONS OF THE SINE RULE

Just as in the case of right-angled triangles, the sine rule becomes very useful. In particular, it means that previous problems that required the partitioning of a non right-angled triangle into two (or more) right-angled triangles can be solved using the sine rule.

We start by considering Q.13., Exercise 9.2:

EXAMPLE 16.18

A surveying team are trying to find the height of a hill. They take a 'sight' on the top of the hill and find that the angle of elevation is 23°27′. They move a distance of 250 metres on level ground directly away from the hill and take a second 'sight'. From this point, the angle of elevation is 19°46′. Find the height of the hill, correct to the nearest metre.

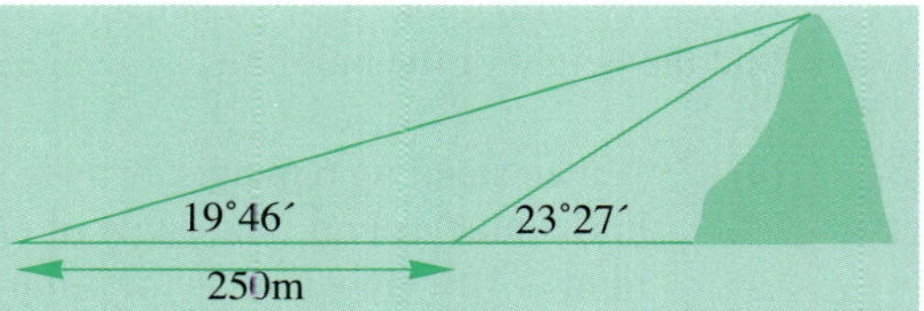

Solution

Labelling the given diagram using the standard notation we have:

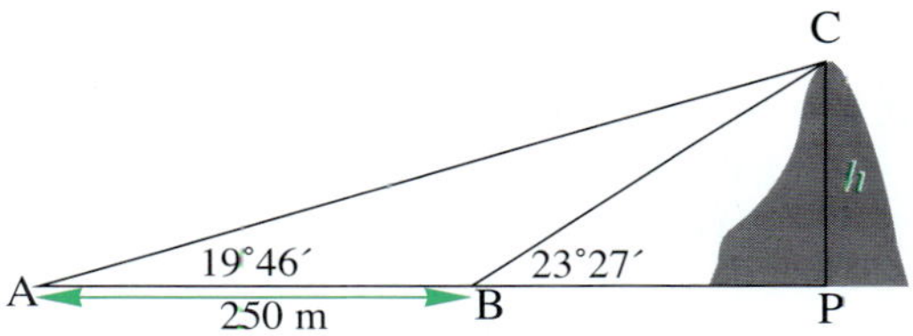

With $\beta = 180 - 23°27' = 156°33'$

and $\gamma = 180 - 19°46' - 156°33' = 3°41'$

Then, using the sine rule,

$$\frac{b}{\sin 156°33'} = \frac{250}{\sin 3°41'}$$

$$\Leftrightarrow b = \frac{250 \sin 156°33'}{\sin 3°41'}$$

$$= 1548.63\ldots$$

```
250*sin(156°33')
/sin(3°41')
         1548.63051
Ans*sin(19°46')
        523.7320966
```

Then, using ΔACP we have,

$$\sin 19°46' = \frac{h}{b} \Leftrightarrow h = b \sin 19°46'$$

$$= 523.73$$

So, the hill is 524 m high (to nearest metre).

A much neater solution (as opposed to solving simultaneous equations – as was required previously).

EXERCISES 16.5.3

1. A short course biathlon meet requires the competitors to run in the direction S60°W to their bikes and then ride S40°E to the finish line, situated 20 km due South of the starting point. What is the distance of this course?

2. A pole is slanting towards the sun and is making an angle of 10° to the vertical. It casts a shadow 7 metres long along the horizontal ground. The angle of elevation of the top of the pole to the tip of its shadow is 30°. Find the length of the pole, giving your answer to 2 d.p.

3. A statue A, is observed from two other statues B and C which are 330 m apart. The angle between the lines of sight AB and BC is 63° and the angle between the lines of sight AC and CB is 75°. How far is statue A from statue B?

4. Town A is 12 km from town B and its bearing is 132°T from B. Town C is 17 km from A and its bearing is 063°T from B. Find the bearing of A from C.

5. The angle of elevation of the top of a building from a park bench on level ground is 18°. The angle of elevation from a second park bench, 300 m closer to the base of the building is 30°. Assuming that the two benches and the building all lie on the same vertical plane, find the height of the building.

6. (a) A man standing 6 metres away from a lamp post casts a shadow10 metres long on a horizontal ground. The angle of elevation from the tip of the shadow to the lamp light is 12°. How high is the lamp light?

(b) If the shadow is cast onto a road sloping at 30° upwards, how long would the shadow be if the man is standing at the foot of the sloping road and 6 metres from the lamp post?

7. At noon the angle of elevation of the sun is 72° and is such that a three metre wall AC, facing the sun, is just in the shadow due to the overhang AB.

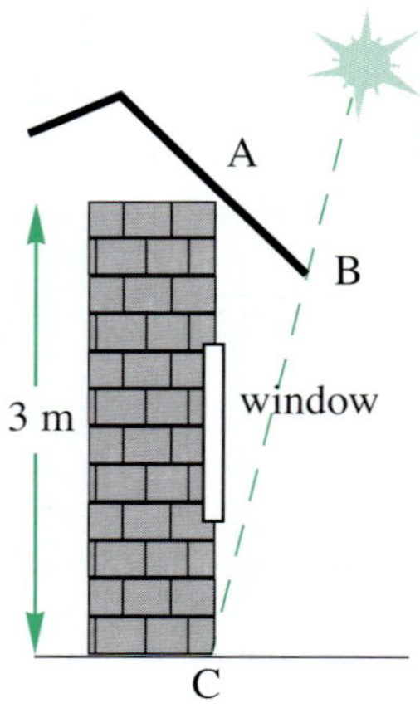

The angle that the overhang makes with the vertical wall is 50°.

(a) Copy and illustrate this information on the diagram shown.
(b) Find the length of the overhang.

At 4 p.m. the angle of elevation of the sun is 40° and the shadow due to the overhang just reaches the base of the window.

(c) How far from the ground is the window?

8. The lookout on a ship sailing due East at 25 km/h observes a reef N62°E at a distance of 30 km.
(a) How long will it be before the ship is 15 km from the reef, assuming that it continues on its easterly course.
(b) How long is it before it is again 15 km from the reef?
(c) What is the closest that the ship will get to the reef?

9. The framework for an experimental design for a kite is shown.
Material for the kite costs \$12 per cm^2.
How much will it cost for the material if it is to cover the framework of the kite.

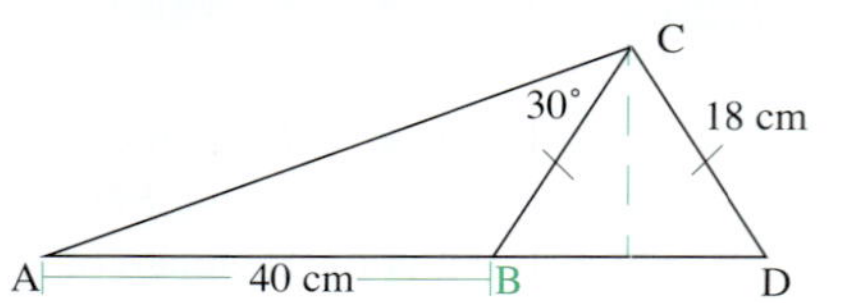

10. A boy walking along a straight road notices the top of a tower at a bearing of 284°T. After walking a further 1.5 km he notices that the top of the tower is at a bearing of 293°T. How far from the road is the tower?

16.5.4 THE COSINE RULE

Sometimes the sine rule is not enough to help us solve for a non right-angled triangle. For example, in the triangle shown, we do not have enough information to use the sine rule. That is, the sine rule only provided the following:

$$\frac{a}{\sin 30^\circ} = \frac{14}{\sin B} = \frac{18}{\sin C}$$

where there are too many unknowns.

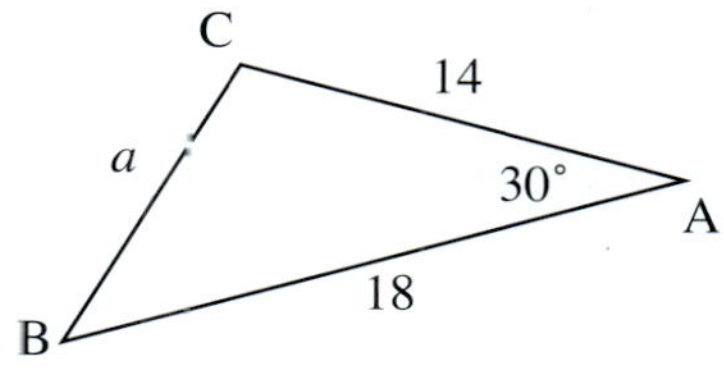

For this reason we derive another useful result, known as the cosine rule. The cosine rule may be used when

1. **two sides and an included angle are given**:
This means that the third side can be determined and then we can make use of the sine rule (or the cosine rule again).

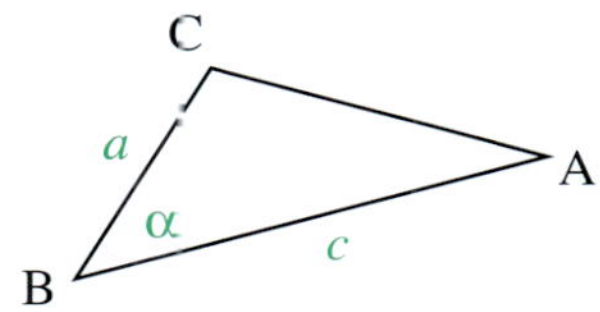

2. **three sides are given**:
This means we could then determine any of the angles.

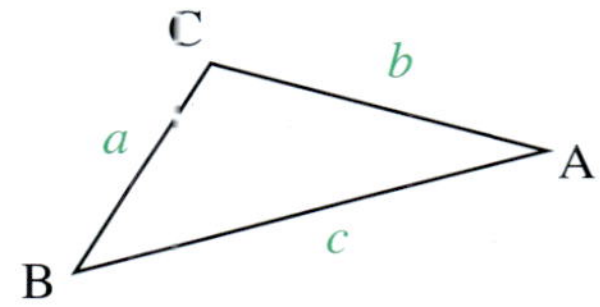

The cosine rule, with the standard labelling of the triangle has three versions:

$$a^2 = b^2 + c^2 - 2bc\cos A$$
$$b^2 = a^2 + c^2 - 2ac\cos B$$
$$c^2 = a^2 + b^2 - 2ab\cos C$$

The cosine rule can be remembered as a version of Pythagoras's Theorem with a correction factor. We now show why this works.

Consider the case where there is an acute angle at A. Draw a perpendicular from C to N as shown in the diagram.

In ΔANC we have $b^2 = h^2 + x^2$

$$\Leftrightarrow h^2 = b^2 - x^2 \quad -(1)$$

In ΔBNC we have $a^2 = h^2 + (c-x)^2$

$$\Leftrightarrow h^2 = a^2 - (c-x)^2 \quad -(2)$$

Equating (1) and (2) we have,

$$a^2 - (c-x)^2 = b^2 - x^2$$
$$\Leftrightarrow a^2 - c^2 + 2cx - x^2 = b^2 - x^2$$
$$\Leftrightarrow a^2 = b^2 + c^2 - 2cx$$

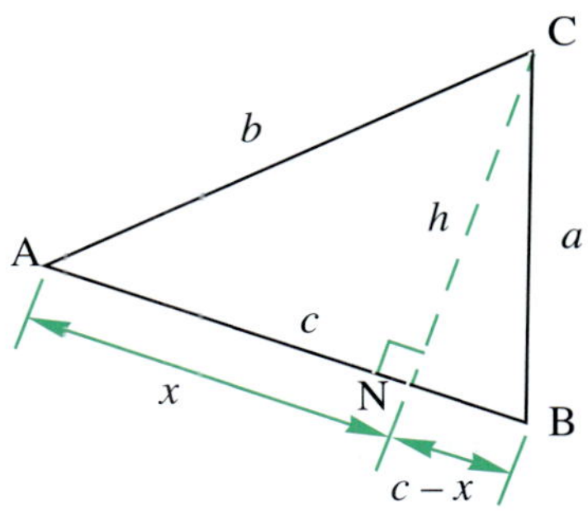

However, from ΔANC we have that $\cos A = \frac{x}{b} \Leftrightarrow x = b\cos A$

Substituting this result for x, we have

$$a^2 = b^2 + c^2 - 2cb\cos A$$

Although we have shown the result for an acute angle at A, the same rule applies if A is obtuse.

EXAMPLE 16.19

Solve the following triangles giving the lengths of the sides in centimetres, correct to one decimal place and angles correct to the nearest degree.

(a)

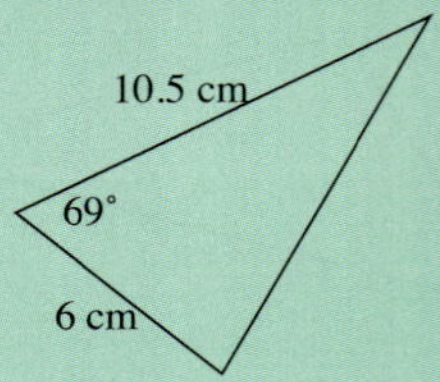

(b)

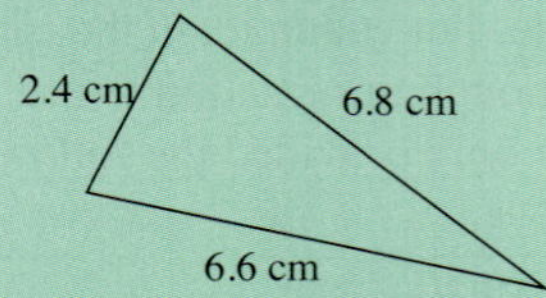

Solution

(a) The data does not include an angle and the opposite side so the sine rule cannot be used. The first step, as with the sine rule, is to label the sides of the triangle. Once the triangle has been labelled, the correct version of the cosine rule must be chosen. In this case, the solution is:

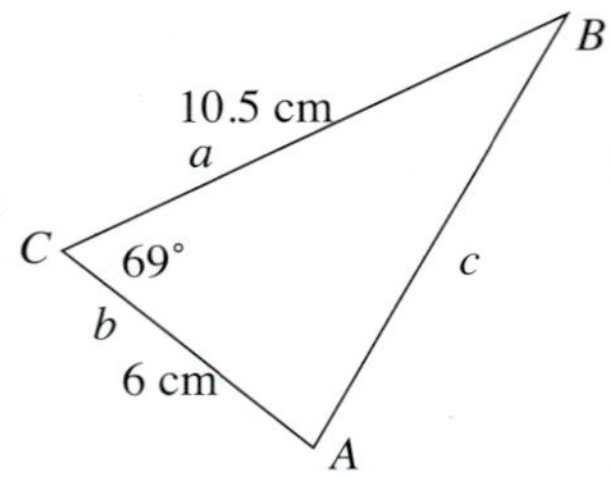

$$\begin{aligned} c^2 &= a^2 + b^2 - 2ab\cos C \\ c^2 &= 10.5^2 + 6^2 - 2 \times 10.5 \times 6 \times \cos 69° \\ &= 101.0956 \\ a &= 10.1 \end{aligned}$$

The remaining angles can be calculated using the sine rule. Again, it is important to carry a high accuracy for the value of c to the remaining problem:

$$\frac{\sin B}{b} = \frac{\sin C}{c} \Leftrightarrow \sin B = \frac{6 \times \sin 69°}{10.0546} \quad \therefore B = 34°$$

Finally, $A = 180° - 34° - 69° = 77°$

(b) In this case, there are no angles given. The cosine rule can be used to solve this problem as follows:

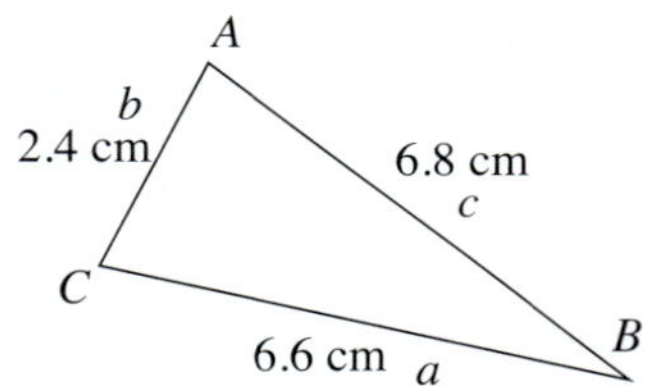

$$\begin{aligned} a^2 &= b^2 + c^2 - 2bc\cos A \\ 6.6^2 &= 2.4^2 + 6.8^2 - 2 \times 2.4 \times 6.8 \times \cos A \\ 2 \times 2.4 \times 6.8 \times \cos A &= 2.4^2 + 6.8^2 - 6.6^2 \\ \cos A &= \frac{2.4^2 + 6.8^2 - 6.6^2}{2 \times 2.4 \times 6.8} \\ &= 0.25858 \\ A &= 75.014° \\ &= 75°1' \end{aligned}$$

Next, use the sine rule: $\frac{\sin B}{b} = \frac{\sin A}{a} \Leftrightarrow \sin B = \frac{2.4 \times \sin 75}{6.6} \quad \therefore B = 20°34'$

So that $C = 180° - 75° - 21° = 84°$

The three angles, correct to the nearest degree are $A = 75°$, $B = 21°$ & $C = 84°$.

EXERCISES 16.5.4

Solve the following triangles

	a cm	*b* cm	*c* cm	*A*	*B*	*C*
1.	13.5		16.7		36°	
2.	8.9	10.8				101°
3.	22.8		12.8		87°	
4.	21.1	4.4				83°
5.		10.6	15.1	74°		
6.		13.6	20.3	20°		
7.	9.2		13.2		46°	
8.	23.4	62.5				69°
9.		9.6	15.7	41°		
10.	21.7	36.0	36.2			
11.	7.6	3.4	9.4			
12.	7.2	15.2	14.3			
13.	9.1		15.8		52°	
14.	14.9	11.2	16.2	63°	42°	75°
15.	2.0	0.7	2.5			
16.	7.6	3.7	9.0			
17.	18.5	9.8	24.1			
18.	20.7	16.3	13.6			
19.		22.4	29.9	28°		
20.	7.0		9.9		42°	
21.	21.8	20.8	23.8			
22.	1.1		1.3		89°	
23.		1.2	0.4	85°		
24.	23.7	27.2				71°
25.	3.4	4.6	5.2			

16.5.5 APPLICATIONS OF THE COSINE RULE

EXAMPLE 16.20

A cyclist rode her bike for 22 km on a straight road heading in a westerly direction towards a junction. Upon reaching the junction, she headed down another straight road bearing 200°T for a distance of 15 km. How far is the cyclist from her starting position?

Solution

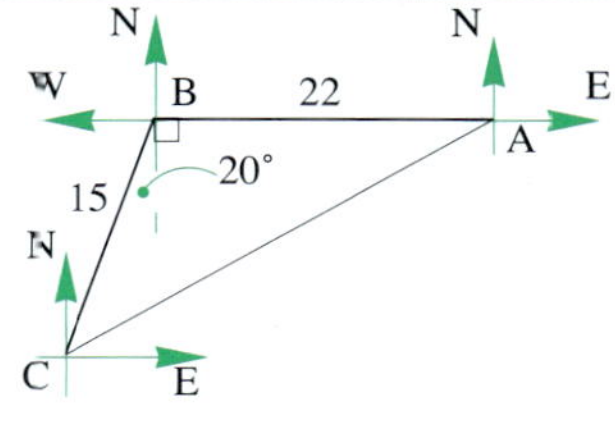

We start with a diagram:

Note that $\angle ABC = 90° + 20° = 110°$

Using the cosine rule we have,

$$AC^2 = 15^2 + 22^2 - 2 \times 15 \times 22 \cos 110°$$

$$\Rightarrow AC = \sqrt{225 + 484 - 660 \times (-0.3420\ldots)}$$

$$\therefore AC = 30.5734\ldots$$

That is, she is (approximately) 30.57 km from her starting point.

EXAMPLE 16.21

A yacht starts from a harbour and sails for a distance of 11 km in a straight line. The yacht then makes a turn to port (left) of 38° and sails for 7 km in a straight line in this new direction until it arrives at a small island. Draw a diagram that shows the path taken by the yacht and calculate the distance from the harbour to the island.

Solution

The question does not give the bearing of the first leg of the trip so the diagram can show this in any direction. H is the harbour, I the island and T the point where the yacht makes its turn.

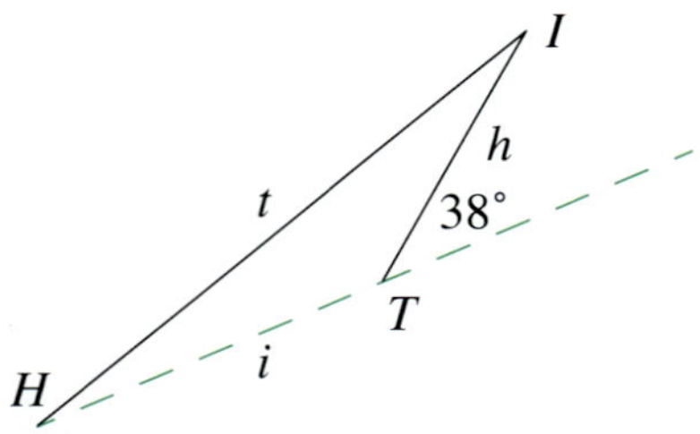

The angle in the triangle at T is 180° – 38° = 142°. The problem does not contain an angle and the opposite side and so must be solved using the cosine rule.

$$\begin{aligned} t^2 &= h^2 + i^2 - 2hi\cos T \\ &= 7^2 + 11^2 - 2 \times 7 \times 11 \times \cos 142° \\ &= 291.354 \\ \therefore t &= 17.1 \end{aligned}$$

That is, distance from the harbour to the port is 17.1 km (to one d.p)

EXAMPLE 16.22

A triangular sandpit having side lengths 5 m, 4 m and 8 m is to be constructed to a depth of 20 cm. Find the volume of sand required to fill this sandpit.

Solution

We will need to find an angle. In this case we determine the largest angle, which will be the angle opposite the longest side.

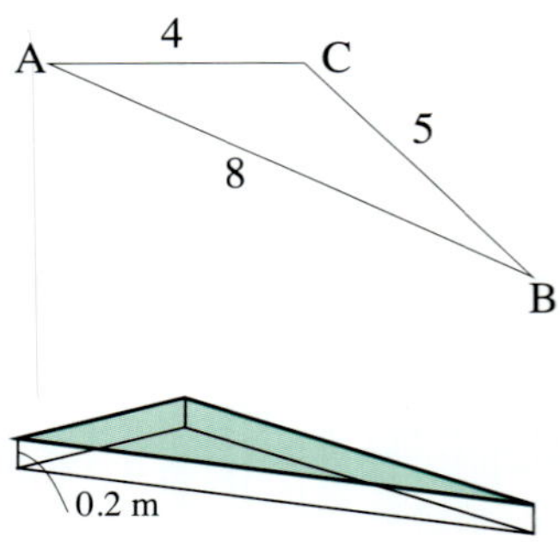

From our diagram we have

$$\begin{aligned} 8^2 &= 4^2 + 5^2 - 2 \times 4 \times 5\cos C \\ \therefore 64 &= 16 + 25 - 40\cos C \\ \Leftrightarrow \cos C &= \frac{16 + 25 - 64}{40} \\ &= -\frac{23}{40} \\ \therefore C &= 125°6' \end{aligned}$$

To find the volume of sand we first need to find the surface area of the sandpit.

$$\text{Area} = \frac{1}{2}ab\sin C = \frac{1}{2} \times 4 \times 5 \times \sin(125°6') = 8.1815 \ \text{m}^2.$$

Therefore, volume of sand required is $0.2 \times 8.1815 = 1.64 \ \text{m}^3$.

EXERCISES 16.5.5

1. Thomas has just walked 5 km in a direction N70°E when he realises that he needs to walk a further 8 km in a direction E60°S.
   (a) How far from the starting point will Thomas have travelled?
   (b) What is his final bearing from his starting point?

2. Two poles, 8 m apart are facing a rugby player who is 45 m from the left pole and 50 from the right one. Find the angle that the player makes with the goal mouth.

3. The lengths of the adjacent sides of a parallelogram are 4.80 cm and 6.40 cm. If these sides have an inclusive angle of 40°, find the length of the shorter diagonal.

4. During an orienteering venture, Patricia notices two rabbit holes and estimates them to be 50 m and 70 m away from her. She measures the angle between the line of sight of the two holes as 54°. How far apart are the two rabbit holes?

5. To measure the length of a lake, a surveyor chooses three points. Starting at one end of the lake she walks in a straight line for 223.25 m to some point X, away from the lake. She then heads towards the other end of the lake in a straight line and measures the distance covered to be 254.35 m. If the angle between the paths she takes is 82°25', find the width of the lake.

6. A light airplane flying N87°W for a distance of 155 km, suddenly needs to alter its course and heads S 34°E for 82 km to land on an empty field.
   (a) How far from its starting point did the plane land.
   (b) What was the plane's final bearing from its starting point?

16.6 MISCELLANEOUS EXERCISES

1. The diagram shows a triangular building plot. The distances are given in metres. Find the length of the two remaining sides of the plot giving your answers correct to the nearest hundredth of a metre.

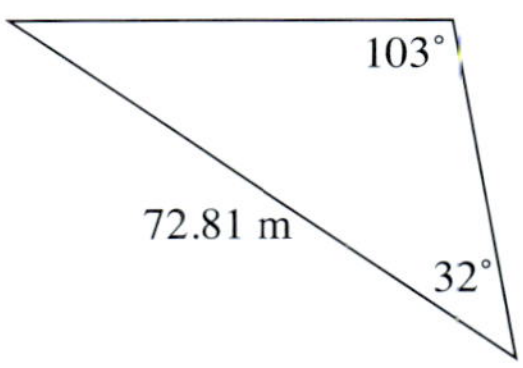

2. Xiang is standing on level ground. Directly in front of him and 32 metres away is a flagpole. If Xiang turns 61° to his right, he sees a post box 26.8 metres in front of him. Find the distance between the flagpole and the post box.

3. A triangular metal brace is part of the structure of a bridge. The lengths of the three parts are shown in metres. Find the angles of the brace.

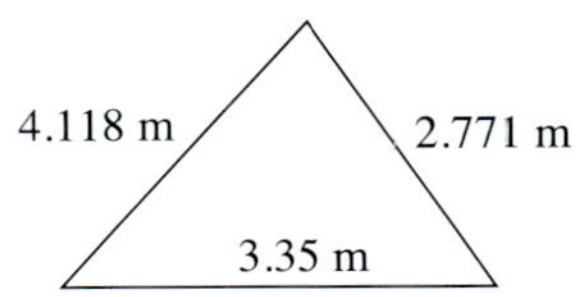

4. Find the smallest angle in the triangle whose sides have length 35.6 cm, 58.43 cm and 52.23 cm.

5. Ayton is directly north of Byford. A third town, Canfield, is 9.93km from Ayton on a bearing of 128° true. The distance from Byford to Canfield is 16.49km. Find the bearing of Canfield from Byford.

6. A parallelogram has sides of length 21.90 cm and 95.18 cm. The angle between these sides is 121°. Find the length of the long diagonal of the parallelogram.

7. A town clock has 'hands' that are of length 62cm and 85cm.
(a) Find the angle between the hands at half past ten.
(b) Find the distance between the tips of the hands at half past ten.

8. A shop sign is to be made in the shape of a triangle. The lengths of the edges are shown. Find the angles at the vertices of the sign.

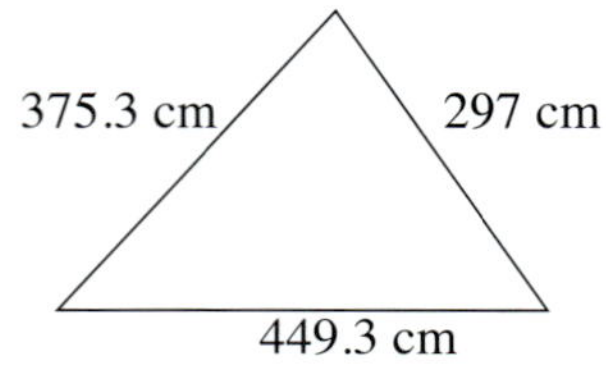

9. An aircraft takes off from an airstrip and then flies for 16.2 km on a bearing of 066° true. The pilot then makes a left turn of 88° and flies for a further 39.51 km on this course before deciding to return to the airstrip.
(a) Through what angle must the pilot turn to return to the airstrip?
(b) How far will the pilot have to fly to return to the airstrip?

10. A golfer hits two shots from the tee to the green.
How far is the tee from the green?

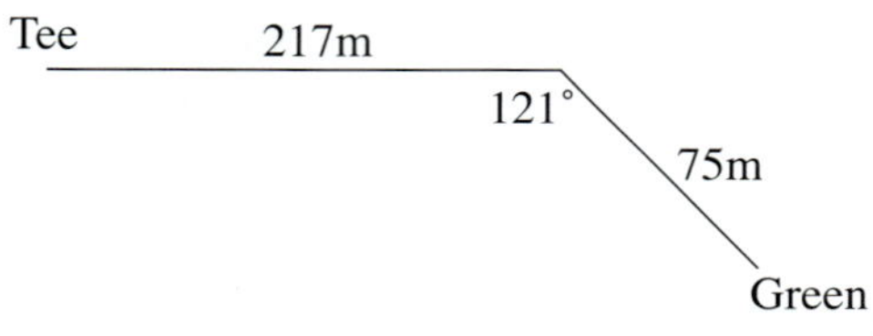

11. The diagram shows a parallelogram. Find the length of the longer of the two diagonals.

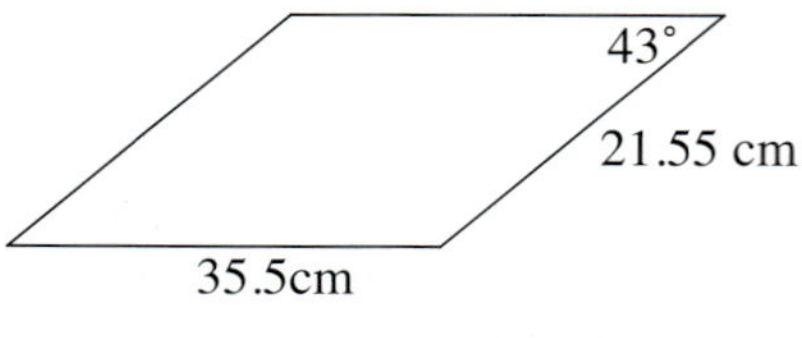

12. A triangle has angles 64°, 15° and 101°. The shortest side is 49 metres long. What is the length of the longest side?

13. The diagram shows a part of the support structure for a tower. The main parts are two identical triangles, ABC and ADE.
AC = DE = 27.4cm and BC = AE = 23.91cm
The angles ACB and AED are 58°.
Find the distance BD.

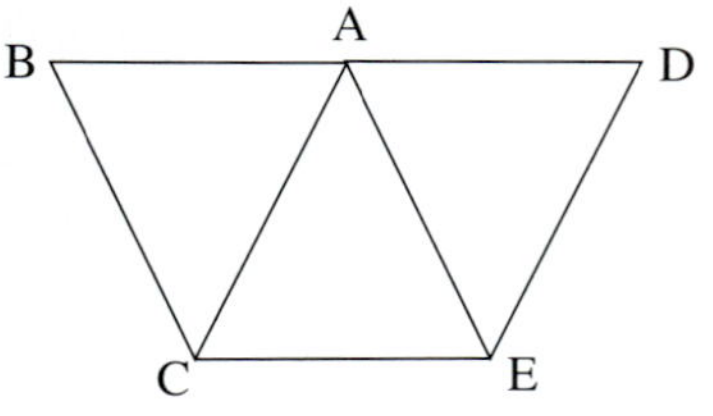

14. The diagram shows a design for the frame of a piece of jewellery. The frame is made of wire.

Find the length of wire needed to make the frame.

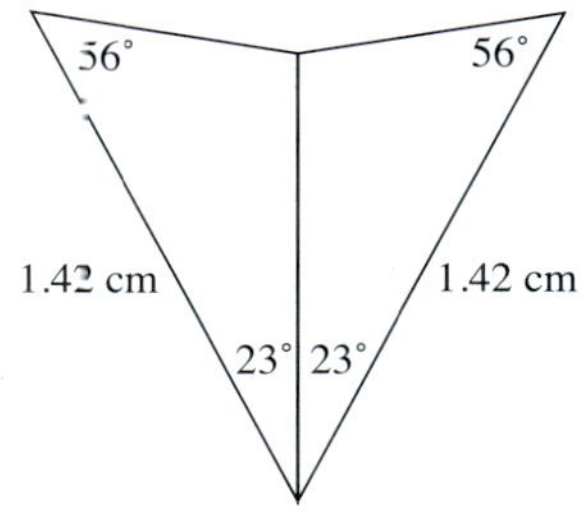

15. A triangular cross-country running track begins with the runners running North for 2050 metres. The runners then turn right and run for 5341 metres on a bearing of 083° true. Finally, the runners make a turn to the right and run directly back to the starting point.

(a) Find the length of the final leg of the run.

(b) Find the total distance of the run.

(c) What is the angle through which the runners must turn to start the final leg of the race?

(d) Find the bearing that the runners must take on the final leg of the race.

16. Show that for any standard triangle ABC, $\frac{\cos A}{a} + \frac{\cos B}{b} + \frac{\cos C}{c} = \frac{a^2 + b^2 + c^2}{2abc}$.

17. A sandpit in the shape of a pentagon ABCDE is to be built in such a way that each of its sides are of equal length, but its angles are not all equal.
The pentagon is symmetrical about DX, where X is the midpoint of AB.
The angle AXE and BXC are both 45° and each side is 2 m long.

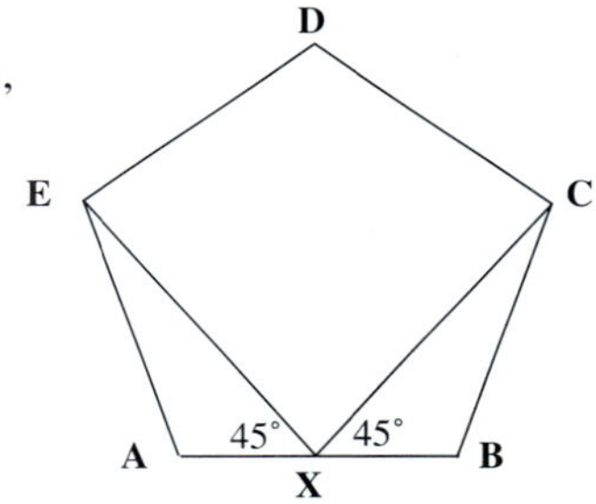

(a) Find $\angle XEA$.

(b) Find the length of EX.

(c) How much sand is required if the sandpit is 30 cm deep? Give your answer to three decimal place.

18. A triangular region has been set aside for a housing development which is to be divided into two sections. Two adjacent street frontages, AB and AC measuring 100 m and 120 m respectively. With the 100 m frontage running in an easterly direction, while the 120 m frontage runs in a north-east direction. A plan for this development is shown alongside. Give all answers to the nearest metre.

(a) Find the area covered by the housing development.

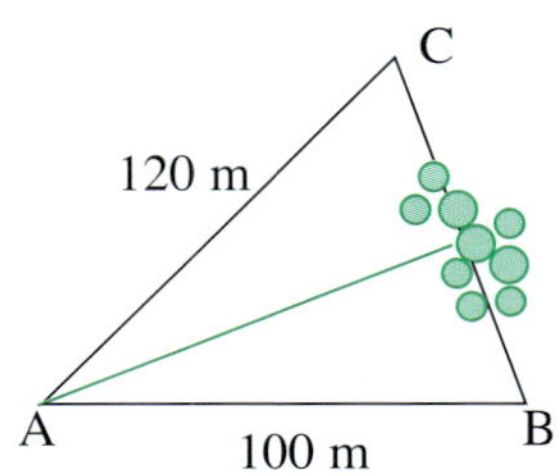

During the development stages, an environmental group specified that existing trees were not to be removed from the third frontage. This made it difficult for the surveyors to measure the length of the third frontage.

(b) Calculate the length of the third frontage, BC.

The estate is to be divided into 2 regions, by bisecting the angle at A with a stepping wall running from A to the frontage BC.

(c) How long is this stepping wall?

19. A cube of edge length 20 cm is shown in the diagram. Find:

(a) the length of the segment $[AC]$.

(b) the length of the segment $[AG]$.

(c) the angle that the segment $[AG]$ makes with the plane $ABCD$.

(d) the angle that the plane $AFGD$ makes with the plane $ABCD$.

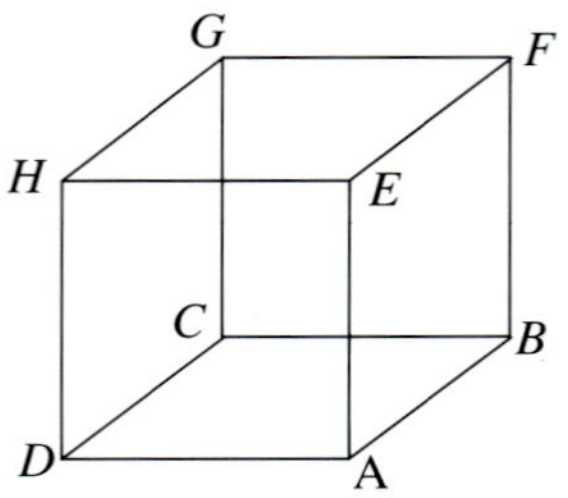

20. The figure shows a right pyramid with a rectangular base. Find

(a) the height of the pyramid.

(b) the angle that the plane BCV makes with the plane $ABCD$.

(c) the angle that the edge $[BV]$ makes with the base $ABCD$.

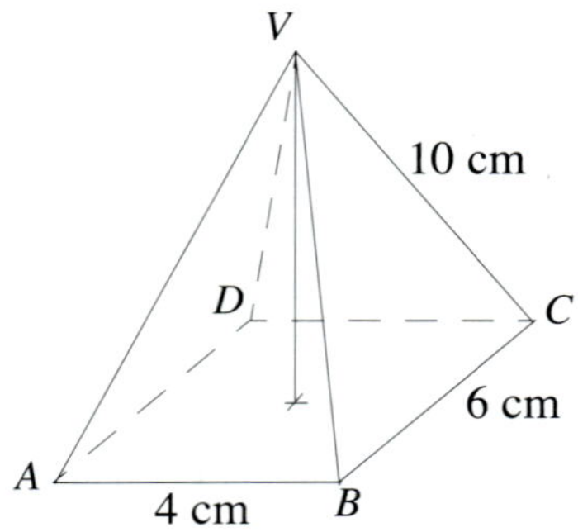

21. From a point A, 150m due south of a tower, the angle of elevation of the top of the tower is $30°$. From a point B, due east of the tower, the angle of elevation of the top of the tower is $40°$.

(a) Illustrate this information on a diagram.

(b) How far apart are the points A and B?

22. In the figure shown, find

(a) the length of $[AC]$.

(b) the length of $[BC]$.

(c) the angle between the edges $[AC]$ and $[BC]$

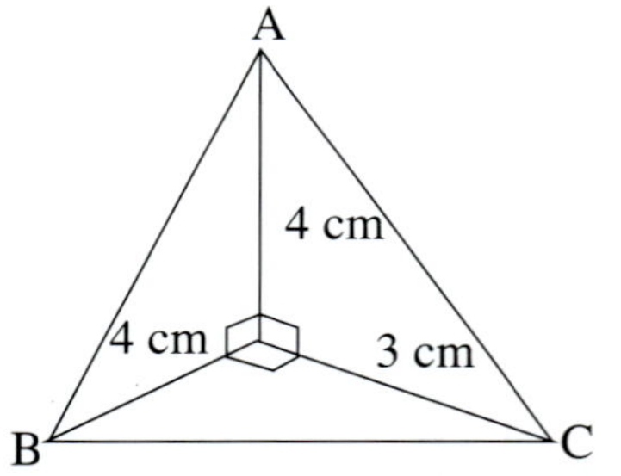

23. A right square pyramid has a height of 16 cm and an edge length of 20 cm.

(a) Illustrate this information on a diagram.

(b) Find
 i. the length of the sides of the base.
 ii. the angle between the sloping face and the base.

24. For the rectangular box shown find the angle which the diagonal AG makes with

(a) the base.

(b) a longer edge of the box.

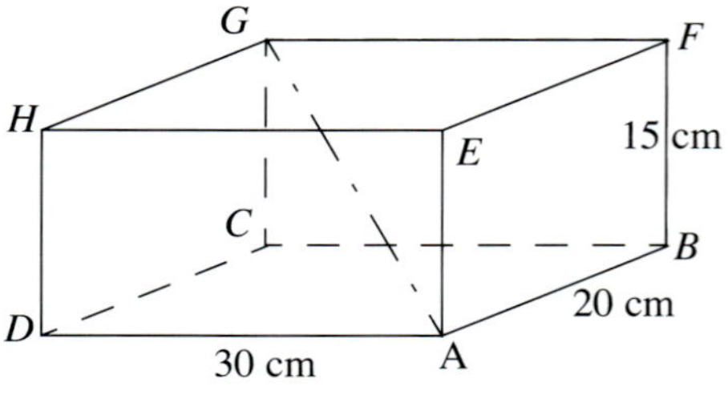

16.7 GRADED REVISION EXERCISES

LEVEL 1

Find the values of the pronumerals in each of the following diagrams:

1.

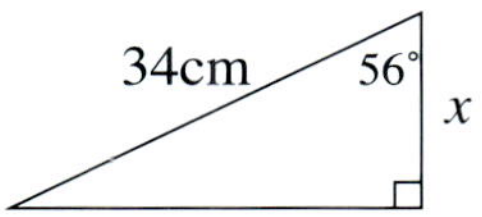

2.

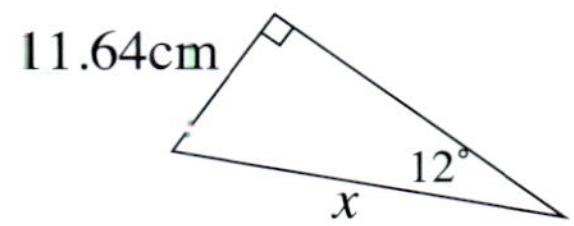

3.

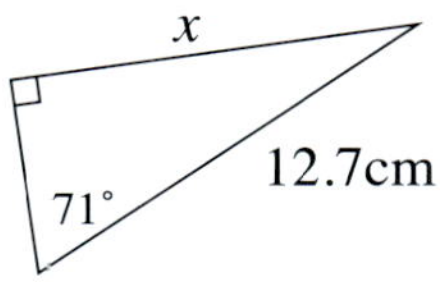

4.

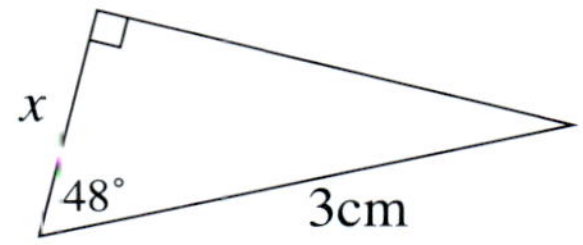

5.

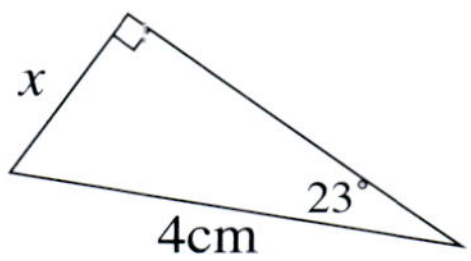

6.

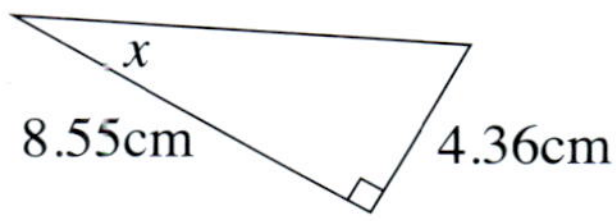

7.

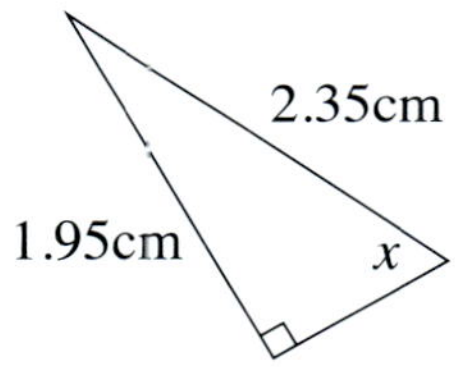

8.

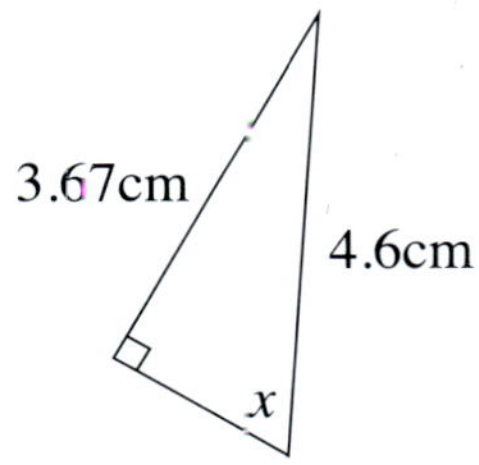

7. For the cube of edge length 8 cm, find

i. the length of DB.

ii. the inclination of the diagonal BH.

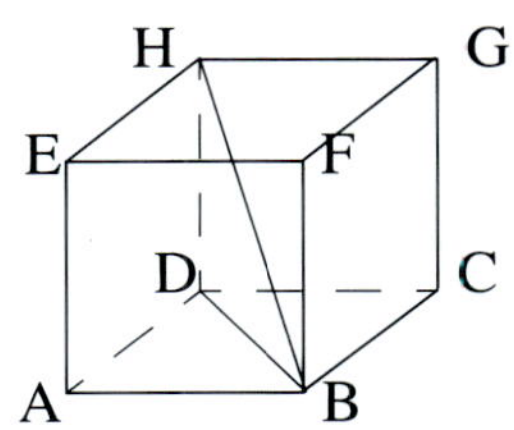

8. The right pyramid shown of vertical height 15 cm stands on a square base of edge length 5cm. Find

i. the length AC.

ii. the inclination of the edge AV.

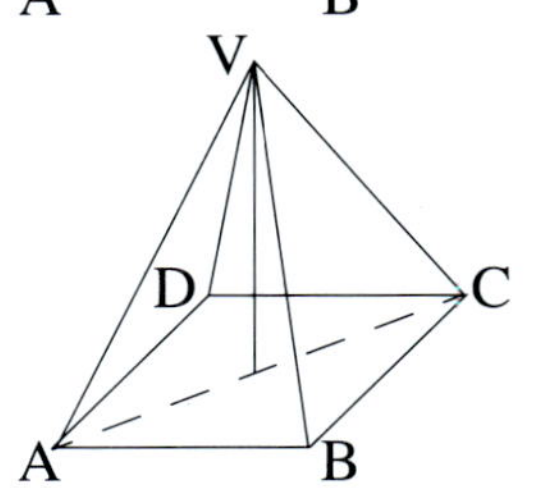

LEVEL 2

Find the values of the pronumerals in each of the following diagrams:

1.

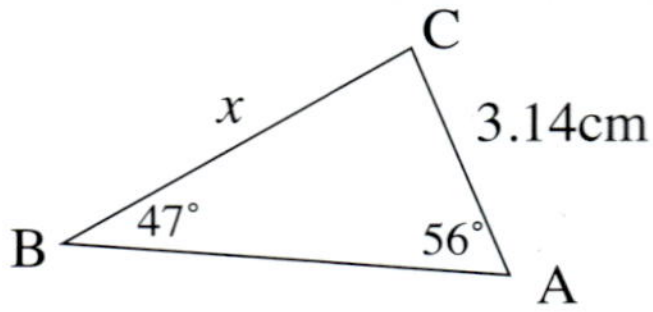

2.

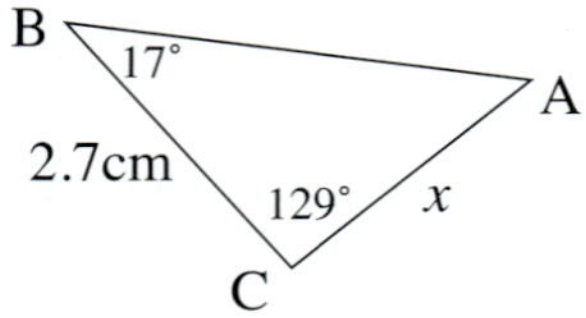

3.

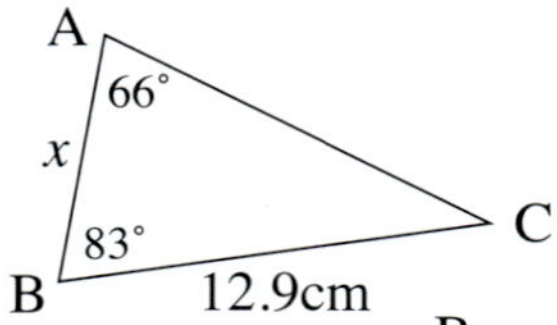

4.

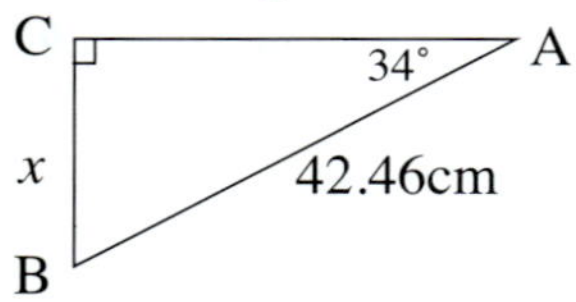

5.

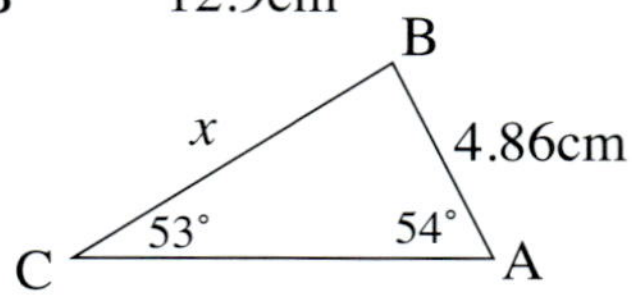

6.

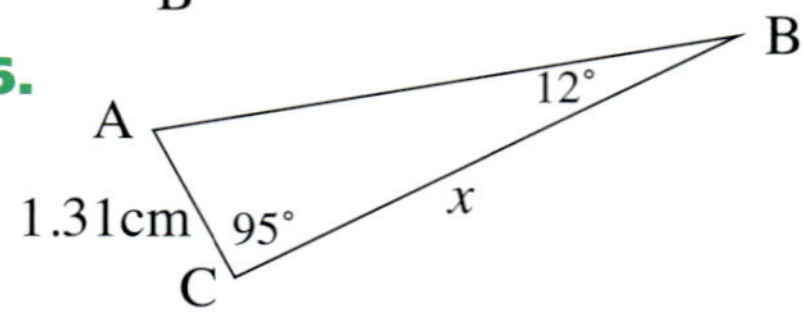

7.

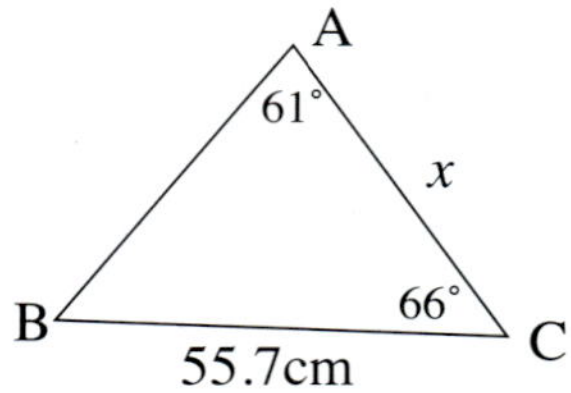

8.

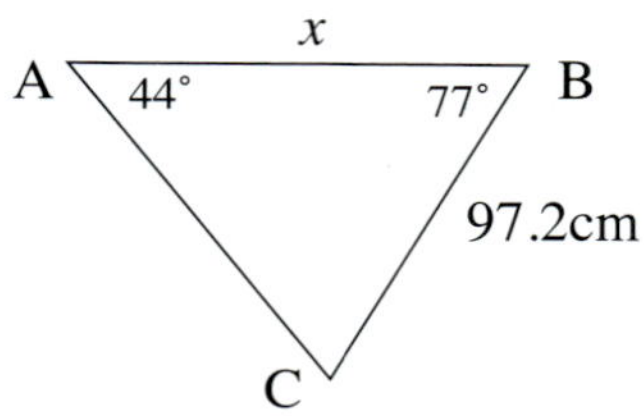

9. For the wedge shown alongside, find

i. the length of $[BE]$.

ii. the inclination of $[AE]$ with the base.

iii. the $\angle AED$.

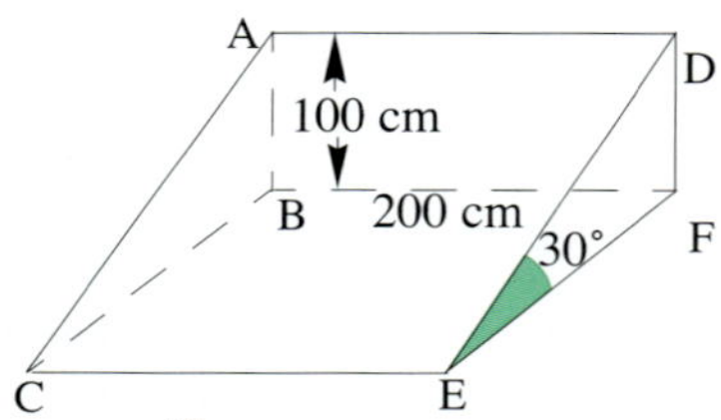

10. In the diagram alongside X is the midpoint of FG, where AB = 10 cm, BC = 6 cm and CG = 5 cm. Find the angle that AX makes with the base.

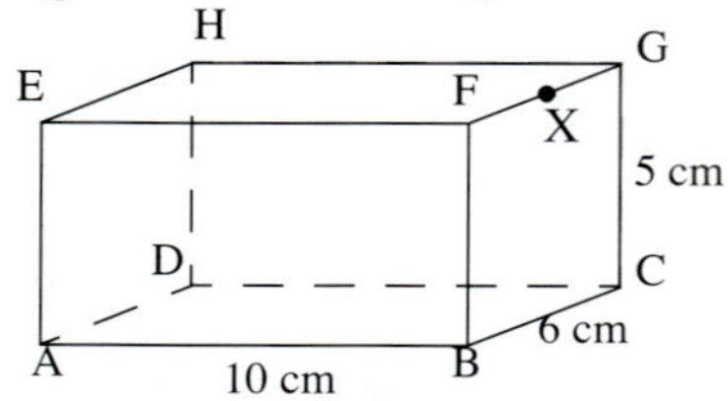

11. A square pyramid stands on a square base of side length 8cm. Each triangular face makes an angle of $60°$ with the base. Find the height of the pyramid.

12. A hemispherical bowl of diameter 12 cm is filled with water to a depth of 5 cm. Find the radius of the surface of the water level.

13. A book 17.6 cm wide and 24.1 cm long is opened at an angle of $30°$.

i. Find the shortest distance between the edge BC and edge EF.

ii. Find $\angle ECF$.

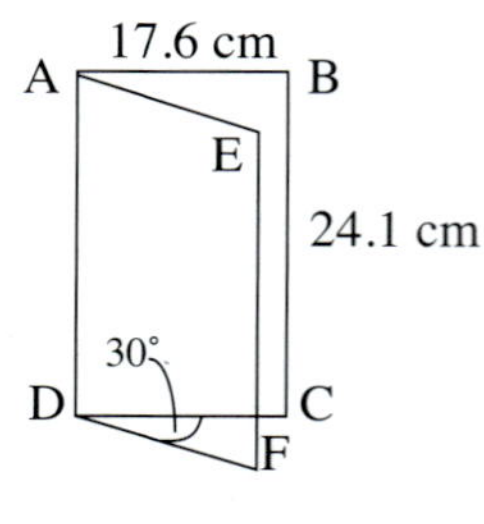

LEVEL 3

1. For the ΔABC shown alongside, where $AB = AC = BC = DC = 1$, find x.

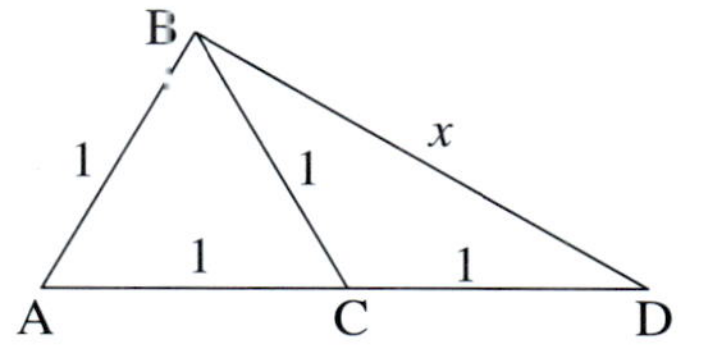

2. A triangle has the measurements shown. Find the two possible values of θ.

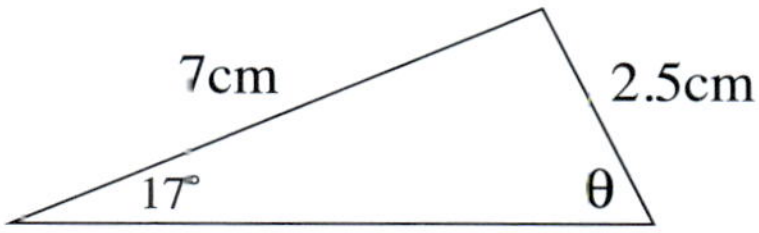

3. Solve the triangle in which $a = 2, b = 1.7, C = 63°$.

4. In the right-angled ΔABC, AD bisects $\angle BAC$. Find the ratio $BD:DC$.

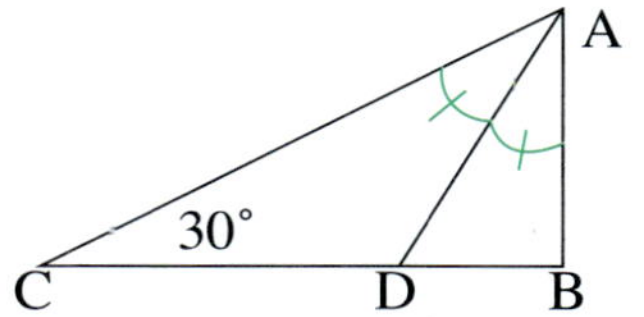

5. Find the perimeter of ΔPQR giving your answer in terms of a.

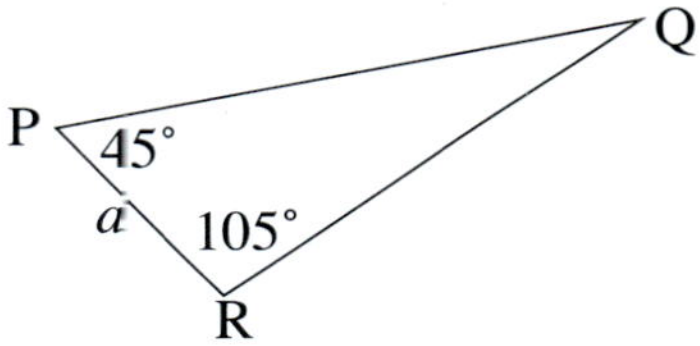

6. A door swings through an angle of 40° from its closed position.
The door is 2m high and 1 m wide.
Find α, $(\angle YXZ)$.

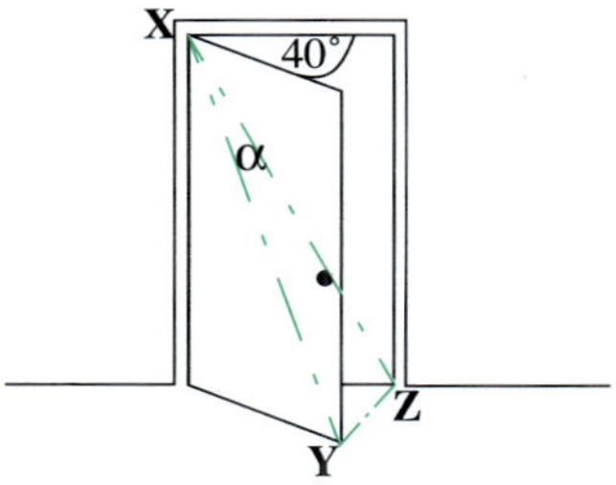

7. i. A solid sphere of radius r rests on top of a solid cylinder of radius r and height $3r$. How high from the base of the cylinder is the top of the sphere?
ii. A solid sphere rests on top of a vertical hollow cylinder having a diameter of 10 cm which is open at the top. The sphere projects 10 cm above the top of the cylinder. What is the length of the diameter of the sphere?

8. A tower stands vertically on a horizontal plane. The angles of elevations from two observers X and Y, 100m apart to the top of the tower are 45° and 30° respectively. X lies East of the tower and Y lies South of X. Find the height of the tower (correct to 2 d.p).

9. A sphere of radius 4 cm rests in a hollow inverted cone with a semi-vertical angle of 20°.
Find i. the distance from the centre of the sphere to the vertex of the cone.
ii. how close the sphere gets to the vertex of the cone.

10. A camera is resting on a tripod. The legs of the tripod rest on a plane and even surface and form a regular tetrahedron with side lengths 1.2 m.
i. How high is the camera from the ground?
ii. Find the angles that the legs of the tripod make with the ground.

LEVEL 4

1. Andrea, Basie and Chai are standing on a sports field. Basie is standing 55 metres north of Andrea, Chai is south east of Basie and Andrea is 70 metres from Chai. Find the true bearing of Chai from Andrea.

2. A helicopter is travelling east at an altitude of 180 metres. An observer sees the helicopter east of her at an angle of elevation of 79°. A minute later, the helicopter is still east of the observer but now has an angle of elevation of 44°. Find the speed of the helicopter in metres per second correct to three significant figures.

3. Two beacons, x m apart, provide a reference point for a helicopter flying vertically above the line joining two stations A and B.

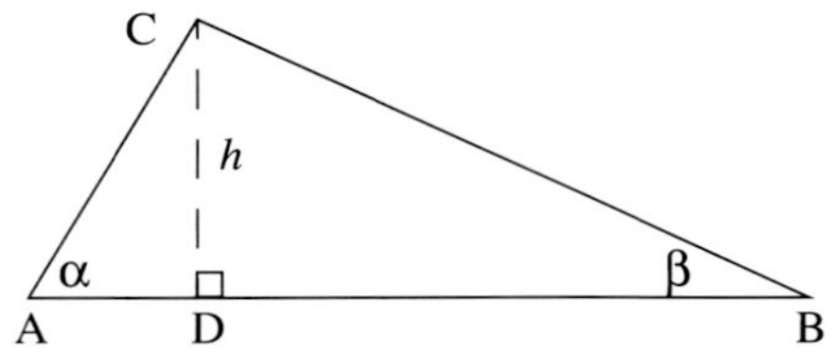

A: beacon 1
B: beacon 2
C: helicopter

The angles of elevation from A and B to C are α and β respectively. The helicopter maintains a constant height h m above ground level.

(a) Express AD in terms of h and α only.

(b) Using ΔABC express AD in terms of x, h and β.

(c) Hence, find an expression for h in terms of x, α and β.

4. Calculate the angle between two adjacent faces of a regular tetrahedron.

5. A mast is erected at the corner A of a rectangular field ABCD. The angle of elevation of the top of the mast from C is $\alpha°$ and the side length of the rectangle are AB = a m and BC = b m. Find the angle of elevation of the top of the mast from B and D.

6. Four identical spheres of radius r are placed together on table so that their centers form a square parallel to the table. A fifth identical sphere is now placed on top of the other four so that it makes contact with all four spheres. How high is the centre of the fifth sphere from the table?

16.8 TOPIC TEST

1. Helen stands 120 metres from the base of a tower. From ground level where Helen is standing, the angle of elevation of the top of the tower is 31°. Find the height of the tower correct to the nearest tenth of a metre.

[2 marks]

2. Find the value of a in the triangle shown, giving your answer correct to five significant figures.

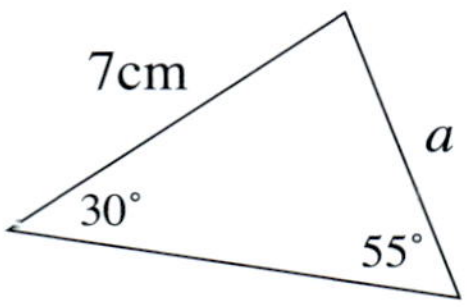

[2 marks]

3. Find the area of the triangle in the diagram, correct to four significant figures.

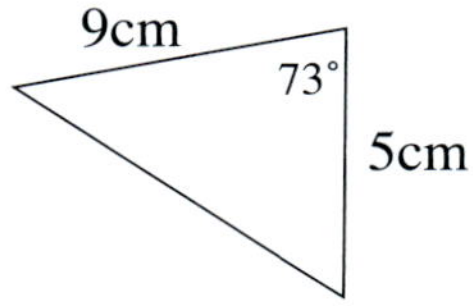

[2 marks]

4. Find, correct to the nearest degree, the largest angle in a triangle with sides of length 11cm, 12cm and 13cm.

[2 marks]

5. The bearing of D from X is N65°W and from Y it is N42°W. The point D is 50 km from the line joining XY and in a direction perpendicular to the line XY.

(a) Draw the diagram displaying the information given.

(b) Find the distance from X to Y.

[6 marks]

6. A triangle ABC rests on a horizontal plane with a third point X vertically above the point B.
Find the length of CX.

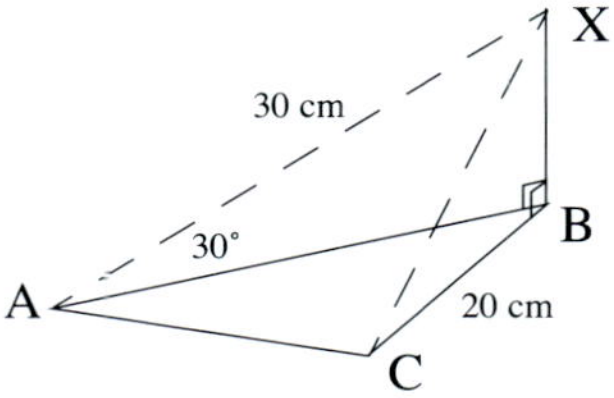

[3 marks]

7. The right pyramid, ABCDV has its rectangular base ABCD on a horizontal plane and its vertex V vertically above ABCD, with AV = BV = CV = DV.
Using the given dimensions, find the angle

i. between the face ABV and the base.

ii. between the edge AV and the base.

iii. the plane ABV and the plane DCV.

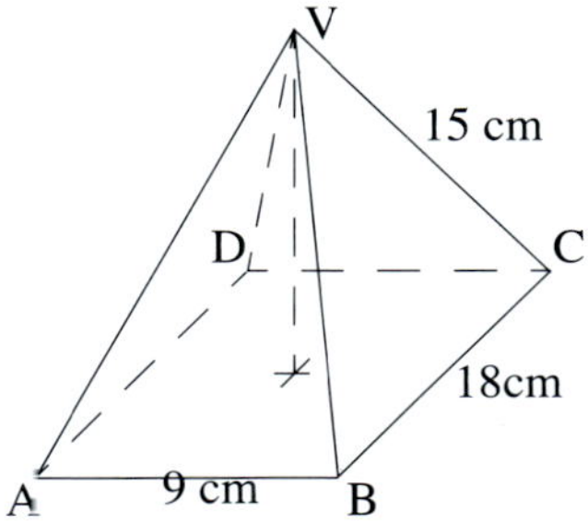

[9 marks]
Total 26 marks

16.9 TOPIC SUGGESTIONS

How did Eratosthenes (~200 BC) measure the size of the Earth? This is an experiment that you might try to repeat. You will need to contact a person who lives a good distance away, ideally on the same line of longitude. A student at another IB school, relative, friend etc. would be ideal. There is a need to coordinate the necessary measurements. Email has now made this a lot easier prospect than it was in 200 BC!

A project that uses trigonometry to make a map, plan or navigation route would seem to offer extensive possibilities in this area of the course.

The task of measuring the height of a hill (or some object whose base cannot be reached) presents challenges that can result in an excellent project, particularly if you analyse the errors in your measurements and the effects that these have on the accuracy of your answers.

17.0 NAVIGATING

Navigation (from Latin words meaning 'ship driving') is the science of knowing where you are and where you are going on the surface of the Earth (and a bit above and below it). We do this by laying imaginary lines of latitude and longitude across the surface of the Earth. The mathematics that underlies this important science is the subject of this chapter.

If you have studied history you may know that the crews of the various expeditions (1490-1500) to the Americas led by Christopher Columbus feared that the Earth was flat and that they would fall off the edge.

From this you might imagine that navigation is a comparatively modern invention. This is not so. To begin with, many animals are exceptional navigators and are known to use the stars and sun as well as landmarks to complete their migrations. The photograph shows a baby albatross on the Galapagos islands. When she grows up and matches the size of her formidable beak (each of her wings will be longer than you are), she will spend years at sea, seldom visiting land. During this time, the albatross will cross tens of thousands of kilometres of the Great Southern Ocean. Then, when the time comes to breed, she will navigate her way back to this island to find a mate. This amazing achievement is matched by the migrations of many other animal species.

Human navigation also has a lengthy history. The Ancient Greeks were certainly aware that the World was spherical and had a good idea of its size. The Phoenicians are known to have traded throughout the Mediterranean, England and perhaps as far as Iceland around 600 BC. The Vikings were also accomplished travellers and are now known to have reached North America at some time between the 9th and the 11th century AD. On the other side of the World, around 2,000 years ago, Polynesian people colonised almost every island in the vast expanse of the Pacific Ocean sometimes crossing thousands of kilometres of empty sea.

The loss of this wisdom for over a thousand years is a warning that progress is not always forwards!

17.1 COORDINATE GEOMETRY

17.1.1 TWO DIMENSIONAL COORDINATE GEOMETRY

In this section we review and extend some of the work that was carried out in Chapter 6. However, our main focus will be with the geometric properties of figures lying on the Cartesian plane. Mostly we shall use coordinates to study geometry – a technique that was first introduced in 1637 by Descartes, a French mathematician and philosopher – who made it possible to use algebra in the study of geometry.

We will discuss geometry in two dimensions, by making reference to the Cartesian plane, i.e., the X–Y plane. In doing so, much of what follows will be based on arbitrary points (x_1, y_1) , (x_2, y_2) and so on, that lie on this plane.

A Cartesian plane is made up of a two dimensional (flat) surface that is divided into four quadrants. These quadrants are the result of a plane divided by two straight lines intersecting at right angles and meeting at a point called the origin, O. The horizontal axis is labelled as the *x*–axis and the vertical axis as the *y*–axis (although other labels can also be used).

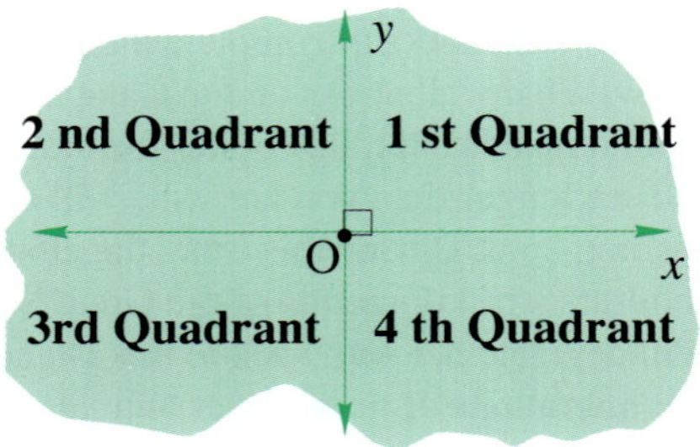

17.1.2 DISTANCE BETWEEN TWO POINTS

The distance between two points on a plane is given by $d = \sqrt{(x_2 - x_1)^2 + (y_2 - y_1)^2}$

This formula can be derived by using Pythagoras's theorem:

Consider the two points $P(x_1, y_1)$ and $Q(x_2, y_2)$.

Constructing a right–angled triangle (as shown) we can apply Pythagoras's theorem ($a^2 = b^2 + c^2$) to obtain the results that

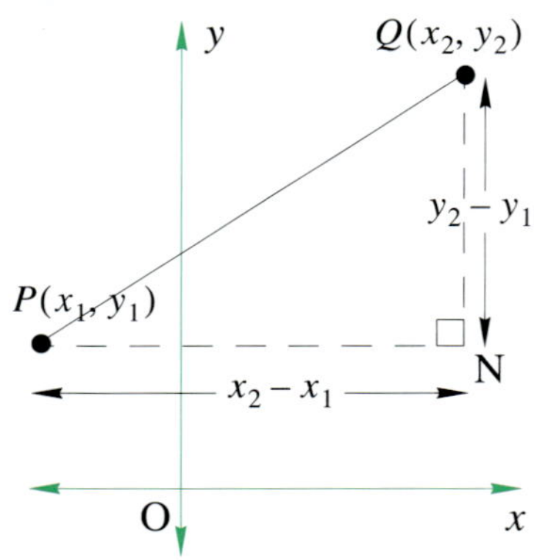

$$PQ^2 = PN^2 + QN^2$$
$$= (x_2 - x_1)^2 + (y_2 - y_1)^2$$
$$\therefore PQ = \sqrt{(x_2 - x_1)^2 + (y_2 - y_1)^2}$$

EXAMPLE 17.1

Find the distances between these pairs of points

(a) (2, 4) and (5, 8)

(b) (–2, 5) and (–4, –6)

Solution

We solve these problems by first making use of Pythagoras's theorem and then using the distance formula.

(a) It is always a good idea to sketch a diagram to illustrate the information:

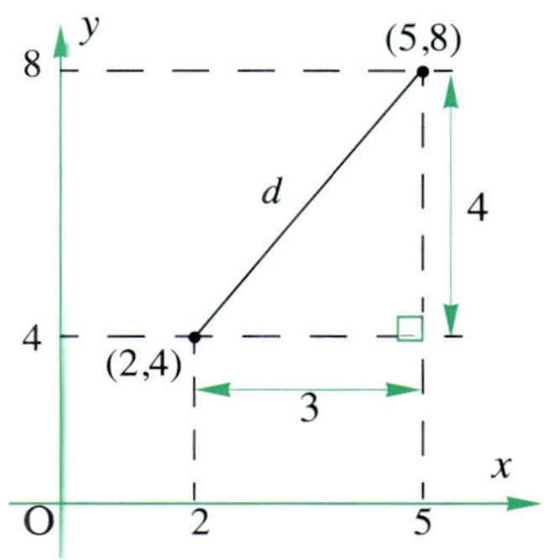

The **vertical distance** between the points is

$8 - 4 = 4$ units

and the **horizontal distance** is

$5 - 2 = 3$ units.

The distance between the points is now found using Pythagoras's theorem:

$$d = \sqrt{3^2 + 4^2} = \sqrt{25} = 5 .$$

Using the formula $d = \sqrt{(x_1 - x_2)^2 + (y_1 - y_2)^2}$ with $x_1 = 2, x_2 = 5$ and $y_1 = 4, y_2 = 8$,

we have $$\begin{aligned} d &= \sqrt{(5-2)^2 + (8-4)^2} \\ &= \sqrt{3^2 + 4^2} \\ &= \sqrt{25} \\ &= 5 \end{aligned}$$

(b) The **vertical distance** between the points is

$5 - (-6) = 11$ units

and the **horizontal distance** is

$(-2) - (-4) = 2$ units.

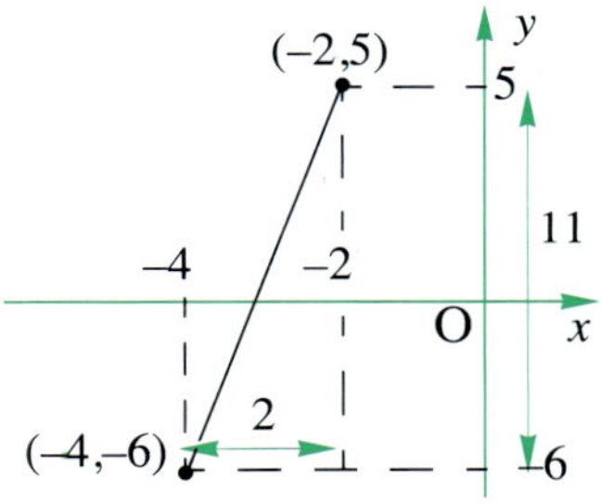

Using Pythagoras's theorem, we have:

$$d = \sqrt{2^2 + 11^2} = \sqrt{125} .$$
$$= 5\sqrt{5}$$

Using the formula $d = \sqrt{(x_1 - x_2)^2 + (y_1 - y_2)^2}$ with $x_1 = -4, x_2 = -2$ and $y_1 = -6, y_2 = 5$,

we have that $d = \sqrt{((-2) - (-4))^2 + (5 - (-6))^2} = \sqrt{2^2 + 11^2}$

$$= \sqrt{125}$$
$$= 5\sqrt{5}$$

EXAMPLE 17.2 A triangle has its vertices at the coordinates A(1, 2), B(3, 6) and C(7, 4). Find the perimeter of the triangle.

Solution

The perimeter is given by AB + BC + CA.
So, all that is required is to determine each distance:

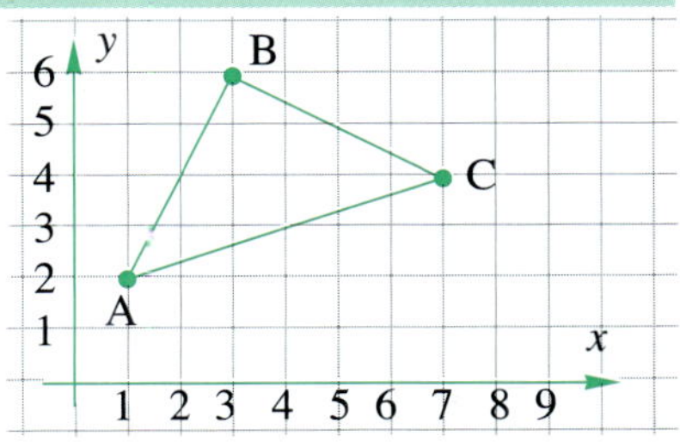

$$AB = \sqrt{(3-1)^2 + (6-2)^2} = \sqrt{4+16} = \sqrt{20}$$
$$BC = \sqrt{(7-3)^2 + (4-6)^2} = \sqrt{16-4} = \sqrt{20}$$
$$CA = \sqrt{(7-1)^2 + (4-2)^2} = \sqrt{36+4} = \sqrt{40}$$

Therefore, perimeter $= \sqrt{20} + \sqrt{20} + \sqrt{40} \approx 15.27$

17.1.3 MID-POINT BETWEEN TWO POINTS

The mid–point between two points has coordinates that are mid–way between the coordinates of the two points.

That is, the coordinates of the mid–point, (x, y) of the line segment $[AB]$, joining the two points $A(x_1, y_1)$ and $B(x_2, y_2)$ are given by

$$x = \frac{x_1 + x_2}{2}, y = \frac{y_1 + y_2}{2}$$

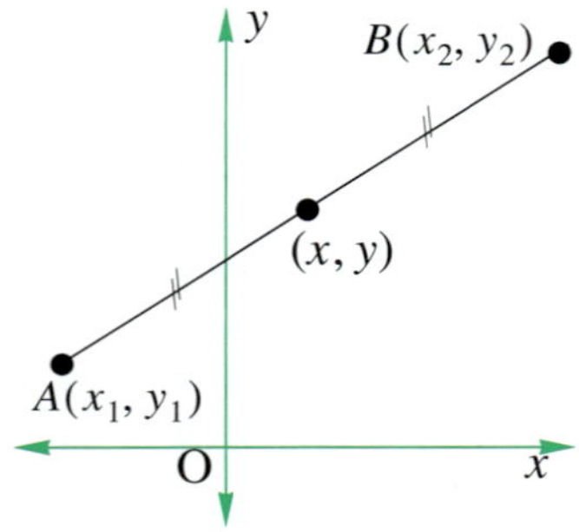

EXAMPLE 17.3

Find the mid-point between these pairs of points:
(a) (3,4) and (7,10) (b) (–3,7) and (5,–2)

Solution

(a) As in the last example, we illustrate the information on the Cartesian plane:

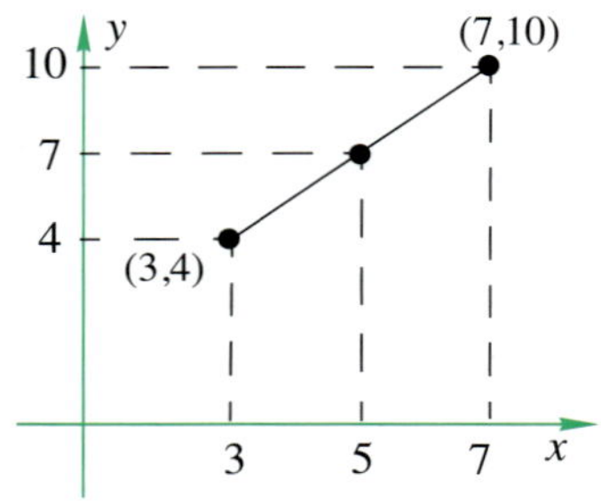

The mid-point between (3,4) and (7,10) is found by finding the x-coordinate and the y-coordinate that lies mid-way between the x- and y- coordinates of the two points.
The x-coordinate is mid-way between 3 & 7.

This is given by $\frac{3+7}{2} = 5$.

The y-coordinate is mid-way between 4 & 10.

This is given by $\frac{4+10}{2} = 7$. Therefore, the mid-point is (5,7)

Using the mid–point formula: $\left(\frac{x_1 + x_2}{2}, \frac{y_1 + y_2}{2}\right)$ with $x_1 = 3, x_2 = 7$ and $y_1 = 4, y_2 = 10$,

we have that the mid–point, is given by $\left(\frac{3+7}{2}, \frac{4+10}{2}\right) = (5, 7)$.

(b) The x-coordinate of the mid-point is mid-way between -3 & 5, i.e., $\frac{-3+5}{2} = 1$.

The y-coordinate of the mid-point is mid-way between 7 & -2, i.e., $\frac{7+(-2)}{2} = 2.5$.

Therefore, the mid-point is $(1, 2.5)$

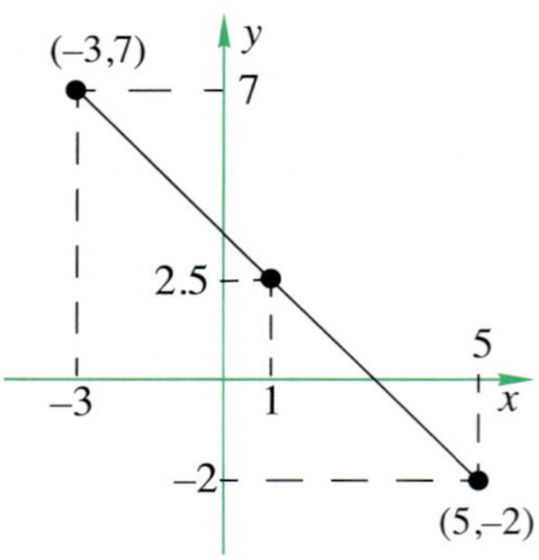

Using the mid–point formula: $\left(\frac{x_1 + x_2}{2}, \frac{y_1 + y_2}{2}\right)$, where

$x_1 = -3, x_2 = 5$ and $y_1 = 7, y_2 = -2$, we have $x = \frac{-3+5}{2}, y = \frac{7+(-2)}{2}$

i.e., $(x, y) = (1, 2.5)$.

EXERCISES 17.1

1. Find the distances between these pairs of points, giving exact answers:

(i)	(2,4) & (6,9)	(ii)	(–8,–3) & (–2,3)	(iii)	(–7,6) & (8,6)
(iv)	(3,8) & (–2,–7)	(v)	(1,–1) & (8,5)	(vi)	(–5,0) & (8,2)
(vii)	(–4,6) & (0,1)	(viii)	(8,–6) & (0,–1)	(ix)	(–9,6) & (6,5)
(x)	(4,–1) & (–5,4)	(xi)	(–7,0) & (7,–3)	(xii)	(–9,6) & (–3,–6)
(xiii)	(–9,–7) & (7,5)	(xiv)	(6,7) & (–6,6)	(xv)	(–7,0) & (–3,0)

2. Find the mid-point of the line segments joining these pairs of points:

(i)	(–7,0) & (–3,0)	(ii)	(6,7) & (–6,6)	(iii)	(–9,–7) & (7,5)
(iv)	(–7,–7) & (2,–4)	(v)	(–7,6) & (–4,–5)	(vi)	(1,–1) & (4,3)
(vii)	(–9,2) & (4,–9)	(viii)	(3,5) & (5,0)	(ix)	(–6,8) & (3,0)
(x)	(–2,6) & (6,3)	(xi)	(9,1) & (8,9)	(xii)	(–6,3) & (4,4)
(xiii)	(4,–1) & (–5,4)	(xiv)	(–9,6) & (–3,–6)	(xv)	(3,8) & (–2,–7)

17.2 LINEAR EQUATIONS

17.2.1 STANDARD FORMS OF LINEAR EQUATIONS

In Chapter 6 we looked at the following forms of a straight line.

1. $y = mx + c$

There are several forms that linear relations can take. Probably the most useful is known as the gradient intercept form:

$$y = mx + c$$

Here, the gradient of the line is equal to m and the y intercept is equal to c.

2. $ax + by + c = 0$

Another common form for a linear equation is:

$$ax + by + c = 0$$

This form of the linear equation does not (immediately) give the gradient and y-intercept of the line. When sketching or plotting linear equations in this form it is best to calculate the intercepts on both axes. This is done as follows:

For x–intercept, let $y = 0$, so that $ax + c = 0 \Rightarrow x = -\frac{c}{a}$

For y–intercept, let $x = 0$, so that $by + c = 0 \Rightarrow y = -\frac{c}{b}$

17.2.2 PROPERTIES OF STRAIGHT LINES

1. Gradient of a line

The gradient, m, of the line through two points (x_1, y_1) and (x_2, y_2) is given by $m = \frac{y_2 - y_1}{x_2 - x_1}$.

Which is also given by $m = \frac{\text{Rise}}{\text{Run}}$

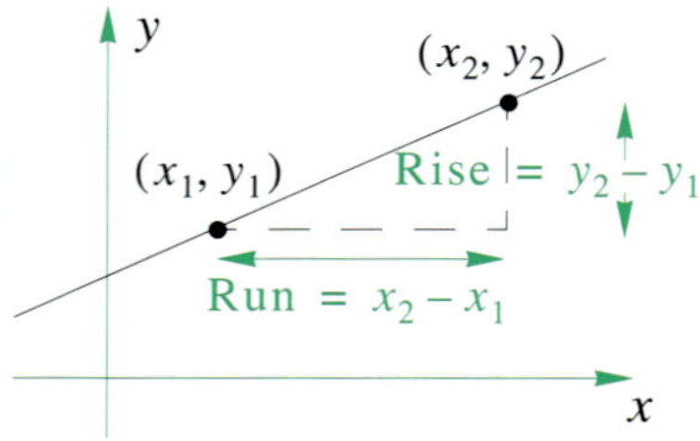

From this we can obtain the point-gradient form of a line. That is, if (x, y) is any point on a straight line having a gradient m, and (x_1, y_1) is another fixed point on that line then the equation of that line is given by

$$y - y_1 = m(x - x_1)$$

2. Parallel lines

Two straight lines
are parallel if they have the same gradient
Conversely,
two straight lines with the same gradient are parallel.

That is, the straight line l_1 with gradient m_1 is parallel to the straight line l_2 with gradient m_2 if and only if $m_1 = m_2$.

That is, $l_1 \parallel l_2$ iff $m_1 = m_2$

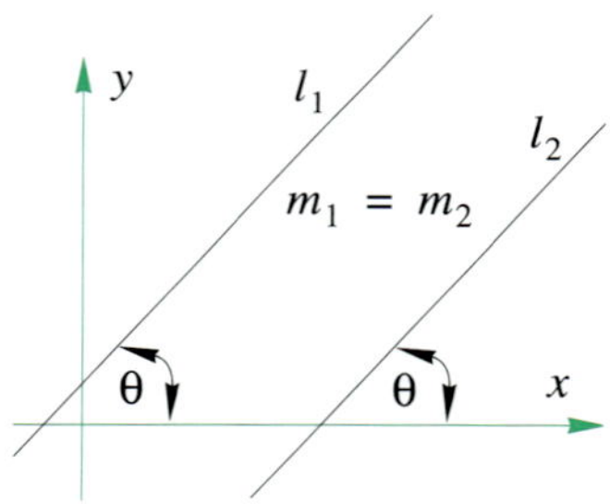

Notice that if the two lines are parallel, they also make equal angles with the x–axis.

4. Perpendicular lines

If two lines are **perpendicular**, then the product of their gradients is –1.

That is, the straight line l_1 with gradient m_1 is perpendicular to the straight line l_2 with gradient m_2 if and only if $m_1 \times m_2 = -1$.
That is,

$$l_1 \perp l_2 \text{ iff } m_1 \times m_2 = -1 \text{ or } m_1 = -\frac{1}{m_2}.$$

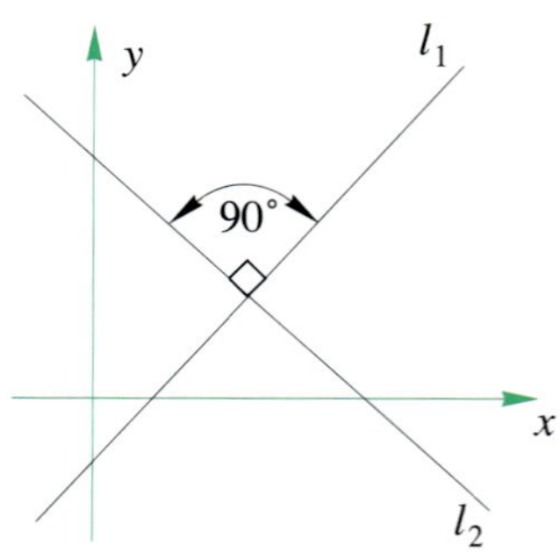

5. Two special cases

Case 1 $y = c$ (i.e., $m = 0$)

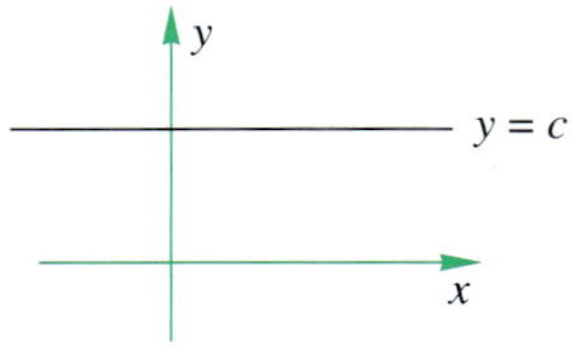

As there is no rise (i.e., rise = 0),

then, $m = \frac{0}{\text{run}} = 0$.

Case 2 $x = k$ (i.e., m is undefined)

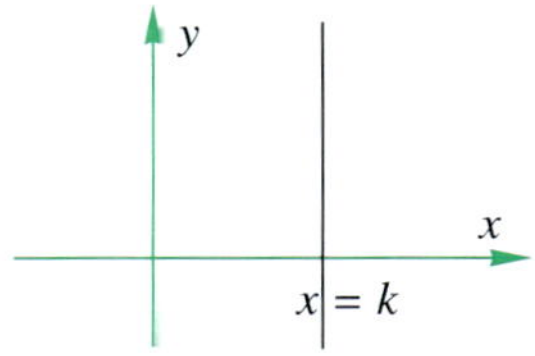

As there is no run (i.e., run = 0),

then, $m = \frac{\text{rise}}{0}$ which is undefined.

(Division by zero is not possible).

EXAMPLE 17.4

Find the equation of the line that passes through the point (–1,3) and is parallel to the line with equation $2x - y + 7 = 0$.

Solution

The gradient of the line $2x - y + 7 = 0$ is found by rearranging it into the form $y = mx + c$ to get: $y = 2x + 7$.

The gradient is 2 and so all the lines parallel to this line will also have a gradient of 2.

Therefore, the equation of the required line is of the form $y = 2x + c$.

The value of the constant c can be found by using the fact that the line passes through the point (–1,3). So,

$$\begin{aligned} y &= 2x + c \\ 3 &= 2 \times -1 + c \\ c &= 5 \\ \therefore y &= 2x + 5 \end{aligned}$$

EXAMPLE 17.5

Find the equation of the line which passes through the point (–1,4) and is perpendicular to the line with equation $2x + 5y + 2 = 0$.

Solution

The gradient form of $2x + 5y + 2 = 0$ is $5y = -2x - 2 \Rightarrow y = -\frac{2}{5}x - \frac{2}{5}$ so the gradient is $-\frac{2}{5}$.

The gradient of all lines that are perpendicular to this is found by using the fact that the product of the gradients is –1: $\left(-\frac{2}{5}\right)m = -1 \Rightarrow m = \frac{5}{2}$.

Therefore, the equation of the required line is of the form $y = \frac{5}{2}x + c$.

The constant c is found in the same way as the previous example:

$$y = \frac{5}{2}x + c$$

Using the point (–1, 4), i.e., $x = -1$ and $y = 4$ we have: $4 = \frac{5}{2} \times -1 + c$

so that $c = 6.5$

That is, the line has equation $y = \frac{5}{2}x + 6\frac{1}{2}$

EXERCISES 17.2

1. Find the gradient of the line joining the points

(i)	(3, 2) and (5, 6)	(ii)	(4, 5) and (6, 11)
(iii)	(–1, 3) and (2, 8)	(iv)	(–5, –7) and (–3, 9)
(v)	(3, 5) and (9, 5)	(vi)	(6, 3) and (4, –1)
(vii)	(2, 4) and (5, –2)	(viii)	(4, 6) and (4, 12)

2. Use the gradient–point form equation to find the equation of the straight line if

(a) it passes through the point (1, 1) and has a gradient of 2.
(b) it passes through the point (–2, 3) and has a gradient of 3.
(c) it passes through the point (3, –4) and has a gradient of –1.

3. Find the gradient of the straight line that is perpendicular to the straight line with gradient equal to

(a) 2. (b) – 3. (c) $-\frac{2}{3}$. (d) $\frac{5}{4}$

4. Find the equation of the straight line that passes through the origin and the point (2,4).

5. Find the equation of the straight line that passes through the points (–1,2) and (0,1).

6. A straight line passes through the point (4,3) and is perpendicular to the line joining the points (–1,3) and (1,–1).
Find the equation of this line, giving the answer in the form $ax + by + c = 0$.

7. A triangle is made from the coordinate axes and the line $3x + 4y + p = 0$. The area of the triangle is 6 square units.
What is the value of p?

8. The lines $px + 4y - 2 = 0$ and $2x - y + p = 0$ are perpendicular. Find the value of p.

9. The lines $x + 2y + 12 = 0$, $x - 2y + 3 = 0$, $2y + x - 8 = 0$ and $x - 2y - 2 = 0$ form a quadrilateral.
What is the best description for this quadrilateral?

10. Find equations for each of the following lines.
Give answers in the gradient/intercept form:

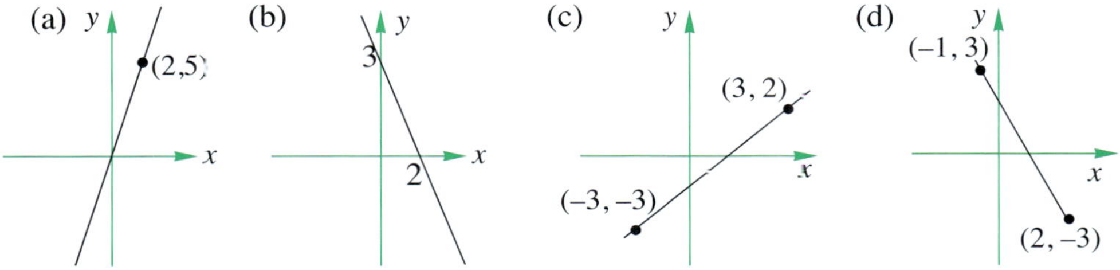

17.3 APPLICATIONS IN 2-D

EXAMPLE 17.6

A triangle is formed by the lines with equations:

$$3x + y - 12 = 0$$
$$x - 3y + 6 = 0$$
$$x + 7y - 44 = 0$$

(a) Prove that the triangle is right angled.
(b) Find the coordinates of the vertices of the triangle.
(c) Find the area of the triangle thus formed.
(d) Prove that the triangle formed by joining the mid-points of the sides is also right angled.

Solution

(a) The gradients of the three sides are:

$$3x + y - 12 = 0 \Rightarrow y = -3x + 12 \Rightarrow \text{Gradient} = -3, [A]$$

$$x - 3y + 6 = 0 \Rightarrow y = \frac{1}{3}x + 2 \Rightarrow \text{Gradient} = \frac{1}{3}, [B]$$

$$x + 7y - 44 = 0 \Rightarrow y = -\frac{1}{7}x + 6\frac{2}{7} \Rightarrow \text{Gradient} = -\frac{1}{7}, [C]$$

Since the product of the gradients of [A] and [B] is $-3 \times \frac{1}{3} = -1$, the lines [A] and [B] are perpendicular and the triangle must be right angled.

(b) The coordinates of the vertices are found by solving the equations of the lines simultaneously. If the gradient/intercept forms of these lines are used, then y can be eliminated immediately:

[A] and [B]: $-3x + 12 = \frac{1}{3}x + 2 \therefore \frac{10}{3}x = 10 \Rightarrow x = 3$

The y-value is found by substituting back into either equation. In this instance we use $y = -3x + 12$: $y = -3 \times 3 + 12 = 3$

[A] and [C]: $-3x + 12 = -\frac{1}{7}x + \frac{44}{7} \therefore -21x + 84 = -x + 44 \Rightarrow 20x = 40$. i.e., $x = 2$.

Using the equation $y = -3x + 12$ we have $y = -3 \times 2 + 12 = 6$

[B] and [C]: $-\frac{1}{7}x + \frac{44}{7} = \frac{1}{3}x + 2 \therefore -3x + 132 = 7x + 42 \Rightarrow 10x = 90$. i.e., $x = 9$.

Using the equation $y = \frac{1}{3}x + 2$ we have $y = \frac{1}{3} \times 9 + 2 = 5$

(c) The coordinates of the vertices of the triangle are (3,3), (2,6) and (9,5). The two perpendicular sides of the triangle are those joining (3,3) to (2,6) and (3,3) to (9,5). These can be used as the base and height of the triangle.
Their lengths are:

$$(3,3) \text{ to } (2,6): h = \sqrt{(3-2)^2+(3-6)^2} = \sqrt{1+9} = \sqrt{10}$$

$$(3,3) \text{ to } (9,5): b = \sqrt{(9-3)^2+(5-3)^2} = \sqrt{36+4} = \sqrt{40}$$

The area of the triangle is: $\frac{1}{2} \times b \times h = \frac{1}{2} \times \sqrt{40} \times \sqrt{10} = \frac{1}{2} \times \sqrt{40 \times 10} = 10$ sq. units.

(d) The midpoints of the three sides have coordinates:

$$(3,3) \text{ to } (2,6) \text{ mid-point is } \left(\frac{3+2}{2}, \frac{3+6}{2}\right) = \left(\frac{5}{2}, \frac{9}{2}\right) \ [P]$$

$$(3,3) \text{ to } (9,5) \text{ mid-point is } \left(\frac{3+9}{2}, \frac{3+5}{2}\right) = (6, 4) \ [Q]$$

$$(9,5) \text{ to } (2,6) \text{ mid-point is } \left(\frac{9+2}{2}, \frac{5+6}{2}\right) = \left(\frac{11}{2}, \frac{11}{2}\right) \ [R]$$

The gradients of the lines joining these points are:

$[PQ]$ Gradient $= \dfrac{\frac{9}{2}-4}{\frac{5}{2}-6} = -\dfrac{1}{7}$, $[PR]$ Gradient $= \dfrac{\frac{9}{2}-\frac{11}{2}}{\frac{5}{2}-\frac{11}{2}} = \dfrac{1}{3}$ $[QR]$ Gradient $= \dfrac{4-\frac{11}{2}}{6-\frac{11}{2}} = -3$

The product of the gradients of the lines $[PR]$ and $[QR]$ is $\frac{1}{3} \times -3 = -1$ so the new triangle is also right angled.

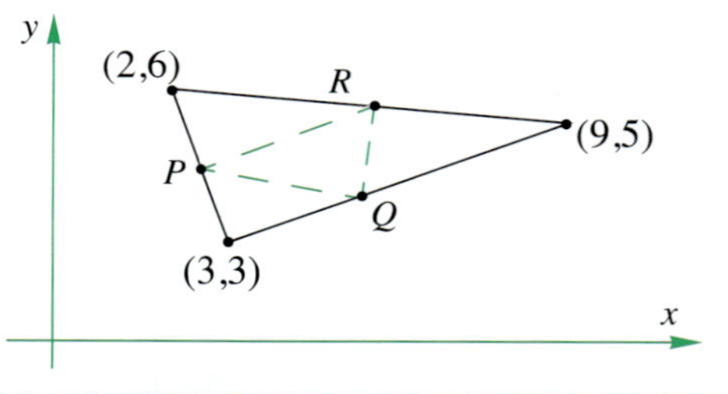

EXERCISES 17.3

1. A quadrilateral is formed by joining the points (–1,–2) to (1,4) to (5,6) to (3,0) and finally back to (–1,–2).
   (a) Prove that the quadrilateral is a parallelogram.
   (b) Find the mid-points of the sides of the parallelogram and prove that these can be joined to form a second parallelogram.

2. The lines with equations $x + 2y - 12 = 0$, $x - 3y + 3 = 0$, $2x - y + 1 = 0$ form a triangle.
   (a) Find the coordinates of the vertices of this triangle.
   (b) The mid-points of the sides of this triangle are joined to make a second triangle. Prove that this second triangle is similar to the first.

3. A straight line has equation $x + 3y = 6$.

(a) Find the equations of the set of lines that are perpendicular to this line.

(b) Find the equation of the line that is perpendicular to $x + 3y = 6$ and which also passes through the point (5,7).

(c) Find the coordinates of the point of intersection of the line found in (b) with $x + 3y = 6$.

(d) Hence find the shortest distance between the point (5,7) and the line with equation $x + 3y = 6$.

4. Two lines have equations $y = 2.5x - 4$ and $y = 2.3x + 7$. The gradients of the two lines are known to 2 significant figures and the intercepts are known exactly. Find the range of possible points of intersection of these lines.

17.4 GEOMETRY OF 2–D & 3-D SHAPES

In Chapter 16 we looked at problems in three dimensions where we calculated lengths and angles. In this section we look at areas and volumes of basic geometric shapes.

17.4.1 PERIMETER AND AREA

Shape	Perimeter	Surface Area	Example
Rectangle 	$P = 2l + 2w$ $= 2(l + w)$	$A = lw$	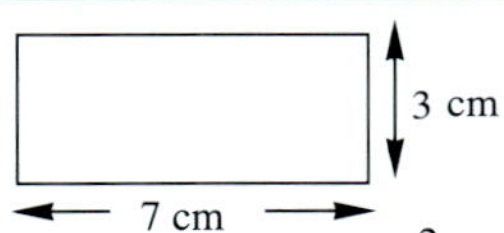 $A = 7 \times 3 = 21$ cm^2 $P = 2(7 + 3) = 20$ cm
Parallelogram 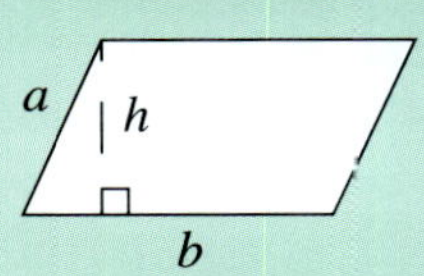	$P = 2a + 2b$ $= 2(a + b)$	$A = bh$	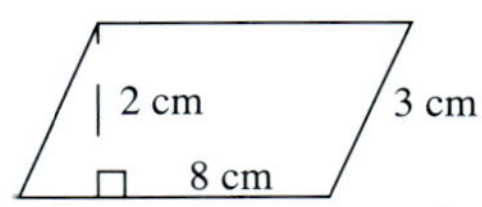 $A = 8 \times 2 = 16$ cm^2 $P = 2(8 + 3) = 22$ cm
Triangle 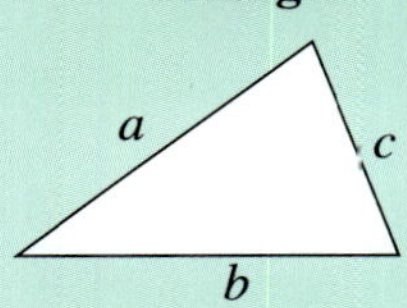	$P = a + b + c$	$A = \frac{1}{2}bh$	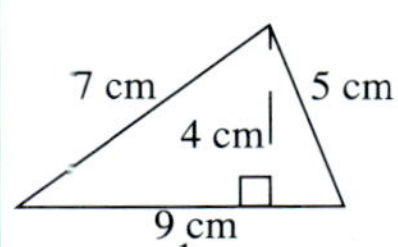 $A = \frac{1}{2} \times 9 \times 4 = 18$ cm^2 $P = 9 + 5 + 7 = 21$ cm
Trapezium 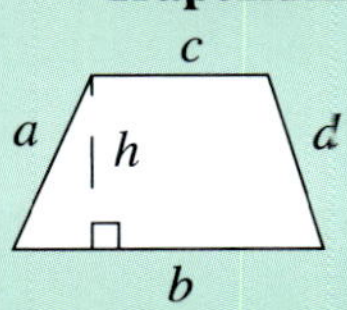	$P = a + b + c + d$	$A = \frac{1}{2}(a + b)h$	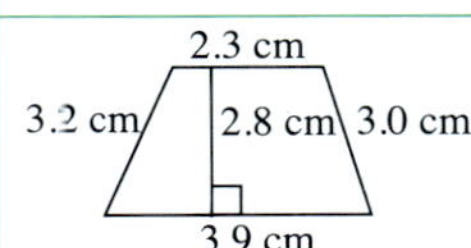 $A = \frac{1}{2}(2.3 + 3.9) \times 2.8 = 8.68$ cm^2 $P = 2.3 + 3.2 + 3.9 + 3.0 = 12.4$ cm

A shape that does not consist of straight lines is the circle:

Shape	Perimeter (circumference)	Surface Area	Example
Circle 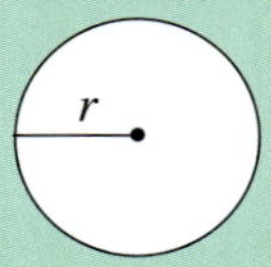	$P = 2\pi r$	$A = \pi r^2$	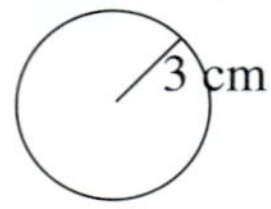 $A = \pi \times 3^2 = 9\pi \text{ cm}^2$ $P = 2\pi \times 3 = 6\pi \text{ cm}$

17.4.2 SURFACE AREA OF PRISMS

When determining the surface area of prisms, the approach is to break the shape up into its net, calculate the area of each side and then add the results.

Shape	Net	Surface Area
Rectangular block 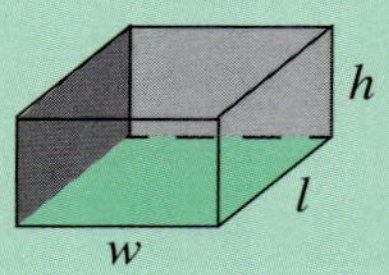	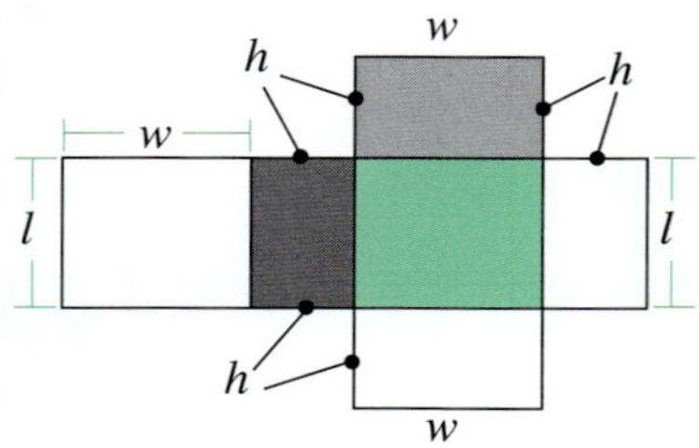	$S_A = 2lw + 2(l + w)h$
Prism 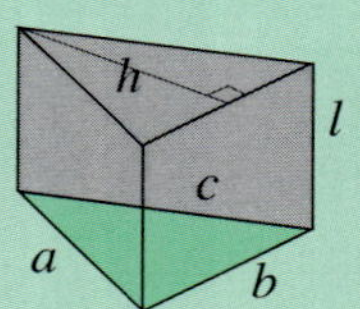	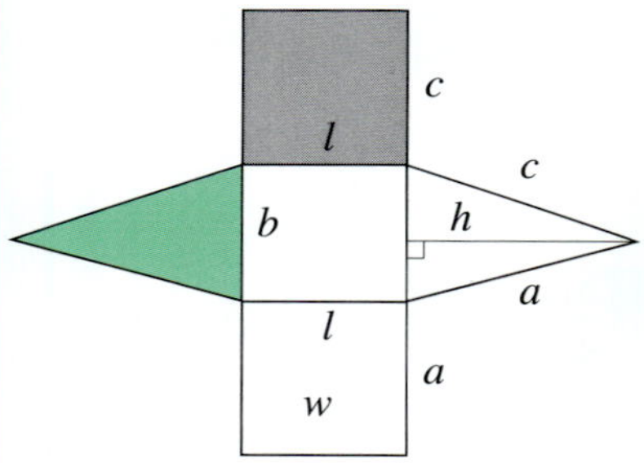	$S_A = 2\left(\frac{bh}{2}\right) + (a + b + c)l$ $= bh + (a + b + c)l$
Square pyramid 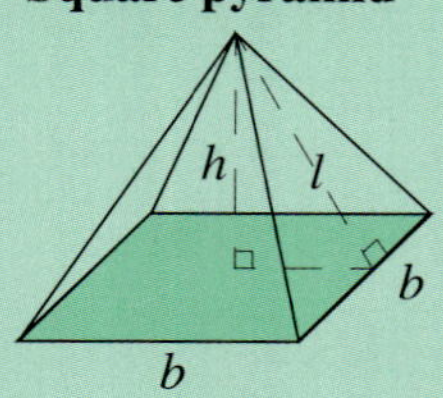	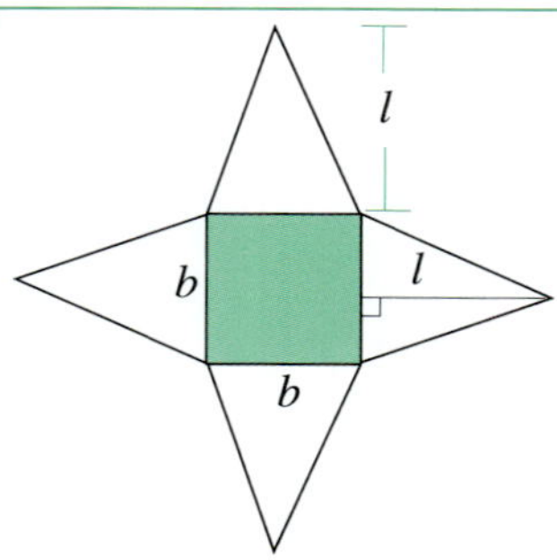	$S_A = b \times b + 4 \times \frac{1}{2}lb$ $= b^2 + 2lb$

Shape	Net	Surface Area
Circular Cylinder 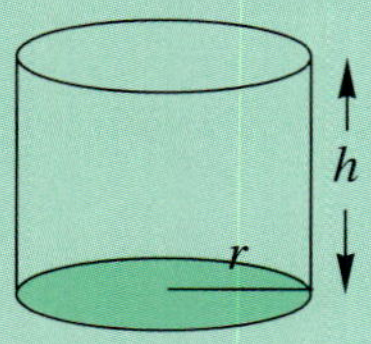	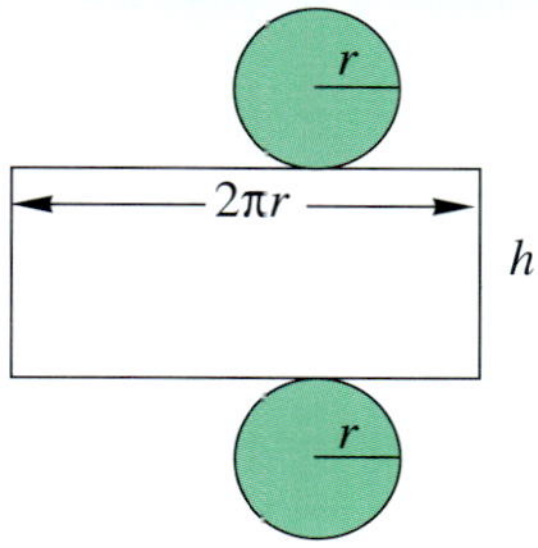	$S_A = 2 \times \pi r^2 + h \times 2\pi r$
Circular Cone 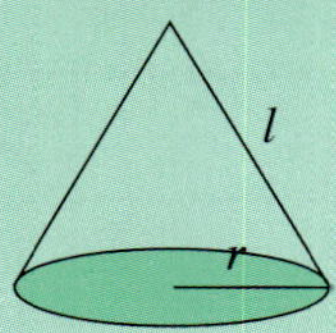	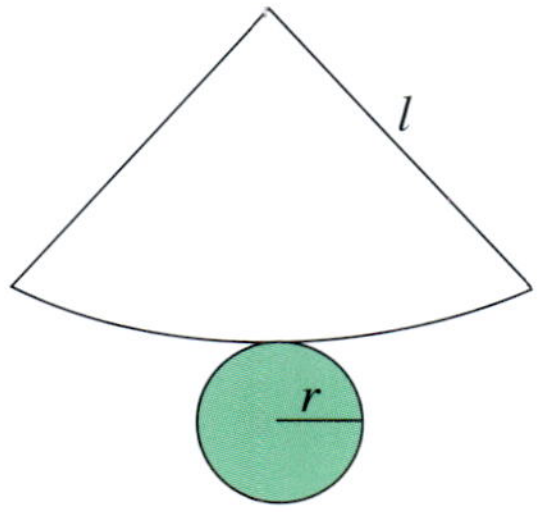	$S_A = \pi r^2 + \pi r l$
Sphere 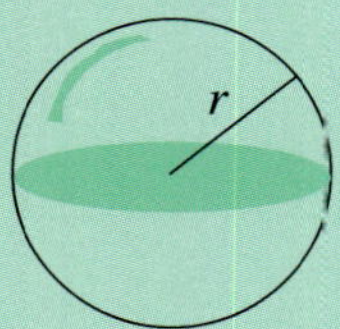	*No net diagram for a sphere*	$S_A = 4\pi r^2$

EXERCISES 17.4

1. Find, where possible, (a) the perimeter (b) the area of the following shaded regions

i.
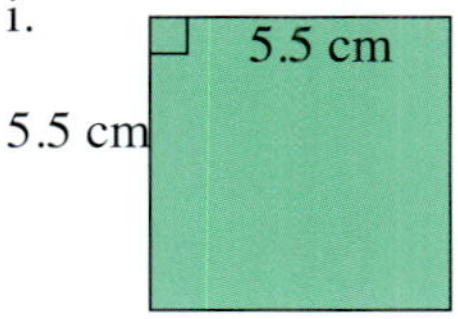

ii.
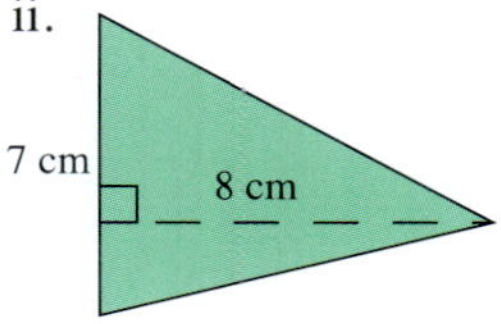

iii.
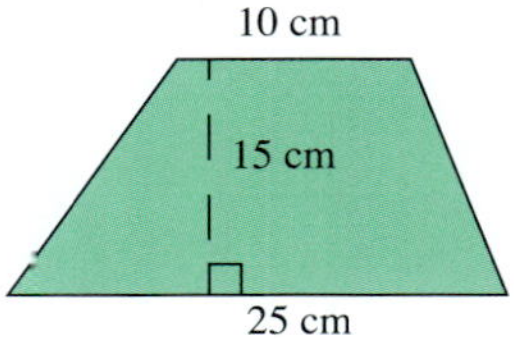

iv.
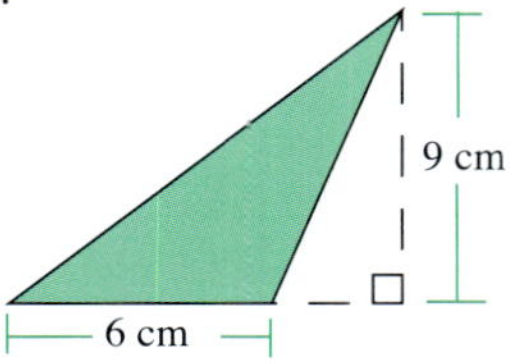

v.
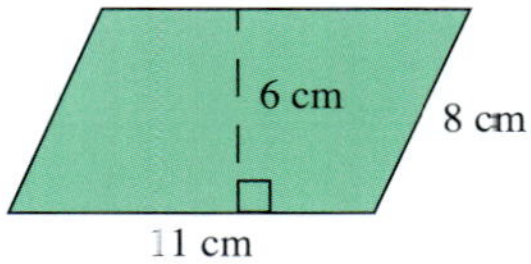

vi.
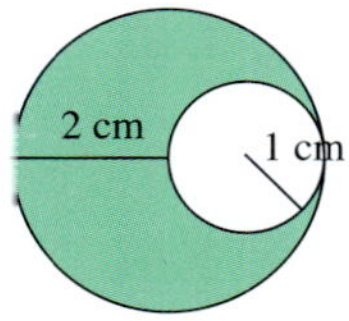

2. Find the area of the shaded region of the following figures

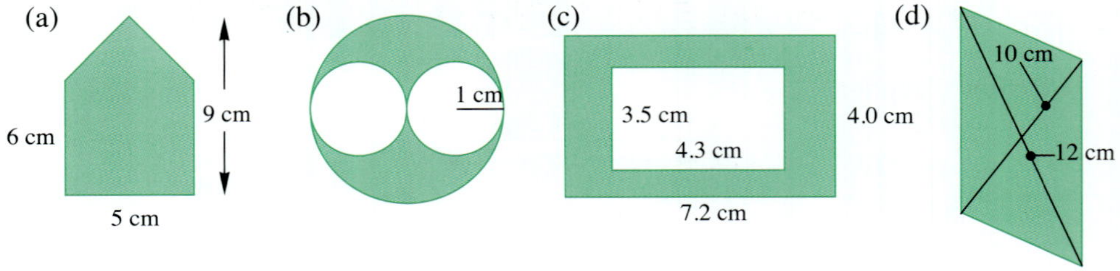

3. Find the surface area of the following figures

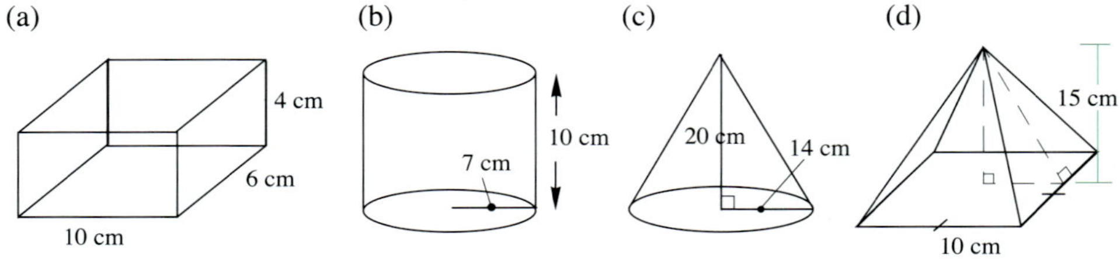

4. Find the surface area of the following figures

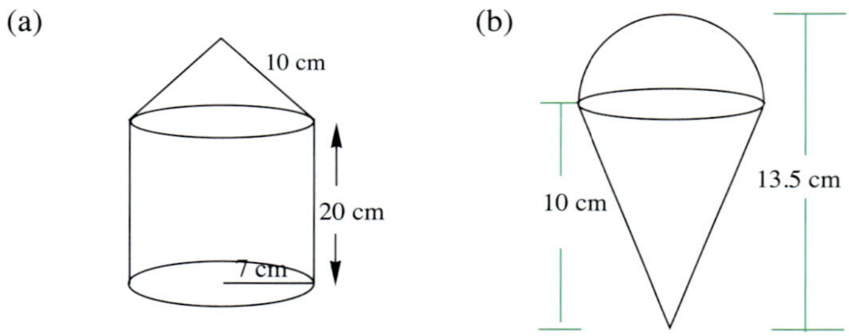

5. Find the volume of the following figures

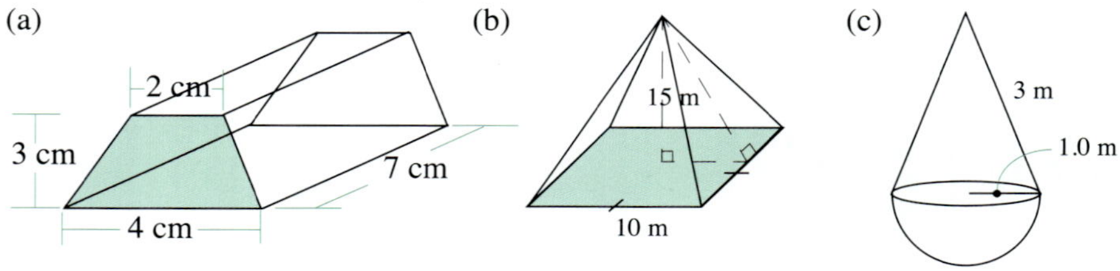

6. Find the volume of the following figures.

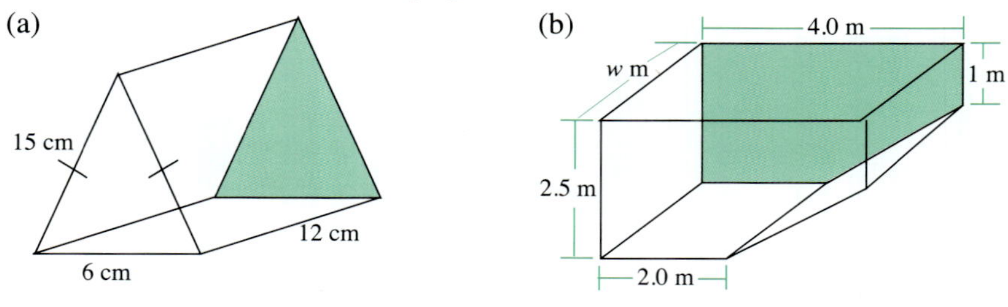

17.5 MISCELLANEOUS EXERCISES

1. The equation of the line l is $y + 3x = 6$.

(a) Find the gradient (slope) of l.

(b) Find the x–intercept and y–intercept of l.

(c) Find the gradient of a line perpendicular to l.

(d) Find the equation of the line in (c) if it passes through the point $(-3, 0)$.

2. What volume of gold will be required to produce the gold pendant shown in the figure.

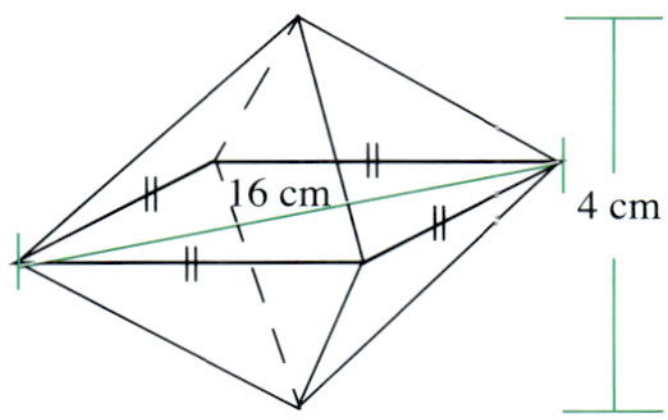

3. A straight line passes through the points $(4, 0)$ and $(0, 8)$.

(a) Find the gradient of this line.

(b) Find the equation of this line.

(c) The line passes through the points $(2, a)$ and $(b, 2)$. Find a and b.

4. Find the equation of the straight line, l, passing through the points $(-4, -2)$ and $(3, 4)$. Hence, find the equation of the straight line perpendicular to l and passing through the point $(0, 4)$

5. Find the equation of each of the following straight lines, given that it passes through the point $(2, -1)$

(a) and has a gradient of 4.

(b) and passes through the point $(-1, 2)$.

(c) and has a y–intercept of $(0, -2)$.

(d) and is perpendicular to the line with equation $y + 2x = 5$.

6. Find the equation of the line passing through $(3, -1)$ and:

(a) perpendicular to (b) parallel to

the line with equation $x - 2y + 1 = 0$

7. Find the point of intersection of the lines with equations $2x - 4y + 1 = 0$ and $x + y - 2 = 0$.

8. Find the area of the figure shown if it has a perimeter of $24 + 6\pi$ cm.

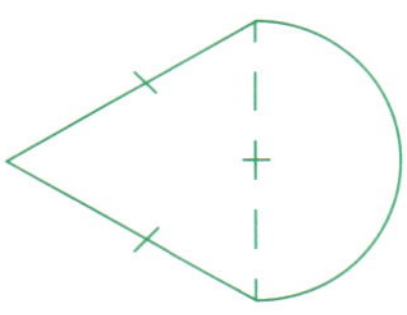

9. The lines $ax - y + 1 = 0$ and $x + 2y + 2 = 0$ intersect at $(1, b)$. Find a and b.

10. Find the equation of the straight line through the point (–1, 2) that is parallel to the line with equation $2y - 4x + 3 = 0$ giving your answer in the form $y = mx + c$.

11. Find the equation of the straight line perpendicular to the line with equation $y = -\frac{2}{3}x + \frac{3}{7}$ which passes through the point (2,4), giving your answer in the form $ax + by + c = 0$.

12. Find the mid-point of the line segment joining the points (–3,7) and (5,1)

13. Find the length of the line segment joining the points (–3, 5) to (2,–3):
(i) exactly
(ii) correct to 3 significant figures.

14. Find the equation of the perpendicular bisector of the line segment joining the points (1,2) and (3,1), giving the answer in the form $y = mx + c$.

15. A pyramid ABCF is sliced off from a rectangular block as shown in the diagram alongside. If the block has an initial volume of V cm^3, how much of the volume is sliced off?

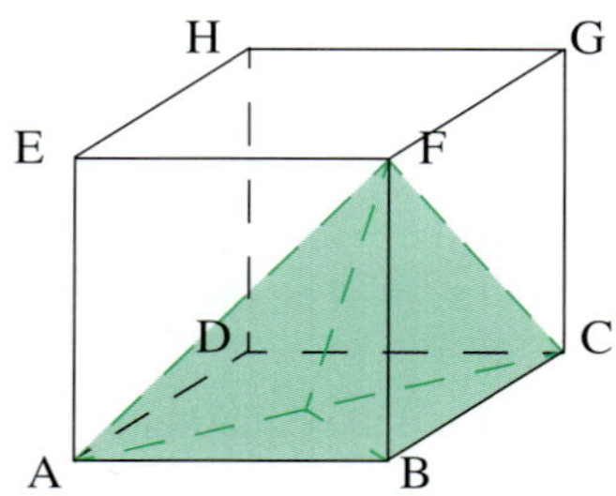

16. Find the ratio of the area of the smaller circle to the area of the larger circle, given that the smaller circle is inscribed in the square and the larger cicle circumscribes the square.

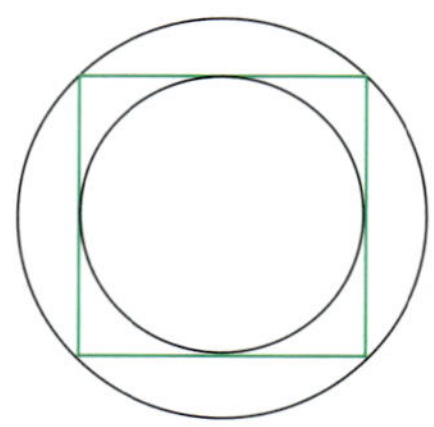

17. A concrete monument is in the shape of a rectangular block surmounted by a pyramid as shown in the diagram. Calculate the volume of concrete required to build the monument.

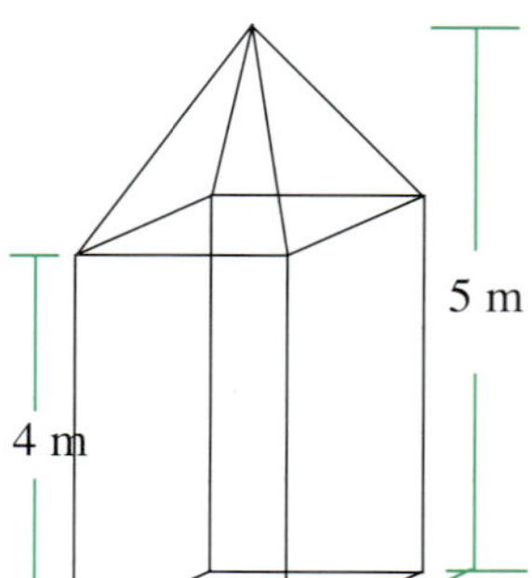

17.6 GRADED REVISION EXERCISES

LEVEL 1

1. For the following equations, state their gradient and y–intercept.

(a) $y = 3x - 1$ (b) $y = -(x+1)$ (c) $2y = x + 4$

2. For the following set of points

(a) (3, 4) & (2,4) (b) (5, 8) & (2, –2) (c) (2, 3) & (–2, –3), find

i. the measure of the distance between them.

ii. the gradient of the line joining them.

iii. the coordinates of the mid-point of the line segment joining them.

3. Find the area of the shaded region in the figures shown below

(a)

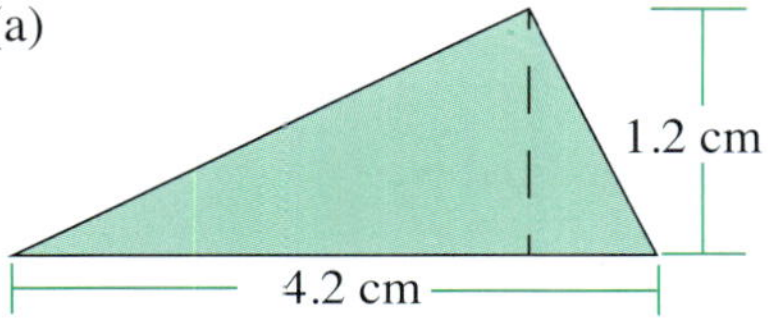

(b)

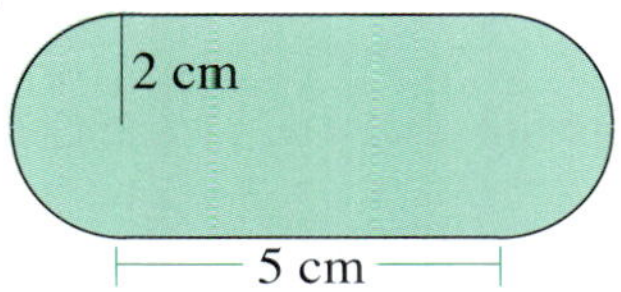

LEVEL 2

1. Find the gradient of the straight line perpendicular to each of the following.

i. $y = 2x - 1$ ii. $y = 3 - \frac{1}{3}x$ iii. $5y = 2x + 10$

2. Find the equation of the straight line passing through (2, 3) with gradient 3.

3. Find the equation of the line passing through (1, 2) and perpendicular to the line with equation $2x + 3y + 1 = 0$.

4. Find the equation of the line passing through (1, 2) and parallel to the line with equation $x - 2y - 1 = 0$.

5. Find the equation of the straight lines shown.

(a)

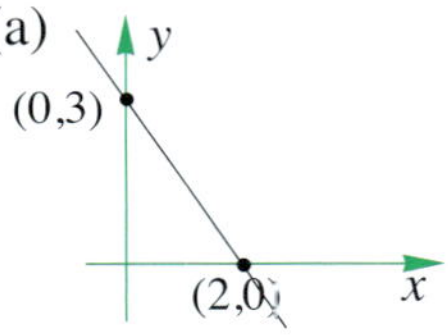

(b)

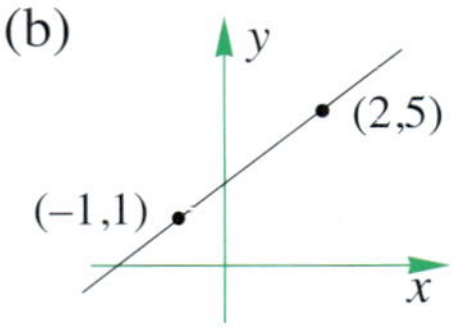

(c)

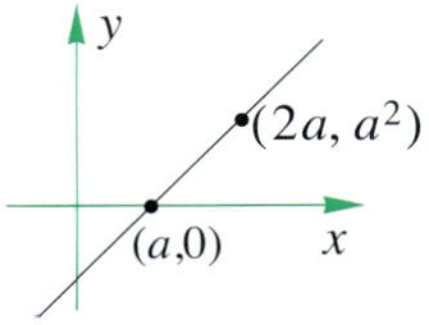

6. Find the surface area of the following figure

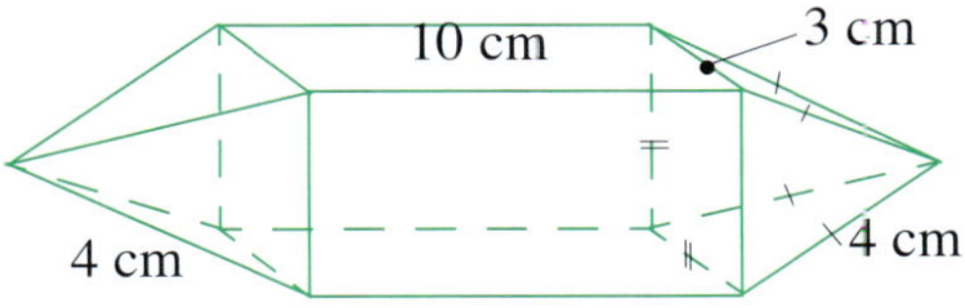

LEVEL 3

1. For what value of k will the lines

(a) $kx - y = 9$ (b) $2x + ky = 4$

be i. parallel ii. perpendicular

to the line with equation $y = 3x + 5$?

2. Given that $4a - 5b = 2c$ and $3a + b = 3c$, find the ratio $a:b$.

3. Find the coordinates of A on the diameter AB of a circle, if B has the coordinates (–2, 4) and the centre of the circle has coordinates (1, 2).

4. The line $ax + by + 10 = 0$ passes through the point (–1, 2) and is perpendicular to the line with equation $8x - 12y + 16 = 0$. Find the values of a and b.

LEVEL 4

1. The points A(3, 2), B$(m, 1)$ and C(1, –4) are collinear. Find m.

2. The vertices A, B and C of a triangle are (0, 6), (–2, 3) and (3, 4) respectively.
- i. Draw the triangle on a set of axes.
- ii. Show that the triangle ABC is an isosceles right-angled triangle.
- iii. Find the coordinates of a fourth point, D, if ABCD is to be a square.

3.
- i. (a) Sketch the graph of $y = 2x^2$.
  (b) Find the gradient of this curve when $x = k, k > 0$.
- ii. Find the equation of the straight line which touches this curve when $x = k$.

This straight line is perpendicular to the line with equation $4y + x = 16$.
- iii. Find the value of k.
- iv. Find the area of the triangle enclosed by the straight lines and the x-axis.

4. Find the coordinates of the point of intersection of the lines $bx + y = ab$ and $ax - by = a^2$, where a and b are real numbers.

5. Find the value of a if the lines $ax + 3y = 5$ and $2x + (a + 1)y = 6$ are

(a) parallel. (b) perpendicular.

6. Let ABC be a triangle with vertices at $A(-2a, 0)$, $B(2a, 0)$ and $C(2b, 2c)$.
- i. Draw the triangle on a set of axes.
- ii. Find the coordinates of the midpoints of AB, AC and BC and mark these on your diagram.
- iii. Show that the equation of the median through
  (a) C is $y = \frac{c}{b}x$.
  (b) A is $(3a + b)y - cx - 2ac = 0$.
- iv. Prove that the medians of a triangle are concurrent.

17.7 TOPIC TEST

1. On the same set of axes, sketch the graph of the straight lines $2y - x = 8$ and $y = 4 - x$. The lines intersect the x–axis at A and B, how far apart are A and B?

[4 + 2 marks]

2. Using the points with coordinates (3, 5) and (5, 1) find

i. the measure of the distance between.
ii. the gradient of the line joining.
iii. the coordinates of their mid-point.

[2 + 2 + 2 marks]

3. Solve the system of linear equations $6x = 5y + 3$ and $8y = 12x + 6$.

[3 marks]

4. Find the relationship between p and q if the simultaneous equations $px - 3y = 8$ and $qx + y = 4$

(a) are perpendicular.
(b) have no real solution.

[4 marks]

5. A triangle ABC is drawn on the set of axis as shown.

The slope of BX is $-\frac{1}{3}$ and $BX \perp AC$

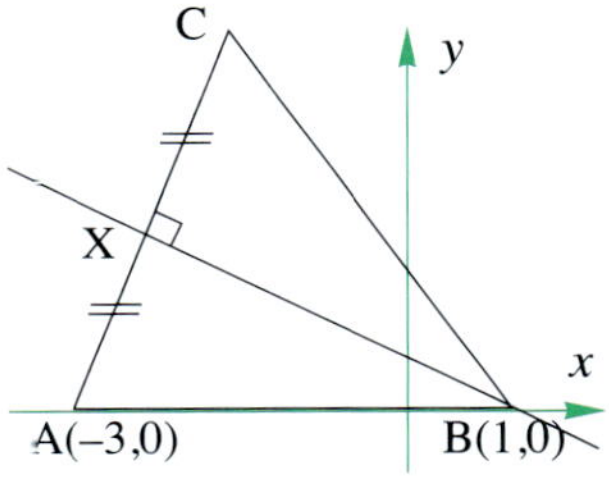

i. Find the equation of
   (a) BX (b) AC
ii. Find the coordinates of X.
iii. Find the area of the triangle ABC.

[2 + 2 + 2 + 2 marks]

6. OABC is a parallelogram as shown on the set of axes.

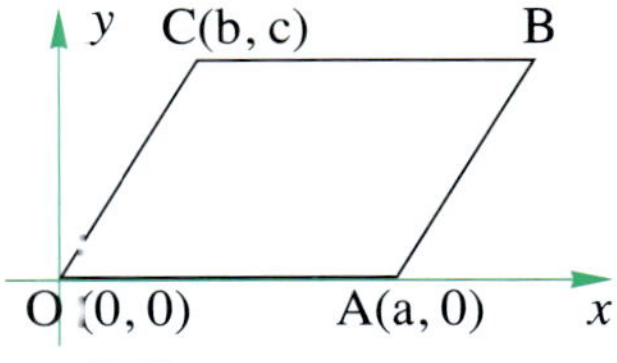

i. Find the coordinates of B.
ii. Find the midpoint of AC and OB.
iii. What can you deduce about the diagonals of a parallelogram?
iv. If OABC is now a rhombus, what are the coordinates of B?
v. Find the gradients of OB and AC for this rhombus.
vi. What can you deduce about the diagonals of a rhombus?

[2 + 4 + 1 + 1 + 3 + 3 marks]

7. A sphere of radius r cm fits tightly into a cylinder.

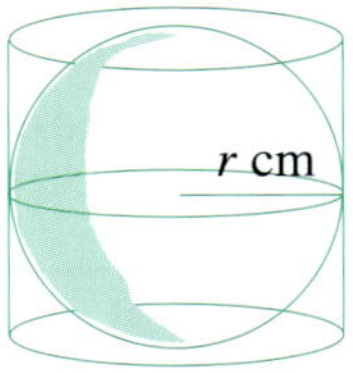

(a) Find, in terms of r, the
   i. height of the cylinder.
   ii. curved surface area of the cylinder.
(b) What do you conclude about the surface area of the sphere and that of the curved surface area of the cylinder?

[1 + 3 + 1 marks]
Total 46 marks

17.8 PROJECT SUGGESTIONS

How did Eratosthenes (~200 BC) measure the size of the Earth?

This is an experiment that you might try to repeat. You will need to contact a person who lives a good distance away, ideally on the same line of longitude. A student at another IB school, relative, friend etc. would be ideal. There is a need to coordinate the necessary measurements. Email has now made this a lot easier prospect than it was in 200 BC!

Most of the mathematics in this chapter referred to flat surfaces (2-dimensions).

How do these results generalise to 3-dimensional space?

Does it make any sense to talk of mathematics in four or more dimensions?

REVISION SET C - PAPER 1 & PAPER 2 STYLE QUESTIONS

1. (a) A is the point (–2, 3) and B is the point (4, –2). Find the length of [AB].
(b) Find the equation of the line that passes through the origin and is perpendicular to the line $x + 3y + 4 = 0$.

2. The line $\frac{x}{15} + \frac{y}{8} = 1$ crosses the x-axis at M and the y-axis at N.

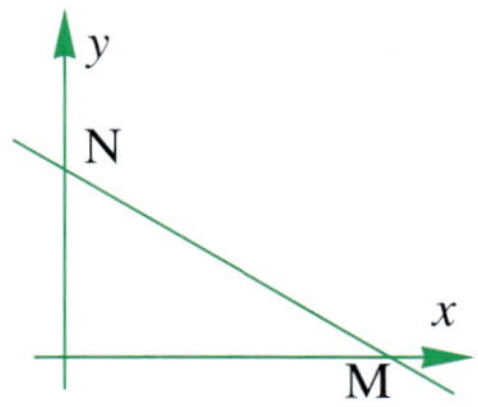

(a) Find the coordinates of i. M
ii. N

(b) Find distance from M to N.

3. (a) Find i. $\{x|x^2 - 11x + 24 = 0\}$.
ii. $\{y|y^2 - 10y + 24 = 0\}$.

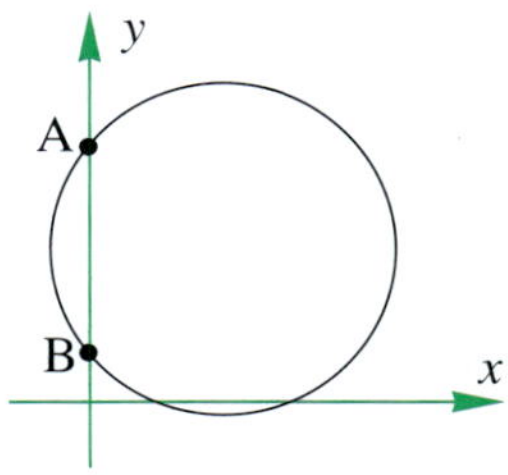

(b) The circle shown alongside is defined by the relation
$$x^2 + y^2 - 11x - 10y + 24 = 0.$$
The circle cuts the y-axis at A and B. Find the distance from A to B.

4. A bush walker walks due east for 5.0 km from her campsite. She then walks 12 km due north. In what direction and how far would she need to walk to return to her campsite?

5. Consider the cuboid shown where AB = 4 cm, BC = 6 cm and BF = 5 cm. Give answers to one decimal place.

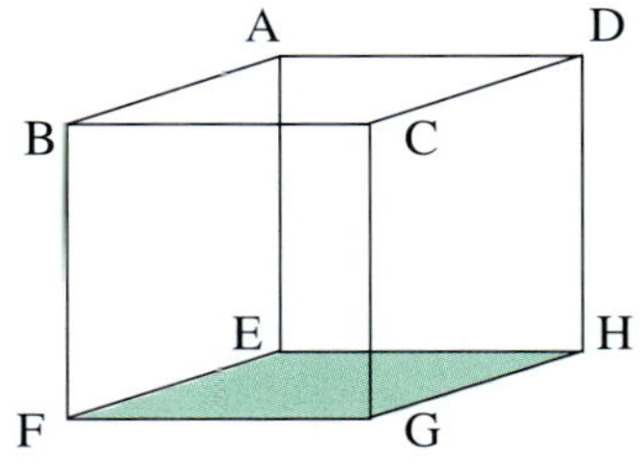

(a) Find the length of
i. [EG].
ii. [AG].
(b) Find the angle
i. $A\hat{B}D$.
ii. $A\hat{B}H$.

6. The points A(–2, –1) and B(3, 2) are shown on the set of axes alongside.

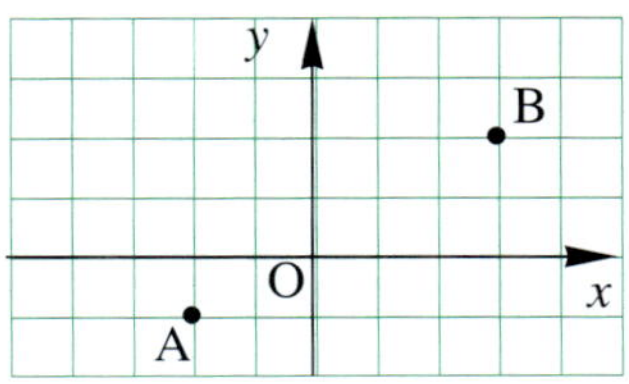

(a) Find
i. the mid-point of [AB].
ii. the length, AB.
(b) Find
i. the equation of the straight line through A and B.
ii. the equation of the line perpendicular to [AB] and passing through O.

7. The sides of a triangle have lengths 7cm, 8cm and 13cm respectively. Find
(a) the size of its largest angle,
(b) the area of the triangle.

8. A photographer's tripod stands on a horizontal floor, with its feet A, B and C at the vertices of an equilateral triangle of side length 0.6 metres. Each leg is 1.2 metres long.

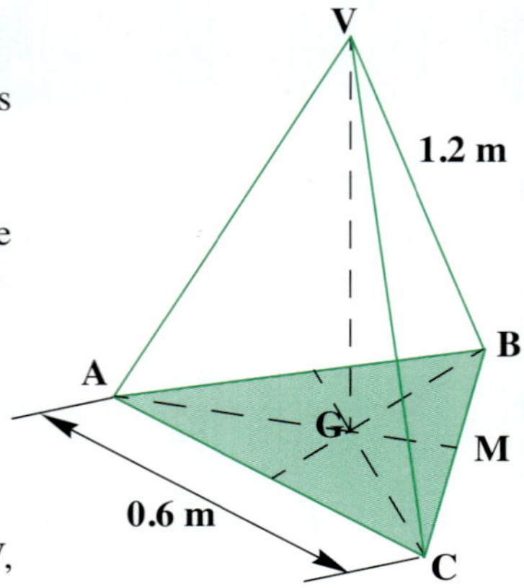

The line [AM] meets the side [BC] at right angles where M is the mid-point of [BC]. The point G is the centroid of triangle ABC, that is, $AG = \frac{2}{3}AM$.

(a) Calculate i. the length of [AM]
ii. the length of [AG]

(b) Calculate the height above the floor of the top, V, of the tripod.

(c) Calculate the angle which a leg makes with the floor.

9. (a) The distance from the point P(1, 1) to the point Q(x, 4) is 5 units. Find x.

(b) Find the gradient of the line perpendicular to the line $4x + 2y = 10$.

(c) Two lines, l_1 and l_2 intersect at the point (2, 8). The gradient of l_1 is twice that of l_2 and line l_1 passes through the (0,0). Find where the line l_2 intersects the y-axis.

10. A building situated on an open flat plain is in the form of a large hemisphere. From a point on the plain which is 80 metres from the nearest point at the base of the building, the angle of elevation of the apparent top of the building is 12°. Find the height of the building in metres (correct to 1 decinal place).

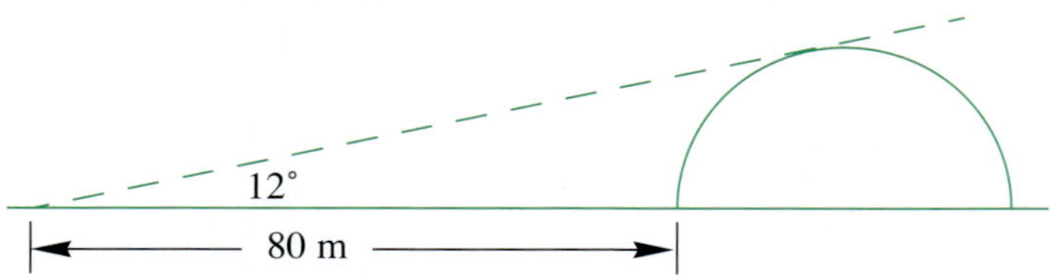

11. The points P(4, 0) and Q(6, 2) are shown on the set of axes alongside.

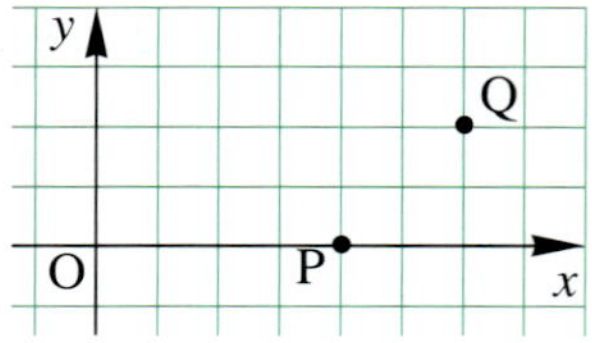

(a) Find the gradient of the line joining P and Q.

(b) R is a point such that OPQR is a parallelogram. Find the coordinates of R.

(c) Find the length of [OQ].

(d) Find the equation of the line which passes through P and Q.

(e) Find the point of intersection of the lines (OQ) and (RP).

12. A rectangular pyramid VABCD has a base measuring 24 cm. by 18 cm. Each slant edge measures 17 cm. Find;

(a) the height of the pyramid,

(b) the angle which a slant edge makes with the base.

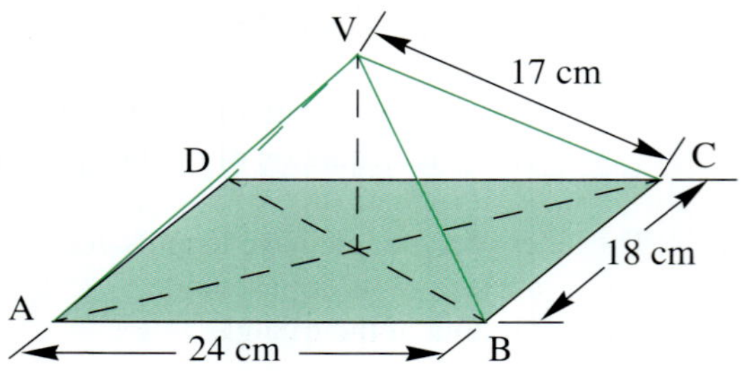

13. A, P and Q are three points in that order in a straight line. The bearing of P and Q from A is 310°. From another point B, 1 kilometre due west of A, the bearing of P is 053°; and Q is due north of B.

(a) Copy and complete the diagram below, clearly showing the position of P and Q and all relevant angles and distances.

(b) Calculate, to the nearest metre,

i. the distance from point P from point B?

ii. distance from P to Q,

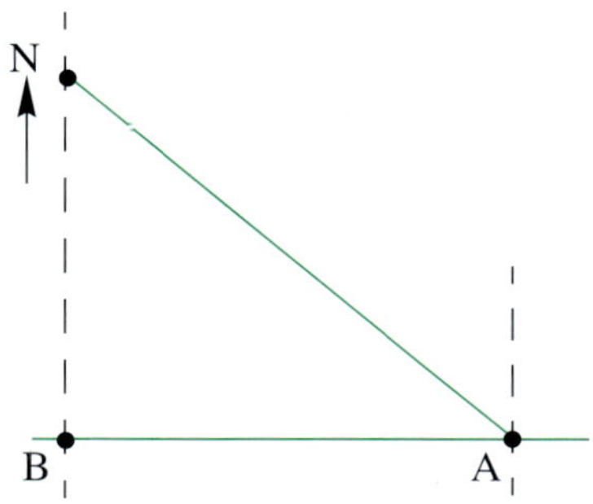

14. A girl walking due east along a straight horizontal road observes a church spire on a true bearing of 076°. After walking 1500 metres further she observes the spire on a true bearing of 067°.

(a) Draw a diagram for this situation.

(b) How far is the church from the road (to the nearest metre)?

15. A monument consists of a truncated square pyramid ABCD, EFGH, of height 2.2 m, surmounted by a square pyramid V ABCD, of height 0.2 m, as shown in the diagram.

The square ABCD has edge 0.5 metres and the square EFGH has edge 0.8 metres.

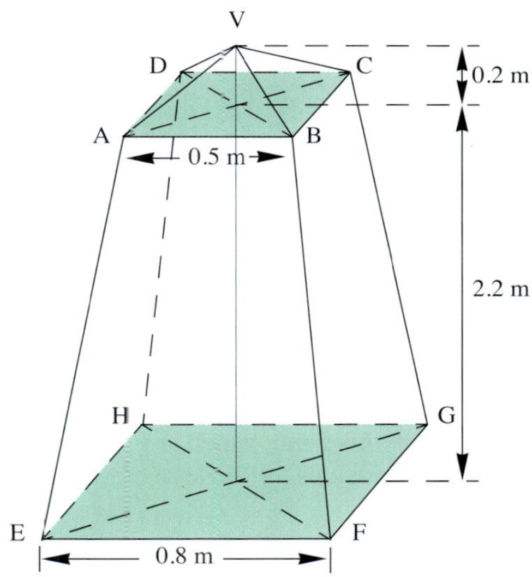

(a) Find the inclination of a sloping triangular face to the base.

(b) Find the surface area of one of the sloping triangular faces.

(c) Find the total surface area of all the sloping faces of the monument.

(d) The monument needs to be rendered with two coats of a cement mix. The cost per square metre for this cement mix is set at $32.00. How much will it cost to render the monument if labour costs will total $300.00?

16. The cross section of a rough, uncut diamond is shown on the set of axes below.

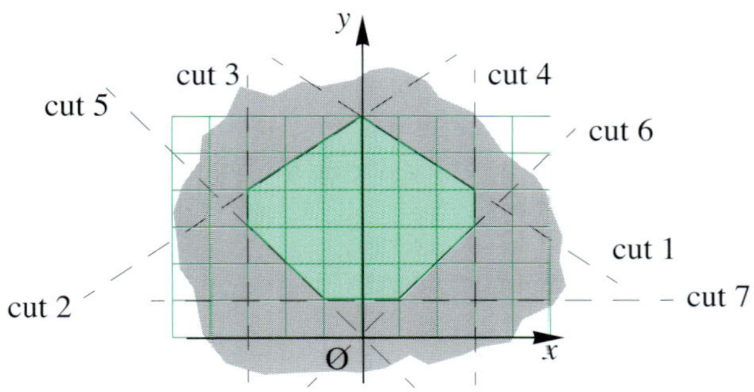

scale: 1 unit = 2 mm

cut	1	2	3	4	5	6	7
1		A		B			
2	A		C				
3		C			D		
4	B					E	
5			D				F
6				E			G
7					F	G	

Also shown are the cuts that a master craftman is intending for the diamond. This will result in the diamond having the shape of a prism of uniform cross-section. The table identifies the points of intersection (on the diamond) along the different cuts.

(a) Using unit measures, state the coordinates of the 7 points, A, B, C, D, E, F and G.
(b) i. Find the equation of the straight line representing cut 6.
    ii. Hence, state the equation of the straight line representing cut 5.
(c) Find the equation of the straight line representing cut 1.
(d) Find the distance from G to C.
(e) Find the angles i. $D\hat{F}G$ ii. $C\hat{A}B$
(f) The diamond will end up having a length of 8 mm. Find its volume in mm^3.

17. A block of land has measurements as shown in the diagram. Calculate its area in hectares, correct to 3 decimal places. (1 hectare = 10000 square metres)

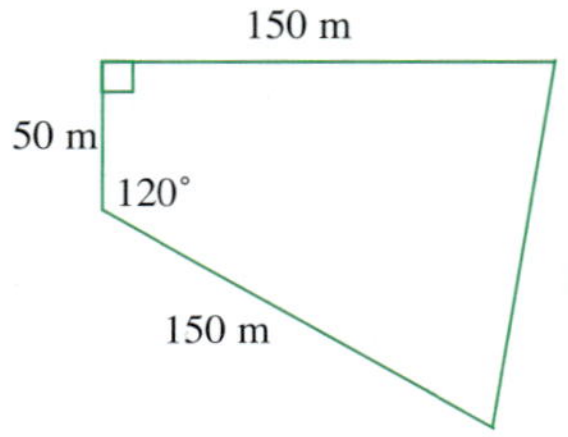

18. From an observation point A, the true bearing of a landmark C is 051° and the bearing of another landmark D is 284°. From another observation point B, 1000 metres due north of A, the bearing of C is 115° and the bearing of D is 224°.
Find the distance between C and D, correct to the nearest metre.

19. A surveyor measures the angle of elevation of the tope of a mountain from a point at sea level as 20°. She then travels 1000 m along a road which slopes uniformily uphill towards the mountain. From this point, which is 100 metres above sea level, she measures the angle of elevation as 23°.

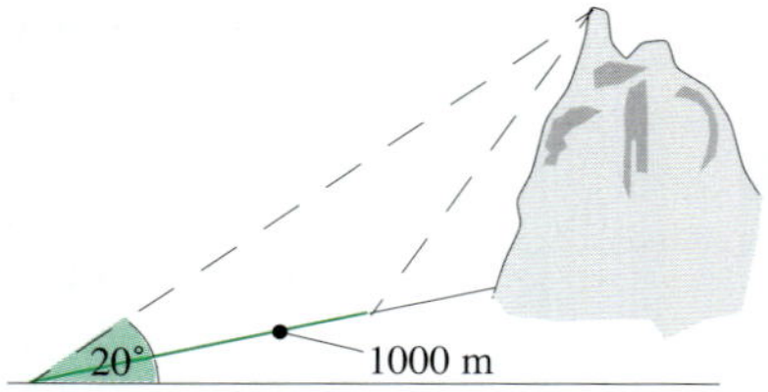

(a) Copy and complete the diagram.
(b) Find the height of the mountain above sea level, correct to the nearest metre.

20. (a) An observer on the balcony at the top of a lighthouse, 130 metres above sea level, observes a ship at sea. The ship is due south of the lighthouse and the angle of depression of the ship **from** the observer is $6°20'$.
Find, to the nearest metre, the horizontal distance of the ship from the lighthouse.

(b) A quarter of an hour later, the observer again notes the position of the ship, which has been travelling on a constant course. The ship is now south east of the lighthouse and the angle of depression of the ship is now $3°30'$.
Find how many kilometres the ship has travelled during the quarter hour, giving your answer correct to two decimal places.

21. V, ABC is a tetrahedron. The equilateral triangular base ABC has edges each of length 6 cm. The slant edges are each of length 8 cm.
Find
(a) the length of the altitude of the tetrahedron that passes through V.
(b) the magnitude of angle VAB.

22. (a) On a set of axes, draw the line joing the points A(0, –3) and B(4, 5).
(b) Calculate the gradient of this line.
(c) Determine the equation of this line.
(d) Find the equation of the line which is perpendicular to [AB] and which cuts the x-axis at the same point as [AB].

23. A right pyramid on a square base of side measure 12 cm has slant edges of 16 cm. Find

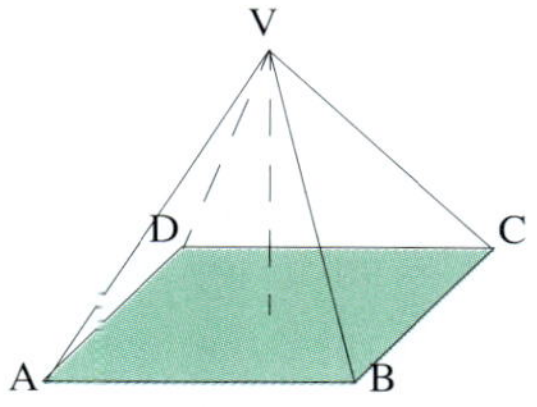

(a) the length of [AC]
(b) the measure of the height of the pyramid.
(c) the measure of the angle between
i. a sloping face and the base.
ii. the lines [VD] and [VB]

Give all answers to 2 decimal places.

24. Two planes, one of which was flying at 300 km per hour and the other at 450 km per hour, left an airport at the same time. Three hours later they were 1200 km apart.
What was the measure of the angle between their flight paths (assuming that they both travelled in a straight line).

25. A has the coordinates (–5, –6) and B has the coordinates (3, 6). The line (AB) cuts the x-axis at P and the y-axis at Q. Find:
(a) the length of [AB].
(b) the gradient if the line (AB).
(c) the angle which (AB) makes with the x-axis.
(d) the equation of the line (AB).
(e) the coordinates i. P ii. Q
(f) the equation of the line through the origin parallel to (AB).

26. A and B are the points (1, –3) and (3, –1) respectively.
(a) Show the points A and B on a set of axes.
(b) i. Find the equation of the perpendicular bisector of [AB].
ii. Draw this perpendicular bisector on the set of axes from part (a).
(c) Find the coordinates of the point on the line $2x + y - 5 = 0$ which is equidistant from A and B.

27. (a) On the same set of axes, draw
i. the straight line with equation $y = 3x - 2$.
ii. the parabola with equation $y = x^2 - 4x + 4$.
(b) The straight line cuts the parabola at the points A and B, find the length of [AB].

28. (a) Find the magnitude of the largest angle of a triangle whose sides have lengths in the ratio 3 : 5 : 7 .

(b) From two points A and B, 100m apart on horizontal ground, the angles of elevation of the top of a vertical tower are respectively 45° and 30°.
If A is due east of the tower, and B is due south of the tower, calculate the height of the tower.

29. (a) The storage tanks of a large oil company are cylindrical in shape.

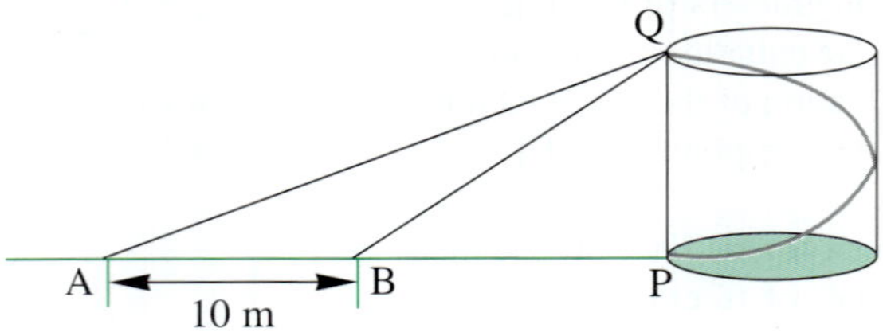

From a point A on the horizontal ground near one such tank, the angle of elevation of its top is $36°52'$; from another point B, 10m closer to the tank, the angle of elevation is $63°26'$.
Calculate the height of the tank.

(b) Starting from a point P at the base of the tank, a ladder spirals up and around the tank exactly once, at a constant slope, reaching the top at a point Q directly above P.
Given that the ladder is exactly 37m long, calculate the outer circumference of the tank.

(c) If the inner circumference is 2m less than the outer circumference, find:
i. the thickness of the walls;
ii. the volume of oil able to be stored in the tank. (Use $\pi = \frac{22}{7}$)

30. A rectangular box has a base measuring 30 cm by 20 cm and a height of 15 cm as shown in the diagram below.

(a) Find the angle which the diagonal [AB] of the box makes with
i. the base;
ii. a longer edge of the box

(b) Point C is the mid-point of the edge as shown in the diagram.
Find the angle $A\hat{C}B$.

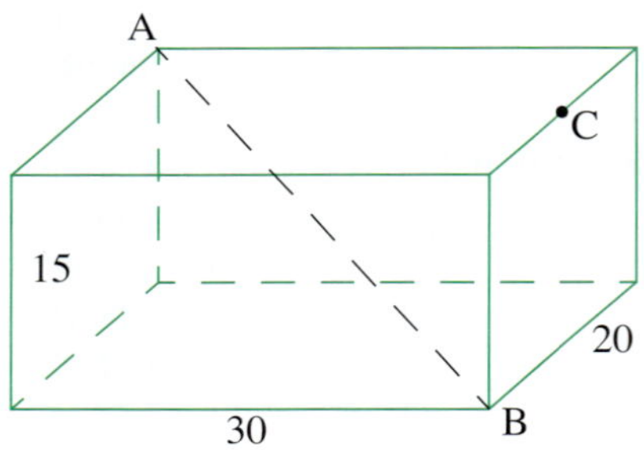

31. The set of axes alongside shows the curve with equation $y = a(x-2)^2 + 5$. The points P and Q have coordinates (0, 6) and (6, 9) respectively.

(a) Show that $a = \frac{1}{4}$.

(b) Show that the distance from A(x, 0) to
i. P is given by $\sqrt{x^2 + 36}$.
ii. Q is given by $\sqrt{x^2 - 12x + 117}$.

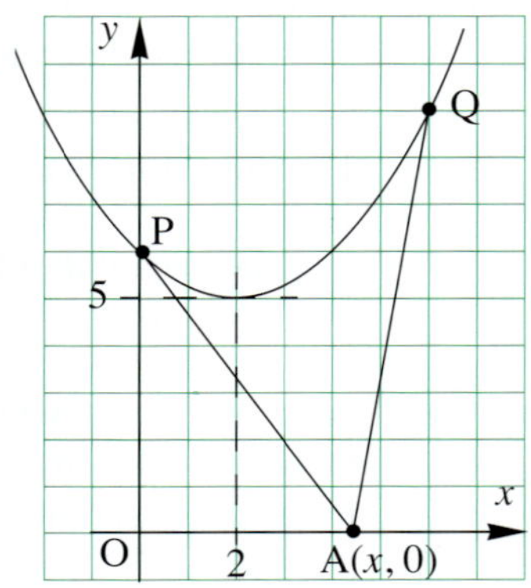

(c) Find the x–value of point A that make AP + AQ a minimum.
Give the value of x to 2 decimal places.

18.0 RUNWAYS

You are in a light aircraft about to land at Harry Hawker aerodrome in Australia. The pilot aims for the 'piano keys' during the descent.

Seconds later the aeroplane is over these 'piano keys' on runway '17R' - see the label - in preparation for touchdown.

The label '17R' is read "one seven right". These labels on runways are not arbitrary. They give information to pilots about the direction of the runway. 17 means that the runway has a bearing of 170° (magnetic) which is 10° east of south. "Right" means that there are two parallel runways in the same direction. The other is '17L' or "one seven left". The runways can be used in either direction, though not at the same time! At the other end, '17R' is labelled '35L' meaning that it runs on a bearing 350° (10° west of north and exactly the opposite of 170°) and is the left of the pair. These directions are rounded to the nearest 10° so that a runway in the direction 157° is called ''16' and a runway in the direction 38° is called '04'.

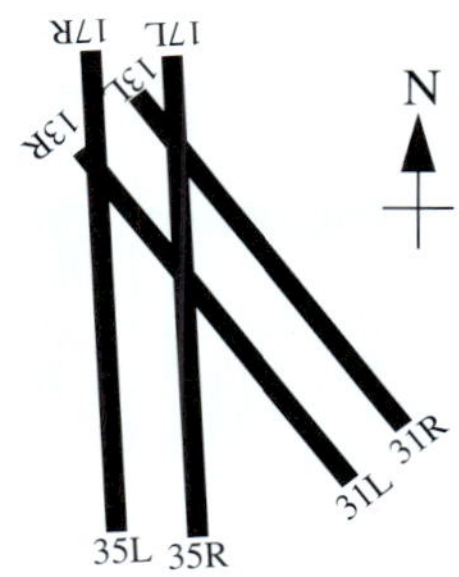

Most aerodromes have more than one runway. This is because it is much easier to land and take off into the wind. The 'duty runway' is chosen to be as much as possible into the wind. The runways at Harry Hawker are shown in the diagram.

These runway directions are not the same at aerodromes in other places. Paraburdoo has one runway labelled '06' at one end and '24' at the other. In which direction does this run?

Aerodromes in the UK often run east-west. How are these directions decided?

Clearly it is better if the runways point in the direction of the most common wind direction(s) at the location.

Thus, a first step in the construction of a new aerodrome should be to collect information about the most common wind directions that are found at the site.

Such data collection and the subsequent analysis is called STATISTICS and is the subject of this chapter.

You might consider a project looking at what runway directions you would choose if you were in charge of building a new aerodrome in your neighbourhood.

18.1 DESCRIBING DATA

18.1.1 DATA COLLECTION

Statistics is the science of getting 'facts from figures' and involves the collection, organisation and **analysis** of sets of observations called **data**. The data represents individual observations to which we can assign some numerical value or categorise in some way that identifies an **attribute**.

The set from which the data is collected is known as the **population**. The population can be finite or infinite, however, the data collected will be a subset from this population. Such a subset is called a **sample**. The process of gathering the data is known as **sampling**. Once we have our sample, we use characteristics of the sample to draw conclusions about the population. This is known as making **statistical inferences**. Note that statistical inference is quite different from simply collecting data and then displaying or summarising it as a 'diagram' – which is known as **descriptive statistics**.

The method that is used in collecting the sample affects the validity of the inferences that can or should be made. The aim then is to obtain a sample that is representative of the population.

This concept can be represented as follows:

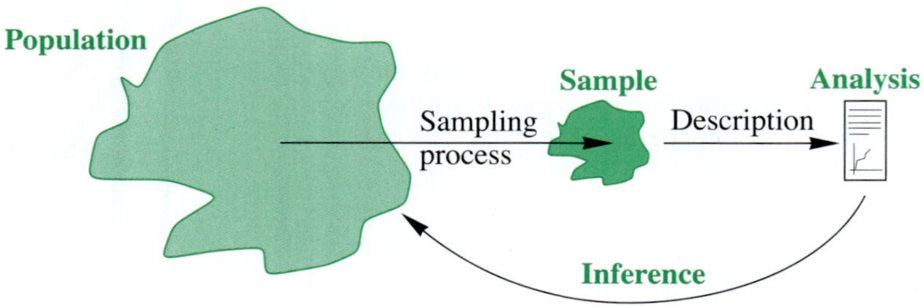

One approach to reduce bias in the sample we acquire is to use a **random sampling process**. By doing this we stand a better chance of obtaining samples that reflect the population as a whole.

18.1.2 TYPES OF DATA

Data can be classified as numerical or categorical –

1. **Numerical data**: These are made up of observations that are quantative and so have a numerical value associated with them.
For example, if the data is to represent heights, then the data would be collected as numerical values, e.g., 172 cm, 165 cm, etc.

2. **Categorical data**: These are made up of observations that are qualitative (which are sometimes also known as nominal data).
For example, if the data is to represent hair colour, then the data would be collected as qualitative data e.g., black, brown, blue, etc.

18.1.3 DISCRETE AND CONTINUOUS DATA

As a rule of thumb, **discrete data** are sets of data that can be **counted**, which distinguishes it from **continuous data** that are sets of data that are **measured**.

EXERCISES 18.1

1. Micro Inc. produces 14,500 electrical components each month. Of these, 2000 are randomly selected and tested.
The test reveals 42 defective components.
(a) What is i. the population size.
ii. the sample size.
(b) Give an estimate of the number of defectives produced during that month.

2. A salmon farm is attempting to determine the number of salmon in its reservoir. On Monday 300 salmon were caught, tagged and then released back into the reservoir. The following Monday 200 salmon were caught and of these 12 were tagged.
(a) Comment on the sampling procedure. Is the sample size large enough? Is there a bias involved?
(b) Estimate the the number of salmon in the reservoir.

3. A manufacturer wishes to investigate the quality of his product – a measuring instrument that is calibrated to within a 0.01 mm reading accuracy. The manufacturer randomly selects 120 of these instruments during one production cycle. She finds that 8 of the instruments are outside the accepted measuring range. One production cycle produces 1500 of these measuring instruments.
(a) What is i. the population size.
ii. the sample size.
(b) Give an estimate of the number of unacceptable instruments produced during a complete production cycle.
(c) In any given week there are 10 production cycles. How many unacceptable instruments can the manufacturer expect at the end of a week. Comment on your result.

4. Classify the following as categorical or numerical data.

(a) The winning margin in a soccer game.
(b) The eye colour of a person.
(c) The number of diagrams in a magazine.
(d) The breed of a cat.
(e) The fire-hazard levels during summer.

5. Classify the following as discrete or continuous data.

(a) The number of cats in a town.
(b) The length of a piece of string.
(c) The time to run 100 metres.
(d) The number of flaws in a piece of glass.
(e) The volume of water in a one litre bottle.

18.2 FREQUENCY DIAGRAMS

18.2.1 PRODUCING A FREQUENCY DIAGRAM

The following figures are the heights (in centimetres) of a group of students:

156	172	168	153	170	160	170	156	160	160	172	174
150	160	163	152	157	158	162	154	159	163	157	160
153	154	152	155	150	150	152	152	154	151	151	154

These figures alone do not give us much information about the heights of this group of people. One of the first things that is usually done in undertaking an analysis is to make a frequency table. In this case as there are a large number of different heights, it is a good idea to group the height data into the categories (or classes) 148-150, 151-153, 154-156, etc. before making a tally.

Heights	Tally	Frequency
148-150	√√√	3
151-153	√√√√√√√√	8
154-156	√√√√√√√	7
157-159	√√√√	4
160-162	√√√√√√	6
163-165	√√	2
166-168	√	1
169-171	√√	2
172-174	√√√	3

Each height is recorded in the appropriate row of the tally column. Finally, the frequency is the number of tally marks in each row. As a check, the total of the frequency column should equal the count of the number of data items. In this case there are 36 heights.

The choice of class interval in making such a frequency table is generally made so that there are about ten classes. This is not inevitably the case and it is true to say that this choice is an art rather than a science. The objective is to show the distribution of the data as clearly as possible. This can best be seen when the data is shown graphically. There are a number of ways in which this can be done. In the present example, we are dealing with heights. Since heights vary continuously, we would most usually use a histogram to display the distribution.

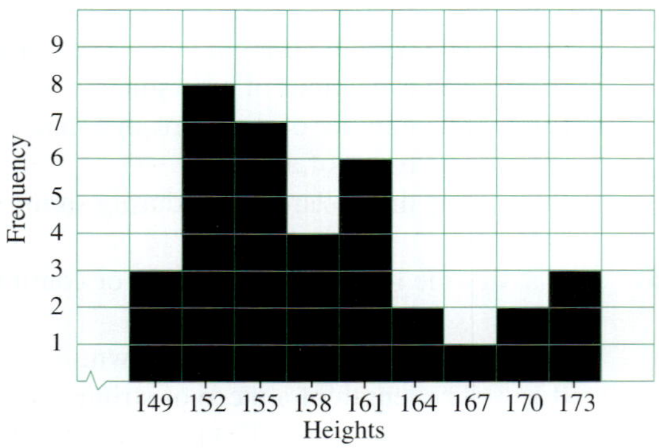

There are two details connected with the construction of histograms that you should not ignore. Firstly, as far as the horizontal scales are concerned, we are representing the continuous variable 'height'. The first class interval represents all the people with heights in the range 148 to 150cm. Since these have been rounded to the nearest whole centimetre, anyone with a height from 147.5 to 150.5cm, or [147.5,150.5), will have been placed in this class. Similarly, anyone with a height in the range [150.5,153.5) will be categorised in the class 151-153cm. If you want to label the divisions between the blocks on the histogram, technically these should be 147.5, 150.5 etc. Secondly, in a histogram, it is the area of the bars and not their height that represents the number of data items in each class. To be completely correct, we should give the area as a measure of the vertical scale. This definition allows us to draw histograms with class intervals of varying widths. This is sometimes done to smooth out the variations at the extremes of a distribution where the data is sparse. This aspect will not be considered in this chapter.

Once we have drawn a histogram, it should be possible to see any patterns that exist in the data. In this case, there is a big group of students with heights from about 150 to 160cm. There are also quite a few students with heights significantly larger than this and very few with heights below the main group. The distribution has a much larger 'tail' at the positive end than at the negative end and is said to be positively skewed. Such patterns can also be seen using other graphical devices. Such as a line graph:

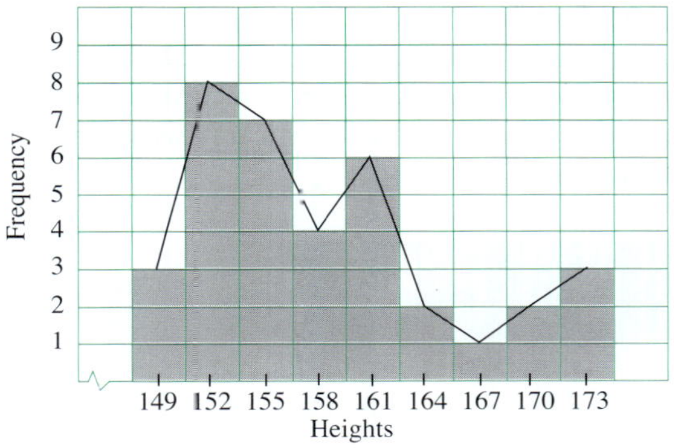

The same patterns are evident from this diagram as were seen from the histogram.

18.2.2 USING A GRAPHICS CALCULATOR

Data can be entered on the calculator either as separate figures from the original data or as a frequency table. In both cases, the data is entered as a list.

To enter the original data, press the **STAT** key and choose **EDIT** from the screen menus. If necessary, press 4 followed by the keys **L1** (2nd 1), **L2** (2nd 2) etc. **ENTER** to clear any previous lists. Next select **STAT EDIT**. The data can now be entered as a column under **L1**.

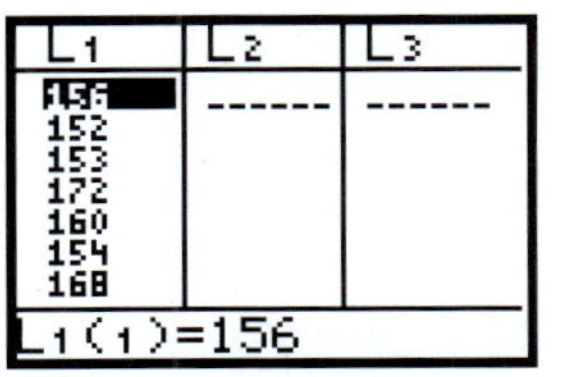

The data can now be displayed as a statistical graph. As with other types of graph, the appropriate window must be set. In the present case, the x data range should be set at 145 to 175. The **Xscl** setting determines the width of class interval that will be used.

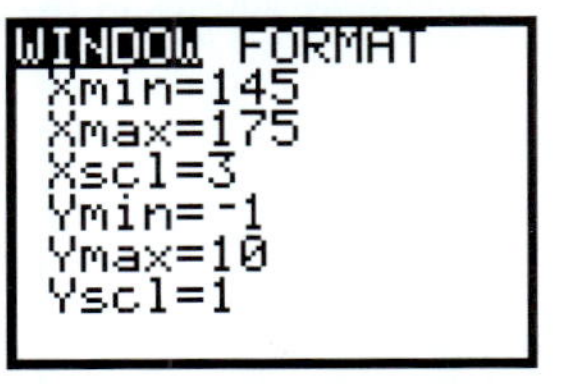

Next, any Cartesian graphs must be cleared. Press **Y=** and clear any rule that you see. The statistical plotting facility must now be activated. Press **2nd STAT PLOT**. Choose plot 1 and turn it on by using the arrows to the word **On** and then press **ENTER**. Also select the histogram symbol from the list of available plot types. We entered the data as **L1** so we must select this as the source of the data. Finally,

because each height was entered separately, the frequency must be defined as 1. Pressing **GRAPH** should now display the histogram. This should be similar to that produced earlier.

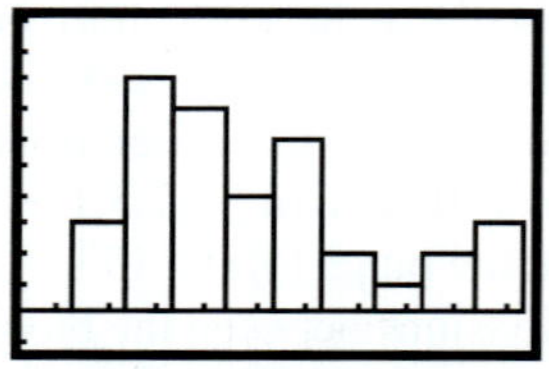

One advantage of using a calculator to produce such plots is that, once the data has been stored, the conditions of the plot can be varied rapidly. For example, if the **Xscl** is changed to 1, the class interval of the frequency table becomes 1 and the histogram is as shown.

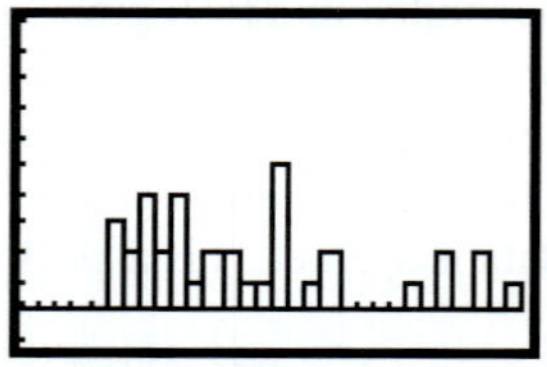

If the data is presented as a grouped list (frequency table), the x values should be entered as **L1** and the frequencies as **L2**. In the case of the height data, the mid-point of each class interval (149, 152, 155 etc.) should be entered in **L1**. The **STAT PLOT** instructions must also be set to record the fact that the frequencies are stored in **L2** before statistical plots or calculations will be successful.

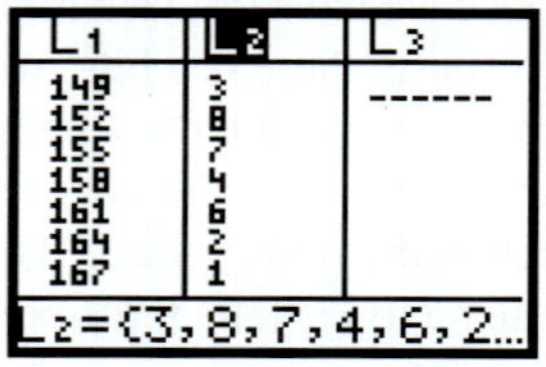

EXERCISES 18.2

1. The following figures are the weights (in grams) of a group of fish sampled from a reservoir:

226	233	233	244	224	235	238	244
222	239	233	243	221	230	237	240
225	230	236	242	222	235	237	240
220	235	238	243	222	232	232	242
229	231	234	241	228	237	237	245
229	231	237	244	225	236	235	240

(a) Construct a frequency table using the following class intervals: 218-220, 221-223, 224-226 etc. grams.

(b) Draw a histogram showing this distribution.

2. In a study of the weights of a sample of semi-precious gem-stones, the following results were obtained (grams):

1.33	1.59	1.82	1.92	1.46	1.57	1.82	2.06
1.59	1.70	1.81	2.02	1.24	1.53	1.69	2.01
1.57	1.62	1.61	1.93	1.11	1.90	1.79	1.91
1.19	1.53	1.90	1.90	1.17	1.97	1.92	2.06
1.41	1.64	1.83	1.90	1.11	1.81	1.83	1.90
1.15	1.68	1.82	1.98	1.39	1.54	1.92	2.04

(a) Construct a frequency table using the following class intervals: [1.1,1.2), [1.2,1.3), [1.3,1.4) etc. grams.

(b) Draw a histogram showing this distribution.

3. Make a frequency table and draw a histrogram for the following sets of data.

Set A:

21.1	28	26.9	31.9	23.7	28.8	27.9	31.3
21.5	26.8	27.4	31.2	21.4	29.9	29.4	31.5
20.4	25.1	25.8	33.6	23.7	25.6	29.1	30.3
21.5	28.2	28.2	31.3	22.4	25.7	25.1	30.3
21.9	29.1	28.7	30.1	21.8	27.8	29.1	34.3
22.5	25.2	25.5	32.9	22.3	29	27.2	33.3

Set B

7	6	5	70	9	9	25	72
7	7	4	72	8	9	28	73
9	9	9	72	6	7	27	71
7	7	9	70	6	8	27	73
8	5	26	73	5	6	26	70
9	9	28	73	5	8	26	71

Compare the two data sets.

18.3 STATISTICAL MEASURES 1

18.3.1 MEASURE OF CENTRAL TENDENCY

After using a graphical presentation of some sort to look at the general pattern of the data, we would usually calculate some representative 'statistics'. The aim of producing these is to reduce the amount of data to a small number of figures that represent the data as well as possible. In the case of the height data we have been studying, we have already observed that the heights group around the range 150-160 cm. This is sometimes known as a 'central tendency' and we have several ways in which we measure this:

18.3.2 MODE

This is the most frequent class of data. In the present case there were more students in the 151-153 cm class than any other so we would give this class as the mode. It is possible for some data to have more than one mode. We describe this as being bimodal, trimodal etc. The mode tends only to be used when there is no alternative such as when we are collecting data on the television stations that people like best.

18.3.3 MEAN

This is the measure commonly (and incorrectly) called average. Numeric data is added and the result is divided by the number of items of data that we have.

Notation: The notation used for the mean depends on whether or not we are claiming to have the mean of all (the **population**) or part (a **sample**) of the possible data set.

In the case of the students, we appear to have a small group of 36 selected from all the possible students in this age group and so we are looking at a sample. It is generally quite clear whether any set of data refers to a population (such as a **census**) or a sample (such as a **poll**).

The **population mean** is denoted by μ and a **sample mean** by $\bar{x}$.

For a data set x, with n items both means are calculated in the same way:

$$\text{Mean} = \frac{\sum x}{n} = \frac{1}{n}\sum x$$

The symbol Σ means 'add all the following'.

If the data is presented in the form of a frequency table in which each item of data x_i is present with a frequency of f_i, then the formula becomes:

$$\text{Mean} = \frac{\sum f_i x_i}{\sum f_i}$$

For the height data, we have two ways of approaching this calculation. One way is to return to the original data and add it all up.

The total is 5694. There are 36 measurements so: Mean $= \frac{5694}{36} = 158.16667$

Alternatively we can use the grouped data formula. There is a convenient way of doing this if we add an extra column to the orginal frequency table:

Heights	Mid-height	Frequency	$f \times h$
148-150	149	3	447
151-153	152	8	1216
154-156	155	7	1085
157-159	158	4	632
160-162	161	6	966
163-165	164	2	328
166-168	167	1	167
169-171	170	2	340
172-174	173	3	519
	Totals:	36	5700

From the table:

$$\sum f_i = 36 \text{ and } \sum f_i \times h_i = 5700 \text{ so Mean } = \frac{\sum f_i \times h_i}{\sum f_i} = \frac{5700}{36} = 158.33333 .$$

This method of calculating the mean will not necessarily give exactly the same answer as the mean calculated from the original data as we have made the assumption that all the students with heights in the range 148-150 cm had a height of 149 cm. This will not generally be a seriously inaccurate assumption as the students with heights below this figure (148 cm) will be balanced by those with heights above this (150 cm). In this case, the difference is quite small.

18.3.4 MEDIAN

The median is found by arranging all the data in order of size and selecting the middle item. For the heights data, there is an even number of figures and so there is not a middle number. In this situation, we take the mean of the middle two data items.

Order	1	2	3	4	5	6	7	8	9
Ht:	150	150	150	151	151	152	152	152	152
	10	11	12	13	14	15	16	17	18
	153	153	154	154	154	154	155	156	156
	19	20	21	22	23	24	25	26	27
	157	157	158	159	160	160	160	160	160
	28	29	30	31	32	33	34	35	36
	162	163	163	168	170	170	172	172	174

The middle heights are the 18th and 19th (156 and 157cm) so the median is 156.5cm.

It is usual to take the mean of the two numbers to give an answer to represent the median, however, there are a number of interpolations that can be used. For our purposes however, we will continue to use the mean of the two observations.

When there are $2n+1$ observation, i.e., there is an odd number of observations, the median corresponds to the $\frac{(2n+1)+1}{2}$ th observation (after they have been placed in order from lowest to highest).

e.g For the data set $\{2, 4, 12, 7, 9\}$ we first list the data from lowest to largest: 2, 4, 7, 9, 12. Here $n = 5$ and so the middle observation is the $\frac{5+1}{2} = 3$ rd observation. i.e., 7.

EXERCISES 18.3

1. For the data set of Exercise 18.2 Q.1., find the mode, mean and median weights.

2. For the data set of Exercise 18.2 Q.2., find the mode, mean and median weights.

3. For the data sets of Exercise 18.2 Q.3., find the mode, mean and median weights.

4. The following numbers represent the annual salaries of the emplyoyees of a small company.

\$20910	\$20110	\$20390	\$20170	\$20060	\$20350
\$21410	\$21130	\$21340	\$21360	\$21360	\$21410
\$20350	\$20990	\$20690	\$20760	\$20880	\$20960
\$21240	\$21060	\$21190	\$21400	\$76000	\$125000

(a) Find the mean salary.
(b) Find the median salary.
(c) Which of the two figures is the better representative measure of salary?

5. The selling prices for the properties in a suburb over June 2004 were:

\$191000	\$152000	\$152000	\$181000
\$180000	\$163000	\$169000	\$189000
\$184000	\$169000	\$167000	\$172000
\$190000	\$169000	\$159000	\$172000
\$202000	\$162000	\$160000	\$154000
\$181000	\$166000	\$163000	\$196000
\$201000	\$154000	\$166000	\$154000
\$178000	\$164000	\$157000	\$185000
\$177000	\$169000	\$157000	\$172000
\$195000	\$150000	\$163000	\$1150000
\$186000	\$166000	\$151000	\$1155000
\$185000	\$151000	\$168000	\$1200000

(a) Find the mean selling price.

(b) Find the median selling price.

(c) Which of the two figures is the better representative measure of selling price?

6. For the figures given below, calculate the mean from the original data.

5	16	15	17	9	16	19	15
6	17	10	16	8	13	13	19
7	16	18	18	8	18	19	18
6	17	19	16	7	13	17	19
9	14	17	19	9	16	17	19
8	18	16	15	8	18	16	15

(a) Use the frequency table method with class intervals 4-6, 7-9 etc. to calculate the mean of the data.

(b) Use the frequency table method with class intervals 1-5, 6-10 etc. to calculate the mean of the data.

18.4 STATISTICAL MEASURES 2

18.4.1 MEASURES OF SPREAD

So far we have only looked at ways of measuring the central tendency of a set of data. This is not necessarily the only feature of a data set that may be important. The following sets of data are test results obtained by a group of students in two tests in which the maximum mark was 20.

Test 1.

4	12	11	10	5	10	12	12	6	8	19	13	3
7	11	13	4	9	12	10	6	13	19	11	3	12
14	11	6	13	16	11	5	10	12	13	7	8	13
14	6	10	12	10	7	10	12	10				

Test 2.

9	8	10	10	8	9	10	11	8	8	11	10	9
8	11	10	9	8	10	11	8	9	11	10	9	8
11	11	9	9	11	10	8	9	11	10	8	9	11
11	8	8	11	10	8	9	10	10				

The means of the two data sets are fairly close to one another (Test 1, 10.1, Test 2, 9.5). However, there is a substantial difference between the two sets which can be seen from the frequency tables.

Test 1

Mark	3	4	5	6	7	8	9	10	11	12	13	14	15	16	17	18	19
Freq.	2	2	2	4	3	2	1	8	5	8	6	2	0	1	0	0	2

Test 2

Mark	3	4	5	6	7	8	9	10	11	12	13	14	15	16	17	18	19
Freq.	0	0	0	0	0	13	11	12	12	0	0	0	0	0	0	0	0

The marks for Test 1 are quite well spread out across the available scores whereas those for Test 2 are concentrated around 9, 10 & 11. This may be important as the usual reason for setting tests is to rank the students in order of their performance. Test 2 is less effective at this than Test 1 because the marks have a very small spread. In fact, when teachers and examiners set a test, they are more interested in getting a good spread of marks than they are in getting a particular value for the mean. By contrast, manufacturers of precision engineering products want a small spread on the dimensions of the articles that they make. Either way, it is necessary to have a way of calculating a numerical measure of the spread of data. The most commonly used measures are variance, standard deviation and interquartile range.

18.4.2 VARIANCE AND STANDARD DEVIATION

To calculate the variance of a set of data, the frequency table can be extended as follows:
Test 1

Mark (M)	Frequency	$M-\mu$	$f(M-\mu)^2$
3	2	–7.10	100.82
4	2	–6.10	74.42
5	2	–5.10	52.02
6	4	–4.10	67.24
7	3	–3.10	28.83
8	2	–2.10	8.82
9	1	–1.10	1.21
10	8	–0.10	0.08
11	5	0.90	4.05
12	8	1.90	28.88
13	6	2.90	50.46
14	2	3.90	30.42
15	0	4.90	0.00
16	1	5.90	34.81
17	0	6.90	0.00
18	0	7.90	0.00
19	2	8.90	158.42
		Total:	640.48

The third column in this table measures the amount that each mark **deviates from the mean** mark of 10.10. Because some of these marks are larger than the mean and some are smaller, some of these deviations are positive and some are negative. If we try to calculate an average deviation using these results, the negative deviations will cancel out the positive deviations. To correct this problem, one method is to square the deviations. Finally, this result is multiplied by the frequency to produce the results in the fourth column.

The last row is calculated: $2 \times (3 - 10.10)^2 = 2 \times 50.41 = 100.82$.

The total of the fourth column is divided by the number of data items (48) to obtain the variance of the marks: Variance $= \dfrac{640.48}{48} = 13.34$

The measure most commonly used is the square root of the variance (remember that we squared the deviations). A measure that is known as the **standard deviation** of the marks. In the previous case: Standard deviation $= \sqrt{13.34} = 3.65$

Repeating this calculation for the second set of marks:

Mark (M)	Frequency	$M - m$	$f(M-m)^2$
8	13	–1.48	28.475
9	11	–0.48	2.534
10	12	0.52	3.245
11	12	1.52	27.725
		Total:	61.979

Variance $= \dfrac{61.979}{48} = 1.291$

Standard deviation $= \sqrt{1.291} = 1.136$

This figure is about one third of the figure calculated for Test 1. This reflects the fact that Test 2 has not spread the students very well.

In summary, the variance and population standard deviation are calculated using the formulas:

Variance $= \dfrac{\sum f_i(x_i - \mu)^2}{\sum f_i}$ Population Standard Deviation $= \sqrt{\dfrac{\sum f_i(x_i - \mu)^2}{\sum f_i}}$

Of course, all of this type of mechanical work is very tedious, and seldom is it the case that we are required to work our way through these calculations. Statisticians (or anyone required to produce statistics such as these) would generally use a statistical package of some sort, where data could be easily transferred and then manipulated by any number of different commands. However, it is not necessary to have access to high powered statistical packages. Spreadsheets are widely used to produce the same results .

In many areas of this course we have seen (and will continue to see) how powerful the graphics calculator is. In the area of statistics, the TI-83 comes into its own. Once the data is entered as a list, it becomes a very powerful tool. And so, as we have said many a time before, make sure that you become familiar with all that your graphics calculator has to offer.

18.4.3 USING A GRAPHICS CALCULATOR

Standard deviation can be calculated directly by first entering the data as a list.

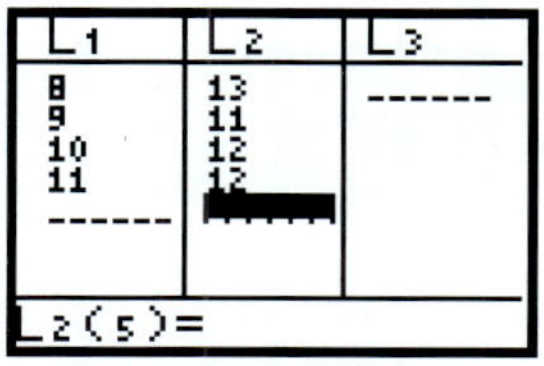

For Test 2, this is best done with the marks as list 1 and the frequencies as list 2.

Next, use **STAT CALC** to access the **1-VarStats** menu. Next, you must nominate the two lists that contain the data. List 1 is **2nd 1** and list 2 is **2nd 2**. The two list names must be separated by a comma.

Finally, press **ENTER**. This screen gives the mean, $(\bar{x})$, the sum of the data, Σx, and the sum of the squares of the data, Σx^2.

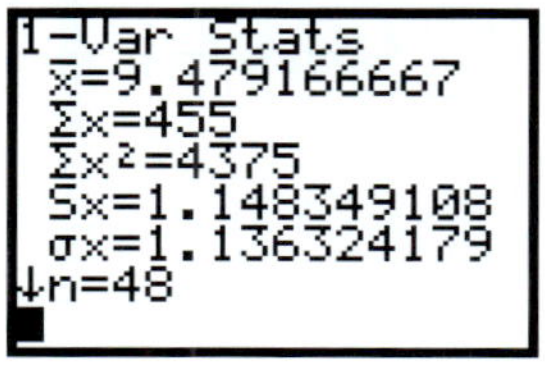

Sx is known as the **sample standard deviation**. This is the same as the standard deviation discussed above but with one less than the number of data items in the denominator (47 in this case).
σx is the **population standard deviation** discussed above.

Sample standard deviation? Population standard deviation? What's it all about? Unfortunately there are regional variations (as well as in textbooks) in the notation and the language that is used to define these terms.

When we refer to the *sample variance*, it suggests that we are finding the variance of a sample and by default, the sample is a subset of a population and so we are in fact finding an estimate of the population variance. This estimate is known as the *unbiased estimate* of the population variance.

The **unbiased estimate** of the population variance, σ^2, is given by $s_{n-1}^2 = \frac{1}{n-1}\sum_{i=1}^{n} f_i(x_i - \bar{x})^2$.

The standard deviation of the sample is given by the square root of s_{n-1}^2, i.e., $\sqrt{s_{n-1}^2}$, which corresponds to the value **Sx** that is produced by the TI-83.

The **variance of a population**, σ^2, is given by $\sigma^2 = \frac{1}{n}\sum_{i=1}^{n} f_i(x_i - \bar{x})^2$. The standard deviation then is $\sigma = \sqrt{\sigma^2}$.

To differentiate between division by n and division by $n-1$ we use s_n^2 for division by n and s_{n-1}^2 for division by $n-1$. Giving the relationship $s_{n-1}^2 = \frac{n}{n-1}s_n^2$.

Then, as the population variance, σ^2, is generally unknown, s_{n-1}^2 serves as an **estimate** of σ^2.

On the TI–83 we have that **Sx** $= s_{n-1}$ and **σx** $= s_n$.

It is therefore important that you are familiar with the notation that your calculator uses for sample standard deviation (unbiased) and population standard deviation.

EXERCISES 18.4

1. The weights (kg) of two samples of bagged sugar taken from a production line

Sample from machine A

1.95	1.94	2.02	1.94	2.07	1.95	2.02	2.06
2.09	2.09	1.94	2.01	2.07	2.05	2.04	1.91
1.91	2.02	1.92	1.99	1.98	2.09	2.05	2.05
1.99	1.97	1.97	1.95	1.93	2.03	2.02	1.90
1.93	1.91	2.00	2.03	1.94	2.00	2.02	2.02
2.03	1.96	2.04	1.92	1.95	1.97	1.97	2.07

Sample from machine B

1.77	2.07	1.97	2.22	1.60	1.96	1.95	2.23
1.79	1.98	2.07	2.32	1.66	1.96	2.05	2.32
1.80	1.96	2.06	1.80	1.93	1.91	1.93	2.25
1.63	1.97	2.08	2.32	1.94	1.93	1.94	2.22
1.76	2.06	1.91	2.39	1.98	2.06	2.02	2.23
1.75	1.95	1.96	1.80	1.95	2.09	2.08	2.29

(a) Find the mean weights of the bags in each sample.

(b) Use the formula $S_x = \sqrt{\frac{\sum f_i(x_i - \bar{x})^2}{\sum f_i - 1}}$ to calculate the sample standard deviations of each sample.

(c) Use the formula $\sigma_x = \sqrt{\frac{\sum f_i(x_i - \mu)^2}{\sum f_i}}$ to calculate the population standard deviations of each sample.

2. The following frequency table gives the numbers of passengers using a bus service over a week long period.

Passengers	0-4	5-9	10-14	15-19	20-24	25-29
Frequency	3	5	11	15	10	7

(a) Find the mean number of passengers carried per trip.
(b) Find the population standard deviation of the number of passengers carried per trip.

3. The number of matches per box in a sample of boxes taken from a packing machine was:

Matches	47	48	49	50	51	52
Frequency	3	6	11	19	12	9

Find the mean and sample standard deviation of the number of matches per box.

18.5 STATISTICAL MEASURES 3

18.5.1 QUARTILES

We have already seen that there is more than one way of measuring central tendency and the same is true of measures of spread. The median is, in some circumstances, a good measure of central tendency. Following on from this, we can define quartiles, the data items that are one quarter and three quarters of the way through a list.

The following data represent the number of employees absent from work over a nine day period:

$$2, 6, 5, 4, 7, 1, 0, 5, 2.$$

Firstly, we order the data to get $0, 1, 2, 2, 4, 5, 5, 6, 7.$
The median is the middle figure:

0 1 2 2 4 5 5 6 7

Median

Lower quartile (Q1) Upper quartile (Q3)

The median divides the distribution into an upper and lower group. The lower quartile is the middle figure of the lower group and the upper quartile is the middle figure of the upper group. As with finding the median, it may be necessary when dealing with a group with an even number of data items to take the mid point between two numbers. This is the case with the current data set. The lower quartile is 1.5 and the upper quartile is 5.5.

When dealing with large data sets or grouped data, there is an alternative method of finding the median and quartiles based on **cumulative frequency**.
This is calculated as follows:

These figures represent the numbers of customers in a small cinema:

Customers	Frequency	Cumulative Frequency
0-9	1	1
10-19	4	5
20-29	9	14
30-39	11	25
40-49	32	57
50-59	23	80
60-69	10	90
70-79	9	99
80-89	1	100

The cumulative frequency is calculated by 'accumulating' the frequencies as we move down the table. Thus the figure 25 in the shaded box means that on 25 occasions there were fewer than 40 customers.

Cumulative frequencies can now be used to produce a **cumulative frequency curve**:

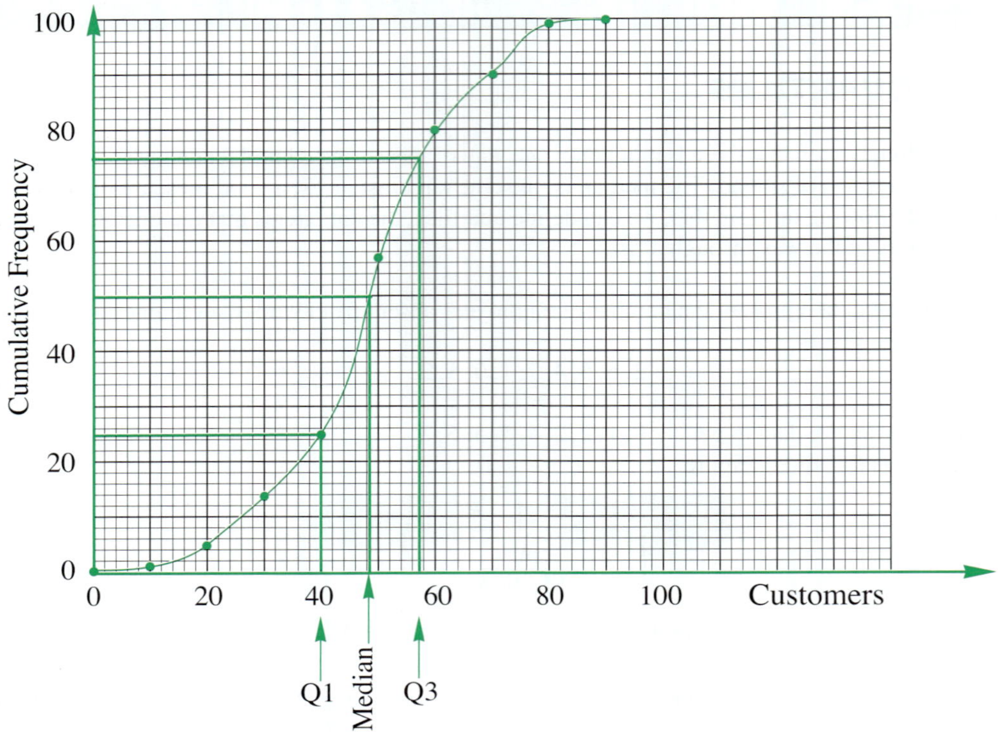

The cumulative frequency curve or **ogive** has effectively placed the data in order. This now enables us to read off estimates of the median and quartiles. The median is half way along the list of data. Since there are 100 figures, the median point is at 50. Technically this should be figure number 51, however, this method only produces an approximate figure and we seldom worry about this distinction. Reading across from 50 and down to the'customers' scale gives a figure of about 48 customers as the median. Similarly, the lower quartile can be found at a cumulative frequency of 25. Reading across from this figure to the graph and then to the horizontal axis gives a lower quartile of approximately 40 customers. Similarly, the upper quartile is about 57 customers.

The difference between the two quartiles is known as the **inter-quartile range**. In this case, the inter-quartile range is 57 – 40 = 17 customers. This is, like the standard deviation, a measure of the spread of the data. For these cinema attendance figures, the standard deviation is about 16 customers. It is not necessarily the case that these two measures of spread give similar answers. When comparing two data sets, choose which measure of spread you wish to use and use that measure throughout the analysis. Do not try to compare the inter-quartile range of one data set with the standard deviation of another.

In choosing which measure of spread to use, we generally use the quartiles and the median for a data set that contains a very few numbers that are very unusual. Such data are known as **outliers**.The data sets in Exercise 13.3 Questions 4 & 5 are examples of this type of data containing outliers. The standard deviation and mean are much more sensitive to outliers than are the median and inter-quartile range. Of course, you will need to look at a data set that has outliers and decide whether or not you want to minimise their effect on the representative statistics that you calculate.

18.5.2 BOX PLOT

Using a graphics calculator

The median and inter-quartile range of a data set can be found directly using a graphics calculator. The data can either be entered as a list or as a frequency table using two lists.

The data set: 5.7, 4.2, 7.9, 3.1, 9.4, 4.2, 7.7 & 8.0 can be entered as a single list.

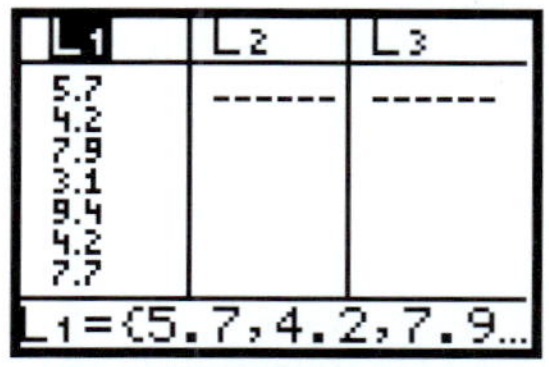

The statistics of this set can be calculated in the same way as the mean and standard deviation were calculated previously. The appropriate figures are found on the second screen (use the down arrow). This screen gives the **five figure summary** of the data: Minimum = 3.1, lower quartile = 4.2, median = 6.7, upper quartile = 7.95 and maximum = 9.4.

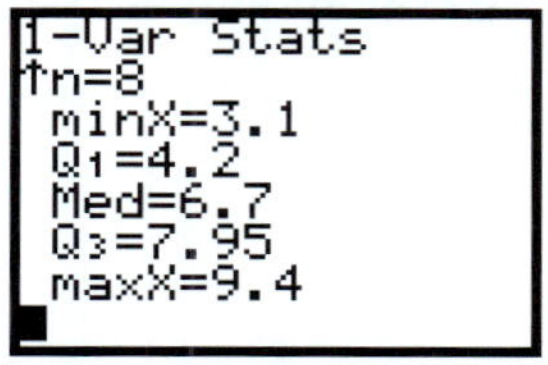

The five figure summary can be displayed using a graph known as a box-plot. This can be displayed on the TI-83 by choosing **2nd STAT PLOT**, turning on plot 1 and selecting the box-plot icon. An appropriate viewing window will also need to be set. The **TRACE** function can be used to identify the five figure summary. The diagram shows the median.

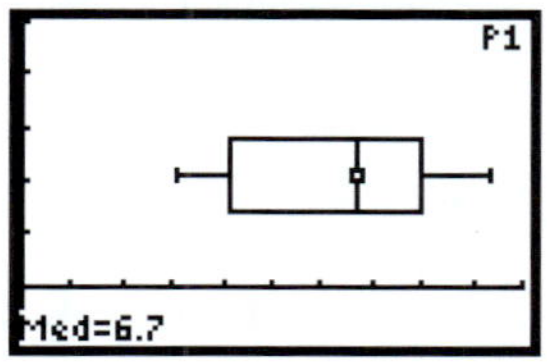

NB: The **first box-plot icon** produces a box plot showing any **outliers**.
The **second box-plot icon** produces a box plot with the **whiskers** running to the outliers.

To construct a **box-plot** (also known as a **box and whisker plot**), draw an appropriate horizontal scale. For this data set, a horizontal scale of 1 to 10 with marks every unit is appropriate. The box plot can now be drawn:

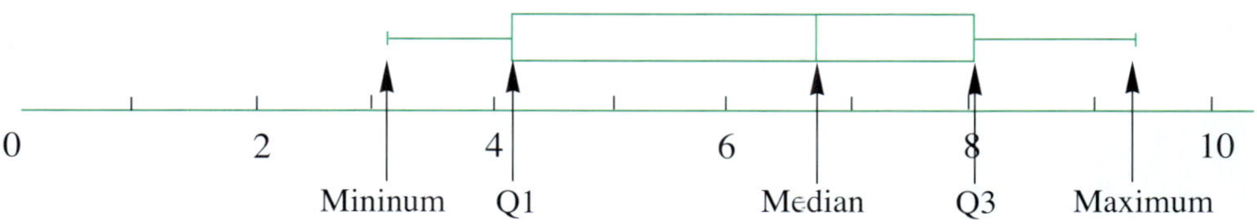

In the same way, the five figure summary for grouped data can be found by entering the data as list 1 and the frequencies as list 2. This frequency table gives the numbers of goals scored in 20 soccer matches:

Goals	0	1	2	3	4	5
Frequency	3	5	5	4	2	1

Enter the goals as **L1** and the frequencies as **L2**.

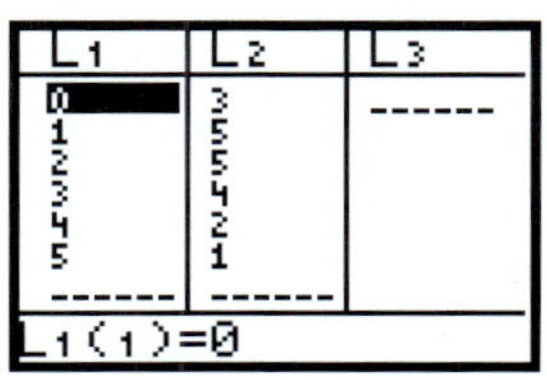

Next use the **STAT CALC** menu and identify the two lists that contain the data.

Press enter and use the down arrow to access the median part of the display.

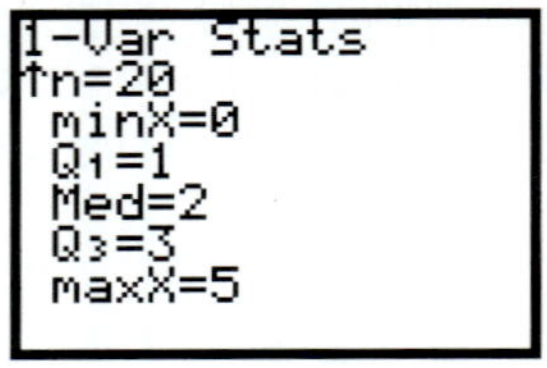

It is also worthwhile comparing a box-plot to their corresponding histogram.

Symmetric

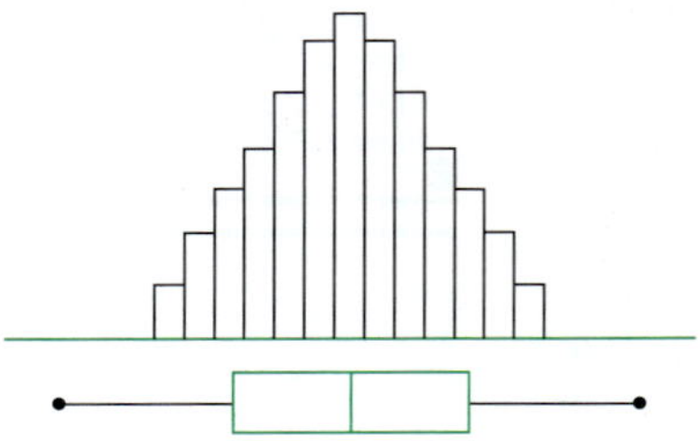

Positively skewed

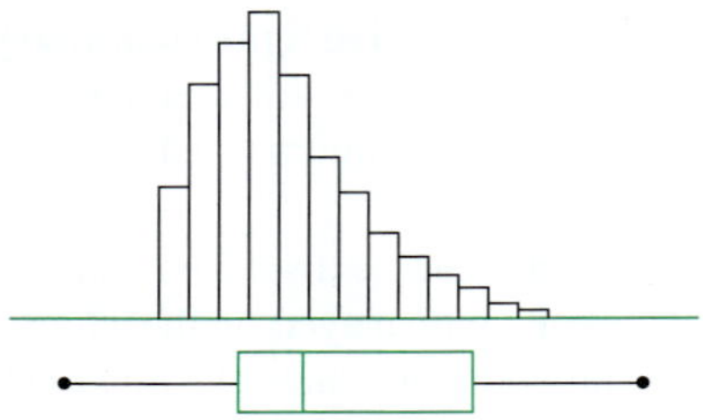

Negatively skewed

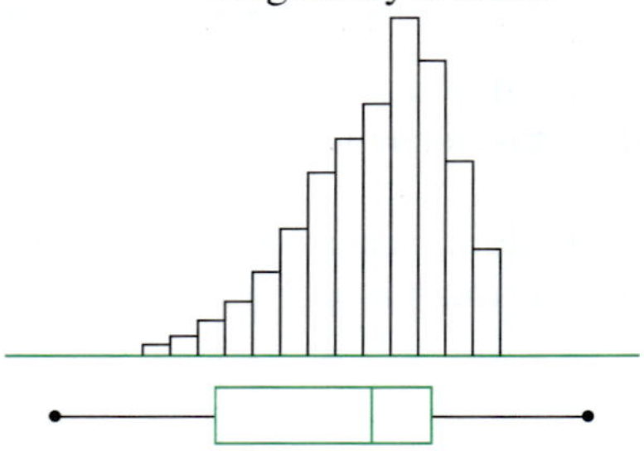

EXERCISES 18.5

1. Find the median, quartiles and inter-quartile range of these data sets:

(a) 5, 6, 2, 8, 9, 2, 7, 0, 5, 3
(b) 2.8, 4.9, 2.8, 0.9, 3.3, 5.8, 2.9, 3.7, 6.9, 3.3, 5.1
(c) 142, 167, 143, 126, 182, 199, 172, 164, 144, 163, 192, 101, 183, 153
(d) 0.02, 0.25, 1.72, 0.93, 0.99, 1.62, 0.67, 1.42, 1.75, 0.04, 1.12, 1.93
(e) 1200, 2046, 5035, 4512, 7242, 6252, 5252, 8352, 6242, 1232

2. Find the median, quartiles and inter-quartile range of these grouped data sets:
(a)

x	0	1	2	3	4	5
Frequency	1	3	6	6	7	1

(b)

x	10	11	12	13	14	15
Frequency	12	45	56	78	42	16

(c)

x	1.0	1.5	2.0	2.5	3.0	3.5
Frequency	2	4	9	9	2	1

(d)

x	10	20	30	40	50	60
Frequency	4	8	15	19	20	5

(e)

x	0	5	10	15	20	25
Frequency	0	3	0	6	7	5

3. The weekly expenses paid to a group of employees of a small company were

\$25	\$0	\$10	\$10
\$55	\$0	\$12	\$375
\$75	\$445	\$7	\$2

(a) Find the mean weekly expense.
(b) Find the population standard deviation of the expenses.
(c) Find the median weekly expense.
(d) Find the quartiles and the inter-quartile range of the expenses.
(e) Which of these statistics are the best representatives of the expenses data?

4. The table shows the numbers of cars per week sold by a dealership over a year.

Cars sold	0	1	2	3	4	5
Number of weeks	2	13	15	12	7	3

(a) Find the mean weekly sales.
(b) Find the population standard deviation of the sales.
(c) Find the median weekly sales.
(d) Find the quartiles and the inter-quartile range of the sales.

5. The table shows the weekly turnover of a small shop over a period during Spring and Summer.

Sales (\$)	\$0-\$99	\$100-\$199	\$200-\$299	\$300-\$399
Number of weeks	2	9	15	7

(a) Find the mean weekly sales.
(b) Find the population standard deviation of the sales.
(c) Construct a cumulative frequency table and draw the cumulative frequency curve.
(d) Find the median weekly sales from your graph.
(e) Find the quartiles and the inter-quartile range of the sales from your graph.

6. Plot the cumulative frequency curves for these data and hence estimate the median, quartiles and inter-quartile range of the data.

x	0-4	5-9	10-14	15-19	20-24	25-29
Frequency	2	5	11	9	7	2

18.6 STATISTICAL MEASURES 4

18.6.1 STEM AND LEAF DIAGRAM

Another format for displaying data is known as a stem and leaf diagram – or a stem plot for short. This form of data display is a sort of compromise between a histogram and a list of values. This is because the diagram will still display all of the recorded data, but it will do so in such a way that it will resemble a histogram. This form of display is best suited to data consisting of up to 50 observations (or so).

Such a diagram is constructed by 'breaking up' the numbers in the data set into two parts – the stem and the leaf.

For example the following numbers (from different data sets) would be broken up as follows:

25	Stem = 2, leaf = 5
35.2	Stem = 35, leaf = 2
1.34	Stem = 1.3, leaf = 4

The break-up 'rule' will always be indicated by a key beside the display. Also, notice that the leaf is always made up of a single digit.

An example of a stem plot is shown alongside.

The key informs us that the data reads as follows

11 12 21 22 31 32 35 36
42 42 47 48 49 50 51 58
62 64 70

```
1 | 1 2
2 | 1 2
3 | 1 2 5 6
4 | 2 2 7 8 9
5 | 0 1 8
6 | 2 4          key: 3|1 = 31 fish
7 | 0
```

There are many advantages of having a stem plot – we can calculate the mean, mode, median and so on using the actual values and so there is no loss of accuracy.

For example, to find the median of the data set displayed by the stem plot above, we determine the middle value by ordering the data (which, in this case are already ordered) and then selecting the middle value. There are 19 observations and so the middle value is the 10th observation:

```
1 | 1 2
2 | 1 2
3 | 1 2 5 6
4 | 2 (2) 7 8 9 ——— 10th observation, therefore, median = 2
5 | 0 1 8
6 | 2 4          key: 3|1 = 31 fish
7 | 0
```

Similarly, we can determine the interquartile range, the mean the mode and so forth.

Of course, not only do you need to be able to read a stem and leaf diagram, you must also be able to construct a stem and leaf diagram.

Construct a stem and leaf diagram for the following data:

195	194	202	194	207	195	202	206
209	209	194	201	207	205	204	191
191	202	192	199	198	209	205	205
199	197	197	195	193	203	202	190
193	191	200	203	194	200	202	202
203	196	204	192	195	197	197	204

Step 1: Make a decision on the stem:
In this instance we use the key 19 | 2 = 192

Step 2: Construct the stem plot without regard to order:

19	5 4 4 5 9 1 1 2 9 8 9 7 7 5 3 0 3 1 4 6 2 5 7 7
20	2 7 2 6 9 9 1 7 5 4 2 9 5 5 3 2 0 3 0 2 2 3 4 4

Step 3: Place the numbers in order:

19	0 1 1 1 2 2 3 3 4 4 4 5 5 5 5 6 7 7 7 7 8 9 9 9
20	0 0 1 2 2 2 2 2 2 3 3 3 4 4 4 5 5 5 6 7 7 7 9 9 9

key: 19 | 2 = 192

Now, although we have constructed the stem plot, we can improve it by using a different stem (i.e., a different class size). We do this by using class sizes of 5 rather than 10. This is done by writing the stems as 19, 19*, 20, 20*:

19	0 1 1 1 2 2 3 3 4 4 4
19*	5 5 5 5 6 7 7 7 7 8 9 9 9
20	0 0 1 2 2 2 2 2 2 3 3 3 4 4 4
20*	5 5 5 6 7 7 9 9 9

key: 19 | 2 = 192

This is an improvement on the previous stem plot.

EXERCISES 18.6

1. In each of the following stem plots, write down the actual data displayed

(a)

0	1 1 2 3
1	0 1 1 1 4 5
2	4 4 5 6
3	3 5
4	6

key 3 | 5 = 35

(b)

60	1 2 4
60*	5 6 7 8 9
70	2 4 4
70*	3

key 60 | 6 = 66

(c)

10	2 3 5 6
20	1 3 4 8 9 0
30	2 4 5
40	1 2
50	0

key 20 | 1 = 201

2. For each of the stem plots in Q. 1
(a) find the median
(b) draw the corresponding box-plot

3. Construct a stem and leaf diagram for the set of data

185	230	205	215	217	206	192	187	207	245	205	181
216	227	239	214	242	248	193	222	217	219	204	234
227	236	234	217	186	236						

4. (a) Construct a stem and leaf diagram for the set of data

1.77	2.07	1.97	2.22	1.60	1.96	1.95	2.23
1.79	1.98	2.07	2.32	1.66	1.96	2.05	2.32
1.80	1.96	2.06	1.80	1.93	1.91	1.93	2.25
1.63	1.97	2.08	2.32	1.94	1.93	1.94	2.22
1.76	2.06	1.91	2.39	1.98	2.06	2.02	2.23
1.75	1.95	1.96	1.80	1.95	2.09	2.08	2.29

(b) Describe the shape of the distribution.

(c) Find

i. the median

ii. interquartile range

5. The times for young atheletes running the 200 m race was recorded as follows

21.1	28	26.9	31.9	23.7	28.8	27.9	31.3
21.5	26.8	27.4	31.2	21.4	29.9	29.4	31.5
20.4	25.1	25.8	33.6	23.7	25.6	29.1	30.3
21.5	28.2	28.2	31.3	22.4	25.7	25.1	30.3
21.9	29.1	28.7	30.1	21.8	27.8	29.1	34.3
22.5	25.2	25.5	32.9	22.3	29	27.2	33.3

(a) Construct a stem and leaf diagram for the set of data

(b) Use you stem plot to identify the median.

18.7 MISCELLANEOUS EXERCISES

1. Identify the sample and the population in each of these situations.

(a) The University of Arkansas tests a new AIDS drug on 100 randomly chosen patients suffering from the disease. After a year of treatment, 42 patients had shown positive signs of recovery and 10 patients had died. The remainder appeared unchanged by the treatment.

(b) The Australian Government surveys 1000 working aged people in NSW to determine unemployment figures in that state.

(c) John asks all members of his IB higher maths class what theme they would like for the Nappa Valley High School senior dance.

2. Classify each of the following as discrete or continuous variables:

(a) The grade received by a student.
(b) The number of people in a passing car.
(c) The playing time of a movie.
(d) Cups of coffee drunk by your maths teacher in a day.
(e) Rainfall in Bali in a month.
(f) The length of a king cobra.
(g) The mass of silica in a sample of 1 kg of earth.

3. Two tetrahedral dice are rolled 60 times, and the sum of their scores recorded:

2, 3, 4, 5, 6, 7, 8, 4, 4, 6, 3, 5, 5, 6, 5, 8, 5, 5,4, 4, 2, 7, 6, 4, 5, 5, 3, 8, 7, 5, 5, 4, 4, 3, 7, 6, 6, 5, 5, 5, 6, 4, 5, 6, 3, 4, 5, 6, 3, 4, 5, 6, 2, 3, 4, 5, 6, 7, 8, 7

(a) Construct a frequency table from this data and draw the histogram.
(b) Complete the cumulative frequency distribution and draw a cumulative frequency graph

4. Choose suitable classes to group the data in each of these situations:

(a) About 15 classes are to be used to group 300 scores which range from 207 to 592.
(b) About 18 classes are to be used to group 900 scores which range from 112 to 418.
(c) SAT scores from 500 students are to be grouped. The lowest score is 458 and the highest score is 784. Use approximately 15 classes.

5. The weights of 50 year 9 students are measured to the nearest kg and recorded. Construct a frequency table using class intervals of 5kg and draw the histogram. Complete the cumulative frequency distribution and draw a cumulative frequency graph.

41	54	37	55	52	60	45	56	56	47
54	51	64	53	64	57	65	40	73	53
57	46	76	56	46	59	55	63	48	43
47	72	41	51	67	44	53	63	58	63
50	49	56	57	48	55	55	53	63	35

6. A group of 100 IB students was given a maths test that was graded out of 20. The following is the distribution of the marks obtained:

mark	9	10	11	12	13	14	15	16	17	18	19
number of students	1	1	3	5	8	13	19	24	14	10	2

(a) Write down the mode.
(b) Draw a cumulative frequency graph
(c) Calculate the mean.
(d) Find the median.
(e) Find the upper and lower quartiles.
(f) Another group of 50 students had a mean mark of 17.16 on the same test. Calculate the mean of the entire group of 150 students.

7. A biologist measures the lengths of 60 mature fern fronds to the nearest centimeter. The results are summarized in the table below:

Frond Length (cm)	Frequency
10 – 14	2
15 – 19	6
20 – 24	8
25 – 29	10
30 – 34	15
35 – 39	9
40 – 44	6
45 – 49	4

(a) Write down the modal class.
(b) Draw a cumulative frequency graph
(c) Calculate the mean.
(d) Find the median.

8. For a mathematics project, Eun-Kee timed the length of 30 popular Korean songs to the nearest second. His raw data is presented below:

185 230 205 215 217 206 192 187 207 245 205 181
216 227 239 214 242 248 193 222 217 219 204 234
227 236 234 217 186 236

(a) Complete the frequency and cumulative frequency distribution table using class intervals of 10 seconds.
(b) Use your table to calculate the mean.
(c) Re-calculate the mean using the raw data.

9. The masses of a sample of new potatoes were measured to the nearest gram and are summarized below. Calculate the mean of the data.

Mass(g)	Frequency
10 – 19	2
20 – 29	14
30 – 39	21
40 – 49	73
50 – 59	42
60 – 69	13
70 – 79	9
80 – 89	4
90 – 99	2

10. Foodcity supermarket recorded the length of time, to the nearest minute, that a sample of 200 cars was parked in their car park. The results were:

Time (minutes)	Frequency
0 – 14	13
15 – 29	23
30 – 44	32
45 – 59	33
60 – 74	27
75 – 89	20
90 – 104	12
105 – 119	11
120 – 134	10
135 – 149	11
150 – 164	8

(a) Draw the cumulative frequency curve and use it to estimate the upper and lower quartiles.
(b) Estimate the 80th percentile.
(c) Estimate the percentage of cars parking for more than 50 minutes.
(d) Calculate the mean time parked for the sample of cars

11. The heights of 10 students, to the nearest centimeter, are as follows. State the range and use a table to calculate the standard deviation.

172 169 163 175 182 170 165 176 177 169

12. The scores of 25 students in a quiz out of 10 are presented below. Use a table to calculate the mean and standard deviation.

7	6	7	4	5	6	4	8	7	6
8	4	5	5	6	7	5	7	4	3
6	4	9	5	7					

Verify your results by using the statistical functions found on your calculator.

13. A scientific researcher weighs a random sample of 30 lizards and records their weights to the nearest gram. Use your calculator to find the sample mean and standard deviation. Provide an unbiased estimate of the population standard deviation.

21	18	15	20	18	17	12	23	19	19
17	20	13	15	17	21	18	14	13	18
22	17	15	12	25	15	16	18	16	17

14. The scores of 100 students in a test out of 10 were:

score	3	4	5	6	7	8	9	10
frequency	5	11	19	24	21	12	6	2

(a) Find the median.
(b) State the range.
(c) Find the upper and lower quartiles.
(d) State the inter-quartile range.
(e) Calculate the mean.
(f) Calculate the standard deviation.

15. For a mathematics project, Eun-Kee timed the length of 30 popular Korean songs to the nearest second. His raw data is presented in Q.8 and below:

185	230	205	215	217	206	192	187	207	245	205	181
216	227	239	214	242	248	193	222	217	219	204	234
227	236	234	217	186	236						

Use your calculator to find the standard deviation, and provide an estimate of the standard deviation of all Korean popular songs.

16. The masses of a sample of new potatoes were measured to the nearest gram and are summarized in Q.9 and below. Estimate the standard deviation of weights of new potatoes.

Mass(g)	Frequency
10 – 19	2
20 – 29	14
30 – 39	21
40 – 49	73
50 – 59	42
60 – 69	13
70 – 79	9
80 – 89	4
90 – 99	2

18.8 GRADED REVISION EXERCISES

LEVEL 1

1. The diagram shows the number of children in ten families:

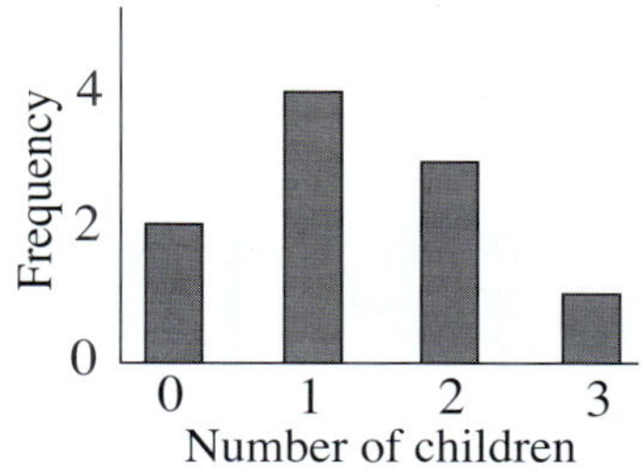

i. What is the modal number of children?
ii. What is the total number of children?
iii. What is the mean number of children per family?
iv. What percentage of the families have one child?

2. Tally the following data into groups 0-4, 5-9, 10-14 etc.

12 14 24 22 18 7 4 8 19 21 13

i. What is the frequency of the 0-4 group?
ii. What is the frequency of the 20-24 group?
iii. What is/are the modal groups?

3. The pie chart shows the proportions of a group of 200 students who play various musical instruments.

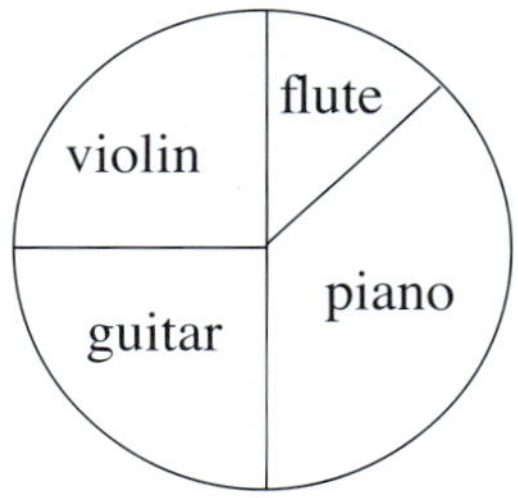

How many students play the piano?

LEVEL 2

1. Find the means of these data sets correct to three significant figures:

i.	4	3	2	5	2	1	0	0	2	4	0
ii.	11	9	4	8	22	31	6	3	22	19	16
iii.	1.4	1.9	2.7	3.8	4.1	3.2	6.8	5.5	3.9	8.5	2.7
iv.	12.8	9	14	2.6	7.2	6.8	6.8	5.9	12.8	4.8	19.4
v.	33.9	45.2	98.7	45.7	2.9	53.8	45.1	84.2	64.7	48.2	55.9

2. Find the modes of these data sets:

i.	3	5	6	2	6	5	7	4	4	3	4
ii.	5	4	3	3	0	0	2	1	4	3	1
iii.	11	14	12	12	13	14	15	11	10	12	12
iv.	10	10	14	13	13	12	11	11	13	12	14
v.	20	20	19	17	16	15	16	17	17	19	16

3. Find the medians of these data sets:

i.	20	17	19	17	16	15	11	17	17	19	16
ii.	31	62	73	11	4	15	88	45	12	6	71
iii.	5.8	3.1	1.9	2.6	2.9	4.2	9.8	6.8	4.2	8.9	3.5
iv.	12.3	54.9	4.6	28.5	83.5	66.7	2.1	6.7	55.9	49.1	84.8
v.	0.1	0.8	0.4	0.7	1.9	4.1	0.4	1.1	8.2	0.4	1.4

LEVEL 3

1. Find the quartiles of these data sets:

i.	3	7	2	8	3	5	9	1	5	2	0
ii.	19	43	91	59	83	62	77	14	82	53	22
iii.	23.5	71.4	55.8	11.8	84.6	54.9	92.6	84.2	17.9	34.8	33.8
iv.	25	11	4	5	13	19	21	6	10		
v.	5.4	6.8	3.4	8.9	3.6	2.9					

2. Find the standard deviations of these data sets, correct to three significant figures:

i.	4	6	2	5	8	1	9	4
ii.	11	75	62	89	71	82	4	68
iii.	3.4	7.9	10.1	5.1	0.3	7.4	8.2	5.3
iv.	124	638	342	836	532	539	842	44
v.	0.23	0.93	0.35	0.02	0.87	0.55	0.63	0.71

LEVEL 4

1. House prices in a particular suburb are generally in the region of \$120 000 to \$150 000. There are, however, a very small number of luxury homes whose values exceed \$2 million. Which measure of average would you use to find a representative house value for this suburb and why?

2. The following data represents the number of eggs laid each year by a species of sea bird.

6	11	4	10	5	5	5	9	2	8
2	4	7	2	5	5	9	7	4	9
9	6	7	5	6	11	9	3	11	9
10	2	8	7	11	4	2	5	5	4

i. Tally the data using class intervals of one.
ii. What is the probability that a randomly chosen bird will lay more than 9 eggs?
iii. Find the mean number of eggs laid.

3. The diagram shows the distribution of scores obtained by a group of 200 trainees, in an aptitude test (scored out of 10).

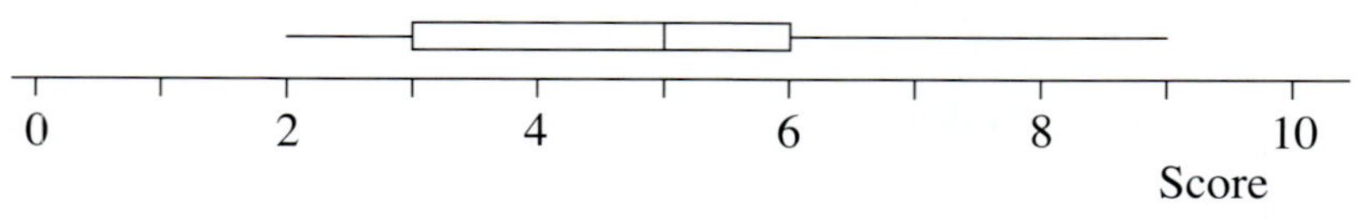

i. Find the number of students who scored more than 6 marks.
ii. Find the median mark.
iii. Find the least mark obtained by the group.
iv. Estimate the mean score.
v. Estimate the standard deviation of the distribution.

18.9 TOPIC TEST

1. The figures below are the heights (in cm, correct to the nearest cm) of a group of students:

175	179	180	179	178	181	180	177	178	180
177	180	181	178	179	179	180	178	177	181

i. Tally these data using no class intervals.

[2 marks]

ii. Show the results on an appropriate statistical graph.

[2 marks]

iii. Tally the data using class intervals of width 5cm. Which of the two tallies is the best for displaying the data?

[1 mark]

iv. Find the mode, mean and median of the data.

[3 marks]

v. Find the standard deviation and interquartile range of these data.

[4 marks]

vi. Construct a stem and leaf diagram for the data set.

[3 marks]

2. The diagram shows the distribution of weekly income of the employees of a small company.

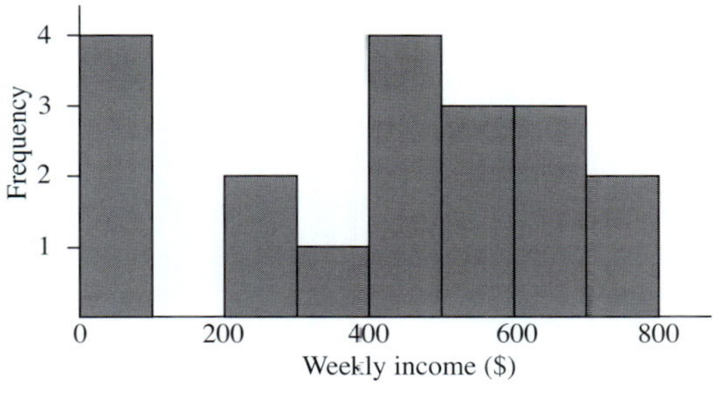

i. How many employees are recorded?
ii. Estimate the mean weekly income.
iii. Estimate the median wage.
iv. Explain the comparatively large number of employees that recive a low weekly wage.
v. If the owner of the company, who receives $1680 per week, is added to the data, which of these measures will be affected most; median or mean?

[8 marks]

Total 20 marks

18.10 PROJECT SUGGESTIONS

Another group of people who are interested in statistics on wind directions are those who have to plan waste tips and chemical factories. Why?

> If you had to plan a new chemical plant in your country, where would you choose to build it and why?

> What would you consider in planning a holiday? Light winds? Nice temperature? No rain?

If you are a surfer, you might like high winds. If you want to ski, you will want low temperatures. Whatever your preferences, weather statistics will be needed to make a rational decision.

> Choose a holiday (scuba diving in Fiji, hiking in Alaska, lazing on a Sri Lankan beach - whatever you dream of) and decide when is the best time to go, supporting your decision with statistical evidence.

19.0 DRUGS AND HEALTH

The World Health Organisation has collected the following figures for the consumption of alcoholic drinks in a South American country in recent years.

Year	Total Adult Per Capita (litres)	Beer	Wine	Spirits
1999	9.59	2.26	6.92	0.42
1998	10.23	2.25	7.56	0.42
1997	10.49	2.2	7.88	0.42
1996	10.33	1.85	8.06	0.42
1995	10.28	1.91	8.09	0.28
1994	11.07	2.1	8.55	0.42
1993	11.21	1.95	8.83	0.43
1992	12.02	1.83	9.9	0.29
1991	12.38	1.56	10.53	0.29
1990	12.29	1.21	10.94	0.14
1989	12.37	1.21	10.87	0.29
1988	13.11	1.06	11.33	0.72
1987	14.44	1.21	11.79	1.45
1986	14.6	1.14	12.01	1.45
1985	14.49	0.85	12.19	1.45
1984	15.76	0.86	13.45	1.45
1983	16.55	0.69	14.41	1.45
1982	16.7	0.5	14.9	1.3
1981	17.43	0.47	15.09	1.88
1980	18.79	0.54	15.37	2.88
1979	18.26	0.5	15.32	2.44
1978	18.03	0.49	16.26	1.29
1977	19.44	0.61	17.4	1.42
1976	19.65	0.66	16.58	2.41
1975	19.87	1.02	16.36	2.49
1974	18.25	1.16	15.11	1.99
1973	17.05	0.79	14.22	2.04
1972	18.35	0.76	15.57	2.02
1971	19.42	0.79	16.82	1.82
1970	21.01	0.93	18.19	1.89
1969	20.35	0.8	17.58	1.97
1968	20.14	0.74	17.42	1.98
1967	19.33	0.66	16.52	2.14
1966	18.88	0.71	16.05	2.13
1965	20.23	0.72	17.21	2.3
1964	20.25	0.5	17.45	2.3
1963	20.07	0.37	17.32	2.38
1962	19.07	0.6	16.36	2.11
1961	18.66	0.75	15.88	2.03

Source: The Global Alcohol Database (World Health Organisation).

These figures represent the actual alcohol consumed per adult. Thus, in 1961, the 'beer' figure was 0.75 litres and the 'wine' figure 15.88 litres. Beer is approximately 5% alcohol and wine approximately 12% alcohol.

This means that the actual volume of beer drunk was $\frac{0.75}{0.05} = 15$ litres per year and for wine it was $\frac{15.88}{0.12}$ = 132.3 litres per year. However, the issue is that this mass of figures serves more to confuse than to inform.

What trends can you see by just looking at the table? Probably very few. The mathematics that helps us analyse such data is the subject of this chapter.

Here is a graph that will give you a taste of this topic. It shows a dot for each year with the dot placed to represent the beer and wine consumption for that year. The dot at the extreme left represents the data for 1963.
Can you see why?

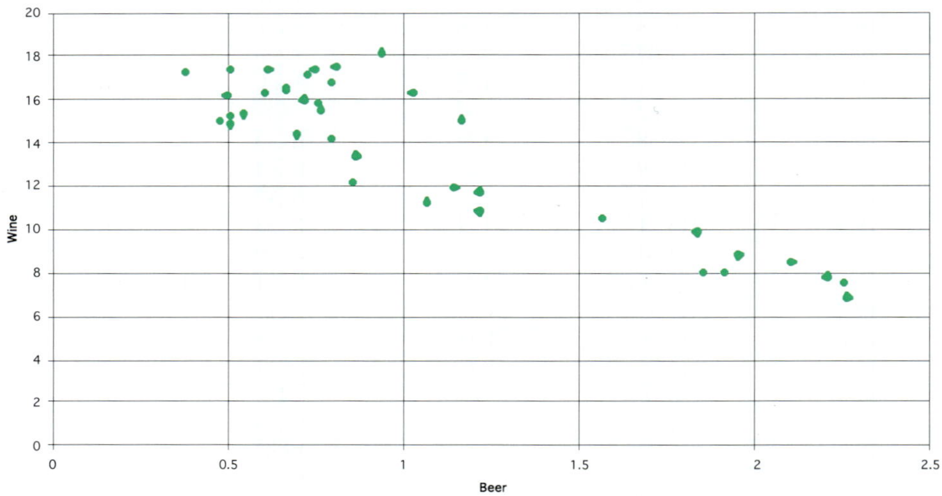

The pattern of dots shows that years in which a lot of wine was drunk were bad years for the brewers of beer and vice versa.

Using this graphical representation of the data it is possible to make valid statements about the relationship between the amounts of beer and wine that are consumed. Of course we need to be careful about what it is that we say.

As we are trying to determine how the two sets of data relate to each other, we say that we have bivariate data, which requires statistical analysis known as bivariate analysis.

Bivariate analysis will require the use of new statistical terms that will enable us to accurately describe the existence or non-existence of a relationship between the two sets of data.

19.1 CORRELATION

19.1.1 INTRODUCTION

This chapter deals with the study of **bivariate data**. In particular we will be looking at the use of **scatter diagrams** as an initial visual aid in describing any relationship that might exist between two variables. Next we look at measuring the strength of such a relationship (if one exists) and finally we consider the issue of **regression analysis**. This will help in obtaining equations that can be used to predict (or explain) the value of one variable (the dependent variable) based on the value of a second observation (the independent variable). In particular, we will only be considering **linear relationships**, as such, the regression lines will take on the form $y = a + bx$.

19.1.2 SCATTER DIAGRAM

A scatter diagram is a method by which we can obtain a very quick **visual appreciation** of how two variables are related. Such diagrams are obtained by plotting a set of points that correspond to the bivariate data. Usually the independent variable runs along the horizontal axis, while the dependent variable runs along the vertical axis. Once the data has been plotted we are interested in giving some indication of the association between the two variables. One such measure is the **correlation**. Qualitative descriptors that are useful include: **direction**, **form** and **strength of relationship**.

DIRECTION OF RELATIONSHIP

If the dependent variable tends to **increase** as the independent variable increases, we say that there is a **positive association** (or relationship) between the variables.

If the dependent variable tends to **decrease** as the independent variable increases, we say that there is a **negative association** (or relationship) between the variables.

Positive association

A. Profit

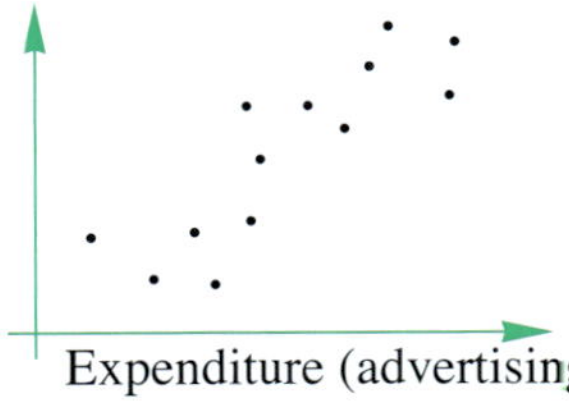

Expenditure (advertising)

Negative association

B. Productivity

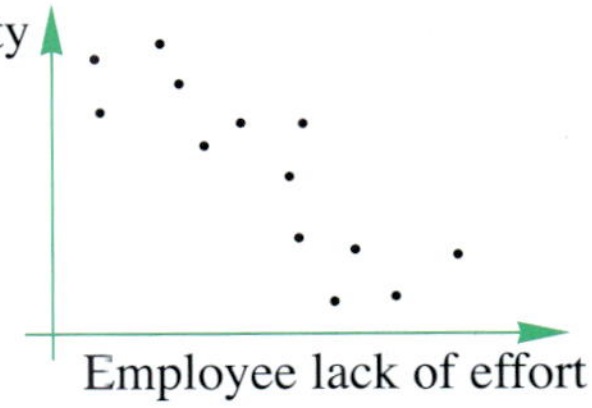

Employee lack of effort

From Graph A., it can then be said that there appears to be a strong suggestion that as more money is spent on advertising, the profit made by the company is also increasing. That is, there is a positive association between increased advertising expenditure and profit.

From Graph B., it can be said that as employees reduce their effort (i.e., the lack of effort increases), their productivity decreases. In this case, we say that there is a negative association between lack of effort and productivity.

FORM OF RELATIONSHIP

The **form** depends on the general shape of the scatter plot. The examples below indicate some types of forms that can be observed. However, we will only consider **linear forms** in this course.

Linear form

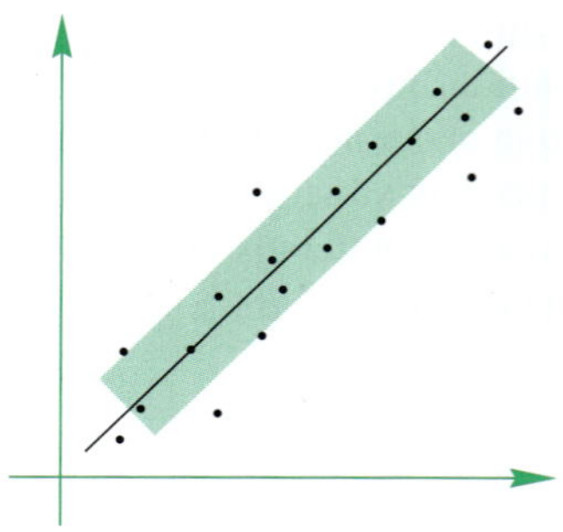

Non–linear form

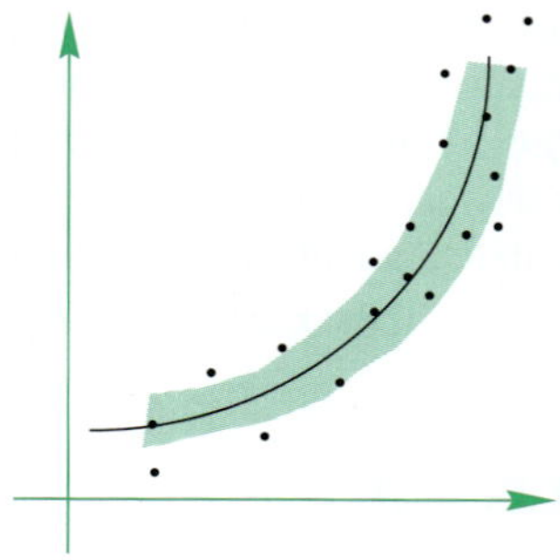

Non–linear form

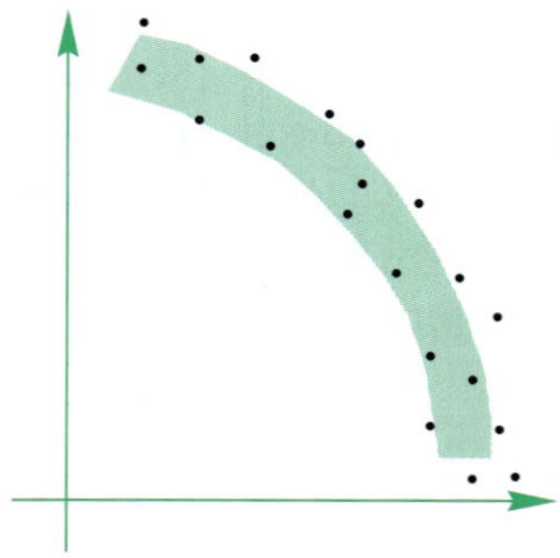

STRENGTH OF RELATIONSHIP

The strength of a linear relationship gives an indication of **how closely** the points in the scatter diagram **fit a straight line**.

> The measure of the **strength of a linear relationship** is determined by the **correlation coefficient**, ***r***.

The following graphs give some indications of the strength (and direction) of the linear relationships that exists between the variables y and x.

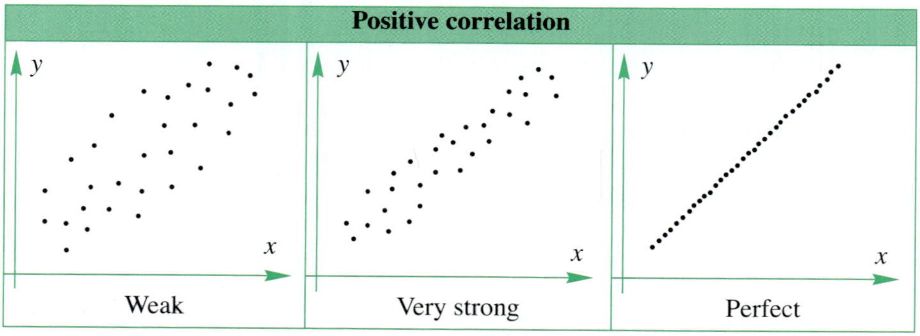

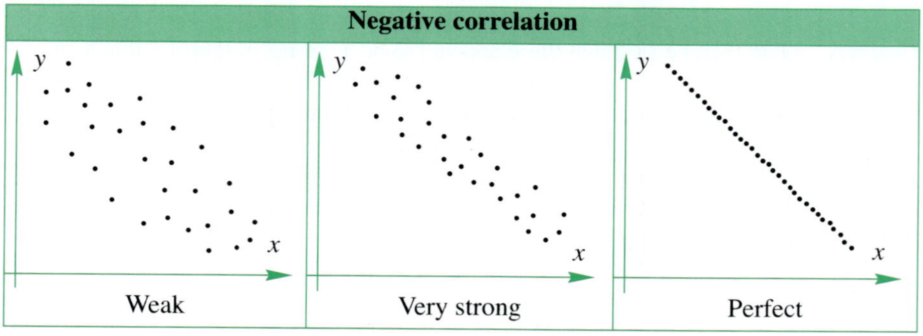

Of course, there also exists the possibility that no relationship exists. A scatter diagram indicating that there is **no linear relationship** between the variables is shown alongside.

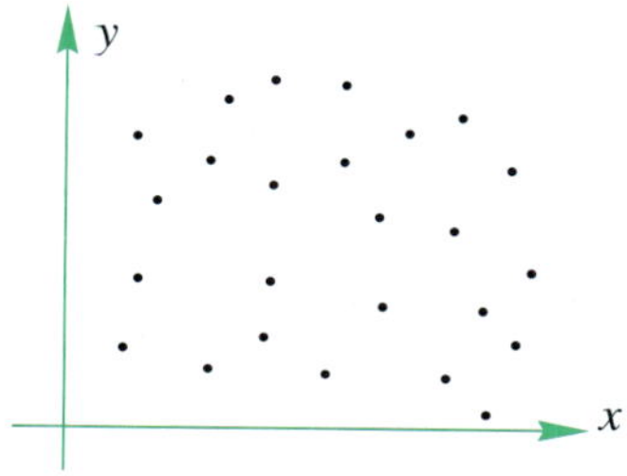

In such a case we say that there is **no correlation** between the variables.

19.1.3 Using the Graphics Calculator

One of the great features of a graphics calculator is its ability to provide a visual display of bivariate data. To use the TI–83 to construct a scatter plot, we first need to make sure that the calculator has been set up correctly. We do this as follows:

1. Press **2nd**, **Y=** (i.e., call up **STAT PLOT**)
2. then press **ENTER**, **ENTER** (this will turn the STAT PLOT function on)
3. then use the arrow keys to highlight the scatter graph option and press **ENTER**
4. then press **2nd MODE** (to **QUIT**).

We next examine examples using two different methods to create a list of data to be used for a scatter graph.

EXAMPLE 19.1

Determine if the data has a linear relationship, stating the direction and strength.

x	2	4	5	7	9	10	11	15
y	3	4	6	6	7	9	10	11

Solution

Having set the TI–83 in the appropriate statistical mode, we next enter the data as two lists, $x \leftrightarrow L_1$ and $y \leftrightarrow L_2$.

For L_1: Press **2nd (** , enter the data and finish off with **2nd)**, then press **STO**➧. **2nd**, **1**.

Similarly for L_2, but this time use type **2nd**, **2**.

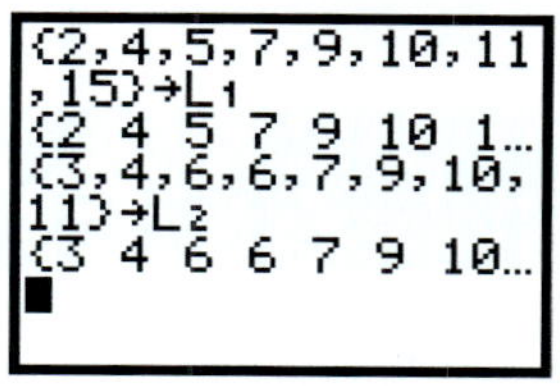

Once we have the data we can plot the scatter diagram.
Set the window to [0,16] by [–1,12] and then press **GRAPH**:

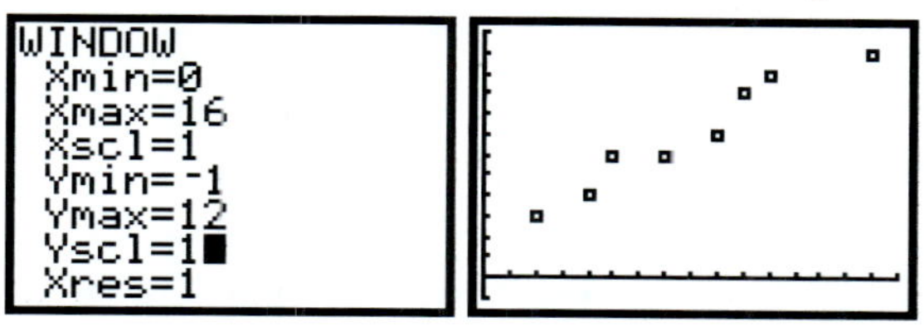

It is always a good idea to check your data list. To do this call up the table of values by pressing **2nd GRAPH**:

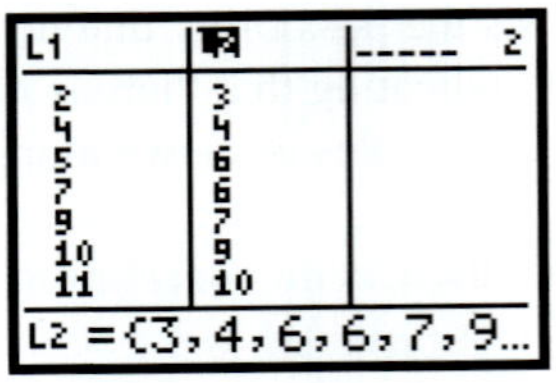

To clear the data lists, call up the **STAT** menu and choose **option 4**, or if you need to, edit your data list by selecting **option1**.

EXAMPLE 19.2

Determine if the data has a linear relationship, stating the direction and strength.

x	14	3	17	2	32	9	17	21
y	125	253	61	190	17	182	134	11

Solution

This time we enter the data directly into the table. To do this, first clear any existing lists. To clear the lists, select the **STAT** menu and select **ClrList L1**, **L2**, . . .
Once this is done, select the **STAT** menu and select **Edit**. At this stage you should have an empty table. Then, enter the data, one at a time, then use an appropriate window setting of [0, 35] by [0, 260] (say) and press **GRAPH**:

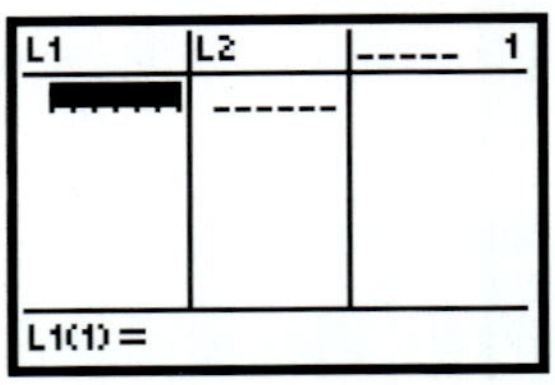

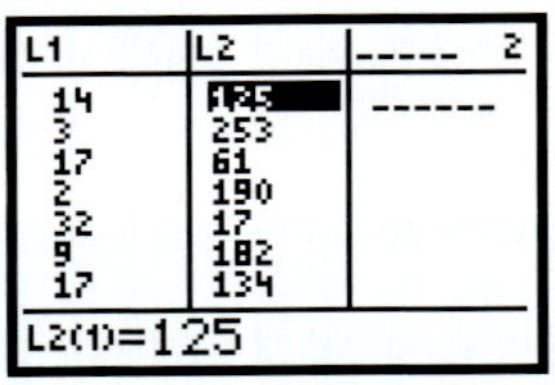

At this stage, the best that can be said about the relationship is that it is a moderate linear relationship with a negative correlation.

EXERCISES 19.1

1. For each of the following, give a statement about
(a) the direction (b) the form (c) strength of the relationship

i.

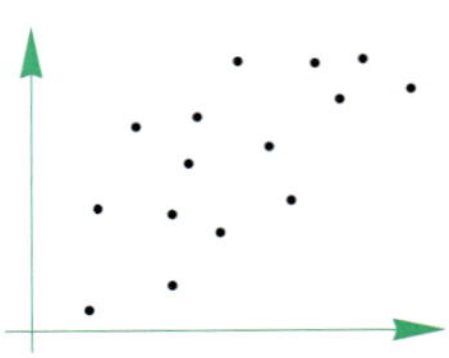

ii.

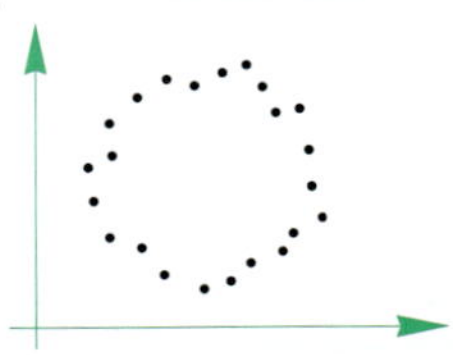

iii.

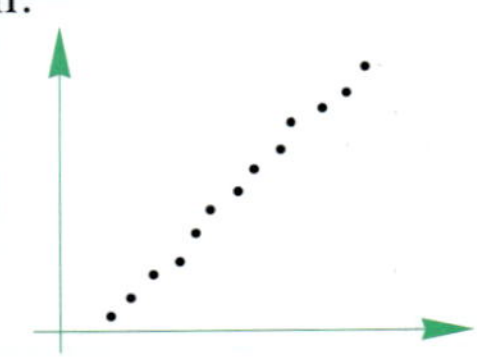

iv.

v.

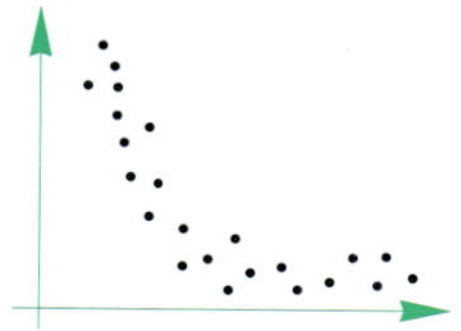

vi.

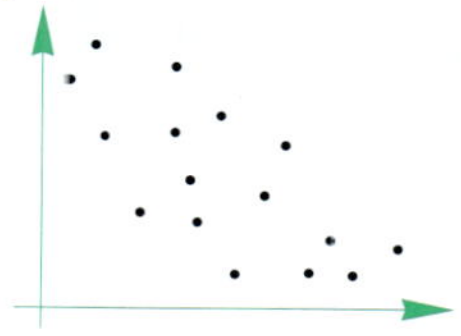

2. A group of students had their Science and Maths results tabulated:

Student	1	2	3	4	5	6	7	8	9	10
Science	55	70	40	67	80	80	55	60	20	84
Maths	60	78	39	65	82	90	50	71	18	79

(a) Plot this data on a scatter diagram.

(b) Describe the direction, form and strength of the relationship between Science marks and Maths marks.

3. The data in the table below shows students' reading–test scores and their corresponding I.Q scores

Student	1	2	3	4	5	6	7	8	9	10
Reading score	50	73	74	62	70	57	60	62	70	65
I.Q scores	99	118	131	111	113	101	105	113	121	118

(a) Plot this data on a scatter diagram.

(b) Describe the direction, form and strength of the relationship between reading scores and IQ scores.

4. The Department of the Environment decided to carry out an investigation into the amount of lead content, due to traffic flow, deposited on the barks of trees running along a stretch of road. The results produced the following table of values.

Traffic flow (in thousands)	32	35	70	73	119	121	125	194	193	204
Lead content (μg/g dry weight)	29	110	164	349	442	337	530	743	540	557

Plot a scatter diagram of this data and use it to comment on these results.

5. The number of industrial accidents in a particular workplace, from 1994 to 2003 were as follows:

Year	1994	1995	1996	1997	1998	1999	2000	2001	2002	2003
Number of accidents	166	131	123	162	160	130	91	82	65	53

(a) How would you rate the work safety policy that the company implemented since 1994?

(b) Plot a scatter diagram and use it to comment on this data

19.2 CORRELATION COEFFICIENT

19.2.1 STRENGTH OF A LINEAR RELATIONSHIP

So far we have given qualitative measures of the strength of a linear relationship, i.e., we have used expressions such as *strong* and *weak*, etc. However, as in all aspects of good statistical analysis, it is important that we provide a **quantitative measure** to describe our observations. Such measures are crucial when comparing sets of data.

The **strength of a linear relationship** is an indication of how closely the points in the scatter diagram fit a straight line. A measure of the strength of a **linear relationship** is given by a correlation coefficient. There are a number of ways that this correlation coefficient can be found. There is the q–correlation coefficient, the Spearman rank correlation coefficient as well as some other dubious correlation coefficients. In this course however, we will only be using the **Pearson's product-moment correlation coefficient** (or simply **Pearson's correlation coefficient**) which is denoted by $\boldsymbol{r}$.

Before we determine how to calculate these values of r, we make the following remarks concerning Pearson's correlation coefficient:

19.2.2 PROPERTIES OF r

1. The value of r does not depend on the units or which variable is chosen as x or y.

2. The value of r always lies in the range $-1 \le r \le 1$. A positive r indicates a positive association between the variables while a negative r indicates a negative association.

3. Perfect linear association, when, scatter plot points lie on a straight line, occurs if $r = \pm 1$.

4. r measures only the strength of a ***linear*** association between two variables.

The last point is particularly important – that is, it is only of use when there is strong evidence that a linear relationship does indeed exist.

A famous example was provided by Anscombe in 1973 where four radically different scatter plots were produced with a contrived value of $r = 0.82$ in each case. These are shown below:

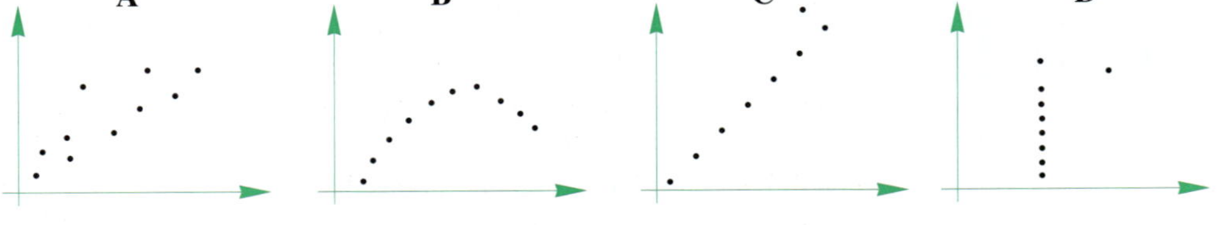

Data set A is a reasonably standard bivariate set for which the linear model seems appropriate and r meaningful. Data set B is a good curvilinear relation so a linear model is unsuitable and r is not valuable. Data set C has an outlier ruining a perfect relationship which may signal an error in data. The correlation probably understates the true relationship. Data set D is anomalous in the extreme - all points are the same except one and this exerts a large influence on the value of r. No meaningful conclusion could be drawn from this set.

19.2.3 SCATTER PLOT AND CORRESPONDING r VALUES

The properties of r and corresponding scatter plots can be summarised as follows:

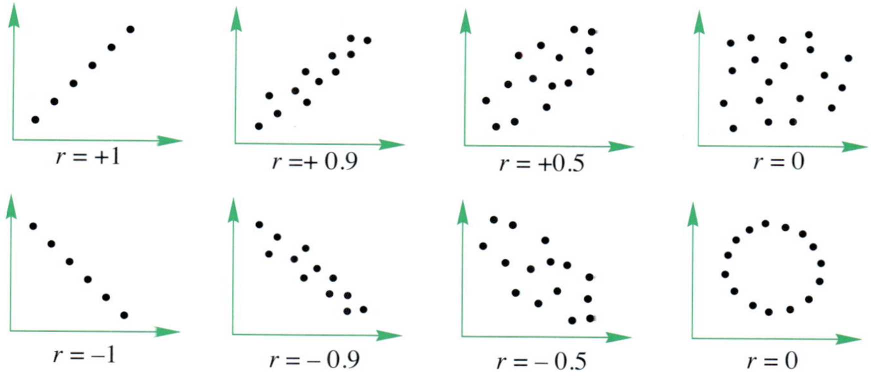

This table provides a good indication of the qualitative description of the strength of the linear relationship and the qualitative value of r:

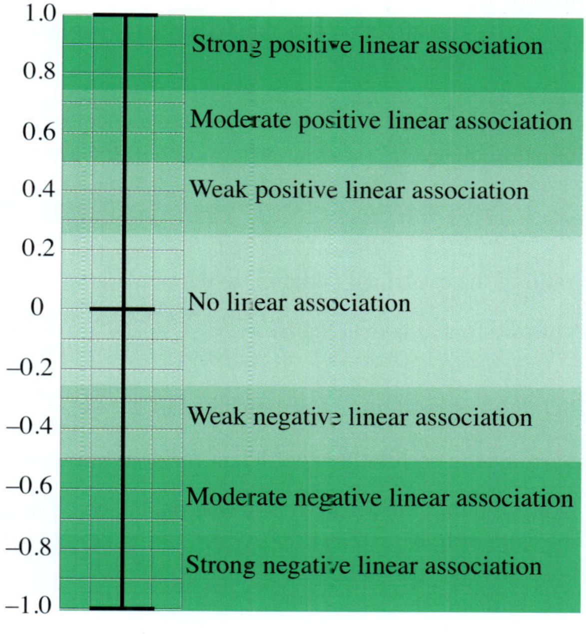

19.2.4 CAUSE AND ASSOCIATION

It can be risky to confuse a relationship with a **cause**. Just because two variables are highly correlated, it does not mean that one necessarily causes the other. For example, the degree of fatigue you experience during summer may be influenced by the temperature of the day (i.e., your fatigue depends on the temperature level), however, if you happen to reach a certain level of fatigue on some other day, this will not in turn indicate the temperature level of the day! That is, the temperature is independent of your fatigue level, and so, a rise in your fatigue level will not cause the temperature to rise.

Some 'relationships' suggested by correlation are spurious. For example, we have:
– Damage caused by fire and the number of firemen fighting it.
– Weight and height in individuals.
– Smoking rates and lung cancer deaths.

Some of these may represent causal relationships, others may seem ridiculous, but too often people jump to unjustified conclusions on the basis of high correlations alone.

Two variables x and y may exhibit a strong link for a number of reasons. These include:

Causation: Changes in x cause changes in y – for example, a change in outdoor temperature causes change in ice cream consumption. In cases where we have control over one variable, if we can change x, we can bring about a change in y. If smoking causes lung cancer, then reducing the incidence of smoking should reduce the prevalence of lung cancer.

Common response:
Both x and y respond to changes in other hidden variables – for example, both the degree of damage caused by a fire and the number of firemen fighting it are related to a third variable, the size of the fire! In this case although x can often be used to predict y, intervening to change x would not bring about a change in y.

Confounding:
The effect on y of the explanatory variable x is mixed up with the effects on y of other variables.

When experiments are not possible, good evidence for causation is less direct and requires a combination of several factors – where each of the above is adequately addressed.

19.2.5 DETERMINING THE VALUE OF r FROM A DATA SET

The ratio of the explained variation to the total variation is called the **coefficient of determination** and is denoted by $r^2 = \frac{\text{Explained variation}}{\text{Total variation}}$. Notice then, that $|r| \leq 1$. The **coefficient of correlation**, r, is then equal to $\pm\sqrt{\frac{\text{Explained variation}}{\text{Total variation}}}$. This definition can be used to determine the correlation coefficient for non–linear, as well as linear, relationships

Note then that **for linear relationships**, the coefficient of determination provides a measure of how well the linear rule linking the variables x and y predicts the value of y based on a given value of x.

As we will only be dealing with linear relationships we can use formulae that can be applied to linear relationships.

Pearson's correlation coefficient, r, gives a numerical measure of the degree to which the points in the scatter diagram behave linearly. To compute this value, we can make use of the formula

$$r = \frac{\sum(x-\bar{x})(y-\bar{y})}{\sqrt{\sum(x-\bar{x})^2\sum(y-\bar{y})^2}} \text{ or } r = \frac{s_{xy}}{s_x s_y}, \text{ or} \ldots$$

where $\bar{x}$ and $\bar{y}$ represent the mean of the x and y scores respectively and s_x and s_y represent the standard deviation of the x and y scores respectively and s_{xy} is the covariance of X and Y. We next look at these formulae in more detail.

The formulae $r = \dfrac{\sum(x-\bar{x})(y-\bar{y})}{\sqrt{\sum(x-\bar{x})^2\sum(y-\bar{y})^2}}$ and $r = \dfrac{\sum xy - \dfrac{\sum x \sum y}{n}}{\sqrt{\left(\sum x^2 - \dfrac{(\sum x)^2}{n}\right)\left(\sum y^2 - \dfrac{(\sum y)^2}{n}\right)}}$

are very useful when manual computation is necessary. The following examples show how these work expressions can be evaluated.

EXAMPLE 19.3

Find the Pearson correlation coefficient for the following set of data.

x	2	3	4	6	8	9	10
y	20	18	17	16	14	12	11

Solution

We will make use of the second formula shown above, meaning that we need

$\sum x = 2+3+4+6+8+9+10 = 42$

$\sum y = 20+18+17+16+14+12+11 = 108$

$\sum xy = (2\times 20)+(3\times 18)+(4\times 17)+(6\times 16)+(8\times 14)+(9\times 12)+(10\times 11)$
$= 588$

$\sum x^2 = 2^2+3^2+4^2+6^2+8^2+9^2+10^2 = 310$

$\sum y^2 = 20^2+18^2+17^2+16^2+14^2+12^2+11^2 = 1730$

Therefore, we have that $r = \dfrac{588 - \dfrac{42\times 108}{7}}{\sqrt{\left(310-\dfrac{(42)^2}{7}\right)\left(1730-\dfrac{(108)^2}{7}\right)}} = \dfrac{-60}{\sqrt{58\times\dfrac{446}{7}}} = -0.9870$

Of course, we should have first determined if it was appropriate to use Pearson's correlation coefficient. That is, we should have established (through the use of a scatter plot) if the data did indicate a linear relationship. As the scatter plot below shows, calculating the r value was appropriate.

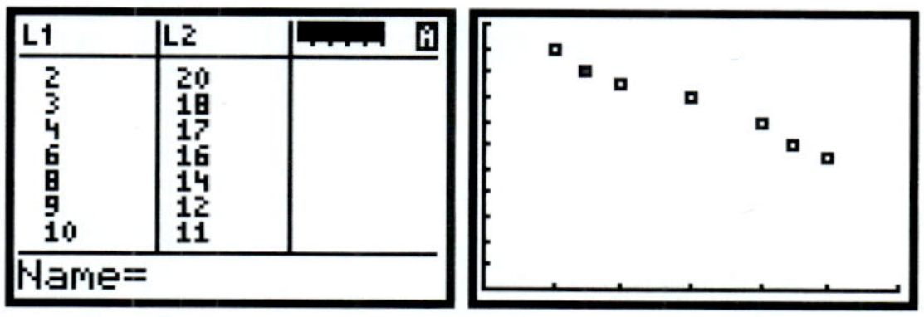

The formula used above is probably the most useful form for use in calculations. However, most calculators that have the facility to deal with bivariate data will be able to produce a value of r with the push of a few buttons. In particular, the TI–83 is excellent for this.

As we have mentioned, one form of the correlation coefficient which can be derived and easy to use, is $r = \frac{s_{xy}}{s_x s_y}$ where $s_{xy} = \frac{1}{n}(\sum xy - n\bar{x}\bar{y})$. The terms s_x and s_y can be recognised as the standard deviations of the random variables *X* and *Y*, while the new quantity, s_{xy}, is called the **covariance of *X* and *Y*.**

Before we look a little closer at how to interpret the value of *r* we proceed with another example of how to calculate the value of *r*, as well as how to use the TI–83 when dealing with bivariate data.

EXAMPLE 19.4

Assuming that the data has a linear relationship, find the coefficient of correlation for this set of data.

x	2	4	5	7	9	10	11	15
y	3	4	6	6	7	9	10	11

Solution

As we are going to use the formula $r = \dfrac{\sum xy - \frac{\sum x \sum y}{n}}{\sqrt{\left(\sum x^2 - \frac{(\sum x)^2}{n}\right)\left(\sum y^2 - \frac{(\sum y)^2}{n}\right)}}$, we need:

$\sum x = 63$, $\sum y = 56$

$\sum xy = 2 \times 3 + 4 \times 4 + 5 \times 6 + \ldots + 15 \times 11 = 522$.

$\sum x^2 = 2^2 + 4^2 + \ldots + 15^2 = 621$, $\sum y^2 = 3^2 + 4^2 + \ldots + 11^2 = 448$

So that $r = \dfrac{522 - \frac{63 \times 56}{8}}{\sqrt{\left(621 - \frac{(63)^2}{8}\right)\left(448 - \frac{(56)^2}{8}\right)}} = 0.9686$.

We now make use of the TI–83:

1. Enter data as two lists, $x \leftrightarrow L_1$ and $y \leftrightarrow L_2$.
2. Check to see if in fact there is a linear relationship.
3. Press, **STAT** then **CALC** and then choose **2:2–Var Stats** and enter L_1, L_2.

1.
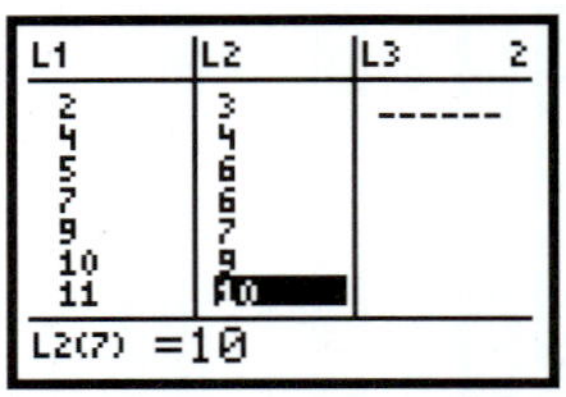

2.
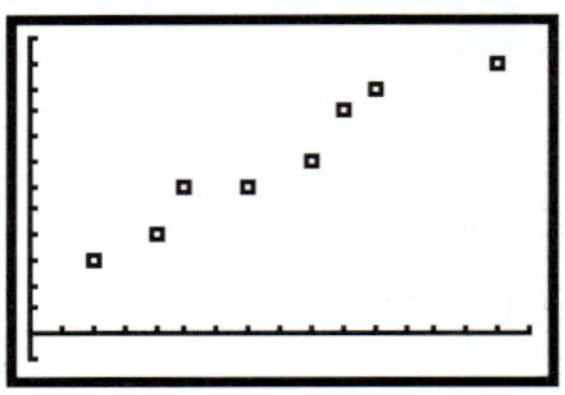

3.
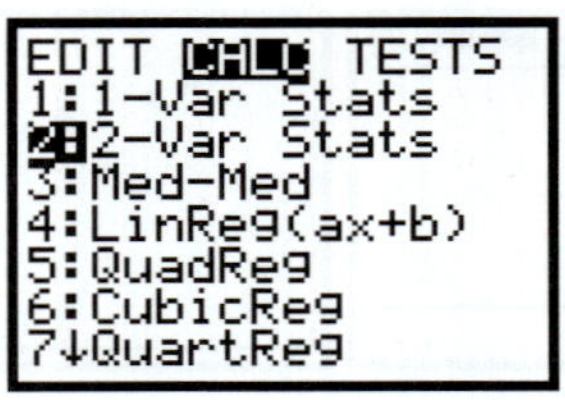

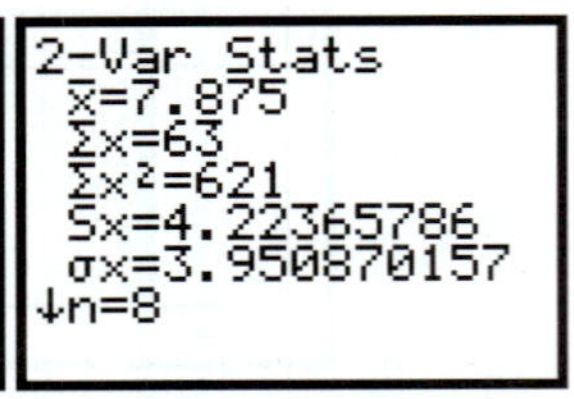

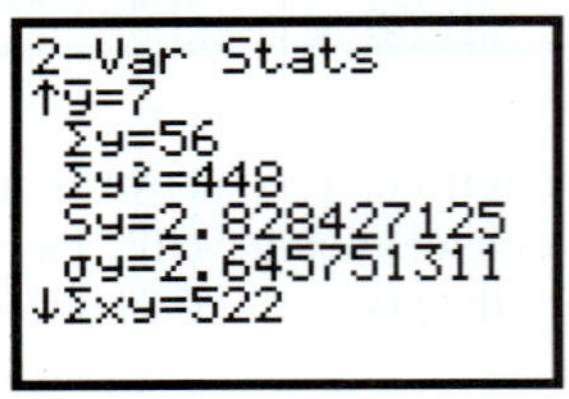

We can now use these results to calculate the correlation coefficient.

However, we can also obtain the correlation coefficient without using the **2–Var Stats** option.

First make sure that the TI–83 has **DiagnosticOn**.
That is, starting from the HOME screen, press **2nd MODE**.

Next, press **2nd 0** (this enables you to access the **CATALOGUE** menu)
Locate the **DiagnosticOn** option (press **APPS** and use the down arrow key)
Then press **ENTER ENTER**.

We can now find the correlation coefficient:

Press **STAT** choose the **CALC** option and then select **4:LinReg(a + bx)** and enter L_1, L_2.

Notice that the screen now displays the values of r, r^2 and the equation of a straight line.
This straight line is in fact the line of best fit using the method of least squares (which we shall look at in the next section).

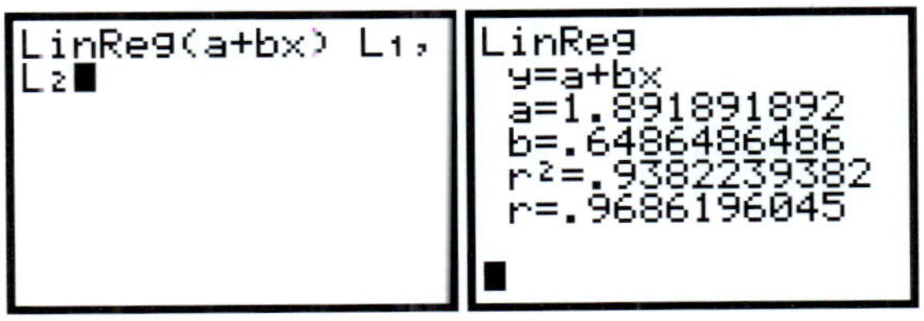

We see that $r = 0.9686 \approx 0.97$. Indicating a very strong positive relationship between x and y.
We also have a value for r^2. The question arises, how do we interpret the value of r^2?

19.2.6 INTERPRETING r AND r^2

Recall that by definition $r^2 = \dfrac{\text{Explained variation}}{\text{Total variation}}$ so that in fact, r^2 is a proportion whereas r is the square root of a proportion. As such, a coefficient of 0.8 does not represent a degree of relationship that is twice as great as a coefficient of 0.4. Also the difference between coefficients of 0.6 and 0.7 is not equal to the difference between coefficients of 0.7 and 0.8.

In general, when interpreting the magnitude of the relation between two variables, regardless of directionality, r^2, the **coefficient of determination**, is more informative. So for two linearly related variables, this value provides the proportion of variation in one variable that can be explained by the variation in the other variable.

In our example, we had $r^2 = 0.938$ or 93.8%, meaning that approximately 94% of the variation in the variable y can be explained by the variation in the variable x. The higher this value is, the better.

Notice that all of a sudden, a value of $r = 0.6$ is not all that impressive! Why? Well, if $r = 0.6$ then $r^2 = 0.36$, meaning that only 36% of the variation in one variable is explained by the variation in the other variable.

One way of visualising the meaning of the coefficient of determination, r^2, is to consider a perfect positive correlation and then observe what happens for a small value of r.

Case 1:

Consider x and y to be standard normal random variables having a perfect correlation, so that individual values of x close to 1.5, say, will have values of y also close to 1.5. In this case, the variation in y can be readily explained by the variation in x. This is shown in the figure alongside.

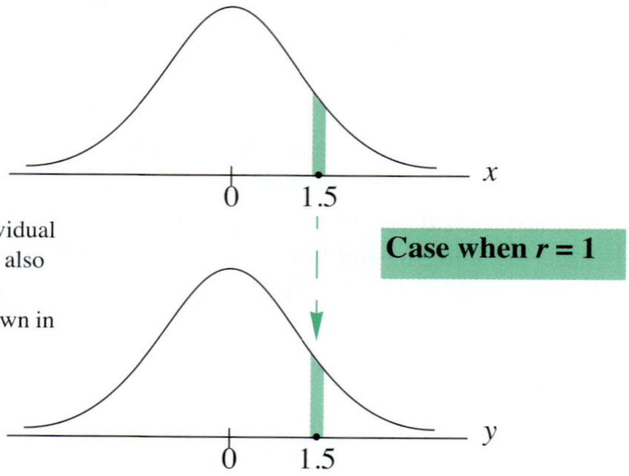

Case 2:

This time, for a small (positive) value of r we see that individuals with values of x close to 1.5 will have values of y widely scattered. In this case the variation in y cannot be attributed accurately to the variation in x.

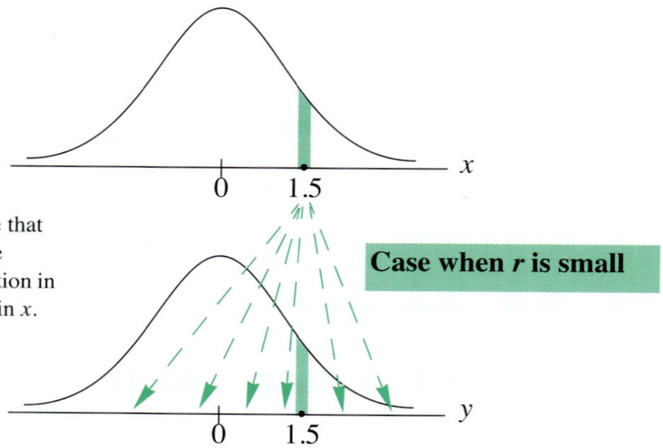

EXAMPLE 19.5

As part of the Mathematical Studies course, a teacher is required to submit marks to the I.B.O for the internal assessment part of the course. The internal marks and the final awarded marks for 10 students in a particular class were as follows:

(a) Draw a scatter diagram for this data.

(b) Would the use of the Pearson correlation coefficient be appropriate in this case? If so, calculate it.

(c) What conclusions can you make about the relationship between internally assessed marks and the final marks awarded to students?

Student	Internal mark	Final mark
Peter	65	72
Mark	80	87
John	78	65
Loise	60	73
Jane	92	95
Tom	81	90
Fiona	50	45
Becci	55	65
Loui	50	51
Sam	90	93

Solution

(a) The first thing to do is enter the data into the calculator: Let the first list correspond to the internal mark and the second list correspond to the final mark:

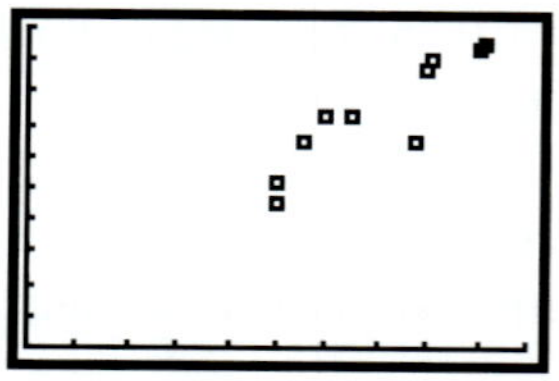

(b) From the scatter diagram it appears reasonable to assume that a linear relationship exists – even though one point might seem not to fit the 'trend'.

As we have concluded that a linear relationship does appear to exist, calculating the value of r is appropriate. We will not make use of the formula to calculate the Spearman correlation coefficient, but rather make use of the TI–83:

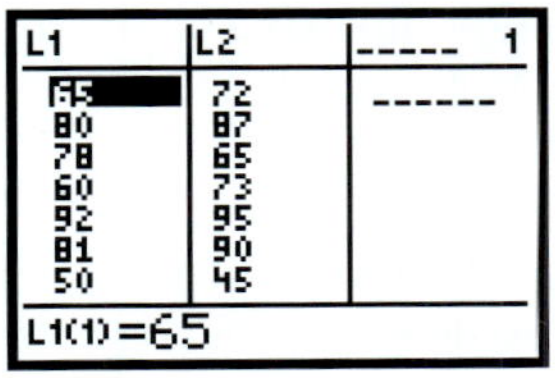

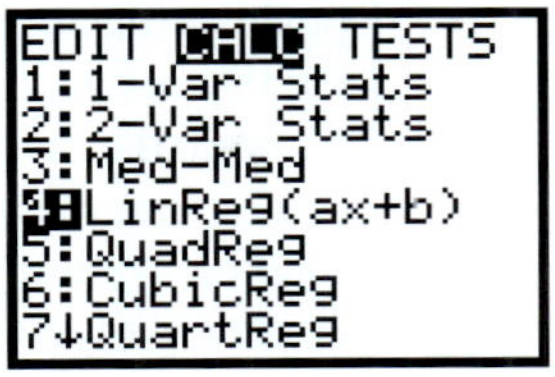

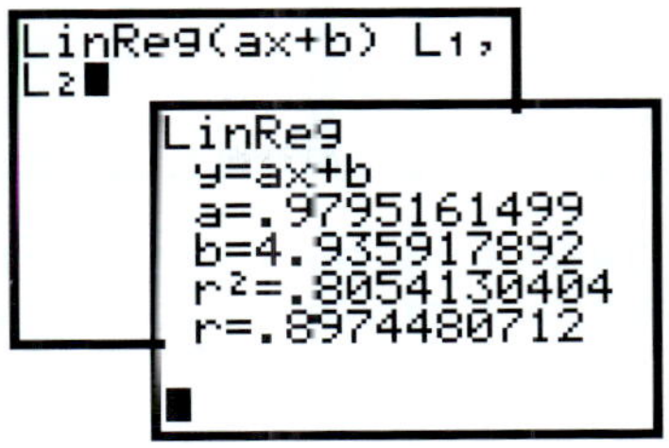

From the TI–83 we find this to be 0.8974.

(c) There exists a positive linear relationship between internal marks and the final mark awarded to the students in this class. As the value of $r = 0.8974$ we can say that there is a strong positive correlation between the internal mark awarded and the final mark awarded. Also, the value of $r^2 = 0.8054$, telling us that 80.54% of the variation in the final mark awarded can be accounted for by the variation in the internal mark awarded. That is, only 20% (approx.) of the variation is attributed to other factors.

WARNING!

The following example makes use of the fact that the value of the covariance, s_{xy} is already provided so that calculating the correlation coefficient, r, is more readily accessible via the formula $r = \frac{s_{xy}}{s_x s_y}$.

It is therefore important that we keep in mind that the **IBO notation may differ** from the notation found on your calculator. The next example deals with this issue when using the TI–83.

EXAMPLE 19.4

Ten students sat for Biology and Mathematics tests. The results were recorded and tabulated as follows:

Student	A	B	C	D	E	F	G	H	I	J
Maths (x)	56	91	84	63	10	63	28	35	91	63
Biology (y)	66	100	60	96	24	46	35	36	72	80

Given that the covariance, $s_{xy} = 493.78$ calculate, correct to 2 d.p. the product moment correlation coefficient, r. What comments can be made based on the value of r.

Solution

We first proceed with the question and then attend to the issue of different notations.

Using the formula $s_{xy} = \frac{1}{n}(\sum xy - n\bar{x}\bar{y})$ we can verify that the value of s_{xy} is 493.78.

Then, using the formula $r = \frac{s_{xy}}{s_x s_y}$ we have that $r = \frac{493.78}{(25.81549)(24.79213)} = 0.7715$.

Before we proceed with making any statement based on the value of r, we need to first establish that there is a linear relationship between the scores on the Mathematics test and the scores on the Biology test. To do this we use a scatter diagram:

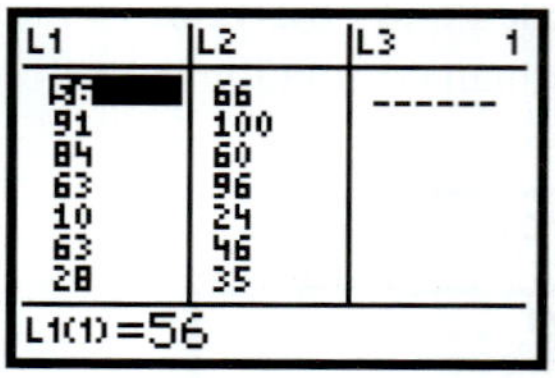

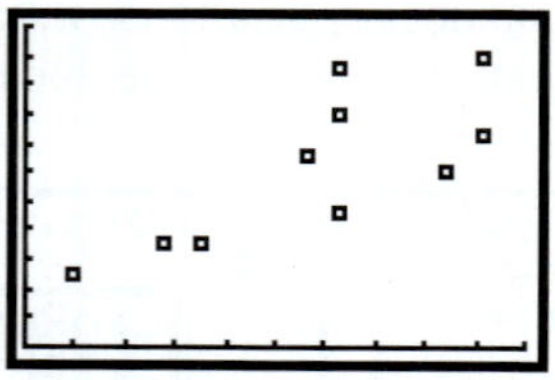

Having established that a linear relationship exists, we can now make comments on the data set. There is strong evidence that a positive linear relationship exists between the test results for Mathematics and Biology. The value of the correlation coefficient, $r = 0.77$ indicates that there is a strong correlation between Mathematics scores and Biology scores, meaning that those students that do well in mathematics, are likely to also do well in biology.

Having said that, it must be understood that based on the value of $r^2 = 0.60$, only 60% of the variation in the Biology scores can be attributed to the variation in the Mathematics scores. Nonetheless, based on the three results:, scatter diagram, value of r and r^2, our comments stand firm.

So, what seems to be the problem you may ask? The 'problem' lies in the values of s_x and s_y that were used in the expression $r = \frac{s_{xy}}{s_x s_y}$. Having entered the data sets as two lists, Mathematics as List 1 and Biology as List 2 we obtain the following results on the TI–83:

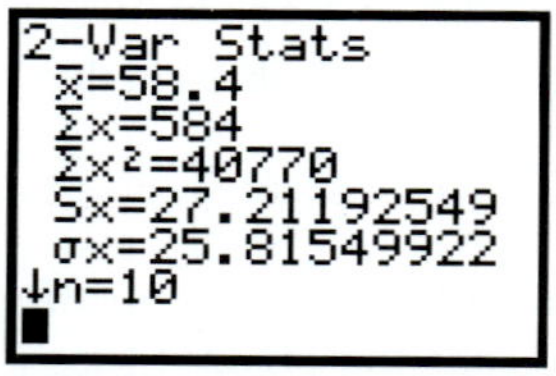

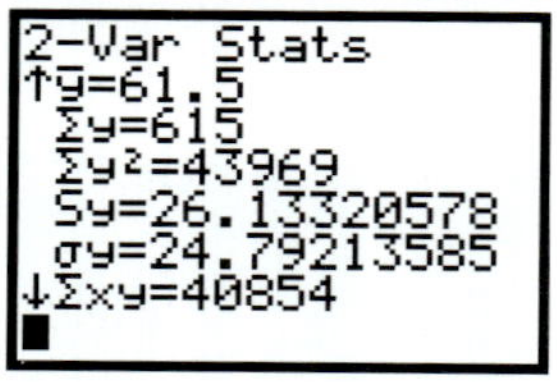

The windows show that $s_x = 27.2119$ and $s_y = 26.1332$. However, we used the values of $\sigma_x = 25.8154$ and $\sigma_y = 24.7921$ in the expression $r = \frac{s_{xy}}{s_x s_y}$.

That is, the IBO use the notation $s_x = \sqrt{\frac{1}{n}\sum_{i=1}^{n}(x_i - \bar{x})^2 f_i}$, whereas (as we saw in Chapter 18) the TI–83 uses the notation $\sigma_x = \sqrt{\frac{1}{n}\sum_{i=1}^{n}(x_i - \bar{x})^2 f_i}$.

This means that you must be aware of these differences – that's all. And, when dealing with questions that involve statistical calculations, just keep these differences in mind. We have used only the TI–83 to demonstrate the differences that can exists between notations. If you are not using a TI graphics calculator, make sure that you check the notation used. A good idea would be to use the above example as a trial run.

EXERCISES 19.2

1. (a) Assuming that the data has a linear relationship, find the coefficient of correlation for this set of data.

x	4	6	7	9	11	12	13	17
y	8	9	11	11	12	14	15	16

(b) Draw a scatter diagram for the given data.

2. (a) Draw a scatter diagram for the given data.

x	1	5	6	6	2	3	4	4
y	2	4	5	3	1	2	5	4

(b) Find the coefficient of correlation for this set of data. What assumption have you made in determining this value.

3. For the set of paired data, find the correlation between x and y. Is this an appropriate use of the correlation coefficient?

x	1	2	3	4	5	6	7
y	4	3	2	1	2	3	4

4. Would it be appropriate to calculate the coefficient of correlation for the data shown below?

x	1	2	3	4	5	6
y	3	2	1	1	2	3

5. Calculate the proportion of the variance of Y which

(a) can be predicted from (or explained by) the variance of X if

i. $r = 0.8$. ii. $r = -0.9$

(b) cannot be predicted from (or explained by) the variance of X if

i. $r = 0.7$. ii. $r = -0.6$.

6. The data shown below represents entrance examination marks (x) and first year average test marks (y) for a group of 10 students.

x	55	59	62	80	92	63	69	84	62	55
y	61	69	52	61	90	85	70	67	72	60

(a) Draw a scatter diagram for the data

(b) Determine the correlation coefficient between x and y.

7. How many times is a difference in predictive capacity between correlations of 0.70 and 0.80 greater than between correlations of 0.20 and 0.30?

8. What correlation between X and Y is required in order to assert that 85% of the variance of X depends on the variance of Y.

9. For the data shown below, calculate the proportion of the variance of y which can be explained by the variance of x

x	3	4	6	7	9	12
y	20	14	12	10	9	7

10. Ten students sat for a Biology and Mathematics test. The results were recorded and tabulated as follows:

Student	A	B	C	D	E	F	G	H	I	J
Maths	5.6	9.1	8.4	6.3	1.0	6.3	2.8	3.5	9.1	6.3
Biology	6.6	10.0	6.0	9.6	2.4	4.6	3.5	3.6	7.2	8.0

(a) Verify that the covariance, $s_{xy} = 4.938$ by using the expression

i. $s_{xy} = \frac{1}{n}(\sum xy - n\bar{x}\bar{y})$

ii. $s_{xy} = r \times s_x s_y$

(b) Using the product moment correlation coefficient, r, comment on the statement:
'The students who do well in Mathematics also do well in Biology.'

(c) What do you notice between the values in this table and those in Example 19.6? What conclusion can you make about two sets of bivariate data:
Set 1: (x_i, y_i) and Set 2: (kx_i, ky_i), where k is a constant

19.3 LINE OF BEST FIT

Having established that a linear relation exists between two variables x and y (say), we can then search for a line of best fit. That is, a line that will best represent the data on the scatter diagram. There are a number of ways this can be done. Some possibilities are:

1. Drawing a line 'by eye'
2. Using the locus of means
3. Using the median-median line of best fit
4. Using the least squares regression equation.

We will consider options 1., 2., and 4

19.3.1 LINE OF BEST FIT - BY EYE

If the scatter plot signals that a linear model is reasonable, we may attempt to model the data using a line of best fit. The choice of a suitable linear model may be done informally or by other formal methods which are usually easier to produce by hand. The problem with drawing a line 'by eye' is that many lines seem equally suitable. Consider the scatter plot shown alongside. Which of the lines, (1), (2) or (3) seems best and are there better lines we could fit?

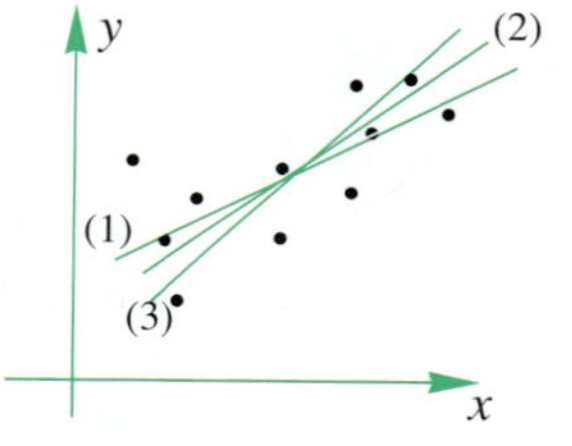

Once a line is decided on, we can then use the methods of Chapter 6 to determine the equation of this line. When we have an equation for the linear model, we can use it for predicting the values of y we may expect for a given value of x.

However, this method relies too much on individual preferences. So, to help in this endeavour, we have two methods:

1. balancing the number of points above and below the line
2. balancing the errors from the line to the data points.

The second of the above uses a method that balances the error on either side of the straight line. However, this method is more suitable when the data points are more scattered and fitting the straight line is not as obvious. Errors are based on the vertical distances between the data points and the fitted straight line, with those above the straight line being positive while those below the straight line being negative.

In this instance we will only consider method 1.

EXAMPLE 19.5

For the data shown below, find the equation of the line of best fit by eye.

x	3	4	6	7	9	12
y	7	9	10	12	14	20

Solution

We start by drawing a scatter plot of the data:

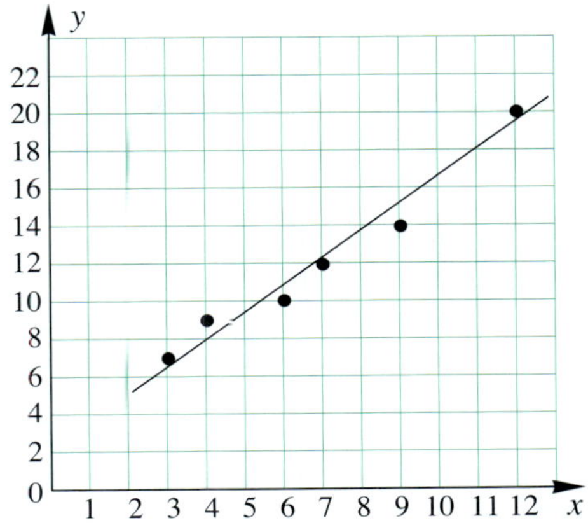

Having fitted a line so that there are three points 'above' the line and three points 'below' the line, we proceed to determine the equation of the straight line.

The line passes through the points with coordinates (4, 8) and (11, 18).
Finding the gradient, we have:

$$m = \frac{18-8}{11-1} = \frac{10}{10} = 1$$

Using the gradient-point form of a straight line we have:

$$y - 8 = 1(x-4)$$

That is, $y = x + 4$.

We can now compare the original data with the results based on the equation $y = x + 4$:

x	3	4	6	7	9	12
y	7	9	10	12	14	20

and

x	3	4	6	7	9	12
$y = x + 4$	7	8	10	11	13	16

The results seem quite good. There is an element of confidence then in using this equation to predict the y-value when $x = 8$ (say).

19.3.2 LINE OF BEST FIT - USING LOCUS OF MEANS

A line can be fitted between any two distinct points. This is the basis of a simple linear model for fitting data where we may divide the x-values and y-values into two sets of upper and lower values and find the means of each of the x and y data in both of the upper (u) and lower (l) groups.

We use the mean value coordinates $(\bar{x}_u, \bar{y}_u)$ and $(\bar{x}_l, \bar{y}_l)$ to define the trend line called the **locus of means**.

The points enable us to find a gradient $m = \dfrac{\bar{y}_u - \bar{y}_l}{\bar{x}_u - \bar{x}_l}$ and then we use either point to produce the equation of a line by a gradient-point form, for example, $y - \bar{y}_u = m(x - \bar{x}_u)$.

If there are an even number of data points then the two groups will divide naturally. For an odd number of data values one point will be in common with both upper and lower groups. We now illustrate this process by an example.

EXAMPLE 19.6

For the data set shown, find the line defined by the locus of two means of these positions.

x	1	2	2	3	4	5	5	6
y	7	9	6	13	18	17	21	24

Solution

We start by plotting the data set:
Inspection of the scatter diagram suggests a linear model may be appropriate as there is a suggestion of a strong correlation.

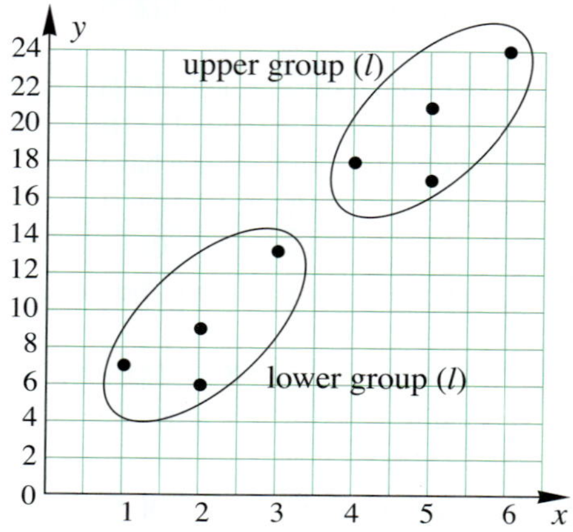

We now determine the means of the x and y values for the upper and lower sets.

Lower level:

$$\bar{x}_l = \frac{1}{4}(1+2+2+3) = 2$$

$$\bar{y}_l = \frac{1}{4}(7+6+9+13) = \frac{35}{4}$$

Therefore we have $(\bar{x}_l, \bar{y}_l) = \left(2, \frac{35}{4}\right)$

Upper level:

$$\bar{x}_u = \frac{1}{4}(4+5+5+6) = 5$$

$$\bar{y}_u = \frac{1}{4}(18+17+21+24) = 20$$

Therefore we have $(\bar{x}_u, \bar{y}_u) = (5, 20)$

The two points defining our line are $(2, 8.75)$ and $(5, 20)$.
The gradient of the line is given by

$$m = \frac{20-8.75}{5-2} = \frac{15}{4}.$$

Hence, the line has equation:

$$y - 20 = \frac{15}{4}(x-5)$$

or

$$y = \frac{15}{4}x + \frac{5}{4}$$

The line can now be added to the scatter plot to see the fit.

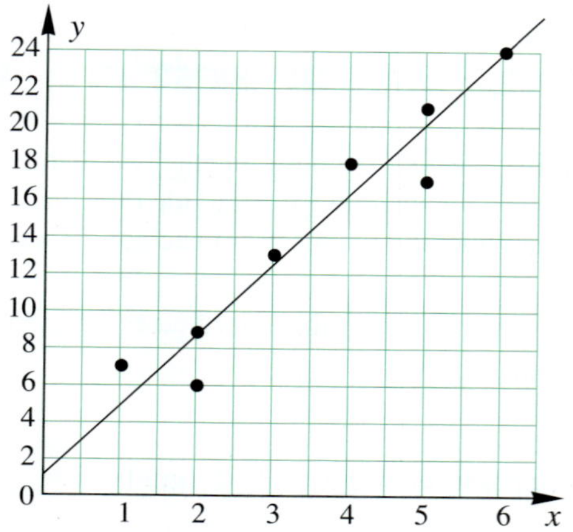

19.3.3 LINE OF BEST FIT - METHODS OF LEAST SQUARES

We have considered a number of alternatives to determine the line of best fit, so far they have all had an element of 'guess' work. Obtaining a line of best fit that adheres to sound statistical procedures is done by using a **simple linear regression** method.

There are a number of different ways this can be achieved, however, we will consider the method known as '**the method of least squares**'. The least squares regression line is the line which makes the sum of the squares of the vertical deviations of the data from the line as small as possible:

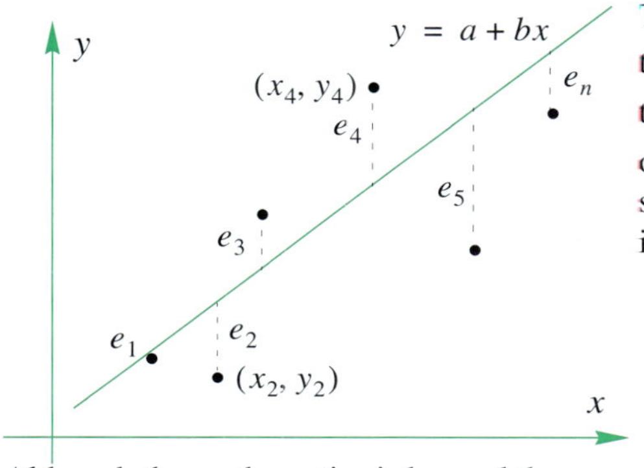

That is, having drawn the scatter diagram, we want to determine the equation $y = a + bx$ for which the sum of the squares of the errors, e_i, between observed values and predicted values (based on the straight line) is minimal.
i.e., we want to minimise the expression

$$\Delta = \sum_{i=1}^{n} e_i^2 = \sum_{i=1}^{n} (y_i - a - bx_i)^2 .$$

Although the mathematics is beyond the scope of this course, we can obtain a rather neat expression that will enable us to determine the value of a and b that give the line of best fit. We quote this result now.

The line of best fit using the method of least squares on n pairs (x_i, y_i) is called the **line of regression of y on x** and is given by $y = a + bx$, where a and b are determined by

$$a = \bar{y} - b\bar{x} \quad \text{and} \quad b = \frac{\sum xy - \frac{1}{n}(\sum x)(\sum y)}{\sum x^2 - \frac{1}{n}(\sum x)^2}$$

Again, it can be seen from these formulae that manual calculations are very tedious, so, it is highly recommended that graphics calculators are used (or any calculator that handles bivariate data).

Once we obtain the line of regression of y on x, it is important that we not only use it to predict values but also to interpret the intercept and the slope of the line in terms of the problem being discussed.

When dealing with regression lines, predictions take on the form of **interpolation** and **extrapolation**. Interpolation is the process of using a regression line to make **predictions within the range of data** that was used to derive the equation. Whereas extrapolation is the process of using a regression line to make **predictions outside the range of data** that was used to derive the equation.

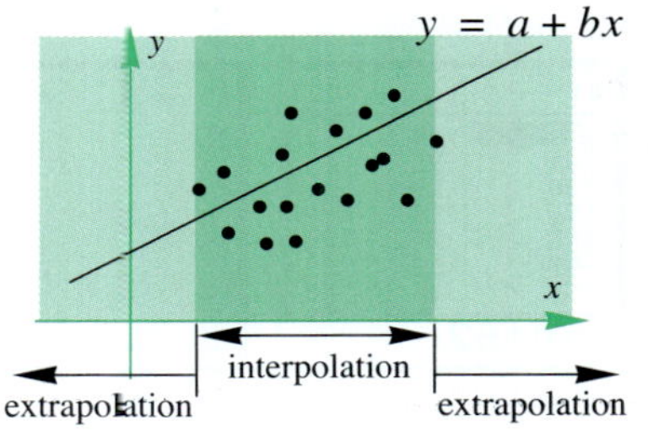

We consider one more time the data set of Example 21.4.

EXAMPLE 19.7 Assuming that the data has a linear relationship, find the regression line of y on x.

x	2	4	5	7	9	10	11	15
y	3	4	6	6	7	9	10	11

Solution

Entering the data for x as **List 1** and for y as **List 2**, we have:

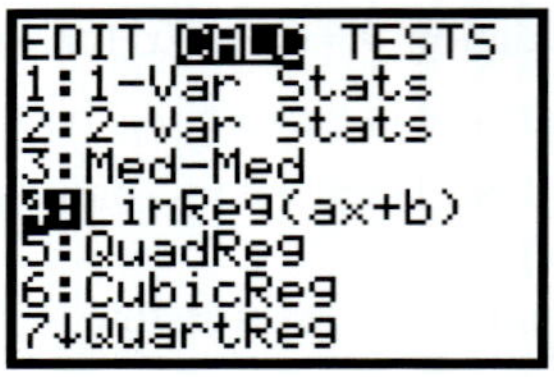

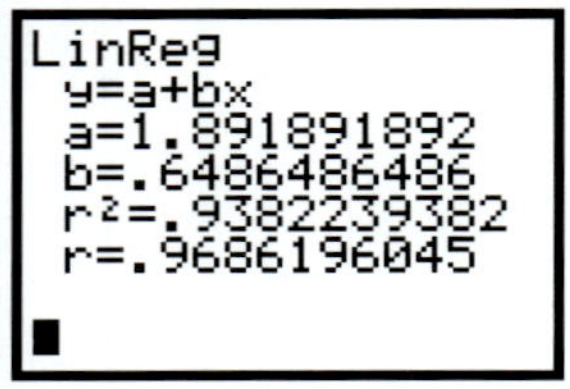

That is, we obtain the least squares regression line: $y = 1.89 + 0.65x$.

From this equation we can then make predictions about other data points that are not in the table.

Now, the values of a and b can also be obtained manually (as follows):

$$b = \frac{\sum xy - \frac{1}{n}(\sum x)(\sum y)}{\sum x^2 - \frac{1}{n}(\sum x)^2} = \frac{522 - \frac{1}{8} \times 63 \times 56}{621 - \frac{1}{8}(63)^2} = \frac{81}{124.875} = 0.6486$$

Then, $a = \bar{y} - b\bar{x} = \frac{56}{8} - \frac{81}{124.875} \times \frac{63}{8} = 1.8918$.

EXAMPLE 19.8 The data shown represents observations made on the rate of cricket sounds by the number y chirps per 15 seconds at different temperatures x in degrees Fahrenheit.

x	62	61	60	59	58	55	53	52	50
y	24	21	19	18	19	15	15	12	11

(a) Plot these points on a scatter diagram.
(b) Determine the line of regression of y on x and draw it on your scatter diagram.
(c) Estimate the cricket's rate when the temperature is 65 degrees Fahrenheit.

Solution

(a) We first enter the data into the TI–83 and then proceed to answer the questions.

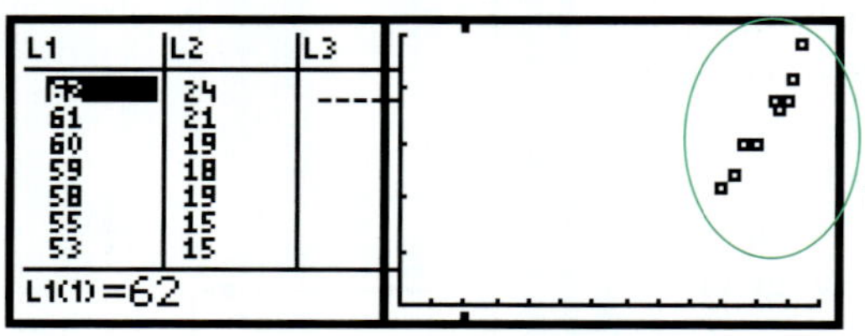

Using different scale

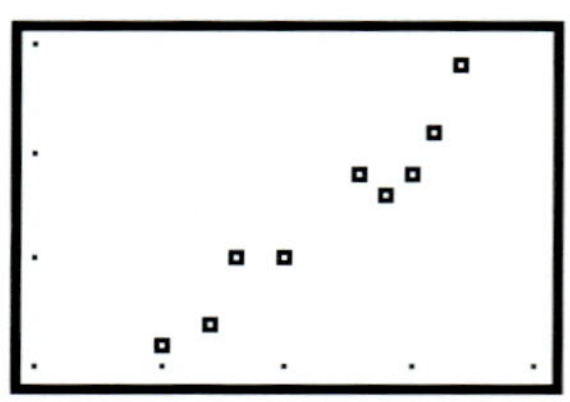

Clearly there is a strong positive relationship

(b) The regression equation is given by $y = -36.62 + 0.95x$ (see display below).

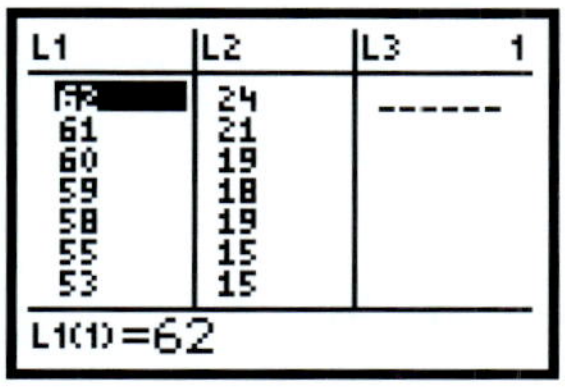

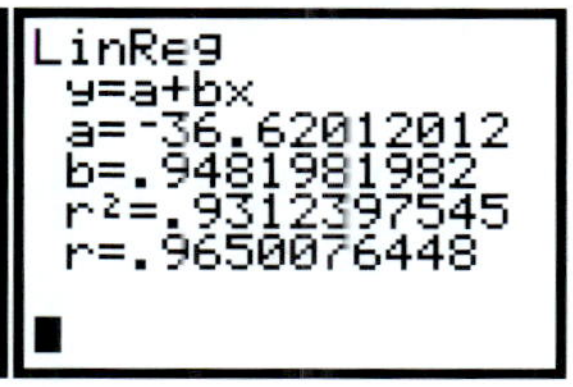

The line of regression of y on x:

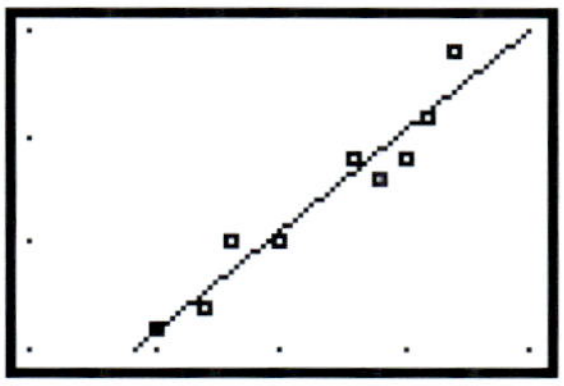

(c) To do this we used the **CALC** function and select **1:value**. So we estimate that the crickets' rate will be 25 when the temperature is 65 degrees Fahrenheit.

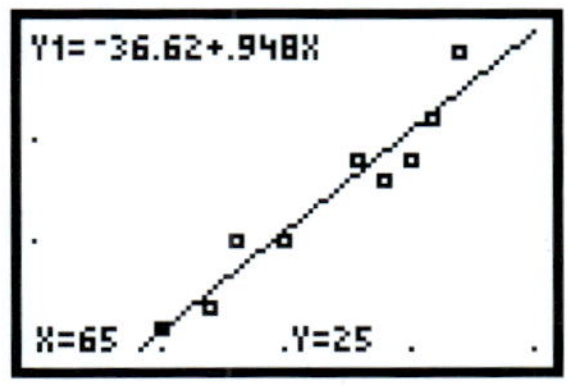

Because this value lies outside our data set, we are in fact **extrapolating** and so we are relying on the assumption that the linear trend based on our regression equation will continue.

19.3.4 ANOTHER FORM OF THE REGRESSION LINE

The regression equation can also be obtained by using the expression

$$(y - \bar{y}) = \frac{s_{xy}}{s_x^2}(x - \bar{x}) \quad \text{where} \quad \frac{s_{xy}}{s_x^2} = r \times \frac{s_y}{s_x}$$

We make use of Example 21.8 to illustrate this approach. We already have that $r = 0.9686$ (using the TI–83 output), then, using the **2-Var Stats** option from the **STAT** menu we have:

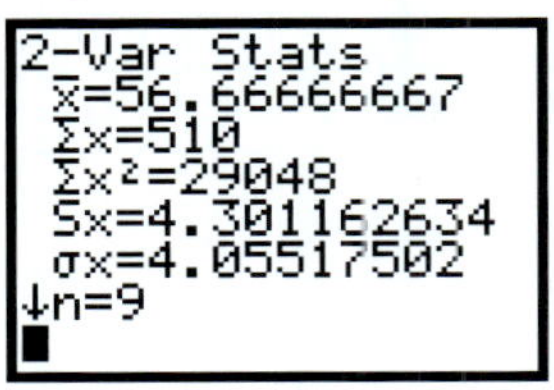

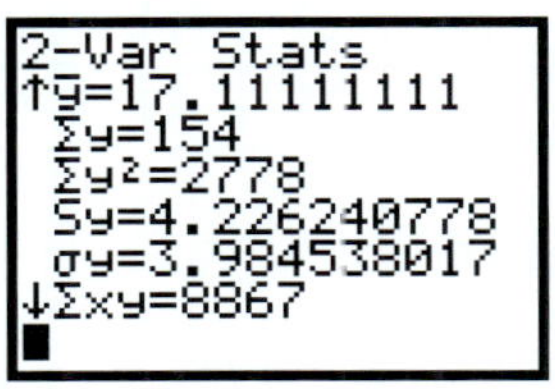

$s_x = 4.0551$, $s_y = 3.9845$,

$\bar{x} = 56.6667$ and $\bar{y} = 17.1111$.

Substituting these values into the equation we have that

$$(y - 17.111) = \frac{0.9686 \times 3.9845}{4.055}(x - 56.6667)$$

So that, $y - 17.111 = 0.9482(x - 56.6667)$

That is, $y = 0.9482x - 36.6215$

Giving us the same result to 2 d.p. (as expected).

Make sure that for examination purposes, you can use either formula.

EXERCISES 19.3

1. For each of the following sets of data
(a) draw a scatter diagram.
(b) make use of the locus of means approach to find the line of best fit.

i.

x	3	4	6	7	9	12
y	20	14	12	10	9	7

ii.

x	15	13	17	14	18	12	20	16	18	17	19
y	18	16	18	15	19	16	18	15	21	17	18

2. For the sets of data shown below,
(a) draw a scatter diagram.
(b) determine the least squares regression line.
(c) draw the regression line on your scatter diagram.

i.

x	3	4	6	7	9	12
y	20	14	12	10	9	7

ii.

x	2	1	4	5	3
y	4	2	6	5	3

iii.

x	11	5	4	5	2	3
y	52	31	30	34	20	25

iv.

x	1	2	3	4	5
y	2	1	3	5	4

3. The following table shows the income (in thousands of dollars) and the annual expenditure, in hundreds of dollars for ten single working people aged 20 – 24 yrs.

Income	22	14	16	18	20	19	16	18	19	18
Expenditure	75	59	67	69	75	73	62	64	70	71

(a) Plot the data on a scatter diagram.
(b) Find the correlation coefficient.
(c) Calculate the proportion of the variance of *Expenditure* which can be explained by the variance of the *Income*.
(d) Find the least squares equation of the regression line.
(e) On the scatter diagram from (a), sketch the regression line.

4. The result of the first two tests given to a group of Mathematics students is shown in the table below:

Test 1 (x)	60	50	80	80	70	60	100	40	90	70
Test 2 (y)	80	70	70	100	50	80	100	60	80	60

(a) Draw a scatter diagram for this data.
(b) Find the coefficient of correlation.
(c) Find the least squares regression line of
i. y on x. ii. x on y.

5. A cafe owner wishes to improve the efficiency of his cafe. One aspect that needs to be looked into is that of the rate at which customers are being served by the staff. The table below shows the number of weeks that eight employees have been working at the cafe and the average number of customers that each served **per hour**.

Weeks at cafe	8	5	15	3	10	2	13	6
Customers served	18.4	12.2	32.3	10	21.0	8.2	28.1	16.5

(a) Draw a scatter diagram for the given set of data.

Define the variable C to represent the average number of customers an employee served per hour and the variable w to represent the number of weeks that employee has been working at the cafe.

(b) The owner decides to use a straight line to model the data. Is the owner justified? Give a reason for your answer.

(c) i. Calculate the correlation coefficient for the given data set.
ii. Use the method of least squares to determine the line of best fit.
iii. Graph the regression line on the scatter diagram in part a.

(d) Estimate how many customers employees should be able to serve in one hour if they have been working at the cafe for
i. 9 weeks.
ii. 50 weeks.
iii. What constraints can you see this model having?

6. The table below shows the results of measurements taken for systolic blood pressures (y) of 8 women and their respective ages (x).

Age (x)	60	42	68	72	42	36	55	49
Blood pressure (y)	155	140	152	160	125	118	155	145

(a) Draw a scatter diagram for the given set of data.

(b) Calculate the correlation coefficient for the given data set. Is this an appropriate statistic to calculate for this data set?

(c) i. Use the method of least squares to determine the line of best fit.
ii. Graph the regression line on the scatter diagram in part a.

(d) Based on your line of best fit, determine the level of systolic blood pressure for a women aged
i. 45.
ii. 85.
iii. What is the difference in using the line of best fit when answering i., and ii.?

7. The yield, y kilograms, of a vegetable, obtained by using x kilograms of a new fertiliser produced the following results:

x	1.4	3.3	5.9	8.8	7.3	5.1
y	5.0	7.5	7.7	10	9	8.3

(a) Draw a scatter diagram for the given set of data.

(b) Calculate the correlation coefficient for the given data set. Is this an appropriate statistic to calculate for this data set?

(c) i. Use the method of least squares to determine the line of best fit.
ii. Graph the regression line on the scatter diagram in part a.

(d) Based on your line of best fit, determine the yield if
i. 6.5 kg of fertiliser were to be used.
ii. 10 kg of fertiliser were to be used.

8. The expected yield, y kilogram per unit area of a crop is related to the amount of fertiliser x kilogram per unit area. The data below give observed yields for various values of x:

x	2	12	3	16	5	6	20	8	13	10
y	20	50	20	57	29	38	67	44	59	39

(a) Draw a scatter diagram for the given set of data.

(b) Calculate the correlation coefficient for the given data set. Is this an appropriate statistic to calculate for this data set?

(c) i. Use the method of least squares to determine the line of best fit.
ii. Graph the regression line on the scatter diagram in part a.

(d) Based on your line of best fit, determine the yield if
i. 4 kg of fertiliser were to be used.
ii. 15 kg of fertiliser were to be used.

9. A firm which produces fungicides notices that the sales in its region appear to depend on rainfall during the previous month. The following data was collected from six months (that are non–consecutive). Unfortunately, some of the data has been smudged and cannot be read.

Rainfall x mm	11	[illegible]	25	27	48	[illegible]
Demand y (1000kg)	[illegible]	38	28	[illegible]	51	81

The following results were calculated just prior to the data smudging.

$\sum x = 180$, $\sum y = 270$, $\sum(x-\bar{x})^2 = 1240$, $\sum(y-\bar{y})^2 = 2554$, $\sum xy = 9592$

(a) Find the correlation coefficient for the given data set.

(b) Calculate the proportion of the variance of y which can be explained by the variance of x.

(c) Find the least square linear regression line, $y = a + bx$.

(d) Based on your regression line, predict the values that should be placed where the smudges occurred. Are there 'problems' in predicting some values?

10. The relationship between the temperature, T °C of water and the mass, M kg of a chemical substance dissolving in the water has been tabulated:

T °C	10	20	30	40	50	60	70	80	90
M kg	50	53	56	61	66	69	70	71	72

(a) Draw a scatter diagram for the given data set.
(b) Find i. the coefficient of correlation.
ii. the covariance.

(c) Find the least squares regression line which best fits the data by evaluating the constants a and b in the equation $M = a + bT$

11. The data set shown below, has a least squares regression line $y = a + bx$. By adding a constant, m to each x_i and to each y_i in the data set shown, how will

x	x_1	x_2	x_3	x_4	x_5
y	y_1	y_2	y_3	y_4	y_5

this affect the coefficient of correlation?

12. The content of sand found in soil at different depths from a particular area recorded and tabulated (as shown below).

x (depth in cm)	0	12	24	36	48	60	72	84	96
y (% sand)	80	63	64	62	57	59	41	47	38

(a) Draw a scatter diagram for the given set of data.

(b) Calculate the correlation coefficient for the given data set. Is this an appropriate statistic to calculate for this data set?

(c) i. Use the method of least squares to determine the line of best fit.
ii. Graph the regression line on the scatter diagram in part a.

(d) Using your line of best fit, determine the percentage of sand in the soil at a depth
i. of 40 cm. Would this be considered as extrapolation or interpolation?
ii. of 120 cm. Would this be considered as extrapolation or interpolation?

19.4 MISCELLANEOUS EXERCISES

1. The school nurse at Queens Hill High School is conducting a study into the relationship between the height and weight of school students. Everyday she weighs and measures students that attend her room. One particular day, 8 students need her attention. She carries out her study with the following results:

Height (cm)	130	120	155	160	140	140	165	135	150
Weight (kg)	45	40	60	61	60	54	70	50	58

(a) Plot the scatter diagram for this data set.

(b) Use the method locus of means to find the line of best fit for this data and draw this line on your scatter diagram.

(c) Calculate the product moment correlation coefficient for this data. What can you conclude from this value?

2. The reaction time of drivers is being measured against their BAC (Blood Alcohol Content) by a policewoman. She plots a scattergram of BAC on the x–axis against reaction time (measured in seconds) on the y-axis. She decides to obtain a linear fit to the data by using the locus of means method. The lower half mean coordinate is $(0.03, 0.17)$ and the upper half mean coordinate is $(0.18, 0.62)$. By studying road accident and medical data she concludes that people with reaction times greater than 0.25 seconds constitute a driving risk. From her model, estimate the minimum BAC which would put people at risk due to increased reaction times

3. Which bivariate sets X and Y would be expected to display a Pearson's correlation coefficient closest to –1?

A. X: the number of words in a book, Y: the number of pages in a book
B. X: the time taken to cycle 10 km, Y: the average speed of the cyclist
C. X: the face up value of a die
Y: the face up value of a die Y in a set of trials where the two dice X and Y are tossed 100 times
D. X: students test marks in Chemistry, Y: students test marks in Mathematics
E. X: Maximum daily temperature Wales Y: Minimum daily temperature in Wales

4. The equation of least squares regression line of y on x is $y = 2x + 7$. The standard deviation of the **Y** data set is $s_Y = 5$ and for the **X** data set it is $s_X = 2$.

(a) Find the product moment correlation coefficient.
(b) Make a brief statement that best describes the level of correlation between **Y** and **X**.

5. For x-values, with corresponding y-values, the results were ordered for x, then the means for x and y for the upper and lower half of the data were calculated. The points were $(5, 3)$ and $(3, 5)$. Using this information, find the predicted value of y when $x = 6.5$.

6. The data in the following table shows the percentage marks in Mathematics (M), English (E) and Physics (P) for a group of students in 2004.

Student	1	2	3	4	5	6	7	8	9	10	11	12
M	100	72	86	88	98	91	72	98	100	75	75	84
E	88	75	88	72	86	68	63	91	91	64	63	82
P	88	65	88	79	94	78	67	100	100	62	65	73

(a) Calculate, correct to two decimal places, the product moment correlation coefficient between Maths marks and English marks.
[You may use the following information:

$\sum M = 1039$ $\quad \sum M^2 = 91303$ $\quad \sum ME = 81687$

$\sum E = 931$ $\quad \sum E^2 = 73637$]

(b) i. Find the equation of the two mean regression line of P on M, giving all constants correct to two decimal places.

ii. Predict the Physics mark for a student who scored 80 in Mathematics.

(c) i. Find the least squares regression equation of M on E.

ii. Predict the Maths mark for a student who scored 95 in English.

iii. In part ii., did you interpolate or extrapolate? How confident are you in your prediction?

(d) If you wished to predict the English marks for a student who had scored 90 in Mathematics, how would you go about doing this.

7. For the data below describe the type of relationship that exists between the variables X and Y, including
i. the direction
ii. the form
iii. the strength of relationship

x	1	9	7.5	10	2.5	5.0	11	5.5	14	6
y	10	47	45	60	20	35	65	23	68	35

8. The following summary was based on 7 ordered pairs of data (x, y):
$\sum x = 144$, $\sum y = 223$, $\sum xy = 4960$, $s_x = 6.08$, $s_y = 10.59$, $s_{xy} = 62.10$.

(a) What is the mean of x and y?
(b) Determine the product–moment correlation coefficient.
(c) Find the least squares regression line of y on x.

9. The results of a set of 10 paired data (x, y) is summarised as follows:
$\sum x = 224.4$, $\sum x^2 = 5099.12$, $\sum y = 254.3$, $\sum y^2 = 6706.91$, $\sum xy = 5595.30$

(a) Determine the equation of the least squares regression line, $y = a + bx$.
(b) Predict the value of x when $y = 25$.
(c) Determine the covariance, $\text{Cov}(X, Y)$.

10. The table shows the income (in thousands of dollars) and the annual expenditure, in hundreds of dollars for ten single male workers aged 30 – 40 yrs.

Income	42	34	36	38	40	39	36	38	39	38
Expenditure	55	39	47	49	55	53	42	44	50	51

(a) i. Plot the data on a scatter diagram.
ii. Find the correlation coefficient.

(b) Calculate the proportion of the variance of *Expenditure* which can be explained by the variance of the *Income*.

(c) Find the least squares equation of the regression line.

(d) On the scatter diagram from (a), sketch the regression line.

(e) Estimate the expenditure by a single working person aged 30–40 yrs if their annual income is $37 000.

11. The relationship between the amount of chemical, x gm, in an item and the durability, y days of the item is thought to be linear. Samples were obtained from 35 randomly selected items. The results of this sample are:

$$\sum x = 154, \sum y = 492, \sum xy = 2235, \sum x^2 = 821, \sum y^2 = 7214$$

(a) Determine the mean and the standard deviation of
i. the amount of chemical in the items.
ii. the durability of the item.

The least squares y on x regression line for this set of data is given by $y - \bar{y} = b(x - \bar{x})$.

(b) Find the value of b, and **use this** to find the product moment correlation coefficient.
(c) Calculate the proportion of the variance of *durability* which can be explained by the variance of the *chemical content*.
(d) Estimate the durability of an item containing 3.5 gm of the chemical.

12. A farmer is trying to establish a relationship between the final weight, x lbs and the carcass weight, y lbs, for a particular type of bull. From a pen of 10 such bulls, she obtained the following weights.

x	1030	1000	1060	980	995	1025	1055	1035	1380	1085
y	614	577	654	594	593	589	629	650	834	691

(a) i. Plot the data on a scatter diagram.
ii. Find the correlation coefficient.
(b) Find the least squares equation of the regression line.
(c) On the scatter diagram from (a), sketch the regression line.
(d) Estimate the carcass weight of a bull if its final weight is known to be 1040 lbs.

19.5 GRADED REVISION EXERCISES

LEVEL 1

1. (a) Draw a scatter plot for the data in the table below.

x	26	28	34	35	38	39	45	45	50	60
y	8	6	12	12	15	10	16	20	19	22

(b) Describe the direction, form and strength of the relationship between x & y.

2. For the set of data shown plot a scatter diagram.

x	5	6	8	8	6	9
y	8	6	1	3	6	0

Describe the direction, form and strength of the relationship.

LEVEL 2

1. For question 1, Level 1, determine the correlation coefficient.

2. Find the coefficient of correlation, r, for the set of data below.

x	12	14	15	17	19	20	21	25
y	5	6	8	8	9	11	12	13

LEVEL 3

1. Interpret the values of r and r^2 obtained in question 2, Level 2.

2. The grades obtained in two assessment tests in a small class studying Latin were:

Test 1 (x)	10	10	13	15	18	19	21	22
Test 2 (y)	15	16	13	13	9	10	11	9

(a) Plot these points on a scatter diagram.
(b) Determine the line of regression of y on x and draw it on your scatter diagram.
(c) Estimate the score on Test 2 if a score of 14 was obtained on Test 1.

3. Ten observations on, x and y, show a linear relationship with the following results:

$$\sum x = 900, \sum y = 140, \sum x^2 = 81956, \sum y^2 = 2214, \sum xy = 13054.$$

The least squares y on x regression line for this set of data is given by

$$y - \bar{y} = b(x - \bar{x})$$

(a) Find the value of b, and **use this** to find the product moment correlation coefficient.
(b) Estimate the value of y when $x = 40$.

4. What correlation between X and Y is required in order to assert that 90% of the variance of X depends on the variance of Y?

LEVEL 4

1. In a class of 5 students, an analysis of their final examination scores (y) compared to their total test scores (x) revealed the following results.:

Test (x)	450	560	355	670	825	910	620	715
Exam (y)	52	60	45	75	90	99	65	80

(a) Plot a scatter diagram for this set of data.

(b) Is a straight line fit an appropriate model for this data set? Why?

(c) i. Find the least squares regression line of y on x.
ii. Graph the regression line on the scatter diagram in part (a).

(d) Use the value of b, in the regression equation, $y = bx + c$, to determine the product moment correlation coefficient.

(e) Estimate the final exam score if a student obtains a total test score of 600.

(f) Is it possible to estimate the total test score based on a final exam score?

How would you go about estimating a test score if a student's final exam score was 56?

2. A model to determine the population density, d, in relation to its radial distance, r, from the city centre, is thought to be of the form $d = k \times e^{-\alpha r}$. The following table displays a set of data taken at 8 different places.

r	0	2	4	6	8	10	12	14	20
d	150	54	20	40	16	6	2	7	1

(a) Plot a scatter diagram of d versus r.

In order to fit the model, a logarithmic transformation is used:

$$\log_e d = \log_e k - \alpha r$$

(b) Using the least squares $\log_e d$ on r regression line derive an estimate of k and α.

(c) i. Plot the fitted curve on the scatter diagram.
ii. Estimate the population density at a radial distance of 16 km from the city centre.

19.6 TOPIC TEST

1. (a) Plot a scatter diagram of the data set tabulated below.

(b) Describe the type of relationship that exists between the variables a and b, including
- i. the direction
- ii. the form
- iii. the strength of relationship

a	2	4	6	6	10	12
b	40.8	32.2	24.2	26.0	14.5	9.2

(c) Use the two means method to find the line of best fit for this data set.

[11 marks]

2. The results of a set of 25 paired data (x, y) is summarised as follows:

$$\sum x = 150, \sum x^2 = 1080, \sum y = 225, \sum y^2 = 2065, \sum xy = 1423$$

(a) Determine the equation of the least squares regression line, $y = a + bx$.

(b) Predict the value of y when $x = 8$.

(c) Find the value of S_{xy}.

[5 marks]

3. What correlation between X and Y is required in order to assert that 80% of the variance of X depends on the variance of Y?

[2 marks]

4. The amount of potassium bromide, x gm, that would dissolve in 100ml of water was recorded for various temperatures, T °C, with the following results:

T	0	10	20	30	40	50
x	54	59	64	73	76	82

(a) Plot a scatter diagram of the data set.

The least squares regression line, $x = a + bT$ is obtained by solving the two simultaneous equations in a and b:

$$na + (\sum T)b = \sum x \quad \text{and} \quad (\sum T)a + (\sum T^2)b = \sum xT$$

(b) Find the least squares regression line of T on x using the simultaneous equations given above.

(c) On your scatter diagram, draw the least squares regression line.

(d) How much potassium bromide do you predict will dissolve in 100ml of water at 25°C?

(e) i. What would you need to do in order to try and predict the temperature required in order to dissolve 70 gm of potassium bromide?

ii. Predict the required temperature in (e) i ,.

[2 + 3 + 2 + 1 + 1 + 3 marks]

Total 30 marks

19.7 PROJECT SUGGESTIONS

The chapter introduction covered some data that might be of commercial interest to owners of vinyards, distilleries and breweries.

Large amounts of genuine data of this sort is now freely available on the internet. As these websites change their addresses, you are advised to type your query into a search engine and follow from there. Try to choose a topic that interests you and is relevant to other people!

How does cigarette smoking affect life expectancy?

What is the connection between drug use and road accidents?

How does family income impact on life expectancy?

How does national income affect life expectancy?

Throughout this chapter we have only considered one measure of correlation, i.e., Pearson's product moment correlation coefficient. However, there are a number of other measures of correlation, for example, the q-correlation coefficient or the Spearman's rank correlation coefficient. Equally, instead of using the two mean regression equation, you could consider the three median regression equation.

A possible investigation could be using these other methods and comparing the results with the methods discussed in this chapter.

Such an investigation should include the investigation of data sensitivity issues and which approach is most robust.

20.0 LONG NECKS AND ISLANDS

The photograph shows a large (over 1 metre long) land tortoise on the Pacific island of Santa Cruz. As you can see, this animal has quite a long neck. Tortoises on other islands in the Galapagos chain have necks of different lengths, some longer and some shorter. You might say that this does not matter. After all, some humans have longer necks than others and no-one worries about that.

However, it was this very question (and a similar one about the lengths of the beaks of the finches that live on the islands) that led to one of the most important discoveries in the history of science.

In 1831, and following an undistinguished period at school and university, Charles Darwin secured the post of naturalist aboard the tiny sailing brig *HMS Beagle* which was about to leave on a round the World trip of exploration. The trip lasted five years during which time Darwin suffered from seasickness and tropical disease. In 1835 the Beagle reached the Galapagos islands, a desolate series of volcanic rocks that were used as a penal colony. It was apparently the prison governor who told Darwin that he could tell which island a tortoise came from by looking at its carapace (shell) and its neck.

On his return to England (with a copious diary and thousands of samples), Darwin married and moved to the village of Downe in Kent (UK) where he spent hours in his garden (see left) thinking about what he had seen and collected on his voyage. After much worry (Darwin was concerned about how his revolutionary ideas would be received), in 1859 he published the *Origin of Species by Means of Natural Selection*. The idea of evolution was born.

Darwin was right to be concerned about how his book would be received. He was vigorously attacked. Darwin died in 1882 at his home in Downe. Today, he is commemorated in many ways. Cities have been named after him as have a finch and a frog. The Darwin Foundation, based in the Galapagos Islands, still researches evolution and works to preserve the islands. His true memorial is, however, an idea.

So what does this have to do with mathematics?

We began with the observation that a tortoise might have an unusually long neck.

Just how do we decide if this variation is normal or due to some significant factor that is worth investigating? The topic of 'significance' forms the subject of this chapter.

20.1 CONTINGENCY TABLES

Consider the following table listing frequencies based on a two-way classification – i.e., classification of 400 individuals that are or are not smokers and that have or do not have lung cancer:

		Classification B		
		Cancer	No cancer	Total
Classification A	Smoker	32	188	220
	Non-smoker	4	176	180
	Total	36	364	400

As there are two divisions for category A and two divisions for category B, we say that we have a 2×2 contingency table.

The question of most interest in this case is whether there exists a relationship between the classifications A and B. In this instance, we would like to determine if there exists a relationship between smoking and lung cancer.

If there is **no relationship** then we would conclude that the **classifications are independent**.

We need to consider the problem of testing for the independence of two factors from the same sample.

To do this we look at the difference between **observed values** and **expected values** (under the hypothesis, H_0, known as the **null hypothesis**). If we reject the null hypothesis H_0, then we accept the alternative hypothesis, denoted by H_1 – indicating that there does exist a relation between the two factors.

The basis for such a statistic, i.e., one that involves a comparison between observed values and expected values, is that if the observed values are in fact very different from their expected values then it would seem reasonable that there was some influencing factor involved to create such a discrepancy.

How do we carry out the test to determine the independence between the two factors?

We consider the following table of observed values:

Observed frequencies:

	B1	B2	
A1	a	b	$a+b$
A2	c	d	$c+d$
	$a+c$	$b+d$	N

If the Bs and As are independent, then $p(A1 \cap B1) = p(A1) \times p(B1) = \frac{a+b}{N} \times \frac{a+c}{N}$

Therefore, under the assumption that the factors are independent, the expected observation in cell

$A1 \cap B1$ is given by $\left(\frac{a+b}{N} \times \frac{a+c}{N}\right) \times N = \frac{(a+b)(a+c)}{N}$.

We can then place this figure in the table of expected values corresponding to $A1 \cap B1$:

Expected frequencies

	B1	B2	
A1	$\frac{(a+b)(a+c)}{N}$		
A2			
			N

That is, we multiply the sum in the first row with the sum in the first column and divide by the total N:

Observed frequencies

	B1	B2	
A1	a	b	$a+b$
A2	c	d	$c+d$
	$a+c$	$b+d$	N

→

Expected frequencies

	B1	B2	
A1	$\frac{(a+b)(a+c)}{N}$		
A2			
			N

In our example, this means that for the observed value 32, we would have a corresponding expected value of $\frac{36 \times 220}{400} = 19.8$.

Similarly, to find the expected value in cell $A1 \cap B2$, we multiply the sum in the first row with the sum in the second column and divide by the total N:

Observed frequencies

	B1	B2	
A1	a	b	$a+b$
A2	c	d	$c+d$
	$a+c$	$b+d$	N

→

Expected frequencies

	B1	B2	
A1		$\frac{(a+b)(b+d)}{N}$	
A2			
			N

In our example, this means that for the observed value 32, we would have a corresponding expected value of $\frac{36 \times 220}{400} = 19.80$. Continuing in this manner we have:

Observed frequencies

	B1	B2	
A1	32	188	220
A2	4	176	180
	36	364	400

Expected frequencies

	B1	B2	
A1	19.80	200.20	220
A2	16.20	163.80	180
	36	364	400

Now that we have the table of observed and expected values, how do we compare them?

It turns out that the statistic $U = \sum\frac{(\text{Obs} - \text{Exp})^2}{\text{Exp}}$, where **Obs** stands for the **observed** value and **Exp** stands for the **expected value**, can be approximated by a distribution known as the **chi squared distribution**, denoted by χ^2, which depends only on the number of **degrees of freedom**, υ. At this stage we leave out the formal definition of degrees of freedom, in preference for stating how to calculate the number of degrees of freedom for contingency tables.

For contingency tables made up of $r \times c$ cells, the number of degrees of freedom is given by $\upsilon = (r-1)\times(c-1)$.

The notation $\chi^2(\upsilon)$ is used to indicate the number of degrees of freedom explicitly.

For our situation, we have that $r = 2$ and $c = 2$, therefore, $\upsilon = (2-1)\times(2-1) = 1$.

In summary, we have:

the χ^2 – test of independence in an $r \times c$ contingency table is based on the result that

$$U = \sum\frac{(\text{Obs} - \text{Exp})^2}{\text{Exp}} \sim \chi^2(\upsilon)\text{, where } \upsilon = (r-1)\times(c-1).$$

We are now in a position to calculate the value U from our tables of expected and observed values:

$$\begin{aligned} U &= \frac{(32-19.8)^2}{19.8} + \frac{(188-200.2)^2}{200.2} + \frac{(4-16.2)^2}{16.2} + \frac{(176-163.8)^2}{163.8} \\ &= 7.5171 + 0.7435 + 9.1877 + 0.9087 \\ &= 18.36 \end{aligned}$$

Now that we have the value of U, what conclusion can we make?

As we have already mentioned, if H_0 is true (i.e., factors are independent), then U is inclined to be small, since the observed values should be near the expected values. On the other hand, if H_0 is not true, then U will be large since the observed values are unlikely to be near the expected values, so that each term $(\text{Obs} - \text{Exp})^2$ is large.

Decision making process:

We **reject** H_0 if $U > c_{(1-\alpha)}\chi^2(\upsilon)$ using an $\alpha\%$ level of significance,

where the level of significance indicates the level or 'error' we are willing to accept in making our conclusion. Typically, we use levels of 1%, 5% and 10%.

What remains then, is to determine what this 'critical value' $c_{(1-\alpha)}\chi^2(\upsilon)$ is (as it will determine whether we accept or reject the null hypothesis).

A list of $c_{0.95}\chi^2(\upsilon)$ values for different values of υ is provided:

υ	**1**	**2**	**3**	**4**	**5**	**6**	**7**
$c_{0.95}\chi^2(\upsilon)$	3.841	5.991	7.815	9.488	11.07	12.59	14.07

These are the critical values that determine if we reject H_0 at the 5% level of significance. So that if we calculate a value of U greater than 9.488 (say) when dealing with 4 degrees of freedom we would reject H_0.

For our example, we have $U = 18.36 > c_{0.95}\chi^2(1)$ $(= 3.841)$ so we would reject H_0 at the 5% level of significance.

We can summarise the process as follows:

Step 1: From the table of observed values, construct the corresponding table of expected values.

Step 2: Calculate the value of $U = \sum \frac{(\text{Obs} - \text{Exp})^2}{\text{Exp}}$.

Step 3: Calculate the number of degrees of freedom. For an $r \times c$ contingency table, the number of degrees of freedom, υ, is given by $\upsilon = (r-1) \times (c-1)$.

Step 4: Decide on a level of significance, $\alpha\%$, (i.e., level of error you are willing to accept in making your decision).

Step 5: Look up the critical value, $c_{(1-\alpha)}\chi^2(\upsilon)$

step 6: If $U > c_{(1-\alpha)}\chi^2(\upsilon)$, we reject H_0 (i.e., reject that the factors are independent) in preference for H_1 (that there exists a relation between the two categories).

EXAMPLE 20.1

We want to decide if hair colour is dependent on eye colour. A sample of 200 randomly selected people had their eye and hair colour noted, the results are shown in the frequency table below:

	Dark hair	Fair hair	
Brown eyes	54	32	86
Blue eyes	42	72	114
	96	104	200

Solution

Step 1: Construct the corresponding table of expected values.

	Dark hair	Fair hair	
Brown eyes	$\frac{86 \times 96}{200} = 41.28$	$\frac{86 \times 104}{200} = 44.72$	86
Blue eyes	$\frac{96 \times 114}{200} = 54.76$	$\frac{114 \times 104}{200} = 59.28$	114
	96	104	200

Step 2:

$$U = \frac{(54-41.28)^2}{41.28} + \frac{(32-44.72)^2}{44.72} + \frac{(42-54.76)^2}{54.76} + \frac{(72-59.28)^2}{59.28} = 13.24$$

Step 3:

We have a 2×2 contingency table, therefore the number of degrees of freedom = 1 [i.e., $(2-1) \times (2-1) = 1 \times 1 = 1$]

Step 4:

We will use a 5% level of significance.

Step 5:

$c_{(1-\alpha)}\chi^2(\upsilon)$ with $\alpha = 5\%$ gives a value of 3.841.

Step 6:

As $U = 13.24 > 3.841$ we would clearly reject H_0 at the 5% level of significance.

Using a Graphics Calculator

This problem can also be solved using the TI–83:

First we need to set up the contingency table for the observed values. This is done using matrices:

Step 1: Press **MATRX** and move cursor to **EDIT**

Step 2: Specify the number of rows and columns and then enter the observed values in the matrix cells.

Step 3: Press **STAT**. Move cursor to **TESTS**. Press **ENTER** and use the down arrow keys to select option **C: χ^2 – Test**

Step 4: The expected values are automatically calculated and placed in Matrix B. All that remains is to select **Calculate**. Move cursor over **Calculate** and press **ENTER**.

Step 5: Read from the screen, the calculated value $\chi^2 = U$ and make your conclusion.

Step 1 & 2:

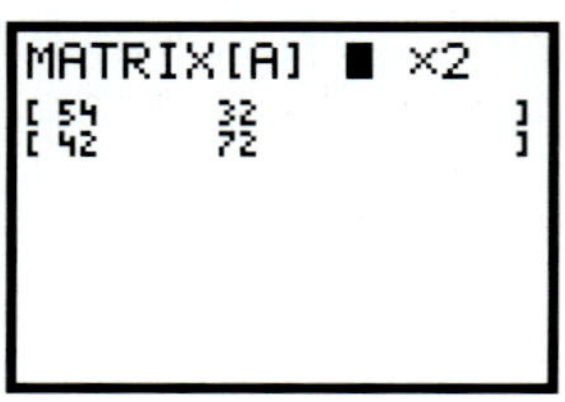

Step 3:

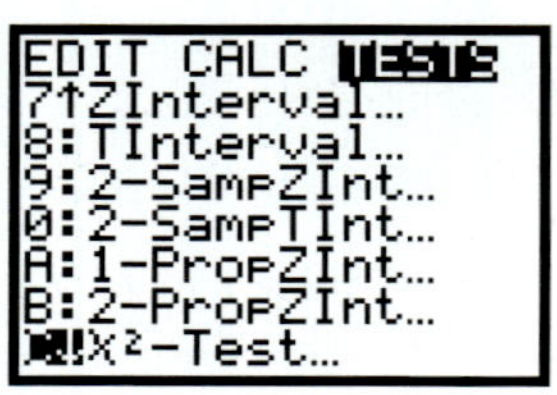

Step 4:

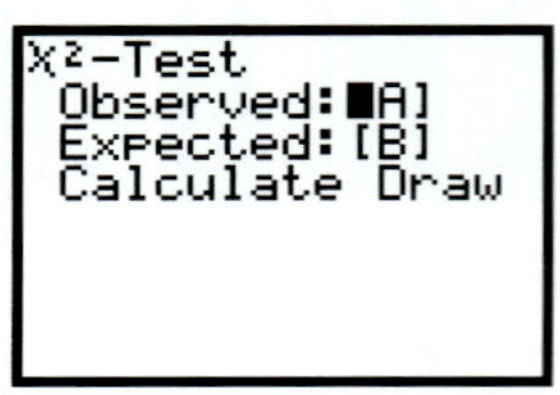

Step 5:

```
X²-Test
 X²=13.2238019
 P=2.7641636E-4
 df=1
```

In this case, as the calculated value $U = 13.22 > 3.841$, we would reject H_0, i.e., we conclude that 'The colour of a person's eyes and the colour of a person's hair are not independent.'

Notice that the display gives no indication of the critical value, $c_{(1-\alpha)}\chi^2(\upsilon)$, and as such, without it, we are unable to make a conclusion about H_0. This means that we would need to have access to a table of values like the one we provided earlier – but even that table only caters for up to 7 degrees of freedom and only for a level of significance of size 5%.

On the otherhand, the TI produces a '*p-value*'. This '*p-value*' gives the probability of calculating a '*value of U as or more extreme than the one obtained if* H_0 *was assumed to be true*'.

In this case, the '*p*–value' is very small, 0.000276, meaning the the probability of obtaining a value of $U = 13.22$ under the assumption that H_0 is true is very small and so we would reject H_0.

If we do not have the critical values that will determine if we reject H_0, we can use the *p*–value.

The corresponding *p-values* for set levels of significance, $\alpha\%$, are as follows:

If $p < 0.01$, then reject H_0 at the 1% level of significance.

If $p < 0.05$, then reject H_0 at the 5% level of significance.

If $p < 0.10$, then reject H_0 at the 10% level of significance.

A table of critical χ^2 values for $\alpha = 1\%, 5\%$ and 10% and degrees of freedom (*d.f*) from 1 to 16 is given below. Use this table if your calculator does not produce *p*–values (or χ^2 values).

d.f	1	2	3	4	5	6	7	8	9	10	11	12	13	14	15	16
1%	6.63	9.21	11.34	13.28	15.09	16.81	18.48	20.09	21.67	23.21	24.73	26.22	27.69	29.14	30.58	32.00
5%	3.84	5.99	7.81	9.49	11.07	12.59	14.07	15.51	16.92	18.31	19.68	21.03	22.36	23.68	25.00	26.30
10%	2.71	4.61	6.25	7.78	9.24	10.64	12.02	13.36	14.68	15.99	17.28	18.55	19.81	21.06	22.31	23.54

EXAMPLE 20.2

It is decided to test the hypothesis that there is no relationship between smoking and gender. A sample of 30 males and 40 females has been randomly selected from a population. The results are:

	Male	7 Smokers (S), 23 Non–smokers(N)
	Female	13 Smokers (S), 27 Non–smokers(N).

Is there a significant relationship between smoking and gender?

Solution

The first thing to do is set up a 2×2 contingency table.
We use M to denote Male, F for Female and S and N as described in question.

Observed frequencies

	S	N	
M	7	23	30
F	13	27	40
	20	50	70

Expected frequencies

	S	N	
M	8.57	21.43	30
F	11.43	28.57	40
	20	50	70

Using the test statistic U with $\upsilon = (2-1) \times (2-1) = 1$ degree of freedom, we have

$$U = \frac{(7-8.57)^2}{8.57} + \frac{(23-21.43)^2}{21.43} + \frac{(13-11.43)^2}{11.43} + \frac{(27-28.57)^2}{28.57} = 0.705 .$$

We now make use of the TI–83:

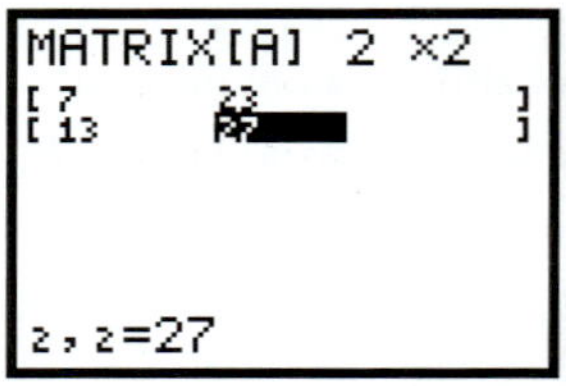

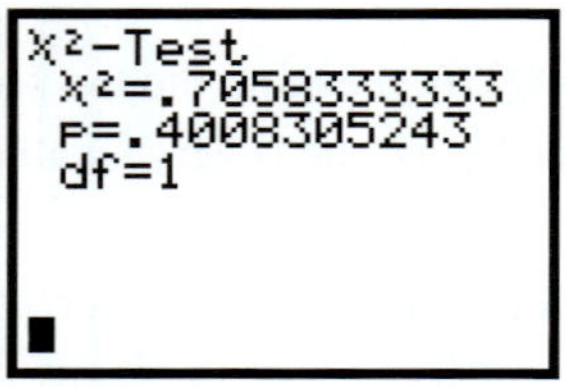

We can also check the expected values:

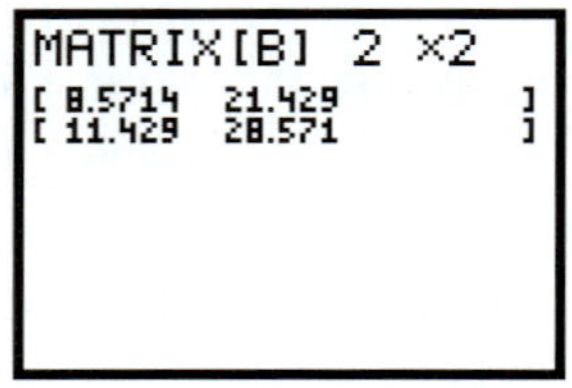

Based on the p–value of 0.40 (>0.05), we accept H_0 at the 5% level of significance.

Also, note that $U = 0.7058 < c_{0.95}\chi^2(1)\ (= 3.841)$ and so we accept H_0 (concurring with the previous result).

The following example includes 'tags' as a guide to the steps described earlier in our summary of how to carry out an analysis of data that involves contingency tables.

EXAMPLE 20.3

A serum is administrated to one of two groups of adults who all have a disease. In group A (the group who were given the serum), 76 recover from the disease and in group B (known as the control group) 64 of them recover from the disease. There are 100 people in each group. Does this serum help to cure the disease?

Solution

We set up our test as follows:

Hypothesis: H_0 : Serum has no effect versus H_1 : Serum has an effect
That is, recovery is independent of the use of the serum.

Test statistic: $U = \sum \frac{(\text{Obs} - \text{Exp})^2}{\text{Exp}} \sim \chi^2(1)$

Level: Use 5% level of significance.
Therefore reject H_0 if $U > c_{0.95}\chi^2(1) = 3.841$.

Next, set up the tables of observed values and expected values:

Observed frequencies

	R	$\bar{R}$	
A	76	24	100
B	64	36	100
	140	60	200

Expected frequencies

	R	$\bar{R}$	
A	70	30	100
B	70	30	100
	140	60	200

Our test statistic then has a value $U = \frac{(76-70)^2}{70} + \frac{(24-30)^2}{30} + \frac{(64-70)^2}{70} + \frac{(36-30)^2}{30}$

$= 3.4287$

Therefore, as our value of U is less than 3.841 we accept H_0 at the 5% level of significance. That is, we accept the hypothesis that the serum has no effect.

Again, using the TI–83, we have:

Observed values

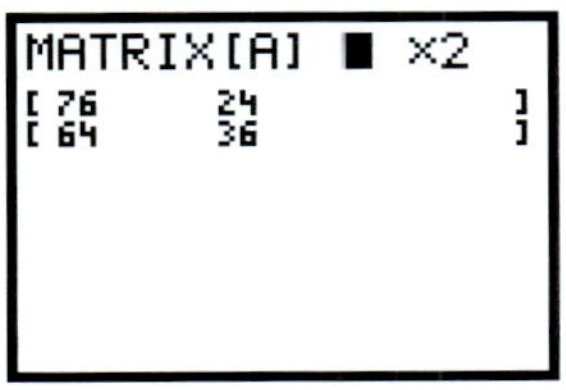

Expected values

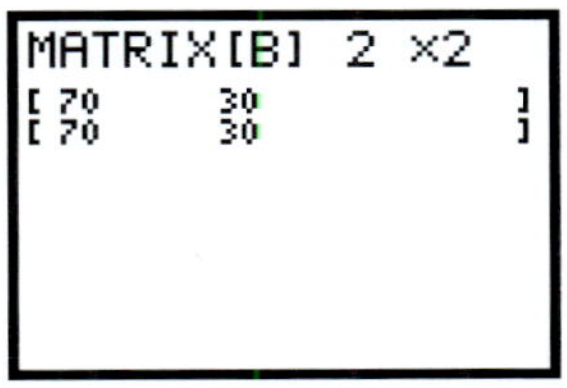

Results:

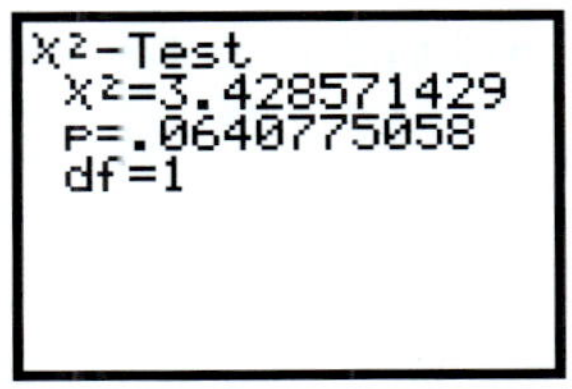

Similarly seeing as $p > 0.05$, we accept the null hypothesis H_0.

Although we have only looked at 2×2 contingency tables, there is no reason why we cannot have 2×3 or 4×3 contingency tables. Next we consider a contigency table that has at least 2 categories per classification.

EXAMPLE 20.4

The issue of capital punishment is raised in a particular state. Three hundred and sixty individuals were randomly selected and grouped by age. Their responses to the question 'Should there be capital punishment in their state' is summarised in the table below. Is there evidence to support that capital punishment opinion and age grouping is dependent?

		Age group			
		18 - 32	33 - 45	46 - 60	
Capital punishment	Yes	22	25	16	63
	No	98	115	84	297
		120	140	100	360

Solution

Based on the observed values, we can calculate the expected values:

Expected values:

		Age group			
		18 - 32	33 - 45	46 - 60	
Capital punishment	Yes	21	24.5	17.5	63
	No	99	115.5	82.5	297
		120	140	100	360

To use the TI–83, first clear all existing matrices:

Press **2nd +**, select **2:Mem Mgmt/Del. . .**, press **ENTER**, select **5:Matrix. . .**, press **ENTER** to select the matrices that need to be deleted (the * symbol will appear) then press **DEL** and then select option **2:Yes** and press **ENTER**.

Once this is done we then enter the observed values into matrix A, and proceed as we have done previously.

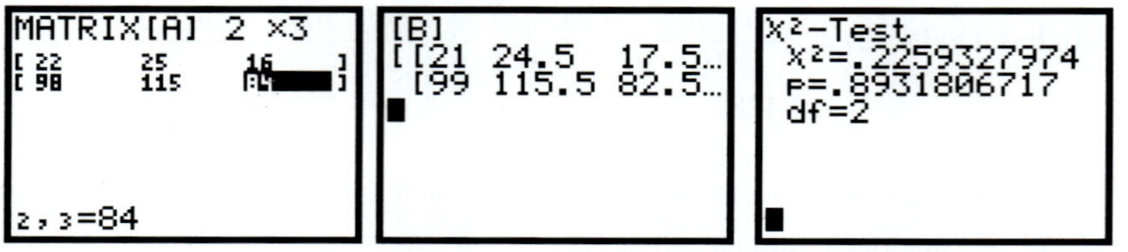

From the p-value, we see that we do not reject H_0 at the 5% (or even 10%) level of significance. That is, we conclude that in fact, opinion on capital punishment is independent of age group.

Equally, we could have calculated the U - value:

$$U = \sum\frac{(\text{Obs} - \text{Exp})^2}{\text{Exp}} = \frac{(22-21)^2}{21} + \frac{(25-24.5)^2}{24.5} + \frac{(16-17.5)^2}{17.5} + \ldots + \frac{(84-82.5)^2}{82.5}$$

$$= 0.2160$$

[Notice that this value differs slightly from that displayed on the screen i.e., 0.2259. This is due to rounding off errors.]

Now, the degrees of freedom for a 2×3 contingency table is given by $(2-1) \times (3-1) = 1 \times 2 = 2$. Therefore, we can look up the table of critical χ^2 values at a level of 5% and 2 degrees of freedom. The tables gives a figure of 5.99. Then as our value $U < 5.99$ we accept H_0 at the 5% level of significance – i.e., opinion on capital punishment is independent of age group.

So, apart from having more categories than a 2×2 contingency table, $r \times c$ contingency tables only require that we calculate extra expected values and look up critical values of χ^2 that correspond to $(r-1) \times (c-1)$ degrees of freedom.

It should also be pointed out at this stage that the expression $U = \sum\frac{(\text{Obs} - \text{Exp})^2}{\text{Exp}}$ can appear in different forms. The form adopted by the I.B.O and hence used in your examination is

$$\chi^2_{calc} = \sum\frac{(f_o - f_e)^2}{f_e},$$

where the observed values, $\text{Obs} = f_o$,

the expected values, $\text{Exp} = f_e$ and

the calculated chi-squared value, $U = \chi^2_{calc}$.

This is only a relabelling of the expression, nothing more. Nonetheless, become familiar with both expressions.

EXERCISES 20.1

1. For the tables shown below, test (at a 5% level of significance) if Factor A is independent of Factor B.

(a)

FACTOR B / FACTOR A	A1	A2	
B1	72	418	490
B2	38	92	130
	110	510	620

(b)

FACTOR B / FACTOR A	A1	A2	
B1	40	12	52
B2	11	9	20
	51	21	72

2. The following table shows the results from a random sample carried out so that the question about the relationship between education and job satisfaction could be analysed.

Have you completed University studies? / *Are you satisfied with your job?*	YES	NO	
YES	272	618	890
NO	238	292	530
	510	910	1420

The questions were: '*Have you completed University studies?*' and
'*Are you satisfied with your job?*'
Test the hypothesis, using a 5% level of confidence, that the responses are independent.

3. The issue of capital punishment was raised in a particular country. There was some thought that one of the sexes was prone to agree with the use of capital punishment for some crimes. The responses of 1000 people were recorded as follows:

	FEMALE	MALE
YES	272	303
NO	188	237

Test, at the 5% level of significance, whether a person's view on capital punishment is independent of their gender.

4. A new therapy is to be tested on patients at a local hospital. The following data relates to patients that attended this local hospital.

In particular, it is desirable to determine if the method of therapy is independent of the patient's level of improvement. The data from this test is as follows:

Therapy 1: 21 registered improvement   33 did not register improvement
Therapy 2: 14 registered improvement   42 did not register improvement

Is the level of improvement independent of the method of therapy?

5. Test, at a 5% level of significance, for the independence between factor A and factor B in the following tables.

(a)

FACTOR A (columns), FACTOR B (rows)

	A1	A2	A3
B1	16	30	40
B2	20	12	12

(b)

FACTOR A (columns), FACTOR B (rows)

	A1	A2
B1	50	30
B2	30	60
B3	10	20

6. Researchers say that alcoholism is linked to marital status. A random sample of people from a particular state had the following results:

	Type 1	Type 2	Type 3
Married	31	47	68
Not married	69	73	52

A Type 1 result stands for a person being diagnosed as an alcoholic. A Type 2 result stands for an undiagnosed alcoholic and Type 3 stands for a non–alcoholic.

Use a 5% level of significance to determine if there is a relationship between the marital status of a person and their alcoholic classification.

7. A study is carried out to determine the 'relationship' between parent and child obesity level. A random sample of 100 obese and 100 non-obese children was taken and then the obesity statuses of their parents were recorded. The results are shown below:

		Children	
		Obese	Not obese
Parent	Obese	68	58
	Not obese	32	42

Using a 5% level of significance, determine if child obesity is dependent on parental obesity.

8. Social researchers believe that alcohol consumption amongst teenagers is dependent on family class. The results of a random sample of 230 teenagers produced the following information:

		Children		
		Frequently	Occasional	None
Parent	Lower class	10	21	9
	Middle class	24	90	24
	Upper class	13	19	7

Use a 5% level of significance to test this hypothesis.

9. We want to decide if hair colour is dependent on eye colour. A sample of 200 randomly selected people had their eye and hair colour noted, the results are shown in the frequency table below:

	Dark hair	Fair hair	
Brown eyes	54	32	86
Blue eyes	42	72	114
	96	104	200

Based on these figures, what conclusion can you make?

10. (a) In an experiment with a flu injection, the following data, representing the number of patients was obtained:

	Treated	Not treated
Developed flu	20	60
Did not develop flu	80	40

Use an appropriate test for the significance of the effect of the flu injection and describe the conclusions from the analysis .

(b) If on the other hand, the data was

	Treated	Not treated
Developed flu	2	6
Did not develop flu	8	4

How are the values of U (χ^2_{calc}) , and the significance of the results affected by this change? Would the significance test used in (a) have been accurate?

11. The following table gives a summary of the infant mortality and overcrowding numbers in 200 districts. The figures are divided into 'high' and 'low' figures.

		Overcrowding	
		High	Low
Infant mortality	High	44	30
	Low	28	98

Is the high infant mortality rate associated with high overcrowding? Use a 1% level of significance.

12. In some remote regions, certain types of surgical operations can be performed either with a local anaesthetic or with a general anaesthetic. The results of these operations over the course of two years is shown below.

	Alive	Died
Local	515	20
General	170	25

Is there an association between the type of anaesthetic used and mortality rates for this operation? Use a 1% level of significance.

20.2 MISCELLANEOUS EXERCISES

1. The following table shows semester 'results' for 400 students who studied Mathematical Studies and Biology at a new I.B. school.

		Biology		
		Excellent	Good	Fail
Maths Studies	Excellent	32	26	20
	Good	40	66	32
	Fail	20	56	108

(a) Construct a table of expected values for the table of observed values given above.

(b) State the null hypothesis.

(c) Calculate the value, $\chi^2_{calc} = \sum \frac{(f_o - f_e)^2}{f_e}$.

(d) i. How many degrees of freedom are associated with this contingency table?

ii. What is the critical χ^2 value, for this contingency table, using a 5% level of significance?

iii. What conclusion can be made about the association between Mathematical Studies results and Biology results?

2. In some developing countries, the outbreak of diseases can often be fatal. International medical teams have worked in such a region on and off over the last 50 years with limited medical resources. The table below shows the results of recorded data from their recent visit:

		Severity of smallpox attack			
		Extreme	Very	Moderate	Light
Year since vaccination took place	0 – 20	45	125	175	150
	20 – 40	185	300	270	180
	Over 40 or unvaccinated	110	90	40	30

(a) Construct a table of expected values for the table of observed values given above.

(b) State the null hypothesis.

(c) i. How many degrees of freedom are associated with this contingency table?

iii. What conclusion can be made about the association between severity of smallpox attack and the number of years that have elapsed since vaccination?

3. Two teams, A and B are given the task of producing electrical components. Each component was then tested. The results of this test are shown in the table below.

Number of electical components

	Passed	Failed
Team A	30	10
Team B	20	20

Can a conclusion be made about which team had the better results?

4. 100 males and 100 females were asked to state if they were for or against a referendum question. The results are given in the table below.

Referendum question

	For	Against
Male	60	40
Female	30	70

(a) Construct a table of expected values for the table of observed values given above.

(b) State, in words, the null hypothesis.

(c) i. Calculate the value, $\chi^2_{calc} = \sum \frac{(f_o - f_e)^2}{f_e}$.

ii. What conclusions can you draw at a 5% level of significance?

5. A survey was conducted to determine how occupation is related to education. Based on a random sample of 500 employed men and women, their occupation and education background was recorded and tabulated as shown below.

		Occupation			
		White collar	Blue collar	Services	Farming
Education (years of schooling)	At least 12 yrs	195	145	30	7
	Less than 12 yrs	20	74	20	9

(a) Based on the table of observed values, above, find the values of a, b, c and d in the table of expected values, shown below.

		Occupation				
		White collar	Blue collar	Services	Farming	Totals
Education (years of schooling)	At least 12 yrs	162.1	b	37.7	12.06	377
	Less than 12 yrs	a	c	12.3	d	123
	Totals	215	219	50	16	500

(b) Write a suitable null hypothesis and alternate hypothesis for this data.

(c) i. Calculate the number of degrees of freedom, and write down the critical value of χ^2 at the 5% significance level.

ii. Calculate the value, χ^2_{calc} to 3 significant figures.

iii. What conclusion can you draw regarding education and occupation?

20.3 GRADED REVISION EXERCISES

LEVEL 1

1. The contingency tables below show the observed 'A*i*' and B*i*' values obtained. For each table, find the expected frequencies (in table form).

(a)

	A1	A2
B1	35	15
B2	30	40

(b)

	A1	A2	A3
B1	19	20	11
B2	9	16	5

LEVEL 2

1. Use the χ^2 test at the 1% level of significance to investigate if there is an association between categories A and B based on the results of the contingency tables (a) and (b) in question 1, Level 1.

LEVEL 3

1. The table shows the absentee records of a sample of unskilled workers from two companies, A and B:

Absentee rocords

		Good	Poor
Company	A	14	16
	B	6	64

(a) Produce an expected frequency table corresponding to the given table.

(b) Use an appropriate test, at the 1% level of significance to determine if there is a significant difference in the absenteeism in the two companies.

Level 4

1. The self-esteem levels of non-smoking and smoking males were recorded as being High, Medium or Low. The results are tabulated below:

Table 1 Level of self-esteem

	High	Medium	Low
Smoker	25	22	22
Non–smoker	25	28	38

Table 2 Level of self-esteem

	High	Low
Smoker	36	33
Non–smoker	39	52

(a) Construct a table of expected frequencies using Table 1. What assumption have you made in constructing your table?

(b) Calculate the $\chi^2{}_{calc}$ – value for this set of data.

(c) Based on this data is there a significant difference between smokers and non–smokers and their level of self–esteem?

Because of the uncertainty in identifying a medium level of self esteem, the results were retabulated as shown in Table 2.

(d) Based on this new set of data, is there a significant difference between smokers and non–smokers and their level of self–esteem?

20.4 TOPIC TEST

1. (a) Using the table of observed values shown below, complete the table of expected values shown next to it.

FACTOR A (rows: FACTOR B)

	A1	A2	
B1	40	60	100
B2	20	80	100
	60	140	200

FACTOR A (rows: FACTOR B)

	A1	A2	
B1	a	b	
B2	c	d	
			200

(b) Using the tables above, calculate the χ^2 – value.

[6 marks]

2. In a statewide survey, "Best Bake', bread–chain outlet, wanted to determine if there was an association between the different types of cakes they produced and their customers' opinions. The table gives the responses of a random sample of 400 customers from its various outlets:

Type of bread

Opinion		W	X	Y	Z
	Satisfied	82	66	56	56
	Dissatisfied	23	20	17	20

Use the χ^2 test at the 5% level of significance to investigate whether there is any significant difference between the different types of breads and the customers' opinions. What initial assumption did you make in your calculations?

[6 marks]

3. The local medical board wish to investigate the effectiveness of a vaccine in combating an infectious diseases. A sample of 190 people who have contracted the disease is obtained and a record is kept informing the board whether or not they have been inoculated with the vaccine within the last 12 months and the level of severity of their attack. The board is then presented with the following table of results.

Table 1 Level of severity

	High	Medium	Low
Vaccinated	30	27	27
Not vaccinated	30	33	43

(a) Write a suitable null hypothesis and alternate hypothesis for this data.

(b) Construct a table of expected frequencies based on Table 1.

(c) i. Calculate the number of degrees of freedom, and write down the critical value of χ^2 at the 5% significance level.

ii. Calculate the value, χ^2_{calc} to 3 significant figures.

iii. What conclusion can the board draw based on this data?

[13 marks]
Total 25 marks

20.5 PROJECT SUGGESTIONS

As with the suggestions we have made elsewhere, try to pick an issue that is relevant and for which you can collect real data. The internet is an invaluable resource for collecting such data.

Women have played a very important role in developments in medicine. They were instrumental in improving medical care for armies in the field, often at great personal risk.

> Did wounded soldiers who received nursing care stand a better chance of living than those who did not. Remember, we are not after a commonsense answer here.
>
> What does the EVIDENCE tell us?

In genetic theory, children having one biological parent of blood-type M and the other biological parent of blood type-N will always be one of three types, M, MN or N and the proportions of these types will be, on average, 1:2:1.

A small hospital reported that of 200 children having one biological parent of blood type-M and one biological parent of blood type-N, 28 % were found to be of type M, 42% of type MN and the remainder of type N.

> How could the work done in this chapter demonstrate the truth of the genetic theory based on these results?
>
> Is it true that brothers and sisters are more alike than cousins?

Too good to be true. All students that study at least one of biology, chemistry or physics will, at some stage, have carried out an experiment. These experiments will, more often than not, require the collection of data and then, based on their collected data, assert that it fits an existing model to within some level of error.

> How can the work in this chapter be used to determine if the data 'recorded' by the student is in fact based on the experiment that was carried out?
>
> Could it be that a student's results are too good to be true?
>
> Would you be able to determine, with some level of certainty, that the results were in fact contrived?

21.0 MONEY

Imagine a world in which you have to do EVERYTHING for yourself. Grow all your own food. Mine and refine the metal with which you make your bicycle. Weave your own shirts. Dig your own well, etc. How long would it take you to make an object as simple as a plastic box if you had to make it all? Probably your entire life in the unlikely event that you managed it at all.

As a species, we are not alone in seeing that things get done better if we collaborate. Pack hunting, termite mounds, beaver dams, flocks of birds in V shapes are all examples of organised behaviour. Humans are not particularly gifted with good sight, hearing, sense of smell, claws, durable teeth etc., but we do collaborate very extensively. Your education is a part of that.

Many societies collaborate effectively using the medium of 'barter'. Often, when the harvest needs to be collected, everyone helps with that. If a new house is needed a group of people work on that. If you have more bread than you need, you 'barter' (exchange) it for another 'good' such as eggs with a person who has too many of these. These Peruvians have travelled to a market in the high Andes to exchange food they have grown for other goods. Sometimes they use barter and sometimes money. Judging by the range and quality of the goods on offer, their system works well!

However, technological societies require people to specialise. You probably live in a country where people spend their working lives as doctors, lathe operators, farmers etc. Exchanging the products of these people's work requires a medium of exchange. Different societies have used different forms of exchange. In the early days, this was almost always gold. This rare metal has very little actual value but achieved the status that a small amount of it could be exchanged for wagon loads of food.

Later, as the supply of goods vastly exceeded the supply of gold, coins (initially made from silver and gold) became the means of exchange. Modern coins are made from non-precious metals. Modern currency is mainly paper (in some countries, plastic). This is rapidly moving towards 'virtual money'. Workers are paid by electronic transfer into their bank accounts. They spend their money using charge and credit cards. The transactions are followed by mathematical calculations. These are the subject of this chapter.

21.1 CURRENCY CONVERSIONS

21.1.1 INTRODUCTION

Most currencies in the modern world are decimalised. This usually means that there is a unit of currency such as the dollar, pound, kroner etc. This basic unit is often divided into smaller parts (often one hundredth of the main unit) such as the cent, penny, ore etc. This means that most currency calculations can be carried out using normal decimal arithmetic and normal calculators and computers. This was not always the case. For example, many countries used a system derived from that of the Roman Empire in which the main unit was the pound. This was divided into twenty shillings which were, in turn, each divided into twelve pennies (originally denarii). This was a system in which it was difficult to perform even simple calculations such as addition. Such non-decimal currency systems are now rare, which is fortunate for accountants! The remainder of this unit will assume a decimalised currency.

21.1.2 BASIC CONVERSIONS

The currencies of different countries are not all of equivalent value. This means that one German Mark does not have the same buying power as, for example, one Japanese Yen. The conversion rate between these currencies is decided by a complex set of market forces. The most powerful of these are connected with currency trading in which dealers buy and sell investments in various countries. Also, the general strength and success of a country's economy will have an effect on the value of its currency. Econonomically successful countries generally have strong currencies. There is, however, a control mechanism that prevents the currencies of economically strong countries from growing in value without limit. The more a country's currency grows in value, the more expensive an export item becomes in other countries. This means that the economy will not grow as rapidly, tending to reduce the value of the currency. It is not important to understand all these forces for this course. That is the area covered by economics.

In simple terms, whilst the values of currencies might fluctuate, it is generally the case that if a kettle costs \$40 in Utopia and 10 Marks in Floria, the same ratio (4:1) will operate for other items so a \$600 mower might be expected to cost about $\frac{600}{4} = 150$ Marks:

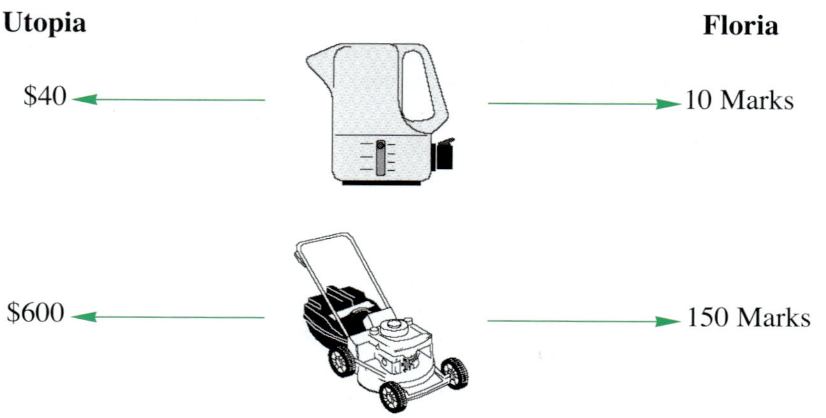

This is a very general observation and is only approximately true in valuing items in different countries.

In the example just discussed, we might say that 1 Florian Mark is worth \$4 in the Utopian currency. This is said to be the **exchange rate** between the two currencies. Such exchange rates, as we have already observed, are determined by many factors, but can generally be found in the business sections of daily papers. They are also displayed in banks and other financial institutions.

EXAMPLE 21.1

The currency conversion rates between five countries are shown.

US	Aus	Sterling	Yen	Mark
1	1.633	0.610	139.490	1.782

Convert:

(a) \$50 US to Japanese Yen
(b) £15 Sterling to \$US
(c) \$1500 Australian to Marks
(d) 125 Yen to Sterling.

Solution

(a) From the table, \$1 US is equivalent to 139.490 Yen.
Thus \$50 US are worth $50 \times 139.490 = 6974.5$ Yen.
The diagram shows how this calculation should look on a TI-83 graphics calculator.

(b) \$1 US is equivalent to £0.61 Sterling. It follows that £1 Sterling must be equivalent to $\frac{1}{0.61} = 1.6393443$ dollars.

So, £15 Sterling is equivalent to $15 \times 1.6393443 \approx 24.59016$ or \$24.59.

A simple alternative is $\frac{15}{0.61} \approx 24.59016$ which gives the answer in one step. Note the use of the automatic answer memory in this example.

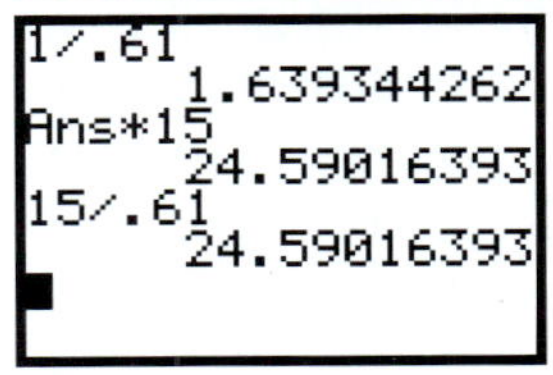

(c) \$1500 Australian to Marks requires us to make the appropriate conversion. From the table, \$1.633 Australian is equivalent to 1.782 marks. Then, \$1 is equivalent to $\frac{1.782}{1.633} \approx 1.0912431$.

So, \$1500 is equivalent to $1500 \times 1.0912431 \approx 1636.86$ Marks.

In one step, this is $\frac{1.782}{1.633} \times 1500 \approx 1636.86$

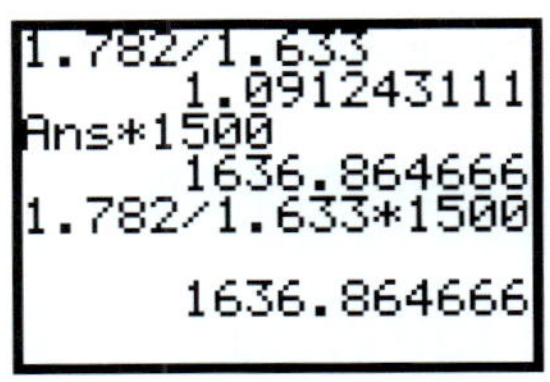

(d) 125 Yen to Sterling needs the value of 1 Yen to be expressed in £ Sterling. This is $\frac{0.610}{139.490} \approx 0.00437307$.

Thus 125 Yen is equivalent to $0.00437307 \times 125 \approx 0.5466$ or $\frac{0.610}{139.490} \times 125 \approx 0.5466$. This is about £0.55

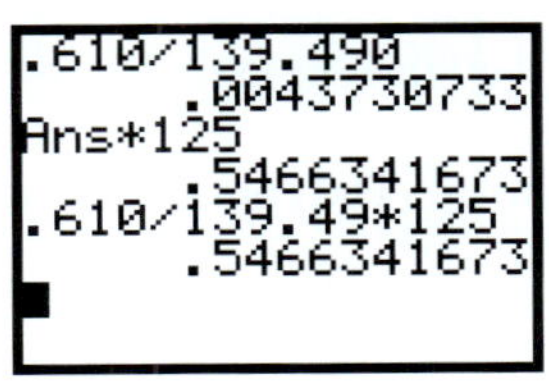

EXERCISES 21.1.1

1. $1 Canadian is equivalent to 4.092 French francs. Convert the following Canadian dollar amounts to French francs, giving your answers to the nearest franc.

(a)	$205	(b)	$138	(c)	$126
(d)	$100	(e)	$183	(f)	$95

2. 1 Swiss franc is equivalent to 3.473 South African rand. Convert the following Swiss franc amounts to South African rand, giving your answers to the nearest rand.

(a)	366	(b)	165	(c)	320
(d)	474	(e)	144	(f)	140

3. 1 German mark is equivalent to 23.261 Indian rupees. Convert the following German mark amounts to Indian rupees, giving your answers to the nearest rupee.

(a)	75	(b)	17	(c)	77
(d)	74	(e)	56	(f)	79

4. $1 US is equivalent to 1740 Italian lire. Convert the following **Italian lire** amounts to $US, giving your answers to the nearest US cent.

(a)	54329	(b)	49550	(c)	988
(d)	34902	(e)	47351	(f)	59876

5. $1 Australian is equivalent to 3.5138 French francs and to 3.0274 South African rand. Convert the following amounts in South African rand to French francs, giving your answers to the nearest franc.

(a)	77	(b)	18	(c)	33
(d)	103	(e)	144	(f)	63

21.1.2 COMMISSION

It is generally the case that agents who change money between currencies will charge for this service. These charges are often called **commission**. There are two main ways in which agents charge such a commision. These are illustrated in the following two examples:

EXAMPLE 21.2

An agent offers to exchange $US to other currencies at the published daily rate. Their commission is $5 per transaction or 1%, whichever is the greater with the commission being paid in $US. Three customers wish to convert the following amounts of $US to Italian lire on a day when the exchange rate was 1760 lire to the dollar:

(a)	$20	(b)	$200	(c)	$20 000

Find the numbers of lire that each will receive after the commission has been charged.

Solution

(a) The commission is either $5 or 1% of the sum to be exchanged. In this case the 1% commission would be 1% of $20 = $\frac{1}{100} \times 20 = 0.20$ or 20 cents.

This is much less than the $5 minimum commission.
The exchange will, therefore, attract a commission of $5.
This means that the customer will only have $20 – $5 = $15 to exchange.

This will yield $1760 \times 15 = 26400$ lire.
In this case, the customer has paid a comparatively large proportion of the money (one quarter) in commission. This is a thing to look out for when you are travelling!

(b) Again the commission is \$5 as 1% of the amount to be exchanged (\$200) is \$2 and this is the smaller amount.
This leaves \$195 to be exchanged.
The amount that the customer will receive is $1760 \times 195 = 343200$ lire.
This customer has paid a much smaller proportion of the money in commission than the first.

(c) If \$20 000 is to be exchanged, the 1% commision is 1% of \$20 000 = \$200.
This leaves \$20 000 – \$200 = \$19 800.
When converted to lire this becomes $1760 \times 19800 = 34848000$ lire.

The second way in which commissions can be charged is to use a slightly different rate when buying or selling a different currency.

EXAMPLE 21.3

A bank offers the following exchange rates for \$1 Australian in relation to the French franc: 'We buy: 3.5959, we sell: 3.5138'. A customer wishes to exchange \$1200 Australian for francs. How many francs will the customer receive? If the customer then immediately exchanges these francs for Australian dollars, how much will she receive? What is the effective exchange rate? Assume that all currency amounts are rounded to the nearest whole number to complete each transaction.

Solution

The bank is selling francs to the customer so the rate for the first transaction is the 'selling rate' of 3.5138.

The customer receives: $3.5138 \times 1200 = 4216.56$ or 4217 (rounded to the nearest franc).

If the customer immediately reconverts this amount back to Australian dollars, the bank will use the 'buying rate' of 3.5959 and she will receive: $\frac{4217}{3.5959} = 1172.7245$.

This amount rounds to \$1173 Australian.
This amounts to a commission of 1200 – 1173 = \$27 on the two transactions.

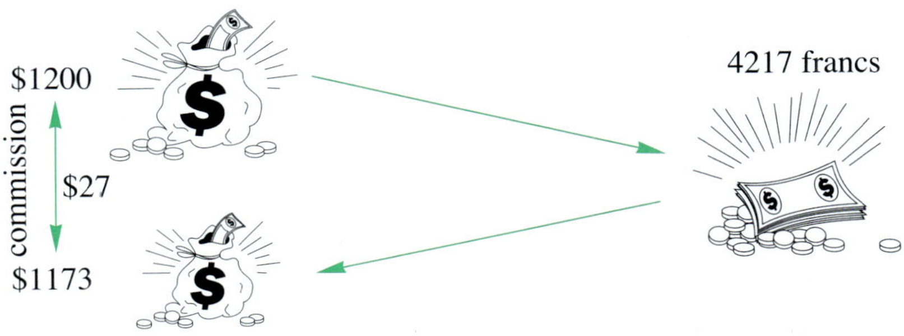

EXERCISES 21.1.2

1. A bank offers the following exchange rates for $1 US in relation to the Japanese yen: 'We buy: 145.7, we sell: 143.1'.
If a customer wishes to exchange the following amounts, find:
i. the number of yen the customer receives, correct to the nearest whole number.
ii. the number of $US that will result if the amount in i. is immediately returned to $US.
iii. the effective commission on the two transactions in $US.

(a)	1000	(b)	1281	(c)	1513
(d)	1356	(e)	1283	(f)	1595

2. A bank offers the following exchange rates for 1 German mark in relation to the Italian lire: 'We buy: 1033, we sell: 1019'.
If a customer wishes to exchange the following amounts, find:
i. the number of Italian lire the customer receives, correct to the nearest whole number.
ii. the number of German marks that will result if the amount in i. is immediately returned to German marks.
iii. the effective commission on the two transactions in German marks.

(a)	1000	(b)	1608	(c)	1359
(d)	1468	(e)	1606	(f)	1217

3. A bank offers the following exchange rates for Swiss francs in relation to the Indian rupee: 'We buy: 33.7, we sell: 31.9'.
If a customer wishes to exchange the following amounts, find:
i. the number of Indian rupees the customer receives, correct to the nearest whole number.
ii. the number of Swiss francs that will result if the amount in i. is immediately returned to Swiss francs.
iii. the effective commission on the two transactions in Swiss francs.

(a)	900	(b)	1433	(c)	1404
(d)	1390	(e)	1334	(f)	1488

4. A bank offers the following exchange rates for $1 US in relation to the South African rand: 'We buy: 6.26, we sell: 6.09'.
If a customer wishes to exchange the following amounts, find:
i. the number of South African rand the customer receives, correct to the nearest whole number.
ii. the number of $US that will result if the amount in i. is immediately returned to $US.
iii. the effective commission on the two transactions in $US.

(a)	6627	(b)	7491	(c)	7358
(d)	7472	(e)	6201	(f)	7030

21.2 INTEREST

When money is borrowed or lent, the borrower usually pays the lender for the service. The amount charged is generally called **interest**. There are two common methods of calculating this interest: simple interest and compound interest.

21.2.1 SIMPLE INTEREST

Simple interest is calculated as a percentage of the amount borrowed (the **principal**).

EXAMPLE 21.4

A family invests \$1250 in an account that pays 6% annual simple interest for 7 years. Find the amount of interest paid over this period.

Solution

The interest is 6% of the principal (\$1250) each year.

This amount is $\frac{6}{100} \times 1250 = \$75$.

Since this is paid in each of the 7 years, the total amount of interest is: $7 \times 75 = \$525$.

The above example could have been solved using the simple interest formula:

$$I = \frac{Crn}{100}$$

I = interest paid,
C = the amount invested (or borrowed), known as the principal or capital,
r = the interest rate (%) and
n is the number of time periods that the loan or investment lasts.

r and n **must refer to the same period of time**. For example, if the time period is a year, the interest rate must be 'per year' (also known as **per annum**).

It is also possible to transform the formula to use it to solve problems in which quantities other than the interest are unknown.

EXAMPLE 21.5

(a) If \$600 is invested in an account and earns \$136.50 simple interest over a 3.5 year period, find the annual interest rate.

(b) If 1450 crowns are invested at a monthly interest rate of 0.08% and earn 8.12 crowns simple interest, find the number of months of the investment.

Solution

(a) The unknown in this question is r. If the subject of the formula is changed to r it becomes:

$$I = \frac{Crn}{100} \Rightarrow r = \frac{100I}{Cn}.$$

Substitution gives: $r = \frac{100 \times 136.50}{600 \times 3.5} = 6.5\,\%$.

(b) Changing the subject of the formula to n, the unknown of this problem gives: $n = \frac{100I}{Cr}$.

The rate in this question is 'per month' and so is the time. If the time periods are not the same, then they must be made so.
So, for example, a time of 2 years would have to become 24 months if the rate was quoted in % per month.

Substituting gives: $n = \frac{100 \times 8.12}{1450 \times 0.08} = 7$ months.

EXERCISES 21.2.1

1. For the following loans, calculate the simple interest earned:

	Principal	Rate (% per annum)	Time (years)
(a)	\$1,340.00	10	6
(b)	\$1,562.00	9	4
(c)	\$780.00	4	7
(d)	\$1,550.00	6	3
(e)	\$10,200.00	8	8
(f)	\$6,250.00	6.5	2

2. For the following loans, calculate the simple interest earned, noting that the time periods are given in months:

	Principal	Rate (% per annum)	Time (months)
(a)	\$1,340.00	12	24
(b)	\$1,562.00	11	36
(c)	\$780.00	4	18
(d)	\$1,550.00	5	15
(e)	\$10,200.00	7.5	3
(f)	\$6,250.00	4.5	8

3. For the following loans, calculate the interest rate (% per annum):

	Principal	Interest	Time (years)
(a)	\$120.00	\$50.40	6
(b)	\$568.00	\$51.12	3
(c)	\$890.00	\$569.60	8
(d)	\$1,650.00	\$396.00	2
(e)	\$125.00	\$132.75	9
(f)	\$6,250.00	\$1,743.75	3

4. For the following loans, calculate the principal (in marks):

	Rate (% per annum)	Interest (marks)	Time (years)
(a)	3	63.00	7
(b)	7	165.90	3
(c)	6	960.00	8
(d)	9	65.70	2
(e)	2	245.16	9
(f)	9	813.15	1

5. For the following loans, calculate the time of the loan (in years):

	Principal (crowns)	Interest (crowns)	Rate (% per annum)
(a)	1560	546.00	7
(b)	200	42.00	3
(c)	2570	411.20	8
(d)	2030	324.80	2
(e)	700	693.00	9
(f)	950	34.20	1

6. For the following loans, calculate the time of the loan (in months):

	Principal (dinars)	Interest (dinars)	Rate (% per annum)
(a)	3000	210.00	7
(b)	5600	98.00	3
(c)	1290	17.20	8
(d)	3400	45.33	2
(e)	780	64.35	9
(f)	2700	8.10	1

7. Helen has 6000 Swiss francs. She wishes to invest this money over a period of 5 years. She has a choice:

EITHER:

Invest in a Swiss bank that offers 7.3% annual simple interest.

OR:

Convert the money to Japanese yen at an exchange that offers
'We buy: 155.7, we sell: 153.1'.
Then invest with an institution that offers 7.4% annual simple interest.

After the investment period, find the number of Swiss francs that Helen will get when she re-converts her investment.

21.2.2 COMPOUND INTEREST

The calculation of simple interest described in the previous section implies that everytime interest is paid, it is withdrawn by the investor. This leaves the principal untouched throughout the period of the loan. This may be the case, however, it is much more usual for investors to add (or **compound**) the interest to the principal so that the principal grows during the term of the investment. It is possible to follow the progress of such a loan period by period as in the following example.

EXAMPLE 21.5

Boris invests \$560 at 6% annual interest compounded annually over a period of 10 years. Compare the results of this with the simple interest alternative.

Solution

The calculation of the interest due at the end of each year is done using the simple interest formula: $I = \frac{Crn}{100}$. For the first year, $I = \frac{560 \times 6 \times 1}{100} = 33.6$ or \$33.60.

This amount is compounded so that the new principal that operates over the second year is: \$560 + \$33.60 = \$593.60.

In the second year the interest is: $I = \frac{593.6 \times 6 \times 1}{100} = 35.612$ or \$35.62 rounded to the nearest cent.

This means that at the end of the second year the amount in the account will be \$593.60 + \$35.62 = \$629.22.

In the third year, the interest will be larger again because it is based on a larger principal.

The complete set of results are:

	Year	Principal	Interest
	Start	\$560.00	\$33.60
After year -	1	\$593.60	\$35.62
	2	\$629.22	\$37.75
	3	\$666.97	\$40.02
	4	\$706.99	\$42.42
	5	\$749.41	\$44.96
	6	\$794.37	\$47.66
	7	\$842.03	\$50.52
	8	\$892.55	\$53.55
	9	\$946.11	\$56.77
	10	\$1,002.87	

Thus after ten years of interest the amount in the account is \$1,002.87. The total interest paid is \$1,002.87 – \$560 = \$442.87.

This compares with the simple interest that might be paid of $\frac{560 \times 6 \times 10}{100} = \$336$, a significantly smaller amount.

Step by step calculations of compound interest are best done by using a calculating device. If using a graphics calculator the steps are:

1. Type in the principal and then press **ENTER**. This places the principal in the automatic memory location called 'Ans'. Every time the calculator performs a calculation the answer is placed in this memory.

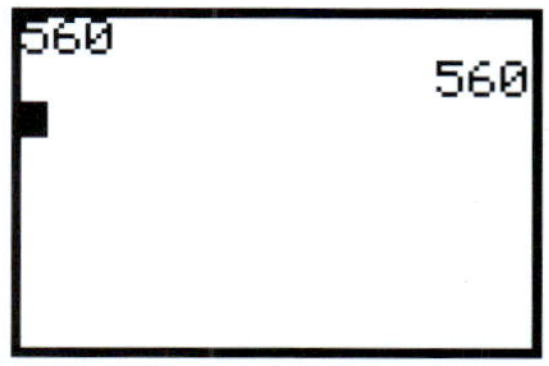

2. Press × 1.06 followed by ENTER. 1.06 is $1 + \frac{6}{100}$ and is the factor that will increase the principal by 6%. Note that this keying sequence will result in the calculator automatically bringing up the 'Ans' memory of the principal.

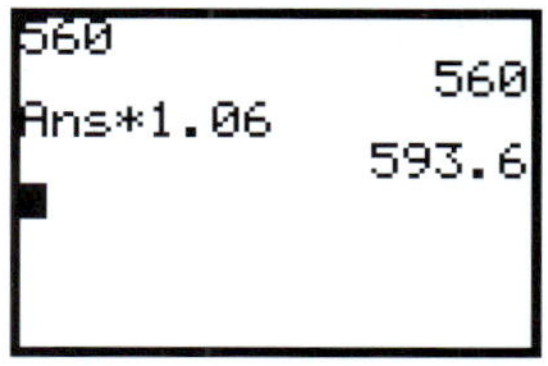

3. Next, press ENTER again. The calculator will repeat the previous calcuation (Ans×1.06) with the new pincipal (593.6). All that is necessary to calculate the amounts in successive years is to continue to press ENTER.

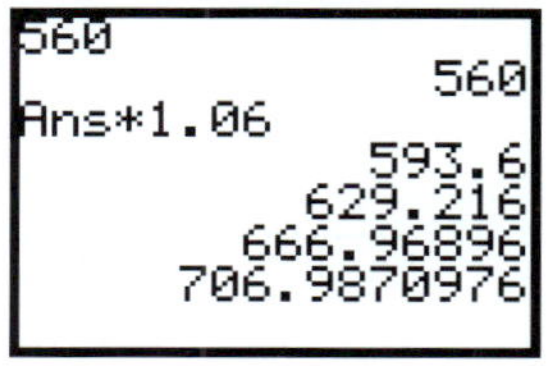

An alternative is to use a spreadsheet. An example of the formulas that could be used is:

A	A	B	C
1	Year	Principal	Interest
2	Start	560	+B2*6/100
3	1	+B2+C2	+B3*6/100
4	2	+B3+C3	+B4*6/100
5	3	+B4+C4	+B5*6/100
6	4	+B5+C5	+B6*6/100
7	5	+B6+C6	+B7*6/100
8	6	+B7+C7	+B8*6/100
9	7	+B8+C8	+B9*6/100

The interest for each year is calculated in column C using the simple interest formula. This is then added to the amount in column B.

In fact, we have already covered this material in Chapter 14 – Geometric Sequences and Series. We have another look at how such a calculation is carried out and then produce a general result.

EXAMPLE 21.6

Find what \$600 amounts to in 20 years if it is invested at 8% p.a. compounding annually.

Solution

We start by finding the value of the investment after it has been left a full year in the investment account.

To do this we first find the interest gained at the end of one year:

We have, $n = 1, r = 8$ and $C = 600$ so that $I = 8\% \times \$600 = 0.08 \times \$600$

Therefore, the total amount in the account at the end of the first year is
$\$600 + 0.08 \times \$600 = \$600(1 + 0.08) = \$600 \times 1.08$

We continue in this manner as shown below:

End of year 1 value $= \$600 + 8\% \times \$600$

$= \$600(1.08)$

End of year 2 value $= \$600(1.08) + 8\% \times \$600(1.08)$

$= \$600(1.08) + 0.08 \times \$600(1.08)$
$= \$600(1.08)[1 + 0.08]$

$= \$600(1.08)^2$

End of year 3 value $= \$600(1.08)^2 + 8\% \times \$600(1.08)^2$

$= \$600(1.08)^2 + 0.08 \times \$600(1.08)^2$
$= \$600(1.08)^2[1 + 0.08]$

$= \$600(1.08)^3$

↓

End of year 20 value $= \$\ 600(1.08)^{20}$

Thus, after 20 years the \$600 amounts to \$2796.57.

Looking closely at the terms of the sequence, they form a G.P:

$$600(1.08), 600(1.08)^2, 600(1.08)^3, \ldots, 600(1.08)^{20}$$

Developing a formula for compound interest

In general, if \$$C$ is invested at $r\%$ p.a. compound interest, it grows according to the sequence

$$C\left(1 + \frac{r}{100}\right), C\left(1 + \frac{r}{100}\right)^2, C\left(1 + \frac{r}{100}\right)^3, \ldots, C\left(1 + \frac{r}{100}\right)^n$$

where $a = C\left(1 + \frac{r}{100}\right)$ and the common ratio 'r' $= \left(1 + \frac{r}{100}\right)$ so that

$$A_n = C\left(1 + \frac{r}{100}\right)^n$$ where A_n is the amount after n time periods.

where C is the capital, r is the interest rate, n is the number of compounding periods and A_n is the amount in the account.

The actual **interest** then, is given by $I = A_n - C = C\left(1 + \frac{r}{100}\right)^n - C$

As with simple interest, the same calculation can be used to follow both loans and investments. We now consider a few examples to see how this works.

EXAMPLE 21.7

$1200 is placed in a savings account that pays 8% annual interest compounded annually. Find the amount in the account and the interest paid after 10 years.

Solution

In this case $C = 1200$, $r = 8$ and $n = 10$.

Using the formula: $A = 1200\left(1 + \frac{8}{100}\right)^{10} = 2590.71$.

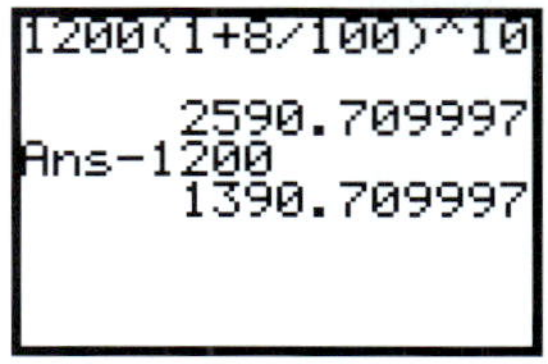

The account will contain $2590.71 after 10 years.

The interest is $2590.71 – $1200 = $ 1390.71

EXAMPLE 21.8

Find the interest earned over a 5 year period if 1500 marks is invested in a savings account that pays 6% annual interest compounded monthly.

Solution

In this case the interest rate is annual but the interest is calculated monthly.
Before using the formula we must express the interest rate as % per month by dividing it by 12 to get $r = 0.5\%$.

The number of interest periods must also be expressed in months ($n = 60$ months).

Using the formula: $A = 1500\left(1 + \frac{0.5}{100}\right)^{60} = 2023.2752$

or 2023 marks.

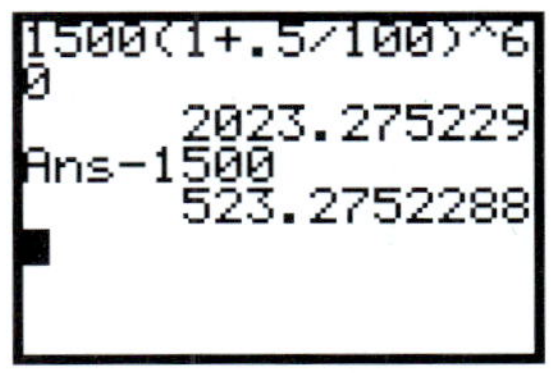

The interest paid is 2023 – 1500 = 523 marks.

EXERCISES 21.2.2

1. Find the amounts that will result if the following principals (in yen) are invested at the rates and for the times given.

	Principal (yen)	Rate % per annum	Time years
(a)	2300	4	12
(b)	600	14	20
(c)	4200	8	20
(d)	1300	4	4
(e)	4300	2	16
(f)	800	12	12

2. Find the amounts that will result if the following principals (in marks) are invested at the rates and for the times given.

	Principal (marks)	**Rate % per annum**	**Time years**
(a)	4800	8	12
(b)	88800	6	4
(c)	79700	2	24
(d)	75300	6	12
(e)	75800	4	24
(f)	26000	4	16

3. Find the amounts that will result if the following principals (in francs) are invested at the rates and for the times given.

	Principal (francs)	**Rate % per annum**	**Time months**
(a)	14900	2	3
(b)	30900	4	2
(c)	29700	8	2
(d)	31200	14	4
(e)	39000	6	1
(f)	27000	14	3

4. Find the amounts that will result if the following principals (in yen) are invested at the rates and for the times given.

	Principal (yen)	**Rate % per annum**	**Time months**
(a)	21700	8	3
(b)	28700	0	5
(c)	23100	4	7
(d)	30000	12	4
(e)	28000	8	6
(f)	1600	4	3

5. If a choice has to be made between investing $1500 for 5 years between the following options, which should the investor choose and why?

Account A: offers 10% per annum calculated annually.

Account B: offers 9.7% per annum calculated monthly.

Account C: offers 9.5% per annum calculated daily.

6. Claudio is searching for a bank so that in two years time, he will have $4000 from his $3000 investment.
What rate, per annum, would a bank need to offer Claudio, if the interest is compounded quarterly?

7. Dean invests $7245 for 5 years. Interest is paid half-yearly and is set at 8.30% p.a. for the first half of the investment period after which it rises to 9.50% p.a for the remaining time. What interest will Dean have earned from this account?

21.2.3 COMPARING SIMPLE & COMPOUND INTEREST

It is useful to compare (at the very least) be aware of the differences between simple interest and compound interest behaviour. We do so by considering an investment of \$1000 in two different accounts; account A returning 8% p.a. simple interest, and account B returning 8% p.a. compounded annually. The results are calculated as follows (with figures rounded up or down):

n (years)	Account A $A = C + \frac{Crn}{100}$	Account B $A = C\left(1 + \frac{r}{100}\right)^n$
0	1000	1000
1	1080	1080
2	1160	1166
3	1240	1260
4	1320	1360
5	1400	1469
6	1480	1587
7	1560	1714
8	1640	1851
9	1720	1999
10	1800	2159
20	2600	4661
30	3400	10 063
40	4200	21 725
50	5000	46 902

These results show that there exists a significant difference in the way simple interest and compound interest behave as the number of periods (n) increases. Graphing these results makes it even more apparent:

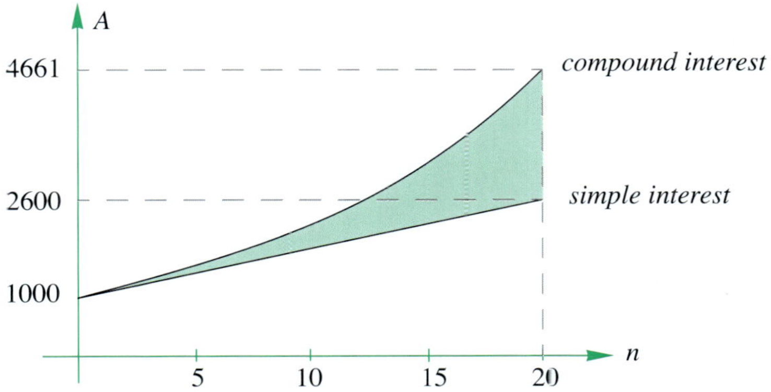

It is therefore important to know what type of interest you are paying or what type of interest is paid to you. For example, the interest charged by a bank on a home loan would be compound interest. But what type of interest does your savings account award you, simple or compound?

Also, is the interest charged/paid yearly, monthly, weekly, daily or *continuously*? How do each of these considerations impact on your decision making?

So, is it better to invest \$1000 at 9% p.a. compounded semiannually or \$1500 at 6.5% simple interest? We consider this problem now.

First we obtain an algebraic expression for both types of interest:

Compound interest: $r = 9$ therefore, semiannually we have 4.5% interest.

If we then have n_1 (semiannual) periods, $A_1 = 1000\left(1+\frac{4.5}{100}\right)^{n_1}$.

Simple interest: $r = 6.5$ therefore, if we have n_2 periods, $A_2 = 1500 + \frac{1500 \times n_2 \times 6.5}{100}$.

Note that after 1 year, $n_1 = 2$ (compounded twice) and $n_2 = 1$. Similarly, after 2 years, $n_1 = 4$ and $n_2 = 2$ and so on.

No. of years	n_1	n_2	A_1	A_2
0	0	0	1000	1500
	1		1045	
1	2	1	1092	1597.5
	3		1141.2	
2	4	2	1192.5	1695
	5		1246.2	
3	6	3	1302.3	1792.5
	7		1360.9	
4	8	4	1422.1	1890
	9		1486.1	
5	10	5	1553	1987.5
	11		1622.9	
6	12	6	1695.9	2085
	13		1772.2	
7	14	7	1851.9	2182.5
	15		1935.3	
8	16	8	2022.4	2280
	17		2113.4	
9	18	9	2208.5	2377.5
	19		2307.9	
10	20	10	2411.7	2475
	21		2520.2	
11	22	11	2633.7	2572.5
	23		2752.2	
12	24	12	2876	2670

We notice that at some point in time, compound interest becomes the better option. Of course, much depends on your needs. In the illustration above, if you require your investment in 10 years time then the simple interest option (with a larger capital deposit) is the better option. However, if you need your money after 11 years, then compound interest is the better option.

A graph of the two available options is shown below.

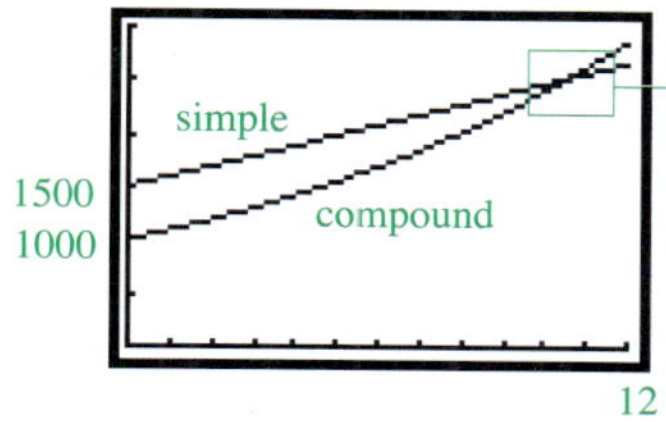

It becomes very useful to know when one option 'overtakes' the other. This can be determined using a graphics calculator.

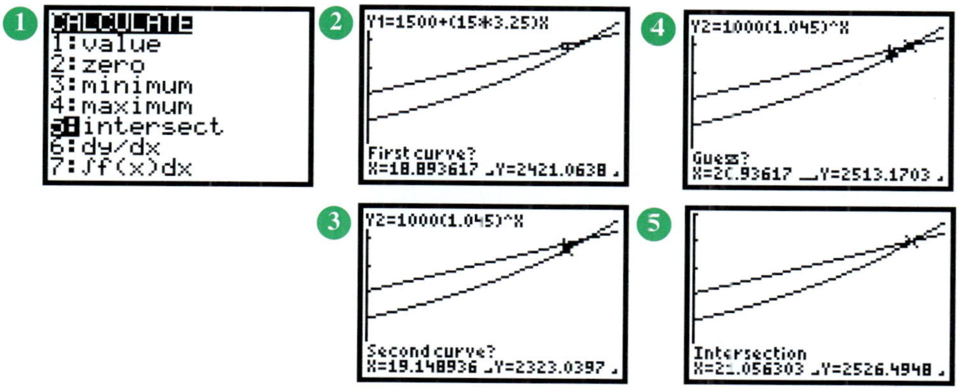

The graphs intersect when $n_1 = 21.0563$, that is, just after the 21st period or just after 10 and a half years.

Of course, we could also have found the intersection by using the solve function:

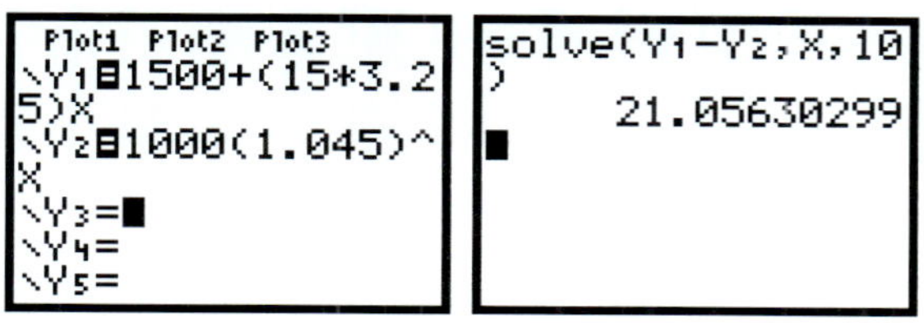

Notice that we have used

$$Y_1 = 1500 + \frac{1500 \times 3.25}{100}X.$$

i.e., we have used an interest rate of 3.25 instead of 6.5. Why is this?

Effective Interest rates

To put different rates and periods of conversion on a comparable basis, we determine the **effective rate**, r_e.

> The effective rate, r_e, is the rate converted annually that will produce the same amount of interest per year as the **nominal** rate r compounded n times a year

For example, if the nominal rate is 8% compounded semiannually, then \$1.00 invested will amount to $1 \times \left(1 + \frac{4}{100}\right)^2 = (1.04)^2 = 1.0816$ [i.e., in one year it is compounded twice at 4%].

That is, the interest on the \$1.00 invested for 1 year is \$0.0816. This is equivalent to an annual rate of 8.16%. That is, 8% compounded semiannually is the same as 8.16% compounded annually.

It is possible to obtain a general expression to find the effective rate.

1. Consider a capital investment, \$C, compounded annually at an effective rate of r_e p.a. At the end of the year this would amount to \$ $C(1+r_e)$.

2. Next consider the same capital invested at a **nominal** rate of r, compounded n times a year. At the end of the year this would amount to \$ $C\left(1+\frac{r}{n}\right)^n$.

Then, for the effective rate to produce the same amount as the compounded rate we need to have

$$C(1+r_e) = C\left(1+\frac{r}{n}\right)^n$$

That is,

$$(1+r_e) = \left(1+\frac{r}{n}\right)^n$$

or,

$$r_e = \left(1+\frac{r}{n}\right)^n - 1$$

In our previous example, $r = 0.08$ and $n = 2$, $\therefore r_e = \left(1+\frac{0.08}{2}\right)^2 - 1 = 1.0816 - 1 = 0.0816$.

When solving financial problems always layout the information in an orderly manner (a list), that way it will be easier for you to apply the rules and not get confused about the correct rates and correct periods. The following examples show one possible approach.

EXAMPLE 21.9 \$2500 is invested at 7.00% p.a., compounded quarterly for 5 years. Find the interest earned.

Solution

We list the known quantities: We are given the rate per annum as 7%. We need to determine the rate per quarter.	**Capital, C**	$= 2500$
Therefore, we divide by 4 (i.e., 4 periods):	**Interest rate, r**	$= \frac{7}{4} = 1.75$
Every year contains 4 periods:	**Number of periods**	$= 5 \times 4 = 20$
To find the interest earned, I, we first need the total amount earned, A.	**Interest earned, I**	$= A - C$

We can now begin our calculations:

$$A = C\left(1+\frac{r}{100}\right)^n = 2500\left(1+\frac{1.75}{100}\right)^{20}$$
$$= 2500(1.0175)^{20}$$
$$= 3536.9455$$

Subtract capital from the total amount, $I = 3536.95 - 2500 = 1036.95$.
That is, the interest earned is \$1036.95

EXAMPLE 21.10

$2000 grows to $3000 over two years when interest is compounded quarterly. Find the annual rate of interest that would allow for this.

Solution

Amount invested $C = 2000$

Amount returned $A = 3000$

Number of periods is 4 per year.

Therefore, over two years there are: $n = 2 \times 4 = 8$ periods

If r is the rate per annum, then, the rate per period is: rate $= \frac{r}{4} = 0.25r$.

So, we have, $3000 = 2000\left(1 + \frac{0.25r}{100}\right)^8$ and we need to solve for r.

Simplifying, we have

$$\frac{3000}{2000} = \left(1 + \frac{0.25r}{100}\right)^8$$

$$\therefore 1.5 = \left(1 + \frac{0.25r}{100}\right)^8$$

$$1.5^{\left(\frac{1}{8}\right)} = 1 + \frac{0.25r}{100}$$

$$1.0519895\ldots = 1 + \frac{0.25r}{100}$$

$$\therefore 0.0519895\ldots = \frac{0.25r}{100}$$

$$\therefore \frac{100}{0.25} \times 0.0519895\ldots = r$$

$$\therefore r = 20.7978\ldots$$

That is, the annual rate would need to be 20.80%

EXERCISES 21.2.3

1. $5000 is invested at 6.5% p.a.. compounded annually for 4 years. find

(a) Find i. the amount of interest earned.

ii. the balance in the account.

(b) What simple interest rate would produce the same balance at the end of 4 years if $5000 was invested? (Give your answer to 3 decimal places).

2. Sally invests $4200 at 6% p.a. compounded quarterly for 5 years.

(a) Find the balance in her account after 5 years.

(b) How much (to the nearest dollar) would Sally need to invest at

i. 6% p.a. simple interest

ii. 8.5% p.a. simple interest

if she was to end up with the same balance at the end of 5 years?

3. \$3000 grows to \$4000 over two years when interest is compounded quarterly. Find the annual rate of interest that would allow for this.

4. Tomiko decides to invest \$4500 for a fixed 5 year period. There are three institutions to choose from. They offer the following options:
Institute A: 12% simple interest.
Institute B: 10.6% p.a. compounded annually.
Institute C: 10 % p.a. compounded monthly.
Which option should Tomiko choose to obtain the best return?

5. \$8000 grows to \$9000 over two years when interest is compounded monthly. Find the annual rate of interest that would allow for this.

6. Janice estimates that she will require \$12 000 in four and a half years time to purchase a new car. She decide to invest money now and allow it to grow for five years. How much does Janice need to invest now if her building society offers an account that pays 8.9% p.a. compounded quarterly?

7. On the first of January 2000 Ross decided to deposit \$2000 into an account paying 5% p.a. compounded semiannually for a period of five years. However, by the first of July 2002 he had to withdraw \$1000 from this account. How much will Ross have in this account at the end of five years?

8. On March 30, 1997, Clive opened an account for his son and deposited \$850 into this account. the acount paid 6% p.a. compounded semiannually. Josh, Clive's son, closed the account on March 30 2004. How much was in the account when Josh closed it?

9. Six years after depositing \$8000 into an investment account, Irma recieves \$12,528.94. During this period the account returned r% p.a. compounded monthly. What was the annual interest rate, r%?

10. Julia has \$10,000 to invest and needs to decided on one of two options.
Option A: 6% p.a. compounded annually for 6 years.
Option B: 5% p.a. compounded annually for the first two years, 6% p.a. compounded annually for the next two years and 7% p.a. compounded annually for the last two years.
(a) Which will give Julia the best return?
(b) By how much?

11. Twelve years ago George invested \$4000. For the first five years the account paid 9% p.a. compounded semiannually. However, for the rest of the time it could only manage 8% p.a. compounded semiannually. How much will George have in the account now?

12. What is the effective rate of interest equivalent to 10% compounded
(a) semiannually (b) quarterly (c) monthly?

13. Twenty two years ago Professor R. McAuliffe bequeathed \$250,000 to the C.H.I.P Foundation for the construction of a building to house part of the new teaching complex. For the first ten years, the money received 12% p.a. compounded quarterly. But for the remainder of the time it only received 10% p.a. compounded semiannually. How much is now available in the fund? Give your answer to the nearest one hundred dollars.

21.2.4 FINDING TIME IN COMPOUND INTEREST

In the previous section we looked at investments that compound over a period of time and determined their value after some given period. However, it might be the case that we wish to determine the period of time for which an investment must be allowed to remain in an account so that it reaches a predetermined value.

That is, from the expression $A = C\left(1 + \frac{r}{100}\right)^n$, given A, C and r, how can we determine n?

Although there is an algebraic method, requiring the use of logarithms, we limit the methods to either a graphical/numerical method or a graphics calculator method. These methods are in fact no different from those already used in Chapter 10, Exponential functions.

EXAMPLE 21.11

How long will it take \$2000 to amount to \$4300 if invested at 8% p.a. with interest compounded annually?

Solution

In this instance we have $C = 2000$, $A = 4300$ and $r = 8$.

From the equation $A = C\left(1 + \frac{r}{100}\right)^n$ we have $4300 = 2000\left(1 + \frac{8}{100}\right)^n$

i.e., $2.15 = (1.08)^n$.

We now have a number of alternatives:

Method 1: Sketch $y = 2.15$ and $y = (1.08)^n$ and then use **CALC 5:intersect**:

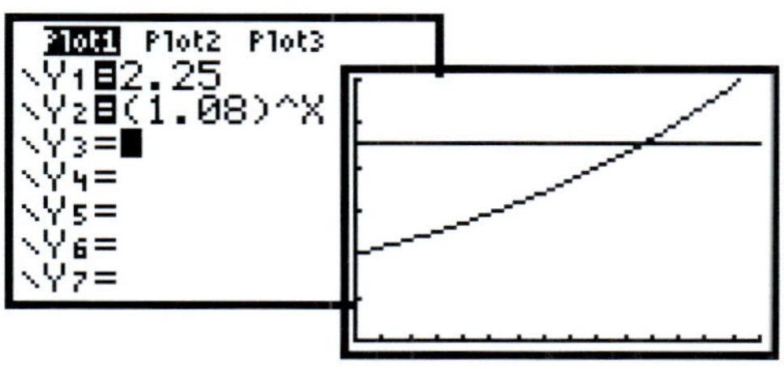

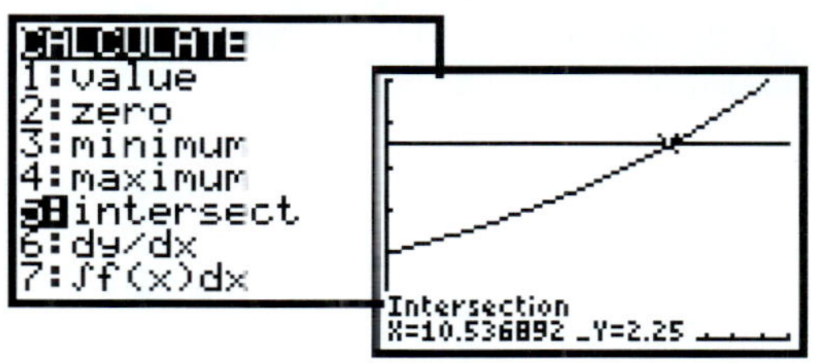

From the last screen dump we see that the graphs intersect at (10.5368, 2.25).
That is, $n = 10.5368$. However, n is an integer (as it is compounded annually) and so we would need to try $n = 10$ or $n = 11$, if we use $n = 10$, we have $A = 2000(1.08)^{10} = 4317.85 > 4300$. Therefore, as we have reached the \$4300 mark by the end of the tenth year, there is no need to try $n = 11$. So it takes 10 years for \$2000 to reach \$4300 at 8% p.a. compounded annually.

Method 2: Use the solver (or solve) options

1. **MATH 0:Solver**:
2. enter equation
3. move cursor over the X
4. press **ALPHA ENTER**

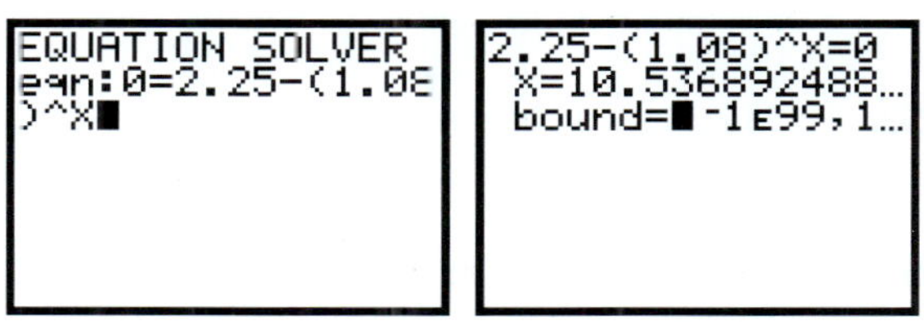

We obtain the same result and would make the same conclusion. That is, it would take 10 years.

EXAMPLE 21.12 How long will it take \$1500 to acrue \$2000 interest if it was invested at 13% p.a. with interest compounding fortnightly?

Solution

This time we are given the *interest amount*! This means that the total amount is \$1500 + \$ 2000 = \$3500.

Also, the 13% is compounded fortnightly, therefore, $r = \frac{13}{26} = 0.5$.

From our formula we now have: $3500 = 1500\left(1 + \frac{0.5}{100}\right)^n$ so that $\frac{7}{3} = (1.005)^n$

Using the graphing option:

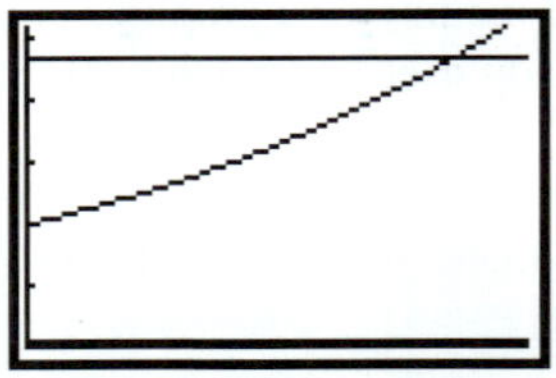

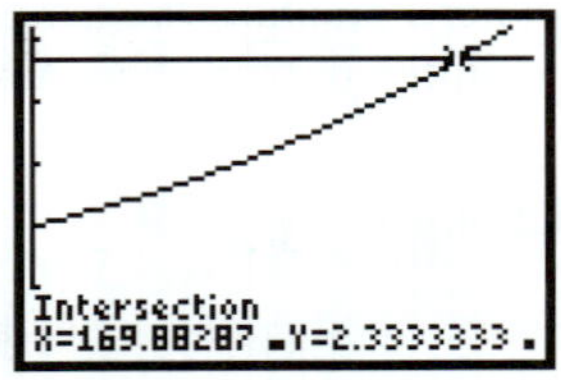

We find that $n = 169.88$

So, when $n = 169$, $A = 1500\left(1 + \frac{0.5}{100}\right)^{169}$

$= 3484.62$

which is less than 3500. Therefore, we need to try $n = 170$:

When $n = 170$, $A = 1500\left(1 + \frac{0.5}{100}\right)^{170} = 3502.04 > 3500$.

Therefore we choose $n = 170$, i.e., 6 years and 14 fortnights (or 6 years and 28 weeks).

EXERCISES 21.2.4

1. How long will it take \$2000 to amount to \$3500 if invested at 8% p.a. with interest compounded
(a) annually? (b) quarterly? (c) daily?

2. Find the number of interest bearing periods, n, for \$3000 to amount to \$4090 when compounding quarterly at a rate of 7% p.a. How long will it take for this to happen?

3. How long will it take for \$1200 to double if invested at 15% p.a. compounded
(a) annually? (b) quarterly?

4. How many fortnights would it take for \$1500 to acrue \$600 interest at 12% p.a. compounded fortnightly?

5. Louise invests \$6350 into an account returning 7.8% p.a compounded monthly. How long will it take for this investment to accumulate to \$8500?

6. Donald invests \$6000 at 8.5% p.a. compounded quarterly so that he can purchase a \$10,000 car in 5 years time.

(a) His parents have told him that they would give him the shortfall if one exists at that time. Is there a shortfall? If so, how much will the parents need to give Donald?

(b) Donald then decides that he wants to purchase a \$12,000 car.

i. Show that the number of quarters, n, that are required satisfy the equation $(1.02125)^n = 2$.

ii. How long does he need to wait until his investment allows him to purchase the car outright and on his own?

7. Kristian and Jørgi both want to save for a holiday. Kristian has \$8000 and invests it into an account returning 8.4% p.a. compounded annually. Jørgi has \$7000 but invests in an account returning 9.2% p.a. compounded quarterly. If the holiday costs \$12,500, who will be the first to go on the holiday? What time difference will there be in their departures?

21.3 DEPRECIATION

21.3.1 INTRODUCTION

Unlike the purchase of a Picasso painting, which will most probably increase in value over the years, items such as cars, computers or pieces of machinery will decrease over the years. For example, one has only to look at the secondhand car market to realise that a car that is two years old is worth a lot less than when purchased new. In fact, as soon as a consumer drives his new car out of the showroom for the first time, the value of the car drops by a significant amount!

The estimated loss in the value of assets is called **depreciation**.

Three terms that are associated with depreciation are

1. Book value
2. Written off value
3. Scrap value

1. Book value:

The book value is the estimated value of an item at any point in time. This is calculated as follows:

Book value after time t = BV_t = Original cost – depreciation after time t.

2. Written off value:

When the book value is equal to zero, the item is said to be written off.

That is, an item is written off when $BV_t = 0$.

3. Scrap value:

At the end of an item's effective (or useful) life, its book value is called its scrap value.

There are a number of ways by which depreciation can be calculated – some, more 'creative' than others, however, we will consider the three primary ways in which depreciation is calculated.

They are:

1. Flat rate (or straight-line) depreciation.
2. Reducing balance depreciation.
3. Unit cost depreciation

We will consider each one in turn in the sections that follow. first, try the following questions.

EXERCISES 21.3.1

1. Patrick purchased a new computer for $4000 two years ago. Each year, the computer depreciates by $400. What is the computer's book value now?

2. Tony purchased a photocopier for $2500. After three years of usage, the machine will be sold. It has depreciated by a total of $1960. What is the scrap value of this machine?

21.3.2 FLAT RATE (STRAIGHT-LINE) DEPRECIATION

If an item depreciates by a **fixed flat rate**, then its value **decreases by a fixed amount** during every period of time (generally the period of time is a year). Because of this constant decrease in the value of the item, we also refer to this as **straight-line depreciation**, that is, the value of the item displays a **linear relationship over time**.

This can be represented as follows:

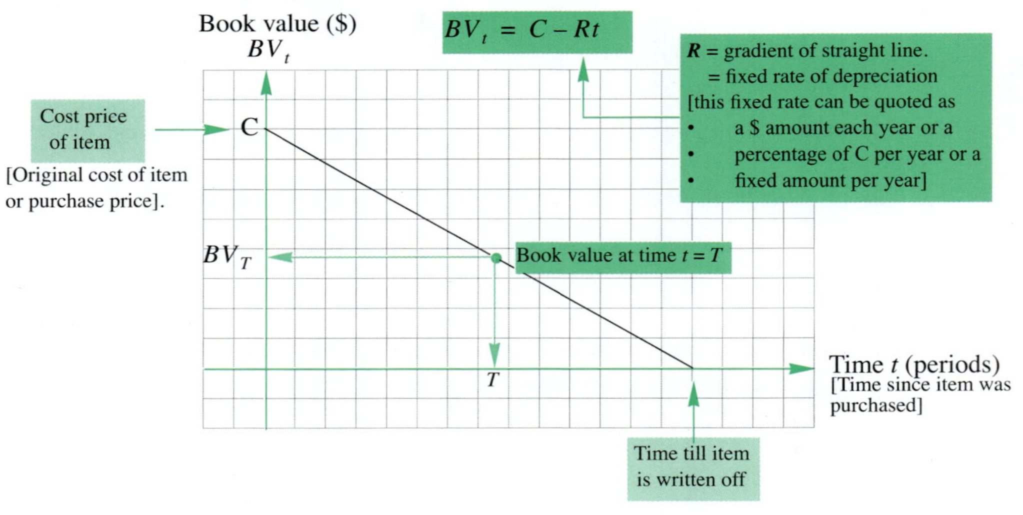

It is often useful to construct a table of values (as was done in Chapter 6 – Linear Graphs). Such a table of values would show the depreciation schedule. The headings most appropriate for such a table would be:

Time, t (periods)	Depreciation 'rate', R	Book value, BV_t
:	:	:
:	:	:

EXAMPLE 21.13

Shannon Books – a book printing company, have just purchased a folding machine for $250,000. Don, the company's chief accountant has informed Mr Cliff Royale, the owner, that they should apply a flat rate depreciation method for the folding machine. The depreciation rate is 18% p.a. of the purchase price.

(a) Find the annual depreciation value.
(b) Set up a depreciation schedule for the machine, showing book values for the first 4 years.
(c) Find the equation of the straight line showing the book value of the machine t years after it was purchased.
(d) Cliff believes that the folding machine only has an effective life of 5 years. Find the scrap value of the folding machine at this point in time.

Solution

(a) We start by identifying the information provided:
Purchase price = $250,000 (= C)

Rate of depreciation (R) = 18% of \$250,000 = $\frac{18}{100} \times \$250000 = \$45000$

That is, every year, the machine's value decreases by $45,000.

(b) We can now set up the depreciation schedule for the machine:

Time, t (periods)	Depreciation 'rate', R	Book value, BV_t
0	0	250,000
1	45,000	205,000
2	45,000	160,000
3	45,000	115,000
4	45,000	70,000

(c) The book value at time t years since it was purchased is given by $BV_t = C - Rt$, and so we have that $BV_t = 250000 - 45000t$.

(d) After 5 years, the scrap value of the machine is given by the book value at $t = 5$.
So, we have that $BV_5 = 250000 - 45000 \times 5$
$= 25,000$
i.e., scrap value after 5 years = $25,000.

Note: When sketching the relationship between the book value and the time since the item was purchased, we restrict our domain to go as far as the time of the item's effective life. For the example above, the graph would be given by:

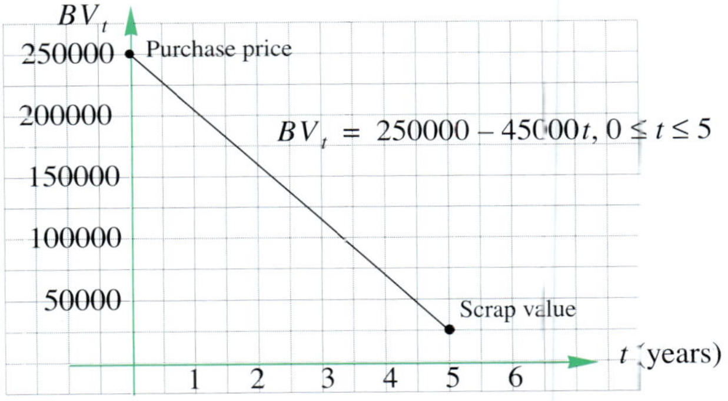

EXAMPLE 21.14 Raoul bought a computer 4 years ago. The current book value of this computer is \$800. If the computer has been depreciating at \$500 p.a. how much did Raoul pay for the computer?

Solution

In this case we have: Purchase price (C) = ?
Rate of depreciation (R) = \$500 per year
Book value at $t = 4$ (BV_4) = \$800

From our equation, $BV_t = C - Rt$, we have:

$$BV_t = C - 500t.$$

Then, when $t = 4$, $BV_4 = 800 \therefore 800 = C - 500 \times 4$

$$\Rightarrow C = 800 + 2000$$
$$= 2800$$

That is, the purchase price was \$2800.00

EXAMPLE 21.15 Mee Yee purchased some office furniture three years ago for \$5000. Today the furniture is worth \$3200. Assuming a flat-rate depreciation model, determine the depreciation rate applied to the furniture.

Solution

In this instance we are given: Purchase price = \$5000 (C)
Book value after 3 years = \$3200 (BV_3)

Using the flat-rate depreciation model, $BV_t = C - Rt$ we have

$$3200 = 5000 - 3 \times R$$
$$\Rightarrow 3R = 5000 - 3200$$
$$\Rightarrow 3R = 1800$$
$$\therefore R = 600$$

That is, the rate of depreciation is \$600 per year.

Another way to look at this is to say that $\text{Rate of depreciation} = \dfrac{\text{Total depreciation}}{\text{number of years}}$

EXERCISES 21.3.2

1. Nonsense Pty Ltd purchased some machinery for \$50,000. Find the book value of the machinery after 5 years if it depreciates at

(a) \$8000 p.a.

(b) 18% of the cost price each year.

2. Chris bought a computer 6 years ago. The current book value of this computer is \$300. The computer has been depreciating at \$550 p.a. How much did Chris pay for it?

3. The graphs below show the depreciation behaviour of different types of items since they were purchased.

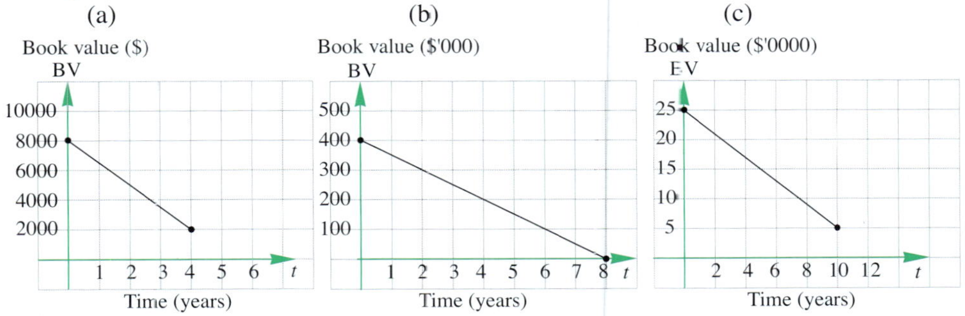

For each item, find

i. the cost price.

ii. the time taken for the item to be written off or reach its scrap value.

iii. the total depreciation of the item for its useful life.

iv. the rate of depreciation.

v. the expression for the linear relationship between its book value, BV, and the time, t years, since it was purchased.

vi. its book value 3 years after it was purchased.

4. Anthony and Juliette have started a mobile dog-washing business. The initial outlay for equipment and van was £10000. They chose to depreciate the equipment and van by the flat rate method at a rate of 15% of the cost price each year. They estimated that the useful life of this equipment would be 5 years.

(a) Set up a depreciation schedule for the life of their equipment and van.

(b) Draw a graph showing the relationship between the book value and the time since the initial outlay, for the equipment and van.

(c) What is the scrap value of the equipment and van at the end of their useful life?

(d) Write the equation relating the book value, BV_t, and the time, t, since the initial outlay for the equipment and van.

5. Jack's Tree Removing Services buys three new removal trucks for \$35,000 each. The owner, Jackie, estimates that the trucks will last 6 years, at which point, their combined scrap value will be \$7000. The firm's accountant decided to use a flat rate depreciation method.

(a) What is the combined annual depreciation for the trucks?

(b) Set up a depreciation schedule for the combined life of the trucks.

(c) Draw a graph showing the relationship between the combined book value and the time since the trucks were purchased.

(d) Write the equation relating the combined book value, BV_t, and the time, t, since the trucks were purchased

6. Two printers are purchased at the same time. Printer A costs \$2500 and is depreciated at \$220 p.a. with a scrap value of \$350, while printer B costs \$3300 and is depreciated at \$300 p.a. with a scrap value of \$450. Which printer will need to be replaced first and what is the time difference between their replacements?

21.3.3 REDUCING BALANCE DEPRECIATION

In business terms, when an asset (in particular, machinery) is purchased and is used for production purposes, more often than not, this asset will 'wear out' very quickly in its first few years of use and then become costly to maintain as it gets older. This is why many businesses replace their equipment frequently. As a consequence, unlike in the flat rate method of the previous section, this asset's book value depreciates more quickly in the first few years than it does in later years. This means that the relationship between the book value and the time since the asset was purchased is no longer linear.

This method of depreciation is known as **reducing balance depreciation**. You will find that most accountants use this method to calculate the annual depreciation of an asset.

So, how does it work? Well, it is pretty much the same as working out the accumulated growth using compound interest, except this time the initial amount is reducing not growing!

For example, let us assume that a new piece of machinery costs \$10,000. Using the **reducing balance depreciation** method we would first need a fixed percentage rate, say we use 15% p.a. for this example. Then, the next step is to find the book value of this piece of machinery at the end of the first year.

We have,

Book value = \$10,000 – 15% of \$10,000
= \$10,000 – \$1500
= \$ 8500

So, at the end of the first year, the piece of machinery has a book value of \$8500.

Next, we determine the book value at the end of another year, but this time, rather than using 15% of the purchase price, we use 15% of the book value at the end of the first year.

So,

Book value = \$8500 – 15% of \$8500
= \$8500 – \$1275
= \$7225

we can then continue in the same manner for the next year:

Book value = \$7225 – 15% of \$7225
= \$7225 – \$1083.75
= \$6141.25

And so on.

Again, it is possible to construct a depreciation schedule:

Time, t (periods)	Book value at start of period	Book value at end of period
0	10,000	8500
1	8500	7225
2	7225	6141.25
3	6141.25	5220.06
4	5220.06	4437.05
:	:	:

Notice that the value at the end of one period becomes the value at the start of the next period.

In fact, reducing the amount by 15% each time is the same as say retaining 85% of the value each time.

That is, rather than working out $10,000 – 15% of $10,000 = $8,500
We could have used: 85% of $10,000 = $8,500

Similarly, instead of using $8500 – 15% of $8500 = $7225
We could have used: 85% of $8500 = $7225

To see how different the behaviour of the reducing balance method is to the flat rate method, we can sketch their graphs.

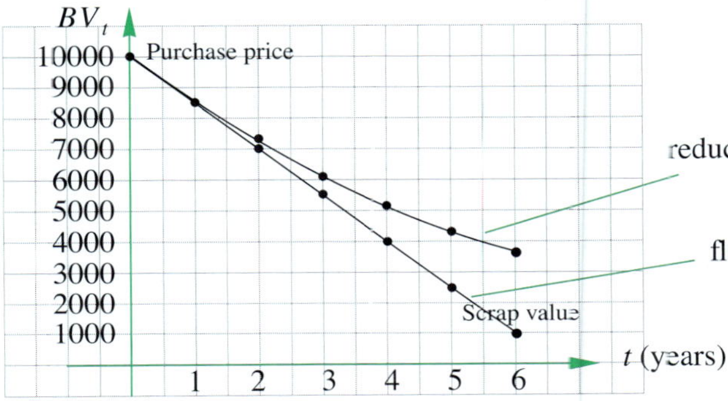

In both methods we use a fixed rate of 15%. The difference between the two methods is what the 15% is used on. For the flate rate method, it is always 15% of the pruchase price. Whereas for the reducing balance method, it is 15% of the book value (which is always changing). Because of this, reducing balance depreciation is also known as **diminishing value depreciation**. This is an example of an exponential **decay** function, and from the graph above, we see that to be the case.

We can generalise the previous example for any rate r% as follows:

$t = 0$ Book value $= \$10000$

$t = 1$ Book value $= \$10000 - \frac{r}{100} \times 10000$

$$= 10000\left(1 - \frac{r}{100}\right)$$

$t = 2$ Book value $= 10000\left(1 - \frac{r}{100}\right) - \frac{r}{100} \times 10000\left(1 - \frac{r}{100}\right)$

$$= 10000\left(1 - \frac{r}{100}\right)\left(1 - \frac{r}{100}\right)$$

$$= 10000\left(1 - \frac{r}{100}\right)^2$$

$t = 3$ Book value $= 10000\left(1 - \frac{r}{100}\right)^2 - \frac{r}{100} \times 10000\left(1 - \frac{r}{100}\right)^2$

$$= 10000\left(1 - \frac{r}{100}\right)^2\left(1 - \frac{r}{100}\right)$$

$$= 10000\left(1 - \frac{r}{100}\right)^3$$

Therefore, continuing in this manner we have that after t periods, the book value, BV, of the item, using the reducing balance depreciation method is given by $BV = 10000\left(1-\frac{r}{100}\right)^t$.

Compare the term $\left(1-\frac{r}{100}\right)^t$ (for depreciation) with that of $\left(1+\frac{r}{100}\right)^t$ (for accumulation) – depreciation contains the decay factor whereas accumulation contains the growth factor! We can summarise our finding as follows:

The book value, BV_t, when applying the reducing balance depreciation method is given by $BV_t = C\left(1-\frac{r}{100}\right)^t$,

where, BV_t = book value after time t
C = purchase price
r = fixed % rate of depreciation
t = time since purchase

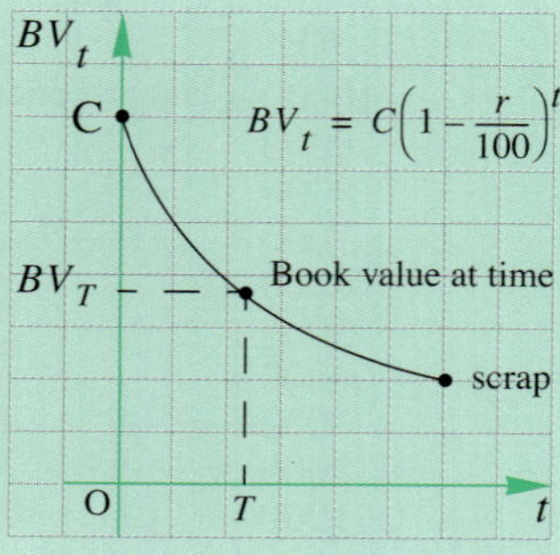

EXAMPLE 21.16

Jack's Tree Removing Services bought two new removal trucks for \$35,000 each. The owner, Jackie, decides that one of the trucks will be depreciated at the flat rate of 15% of the purchase price, while the other will be depreciated at 20% using a reducing balance method. It is estimated that the trucks will have an effective life of 6 years.

(a) Construct a depreciation schedule for both trucks.
(b) Sketch the graphs showing the relationship between the trucks' book values and the time since they were purchased.
(c) At the end of six years, Jackie analyses the depreciation behaviour for both methods. Is there a preference as to which method should be used in the future?

Solution

(a) We start by determine the equations for the book value under each method.

Flat rate method:

Rate of depreciation is 15% of purchase price, i.e., 15% of \$35,000

Therefore, each year the value of the truck reduces by $\frac{15}{100} \times 35000$.

That is, the depreciation rate is \$5250 per year.

Therefore, the book value is given by: $BV_t = 35000 - 5250t,\ 0 \le t \le 6$

Reducing balance method:

Rate of depreciation is 20% of previous book value, i.e., $r = 20$.

Therefore, the book value is given by: $BV_t = 35000\left(1-\frac{20}{100}\right)^t,\ 0 \le t \le 6$

i.e., $BV_t = 35000(0.8)^t,\ 0 \le t \le 6$

Now that we have both equations, we can enter these into the graphics calculator and produce the required tables of values:

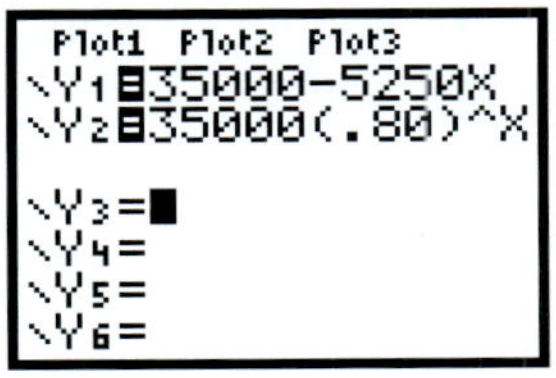

X	Y1	Y2
0	35000	35000
1	29750	28000
2	24500	22400
3	19250	17920
4	14000	14336
5	8750	11469
6	3500	9175

X=6

Y_1 (flat rate)
Y_2 (reducing balance)

← scrap value for each truck

(b)

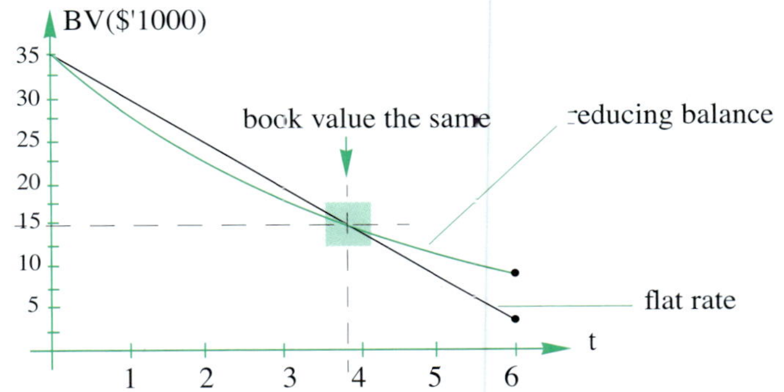

(c) From the graph we see that there will be a time, near $t = 4$, where the book value of both trucks will be the same.
For $t = 1, 2$ and 3, the book value using the flat rate will be geater than that using the reducing balance method.
For $t = 4, 5$ and 6 the reverse happens. The book value using the flat rate will be less than that using the reducing balance method.
Because most businesses want a lower book value of their assets early on, then they should use the reducing balance method. In particular, Jackie should look at reducing the useful lifetime of the trucks to 4 years – the value of the trucks will have depreciated the fastest over the first four years.

EXERCISES 21.3.3

1. A piece of machinery was purchased for $20,000 four years ago. It depreciates at a rate of 20% p.a. by the reducing balance method. What is the book value (to the nearest dollar) of this machine (a) now?
(b) in two years time?

2. Mirko is setting up his dental laboratory and needs to purchase equipment to the value of $250,000. Graham, his accountant suggests that the equipment should be depreciated at a fixed rate of 25% p.a. using the reducing balance method. The equipment will last 8 years.
(a) Construct a depreciation schedule for the equipment.
(b) Draw a graph showing the relationship between the book value of the equipment and the time since it was purchased.

3. Sandro has just resigned as Global cards manager for the National Australian Bank and has decided to set up his own pizza shop – The Zizza Pizza Shop. His outlay for large pizza ovens is $90,000. Sandro decided that he will remain in the business for 10 years. The ovens will be depreciated at 22% p.a. using the reducing balance method.

(a) Find the book value of the ovens after i. one year.
ii. five years.

(b) Find the equation for the book value, $\$BV_t$, of the ovens, in terms of the time, t years, since they were purchased.

(c) What is the scrap value of the ovens?

(d) Draw a graph showing the relationship between the book value and the time since purchase of the ovens.

4. A car was purchased for $15,000 at the start of the new financial year and will be depreciated by 20% p.a. of the previous end of financial year's book value. It is expected that the car will have an effective life of 4 years.

(a) Construct a depreciation schedule for this car.

(b) Draw a graph showing the relationship between the car's book value and the time since it was purchased.

(c) Find the equation relating the book value, $\$BV_t$, of the car and the time, t years, since it was purchased.

(d) What is the car's book value i. after 2 years?
ii. after 3 years and 6 months?

(e) What is the car's scrap value?

5. A piece of equipment has a scrap value of $2000. How much did the equipment cost originally if a diminshing value depreciation schedule has been applied at a fixed rate of 15% p.a. for the last 5 years?

6. Jack's Tree Removing Services bought two new trucks for $50,000 each. The owner, Jackie, decides that truck 1 will be depreciated at the flat rate of 15% of the purchase price, while truck 2 will be depreciated at 22% using a reducing balance method. It is estimated that the trucks will have an effective life of 6 years.

(a) Construct a depreciation schedule for both trucks.

(b) Sketch the graphs showing the relationship between the trucks' book value and the time since they were purchased.

(c) During which year will the book value of truck 1 be less than that of truck 2?

(d) Find the scrap value of both trucks.

7. How many years will it take an asset valued at $62,000 to have a book value of $6210 if it has been depreciating at a fixed rate of 25% p.a. using the reducing balance method?

8. Keystrokes Pty Ltd bought a new delivery van for $32,000. They decided to keep the van until its scrap value (to the nearest one hundred dollars) was $6700. They estimated that at that time they would need to purchase a new van for $35,000. The director, Rory, decided that they should invest a lump sum into an account returning 12% p.a. compounded quarterly. The van, purchased for $32,000 was also to depreciate at a fixed rate of 20% p.a. using the diminishing value method.

(a) How long was it before Rory had to replace the van?

(b) How much should Rory have invested so that the new van could be purchased without Keystrokes Pty Ltd outlaying any other money?

21.3.4 UNIT COST DEPRECIATION

The previous two methods of depreciation, the flat rate method and reducing balance method, are based on a time calculation. That is, they are based on the age of the item. However, consider a bakery that has just purchased an oven and an automated dough-rolling machine. This bakery specialises in making croissants. The time of usage (or productivity) of the different items is very different. While the oven will be running for about 8 to 9 hours, the dough-rolling machine will only be operating for 1 to 2 hours. In all, both items will have 'aged' for the same amount, but will have been in actual use for different amounts.

The unit cost method of depreciation recognises that the life of an asset can be estimated by its actual usage (or productivity or output) rather than how old it is. This, as in the case of the bakery, takes into account the fact that an asset may sit idle and so not depreciate as fast as another asset that is used much more frequently. Another example of this is when a company purchases a computer and a printer. Will they both be in actual use for the same amount of time?

The difficulty in this method is to come up with a measure of usage of the asset. Some standard measures are, number of kilometres for vehicles, number of pages printed for printers, number of hours of usage for an oven, and so on. Whatever measure of usage is considered, to calculate depreciation, some estimate of the usage needs to be made. This is calculated as follows:

Amount of depreciation = amount of actual use × rate of depreciation

Where the rate of depreciation is given by $\frac{\text{Total depreciation}}{\text{Estimated total usage}}$

Notice that Total depreciation = Purchase price – scrap value

EXAMPLE 21.17

A photocopy machine was purchased for \$12,000 and has an estimated life span of 1,000,000 copies at which point its scrap value will be \$1500. During its first year of operation, 110,000 copies are made. Find its book value at the end of the first year.

Solution

Book value (after 110,000 copies) = Purchase price – depreciation (after 110,000 copies). The purchase price is \$12,000 so all that remains is to determine the amount of depreciation.

As we are given the actual usage, we can make use of the unit cost depreciation method. The infomation is as follows:

Amount of actual use = 110,000 (copies)
Estimated total usage = 1,000, 000 (copies)
Total depreciation = \$(12,000 – 1500)
= \$10500

$$\text{Therefore, depreciation (after the first 110,000 copies)} = \$110,000 \times \frac{10500}{1000000}$$
$$= \$1155$$

And so, the book value (after 110,000 copies) = \$12,000 – 1155
= \$10,845

Notice that by rearranging our 'formula'

$$\text{Amount of depreciation} = \text{amount of actual use} \times \text{rate of depreciation}$$
$$= \text{amount of actual use} \times \frac{\text{Total depreciation}}{\text{Estimated total usage}}$$

to

$$\text{Amount of depreciation} = \frac{\text{Amount of actual use}}{\text{Estimated total usage}} \times \text{Total depreciation}$$

The term $\frac{\text{Amount of actual use}}{\text{Estimated total usage}}$ represents the **fraction of the total usage** of the asset. This version of the formula is more in line with the definition of the unit cost method of depreciation.

EXAMPLE 21.18

A car was purchased for \$30,000 with an estimated scrap value of \$4500 once the car reaches 90,000 km.

(a) Find the depreciation rate in dollars per km.
(b) Find the book value of the car after reaching 50,000 km.
(c) Construct a depreciation schedule for every 10,000 km travelled.

Solution

(a) To find the depreciation rate we first need to find the total depreciation:
Total depreciation = Purchase price – scrap value
= \$30,000 – \$4500
= \$25500
Therefore, the rate is given by 255000/90000 = 0.283 i.e., \$0.28 per km

(b) Book value (at 50,000 km) = Purchase price – depreciation (at 50,000 km)

To find the depreciation (at 50,000 km) we can first determine the fraction of usage that 50,000 km makes up.

In this case it is $\frac{50000}{90000} = \frac{5}{9}$.

Then, the depreciation (at 50,000 km) = $\$\frac{5}{9} \times \text{Total depreciation}$
$= \$\frac{5}{9} \times 25500$
$= \$\ 1416.67$

Therefore, the book value (at 50,000 km) = \$ 30,000 – \$ 1416.67
= \$ 15833.33

(c) Every 10,000 km, the fraction of usage is $\frac{10000}{90000} = \frac{1}{9}$.

Therefore, every 10,000km the car depreciates by $\$\frac{1}{9} \times 25500 = \frac{25500}{9}$.

We can then use this value to construct the depreciation schedule:

Where the X column is measured in 10,000 km and the Y_1 column is the book value in dollars.

X	Y_1	X	Y_1
0	30000	3	21500
1	27167	4	18667
2	24333	5	15833
3	21500	6	13000
4	18667	7	10167
5	15833	8	7333.3
6	13000	9	4500

X=0 X=9

It is also possible to produce an equation showing the relationship between the book value and the number of kilometres for part (c):

If n is the number of successive 10,000 km readings, the book value can be determined using

$$BV_n = 30000 - \frac{25500}{9}n, n = 0, 1, 2, \ldots, 9.$$

EXERCISES 21.3.4

1. Jones Inc. purchases a car for $42,000 which depreciates at a rate of 24 cents per km. In its first year its odometer reads 18,660 km. It is estimated that the car will have a scrap value of $6000.
(a) By how much has the car depreciated in the first year of usage?
(b) What is the car's effective life?

2. Print 'R' Ass., purchased a colour digital printer for $18,000. Depreciation is calculated at a rate of 20 cents per 100 copies made. In the first year of usage, the counter reads 812,000 copies. Thereafter the machine is set to run at 900,000 copies per year.
(a) By how much has the printer depreciated by the first year of usage?
(b) What is the printer's book value at the end of the third year.
(c) When will the printer be written off?

3. A car purchased for $25,000 has its scrap value estimated at $3000, after having an effective life covering 85,000 km. In its first year, the car travels 22,000 km. Find the car's book value at the end of the first year.

4. During its first year, a car depreciates by $3600. The car was purchased for $30,000 and has its scrap value estimated at $2300, after having an effective life covering 90,000 km.
(a) Find the car's book value at the end of the first year.
(b) How many kilometres did the car travel during its first year?

5. The machinery in an automated car-wash cost $750,000 when new. It is estimated that this car wash will have a life of 15,000 cars passing through it after which it has to be completely replaced. The table below shows the history of the car-wash for its first four years of operation.

Number of years of operation	Number of cars washed
1	2200
2	2650
3	3000
4	2840

(a) Find the car-wash's depreciation for each year of operation shown above.
(b) After its 4th year of operation water restrictions have been put in place and the car-wash can only allow a maximum of N cars per year to be washed. How long will it be before the car-wash needs to be completely replaced?

6. A farmer purchased a machine for \$44,000. His accountant will depreciate the machine based on the number of outputs recorded. The scrap value for the machine is estimated to be \$5700 once the counter reaches 800,000.

(a) What is the machine's total depreciation value at the end of its effective life?

(b) During the first year of usage, the counter on the machine reads 80,000. Find the book value of the machine at the end of the first year.

(c) What should the yearly output be, assuming it to remain constant, from the start of the second year so that the machine lasts another 11 years?

7. Print 'n' Vision have bought printing equipment to the value of \$668,000, with an estimated scrap value of \$82,000. The machine is good for 2,000,000 runs and the company will be carrying out 390,000 runs per year. The accountant is trying to decide on the 'best' method of depreciating the equipment. The options available to him are:

Option A. Flat rate depreciation at a fixed rate of 12% p.a.
Option B. Fixed rate of 18% p.a. by diminishing value method.
Option C. Using a unit cost depreciation as per the annual printing schedule.

(a) Complete the depreciation schedule for each option using the table below.

years	Option A	Option B	Option C
0	668000	668000	668000
1	587840	547760	553730
2			
3			

(b) Show that after t years, the book value, BV, for each option is as follows

Option A: $BV_A = 668000 - 80160t$.

Option B: $BV_B = 668000 \times (0.82)^t$.

Option C: $BV_C = 668000 - 114270t$.

(c) On the same set of axes, sketch the three graphs representing the different options to calculate the machine's book value, t years after it was purchased.

(d) Which option will provide the
  i. greatest amount of depreciation over the first three years?
  ii. least amount of depreciation over the first three years?

(e) What is the scrap value of the machine after 5 years under
  i. Option A?
  ii. Option B?
  iii. Option C?

(f) i. Under which option will the machine be written off first?
  ii. When will this occur?

21.4 FINANCIAL TABLES

There are a number of financial calculations that are often performed using tables. These are generally used to save calculation.

21.4.1 INVESTMENTS AND SAVINGS

The following table shows the compound interest amounts produced by \$1000 compounded monthly over various periods.

Rate(%) Month	2	2.5	3	3.5	4	4.5	5	5.5
1	\$1,001.67	\$1,002.08	\$1,002.50	\$1,002.92	\$1,003.33	\$1,003.75	\$1,004.17	\$1,004.58
2	\$1,003.34	\$1,004.17	\$1,005.01	\$1,005.84	\$1,006.68	\$1,007.51	\$1,008.35	\$1,009.19
3	\$1,005.01	\$1,006.26	\$1,007.52	\$1,008.78	\$1,010.03	\$1,011.29	\$1,012.55	\$1,013.81
4	\$1,006.68	\$1,008.36	\$1,010.04	\$1,011.72	\$1,013.40	\$1,015.08	\$1,016.77	\$1,018.46
5	\$1,008.36	\$1,010.46	\$1,012.56	\$1,014.67	\$1,016.78	\$1,018.89	\$1,021.01	\$1,023.13
6	\$1,010.04	\$1,012.57	\$1,015.09	\$1,017.63	\$1,020.17	\$1,022.71	\$1,025.26	\$1,027.82
7	\$1,011.73	\$1,014.67	\$1,017.63	\$1,020.60	\$1,023.57	\$1,026.55	\$1,029.53	\$1,032.53
8	\$1,013.41	\$1,016.79	\$1,020.18	\$1,023.57	\$1,026.98	\$1,030.40	\$1,033.82	\$1,037.26
9	\$1,015.10	\$1,018.91	\$1,022.73	\$1,026.56	\$1,030.40	\$1,034.26	\$1,038.13	\$1,042.01
10	\$1,016.79	\$1,021.03	\$1,025.28	\$1,029.55	\$1,033.84	\$1,038.14	\$1,042.46	\$1,046.79
11	\$1,018.49	\$1,023.16	\$1,027.85	\$1,032.56	\$1,037.28	\$1,042.03	\$1,046.80	\$1,051.59
12	\$1,020.18	\$1,025.29	\$1,030.42	\$1,035.57	\$1,040.74	\$1,045.94	\$1,051.16	\$1,056.41
13	\$1,021.88	\$1,027.42	\$1,032.99	\$1,038.59	\$1,044.21	\$1,049.86	\$1,055.54	\$1,061.25
14	\$1,023.59	\$1,029.56	\$1,035.57	\$1,041.62	\$1,047.69	\$1,053.80	\$1,059.94	\$1,066.11
15	\$1,025.29	\$1,031.71	\$1,038.16	\$1,044.65	\$1,051.18	\$1,057.75	\$1,064.36	\$1,071.00
16	\$1,027.00	\$1,033.86	\$1,040.76	\$1,047.70	\$1,054.59	\$1,061.72	\$1,068.79	\$1,075.91
17	\$1,028.71	\$1,036.01	\$1,043.36	\$1,050.76	\$1,058.20	\$1,065.70	\$1,073.24	\$1,080.84
18	\$1,030.43	\$1,038.17	\$1,045.97	\$1,053.82	\$1,061.73	\$1,069.70	\$1,077.72	\$1,085.79
19	\$1,032.15	\$1,040.33	\$1,048.58	\$1,056.90	\$1,065.27	\$1,073.71	\$1,082.21	\$1,090.77
20	\$1,033.87	\$1,042.50	\$1,051.21	\$1,059.98	\$1,068.82	\$1,077.73	\$1,086.72	\$1,095.77
21	\$1,035.59	\$1,044.67	\$1,053.83	\$1,063.07	\$1,072.38	\$1,081.77	\$1,091.24	\$1,100.79
22	\$1,037.32	\$1,046.85	\$1,056.47	\$1,066.17	\$1,075.96	\$1,085.83	\$1,095.79	\$1,105.84
23	\$1,039.04	\$1,049.03	\$1,059.11	\$1,069.28	\$1,079.54	\$1,089.90	\$1,100.36	\$1,110.91
24	\$1,040.78	\$1,051.22	\$1,061.76	\$1,072.40	\$1,083.14	\$1,093.99	\$1,104.94	\$1,116.00

This table can be used directly to find the amounts that will result from investing \$1000. It can also be used to find the results of investing other amounts. Of course, a table such as this one was of great significance at a time when no electronic calculating device was available and all calculations were done by hand – imagine that! In fact, calculations such as multiplication, division, finding values of trigonometric ratios and so on were all tabulated. You will find that at the end of many 'old' books there existed sets of tables for all sorts of calculations.

EXAMPLE 21.19 \$1000 is placed in a savings account that pays 4.5% annual interest compounded monthly. Find the amount in the account and the interest paid after 8 months.

Solution

We concentrate on the relevant row and column of our table ($n = 8$, $r = 4.5$)

Rate(%) Month	2	2.5	3	3.5	4	4.5	5	5.5
6	\$1,010.04	\$1,012.57	\$1,015.09	\$1,017.63	\$1,020.17	\$1,022.71	\$1,025.26	\$1,027.82
7	\$1,011.73	\$1,014.67	\$1,017.63	\$1,020.60	\$1,023.57	\$1,026.55	\$1,029.53	\$1,032.53
8	\$1,013.41	\$1,016.79	\$1,020.18	\$1,023.57	\$1,026.98	\$1,030.40	\$1,033.82	\$1,037.26
9	\$1,015.10	\$1,018.91	\$1,022.73	\$1,026.56	\$1,030.40	\$1,034.26	\$1,038.13	\$1,042.01

Then, reading directly from the table, the amount is \$1030.40 and the interest is \$30.40

EXAMPLE 21.21

Find the interest earned over a 1 year period if 1500 marks is invested in a savings account that pays 3% annual interest compounded monthly.

Solution

Again from the the table:

Rate(%) Month	2	2.5	3	3.5	4	4.5	5	5.5
....								
11	\$1,018.49	\$1,023.16	\$1,027.85	\$1,032.56	\$1,037.28	\$1,042.03	\$1,046.80	\$1,051.59
12	\$1,020.18	\$1,025.29	\$1,030.42	\$1,035.57	\$1,040.74	\$1,045.94	\$1,051.16	\$1,056.41
13	\$1,021.88	\$1,027.42	\$1,032.99	\$1,038.59	\$1,044.21	\$1,049.86	\$1,055.54	\$1,061.25
....								

1000 marks will amount to 1030.42.

1500 marks will amount to $\frac{1500}{1000} \times 1030.42 = \$1545.63$

EXERCISES 21.4.1

1. Use the table in § 21.4.1 to find the amount that 1000 yen will amount to after 10 months in an account that pays 4% annual interest compounded monthly.

2. Use the table in § 21.4.1 to find the amount that 20 000 rupees will amount to after 18 months in an account that pays 3.5% annual interest compounded monthly.

3. Use the table in § 21.4.11 to find the amount that \$150 will amount to after 16 months in an account that pays 4.5% annual interest compounded monthly.

4. Use the table in § 21.4.1 to find the amount that 120 000 lire will amount to after 20 months in an account that pays 2.5% annual interest compounded monthly.

5. Use the table in § 21.4.1 to find the amount that 250 000 crowns will amount to after 2 months in an account that pays 5.5% annual interest compounded monthly.

21.4.2 LOAN REPAYMENTS

When a consumer takes out a loan it is common for the interest on the loan to be calculated on the balance at each stage of the loan. As the debt decreases, so does the interest. This is known as **decreasing balance** interest. This makes the calculation more difficult to follow than the simple interest method.

A good way to follow such a loan is to use a spreadsheet. The following example shows this.

EXAMPLE 21.22

Frieda borrows 12 000 Swiss francs to help her buy a car. The company that issues the loan charges 9% annual interest calculated monthly. The repayments are monthly and the loan is to last 5 years. Write a spreadsheet that will track this loan and find the repayment that will discharge the loan in 5 years.

Solution

The diagram shows the result of fixing the payment at 300 Swiss francs.

A	A	B	C	D	E	F
1	Ann rate	9				
2	Principal	12000				
3	Payment	300			Final bal:	-3839
4						
5	Month	Loan	Interest	Payment		
6	0	12000	90	300		
7	1	11790	88	300		
8	2	11578	87	300		
9	3	11365	85	300		

The sheet extends below this diagram. The formulas used are:

A	A	B	C	D	E	F
1	Ann rate	9				
2	Principal	12000				
3	Payment	300			Final bal:	+B66
4						
5	Month	Loan	Interest	Payment		
6	0	+B2	+B6*$B$1/1200	+$B$3		
7	+A6+1	+B6+C6-D6	+B7*$B$1/1200	+$B$3		
8	+A7+1	+B7+C7-D7	+B8*$B$1/1200	+$B$3		
9	+A8+1	+B8+C8-D8	+B9*$B$1/1200	+$B$3		

Column A uses a formula to write a column of sequential numbers to represent the months of the loan.
Column C calculates the simple interest due on the balance of the loan at the beginning of each month. The denominator of the interest formula is 1200, not 100, because the quoted interest rate is annual but the interest is charged monthly. The formula that calculates this uses an 'absolute cell reference'. $B$1. The effect of the '$' signs is to make sure that this part of the formula is reproduced exactly when the formula is copied. Cell references that do not have the '$' sign are updated when they are copied.

Column B calculates the new debt as the old debt + interest – payment.

Cell F3 records the debt after 60 months (5 years) so that the user of the spreadsheet can change the key parameters of the problem: the interest rate in B1 the principal in B2 and the monthly payment in B3. The final balance is in cell B66, well off the home screen. To enable a user to change the main parameters of the problem (B1 to B3) and see the final debt immediately.

After a some trial and error, you should be able to arrive at the solution to the problem:

A	A	B	C	D	E	F
1	Ann rate	9				
2	Principal	12000				
3	Payment	249			Final bal:	8
4						
5	Month	Loan	Interest	Payment		
6	0	12000	90	249		
7	1	11841	89	249		
8	2	11681	88	249		

This shows that if the monthly repayment is set at 249 Swiss francs, the final debt will be 8 Swiss francs. This amounts to an almost complete discharge of the loan.

The above 'trial and error' method should be used to solve the following problems.

EXERCISES 21.4.2

1. Find the monthly payments necessary to discharge the following loans at the interest rates given in the periods specified.

	Loan (Crowns)	Rate	Period (months)
(a)	3000	5	60
(b)	6700	6	12
(c)	2300	7	80
(d)	7000	4	50
(e)	3900	8	60
(f)	12000	7	48
(g)	3400	9	120
(h)	15000	3	120

Rather than using 'trial and error', it is possible to use tables to find appropriate payments for reducing balance loans.

The following table shows the monthly payments needed to repay a loan of $10 000 for various periods of time.

Rate Months	2	3	4	5	6	7	8	9
6	$1,676.40	$1,681.28	$1,686.17	$1,691.06	$1,695.95	$1,700.86	$1,705.77	$1,710.69
12	$842.39	$846.94	$851.50	$856.07	$860.66	$865.27	$869.88	$874.51
18	$564.39	$568.84	$573.31	$577.81	$582.32	$586.85	$591.40	$595.98
24	$425.40	$429.81	$434.25	$438.71	$443.21	$447.73	$452.27	$456.85
30	$342.01	$346.41	$350.83	$355.29	$359.79	$364.32	$368.88	$373.48
36	$286.43	$290.81	$295.24	$299.71	$304.22	$308.77	$313.36	$318.00
42	$246.72	$251.11	$255.55	$260.03	$264.56	$269.14	$273.77	$278.45
48	$216.95	$221.34	$225.79	$230.29	$234.85	$239.46	$244.13	$248.85
54	$193.80	$198.20	$202.66	$207.18	$211.77	$216.42	$221.12	$225.89
60	$175.28	$179.69	$184.17	$188.71	$193.33	$198.01	$202.76	$207.58
66	$160.13	$164.55	$169.04	$173.62	$178.26	$182.98	$187.78	$192.65
72	$147.50	$151.94	$156.45	$161.05	$165.73	$170.49	$175.33	$180.26
78	$136.83	$141.27	$145.81	$150.43	$155.14	$159.94	$164.83	$169.81
84	$127.67	$132.13	$136.69	$141.34	$146.09	$150.93	$155.86	$160.89
90	$119.75	$124.22	$128.79	$133.47	$138.25	$143.13	$148.12	$153.20
96	$112.81	$117.30	$121.89	$126.60	$131.41	$136.34	$141.37	$146.50
102	$106.69	$111.19	$115.81	$120.54	$125.39	$130.36	$135.44	$140.62
108	$101.25	$105.77	$110.41	$115.17	$120.06	$125.06	$130.19	$135.43
114	$96.39	$100.92	$105.58	$110.38	$115.29	$120.34	$125.51	$130.81
120	$92.01	$96.56	$101.25	$106.07	$111.02	$116.11	$121.33	$126.68
126	$88.06	$92.62	$97.33	$102.17	$107.16	$112.29	$117.56	$122.96
132	$84.46	$89.04	$93.77	$98.64	$103.67	$108.84	$114.15	$119.61
138	$81.18	$85.77	$90.52	$95.43	$100.49	$105.70	$111.06	$116.57
144	$78.17	$82.78	$87.55	$92.49	$97.59	$102.84	$108.25	$113.80
150	$75.40	$80.03	$84.83	$89.79	$94.92	$100.22	$105.67	$111.28

EXAMPLE 21.23

Find the monthly repayment needed to repay a loan of 200 000 marks over a 5 year period with 5% annual interest paid monthly on the reducing balance of the loan.

Solution

The period is 60 months. From the table:

Rate / Months	2	3	4	5	6	7	8	9
. . .								
54	$193.80	$198.20	$202.66	$207.18	$211.77	$216.42	$221.12	$225.89
60	$175.28	$179.69	$184.17	$188.71	$193.33	$198.01	$202.76	$207.58
66	$160.13	$164.55	$169.04	$173.62	$178.26	$182.98	$187.78	$192.65
. . .								

The repayment for a loan of 10 000 marks (the fact that the table refers to dollars does not matter) is $188.71. The repayment for a loan of 200 000 is 20×188.71 = 3774 marks

EXERCISES 21.4.2 – continued . . .

2. Find the monthly payments needed to pay off the following loans (in francs) in the periods given.

	Loan	Rate (% p.a.)	Period (months)
(a)	9700	4	12
(b)	4100	4	24
(c)	5100	3	78
(d)	2600	2	18
(e)	1300	5	6
(f)	6800	8	18
(g)	5300	3	102
(h)	5100	5	60
(i)	8200	8	12
(j)	8800	4	108
(k)	1000	4	90
(l)	8200	5	24
(m)	4500	3	24
(n)	6900	2	78
(o)	400	5	84
(p)	8200	7	30
(q)	3000	7	30
(r)	900	5	24
(s)	2900	7	84
(t)	3700	7	84

21.4.3 INFLATION

Over recent years it has been generally true that wages and prices have risen steadily. This effect, known as **inflation**, does not generally make people more wealthy because rises in wages are offset by rises in prices. Such rises are often expressed as percentage rates

EXAMPLE 21.24

In a period when inflation is running at 5%, find the price of a television that originally cost $450, 4 years later.

Solution

The price rises are cumulative (like compound interest).

Price after 4 years: $450\left(1+\frac{5}{100}\right)^4 = 546.97781$ or $546.98

An alternative to the calculation above is to use an inflation table:

The table shows the price of a $100 article at various inflation rates after various periods.

Rate / Year	2	2.5	3	3.5	4	4.5	5
1	$102.00	$102.50	$103.00	$103.50	$104.00	$104.50	$105.00
2	$104.04	$105.06	$106.09	$107.12	$108.16	$109.20	$110.25
3	$106.12	$107.69	$109.27	$110.87	$112.49	$114.12	$115.76
4	$108.24	$110.38	$112.55	$114.75	$116.99	$119.25	**$121.55**
5	$110.41	$113.14	$115.93	$118.77	$121.67	$124.62	$127.63
6	$112.62	$115.97	$119.41	$122.93	$126.53	$130.23	$134.01
7	$114.87	$118.87	$122.99	$127.23	$131.59	$136.09	$140.71
8	$117.17	$121.84	$126.68	$131.68	$136.86	$142.21	$147.75
9	$119.51	$124.89	$130.48	$136.29	$142.33	$148.61	$155.13
10	$121.90	$128.01	$134.39	$141.06	$148.02	$155.30	$162.89
11	$124.34	$131.21	$138.42	$146.00	$153.95	$162.29	$171.03
12	$126.82	$134.49	$142.58	$151.11	$160.10	$169.59	$179.59
13	$129.36	$137.85	$146.85	$156.40	$166.51	$177.22	$188.56
14	$131.95	$141.30	$151.26	$161.87	$173.17	$185.19	$197.99
15	$134.59	$144.83	$155.80	$167.53	$180.09	$193.53	$207.89

The table can be used to solve the previous example by selecting the entry in the 5% column and the 4 year row (**bold**).

The price of a $450 television after 4 years at 5% inflation is $\frac{450}{100}(121.55) = 546.975$ or $546.98. Use of a table may result in small rounding errors over a calculator based solution.

EXERCISES 21.4.3

Find the prices of the following items at the inflation rates given after the periods indicated:

	Original price	Rate	Years
1.	$665	8	3
2.	$1282	5	8
3.	$983	1	4
4.	$548	10	1
5.	$204	4	13
6.	$197	10	13
7.	$911	10	5
8.	$90	9	6
9.	$499	6	2
10.	$1350	11	10
11.	$313	5	9
12.	$511	6	1.5
13.	$316	1	10
14.	$569	8	2.75
15.	$1487	6	13

21.5 MISCELLANEOUS EXERCISES

1. A computer is bought at a cost of $2500 and depreciates in value at a rate of 15% p.a. Find, to the nearest dollar, the value of the computer after four years.

2. An account receives 8% p.a. interest paid quarterly. The account begins the year with $1000. Find, to the nearest dollar, the amount of money in the account one year later.

3. An item of clothing costs $50 at a sale. The item was said to have been reduced by 25%. What was the original price of the item. Give your answer to the nearest dollar.

4. House values in a particular area appreciate on average by 2.5% p.a. Find the least number of years before a $100,000 house would have a value of $125,000.

5. A sum of $10,000 is invested for 3 years at 14% p.a. compounded annually. What is the amount in the account after 3 years?

6. A sum of $100 is invested on January 1st for three years at the rate of 12% p.a. compounded annually. One year later, another $100 is invested on January 1st for 2 years at the same rate. How much interest has acrued after three years?

7. A new car is valued at $24,000. It loses 15% of its value in the first year, 10% in the second year and 5% in the third year. What is the car's value at the end of the third year?

8. On the 1st of January each year I bank $1000 into an account. On the 31st December each year I am paid 15% p.a. interest on my balance. What is the total amount of money in the account, to the nearest dollar, just after I have banked my third amount of $1000?

9. If an investment, in a fixed deposit account, doubles in value in five years, find to one decimal place, the rate of interest per annum, compounded annually.

10. A car purchased for \$30,000 depreciates by 12% in the first year and then 8% p.a. in each successive year using the reducing balance method. Find, to the nearest \$100, the value of the car after 4 years.

11. Janette invests \$8500 which 5 years later has increased to \$12,500.
(a) What was the interest from this investment
(b) i. Find, the simple interest rate, % per annum, for this investment.
ii. Find, the compound interest rate, % per annum, for this investment.

12. A car purchased for \$45,000 depreciates at 12% in the first year, then 8% p.a. in each successive year using the reducing balance method. How many years from purchase will it take to drop below 50% of the purchase price?

13. A person investing \$10,000 is offered 5% per annum compounded quarterly by one institution but 5% per annum compounded daily by another institution. What is the difference that the person can expect to receive. Give your answer to the nearest dollar.

14. A person invests \$1200 each year in an account where compound interest of 5% p.a. is calculated at the end of each year. What is the value of this investment at the end of ten years. Give your answer to the nearest \$1000.

15. How many years will it take to earn £5520 interest if £8000 is invested at a simple interest rate of 13.8% p.a.

16. To the nearest dollar, what will \$15,000 invested for 3 years at an interest rate of 16% p.a. compounded quarterly amount to?

17. A construction company has a machine worth £195,000. The company's accountant estimates that its value will depreciate at the rate of 12% p.a. reducing balance. What will its book value be after five years?

18. A new car costing \$18,500 can be purchased for a deposit of \$2000 and monthly payments, over a period of 4 years, at a flat rate of 18% p.a. What should the monthly repayments be? Give your answer to the nearest dollar.

19. A person planning for retirement invests, on the first day of each year, an amount of £2500. Interest is compounded annually at 12.5% p.a. Find, to the nearest dollar, the total amount of money in the account immediately after the investment of the third amount of £2500.

20. Wendy and Frank borrowed \$20,000 as a loan to finance an overseas trip. They agreed to repay the loan at \$3600 per year with interest charged at 14% p.a. on the unpaid balance at the end of each year. Find the amount still owing, to the nearest dollar, immediately after they made their third payment.

21. (a) £4000 is placed into a savings account returning 4% p.a. simple interest.
i. How much interest is payable over ten years?
ii. Find the amount in the account after ten years.
(b) How long would it take \$9000 to earn \$8000 interest at 15% p.a. simple interest?
(c) What simple rate of interest is required over 3 years if \$400 accumulates to \$510?
(d) \$30 in simple interest was earned over 2 years at 12% p.a. What was the capital?

22. The currency conversion rates between five countries are shown.

US	Aus	Sterling	Yen	Mark
1	1.633	0.610	139.490	1.782

Convert:

(a) \$100 US to Japanese Yen
(b) £150 Sterling to \$US
(c) \$150 Australian to Marks
(d) 1250 Yen to Sterling.

23. An agent offers to exchange \$US to other currencies at the published daily rate. Her commission is \$5 per transaction or 1%, whichever is the greater with the commission being paid in \$US. Three customers wish to convert the following amounts of \$US to Italian lire on a day when the exchange rate is 1760 lire to the dollar:

(a) \$200 (b) \$20 (c) \$2000

Find the numbers of lire that each will receive after the commission has been charged.

24. \$1 Australian is equivalent to 3.5138 French francs and to 3.0274 South African rand. Convert the following amounts in South African rand to French francs, giving your answers to the nearest franc.

(a) 770 (b) 180 (c) 330

25. A bank offers the following exchange rates for \$1 Australian in relation to the French franc: 'We buy: 3.5959, we sell: 3.5138'. A customer wishes to exchange \$2400 Australian for francs. How many francs will the customer receive? If the customer then immediately exchanges these francs for Australian dollars, how much will she receive? What is the effective exchange rate? Assume that all currency amounts are rounded to the nearest whole number to complete each transaction.

26. A family invests \$2500 in an account that pays 8% annual simple interest for 9 years. Find the amount of interest paid over this period.

27. (a) If \$800 is invested in an account and earns \$116.50 simple interest over a 3.5 year period, find the annual interest rate.

(b) If 2900 crowns are invested at a monthly interest rate of 0.10% and earn 10.12 crowns simple interest, find the number of months of the investment.

28. Boris invests \$660 at 8% annual interest compounded annually over a period of 10 years. Compare the results of this with the simple interest alternative.

29. Find what \$900 amounts to in 20 years if it is invested at 9% p.a. compounding annually.

30. \$2400 is placed in a savings account that pays 7% annual interest compounded annually. Find the amount in the account and the interest paid after 12 years.

31. Find the interest earned over a 6 year period if 3000 marks is invested in a savings account that pays 6.24% annual interest compounded monthly.

32. \$1500 is invested at 9.00% p.a., compounded quarterly for 6 years. Find the interest earned.

33. \$3000 grows to \$4000 over two years when interest is compounded quarterly. Find the annual rate of interest that would allow for this.

34. Danielle decides to invest \$5000 for a fixed 6 year period. There are three banks to choose from, offering the following options:
Bank A: 13% simple interest.
Bank B: 11.6% p.a. compounded annually
Bank C: 11 % p.a. compounded monthly.
With which bank should Danielle invest her money to obtain the best return?

35. Ten years ago Kirk invested \$8000. For the first five years the account paid 10% p.a. compounded semiannually. However, for the rest of the time it could only manage 7% p.a. compounded semiannually. How much will Kirk have in the account now?

36. How long will it take \$3000 to amount to \$4300 if invested at 12% p.a. with interest compounded annually?

37. How long will it take \$3000 to accrue \$4000 interest if was invested at 15% p.a. with interest compounded fortnighly?

38. Lucille invests \$8000 at 7.5% p.a. compounded quarterly so that she can purchase a \$12,000 car in 4 years time.
(a) Her parents have told her that they would give her the shortfall if one exists at that time. Is there a shortfall? If so, how much will the parents need to give Lucille?
(b) Lucille then decides that she wants to purchase a \$16,000 car.
i. Show that the number of quarters, n, that are required satisfy the equation $(1.01875)^n = 2$.
ii. How long does she need to wait until her investment allows her to purchase the car outright and on her own?

39. McPherson Printers – a book printing company, have just purchased a printing machine for \$350,000. The company's accountant believes that a flat rate depreciation method should apply for this machine. The depreciation rate is 20% p.a. of the purchase price.
(a) Find the annual depreciation value.
(b) Set up a depreciation schedule for the machine, showing book values for the first 4 years.
(c) Find the equation of the straight line showing the book value of the machine t years after it was purchased.
(d) The machine has an effective life of 5 years. Find the scrap value of the printing machine at that point in time.

40. Veronique bought a computer 5 years ago. The current book value of this computer is \$700. If the computer has been depreciating at \$550 p.a. how much did Veronique pay for the computer?

41. Cheng Tek purchased some office furniture four years ago for \$6000. Today the furniture is worth \$2200. Assuming a flat-rate depreciation model, determine the depreciation rate applied to the furniture.

42. A piece of machinery was purchased for \$40,000 five years ago. It depreciates at a rate of 22% p.a. by the reducing balance method. What is the book value (to the nearest dollar) of this machine (a) now? (b) in three years time?

43. Tim's earth moving company bought two new trucks for \$45,000 each. Tim decides that one of the trucks will be depreciated at the flat rate of 18% of the purchase price, while the other will be depreciated at 23% using a reducing balance method. It is estimated that the trucks will have an effective life of 5 years.

(a) Construct a depreciation schedule for both trucks.
(b) Sketch the graphs showing the relationship between the truck's book value and the time since they were purchased.
(c) Compare the behaviour of both methods of depreciation.

44. A car was purchased for \$22,000 at the start of the new financial year and will be depreciated by 18% p.a. of the previous end of financial year's book value. It is expected that the car will have an effective life of 5 years.

(a) Construct a depreciation schedule for this car.
(b) Draw a graph showing the relationship between the car's book value and time since it was purchased.
(c) Find the equation relating the book value, BV_t, of the car and the time, t years, since it was purchased.
(d) What is the car's book value i. after 2 years?
ii. after 3 years and 9 months?
(e) What is the car's scrap value?

45. A car was purchased for \$35,000 with an estimated scrap value of \$5500 once the car reaches 100,000 km.

(a) Find the depreciation rate in dollars per km.
(b) Find the book value of the car after reaching 60,000 km.
(c) Construct a depreciation schedule for every 10,000 km travelled.

46. A printing company purchased a binding machine for \$300,000 and decided to depreciate its value at 18% p.a. using a flat rate method.

(a) i. Calculate its book value at the end of three years.
ii. By how much has the machine depreciated after five years of operation?
(b) Construct a depreciation schedule for the binding machine.
(c) The binding machine's book value, $\$BV_t$, is given by the equation $a - bt$, where t is the time in years since it was purchased. Determine the values of a and b.
(d) Draw a graph depicting the book value of the machine since it was purchased.
(e) How long will it take for the machine to be written off?

47. A photocopy machine was purchased for \$20,000 and has an estimated life span of 2,000,000 copies at which point its scrap value will be \$2500. During its first year of operation, 210,000 copies are made. Find its book value at the end of the first year.

48. Solve the following problems using tables.

(a) \$1000 is placed in a savings account that pays 5.5% annual interest compounded monthly. Find the amount in the account and the interest paid after 9 months.

(b) Find the interest earned over a 1 year period if 3000 marks is invested in a savings account that pays 4% annual interest compounded monthly.

(c) In a period when inflation is running at 6%, find the price of a television that originally cost \$550, 5 years later.

21.6 GRADED REVISION EXERCISES

LEVEL 1

1. The currency conversion rates between four countries are shown:

US \$1.00 Aus \$1.92 Sterling 0.58 Yen 152.35

Convert

i. \$25 US to Yen.

ii. £50 Sterling to \$Australian.

iii. ¥150 Yen to \$Australian.

2. Tom invests \$745 in an account paying 7% simple interest per annum for 5 years. How much interest has been paid over this period?

LEVEL 2

1. Sylvia receives a flat salary of \$25,000 and a 6% commission on sales made. If she sells \$30,000 worth of products a year, what is her income?

2. Find the interest charged for a loan of \$4000 if it is levied at 6% compound interest over four years compounding annually.

LEVEL 3

1. Jack invests £650 at 5% annual interest compounded annually over a period of 15 years. Find the total interest paid. What simple interest rate would be needed to achieve the same result after 15 years?

2. Joanne invested GBP 25,000 in bonds, which return monthly interest at the simple rate of 12.0% per annum.

i. What is the monthly interest rate?

ii. How much does Joanne receive each (a) year? (b) month?

iii. How long did it take for Joanne to receive \$9000 interest?

3. Yuriko invested ¥1000 at 9% per annum compounded monthly. After t years, her return was ¥892.

i. Show that $1.892 = 1.0075^{12t}$.

ii. For how long did Yuriko invest her money?

LEVEL 4

1. On the same set of axes, plot a graph of the following investments:

i. \$1000 at 9% simple interest over a 10 year period.

ii. \$1000 at 7% compounded monthly for a period of 10 years.

Discuss the progress of the investment during the 10 year period.

2. A good way to compare investments is to determine their effective rates – that is the simple interest rates that would produce the same return in 1 year if the same principal had been invested at simple interest without compounding.

Bond A pays 16% compounded monthly, while bond B pays 16.2% compounded semiannually. Which is the better investment?

21.7 TOPIC TEST

1. 1 Singapore dollar is worth US $0.70. Convert

i. 50 Singapore dollars to US dollars.

ii. $90 US into Singapore dollars.

[3 marks]

2. Find the interest on 2500 crowns if they are they invested at

i. 4.8% p.a. simple interest for an eight year period.

ii. 4.8% p.a. compounded monthly for an eight year period.

[4 marks]

3. A bank offers the following exchange rates for $1 Australian in relation to the French franc: 'We buy: 3.7979, we sell: 3.6848'.
A customer wishes to exchange $1500 Australian for francs. How many francs will the customer receive? If the customer then immediately exchanges these francs for Australian dollars, how much will she receive? What is the effective commission? Round your answers to the nearest dollar.

[4 marks]

4. How long will it take for $12,000 to grow to $14,000 if it is invested at 8% p.a. compounded quarterly?

[3 marks]

5. Which is the better investment?
Option A: Buying notes paying 8% compounded monthly.
Option B: Buying notes paying 8.1% compounded semiannually.

[5 marks]

6. The graph below shows the depreciation behaviour of an item since it was purchased. Find

(a) the cost price.

(b) the time taken for the item to reach its scrap value.

(c) the total depreciation of the item for its useful life.

(d) the rate of depreciation.

(e) the expression for the linear relationship between its book value, BV, and the time, t years, since it was purchased.

(f) its book value 3 years and 4 months after it was purchased.

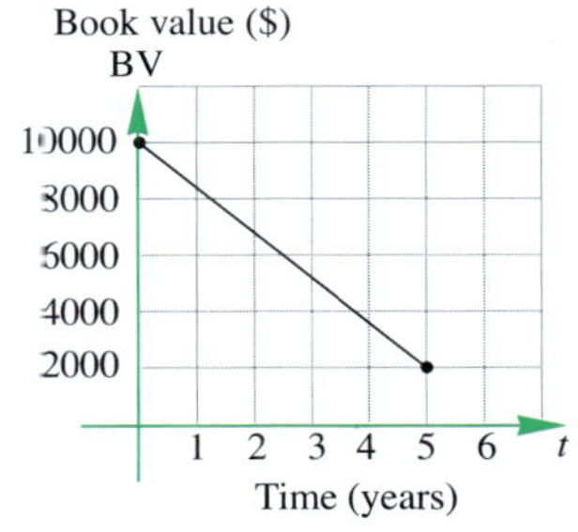

[10 marks]

7. A piece of machinery was purchased for $50,000 five years ago. It depreciates at a rate of 25% p.a. by the reducing balance method.

(a) What is the book value (to the nearest dollar) of this machine

i. now?

ii. in three years time?

(b) The book value, $\$BV_t$, of this machine, t years from the day it was purchased can be calculated using the equation, $BV_t = a \times b^t, 0 \le t \le 9$.

i. Find the values of a and b.

ii. Calculate its scrap value.

[6 marks]
Total 35 marks

21.8 PROJECT SUGGESTIONS

There are many good financial starting points for projects.

Some advertisements offer 'terms' (10% down followed by 24 monthly payments of ...). Comparisons between these can provide you with some interesting investigations. What would happen if you used a credit card, bank loan etc.

Home loans provide an excellent starting point for a project.

Consider the following situation:
Linda borrows $300000 at 1% per month reducible interest. If she repays the loan in equal monthly instalments over 20 years, how much is each instalment?

Using Linda's basic approach, introduce variables into her schedule. For example, what if Linda decides to marry 5 years after she started her loan and started to have children two years after that.

What are her options? Does she want to continue working? If not, how does this impact on what was once a two wage family? If she does return to work, what new financial commitments will there be? Such questions are just the tip of the iceberg – analysing a family's financial situation is full of mathematics that involves careful planning!

Superannuation is an area rich in financial mathematics and decision making.

Consider the relatively simple problem:
A woman invests $1000 at the beginning of each year in a superannuation scheme. If the interest is paid at the rate of 12% p.a. on the investment (compounding annually), how much will her investment be worth after 20 years?

However, depending when you start contributing to a superannuation fund, you will be able to determine the risks that you are prepared to take. For example, early on in your contributing life you may want the funds in your portfolio to be partitioned so that 80% of your funds are invested in high yield-high risk investments and 20% in low yield-secure investments. However, as you get older, say 5 years before retirement, you may decide that your portfolio should be 5% high yield-high risk while the other 95% should be low yield-secure investments.
Risk analysis such as this can make a big difference in the amount of money in your superannuation fund!
How often should you restructure your fund to give the best return?

Insurance is also an excellent topic as insurance premiums enable you to link your investigation to probability.
Seek out an actuary from an insurance company and they will have a wealth of problems that can start you off on a project.

Topics like: Estimation of outstanding claim provisions and
Reserve funding and risk analysis
will cover a wide range of financial and statistical applications.

REVISION SET D - PAPER 1 & PAPER 2 STYLE QUESTIONS

1. Using the box-plot shown, determine
(a) the first quartile.
(b) the interquartile range.
(c) the median.
(d) the range.

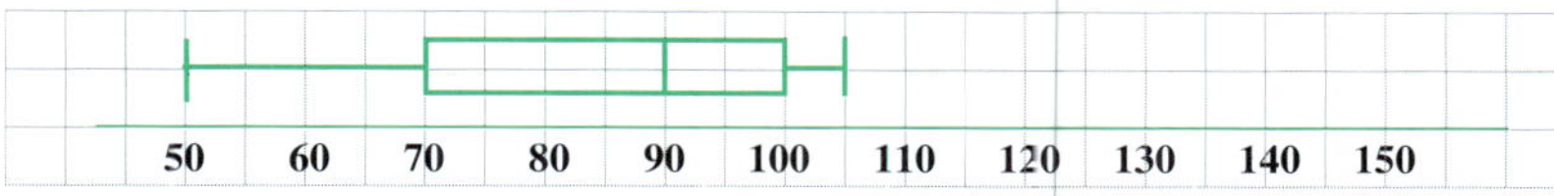

2. Consider the following stem and leaf diagram of test results for 45 students

Stem	Leaf
0	3 3 4
0*	5 5 6 6 6 7 8 9 9
1	0 0 0 0 1 1 1 1 1 1 2 2 2 2 2 2 2 3 3 3 3 3 4
1*	5 5 5 6 6 7 7 8 9
2	0

key 1 | 1 = 11

(a) State i. the highest score. ii. the lowest score.
(b) Determine
i. the median. ii. the mode.

3. 1 German mark is equivalent to 23.261 Indian rupees.
i. Convert 150 marks to rupees.
ii. Convert 1250 rupees to marks.

4. \$1 New Zealand is worth 60 US cents. An exchange service charges a 3% fee to exchange money. How much will a traveller who wants to change \$500 NZ to \$US get?

5. Find the simple interest on 60 000 lire invested at 5.6% p.a. simple interest for 6 years.

6. The yield, y kilograms, of potato, obtained by using x kilograms of no–name brand fertiliser produced the following results:

x	10	49	54	14	69	77	60
y	500	710	770	500	780	830	740

(a) Draw a scatter diagram for the given set of data.
(b) Calculate the correlation coefficient for the given data set. Is this an appropriate statistic to calculate for this data set?
(c) Calculate the proportion of the variance of y which can be explained by the variance of x.
(d) i. Use the method of least squares to find the regression equation of y on x.
ii. Graph the regression line on the scatter diagram in part a.
(e) Based on your line of best fit, determine the yield if 50 kg of fertiliser were used.
(f) A yield of 600 kg of this potatoes were produced. How much fertiliser would have been used?

7. Process PLC is a small computer business. The main assets of this company are:
A deposit account that earns 5% annual interest calculated daily on the closing balance and credited at the end of each month.

A computer, worth \$4500 at the end of May 1997, which depreciates at the rate of 22% per annum.

A current account. The charges are 70c per debit and 55c per credit on all transactions except bank charges.

A car worth \$18000 at the end of May 1997, depreciating at 15% per annum.
The liabilities are:
A \$15000 loan for the car (as at the end of May 1997). This loan charges 9% annual interest calculated monthly on the diminishing balance. The repayment is \$360 per month.
A debt of \$1500 owed to a supplier and which does not need to be paid until the end of June.
The tables show the activity in the two bank accounts over one month.

Current account:

Date	Detail	Debit	Credit	Balance
1/5/97	Carried			\$1263.65
5/5/97	Cheques		\$647.45	
12/5/97	162345	\$56.70		
17/5/97	162344	\$126.30		
25/5/97	Transfer		\$2300.00	
26/5/97	162347	\$534.55		
31/5/97	Charges			

Deposit account:

Date	Detail	Debit	Credit	Balance
1/5/97	Carried			\$6500.45
10/5/97	Cheques		\$1245.50	
31/5/97	Interest			

(a) i. Complete the balance column for the current account, including the charges at the end of the month.
ii. Complete the balance column for the deposit account, including the interest credit at the end of the month.
iii. A company's equity is defined as the difference between its assets and liabilities. What is this company's equity at the end of May 1997?

(b) i. The computer's value changes with time. *Sketch* the graph of the computer's value against time.
ii. Find the computer's value at the end of May 2000.
iii. Process PLC intend to replace the computer as soon as it falls to half of its value at the end of May 1997. In which month will the computer be replaced?

8. Parents at Zone One High School are concerned with the results of that school's common test obtained by students. Some parents suggests that the discrepancy could be due to the different teaching styles adopted by the three teachers teaching these classes.

The table below gives the results for the number of students that passed and failed from classes taken by the three teachers, A, B and C:

	A	B	C	
Passed	60	57	66	183
Failed	15	24	18	57
	75	81	84	240

The table below shows the expected number of students that passed and failed, under the assumption that the results are independent of teaching style.

	A	B	C	
Passed	a	b	64.05	183
Failed	c	d	19.95	57
	75	81	84	240

(a) Find the values of a, b, c and d.

(b) i. Write a suitable null hypothesis.

ii. Write a suitable alternate hypothesis.

(c) Calculate the value, $\chi^2_{calc} = \sum \frac{(f_o - f_e)^2}{f_e}$.

(d) i. How many degrees of freedom are associated with this contingency table?

ii. What is the critical χ^2 value, for this contingency table, using a 5% level of significance?

iii. What conclusion can be made about the association between teaching style and success in common-tests for students at this school?

9. Find the amount the 3500 Swiss francs will amount to after 5 years invested at 6.8% annual interest compounded annually.

10. Find the amount the 1250 crowns will amount to after 7 years invested at 9% annual interest compounded monthly.

11. Consider the frequency table below which displays the results of a test taken by 30 students. The test is out of 10 marks

score	0	1	2	3	4	5	6	7	8	9	10
frequency	0	1	1	1	2	4	8	6	3	2	2

Which of the statement(s) below is/are correct

A. the mean mark is less than the median mark.

B. the median and mode marks have the same value.

C. the mode mark is greater than the mean mark.

D. the mean mark is equal in value to the median.

E. all three measures of central tendency have equal values.

12. A small business employing 52 staff has an annual labour cost structure summarised in the table below (where \$K means \$1000):

Income \$K	26	29	32	35	40	45	55
Frequency	8	15	10	8	6	3	2

(a) Find the mean, median and mode salaries for the company.
Comment on the results of your calculations.

(b) Find the population variance and hence population standard deviation for the salaries of workers in this company.

A group of workers write to the senior management requesting a pay increase. They suggest either a \$1000 flat increase or a 3% increase.

(c) Which option financially benefits which workers? Give detailed calculations.

(d) Which option is management likely to accept? Why?

(e) Which option increases the range of salaries in the company? By how much?

(f) What flat increase would cost the company the same as a 3% increase?

(g) What percentage increase would cost the company the same as a flat \$1000 increase?

(h) A committee of two people is to be selected at random from the group of 52 workers to form a conflict resolution working party.
What is the probability that the combined salaries of these two workers will exceed twice the mean salary for the company?

13. The table below displays a currency conversion between GBP, USD and AUD.

	GBP	USD	AUD
GBP	1	1.42	x
USD	y	1	
AUD	0.38	z	1

Note: all answers are to be given to two decimal places.

(a) Find the values of i. x ii. y iii. z

(b) Debbie decides to exchange GBP 500 at a money exchange outlet.
How many AUD will Debbie receive if

i. there is no commission?

ii. the money exchanger charges 2% commission?

(c) Debbie then decides to invest 50% of her AUD (from (b) ii.) in bonds, with an annual rate of return set at 6.5% compounded annually.

i. How much interest will she earn after 5 years?

ii. She decides to cash in her bonds after she receives in interest, twice the initial amount she invested. How long will Debbie have to wait?

14. A random sample of 1000 people was selected and tested for hearing difficulties. The results of this survey were recorded and tabulated as shown below.

	No hearing difficulty	Hearing difficulty
Male	320	80
Female	540	60

Do these figures suggest that there exists a relationship between gender and hearing difficulties at the i. 5% level of significance? ii. 1% level of significance?

15. John wishes to start a luxury car dealership. However, he is unsure as to where the showroom should be located. He obtainted the following data, for the number of luxury cars sold during the first three months of the year.

		Type of car		
		Luxury	Non-luxury	Totals
Location of suburb	Northern	57	28	85
	Southern	26	14	40
	Western	17	38	55
	Eastern	30	10	40
	Totals	130	90	220

The table below shows the expected number of luxury and non-luxury cars sold in each suburb, if the categories were in fact independent.

Expected values:

		Type of car		
		Luxury	Non-luxury	Totals
Location of suburb	Northern	a	b	85
	Southern	c	d	40
	Western	e	32	55
	Eastern	30	10	40
	Totals	130	90	220

(a) i. Show that the number of expected luxury cars sold in the Northern suburb is given by 50.23.

ii. Hence find the values of b, c, d, and e.

(b) i. Write a suitable null hypothesis.
ii. Write a suitable alternate hypothesis.

(c) i. What type of test should be used to address your null hypothesis?
ii. Carry out your test.
iii. What conclusions can you draw from your results?

16. The data below are the weights in kilograms of a sample of a species of small rodents.

3.5	5.2	6.9	2.8	8.4	12.1	9.4	5.7	1.9	6.2
4.4	5.7	4.9	8.5	9.2	9.9	4.2	7.9	3.4	6.8

i. Make a frequency table of these weights using the class intervals 0-0.9, 1.0-1.9 etc.
ii. Find the mean of these weights.
iii. Find the standard deviation of the weights correct to four significant figures.
iv. Draw a stem and leaf diagram for the data.

17. For a bivariate data set of 26 pairs, the following statistics were calculated:
$\bar{x} = 90, \bar{y} = 80, s_{xy} = 100, s_x = 20$ and $s_y = 10$.

(a) Find the product moment correlation coefficient.
(b) The equation of the least squares regression line of y on x.
(c) Which one of the following statements is not true about the least squares regression line of y on x?

A. It includes the point $(\bar{x}, \bar{y})$.

B. It has a gradient given by $\frac{s_{xy}}{s_x^2}$.

C. It minimises the sum of the squares of the vertical distances from each point on the scatter plot to the line.

D. It has a negative gradient if the product moment correlation is negative.

E. It is the same as the least squares regression line of x on y.

(d) Find the two mean regression line of y on x for the following bivariate data set.

x	0	1	2	3	4	5
y	1	2	4	9	16	25

18. A sample of thousand electric lamps were operated continuously to determine their reliability. The result of this quality control process is shown in the table below:

Life length (hours)	frequency
[0, 200[	1
[200, 400[	9
[400, 600[	36
[600, 800[	52
[800, 1000[	125
[1000, 1200[	165
[1200, 1400[	203
[1400, 1600[	179
[1600, 1800[	121
[1800, 2000[	68
[2000, 2200[	31
[2200, 2400[	8
[2400, 2600[	2

(a) Draw a histogram to represent the frequency distribution.
(b) What is the modal class?
(c) Construct a relative frequency polygon and use it to find the median.
(d) Calculate the mean and the unbiased sample standard deviation for this data.
(e) Draw a box plot of the data.

22.0 ATHLETICS

We are all familiar with the idea of change. The World around us just does not stand still. It is odd, therefore, that up until a few hundred years ago we had no method of analysing situations in which change played a part. The movements of the planets were a mystery, as were the changes that happen in the human body.

Will a woman ever run the Marathon faster than the fastest man? The Marathon is a race named after the feat of a Greek soldier who ran from Marathon to Athens to report a victory against an invading army, a distance of about 40 km.

The distance of the race was standardised at 42,195 metres, or 26 miles 385 yards at the 1908 Olympics in London. The race started at Windsor Castle and finished in front of the royal box in the stadium in London - so, of course, that fixed the distance for ever.

If we look at the World record times over the last three decades, they are:

Women's World Record

Aug 1, 1971 - Adrienne Beames (Australia) - 2:46:30 in Australia.
Apr 13, 2003 - Paula Radcliffe (Great Britain) - 2:15:25, London Marathon, London, UK.

Men's World Record

Dec 3, 1967 - Derek Clayton (Australia) - 2:09:36, Fukuoka Marathon, Fukuoka, Japan.
Sep 28, 2003 - Paul Tergat (Kenya) - 2:04:55. Berlin Marathon, Berlin, Germany.

As you can see, during this period, the men's record has always been faster than the women's. But if we take into account the fact that the records have changed, a different picture emerges.

The women's record has improved by 46:30 – 15:25 or 31 minutes and 5 seconds over the period 1971 to 2003 (32 years). The men's record has improved by 9:36 – 4:55 or 4 minutes 41 seconds over the period 1967 to 2003 (36 years).

The women have improved by 1865 seconds over 32 years and the men by 281 seconds over 36 years. What matters here is the RATE at which the records are being improved.

The women are improving at $\frac{1865}{32} \approx 58$ seconds per yr and the men at $\frac{281}{36} \approx 8$ seconds per yr.

So what will the Marathon records be in 2020?

For both women and men, this is 2020 – 2003 = 17 years in the future. The women's record can be expected to improve by $58 \times 17 = 986$ seconds and the men's by $8 \times 17 = 136$ seconds.

A little bit of calculation with the times will tell you that the women's record will have improved by 16 minutes, 26 seconds to 1 hour, 58 minutes and 59 seconds. The men's by 2 minutes and 16 seconds to 2 hours, 2 minutes and 39 seconds. It appears that in 2020 we can expect that women will be running Marathons much faster than men. Is this a realistic suggestion? The data on which it is based suggests that at the next millennium, women will be running the Marathon in under half an hour. Is this sensible?

These ideas are based on the principle of 'rate of change', the subject of this chapter.

22.1 QUANTITATIVE MEASURE

22.1.1 FUNCTIONAL DEPENDENCE

The notion of functional dependence of a function $f(x)$ on the variable x has been dealt with in Chapter 5. However, apart from this algebraic representation, sometimes it is desirable to describe a graphical representation using a qualitative rather than quantitative description. In doing so, there are a number of key words that are often used.

Words to be kept in mind are:

Rate of change (slow, fast, zero)	Increasing, decreasing
Positive, negative	Maximum, minimum
Average	Instantaneous
Stationary	Initial, final
Continuous, discontinuous	Range, domain

Such terms enable us to describe many situations that are presented in graphical form. There is one crucial point to be careful of when describing the graphical representation of a given situation. Graphs that look identical could very well be describing completely different scenarios. Not only must you consider the behaviour (shape) of the graph itself, but also take into account the variables involved.

Consider the two graphs below. Although identical in form, they tell two completely different stories. We describe what happens in the first five minutes of motion:

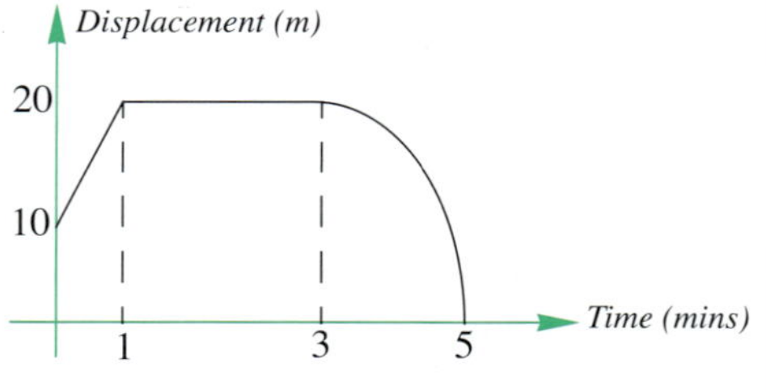

An object is moving in such a way that its displacement is increasing at a constant rate, that is, the object maintains a constant velocity (or zero acceleration) for the first minute. During the next two minutes the object remains stationary, that is, it maintains its displacement of 20 metres (meaning that it doesn't move any further from its starting position). Finally the particle returns to the origin.

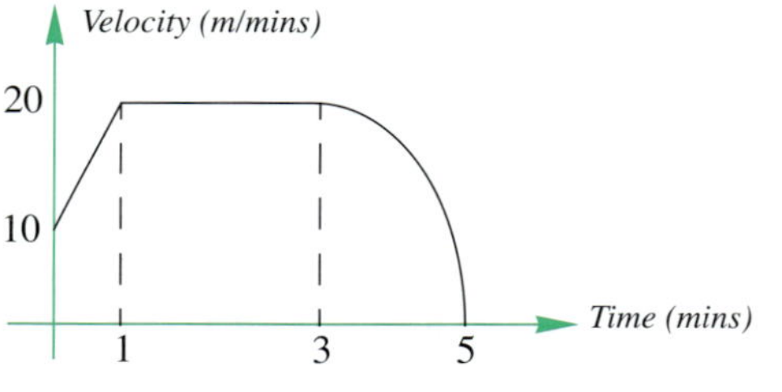

An object is moving at 10 m/min and keeps increasing its velocity at a constant rate until it reaches a velocity of 20 m/min, that is, it maintains a constant acceleration for the first minute. During the next two minutes the object is moving at a constant velocity of 20 m/mins (meaning that it is moving further away from its starting position). Finally, the particle slows to rest, far from the origin.

Although the shape of the graphs are identical, two completely different situations have been described!

22.1.2 QUANTITATIVE ASPECTS OF CHANGE

When dealing with the issue of rates of change, we need to consider two types of rates:

1. that of the **average rate** of change and
2. that of the **instantaneous rate** of change.

We start by considering the first of these terms, the average, and then see how the second, the instantaneous rate, is related to the first.

22.1.3 AVERAGE RATE OF CHANGE.

The average rate of change can be best described as an 'overview' of the effect that one variable (the independent variable) has on a second variable (the dependent variable). Consider the graph below, we can describe the change in the y-values (relative to the change in the x-values) as follows:

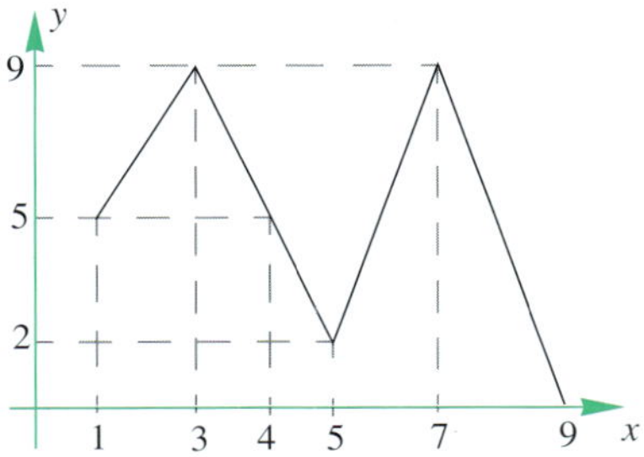

For $x \in [1, 3]$:

There is a **constant** increase from $y = 5$ to $y = 9$ as x increases from 1 to 3. So that an increase of 2 units in x has produced an increase of 4 units in y.
We say that the **average rate** of change of y with respect to x is $\frac{4}{2} = 2$

For $x \in [1, 4]$:

This time, the overall change in y is 0.
That is, although y increases from 5 to 9, it then decreases back to 5. So from its initial value of 5, because it is still at 5 as x increases from 1 to 4, the overall change in y is 0. This time the average rate of change is $\frac{0}{3} = 0$.

For $x \in [1, 5]$:

As x now increases from 1 to 5 we observe that there is an overall decrease in the value of y, i.e., there is an overall decrease of 3 units (y: $5 \to 9 \to 5 \to 2$). In this instance we say that the average rate of change is $-\frac{3}{4} = -0.75$.

Notice that we have included a negative sign to indicate that there was an overall decrease in the y values (as x has increased by 4).

Similarly for the rest of the graph. Note that we need not start at $x = 1$. We could just as easily have found the change in y for $x \in [3, 5]$. Here, the average rate of change is $-\frac{7}{2} = -3.5$.

The question then remains, is there a simple way to find these average rates of change and will it work for the case where we have non–linear sections? As we shall see in the next sections, the answer is 'yes'.

22.1.4 DETERMINING THE AVERAGE RATE OF CHANGE

To find the average rate of change in y it is necessary to have an initial point and an end point, as x increases from x_1 to x_2.

At A $x = x_1, y = y_1$ and at B $x = x_2, y = y_2$.

To obtain a numerical value, we find the gradient of the straight line joining these two points.

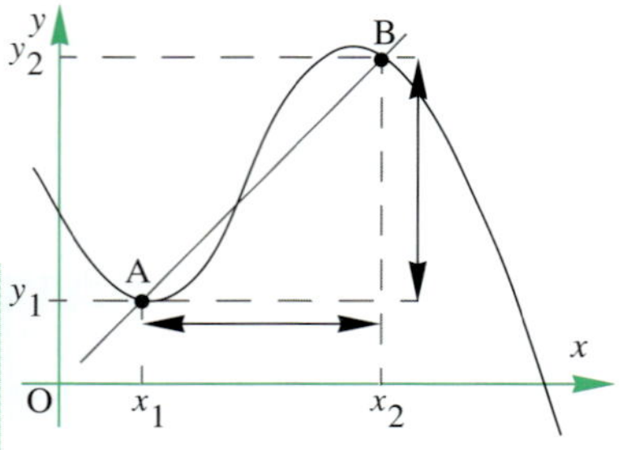

Average rate of change from A to B = gradient from A to B

$$= \frac{y_2 - y_1}{x_2 - x_1}$$

EXAMPLE 22.1

For each of the graphs below, find the average rate of change of y with respect to x over the interval specified (i.e., over the domain L).

(a)

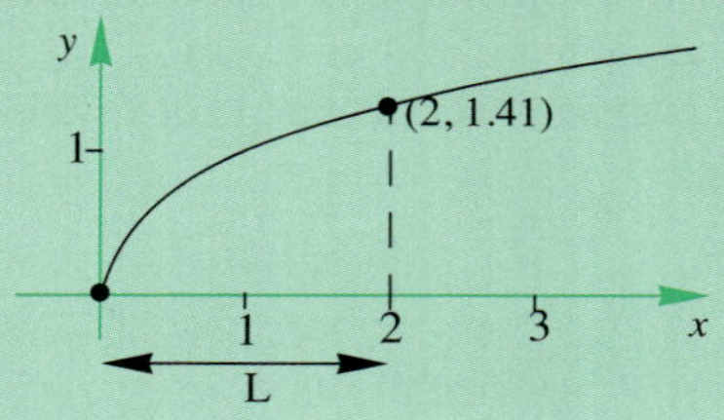

(b)

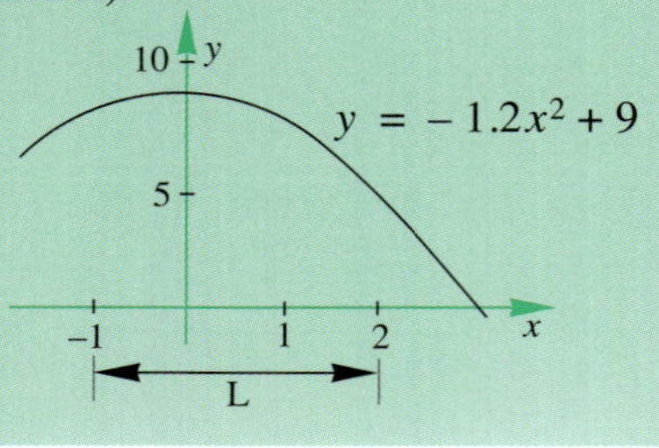

Solution

(a) For this case we have the 'starting point' at the origin (with coordinates (0,0)) and the 'end' point with coordinates (2,1.41).
This means that the average rate of change of y with respect to x, over the domain L is given by $\frac{y_2 - y_1}{x_2 - x_1} = \frac{1.41 - 0}{2 - 0} = 0.705$.

(b) This time we will need to first determine the coordinates of the extreme points:
For $x = -1$, $y = -1.2 \times (-1)^2 + 9 = 7.8$ and for $x = 2$, $y = -1.2 \times (2)^2 + 9 = 4.2$.

Therefore, the average rate of change is equal to $\frac{y_2 - y_1}{x_2 - x_1} = \frac{4.2 - 7.8}{2 - (-1)} = -1.2$.

It is not always necessary to have a graph in order to find the average rate of change. Often we are given information in the form of a table.

EXAMPLE 22.2

The table below shows the number of bacteria, N, present in an enclosed environment. Find the average growth rate of the population size over the first 4 hours

Time (hrs)	0	1	2	3	4	5	6	7	9
N	30	36	43	52	62	75	90	107	129

Solution

This time we need to consider the time interval $t = 0$ to $t = 4$. From the table we observe that the coordinates corresponding to these values are; (0,30) and (4,62).
Therefore, the average rate of growth of the number of bacteria over the first 4 hours is equal to $\frac{62-30}{4-0} = \frac{32}{4} = 8$.

This means that during the first 4 hours, the number of bacteria was increasing (on average) at a rate of 8 every hour.

Notice that in the 1st hour, the average rate was $\frac{36-30}{1-0} = \frac{6}{1} = 6$ (< 8), whereas in the 4th hour the average rate of increase was $\frac{62-52}{4-3} = \frac{10}{1} = 10$ (> 8).

22.1.5 VELOCITY AS A MEASURE OF THE RATE OF CHANGE OF DISPLACEMENT

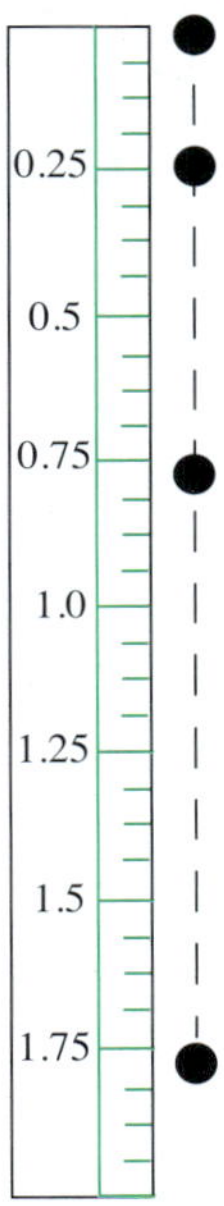

Consider a marble that is allowed to free fall from a height of 2 metres (see diagram). As the marble is falling, photographs are taken of its fall at regular intervals of 0.2 second.

From its motion, we can tell that the rate at which the marble is falling is increasing (i.e., its velocity is increasing).

What is its average velocity over the first 0.6 second?

Reading from the diagram, we see that the marble has fallen a total distance of 1.75 (approximately), therefore, the average velocity v_{ave} of the marble, given by the rate at which its displacement increases (or decreases), is given by

$$v_{ave} = \frac{1.75-0}{0.6-0} \approx 2.92 \text{ m/sec}$$

EXAMPLE 22.3

The displacement, x m, of an object, t seconds after it is dropped from the roof of a building is given by $x = 4.9t^2$ m.

(a) What is the object's displacement after 4 seconds?
(b) What is the average velocity of the object over the first 4 seconds of its motion?

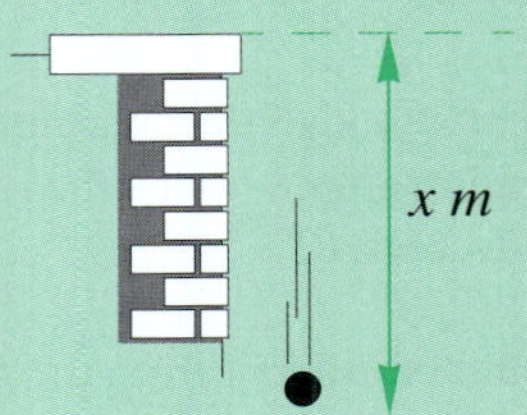

Solution

(a) After 4 seconds of free fall, the object's displacement will be $4.9(4)^2 = 78.4$ m. We obtained this result by substituting the value of $t = 4$ into the equation for the displacement $x = 4.9t^2$.
(b) The average velocity is given by the average rate of change of displacement, x m, with respect to the time t seconds.

Once we have the starting position and the end position we can determine the average velocity using: $v_{ave} = \frac{x_2 - x_1}{t_2 - t_1} = \frac{78.4 - 0}{4 - 0} = 19.6$.

That is, the object's average velocity over the first 4 seconds is 19.6 m/s.

EXAMPLE 22.4

The concentration of a drug, in milligrams per millilitre, in a patient's bloodstream, t hours after an injection is approximately modelled by the function

$$t \mapsto \frac{2t}{8 + t^3}, t \geq 0.$$

Find the average rate of change in the concentration of the drug present in a patient's bloodstream;
(a) during the first hour.
(b) during the first two hours.
(c) during the period $t = 2$ to $t = 4$.

Solution

To help us visualise the behaviour of this function we will make use of the TI–83. Begin by introducing the variable C, to denote the concentration of the drug in the patient's bloodstream t hours after it is administered. So that $C(t) = \frac{2t}{8 + t^3}, t \geq 0$.

(a) Initially the concentration is 0 milligrams per millilitre, the concentration after 1hr is given by

$$C(1) = \frac{2 \times 1}{8 + 1^3} = \frac{2}{9} \approx 0.22.$$

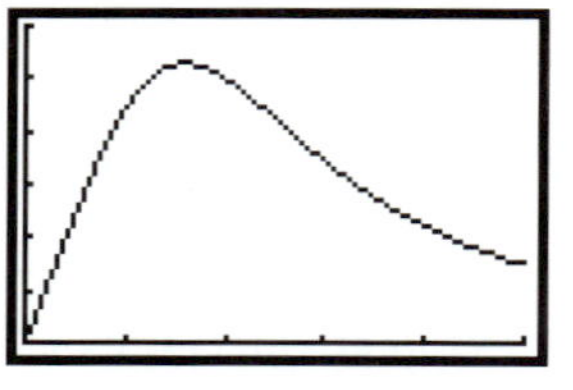

Therefore, the average rate of change in concentration (C_{ave}) during the first hour is given by $C_{ave} = \frac{0.22 - 0}{1 - 0} = 0.22$.

Note: the units are *mg/ml/hr*.

(b) The concentration 2 hours after the drug has been administered is $C(2) = \frac{2 \times 2}{8 + 2^3} = 0.25$.

That is, 0.25 *mg/ml*.
Therefore, the average rate of change in concentration with respect to time is

$$C_{ave} = \frac{0.25 - 0}{2 - 0} = 0.125.$$

Notice that although the concentration has increased (compared to the concentration after 1 hour), the rate of change in the concentration has actually decreased!

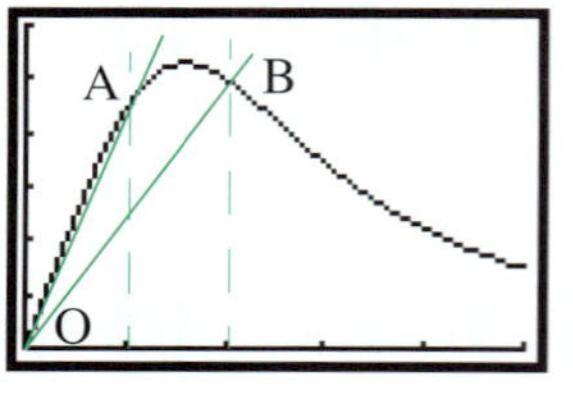

This should be evident from the graph of $C(t)$ versus t.
The slope of the straight line from the origin to A(1, 0.22), m_{OA}, is greater than the slope from the origin O to the point B(2, 0.25), m_{OB}.
That is $m_{OA} > m_{OB}$.

(c) The average rate of change in concentration from $t = 2$ to $t = 4$ is given by $\frac{C(4) - C(2)}{4 - 2}$.

Now, $\dfrac{C(4)-C(2)}{4-2} = \dfrac{\frac{2\times 4}{8+4^3}-0.250}{4-2} \approx \dfrac{0.111-0.250}{2} = -0.0694$

Therefore, the average rate of change of concentration is –0.070 *mg/ml/hr*.
Meaning that the overall amount of drug in the patient's bloodstream is decreasing during the time interval $2 \le t \le 4$.

EXERCISES 22.1

1. For each of the following graphs determine the average rate of change over the specified domain.

(a) $x \in [4, 8]$

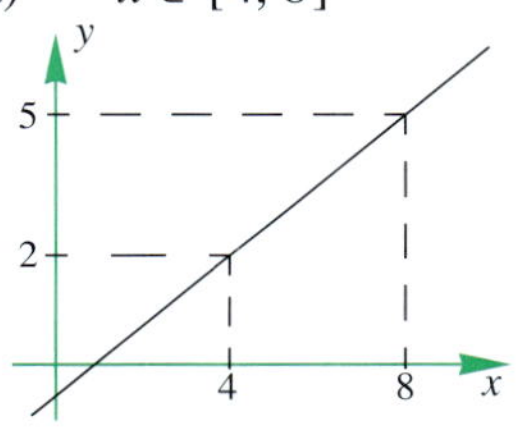

(b) $x \in [-b, 3b]$

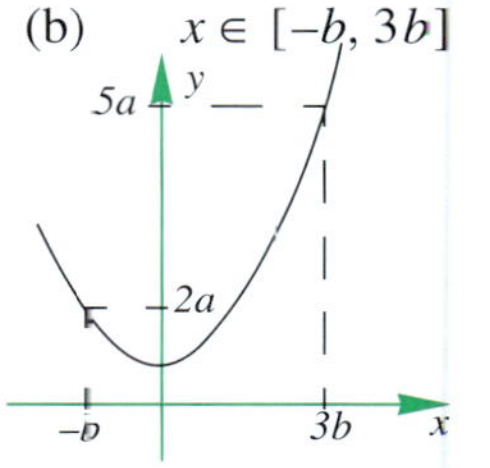

(c) $x \in [-2, 2]$

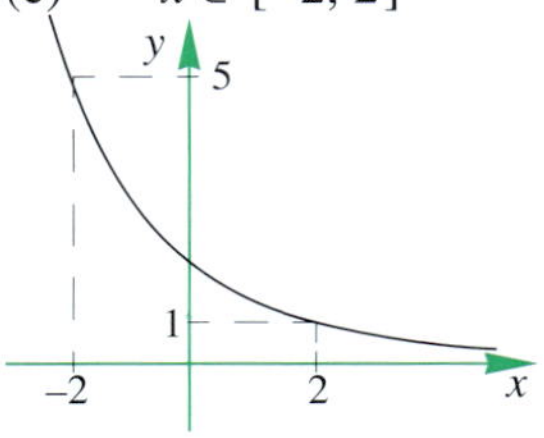

(d) $x \in [-3, 2]$

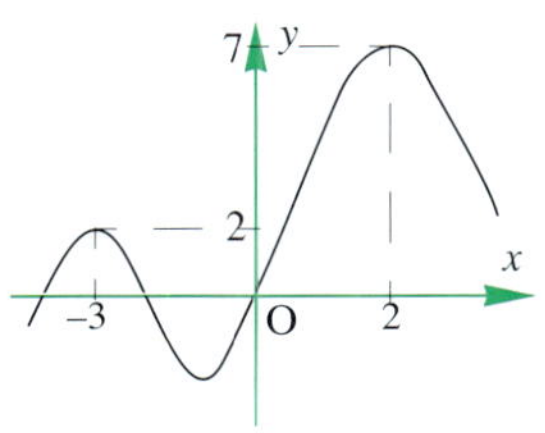

(e) $x \in [-2, 6]$

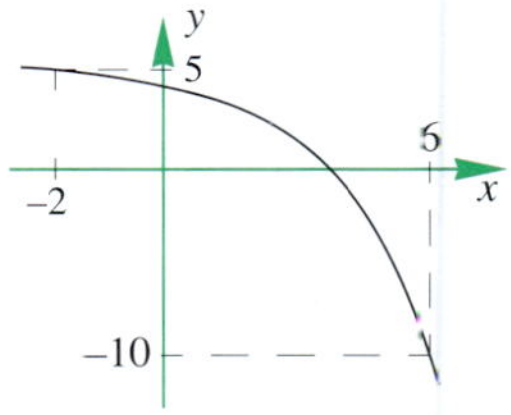

(f) $x \in [-1, 1]$

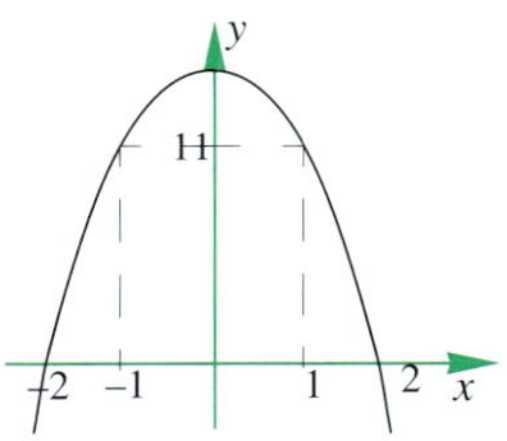

2. For each of the following functions, find the average rate of change over the given domain.

(a) $x \mapsto x^2 + 2x - 1,\ x \in [0, 2]$ (b) $x \mapsto \sqrt{x+1},\ x \in [3, 8]$

(c) $x \mapsto 10 - \dfrac{1}{\sqrt{x}},\ x \in [2, 20]$ (d) $x \mapsto \dfrac{x}{x+1},\ x \in [0.1, 1.1]$

(e) $x \mapsto \dfrac{1}{1+x^2} - 1,\ x \in [0, 100]$ (f) $x \mapsto x\sqrt{400-x},\ x \in [300, 400]$

(g) $x \mapsto 2^x,\ x \in [0, 5]$ (h) $x \mapsto (x-1)(x+3),\ x \in [-3, 2]$

3. The displacement of an object, t seconds into its motion is given by the equation,

$$s(t) = t^3 + 3t^2 + 2t,\ t \ge 0 .$$

Find the average rate of change of displacement during

(a) the first second.
(b) the first 4 seconds.
(c) the interval when $t = 1$ to $t = 1 + h$.

4. The distance s metres that a particle has moved in t seconds is given by the function $s = 4t + 2t^2,\ t \ge 0$. Find the particle's average speed over the first 4 seconds.

5. The distance s metres that a particle has moved in t seconds is given by the function $s = 4t + 2t^2, t \geq 0$. Find the particle's average speed during the time interval from when $t = 1$ to $t = 1 + h$.

6. The temperature T °C of food placed inside cold storage is modelled by the equation

$$T = \frac{720}{t^2 + 2t + 25} \text{ where } t \text{ is measured in hours.}$$

Find the average rate of change of the temperature, T°C, with respect to the time, t hours, during the first 2 hours that the food is placed in the cold storage.

7. The volume of water in a hemispherical bowl of radius r is given by $V = \frac{1}{3}\pi h^2(3r - h)$, where h is the height of the water surface inside the bowl. For the case where r is 20 cm,

(a) find the average rate of increase in the amount of water inside the bowl with respect to its height, h cm, as the water level rises from 2 cm to 5 cm.

(b) Find the average rate of increase in the amount of water inside the bowl with respect to its height, h cm, as the water level rises by

i. 1 cm ii. 0.1 cm iii. 0.01 cm

8. An amount of money is placed in a bank and is accumulating interest on a daily basis. The table below shows the amount of money in the savings account over a period of 600 days.

t (days)	100	200	300	400	500	600	700
$\$D$/day	1600	1709	1823	1942	2065	2194	2328

(a) Plot the graph of $\$D$ versus t (days).

(b) Find the average rate of change in the amount in the account during the period of 100 days to 300 days.

9. The temperature of coffee since it was poured into a cup was recorded and tabulated below.

t min	0	2	4	6	9
T °C	60	50	30	10	5

(a) Plot these points on a set of axes that show the relationship between the temperature of the coffee and the time it has been left in the cup.

(b) Find the average rate of change of temperature of the coffee over the first 4 minutes.

(c) Over what period of time is the coffee cooling the most rapidly?

10. The displacement, d metres, of an object, t seconds after it was set in motion is described by the equation

$$d = 4t + 5t^2, \text{ where } t \geq 0.$$

(a) Find the distance that the object travels in the first 2 seconds of its motion.

(b) Find the average rate of change of distance with respect to time undergone by the object over the first 2 seconds of its motion.

(c) What quantity is being measured when determining the average rate of change of distance with respect to time?

(d) How far does the object travel during the 5th second of motion?

(e) Find the object's average speed during the 5th second.

11. A person invested \$1000 and estimates that on average, the investment will increase each year by 16% of its value at the beginning of the year.
(a) Calculate the value of the investment at the end of each of the first 5 years.
(b) Find the average rate at which the investment has grown over the first 5 years.

22.2 QUALITATIVE MEASURE

22.2.1 QUALITATIVE ASPECTS OF CHANGE

Apart from quantitative measures (i.e., providing numerical values), it is also important to be able to provide qualitative descriptions of the behaviour of graphs. In doing so, many of the key words mentioned at the start of this chapter should be referred to.

22.2.2 DESCRIBING THE BEHAVIOUR OF A GRAPH

Consider the graph shown:

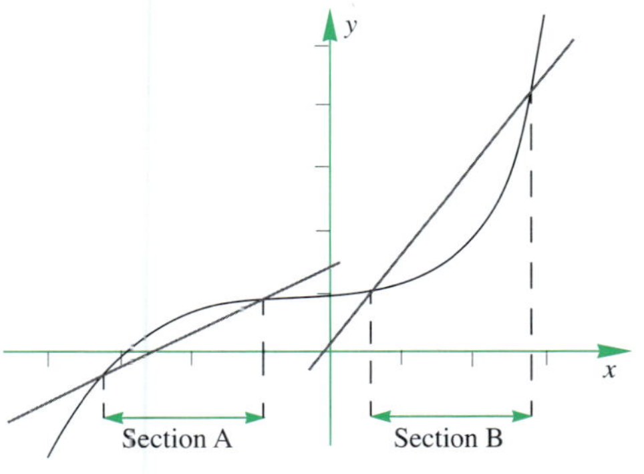

In both Section A and Section B, the gradients of the lines are positive. However, the gradient of the straight line in section B is steeper than that of the line in Section A. We can then say that over Section B the graph is increasing at a faster rate than it is over Section A.

In fact, if we were able to walk along this curve, from left to right, we could describe our 'journey' as follows:

As we walk from the left hand side and towards that part of the graph that lies above Section A, the **function is increasing**, i.e., as the values of x increase, so too do the values of y. As the values of x approach 0 (from the left side of the y–axis) the rate at which the function is increasing is slowing down. That is, I do not need to make as much effort to move as I get closer to the y–axis. Even though the function is still increasing (as we are getting closer to the y–axis), we then have that the rate of change of the function is in fact decreasing! Actually, by the stage when we have reached the y–axis we could almost say that the **function remains stationary**, i.e., it has stopped increasing. In this instance, we would say that the **rate of change of the function is zero**. As we pass the y–axis and keep moving along the curve we find it more difficult to walk along the curve. That is, the effort that we need to make to keep walking is increasing. In this instance the function is increasing but so too is the rate at which it is increasing.

22.2.3 PRODUCING A GRAPH FROM A PHYSICAL SITUATION

In this section we will concentrate on producing a graph to describe the behaviour of the flow of liquid into a container. The importance of such problems is that they enable us to describe how changing one variable will affect a second (related) variable. That is, the effect the independent variable has on the dependent variable. One way to do this is by increasing the independent variable (usually x) and observing the change in the dependent variable (usually y).

EXAMPLE 22.5

A cylindrical vase is placed under a tap and water is allowed to flow into it at a constant rate. Provide a graphical representation of the relationship between the volume of water in the vase and

(a) the time for which water flows into the vase.
(b) the level of water.

Solution

(a) The independent variable in this case is time, t seconds. Consider the volume of water, V cm^3, that flows into the vase in equal time intervals (of 2 seconds say). In these equal time intervals we have equal amounts of water flowing into the vase.
For example, if 10 cm^3 of water flows into the vase every 2 seconds, we could produce the following table of values:

t seconds	0	2	4	6	...
V cm^3	0	10	20	30	...

Based on the results of this table we can produce a graph of V cm^3 versus t seconds:

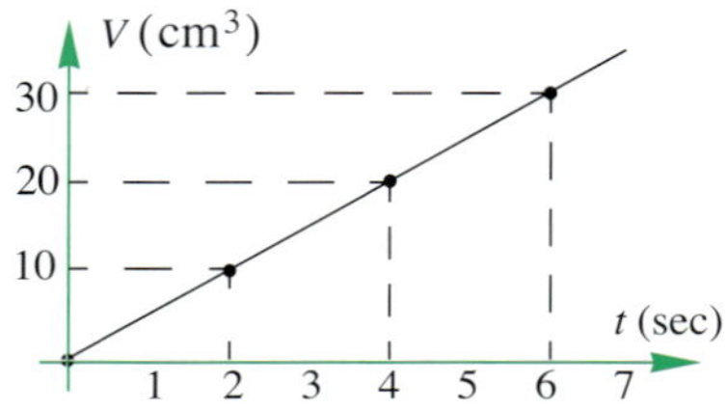

(b) To see how the volume changes with respect to the level of water, we use a different approach— this time we consider a 'frame–by–frame' sequence of the vase as it is filled.

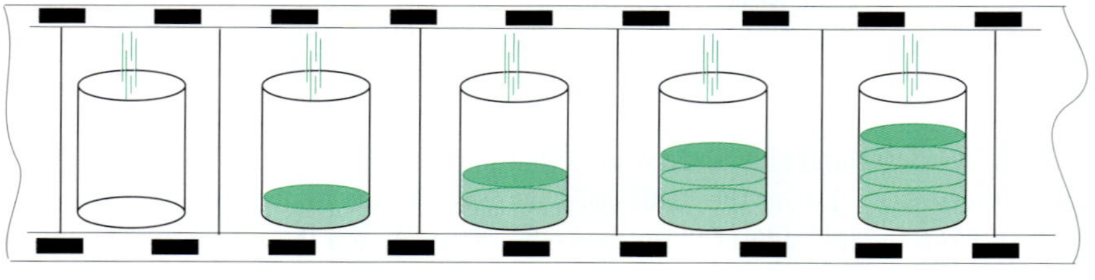

From our 'snap–shots', we see that for equal heights, equal amounts of water flow into the vase. So that every time the water level increases by 1cm, the volume increases by 8 cm^3.

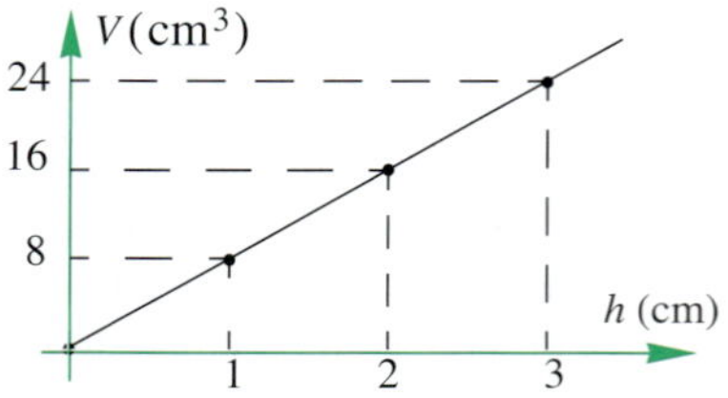

This would imply that the relationship between the volume, V cm^3 of water in the vase and the level of water, h *cm*, is linear.

EXAMPLE 22.6

Sketch a graph showing the relationship between the level of water in a flask and the time for which water has been flowing into the flask.

Solution

2. Let the level of water in the flask be denoted by h cm and the time for which water has been flowing be denoted by t seconds. Again, we use our 'frame–by frame' approach:

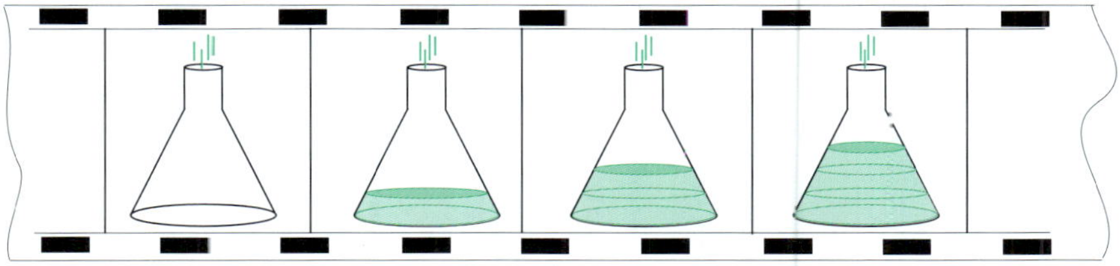

As we consider equal time intervals we see that the same amount of water will flow into the flask during each of these time intervals. However, because the flask becomes narrower as the level rises, then (because we still have the same volume of water flowing into the flask), the height of the space occupied by these equal volumes of water must increase at a faster rate than it had for the 'cone' section. A cross sectional view of the flask (shown above) shows this more clearly.

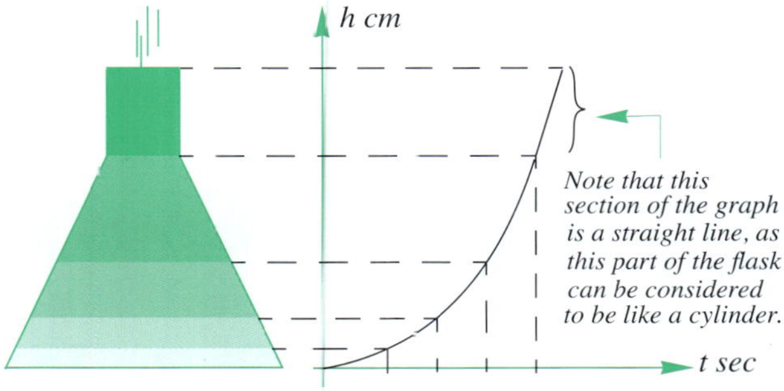

EXERCISES 22.2

1. (a) The cross section of a basin, shown in Figure A, is being filled by water flowing at a constant rate.

Sketch a graph of the relationship between the level of water, h cm, and the time, t sec, that water has been flowing.

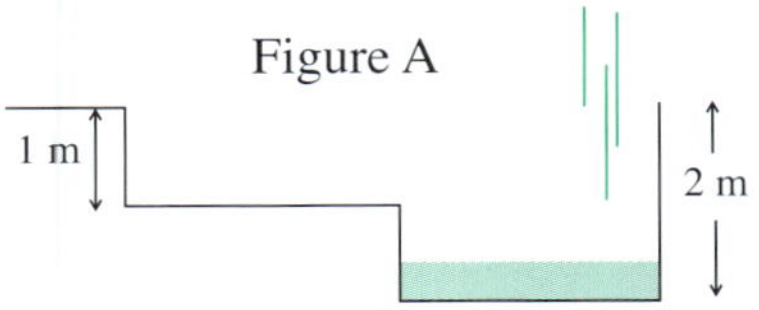

(b) The cross section of a second basin is shown in Figure B. Water is flowing into this basin at the same rate as in (a).

Sketch a graph of the relationship between the level of water, h cm, and the time, t sec, that water has been flowing for this basin.

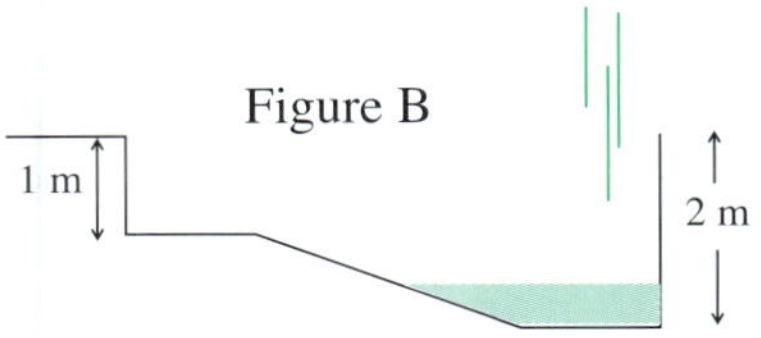

2. For each of the following bottles, sketch the graph that would show the relationship between the level of water, h cm, and the volume of water, V cm^3 in the bottle. That is, sketch a graph of h versus V.

You may assume that water is flowing into each bottle at a constant rate.

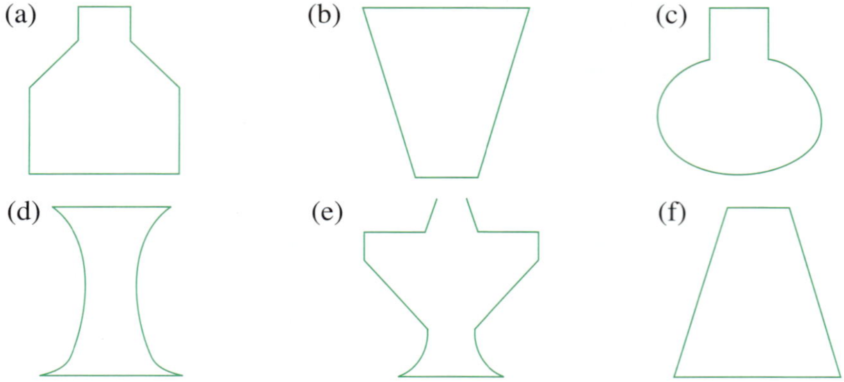

22.3 INSTANTANEOUS RATE OF CHANGE

22.3.1 INFORMAL IDEA OF LIMITS

As already discussed, the average rate of change between two points on a curve is determined by finding the gradient of the straight line joining these two points. However we often need to find the rate of change at a particular instant, and so the method used for finding the average rate of change is no longer appropriate. However, it does provide the foundation that leads to obtaining the instantaneous rate of change. We refine our definition of the average rate of change to incorporate the notion of the instantaneous rate of change. The basic argument revolves around the notion of magnifying near the point where we wish to find the instantaneous rate of change, that is, by repeatedly 'closing in' on a section of a curve. This will give the impression that over a very small section, the curve can be approximated by a straight line. Finding the gradient of that straight line will provide us with a very good approximation of the rate of change of the curve (over the small region under investigation). To obtain the exact rate of change at a particular point on the curve we will then need to use a **limiting** approach.

The process used to determine the rate of change at A is carried out as follows:

1. Start by drawing a secant from A to B, where B is chosen to be close to A. This will provide a reasonable first approximation for the rate of change at A. Then, to obtain a better approximation we move point B closer to point A.

2. Next, **zoom–in** towards point A, again. We move point B closer to point A, whereby a better measure for the rate of change at point A is now obtained. We then repeat step 2, i.e., move B closer to A and zoom in, move point B closer to A and zoom in, and so on.

3. Finally, the zooming–in process has reached the stage whereby the secant is now virtually lying on the curve at A. In fact the secant is now the tangent to the curve at the point A.

Using the process of repeatedly zooming–in to **converge** on a particular region lies at the heart of the **limiting process**. Once we have understood the concepts behind the limiting process, we can move on to the more formal aspect of limits. However, apart from an informal treatment of limits, work on limits is beyond the scope of the core work in HL mathematics.

We now provide a 'visual' representation of steps 1. to 3. described above.

As we magnify, and move point B closer to point A the secant from A to B becomes the tangent at A:

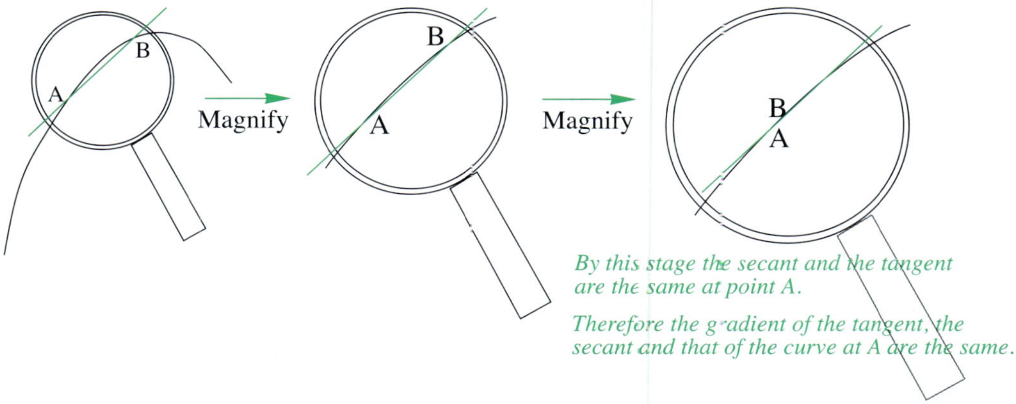

We now investigate this limiting process further through a number of examples. We will still maintain an informal approach to determining limits, but at the same time we will steer our work on limits towards the fundamentals behind the ideas in Chapter 19 – Differential calculus.

EXAMPLE 22.7 An object moves along a straight line. Its position, x metres (from a fixed point O), at time t seconds is given by $x(t) = t - \frac{1}{4}t^2, t \geq 0$. Determine

(a) its average velocity over the interval from $t = 1$ to $t = 2$
(b) its average velocity over the interval $t = 1$ to $t = 1.5$
(c) its average velocity over the interval $t = 1$ to $t = 1.1$
(d) its average velocity over the interval $t = 1$ to $t = 1 + h$, where h is small.

How can the last result help us determine the object's velocity at $t = 1$?

Solution

(a) The average velocity over the required second (from $t = 1$ to $t = 2$) is found by looking at the slope of the secant joining those two points on the graph of $x(t)$.

At $t = 2$, we have $x(2) = 2 - \frac{1}{4}(2)^2 = 1$, and at $t = 1$, $x(1) = 1 - \frac{1}{4}(1)^2 = \frac{3}{4}$.

Therefore, we have that
$$v_{ave} = \frac{x(2) - x(1)}{2 - 1} = \frac{1 - 0.75}{1} = 0.25$$

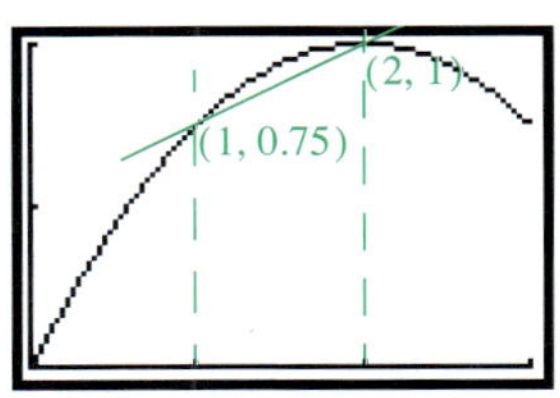

Therefore, the average velocity over the second is 0.75m/s.

(b) For $t = 1$ to $t = 1.5$ we have, $v_{ave} = \frac{x(1.5) - x(1)}{1.5 - 1} = \frac{(1.5 - 0.25 \times 1.5^2) - 0.75}{0.5} = 0.375$

(c) Similarly, for $t = 1$ to $t = 1.1$, we have $v_{ave} = \frac{x(1.1) - x(1)}{1.1 - 1} = 0.475$

(d) We are now in a position to determine the average rate over the interval $t = 1$ to $t = 1 + h$. The average velocity is given by $v_{ave} = \frac{x(1 + h) - x(1)}{1 + h - 1}$

Now, $x(1+h) = (1+h) - 0.25(1+h)^2 = 1 + h - 0.25(1 + 2h + h^2)$

$$= 0.75 + 0.5h - 0.25h^2$$

Therefore, $v_{ave} = \dfrac{0.75 + 0.5h - 0.25h^2 - 0.75}{1 + h - 1} = \dfrac{0.5h - 0.25h^2}{h}$

$$= \frac{h(0.5 - 0.25h)}{h}$$

$$= 0.5 - 0.25h,\ h \neq 0$$

Notice that for ii., (i.e., $t = 1$ to $t = 1.5$) $h = 0.5$, so that substituting $h = 0.5$ into this equation we have, $v_{ave} = 0.5 - 0.25(0.5) = 0.375$, providing the same result as before.

We can set up a table of values and from it determine what happens as we decrease the time difference.

h	v_{ave}
0.1	0.475
0.01	0.4975
0.001	0.4999

We notice that as h becomes very small, the average rate of change from $t = 1$ to $t = 1 + h$ becomes the instantaneous rate of change at t = 1! This is because we are zooming in onto the point where $t = 1$.

This means that the rate of change at $t = 1$ (h '= 0') would therefore be 0.5 m/s. So that the particle would have a velocity of 0.5 m/s after 1 second of motion.

EXAMPLE 22.8

For the graph with equation $f: x \mapsto (x+2)(x-1)(x-4)$,

(a) Find the average rate of change of f over the interval $[-1,2]$.

(b) Find the rate of change of f, where $x = 4$.

Solution

(a) We first find the coordinates of the end points for the interval $[-1,2]$:

$x = -1, y = f(-1) = (-1+2)(-1-1)(-1-4) = 10$.

$x = 2, y = f(2) = (2+2)(2-1)(2-4) = -8$.

Therefore, the average rate of change in y with respect to x over the interval $[-1,2]$ is given by

$$\frac{f(2) - f(-1)}{2 - (-1)} = \frac{-8 - 10}{3} = -6$$

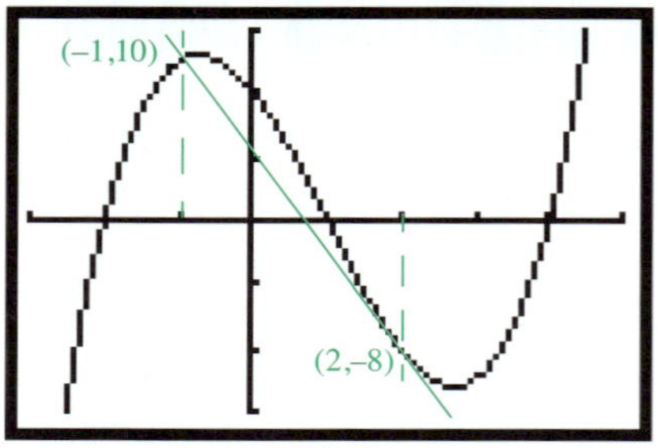

(b) To determine the rate of change at $x = 4$, we choose a second point close to $x = 4$. In this case, we use the point $x = 4 + h$, where h can be considered to be a very small number.

We will look at what happens to the gradient of the secant joining the points $(4, 0)$ and $(4 + h, f(4 + h))$ as h approaches zero.

The gradient of the secant is given by

$$\frac{f(4+h) - f(4)}{(4+h) - 4} = \frac{f(4+h) - f(4)}{h}$$

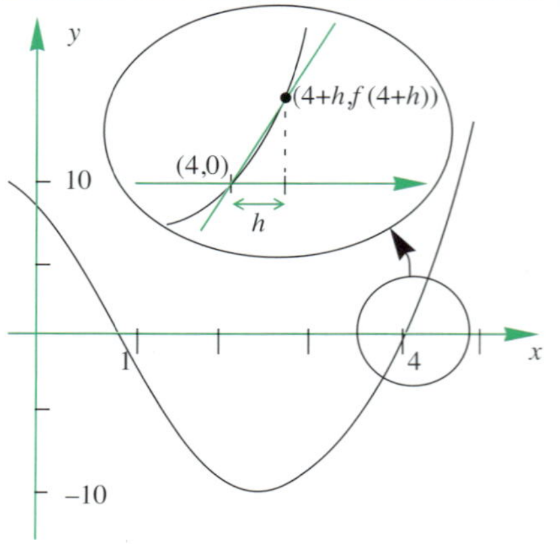

We now need to determine the value of $f(4+h)$ and $f(4)$. However, we already know that $f(4) = 0$.

We can now find values for $f(4+h)$ as h approaches zero.

For $h = 0.1$, $f(4+0.1) = f(4.1) = (4.1+2)(4.1-1)(4.1-4) = 6.1 \times 3.1 \times 0.1 = 1.891$.

Therefore, $\frac{f(4+h)-f(4)}{h} = \frac{1.891-0}{0.1} = 18.91$, for $h = 0.1$.

We can continue in this same manner by making the value of h smaller still.
We do this by setting up a table of values:

h	$\frac{f(4+h)-f(4)}{h}$
0.01	18.09010000
0.001	18.00900100
0.0001	18.00090001

From the table, it appears that as h approaches zero, the gradient of the secant (which becomes the gradient of the tangent at (4,0)) approaches a value of 18.
Therefore, we have that the rate of change of f at (4,0) is 18.

More formally we write this result as $\lim_{h \to 0} \frac{f(4+h)-f(4)}{h} = 18$, which is read as

"The limit as h tends to zero of $\frac{f(4+h)-f(4)}{h}$ is equal to 18."

Which is saying that if we make h as small as we can, then $\frac{f(4+h)-f(4)}{h}$ equals 18.

EXERCISES 22.3

1. For each of the graphs shown, find the gradient of the secant joining the points P and Q.

(a)

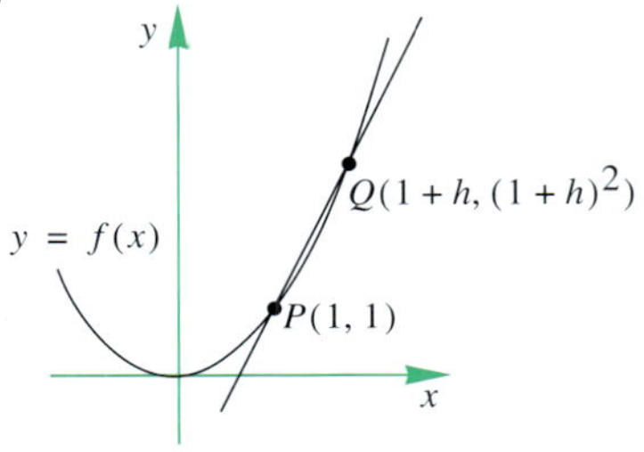

(b)

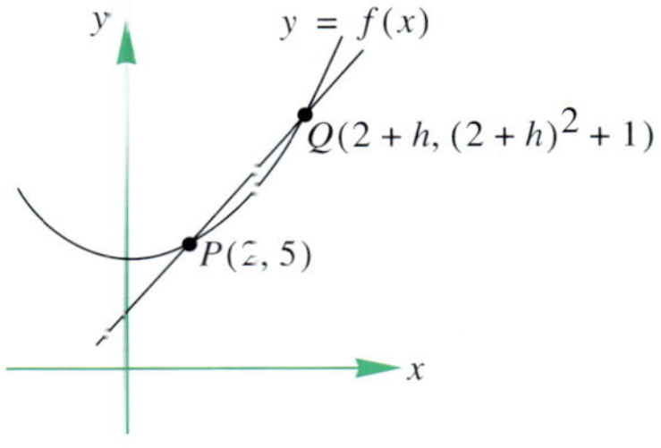

(c)

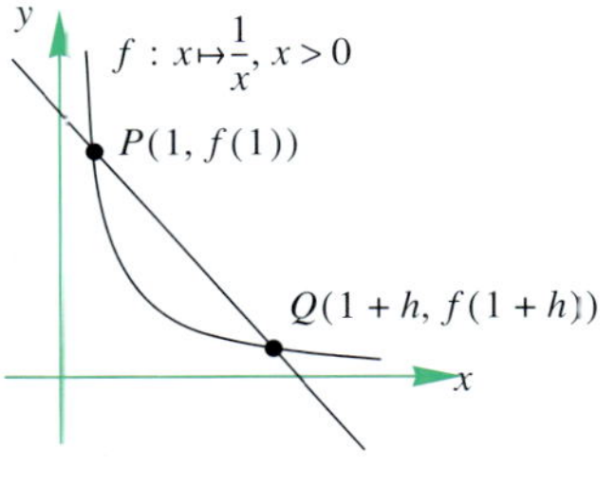

(d)

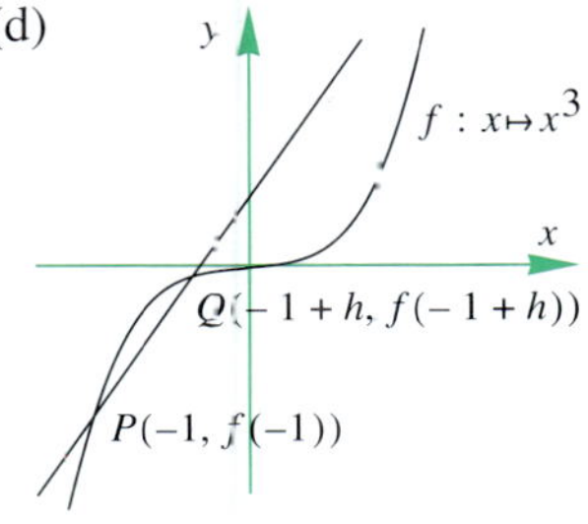

2. For each of the graphs in Q.1., use a limiting argument to deduce the instantaneous rate of change of the given function at the point *P*.

3. For each of the functions, f, given below, find the gradient of the secant joining the points $P(a, f(a))$ & $Q(a+h, f(a+h))$.

(a) $f(x) = 3 + x^2$ (b) $f(x) = 1 - x^2$ (c) $f(x) = (x+1)^2 - 2$

(d) $f(x) = x^3 + x$ (e) $f(x) = 2 - x^3$ (f) $f(x) = x^3 - x^2$

(g) $f(x) = \frac{2}{x}$ (h) $f(x) = \frac{1}{x-1}$ (i) $f(x) = \sqrt{x}$

4. For each of the functions, f, given below, find the gradient of the secant joining the points $P(a, f(a))$ & $Q(a+h, f(a+h))$ and hence deduce the gradient of the tangent drawn at the point *P*.

(a) $f(x) = x$ (b) $f(x) = x^2$ (c) $f(x) = x^3$ (d) $f(x) = x^4$.

Hence deduce the gradient of the tangent drawn at the point $P(a, f(a))$ for the function $f(x) = x^n, n \in N$.

5. An object moves along a straight line. Its position, x metres (from a fixed point O), at time t seconds is given by $x(t) = 2t^2 - 3t + 1, t \geq 0$.

(a) Sketch the graph of its displacement function.

(b) Determine
 i. its average velocity over the interval from $t = 1$ to $t = 2$
 ii. its average velocity over the interval $t = 1$ to $t = 1.5$
 iii. its average velocity over the interval $t = 1$ to $t = 1.1$

(c) Show that its average velocity over the interval $t = 1$ to $t = 1 + h$, where h is small is given by $1 + 2h$.

(d) How can the last result help us determine the objects' velocity at $t = 1$?

(e) Show that its average velocity over any time interval of length h is given by $4t + 2h - 3$. Hence deduce the object's velocity at any time t during its motion.

6. The healing process of a certain type of wound is measured by the decrease in surface area that the wound occupies on the skin. A certain skin wound has its surface area modelled by the equation $S(t) = 20 \times 2^{-0.1t}$ where S sq. cm is the unhealed area t days after the skin received the wound.

(a) Sketch the graph of $S(t) = 20 \times 2^{-0.1t}, t \geq 0$.

(b) i. What area did the wound originally cover?
 ii. What area will the wound occupy after 2 days?
 iii. How much of the wound healed over the two day period?
 iv. Find the average rate at which the wound heals over the first two days.

(c) How much of the wound would heal over a period of h days?

(d) Find the rate at which the wound heals
 i. immediately after it occurs
 ii. one day after it occured.

22.4 DIFFERENTIATION PROCESS

22.4.1 THE DERIVATIVE AND THE GRADIENT FUNCTION

In the previous sections we concentrated our efforts on determining the average rate of change for a function over some fixed inteval. Then proceeded to find the instantaneous rate of change at a particular point (on the curve). We now consider the same process, with the exception that we will discuss the instantaneous rate at any point $P(x, f(x))$. The result will be an expression that will enable us to determine the instantaneous rate of change of the function at any point on the curve. Because the instantaneous rate of change at a point on a curve is simply a measure of the gradient of the curve at that point, our newly found result will be known as the **gradient function** (otherwise known as the derivative of the function).

For a continuous function, $y = f(x)$, we deduced that the instantaneous rate of change at the point $P(a, f(a))$ is given by $\frac{f(a+h)-f(a)}{h}$, where h is taken to be very small (in fact we say that h approaches or tends to zero).

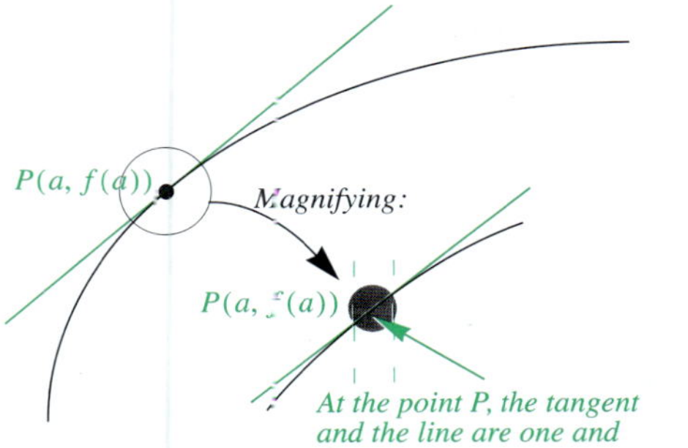

So, to determine the rate at which a graph changes at a single point, we need to find the slope of the tangent line at that point.

This becomes obvious if we look back at our 'zooming in process—where the tangent line to the function at the point $P(a, f(a))$ is the line that best approximates the graph at that point.

Rather than considering a fixed point $P(a, f(a))$, we now consider any point $P(x, f(x))$ on the curve with equation $y = f(x)$:

The rate of change of the function f at $P(x, f(x))$ is therefore given by the gradient of the tangent to the curve at P.

If point Q comes as close as possible to the point P, so that h approaches zero, then, the gradient of the tangent at P is given by the gradient of the secant joining the points $P(x, f(x))$ and $Q(x+h, f(x+h))$ as $h \to 0$.

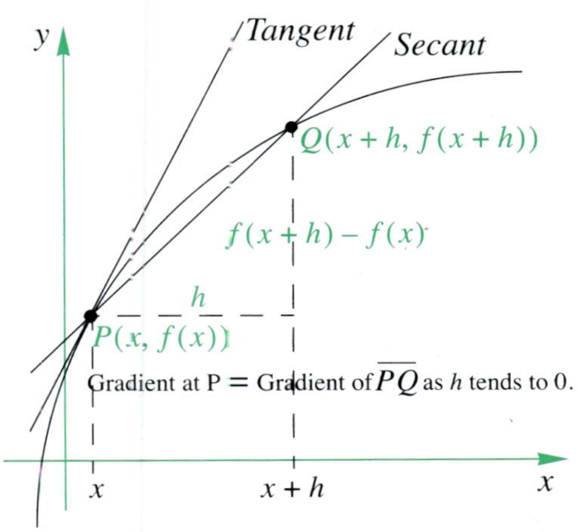

In mathematical notation we have:

$$\text{Rate of change at P} = \lim_{h \to 0} \frac{f(x+h)-f(x)}{h}$$

22.4.2 NOTATION AND LANGUAGE

We now introduce the term **derivative of a function**:

The **rate of change** of $f(x)$ at $P(x, f(x))$ = **Gradient function** of $f(x)$ at $P(x, f(x))$

= **The derivative** of $f(x)$

$$= \lim_{h \to 0} \frac{f(x+h) - f(x)}{h}$$

The derivative of a function $f(x)$ is denoted by $f'(x)$ and is read as "*f* dash of *x*".

That is, $$f'(x) = \lim_{h \to 0} \frac{f(x+h) - f(x)}{h}$$

Finding the derivative of a function using this approach is referred to as *finding the derivative of f from first principles*.*

It is important to realise that in finding $f'(x)$ we have a new function—called the **gradient function**, because the expression $f'(x)$ will give the gradient anywhere on the curve of $f(x)$. So that if we want the gradient of the function $f(x)$ at $x = 5$, we first determine $f'(x)$ and then substitute the value of $x = 5$ into the equation of $f'(x)$.

EXAMPLE 22.9

Using the first principles method, find the derivative (*or the gradient function*) of the function $f(x) = 3x^2 + 4$. Hence, find the gradient of the function at $x = 3$.

Solution

Using the first principles method means that we must make use of the expression

$$f'(x) = \lim_{h \to 0} \frac{f(x+h) - f(x)}{h} \quad -(1)$$

We start by first evaluating the expression $f(x+h) - f(x)$:

That is:

$$\begin{aligned} f(x+h) - f(x) = 3(x+h)^2 + 4 - [3x^2 + 4] &= 3(x^2 + 2xh + h^2) + 4 - 3x^2 - 4 \\ &= 3x^2 + 6xh + 3h^2 - 3x^2 \\ &= 6xh + 3h^2 \end{aligned}$$

Substituting this result into (1):

$$\begin{aligned} \lim_{h \to 0} \frac{f(x+h) - f(x)}{h} &= \lim_{h \to 0} \frac{6xh + 3h^2}{h} \\ &= \lim_{h \to 0} \frac{h(6x + 3h)}{h} \\ &= \lim_{h \to 0} (6x + 3h), h \neq 0 \\ &= 6x \end{aligned}$$

That is, we now have the gradient function $f'(x) = 6x$.

To determine the gradient of the function at $x = 3$, we need to substitute the value $x = 3$ into the gradient function. That is, $f'(3) = 6 \times 3 = 18$.

*In examinations: questions on differentiation from first principles will not be set.

Using the TI–83 we can determine the derivative at $x = 3$ by using the '**nDeriv(**' command from the **MATH** menu:

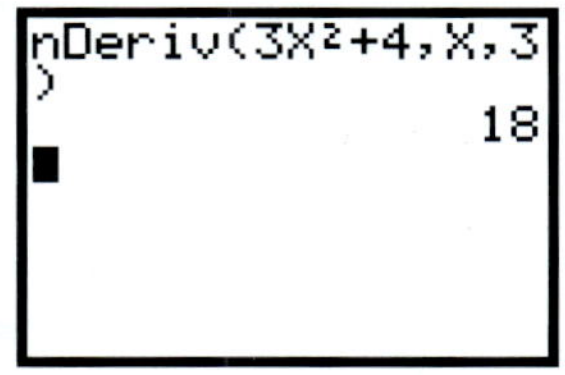

EXAMPLE 22.10 A particle moving along a straight line has its position at time t seconds governed by the equation $p(t) = t - 0.2t^2, t \geq 0$, where $p(t)$ is its position in metres from a fixed point O.

(a) Find the particle's velocity after it has been in motion for 1 second.

(b) Find the particle's velocity at time $t = a$, $a > 0$.

Solution

(a) This part is readily done by making use of the TI–83:
So, we have that the particle's velocity is 0.6 m/s

(b) Let us again make use of the TI–83:
We enter the function $p(t) = t - 0.2t^2, t \geq 0$ as $\mathbf{Y}_1$ and then enter $\mathbf{Y}_2$ as **nDeriv(**$\mathbf{Y}_1$**, X, X)**. Once this is done we sketch both $\mathbf{Y}_1$ and $\mathbf{Y}_2$.
The graph of $\mathbf{Y}_2$ represents the gradient function of $\mathbf{Y}_1$, i.e., the derivative of $\mathbf{Y}_1$.

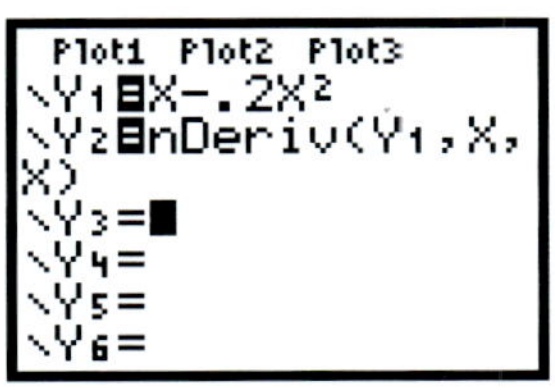

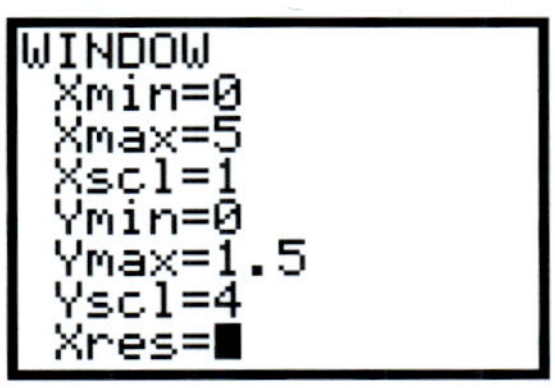

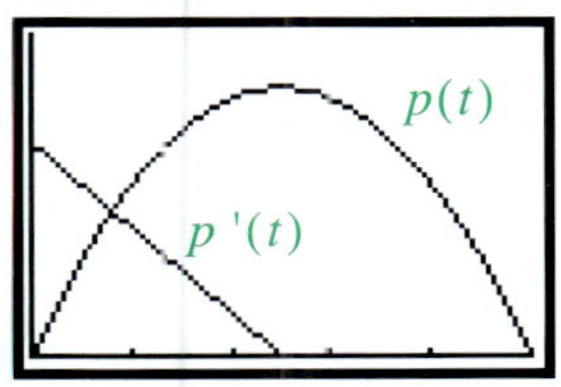

The graph of $p'(t)$ appears to be a straight line passing through the points (0,1) and (2.5, 0) and so we can obtain the equation of this straight line:

gradient $= m = \dfrac{0-1}{2.5-0} = -0.4$. Then, using $y - 1 = (-0.4)(x-0)$ we have, $y = -0.4x + 1$.

Changing to the appropriate variables we have, $v(t) = -0.4t + 1$.

Therefore at $t = a$, $v(a) = -0.4a + 1$.

Obviously, this method relied on our ability to spot the type of function **nDeriv(**$\mathbf{Y}_1$**, X, X)** produced. The fact that it could then be readily determined made it all fairly straight forward. However, more often than not, this will not be the case. So we make use of the definition of differentiation to obtain the velocity as the derivative of the position function.

$$v(t) = p'(t) = \lim_{h \to 0} \frac{p(t+h) - p(t)}{h}$$

We start by determining $p(t+h) - p(t) = (t+h) - 0.2(t+h)^2 - [t - 0.2t^2]$

$$= t + h - 0.2(t^2 + 2th + h^2) - t + 0.2t^2$$

$$= -0.4th + h - 0.2h^2$$

Therefore, $v(t) = \lim_{h \to 0} \frac{-0.4th + h - 0.2h^2}{h} = \lim_{h \to 0}(-0.4t + 1 - 0.2h) = -0.4t + 1$

And so, $v(a) = -0.4a + 1$.

EXERCISES 22.4

1. Use a limiting process to find the gradients of these curves at the points indicated:

(a) $x \mapsto x^3$ at $x = 1$ (b) $v = 2t^2 - 1$ at $t = 2$

(c) $f(x) = \frac{1}{x}$ at $x = 3$ (d) $x \mapsto 2^x$ at $x = 1$

(e) $f = t^2 - 2t + 3$ at $t = 0.5$ (f) $t \mapsto \frac{t^2 - 1}{t}$ at $t = 4$

2. An object is dropped from a high building. The distance, d metres, that the object has fallen, t seconds after it is released, is given by the formula $d = 4.9t^2, 0 \leq t \leq 3$.

(a) Find the distance fallen during the first second.
(b) Find the distance fallen between $t = 1$ and $t = h + 1$ seconds.
(c) Hence, find the speed of the object 1 second after it is released.

3. Find, from first principles, the gradient function, f' , of the following

(a) $f:x \mapsto 4x^2$ (b) $f:x \mapsto 5x^2$ (c) $f:x \mapsto 4x^3$

(d) $f:x \mapsto 5x^3$ (e) $f:x \mapsto 4x^4$ (f) $f:x \mapsto 5x^4$

Can you see a pattern in your results?

4. Find, from first principles, the derivative of the following functions

(a) $f(x) = 2x^2 - 5$ (b) $g(x) = 2 - x$ (c) $g(x) = 2 - x + x^3$

5. A particle moving along a straight line has its position at time t seconds governed by the equation $x(t) = 2t - 0.5t^2, t \geq 0$, where $x(t)$ is its position in metres from the origin O.

(a) Find the particle's velocity after it has been in motion for 1 second.
(b) Find the particle's velocity at time $t = a$, $a > 0$.

6. A particle moving along a straight line has its position at time t seconds governed by the equation $x(t) = 4t^2 - t^3, t \geq 0$, where $x(t)$ is its position in metres from the origin O.

(a) Sketch the displacement-time graph of the motion over the first five seconds
(b) Find the particle's velocity at time i. $t = 1$
ii. $t = 2$
(c) Find the particle's velocity at any time $t, t \geq 0$.
(d) When will the particle first come to rest?

22.5 MISCELLANEOUS EXERCISES

1. For each of the graphs below, find the average rate of change of y with respect to x over the interval specified (i.e., over the domain L).

(a)

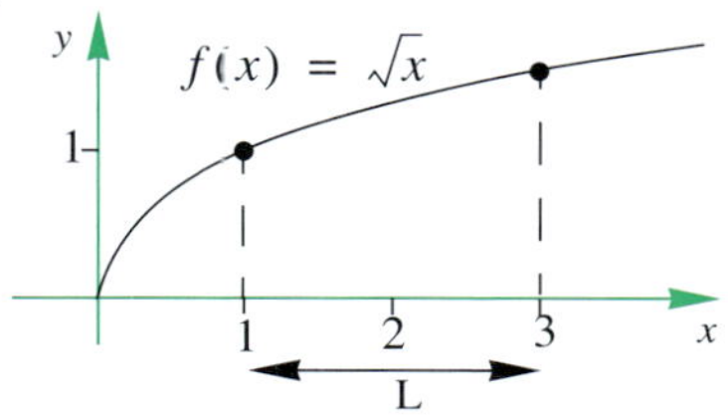

(b)

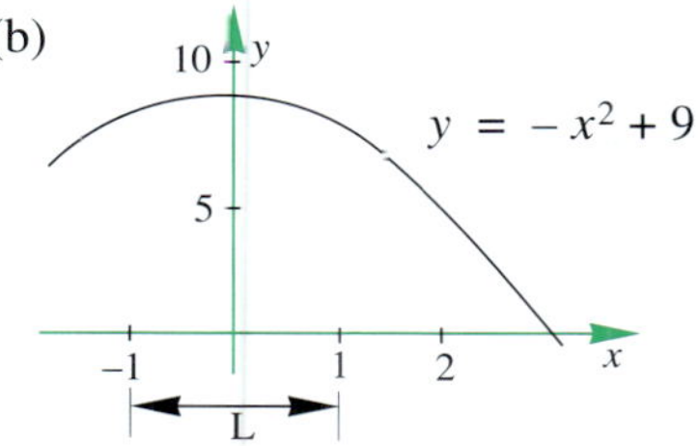

2. The table below shows the number of bacteria, N, present in an enclosed environment. Find the average growth rate of the population size over the last 4 hours

Time (hrs)	0	1	2	3	4	5	6	7	9
N	30	36	43	52	62	75	90	107	129

3. The displacement, x m, of an object, t seconds after it is dropped from the roof of a building is given by $x = 4.9t^2$ m.

(a) What is the object's displacement after 5 seconds?

(b) What is the average velocity of the object over the first 5 seconds of its motion?

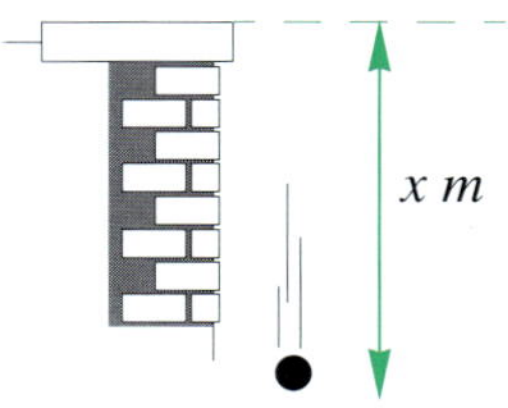

4. The concentration of a drug, in milligrams per millilitre, in a patient's bloodstream, t hours after an injection is approximately modelled by the function

$$t \mapsto \frac{2t}{8+t^3},\ t \geq 0 .$$

Find the average rate of change in the concentration of the drug present in a patient's bloodstream;

(a) during the first hour.

(b) during the first two hours.

(c) during the period $t = 2$ to $t = 4$.

5. A cylindrical vase is placed under a tap and water is allowed to flow into it at a constant rate. Provide a graphical representation of the relationship between the volume of water in the vase and

(a) the time for which water flows into the vase.

(b) the level of water.

6. Each graph shows the relationship between the height of water and the volume that has been poured into it. Draw the profile of the vase that corresponds to each graph.

(a)

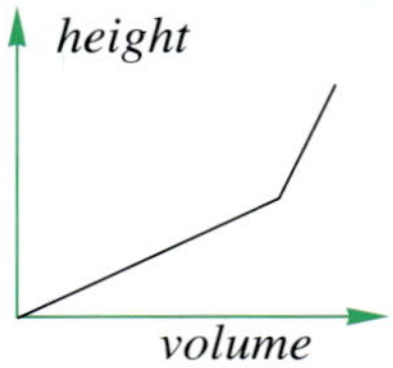

(b)

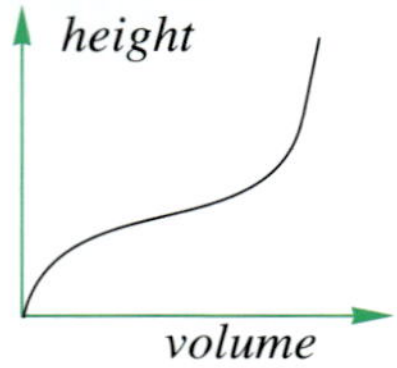

(c)

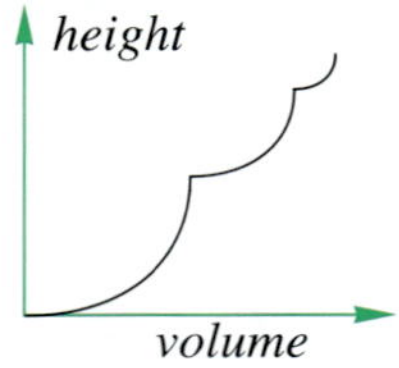

7. An object moves along a straight line. Its position, x metres (from a fixed point O), at time t seconds is given by $x(t) = t - \frac{1}{16}t^2, t \geq 0$. Determine

(a) its average velocity over the interval from $t = 2$ to $t = 3$

(b) its average velocity over the interval $t = 2$ to $t = 2.5$

(c) its average velocity over the interval $t = 2$ to $t = 2.1$

(d) its average velocity over the interval $t = 2$ to $t = 2 + h$, where h is small.

How can the last result help us determine the object's velocity at $t = 2$?

8. For the graph with equation $f: x \mapsto (x+1)(x-2)(x-1)$.

(a) Find the average rate of change of f over the interval $[0,2]$.

(b) Find the rate of change of f, where $x = 0$.

9. Consider the function $f: x \mapsto x^2 + 3$.

(a) Sketch the graph of f.

(b) i. Find an expression for $f(x+h) - f(x)$.

ii. Simplify $\frac{f(x+h) - f(x)}{h}, h \neq 0$.

(c) Using your results of part (b), deduce the gradient function of f.

10. Consider the position-time graphs shown below.

i.

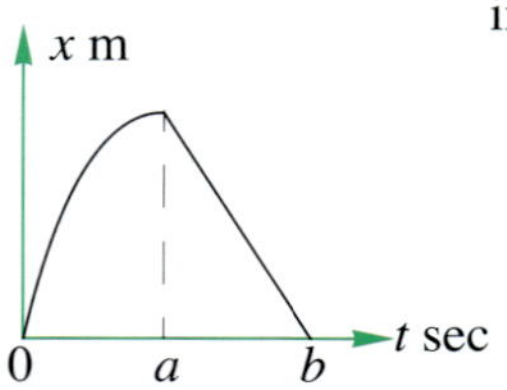

ii.

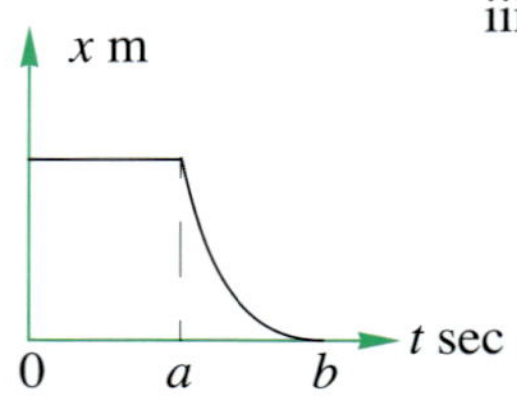

iii.

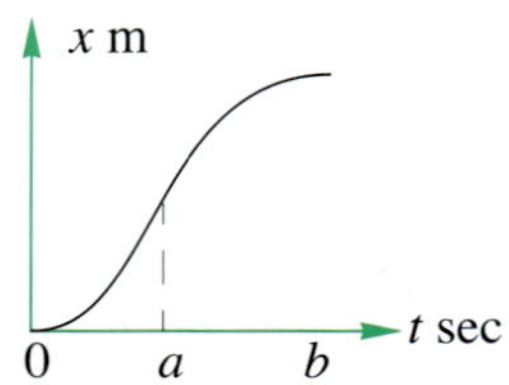

(a) Match the position-time graphs with the velocity-time graphs shown below. For the case where there is no match, sketch the corresponding velocity-time graph.

α.

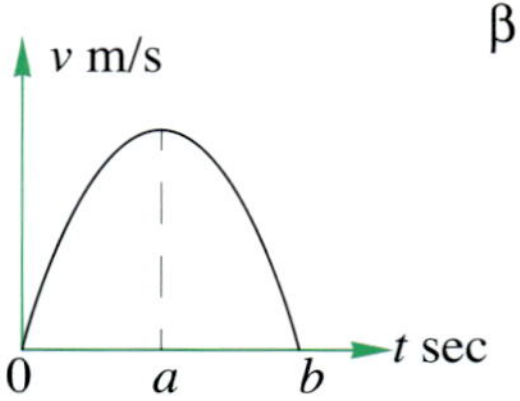

β.

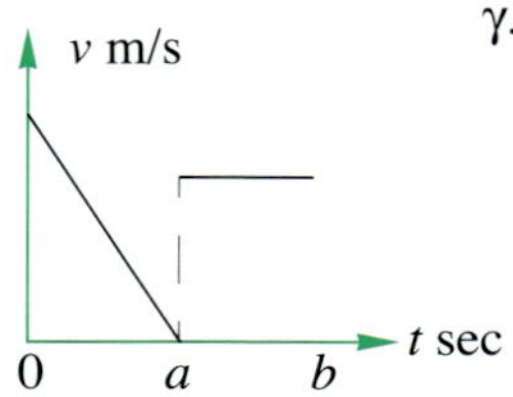

γ.

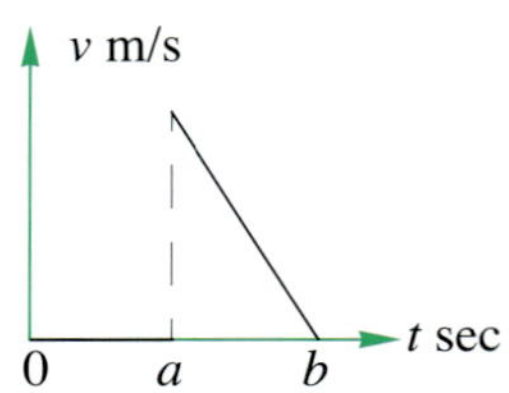

11. A body is moving along a straight line and its distance, x m, from a fixed point O on the line is given by $x(t) = 3 - 2t^2, t \geq 0$.

(a) Find i. $x(3)$

ii. $x(4)$

iii. the average velocity over the interval $t = 3$ to $t = 4$.

(c) Find i. $x(3 + h)$

ii. $x(3 + h) - x(3)$

iii. the average velocity over the interval $t = 3$ to $t = 4$.

iv. the velocity at $t = 3$.

12. The temperature, T°C, of a kettle at time t minutes after removing it from a heat source was tabulated:

t	0	1	2	3	4	5	6
T	60	50	30	15	10	8	2

(a) Use the table to accurately plot and sketch the graph of T °C versus t minutes.

(b) Find the average rate of change of temperature between

i. $t = 2$ and $t = 6$.

ii. $t = 3$ and $t = 3$.

iii. $t = 3$ and $t = 3 + h$, where h is a small positive number.

(c) Using part (b) find the rate of change of temperature when $t = 3$.

(d) i. Draw the tangent on the curve when $t = 3$.

ii. Calculate the gradient of the tangent to the curve at t = 3.

(e) What conclusion can you draw from parts (c) and (d)?

13. The volume of water in a container is shown in the graph below.

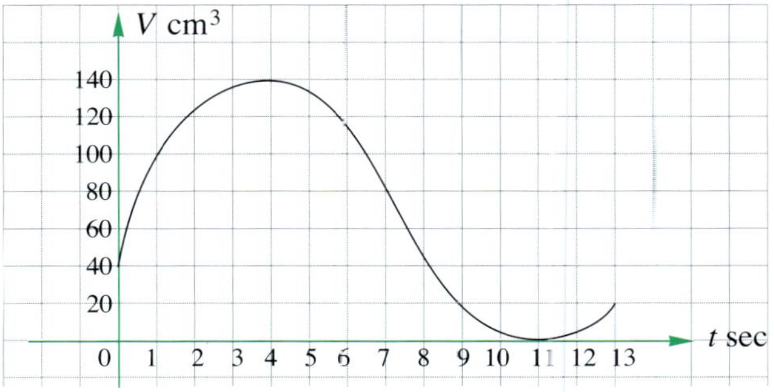

(a) What amount of water is in the container initially?

(b) What is the average rate of change of the volume of water in the container over the first 13 seconds?

(c) What is i. the maximum volume of water in the container?

ii. the minimum volume of water in the container?

(d) At what time will the volume of water be decreasing fastest?

(e) Find the rate of change of volume of water in the container when

i. $t = 1$

ii. $t = 4$

22.6 GRADED REVISON EXERCISES

LEVEL 1

1. State the rate of change of y with respect to x if:

i. $y = 2x - 1$ ii. $y = 3 - x$ iii. $2y = -x + 4$

2. Determine the average rate of change over the given interval, **I**, for the following:

(a)

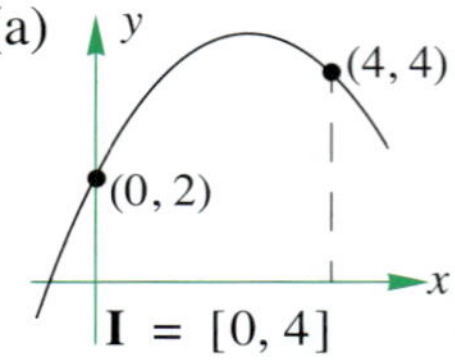

(b)

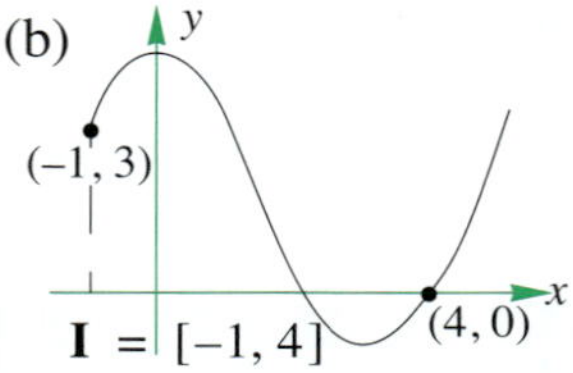

(c)

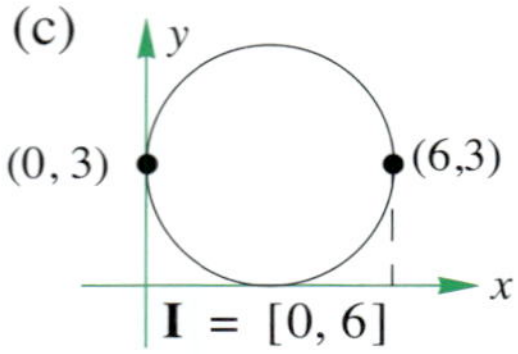

3. Over which section(s) of the graph shown below is the gradient

i. positive?
ii. negative?
iii. increasing?
iv. decreasing?

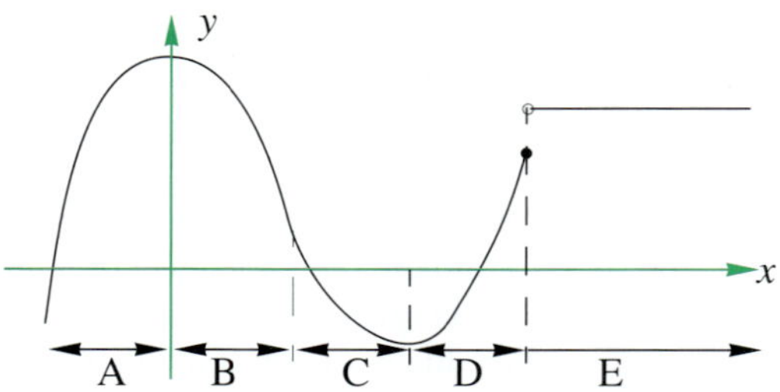

LEVEL 2

1. For each of the following, find the average rate of change over the domain **I**.

(a) $f(x) = 5 - x^2, \mathbf{I} = [1,3]$ (b) $h(t) = \frac{20}{t} + 2, \mathbf{I} = [2,5]$

2. Given that $f(x) = \sqrt{x+1} - 1$, find the slope of the secant line through the points $(3, f(3))$ and $(8, f(8))$. What does this slope measure?

3. The temperature, T °C, of hot tea, t minutes after it has been poured into a cup is given by

$$T = 30 + \frac{60}{t+1}, t \geq 0.$$

Find the average rate of change in the temperature of the tea (after been poured)

i. over the first 9 minutes.
ii. over the last 10 minutes of the first hour.

4. If $f(x) = x^2$, find i. $f(x+h) - f(x)$. ii. $\frac{1}{h}(f(x+h) - f(x))$.

5. A vase is being filled with water at a constant rate.
The cross section of the vase is shown alongside.

Sketch a graph showing the variation in the height of the water level with respect to time.

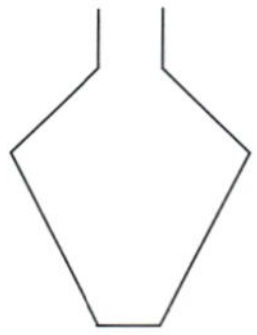

LEVEL 3

1. Find the gradient of the secants passing through the points A and B:

(a) (b)

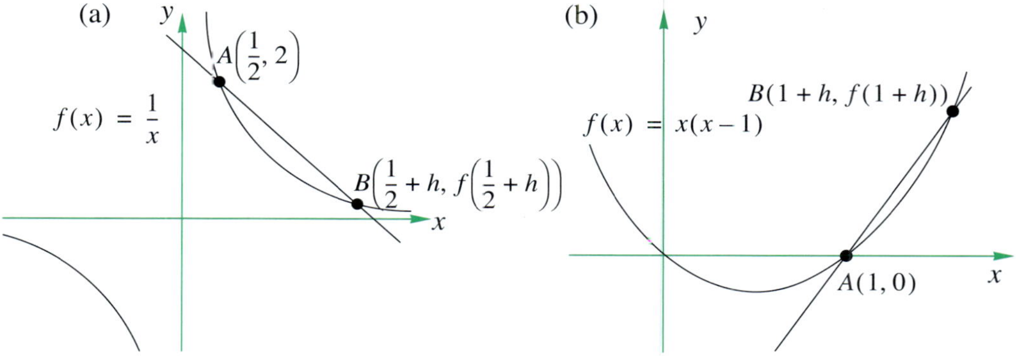

2. For the functions in question 1., find the gradient of the curve at A using a limiting argument.

3. Find the gradient of the secant joining the points P at $x = a$ and Q at $x = a + h$, for the function $h(x) = a - x^3$. Hence, deduce the gradient of $h(x)$ at $x = a$.

4. Find from first principles, the derivative of the following:

(a) $f(x) = x^3$. (b) $g(x) = \frac{1}{x} + 2$.

LEVEL 4

1. The revenue, \$ $R(x)$, from the sale of x wooden planter boxes is given by

$$R(x) = 30x - 0.03x^2.$$

(a) i. For what value(s) of x will the revenue remain positive?
  ii. Sketch the graph of the revenue function.

(b) Find the change in revenue when production increases from
  i. 100 to 400 planters. ii. 100 to 900 planters.

(c) The initial production is set at 200 planters. What is the largest increase in production allowed if the average rate of change in revenue must remain positive?

(d) Using the points P, $x = 200$ and Q, $x = 200 + h$, where h is small, find the rate of change in revenue when 200 wooden boxes are manufactured.

2. The x–coordinate of a particle after t seconds is given by $x(t) = 10t - 2t^2, t \geq 0$.

(a) i. Find an expression for $x(t+h) - x(t)$.
  ii. Hence find the velocity of the particle 2 seconds into its motion.

(b) Find the particle's acceleration when it comes to a stop.

3. Find from first principles, the derivative of $f(x) = 2^x$ when $x = 1.4$. Verify this result using a graphics calculator.

22.7 TOPIC TEST

1. A vase is being filled with water at a constant rate. The profile of the vase is shown alongside. Sketch a graph of how the water level changes with height.

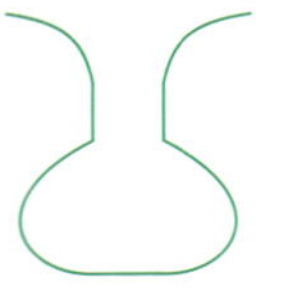

[4 marks]

2. Use a limiting process to find the gradient of $g(t) = 2^t$ at $t = 2$ (ans to 2 dec.pl).

[4 marks]

3. A particle's position along the x–axis is given by $x(t) = 2t^2 - 4t$, $t \geq 0$. Find its average velocity during the first 2 seconds of motion.

[3 marks]

4. The number, N, of infant deaths per 100,000 births in a particular country since1960 has been tabulated below:

t (years)	0	10	20	30	40
N	52.4	38.2	26.6	17.6	11.2

(a) Plot and sketch a graph of the above data on an appropriate set of axes.

(b) Find the average rate of change in the number of infant deaths per 100,000 during the period 1970 to 1990.

(c) During which decade is the rate of change least?

It is suggested that a model for the data has the form $N(t) = at^2 - bt + c$, $t \geq 0$.

(d) i. Show that $a = 0.013$, $b = 1.55$ and $c = 52.4$.

ii. Based on this model, how many deaths can be expected in 2010?

(e) i. Find an expression for $N(t+h) - N(t)$.

ii. Hence, find the rate of change in the number of deaths per 100,000 in 1995.

[16 marks]

5. The temperature, T °C, of a kettle t minutes after it is switched off is modelled by the function $T(t) = 30 + \dfrac{60}{t+1}$, $t \geq 0$.

(a) Construct a table of values showing the trend in the kettle's temperature during the first 10 minutes after it is switched off .

(b) Find the average rate of change in its temperature during the first 2 minutes after it was switched off.

(c) i. Find an expression in terms of h, for the average rate of change in temperature after it is switched off, in the time interval $[2, 2+h]$.

ii. Hence, find the rate of change in temperature 2 minutes after it was switched off.

(d) i. Sketch the graph of the kettle's temperature t minutes after it is switched off.

ii. From your graph deduce the kettle's temperature over the long term.

[13 marks]

Total 40 marks

22.8 PROJECT SUGGESTIONS

The discussion at the start of this chapter was based on the records for the Marathon. If you look at the complete set of records, a different picture emerges. At the time of writing, these can be found on the website: http://www.marathonguide.com

There are other sports at which women perform as well as, or better than, men. An example is the extreme sport of 'free diving' (diving to great depth without an air source).

Which sports are these and what is it about them that makes them differ from a sport such as weight-lifting where the records for men and women are quite different?

Chapter 3 used a case study based on rates of change. There are suggestions for experimentally based projects there.

23.1 DIFFERENTIATION

Optimisation is a key factor in the operation of any business. At the end of the day, it is the bottom line that matters – all you need to do is attend an annual general meeting of investors to realise the pressure that is placed on the CEO to provide the investors with the largest possible return on their investment.

> **W**hat profit has been generated?
> **W**hat are the conditions that will provide a maximum profit?

These two basic questions are the driving force behind many (but not all) decisions that are made in the path to maximising a company's profit.

The annual operating cost of a truck to travel x thousand kilometres in a year, where $0 \le x \le 100$, consists of a standing cost plus a running cost which varies with x. For a given truck, for which the standing cost is \$2000, and the running cost is $\$(200x - 0.4x^2)$, the annual operating cost, $\$B(x)$, is therefore given by $B(x) = 2000 + 200x - 0.4x^2, 0 \le x \le 100$.

This is a simplified version of a standard problem that many truck companies face when considering their operating cost. What need to be done is to determine their annual income for a truck to travel x thousand kilometres in a year. Sometimes that information will be available at the start of the year and sometimes the income might have to be estimated using some form of model.

Once an annual profit model is created, $\$P(x)$, it is then possible to determine the maximum profit that a truck will generate.

Determining the maximum value that $P(x)$ can have is made easy when calculus is used, in particular, differential calculus.

In this chapter we will consider the tools that will enable you to investigate problems such as the one just discussed.

23.1 DIFFERENTIATION

23.1.1 REVIEW

1. Rate of change, gradient and the derivative

The **rate of change** of a curve at a point gives a **measure of the gradient** of the curve at that point. When finding the **derivative** of the equation of a curve we obtain the **gradient function**. As the name suggests, the gradient function enables us to find the gradient at any point on the curve.

2. Differentiation

Differentiation is the process of finding the derivative of a function. The derivative of a function is often called its **derived function**.

3. Language and notation

The **derivative of $f(x)$ with respect to x** is usually written as $f'(x)$ (read as "*f* dash of *x*") or $\frac{d}{dx}(f(x))$ (read as "dee–dee–*x* of $f(x)$").

The **derivative of y with respect to x** is usually written as $\frac{dy}{dx}$ (read as "dee–*y*–dee–*x*") or y' (read as "*y*–dash of *x*").

4. Average rate $\rightarrow$ instantaneous rate

The **average rate** of change of the function $f(x)$ over the interval $x = x_1$ to $x = x_2$ is graphically represented by the **gradient of the secant** passing through the two points $(x_1, f(x_1))$ and $(x_2, f(x_2))$ on the curve $y = f(x)$.

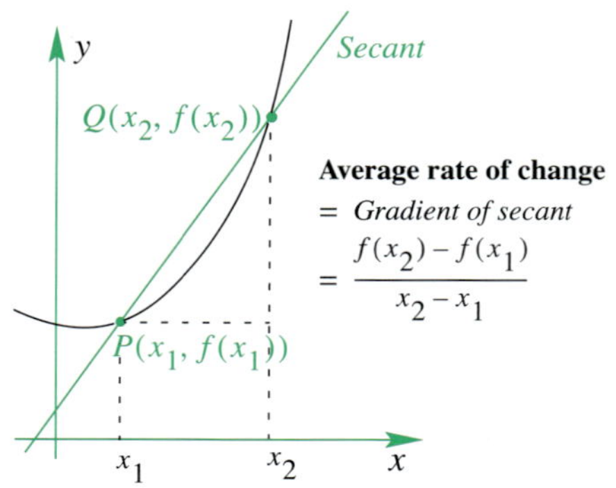

Notice that when we talk about the average rate, we require two points on a curve.
i.e., ". . *over the interval* . ." or ". . *from* . . *to* . ."

Whereas the **instantaneous rate** of change of $f(x)$ at the point $x = x_1$ is graphically represented by the **gradient of the tangent** at the point $(x_1, f(x_1))$ on the curve $y = f(x)$.

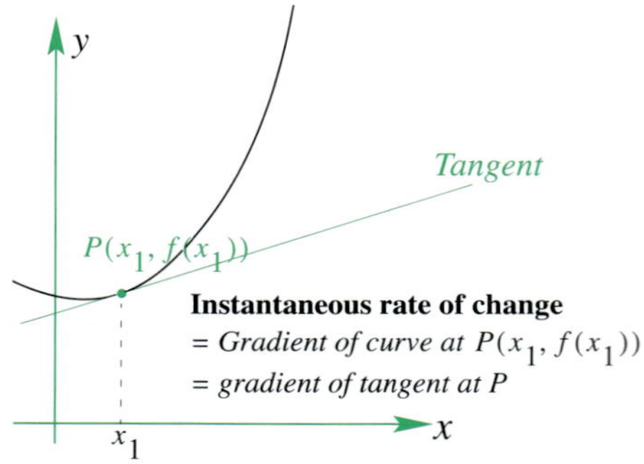

Notice that when we talk about the instantaneous rate, we refer to only one point,
i.e., ". . *at the point* . ."

5. Different notation

The basic difference between the two can be seen in the diagram below:

Functional Notation	**Delta Notation**

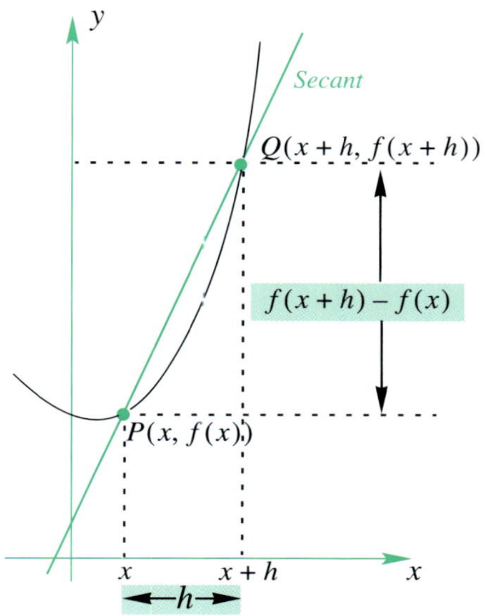

Gradient of secant = $\dfrac{f(x+h)-f(x)}{h}$

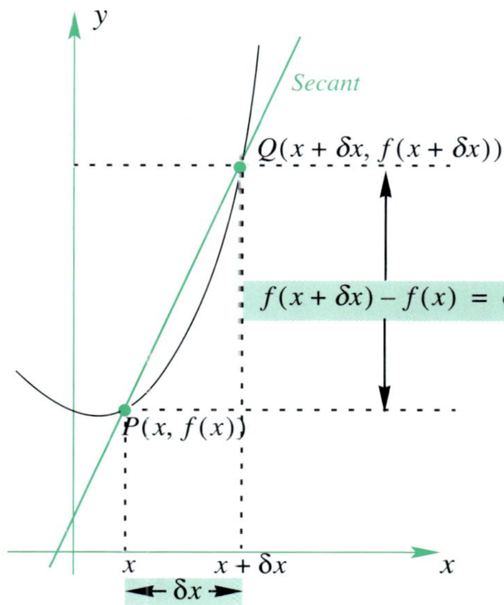

Gradient of secant = $\dfrac{f(x+\delta x)-f(x)}{\delta x} = \dfrac{\delta y}{\delta x}$

To find the gradient at some point *P*, that is, to find the derivative of the function at any point *P* on the curve defined by the equation $y = f(x)$, we use the method of first principles:

$$f'(x) = \lim_{h \to 0}\frac{f(x+h)-f(x)}{h}, h \neq 0$$

$$\frac{dy}{dx} = \lim_{\delta x \to 0}\frac{\delta y}{\delta x}, \delta x \neq 0$$

Using the delta notation, we read $\frac{dy}{dx} = \lim_{\delta x \to 0}\frac{\delta y}{\delta x}, \delta x \neq 0$ as "dee–*y*–dee–*x* is equal to the limit as delta *x* (δx) tends to zero of delta *y* (δy) on delta *x* (δx)". This is in part where the expression "The derivative of *y* with respect to *x*" stems from. The notation $f'(x)$ is due to one of the greatest eighteenth century mathematicians, Joseph Louis Lagrange (1736–1813), whereas the notation $\frac{dy}{dx}$ is attributed to that other great mathematician, Gottfried Wilhelm Leibniz (1646–1716).

As we now have the definition of the derivative, given by the expression $\frac{dy}{dx} = \lim_{\delta x \to 0}\frac{\delta y}{\delta x}, \delta x \neq 0$ or $f'(x) = \lim_{h \to 0}\frac{f(x+h)-f(x)}{h}, h \neq 0$, then to differentiate a function $y = f(x)$ one of these expressions is used. Notice that the expression $f'(x)$ is itself a function and for this reason we also refer to the derivative as the **gradient function** of $y = f(x)$.

EXAMPLE 23.1

Differentiate the function $f(x) = 2 - x^2$.

Solution

As in the previous example, we first simplify the expression $f(x+h) - f(x)$:

$$\begin{aligned} f(x+h) - f(x) &= 2 - (x+h)^2 - [2 - x^2] \\ &= 2 - (x^2 + 2xh + h^2) - 2 + x^2. \\ &= -2xh - h^2 \end{aligned}$$

Next we simplify the expression $\frac{f(x+h) - f(x)}{h}$:

$$\begin{aligned} \frac{f(x+h) - f(x)}{h} &= \frac{-2xh - h^2}{h} \\ &= \frac{(-2x - h)h}{h} \\ &= -2x - h \end{aligned}$$

Finally we using the definition $f'(x) = \lim_{h \to 0} \frac{f(x+h) - f(x)}{h}$:

$$\begin{aligned} f'(x) &= \lim_{h \to 0} \frac{f(x+h) - f(x)}{h} \\ &= \lim_{h \to 0} (-2x - h) \\ &= -2x \end{aligned}$$

EXAMPLE 23.2

Find the gradient function of $f(x) = x^3 + 1$.

Solution

By definition, we have that $f'(x) = \lim_{h \to 0} \frac{f(x+h) - f(x)}{h}, h \neq 0$, and so, we start by simplifying the expression $f(x+h) - f(x)$:

$$\begin{aligned} f(x+h) - f(x) &= (x+h)^3 + 1 - (x^3 + 1) \\ &= x^3 + 3x^2h + 3xh^2 + h^3 + 1 - x^3 - 1 \\ &= 3x^2h + 3xh^2 + h^3 \\ &= h(3x^2 + 3xh + h^2) \end{aligned}$$

Therefore we have $f'(x) = \lim_{h \to 0} \frac{f(x+h) - f(x)}{h}, h \neq 0 = \lim_{h \to 0} \frac{h(3x^2 + 3xh + h^2)}{h}, h \neq 0$

$$\begin{aligned} &= \lim_{h \to 0} (3x^2 + 3xh + h^2), h \neq 0 \\ &= 3x^2 \end{aligned}$$

That is, $f(x) = x^3 + 1 \Rightarrow f'(x) = 3x^2$.

23.1.2 POWER RULE FOR DIFFERENTIATION

Finding the derivative from first principles can be tedious. The previous two examples clearly show this. However, using the first principles approach, produces the results shown in the table below:

Function $y = f(x)$	x^4	x^3	x^2	x^1	x^{-1}	x^{-2}
Derivative $\frac{dy}{dx} = f'(x)$	$4x^3$	$3x^2$	$2x^1$	$1x^0$	$-1x^{-2}$	$-2x^{-3}$

Based on these results and following the general pattern, it is reasonable to assume the general result that if

$$y = x^n, n \in Z, \text{ then } \frac{dy}{dx} = nx^{n-1}.$$

In fact this rule is true for any exponent $n \in \mathbb{R}$, i.e., for any real number n.
This result is known as the **power rule for differentiation**.

Notice that for the case $n = 0$, then $y = x^0$ and so we have that $\frac{dy}{dx} = 0x^{0-1} = 0$.

Note: The function $y = x^0$ represents the horizontal straight line $y = 1$, and so its gradient will always be 0. In fact, for the case where $y = k$ (a real constant) $y = kx^0 \Rightarrow \frac{dy}{dx} = k \times 0x^{0-1} = 0$.
As the function $y = k$ represents a horizontal straight line its gradient will always be 0. We therefore have the power rule:

The function $f:x \mapsto x^n$, has a gradient function (or derivative) given by $f':x \mapsto nx^{n-1}$.
This can also be written as:

If $y = x^n$ then $\frac{dy}{dx} = nx^{n-1}$ or If $f(x) = x^n$ then $f'(x) = nx^{n-1}$

It is also possible to find the derivative of the derivative, that is, $f''(x)$. This is known as the second derivative. To find $f''(x)$, differentiate $f(x)$ once, and then differentiate again.

EXAMPLE 23.3

Use the power rule to find (a) $f'(x)$ (b) $f''(x)$ of
i. $f(x) = x^6$ ii. $f(x) = x^{10}$

Solution

(a) i. $f(x) = x^6 \Rightarrow f'(x) = 6x^{6-1} = 6x^5$
ii. $f''(x) = 6 \times 5x^4 = 30x^4$

(b) i. $f(x) = x^{10} \Rightarrow f'(x) = 10x^{10-1} = 10x^9$
ii. $f''(x) = 10 \times 9x^8 = 90x^8$

An extension of the power rule is the derivative of $f:x \mapsto ax^n$, where a is a real constant. In this case we have the general result:

If $f:x \mapsto ax^n$ then $f':x \mapsto anx^{n-1}$

EXAMPLE 23.4

Differentiate the following functions:

(a) $12x^3$ (b) $-4x^5$ (c) $\frac{x^2}{7}$ (d) $3x^{-3}$ (e) $\frac{3}{x^2}$

Solution

(a) Let $y = 12x^3 \Rightarrow \frac{dy}{dx} = 12 \times 3x^{3-1}$

$= 36x^2$

(b) Let $f(x) = -4x^5 \therefore f'(x) = -4 \times 5x^{5-1}$

$= -20x^4$.

(c) Let $f(x) = \frac{x^2}{7}$, that is, $f(x) = \frac{1}{7}x^2 \Rightarrow f'(x) = \frac{1}{7} \times 2x^{2-1} \therefore f'(x) = \frac{2}{7}x$

(d) Let $y = 3x^{-3} \therefore \frac{dy}{dx} = 3 \times -3x^{-3-1} = -9x^{-4}$

(e) Let $y = \frac{3}{x^2} = 3x^{-2} \therefore \frac{dy}{dx} = -6x^{-2-1} = -\frac{6}{x^3}$

23.1.3 DERIVATIVE OF A SUM OR DIFFERENCE

This rule states that the derivative of a sum (or a difference) is equal to the sum (or the difference) of the derivatives. That is,

If $y = f(x) \pm g(x)$ then $\frac{dy}{dx} = f'(x) \pm g'(x)$

EXAMPLE 23.5

Differentiate the following functions

(a) $2x^3 + 5x - 9$ (b) $x - 5x^2 + 2$ (c) $x(x^3 - 3) + 5x^5$

Solution

(a) Let $y = 2x^3 + 5x - 9 \Rightarrow \frac{dy}{dx} = \frac{d}{dx}(2x^3 + 5x - 9) = \frac{d}{dx}(2x^3) + \frac{d}{dx}(5x) - \frac{d}{dx}(9)$

$= 6x^2 + 5$

Notice we have used a different notation, namely that $f'(x) = \frac{d}{dx}(f(x))$. We can think of $\frac{d}{dx}$ as the differentiation operator, so that $\frac{d}{dx}(f(x))$ or $\frac{d}{dx}(y)$ is an operation of differentiation done on $f(x)$ or y respectively.

(b) Let $f(x) = x - 5x^2 + 2 \therefore f'(x) = 1x^{1-1} - 5 \times 2x^{2-1} + 0$

$= x - 10x$

(c) We need to first expand and simplify the expression:

Let $y = x(x^3 - 3) + 5x^5$

$= x^4 - 3x + 5x^5$

$\therefore \frac{dy}{dx} = 4x^3 - 3 + 25x^4$

EXAMPLE 23.6

Differentiate the following (a) $\frac{5x^2 + 4x - 3x^3}{x}$ (b) $(x^3 + 2)^2$

Solution

(a) $\frac{d}{dx}\left(\frac{5x^2 + 4x - 3^3}{x}\right) = \frac{d}{dx}\left(\frac{5x^2}{x} + \frac{4x}{x} - \frac{3x^3}{x}\right) = \frac{d}{dx}(5x + 4 - 3x^2) = 5 - 6x$.

(b) The first step is to expand the bracket so that each term is in the form of ax^n.

$\frac{d}{dx}((x^3 + 2)^2) = \frac{d}{dx}(x^6 + 4x^3 + 4) = 6x^5 + 12x^2$

EXERCISES 23.1

1. Find the derivative of each of the following

(a)	x^5	(b)	x^9	(c)	x^{25}
(d)	$9x^3$	(e)	$-4x^7$	(f)	$\frac{1}{4}x^8$
(g)	$x^2 + 8$	(h)	$5x^4 + 2x - 1$	(i)	$-3x^5 + 6x^3 - x$
(j)	$20 - \frac{1}{3}x^4 + 10x$	(k)	$3x^3 - 6x^2 + 8$	(l)	$3x - 1 + \frac{x^2}{5} + x^4$

2. Find the derivative of each of the following

(a)	$x(x + 2)$	(b)	$(x + 1)(x^3 - 1)$	(c)	$x\left(x^2 + 1 - \frac{1}{x}\right), x \neq 0$
(d)	$\frac{2x - 1}{x}, x \neq 0$	(e)	$(x^2 - 2x)^2$	(f)	$\frac{x^2 - x + x^3}{2x}, x \neq 0$

3. Show that if $f(x) = x^2 - x$, then $f'(x) = 1 + \frac{2f(x)}{x}$.

4. Show that if $y = ax^n$ where a is real and $n \in N$, then $\frac{dy}{dx} = \frac{ny}{x}, x \neq 0$.

5. (a) Factorise i. $6x - 2$ ii. $81x^2 - 54x + 9$

(b) Differentiate i. $3x-1$ ii. $(3x-1)^2$ iii. $(3x-1)^3$

(c) Deduce an expression for the derivative of $(3x-1)^n$

6. Given that $f'(x) = a(x-b)^2+8$, find a and b if $f(x) = \frac{1}{3}x^3 - x^2 + 9x + 10$.

7. Find (a) the first derivative (b) the second derivative of the following

i. $\left(x-\frac{2}{x}\right)^2$ ii. $\frac{x^3-x+2}{x^2}$ iii. $\frac{1}{x}\left(\sqrt{x}-\frac{3}{\sqrt{x}}\right)^2$

iv. $\left(x-\frac{2}{x^4}\right)$ v. $\frac{2x-3x^2+5}{x^3}$ vi. $\left(x-\frac{2}{x^2}\right)\left(x^2+\frac{3}{x^3}\right)$

23.2 GRAPHICAL INTERPRETATION OF THE DERIVATIVE

23.2.1 THE VALUE OF THE DERIVATIVE AT A PARTICULAR POINT ON A CURVE

So far in this chapter we have looked at the **gradient function** of $f(x)$, namely $f'(x)$. The function $f'(x)$ represents the gradient at any point on the curve of the original function $f(x)$. In order to determine the gradient at a particular point $x = a$ (say), we revisit the definition of the derivative, $f'(x) = \lim_{h \to 0} \frac{f(x+h)-f(x)}{h}$. Rather than finding the derivative at 'any' point 'x', we determine the derivative at a particular point '$x = a$'. Note we are assuming that the derivative exists at the point '$x = a$'.

If the function $f(x)$ can be differentiated (i.e., is differentiable) at $x = a$, then

$$f'(a) = \lim_{h \to 0} \frac{f(a+h)-f(a)}{h}$$

Since $f'(a)$ gives the gradient of the tangent at $x = a$, it also gives the gradient of the graph of $y = f(x)$ and so the derivative may be used to determine gradients or to find where on a curve a particular gradient exists.

Another form of the derivative at $x = a$ is given by $f'(a) = \lim_{x \to a} \frac{f(x)-f(a)}{x-a}$. This result follows directly from our previous definition by replacing h with $x-a$. This can be seen by using the following 'replacements':

1. If $h = x-a$ then $h \to 0 \Rightarrow x-a \to 0 \Rightarrow x \to a$ and so we replace $\lim_{h \to 0}$ with $\lim_{x \to a}$.

2. If $h = x-a$ then $h+a = x$.

We first look at an example that uses a 'first principle' approach to determining the gradient of a curve at a particular point.

EXAMPLE 23.7

Let P and Q be points on the curve $y = x^2 - 2x$ for which $x = 2$ and $x = 2 + h$ respectively. Express the gradient of PQ in terms of h and hence find the gradient of the curve $y = x^2 - 2x$ at $x = 2$.

Solution

We begin by sketching the relevant curve and placing all the information on our diagram.

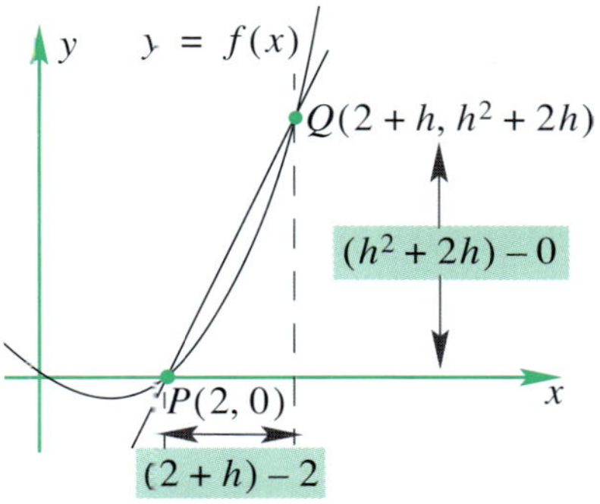

At P, $x = 2$ and so $y = 2^2 - 2(2) = 0$.

At Q, $x = 2 + h$ and so, $y = (2+h)^2 - 2(2+h) = h^2 + 2h$

Next, we find the gradient of PQ, $m_{PQ} = \frac{rise}{run} = \frac{y_2 - y_1}{x_2 - x_1}$.

So, $m_{PQ} = \frac{(h^2 + 2h) - 0}{(2+h) - 2} = \frac{h^2 + 2h}{h} = \frac{h(h+2)}{h}$

After cancelling the 'h' we have: $m_{PQ} = h + 2, h \neq 0$

Note the additional condition, $h \neq 0$. Why is this?

We are now in a position to determine the gradient of the curve at $x = 2$.

Gradient at $P = \lim_{h \to 0} m_{PQ} = \lim_{h \to 0}(h + 2) = 2$.

EXAMPLE 23.8

Find the gradient of the curve $y = \frac{1}{x-1}$ at the point (2, 1).

Solution

As in the previous example, we choose two points P and Q where $x = 2$ and $x = 2 + h$ respectively.

When $x = 2$, $y = \frac{1}{2-1} = 1$ and when $x = 2 + h$, $y = \frac{1}{2+h-1} = \frac{1}{h+1}$.

Meaning that P has coordinates $(2, 1)$ and

Q has coordinates $\left(2 + h, \frac{1}{1+h}\right)$.

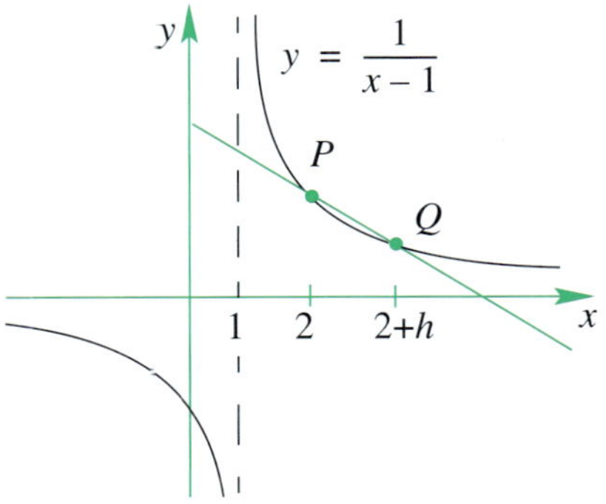

$$m_{PQ} = \frac{\frac{1}{1+h} - 1}{(2+h) - 2} = \frac{\frac{1-(1+h)}{1+h}}{h} = \frac{-\frac{h}{1+h}}{h} = -\frac{1}{1+h}$$

Then, as $h \to 0$, $m_{PQ} \to -\frac{1}{1}$.

That is, gradient at (2, 1) is given by $\lim_{h \to 0} -\frac{1}{1+h} = -1$.

As can be seen from the last two examples, finding the gradient of a curve at a particular point using a 'first principles approach' is rather lengthy. In fact, the process for finding the value of the derivative or the gradient at a particular point on a curve is rather straight forward. This process requires the use of two steps:

Step 1: Find the gradient function (i.e., the derivative). e.g., if $y = f(x)$ find $\frac{dy}{dx} = f'(x)$

Step 2: Substitute the x–value of the point in question into the equation of the derivative. i.e., if we want the gradient at $x = a$, determine $f'(a)$.

EXAMPLE 23.9

Find $f'(3)$ given that $f(x) = x^3 - 2x^2 + 10$.

Solution

Using the power rule for differentiation we have

$$f(x) = x^3 - 2x^2 + 10 \Rightarrow f'(x) = 3x^2 - 4x$$
$$\therefore f'(3) = 3 \times 3^2 - 4 \times 3$$
$$= 15$$

Notice how much more efficient this is compared to using $f'(3) = \lim_{h \to 0} \frac{f(3+h) - f(3)}{h}$

EXAMPLE 23.10

Find the gradient of the curve with equation $y = 9x - x^3$ at the point (2, 10).

Solution

First, find the equation that gives the gradient at any point on the curve, that is, find $\frac{dy}{dx}$.

Using the power rule we have, $y = 9x - x^3 \Rightarrow \frac{dy}{dx} = 9 - 3x^2$. Substituting $x = 2$ into the derivative equation, $\frac{dy}{dx} = 9 - 3(2)^2 = -3$. i.e., the gradient at the point (2, 10) is –3.

EXAMPLE 23.11

Determine the coordinate(s) on the curve $x \mapsto x^3 - x + 2$ where the gradient is 11.

Solution

Let $f(x) = x^3 - x + 2$, we need to find values of x for which $f'(x) = 11$:

We have that $f'(x) = 3x^2 - 1$, so that $3x^2 - 1 = 11 \Leftrightarrow 3x^2 - 12 = 0$

$$\Leftrightarrow 3(x^2 - 4) = 0$$
$$\Leftrightarrow 3(x-2)(x+2) = 0$$
$$x = 2 \text{ or } x = -2$$

For $x = 2$, $f(2) = 8$ and for $x = -2$, $f(-2) = -4$.

Therefore, the required coordinates are (2, 8) and (–2,–4).

Notice that it is possible for a curve to have the same gradient at different points.

EXAMPLE 23.12

Given that $f(x) = x^3 - x^2 - x + 1$, find the coordinates of all points for which the curve with equation $y = f(x)$ has a horizontal tangent.

Solution

If the tangent at a point on the curve is horizontal then the gradient of the curve at that point must be zero. So, to find those values of x where $f'(x) = 0$.

Now, $f'(x) = 3x^2 - 2x - 1 \therefore f'(x) = 0 \Leftrightarrow 3x^2 - 2x - 1 = 0$

$$\Leftrightarrow (3x+1)(x-1) = 0$$

$$\Leftrightarrow x = -\frac{1}{3} \text{ or } x = 1$$

For $x = -\frac{1}{3}$, $y = f\left(-\frac{1}{3}\right) = -\frac{1}{27} - \frac{1}{9} + \frac{1}{3} + 1 = \frac{32}{27}$ and for $x = 1$, $y = 1 - 1 - 1 + 1 = 0$.

So, the relevant points are $\left(-\frac{1}{3}, \frac{32}{27}\right)$ and $(1, 0)$.

We now make use of a graphics calculator to find the gradient at a particular point on a curve.

EXAMPLE 23.13

Use a graphics calculator to solve Example 24.10.

Solution

Using the TI–83 we first enter the equation, $y = 9x - x^3$ into the equation screen. Then we **QUIT** and have a blank screen.

At this stage we select **MATH** and use the down arrow to select option **8:nDeriv(**, we now call up the **VARS** screen.

Using the arrows, select **Y–VARS**. Next select option **1:Function** followed by option **1:Y1** and then press **ENTER**.

At this stage we have a screen displaying **nDeriv(Y1**. Next we enter our parameters, **,X, 2)** – dont forget the commas. The '**X**' informs the calculator that the variable in question is '**X**' and the '**2**' informs the calculator that we wish to evaluate the derivative at '$x = 2$'.

The screen sequence is now displayed:

Notice that the answer we have obtained is –3.000001 (which, for all intended purposes is –3). The reason lies in our original 'accuracy' settings on the calculator – see your graphics calculator handbook for more details on this.

We could also have entered the information on the one screen as follows:

That is, type in the equation into the **nDeriv(** function as opposed to using the **VARS** approach.

However, you should spend some time in considering when one approach is more beneficial than the other.

EXERCISES 23.2.1

1. Let P and Q be points on the curve $y = x^2 - 4x$ for which $x = 4$ and $x = 4 + h$ respectively. Express the gradient of PQ in terms of h and hence find the gradient of the curve $y = x^2 - 4x$ at $x = 4$.

2. For a curve with equation $y = \frac{2}{x+1}$ determine the coordinates of the points P and Q where $x = 1$ and $x = 1 + h$ respectively. Express the gradient of the line PQ in terms of h and hence find the gradient of the curve at $x = 1$.

3. Using a first principles method find the gradient of the curve with equation $y = 6 - x^3$ at the point where $x = 2$.

4. Find the gradient of the function at the indicated point.

(a) $f(x) = x^3 - 2$ at $(1, -1)$ (b) $f(x) = \frac{1}{x}$ at $(2, 0.5)$

(c) $f(x) = (2x-1)^2$ at $(2, 9)$ (d) $y = (2x+1)^2$ at $(0, 1)$

5. Find the value(s) of x, so that $f'(x) = 0$ given that $f(x) = x^3 - 8x$.

6. For the curve with equation $y = x^2 - 12x$, find:

(a) $\frac{dy}{dx}$ (b) the gradient where $x = -3$

(c) the coordinates of the point where the gradient is 4.

7. For the curve with equation $y = -x^3 + 3x$, find:

(a) $\frac{dy}{dx}$ (b) the gradient where $x = 1$

(c) the coordinates of the point where the gradient is –3.

8. For the curve with equation $f(x) = \frac{1}{4}x^2(x^2 - 1)$, find:

(a) the coordinates where its gradient is zero.
(b) the set of values of x for which its gradient is positive.

9. Determine those values of x for which the curve with equation $y = 8 - x^2$ will have the same gradient as the curve with equation $y = x^3 - x$.

10. Find the gradient of the function $x: \mapsto x^3 + x^2 - 2x$ at the points where
(a) it crosses the x–axis, (b) it cuts the y–axis.

11. The curve with equation $y = ax + bx^2$ passes through the point $(2, 0)$, where its gradient is found to be 3. Determine the values of a and b.

12. Given that $g(x) = x^2$, show that $f'(a + b) = f'(a) + f'(b)$.

13. Given that $h(x) = 4x - 2x^2, x \geq 0$, find (a) $f(a^2)$ (b) $f'(a^2)$.

14. The function $f(x) = ax^3 - bx^2$ has a gradient of 2 at the point $(1, 6)$. Find its gradient when $x = 2$.

23.2.2 GRADIENT FUNCTION FROM A GRAPH

We now have a brief look at some geometrical properties of the gradient value $f'(a)$ at a point $(a, f(a))$ of the graph $y = f(x)$.

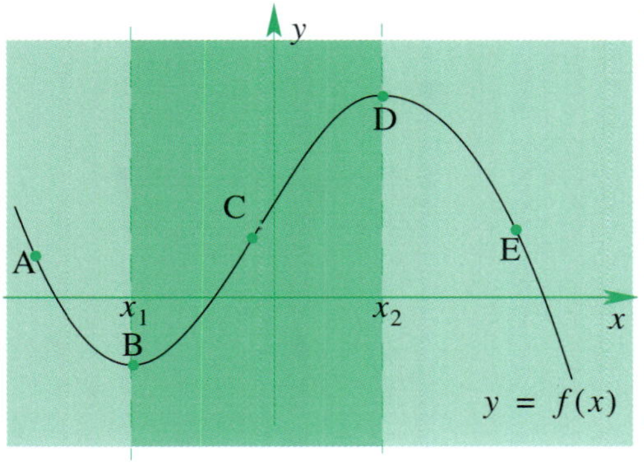

1. At A, tangent has a negative slope.
2. For $x < x_1$ gradient is < 0, i.e., $g'(x) < 0$.
3. At B, tangent is a horizontal line. Meaning that it has a gradient of zero.
4. At $x = x_1$ gradient is zero, i.e., $g'(x_1) = 0$.
5. At C, tangent has a positive slope.
6. For $x_1 < x < x_2$ gradient is positive.
7. At D, tangent is a horizontal line. Meaning that it has a gradient of zero.
8. At $x = x_2$ gradient is zero, i.e., $g'(x_2) = 0$.
9. For $x > x_2$ gradient is negative.
10. At E, tangent has a negative slope.

These properties are extremely useful in helping us 'find' and/or 'sketch' the gradient function of a graph when the equation of the function is not provided.

EXAMPLE 23.14

Sketch the graph of the gradient function of the curves shown below.

(a)

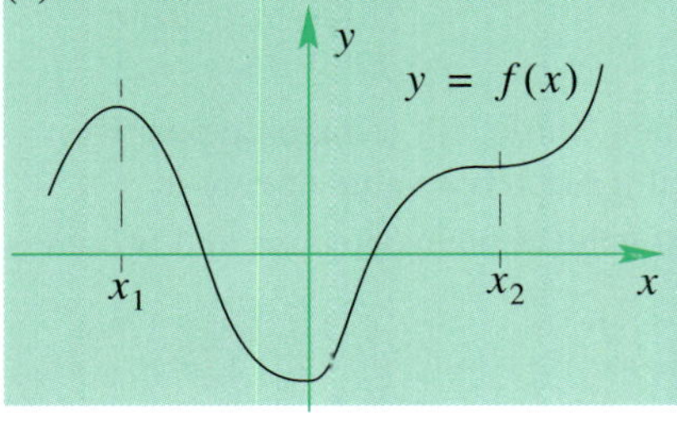

(b)

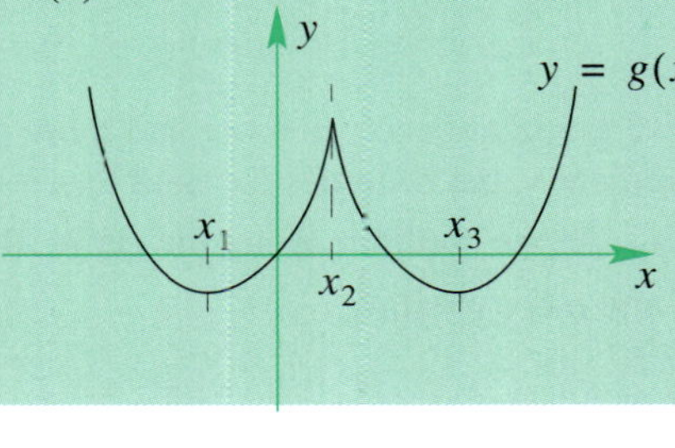

Solution

A general approach to these types of problems is to first locate where a horizontal tangent line would occur on the curve. Once the point(s) have been located we can then 'break up' the remainder of the curve into appropriate regions. e.g., regions of positive gradients, regions of negative gradients and so on.

(a) In this case, our 'key' points occur when $x = x_1$, $x = x_2$ and $x = 0$. At each of these points the gradient is zero. We observe that for $x < x_1$, $0 < x < x_2$ and $x > x_2$ the gradient is positive. Also, for $x_1 < x < 0$ the gradient is negative. We start by identifying these regions on the set of axes defining $f'(x)$ versus x – this will show us where we can sketch the gradient function.

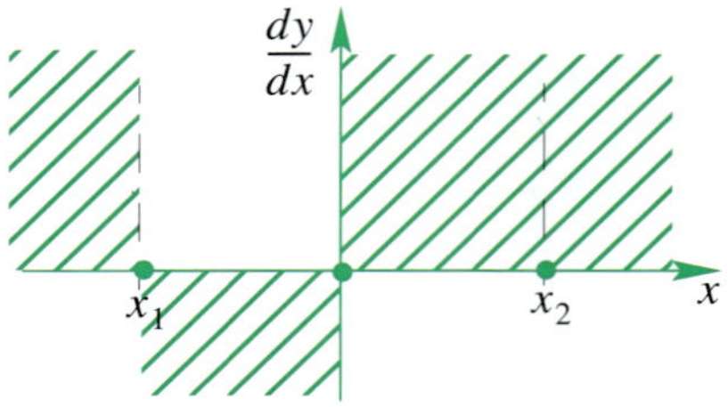

All that remains is to 'identify' the gradient values along the curve. By 'identify' we mean determine the relative gradient values. That is, the region where the gradient has been identified as being either positive or negative. As we move along the curve (from left to right) are the gradient values increasing or decreasing? As you move along the curve, keep asking yourself questions like, "Is the gradient value becoming more and more positive?", "Is the gradient value becoming more and more negative?", "Is the gradient becoming smaller and smaller?" and so on.

These questions can be more easily answered if you use a ruler or pencil and run it along the curve from left to right.

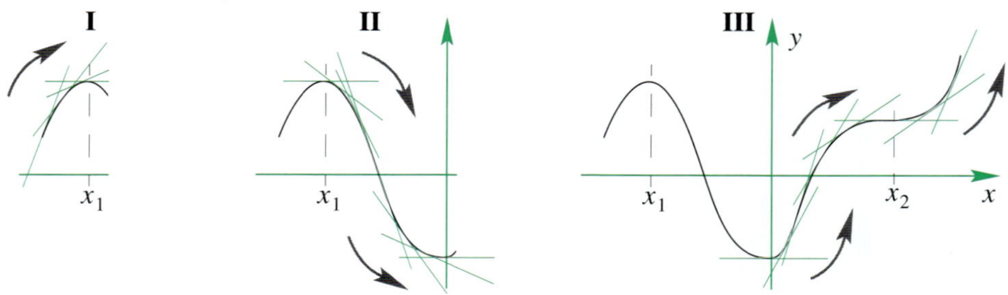

Section **I** ($x < x_1$):

Positive gradient that is decreasing in value as we get closer to x_1.

Section **II** ($x_1 < x < 0$):

Negative gradient becomes more and more negative, then, while still remaining negative, the gradient value becomes less and less negative until it reaches a value of zero at $x = 0$.

Section **III** ($x > 0$):

Positive gradient becomes more and more positive, then, while still remaining positive, the gradient value becomes less and less positive until it reaches a value of zero at $x = x_2$. As we continue along the curve, the gradient value remains positive and becomes more and more positive.

Combining all of our findings we can produce the following sketch of the gradient function:

Realise that we are not so much interested in the numerical values of the gradient function but rather the general shape of the gradient function.

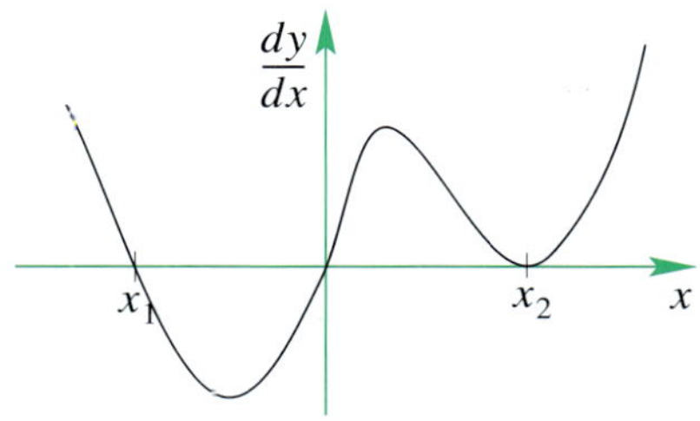

(b) Using the same approach we first determine the 'regions' where the gradient function lies (i.e., where it is positive and where it is negative).

I. For $x < x_1$, $g'(x) < 0$.

II. At $x = x_1$, $g'(x) = 0$.

III. For $x_1 < x < x_2$, $g'(x) > 0$.

IV. At $x = x_2$ the gradient value cannot be found.

V. For $x_2 < x < x_3$, $g'(x) < 0$.

VI. At $x = x_3$, $g'(x) = 0$.

VI. For $x > x_3$, $g'(x) > 0$.

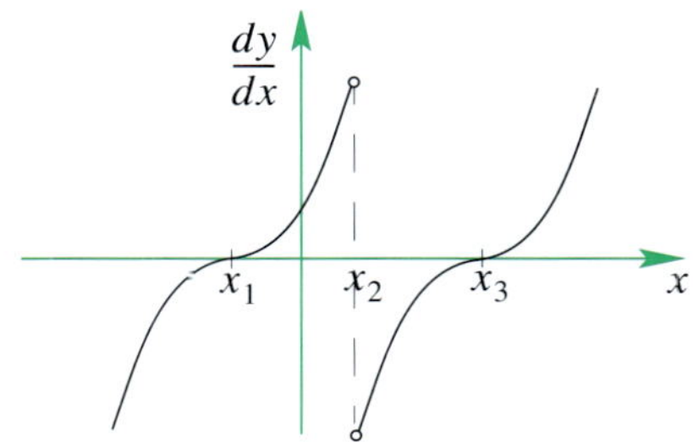

Notice that at $x = x_2$ the curve finishes in a peak. Trying to place a tangent at a point such as this is not possible – i.e., approaching $x = x_2$ from the left and from the right provides two different tangents that will not coincide as we get closer and closer to $x = x_2$. In order for a gradient-value to exist at $x = a$ (say) one of the conditions is that the tangents (as we approach $x = a$) from the left and from the right must coincide. This reinforces the need to understand that

$\lim_{h \to 0} \frac{f(a+h) - f(a)}{h}$ exists if and only if $\lim_{h \to 0^+} \frac{f(a+h) - f(a)}{h} = \lim_{h \to 0^-} \frac{f(a+h) - f(a)}{h}$,

i.e., the right-hand side limit equals the left-hand side limit.

EXERCISES 23.2.2

1. For each of the following functions, sketch the corresponding gradient function.

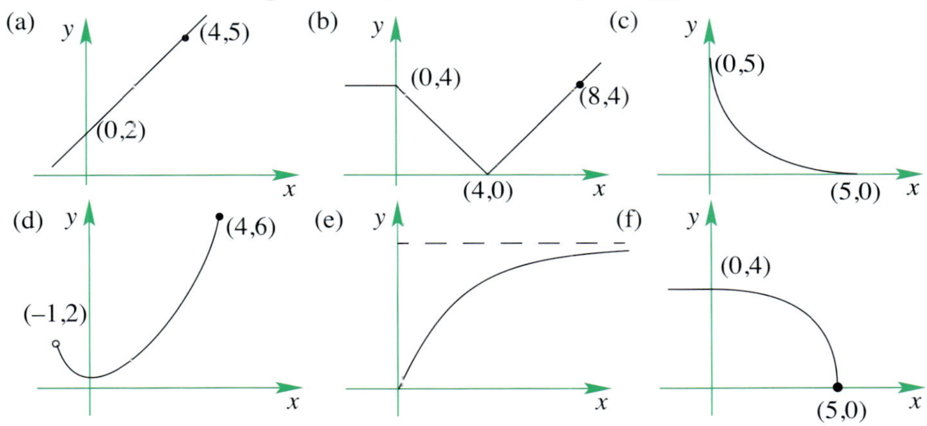

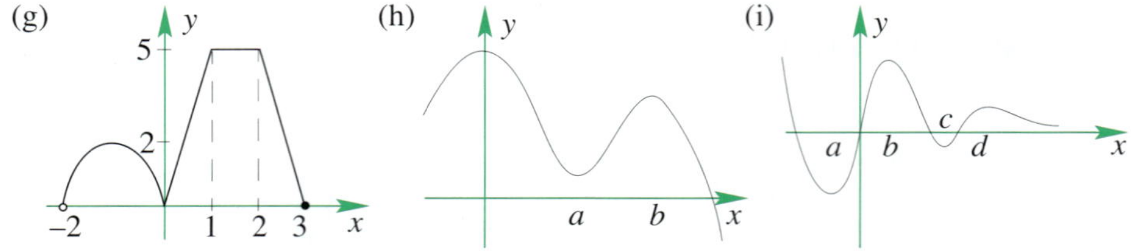

2. Sketch the graph of the function $f(x)$ given that $f(1) = 0, f'(x) = -1$ for all real x.

3. Sketch the graph of $f:\mathbb{R}\mapsto\mathbb{R}$ given that $f(2) = 2, f'(2) = 0, f(1) = 4, f'(1) = 0$, $f'(x) > 0$ for $x > 2$ and $x < 1$ and $f'(x) < 0$ for $1 < x < 2$.

23.3 TANGENTS

23.3.1 EQUATION OF TANGENT

The gradient of a curve $y = f(x)$ at any point (x_1, y_1) is equal to the gradient of the tangent to the curve at that point. To find the gradient at (x_1, y_1), you need to

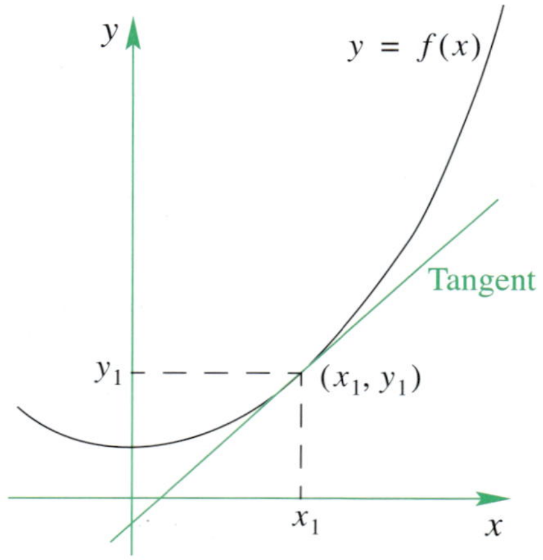

1. Find the gradient function of $y = f(x)$, that is, find the derivative $\frac{dy}{dx} = f'(x)$.

2. Find the gradient at (x_1, y_1), that is, find $f'(x_1)$.

This gives the gradient of the tangent, m, at the point (x_1, y_1).

Finally, to find the **equation of the tangent**, you need to use the straight line equation

$$y - y_1 = m(x - x_1), \text{ where } m = f'(x_1)$$

EXAMPLE 23.15

Find the equation of the tangent to the curve $y = 5 - x^2$ at the point (1,4).

Solution

Given that $y = 5 - x^2 \Rightarrow \frac{dy}{dx} = -2x$.

Then, for $x = 1$, we have $\frac{dy}{dx} = -2(1) = -2$.

Therefore, using $y - y_1 = m(x - x_1)$, with $m = -2$ and $(x_1, y_1) \equiv (1, 4)$, we have the equation of the tangent given by

$$y - 4 = (-2)(x - 1) \Leftrightarrow y - 4 = -2x + 2$$

That is, $y = -2x + 6$

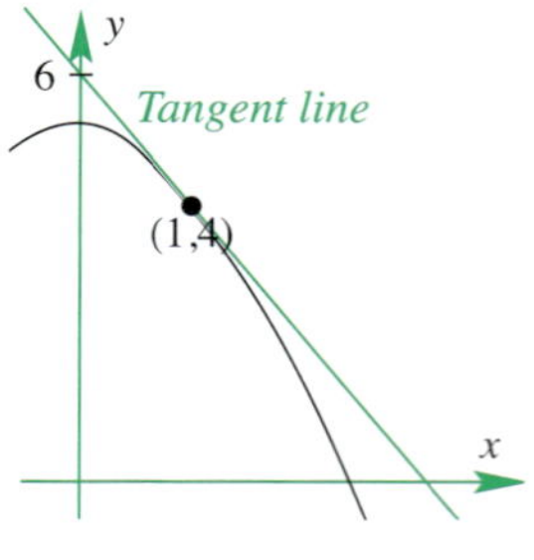

EXAMPLE 23.16

Find the equation of the tangent to the curve $y = x^3 - 8$ where $x = 2$.

Solution

Given that $y = x^3 - 8 \Rightarrow y' = 3x^2$. Then, for $x = 2$, $y' = 3 \times 2^2 = 12$. i.e., $m = 12$.
In order to use the equation $y - y_1 = m(x - x_1)$ we need both x- and y-values. As we are only given the x-value, we now determine the corresponding y-value, i.e., $x = 2 \Rightarrow y = 2^3 - 8 = 0$.
With $(x_1, y_1) \equiv (2, 0)$ the equation of the tangent is: $(y - 0) = 12(x - 2) \Leftrightarrow y = 12x - 24$.

EXAMPLE 23.17

Find the equation of the tangent to the curve $y = x^3 - x$ where $x = 1$.

Solution

We solve this problem using the TI–83:

Step 1: Enter equation, $y = x^3 - x$
Step 2: Set window range
Step 3: Graph the function

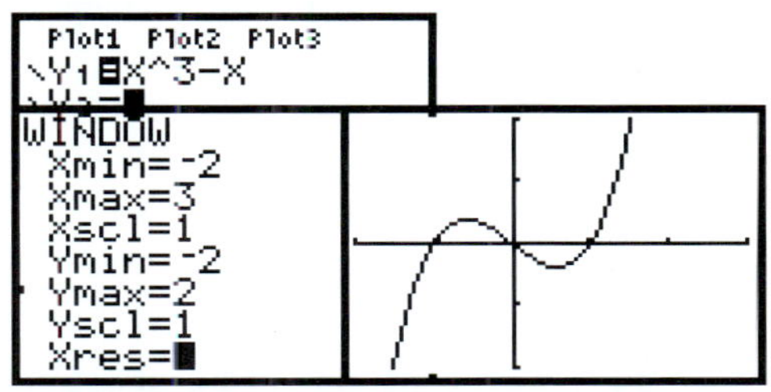

Step 4: Call up the DRAW tool (press 2nd PRGM)
Step 5: Enter the value 1 and press ENTER

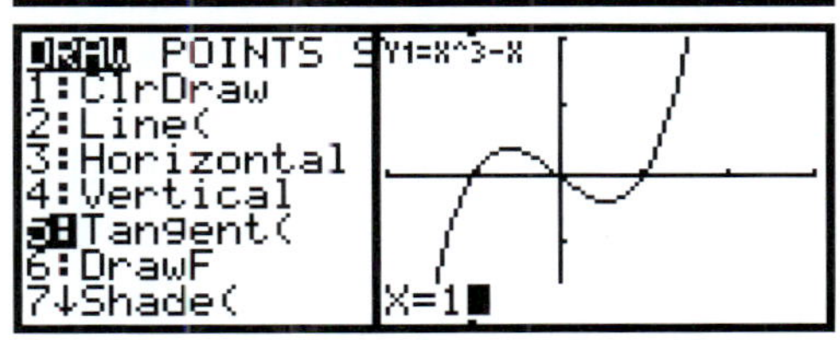

Notice that you are automatically provided with the equation of the tangent at the bottom of the screen.

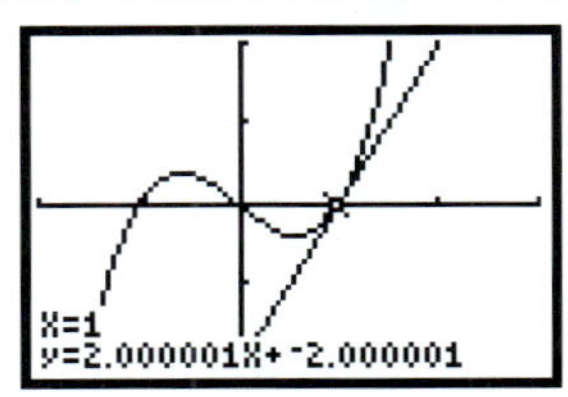

Equation of tangent is $y = 2x - 2$

EXERCISES 23.3

1. Find the equations of the tangents to the following curves at the points indicated

(a) $y = x^3 - x^2 - x + 2$ at (2,4)
(b) $y = x^4 - 4x^2 + 3$ at $(1, 0)$
(c) $y = x^2 - 7$ at $(3, 2)$
(d) $y = (x - 4)^2$ at $(5, 1)$
(e) $f(x) = x^3 - 9x$ at $(-1, 8)$
(f) $f(x) = \frac{1}{2}x - x^2 + 3$ at $(2, 0)$
(g) $x \mapsto x(x^3 - 4)$ at $(2, 8)$
(h) $x \mapsto x^2(x - 1)$ at $(2, 4)$

2. Find the equation of the tangent to the curve $y = x^2(x^2 - 1)$ at the point A(2, 12).
The tangent at a second point, B(–2,12), intersects the tangent at A at the point C.
Determine the type of triangle enclosed by the points A, B and C.
Show that the tangents drawn at the points X and Y, where $x = a$ and $x = -a$ respectively will always meet at a third point Z which will lie on the y–axis.

3. Find the equation of the tangent to the curve $y = x^2 - 2x$ that is parallel to the line with equation $y = 4x + 2$.

4. Find the equations of the tangents to the curve $y = x^3 - 12x + 3$ at the point where the gradient is 0.

5. The straight line $y = -x + 4$ cuts the parabola with equation $y = 16 - x^2$ at the points A and B.
(a) Find the coordinates of A and B.
(b) Find the equation of the tangents at A and B, and hence determine where the two tangents meet.

6. The line L and the curve C are defined as follows,

$$L: y = 4x - 2 \text{ and } C: y = mx^3 + nx^2 - 1$$

The line L is a tangent to the curve C at $x = 1$.
(a) Using the fact that L and C meet at $x = 1$, show that $m + n = 3$.
(b) Given that L is a tangent to C at $x = 1$, show that $3m + 2n = 4$.
(c) Hence, solve for m and n.

7. (a) Sketch the graph of the function $f(x) = x(x-1)(x+2)$.
(b) Find the equation of the tangent to the curve
i. at the point (0, 0). ii. at the point (–2, 0).
(c) Find the area of the region enclosed by the x-axis and the two tangents.

8. The functions $f(x) = x + 2$ and $g(x) = x^2$ are shown in the diagram alongside.
(a) The curves intersect at A and B, find the values of a, b and c.
The tangents to the curve at A and B intersect at T.
(b) Find the coordinates of T.
(c) Find the perimeter of ΔABT giving your answer to 2 decimal places.

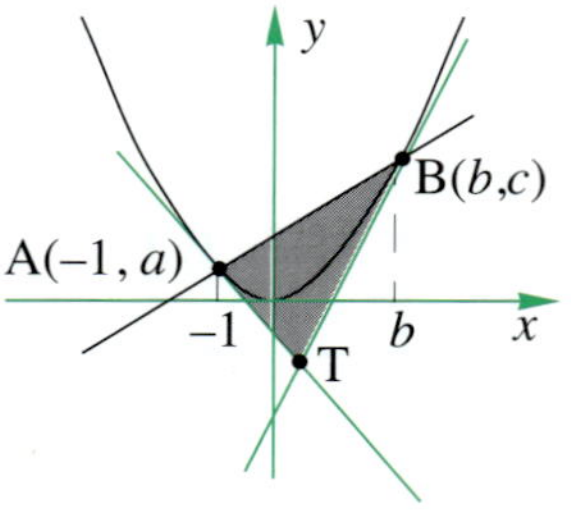

9. Find the equations of the tangents to the curve $y = \frac{1}{3}x^3 - x^2 - x + 1$ at the points on the curve where the tangent is parallel to the line $y = 7x - 2$.

10. Find the equation of the tangent to the parabola $y = (x-1)^2 + 3$ at the points where
(a) it is horizontal.
(b) it is perpendicular to the straight line $2y + x = 1$

11. The tangent to the curve $y = x^2(x - 1)$ at the point where the curve crosses the x-axis meets the y-axis at the point Y(0, a). Find the value of a.

23.4 CURVE SKETCHING

23.4.1 INCREASING AND DECREASING FUNCTIONS

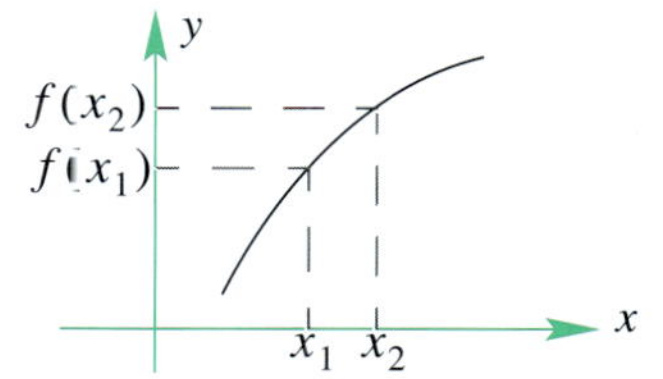

A function f is said to be **increasing** if its **graph rises** as it is sketched from left to right.

That is, if $x_2 > x_1 \Rightarrow f(x_2) > f(x_1)$

(i.e., the y–values increase as the x–values increase).

Similarly:

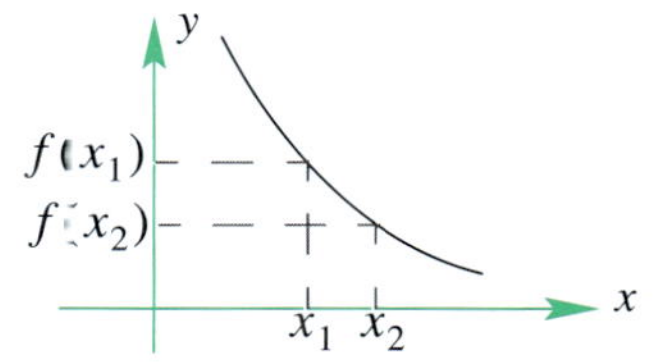

A function f is said to be **decreasing** if its **graph falls** as it is sketched from left to right.

That is, if $x_2 > x_1 \Rightarrow f(x_2) < f(x_1)$

(i.e., the y–values decrease as the x–values increase).

. . . A calculus point of view

The derivative can be used to determine whether a function is increasing or decreasing and so it can be used to help find those values of x for which the function is increasing or decreasing.

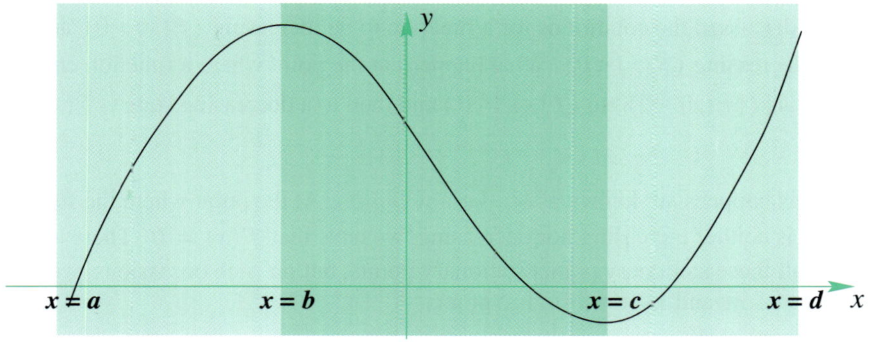

Any tangent line drawn in the interval $a < x < b$ will always have a positive gradient. This means that $\frac{dy}{dx} = f'(x) > 0$ in this region.

Any tangent line drawn in the interval $b < x < c$ will always have a negative gradient. This means that $\frac{dy}{dx} = f'(x) < 0$ in this region.

Any tangent line drawn in the interval $c < x < d$ will always have a positive gradient. This means that $\frac{dy}{dx} = f'(x) > 0$ in this region.

This means that to determine where a function is increasing or decreasing, the values of x for which $f'(x) > 0$ and $f'(x) < 0$ respectively need to be found.

EXAMPLE 23.18

Find the values of x for which the function $f(x) = 1 + 4x - x^2$ is increasing.

Solution

By definition, a function is increasing for those values of x for which $f'(x) > 0$.

Therefore: **1.** find $f'(x)$

2. find the values of x such that $f'(x) > 0$

Now, $f(x) = 1 + 4x - x^2 \Rightarrow f'(x) = 4 - 2x$

Then, $f'(x) > 0 \Leftrightarrow 4 - 2x > 0$

$\Leftrightarrow 4 > 2x$

$\Leftrightarrow x < 2$

We could also have determined this by sketching the graph of $f(x) = 1 + 4x - x^2$. The turning point can be determined by completing the square. i.e., $f(x) = -(x-2)^2 + 5$ giving the axis of symmetry as $x = 2$.

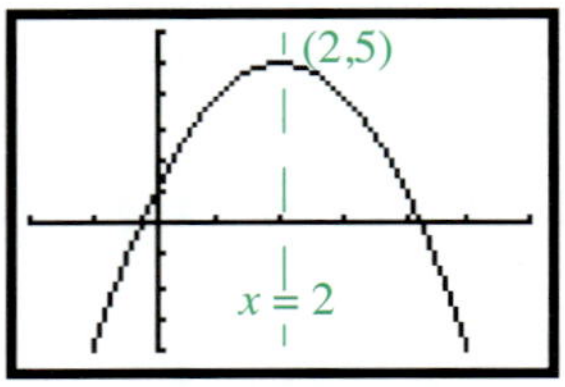

23.4.2 STATIONARY POINTS

So far we have discussed the conditions for a function to be increasing $(f'(x) > 0)$ and for a function to be decreasing $(f'(x) < 0)$. What happens at the point where a function changes from an increasing state $((f'(x) > 0))$ to $(f'(x) = 0)$ and then to a decreasing state $((f'(x) < 0))$ or vice–versa?

Points where this happens are known as **stationary points.** At the point where the function is in a state where it is neither increasing nor decreasing, we have that $f'(x) = 0$. There are times when we can call these stationary points stationary points, but on such occassions, we prefer the terms **local maximum** and **local minimum** points.

At the point(s) where $\frac{dy}{dx} = f'(x) = 0$ we have a **stationary point**.

There are three types of stationary points, namely;

local maximum point,

local minimum point and

stationary point of inflection.

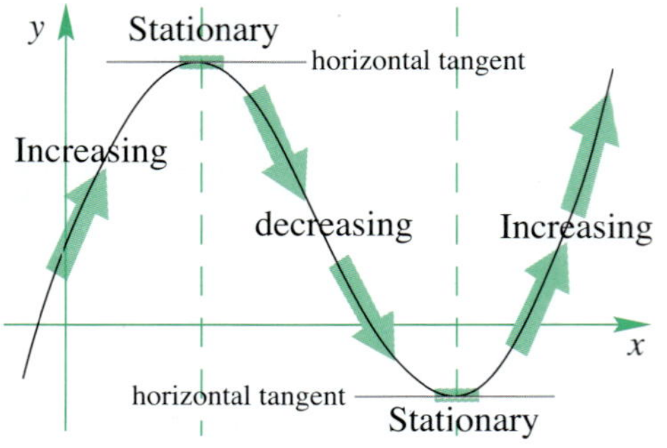

1. Local Maximum

When sketching a curve, if the following properties hold:

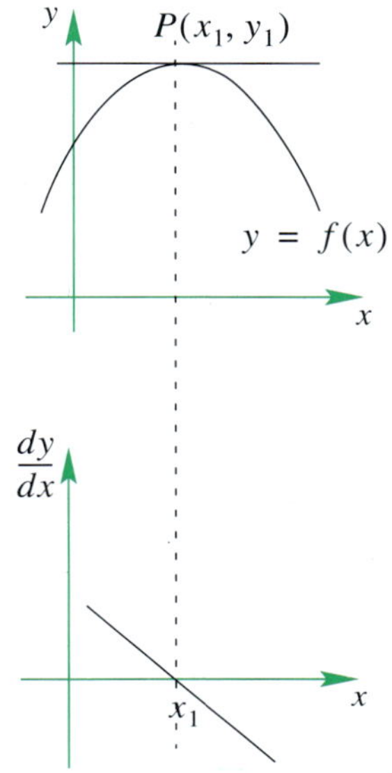

i. At $P(x_1, y_1)$, $\frac{dy}{dx} = f'(x) = 0$ that is $f'(x_1) = 0$.

ii. For $x < x_1$ then $\frac{dy}{dx} > 0$

$x > x_1$ then $\frac{dy}{dx} < 0$

where the two chosen values of x are such that one is just slightly less than x_1 and the other is just slightly greater than x_1. Then,

$y = f(x)$ has a local maximum point (also known as a **relative maximum**) at the point $P(x_1, y_1)$.

iii. Graph of the gradient function:

Notice that the values of $\frac{dy}{dx}$ are changing from **positive** to **negative**. Sometimes this is referred to as the sign of the first derivative.

At this stage, it isn't so much the magnitude of the derivative that is important, but that there is a change in the sign of the derivative near $x = x_1$.

In this instance the sign of the derivative changes from positive to negative. This change in sign is sometimes represented via the diagram alongside, which is referred to as a sign diagram of the first derivative. Such diagrams are used to confirm the nature of stationary points (in this case, that a local maximum occurs at $x = x_1$).

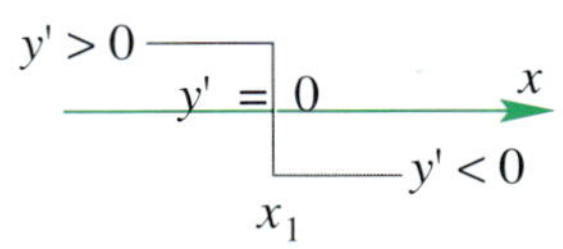

EXAMPLE 23.19 Find the local maximum value of the function whose equation is $f(x) = -3 + 4x - x^2$.

Solution

First we differentiate: $f(x) = -3 + 4x - x^2 \Rightarrow f'(x) = 4 - 2x$

Next, equate $f'(x)$ to 0 and solve for x: $0 = 4 - 2x$

$\Leftrightarrow x = 2$

To ensure that we have obtained a local maximum we choose values of x slightly less than 2 and slightly greater than 2, for example, choose $x = 1.9$ and $x = 2.1$.

For $x = 1.9$, we have that $f'(1.9) = 4 - 2(1.9) = 0.2$.

For $x = 2.1$ we have that $f'(2.1) = 4 - 2(2.1) = -0.2$.

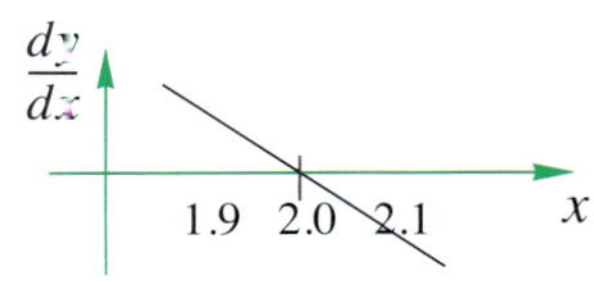

Using the graph of the gradient function, $\frac{dy}{dx}$, confirms that there is a local maximum at $x = 2.0$.

The local maximum value of $f(x)$, is found by substituting $x = 2$ into the given equation: $f(2) = -3 + 4(2) - (2)^2 = 1$. That is, the local maximum occurs at the point (2,1). This process can also be carried out using the TI–83:

1. Enter equation

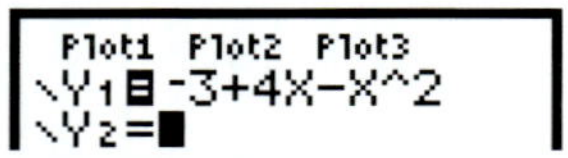

2. Use the **CALC** function, and then select **4:maximum**

3. Press **ENTER**. This will prompt you to choose a lower (left) bound then an upper (right) bound and finally to make a guess

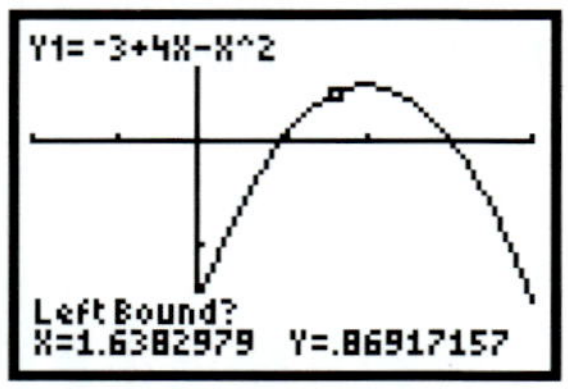

3.1

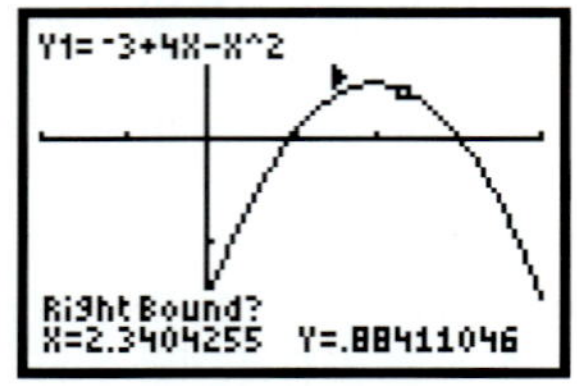

3.2

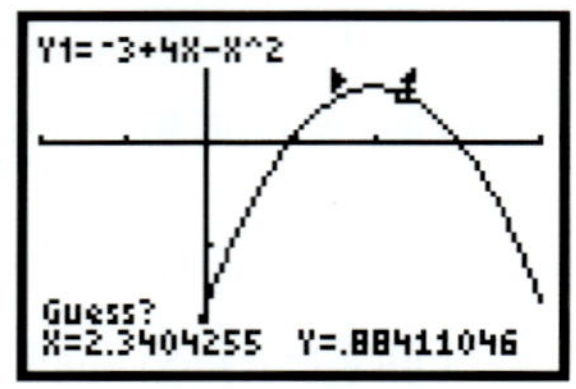

Move the cursor to a point between the two arrows and press **ENTER**.

4. Press **ENTER** once more. This will provide both x– and y– values of the stationary point.

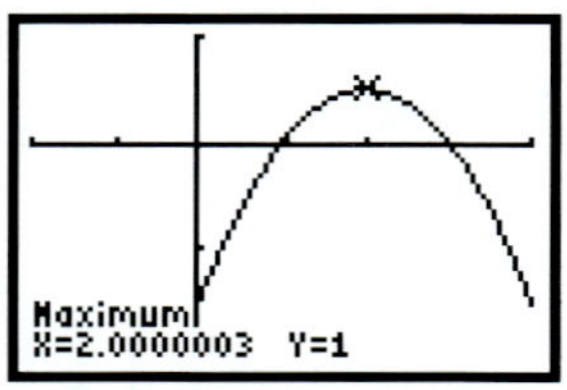

The other option is to use the **fMax** command from the **MATH** screen.

2. Local Minimum

When sketching a curve, if the following properties hold:

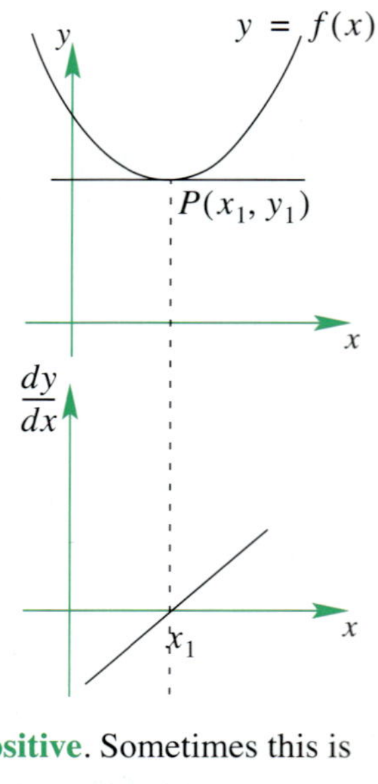

i. At $P(x_1, y_1)$, $\frac{dy}{dx} = f'(x) = 0$ that is $f'(x_1) = 0$.

ii. For $x > x_1$ then $\frac{dy}{dx} > 0$

$x < x_1$ then $\frac{dy}{dx} < 0$

where the two chosen values of x are such that one is just slightly greater than x_1 and the other is just slightly less than x_1. Then,

$y = f(x)$ has a local minimum point (also known as a **relative minimum**) at the point $P(x_1, y_1)$.

iii. Graph of the gradient function:

Notice that the values of $\frac{dy}{dx}$ are changing from **negative** to **positive**. Sometimes this is referred to as the sign of the first derivative.

Again we can represent the change in the sign of the first derivative via the diagram alongside, which is referred to as a sign diagram of the first derivative. Such diagrams are used to confirm the nature of stationary points (in this case, that a local minimum occurs at $x = x_1$).

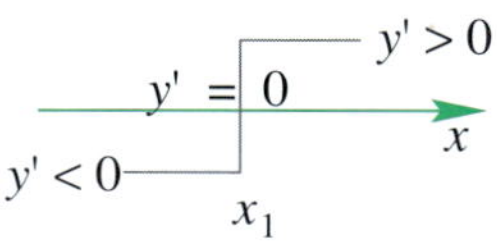

EXAMPLE 23.20

Find the minimum value of $y = x^4 - 4x^3$.

Solution

We first need to determine where the gradient is zero.

Therefore, we solve for $\frac{dy}{dx} = 0$

Now, $\frac{dy}{dx} = 4x^3 - 12x^2 \therefore \frac{dy}{dx} = 0$

$$\Leftrightarrow 4x^2(x-3) = 0$$

Therefore, $x = 0$ or $x = 3$.

From the graph we see that there is a local minimum at $x = 3$.

A quick sketch of the function indicates that a point of inflection occurs at $x = 2$:

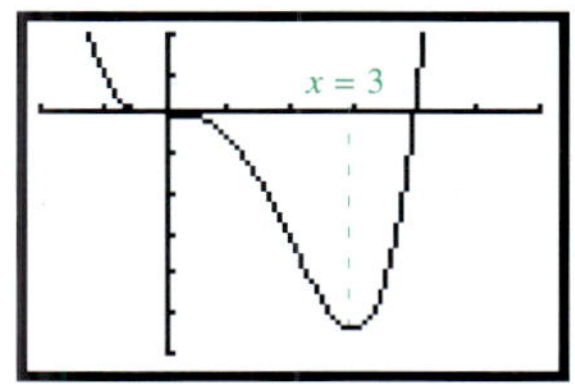

We can verify this by using the sign of the first derivative:

For $x = 3.1$, $\frac{dy}{dx} = -3.84$.

For $x = 2.9$, $\frac{dy}{dx} = -3.36$

Sign diagram:

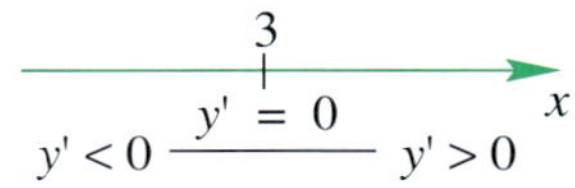

Therefore there is a local minimum at $x = 3$.

The minimum value is then given by $y = (3)^4 - 4 \times (3)^3 = -27$

Again, we see that the graphics calculator can be used to solve this problem directly:

Step 1: call up CALC
Step 2: selecting option 3: minimum
Step 3: select a left bound, then a right bound and finally make a guess.

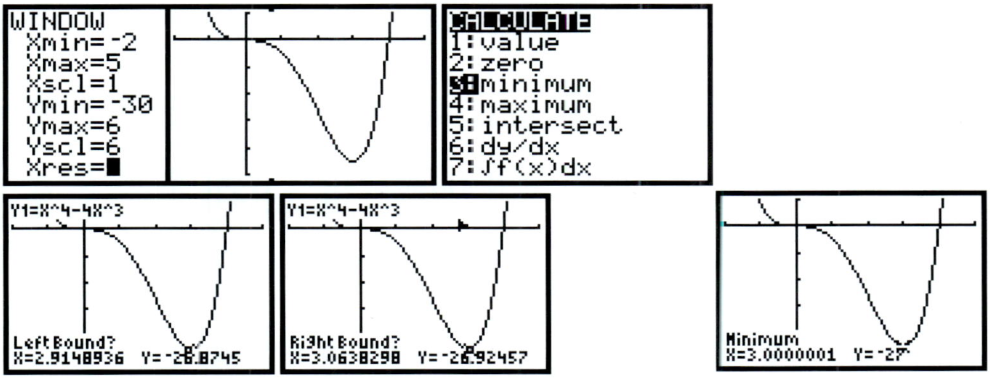

The last screen dump provides where the minimum occurs.

3. Points of Inflection

There are two types: **A.** **Stationary points of inflection**
B. **Non–stationary points of inflection.**

A. Stationary point of inflection

The following properties hold at a stationary point of inflection.

i. At $P(x_1, y_1)$, $f'(x) = 0$. That is $f'(x_1) = 0$.

ii. For $x < x_1$, $f'(x) > 0$ and for $x > x_1$, $f'(x) > 0$.

Similarly,

At $P(x_2, y_2)$, $f'(x) = 0$. That is $f'(x_2) = 0$
and for $x < x_2$, $f'(x) < 0$ and for $x > x_2$, $f'(x) < 0$.

That is, the gradient of the curve on either side of x_1 (or x_2) has the same sign.

iii. Graph of the **gradient function**, $y = f'(x)$:
Notice that the values of $f'(x)$ have the same sign on either side of $x = x_1$.

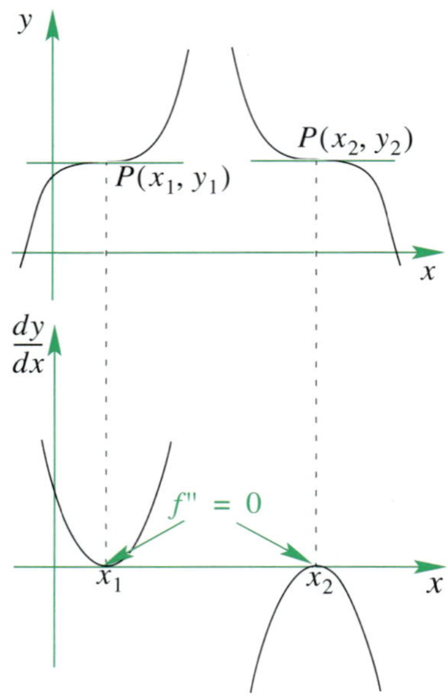

Notice that at $x = x_1$, the gradient of $f'(x)$ is also equal to zero. That is, the derivative of the derivative is equal to zero.

EXAMPLE 23.21

Find the stationary point of inflection for the graph with equation $y = (x-1)^3(x+2)$.

Solution

We solve this problem using the graphics calculator.
We start by sketching the graph of $y = (x-1)^3(x+2)$.
From the sketch it appears that there is a stationary point of inflection at $x = 1$.

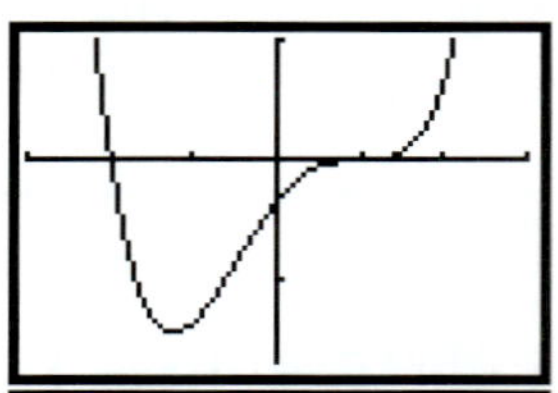

We do this by determining the gradient at $x = 1$.
We can make use of the **MATH** menu and selecting option **8: nDeriv(**

We obtain a derivative value of 3×10^{-6}, which can be considered to be zero.

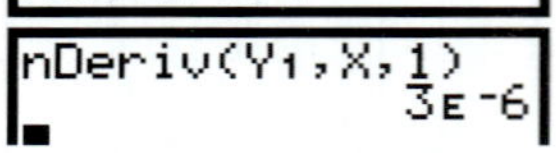

Then, as $\frac{dy}{dx} = 0$ at $x = 1$ and $\frac{dy}{dx} > 0$ on either side of $x = 1$ there is a stationary point of inflection at $x = 1$.

EXAMPLE 23.22

Find and classify all stationary points of $f(x) = x^3 - 3x^2 - 9x + 1$.

Solution

Now, $f(x) = x^3 - 3x^2 - 9x + 1 \Rightarrow f'(x) = 3x^2 - 6x - 9$

Solving for stationary points we have, $3x^2 - 6x - 9 = 0 \Leftrightarrow 3(x-3)(x+1) = 0$

$\Leftrightarrow x = 3$ or $x = -1$

So that $f(-1) = 6$ $f(3) = -26$ and.

Using the sign of the first derivative, we have:

At $x = 3$:

For $x < 3$ $(x = 2.9)$ $f'(2.9) < 0$ and

for $x > 3$ $(x = 3.1)$ $f'(3.1) > 0$.

Therefore, there is a local minimum at $(3,-26)$.

At $x = -1$:

For $x < -1$ $(x = -1.1)$ $f'(-1.1) > 0$ and

for $x > -1$ $(x = -0.9)$ $f'(-0.9) < 0$.

Therefore, there is a local maximum at $(-1,6)$.

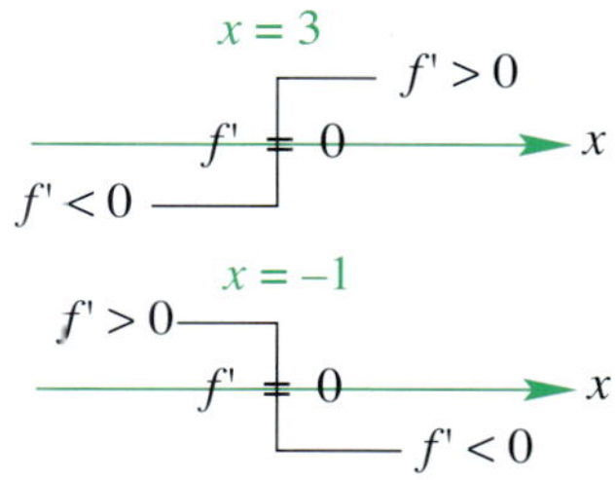

EXAMPLE 23.23

Sketch the graph of the function $f(x) = x^4 - 2x^2$, clearly marking any stationary points.

Solution

We first find the **stationary points** (if any).

This means that we must solve for $f'(x) = 0$:

Now, $f'(x) = 4x^3 - 4x = 4x(x^2 - 1) \therefore 4x(x^2 - 1) = 0 \Leftrightarrow 4x(x+1)(x-1) = 0$

$\Leftrightarrow x = -1$ or $x = 1$ or $x = 0$

We now check for the nature of each point

At $x = 0$: $x = -0.1$, $f'(-0.1) = 0.396 > 0$

$x = 0.1$, $f'(0.1) = -0.396 < 0$

Therefore, we have a **local maximum** point at $(0,0)$.

At $x = 1$:

$x = 0.9$, $f'(0.9) = -0.684 < 0$

$x = 1.1$, $f'(1.1) = 0.924 > 0$

Therefore, we have a **local minimum** point at $(1,-1)$.

At $x = -1$:

$x = -0.9$, $f'(-0.9) = 0.684 > 0$

$x = -1.1$, $f'(-1.1) = -0.924 < 0$

Hence, we have a **local minimum** point at $(-1,-1)$.

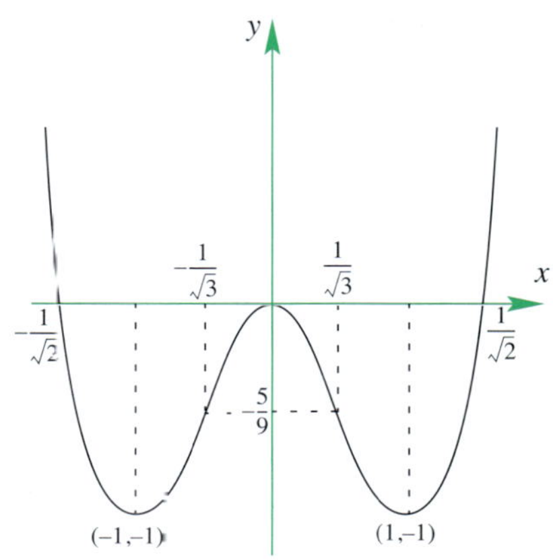

23.4.3 GLOBAL MAXIMA AND MINIMA

Until now we have only considered locating the **local maxima** or the **local minima**. The process has been straightforward enough in the sense that there exists a procedure for locating these points, i.e., find the derivative, equating it to zero, solve and then use a sign diagram of the first derivative to identify the nature of the stationary point.

Consider the function $f(x) = 4x^3 + 9x^2 - 12x + 10$. Assuming that we have been able to find the stationary points we can sketch the graph of $f(x)$:

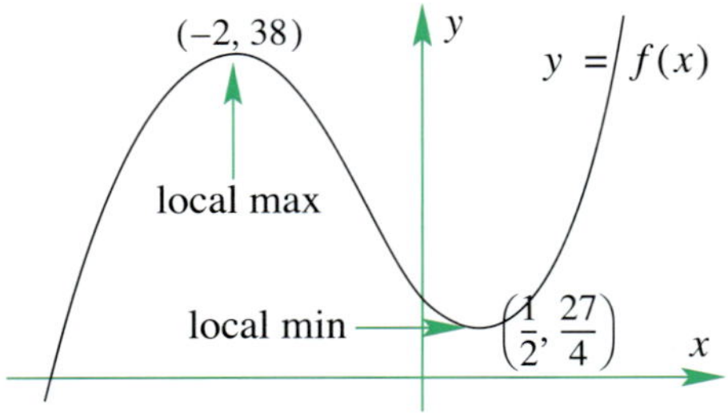

Going through our standard process we find that the curve has a

local maximum at (–2, 38)

and a

local minimum at $\left(\frac{1}{2}, \frac{27}{4}\right)$.

In sketching this curve we have (correctly) assumed that $x \in \mathbb{R}$ and as such the graph extends indefinitely and as such, no overall minimum or maximum can be given. However, what if we wished to find the maximum value of this function but this time have a restriction on the domain, e.g., $f(x) = 4x^3 + 9x^2 - 12x + 10, -3 \le x \le 2$.

In this instance, proceeding with our standard approach, i.e., finding the derivative and so forth gives the same results as above. However, we should now sketch the graph over the given domain. A sketch of this graph over the given domain is now produced:

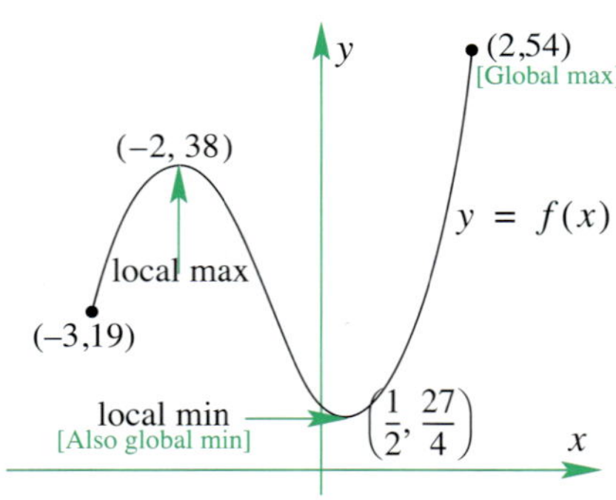

From the sketch we can see that although the point (–2, 38) is still a stationary point, it only identifies the local maximum.

Over the new domain, the maximum value of $f(x)$ is in fact 54. We say that $f(x)$ has an endpoint maximum. This **endpoint** maximum occurs at (2, 54). Then, as there is no other value greater than 54, we also say that the **global maximum** or the **absolute maximum** is 54.

Notice too that the point $\left(\frac{1}{2}, \frac{27}{4}\right)$ is both a local minimum **and** a **global minimum**, because for this domain the minimum value of $f(x)$ is $\frac{27}{4}$, which happens to coincide with the local miminum. Had the domain been $-4 \le x \le 2$, then the **absolute minimum** would have occured at $x = -4$ with a value of $f(-4) = -54$.

So, when using the term '**local**' we are in fact referring to points that are in the **immediate vicinity** (or **neighbourhood**) of the where the gradient is zero.

Notice also that the derivative, $f'(x)$, does not exist at (2, 54) [or (–3, 19) for that matter] – however, we still have a maximum at that point. Why is it that the derivatives does not exist at these end point?

23.4.4 SUMMARY

We now provide a visual summary of the definitions we have covered to date.

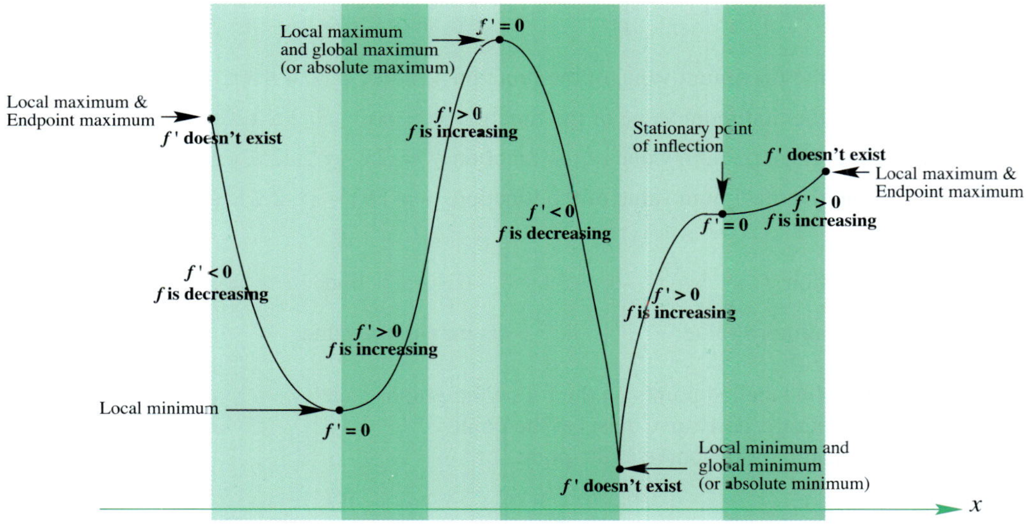

EXERCISES 23.4

1. Draw a sketch of a graph of the function $f(x), x \in \mathbb{R}$, where,

(a) $f(1) = 2, f'(1) = 0, f(3) = -2, f'(3) = 0, f'(x) < 0$ for $1 < x < 3$ and $f'(x) > 0$ for $x > 3$ and $x < 1$.

(b) $f'(2) = 0, f(2) = 0, f'(x) > 0$ for $0 < x < 2$ and $x > 2$, $f'(x) < 0$ for $x < 0$ and $f(0) = -4$.

(c) $f(4) = f(0) = 0, f'(0) = f'(3) = 0, f'(x) > 0$ for $x > 3$ and $f'(x) < 0$ for $x < 0$ and $0 < x < 3$.

2. Find the coordinates and nature of the stationary points for the following:

(a) $y = 3 + 2x - x^2$ (b) $y = x^2 + 9x$

(c) $y = x^3 - 27x + 9$ (d) $f(x) = x^3 - 6x^2 + 8$

(e) $f(x) = 3 + 9x - 3x^2 - x^3$ (f) $y = (x-1)(x^2-4)$

(g) $y = (x-1)^2(x+1)$ (h) $g(x) = x^4 - 8x^2 + 16$

3. Sketch the following functions:

(a) $y = 5 - 3x - x^2$ (b) $f(x) = x^2 + \frac{1}{2}x + \frac{3}{4}$

(c) $f(x) = x^3 + 6x^2 + 9x + 4$ (d) $f(x) = x^3 - 4x$

(e) $f(x) = \frac{1}{3}x^3 - x^2 + 4$ (f) $y = 4x^3 - x^4$

(g) $y = x^3 - 8$ (h) $y = x^4 - 16$

4. Find and describe the nature of all stationary points and points of inflection for the function $f(x) = x^3 + 3x^2 - 9x + 2$.

5. Sketch the graph of $x \mapsto x^4 - 4x^2$.

6. (a) Find the maximum value of the function $y = 6x - x^2, 4 \le x \le 7$.

(b) Find the minimum value of the function $y = 6x - x^2, 2 \le x \le 6$.

(c) Find the maximum value of the function $y = 2x - x^3, -2 \le x \le 6$.

(d) Find the maximum value of the function $y = 36x - x^4, 2 \le x \le 3$.

7. For the function $f(x) = \frac{1}{3}x^3 - x^2 - 3x + 8, -6 \le x \le 6$ find;

i. its minimum value, ii. its maximum value.

8. For each of the labelled points on the following graphs state

i. whether the derivative exists at the point.

ii. the nature of the curve at the point.

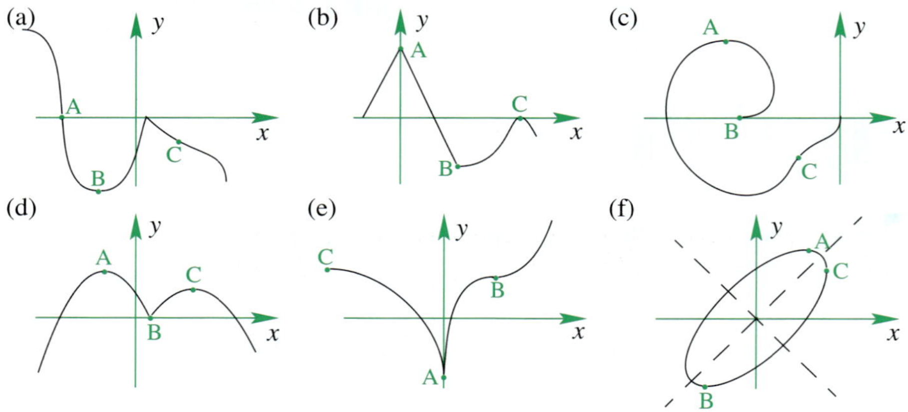

9. Identify which graph corresponds to

i. $f(x)$ ii. $f'(x)$ iii. the gradient function of $f'(x)$

(a) (b)

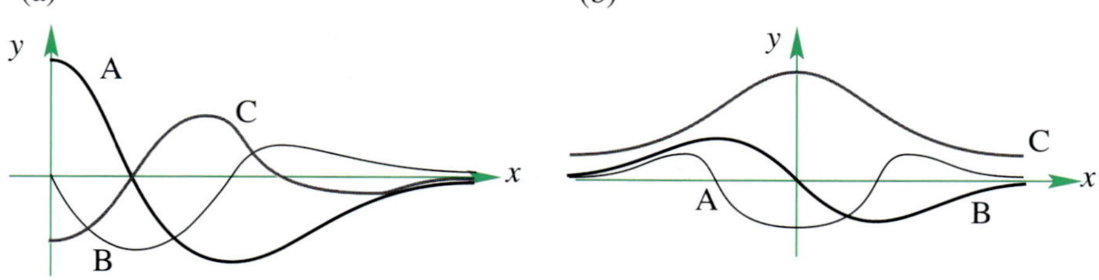

10. For each of the functions, $f(x)$, sketch $f'(x)$.

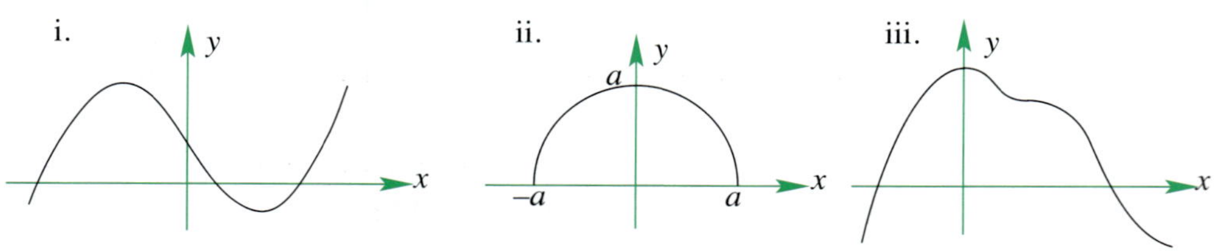

11. The curve with equation $y = ax^3 + bx^2 + cx + d$ has a local maximum where $x = -3$ and a local minimum where $x = -1$. If the curve passes through the points (0,4) and (1, 20) sketch the curve for $x \in \mathbb{R}$.

12. The function $f(x) = ax^3 + bx^2 + cx + d$ has turning points at $\left(-1, -\frac{13}{3}\right)$ and (3, –15). Sketch the graph of the curve $y = f(x)$.

13. The function $f: x \mapsto \mathbb{R}$, where $f(x) = ax^5 + bx^3 + cx$ has stationary points at (–2, 64) (2, –64) and (0, 0). Find the values of a, b and c and hence sketch the graph of f.

14. Find m and n so that $f'(1)$ exists for the function $f(x) = \begin{cases} mx^2 + n & \text{if } x \le 1 \\ \frac{1}{x} & \text{if } x > 1 \end{cases}$.

15. The curve with equation $y = ax^3 + bx^2 + cx + 5$ has a stationary point at (0, 5), an x–intercept at $x = -1$ and an inflection point where $x = 0.5$. Find the values of a, b and c.

16. The curve with equation $y = ax^3 + bx^2 + cx + d$ intersects the x-axis at $x = 1$ and cuts the y-axis at (0, –34). Given that the curve has turning points at $x = 3$ and $x = 5$, determine the values of a, b, c and d. Sketch this curve.

23.5 MISCELLANEOUS EXERCISES

1. Find the local maximum point on the curve $f(x) = 2x^3 + 3x^2 - 12x + 1$.

2. Sketch the graph of the function $y = 5x^3 - x^2$, clearly showing all intercepts with the axes as well as stationary points.

3. Find the coordinates on the curve of the function $f(x) = \frac{1}{3}x^3 - x^2 + 3x$ where its gradient is 6.

4. Find the values of k for which $f'(k) = 0$ where $f(x) = x^3 - 3x^2 - 24x + 1$

5. Find the equation of the tangent to the curve $y = x^3 - 3x^2 - 3x$ at the points where the curve has a gradient of 6.

6. Sketch the curve of the function $f(x)$ given that $f(2) = 4$, $f(4) = -2$, $f(6) = 0$, $f(0) = 2$, $f'(2) = 0 = f'(4)$, $f'(x) > 0$ for $x < 2, x > 4$ and $f'(x) < 0$ for $2 < x < 4$.

7. (a) Find the coordinates of all stationary points on the curve $y = 3x^4 - 4x^3 - 12x^2$.

(b) Sketch the graph of the curve $y = 3x^4 - 4x^3 - 12x^2$.

8. For the function $f(x) = x^3 - x^2 - x + 2$, find the values of x for which

(a) $f'(x) = 0$

(b) $f'(x) < 0$

(c) $f'(x) > 0$

9. (a) Sketch the graph of the function $g(x) = x^3 - 12x + 16$, $-4 \le x \le 3$.

(b) Determine

i. the maximum value of $x^3 - 12x + 16$, $-4 \le x \le 3$

ii. the minimum value of $x^3 - 12x + 16$, $-4 \le x \le 3$

(c) For what value of k will the equation $x^3 - 12x + 16 = k$, $-4 \le x \le 3$ have

i. one solution. ii. two solutions iii. three solutions.

10. (a) For what values of x will the curve $y = 3x^2 - x^3$ have a gradient of zero?

(b) Find the equation of the tangent to the curve $y = 3x^2 - x^3$ at the point

i. $x = 3$ ii. $x = 2$.

(c) Find the point of intersection of the two tangents in (b).

11. Find the values of x for which the function $f(x) = x^3 + x^2 - 5x + 2$

(a) is increasing.

(b) is decreasing.

(c) changes from increasing to decreasing.

12. (a) Sketch the graph of the function $f(x) = x^3(4 - x)$ clearly identifying turning points and stationary points of inflection.

(b) Determine the equation of the tangent to the curve in (a) when $x = 2$.

13. Consider the curve with equation $y = x^3 + 3x^2 + 3x + 6$

(a) Find i. $\dfrac{dy}{dx}$

ii. the gradient of the curve when $x = 3$.

(b) i. Is the function y decreasing when $x = -2$?

ii. For what values of x is the function y increasing?

(c) Find the coordinates of

i. the local maximum of the function y.

ii. the local minimum of the function y.

14. Find the values of a and b given that the function $f(x) = ax^3 + x^2 - bx + 1$ has a gradient of 12 at the point $(1, 4)$.

15. (a) Find the equation of the tangent to the curve $y = \dfrac{3}{x}$ at the point when $x = \dfrac{1}{2}$.

(b) Find the equation of the tangent to the curve $y = 3x^4 - 2x^{-2}$ when $x = 1$.

16. The function $y = ax + \dfrac{b}{x^2}$ passes through the point $(2, 0)$, where its gradient is found to be 3. Determine the values of a and b.

23.6 GRADED REVISION EXERCISES

LEVEL 1

1. Find the derivative of:

i. x^7 ii. $6x^2 + 1$ iii. $12 - x$

2. Find the derived functions of the following.

i. $(x+1)(x-1)$ ii. $(2x+1)(3x-2)$ iii. $x\left(x - \frac{1}{x}\right)$

3. Differentiate the following.

i. $x^2(x-1)$ ii. $\frac{2}{x} - x$ iii. $(x^2-3)^2$

iv. $\frac{(x+1)^2}{x}$ v. $x(9-x)^2$ vi. $x^3 - \frac{1}{x^3} + 4$

LEVEL 2

1. Find the gradient of the function at the point indicated.

(a) $y = x - \frac{1}{x^2} + 1$ at (1,1) (b) $f(x) = 27 - x^3$ at (3,0)

2. Find the coordinate(s) on the curve $y = x^4 - 4x^2$ where the gradient is zero.

3. Find the equation of the tangent to the curve $y = x^3 - 4x^2 + x$ at the point where the curve cuts the y-axis.

4. Find the slope of the curve $y = x^3 - x + 2$ at $x = -1$.

5. Sketch the graph of the gradient function corresponding to each of the folowing.

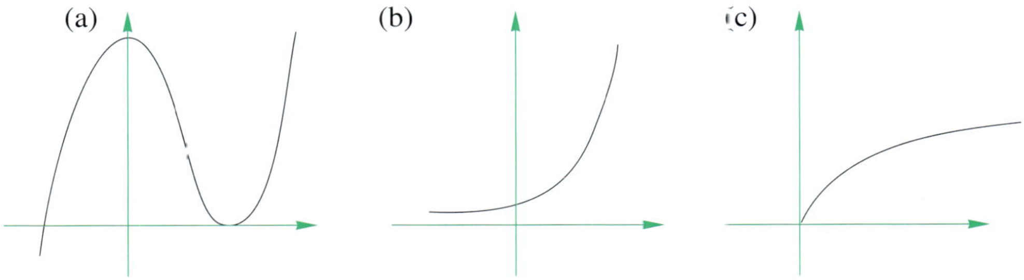

6. Find the gradient of the curve with equation $y = x^2 - 3x + 2$ at the points where it cuts the x-axis.

7. Find the equation of the tangent to the curve $y = \frac{1}{x^2} + 3$ at the point where $x = 2$.

LEVEL 3

1. For the curve with equation $y = x^3 - 3x$, find

(a) $\frac{dy}{dx}$. (b) its gradient at the origin.

(c) the coordinates where the gradient is zero.

2. Find the equations of the tangents to the curve $y = \frac{1}{x^2} + 3$ at the point where $y = 4$.

3. Use first principles to find $f'(x)$, where $f(x) = 2 - 3x^{-2}$.

4. Find a and b, given that $f(x) = ax^2 + 2bx + 4$ and $f(1) = 10$ and $f'(2) = 0$.

5. Given that $f(x) = x^2(x-3)$, find the value(s) of x for which $f'(x) = 0$.

6. (a) A curve has the equation $y = \frac{(x-2)^2}{x}$. Find the equation of the tangent to the curve at the point (1, 1).

(b) Find the equation of a second tangent to the curve that is parallel to the tangent in (a).

LEVEL 4

1. Let the function f be defined by $f(x) = x - 3 + \frac{4}{x^2}, x \neq 0$

(a) Find $f'(x)$

(b) i. Find the values of x for which $f(x) \geq 0$.

ii. Find the values of x for which $f'(x) \geq 0$.

iii. Sketch the graph of f, show clearly any stationary points and asymptotes.

2. A curve has equation $y = 3x^4 - 4x^3 - 6ax^2 + 12ax$, where a is a positive constant.

(a) Expand and simplify $(x^2 - a)(x - 1)$.

(b) Find $\frac{dy}{dx}$.

(c) For what values of x does the curve have a horizontal tangent?

(d) Determine the nature of all stationary points if

i. $0 < a < 1$

ii. $a = 1$

(e) Sketch the curve when $a = 1$. State the coordinates of all stationary points but make no attempt to determine exactly the x-coordinates of any points (other than the origin) at which the curve crosses the x-axis.

3. Consider the graph $y = x^2 + \frac{2}{x}$.

(a) Find the coordinates of any turning points.

(b) i. Find the equation of the tangent to the graph at the point where $x = 2$.

ii. Find the coordinates of the other point which is common to the tangent in i., and the graph.

23.7 TOPIC TEST

1. i. For the function $f(x) = 2x^2 + 1$ find (a) $f(2)$ (b) $f(2+h)$

ii. Find the average gradient of $f(x)$ from $x = 2$ to $x = 2 + h$.

iii. Hence, deduce the gradient at $x = 2$.

[6 marks]

2. For the curve with equation $y = -2x^3 + 24x - 32$ find:

(a) its gradient function.

(b) its gradient when $x = a$.

(c) the coordinates where its gradient is 18.

[1 + 1 + 2 marks]

3. The curve with equation $f(x) = x^3 - bx^2 - 5x + 1$ has a local maximum when $x = -1$. Find b.

[3 marks]

4. i. Sketch the graph of the curve with equation $y = \frac{1}{4}x^2(3-x)$, clearly showing all points of intersection with the axes and the coordinates of its turning point(s).

ii. Hence, state the maximum value of $kx^2(3-x)$, for $x > 0$.

[7 marks]

5. Find $f''(x)$ where $f(x) = x^3 - 4x^2 + x$.

[2 marks]

6. Find the equation of the tangent to the curve $y = 2x - \frac{4}{x}$ at the point where $x = 1$.

[4 marks]

7. i. Consider the graph of $y = f(x)$ as shown.

(a) Identify the nature of the turning point for $x = a$.

(b) Identify those regions for which the gradient is positive.

(c) Sketch the graph of the gradient function.

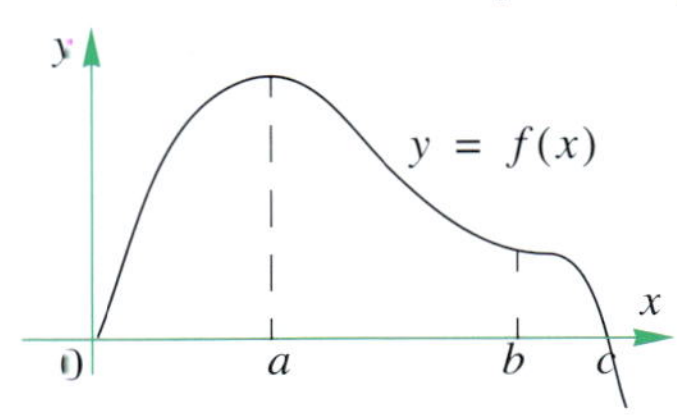

ii. Part of the graph in i., is shown again. For $a < x < b$, the curve can be approximated by the equation, $y = 1 - (x-2)^3$.

(a) Copy and complete the table below.

h	0.1	0.01	0.001
PR			

(b) Using the table in (a), deduce the gradient of the curve with equation $y = 1 - (x-2)^3$ at $(1, 2)$.

(c) Certify your results in (b).

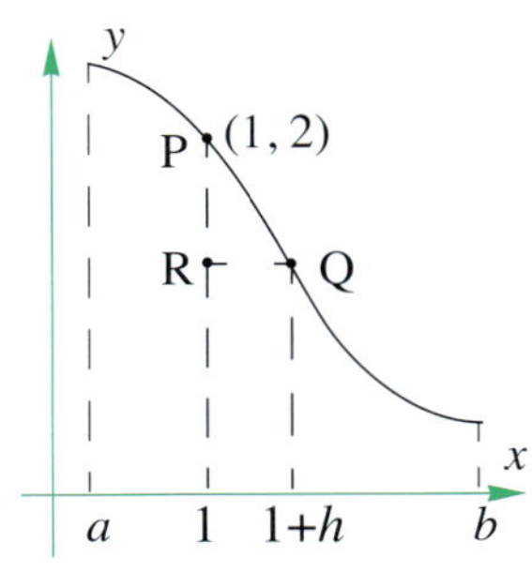

[1 + 2 + 3 + 2 + 2 + 2 marks]
Total 38 marks

23.8 PROJECT SUGGESTIONS

Although we have not looked at optimisation problems as such, you now have the skills that will enable you to tackle problems such as the one presented in the introduction of this chapter.

Combining the modelling skills you have developed throughout the course, in particular, those from Chapters 3 and 9, you will now be able to attempt a wider range of problems, many of which would make a excellent project.

Many things we buy are stored and sold in steel cans. Visit a factory that uses cylindrical cans to store their product.

What is the 'best' height and radius of such cans?

This seemingly straight forward task can be opened up into a very interesting project.

There is no need to restrict the task to one involving cylindrical cans, an investigation of any storage container would make an interesting project.

REVISION SET E - PAPER 1 & PAPER 2 STYLE QUESTIONS

1. Let P and Q be the points on the curve $y = \frac{1}{x^2}$ at which $x = 1$ and $x = 1 + h$ respectively.

(a) Express the gradient of the line [PQ] in terms of h.

(b) Hence find the gradient of the tangent to the curve at $x = 1$.

2. (a) Sketch the graph of $y = 4(x^2 - x^4)$, locating the stationary points and giving their coordinates.

(b) Sketch on the same set of axes the graph of $y = \frac{1}{4(x^2 - x^4)}$

Label both graphs clearly.

3. Identify all intercepts with the axes and all stationary points, state their coordinates and the nature of the stationary point for the curve with equation $y = \frac{1}{9}(4x^3 - x^4)$.

4. Let $f(x) = 1 - \frac{3}{x}, x \neq 0$ and $g(x) = x^2 - 3x$

(a) Find the coordinates of the points of intersection of the graphs of f and g.

(b) Sketch on the same set of axes the graphs of f and g.

(c) Find the equation of the tangents to each curve at the point(s) of intersection where $x > 0$. having

5. The graph of the function $y = f(x)$ is shown below.

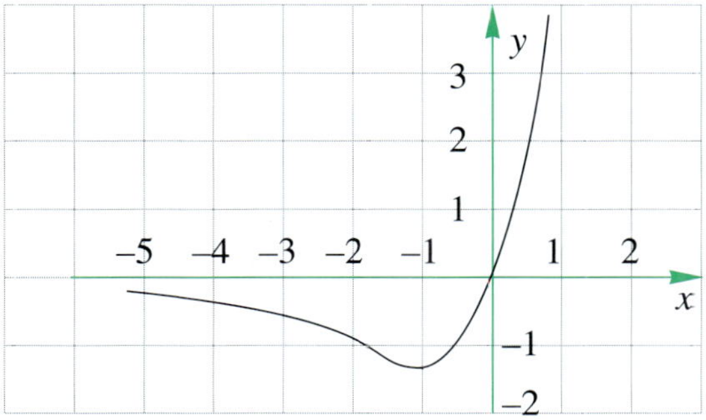

On different sets of axes, sketch the graphs of

i. $y = f'(x)$.

ii. $f = f''(x)$.

6. (a) Find the gradient of the graph with equation $y = x^2 + 3$ at the point where $x = 3$.

(b) Find the equation of the tangent to this curve at the point where $x = 3$.

7. (a) Find i. $f'(x)$ ii. $f''(x)$ if $f(x) = x^3 + 5x^2 + 7x$

(b) For what values of x is the function $f(x) = x^3 + 5x^2 + 7x$ increasing?

8. Let P and Q be the points on the curve $y = x^3$ at which $x = 2$ and $x = 2 + h$, respectively.

(a) Express the gradient of the line [PQ] in terms of h.

(b) Hence find the gradient of the tangent to the curve $y = x^3$ at $x = 2$.

9. The volume, $V(t)$, of water in a reservoir at time t is given by $V(t) = 3 + 2\sin\frac{1}{4}t$.

(a) Find the volume of the water in the reservoir at time $t = 10$.

(b) i. Find $V'(t)$.

ii. Find the rate of change of the volume of water in the reservoir at time $t = 10$. (Give your answers correct to two decimal places.)

10. The graph of $y = ax^3 + bx^2 + cx$ passes through the point (–1, 16) and has a stationary point at (1, –4). Find the values of a, b and c.

11. When a bus travels along a straight road from one stop to the next, its velocity-time function for the trip is described by $v(t) = \frac{1}{10}t(5 - t)$.

(a) State the domain of this function.

(b) Sketch the graph of the function.

(c) Find the maximum value of v.

(d) Find a rule for the acceleration of the bus between the stops.

(e) Sketch the acceleration-time graph for this bus.

12. $f(t) = \frac{(1+t)^2}{t}$ Find (a) the derivavtive of $f(t)$

(b) the equation of the tangent at its t-intercept.

13. Find $\frac{dy}{dx}$ where $y = 3$, for the curve with equation $y = \frac{2x+5}{x}$.

14. Consider the function $f(x) = 2x^4 + 2x^3$ where x is any real number.

(a) Find the coordinates of its stationary points and determine their nature.

(b) Find the x–intercepts.

(c) Sketch the graph of $f(x) = 2x^4 + 2x^3$.

(d) For what values of x is $f(x)$ increasing?

15. Two straight roads ACB and DCE cross at right angles at C. At 9:00 a.m a boy starts from a point on AC, 20 km from C, walking at a constant speed of 3 kmh^{-1} towards C. Also at 9:00 a.m a girl starts to walk towards C at 4 kmh^{-1} from a point on DC which is 10 km from C.

(a) Draw the information on a set of axes.

(b) If the boys position after t hours is $(20 - 3t, 0)$, the girl's position at the same time is given by which one of the following options

A. (10 – 4t, 0) **B.** (10 + 4t, 0) **C.** (0, 10 – 4t) **D.** (0, 10 + 4t)

(c) Show that the length, d km, of the straight line joining the positions of the boy and the girl is given by $d^2 = 25t^2 - kt + 500$

(d) State the value of k.

(e) Calculate when the boy and girl are nearest to each other.

16. Find the maximum value of N, given that $N = 1 + x - 3x^3, x > 0$

17. Consider the function $g(x) = 4x^3 - 6x^2 - 24x$.

(a) Find all values of x for which $g(x) = 0$.
(b) Determine the location and the nature of the stationary points.
(c) Sketch the graph of $g(x)$.
(d) Find the equation of the tangent at the point $(1, -26)$.
(e) Find the area of the triangular region enclosed by the tangent and both axes.

18. For a particular cubic function:

$f'(x) < 0$ for $x < -1$ and $x > 1$
$f'(-1) = 0 = f'(1)$
$f'(x) > 0$ for $-1 < x < 1$

Identify the nature of the stationary points of $y = f(x)$

19. Consider the function $f : \mathbb{R} \mapsto \mathbb{R}$, where $f(x) = \frac{1}{3}x^3 - x^2 + x$.

(a) Sketch the graph of $y = f'(x)$.
(b) i. Find the gradient of the function at the point $x = 0$.
ii. Find the equation of the tangent to the curve at the point where $x = 0$.
(c) Find the coordinates of the point A where the tangent to the curve at $x = 0$ meets the function for a second time.
(d) Find the area of the triangular region enclosed by the y-axis and the tangents to the curve at $x = 0$ and at A.

20. An object is thrown vertically upwards so that its height above the ground, h m, at time t seconds, $0 \le t \le 10$ is given by $h = 100t - 10t^2$. Calculate

(a) the average rate at which the height is increasing over the first five seconds.
(b) the initial rate at which the height is increasing, at the instant the object is thrown.

21. The graph below represents the derivative function $y = f'(x)$. Sketch the graph of the function $y = f(x)$ given that $f(0) = 0$ and $f(-4) = -4$.

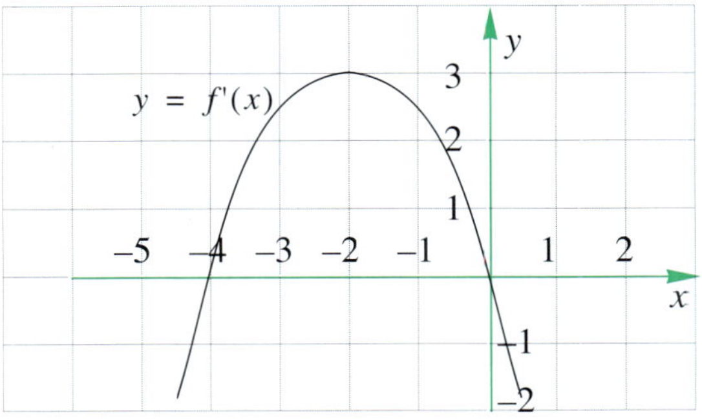

22. A tangent to the curve $y = -2x^2 + 12x - 11$ has a gradient of –4. Find the equation of this tangent.

23. Find the coordinates of the points on the curve $y = 4x - \frac{1}{x}$ where the gradient is 8.

24. Consider the functions $g(x) = x^2 + 4x + 1$ and $f(x) = 5x + 3$.

(a) State, in surd form, the x-values of the points where the parabola crosses the x-axis.
(b) Show, algebraically, that the graphs of the two functions meet at $x = -1$ and $x = 2$.
(c) Find the gradients of the tangents to the parabola at the points of intersection.
(d) Give the equation of the tangent to the parabola where $x = -1$.

25. (a) The graph shows sections of two curves, $y = f(x)$ and $y = g(x)$ for $x \geq 0$.

The coordinates of the points shown are as follows:
A = (0.3, 10.12), B = (0.4, 11.08)
C = (0.3, 5.307), D = (0.4, 6.784)

Compare the average rate of change of $y = f(x)$ with the average rate of change of $y = g(x)$, between $x = 0.3$ and $x = 0.4$.

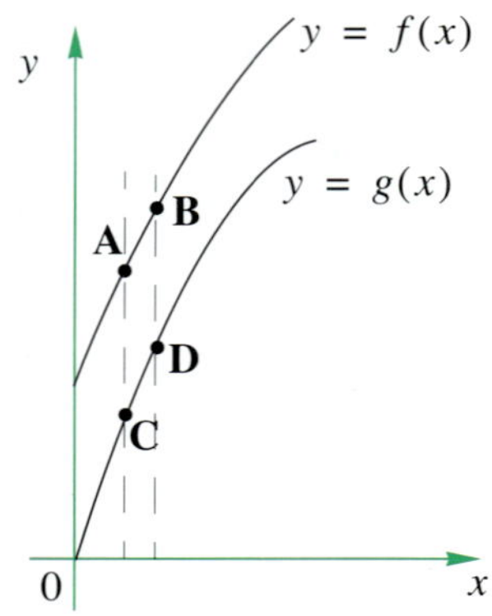

(b) In fact, the graph in (a) is a section of the graph shown below, where $f(x) = 7 + 11x - 2x^2, x \geq 0$ and $g(x) = x^3 - 8x^2 + 20x, x \geq 0$.

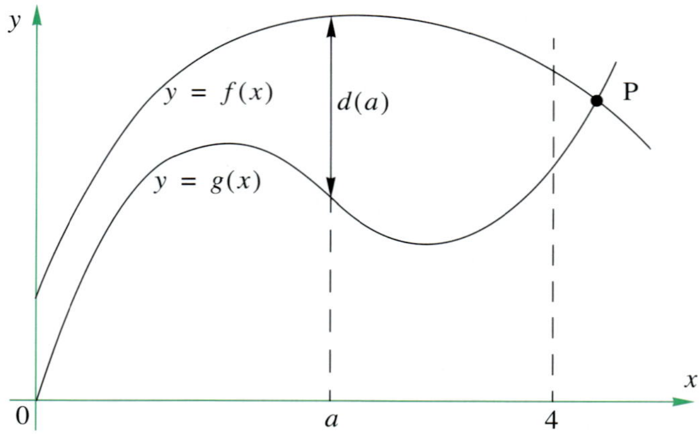

The vertical distance between the curves is given by the function $d(x)$. One such value of $d(x)$ is illustrated when $x = a$.

i. Find the rule for the function $d(x)$ when $0 \leq x \leq 4$.
ii. Find the maximum distance between the curves for $0 \leq x \leq 4$ and the value(s) of x at which it occurs.
iii. Find the minimum distance between the curves for $0 \leq x \leq 4$ and the value(s) of x at which it occurs.

(c) P, the point of intersection of the curves in (b), cannot be easily found. However, an approximation for the x-coordinate of P can be found by solving (approximately) the equation $d(x) = 0$. A section of the graph of $y = d(x)$ (not to scale) is shown. QR is the tangent at the point where $x = 4$ and $R = (\alpha, 0)$.

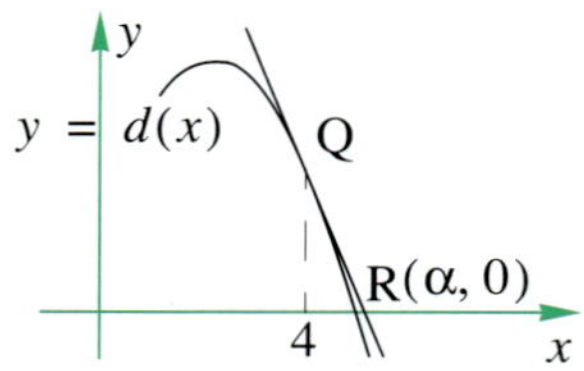

i. Find the equation of QR and use it to find α.
ii. What does the value of α represent?

Note: Answers to the Revision Sets – Paper I and Paper II Style Questions will be available in the Solutions Manual – which includes fully worked solutions to every question in the textbook. Check the ibid press website for updates and availability.

www.ibid.com.au

or contact fabio@ibid.com.au

ANSWERS TO CHAPTER 4

EXERCISES 4.2

1. (a) 4 (b) 5 (c) 2 (d) 4 (e) 3 (f) 3 (g) 5 (h) 3 (i) 6 (j) 4 (k) 6 (l) 5 **2.** (a) 2.5 (b) 24700 (c) 0.35 (d) 45630 (e) 0.45 (f) 4 (g) 57000 (h) 0.0454 (i) 0.0045 (j) 346000 (k) 0.045 (l) 90000 **3.** 8.2 cm **4.** 11cm **5.** 44m **6.** 1040mm **7.** 30 terms **8.** 3 S.F. **9.** 19

EXERCISES 4.3

1. (a) 2.2×10^1 (b) 1.2×10^{-12} (c) 4.8×10^9 **2.** $r = 4.72 \times 10^9 m$ $C = 2.97 \times 10^{10} m$, Vol = $4.40 \times 10^{29} m^3$ **3.** $7.48 \times 10^{-6} m^3$ **4.** 280km **5.** 4.494×10^{15} kgm^{2}s^{-2} **6.** 3.4×10^9 **7.** 74 generations, 1.9×10^{22} relations. **8.** 1.25×10^8 (to 3 sig fig) or 1.3×10^8 (to 2 sig. fig.) **9.** 3.1×10^{-24} gms **10.** 4.3×10^6

EXERCISES 4.4

1. 0.0033333 (etc.), 0.01, 1% **2.** 0.0144, 0.0053, 0.53% **3.** The absolute errors are all 0.2kg. The relative and percentage errors are: i. 0.027 (2.7%) ii. 0.0102 (1.02%) iii. 0.00893 (0.893%) iv. 0.00352 (0.352%) v. 0.00202 (0.202%). **4.** Absolute errors are all 0.5 mm i. 0.1 (10%) ii. 0.05 (5%) iii. 0.00417 (0.417%) iv. 0.002 (0.2%) **5.** i. 0.018% ii. 0.0014% iii. 0.273% iv. 0.0030% **6.** i. 0.45% ii. 0.073% iii. 0.020% iv. 0.0022% **7.** 6.6×10^{-4} % which is plainly far more accuracy than can be justified in the times quoted for this flight.

EXERCISES 4.5

1. (a) 0.209 [0.204, 0.215] (b) 3384 [1726, 64752] (c) 194.9 [206.3, 184.85] **2.** 516577 mm^2 [514157, 519000] 0.43% **3.** 164553 cm^2 [150456, 179503] 9.1% **4.** 17961 [17936, 17986] **5.** 28.24m [27.39, 29.11] 3.1% accuracy. **6.** 105.6m [104.9, 106.2] 0.66% **7.** $a \approx b$ **8.** 0.25kg **9.** 15.42cm [14.90, 15.97] 3.6% **10.** 12096 [10646, 13672] 13% **11.** 2208 [2076, 2344] 6.2% **12.** 0.00069%

EXERCISES 4.6 – MISCELLANEOUS EXERCISES

1. (a) 0.380568 (b) 0.381 (c) 0.38 (d) 3.80568×10^{-1} **2.** (a) 24548.616 (b) 24500 (c) 24548.62 (d) 24549 (e) 2.4548616×10^4 **3.** (a) i. 75.0 ii. 74.97 iii. 75 (b) 7.5×10^1 **4.** (a) 790 (b) 790.02 (c) 790 **5.** $\frac{1}{7}$ = 0.14286 to 5 SF, absolute error = 2.86×10^{-6} % error = 0.0020 **6.** ≈ 0.20467227 or 0.205 to 3SF **7.** (a) 3 (b) 3 (c) 2 **8.** (a) 4.51×10^{21} (b) 6.01×10^{-17} (c) 2.12×10^{-1} **9.** (a) 34000 (b) 0.000705 **10.** (a) 19.9 ± 0.8 (b) 45.44 or 41.54 to 49.5 (c) ≈ 1.429, 0.769 to 10 **11.** Series gives 2.96339 error (–)0.1782 (5.7%)

EXERCISES 4.7 – GRADED REVISION EXERCISES

LEVEL 1

1. i., ii., iii., v., **2.** i. 3.4×10^1 ii. 5.67×10^4 iii. 6.00056×10^5 iv. 4×10^{-1} v. 4.38×10^{-3} vi. 4.0456×10^0 vii. 2.0×10^{-3} viii. 7.5×10^{-1} ix. 5.4×10^{-6} x. 1×10^4 xi. 6×10^{-2} xii. 7.5×10^{-2}

LEVEL 2

1. i. 7.92×10^3 ii. 1.96×10^2 iii. 3.15×10^{-2} iv. 4.83×10^1 v. 2.45×10^{-6} **2.** i. 3.3%

ii. 11% iii. 6.9% iv. 11% v. 11% **3.** i. $\frac{13}{10}$ ii. $\frac{1}{4}$ iii. $\frac{1}{3}$ iv. $\frac{5}{9}$ v. $\frac{5}{6}$ vi. $\frac{83}{99}$

LEVEL 3

1. i. $v = \frac{uf}{u-f}$ ii. Smallest value =0.46735765, Largest value = 0.54967293 iii. 8.12 %

2. i. 1.8×10^{16} kg m^2 sec^{-2}. ii. 0.0007515 J **3.** i. 76.4 cm^3 ii. 76.43 cm^3 iii. 76 cm^3 .

LEVEL 4

1. $k = m(2+\sqrt{2}), m \in \mathbb{Q}$. **2.** 6.6 %. **3.** The percentage error is $\frac{150028}{14797665}(100) = 1.0138626$ or about 1%.

EXERCISES 4.8 – TOPIC TEST

1. i. 1.45×10^2 ii. 1.08×10^{11} iii. 1.59×10^{-6} . **2.** i. 4010 ii. 0.00506 iii. 109000**3.** $\frac{1}{3}\% \approx 0.33\%$.

4. (a) 0.319024 (b) 0.319 (c) 0.32 (d) 3.19024×10^{-1} **5.** $2.\dot{2}\dot{3} = \frac{221}{99}$. Therefore, $2.\dot{2}\dot{3} \in \mathbb{Q}$

6. i. $9.41 \le A \le 10.05$ ii. % error = 3.275% ≈ 3.3%

ANSWERS TO CHAPTER 5

EXERCISES 5.1

1. (a) (b) (c) (d) (e) (f)

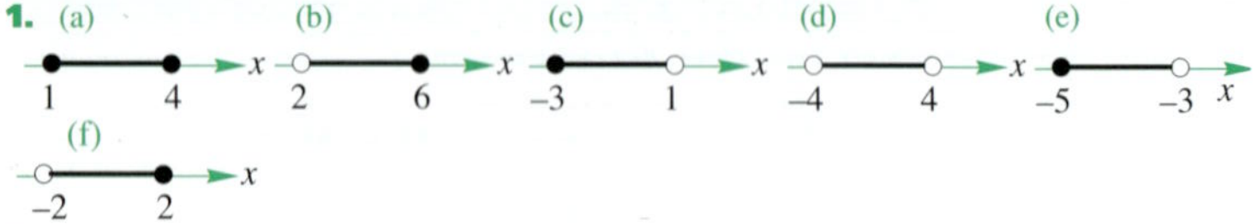

2. (a) (b) (c) (d) (e) (f)

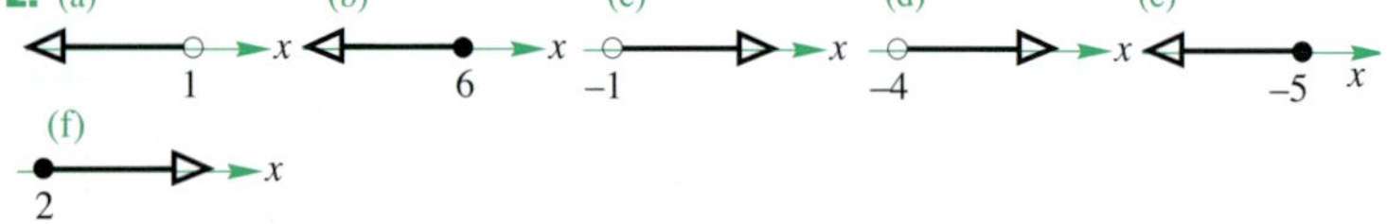

3. (a)]2, 8] (b) [4, ∞[(c) [0, 6] (d)]–∞, 10] (e)]–∞, 12[(f) [3, 5[(g)]–2, ∞[(h)]–3, 1] (i)]–∞, –7[**4.** (a) $\{x \mid 4 \le x \le 6\}$ (b) $\{x \mid -2 < x < 5\}$ (c) $\{x \mid 8 < x \le 12\}$ (d) $\{x \mid 12 \le x < 15\}$ (e) $\{x \mid 5 \le x < \infty\}$ (f) $\{x \mid -\infty < x < 8\}$ (g) $\{x \mid 2 < x < \infty\}$ (h) $\{x \mid -\infty < x \le 3\}$ (i) $\{x \mid -2 < x < \infty\}$

5. (a) (b) (c) (d) (e) (f) (g) (h) (i)

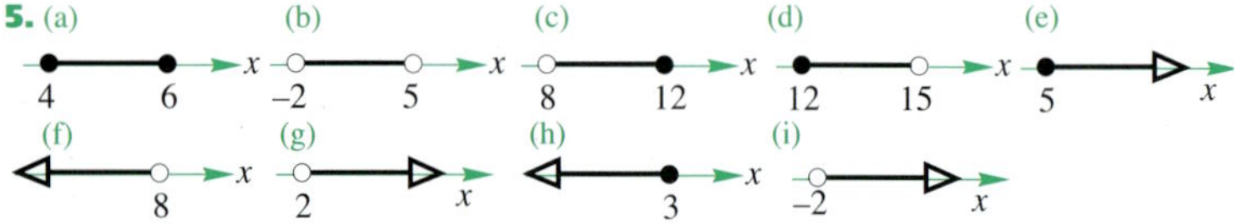

EXERCISES 5.2

1. (a) 5 (b) 5 (c) –3 (d) 21 (e) $\frac{28}{3}$ (f) $-\frac{19}{3}$ (g) 11 (h) 11 (i) 4 **3.** (a) 9 (b) 20 (c) 10 (d) $\frac{4}{3}$ (e) –9 (f) 15 **5.** (a) –1 (b) 2 (c) –5 (d) –1 (e) 1 (f) $\frac{11}{13}$ **7.** (a) $-\frac{13}{7}$ (b) $-\frac{7}{4}$ (c) 22 (d) $\frac{1}{31}$ (e) –19 (f) $\frac{16}{7}$ (g) $-\frac{9}{2}$ (h) 24.5 **9.** (a) $\frac{5}{6}$ (b) $-\frac{4}{3}$ (c) $-\frac{2}{3}$ (d) $\frac{11}{4}$ (e) $-\frac{1}{2}$ (f) $\frac{1}{4}$ (g) $\frac{7}{11}$ (h) $\frac{6}{5}$ (i) $\frac{4}{3}$ (j) $\frac{1}{2}$ (k) $\frac{5}{2}$ (l) $\frac{7}{6}$ **11.** (a) –0.5 (b) –1 (c) $\frac{13}{4}$ (d) $\frac{73}{11}$ (e) $\frac{9}{8}$ (f) –13 (g) $\frac{7}{3}$ (h) $-\frac{23}{2}$ (i) $\frac{1849}{150}$ (j) $-\frac{4}{3}$ (k) $\frac{3}{2}$ (l) $\frac{8}{5}$ **13.** (a) $\frac{a-2}{a}$ (b) $\frac{1+b}{b}$ (c) $-\frac{2}{a}$ (d) $\frac{5}{a}$ (e) $-\frac{2}{a}$ (f) $2a$ (g) $\frac{5}{4}(3-2a)$ (h) $1-a$ (j) $\frac{2}{3}\left(\frac{a-1}{a+1}\right)$ **14.** (a) $\frac{5}{a-b}$ (b) $\frac{a+b}{a-b}$ (c) 1 (d) –1 (e) $1-a$ (f) $2a-a^2$ (g) $\frac{ab}{a+b}$ (h) $\frac{a^2-b^2}{a-b} = a+b$ (i) 1

EXERCISES 5.3

1. (a) $12 - y = 5$ (b) 7 **2.** (a) $4x + 2 = 30$ (b) 7 **3.** $2x - 5 = 19, x = 12$ **4.** 12, 13 **5.** 31, 32, 33 **6.** $x + 3x + 2x = 84$, 14, 28, 42 **7.** 20 **8.** 84 **9.** 200 **10.** $3388.9 **11.** $\frac{4}{3}$ **12.** $13,739 **13.** (a) 7 ml (b) 23.33% **14.** Tim–8yrs, Claire–16yrs, Thomas–10yrs **15.** 3.75 km **16.** 22 **17.** 5% **18.** (a) 144 km (b) 40 kmh^{-1} **19.** Ab –$12,500, Ba–$5,500 **20.** 83.33% **21.** (a) $\frac{(2m+1)}{m(m+1)}d$ hr (b) $\frac{2m(m+1)}{2m+1}$ kmh^{-1} **22.** 32% **23.** (a) i. $10x + 4$ ii. $400 + x$ (b) 364 **24.** 60 kg

EXERCISES 5.4

1. (a) $x > 4$ (b) $x \le 9$ (c) $x < 4$ (d) $x < 1$ (e) $x \le 6$ (f) $x \le -1$ (g) $x \ge 5$ (h) $x < 5$ (i) $x \le -3$ (j) $x > 21$ (k) $x \ge \frac{28}{3}$ (l) $x > -\frac{19}{3}$ (m) $x \le 11$ (n) $x < 11$ (o) $x < -\frac{3}{2}$

2. (a) (b) (c) (d) (e) (f) (g) (h) (i) (j) (k) (l) (m) (n) (o)

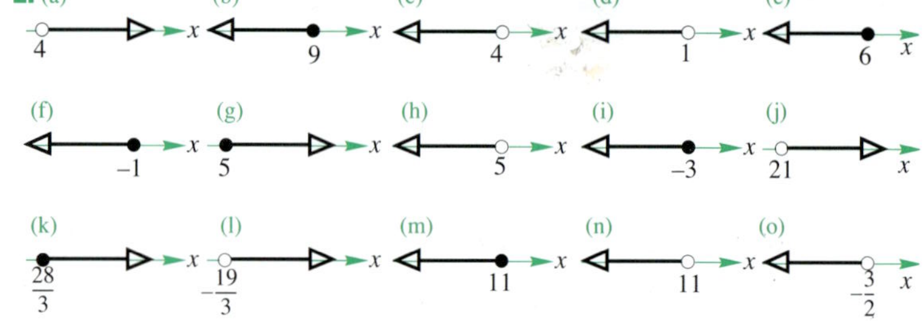

3. (a) $x < 9$ (b) $x \ge 20$ (c) $x > 10$ (d) $x \le \frac{4}{3}$ (e) $x > -9$ (f) $x \ge 15$

4. (a) (b) (c) (d) (e) (f)

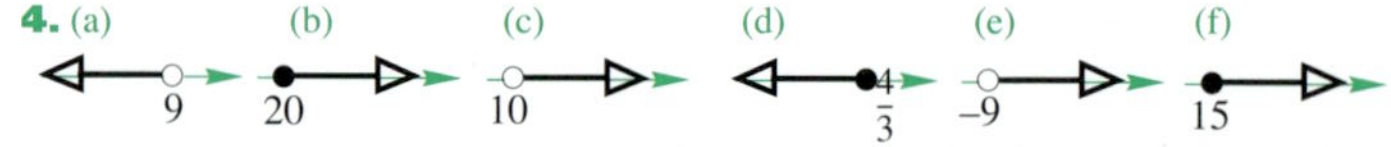

5. (a) $x > -1$ (b) $y < 2$ (c) $x \le -5$ (d) $x \ge -1$ (e) $x < 1$ (f) $x < \frac{11}{13}$

6. (a) (b) (c) (d) (e) (f)

7. (a) $x < -\frac{13}{7}$ (b) $x \le -\frac{7}{4}$ (c) $x \ge 22$ (d) $y \ge \frac{1}{31}$ (e) $x \le -19$ (f) $w > \frac{16}{7}$

8. (a) (b) (c) (d) (e) (f)

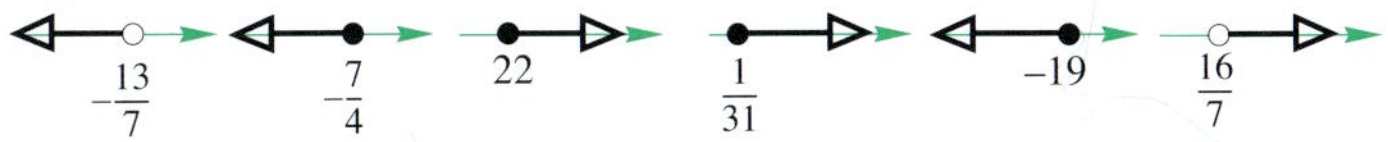

9. (a) $x < -27$ i.e., $]-\infty, -27[$ (b) $x \ge -12$ i.e., $[-12, \infty[$ (c) $x < \frac{180}{29}$ i.e., $\left]-\infty, \frac{180}{29}\right[$

(d) $x \ge -\frac{1}{9}$ i.e., $\left[-\frac{1}{9}, \infty\right[$ **10.** –5, –4, –3, –2, –1

EXERCISES 5.5

1. $x \ge 15$ **2.** $3 \le x < 20$ **3.** $38 < x < 50$ **4.** 40.5 kg **5.** (a) $x > 5$ (b) $x > 4$ **6.** >\$2000
7. $>25\,\text{kmh}^{-1}$ **8.** > 93 (or ≥ 94)

EXERCISES 5.6 – MISCELLANEOUS EXERCISES

1. $x = \frac{5}{2}$ **2.** (a) $x = -\frac{15}{4}$ (b) $x < \frac{5}{4}$ **3.** (a) $x \le -\frac{9}{2}$ 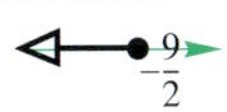 (b) $x < 1$ 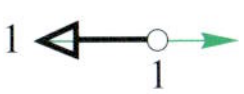

4. (a) $\frac{27}{2}$ (b) 1 **5.** (a) $x > -5$ **6.** (a) $-\frac{5}{2}$ (b) $\frac{19}{5}$ **7.** 23 **8.** 26 five cent coins.

9. (a) $a = 7$ (b) $y = -2.4$ (c) $x = \frac{5}{3}$ **10.** (a) $x = 6$ (b) $x < -\frac{13}{5}$.

11. (a) $-2 \le x$ (b) $-54 > y$

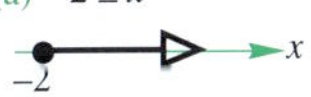

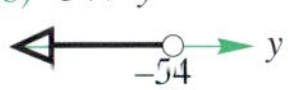

12. (a) $x = b + 1$ (b) $x = ab$ **13.** 3 litres

EXERCISES 5.7 – GRADED REVISION EXERCISES

LEVEL 1

1. (a) i. $\{x | 4 \le x \le 9\}$ ii. $[4, 9]$ (b) i. $\{x | x > 6\}$ ii. $]6, \infty[$ (c) i. $\{x | x < -3\}$ ii. $]-\infty, -3[$

2. i.  ii. 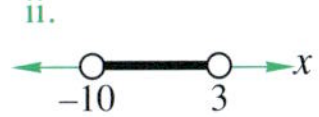 iii. 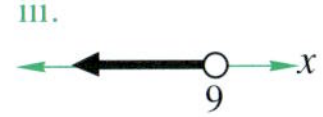

LEVEL 2

1. i. $x = 4$ ii. $s = 3$ iii. $x = 1$ iv. $t = 4$ v. $x = 4.5$ vi. $y = -\frac{10}{3}$

2. i. $x > -1$ ii. $x \ge -4$ iii. $y < -6$ iv. $x > 2$ v. $x < 7$ vi. $x > 0$.

LEVEL 3

1. i. $x = 4$ ii. $x = 24.5$ iii. $s > 0.8$ iv. $y > \frac{17}{7}$ v. $x = \frac{3}{2}$ vi. $a \ge -\frac{5}{11}$ **2.** Joseph is 22.

LEVEL 4

1. i. $x = b - 1$ ii. If $ab > 0 \Rightarrow x > \frac{ab(a+b)}{a^2 + b^2}$, if $ab < 0 \Rightarrow x < \frac{ab(a+b)}{a^2+b^2}$.

2. 400 km. We have assumed that they travel in the same direction.

EXERCISES 5.8 – TOPIC TEST

1.  **2.** i. 2 ii. 5 iii. 4 **3.** (a) $x < 5$ (b) 
4. i. –13 ii. $x \le \frac{1}{23}$

5. i. $10 - x$ ii. $x = \frac{10}{3}$ **6.** ii. $y = \frac{100}{31}$ iii. $x = 6200$ **7.** $w = 83\frac{1}{3}$

ANSWERS TO CHAPTER 6

EXERCISES 6.1

1. i.

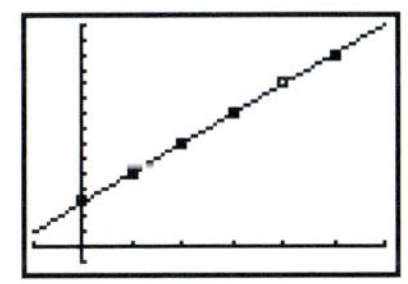

ii.

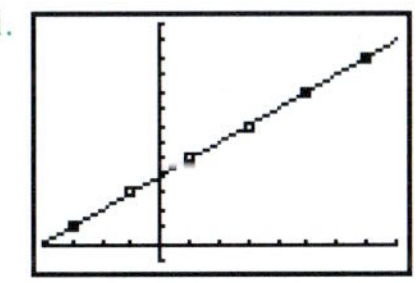

iii.

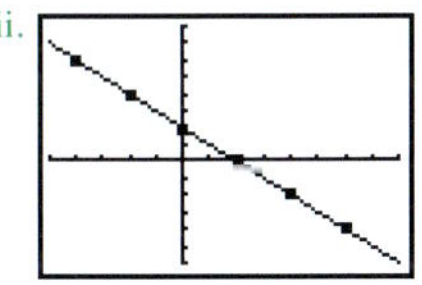

iv.

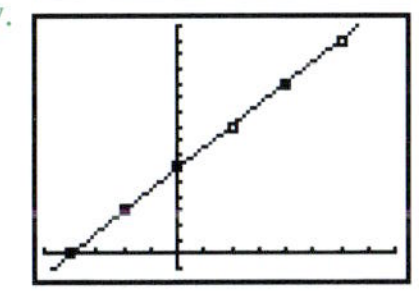

2. (a) [–5, 5] by [–5, 5]

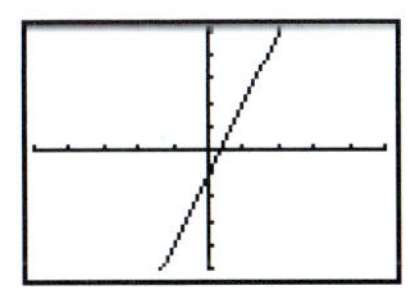

(b) [–5, 5] by [–5, 5]

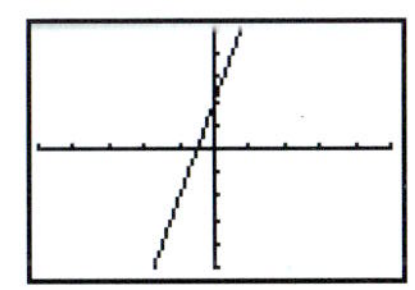

(c) [–5, 5] by [–5, 5]

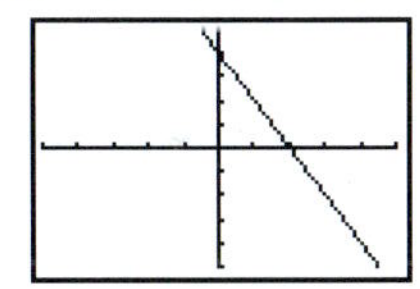

(d) [–5, 8] by [–5, 5]

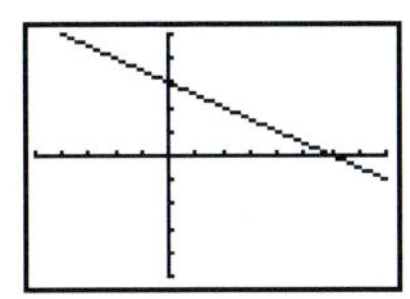

(e) [–5, 8] by [–5, 5]

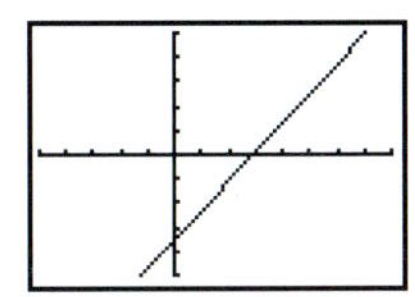

(f) [–5, 8] by [–5, 5]

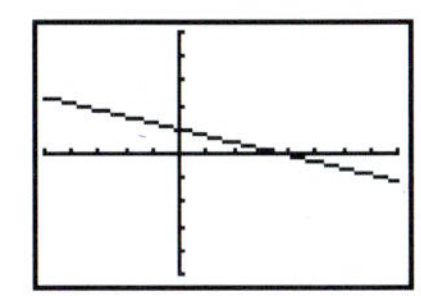

3. (a) $m = -1, c = 1$ (b) $m = 1, c = -2$ (c) $m = -\frac{2}{3}, c = 4$ (d) $m = -3, c = 6$ (e) $m = \frac{1}{3}, c = \frac{1}{3}$

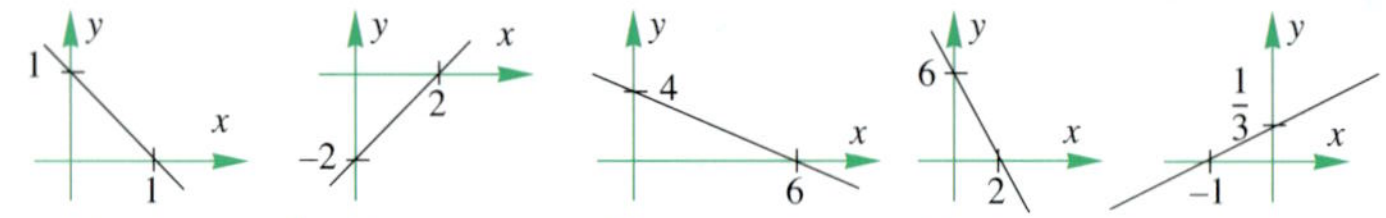

(f) $m = 4, c = 2$ **4.** (a) (b) (c)

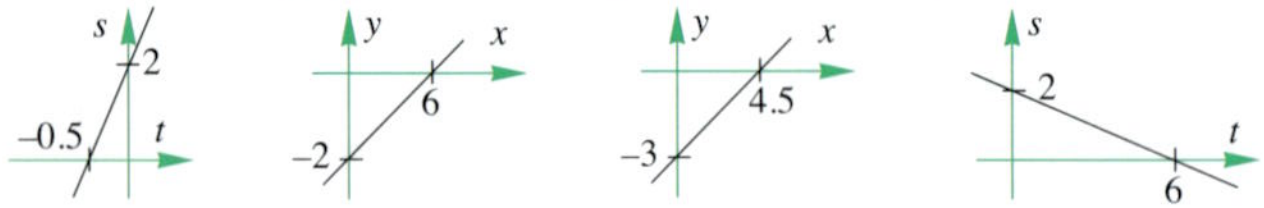

5. (a) (b) (c) **6.** (a) (b) (c)

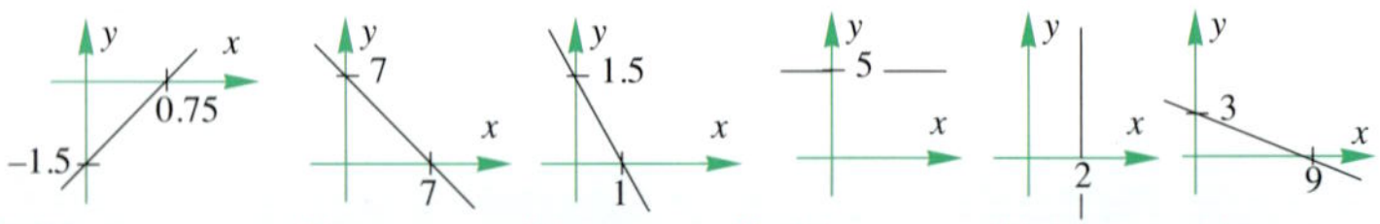

7. (a) (b) (c) **8.** (a) (b) (c)

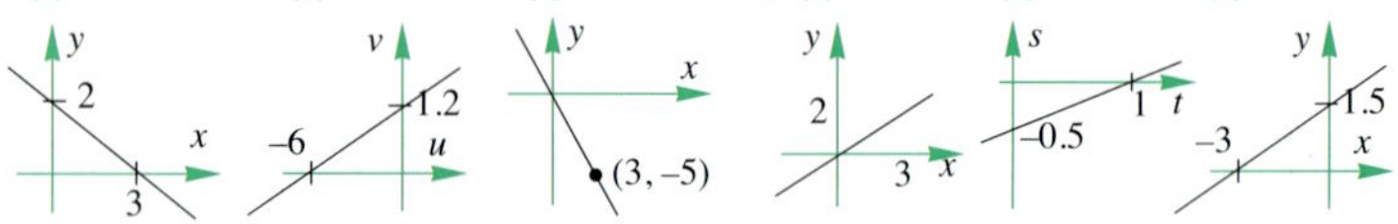

9. Check your graphs against the graphs with equations (a) $y = 2x + 5$ (b) $y = -x + 3$ (c) $y = -2x + 8$ (d) $y = -0.75x + 3$ (e) $y = 2x - 2$

10. (a)

t (years)	0	1	2	3	4	5	6
V (\$)	900	750	600	450	300	150	0

(b)

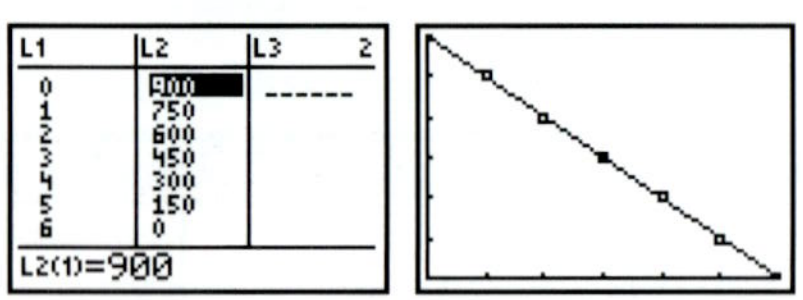

(c) Value of computer ≥ \$0.

11. (a)

t (hours)	0	1	2	3	4	5	6	7	8	9	10
V (Litres)	250	225	200	175	150	125	100	75	50	25	0

(b), (c)

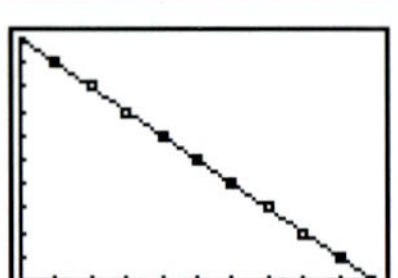

(d) $V \geq 0, 0 \leq t \leq 10$.

12. (a)

n	0	1	2	3	4	5	10	20	30	40	...
I	750	825	900	975	1050	1125	1500	2250	3000	3750	...

(b)

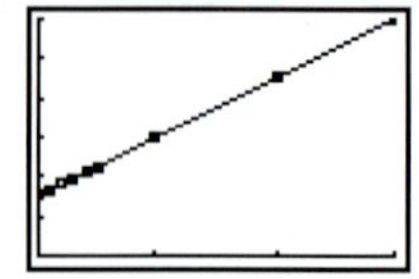

(c) Same as (b) – note: this is a discrete linear relation

13. (a) An appropriate range for the father's height would be:

H	140	145	150	155	160	165	170	175	180	185	190
h	157	159.5	162	164.5	167	169.5	172	174.5	177	179.5	182

(b)

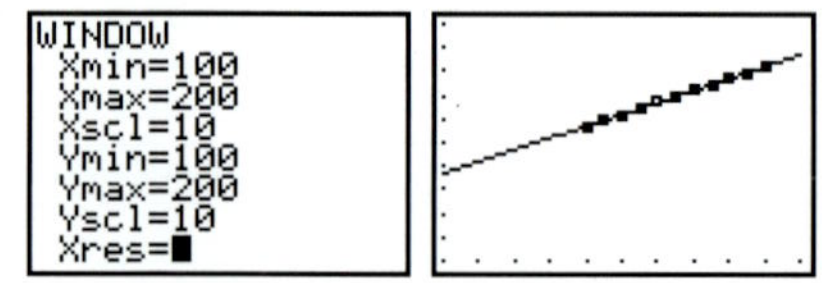

(c) 180.5 cm

EXERCISES 6.2

1. (a) 1.5 (b) 1 (c) 2 (d) $\frac{8}{3}$ (e) $-\frac{9}{7}$ (f) 0.6 (g) –0.25 (h) 3 (i) a (j) –2 **2.** (a) 1.5 (b) –0.75 (c) 2.5 (d) 1.25 (e) 0 (f) undefined **3.** (a) $y = 2x + 5$ (b) $y = -x + 3$ (c) $y = -2x + 8$ (d) $y = -0.5x + 3$ (e) $3y = 2x$ **4.** (a) $y = -\frac{3}{4}x + 3$ (b) $y = 2x - 2$ (c) $y = 2x - 5$ (d) $y = -x - 4$ (e) $y = -3x + 10$ **5.** (a) $2y = 3x$ (b) $3x + 4y = 12$ (c) $2y - 5x + 10 = 0$ (d) $4y - 5x = 20$ (e) $y = 3$ (f) $x = 2$ **6.** Straight line should be drawn in such a way that it passes through the two intercepts: (a) *x-int*: (4, 0), *y-int*: (0, 2) (b) *x-int*: (–6, 0), *y-int*: (0, 2) (c) *x-int*: (–2.25, 0), *y-int*: (0, 9) (d) *x-int*: (–2, 0), *y-int*: (0, 2) (e) *x-int*: (–2, 0), *y-int*: (0, 3) (f) *x-int*: (–1, 0), *y-int*: (0, –2.5) **7.** Straight line should be drawn in such a way that it passes through the two intercepts: (a) *x-int*: (2, 0), *y-int*: $\left(0, -\frac{4}{3}\right)$ (b) *x-int*: (–10, 0), *y-int*: (0, –9) (c) *x-int*: (7, 0), *y-int*: (0, 14)

8. (a) $C = 4 + 0.02n, n = 0, 1, 2, 3...$ (c) \$44

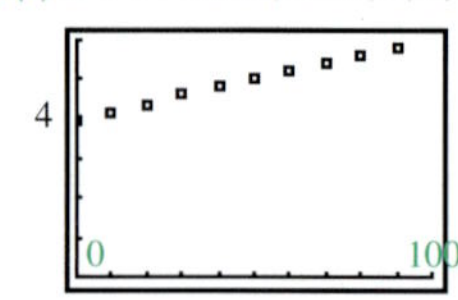

9. (a) $\mathbf{B} = 44 + 0.22n, n = 0, 1, 2, ...$ (c) \$80.96

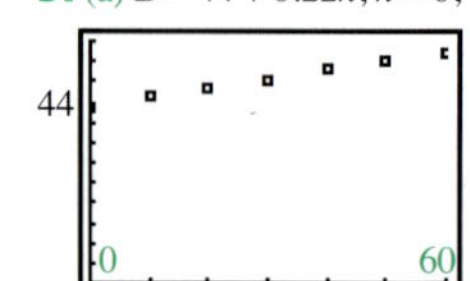

(d) 192

10. (a) i. $P = 0.35k$ ii. $G = 0.27k$ (c) 12500 km

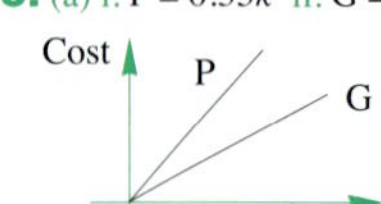

EXERCISES 6.3.1

1. (a) $(3, 3)$ (b) $(2, -1)$ (c) $(-1.5, -1.5)$ (d) $(-1, 1)$ **2.** (a) $(2, 2)$ (b) $\left(\frac{1}{3}, \frac{5}{3}\right)$ (c) $(-10, 8)$

3. (a) $\left(\frac{5}{2}, \frac{15}{2}\right)$ (b) $\left(\frac{9}{5}, \frac{51}{5}\right)$ (c) $(-3, 1)$ **4.** (a) $x = -3, y = -0.5$ (b) $u = -0.75, v = 4.25$ (c) $m = -1, n = 5$ (d) $a = 2, b = -2$ (e) no solution (f) no solution **5.** (a) $x = 1, y = 4$ (b) $x = 3, y = 2$ (c) $x = 2, y = -1$ (d) $x = 4, y = 5$ (e) $x = 4, y = -3$ (f) $x = -3, y = 2$

EXERCISES 6.3.2

1. (a) $x = 1, y = 2$ (b) $x = 3, y = 5$ (c) $x = -1, y = 2$ (d) $x = 0, y = 1$ (e) $x = -2, y = -3$ (f) $x = -5, y = 1$ **2.** (a) $x = \frac{13}{11}, y = \frac{17}{11}$ (b) $x = \frac{9}{14}, y = \frac{3}{14}$ (c) $x = 0, y = 0$ (d) $x = \frac{4}{17}, y = -\frac{22}{17}$ (e) $x = -\frac{16}{7}, y = \frac{78}{7}$ (f) $x = \frac{5}{42}, y = -\frac{3}{28}$ **3.** (a) $(3, -3)$ (b) $(2, -1)$ (c) $(3, 1)$ (d) $(3, -1)$ (e) $(-1, -4)$ (f) $\left(\frac{62}{15}, \frac{19}{15}\right)$ **4.** (a) $x = a, y = 0$ (b) $x = \frac{1}{2}(a+b), y = \frac{1}{2}(a-b)$ (c) $x = \frac{a}{2}, y = -\frac{a}{2}$

5. (a) Lines are parallel with different y-intercepts, so no solution. (b) Lines are parallel with different y-intercepts, so no solution. (c) Lines are parallel with same y-intercepts (i.e., they are coincident), so there are an infinite number of solutions. **6.** (a) –3 (b) –5 (c) –1.5

7. (a) $m = 2, a = 8$ (b) $m = 10, a = 24$ (c) $m = -6, a = 9$ **8.** (a) $a = 1, b = 1$ (b) $a = \frac{9}{4}, b = 3$

9. (a) $\{(a^{-1}, 0)\}$ (b) $\{(a, b)\}$ (c) $\{(b, 0)\}$ (d) $\{(1, 0)\}$

EXERCISES 6.3.3

1. 36, 21 **2.** CD: \$8.50 video: \$6.00 **3.** shirt: \$9.00 socks: \$2.00 **4.** length 15m, width 12m **5.** maths 72, physics 44 **6.** adult \$5.00, child \$2.00 **7.** brown \$10, yellow \$12 **8.** 21600 **9.** \$26 (the tapes cost \$18). **10.** Adults \$17, children \$9. **11.** (a) $x = 48, y = -70$ (b) $x = 94.67, y = -140$ (c) Graphically, the lines are almost parallel so small changes in the coefficients produce big changes in the solutions.

EXERCISES 6.4 – MISCELLANEOUS EXERCISES

1. (a) $(2, -1)$ (b) $(3, 4)$ (c) $\left(-4, -\frac{11}{2}\right)$ (d) $(2, 3)$ **2.** (a) $r = \frac{16}{5}, s = \frac{2}{5}$ (b) $(3, 1)$ (c) $(-3, 2)$ (d) $(2, 7)$ **3.** (a) $(3, -1)$ (b) $(16, 14)$ (c) $\left(\frac{18}{5}, \frac{3}{5}\right)$ (d) $u = k, v = 4 - 4k, k \in \mathbb{R}$ (*i.e., infinite soln*).

4. (a) $(-2, 1)$ (b) $p = 0.5, q = 4$ (c) $(0, 3)$ (d) $u = 0.5, v = 1.5$ (e) $m = \frac{3}{14}, n = \frac{11}{28}$

5. (a) $\left(\frac{a+b}{2a}, \frac{a-b}{2}\right)$ (b) $p = \frac{(a-1)b}{b-a}, q = \frac{a^2-b}{b-a}$ **6.** $\left(\frac{2ab}{a+b}, \frac{2ab}{a+b}\right)$ **7.** 36, 21 **8.** 12, 9

9. 35, 7 **10.** \$6.20 **11.** 7 two-cent coins, 32 one-cent coins **12.** $i = 5, j = 1$ **13.** $p = 1.36, q = 1.28$ **14.** \$5.00 **15.** $a = 7, b = 3$ **16.** Felicity is 36, Joanne is 43 **17.** 4.5 km h^{-1} and 3.75 km h^{-1} **18.** 72 **19.** A = 336 m^2 dim: 12×28, **20.** (a) 0.75 km h^{-1} (b) 6.75 km h^{-1}

EXERCISES 6.5 – GRADED REVISION EXERCISES

LEVEL 1

1. (a) $m = 2, c = 5$ (b) $m = -0.3, c = -0.4$ (c) $m = -0.4, c = 1.6$ **2.** (a) $(0, 2)$ (b) $(-1.5, -1.5)$ (c) $(-3, 4)$ **3.** (a) 2.2 (b) –2.4 (c) –4

LEVEL 2

1. (a) –4.5 (b) 2 **2.** $(0, 10)$ **3.** (a) (b) (c)

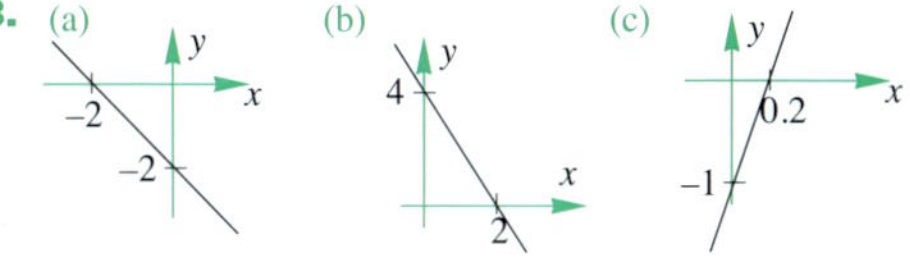

4. (a) $2y - x = 1$
(b) $3y + 4x = 6$
(c) $y = 2$

5. 150 **6.** $(a, 0)$

LEVEL 3

1. (a) $x = 3, y = 4$ (b) $x = 3, y = 5$ (c) $x = 13, y = 5$ **2.** (a) $2y + 9x = 41$ (b) $y + x = 1$ (c) $y - x = 2b$

3. i. (a)

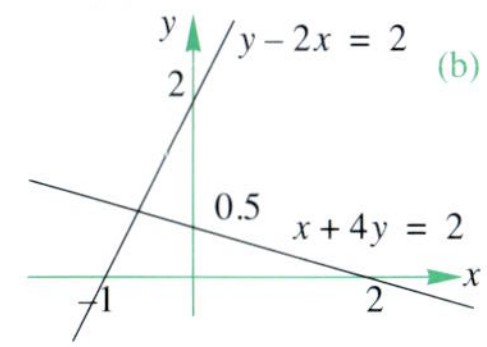

(b) $\left(-\frac{2}{3}, \frac{2}{3}\right)$

ii. (a)

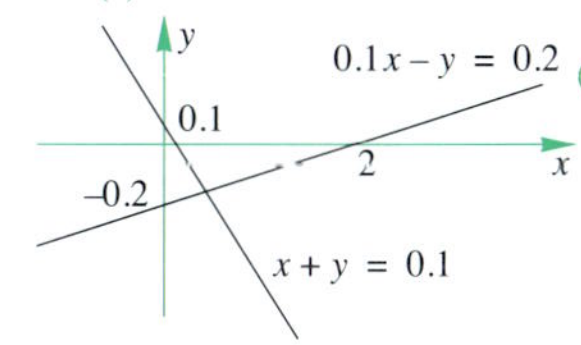

(b) $\left(\frac{3}{11}, -\frac{19}{110}\right)$

4. $\sqrt{41}$ **5.** –13.5 **6.** \$2

LEVEL 4

1. $a = \frac{16}{7}, b = \frac{20}{7}$ **2.** $x = a$, but no finite y. **3.** $x = -2, y = 1, z = 4$ **4.** (a) i. $\left(\frac{k}{2}, 0\right)$ ii. $\left(0, \frac{k}{3}\right)$ (b) need to sketch graph (c) $k = 6$

EXERCISES 6.6 – TOPIC TEST

1. $m = -3, (0, 7)$ **2.** $m = 2$ **3.** (a) $y = -\frac{2}{3}x + 2$ (b) $a = -\frac{3}{2}$ **4.** $(3, 3)$ **5.** $\left(\frac{5}{4}, \frac{7}{4}\right)$

6. adult \$40, teen \$8 **7.** (a)

n	0	1	2	3	4	5	10
Total income (\$$T$)	200	240	280	320	360	400	600

(b)

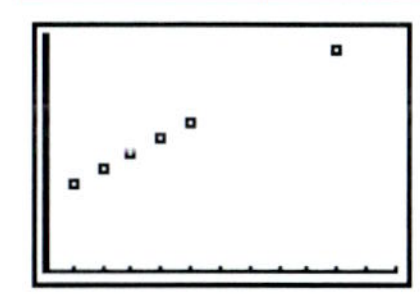

(c) i. 520 ii. 13
(d) i. positive integers including zero.
ii. $T = 200 + 40n, n = 0, 1, 2, \ldots$

ANSWERS TO CHAPTER 7

EXERCISES 7.1

1. (a) $2(ax + 2)$ (b) $y(9 - a)$ (c) $t(4 + s)$ (d) $x^2(1 + x)$ (e) $xy(3 - y)$ (f) $rs(r^2 + s^2)$ (g) $z(2z - 3y)$ (h) $ab^2(1 - a^3b)$ (i) $xy(3xy - 8)$ (j) $3y^2(xy + 3)$ (k) $2wt(1 + 4w)$ (l) $3ps(1 - 4ps)$

2. (a) $5a(x + 2y - xy)$ (b) $xu(2x - 4u + 1)$ (c) $mn(3 + 4n)$ (d) $ab^2(c - a^2 + abc)$ (e) $3xy(y^2 + 2x - 4)$ (f) $3uv(4 - 3u - 2uv)$ **3.** (a) $(2 + z)(x + y)$ (b) $(3 - r)(t - s)$ (c) $(x - y)(x + 2)$ (d) $(st + 2)(a - b)$ (e) $(r^2 + 4)(xy + 1)$ (f) $(y^3 + 5)(2 + z)$ **4.** (a) $(4 + x)(s + 1)$ (b) $(a + b)(2 - c)$ (c) $(2 + t)(5x + y)$ (d) $(3 - 5k)(x + 2)$ (e) $(b + 4)(ab - 1)$ (f) $(y - z)(a + b)$

5. (a) $(a^3+4)(a-1)$ (b) $(1-yz)(x+1)$ (c) $(m-1)(n-1)$ (d) $(y^2+3)(y-1)$ (e) $(x-b)(x-b+1)$ (f) $(a+b)(a+b-1)$ (g) $(2-x)(a-b)$ (h) $(z+3)(x-y)$

EXERCISES 7.2.1

1. (a) $(x+8)(x+1)$ (b) $(x+10)(x+1)$ (c) $(x+12)(x+1)$ (d) $(x+6)(x+4)$ (e) $(x+3)(x+5)$ (f) $(x+9)(x+4)$ (g) $(x+8)(x+2)$ (h) $(x+8)(x+3)$ (i) $(x+5)(x+5) = (x+5)^2$ (j) $(x+6)(x+6) = (x+6)^2$ (k) $(x+11)(x+3)$ (l) $(x+13)(x+6)$ **2.** (a) $(x-8)(x-1)$ (b) $(x-10)(x-1)$ (c) $(x-12)(x-1)$ (d) $(x-8)(x-2)$ (e) $(x-5)(x-4)$ (f) $(x-8)(x-4)$ (g) $(x-9)(x-4)$ (h) $(x-10)(x-11)$ (i) $(x-7)(x-8)$ (j) $(x-9)(x-3)$ (k) $(x-7)(x-5)$ (l) $(x-6)(x-6) = (x-6)^2$ **3.** (a) $(x+7)(x-4)$ (b) $(x+9)(x-3)$ (c) $(x+6)(x-2)$ (d) $(x+12)(x-4)$ (e) $(x+8)(x-5)$ (f) $(x+12)(x-6)$ (g) $(x+14)(x-3)$ (h) $(x+15)(x-5)$ (i) $(x+11)(x-9)$ (j) $(x+9)(x-8)$ (k) $(x+13)(x-3)$ (l) $(x+13)(x-4)$ **4.** (a) $(x+3)(x-7)$ (b) $(x-8)(x+2)$ (c) $(x-9)(x+4)$ (d) $(x-12)(x+2)$ (e) $(x-14)(x+4)$ (f) $(x-12)(x+6)$ (g) $(x-15)(x+5)$ (h) $(x-15)(x+7)$ (i) $(x-13)(x+4)$ (j) $(x-13)(x+7)$ (k) $(x-16)(x+4)$ (l) $(x-7)(x+6)$

EXERCISES 7.2.2

1. (a) $(2x+1)(x+1)$ (b) $(2x+1)(x+2)$ (c) $(2x+1)(x+3)$ (d) $(3x+1)(x+1)$ (e) $(3x+1)(x+2)$ (f) $(3x+1)(x+3)$ **2.** (a) $(2x-1)(x-1)$ (b) $(2x-1)(x-2)$ (c) $(2x-1)(x-3)$ (d) $(3x-1)(x-1)$ (e) $(3x-1)(x-2)$ (f) $(3x-1)(x-3)$ **3.** (a) $(2x+1)(x-1)$ (b) $(2x+1)(x-2)$ (c) $(2x+1)(x-3)$ (d) $(3x+1)(x-1)$ (e) $(3x+1)(x-2)$ (f) $(3x+1)(x-3)$ (g) $(2x+1)(3x-1)$ (h) $(2x-1)(3x+1)$ (i) $(2x-1)(2x-1) = (2x-1)^2$ **4.** (a) $-3(x+2)(x-1)$ (b) $3(x+2)(x+1)$ (c) $(5x-2)(x-1)$ (d) $(5x+2)(x-3)$ (e) $2(1-2x)(x+3)$ (f) $-(3x-1)(x+4)$ (g) $3(3x+2)(x+1)$ (h) $2(3-x)(x+2)$ (i) $(5x+2)(2x+5)$ (j) $(2x+1)(4x-3)$ (k) $(2-3x)(2x+3)$ (l) $(9x+1)(2-x)$

EXERCISES 7.2.3

1. (a) $(x+2)^2$ (b) $(x+7)^2$ (c) $(y+8)^2$ (d) $(4x+1)^2$ (e) $(5a+1)^2$ (f) $(6z+1)^2$ (g) $(2x+3)^2$ (h) $(3x+4)^2$ (i) $(5x+2)^2$ (j) $(x+1)^2$ (k) $(x+\frac{1}{2})^2$ (l) $(x+\frac{3}{2})^2$

2. (a) $(x-2)^2$ (b) $(x-7)^2$ (c) $(y-8)^2$ (d) $(2u-3)^2$ (e) $(3a-4)^2$ (f) $(5v-2)^2$ (g) $(y-\frac{1}{2})^2$ (h) $(a-\frac{5}{2})^2$ (i) $(x-\frac{7}{2})^2$ **3.** (a) $(a-y)(a+y)$ (b) $(y-1)(y+1)$ (c) $(10-a)(10+a)$ (d) $(8-b)(8+b)$ (e) $(9-z)(9+z)$ (f) $(r-7)(r+7)$ (g) $(4x-7)(4x+7)$ (h) $(5-2a)(5+2a)$ (i) $(1-3y)(1+3y)$ (j) $4(a-3)(a+3)$ (k) $2(m-2)(m+2)$ (l) $3(y-3)(y+3)$ (m) $(x+5)(x-1)$ (n) $(ab-a-b)(ab+a+b)$ (o) $-3y(2x-y)$ **4.** (a) $(x-\sqrt{3})(x+\sqrt{3})$ (b) $(y-\sqrt{5})(y+\sqrt{5})$ (c) $(b-\sqrt{7})(b+\sqrt{7})$ (d) $(a+b-\sqrt{5})(a+b+\sqrt{5})$ (e) $(a-b\sqrt{10})(a+b\sqrt{10})$ (f) $(x+y-\sqrt{2})(x+y+\sqrt{2})$ (g) $2(x+\sqrt{2}(x-y))(x-\sqrt{2}(x-y))$ (h) $4(x-\sqrt{5})(x+\sqrt{5})$ (i) $18(\sqrt{2}a-b)(\sqrt{2}a+b)$ (j) $(a+2-\sqrt{10})(a+2+\sqrt{10})$ (k) $(x-3-\sqrt{6})(x-3+\sqrt{6})$ (l) $(x-y)(2-x+y)(2+x-y)$

EXERCISES 7.2.4

1. (a) $(x+1+\sqrt{3})(x+1-\sqrt{3})$ (b) $(x+2-\sqrt{6})(x+2+\sqrt{6})$ (c) $(x-3-\sqrt{7})(x-3+\sqrt{7})$ (d) $(x-4-\sqrt{19})(x-4+\sqrt{19})$ (e) $(x-1-\sqrt{2})(x-1+\sqrt{2})$ (f) $(x-2-2\sqrt{2})(x-2+2\sqrt{2})$ (g) $2\left(x-\frac{3}{2}+\frac{\sqrt{13}}{2}\right)\left(x-\frac{3}{2}-\frac{\sqrt{13}}{2}\right)$ (h) $3(x+2+\sqrt{7})(x+2-\sqrt{7})$ (i) $2\left(x+\frac{1}{4}+\frac{\sqrt{17}}{4}\right)\left(x+\frac{1}{4}-\frac{\sqrt{17}}{4}\right)$

EXERCISES 7.2.5 – MISCELLANEOUS QUESTIONS

1. (a) $(x+1)(x+2)$ (b) $(x+6)(x+1)$ (c) $(x+4)(x+2)$ (d) $(x+4)(x+5)$ (e) $(z+6)(z+3)$ (f) $(x+2)(x+5)$ **2.** (a) $(x-2)(x-1)$ (b) $(x-6)(x-1)$ (c) $(x-4)(x-2)$ (d) $(x-4)(x-5)$ (e) $(z-6)(z-3)$ (f) $(x-2)(x-5)$ **3.** (a) $(x-3)(x+1)$ (b) $(y+5)(y-2)$ (c) $(s-5)(s+2)$ (d) $(x-4)(x+3)$ (e) $(y-7)(y+2)$ (f) $(r-9)(r+5)$ **4.** (a) $2(x-2)(x-1)$ (b) $3(x+2)(x+5)$ (c) $4(s-5)(s+2)$ (d) $3(y+5)(y-2)$ (e) $5(x+4)(x+2)$ (f) $6(x+1)(x-3)$ **5.** (a) $(y-15x)(y-x)$ (b) $(z+7w)(z-6w)$ (c) $(a+7b)(a-5b)$ (d) $2(y+6x)(y+3x)$ (e) $3(x-4y)(x+3y)$ (f) $5(a-2b)(a-3b)$ **6.** (a) $(2x+1)(x+2)$ (b) $(3x+2)(x+1)$ (c) $(3x+1)(x+2)$ (d) $(2x-1)(x-2)$ (e) $(2x-1)(x+2)$ (f) $(7s-5)(s+1)$ (g) $(5x-3)(x+1)$ (h) $(7x-1)(x-5)$ (i) $(3y+1)(y-2)$ (j) $(3z-4)(z-2)$ (k) $(5w+4)(w-1)$ (l) $(2x+3)(x-5)$ (m) $(2y+3)(y+2)$ (n) $(5x+4)(x+1)$ (o) $(3z+2)(z+2)$ (p) $2(1-4x)(1+x)$ (q) $(3-5x)(1+x)$ (r) $(4+5x)(1-x)$ **7.** (a) $(x+4)^2$ (b) $(y+5)^2$ (c) $(z+3)^2$ (d) $(x-5)^2$ (e) $(b-6)^2$ (f) $(x-7)^2$ (g) $(y-5)(y+5)$ (h) $(x-6)(x+6)$ (i) $(z-4)(z+4)$ (j) $4x(x-3)(x+3)$ (k) $3(s-4)(s+4)$ (l) $(2x-3y)(2x+3y)$ (m) $(2x+1)^2$ (n) $(3z+2)^2$ (o) $(3z+1)^2$ **8.** (a) $(x+1-\sqrt{5})(x+1+\sqrt{5})$ (b) $(x-1-\sqrt{3})(x-1+\sqrt{3})$ (c) $(x+2-\sqrt{2})(x+2+\sqrt{2})$ (d) $(x+2-\sqrt{7})(x+2+\sqrt{7})$ (e) $(x-3-\sqrt{6})(x-3+\sqrt{6})$ (f) $(x+3-\sqrt{7})(x+3+\sqrt{7})$ (g) $(x+4-\sqrt{3})(x+4+\sqrt{3})$ (h) $(x-4-\sqrt{2})(x-4+\sqrt{2})$ (i) $(z+5-\sqrt{5})(z+5+\sqrt{5})$ (j) $(x-5-\sqrt{5})(x-5+\sqrt{5})$ (k) $(z-2-\sqrt{3})(z-2+\sqrt{3})$ (l) $(a+7-\sqrt{19})(a+7+\sqrt{19})$

EXERCISES 7.3.1

1. (a) 0, 7 (b) $\pm\sqrt{7}$ (c) 0, –3 (d) no real solutions. (e) –4, 3 (f) 0, 2 (g) –4, 2 (h) –2, 5 (i) –2, 5 (j) –6, 5 (k) –6, 1 (l) –4, 7 (m) –4, 3.5 (n) 4, 7 (o) 3, 9 (p) –3, 9 (q) –4, –3 (r) –3, –4/3 (s) –7, –2 (t) –0.5, 4 (u) –1/3, 4 (v) 0.5, 5 (w) 0.5, 5/3 (x) 1/4, 5/3 (y) 1/3, 2/5 (z) 1/4, 2

EXERCISES 7.3.2

1. (a) $\frac{-5\pm\sqrt{17}}{2}$ (b) $-2\pm\sqrt{7}$ (c) $1\pm\sqrt{6}$ (d) N.R.S (e) $\frac{-5\pm\sqrt{53}}{2}$ (f) $\frac{-7\pm\sqrt{85}}{2}$ (g) $\frac{-11\pm\sqrt{157}}{2}$ (h) $\frac{-3\pm\sqrt{57}}{2}$ (i) $\frac{-5\pm\sqrt{73}}{2}$ (j) N.R.S (k) $2\pm\sqrt{11}$ (l) $3\pm2\sqrt{5}$ (m) $-2\pm2\sqrt{6}$ (n) N.R.S (o) $1\pm\frac{\sqrt{22}}{2}$ (p) $-\frac{1}{2}, 3$ (q) $-\frac{1}{2}, 4$ (r) $\frac{3\pm\sqrt{41}}{4}$ (s) $\frac{3\pm\sqrt{57}}{6}$ (t) $\frac{5\pm\sqrt{73}}{6}$

EXERCISES 7.3.3

1. (a) $\frac{3\pm\sqrt{37}}{2}$ (b) $\frac{5\pm\sqrt{33}}{2}$ (c) $\frac{3\pm\sqrt{33}}{2}$ (d) $\frac{7\pm\sqrt{57}}{2}$ (e) $\frac{-7\pm\sqrt{65}}{2}$ (f) –4, 2 (g) $-1\pm2\sqrt{2}$

(h) $\frac{-5 \pm \sqrt{53}}{2}$ (i) $\frac{3 \pm \sqrt{37}}{2}$ (j) no real solutions (k) $4 \pm \sqrt{7}$ (l) no real solutions (m) $\frac{2 \pm \sqrt{13}}{2}$

(n) $\frac{3 \pm 2\sqrt{11}}{5}$ (o) $\frac{6 \pm \sqrt{31}}{5}$ (p) $\frac{6 \pm \sqrt{29}}{7}$

EXERCISES 7.4.1

1.

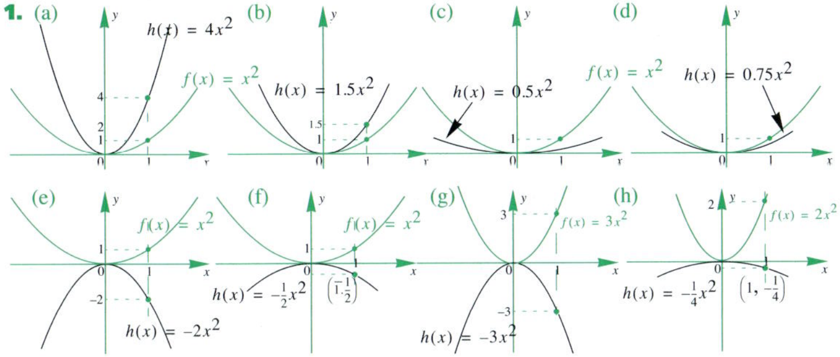

2.

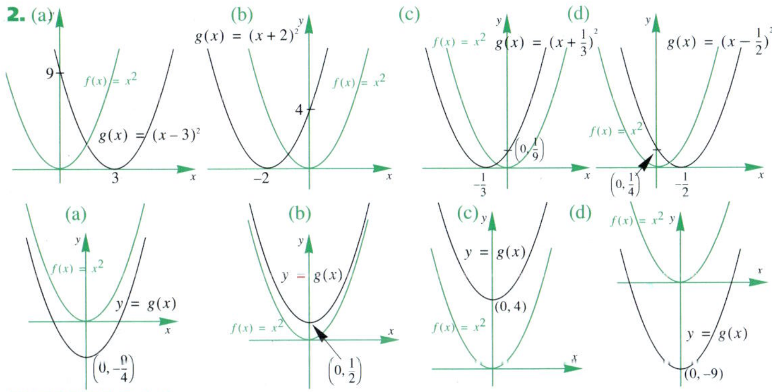

EXERCISE 7.4.2

1.

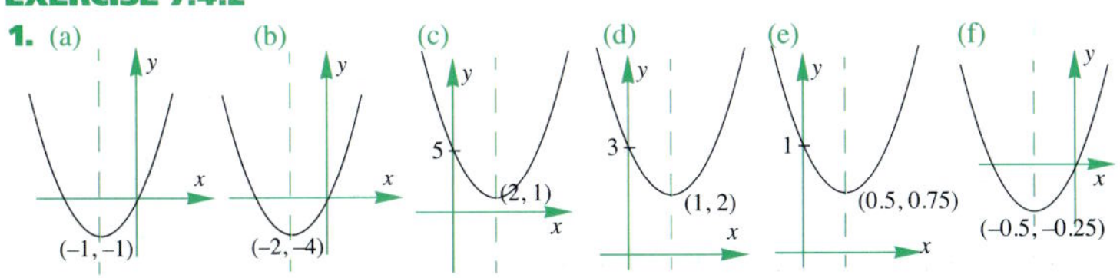

2.

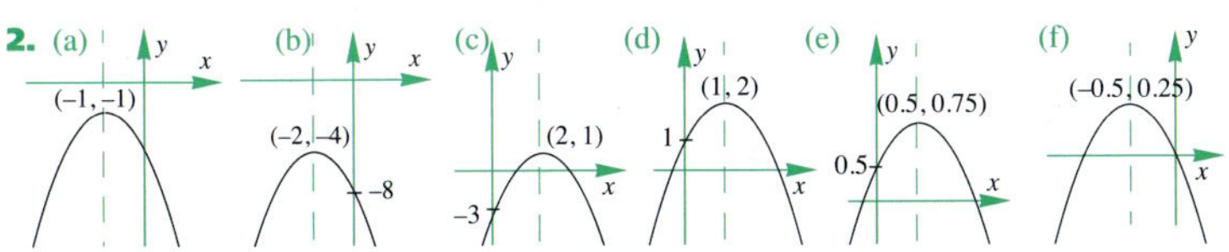

3.

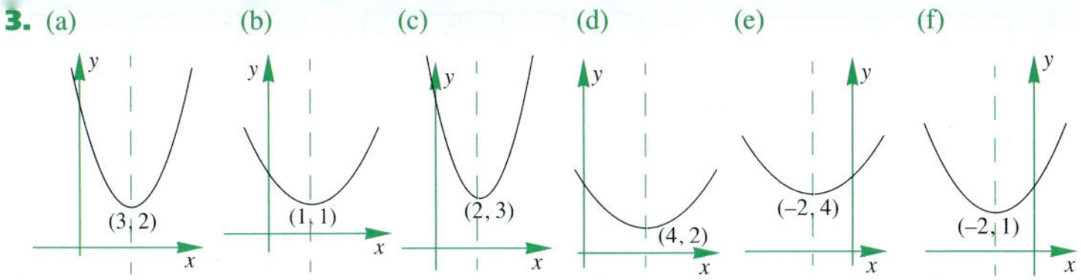

4.

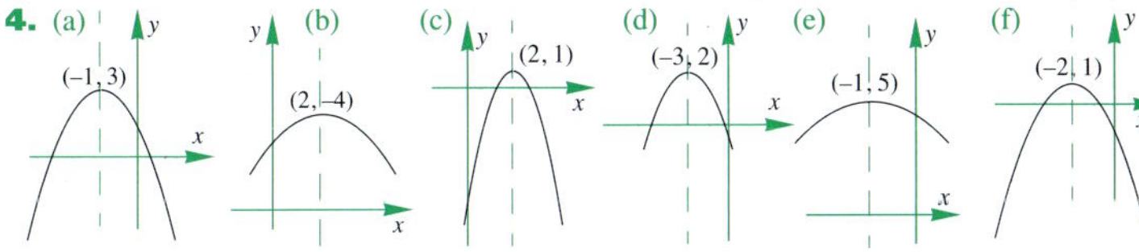

EXERCISES 7.4.3

1. Graphs are shown using the **ZOOM4** viewing window:

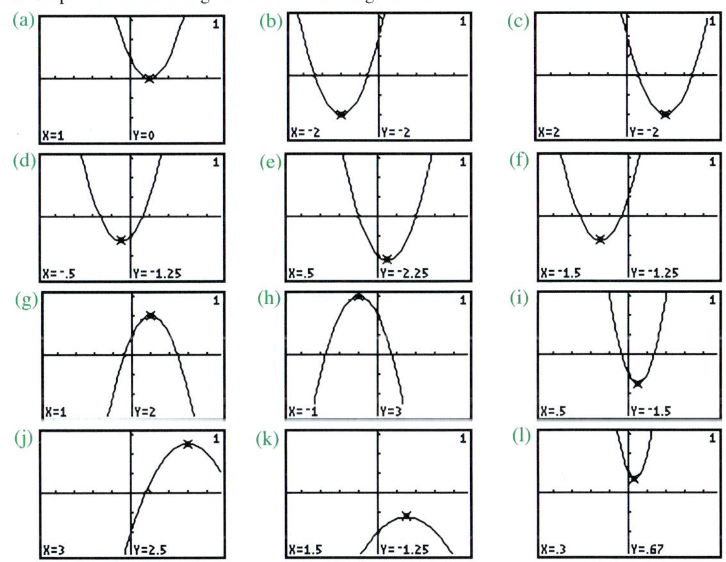

EXERCISES 7.4.4

1. Graphs are shown using the **ZOOM6** viewing window:

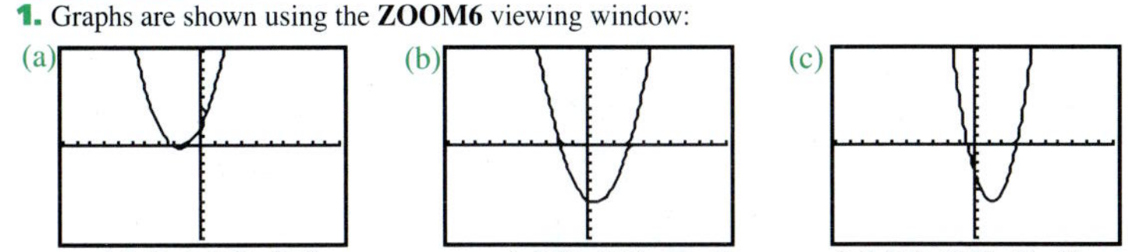

(a) (–2,0), (–1,0), (0,2) (b) (–2,0), (3,0), (0,–6) (c) (–0.5,0), (3,0), (0,–3)

(d)
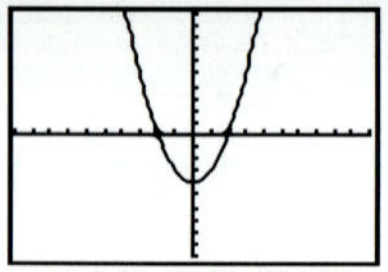
(−2,0), (2,0), (0,−4)

(e)
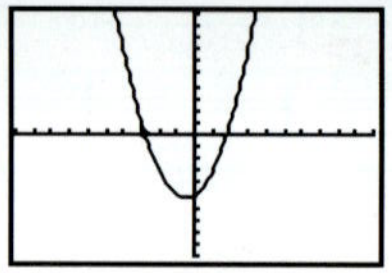
(−2.79,0), (1.79,0), (0,−5)

(f)
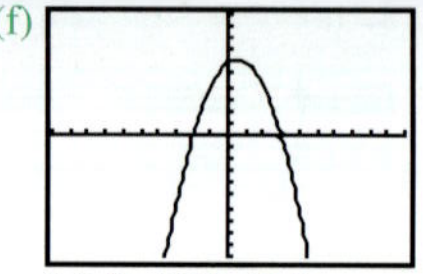
(−2,0), (3,0), (0,6)

(g)
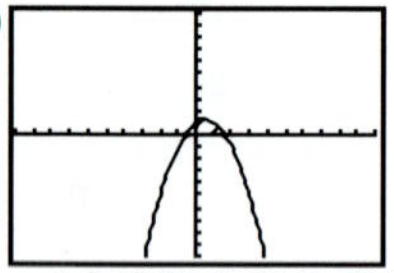
(−0.62,0), (1.62,0), (0,1)

(h)
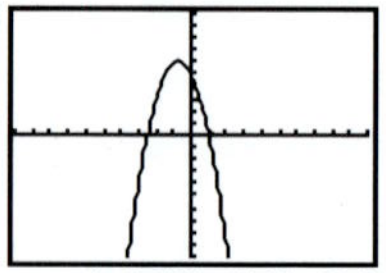
(−2.5,0), (1,0), (0,5)

(i)
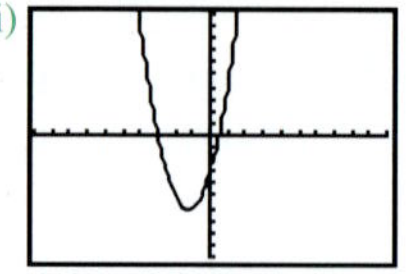
(−3,0), (0.5,0), (0,−3)

(j)
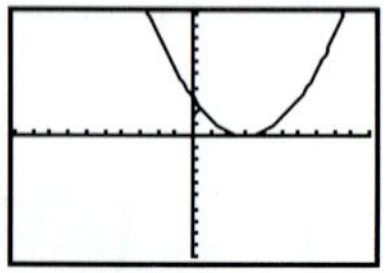
(3,0), (0,3)

(k)
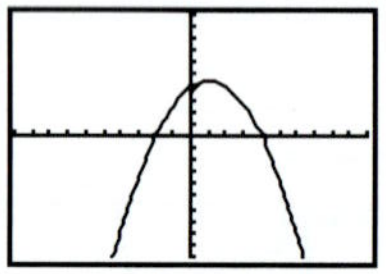
(−2,0), (4,0), (0,4)

(l)
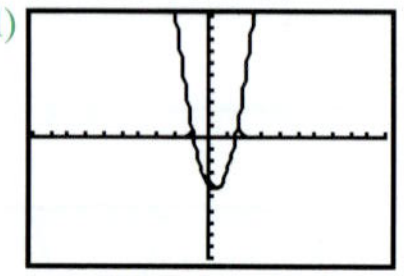
(−0.87,0), (1.54,0), (0,−4)

EXERCISES 7.5

1. (a) $(x-2)(x-1)$ (b) $(x-6)(x-1)$ (c) $(x-4)(x-2)$ (d) $(x-4)(x-5)$ (e) $(z-6)(z-3)$ (f) $(x-2)(x-5)$ (g) $(x-3)(x+1)$ (h) $(y+5)(y-2)$ (i) $(s-5)(s+2)$ (j) $(x-4)(x+3)$ (k) $(y-7)(y+2)$ (l) $(r-9)(r+5)$ (m) $2(x-1)(x-2)$ (n) $3(x+5)(x+2)$ (o) $4(s-5)(s+2)$ (p) $3(y+5)(y-2)$ (q) $5(x+4)(x+2)$ (r) $6(x-3)(x+1)$

EXERCISES 7.6.1

1. $x=1, 5$ **2.** $x=-2, 3$ **3.** $x=-2, 2$ **4.** $x=-1, 3$ **5.** $x=-2, 2$
6. (a) (−2, 4); (3, 9) (b) (2, 5); (−2, 5) (c) (1, 2) (d) (−2, 0); (3, 5) (e) (−3, 7); (2, 2) (f) (3, 9)
7. $(a, 3a^2)$; $(-2a, -3a^2)$.

EXERCISES 7.6.2

1. (b) (−2, 4); (2, 4) **2.** (−1, 2); (1, 2) **3.** (−1, 3); (2, 0) **4.** (a) (3, 17); (−2, 2) (b) (1, 6); (2, 13) (c) (−1, 1); (1, 9) (d) (0.5, 2.25); (−2, 1) (e) (4, 46); (5, 69) (f) (−3, 29); (−2, 12)

EXERCISES 7.7 – MISCELLANEOUS EXERCISES

1. (a) $4y(x-3y)$ (b) $2ab(3a+2b)$ (c) $3uv(v-4u)(v+u)$ **2.** 23 units. **3.** (2, −16)
4. none **5.** $x=1$, (1, −1) **6.** 25 **7.** (a) 2 (b) 4 **8.** −21 **9.** −2 **10.** (a) $(3+y)(x-4)$ (b) $x=4, y=-3$ **11.** 75 **12.** 2 **13.** 72 **14.** 4, 6 **15.** −1.5, 11 **16.** (a) $x=1.5$ (b) none (c) $x=0$ or $x=6$ (d) all real values **17.** 2 **18.** (a) −8, 3 (b) $\frac{5+\sqrt{57}}{4}, \frac{5-\sqrt{57}}{4}$ **19.** $b=0, c=-7$
20. 2 **21.** (a) $(x+4)(x-2)$ (b) Using the graph of $y=(x+4)(x-2)$ we have $x=-3, -2, -1, 0, 1$.
22. (a) (i) $(x+5)(x+1)$ (ii) $(x-6)(x+1)$
(b) (i)
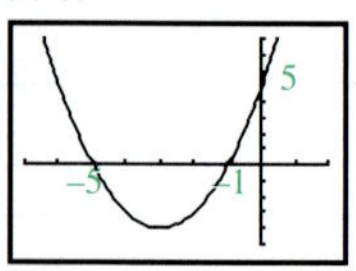

(ii)
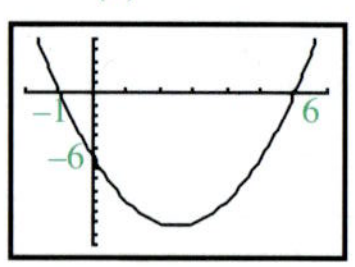

23. (a) $(x-4)(x+3)$ (b) $(3x-2)(2x-3)$ (c) $(x-1-y)(x-1+y)$
24. (a)
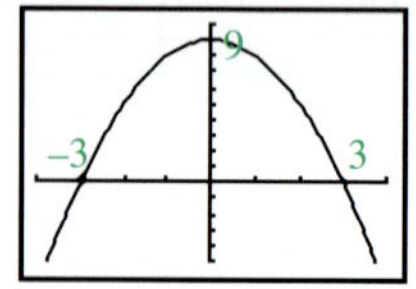

(b)
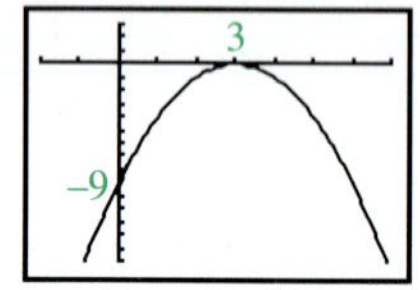

25. (a) $b=16, a=10$ (b) $(x-8)(x-2)$ **26.** $y=\left(x-\frac{3}{2}\right)^2+4\frac{3}{4}$, $\left(\frac{3}{2}, 4\frac{3}{4}\right)$ **27.** $a=-1, b=-3$

28. −3, 0.5 **29.** (0,7), $(1\pm2\sqrt{2}, 0)$ **30.** $b^2-4ac>0$, $b<-6$ or $b>6$ **31.** (a) $a=0.5, b=-3, c=-8$ (b) $\frac{1}{2}(x+2)(x-8)$ (c) i. −1, 0, 1, 2, 3, 4, 5, 6, 7 ii. −12.5 **32.** (a) (−2, −6); $\left(\frac{2}{3}, \frac{26}{9}\right)$

(b) $\left(-\frac{1}{2}, \frac{15}{4}\right)$; (2, 0) (c) (2, −1); (−1, 2) **33.** (a) $X(-2-\sqrt{2}, 0)$ & $Y(-2+\sqrt{2}, 0)$

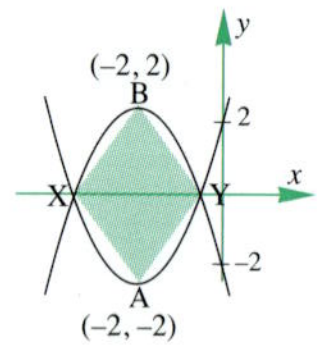

(b) i. 4 units ii. 8 sq. units **34.** mn

EXERCISES 7.8 – GRADED REVISION EXERCISES

LEVEL 1

1. (a) No (b) Yes (c) Yes **2.** (a) $x(x+1)$ (b) $x(2-x)$ (c) $ax(4+x)$ (d) $6y(y+2)$ (e) $3b(3b-1)$ (f) $4ay(3y+1)$ **3.** (a) $(2+a)(a+b)$ (b) $(z-3)(z-1)$ (c) $(x-2)(x+1)$

LEVEL 2

1. (a) $(x+4)(x-1)$ (b) $(y-5)(y-1)$ (c) $(x+7)(x-2)$ (d) $(a-1)^2$ (e) $(a-3)(a+1)$ (f) $(y-9)(y-7)$ **2.** (a) $(2x-1)(x+3)$ (b) $(3x+1)(x+1)$ (c) $(2y+3)(y-1)$ (d) $(3z+1)(2z+1)$ (e) $(4x+3)(3x+4)$ (f) $(5x+1)(x-3)$ **3.** i. −4, 1 ii. 1, 5 iii. −3, −2 iv. 1 v. −7, 2 vi. 7, 9 **4.** i. $\frac{-1\pm\sqrt{29}}{2}$ ii. $\frac{-3\pm\sqrt{41}}{4}$ iii. $-2\pm\sqrt{13}$ iv. $\frac{3\pm\sqrt{89}}{10}$ v. none vi. $\frac{-5\pm\sqrt{133}}{6}$

LEVEL 3

1. i. (3,4) ii. (−1,3) iii. (1,−3) iv. (1,1) v. (0.5,3) vi. (1,0) **2.** i. $y=x^2-x-2$ ii. $y=4-x^2$ iii. $y=x^2-2x-1$ iv. $y=x^2-4x+4$ **3.** (a) $x=\frac{-1\pm\sqrt{17}}{2}$ (b) $x=\frac{1\pm\sqrt{15}}{2}$

(c) $x=3\pm\sqrt{10}$ (d) $x=2\pm\sqrt{3}$ (e) $x=\frac{1\pm\sqrt{17}}{4}$ (f) $x=3\pm\sqrt{7}$

4. (a) (b) (c) (d) (e) (f)

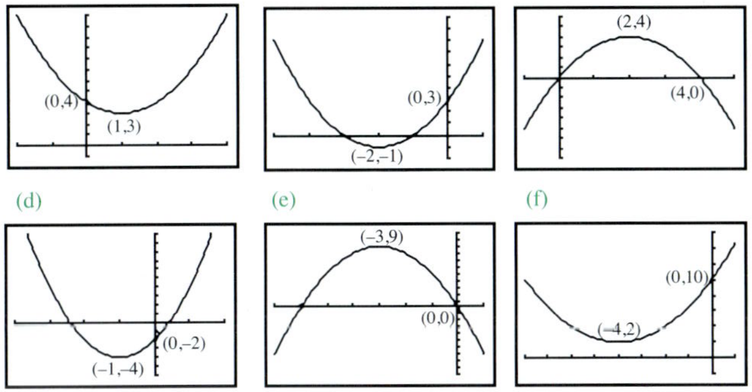

5. (a) (b) (c) (d) (e) (f)

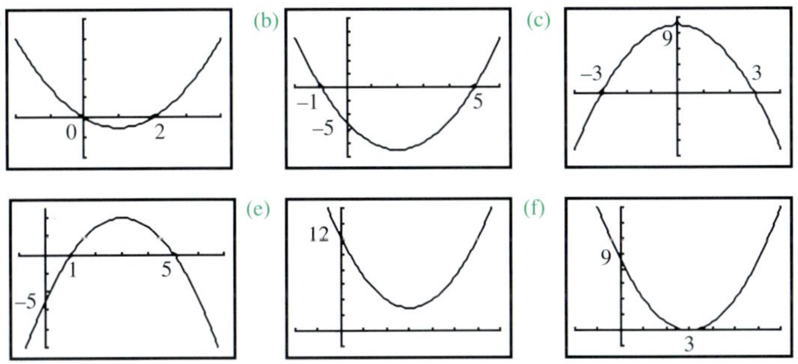

6. (a) (–5, 15); (3, 23) (b) (–4, –6); (2, 6) **7.** (a) $a = 2, k = 4, c = -4$ (b) –0.5, 4 (c) $x = 1.75$ (d) –2.125

LEVEL 4

1. $\left(-\frac{k}{2}, -\frac{k^2}{4}+4\right)$. **2.** $N_{min} = \frac{3}{4}$. **3.** (a)

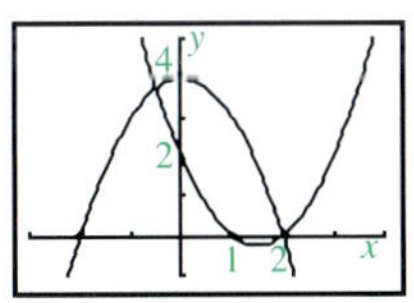

(b) $x = -0.5$ or $x = 2$

(c) (–0.5, 3.75) & (2, 0) **4.** (a) $x = -a$ or $x = -1$ (b) $x = -2$ or $x = -1$.

EXERCISES 7.9 – TOPIC TEST

1. $x = 7$ or $x = -1$. **2.** (a) $(x-8)(x-4)$ (b) $(x-6)^2$ (c) $(x-6-\sqrt{6})(x-6+\sqrt{6})$

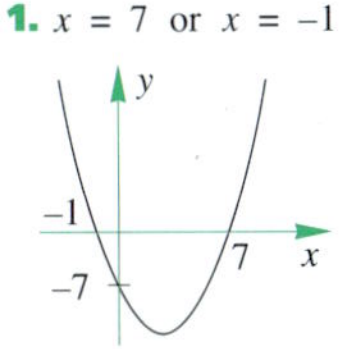

3. 0.884, –1.88 **4.** $y = -x^2+2x+3$, (1,4) **5.** –2, 3 **6.** $k = 3$

7. (a) i. $-(x-10)(x-4)$ ii. $x(8-x)$

$A \equiv (4,0), B \equiv (8,0)$ and $C \equiv (10,0)$ (b) $x = \frac{20}{3}$

(c) $\frac{80}{9}$sq.u **8.** (a) $y = -0.5x+1$ (b) $y = 2x(2-x)$ (c) (0.25, 0.875)

ANSWERS TO CHAPTER 8

EXERCISES 8.1

1. (a) dom = {2, 3, –2}, ran = {4, –9, 9} (b) dom = {1, 2, 3, 5, 7, 9}, ran = {2, 3, 4, 6, 8, 10} (c) dom = {0, 1}, ran = {1, 2} **2.** (a)]1, ∞[(b) [0, ∞[(c)]9, ∞[(d)]–∞, 1] (e) [–3, 3] (f)]–∞, ∞[(g)]–1, 0] (h) [0, 4] (i) [0, ∞[(j) [1, 5] (k)]0, 4[(l)]–∞, –1] ∪ [1, ∞[

3. (a) $r = [-1, \infty[, d = [0, 2[$ (b) $r = \{y : y \geq 0\}\backslash\{4\}$, $\mathbb{R}$ (c) r = [0, ∞[\{3}, d = [–4, ∞[\{0} (d) $r = [-2, 0[, d = [-1, 2[$ (e) $r =]-\infty, \infty[$ $d =]-\infty, -3] \cup [3, \infty[$ (f) $r = [-4,4], d = [0,8]$

4. (a) one to many (b) many to one (c) many to one (d) one to one (e) many to many (f) one to one **5.** (a) $\mathbb{R}\backslash\{-2\}$ (b)]–∞, 9[(c) [–4,4] (d)]–∞, –2] ∪ [2, ∞[(e) $\mathbb{R}\backslash\{0\}$ (f) $\mathbb{R}$

EXERCISES 8.2

Graphs with graphics calculator output have standard viewing window unless otherwise stated.

1. (a) 3, 5 (b) i. $2(x+a)+3$ ii. $2a$ (c) 3 **2.** (a) 0, $\frac{10}{11}$ (b) $-\frac{5}{4}$ (c) $\left[0, \frac{10}{11}\right]$

3. (a) $-0.5x^2 - x + 1.5$, $-0.5x^2 + x + 1.5$ (b) $\pm\sqrt{2}$ (c) no solution

4. (a) $x = 0, 1$ **5.** (a) i. ii.

(b)

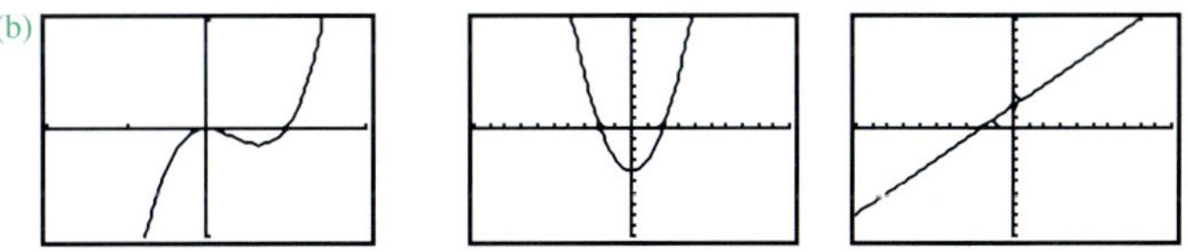

Window [–2,2], [–1,1]
Range: [–12, 4]

(b) i. $\{2\sqrt{2}, -2\sqrt{2}\}$ ii. {3, –2}

6. (b), (c), (d), (e) **8.** (a), (d), (e), (f) **10.** (a) $\{y : y > 1\} \cup \{y : y \leq -1.25\}$ (b) 10

11. (a) [–3, ∞[(b) [–3,0] (c) [3, ∞[

9. (a)

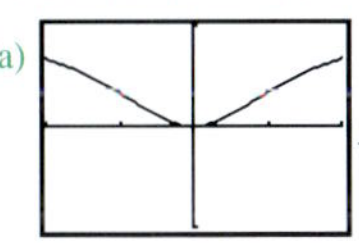

Window [-2,2], [-1,1]
(b) [0, 1[

EXERCISES 8.3.1

1. (a) (b) (c) (d) **2.**

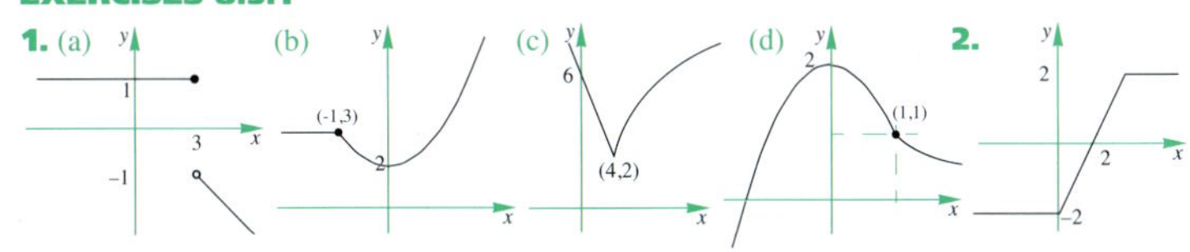

3. (a) (b) (c) (d) **4.**

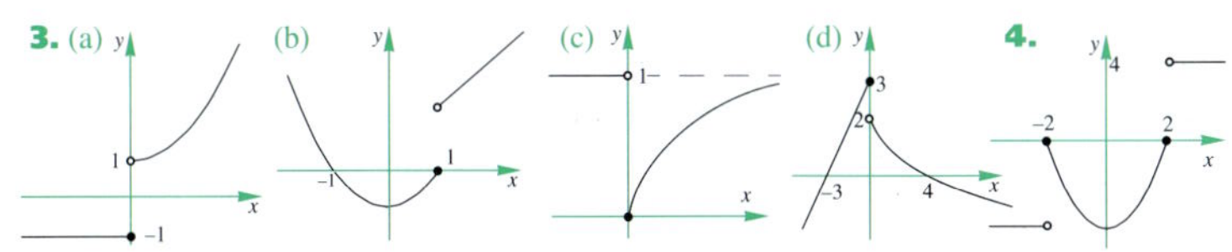

EXERCISES 8.3.2

1. (a) (b) (c) (d) (e) (f) (g) (h) (i)

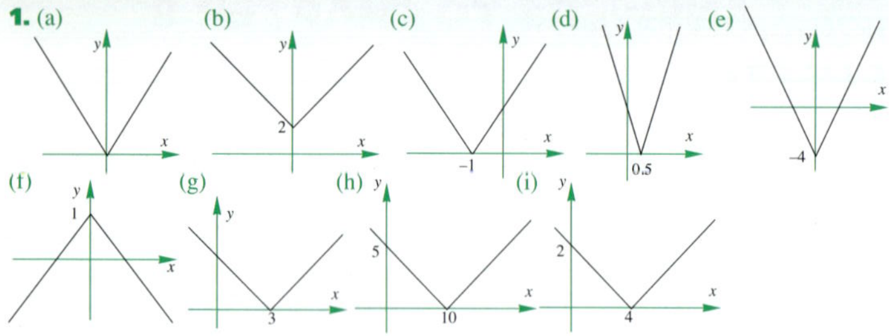

EXERCISES 8.4

1. (a) –2, 1 (b) 1, 2 (c) 1.25 (d) 1.52 (e) 1 (f) 2.04 (g) 1.25, –1.25 (h) 0.34, 8.99 (i) No solution (j) 3 (k) 3.44 (l) –1.67

EXERCISES 8.5 – MISCELLANEOUS EXERCISES

1. (a) Yes (b)

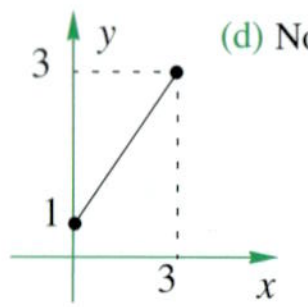

(c) i. $0 \le x \le 3$ ii. $[1, 3]$

(d) None, $x = 12$ lies outside the domain.

2. (a) 2 m (b) 66 m (c)

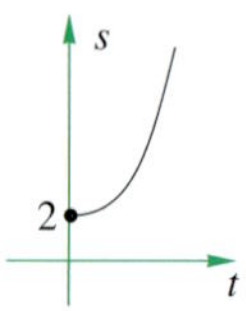

3.

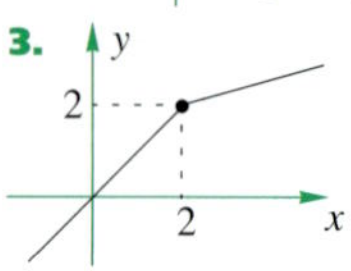

4. (a) $y \ge -2$ (b) i. $(0, 4)$ ii. $(3 \pm \sqrt{3}, 0)$ (c) $b = 2, a = \frac{2}{3}$

5. (a) 60, 44, 36, 32, 30 (b) Use graphics calc. (c) 28 (d) $\frac{16}{3}$ minutes

6. (a) 3, $a = 2, -1$ **7.** (b) $b = 3, c = 2$ **8.** Range $=]-\infty, 7]$; Many:one

9.

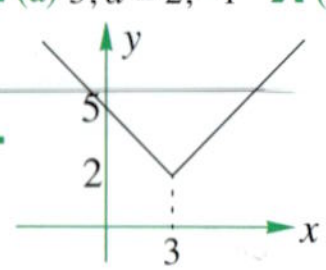

10. 1 **11.** (b) $V = 0$ 60 64 36 0 (c) 67.6

EXERCISES 8.6 – GRADED REVISION EXERCISES

LEVEL 1

1. (a) dom = (–1, 4], ran = [0, 5] (b) dom = [0, 6], ran = [0, 6] (c) dom = $\mathbb{R}\backslash\{2\}$, ran = $\mathbb{R}\backslash\{3\}$.

2. (a) and (c). **3.** (a) $y = 2 + 2 = 4$ (b) 18 (c) 6

LEVEL 2

1. $-\frac{8}{3}$ **2.** (a) $x \le 2$ (b) $x \ge 3$ (c) $x > -1$ **3.** (a) $g(f(x)) = \frac{1}{2x}$ (b) $f(g(x)) = \frac{1-x}{1+x}$

4. (a) i.

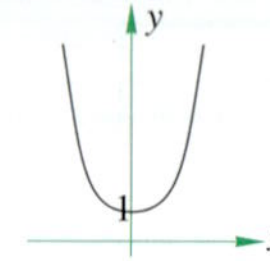

ii.

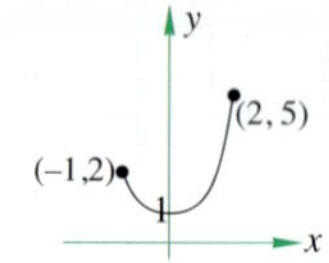

(b) i. ran = $[1, \infty)$ ii. ran = $[1, 5]$

5. (a) –1 (b) $4x^2 + 4x$ (c) $x^2 - 2x$ (d) $4x$

6. (a)

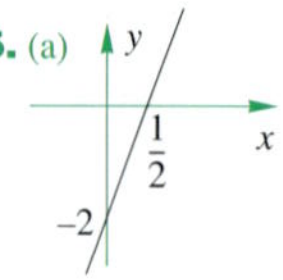

(b)

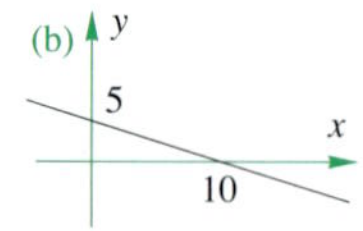

(c)

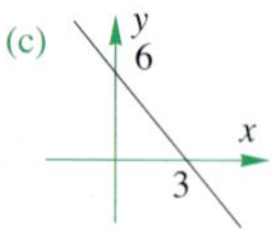

7. (a) i. 0 ii. $h^2 - 3h$ (b) $3h + h^2$

LEVEL 3

1. (a)

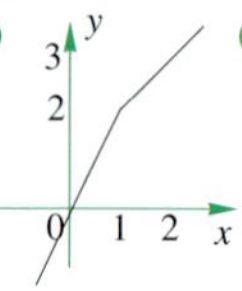

(b)

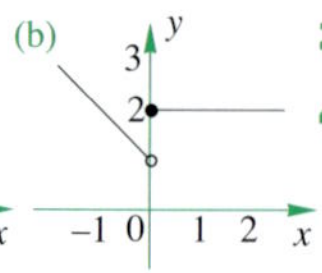

2. i. 4 ii. $3 + x$ **3.** $g(f(x)) = x$; $f(g(x)) = x$

4. Calculator check. **5.** $x = \frac{k}{k-1}$

6.

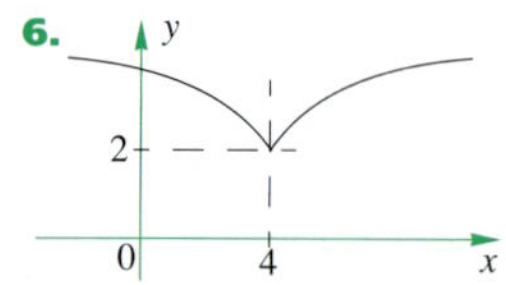

LEVEL 4

1. S = (0, 2) **2.** $\{\pm\sqrt{3}, \pm 1\}$. **3.** $-3 < x < 3$;

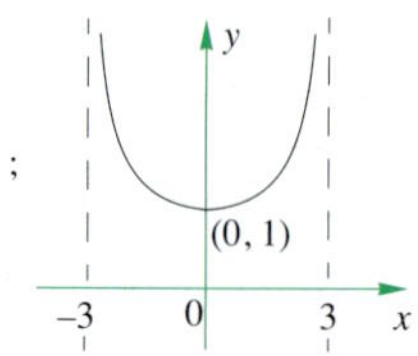

4. (a) $\therefore a = 2, b = 8, c = 6$ (b) –0.75

5. (a)

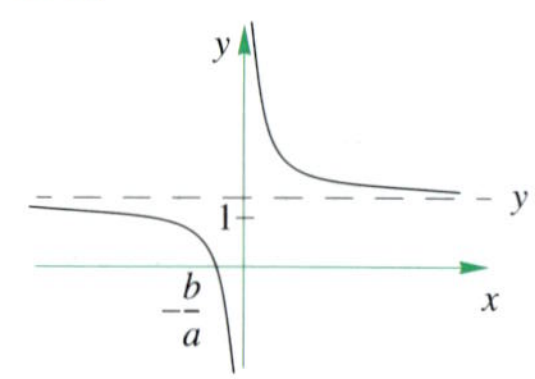

(b) i. & ii.

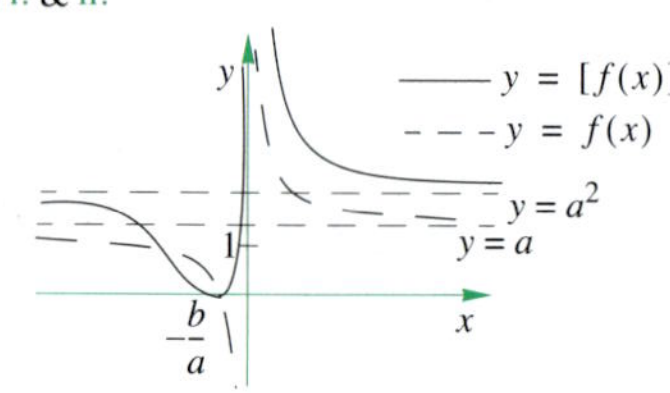

EXERCISES 8.7 – TOPIC TEST

1. i. 1 ii. 1

2.

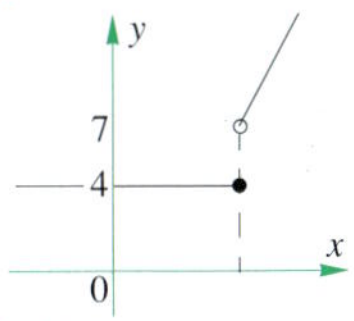

3. (a) range =]–∞, 16] (b) range =]–∞, 16+*k*]

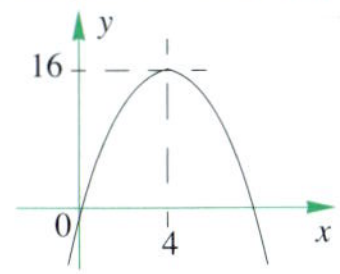

This occurs because all that has happened is that '*k*' has been added to the function in (a) and so the graph in (a) is moved up '*k*' units.

4. (b)

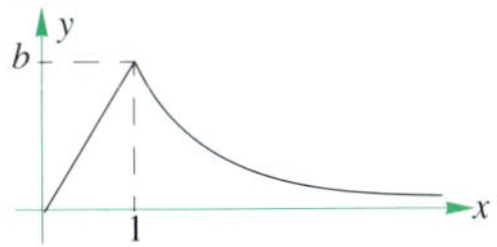

5. (a)

t min	1	3	7	14	29	59
T °C	60	45	37.5	34	32	31

5. (b)

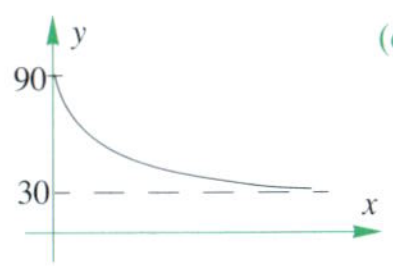

(c) $T \to 30$

(d) 2

6.

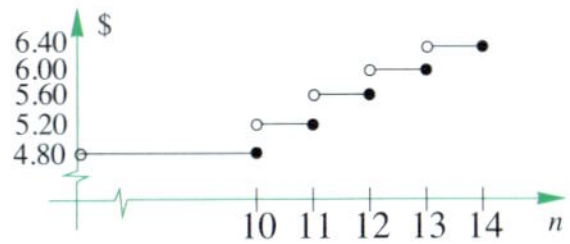

7. (a) $y = \begin{cases} 0.5x+1 & \text{if } -2 \le x < 0 \\ 1-0.5x & \text{if } 0 \le x \le 2 \end{cases}$ (b) $y = \begin{cases} -1 & \text{if } -3 < x \le -1 \\ 0 & \text{if } -1 < x \le 1 \\ 1 & \text{if } 1 < x \le 3 \end{cases}$

ANSWERS TO CHAPTER 9

EXERCISES 9.1.1

1. 8 **2.** 4, 0.25 **3.** 8, 18 **4.** 8 and 11 or –8 and –11 **5.** 6, –10 **6.** 2 m **7.** 51 kmh^{-1} **8.** 11 &13 or –11 & –13 **9.** 25 days **10.** 30 **11.** (a) 30 (b) \$50 each. **12.** 6 kmh^{-1} **13.** 16 **14.** 6 **15.** 3 hours **16.** 9 **17.** (a) 15 hrs (b) 10 hrs **18.** Chair-one: 20; Chair-two: 24 **19.** (a) 2 km/day (b) 2.5 km/day **20.** 7.5 hrs, 10.5hrs

EXERCISES 9.1.2

1. (a) i. $100 - 2x$ ii. $0 < x < 50$ [Nb: if $x = 0$ or 50, $A = 0$ and so there is no enclosure]
(b) i. $A = 2x(50 - x), 0 < x < 50$ ii. 10m by 80m or 40m by 20m iii. 1250 m^2 iv. 25m by 50 m

2. (a) ii. $0 < x < 12$ (b) i. 20 m^2 ii. 32 m^2 iii. 32 m^2 (c) (d) 6 m by 6 m

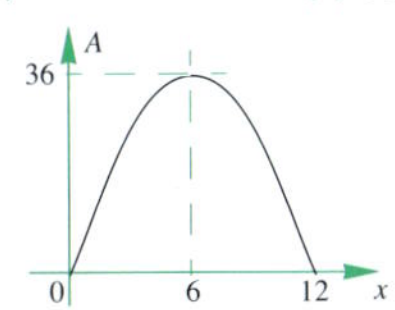

3. (a)

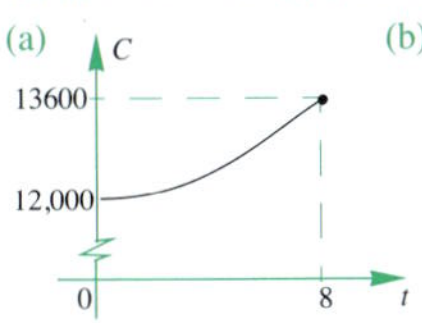

(b) \$12,900

4. (a) $R(x) = xp = x(40 - 0.0004x), 0 \le x \le 100,000$ (b) i. \$960,000 ii. 18377 or 81622 [must be integer values] iii. 1,000,000

5. (a) $y = \frac{4}{3}(50 - x)$ (b) i. $A = \frac{8}{3}x(50 - x)$ ii. $0 < x < 50$ (c) i. $\frac{5000}{3}$ m^2

ii. $x = 25, y = \frac{100}{3}$; dimensions 50 m by $\frac{100}{3}$ m. **6.** 35.83 kmh^{-1} **7.** (a) 100%

(b) $t = 0.229$ (first time) then again at $t = 13.104$ (c) (d) i. 42.26% ii. 1.73 weeks

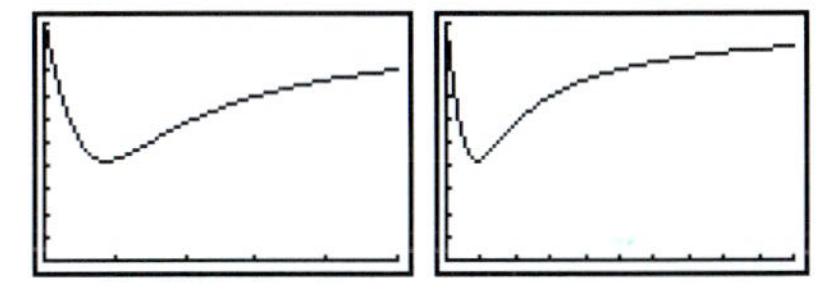

[0, 10] by [0, 1] [0, 20] by [0, 1]

(e) As time increases, oxygen level will be 100%

8. (a) (b) \$95.26

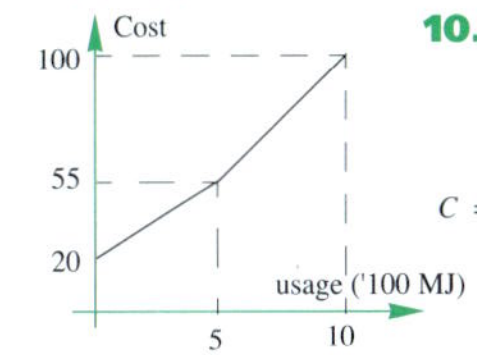

9. B(3254, 1953), C(6146,3687) units in metres

10. (a)

$C = \begin{cases} 0.25, & 0 \le x < 50 \\ 0.65, & 50 \le x < 80 \\ 1.30, & 80 \le x < 160 \\ 1.95, & 160 \le x < 320 \\ 2.50, & 320 \le x < 500 \\ 3.40, & 500 \le x < 1000 \end{cases}$

(b)

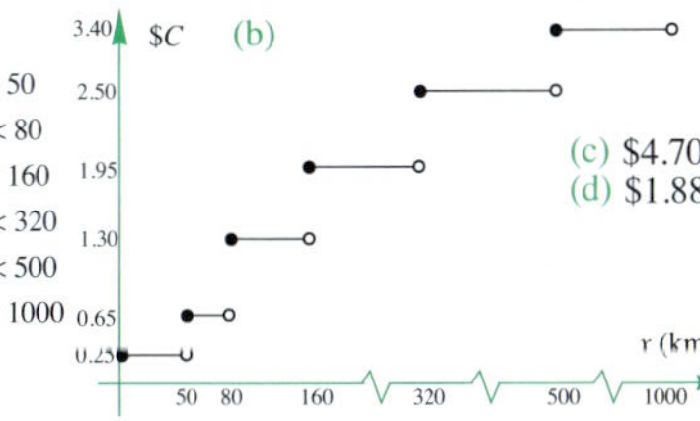

(c) \$4.70
(d) \$1.88

11. (a) $0 < x < 4$ (b) $A(x) = 3x + 0.25x^2, 0 < x < 4$ (c) i. ii. 3.25 iii. $-6 + \sqrt{48}$

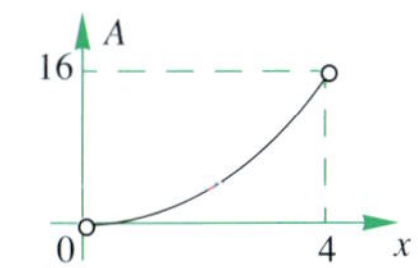

12. (a) i. 200 m ii. 320 m
(c) 12 sec (b) i. 0.34 sec and 11.66 sec ii. 11.31 sec
(d) 360 m

13. (a) 0.53 sec (on the way up) and 9.47 sec (on the way down)
(b) 10 sec (c) 500 m (d) 12.07 sec (e) 750 m

14. (a) \$72500 (b) No. (Loss of \$20000) (c) 2500

15. (a)

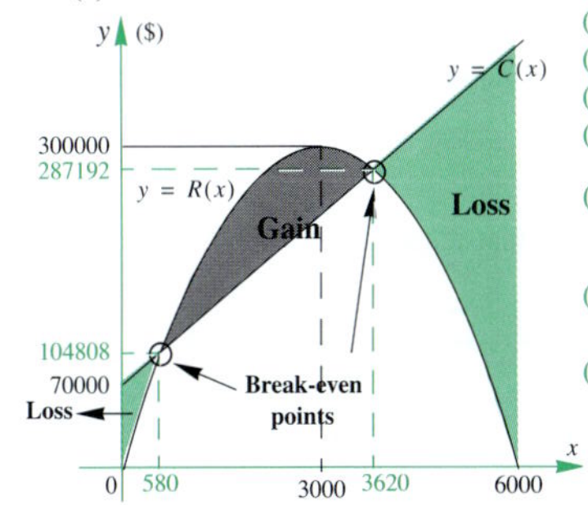

(b) i. \$70,000 ii. \$300,000
(c) Fixed cost (e.g., salary, electricity, ...)
(d) See graph in (a)
(e) \$76667 (to nearest dollar)
(f) i. $P(x) = 140x - \frac{1}{30}x^2 - 70000, 0 \le x \le 6000$
ii. \$77,000 iii. 2100
(g) i. $0 \le x \le 580$ or $3620 \le x \le 6000$
ii. $581 \le x \le 3619$
(h) See graph in (a)

16. (b) i. $P(x) = -\frac{1}{30}x^2 + 194x - 72000$ ii. $0 \le x \le 6000$

(d) The company will break-even at 399 radios and 5421 radios. Provided the company sells between 399 and 5421 radios they will make a profit. (e) 2910

10. (c)

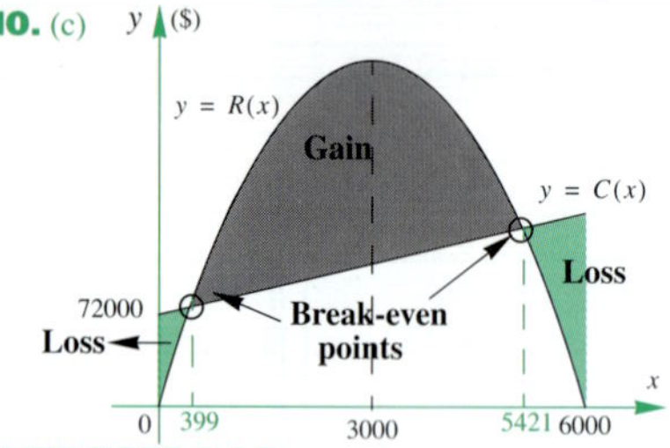

EXERCISES 9.1.3

1. i. (b) $y = 0.4x + 7.2$ ii. (b) $y = 6 - 2x$ iii. (b) $y = 0.5x + 3.2$ **2.** Second difference = 0.64

3. (b) $y = x^2 + 4x + 2$ **4.** (a) & (c)

4. (b) $y = 2x^2 - 3x + 7$ **5.** $y = 2x^2 - x + 3$ **6.**

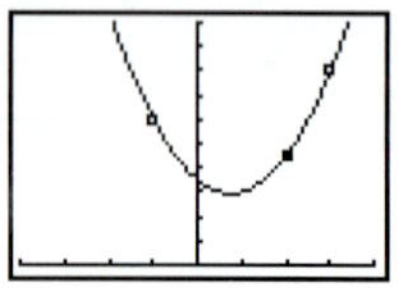

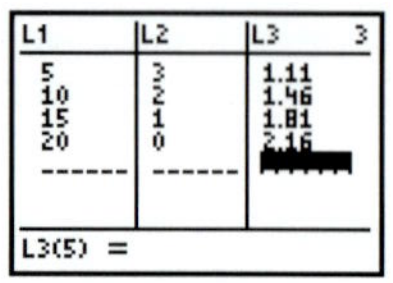

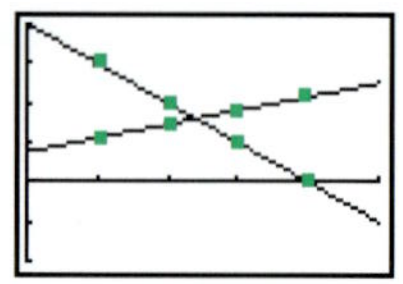

(c) i. $p = -0.2q + 4$ ii. $p = 0.07q + 0.76$

(d) Optimium scenario: demand = supply. This occurs when $p = 1.60$, $q = 12$.

7. (a) Using the coordinates (4,6), (6,11) & (9,9): $y = -0.6333x^2 + 8.8833x - 19.2$

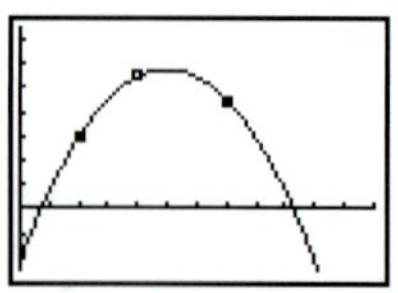

(b) –40°C at 11 pm. The model is not valid outside data range. Therefore extrapolation will not necessarily work.

8. Equation of path: $y = -\frac{31}{2400}x^2 + \frac{49}{48}x + 1$. Greatest height: 21.17 m.

9. (a) (b) Linear (c) i. ii. $M = 0, x = 1$. i.e., 1 m

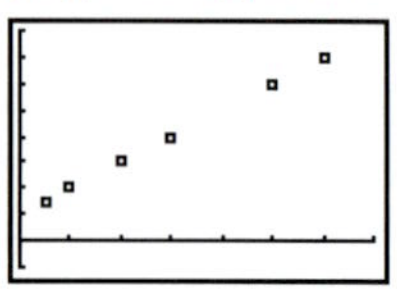

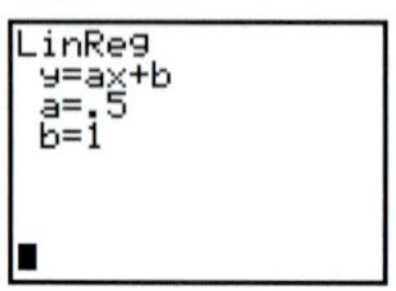

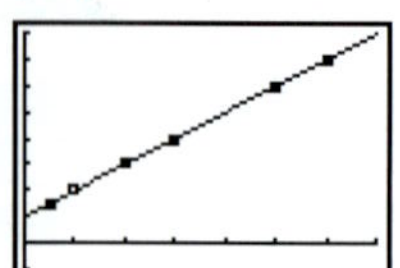

10. (a) (b) (c) i. parabola (d)

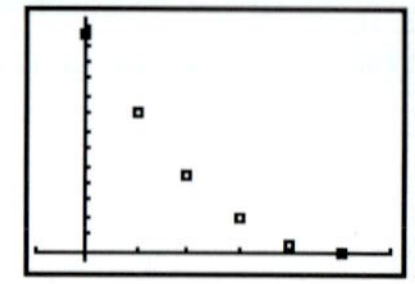

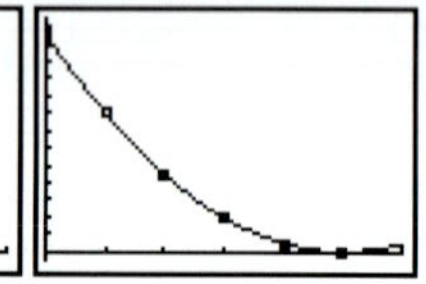

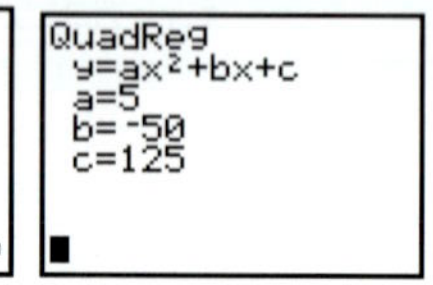

11. (a) $P(x) = -2x^2 + x + 3$ (b) $P(x) = (1-k)x^2 + x + k, x \in \mathbb{R}, k \neq 1$

12. (a) i. $a = 45000, b = 500000$ ii. $k_1 = 0, k_2 = 4$

(b)

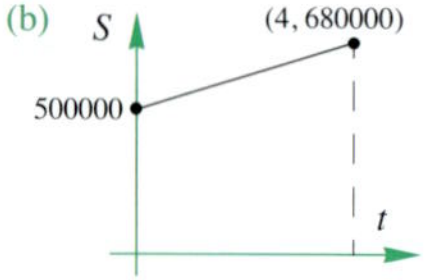

13. (a) (b) i. Parabolic

ii. $h(x) = -0.04694x^2 + 0.96518x + 1.7896$

(c) i. 6.75 m ii. 22.27 m

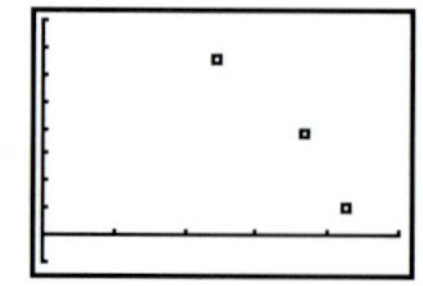

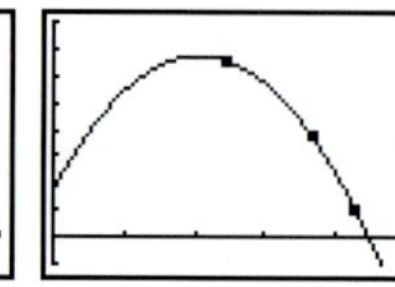

14. (a) $y = -\frac{2}{9}x^2 + \frac{11}{9}x + \frac{9}{2}, 0 \le x \le 5.5$ (b) [BE] : $y = -\frac{11}{9}x + \frac{121}{18}, 1 \le x \le 5.5$, [BO] : $y = 5.5x$, $0 \le x \le 1$ (c) 49°36′

15. (a)

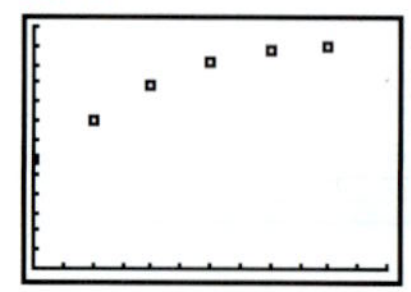

(b) second difference is constant = –50

(c) $y = -0.25x^2 + 25x + 580$ (d) $22500 per car (e) i. $824,750 ii. $19750

16. (a) i. & ii. have a constant gradient and so are linear.

iii. use trial and error to verufy quadratic relation

(b) ii. $p = 10 - 0.001x$, $C(x) = 2x + 7000$,

$R(x) = -0.001x^2 + 10x$.

(c) $P(x) = -0.001x^2 + 8x - 7000$, max profit = $P(4000) = 9000$

ANSWERS TO CHAPTER 10

EXERCISES 10.1

1. (a)–(j)

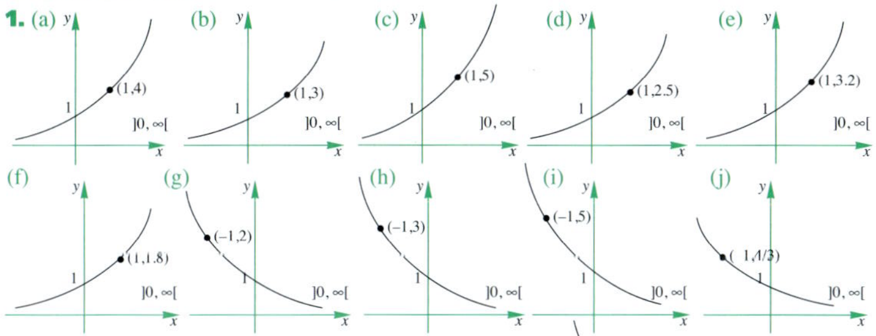

(k), (l)

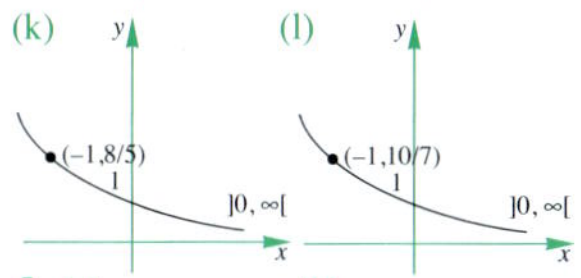

2. (a), (b)

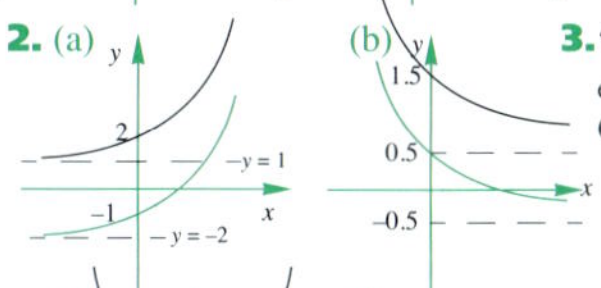

3. 'b' has a dilation effect on $f(x) = a^x$ (along the y axis).

4. (a)–(d)

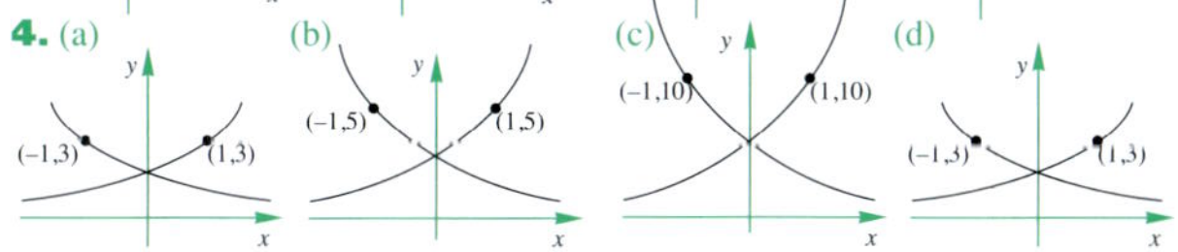

5. (a) [1, 16] (b) [3, 27] (c) [0.25, 16] (d) [0.5, 4] (e) [0.125, 0.25] (f) [0.1, 10]

6. (a) −1, 5
(b)

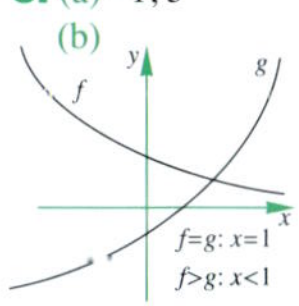

7. (a), (b)

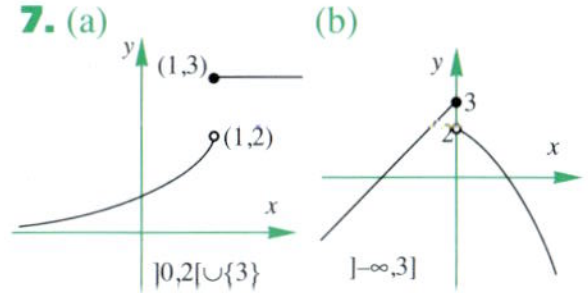

EXERCISES 10.2

1. i. 3 ii. 3 iii. 2 iv. 3.91 v. 2.93 vi. 1.66 vii. 2.40 viii. 1.76 ix. 4.17 x. −1/3 xi. −1/3 xii. −1/2 xiii. −3.32 xiv. 0 xv. −2.32 **2.** i. 1.63 ii. 1.74 iii. 2.32 iv. 1.89

EXERCISES 10.3

1. (a) 444.5 (b) 31.63% (c) 6.02 sec
(d)

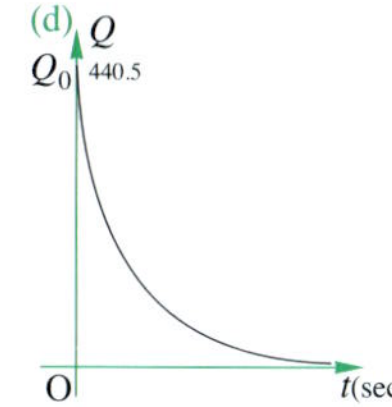

2. (a) 1200 (b) i. 2400 ii. 4800 (c) 10.3 yrs (approx)
(d)

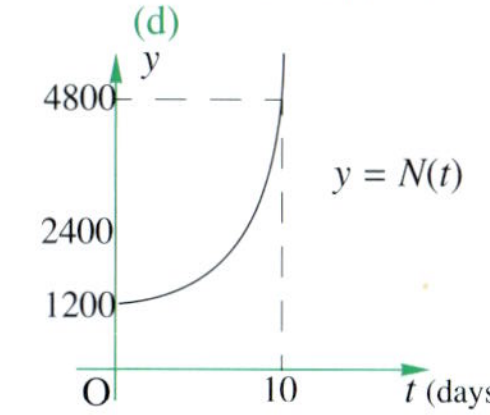

3. (a) 120 (b) 0.1833 (c) i. 750 ii. 900 (d) 11.57 yrs

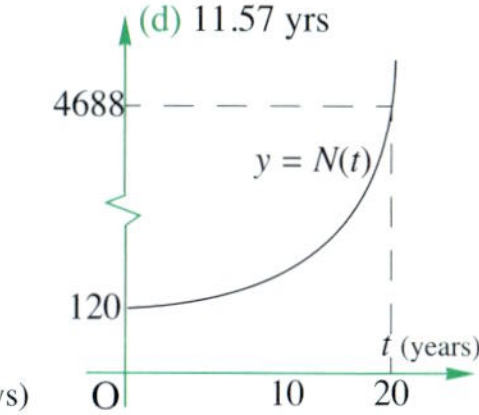

4. (a) 1000 (b) i. 1516 ii. 2000 (c) 10 days **5.** (a) 0.0013 (b) 2.061 kg (c) 231.56 yrs
(d)

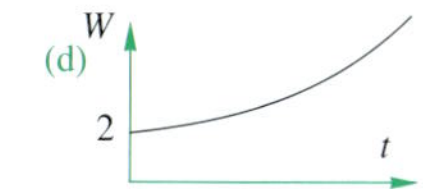

6. (a) i. 157 ii. 165 iii. 191 (b) 14.2 yrs (c) 20.1 yrs (d)

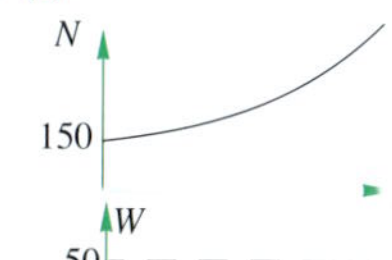

7. (a) 50 (b) 0.0222 (c) 18 kg (d)

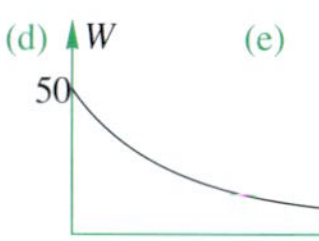

(e)

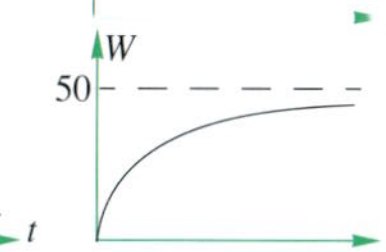

8. (a) 15 000°C (b) i. 11 915°C ii. 1500°C (c) 3.01 million yrs (d)

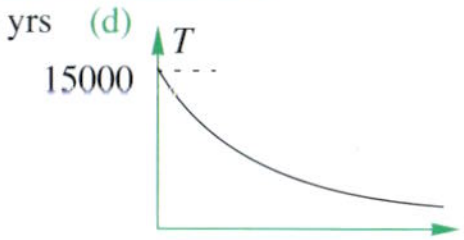

9. (a) 0.0151 (b) 12.5gm (c) 20 years (d)

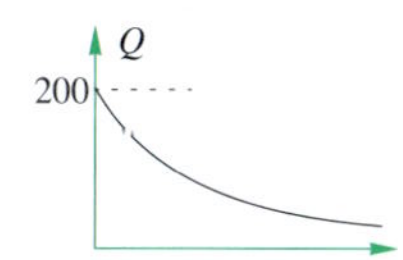

10. (a) \$2 million (b) \$1.589 mil (c) 30.1 years (d)

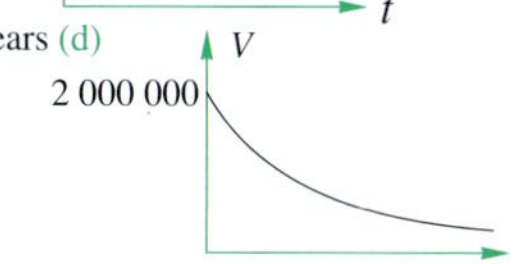

11. (b) 0.01761 (c) 199 230 (d) 22.6 years

EXERCISES 10.4 – MISCELLANEOUS EXERCISES

1. (a) (b) – 0.6 **2.** (a) $c = 1$ (b) 3 **3.** (a) 8 16 32 64 128 256 (b) i. 45 ii. 3.6

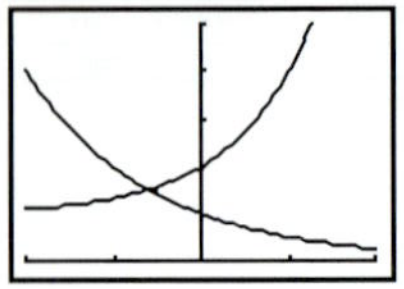

4. (a) 400
(b) i. 705 ii. 1242
(d) 14.2

5. (a)
(b) 0.5 sec

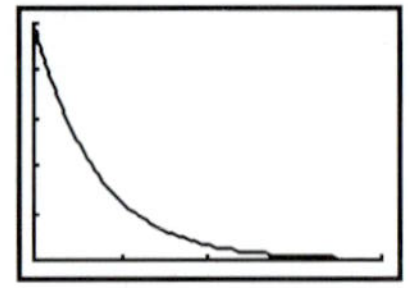

6.

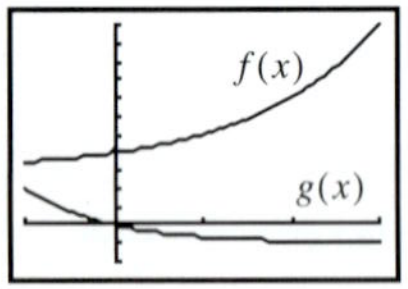

7. (a) $x = 2$ (b) $x = -1$ **8.** $c = -2,\ k = 0.25$
9. (a) 10 20 40 80 160 320 640
(c) i. 56 (or 57) ii. $t = 5.64$

10. (a)

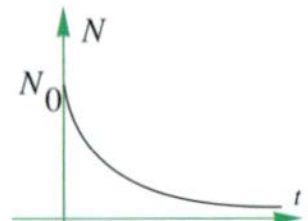

(b) 9.24% of N_0
(c) $t = 3.50$ (years)
(d) 160.43 cm
(e) 72 days old.

(f)

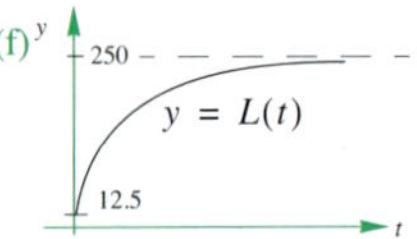

EXERCISES 10.5 – GRADED REVISION EXERCISES

LEVEL 1

1. i. Window : [–2, 3] by [0,8] ii. Window : [–2, 3] by [0,8] iii. Window : [–2, 3] by [0,8]

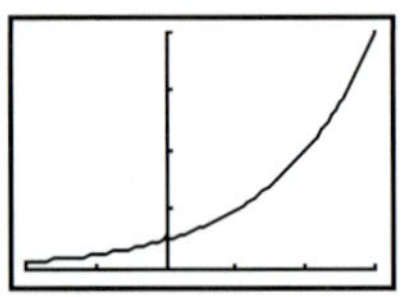

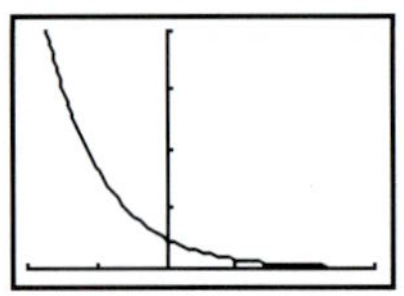

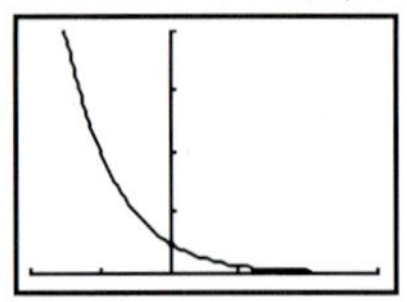

2. i. 17 ii. 5/3

LEVEL 2

1. i. 3.6 ii. 2.5 iii. 1.8

LEVEL 3

1.

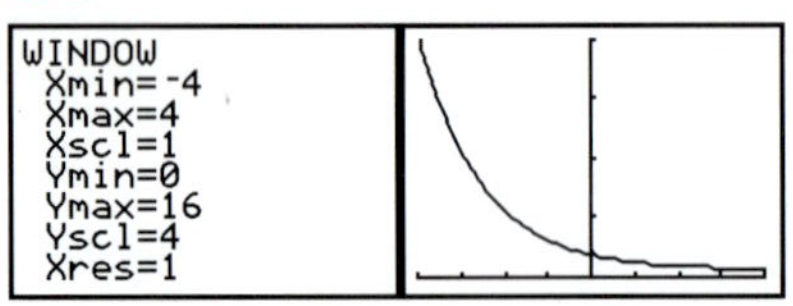

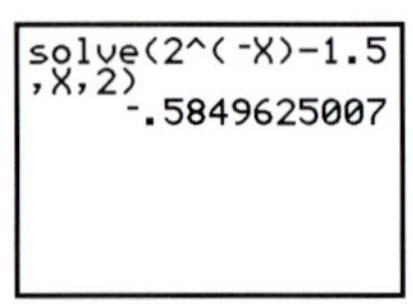

2. $c = 2, k = -2/3$

3. i.

t seconds	0	15	30	45	60
N	2	4	8	16	32

ii.

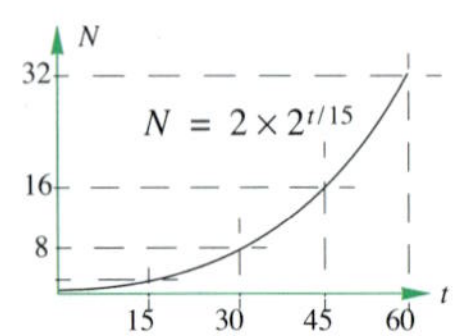

iii. 2097152

LEVEL 4

1. i.

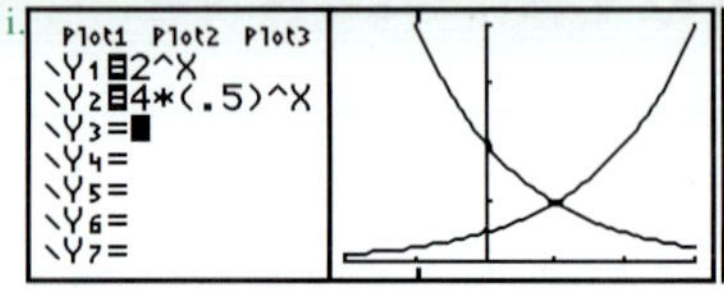

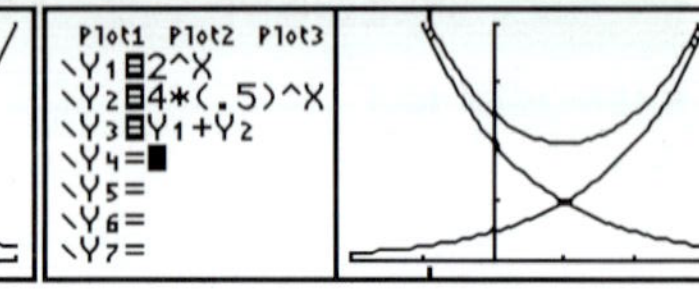

ii. 0, 2 **2.** i. 5 ii. (a)

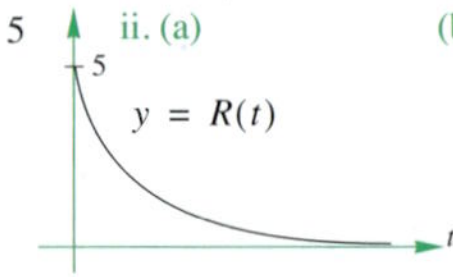

(b)

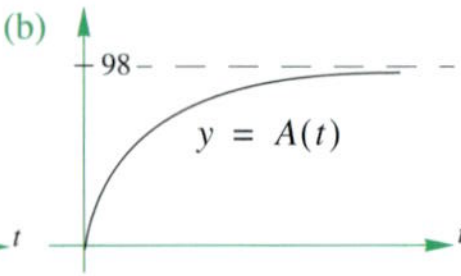

iii. 49 mg.

EXERCISES 10.6 – TOPIC TEST

1.

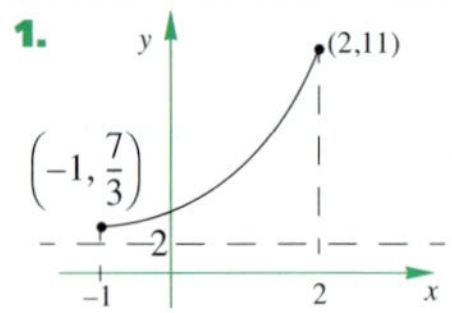

2. $a = 8, k = 4, b = 2$
3. (a) $x = 3.170$ (to 3 dec. pl.)
(b) $x = -0.708$ (to 3 dec. pl.)

4. i. 0.866 ii. approx. 6.4 yrs.

iii. $M = 200 \times (0.75)^{10/2} = 47.46$

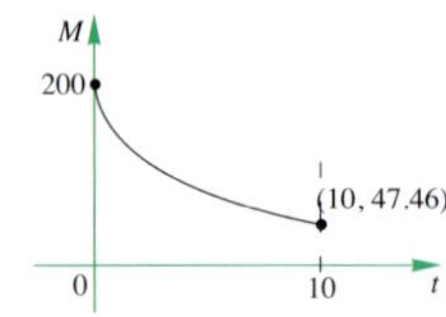

iv. $A = 200 - M = 200 - 200(0.75)^{t/2}$

$= 200\left[1 - \left(\frac{3}{4}\right)^{t/2}\right]$

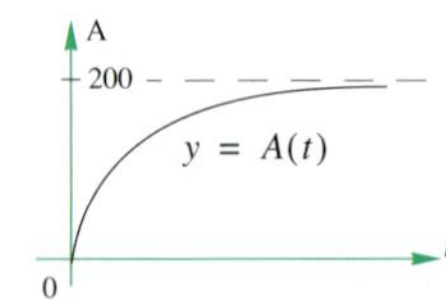

5. From the graphs below, we have that $x \sim 1.9$

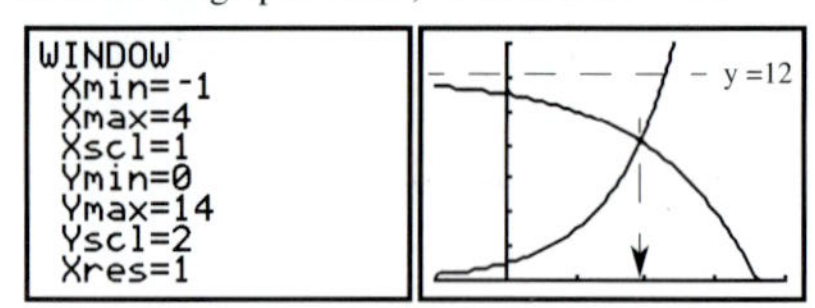

ANSWERS TO CHAPTER 11

EXERCISES 11.1

1. (i) 360°, 2 (ii) 1080°, 3 (iii) 180°, 4 (iv) 180°, 3 (v) 120°, 0.25 (vi) 480°, $\frac{2}{3}$

2. Graphics calculator outputs using a window [-360,360] and [–4,4].

(i)

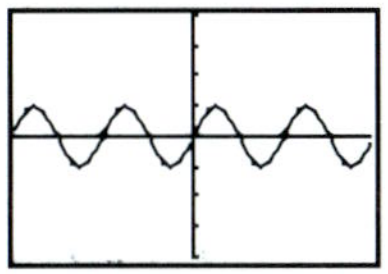

(ii)

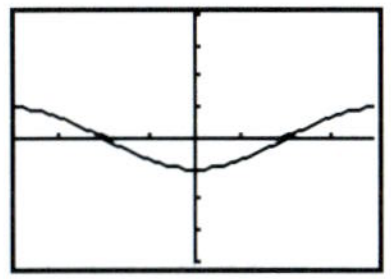

(iii)

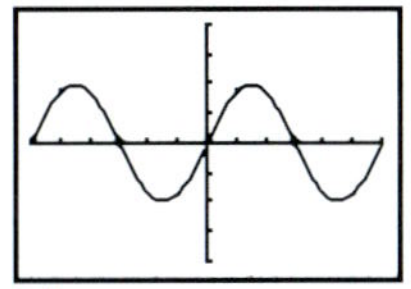

(iv)

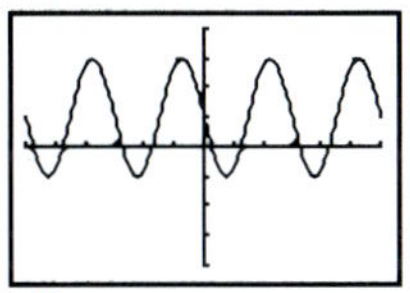

(v)

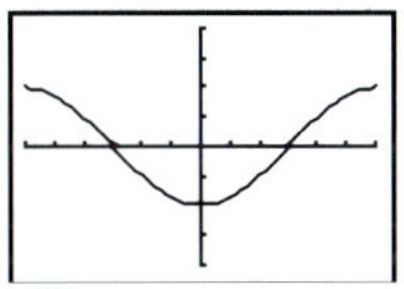

(vi)

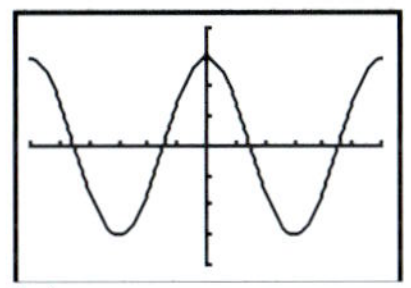

(vii)

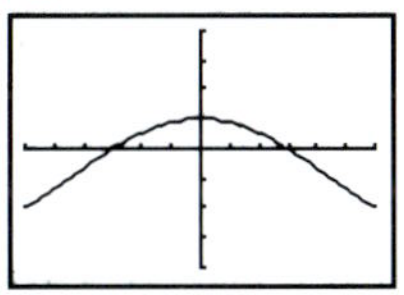

(viii)

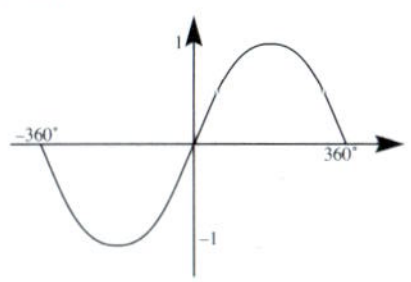

(ix)

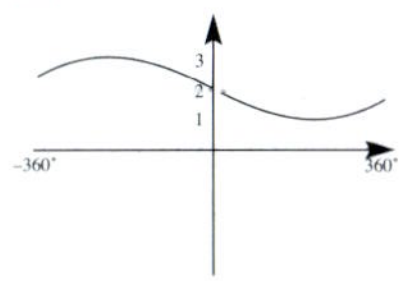

(x)

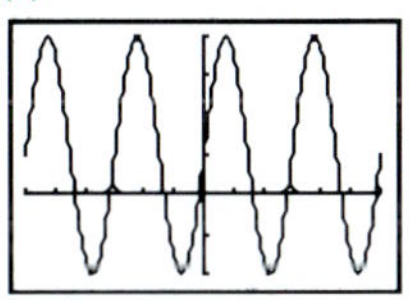

(xi)

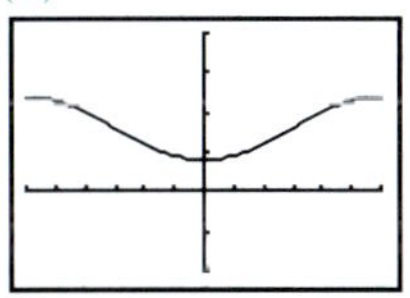

(xii)

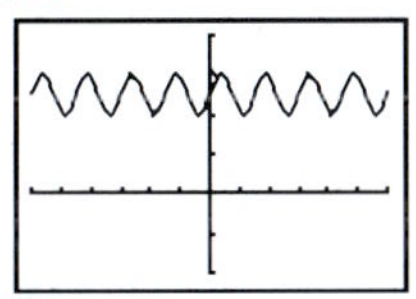

3. (i)

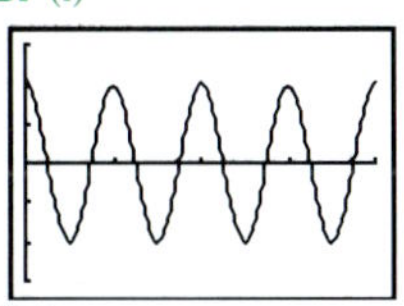

(ii)

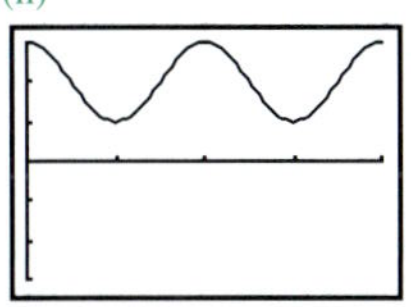

(iii)

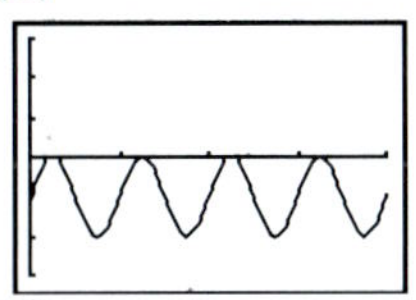

(iv)

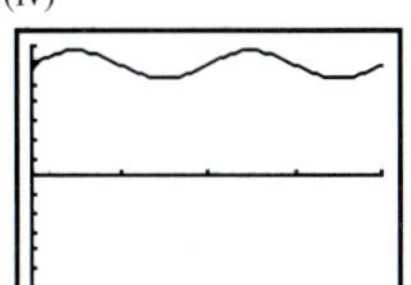

(v)

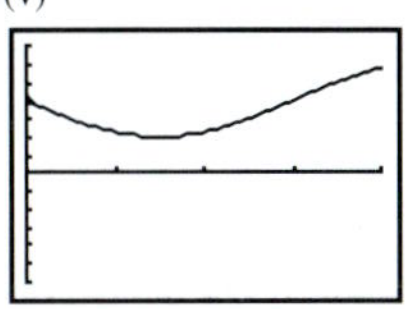

(vi)

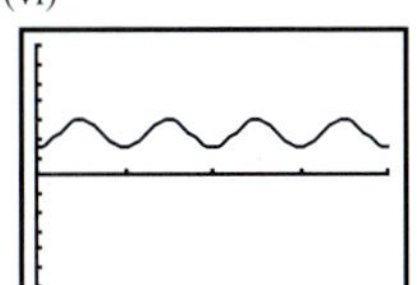

4. (a) $a = -4, b = 1, c = 2$ (b) $a = 0.2, b = 0.5, c = 0$

5. (a) Use graphics calculator to verify your graph. (b) i. 2.5 ii. [0.5, 5.5] (c) Use graphics calculator – to give $f(t) = \cos(45t)°$. (d) i. $k > 5.5$ or $k < 0.5$ ii. $k = 5.5, 0.5$ iii. $0.5 < k < 5.5$

EXERCISES 11.2

1. i. 5, 24, 11.46°, 19 ii. $T = 5\sin(15t - 171.89) + 19$ iii. 23.6° **2.** i. 3. 4.2, 2, 7
ii. $L = 3\sin(85.71t - 171.42) + 7$ i.e., $L = 3\sin 85.71(t-2) + 7$

3. i. 5, 11, 0, 7 ii. $V = 5\sin(32.73t) + 7$

4. i. 1, 11, 1.75, 12 ii. $P = \sin(32.73t - 57.28) + 12$

5. i. 2.7, 7, 2.22°, 6 ii. $S = 2.7\sin(51.43t - 114.17) + 6$

6. i. 0.6, 3.5, 0, 11 ii. $P = 0.6\sin(102.86t) + 11$

7. (a) 3000 (b) 1000, 5000 (c) $\frac{2}{3}$ **8.** (a) 6.5 m, 7.5 m (b) 1.58 sec, 3.42 sec

9. (a) 15000 (b) 12 months (c)

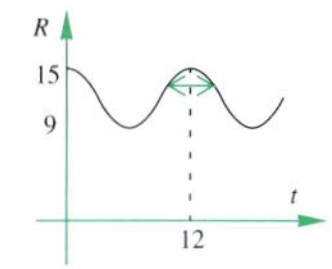

(d) 4 months

10. (a)

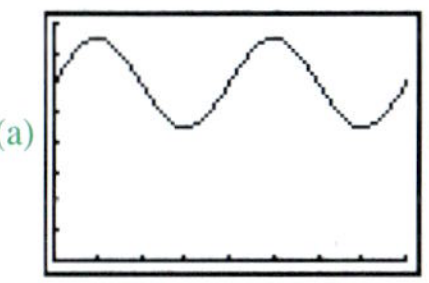

(b) i. 7, 11, 19, 23 ii. $[0, 7] \cup [11, 19] \cup [23, 24]$ (c) 14.90 m

11. (a) amp = 20, period = 1.2 sec (b) 120 mmHg (d) 0.40, 0.80

(c)

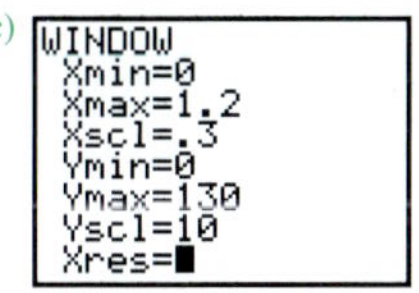

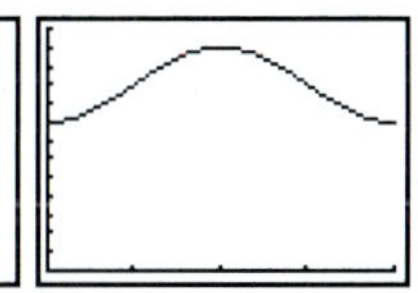

EXERCISES 11.3 – MISCELLANEOUS EXERCISES

1. (a)

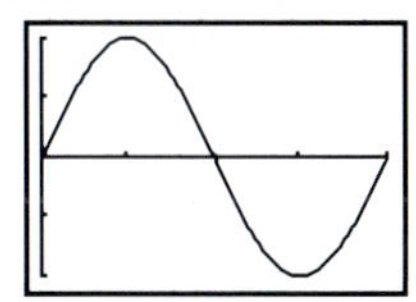

(b)

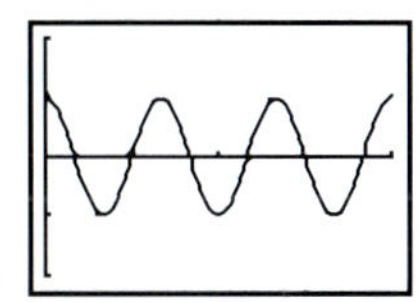

2. (a)

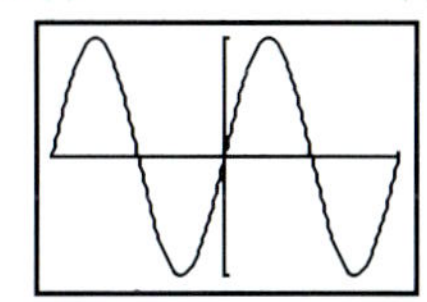

(b) 15°, 75°

3. (a)

t	0	4	8	12	16	20	24
h	5	6.73	3.27	5	6.73	3.27	5

(b)

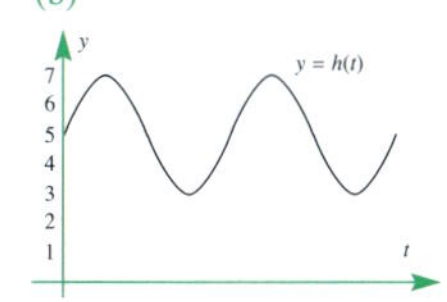

(c) (i) 7 (ii) 5.52
(d) 9, 21

4. (a) 60 sec

(b)

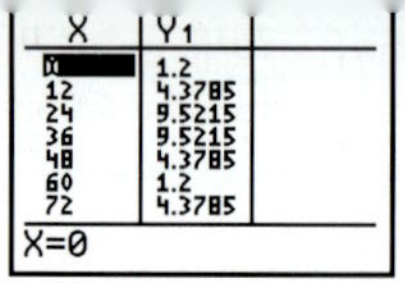

(c)

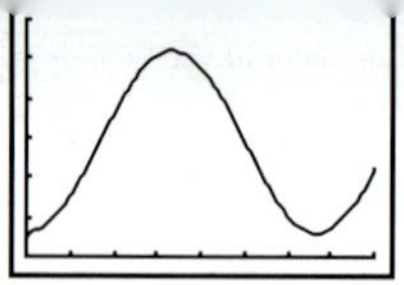

(d) 19.76

5. (a) (i) 5 (ii) 120° (iii) 3 (b) 20°

6. (i) (a) 120° (b) 2 (ii) (a) 720° (b) 4 **7.** 6

8. (i)

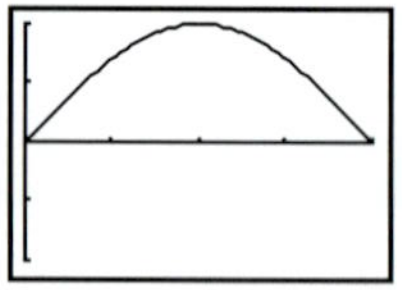

(ii)

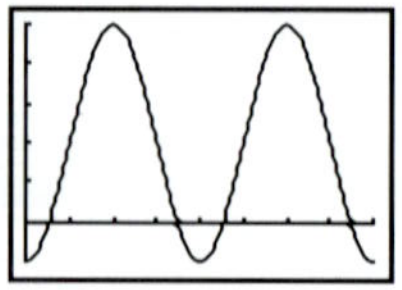

9. (a) Amp = 20 mmHg, period = 1.2 sec (b) 120 mmHg (c) Invert cos function, starts at (0, 80) range [80, 120]

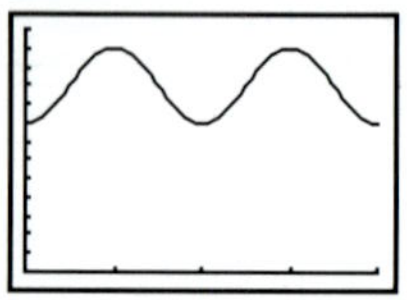

(d) 0.4, 0.8

(e) 33.33%

10. (a) amp = 0.8, period = 4.7, 2.6, 11 (b) $D = 0.8\sin(76.59t - 199.13) + 11$

EXERCISES 11.4 – GRADED REVISION EXERCISES

LEVEL 1

1. i. 0.3907 ii. 0.5736 iii. 0.9848 iv. –0.3420 v. –0.7660 vi. 0.2720

LEVEL 2

1. i. 2 ii. 2 iii. 7 iv. 1 v. 1 vi. 4.

2. i. 360° ii. 720° iii. 120° iv. 360° v. 1080° vi. 480°

3. i. $y = 1$ ii. $y = 2$ iii. $y = -3$

LEVEL 3

1. i.

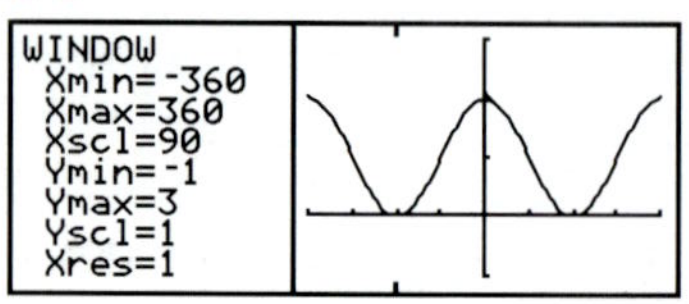

ii.

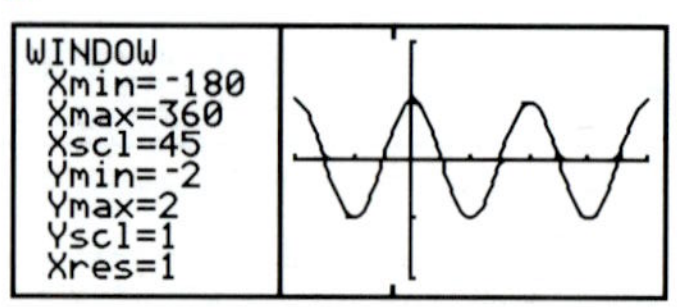

iii.

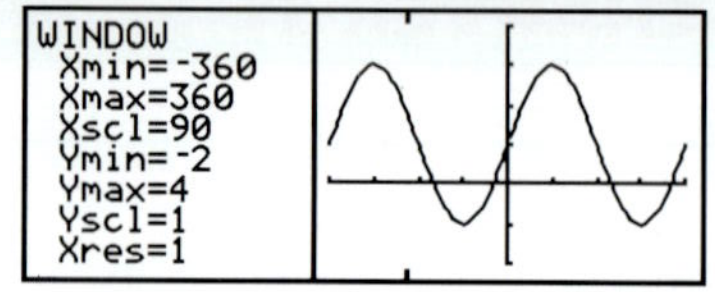

iv.

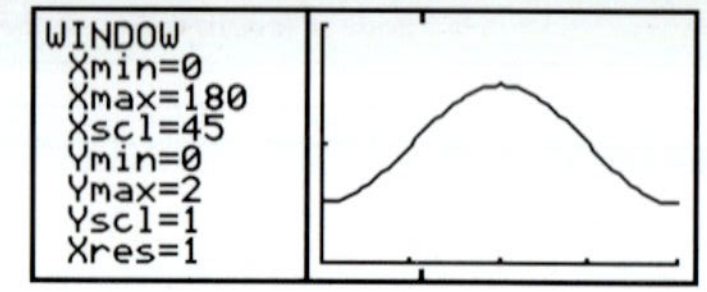

v.

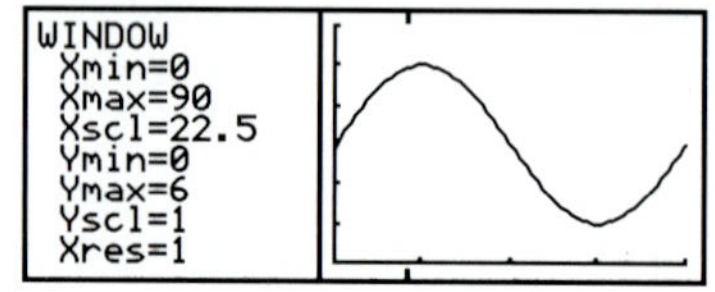

vi.

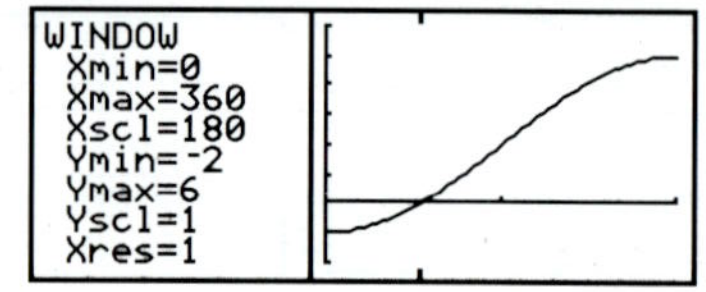

2. i. 45° ii. 60° iii. 270°

LEVEL 4

1. i. $f(x) = 1 - 2\sin x$ ii. $f(x) = 1 + 2\cos 2x$ iii. $f(x) = 2 - \cos 2x$ iv. $f(x) = 1 + \cos 4x$

2. i. (a) 12 (b) 23.2°C. ii. (a) max temp. = 24°C, min temp. 12°C

(b) max. occurs when $t = 12$, min. occurs when $t = 0, 24$.

iii.

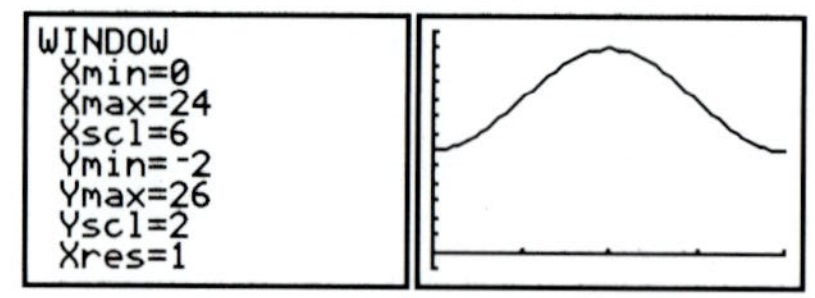

iv. Therefore, between 7:18 am and 4:42pm.

v. Rising most rapidly at t = 6 (i.e., 6 am). Using the gradient of the tangent or graphics calculator we have largest rate = 1.57°C/hr.

3. i. (a) 1 (b) 3.

ii.

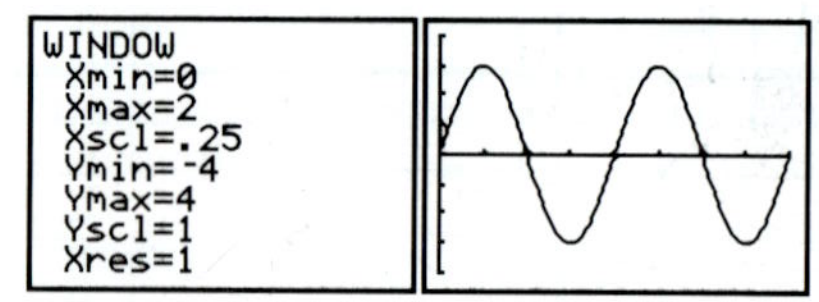

iii. ~ 15.59

4. i.

t hours	0	1	2	3	4	5	6	7	8
h metres	5.5	6.6	7	6.6	5.5	4.4	4	4.4	5.5

ii.

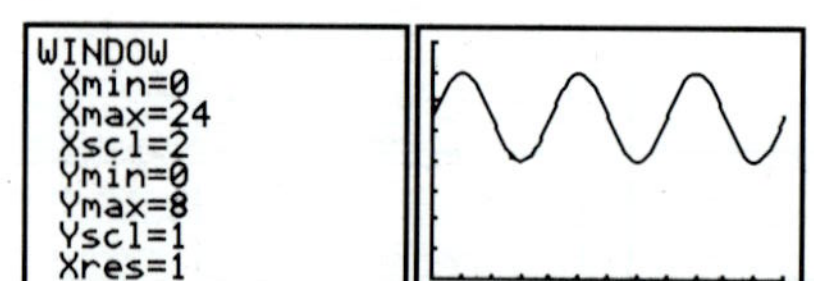

iii.

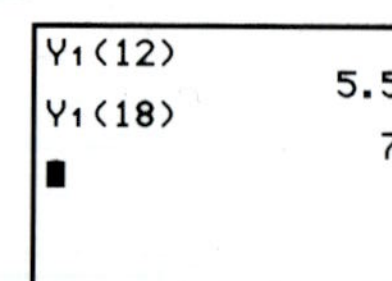

iv. 6, 14, 22

5. i. (a) 5 m (b) 1 m ii. $t = \frac{1}{2}, \frac{5}{2}, \frac{7}{2}, \frac{11}{2}$

iii. iv. 33.33%

```
WINDOW
 Xmin=0
 Xmax=6
 Xscl=.75
 Ymin=0
 Ymax=6
 Yscl=1
 Xres=1
```

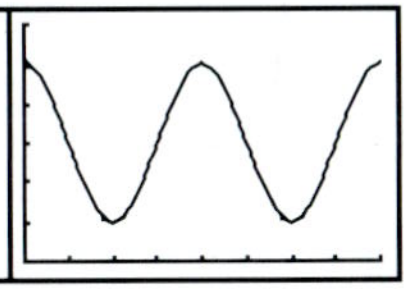

6. i. $k = 30$ ii. By observation, we have: $b = 11$ and $a = -10$.

iii. iv. 4, 8

v. 33.33% of the time.

```
WINDOW
 Xmin=0
 Xmax=12
 Xscl=3
 Ymin=0
 Ymax=22
 Yscl=2
 Xres=1
```

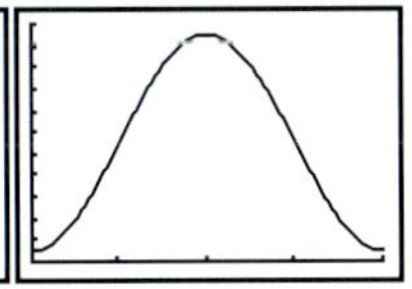

EXERCISES 11.5 – TOPIC TEST

1. i. 1 ii. $-\frac{1}{\sqrt{2}}$ **2.** i. 2 ii. 120° iii. 7 **3.** 10 **4.**

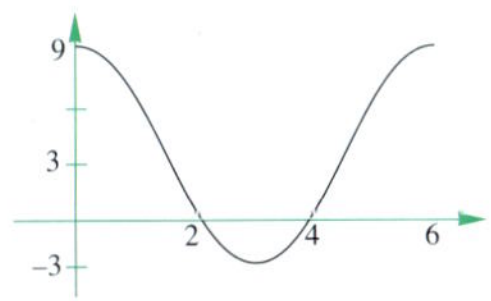

5. $k = 2$, Amp. = 2 transl. = +2, Equation: $f(x) = 2 + 2\sin(2x°)$

6. i. period = 52 weeks. ii. $N_{max} = 7000$ $N_{min} = 2000$

iii.

```
WINDOW
 Xmin=0
 Xmax=143
 Xscl=13
 Ymin=0
 Ymax=8000
 Yscl=1000
 Xres=1
```

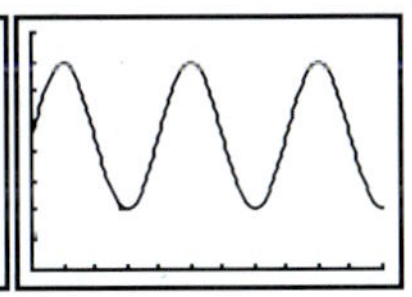

iv. (a) By observation, $t = 13$ (b) $t = 5.33$ v. $t = 31.32, 46.82, 83.32, 98.68$
Now, 31.32 weeks = 219 days, i.e., 9th August. While 46.82 weeks = 328 days, i.e., 24 Nov.
Therefore, population will be at least 3000 from August (9th) to November (24th).

ANSWERS TO CHAPTER 12

EXERCISES 12.1.1

1. i. (b) 4 (c) $t_n = 4n - 2$ ii. (b) –3 (c) $t_n = -3n + 23$ iii. (b) –5 (c) $t_n = -5n + 6$ iv. (b) 0.5 (c) $t_n = 0.5n$ v. (b) 2 (c) $t_n = y + 2n - 1$ vi. (b) –2 (c) $t_n = x - 2n + 4$ **2.** –28 **3.** 9,17 **4.** –43 **5.** 7 **6.** 7 **7.** –5 **8.** 0 **9.** (a) 41 (b) 31st **10.** 2, $\sqrt{3}$ **11.** (a) i. 2 ii. –3 (b) i. 4 ii. 11 **12.** $x - 8y$ **13.** $t_n = 5 + \frac{10}{3}(n-1)$ **14.** (a) –1 (b) 0

EXERCISES 12.1.2

1. (a) 145 (b) 300 (c) –170 **2.** (a) –18 (b) 690 (c) 70.4 **3.** (a) –105 (b) 507 (c) 224 **4.** (a) 126 (b) 3900 (c) 14th week **5.** 855 **6.** (a) 420 (b) –210

EXERCISES 12.1.3

1. 123 **2.** –3, –0.5, 2, 4.5, 7, 9.5, 12 **3.** 3.25 **4.** $a = 3$ $d = -0.05$ **5.** 10 000 **6.** 330 **7.** –20 **8.** 328 **9.** \$725, 37wks **10.** i. \$55 ii. 2750 **11.** (a) (i) 8m (ii) 40m (b) 84m (c) Dist = $2n^2 - 2n = 2n(n-1)$ (d) 8 (e) 26 players, 1300m **12.** (a) 5050 (b) 10200 (c) 4233 **13.** (a) 145 (b) 390 (c) –1845 **14.** (b) $3n - 2$

EXERCISES 12.2.1

1. (a) $r = 2, u_5 = 48, u_n = 3 \times 2^{n-1}$ (b) $r = \frac{1}{3}, u_5 = \frac{1}{27}, u_n = 3 \times \left(\frac{1}{3}\right)^{n-1}$ (c) $r = \frac{1}{5}, u_5 = \frac{2}{625}, u_n = 2 \times \left(\frac{1}{5}\right)^{n-1}$ (d) $r = -4, u_5 = -256, u_n = -1 \times (-4)^{n-1}$ (e) $r = \frac{1}{b}, u_5 = \frac{a}{b^3}, u_n = ab \times \left(\frac{1}{b}\right)^{n-1}$ (f) $r = \frac{b}{a}, u_5 = \frac{b^4}{a^2}, u_n = a^2 \times \left(\frac{b}{a}\right)^{n-1}$ **2.** (a) ±12 (b) $\frac{\pm\sqrt{5}}{2}$ **3.** (a) ±96 (b) 15th **4.** (a) $u_n = 10 \times \left(\frac{5}{6}\right)^{n-1}$ (b) $\frac{15625}{3888} \approx 4.02$ (c) $n = 5$ (4 times) **5.** $-2, \frac{4}{3}$ **6.** (a) i. \$4096 ii. \$2097.15 (b) 6.2 yrs **7.** $u_n = \frac{1000}{169} \times \left(\frac{12}{5}\right)^{n-1}, \frac{1990656}{4225} \approx 471.16$ **8.** 2.5,5,10 or 10,5,2.5 **9.** 53757 **10.** 108 952 **11.** (a) \$56 156 (b) \$299 284

EXERCISES 12.2.2

1. (a) 3 (b) $\frac{1}{3}$ (c) –1 (d) $-\frac{1}{3}$ (e) 1.25 (f) $-\frac{2}{3}$ **2.** (a) 216513 (b) 1.6384×10^{-10} (c) $\frac{256}{729}$ (d) $\frac{729}{2401}$ (e) $-\frac{81}{1024}$ **3.** (a) 11; 354292 (b) 7; 473 (c) 8; 90.90909 (d) 8; 172.778 (e) 5; 2.256 (f) 13; 111.1111111111 **4.** (a) $\frac{127}{128}$ (b) $\frac{63}{8}$ (c) $\frac{130}{81}$ (d) 60 (e) $\frac{63}{64}$ **5.** 4; 118096 **6.** \$2109.50 **7.** 9.28cm **8.** (a) $V_n = V_0 \times 0.7^n$ (b) 7 **9.** 54 **10.** 53.5gm; 50 weeks. **11.** 7 **12.** 9 **13.** –0.5, –0.7997 **14.** $r = 5$, 1.8×10^{10} **15.** \$8407.35 **16.** 1.84×10^{19} ~ 200 billion tonnes.

EXERCISES 12.2.3

1. Term 9 AP = 180, GP = 256. Sum to 11 terms AP = 1650, GP = 2047. **2.** 18. **3.** 12 **4.** 12, 7 **5.** 8 weeks (Ken \$220 & Bo-Youn \$255) **6.** (a) week 8 (b) week 12 **7.** (a) 1.618 (b) 121379 [~121400, depends on rounding errors]

EXERCISES 12.2.4

1. (i) $\frac{81}{2}$ (ii) $\frac{10}{13}$ (iii) 5000 (iv) $\frac{30}{11}$ **2.** $23\frac{23}{99}$ **3.** 6667 fish. [Nb: $t_{43} < 1$. If we use $n = 43$ then ans is 6660 fish]; 20 000 fish. Overfishing means that fewer fish are caught in the long run. **4.** 27 **5.** 48,12,3 or 16,12,9 **6.** (a) $\frac{11}{30}$ (b) $\frac{37}{99}$ (c) $\frac{191}{90}$ **7.** 128 cm **8.** $\frac{121}{9}$ **9.** $2+\frac{4}{3}\sqrt{3}$

EXERCISES 12.2.5

1. 3, –0.2 **2.** $\frac{2560}{93}$ **3.** $\frac{10}{3}$ **4.** $\frac{43}{18}, \frac{458}{99}, \frac{413}{990}$ **5.** 9900 **6.** 3275 **7.** 3 **8.** $t_n = 6n - 14$

9. 6 **10.** $-\frac{1}{6}$ **11.** i. 12 ii. 26 **12.** 9 cm, 12 cm **13.** ±2 **14.** (5, 5, 5) [trivial case], (5, –10, 20)

15. (a) 2, 7 (b) 2, 5, 8 (c) $3n - 1$ **16.** (a) 5 (b) 2 m

EXERCISES 12.3

1. \$2773.08 **2.** \$4377.63 **3.** \$1781.94 **4.** \$12216 **5.** \$35816.95 **6.** \$40349.37 **7.** \$64006.80 **8.** \$276971.93, \$281325.41 **9.** \$63762.25 **10.** \$98.62, \$9467.14, interest \$4467.52. Flat interest = \$6000 **11.** \$134.41, \$3790.44, 0.602% /month (or 7.22% p.a)

EXERCISES 12.4 – MISCELLANEOUS EXERCISES

1. (a) (i) 5 (ii) 3 (b) 5, 8, 11, 14, 17 (c) 62 (d) 100 **2.** 29 **3.** (b) (i) 10, 22, 34, 46 (ii) 238 (c) $d = 12, a = 10$, sum = 640 **4.** (a) 3 (b) 98415 (c) 147620 **5.** (a) \$3027 (b) \$5526.70

6. (a) 0.85 (b) \$6264.10 (c) (i) $u_n = 12000 \times (0.85)^n$ (ii) 8.53 years **7.** (a) 5.78 (b) 13 (c) 51.34

8. 288 **9.** –96 **10.** 7 terms **11.** 1000 numbers. **12.** (a) $1250 \times 1.12^5 \approx \$2203$ (b) 19 years

13. (a) $A_n = 12000 + 200n$ $B_n = 8000 \times 1.05^n$ (b) $n = 13$.

EXERCISES 12.5 – GRADED REVISION EXERCISES

LEVEL 1

1. i. A.P ii. A.P iii. Neither iv. G.P **2.** i. 13 ii. –19 iii. 6 iv. 16 **3.** i. 8 ii. 48. **4.** i. –40 ii. 1.875 iv. 49152 **5.** i. –285 ii. 15 iii. 505 iv. 32769

LEVEL 2

1. 45 **2.** 7. **3.** 4782968 **4.** 22 years **5.** $b = \frac{a+c}{2}$ **6.** \$360,000. **7.** $x = \pm 84$ **8.** $\frac{121}{9}$ **9.** 9

LEVEL 3

1. (a) 9 (b) 93 (c) 48 **2.** (a) 313 (b) 1738 (c) 2 (d) 6141 **3.** 28 **4.** $b = \frac{1}{3}(d + 2a)$.

LEVEL 4

1. 21 **2.** 21 months. **3.** 15% **4.** (a) 4950 (b) 24th day (c) 130 150 **5.** (c) \$5,692.84

EXERCISES 12.6 – TOPIC TEST

1. 28 **2.** 425 **3.** (a) \$26,226.75

(b)

Years since purchase (t)	1	2	3	4
Value of car (\$ V)	28050	26226.75	24522.01	22928.08

(c) $V_t = 30000 \times (0.935)^t, t = 0, 1, 2,$ **4.** (a) 10 (b) $S_n = \frac{n}{2}[2(1) + (n-1)1] = \frac{n^2}{2} + \frac{n}{2}$

(c) 10 rows **5.** (a) 16326 ratchets (b) 70399 ratchets (c) $n = 10.006468$ (d) the 26th year

ANSWERS TO CHAPTER 13

EXERCISES 13.1

1. (a) hammer is an element of set A. (b) axe is an element of set C. (c) Tuesday is an element of set B. (d) Tuesday is not an element of set A. (e) January is an element of set B. (f) Sunday is not an element of set C. **2.** (a) $32 \in C$ (b) $45 \in N$ (c) $Green \notin K$ (d) $Mary \notin P$ (e) $Horse \notin M$ (f) $Banana \in H$ **3.** (a) i. T ii. F (b) i. T ii. T **4.** (a) $A = \{x | x = 2n, n = 1, 2, 3, 4\}$

(b) $B = \{x | 40 < x^2 < 169$, where x^2 is odd$\}$ (c) $C = \{x | 0 < x < 12, x \in \mathbb{N}\}$

(d) $D = \{x | 0 < x < 25$, and x is a prime number$\}$ (e) $E = \{x | x$ are factors of 28$\}$

5. (a) {4, 9, 16, 25, 36} (b) {(1,6), (4,4), (7,2)} (c) { } or ∅ **6.** (a) T (b) T (c) F **7.** (e) Non empty sets: (a) {1, 3}, (b) {2, 4} (c) {92} (d) {3, 4} **8.** (a) T (b) F (c) T (d) F (e) F (f) F

EXERCISES 13.2

1. (a) F (b) T (c) T (d) F (e) F (f) T (g) T (h) F (i) F (j) F (k) F **2.** (a) $\frac{5}{11}$ (b) $\frac{316}{99}$ (c) $\frac{1763}{330}$

(d) $\frac{8441}{660}$ **3.** (a) $x = \frac{8}{5} \therefore x \in \mathbb{Q}$ (b) {(0,5), (1,4), (2,3), (3,2), (4,1), (5,0)}

(c) {(1,2), (1,3), (1,4), (1,5), (2,1), (2,3), (2,4), (2,5), (3,1), (3,2), (3,4), (3,5), (4,1), (4,2), (4,3), (4,4), (5,1), (5,2), (5,3), (0,6), (6,0)} (d) $\{x | 3 < x \le 6\}$

EXERCISES 13.3

1. (a) T (b) F (c) F (d) T (e) T (f) F (g) T (h) T (i) T **2.** (a) 4 (b) 12 (c) 7 (d) 11 **3.** (a) Finite (b) Infinite (c) Finite (d) Finite (e) Infinite (f) Finite **4.** (a) { } (b) {a , e} (c) {Monday, Tuesday, Wednesday, Thursday, Friday, Saturday, Sunday} (d) {4, 5} (e) {0, 1, 2, 3, 4, 5} (f) {–4, –3, –2, –1, 0, 1, 2, 3, 4, 5} (g) Not possible (h) Not possible **5.** (a) 0 (b) 2 (c) 7 (d) 2 (e) 6 (f) 10 (g) Infinite (h) Infinite **6.** G and H **7.** (a) $\{x | x \in \mathbb{N}, 3 < x < 18\}$

(b) $\{x | x \in \mathbb{R}, x < 12\}$ (c) $\{x | x \in \mathbb{Z}, -56 \le x \le 45\}$ (d) $\{x | x \in \mathbb{Q}, -5 \le x \le 5\}$

EXERCISES 13.4

1. (a) T (b) F (c) T (d) T (e) T (f) T (g) F (h) T (i) F **2.** (a) T (b) F (c) T (d) T (e) T (f) T (g) F (h) T (i) F **3.** { }, {α}, {β}, {μ}, {σ}, {α, β}, {α, μ}, {α, σ}, {β, μ}, {β, σ}, {μ, σ}, {α, β, μ}, {α, β, σ}, {β, μ, σ}, {α, μ, σ}, {α, β, μ, σ} **4.** 32 **5.** (a) 31 (b) 15 **6.** (a) F (b) F (c) T (d) T **7.** (a) Yes (b) No (c) Yes (d) Yes **8.** (a) {1} {2} {1,2} { } (b) { } (c) {{1}}, {{1}, {1,2}}, {{1}, 3}, {{1,2},3}, {3}, {{1}, {1,2},3} { } **9.** (a) T (b) T (c) F (d) F (e) T (f) F (g) F (h) T **10.** 6

11. (a) {6, 12, 18} (b) {4, 16} (c) {9} (d) {} (e) {2, 3, 4, 6, 8, 9, 10, 12, 14, 15, 16, 18}
(f) {1, 3, 4, 6, 9, 12, 15, 16, 18} (g) {1, 2, 4, 6, 8, 9, 10, 12, 14, 16, 18}
(h) {1, 2, 3, 4, 6, 8, 9, 10, 12, 14, 15 16, 18} (i) {4, 9, 16} (j) {3, 4, 6, 9, 12, 15, 16, 18}
(k) {2, 4, 6, 8, 9, 10, 12, 14, 16, 18} **12.** (a) 3 (b) 2 (c) 1 (d) 0 (e) 12 (f) 9 (g) 11 (h) 13
13. (a) {–3, –2} (b) {–3, –2} (c) {–3, –2, –1, 0, 1, 2, 3} (d) {–3, –2} (e) {–5, –4, –3, –2, –1, 0, 1, 2, 3} (f) {–3, –2, –1, 0, 1, 2, 3, 4, 5} (g) {–5, –4, –3, –2, –1, 0, 1, 2, 3, 4, 5}
(h) {–5, –4, –3, –2, –1, 0, 1, 2, 3, 4, 5} (i) {–3, –2, –1, 0, 1, 2, 3} (j) {–3, –2, –1, 0, 1, 2, 3}
(k) {–5, –4, –3, –2, –1, 0, 1, 2, 3} **14.** (a) 2 (b) 2 (c) 7 (d) 2 (e) 9 (f) 9 (g) 11 (h) 11
15. (a) $\{x \mid x \in \mathbb{R}, -3 \le x \le -2\}$ (b) $\{x \mid x \in \mathbb{R}, -3 < x \le -2\}$ (c) $\{x \mid x \in \mathbb{R}, -3 < x < 4\}$
(d) $\{x \mid x \in \mathbb{R}, -3 < x \le -2\}$ (e) $\{x \mid x \in \mathbb{R}, -5 \le x < 4\}$ (f) $\{x \mid x \in \mathbb{R}, -3 \le x \le 5\}$
(g) $\{x \mid x \in \mathbb{R}, -5 \le x \le 5\}$ (h) $\{x \mid x \in \mathbb{R}, -5 \le x \le 5\}$ (i) $\{x \mid x \in \mathbb{R}, -3 < x < 4\}$
(j) $\{x \mid x \in \mathbb{R}, -3 \le x < 4\}$ (k) $\{x \mid x \in \mathbb{R}, -5 \le x < 4\}$
16. (a) { b, d, f, g, j, k, l, n, o, p, q, r, u, v, w, x, y, z } (b) { a, c, i, o, u }
(c) { c, e, f, g, h, i, k, q, r, s, t, u, v, w, x, y } (d) { o, u } (e) { u } (f) { o, u } (g) { u }
(h) { b, d, e, f, g, i, j, k, l, n, o, p, q, r, u, v, w, x, y, z }
17. (a) { –10, –9, –8, –7, –6, 3, 4, 5, 6, 7, 8, 9, 10 }
(b) { –10, –9, –8, –7, –6, –5, –4, –3, –2, 7, 8, 9, 10 }
(c) { –10, –9, –8, –7, –6, –5, –4, –3, –2, –1, 0, 3, 4, 5, 6,7,8, 9, 10 }
(d) { –10, –9, –8, –7, –6, 7, 8, 9, 10 (e) { –10, –9, –8, –7, –6, 7, 8, 9, 10 }
(f) {–10, –9, –8, –7, –6, 7, 8, 9, 10 } (g) { –10, –9, –8, –7, –6, 7, 8, 9, 10 }
(h) { –10, –9, –8, –7, –6, –5, –4, –3, –2, 3, 4, 5, 6, 7, 8, 9, 10 }
18. (a) $\{x \mid x \in \mathbb{Z}, x < -5 \text{ or } x > 2\}$ (b) $\{x \mid x \in \mathbb{Z}, x < -1 \text{ or } x > 6\}$
(c) $\{x \mid x \in \mathbb{Z}, x \le 0 \text{ or } x \ge 3\}$ (d) $\{x \mid x \in \mathbb{Z}, x < -5 \text{ or } x > 6\}$ (e) $\{x \mid x \in \mathbb{Z}, x < -5 \text{ or } x > 6\}$
19. (a) $\{x \mid x \in \mathbb{R}, -10 \le x \le -5 \text{ or } -3 < x \le 10\}$ (b) $\{x \mid x \in \mathbb{R}, -10 \le x \le 3\}$
(c) $\{x \mid x \in \mathbb{R}, -10 \le x \le -3, 4 \le x \le 10\}$ (d) $\{x \mid x \in \mathbb{R}, -10 \le x \le -5 \text{ or } -3 < x \le 3\}$
(e) $\{x \mid x \in \mathbb{R}, -10 \le x \le -5\}$ **20.** (a) 28 (b) 19 (c) 25

EXERCISES 13.5

1. (a) i. ii. iii.
(b) i. ii. iii.
(c) i. ii. iii.
(d) i. ii. iii.

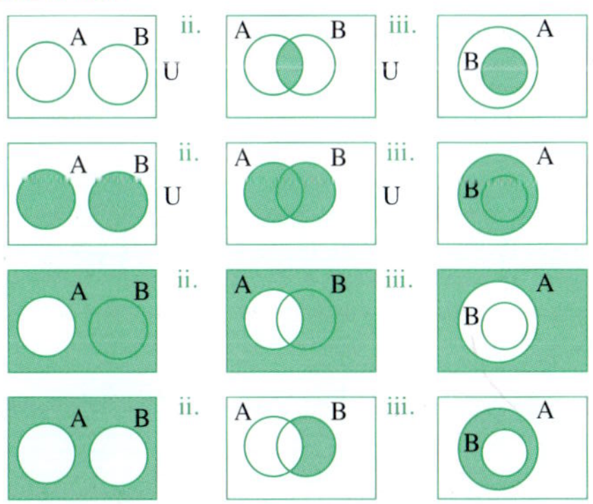

2. (a) A = {3, 6, 9, 12, 15, 18, 21, 24, 27, 30, 33, 36, 39}
B = {4, 8, 12, 16, 20, 24, 28, 32, 36}
(b) Multiples of both 3 and 4, less than 40
(c) (i) 13 (ii) 9 (iii) 3 (iv) 19 (d) 19 = 13 + 9 – 3
(e)

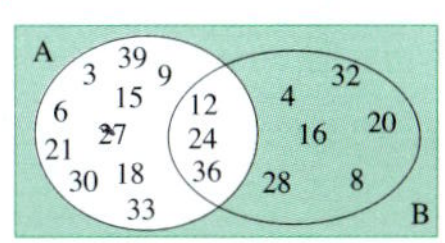

3. (a) A = { 1, 2, 3, 4, 6, 9, 12, 18, 36 }
B = { 1, 2, 3, 4, 5, 6, 10, 12, 15, 20, 30, 60 }
(b) Numbers that are both factors of 36 and 60
(c) (i) 9 (ii) 12 (iii) 6 (iv) 15 (d) 15 = 12 + 9 – 6
(e)

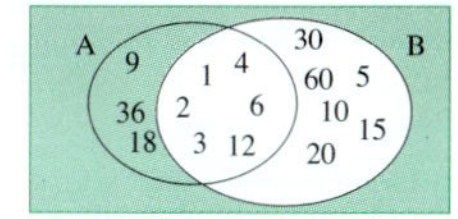

4. (a) A = {2, 3, 5, 7, 11, 13, 17, 19, 23, 29}
B = {1, 3, 5, 7, 9, 11, 13, 15, 17, 19, 21, 23, 25, 27, 29}
(b) Odd numbers less than 30 that are prime numbers
(c) (i) 10 (ii) 15 (iii) 9 (iv) 16 (d) 16 = 10 + 15 – 9
(e)

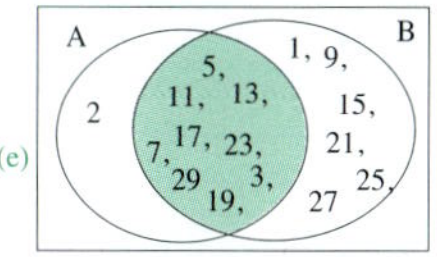

5. (a)

(b) The fruits that are common to both sets
(c) (i) 5 (ii) 3 (iii) 3 (iv) 5 (d) 5 = 5 + 3 – 3

6. (a) A = {1, 2, 3, 4, 6, 9, 12, 18, 36}
B = {6, 12, 18, 24, 30, 36, 42, 48, 54, 60}
C = {2, 4, 6, 8, 10, 12, 14, 16, 18, 20}
(b) Nunbers that are factors of 36, multiples of 6 and even. (c)
(i) 9 (ii) 10 (iii) 10 (iv) 3 (v) 19 (d)

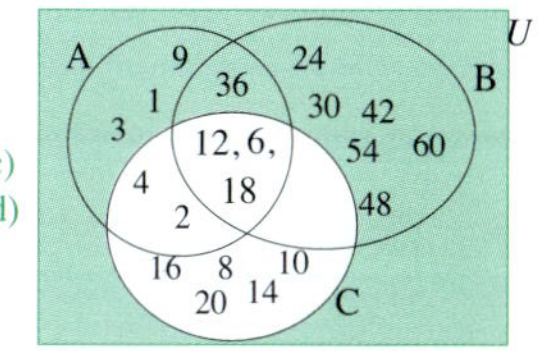

7. (a) (i) (ii) (iii) (iv)
(v) (vi) (vii)

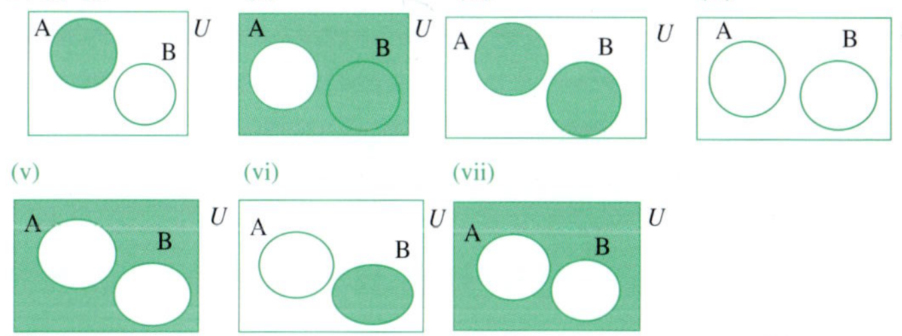

8. (a) (i) (ii) (iii) (iv)
(v) (vi) (vii)

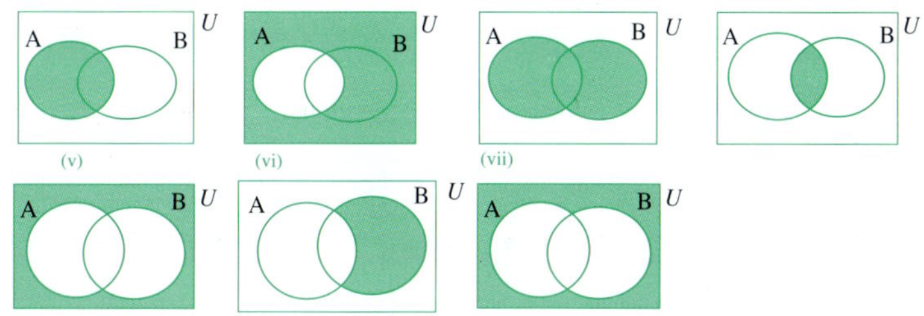

9. (a) (i) (ii) (iii) (iv) (v) (vi) (vii)

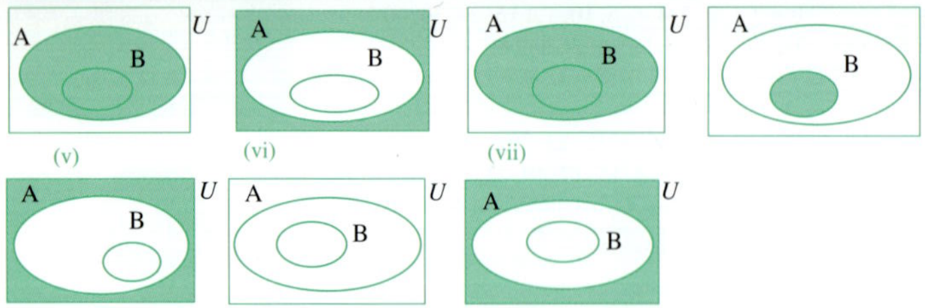

10. (a) (i) (ii) (iii) (iv) (v) (vi) (vii) (viii)

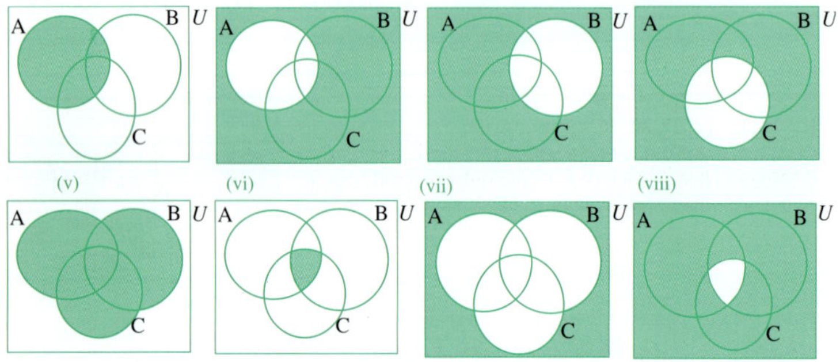

11. (a) A (b) ∅ (c) U (d) A (e) A (f) A (g) A (h) A (i) $A \cup B$ (j) $B \cap A$

EXCERCISES 13.6.1

1. (a) (b) (c) (d) (e)

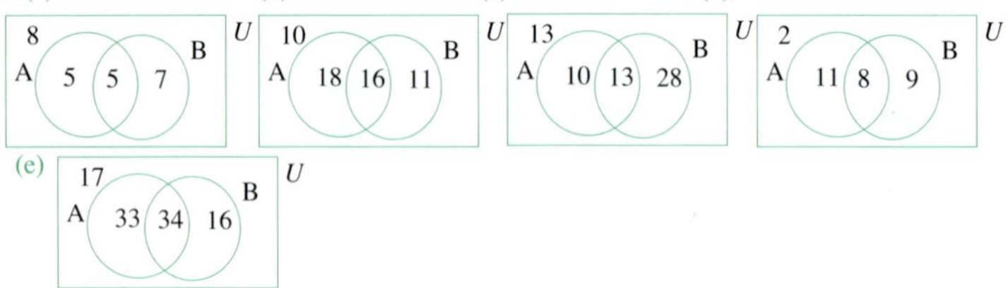

2.

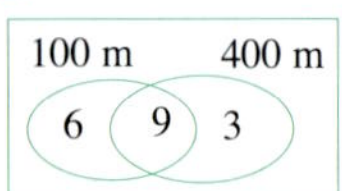

3.

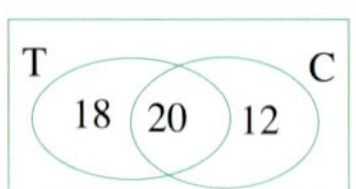

4.

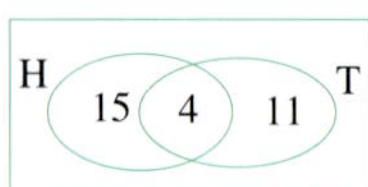

5. $p = 15$

6. Inconsistent information–no Venn diagram possible

7. (a) 65 (b) 95

8. 240

9.

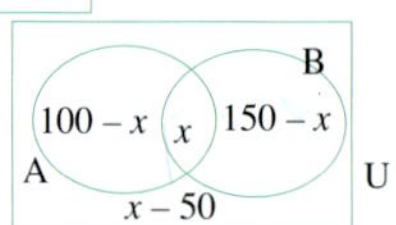

Need: $50 \le x \le 100$

EXERCISES 13.6.2

1. **2.** **3.**

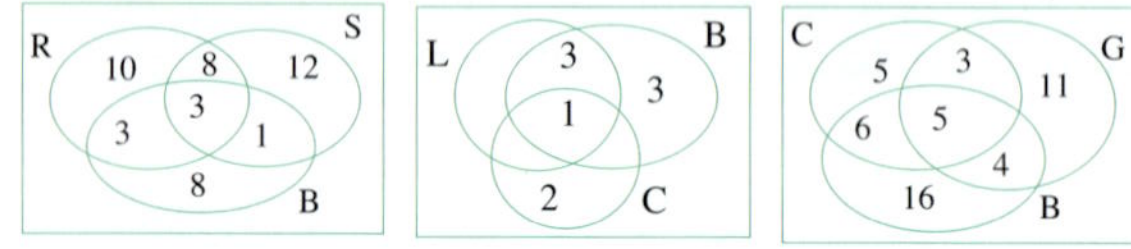

4. (a) 44 (b) 0 **5.** (a) 44 (b) 9 **6.** 18 **7.** (a) 54 (b) 12

EXERCISES 13.7 – MISCELLANEOUS EXERCISES

1. (a) F (b) T (c) F (d) F (e) F (f) T (g) T (h) F (i) T (j) T (k) T (l) T

2. (a) {1, 3, 5} (b) {2, 4} (c) ∅ (d) ∅ **3.** (a) ∅ (b) {a, e, i, o, u, b, c, d, f, g, h} (c) ∅ (d) {a, e, g, h} (e) {g, h} **4.** (a) i.{1, 2} ii. 2 (b) i. {6, 8, 10} ii. 3 (c) i. {5, 9, 13, 17, 21} ii. 5 (d) i. {3, 5, 7, 9, 11, 4, 6, 8, 10, 12, 13, 14, 15, 16} ii. 14 (e) i. {(2,2), (4,4), (6,6)} ii. 3

5. (a) Finite (b) Infinite (c) Finite (d) Finite (e) Infinite **6.** (a) {3, 4, 5 ,6 , . . . } (b) {1, 2, 3, 4,5 ,6 ,7 ,8} (c) {1, 2, 3,4} (d) {1, 2, 3, 4, 5} **7.** (a) No (b) Yes **8.** (a) {8} (b) {±2} (c) {2}

9. (a) i. 4 ii. {0, 4} iii. 2 (b) i. 2 ii. {1.5, 2, 3} iii. ∅ **10.** (a) {1, 2, 3, 4, 5, 6, 7, 9} (b) {3, 4, 7} (c) {9} **11.** $A \cap B = A$ **12.** (ii) & (iv) are correct

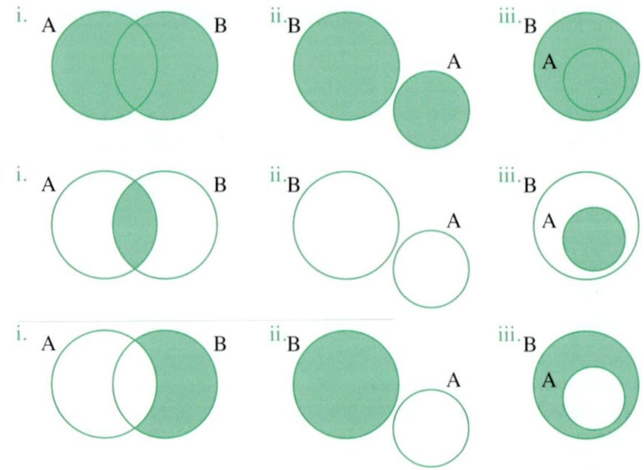

14. (a) 0 (b) 2 (c) 4 (d) 4 **15.** (a) {0, 1, 2, 3, 4, 5, 6} (b) {7, 8} (c) {0, 1, 5, 6, 7, 8} (d) {0, 1, 2, 3, 4} (e) {3, 4, 5, 6, 7, 8} (f) {5, 6} **16.** (a) {1, 2, 3, 4, 5} (b) {1} (c) {3, 5} (d) {6, 7, 8, 9} (e) ∅ (f) ∅

17. (a) (b) (c) (d)

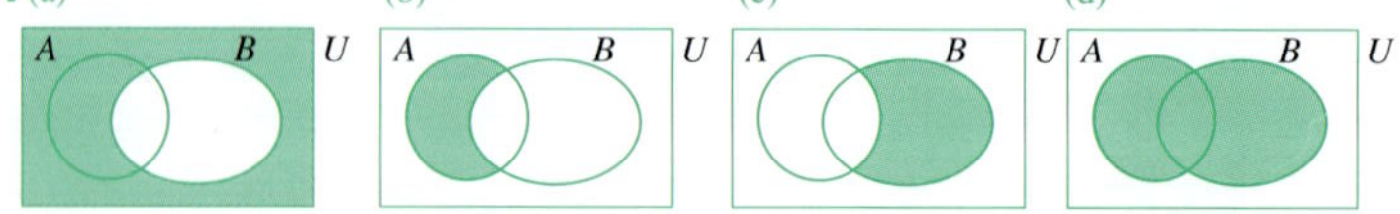

18. (a) F (b) T (c) F **19.** (a) 14 (b) 60 (c) 48 (d) 60 **20.** (a) 20 (b) 10 (c) 10

21.

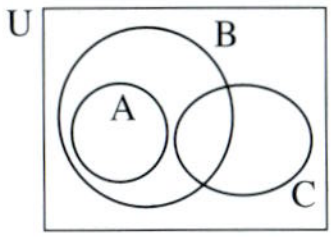

22. (a) i. 22 ii. 5 (b) $A \cap B = \{5\}$, $A \cup B = \{1, 2, 3, 4, 5, 6, 7, 8, 9\}$
23. (a) 7 (b) 49

24. (a) $B \cap C = \{x | x \in A$ where x is a perfect square prime number$\}$

$B \cap D = \{x | x \in A$ where x is a prime number whose first digit is greater than its second digit$\}$

(b) $B \cap D = \{\ \} = \varnothing$, $C \cap D = \{31, 43, 53, 61\}$

25. (a)

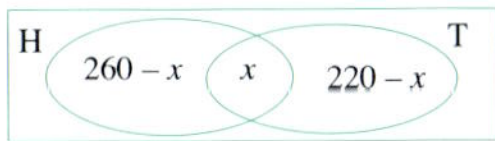

(b) 80 **26.** (a) 22 (b) 3 **27.** (a) 15 (b) 14 (c) 8

EXERCISES 1.83 – GRADED REVISION EXERCISES

LEVEL 1

1. $\{3, 4, 5, 6, 7\}$ **2.** 4

3. i.

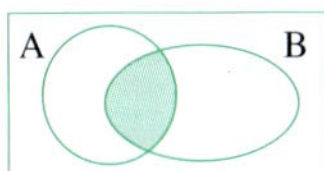

ii.

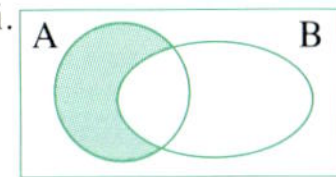

iii.

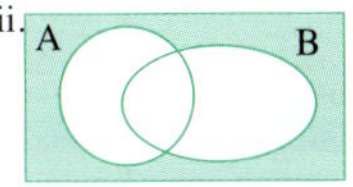

4. $\{\ \}, \{4\}, \{8\}, \{12\}, \{4, 8\}, \{4, 12\}, \{8, 12\}, \{4, 8, 12\}$ **5.** 16

LEVEL 2

2.

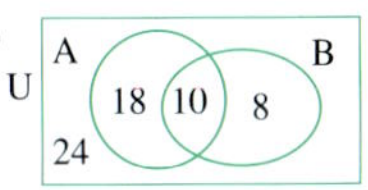

3. (a)

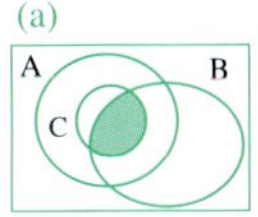

(b)

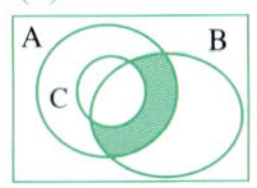

(c)

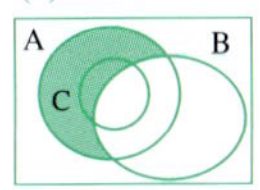

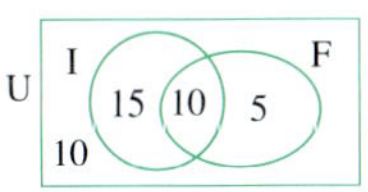

5. (a) 5 (b) $\{3, 4, 5, 6, 8\}$

LEVEL 3

1. (a) 1 or 2 (b) 0 or 1

LEVEL 4

1. $A' \cup B' \cup (A \cap B \cap C') = (A \cap B \cap C)'$

EXERCISES 13.9 – TOPIC TEST

1. i.

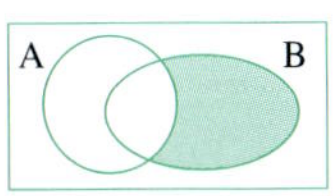

ii.

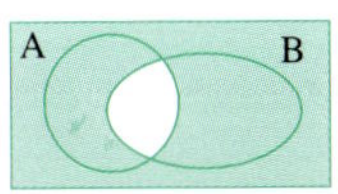

2. (a) i. $\{4, 8, 12\}$ ii. $\{1, 2, 4, 5, 8, 10\}$

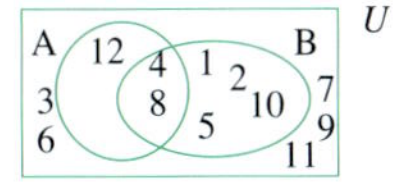

3.

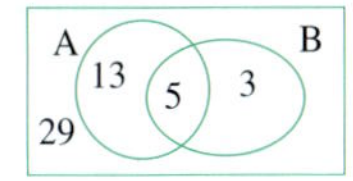

$n(A \cap B') = 13$

4. (a)

	The Fanom is their favourite	The Fanom is not their favourite	Total
Justecia is their favourite	x	12	$12 + x$
	25	$5 - x$	$30 - x$
Total	$25 + x$	$15 - x$	42

(b)

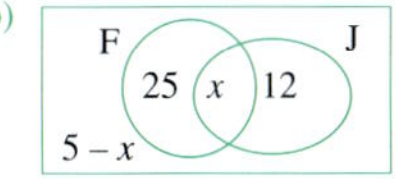

(c) 12

5. i. 12 ii. 17 iii. 3

ANSWERS TO CHAPTER 14

EXERCISES 14.1.1

1. (a) (b) (d) (f) (h) (i) (k) (l)

EXERCISES 14.1.2

1. (a) Brutus is not sick (b) The cup is not empty (c) Monday is not a holiday (d) There are not twelve months in a year (e) 4 is not less than 6 (f) Ismael does not study Geography (g) Yoshi does not play football (h) All roses are not red (i) March does not have 31 days (j) Not all quadrilaterals are rectangles (k) Carlos does not live in Japan

EXERCISES 14.1.3

1. (a) All trees are green and all mammals are warm blooded (b) Josh and Anne study French (c) Abdul plays both football and squash (d) The moon orbits the earth and the earth orbits the sun (e) It is raining and today is Saturday

2. (a) p: Jennifer studies Physics q: Jennifer studies Chemistry (b) p: All mammals are warm blooded q: All roses are red (c) p: Ruth is 16 years old q: Janet is 17 years old (d) p: Today is Sunday q: It is fine (e) p: Ronnie plays football q: Renee plays tennis

3. (a) i. P = $\{4, 8, 12, 16\}$, Q = $\{6, 12, 18\}$ iii. $p \wedge q$ truth set = $\{12\}$ ii.

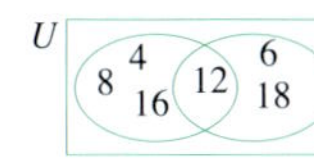

(b) i. P = $\{1, 4, 9, 16, 25, 36, 49, 64, 81\}$ Q = $\{6, 16, 26, 36\}$ ii.
iii. $p \wedge q$ truth set = $\{16, 36\}$

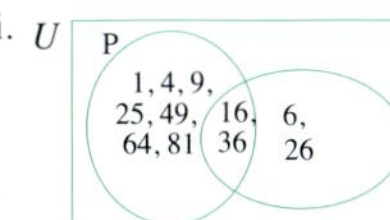

(c) i. P = $\{2, 3, 5, 7, 11, 13, 17, 19\}$ Q = $\{3, 6, 9, 12, 15, 18\}$ ii.
iii. $p \wedge q$ truth set = $\{3\}$

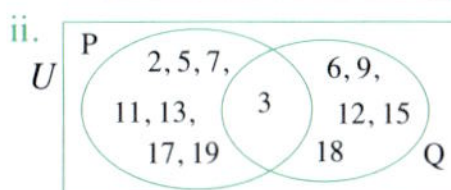

4. (a) All trees are green or all mammals are warm blooded
(b) Josh or Anne study French (c) Abdul plays football or squash (d) The moon orbits the earth or the earth orbits the sun (e) It is raining or today is Saturday

5. (a) p: Josh will buy a TV q: Josh will buy a stereo
(b) p: Yoshi plays tennis q: Mohammed plays squash
(c) p: The moon orbits the earth q: The moon orbits the sun
(d) p: Emus are birds q: Goats are horses
(e) p: Ruth likes watching the theatre q: Ruth likes watching sports

6. (a) i. P = { 4, 8, 12, 16 } Q = { 6, 12, 18 } ii.

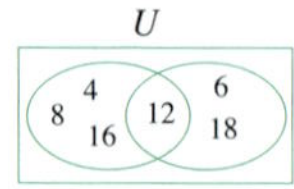

iii. $p \vee q$ truth set = { 4, 6, 8, 12, 16, 18 }

(b) (i) P = {1, 4, 9, 16 25, 36, 49, 64, 81} Q = { 6, 16, 26, 36} ii.

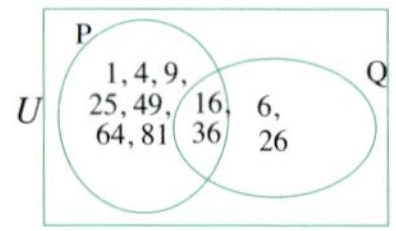

iii. $p \vee q$ truth set = { 1, 4, 6, 9, 16, 25, 26, 36, 49, 64, 81}

(c) i. P ={2, 3, 5, 7, 11, 13, 17, 19} Q = { 3, 6, 9, 12, 15, 18 } ii.

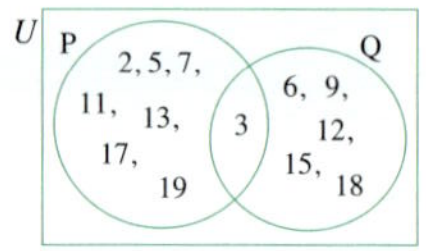

ii. $p \vee q$ truth set = {2, 3, 5, 6, 7, 9, 11, 12, 13, 15, 17, 18, 19}

EXERCISES 14.1.4

1. (a) If x is a multiple of 9 then x is divisible by 3 (b) If Yoshi is sick then Yoshi is not at work (c) If Paris is in France then the sky is blue (d) If this is the right time for an argument then it is the right place for an argument (e) If John works hard then John earns money

2. (a) p: A number is divisible by 10 q: A number ends in zero
(b) p: I will earn more money q: I work hard
(c) p: The flowers will grow q: There is enough rain
(d) p: A number is divisible by 4 q: A number is even
(e) p: I think q: I am

3. (a) i. P = { 16, 25, 49 } Q = { 15, 16, 19, 25, 26, 29, 35, 36, 39, 45, 46, 49 }

ii. 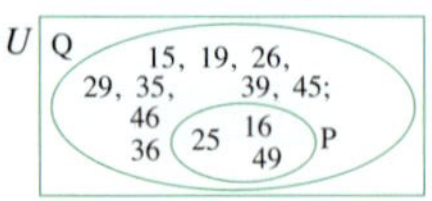

(b) 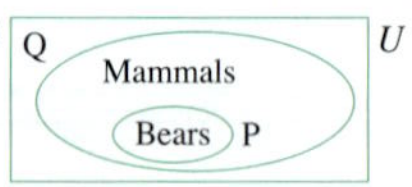

EXERCISES 14.1.5

1. (a) Converse : If a number ends in zero then it is divisible by 10
Inverse : If a number is not divisible by 10 then it does not end in zero
Contrapositive: If a number does not end in zero then it is not divisible by 10
(b) Converse : I work hard only if I will earn more money
Inverse : If I will not earn more money then I will not work hard
Contrapositive : If I do not work hard then I will not earn more money
(c) Converse : There is enough rain only if the flowers will grow
Inverse : The flowers will not grow only if there is not enough rain
Contrapositive : If there is not enough rain then the flowers will not grow
(d) Converse : If a number is even then it is divisible by four
Inverse : If a number is not divisible by four then it is not even
Contrapositive : If a number is not even then it is not divisible by four
(e) Converse : If I am then I think
Inverse : If I do not think then I am not
Contrapositive : If I am not then I do not think

EXERCISES 14.1.6

1. (a) I will paint the house if and only if it is fine (b) We will go to Disneyland if and only if there are exciting rides (c) I will take my umbrella if and only if it is raining (d) John will hit a home run if the pitcher is useless

2. (a)

p	q	$\neg p$	$\neg p \vee q$
T	T	F	T
T	F	F	F
F	T	T	T
F	F	T	T

(b)

p	q	$\neg$p	$\neg$q	$\neg p \vee \neg q$
T	T	F	F	F
T	F	F	T	T
F	T	T	F	T
F	F	T	T	T

(c)

p	q	$\neg$p	$\neg$q	$\neg p \wedge \neg q$
T	T	F	F	F
T	F	F	T	F
F	T	T	F	F
F	F	T	T	T

(d)

p	q	$\neg$p	q	$\neg p \Rightarrow q$
T	T	F	T	T
T	F	F	F	T
F	T	T	T	T
F	F	T	F	F

(e)

p	q	$\neg$q	$p \vee \neg q$	q	$(p \vee \neg q) \Rightarrow q$
T	T	F	T	T	T
T	F	T	T	F	F
F	T	F	F	T	T
F	F	T	T	F	F

(f)

p	q	$\neg$p	$\neg p \Leftrightarrow q$
T	T	F	F
T	F	F	T
F	T	T	T
F	F	T	F

(g)

p	q	¬q	p ∨ ¬q	p ∧ ¬q	(p ∨ ¬q) ⇒ (p ∧ ¬q)
T	T	F	T	F	F
T	F	T	T	T	T
F	T	F	F	F	T
F	F	T	T	F	F

(h)

p	q	¬p	¬p ∧ q	q ∧ p	(¬p ∧ q) ∨ (q ∧ p)
T	T	F	F	T	T
T	F	F	F	F	F
F	T	T	T	F	T
F	F	T	F	F	F

(i)

p	q	¬q	p ⇒ ¬q	¬(p ⇒ ¬q)
T	T	F	F	T
T	F	T	T	F
F	T	F	T	F
F	F	T	T	F

(j)

p	q	p ∧ q	p ∨ q	(p ∧ q) ⇒ (p ∨ q)
T	T	T	T	T
T	F	F	T	T
F	T	F	T	T
F	F	F	F	T

(k)

p	q	p ∧ q	¬(p ∧ q)	q ⇔ p	¬(q ⇔ *p*)	¬(p ∧ q) ∨ ¬(q ⇔ p)
T	T	T	F	T	F	F
T	F	F	T	F	T	T
F	T	F	T	F	T	T
F	F	F	T	T	F	T

EXERCISES 14.1.7

(a)

p	q	r	p ∨ q	r	(p ∨ q) ⇒ r
T	T	T	T	T	T
T	T	F	T	F	F
T	F	T	T	T	T
T	F	F	T	F	F
F	T	T	T	T	T
F	T	F	T	F	F
F	F	T	F	T	T
F	F	F	F	F	T

(b)

p	q	r	¬p	¬q	¬p ∨ ¬q	(¬p ∨ ¬q) ∧ r
T	T	T	F	F	F	F
T	T	F	F	F	F	F
T	F	T	F	T	T	T
T	F	F	F	T	T	F
F	T	T	T	F	T	T
F	T	F	T	F	T	F
F	F	T	T	T	T	T
F	F	F	T	T	T	F

(c)

p	q	r	¬p	¬q	¬r	¬p ∧ ¬q ∧ ¬r
T	T	T	F	F	F	F
T	T	F	F	F	T	F
T	F	T	F	T	F	F
T	F	F	F	T	T	F
F	T	T	T	F	F	F
F	T	F	T	F	T	F
F	F	T	T	T	F	F
F	F	F	T	T	T	T

(d)

p	q	r	¬p	q ∨ r	¬p ⇒ (q ∨ r)
T	T	T	F	T	T
T	T	F	F	T	T
T	F	T	F	T	T
T	F	F	F	F	T
F	T	T	T	T	T
F	T	F	T	T	T
F	F	T	T	T	T
F	F	F	T	F	F

(e)

p	q	r	¬*q*	p ∨ ¬q	r	(p ∨ ¬q) ⇒ r
T	T	T	F	T	T	T
T	T	F	F	T	F	F
T	F	T	T	T	T	T
T	F	F	T	T	F	F
F	T	T	F	F	T	T
F	T	F	F	F	F	T
F	F	T	T	T	T	T
F	F	F	T	T	F	F

(f)

p	q	r	¬p	¬p ⇔ q	(¬p ⇔ q) ∨ r
T	T	T	F	F	T
T	T	F	F	F	F
T	F	T	F	T	T
T	F	F	F	T	T
F	T	T	T	T	T
F	T	F	T	T	T
F	F	T	T	F	T
F	F	F	T	F	F

(g)

p	q	r	¬r	p ∨ ¬r	¬q	r ∧ ¬q	p ∨ ¬r) ⇒ (r ∧ ¬q
T	T	T	F	T	F	F	F
T	T	F	T	T	F	F	F
T	F	T	F	T	T	T	T
T	F	F	T	T	T	F	F
F	T	T	F	F	F	F	T
F	T	F	T	T	F	F	F
F	F	T	F	F	T	T	T
F	F	F	T	T	T	F	F

(h)

p	q	r	¬p	¬p ∧ q	q ∧ r	(¬p ∧ q) ∨ (q ∧ r)
T	T	T	F	F	T	T
T	T	F	F	F	F	F
T	F	T	F	F	F	F
T	F	F	F	F	F	F
F	T	T	T	T	T	T
F	T	F	T	T	F	T
F	F	T	T	F	F	F
F	F	F	T	F	F	F

(i)

p	q	r	p ∧ r	r ∨ q	(p ∧ r) ⇒ (r ∨ q)
T	T	T	T	T	T
T	T	F	F	T	T
T	F	T	T	T	T
T	F	F	F	F	T
F	T	T	F	T	T
F	T	F	F	T	T
F	F	T	F	T	T
F	F	F	F	F	T

(j)

p	q	r	p ∧ q	¬(p ∧ q)	q ⇔ r	¬(q ⇔ r)	¬(p ∧ q) ∨ ¬(q ⇔ r)
T	T	T	T	F	T	F	F
T	T	F	T	F	F	T	T
T	F	T	F	T	F	T	T
T	F	F	F	T	T	F	T
F	T	T	F	T	T	F	T
F	T	F	F	T	F	T	T
F	F	T	F	T	F	T	T
F	F	F	F	T	T	F	T

EXERCISES 14.1.8

1. (b) Invalid (c) Invalid

EXERCISES 14.2 – MISCELLANEOUS EXERCISES

1. (a) Yes (b) No (c) No (d) No (e) Yes (f) Yes

2. (a) Implication : If Fuzzy is a bear then Fuzzy is cute
Converse : If Fuzzy is cute then Fuzzy is a bear
Inverse : If Fuzzy is not a bear then Fuzzy is not cute
Contrapositive : If Fuzzy is not cute then Fuzzy is not a bear

(b) Implication : If Mary lives in Spain then Mary loves fish
Converse : If Mary loves fish then Mary lives in Spain
Inverse : If Mary does not live in Spain then Mary does not love fish
Contrapositive : If Mary does not love fish then Mary does not live in Spain

(c) Implication : If it is fine then we will go to the concert
Converse : We will go to the concert only if it is fine
Inverse : If it is not fine then we will not go to the concert
Contrapositive : We will not go to the concert only if it is not fine

(d) Implication : If I work hard then I will pay taxes
Converse : I will pay taxes only if I work hard
Inverse : If I do not work hard then I will not pay taxes
Contrapositive : I will not pay taxes only if I do not work hard

(e) Implication : If John loves fishing then John lives by the sea
Converse : If John lives by the sea then John loves fishing
Inverse : If John does not love fishing then he does not live by the sea
Contrapositive : If John does not live by the sea then John does not love fishing

3. (a) **4.** row 2 and row 3 **5.** All **6.** Tautology **7.** (a) and (b)

8.

p	q	$p \wedge q$	$\neg p$	$\neg q$	$(\neg p \wedge \neg q)$	$(p \wedge q) \vee (\neg p \wedge \neg q)$
F	F	F	T	T	T	T
F	T	F	T	F	F	F
T	F	F	F	T	F	F
T	T	T	F	F	F	T

9. (a) 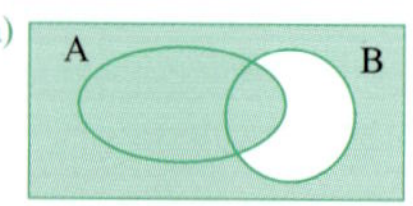

(b) $p \vee \neg q$ **10.** (a) $A \cap B$ (b) $A \cup (B \cup C)$
(c) $A \cap (B \cup C)$

11. $p \vee q$ **12.** (a) p (b) p (c) p (d) q (e) T (Tautology) (f) p **13.** (c)

14. **15.** (e)

p	q	$\neg p$	$\neg p \wedge q$
T	T	F	F
T	F	F	F
F	T	T	T
F	F	T	F

16. (a) p : Today is Tuesday $\neg p$ (b) $p : x$ is a even number $q : x$ is a prime number $p \vee q$
(c) p : Mary studies French q : John studies French $p \wedge q$
(d) p : It is raining q : The concert will be cancelled $p \Rightarrow q$
(e) p : Yoshi will go to the concert q : It is raining $p \Leftrightarrow \neg q$
(f) p : Birgit likes ice cream q : Birgit likes cake $p \wedge q$
(g) p : Jessica will go to the concert q : Mary goes to the concent r : It is raining $p \Leftrightarrow (q \wedge \neg r)$
(h) p : It is fine q : Temperature is between $20°$C and $30°$C
r : Paul will play tennis $(p \wedge q) \Rightarrow r$
(i) p : I work hard q : I will pass my exams $p \Rightarrow q$
(j) p : Bill wins his race q : Bill will make the final $p \Rightarrow q$
(k) p : Bill wins his race q : Bill will make the final $\neg p \Rightarrow \neg q$
17. (a) Invalid (b) Invalid (c) Invalid (d) Valid (e) Invalid
18. (c) **19.** (a) No (b) Yes (c) $q \Rightarrow p$

EXERCISES 14.3 – GRADED REVISON EXERCISES

LEVEL 1

1. If x is not a blet, then x is not a flib **2.** "I will not go swimming." is given by $\neg q$. "If it is hot, then I will not go swimming." is the implication. Therefore, p implies the negative of q. i.e., $p \Rightarrow \neg q$. **3.** (a) If Physics is difficult, then I don't study hard (b) If I don't study hard, then Physics is difficult (c) Physics is difficult if and only if I don't study hard. (d) If I do not study hard, then Physics is not difficult.

LEVEL 2

1. i. $\neg p \Rightarrow q$ is an implication, stating the negative of p implies q. Therefore the converse is $q \Rightarrow \neg p$ ii. "She is not tall" is represented by $\neg x$. "She is not tall and she is beautiful" is the conjunction of $\neg x$ and y. i.e., $\neg x \wedge y$. **2.** (e)

LEVEL 3

1. i.

p	q	$\neg p$	$\neg p \Rightarrow q$	$p \vee q$	$(\neg p \Rightarrow q) \Leftrightarrow (p \vee q)$
T	T	F	T	T	T
T	F	F	T	T	T
F	T	T	T	T	T
F	F	T	F	F	T

ii. Regardless of the truth value of p and q, the result is always true. So, $(\neg p \Rightarrow q) \Leftrightarrow (p \vee q)$ is a tautology. iii. If I do not train, I will get into trouble. $p \vee q$ is the disjunction of p and q, that is, "I train or I get into trouble". So, either p and q are both true, or at least one of them is true.
2. i. $x + xy = x(1+y) = x(1) = x$ ii. $xy' + x'y + xy = x + y$ (Use a Venn diagram). **3.** valid

LEVEL 4

1. Consider the contrapositive approach:

$\neg q$ i.e., the triangle is equilateral

$\neg p$ i.e., the line drawn from the vertex of a triangle to the midpoint of the opposite side does intersect this side at a right angle.

Therefore, we must prove that $\neg q \Rightarrow \neg p$.

Consider the ΔABC as shown in the diagram: Given that $AB = AC$ & $BX = CX$

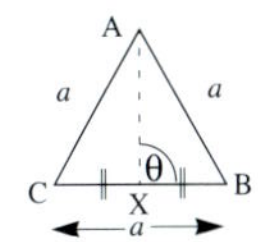

then, as AX is common

we have $\Delta ABX \equiv \Delta ACX$ (SSS)

$\therefore \angle AXB = \angle AXC$

$\therefore \angle AXB = 90°$ (straight line)

Hence, $\neg q \Rightarrow \neg p$ which is equivalent to $p \Rightarrow q$ (as required).

Or very simply: $\neg q \Rightarrow \neg p$ so $\neg(\neg q) \Rightarrow \neg(\neg p)$ or $p \Rightarrow q$
2. Use truth table to show that the last column consists of Ts only.

EXERCISES 14.4 – TOPIC TEST

1. p:The movie is over. q: We go home
2. A: $x \wedge y$ is the conjunction of x and y. It is true only if both x and y are true. Since y is false, then $x \wedge y$ is false.

B: $x \vee y$ is the disjunction of x and y. It is true if either x or y is true. If x is true then $x \vee y$ is true.

C: The statement y is false because $3 \times 4 = 12$ $(\neq 18)$.

D: $x \Rightarrow y$ is the implication, if x is true then y is true. However, x is true but y is false.

3. If x is prime and odd, then $x > 2$. (b) $(p \wedge r) \Rightarrow q$
4. i.

p	q	$p \Rightarrow q$	$\neg q$	$p \wedge \neg q$	$\neg(p \wedge \neg q)$	$(p \Rightarrow q) \Leftrightarrow \neg(p \wedge \neg q)$
T	T	T	F	F	T	T
T	F	F	T	T	F	T
F	T	T	F	F	T	T
F	F	T	T	F	T	T

ii. The statement "If Nora lives in Sydney, then Nora lives in N.S.W" is represented by $p \Rightarrow q$. However, the statement "It is false that Nora lives in Sydney and Nora does not live in N.S.W" is given by $\neg(p \wedge \neg q)$.
5. (a) He does not eat too much (b) He does not eat too much and he is healthy (c) He is healthy or happy (d) If he eats too much then he is not happy (e) If he does not eat too much and he is healthy than he is happy (f) He eats too much if and only if he is not happy (g) He eats too much or he is happy (h) If he does not eat too much or he is healthy than he is happy (i) If he does not eat too much then he is healthy and happy (j) He is happy if and only if he does not eat too much and he is healthy

6. (a) i. 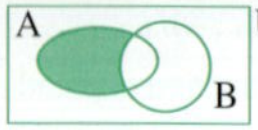

ii. 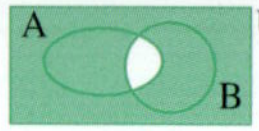

(b) i. $A \cap B'$ ii. $(A \cap B)' \cup (A \cap B)$

7. (a) p is not a sufficient condition for q. (b) q is a sufficient condition for p (c) $q \Rightarrow p$

ANSWERS TO CHAPTER 15

EXERCISES 15.1

1. (a) $\frac{2}{5}$ (b) $\frac{3}{5}$ (c) $\frac{2}{5}$ **2.** (a) $\frac{2}{7}$ (b) $\frac{5}{7}$ **3.** (a) $\frac{5}{26}$ (b) $\frac{21}{26}$ **4.** {HH, HT, TH, TT} (a) $\frac{1}{4}$ (b) $\frac{3}{4}$

5. {HHH,HHT,HTH,THH,TTT,TTH,THT,HTT} (a) $\frac{3}{8}$ (b) $\frac{1}{2}$ (c) $\frac{1}{4}$ **6.** (a) $\frac{2}{9}$ (b) $\frac{2}{9}$ (c) $\frac{2}{3}$ (d) $\frac{1}{3}$

7. (a) $\frac{1}{2}$ (b) $\frac{3}{10}$ (c) $\frac{9}{20}$ **8.** (a) $\frac{11}{36}$ (b) $\frac{1}{18}$ (c) $\frac{1}{6}$ (d) $\frac{5}{36}$ **9.** {GGG, GGB. GBG, BGG, BBB, BBG,

BGB, GBB} (a) $\frac{1}{8}$ (b) $\frac{3}{8}$ (c) $\frac{1}{2}$ **10.** (a) $\frac{1}{2}$ (b) $\frac{1}{4}$ (c) $\frac{1}{4}$ **11.** (a) $\frac{3}{8}$ (b) $\frac{1}{4}$ (c) $\frac{3}{8}$ (d) $\frac{3}{4}$

12. (a){(1, H),(2, H),(3, H),(4,H),(5, H),(6, H),(1, T),(2, T),(3, T),(4, T),(5, T),(6,T)}

(b) $\frac{1}{4}$ **13.** (a) $\frac{1}{216}$ (b) $\frac{1}{8}$ (c) $\frac{3}{8}$

EXERCISES 15.2

1. (a) $\frac{1}{4}$ (b) $\frac{5}{8}$ (c) $\frac{3}{4}$ **2.** (a) $\frac{1}{13}$ (b) $\frac{1}{2}$ (c) $\frac{1}{26}$ (d) $\frac{7}{13}$ **3.** $\frac{9}{26}$ **4.** (a) 1.0 (b) 0.3 (c) 0.5 **5.** (a) 0.65 (b) 0.70 (c) 0.65 **6.** (a) 0.95 (b) 0.05 (c) 0.80 **7.** (a) {TTT,TTH,THT,HTT,HHH,HHT,HTH, THH} (b) $\frac{3}{8}$ (c) $\frac{1}{2}$ (d) $\frac{1}{4}$ (e) $\frac{3}{8}$ **8.** (a) $\frac{6}{25}$ (b) $\frac{6}{25}$ (c) $\frac{13}{25}$ **9.** (b) $\frac{3}{4}$ (c) $\frac{1}{2}$ (d) $\frac{1}{6}$ (e) $\frac{7}{12}$

10. (a) $\frac{1}{4}$ (b) $\frac{1}{2}$ (c) $\frac{8}{13}$ (d) $\frac{7}{13}$ **11.** (a) 0.1399 (b) 0.8797 (c) 0.6 **12.** (b) $\frac{4}{15}$ (c) $\frac{4}{15}$ (d) $\frac{11}{15}$

EXERCISES 15.3

1. (a) 0.7 (b) 0.75 (c) 0.50 (d) 0.5 **2.** (a) 0.5 (b) 0.83 (c) 0.10 (d) 0.90

3. (a) 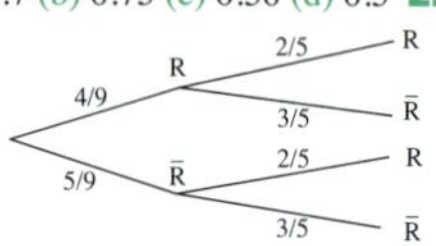

(b) $\frac{8}{45}$ (c) $\frac{22}{45}$ (d) $\frac{6}{11}$ **4.** (a) 0.5 (b) 0.30 (c) 0.25

5. (a) 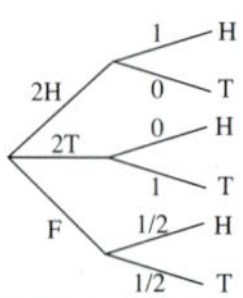

(b) $\frac{1}{2}$ (c) $\frac{2}{3}$ **6.** $\frac{1}{3}$ **7.** (a) 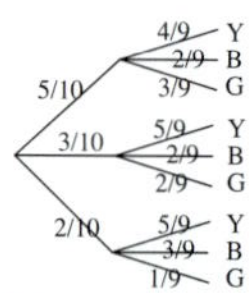

(b) $\frac{31}{45}$ (c) $\frac{2}{9}$ **8.** $\frac{2}{3}$

9. (a) 0.88 (b) 0.42 (c) 0.6 (d) 0.28 **10.** (a) 0.33 (b) 0.49 (c) 0.82 (d) 0.551 **11.** (a) 0.22 (b) 0.985 (c) 0.8629 **12.** (a) 0.44 (b) 0.733 **14.** (a) 0.512 (b) 0.128 (c) 0.8571 **15.** (a) 0.2625 (b) 0.75 (c) 0.4875 (d) 0.7123 **16.** (a) 0.027 (b) 0.441 (c) 0.4532

EXERCISES 15.4 – MISCELLANEOUS EXERCISES

1. (a) 0.16 (b) 0.70 (c) 0.58 **2.** (a) 0.5 (b) 0.7 (c) 0.4 (d) 0.2 **3.** (a) 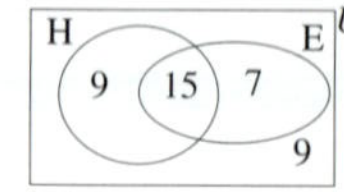

(b) (i) 0.375 (ii) 0.225 (iii) 0.4 (iv) 0.225

4. (a) 0.343 (b) 0.147 (c) 0.8125

5. (i) Use of diagram (e.g. Venn), 19 (ii) $\frac{1}{7}$ (iii) $\frac{19}{37}$ (iv) Dependent.

6. (i) $\frac{1}{9}$ (ii) $\frac{4}{9}$ (iii) $\frac{2}{3}$

8. i.

Number of eggs	2	3	4	5	6	7	8	9	10	11
Frequency	5	1	5	8	3	4	2	6	2	4

ii. $\frac{3}{20}$ iii. 6.325 eggs per bird iv. $\frac{9}{20}$ v. 2.787 eggs vi. The interval is 3.5376 to 9.11236 or in whole numbers 4 to 9. vii. $\frac{7}{10}$

9. i. Each spin is physically separate from the others. The result of the first spin will not affect the second.

ii.

Heads	Probability
0	0.064
1	0.288
2	0.432
3	0.216

iii. 0.36

iv. 0.0040 .

EXERCISES 15.5 – GRADED REVISION EXERCISES

LEVEL 1

1. $\frac{1}{26}$ **2.** $\frac{4}{9}$ **3.** $\frac{1}{8}$ **4.** $\frac{2}{6} = \frac{1}{3}$ **5.** $\frac{1}{4}$ **6.** $\frac{1}{2}$

LEVEL 2

1. 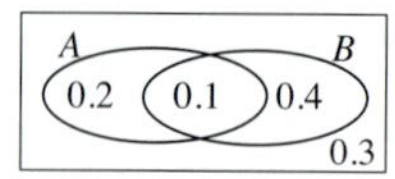

i. 0.7 ii. 0.5 iii. 0.4

2. i. 16 ii. $\frac{1}{16}$ iii. $\frac{3}{8}$ iv. $\frac{5}{16}$ **3.** i. $\frac{1}{81}$ ii. $\frac{1}{72}$

4. i. 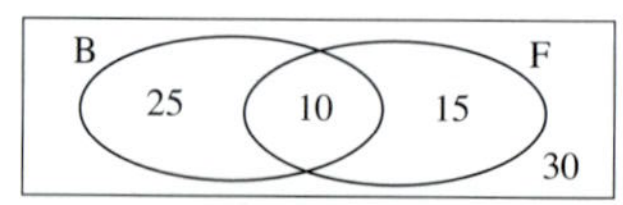

ii. $\frac{1}{8}$ iii. $\frac{5}{16}$ iv. $\frac{3}{8}$

LEVEL 3

1. 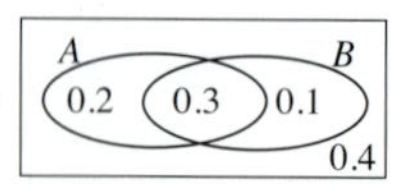

i. $\frac{3}{4}$ ii. $\frac{3}{5}$ iii. 0.2 , 0.3 so the events are not independent.

2. i. $\frac{4}{51}$ ii. No iii. $\frac{1}{13}$ **3.** (a) i. Yes ii. $\frac{9}{64}$ iii. $\frac{1}{4}$ (b) i. No ii. $\frac{3}{28}$ iii. $\frac{3}{14}$.

LEVEL 4

1.

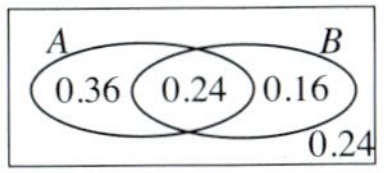

i. 0.76 ii. 0.24 iii. 0.6

2. i.

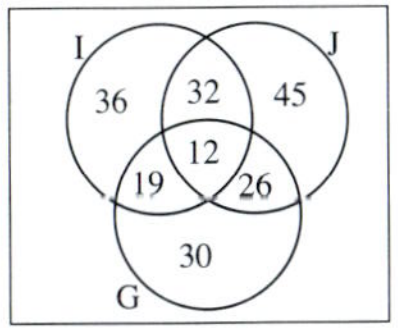

ii. $\frac{3}{50}$ iii. $\frac{77}{200}$ iv. $\frac{44}{115}$

v. No, selections made without replacement are always dependent.

EXERCISES 15.6 – TOPIC TEST

1. i. 0.2 ii. 0.4 **2.** i. 0.25 ii. $\frac{3}{51}$ iii. $\frac{25}{51}$ **3.** i.

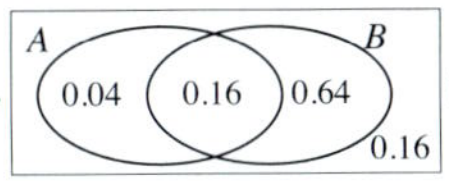

ii. 0.84 iii. 0.16

iv. $\frac{1}{5}$ **4.** i. $\frac{3}{7}$ (the results are independent) ii. $\frac{864}{2401}$ iii. $\frac{9}{49}$

ANSWERS TO CHAPTER 16

EXERCISES 16.1

1.

	a cm	b cm	c cm	A	B	C
1	3.8	4.1	1.6	67°	90°	23°
2	81.5	98.3	55.0	56°	90°	34°
3	32.7	47.1	33.9	44°	90°	46°
4	1.61	30.7	30.7	3°	90°	87°
5	2.3	2.74	1.49	57°	90°	33°
6	48.5	77	59.8	39°	90°	51°
7	44.4	81.6	68.4	33°	90°	57°
8	2.93	13.0	12.7	13°	90°	77°
9	74.4	94.4	58.1	52°	90°	38°
10	71.8	96.5	64.6	48°	90°	42°
11	23.3	34.1	24.9	43°	90°	47°
12	43.1	43.2	2.3	87°	90°	3°
13	71.5	80.2	36.4	63°	90°	27°
14	33.5	34.1	6.5	79°	90°	11°
15	6.1	7.2	3.82	58°	90°	32°
16	29.1	30	7.3	76°	90°	14°
17	29.0	29.1	2.0	86°	90°	4°
18	34.5	88.2	81.2	23°	90°	67°
19	24.0	29.7	17.5	54°	90°	36°
20	41.2	46.2	21.0	63°	90°	27°
21	59.6	72.9	41.8	55°	90°	35°
22	5.43	6.8	4.09	53°	90°	37°
23	13.0	19.8	14.9	41°	90°	49°
24	14.0	21.3	16.1	41°	90°	49°
25	82.4	88.9	33.3	68°	90°	22°

2. (a) $2\sqrt{3}$ (b) $5(1+\sqrt{3})$ (c) 4 (d) $2(1+\sqrt{3})$ (e) $\frac{4}{3}(3+\sqrt{3})$ (f) $\sqrt{106}-5$ **4.** (a) $25(1+\sqrt{3})$

(b) $\frac{40\sqrt{3}}{3}$

EXERCISES 16.2

1. (a) i. 030°T ii. 330°T iii. 195°T iv. 200°T (b) i. N25°E ii. S iii. S40°W iv. N10°W

2. 37.49m **3.** 18.94m **4.** 37° 18' **5.** $\frac{26}{9}$ m/s **6.** N58° 33'W, 37.23 km **7.** 199.82 m **8.** 10.58 m

9. 72.25 m **10.** 25.39 km **11.** 15.76 m **12.** (a) 3.01km N, 3.99km E (b) 2.87km E 0.88km S (c) 6.86km E 2.13km N (d) 7.19km 253°T **13.** 524m

EXERCISES 16.3

1. (a) 39°48' (b) 64°46' **2.** (a) 12.81 cm (b) 61.35 cm (c) 77°57' (d) 60.83 cm (e) 80° 32'
3. (a) 21°48' (b) 42°2' (c) 26°34' **4.** (a) 2274 (b) 12.7° **5.** 251.29 m **6.** (a) 103.52 m (b) 35.26°

(c) 39.23° **7.** (b) 53.43 m (c) 155.16 m (d) 98.37 m **8.** (b) 48.54 m **9.** (a) $\sqrt{(b-c)^2+h^2}$

(b) $\tan^{-1}\left(\frac{h}{a}\right)$ (c) $\tan^{-1}\left(\frac{h}{b-c}\right)$ (d) $2(b+c)\sqrt{h^2+a^2}+2a\sqrt{(b-c)^2+h^2}$ **10.** 82.80 m

11. (a) 40.61 m (b) 49.46 m **12.** (a) 10.61 cm (b) 75° 58' (c) 93° 22' **13.** (a) 1.44 m (b) 73° 13' (c) 62° 11'

EXERCISES 16.4

1. (a) 1999.2 cm^2 (b) 756.8 cm^2 (c) 3854.8 cm^2 (d) 2704.9 cm^2 (e) 538.0 cm^2 (f) 417.5 cm^2 (g) 549.4 cm^2 (h) 14.2 cm^2 (i) 516.2 cm^2 (j) 281.5 cm^2 (k) 918.8 cm^2 (l) 387.2 cm^2 (m) 139.0 cm^2 (n) 853.7 cm^2 (o) 314.6 cm^2 **2.** 69345 m^2 **3.** $100\pi - 6\sqrt{91}$ cm^2 **4.** 17.34 cm

5. (a) 36.77sq units (b) 14.70 sq units (c) 62.53 sq units **6.** 52.16 cm^2 **7.** 27° 2'

8. $\frac{(b + a \times \tan\theta)^2}{2\tan\theta}$ **9.** Area of $\Delta ABC = 61.38$ cm^2, Area of $\Delta ACD = 101.78$ cm^2

EXERCISES 16.5.1

	a cm	b cm	c cm	A	B	C
1	13.3	37.1	48.2	10°	29°	141°
2	2.7	1.2	2.8	74°	25°	81°
3	11.0	0.7	11.3	60°	3°	117°
4	31.9	39.1	51.7	38°	49°	93°
5	18.5	11.4	19.5	68°	35°	77°
6	14.6	15.0	5.3	75°	84°	21°
7	26.0	7.3	26.4	79°	16°	85°
8	21.6	10.1	28.5	39°	17°	124°
9	0.8	0.2	0.8	82°	16°	82°
10	27.7	7.4	33.3	36°	9°	135°
11	16.4	20.7	14.5	52°	84°	44°
12	21.4	45.6	64.3	11°	24°	145°
13	30.9	27.7	22.6	75°	60°	45°

14	29.3	45.6	59.1	29°	49°	102°
15	9.7	9.8	7.9	65°	67°	48°
16	21.5	36.6	54.2	16°	28°	136°
17	14.8	29.3	27.2	30°	83°	67°
18	10.5	0.7	10.9	52°	3°	125°
19	11.2	6.9	17.0	25°	15°	140°
20	25.8	18.5	40.1	30°	21°	129°

EXERCISES 16.5.2

	a	b	c	$A°$	$B°$	$C°$	$c*$	$B*°$	$C*°$
1	7.40	18.10	21.06	20.00	56.78	103.22	12.95	123.22	36.78
2	13.30	19.50	31.36	14.00	20.77	145.23	6.49	159.23	6.77
3	13.50	17.00	25.90	28.00	36.24	115.76	4.12	143.76	8.24
4	10.20	17.00	25.62	15.00	25.55	139.45	7.22	154.45	10.55
5	7.40	15.20	19.55	20.00	44.63	115.37	9.02	135.37	24.63
6	10.70	14.10	21.41	26.00	35.29	118.71	3.94	144.71	9.29
7	11.50	12.60	22.94	17.00	18.68	144.32	1.16	161.32	1.68
8	8.30	13.70	18.67	24.00	42.17	113.83	6.36	137.83	18.17
9	13.70	17.80	30.28	14.00	18.32	147.68	4.27	161.68	4.32
10	13.40	17.80	26.19	28.00	38.58	113.42	5.24	141.42	10.58
11	12.10	16.80	25.63	23.00	32.85	124.15	5.30	147.15	9.85
12	12.00	14.50	24.35	21.00	25.66	133.34	2.72	154.34	4.66
13	12.10	19.20	29.34	16.00	25.94	138.06	7.57	154.06	9.94
14	7.20	13.10	19.01	15.00	28.09	136.91	6.30	151.91	13.09
15	12.20	17.70	23.73	30.00	46.50	103.50	6.93	133.50	16.50
16	9.20	20.90	27.97	14.00	33.34	132.66	12.59	146.66	19.34
17	10.50	13.30	21.96	20.00	25.67	134.33	3.03	154.33	5.67
18	9.20	19.20	26.29	15.00	32.69	132.31	10.80	147.31	17.69
19	7.20	13.30	18.33	19.00	36.97	124.03	6.82	143.03	17.97
20	13.50	20.40	25.96	31.00	51.10	97.90	9.01	128.90	20.10

21. Non of the triangles are possible with the given data.

EXERCISES 16.5.3

1. 30.64 km **2.** 4.57 m **3.** 476.4 m **4.** 201°47'T **5.** 222.9 m **6.** (a) 3.40 m (b) 3.11 m **7.** (b) 1.000 m (c) 1.714 m **8.** (a) 51.19 min (b) 1 hr 15.96 min (c) 14.08 km **9.** $4886 **10.** 906 m

EXERCISES 16.5.4

	a cm	b cm	c cm	A	B	C
1	13.5	9.8	16.7	54°	36°	90°
2	8.9	10.8	15.2	35°	44°	101°
3	22.8	25.6	12.8	63°	87°	30°
4	21.1	4.4	21.0	85°	12°	83°
5	15.9	10.6	15.1	74°	40°	66°
6	8.8	13.6	20.3	20°	32°	128°
7	9.2	9.5	13.2	44°	46°	90°
8	23.4	62.5	58.4	22°	89°	69°
9	10.5	9.6	15.7	41°	37°	102°
10	21.7	36.0	36.2	35°	72°	73°
11	7.6	3.4	9.4	49°	20°	111°
12	7.2	15.2	14.3	28°	83°	69°
13	9.1	12.5	15.8	35°	52°	93°
14	14.9	11.2	16.2	63°	42°	75°
15	2.0	0.7	2.5	38°	13°	129°
16	7.6	3.7	9.0	56°	24°	100°
17	18.5	9.8	24.1	45°	22°	113°
18	20.7	16.3	13.6	87°	52°	41°
19	14.6	22.4	29.9	28°	46°	106°
20	7.0	6.6	9.9	45°	42°	93°
21	21.8	20.8	23.8	58°	54°	68°
22	1.1	1.7	1.3	41°	89°	50°
23	1.2	1.2	0.4	85°	76°	19°
24	23.7	27.2	29.7	49°	60°	71°
25	3.4	4.6	5.2	40°	60°	80°

EXERCISES 16.5.5

1. (a) 10.14 km (b) 121°T **2.** 7° 33' **3.** 4.12 cm **4.** 57.32 m **5.** 315.5 m **6.** (a) 124.3 km (b) W28° 47' S

EXERCISES 16.6 – MISCELLANEOUS EXERCISES

1. 39.60m 52.84m **2.** 30.2m **3.** 54°,42°, 84° **4.** 37° **5.** 028°T. **6.** 108.1cm **7.** (i) 135° (ii) 136.1cm **8.** 41°, 56°, 83° **9.** (i) 158° left (ii) 43.22km **10.** 264m **11.** 53.33cm **12.** 186m **13.** 50.12cm **14.** 5.17cm **15.** (a) 5950m (b) 13341m (c) 160° (d) 243° **17.** (a) 20.70° (b) 2.578 m (c) 1.994 m^3 **18.** (a) 4243 m^2 (b) 86 m (c) 101 m **19.** (a) 28.28 cm (b) 34.64 cm (c) 35°16' (d) 45° **20.** (a) 9.33cm (Approx) (b) 77°54' (c) 68°52' **21.** (b) 182.08 m **22.** (a) 5 (b) 5 (c) 68°54′ **23.** (b) i. $x = \sqrt{288} \approx 16.97$ ii. 62° **24.** (a) 22°35′ (b) 39°48'

EXERCISES 16.7 – GRADED REVISION EXERCISES

LEVEL 1

1. 19.01cm **2.** 56cm **3.** 12.01cm **4.** 2.01cm **5.** 1.56cm **6.** 27° **7.** 56° **8.** 53° **9.** i. $8\sqrt{2} \approx 11.31$ cm ii. $\theta = 35°16'$ **10.** i. $\therefore AC = 5\sqrt{2} = 7.07$ ii. 76°44'

LEVEL 2

1. 3.56cm **2.** 1.41cm **3.** 7.27cm **4.** 23.7cm **5.** 4.92cm **6.** 6.03cm **7.** 50.86cm **8.** 119.94cm **9.** i. 264.58 cm ii. 20°42' iii. 45° **10.** 25° 35' **11.** 6.93 cm **12.** 5.92 cm **13.** i. 9.11 ii. 69°17'

LEVEL 3

1. $\therefore x = \sqrt{3}$ **2.** $\theta = 55°$. The other angle is $= 180° - 55° = 125°$.
3. $c = 1.95$ cm $A = 66°$, $B = 180 - 66 - 63 = 51°$ **4.** 0.5 **5.** $\sim 4.35a$. **6.** 17°36′
7. i. $5r$ ii. 12.5 cm **8.** 70.71 m **9.** i. ~ 11.70 cm ii. 7.70 cm **10.** i. 0.98 m ii. 54°44′

LEVEL 4

1. 101°. This is a true bearing. An alternative answer is east 11° south. **2.** 2.52 m/s to 3 S.F.

3. (a) $AD = \frac{h}{\tan\alpha}$ (b) $\therefore AD = x - \frac{h}{\tan\beta}$ (c) $h = \frac{x\tan\alpha\tan\beta}{\tan\alpha + \tan\beta}$ **4.** $\alpha = 70°32'$

5. From B $\theta = \tan^{-1}\left(\left(\sqrt{1 + \left(\frac{b}{a}\right)^2}\right)\tan\alpha\right)$ From D: $\beta = \tan^{-1}\left(\left(\sqrt{\left(\frac{a}{b}\right)^2 + 1}\right)\tan\alpha\right)$.

6. $r + \sqrt{2}r = (1 + \sqrt{2})r$.

EXERCISES 16.8 – TOPIC TEST

1. 72.1metres **2.** $a = 4.2727$ to 5 S.F. **3.** 21.52cm^2 to 4 S.F. **4.** 69° to the nearest degree
5.

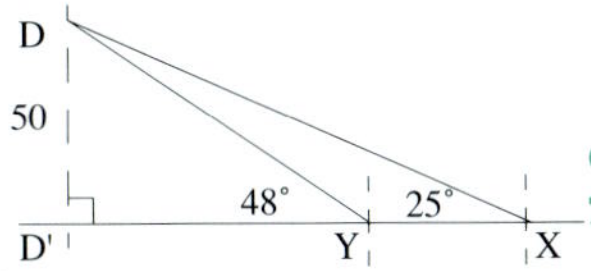

$$XY = D'X - D'Y = 50\left(\frac{1}{\tan 25°} - \frac{1}{\tan 48°}\right) = 62.21$$

6. 25 cm
7. i. 51°1′ ii. 47°52′ iii. 77°57′

ANSWERS TO CHAPTER 17

EXERCISES 17.1

1. i. $\sqrt{41}$ ii. $\sqrt{72}$ iii. $\sqrt{225}=15$ iv. $\sqrt{250}$ v. $\sqrt{85}$ vi. $\sqrt{173}$ vii. $\sqrt{41}$ viii. $\sqrt{89}$ ix. $\sqrt{226}$ x. $\sqrt{106}$ xi. $\sqrt{205}$ xii. $\sqrt{180}$ xiii. $\sqrt{400} = 20$ xiv. $\sqrt{145}$ xv. 4
2. i. (–5,0) ii. (0,6.5) iii. (–1,–1) iv. (–2.5,–5.5) v. (-5.5,0.5) vi. (2.5,1) vii. (-2.5,–3.5) viii. (4,2.5) ix. (–1.5,4) x. (2,4.5) xi. (8.5,5) xii. (–1,3.5) xiii. (–0.5,1.5) xiv. (–6,0) xv. (0.5,0.5)

EXERCISES 17.2

1. i. 2 ii. 3 iii. $\frac{5}{3}$ iv. 8 v. 0 vi. 2 vii. –2 viii. undefined.

2. (a) $y = 2x - 1$ (b) $y = 3x + 9$ (c) $y = -x - 1$ **3.** (a) $-\frac{1}{2}$ (b) $\frac{1}{3}$ (c) $\frac{3}{2}$ (d) $-\frac{4}{5}$ **4.** $y = 2x$

5. $y = -x + 1$ **6.** $2y - x - 2 = 0$ **7.** ±12 **8.** 2. **9.** Parallelogram.

10. (a) $y = \frac{5}{2}x$ (b) $y = -\frac{3}{2}x + 3$ (c) $y = \frac{5}{6}x - \frac{1}{2}$ (d) $y = -2x + 1$

EXERCISES 17.3

1. (b) (0,1), (3,5), (4, 3) & (1,–1) **2.** (a) (0,1), (2,5) & (6,3) **3.** (a) $y = 3x + c$ (b) $y = 3x - 8$ (c) (3,1) (d) $\sqrt{40}$ **4.** (36.67,89.5) to (110,265.5)

EXERCISES 17.4

1. i. (a) 22 cm (b) 30.25 cm^2 ii. (a) not enough info! (b) 28 cm^2 iii. (a) not enough info! (b) 262.5 cm^2 iv. (a) not enough info! (b) 27 cm^2 v. (a) 38 cm (b) 66 cm^2 vi. (a) 6π cm (b) 3π cm^2 **2.** (a) 37.5 cm^2 (b) 2π cm^2 (c) 13.75 cm^2 (d) 60 cm^2 **3.** (a) 248 cm^2 (b) 238π cm^2 (c) $(196 + 14\sqrt{596})\pi$ cm^2 (d) $100(1 + \sqrt{10})$ cm^2 **4.** (a) 399π cm^2 (b) $(24.5 + 3.5\sqrt{112.25})\pi$ cm^2 **5.** (a) 63 cm^3 (b) 500 cm^3 (c) $\frac{5}{3}\pi$ cm^3 **6.** (a) $36\sqrt{216}$ cm^3 (b) Vol = surface area × width = $8.5w$ cm^3, where w is the width of shape.

EXERCISES 17.5 – MISCELLANEOUS EXERCISES

1. (a) –3 (b) (2,0), (6, 0) (c) $\frac{1}{3}$ (d) $3y - x - 3 = 0$ **2.** $\frac{512}{3}$ cm^3 **3.** (a) –2 (b) $y = -2x + 8$ (c) $a = 4, b = 3$ **4.** $7y - 6x - 10 = 0, 6y + 7x - 24 = 0$ **5.** (a) $y = 4x - 9$ (b) $y = -x + 1$ (c) $2y - x + 4 = 0$ (d) $2y - x + 4 = 0$ **6.** (a) $y + 2x - 5 = 0$ (b) $2y - x + 5 = 0$ **7.** $\left(\frac{7}{6}, \frac{5}{6}\right)$ **8.** $r = 6$, Area = 118.90 cm^2 **9.** $b = -1.5, a = -2.5$ **10.** $y = 2x + 4$ **11.** $3x - 2y + 2 = 0$ **12.** (1, 4) **13.** (i) $\sqrt{89}$ (ii) 9.43 **14.** $y = 2x - 2.5$ **15.** $\frac{1}{6}V$ cm^3 **16.** 1:2 **17.** $\frac{13}{4}$ m^3

EXERCISES 17.6 – GRADED REVISION EXERCISES

LEVEL 1

1. (a) $m = 3, c = -1$ (b) $m = -1, c = -1$ (c) $m = 0.5, c = 2$. **2.** i. (a) 1 (b) $\sqrt{109}$ (c) $2\sqrt{13}$ ii. (a) 0 (b) 10/3 (c) 1.5 iii. (a) (2.5, 4) (b) (3.5, 3) (c) (0,0)
3. (a) 2.52 cm^2 (b) $4\pi + 20$ cm^2

LEVEL 2

1. i. –0.5 ii. 3 iii. –2.5 **2.** $y = 3x - 3$ **3.** $2y - 3x - 1 = 0$ **4.** $2y - x - 3 = 0$.
5. (a) $y = -\frac{3}{2}x + 3$ (b) $y = \frac{4}{3}x + \frac{7}{3}$ (c) $y = ax - a^2$ **6.** $120 + 12\sqrt{13.75}$ cm^2 ~ 164.50 cm^2

LEVEL 3

1. (a) i. $k = 3$ ii. $k = -\frac{1}{3}$ (b) i. $k = -\frac{2}{3}$ ii. $k = 6$ **4.** $\frac{a}{b} = \frac{17}{6}$ **5.** $A \equiv (4, 0)$.
6. $b = -20, a = -30$.

LEVEL 4

1. $m = \frac{8}{3}$

2. i.

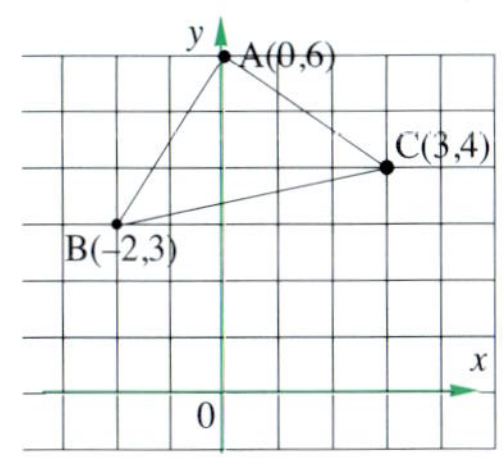

iii. $D \equiv (1, 1)$

3. i. (a)

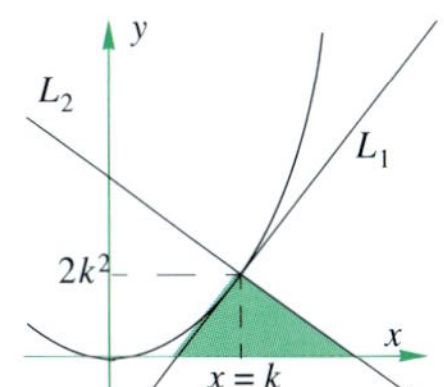

(b) $y = 2x^2 \therefore y' = 4x$. So, at $x = k$, $y' = 4k$. i.e., gradient = $4k$.
At $(k, 2k^2)$, $m = 4k \therefore (y - 2k^2) = 4k(x - k) \Rightarrow y = 4kx - 2k^2$
iii. $4y + x = 16 \Leftrightarrow y = -\frac{1}{4}x + 4 \therefore -\frac{1}{4} \times 4k = -1 \Rightarrow k = 1$.
iv. For L_1 : $y = 0 \Rightarrow 0 = 4kx - 2k^2 \Leftrightarrow x = \frac{2k^2}{4k} = \frac{1}{2}k$.
For L_2 : $y = 0 \Rightarrow 4 \times 0 + x = 16 \Leftrightarrow x = 16$.
$\therefore$ Area $= \frac{1}{2} \times \left(16 - \frac{1}{2}\right) \times 2(1)^2 = \frac{1}{2} \times \frac{31}{2} \times 2 = \frac{31}{2}$ sq. units

4. $(a, 0)$ **5.** (a) $a = 2, -3$ (b) $a = -0.6$

6. i.

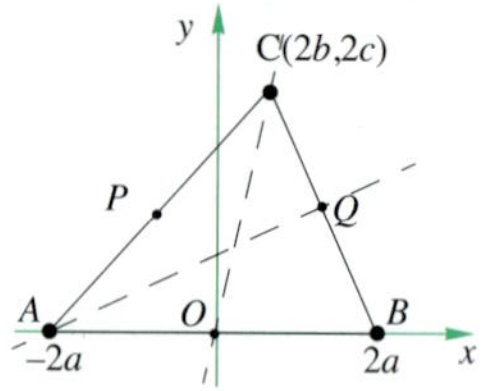

ii. mid-points: $O \equiv (0, 0)$ i.e.,

$$P \equiv \left(\frac{2b + (-2a)}{2}, \frac{2c + 0}{2}\right) \equiv (b - a, c)$$

$$Q \equiv \left(\frac{2a + 2b}{2}, \frac{2c + 0}{2}\right) \equiv (a + b, c)$$

EXERCISES 17.7 – TOPIC TEST

1.

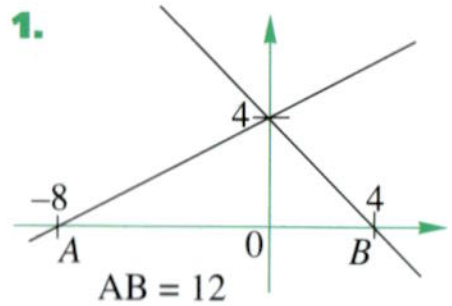

2. i. $2\sqrt{5}$ ii. –2 iii. (4, 3)

3. $x = -4.5, y = -6$

4. (a) If $pq = 3$ (b) $p = -3q$

5. i. (a) $y = -\frac{1}{3}(x - 1)$ (b) $y = 3(x + 3)$ ii. $X \equiv (-2.6, 1.2)$

iii. 4.8 sq. units

6. i. $B \equiv (a + b, 0 + c) \equiv (a + b, c)$ ii. For $\overline{AC}$: $\left(\frac{1}{2}(a + b), \frac{1}{2}c\right)$ $\overline{OB}$: $\left(\frac{1}{2}(a + b), \frac{1}{2}c\right)$

iii. They bisect iv. $B \equiv (b + \sqrt{b^2 + c^2}, c)$ v. $m_{OB} = \dfrac{c}{b + \sqrt{b^2 + c^2}}$, $m_{AC} = \dfrac{-c}{\sqrt{b^2 + c^2} - b}$

vi. $m_{OB} \times m_{AC} = -1$. Therefore, diagonals are perpendicular.

7. (a) i. 2r ii. $4\pi r^2$ (b) They are equal

ANSWERS TO CHAPTER 18

EXERCISES 18.1

1. (a) i. 14 500 ii. 2 000 (b) 305 (304.5) **2.** Sample size is large but may be biassed by factors such as the location of the catch. Population estimate is 5 000. **3.** (a) i. 1500 ii. 120 (b) 100 (c) 1 000 **4.** (a), (c) numerical, (b), (d), (e) categorical **5.** (a), (d) discrete, (b), (c), (e) continuous

EXERCISES 18.2

1.

218-220	221-223	224-226	227-229	230-232	233-235	236-238	239-241	242-244	245-247
1	4	4	3	6	8	9	5	7	1

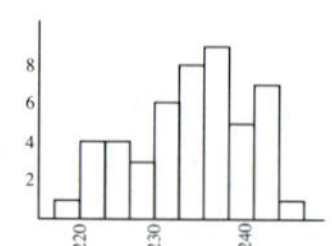

2.

1.1-1.2	1.2-1.3	1.3-1.4	1.4-1.5	1.5-1.6	1.6-1.7	1.7-1.8	1.8-1.9	1.9-2.0	2.0-2.1
5	1	2	2	7	6	1	12	7	5

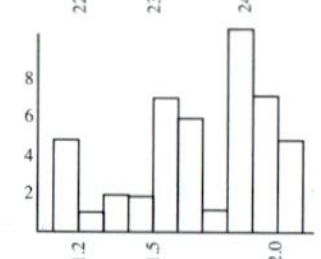

3. Set A Mode = 29.1 Mean = 27.2 Median = 27.85
Set B Mode = 9 Mean = 26.6 Median = 9. Set B is much more spread out than set A and although the two sets have a similar mean, they have very different mode and median.

EXERCISES 18.3

1. Mode = 236-238gms Mean = 234gms Median = 235gms **2.** Mode = 1.8-1.9gms Mean = 1.69gms Median = 1.80gms **3.** Set A Mode = 29.1 Mean = 27.2 Median = 27.85 Set B Mode = 9 Mean = 26.6 Median = 9. **4.** (a) $27522 (b) $21025 (c) Median **5.** i. $233 300 ii. $169 000 iii. Median **6.** (a) 14.375 (b) 14.354

EXERCISES 18.4

1. (a) Sample A Mean = 1.99kg Sample B Mean = 2.00kg (b) Sample A Sample std = 0.0552kg Sample B Sample std = 0.1877kg (c) Sample A Population std = 0.0547kg Sample B Population std = 0.1858kg **2.** (a) 16.41 (b) 6.84 **3.** Mean = 49.97 Std = 1.365

EXERCISES 18.5

1. (a) Med = 5, Q1 = 2, Q3 = 7, IQR = 5 (b) Med = 3.3, Q1 = 2.8, Q3 = 5.1, IQR = 2.3 (c) Med = 163.5, Q1 = 143, Q3 =182, IQR = 39 (d) Med = 1.055, Q1 = 0.46, Q3 = 1.67, IQR = 1.21 (e) Med = 5143.5, Q1 = 2046, Q3 = 6252, IQR = 4206
2. (a) Med = 3, Q1 = 2, Q3 = 4, IQR = 2 (b) Med = 13, Q1 = 12, Q3 = 13, IQR = 1 (c) Med = 2, Q1 = 2, Q3 = 2.5, IQR = 0.5 (d) Med = 40, Q1 = 30, Q3 = 50, IQR = 20 (e) Med = 20, Q1 = 15, Q3 = 22.5, IQR = 7.5
3. (a) $84.67 (b) $147.85 (c) $11 (d) Q1 = $4.50, Q3 = $65 IQR = $60.50 (e) Median & IQR.
4. (a) 2.35 (b) 1.25 (c) 2 (d) Q1 = 1, Q3 = 3, IQR = 2 **5.** (a) $232 (b) $83

(c-e)

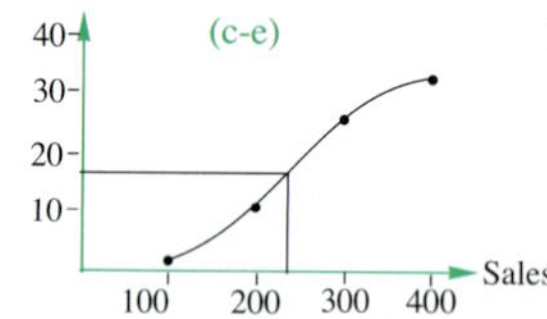

Med = $220
Q1 = $160
Q3 = $310
IQR = $150

6.

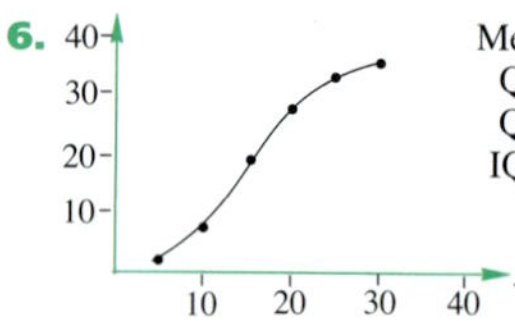

Med = 14
Q1 = 10
Q3 = 19
IQR = 9

EXERCISES 18.6

1. (a) 1, 1, 2, 3, 10, 11, 11, 11, 14, 15, 24, 24, 25, 26, 33, 35, 46 (b) 61, 62, 64, 65, 66, 67, 68, 69, 72, 74, 74, 73 [Note the misplacement of the last data point in the display!] (c) 12, 13, 15, 16, 20, 21, 23, 24, 28, 29, 32, 34, 35, 41, 42, 50

2. (a) (a) 14 (b)

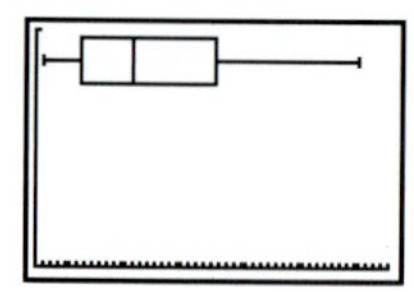

(b) (a) 67.5 (b)

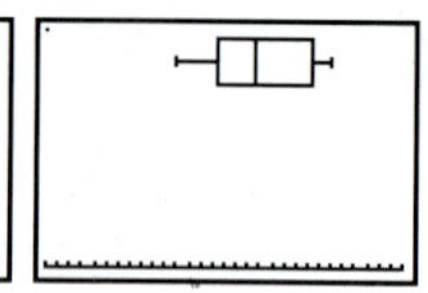

(c) (a) 206 (b)

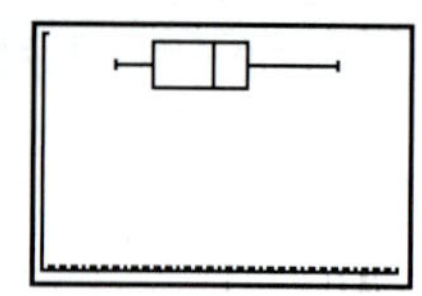

3.

```
18 | 1 5 6 7
19 | 2 3
20 | 4 5 5 6 7
21 | 4 5 6 7 7 7 9
22 | 2 7 7
23 | 0 4 4 6 6 9
24 | 2 5 8

key 21 | 4 = 214
```

4. (a)

```
1.6 | 0 3 6
1.7 | 5 6 7 9
1.8 | 0 0 0
1.9 | 1 1 3 3 3 4 4 5 5 5 6 6 6 6 7 7 8 8
2.0 | 2 5 6 6 6 7 7 8 8 9
2.1 |
2.2 | 2 2 3 3 5 9
2.3 | 2 2 2 9
2.4 |
2.5 |

key 1.9 | 3 = 1.93
```

(b) mildly +vely skeded (c) i. 1.965 ii. 0.16

5. (a) (b) median = 27.85

20	4
21	1 4 5 5 8 9
22	3 4 5
23	7 7
24	
25	1 1 2 5 6 7 8
26	8 9
27	2 4 8 9
28	0 2 2 7 8
29	0 1 1 1 4 9
30	1 3 3
31	2 3 3 5 9
32	9
33	3 6
34	3

key 31 | 2 = 31.2

EXERCISES 18.7 – MISCELLANEOUS EXERCISES

1. (a) Sample–100 randomly selected patients, population – all suffering from AIDS
(b) Sample–1000 working aged people in N.S.W, population – all working aged people in N.S.W
(c) Sample – John's I.B Higher Maths class, population – all seniors at Nappa Valley High School.

2. Discrete: a, b, d; Continuous: c, e, f, g.

3. (b)

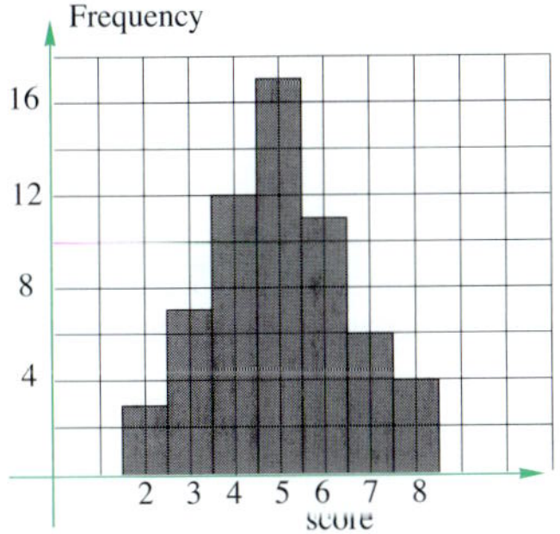

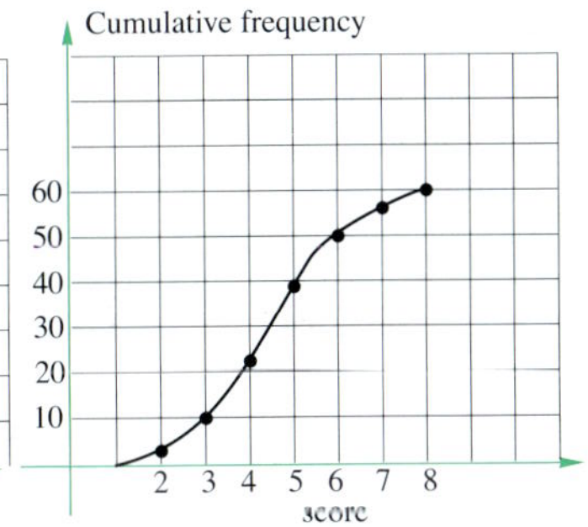

4. suggested answers only: (a) 200–224; 225–249; 250–274; . . . 575–599
(b) 100–119; 120–139; . . . 400–419 (c) 440–459; 460–479; . . . 780–799.

5. Make use of your graphics calculator

6. (a) 16 (b) graphics calculator (c) 15.23
(d) 15.5 (e) Q1 = 14, Q3 = 17 (f) 15.87 (2dec.pl.)

7. (a) 30–34 (b) graphics calculator (c) 30.4
(d) 32 (approx)

8. (b) 215.5 (c) 216.2 **9.** 48.17

10. (a) Q1~ 35, Q3~ 95 (b) ~ 104 (c) 60% (d) 67.15

11. range = 19, s = 5.49 **12.** 5.8; 1.50 **13.** 17.4; s_n = 3.12 s_{n-1} = 3.18

14. (a) 6 (b) 7 (c) Q1 = 5, Q3 = 7
(d) 2 (e) 6.15 (f) 1.61

15. s_n = 18.8 , s_{n-1} = 19.1 **16.** 14.18

EXERCISES 18.8 – GRADED REVISION EXERCISES

LEVEL 1

1. i. 1 ii. 13 iii. 1.3 iv. 40% **2.** i. 1 ii. 3 iii. 10-14 & 20- 24. **3.** 75

LEVEL 2

1. i. 2.09 ii. 13.7 iii. 4.05 iv. 9.28 v. 52.6 **2.** i. 4 ii. 3 iii. 12 iv. 13 v. 16 & 17
3. i. 17 ii. 31 iii. 4.2 iv. 49.1 v. 0.8

LEVEL 3

1. i. 2 & 7 ii. 22 & 82 iii. 23.5 & 84.2 iv. 5.5 & 20 (take the mid points between 5 & 6 and 19 & 21) v. 3.4 & 6.8 **2.** i. 2.57 ii. 30.1 iii. 2.91 iv. 279 v. 0.296

LEVEL 4

1. The median as this is the measure that is least affected by the small number of outliers (the luxury homes).
2. i. Frequency table required ii. mean = 6.35 gm iii. 2.615 gm
3. i. 50 ii. 5 iii. 2 iv. The mean is likely to be about 5 as there is a positive skew.
v. The range is 9 – 2 = 7. A complete distribution usually covers 6 standard deviations. On this basis the standard deviation might be expected to be about 7/6 or 1.2 marks.

EXERCISES 18.9 – TOPIC TEST

1. i.

Height	175	176	177	178	179	180	181
Frequency	1	0	3	4	4	5	3

ii.

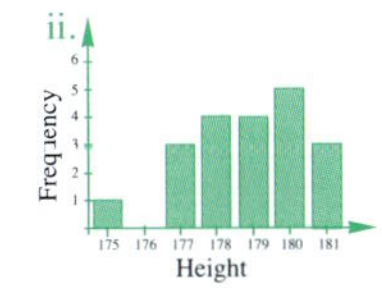

iii.

Height	175-179	180-184
Frequency	12	8

First method is best.

iv. Mode = 180cm, Mean = 178.85, Median = 179cm
v. 1.558cm, Quartiles are 178 & 180cm, Interquartile range = 2cm
vi.

17	5 7 7 7 8 8 8 8 9 9 9 9
18	0 0 0 0 0 1 1 1

key: 17 | 5 = 175

2. i. 19 ii. 418.42105 ≈ \$420 iii. \$450 iv. These are probably part time or casual workers. v. The mean.

ANSWERS TO CHAPTER 19

EXERCISES 19.1

1. (a) i. Increasing, positive ii. approx. linear iii. mild (to weak)
(b) i. No association ii. – iii. 0
(b) i. Increasing, positive ii. linear iii. very strong
(d) i. Increasing, positive ii. square root iii. mild (strength not appropriate as it is a non–linear relationship!)
(e) i. Decreasing, negative ii. exponential iii. mild (stength not appropriate as it is a non–linear relationship!)
(f) i. Decreasing ii. approx. linear iii. mild.

2. (a)

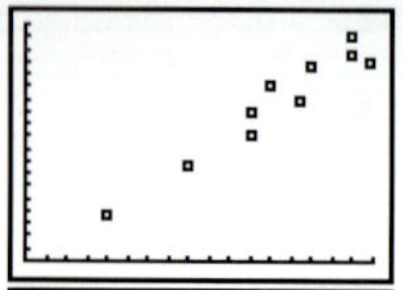

(b) Positive association, linear, strength: very strong

3. (a)

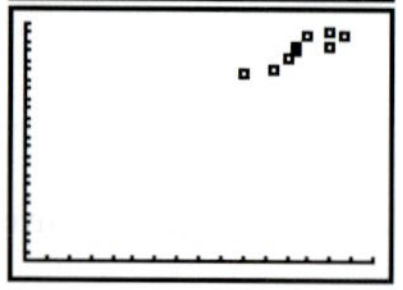

(b) Positive association, linear, strength: very strong

4.

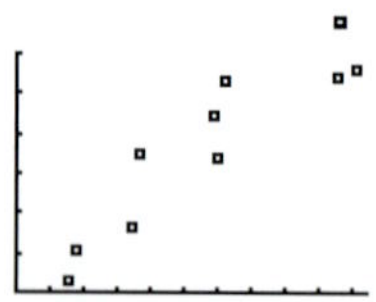

Data displays a strong positive association. Increase in lead content can be attributed to increase in traffic flow.

5.

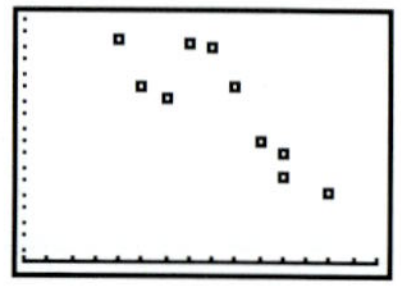

(a) Worksafety policy has had desired effect. i.e., number of accidents has decreased. Data displays a strong negative association.

EXERCISES 19.2

1. (a) $r = 0.96$

(b)

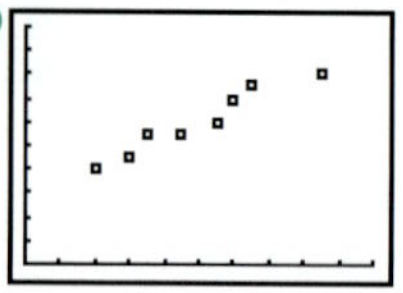

2. (b) $r = 0.70$ (assumed linear)

(a)

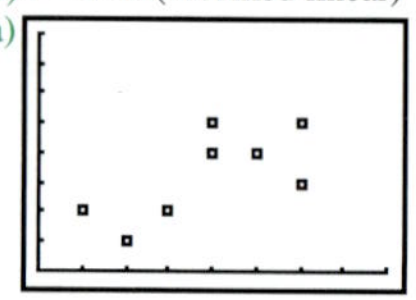

3. $r = 0$. No, not linear!

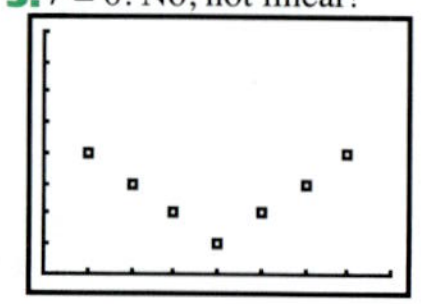

4. No. The relationship is not linear. **5.** (a) (i) 64% (ii) 81% (b) (i) 51% (ii) 64%

6. (a)

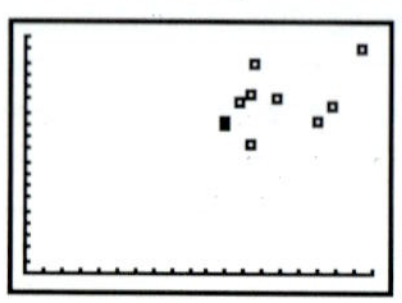

(b) $r = 0.45$ **7.** 3 **8.** ±0.922 **9.** 82%

10. (b) $r = 0.7715$

There is strong evidence to suggest that a student who does well in maths will also do well in biology.

(c) Values of r are the same.

EXERCISES 19.3

1. (a) i. ii. $y = -1.33x + 21.11$

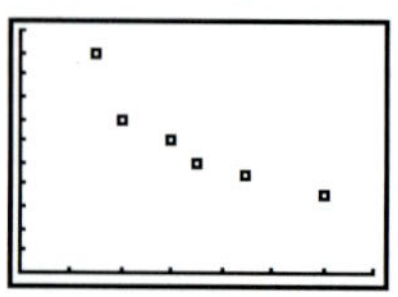

(b) i. ii. $y = 0.64x + 6.94$

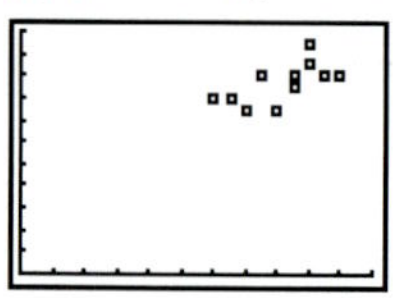

2. (i) (a)

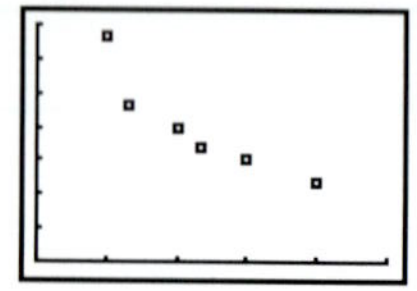

(b) $y = 20.6 - 1.26x$

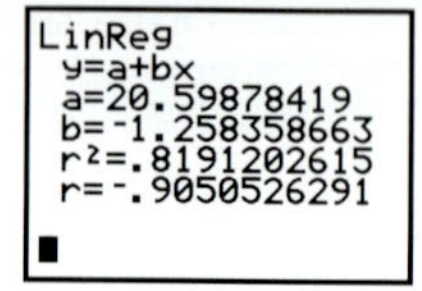

(c)

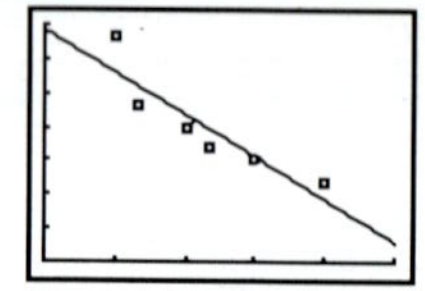

(ii) (a)

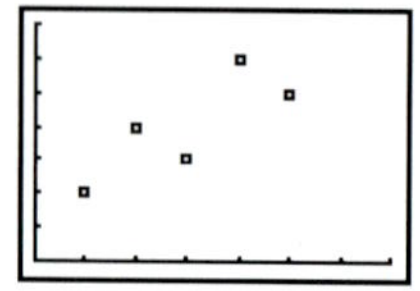

(b) $y = 1.6 + 0.8x$

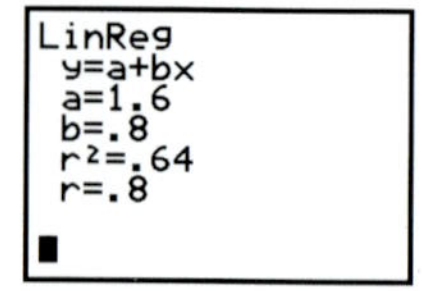

(c)

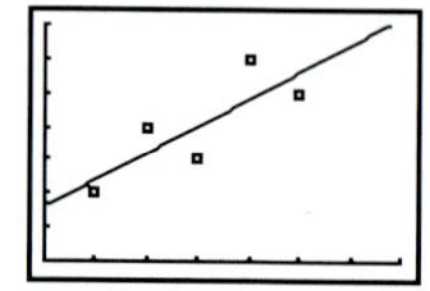

(iii) (a)

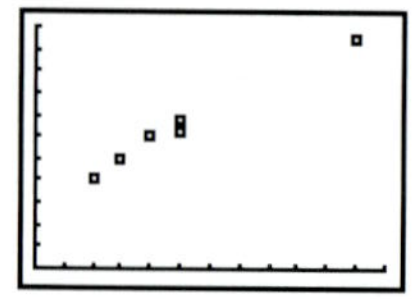

(b) $y = 14.8 + 3.44x$

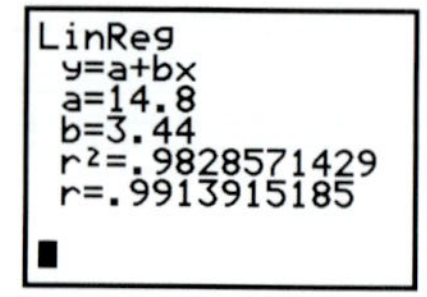

(c)

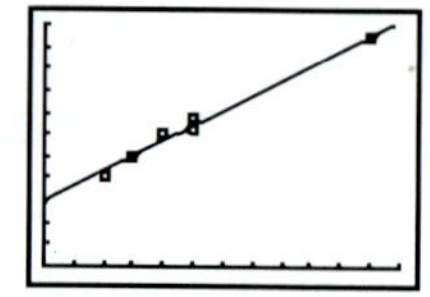

(iv) (a)

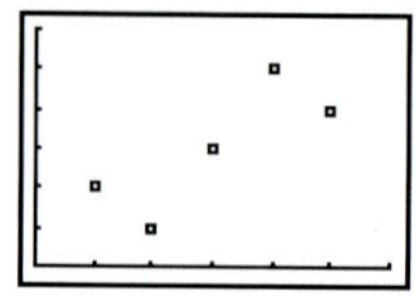

(b) $y = 0.6 + 0.8x$

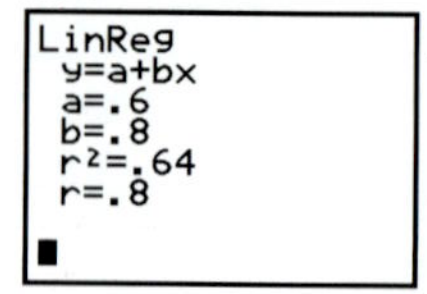

(c)

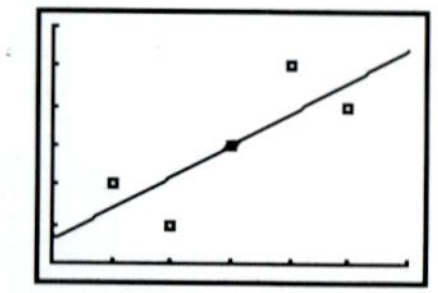

3. (a)

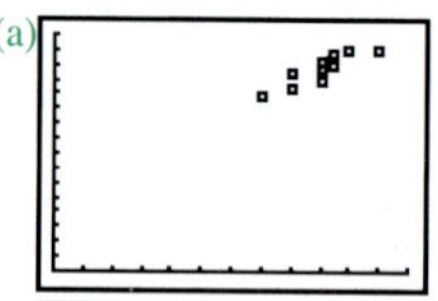

(b) $r = 0.891$
(c) 79.4%
(d) $y = 29.76 + 2.15x$

(e)

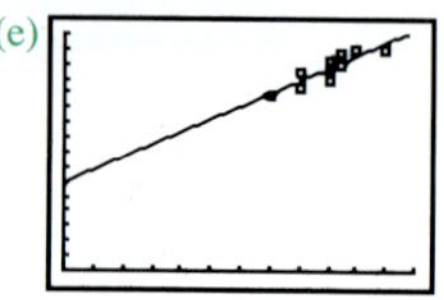

4. (a)

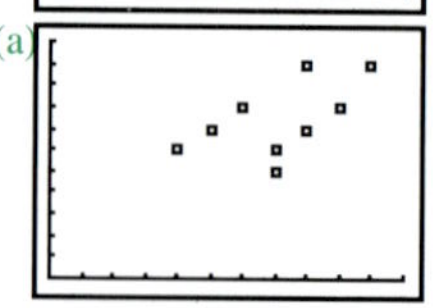

(b) $r = 0.553$
(c) i. $y = 40 + 0.5x$
ii. $x = 24.1 + 0.61y$

5. (a)

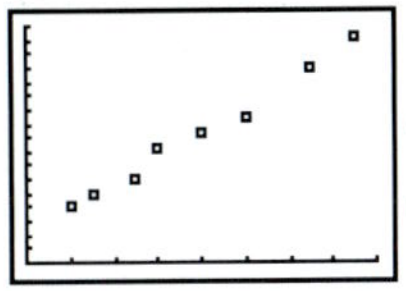

(b) Based on the scatter diagram, there is a definite linear relationship. Therefore, owner is justified.

(c)(i) $r = 0.99$ (ii) $C = 4.19 + 1.82w$ (iii)

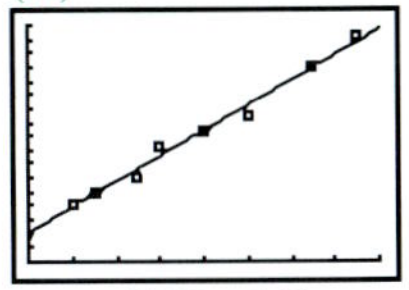

(d)(i) 20.57 i.e., 21
(ii) 95.19 i.e., 95
(iii) From (ii), serving 95 people per hr is unrealistic.

6. (a)

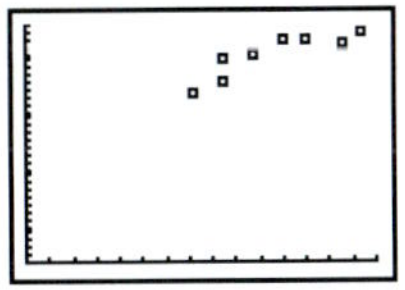

(b) Scatter diagram shows a linear relationship. Therefore statistic is appropriate, $r = 0.877$

(c) (i) $y = 89.50 + 1.02x$
(ii)

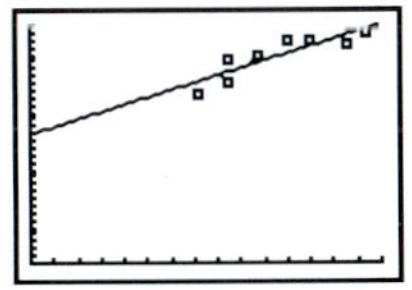

(d)(i) 135.6 (ii) 176.5
(iii) $x = 85$ is a fair way out from the set of values used to obtain the regression line

7.(a)

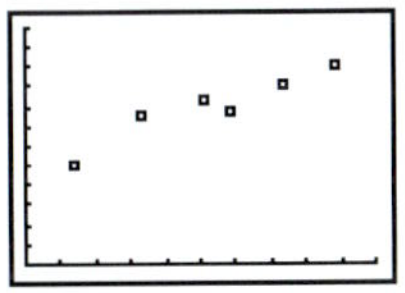

(b) Scatter diagram shows a linear relationship. Therefore statistic is appropriate, $r = 0.945$

(c) (i) $y = 4.74 + 0.6x$
(ii)

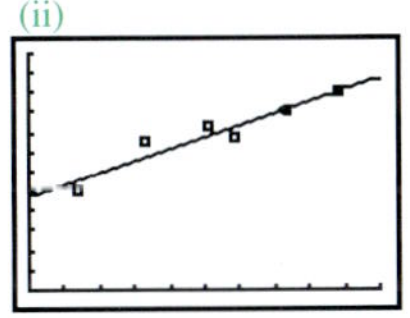

(d) i. 8.63 ii. 10.73

8. (a) (b) Linear trend exists, $r = 0.96$ (c) i. $y = 2.68x + 16.86$ (d) i. 27.57 ii. 57.03

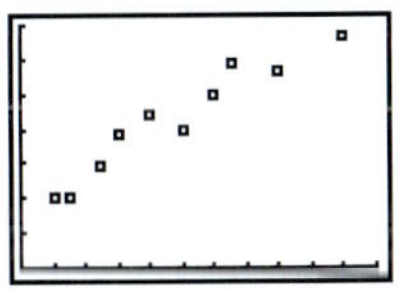

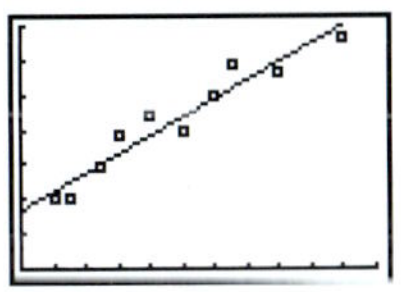

9. (a) $r = 0.8384$ (b) 70.29% (c) $y = 1.20x + 8.9$ (d) (11, 22.1), (24.45, 38), (27, 41.3), (60.08, 81). The equation is used to predict y from x-values, not x from y-values. We would need to find the regression of x on y.

10. (a) (b) i. $r = 0.97$ ii. 222 (c) $M = 0.2967T + 48.28$ **11.** Remains the same

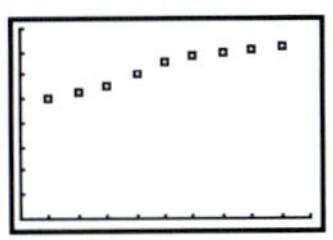

12. (a) (b) –0.93 (c) i. $y = -0.37x + 74.44$ (d) $y(40) = 59.722$ [interpolation] $y(120) = 30.278$ [extrapolation]

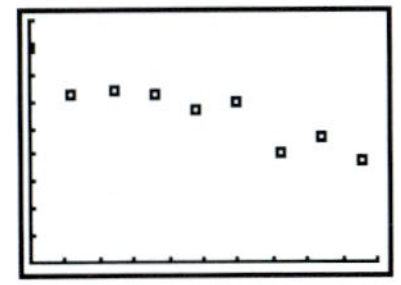

ii.

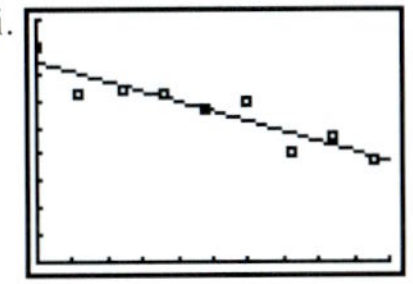

EXERCISES 19.4 – MISCELLANEOUS EXERCISES

1. (a) (b) $y = 0.57x - 26.2$ (c) 0.9388

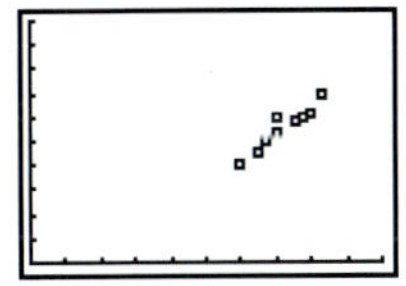

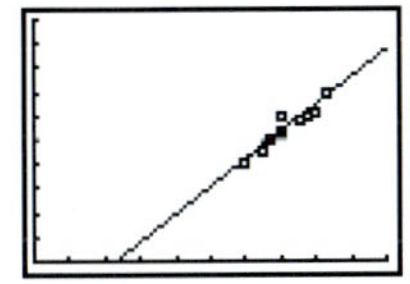

Because of the strong positive linear association, and the high r value, we can say that the taller the student the greater their weight.

2. 0.057 **3.** B **4.** (a) 0.8 (b) strong positive relationship **5.** 1.5

6. (a) 0.78 (b) i. P = 1.07M – 12.91 ii. 73% (c) i. M = 0.77E + 27.14 ii. 100 iii. Extrapolated. Continued linear trend highly likely. Therefore confident (d) Find regression equation of E on M, then use M = 90 into this new equation. **7.** i. positive ii. linear iii. very strong

8. (a) $\bar{x} = 20.57$, $\bar{y} = 31.86$ (b) 0.9645 (c) $y = 1.68x - 2.7$

9. (a) $y = -1.75x + 64.67$ (b) 22.67 (c) –11.12

10. (a) i. ii. $r = 0.8908$ (b) $r^2 = 0.7935$. That is, 79.35% (c) $y = 2.15x - 33.28$

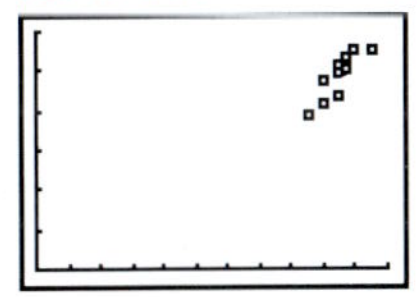

(d)

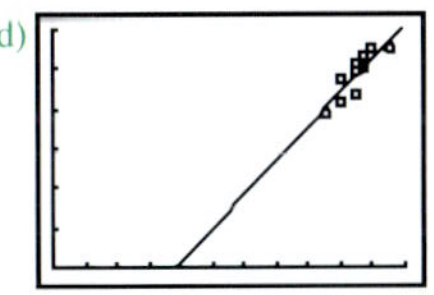

(e) $x = 37$, $y = 46.35$. Expenditure is $4635

11. (a) i. 4.4; 2.02 ii. 14.06; 2.92 (b) $b = 0.4895$; $r = 0.34$ (c) $r^2 = (0.3397)^2 = 0.1154$.
(d) Regression equation is $(y - 14.06) = 0.4895(x - 4.4)$ ∴ when $x = 3.5$, $y = 13.63$

12. (a) i. ii. $r = 0.9629$

(b) $y = 0.635x - 33.815$
(c) Diagram shown alongside:
(d) When $x = 1040$, $y = 626.59$
The carcass weighs 626.59lbs

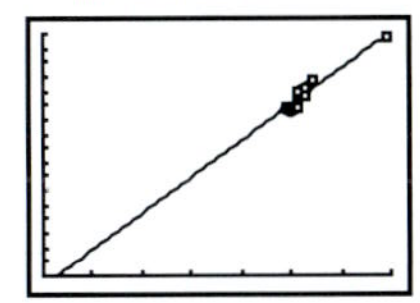

EXERCISES 19.5 – GRADED REVISION EXERCISES

LEVEL 1

1. (a) (b) +ve direction, strong linear relationship. **2.** –ve direction, strong and linear relationship

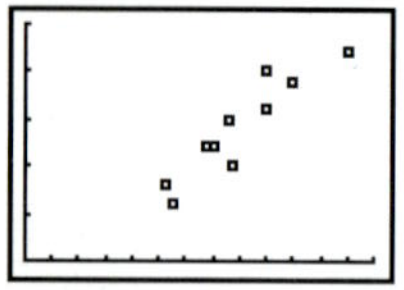

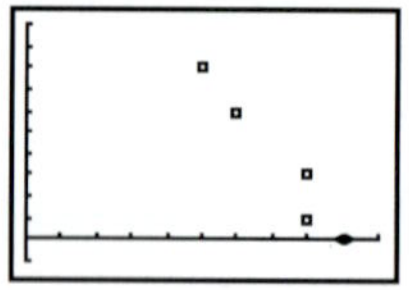

LEVEL 2

1. $r = 0.9213$ **2.** $r = 0.9686$

LEVEL 3

1. As the scatter plot indicates, there exists a strong positive linear relationship, r measures the strength of the relationship, whereas, r^2 represents the proportion of the variation in y which can be attributed to the variation in x (in this case it would be 93.81%).

2. (a) (b) $y = -0.52x + 20.31$ (c) $x = 14, y = 13.03$

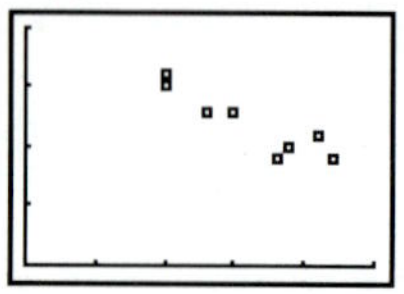

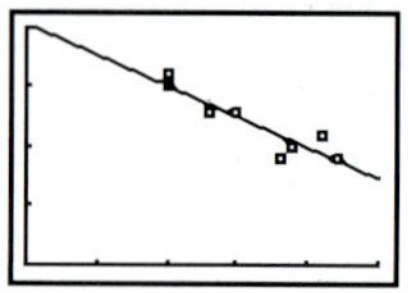

3. (a) $b = 0.4749$ $r = 0.92$ (b) –9.745

4. $r^2 = 0.9 \therefore r = 0.9487$

LEVEL 4

1. (a)

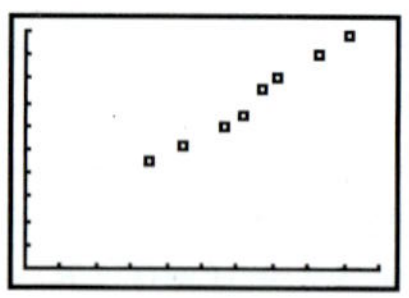

(b) Yes. Scatter diagram shows a strong +ve linear relationship.

(c) i. $y = 0.1005x + 6.638$ ii.

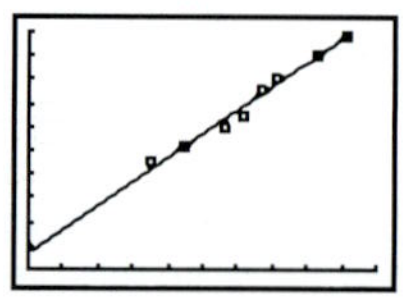

(d) = 0.993

(e) $x = 600, y = 66.94$ (f) Yes. But **not** by using $y = 0.1005x + 6.638$. First you need to find the regression of x on y, i.e., $x = by + c$. Which is given by $x = 9.808y - 55.76$. So, $x(56) = 493$.

2. (a)

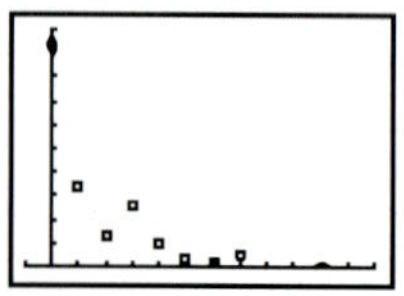

(b) $Y = -\alpha r + C$ with $Y = \ln d$ where, $C = \ln k$, we have $Y = -0.2373r + 4.5471$. $\Rightarrow k = e^{4.5471} = 94.36$

(c) i.

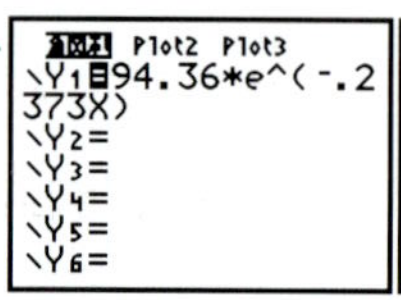

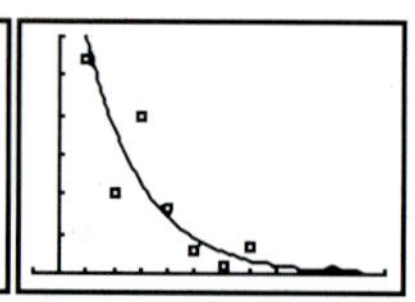

ii. $r = 16, d = 2.117$

EXERCISES 19.6 – TOPIC TEST

1. (a) (b) i. –ve ii. linear iii. strong (c) $y = -2.97x + 44.28$

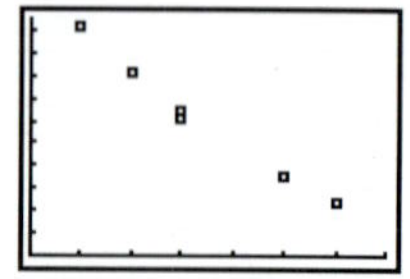

2. (a) $y = 0.4056x + 6.566$ (b) $x = 8, y = 9.81$

(c) $s_{xy} = 2.92$

3. $r^2 = 0.80 \therefore r = 0.8944$

4. (a)

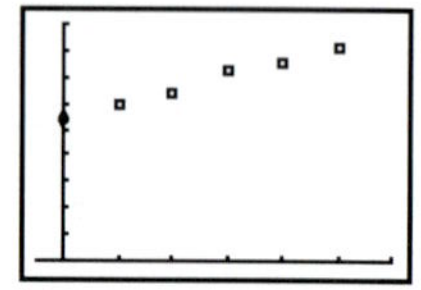

(b) From the given data: $6a + 150b = 408$ —(1)

$150a + 5500b = 11200$ —(2)

$b = \frac{4}{7} \therefore a = \frac{376}{7}$

(c)

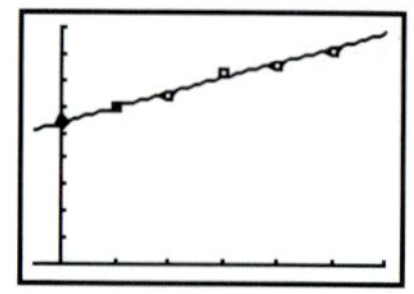

(d) $T = 25, x = 68$

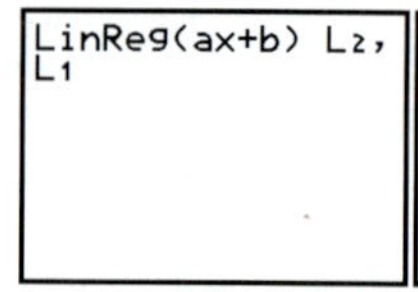

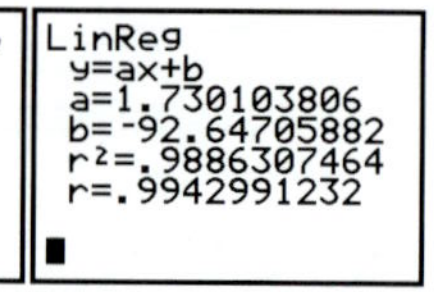

(e) i. First you need to find the regression of T on x, i.e., $T = bx + a$ ii. $T = 1.73x - 92.65$
So, $x = 70, T = 28.45$

ANSWERS TO CHAPTER 20

EXERCISE 20.1

1. (a) Reject independence (i.e., reject H_0) (b) Accept independence (i.e., accept H_0).

2. Reject H_0 **3.** Accept H_0 **4.** Accept H_0 (i.e., independent) **5.** (a) Reject H_0 (b) Reject H_0

6. Reject H_0 **7.** Accept H_0 **8.** Accept H_0 **9.** Reject H_0 at the 5% level of significance.

10. (a) The treatment has a significant effect, that is, of those treated, a significantly smaller proportion developed the flu than those that were not treated.

(b) Results are not significant. The significance test used in (a) is accurate because the sample size is large, but in (b) it is inaccurate because the sample size is too small.

11. Yes, $p = 1.178 \times 10^{-7}$ **12.** Yes, $p = 6.35 \times 10^{-6}$

CHAPTER 20 – MISCELLANEOUS EXERCISES

1. (a)

Expected values:

Maths Studies		Biology		
		Excellent	Good	Fail
	Excellent	17.94	28.86	31.2
	Good	31.74	51.06	55.2
	Fail	42.32	68.08	73.6

(b) Biology results are independent of Mathematical Studies results.

(c) $\chi^2_{calc} = 61.59$ (d) i. 4 ii. 9.488 iii. Reject H_o. That is, there exists a positive relationship between doing well in Mathematical Studies and Biology.

2. (a)

Expected values: Severity of smallpox attack

Year since vaccination took place		Extreme	Very	Moderate	Light
	0 – 20	99	149.96	141.22	104.82
	20 – 40	187	283.25	266.75	198
	Over 40 or unvaccinated	54	81.79	77.03	57.18

(b) The severity of smallpox is independent of when vaccination was administered.

(c) i. 6 ii. $\chi^2_{calc} = 153.46$, reject H_o, i.e., severity of smallpox is dependent on when vaccination was administered.

3. $\chi^2_{calc} = 5.33$, $p = 0.021$. Therefore reject the null hypothesis at 5% level. That is, difference between team performance exists.

4. (a)

Referendum question

	For	Against
Male	45	55
Female	45	55

(b) Preference on referendum issue is independent of gender.

(c) i. 18.18 ii. $p = 2.01 \times 10^{-5}$. Therefore, reject the null hypothesis. That is, gender does play a part in preference on referendum issue.

5. (a) $a = 52.89, b = 165.13, c = 53.87, d = 3.94$ (b) H_o : Occupation is independent of the amount of schooling received. H_1 : Occupation is dependent on the numbers of years spent at school (c) i. d.f = 3, $\chi^2 = 7.81$ ii. $\chi^2_{calc} = 52.1$ iii. Reject H_o.

EXERCISES 20.3 – GRADED REVISION EXERCISES

LEVEL 1

1. (a)

	A1	A2
B1	27.08	22.92
B2	37.92	32.08

(b)

	A1	A2	A3
B1	17.5	22.5	10
B2	10.5	13.5	6

LEVEL 2

1. (a) There exists an association (b) No association

LEVEL 3

1. (a)

	Good	Poor
A	6	24
B	14	56

(b) As $p = 1.27 \times 10^{-5}$ we reject H_o. That is, there is a significant difference between company and absenteeism.

LEVEL 4

1. (a) **Table 1**

Level of self-esteem

	High	Medium	Low
Smoker	21.56	21.56	25.88
Non–smoker	28.44	28.44	34.12

We have assumed that there exists no association between the level of self-esteem and the smoking habits of males.

(b) $\chi^2_{calc} = 2.00$ (c) p - value = 0.37 and so as $p > 0.05$ we accept H_o. That is, there is no significant difference between self-esteem of smokers as compared to non-smokers.

(d) $p = 0.242$, therefore still no significant difference (i.e., we still accept H_o).

EXERCISES 20.4 – TOPIC TEST

1. (a) $a = 30, b = 70, c = 30, d = 70$ (b) $\chi^2_{calc} = 9.52$. **2.** $\chi^2_{calc} = 0.488$, $p = 0.921 > 0.05$, therefore accept H_o. That is, there is no significant difference between the type of bread and opinion. Initial assumption was that the categories were independent (so that the expected values table could be constructed). **3.** (a) H_o : Vaccine has no effect on the severity of infectious disease. H_1 : Level of severity of infectious disease is dependent on the vaccine having been administered.

(b)

	High	Medium	Low
Vaccinated	26.53	26.53	30.94
Not vaccinated	33.47	33.48	39.05

(c) i. d.f = 2, $\chi^2_{critical} = 5.99$ (at 5% sig. level) ii. $\chi^2_{calc} = 1.73$ iii. Accept H_o, i.e., this vaccine has no effect on the severity of the infectious disease.

ANSWERS TO CHAPTER 21

EXERCISES 21.1.1

1. a. 839 b. 565 c. 516 d. 409 e. 749 f. 389 **2.** a. 1271 b. 573 c. 1111 d. 1646 c. 500 f. 486 **3.** a. 1745 b. 395 c. 1791 d. 1721 e. 1303 f. 1838 **4.** a. 31.22 b. 28.48 c. 0.57 d. 20.06 e. 27.21 f. 34.41 **5.** a. 89 b. 21 c. 38 d. 120 e. 167 f. 73

EXERCISES 21.1.2

1. a.i. ¥143100 ii. \$U.S 982.16 iii. \$U.S17.84 b.i. ¥183311 ii. \$U.S1258.14 iii. \$U.S22.86 c.i. ¥216510 ii. \$U.S 1486 iii. \$U.S 27 d. i. ¥194044 ii. \$U.S 1331.80 iii. \$U.S 24.2 e.i. ¥183597 ii. \$U.S 1260.11 iii. \$U.S 22.89 f. i. ¥228245 ii. \$U.S 1566.54 iii. \$U.S28.46 **2.** a.i. 1019000 ii. 986 iii. 14 b.i. 1638552 ii. 1586 iii. 22 c.i. 1384821 ii. 1341 iii. 18 d.i. 1495892 ii. 1448 iii. 20 e.i. 1636514 ii. 1584 iii. 22 f.i. 1240123 ii. 1201 iii. 16 **3.** a.i. 28710 ii. 852 iii. 48 b.i. 45713 ii. 1356 iii. 77 c.i. 44788 ii. 1329 iii. 75 d.i. 44341 ii. 1316 iii. 74 e.i. 42555 ii. 1263 iii. 71 f.i. 47467 ii. 1409 iii. 79 **4.** a.i. 40358 ii. 6447 iii. 180 b.i. 43620 ii. 7288 iii. 203 c.i. 44810 ii. 7158 iii. 200 d.i. 45504 ii. 7269 iii. 203 e.i. 37764 ii. 6033 iii. 168 f.i. 42813 ii. 6839 iii. 191

EXERCISES 21.2.1

1. a. \$804.00 b. \$562.32 c. \$218.40 d. \$279.00 e. \$6,528.00 f. \$812.50 **2.** a. \$321.60 b. \$515.46 c. \$46.80 d. \$96.88 e. \$191.25 f. \$187.50 **3.** a. 7% b. 3% c. 8% d. 12% e. 11.8% f. 9.3% **4.** a. 300 b. 790 c. 2000 d. 365 e. 1362 f. 9035 (all marks) **5.** a. 5 b. 7 c. 2 d. 8 e. 11 f. 3.6 (all years) **6.** a. 12 b. 7 c. 2 d. 8 e. 11 f. 3.6 (months)

7. Option 1: $I = \frac{6000 \times 7.3 \times 5}{100} = 2190$ francs. Total is 8190 francs. Option 2: ¥6000 × 153.1 = ¥918600. Interest is $\frac{918600 \times 7.4 \times 5}{100} = 339882$ and the total = ¥1258482. Conversion back to Swiss francs gives $\frac{1258482}{155.7} = 8082.736$ or 8083. Option 1 is better.

EXERCISES 21.2.2

1. a. 3682 b. 8246 c. 19576 d. 1521 e. 5903 f. 3117 **2.** a. 12087 b. 112108 c. 128192 d. 151518 e. 194298 f. 48698 **3.** a. 14974 b. 31103 c. 30083 d. 32593 e. 39190 f. 27899 **4.** a. 22122 b. 28700 c. 23635 d. 31155 e. 29098 f. 1616 **5.** A gives \$2415.77 B gives \$2431.52 & C gives \$2411.87 so B is best. **6.** 14.65% **7.** \$3952.08

EXERCISES 21.2.3

1. (a) i. \$1432.33 ii. \$6432.33 (b) 7.162% **2.** (a) \$5656.79 (b) i. \$4856 ii. \$3428 **3.** 14.65% **4.** A: \$7200 B: \$7447.12 C: \$7403.89 Therefore choose B **5.** 5.90% **6.** \$8075.20 **7.** \$1428.76 **8.** \$1285.70 **9.** 7.5% **10.** (a) Option A (b) \$2.52 **11.** \$10,756.96 **12.** (a) 10.25% (b) 10.38% (c) 10.47% **13.** \$2 630 100

EXERCISES 21.2.4

1. (a) 7 years 3 months (b) 7.06 years (c) 7 years **2.** $n = 18; 4\frac{1}{2}$ years **3.** (a) 5 years (b) 4 yrs and 9 months (or 19 quarters) **4.** 73 **5.** 3 years and 9 months **6.** (a) yes, \$863.23 (b) ii. 8 years three month **7.** Kristian: $n = 5.53$. Jørgi $n > 25.498$ i.., 6.37 yrs. Kristian goes 10 months before Jørgi. Nb: If they cannot take their money out until the end of the period then Kristian goes 6 months before Jørgi.

EXERCISES 21.3.1

1. \$3200 **2.** \$540

EXERCISES 21.3.2

1. (a) \$10000 (b) \$5000 **2.** \$3600
3. (a) i. \$8000 ii. 4 yrs iii. \$6000 iv. \$1500 per yr v. BV = 8000 – 1500t, $0 \le t \le 4$ vi. \$3500 (b) i. \$400,000 ii. 8 yrs iii. \$400,000 iv. \$50,000 per yr v. BV = 400000 – 50000t, $0 \le t \le 8$. vi. \$250000 (c) i. \$250000 ii. 10 yrs iii. \$200000 iv. \$20000 per yr v. BV = 250000 – 20000t, $0 \le t \le 10$ vi. \$190000
4. (a) Table of values using the expression BV = 10000 – 1500t, $0 \le t \le 5$.

(b) 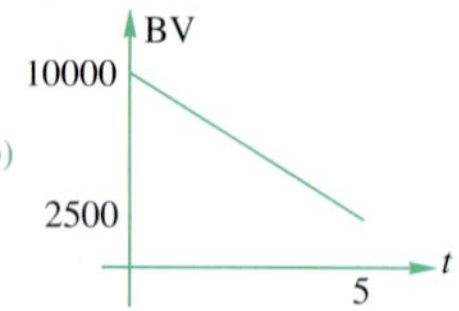

(c) \$2500 (d) $BV_t = 10000 - 1500t, 0 \le t \le 5$

5. (a) \$16333.33 per year (b) Table of values using $BV_t = 105000 - \frac{49000}{3}t, 0 \le t \le 6$

(c) 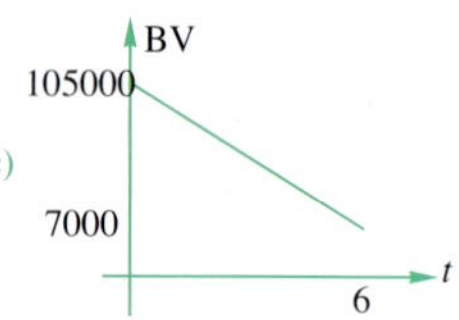

(d) $BV_t = 105000 - \frac{49000}{3}t, 0 \le t \le 6$

6. Printer B by 3 months and 8 days

EXERCISES 21.3.3

1. (a) \$8192 (b) \$5243

2. (a) 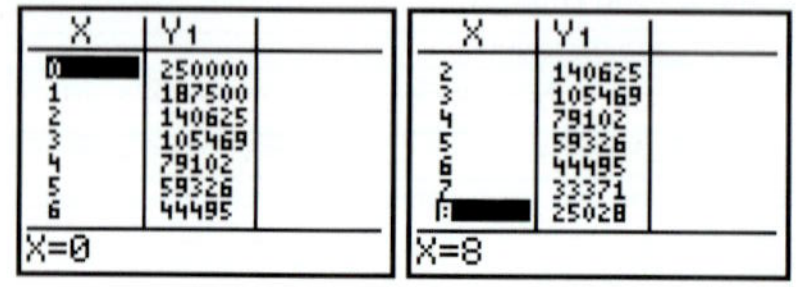

(b) 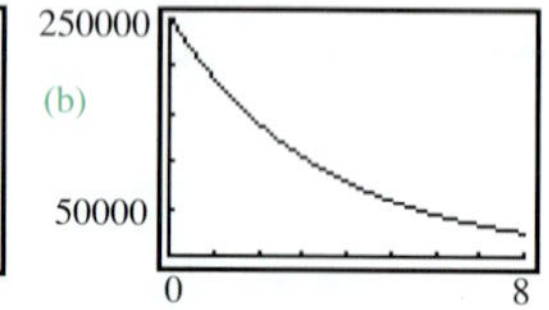

3. (a) i. \$70200 ii. \$25984.56 (b) $BV_t = 90000 \times (0.78)^t, 0 \le t \le 10$ (c) \$7502.20

(d) 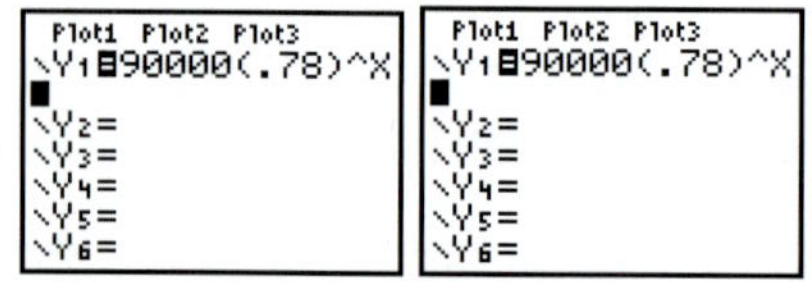

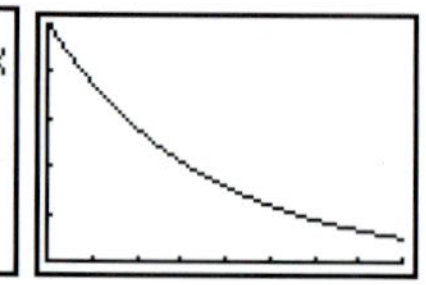

4. (a) 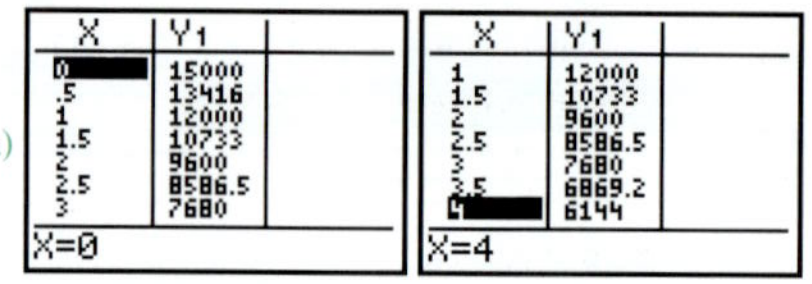

(b) 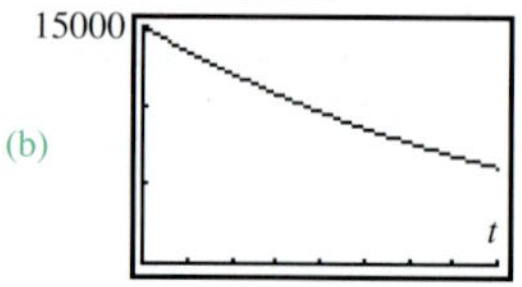

(c) $BV = 15000(0.8)^t, 0 \le t \le 4$ (d) i. \$9600 ii. \$6869.20 (e) \$6144.00 **5.** \$4507.50

6. (a) 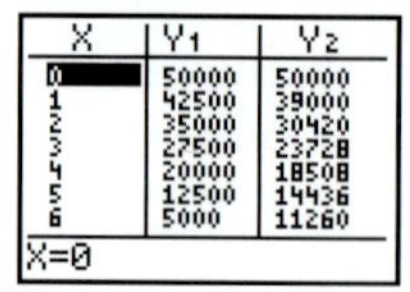

(b) 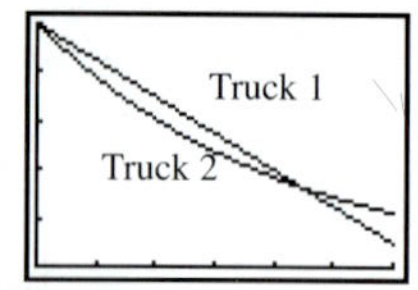

(c) 4th year
(d) Truck 1: \$5000
Truck 2: \$11260

7. 7.9983 ~ 8 years **8.** (a) 7 years (b) \$12801.50

EXERCISES 21.3.4

1. (a) \$37521.60 (b) 150,000 km **2.** (a) \$1624 (b) \$12776 (c) 10 yrs 1 month (approx)
3. \$19305.88 **4.** (a) \$26400 (b) 11696.75 km **5.** (a) \$110000; \$132500; \$150000; \$142000 (b) $\frac{4310}{N}$ yrs **6.** (a) \$38300 (b) \$40170 (c) Approx. 65455

7. (a) Construct a table using the following equations:
A: BV = 668000 – 80160t
B: BV = 668000×$(0.82)^t$
C: BV = 668000 – 114270t

(c) 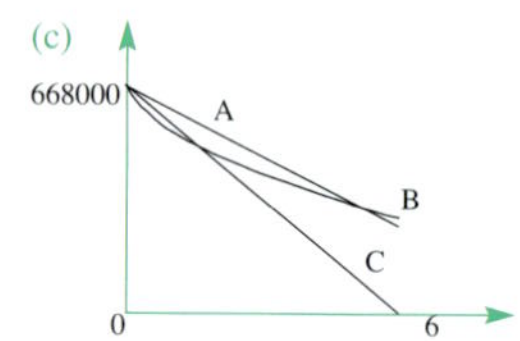

(d) i. C ii. A
(e) i. \$267200
ii. \$247645
iii. \$96650
(f) i. C ii. 5.85 yrs

EXERCISES 21.4.1

1. 1034 **2.** 21076 **3.** \$159.26 **4.** 125100 **5.** 252298

EXERCISES 21.4.2

1. a. 56.61 b. 576.65 c. 36.06 d. 152.22 e. 79.08 f. 287.35 g. 43.07 h. 144.84
2. a. \$825.95 b. \$178.04 c. \$72.05 d. \$146.74 e. \$219.84 f. \$402.15 g. \$58.93

h. $96.24 i. $713.31 j. $97.16 k. $12.88 l. $359.75 m. $193.42 n. $94.41
o. $5.65 p. $298.74 q. $109.30 r. $39.48 s. $43.77 t. $55.84

EXERCISES 21.4.3

1. $837.71 **2.** $1,894.10 **3.** $1,022.91 **4.** $602.80 **5.** $339.67 **6.** $680.10 **7.** $1,467.17 **8.** $150.94 **9.** $560.68 **10.** $3,833.22 **11.** $485.57 **12.** $557.67 **13.** $349.06 **14.** $703.12 **15.** $3,171.66

EXERCISES 21.5 – MISCELLANEOUS EXERCISES

1. $1305 **2.** $1082 **3.** $67 **4.** 9 **5.** $14815.44 **6.** $65.93 **7.** $17,442 **8.** $3473 **9.** 14.9% **10.** $20,600 **11.** (a) $4000 (b) i. 9.41% ii. 8.02% **12.** 8 **13.** $3 **14.** $15,000 **15.** 5 **16.** $24,015 **17.** £102907.72 **18.** $591 **19.** £8477 **20.** $17,248 **21.** (a) i. £1600 ii. £5600 (b) 5 yrs 11 months 3days (c) 9.17% (d) $125 **22.** (a) ¥13949 (b) $245.90 (c) 163.69 marks (d) £5.47 **23.** (a) 343 200 Lire (b) 26400 Lire (c) 3484800 Lire **24.** (a) 893.71 ~ 894 francs (b) 208.92 ~ 209 francs (c) 383.02 ~ 383 francs **25.** 8433 francs; $AU 2345; $55 **26.** $1800 **27.** (a) 4.16% (b) 3.489 months (3 months 15 days) **28.** comp. int = $764.89, simple interest = $528 **29.** $5043.96 ~ $5044 **30.** $5405.26; $3005.26 **31.** $1358.13 **32.** $1058.65 **33.** 14.65% **34.** Opt A: $3900, Opt B: $4659.51, Opt C: $4644.91. Option B is the best option. **35.** $18381.73 **36.** 3.18 yrs **37.** 147.29 fortnights **38.** (a) Shortfall of $1231.08 (b) ii. $n = 37.31$, need 38 quarters, i.e., 9.5 years. **39.** (a) $70,000 (b) $t = 0$, BV = $350000; $t = 1$, BV = $280000; $t = 2$, BV = $210000; $t = 3$, BV = $140000; $t = 4$, BV = $70000; $t = 5$, BV = $0 (c) $BV = 350000 - 70000t, 0 \le t \le 5$ (d) scrap value = $0
40. $3450 **41.** $950 **42.** (a) $11549 (b) $5480
43. (a)

t	0	1	2	3	4	5
Truck 1:	45000	36900	28800	20700	12600	4500
Truck 2:	45000	34650	26680.5	20544	15819	12180

(b)

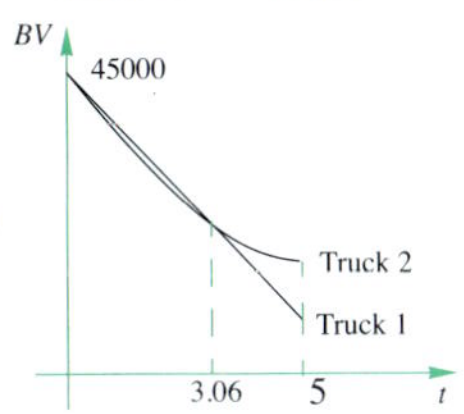

(c) $0 \le t < 3.06, BV_1 > BV_2$
$3.06 < t \le 5, BV_2 > BV_1$
$t = 3.06, BV_1 = BV_2$

44. (a) Use the equation $BV = 22000 \times (0.82)^t, 0 \le t \le 5$ to generate the schedule

(b)

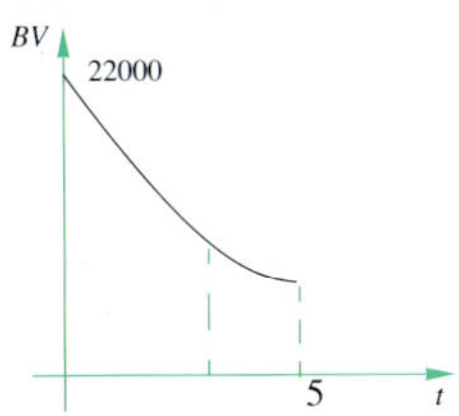

(c) $BV = 22000 \times (0.82)^t, 0 \le t \le 5$
(d) i. $14792.80 ii. $10452.61
(e) $8156.28

45. (a) $ 0.295 per km (b) $17300 (c) Use $BV = 35000 - 0.295N, 0 \le N \le 100000$ to generate the schedule. **46.** (a) i. $138000 ii. $270000 (b) Use the equation $BV = 300000 - 54000t$ to generate the schedule. (c) $a = 300000, b = 54000$

(d)

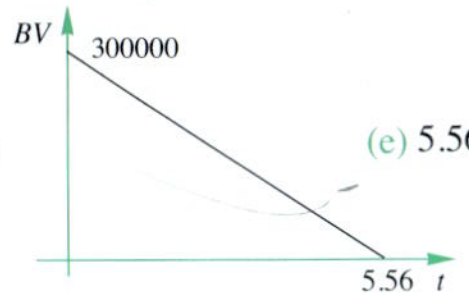

(e) 5.56 yrs

47. (c) $18162.50
48. (a) $1042.01; $42.01 (b) $122.22 (c) $736.01

EXERCISES 21.6 – GRADED REVISION EXERCISES

LEVEL 1

1. i. US $ 25 = ¥ 3808.75 ii. £50 = AUS $ 165.51 iii. ¥150 = AUS $ 1.89 **2.** $260.75

LEVEL 2

1. $26,800 **2.** $1049.91

LEVEL 3

1. Interest = £701.30; equivalent rate, $R = 7.20\%$ p.a **2.** i. 1.0% ii. (a) $3000 (b) $250 iii. 36 months (or 3 years). **3.** 7.113 ~ 7 yrs and 5 weeks.

LEVEL 4

1.

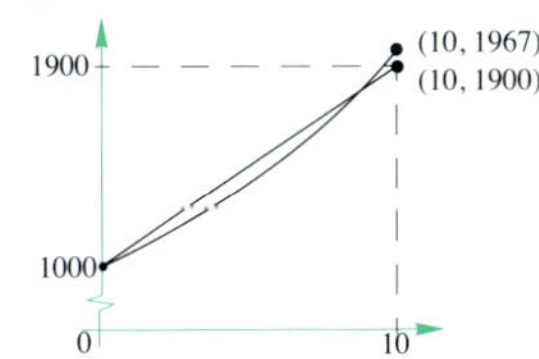

For $0 \le t \le 8.07$ simple interest has a higher return. After 8.07 years, compound interest does better.

2. **Bond A**: 16% p.a. compounded monthly. **Bond B**: 16.2% p.a. compounded semi-annually. Bond A: effective rate is 17.23%. Bond B: effective rate is 16.86%. Bond A is the better investment.

EXERCISES 21.7 – TOPIC TEST

1. i. 35 (USD) ii. 128.57 (SING) **2.** i. 960 ii. 1167.55 crowns **3.** $5527.2; $1455.33; commission is 1500 – 1455 = $45 **4.** 1.946 ~1.95 years **5.** Option A: i.e., 8.299% ; Option B: i.e., 8.264% Therefore, Option A is the better investment. **6.** (a) 10000 (b) 5 years (c) $8000 (d) $1600 per year (e) $BV = 10000 - 1600t, 0 \le t \le 5$ (f) $4666.67
7. (a) i. $11865.23 ii. $5005.65 (b) i. $a = 50000, b = 0.75$ ii. $3754.23

ANSWERS TO CHAPTER 22

EXERCISES 22.1

1. (a) $\frac{3}{4}$ (b) $\frac{3a}{4b}$ (c) –1 (d) 1 (e) $-\frac{15}{8}$ (f) 0 **2.** (a) 4 (b) 0.2 (c) 0.027 (d) 0.433 (e) –0.01 (f) –30 (g) 6.2 (h) 1 **3.** (a) 6 m/s (b) 30 m/s (c) $11 + 6h + h^2$ m/s **4.** 12 m/s **5.** $8 + 2h$ **6.** –3.49 ºC/sec **7.** (a) 127π cm^3/cm (b) i. 19.6667π cm^3/cm ii. 1.9967π cm^3/cm iii. 0.2000π cm^3/cm **8.** 1.115 **9.** (a) –7.5 ºC/min (b) $t = 2$ to $t = 6$ **10.** (a) 28 m (b) 14 m/s (c) average speed (d) 49 m (e) 49 m/s **11.** (a) $1160, $1345.6, $1560.90, $1810.64,$ 2100.34 (b) $220.07 per year

EXERCISES 22.2

1. (a) (b) (c) (d) (e) (f) **2.** (a) (b)

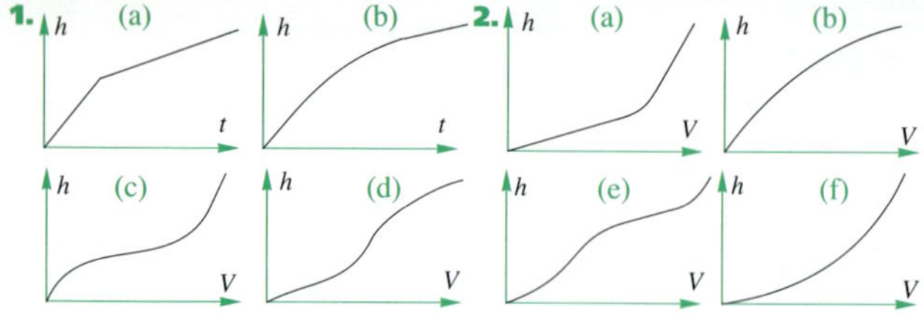

EXERCISES 22.3

1. (a) $h + 2$ (b) $4 + h$ (c) $\frac{-1}{1+h}$ (d) $3 - 3h + h^2$ **2.** (a) 2 (b) 4 (c) –1 (d) 3 **3.** (a) $2a + h$

(b) $-(2a + h)$ (c) $(2a + 2) + h$ (d) $3a^2 + 1 + 3ah + h^2$ (e) $-(3a^2 + 3ah + h^2)$

(f) $3a^2 - 2a + (3a - 1)h + h^2$ (g) $\frac{-2}{a(a+h)}$ (h) $-\frac{1}{(a-1)(a-1+h)}$ (i) $\frac{1}{\sqrt{a+h}+\sqrt{a}}$

4. (a) 1 ; 1 (b) $2a + h$; $2a$ (c) $3a^2 + 3ah + h^2$; $3a^2$ (d) $4a^3 + 6a^2h + 4ah^2 + h^3$; $4a^3$

5. (a) (b) i. 3 ms^{-1} ii. 2 ms^{-1} iii. 1.2 ms^{-1} (d) Find (limit) as $h \to 0$ (e) $4t - 3$

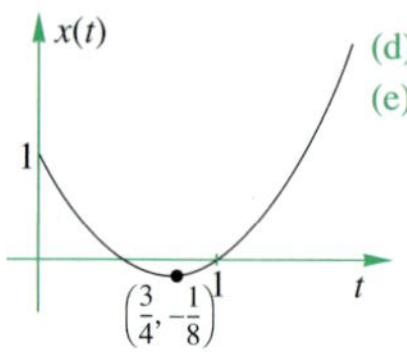

6. (a) (b) i. 20 cm^2 ii. 17.41 cm^2 iii. 2.59 cm^2 iv. –1.29 cm^2 day^{-1} (c) $20(1 - 2^{-0.1h})$ cm^2 day^{-1} (d) i. –1.3863 cm^2 day^{-1} ii. –1.2935 cm^2 day^{-1}

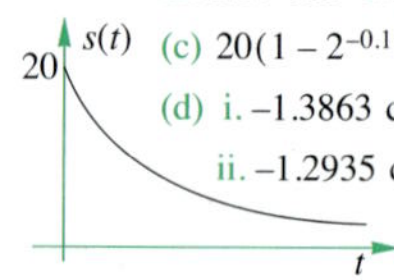

EXERCISES 22.4

1. (a) 3 (b) 8 (c) $-\frac{1}{9}$ (d) 1.39 (e) –1 (f) $\frac{17}{16}$ **2.** 4.9 m (b) $4.9(h^2 + 2h)$ m (c) 9.8 m/s

3. (a) $8x$ (b) $10x$ (c) $12x^2$ (d) $15x^2$ (e) $16x^3$ (f) $20x^3$ **4.** (a) $4x$ (b) –1 (c) $-1 + 3x^2$ **5.** (a) 1 ms^{-1} (b) $(2 - a)$ ms^{-1}

6. (a)

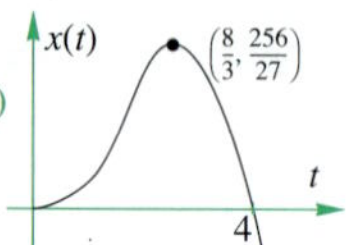

(b) i. 5 ms^{-1} ii. 4 ms^{-1} (c) $8t - 3t^2$ ms^{-1} (d) $\frac{8}{3}$ sec

EXERCISES 22.5 – MISCELLANEOUS EXERCISES

1. (a) $\frac{1}{2}(\sqrt{3} - 1)$ (b) 0 **2.** 13.5 bact. per hour **3.** (a) 122.5 m (b) 24.5 ms^{-1} **4.** (a) $\frac{2}{9}$ mg/ml/hr

(b) $\frac{1}{8}$ mg/ml/hr (c) $-\frac{5}{72}$ mg/ml/hr **5.** (a)

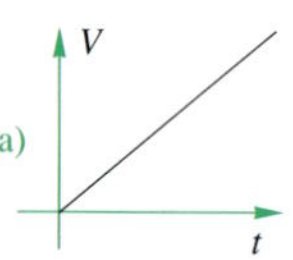

(b)

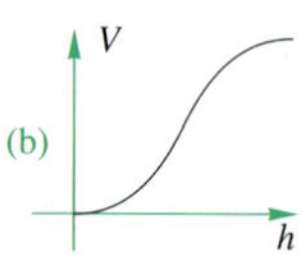

6. (a) (b) (c)

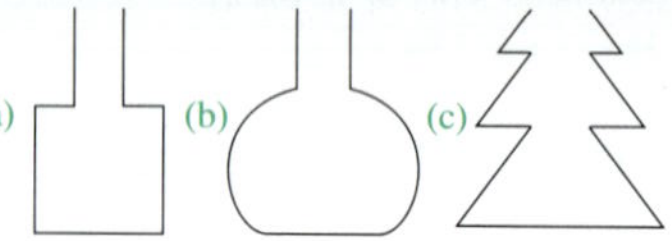

7. (a) 0.6875 (b) 0.71875 (c) 0.74375

(d) $v = 0.75 - 0.0625h$; as $h \to 0$, $v \to 0.75$ **8.** (a) –1 (b) –1 **9.** (a)

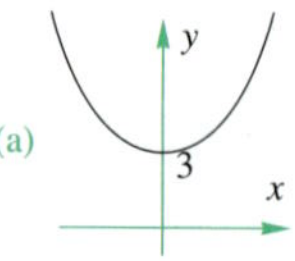

(b) i. $2xh + h^2$ ii. $2x + h$ (c) $2x$ **10.** (a) iii. $\leftrightarrow \alpha$

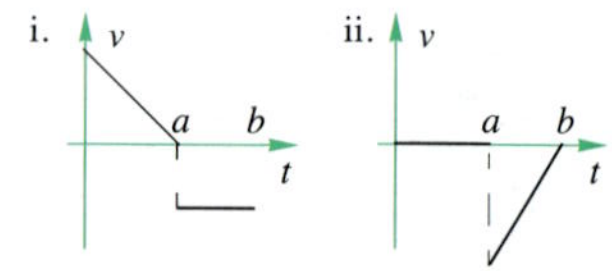

11. (a) i. –15 ii. –29 iii. –14 (b) i. $-15 - 12h - 2h^2$ ii. $-12h - 2h^2$ iii. –14 iv. –12 **12.** (a) Need **accurate** plot (b) i. –7 ii. 0 iii. This answer will depend on the accuracy of your plot, however, you will not be able to obtain an algebraic expression for this answer (unless you can deduce the equation of the function). Again, part (d) will depend on the accuracy of you plot. (e) gradient of tangent = gradient of curve at the same point. **13.** (a) 40 cm^3 (b) –0.7692 (c) i. 140 cm^3 ii. 0 cm^3 (d) $t = 7.5$

(e) i. Answers will vary, depending on the accuracy of your plot. ii. 0

EXERCISES 22.6 – GRADED REVISION EXERCISES

LEVEL 1

1. i. 2 ii. –1 iii. –0.5 **2.** (a) 0.5 (b) –0.6 (c) 0 **3.** i. A, D ii. B, C iii. C, D iv. A, B

LEVEL 2

1. (a) –4 (b) –2 **2.** 0.2, measures average rate of change of $f(x)$ over the interval [3, 8].

3. i. –6 °C/min ii. –0.0193 °C/min **4.** i. $2xh + h^2$ ii. $2x + h$

5.

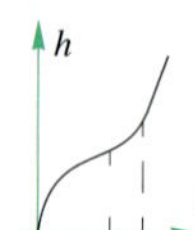

LEVEL 3

1. (a) $-\frac{4}{1+2h}$ (b) $1 + h$ **2.** (a) –4 (b) 1 **3.** $-3a^2 - 3ah$; $-3a^2$ **4.** (a) $3x^2$ (b) $-\frac{1}{x^2}$

LEVEL 4

1. (a) i. $0 < x < 1000$ ii.

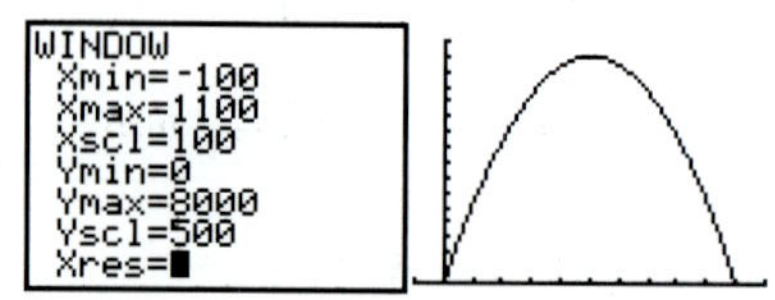

(b) i. 4500 ii. 0

(c) 599 (d) 18 **2.** (a) i. $10h - 4ht - 2h^2$ ii. 2 (b) –4 **3.** 1.829

EXERCISES 22.7 – TOPIC TEST

1. 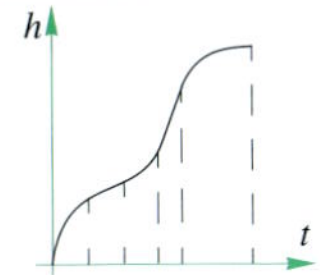

2. 2.77 **3.** 0 **4.** (a) Use graph paper to plot and sketch (b) –1.03

(c) Last decade (d) ii. 7.4 deaths per 100,000 (e) i. $2aht - bh + ah^2$ ii. –0.64

5. (a) 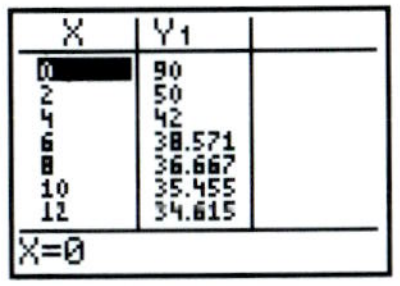

X	Y1
0	90
2	50
4	42
6	38.571
8	36.667
10	35.455
12	34.615

X=0

(b) –20°C/min (c) i. $-\frac{20}{h+3}$ ii. $-\frac{20}{3}$ °C/min

(d) i. 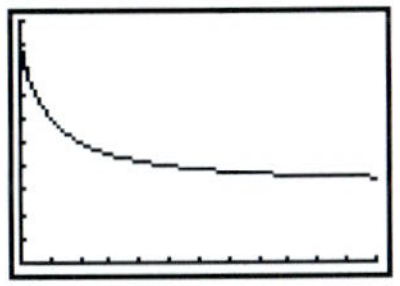 ii. $T \to 30$

ANSWERS TO CHAPTER 23

EXERCISES 23.1

1. (a) $5x^4$ (b) $9x^8$ (c) $25x^{24}$ (d) $27x^2$ (e) $-28x^6$ (f) $2x^7$ (g) $2x$ (h) $20x^3+2$ (i) $-15x^4+18x^2-1$ (j) $-\frac{4}{3}x^3+10$ (k) $9x^2-12x$ (l) $3+\frac{2}{5}x+4x^3$ **2.** (a) $2x+2$ (b) $4x^3+3x^2-1$ (c) $3x^2+1$ (d) $\frac{1}{x^2}$ (e) $4x^3-12x^2+8x$ (f) $\frac{1}{2}+x$ **5.** (a) i. $2(3x-1)$ ii. $9(3x-1)^2$ (b) i. 3 ii. $18x-6$ iii. $9(3x-1)^2$ (c) $3n(3x-1)^{n-1}$ **6.** $a=1, b=1$ **7.** i. (a) $2x-\frac{8}{x^3}$ (b) $2+\frac{24}{x^4}$

ii. (a) $1+\frac{1}{x^2}-\frac{4}{x^3}$ (b) $-\frac{2}{x^3}+\frac{12}{x^4}$ iii. (a) $\frac{6}{x^2}-\frac{18}{x^3}$ (b) $-\frac{12}{x^3}+\frac{54}{x^4}$ iv. (a) $1+\frac{8}{x^5}$ (b) $-\frac{40}{x^6}$

v. (a) $-\frac{4}{x^3}+\frac{3}{x^2}-\frac{15}{x^4}$ (b) $\frac{12}{x^4}-\frac{6}{x^3}+\frac{60}{x^5}$ vi. (a) $3x^2-\frac{6}{x^3}+\frac{30}{x^6}$ (b) $6x+\frac{18}{x^4}-\frac{180}{x^7}$

EXERCISES 23.2.1

1. $m_{PQ} = 4+h$; $\lim_{h\to 0} m_{PQ} = 4$ **2.** $P(1,1), Q\left(1+h, \frac{2}{2+h}\right)$; $m_{PQ} = -\frac{1}{2+h}$; $\lim_{h\to 0} m_{PQ} = -\frac{1}{2}$ **3.** –12 **4.** (a) 3 (b) $-\frac{1}{4}$ (c) 12 (d) 4 **5.** $\pm\sqrt{\frac{8}{3}}$ **6.** (a) $2x-12$ (b) –18 (c) (8, –32)

7. (a) $-3x^2+3$ (b) 0 (c) $(\pm\sqrt{2}, \pm\sqrt{2})$ **8.** (a) $\left(\pm\frac{\sqrt{2}}{2}, -\frac{1}{16}\right)$, (0,0)

(b) $\left\{x : \frac{-1}{\sqrt{2}} < x < 0\right\} \cup \left\{x : x > \frac{1}{\sqrt{2}}\right\}$ **9.** $x = \frac{1}{3}, -1$ **10.** (a) –2, 6, 3 (b) –2 **11.** $a=-3$ $b=1.5$

12. $f'(a+b) = 2(a+b) = 2a+2b$ **13.** (a) $4a^2-2a^4$ (b) $4-4a^2$ **14.** –56

EXERCISES 23.2.2

1. 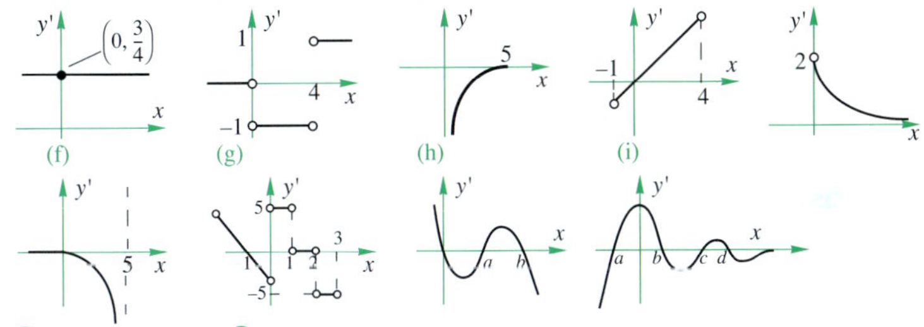

2. 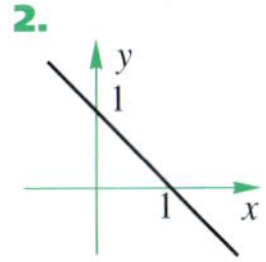

3. 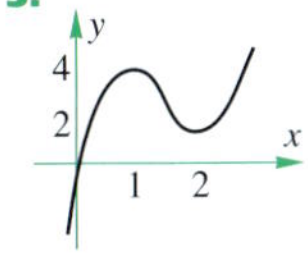

EXERCISES 23.3

1. (a) $y=7x-10$ (b) $y=-4x+4$ (c) $y=6x-16$ (d) $y=2x-9$ (e) $y=-6x+2$ (f) $y=-3.5x+7$ (g) $y=28x-48$ (h) $y=8x-12$ **2.** A: $y=28x-44$, B: $y=-28x-44$, Isosceles. $z \equiv (0, a^2-3a^4)$ **3.** $y=4x-9$ **4.** $y=-13, y=19$ **5.** (a) A(–3,7), B(4,0) (b) A: $y=-8x+32$, B: $y=6x+25$, $\left(\frac{1}{2}, 28\right)$ **6.** (c) $m=-2, n=5$ **7.** (a) 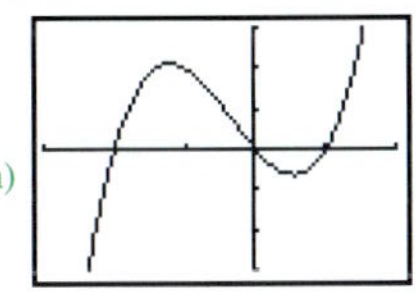 (b) i. $y=-2x$ ii. $y=6(x+2)$

(c) 3 sq. units **8.** (a) $a=1, b=2, c=4$ (b) (0.5, –2) (c) $\sqrt{11.25}+\sqrt{38.25}+\sqrt{18} \approx 13.7814$

9. $3y = 21x+31$; $3y = 21x-77$ **10.** (a) $y=3$ (b) $y=2x$ **11.** $a=-1$

EXERCISES 23.4

1. (a) 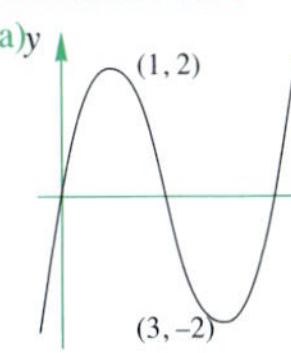

(b) 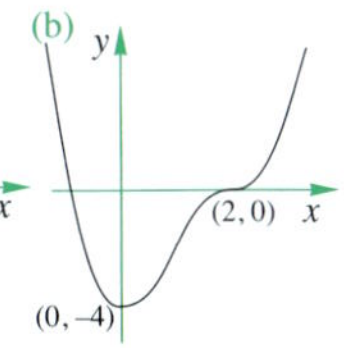

(c) 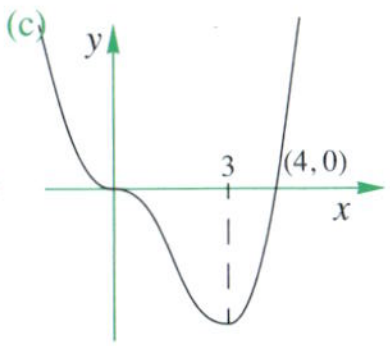

2. (a) max at (1, 4) (b) min at $\left(-\frac{9}{2}, -\frac{81}{4}\right)$ (c) min at (3, –45) max (–3, 63) (d) max at (0, 8), min at (4, –24) (e) max at (1, 8), min at (–3, –24) (f) min at $\left(\frac{1+\sqrt{13}}{3}, \frac{70-26\sqrt{13}}{27}\right)$, max at $\left(\frac{1-\sqrt{13}}{3}, \frac{70+26\sqrt{13}}{27}\right)$ (g) min at $(1, 0)$ max at $\left(-\frac{1}{3}, \frac{32}{27}\right)$ (h) max at $(0, 16)$, min at $(2, 0)$, min at $(-2, 0)$

3. (a) (b) (c) (d) (e) (f) (g) (h)

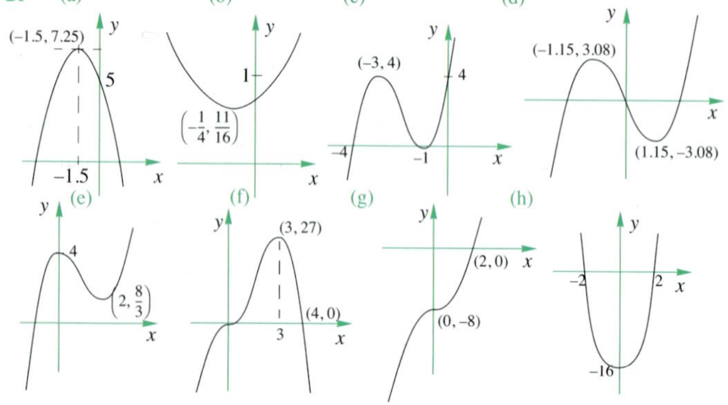

4. min at $(1, -3)$, max at $(-3, 29)$, non-stationary infl $(-1, 13)$ **5.**

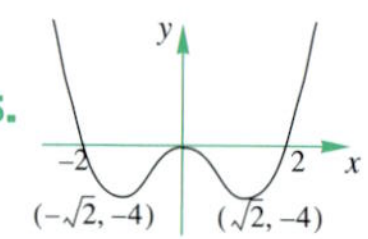

6. (a) 8 (b) 0 (c) 4 (d) $27\sqrt[3]{9} \approx 56.16$ **7.** i. min value –82. ii. max value 26

8. (a) A i. Yes ii. non-stationary pt of inflect. B. i. Yes ii. Stationary point (local/global min) C. i. Yes ii. non-stationary pt of inflect.
(b) A. i. No ii. Local/global max B. i. No. ii. Local/global min C. i. Yes ii. Stationary point (local max)
(c) A. i. Yes ii. Stationary point (local/global max) B. i. No ii. Local min C. i. Yes ii. non-stationary pt of inflect.
(d) A. i. Yes ii. Stationary pt (local/global max) B. i. No ii. Local min C. i. Yes ii. Stationary point (local max)
(e) A. i. No ii. Cusp (local min) B. i. Yes ii. Stationary pt of inflect C. i. No ii. Local max
(f) A. i. Yes ii. Stationary point (local/global max) B. i. Yes ii. Stationary point (local/global min) C. i. No ii. Tangent parallel to y–axis.

9. (a) i. A ii. B iii. C (b) i. C ii. B iii. A

10. i. ii. iii.

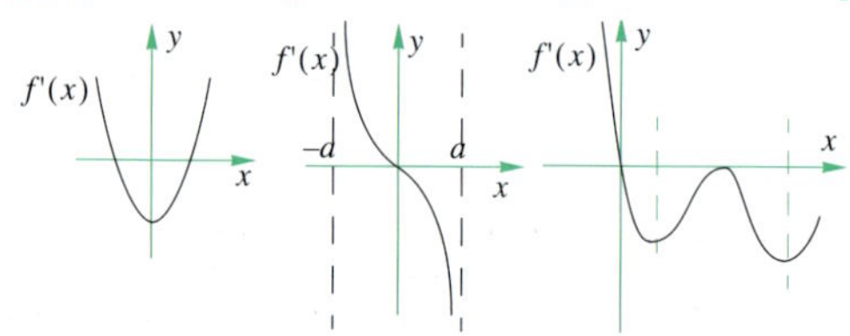

11. $y = x^3 + 6x^2 + 9x + 4$

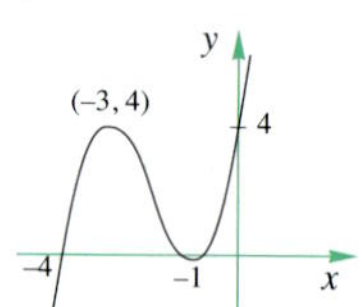

12. $f(x) = \frac{1}{3}x^3 - x^2 - 3x - 6$ **13.** $f(x) = 3x^5 - 20x^3$ **14.** $m = -0.5, n = 1.5$

15. $a = 2, b = -3, c = 0$

16. $a = 1, b = -12, c = 45, d = -34$

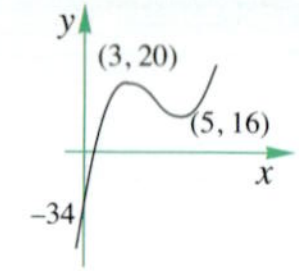

EXERCISES 23.5 – MISCELLANEOUS EXERCISES

1. $(-2, 21)$ **2.**

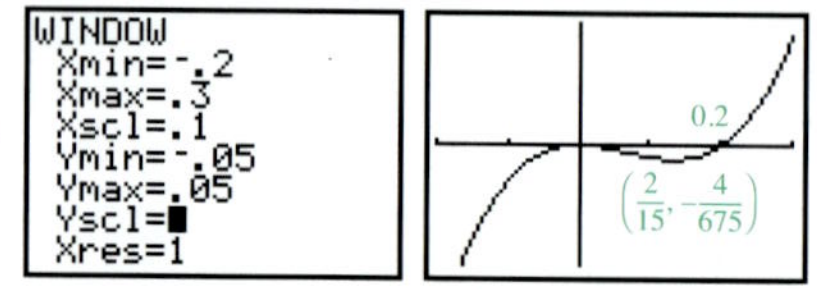

3. $(3, 9); \left(-1, \frac{5}{3}\right)$

4. $k = -2, 4$ **5.** $y = 6x + 5; y = 6x - 27$ **6.**

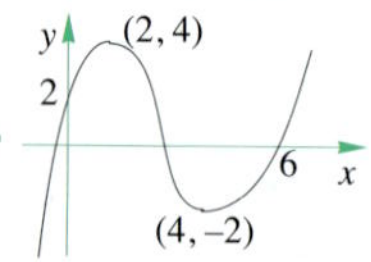

7. (a) $(0, 0); (-1, -5); (2, -32)$ (b)

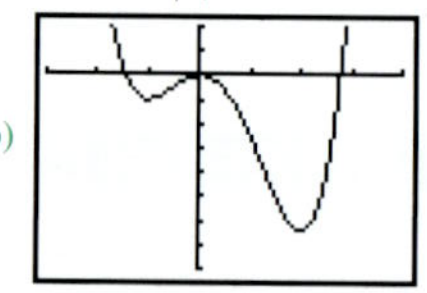

8. (a) $-\frac{1}{3}$ (b) $-\frac{1}{3} < x < 1$

(c) $x < -\frac{1}{3}$ or $x > 1$ **9.** (a)

(b) i. 32 ii. 0 (c) i. $k = 32$ ii. $7 < k < 32, k = 0$ iii. $0 < k \leq 7$ **10.** (a) 0, 2 (b) i. $y = -9x + 27$ ii. $y = 4$ (c) $\left(\frac{23}{9}, 4\right)$

11. (a) $x < -\frac{5}{3}$ or $x > 1$ (b) $-\frac{5}{3} < x < 1$ (c) $x = -\frac{5}{3}$ **12.** (a)

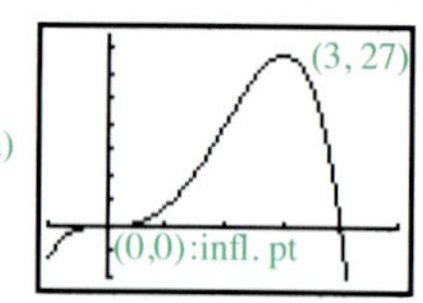

(b) $y = 16x - 16$ **13.** (a) i. $\frac{dy}{dx} = 3x^2 + 6x + 3$ ii. 48 (b) i. no ii. $x < -1$ or $x > -1$ (c) no local max or local min, only one stationary point of inflection at $(-1, 5)$ **14.** $a = 4, b = 2$

15. (a) $y = -12x + 12$ (b) $y = 16x - 15$ **16.** $a = 1, b = -8$

EXERCISES 23.6 – GRADED REVISION EXERCISES
LEVEL 1

1. i. $7x^6$ ii. $12x$ iii. -1 **2.** i. $2x$ ii. $12x-1$ iii. $2x$ **3.** i. $3x^2-2x$ ii. $-\frac{2}{x^2}-1$ iii. $4x^3-12x$ iv. $1-\frac{1}{x^2}$ v. $81-36x^2+3x^2$ vi. $3x^2+\frac{3}{x^4}$

LEVEL 2

1. (a) 3 (b) –27 **2.** $(0,0)$ $(\pm\sqrt{2},-8)$ **3.** $y = x$ **4.** 2

5. 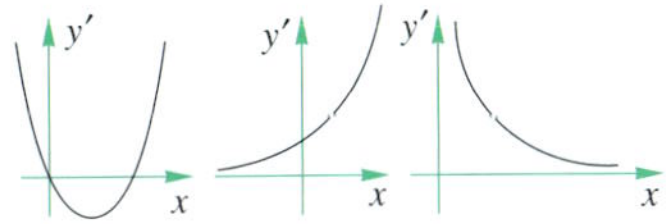

6. $x=1,\ y'=-1$; $x=2,\ y'=1$
7. $4y+x=15$

LEVEL 3

1. (a) $3x^2-3$ (b) –3 (c) $(\pm1, 0)$ **2.** $y=-2x+6$; $y=2x+6$ **3.** $\frac{6}{x^3}$ **4.** $a=-2, b=4$

5. 0, 2 **6.** (a) $y=-3x+4$ (b) $y=-3x-12$

LEVEL 4

1. (a) $f'(x) = 1-\frac{8}{x^3}$ (b) i. $-1 \le x < 0$ or $x > 0$ ii. $x < 0$ or $x \ge 2$ iii. 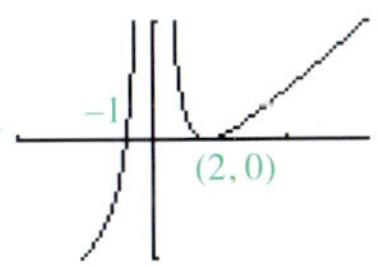

2. (a) x^3-x^2-ax+a (b) $y' = 12(x^3-x^2-ax+a)$ (c) $x=\pm\sqrt{a}$ or $x=1$ (d) i. Local min at $x=-\sqrt{a}$ & $x=1$; Local min at $x=\sqrt{a}$ ii. Stationary point of infl. $x=1$; local min at $x=-1$

(e) $y = 3x^4-4x^3-6x^2+12x$ 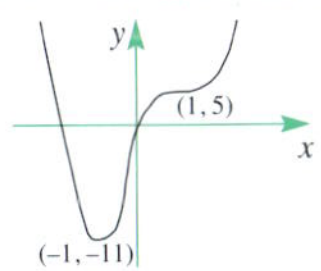

3. (a) (1, 3) (b) i. $y=3.5x-2$ ii. $(-0.5, -3.75)$

EXERCISES 23.7 – TOPIC TEST

1. i. (a) 9 (b) $9+8h+2h^2$ ii. $8+2h$ iii. 8 **2.** (a) $-6x^2+24$ (b) $-6a^2+24$ (c) $(1,-10)$ & $(-1,-54)$ **3.** $b=1$ **4.** i. 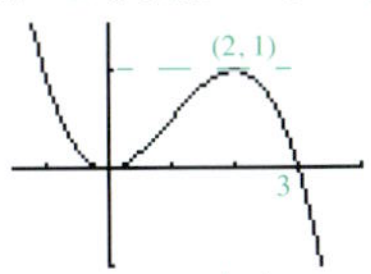

ii. $4k$ **5.** $6x-8$ **6.** $y=6x-8$

7. i. (a) global max. (b) $0 < x < a$ (c) 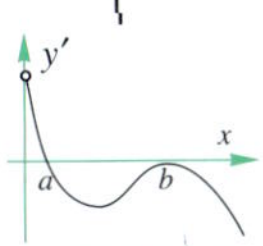

ii. (a) (0.1, 0.271); (0.01, 0.0297); (0.001,0.002997) (b) –3

ANSWERS TO REVISION EXERCISES

REVISION SET A

1. i. No ii. Yes iii. Yes **2.** i. 6.25×10^{-4} ii. 1.0×10^{-2} iii. 8.1×10^{-7} **3.** i. 1.39×10^{0} ii. 1.96×10^{2} iii. 2.48×10^{-2} iv. -6.27×10^{7} v. 9.85×10^{-1} **4.** 13.9681% ~ 14%
6. (c) 21°C hr^{-1} (d) $O(20)_{min}=853, O(20)_{max}=1117$; $O(30)_{min}=997, O(30)_{max}=1393$; 6.6%
7. [19.25, 29.25] **8.** (a) 5% (b) ~ 14.7% (c) 10% **9.** ~ 0.64% **10.** (a) Intercepts:(0, 0) & (4, 0), turning point at (2, 4) (b) I = [2, ∞[(c) ii.]–∞, 4] **11.** (b)]0, 4] (c) $x=2$ **12.** (b) $k=-1$
13. (a) $f(g(x)) = 9-x$ (b) i. [0, ∞[iii.]–∞, 9] **14.** (a) Yes (c) i. $\left[0, \frac{16}{3}\right]$ ii. [0, 4] (d) $\frac{8}{3}$
15. (a) 8 (b) 6 (c) 6 **16.** (a) i. $\frac{4}{5}$ or $-\frac{2}{3}$ ii. 0 or $\frac{5}{9}$ (b) –1.22 or 0.55 **17.** (a) i. (4, –2) ii. 4 iii. min values = –2 (b) 3 or 5 **19.** 5cm by 20cm by 30cm **20.** (a) A: 31.2 ms^{-1}; B: 44.5 ms^{-1} (b) A: 4.6 sec; B: 17.1 sec (c) i. 8.325 sec ii. 40.51 ms^{-1} **21.** (a) i. \$6000 ii. 20% iii. \$26,000 iv. $m=0.3, c=-3000$ (b) i. \$6000 ii. $m_1=0.4, c_1=-7750$; $m_2=0.2, c_2=0$ (c) $30000 < x < 47500$ **22.** i. 4m ii. $h(t) = -5(t-1.2)^2+7.2$, max height = 7.2 m when $t = 1.2$ iii. 0.537 sec & 1.86 sec iv. 2.4 sec **23.** (a) $5x-\frac{1}{2}x^2$ cm^2 (b) 1.25 **24.** (a) $a=12, b=27$ (b) –9 **25.** i. $4(s+10)$ km ii. 87.14 km hr^{-1} **26.** 20 units **27.** i. $50x+10y > 450$ ii. $y=26$ x iii. 5 **28.** (a) $a=0, b=10$ (b) 50 m^2 **29.** (a) 0.5 (b) $F=0.00125x, 0 \le x \le 80$ (c) $y = 0.000625x(x-80)^2, 0 \le x \le 80$ (e) 26.67 m

REVISION SET B

1. (a) 5 (b) 390 **2.** (a) 189 (b) 99 (c) –96 **3.** (a) $A=10$ (b) $k=0.009691$ **4.** 1 **5.** (a) i. 7m ii. 4m iii. 1m (c) 2.35, 9.65, 14.35, 21.65 (d) 9:39 am to 2:21 pm and 9:39 pm to midnight (e) $h(t) = 3\cos\left(\left(\frac{360}{13}t°\right)+4\right)$ **6.** (a) i. \$17680 ii. \$130 000 (b) i. \$19618.12 ii. \$1325999.37 (c) in the 5th year **7.** (b) i. 28 ii. 40 iii. 10 **8.** (a) 0.12 (b) $\frac{14}{23}$ **9.** (a) 0.89 (b) 0.525 (c) 0.4494
10. (a) 0.46 (b) i. 0.60 ii. $\frac{9}{23}$ **11.** (a) 341.99 (approx) (b) 20 terms **12.** (a) ii. 2000 (c) 52 hrs (d) 176995 **13.** (a) 23 (b) 3 (c) 7 (d) 26 **14.** (b) ii. –2.51 **15.** 20 rows
16. (a) P = {3, 6, 9, 12, 15, 18, 21, 24, 27, 30, 33, 36, 39}; R = {1, 2, 3, 4, 6, 9, 12, 18, 36}; Q = { 2, 4, 6, 8, 10, 12, 14, 16, 18, 20, 22, 24, 26, 28, 30, 32, 34, 36, 38}
(b) i. (a) x is not a multiple of 3 between 0 and 40 (b) x is a multiple of 3 and an even number between 0 and 40 (c) x is a multiple of 3 and an even number and a factor of 36 between 0 and 40 (d) x is either a multiple of 3 between 0 and 40 or an even number between 0 and 40 or a factor of 36 (e) x is not a multiple of 3 between 0 and 40 and x is not an even number between 0 and 40 (f) x is not a multiple of 3 between 0 and 40 and x is not an even number between 0 and 40 and x is not a factor of 36 (g) x is not an even factor of 36 that is a multiple of 3 (h) x is an even number or a factor of 36 but is not a multiple of 3.
iv. (a) {1, 2, 4, 5, 7, 8, 10, 11, 13, 14, 16, 17, 19, 20, 22, 23, 25, 26, 28, 29, 31, 32, 34, 35, 37, 38}

24, 26, 27, 28, 30, 32, 33, 34, 36, 38, 39} (e) {1, 5, 7, 11, 13, 17, 19, 23, 25, 29, 31, 35, 37} (f) {5, 7, 11, 13, 17, 19, 23, 25, 29, 31, 35, 37} (g) U\{6, 12, 18, 36} (h) {1, 2, 4, 8, 10, 14, 16, 20, 22, 26, 28, 32, 34, 38} **17.** (a) $\frac{1}{4}$ (b) $\frac{3}{8}$ (c) 0.3169 **18.** $\frac{64}{425}$ **19.** $\frac{10}{21}$ **21.** (b) i. $k = 0, 6$ ii. $k = 3$

22. (a) $(4-\pi)$ cm² (b) $\frac{1}{2}(4-\pi)$ cm² (c) ii. 0.5 iii. $A_n = (4-\pi)\times\left(\frac{1}{2}\right)^{n-1}, n = 1, 2, \ldots$ iv. $\frac{1}{16}(4-\pi)$ cm² (d) i. $\frac{31}{16}(4-\pi)$ cm² ii. $2(4-\pi)$ cm² (e) Geometric sequence

23. (a) – (c) Neither (d) contradiction (e) – (f) Tautology (g) contradiction (h) Neither

24. i. 0.02 ii. 0.06 iii. 0.225 iv. 0.635 **25.** (b) 6 (c) 18 **26.** (b) i. $\frac{7}{3}, \frac{11}{3}$ ii. $a = \frac{7}{3}, b = \frac{11}{3}$ (b) i. $k = -2$ ii. $-2 < k < 2$ **27.** i. 0.8 ii. 0.25 **28.** (a) $\frac{4}{15}$ m (b) 48 m **29.** (b) $a = 1, b = 9, c = -4.5, d = -4$ **30.** (a) i. 1 am & 7 am ii. 0.9 m (b) 15hrs (c) i. $a = 1.5, b = 45, c = 2$ ii. 2m (d) i. 6 times ii. 43.75% **31.** i. $\frac{2}{3}$ ii. $\frac{2}{9}$ iii. Not independent **32.** (a) \$50,000 ii. \$5801 iii. 2yrs 7wks (approx) iv. between 2.8 yrs and 3.7 yrs (b) k = 8839.71

REVISION SET C

1. (a) $\sqrt{61}$ (b) $y = 3x$ **2.** (a) i. (15, 0) ii. (0, 8) (b) 17 **3.** (a) i. {3, 8} ii. {4, 6} (b) 2

4. 13 km S22°37′W **5.** (a) i. 7.2 cm ii. 8.8 cm (b) i. 56.3° ii. 62.9° **6.** (a) i. $\left(\frac{1}{2}, \frac{1}{2}\right)$ ii. $\sqrt{34}$ (b) i. $3x - 5y + 1 = 0$ ii. $5x + 3y = 0$ **7.** (a) 120° (b) 24.25 cm² **8.** (a) i. 0.5196 m ii. 0.3464 m (b) 1.15 m (c) 73°13′ **9.** (a) –3 or 5 (b) 0.5 (c) (0, 4) **10.** 21.0 cm **11.** (a) 1 (b) (2, 2) (c) $\sqrt{40}$ (d) $y = x - 4$ (e) (3, 1) **12.** (a) 8 cm (b) 28°4′ **13.** (b) i. 660 m ii. 688 m **14.** 906 m

15. (a) 38°40′ (b) 0.08004 cm² (c) 6.05 cm² (d) \$493.71

16. (a) A(0, 6); B(3, 4); C(–3, 4); D(–3, 3); E(3,3); F(–1, 1); G(1, 1) (b) i. $y = x$ ii. $y = -x$ (c) $y = -\frac{2}{3}x + 6$ (d) 5 (e) i. 135° ii. 112°38′ **17.** 1.262 ha **18.** 1623 m **19.** 1939 m

20. (a) 1171m (b) 1.54km **21.** (a) 7.2cm (b) 67°58′ **22.** (b) 2 (c) $y = 2x - 3$ (d) $2x + 4y = 3$

23. (a) 16.97 cm (b) 13.56 cm (c) i. 66.14° ii. 66.06° **24.** 60°36′

25. (a) $4\sqrt{13}$ (b) 1.5 (c) 56°19′ (d) $3x - 2y + 3 = 0$ (e) i. P(–1, 0) ii. Q(0, 1.5) (f) $y = 1.5x$

26. (b) i. $y = -x$ (c) (5, –5) **27.** (b) $5\sqrt{10}$ **28.** (a) 120° (b) 50 m **29.** (a) 12 m (b) 35 m (c) i. $\frac{7}{22}$ m ii. 1039.5 m³ **30.** (a) i. 22.59° ii. 39.81° (b) 100.1° **31.** (c) $x = 2.40$

REVISION SET D

1. (a) 70 (b) 30 (c) 90 (d) 55 **2.** (a) i. 20 ii. 3 (b) i. 12 ii. 12 **3.** i. 3489.15 rupees ii. 53.74 marks **4.** \$U.S 291 **5.** 20160 lire **6.** (b) r = 0.9843, yes (c) 96.89% (d) i. $y = 5.1037x + 447.2070$ (e) 702.4 kg (f) 30.46 kg **7.** (a) i. Total charges = \$3.20 ii. Total balance = \$7777.31 iii. Equity = \$17267.66 (b) ii. \$2135.48 iii. 2yrs 9.5 months – mid February 2000 **8.** (a) $a = 57.19, b = 61.76, c = 17.81, d = 19.24$ (b) i. H_0: Student results are independent

ii. 5.991 iii. Accept H_0 **9.** 4863.22 Swiss francs **10.** 2341.50 crowns **11.** B

12. (a) mean = \$33.23K, mode = \$32K, median = \$32K Distribution is positively skewed (b) \$6.74K (c) workers on 35K, 50K, 45K, 55K (d) Managers (e) \$0.87K (f) \$997 per worker (g) 3.009% increase (h) 0.4268 **13.** (a) i. 0.54 ii. 0.70 iii. 0.54 (b) i. \$1315.79 ii. \$1289.47 (c) i. 17.45 yrs **14.** i. Reject H_0 ii. same as i. (but at the 1% level of sig.) **15.** (a) i. 50.23 ii. $b = 34.77, c = 26.77, d = 13.23, e = 23$ (b) i. H_0: Type of car sold is independent of location of showroom. ii. H_1: Type of car sold is dependent of location of showroom. (c) i. Chi-square test for independence. iii. Reject H_0 at the 5% level of significance. **16.** ii. 6.35gm; 2.615gm

17. (a) 0.5 (b) $y = 0.25x + 57.5$ (c) E (d) $y = \frac{43}{9}x - 12$ **18.** (b) [1200, 1400[(c) median = 1340 (d) 1307 hrs; 400.79 hrs

REVISION SET E

1. (a) $-\frac{2+h}{(1+h)^2}$ (b) –2 **2.** (a) Stationary points: (0, 0), $\left(\pm\frac{1}{\sqrt{2}}, 1\right)$ **3.** Intercepts: (0, 0) & (4, 0); Stationary points: (0, 0) & (3, 3) **4.** (a) (1, –2), (–1, 4), (3, 0) (c) At (1, –2) tangent is $y = -x - 1$ and at (3, 0) tangent is $y = 3x - 9$ **6.** (a) 9 (b) $y = 9x - 9$ **7.** (a) i. $3x^2 + 10x + 7$ ii. $6x + 10$ (b) $x < -\frac{7}{3}$ or $x > -1$ **8.** (a) $12 + 6h + h^2, h \neq 0$ (b) 12 **9.** (a) 5.39 (b) i. $\frac{1}{2}\cos\left(\frac{1}{4}t\right)$ ii. –0.40

10. $a = -1, b = 6$ and $c = -9$ **11.** (a) [0, 5] (c) 0.625 (d) $\frac{1}{10}(5 - 2t)$ ms⁻² **12.** (a) $1 - \frac{1}{t^2}$ (b) y = 0 **13.** –0.2 **14.** (a) Stationary point of infl. at (0, 0); local min at $\left(-\frac{3}{4}, -\frac{27}{128}\right)$ (b) 0, –1 (d) $\left\{x: -\frac{3}{4} < x < 0\right\} \cup \{x: x > 0\}$ **15.** (b) $y = 10 - 4t$, i.e., C (c) Boy at $(20 - 3t, 0)$ and girl at $(0, 10 - 4t)$ (d) $k = 200$ (e) $t = 4$ **16.** $\frac{11}{9}$ **17.** (a) 0, $\frac{3 \pm \sqrt{105}}{4}$ (b) local max at (–1, 14); local min at (2, –40) (d) $y = -24x - 2$ (e) $\frac{1}{12}$ sq. units **18.** local min at $x = -1$; local max at $x = 1$

19. (b) i. 1 ii. $y = x$ (c) (3, 3) (d) 36 sq. units **20.** (a) 50 ms⁻¹ (b) 100 ms⁻¹ **22.** $y = -4x + 21$

23. (0.5, 0) & (–0.5, 0) **24.** (a) $-2 \pm \sqrt{3}$ (b) 2 or –1 (c) At $x = 2, m = 8$; at $x = -1, m = 2$ (d) $y = 2x$ **25.** (a) For $f(x)$: 9.6. For $g(x)$: 14.77 (b) i. $d(x) = -x^3 + 6x^2 - 9x + 7, 0 \leq x \leq 4$ ii. 7 where $x = 3$ and $x = 0$ iii. 3 at $x = 1$ and at $x = 4$ (c) i. $y = -9x + 39$ ii. α is an approximation of where $d(x) = 0$